Hazards XX

Process safety and environmental protection
Harnessing knowledge – Challenging complacency

Institution of Chemical Engineers, Rugby, UK

Hazards XX

Process safety and environmental protection
Harnessing knowledge – Challenging complacency

Orders for this publication should be directed as follows:

Institution of Chemical Engineers,
Davis Building,
165–189 Railway Terrace, RUGBY,
Warwickshire CV21 3HQ, UK

Tel: +44 (0)1788 578214
Fax: +44 (0)1788 560833
Website: www.icheme.org/shop

Sponsors

Main event sponsors:
bp plc
ABB Engineering Services
Burgoyne Consultants Ltd

This symposium is supported by the Health and Safety Executive (HSE) and the Environment Agency and sponsored by:
Det Norske Veritas (DNV)
Aker Kvaerner

It is co-sponsored by:
Centre for Chemical Process Safety (AIChE)
Society of Chemical Industry (SCI)
Chemical Industries Association (CIA)
The Royal Society of Chemistry (RSC)
European Process Safety Centre (EPSC)
Safety and Reliability Society
Institution of Occupational Safety and Health (IOSH)
Society of Industrial Emergency Services Officers (SIESO)
IChemE Subject Group for Safety and Loss Prevention (SLPSG)
IChemE Subject Group for Environment (ESG)

Printed by Antony Rowe Ltd, Chippenham, UK

Hazards XX

Process safety and environmental protection
Harnessing knowledge – Challenging complacency

A three-day symposium organized by the Institution of Chemical Engineers (North West Branch) and held at the Weston Building, Manchester Conference Centre, UK, 14–17 April, 2008.

This book contains the papers and poster papers presented at Hazard XX.
There is also an accompanying CD-ROM.

Organizing Committee

M.F. Pantony (Chairman)	Consultant
M.J. Adams	Symposium organizer/secretary
G.R. Astbury	Consultant
C.J. Beale	Ciba Speciality Chemicals plc
S.R. Beattie	Syngenta plc
A. Boyd	AstraZeneca
G.A. Chamberlain	Shell Global Solutions
T. Clayton	Environment Agency
H. Conlin	RPS Risk Management
H.R. Cripps	HRC Consultants Ltd
K. Dixon-Jackson	Ciba Specialty Chemicals plc
M.I. Essa	Health and Safety Executive
R.F. Evans	Health and Safety Executive
B. Fulham	Health and Safety Executive
N. Gibson	Burgoyne Consultants Ltd
S. Hawksworth	Health and Safety Laboratory
P. Hooker	NPIL Pharmaceuticals (UK) Ltd
M. Hoyle	AstraZeneca
I. Kempsell	Sellafield Ltd
T.A. Kletz	Loughborough University
I.F. McConvey	AstraZeneca
G.S. Melville	Phoenix Loss Prevention Ltd
M.J. Pitt	Sheffield University
A. Saimbi	Morgan Professional Services
R.C. Santon	Consultant
G. Sellers	Consultant

Corresponding members of the committee

M.S. Mannan	Mary Kay O'Connor Process Safety Centre, Texas A&M University System
R.L. Rogers	Inburex GmbH

INSTITUTION OF CHEMICAL ENGINEERS
SYMPOSIUM SERIES No. 154
ISBN 978 0 85295 523 9

Preface

It gives me great pleasure in writing this preface for Hazards XX. This is the twentieth symposium since the first was held in March 1960, nearly half a century ago. Hazards' success and its growing popularity help bring about global recognition from process safety and environment professionals, experts and practitioners worldwide. This is so that we may share, harness and adopt knowledge, experience and expertise for the benefit of us all. I passionately believe that, through those who lead, manage and train others, Hazards will help catalyse a dynamic safety culture with a view to reducing risk and harm, help develop harmony in societies and a much cleaner environment, not to mention sustainable and profitable businesses.

Hazards XX offered the opportunity to be involved with two workshops entitled 'Recent advances in Safety Training' presented by IOSH and 'Health and Safety Update' presented by NEBOSH, held on Monday the 14th April 2008. The Symposium and Exhibition stretched over the three subsequent days offered the opportunity to engage in a whole host of high quality technical papers, covering aspects and issues such as: Safety Management, Environmental Protection, Safe Process Design, Compliance and Standards, Transport and Storage, Human Factors and Behaviour etc. The aforementioned themes were covered in several parallel sessions run over the three days. This year the organising committee made special effort to weave into the programme 15-minute slots for those presenting Poster Papers in session 5 on the afternoon of day 1 of the Symposium and Exhibition.

The Plenary Session 1, on Day 1 of the Symposium – Tuesday 15th April 2008, covered Keynote Papers. This session kicked off with The Burgoyne Memorial Lecture from Mr Peter Webb of Basell Polyolefins UK Ltd on 'Performance Indicators a tool to help senior managers show process safety leadership'. In this session we also benefited from a paper from Mr G. Visscher of the US Chemical Safety and Hazard Investigation Board on Expecting the Unexpected: Lessons from CSB accidents investigations. A joint paper from Dr D. Edwards of Granherne, Moscow and Mr C. Brandon on 'Challenging the SHE culture in the Russian Federation' and a paper from Mr P. Bratt and Ms C. Henney of Hammonds, Trinity Court, Manchester on 'Buncefield – Legal Impacts' proved to be interesting. Above all a paper by Professor Trevor Kletz on 'Accidents of the Next 15 years?' helped us address and challenge complacency.

The plenary session (Session 16) on the final day of the Symposium included a paper and DVD presentation involving Centrica Rough 47/3 Bravo Offshore Platform incident of 16th February 2006, covering the catastrophic vessel failure and subsequent hydrocarbon fire. Furthermore, an HSE paper from J. Carter, I. Travers and V. Beckett on the use of process safety performance indicators to ensure the effective management of major accident hazard risk. This paper is timely in the wake of high profile incidents such as the BP's Texas City Refinery incident of October 2005 and Buncefield incident of December 2005.

Sadly, during the organisation of this conference a member of the Hazards Committee, Dr Richard Rogers, died unexpectedly. For some thirty years, Richard has made a significant contribution to the development of process safety technology, its application in industry to produce safe processes, and in the preparation of European standards. Initially Richard was a member and then manager of the UK ICI/AstraZeneca Hazards and Process Studies Group and more recently a consultant with Inburex GmbH, Germany. His personality and contribution to process safety will be missed

Finally, I hope you found your attendance at Hazards XX fruitful, informative and inspiring. The success of Hazards is largely due to dedicated efforts of its committee members who put together the whole symposium programme including the organisation of the Symposium Dinner with excellent steer from the Chairman, Dr Martin Pantony. Regrettably, Martin, after many years of service to hazards, both as the Chairman and as a Committee Member is stepping down as the Chair of the Hazards Committee. I am sure you will all join me in thanking Martin for his outstanding contribution to Hazards and in wishing him well in the future.

M Iqbal Essa
Chairman – NW IChemE Branch
(HM Principal Specialist Inspector – HSE)

Contents

Keynote papers (plenary session 1)

Paper 1 Page 1	The Burgoyne Memorial Lecture Process safety performance indicators – a tool to help senior managers show process safety leadership P. Webb (*Basell Polyolefins UK Ltd, UK*)
Paper 2 Page 16	Selecting risk control measures – why organisations often demonstrate poor risk management C.J. Beale (*Ciba Expert Services, UK*)
Paper 3 Page 34	Some observations about major chemical accidents from recent CSB investigations G. Visscher (*US Chemical Safety Board, USA*)
Paper 4 Page 49	Buncefield – legal impacts P. Bratt and C. Henney (*Hammonds, UK*)
Paper 5 Page 62	Challenging the 'SHE' culture in the Russian Federation C. Brandon (*HSE Consultant*) and D.W. Edwards (*Granherne, Russia*)
Paper 6 Page 73	Accidents of the next 15 years? T.A. Kletz (*Loughborough University, UK*)

Safety management (plenary session 2)

Paper 7 Page 82	Lessons learned equals improved safety culture F.K. Crawley (*University of Strathclyde, UK*)
Paper 8 Page 93	Unlocking safety culture excellence: our behaviour is the key J. Hunter (*GlaxoSmithKline, UK*) and R. Lardner (*The Keil Centre Ltd, UK*)
Paper 9 Page 105	Managing risk competence C. Urwin (*Det Norske Veritas, UK*)
Paper 10 Page 112	Integrity management for the 21st century with 20th century equipment L. Krstin (*ABB Global Consulting*)

Environmental protection (session 3)

Paper 11 Page 121 — Bunding at Buncefield: successes, failures and lessons learned
A. Whitfield and M. Nicholas (*Environment Agency, UK*)

Paper 12 Page 134 — Practical implementation of a PPC site protection and monitoring plan
A. Buchanan (*Aker Kvaerner Consultancy Services, UK*) and J. Holden (*Elementis Chromium, UK*)

Paper 13 Page 146 — A successful regulatory intervention at William Blythe from despair to delivery
L. Murray (*Health and Safety Executive, UK*) and R. Costello (*William Blythe*)

Paper 14 Page 154 — What do we want to sustain and how do we decide?
M.S. Mannan, D. Narayanan and Y. Guo (*Mary Kay O'Connor Process Safety Center, USA*)

Paper 15 Page 170 — The implementation of IPPC under Schedule 1, Section 4.5 part A (1) a), for a small manufacturing enterprise, producing contract chemicals under a multiproduct protocol
S. Hollingworth (*NPIL Pharma (UK) Limited, UK*)

Safe process design (session 4)

Paper 16 Page 182 — Investigation into a fatal fire at Carnauld Metalbox Ltd
M.I. Essa (*Health and Safety Executive, UK*) and A. Thyer (*Health and Safety Laboratory, UK*)

Paper 17 Page 192 — The only good waste is 'dead' waste – WASOP, a methodology for waste minimisation within complex systems
N. Blundell (*Health and Safety Executive, UK*) and D. Shaw (*Aston Business School, UK*)

Paper 18 Page 215 — Impact of emergency shutdown devices on relief system sizing and design
R.K. Goyal and E.G. Al-Ansari (*Bahrain Petroleum Company, The Kingdom of Bahrain*)

Paper 19
Page 236
Background to and experience using the NII's new safety assessment principles – learning for the high hazard sector?
A. Trimble (*HM Nuclear Installations Inspectorate, UK*)

Paper 20
Page 247
Explosion processes and DDT of various flammable gas/air mixtures in long closed pipes containing obstacles
C. Lohrer, C. Drame, D. Arndt, R. Grätz (*Federal Institute for Materials Research and Testing (BAM), Germany*) and A. Schönbucher (*University of Duisburg-Essen, Germany*)

Paper 21
Page 262
Acceptance criteria for damaged and repaired passive fire protection
D. Kerr, D. Willoughby (*Health and Safety Laboratory, UK*), S. Thurlbeck (*MMI Engineering Ltd, UK*) and S. Connolly (*Health and Safety Executive, UK*)

Poster session papers (session 5)

Paper 22
Page 279
Competence assurance in the major hazard industries
M. Bromby and C. Shea (*ESR Technology*)

Paper 23
Page 290
Moodle e-learning environment – an effective tool for a development of a learning culture
V. Siirak (*Tallinn University of Technology, Estonia*)

Paper 24
Page 297
Using ARX approach for modelling and prediction of the dynamics of a reactor-exchanger
Y. Chetouani (*Université de Rouen, France*)

Paper 25
Page 306
Integrating risk into your plant lifecycle – a next generation software architecture for risk based operations
N. Cavanagh, J. Linn and C. Hickey (*DNV Software, UK*)

Paper 26
Page 316
Assessment of high integrity instrumented protective arrangements
A.G. King (*ABB Engineering Services, UK*)

Paper 27
Page 325
Considerations for layer of protection analysis for licensed plant
J. Fearnley (*Aker Kvaerner Consultancy Services, UK*)

Paper 28 Page 333 — Harmfulness and hazard categorisation – impact of emerging technologies on equipment design in the mining industry
D.S. Dolan (*Fluor Australia Pty Ltd*) and J. Frangos (*Toxikos Pty Ltd*)

Paper 29 Page 351 — Designing for safety – how to design better water treatment works
P. Bradley (*United Utilities, UK*)

Paper 30 Page 362 — Guidance on effective workforce involvement in health and safety
D. Pennie, M. Wright, P. Leach (*Greenstreet Berman, UK*) and M. Scanlon (*Energy Institute, UK*)

Paper 31 Page 373 — Development of an efficient safety and learning culture in Romanian small and medium enterprises (SME's) through virtual reality safety tools
S. Kovacs (*INCDPM, Romania*)

Paper 32 Page 385 — Knowledge transfer – critical components in occupational health and safety – an Estonian approach
M. Järvis and P. Tint (*Tallinn Technical University, Estonia*)

Compliance with standards (session 6)

Paper 33 Page 398 — Industrial processing sites – compliance with the new Regulatory Reform (Fire Safety) Order 2005
S.J. Manchester (*BRE Fire and Security, UK*)

Paper 34 Page 410 — The revised EN 13463-1 standard for non-electrical equipment for use in potentially explosive atmospheres
K. Brehm (*Bayer Technology Services GmbH, Germany*) and R.L. Rogers

Paper 35 Page 421 — Implementation experience of ATEX 137 for a petrochemical site
J. Fearnley (*Aker Kvaerner Consultancy Services, UK*) and R. Perbal (*SABIC Europe BV, The Netherlands*)

Paper 36 Page 431 — Practical application of static hazard assessment for DSEAR compliance
G.R. Ellis (*ABB Engineering Services, UK*)

Paper 37 Page 441 — Experiences with the application of IEC 61511 in the process industry in Austria
R. Preiss, K. Findenig and M. Doktor (*TÜV, Austria*)

Paper 38 Page 454 — Independent review of some aspects of IP15 Area classification code for installations handling flammable fluids
P. Nalpanis, S. Yiannoukas, J. Daycock, P. Crossthwaite (*DNV, UK*) and M. Scanlon (*Energy Institute, UK*)

Transport and storage (session 7)

Paper 39 Page 469 — Potential for flashback through pressure/vacuum valves on low-pressure storage tanks synopsis
A. Ennis (*Haztech Consultants Ltd, UK*) and D. Long (*Protego UK*)

Paper 40 Page 479 — Methods of avoiding tank bund overtopping using computational fluid dynamics tool
S.R. Nair (*Aker Kvaerner Consultancy Services, UK*)

Paper 41 Page 496 — Hazards in the maritime transport of bulk materials and containerised products
J.B. Kelman (*CWA International, UK*)

Paper 42 Page 511 — The causes of IBC (international bulk container) leaks at chemical plants – an analysis of operating experience
C.J. Beale (*Ciba Expert Services, UK*)

Paper 43 Page 522 — Liquid dispersal and vapour production during overfilling incidents
G. Atkinson, S. Gant, D. Painter (*Health and Safety Executive, UK*), L. Shirvill and A. Ungut (*Shell Global Solutions, UK*)

Human factors and behaviour (session 8)

Paper 44 Page 537 — Continuous monitoring of risks – people, plant and process
J. Bond (*Consultant, UK*)

Paper 45 Page 549 — The necessity of trust and 'creative mistrust' for developing a safe culture
J. Mitchell (*The Keil Centre, UK*)

Paper 46 Page 557 — Using the best available techniques to change behaviour in the construction industry
M. Worthington (*Morgan Sindall*), S. Hughes (*Morgan Ashurst*), and A. Saimbi (*Morgan Professional Services*)

Paper 47 Page 567 — Implementing and sustaining human reliability programmes of work – a managers' guide
A. Hubbard and J. Henderson (*Human Reliability*)

Paper 48 Page 577 — Improving shift handover and maximising its value to the business
A. Brazier (*Consultant, UK*) and B. Pacitti (*Infotechnics Ltd, UK*)

Paper 49 Page 590 — An investigation into a 'weekend (or bank holiday) effect' on major accidents
N.C. Healey (*Health and Safety Laboratory, UK*) and A.G. Rushton (*Health and Safety Executive, UK*)

Risk assessment and analysis (session 9)

Paper 50 Page 603 — Is HAZOP always the method of choice for identification of major process plant hazards?
A. Verna and G. Stevens (*Arthur D. Little Limited, UK*)

Paper 51 Page 611 — HAZOP for dust handling plants: a useful tool or a sledgehammer to crack a nut?
A. Tyldesley (*Haztech Consultants Ltd, UK*)

Paper 52 Page 617 — A rule-based system for automated batch HAZOP studies
C. Palmer, P.W.H. Chung (*Loughborough University, UK*) and J. Madden (*Hazid Technologies Ltd, UK*)

Paper 53 Page 630 — A consistent approach to the assessment and management of asphyxiation hazards
K.A. Johnson (*Sellafield Ltd, UK*)

Paper 54 Page 641 — Interpretation of the HCR for QRA – and its application beyond the North Sea
S.A. Richardson (*Shell Global Solutions, UK*)

Paper 55 Page 652 — An improved approach to offshore QRA
B. Bain (*DNV Energy, UK*) and A. Falck (*DNV Energy, Norway*)

Chemical reactions (session 10)

Paper 56
Page 671
Handling of reactive chemical wastes – a review
J.C. Etchells, H. James, M. Jones, and
A.J. Summerfield (*Health and Safety Executive, UK*)

Paper 57
Page 682
Dewar scale-up for reactive chemical waste handling
L. Véchot and J. Hare (*Health and Safety Laboratory, UK*)

Paper 58
Page 701
What kind of relationship do you have with your tollers?
C. Williams, M. Luginbuehl and P. Brown (*Syngenta, UK*)

Paper 59
Page 717
Thermal stability at elevated pressure – an investigation using differential scanning calorimetry
I.J.G. Priestley, P. Brown (*Syngenta Huddersfield, UK*), J. Ledru (*University of Aberdeen, UK*), and E. Charsley (*University of Huddersfield, UK*)

Paper 60
Page 729
Case studies in hazards during early process development
A. Boyd, P. Gillespie, M. Hoyle and I. McConvey (*AstraZeneca, UK*)

Issues from Seveso (session 11)

Paper 61
Page 743
Adapting the EU Seveso II Directive for the globally harmonised system of classification and labelling of chemicals (GHS) in terms of acute toxicity to people: initial study into potential effects on UK industry
M. Trainor, A. Rowbotham, J. Wilday, S. Fraser, J.L. Saw (*Health and Safety Laboratory, UK*) and D. Bosworth (*Health and Safety Executive, UK*)

Paper 62
Page 760
Practical experience in radio frequency induced ignition risk assessment for COMAH/DSEAR compliance
I.R. Bradby (*ABB Engineering Services, UK*)

Paper 63
Page 775
Lessons learnt from decommissioning a top tier COMAH site
K. Dixon-Jackson (*Ciba Expert Services, UK*)

Safety management (continued) (session 11)

Paper 64 Page 787 — A safety culture toolkit – and key lessons learned
P. Ackroyd (*Greenstreet Berman Ltd*)

Paper 65 Page 795 — Safety model which integrates human factors, safety management systems and organisational issues applied to chemical major accidents
L.J. Bellamy (*White Queen Safety Strategies, The Netherlands*), T.A.W. Geyer (*Environmental Resources Management Limited, UK*), J.I.H. Oh (*Ministry of Social Affairs and Employment, The Netherlands*), and J. Wilkinson (*Health and Safety Executive, UK*)

Paper 66 Page 809 — Can we still use learnings from past major incidents in non-process industries?
F. Gil (*BP, UK*) and J. Atherton (*Process Safety Consultant, UK*)

Risk assessment and analysis (continued) (session 12)

Paper 67 Page 825 — Liquid mists and sprays flammable below the flash point: the problem of preventative bases of safety
S. Puttick (*Syngenta Huddersfield Manufacturing Centre, UK*)

Paper 68 Page 838 — Modelling of vented dust explosions – empirical foundation and prospects for future validation of CFD codes
T. Skjold (*GexCon AS and University of Bergen, Norway*), K. van Wingerden, O.R. Hansen (*GexCon AS, Norway*) and R.K. Eckhoff (*University of Bergen, Norway*)

Paper 69 Page 851 — Bringing risk assessments to life by integrating with process maps
G. Sellers (*Safety Management Consultant, UK*), A. Webb and C. Thornton (*Biffa Waste Services Ltd, UK*)

Paper 70 Page 858 — Simplified flammable gas volume methods for gas explosion modelling from pressurized gas releases: a comparison with large scale experimental data
V.H.Y. Tam, M. Wang, C.N. Savvides (*EPTG, BP Exploration, UK*), E. Tunc, S. Ferraris and J.X. Wen (*Kingston University, UK*)

Paper 71 Page 869 — Pressurised CO_2 pipeline rupture
H. Mahgerefteh, G. Denton (*University College London, UK*), and Y. Rykov (*Keldysh Institute of Applied Mathematics, Russia*)

Paper 72 Page 880 — Verification and validation of consequence models for accidental releases of hazardous chemicals to the atmosphere
H.W.M. Witlox and A. Oke (*DNV Software, UK*)

Safe process design (continued) (session 13)

Paper 73 Page 893 — Avoidance of ignition sources as a basis of safety – limitations and challenges
S. Puttick (*Syngenta Huddersfield Manufacturing Centre, UK*)

Paper 74 Page 902 — Assessment of flammable gas ingestion and mixing in offshore HVAC ducts: implications for gas detection strategies
C.J. Lea (*Lea CFD Associates Ltd, UK*), M. Deevy (*Health and Safety Laboratory, UK*), and K. O'Donnell (*Health and Safety Executive, UK*)

Paper 75 Page 917 — Guidance on the use of non-certified electrical equipment in laboratory fume cupboards
G.R. Astbury (*formerly Senior Manager of Health and Safety Laboratory, UK*)

Paper 76 Page 931 — Maximise the use of your existing flare structures
N. Prophet, G.A. Melhem, and R.P. Stickles (*ioMosaic Corporation, USA*)

Paper 77 Page 943 — Health and safety in biodiesel manufacture
S.W. Harper, J.C. Etchells, A.J. Summerfield, and A. Cockton (*Health and Safety Executive, UK*)

Paper 78 Page 952 — A methodology to guide industrial explosion safety system design
R.J. Lade (*Kidde Research, UK*) and P.E. Moore (*UTC Fire & Security, UK*)

Safety management (continued) (session 14)

Paper 79 Page 966 — APELL, safer production and corporate social responsibility – linking three initiatives to improve chemical safety in the Thai chemical industry
M. Hailwood (*Germany*)

Paper 80 Page 978 — Demonstrating continuous risk reduction
A. Bird, A. Lyon (*DNV Consulting*) and V. Edwards (*BP Trinidad and Tobago*)

Paper 81 Page 990 — Business continuity and the link to insurance: a pragmatic approach to mitigate principal risks and uncertainties
N.J.L. Gardener (*Elementis plc*)

Paper 82 Page 1006 — A world class approach to process safety management (PSM) after the Texas city disaster
E. Pape (*DNV Energy, UK*)

Paper 83 Page 1021 — Leading indicators for the management of maintenance programmes; a joint industry programme
K. Hart (*Energy Institute*), J. Sharp (*Cranfield University, UK*), J. Wintle (*TWI Ltd*), D. Galbraith (*Poseidon International Limited*), and E. Terry (*Sauf Consulting Limited*)

Risk assessment and general process safety issues (session 15)

Paper 84 Page 1037 — A progressive risk assessment process for a typical chemical company: how to avoid the rush to QRA
R.T. Gowland (*European Process Safety Centre, UK*)

Paper 85 Page 1046 — Managing business risks from major chemical process accidents
M. Bardy, L. Fernando Oliveira (*DNV Energy Solutions South America*) and N. Cavanagh (*DNV Software, UK*)

Paper 86 Page 1062 — Overview of health and safety in China
H. Wei, L. Dang (*Tianjin University, PR China*) and M. Hoyle (*AstraZeneca plc, UK*)

Paper 87 Page 1070 — Analysis of past incidents in the process industries
I.M. Duguid (*Consultant*)

Paper 88 Page 1081 — Hazard of an expert witness – an Australian Experience
R. Ward (*University of New South Wales, Australia*)

Plenary session 16

Paper 89 Page 1093 — 'Incredible'
G. Sibbick (*Centrica Storage Ltd, UK*)

Paper 90 Page 1097 — Dust explosion in sugar silo tower: investigation and lessons learnt
M. Westran (*British Sugar, Peterborough*),
F. Sykes (*Health and Safety Executive, Norwich*),
S. Hawksworth, and G. Eaton (*Health and Safety Laboratory, Buxton*)

Paper 91 Page 1111 — The use of process safety performance indicators to ensure the effective management of major accident hazard risks – the Health and Safety Executive's experience
I. Travers, V. Beckett, and J. Carter (*Health and Safety Executive, UK*)

Page 1120 — Index

PROCESS SAFETY PERFORMANCE INDICATORS – A TOOL TO HELP SENIOR MANAGERS SHOW PROCESS SAFETY LEADERSHIP

Peter Webb
Basell Polyolefins UK Ltd, Carrington Site, Urmston, Manchester, M31 4AJ
E-mail: Peter.Webb@Basell.com

As is the case with many major hazard companies, Basell Polyolefins has achieved a significant reduction in work place injuries over recent years. A key factor in achieving this has been the leadership and engagement of senior managers. They have been helped by having been able to measure their successes and failures by counting the number of injuries. Since 2003, Basell has been developing a process safety performance indicator (PSPI) system which is intended to help senior managers similarly engage in achieving improvements in process safety. While process safety incidents across the industry have not killed or injured as many people as falls from height and moving vehicle accidents, they often have far greater business consequences.

This paper will describe Basell's leading and lagging PSPI framework, and show how it is helping senior managers to get engaged in process safety. This is true not only for managers of a technical but also a non technical background. They now have a tool which will help them show leadership in process safety.

INTRODUCTION

Lack of adequate process safety performance indicators (PSPIs) has been cited as a factor in a number of recent major accidents (Hopkins 2000, HSE 2003, US Chemical Safety and Hazard Investigation Board 2007). Against this background and interest from the regulators (HSE 2006), many companies involved in major hazards are developing PSPI programmes.

Basell's PSPI corporate framework was piloted at the company's Carrington UK site over the period 2003 to 2006, and was adapted and implemented as a company wide initiative in 2007. Basell makes polypropylene and polyethylene in 19 countries and has 6,800 employees.

The framework comprises lagging and leading indicators. A suite of lagging indicators is analogous to the different categories of injuries. These are standard across the whole company, and are reported every month up through the line. There was already a mature incident reporting system in all sites in the company, and these incidents would anyway have been recorded. But what is new is the headlining of these narrowly defined incidents. This offers learning opportunities which have been derived from analysing the information, to identify "big picture" conclusions.

The other part is leading indicators, and the decision was taken that it was not appropriate to try to select a standards set for the whole company. It was decided that it was better for sites to choose their own. While this precludes the possibility to benchmark, it allows the sites to focus on issues which are relevant to them, and also increases buy in. Some successes have also been achieved with leading indicators, although somewhat less tangible.

THE SWORD OF DAMOCLES

Legend has it that Damocles was a young courtier in the court of Dionysius II of Syracuse. And he was a bit of a wag. One day he observed to Dionysius that the job of being king must be a pretty good one, with all the power and privilege. Dionysius responded that he could give it a try if he liked. So Damocles was installed in the throne, and Dionysius arranged for a slap up banquet with wine women and song. At the end of the evening, Dionysius asked Damocles what he had thought of his spell in charge, and Damocles told him that he had had a great time. Dionysius then asked him what he had thought about the sword. He pointed to a large sharp sword which was hanging above the throne by a single thread of horse hair. Damocles swiftly vacated the position. Dionysius explained that the sword was there as a reminder that the position carried responsibilities. Without it, it would be too easy to get busy enjoying the trappings of power, and forget about the important stuff. Damocles had gained a better understanding of the role, and was no longer envious of Dionysius.

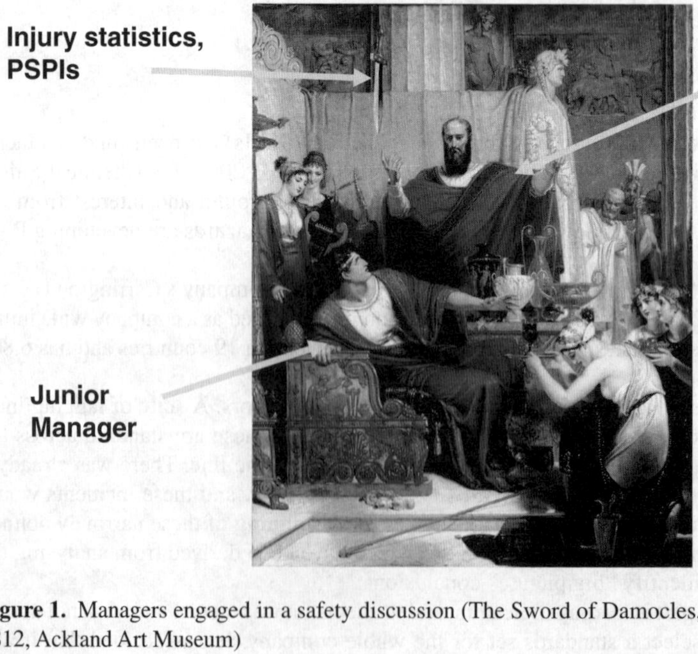

Figure 1. Managers engaged in a safety discussion (The Sword of Damocles, Richard Westall 1812, Ackland Art Museum)

THE HISTORY OF PROCESS SAFETY PERFORMANCE INDICATORS AT BASELL

In the 1990s and early 2000s the measuring and reporting of injury performance as a Total Recordable Rate (TRR) in our company had been an effective Sword of Damocles. It had helped to drive down injuries. Senior managers had played a key role in this. They were interested in safety, and TRR was a simple, easily understood tool which they could use to engage themselves. Senior managers had given site managers clear injury performance objectives, and site managers had found solutions appropriate to their own sites to achieve the desired objectives. TRR had a strong symbolic value.

WHAT INTERESTS MY BOSS FASCINATES ME!

Organisations need objectives. The greatest advantage of management by objectives is that it makes it possible for a manager to control his own performance (Drucker 1968). Defining what needs to be achieved rather than prescribing how it should be done, allows managers to apply a broader vision. This leads to stronger motivation and higher performance. In order for it to work, a manager needs to know not only what his goals are. He also needs to be able to measure his performance and results against the goals. Managers need to be provided with clear measurements or metrics. According to Drucker,

> *"They do not need to be rigidly quantitative; nor do they need to be exact. But they have to be clear simple and rational. They have to be relevant and direct attention and efforts where they should go. They have to be reliable – at least to the point where their margin of error is acknowledged and understood. And they have to be, so to speak, self announcing, understandable without complicated interpretation or philosophical discussion.*
>
> *Each manager should have the information he needs to measure his own performance and should receive it soon enough to make any changes necessary for the desired results."*

In other words, what gets measured gets done. Or as it has been said "What interests my boss fascinates me".

But the shortcomings of TRR started to become apparent. Despite a TRR trend which headed ever downwards, the company at sites around the world had nevertheless had fatalities. Because of the profile which TRR was given, significant management time was spent on sometimes minor injuries. At the same time, senior managers seemed to be less interested in losses of containment and fires.

From the outset we realised that it was essential to think not only about how and what to measure, but also how the tool could be used by senior managers to drive improvements. Hopkins (2007) examined how BP at Texas City had been measuring but not managing process safety performance. In contrast, by tracking and "headlining" the number of injuries, Texas City workplace safety had improved dramatically. Managers' remuneration, even though it was only small amounts of money, was affected by injury performance in their area of control. Injury statistics, like Damocles' Sword, had a powerful symbolic value which had strongly influenced manager behaviour.

So in 2003 at Carrington we set about designing a simple tool comprising leading and lagging indicators, which we hoped would achieve for process safety what TRR had done for workplace safety. We recognised that senior managers at the highest level in our company were interested in safety, and we wanted to channel their interest towards process safety. We had in mind that if it worked at a site level, it could form the basis of a corporate implementation.

There was nothing new about the concept of leading and lagging indicators. OHSAS18001 (BSI 1999) for example talks about proactive and reactive indicators. We were trying to apply the concepts to process safety.

SITE LAGGING INDICATORS

We started by establishing a suite of seven lagging process safety indicators, which we defined in terms of consequence, Appendix 1. In arriving at our definitions we took account of those in RIDDOR for dangerous occurrences (HSE 1995), and a PSPI pilot run in Scotland by the HSE and the SCIA (HSE/SCIA 2003). Additionally we adjusted and calibrated them to suit our own situation. We added a category for releases of substances which could have an impact on the environment. Note that we do not have a category for toxics, as this is not relevant for our business.

To facilitate incident reporting and analysis, the seven categories were set up in the site's computer based HSEQ management system QPulse.

SITE LEADING INDICATORS

In 2004 we moved on to develop a system for leading indicators. We decided that we needed to define the framework in Table 1 for each indicator.

WHICH RISK CONTROL SYSTEMS TO MEASURE

In planning our approach we struggled with items 1 and 2 in Table 1- how to decide which process safety issues to tackle. Of course we could think of lots of things to measure,

Table 1. Framework for leading PSPIs

1. Which risk control system is weak and needs to be reinforced.
2. What specifically is the issue; what does the weakness relate to.
3. Definition of a metric.
4. How will the data be collected, and who will do it.
5. How will the results be monitored, and corrective actions identified.
6. What target would we aim for.

but we wanted to work with only a handful of indicators, five or six. We could have locked the management team in a smoke filled room until they came up with the five. But we decided to use a team of people who represented horizontal and vertical cross sections of the work force. We asked the team to brain storm around the question "**As you go about your work, what are the things which make you feel uncomfortable about process safety**". The team used a diagrammatic representation of our HSE management system as a guideword list of risk control systems, to prompt thought about which systems were weak. The only constraint we put on the brainstorming was that, for a weakness to be added to the list, the proposer had to give an example of how it had in his experience given rise to a risk. We got a list of 14 items, and used a risk ranking tool to prioritise them.

DEFINING THE METRICS AND MEASUREMENT SYSTEM

We then looked at trying to define a metric for each of the top five issues, and how we would collect the data. From our investigations, we had decided that the metrics should have the attributes listed in Table 2.

We were not able to find a metric which met these criteria for every one of the top five issues. For example, the team had identified shift handover as a potentially weak area, but we were not able to find a suitable metric. So we concluded that although shift handover deserved attention, it would have to be via a means other than as a leading PSPI. We picked another weakness to make up the list of five. As an aside, in the meantime we have given further thought to how we could measure shift handover. We have found some criteria which we could use as the basis for periodic audits, but have some concerns about whether this will be cost effective in terms of data collection; unless we can get the teams to audit themselves, somebody probably from the HSEQ Department will have to work unsociable hours to do the audits!

MONITORING

In planning how we would monitor the data we were mostly able to use existing arrangements, e.g. the site HSE Committee, or the periodic inspection review meeting. But we

Table 2. Desirable attributes of leading PSPI metrics

- Support continual improvement
- Drive appropriate behaviour
- Emphasise achievements rather than failures
- Be precise and accurate
- Be difficult to manipulate
- Be owned and accepted by the people involved in related work activities and those using the metrics
- Be easily understood
- Be cost-effective in terms of data collection

recognised it was important that people with the power to make decisions were part of the monitoring process.

The leading metrics we settled on and their associated framework are given in Appendix 2.

CONCERNS PEOPLE RAISED

"Oh no! Not another report for management!"

A valid concern, but we sought to impress on people that it is not just about reporting, it is about creating a simple tool which managers can use to focus on and improve process safety.

"But all incidents already get reported"

This is true, but the process safety incidents get buried, and we do not see the process safety big picture.

"We do see the big picture, because all process safety incidents get a full investigation"

We calibrated the lagging incident definitions so that we get about twenty incidents a year, on a site of about 150 people and three plants. We have to be somewhat selective on which incidents to carry out a full in depth root cause analysis, and do not consider it a good use of resources to do twenty of those a year.

"We should take account of risk in deciding which incidents to count"

As mentioned in Table 2, good indicators are precise and accurate, difficult to manipulate, and easily understood. Risk is none of these. We based our definitions on pure consequence, which means that a 49 kg release of a flammable material in the open air would not count as a PSPI, while a 50 kg would. This does not mean the 49 kg release is unimportant, but you have to draw the line somewhere. And if the 49 kg release was of particular interest, we would anyway investigate it to get root causes and learnings.

Perhaps a more fundamental objection to using risk criteria is that recent work appears to indicate that high risk near misses are a poor predictor of actual incidents. The things which we perceive as high risk are apparently not the things which cause the actual incidents. The Institute of Petroleum (2005) analysed the causes of some 600 potential and 800 actual major hazard incidents. The analysis considered 31 causes. There were substantial differences in the frequency of causes of actual and potential incidents. For example "Leak from flanged joint or coupling" was the second most frequently occurring potential incident cause, but only ranked seventh in the actual incidents. A PSPI programme which identified this cause for attention based on counting potential incidents, might divert resources away from work which could reduce the number of actual incidents.

"The iceberg theory tells us that our existing near miss reporting and behavioural programmes will prevent major accidents"

Groeneweg (2006) analysed the Oil & Gas Producers Association safety performance indicators over the years 1997 to 2004. This represented 12.3 billion hours of data. He looked for correlations between companies for the different severities of injuries. He wanted to find out if the companies which had the highest fatal accident rate (FAR) also had the highest lost time injury frequency (LTIF). He found no correlation. He also looked for correlations between other indicators: total recordable injury frequency, and near misses. He found no (or hardly any) correlation. His conclusion is that the iceberg theory does not hold. He suggests that injuries which affect one or a few people are related to behaviour, but the injuries which result in large numbers of fatalities are due to failures in the "primary processes". Put another way, if you want to prevent fatalities, you need to target your safety programme on the things which cause fatalities. Likewise for major accidents. Kletz (1993) previously suggested that injury statistics have been over valued by industry, and that we should measure the things we want to control.

THINGS WE HAVE LEARNED FROM THE SITE IMPLEMENTATION

THE NEED FOR A CHAMPION

Somebody needs to be given the job of overseeing the programme. This is not complicated stuff, and many of the elements of it use existing systems. But it does need to be tended.

SENIOR MANAGEMENT INVOLVEMENT

The Site Manager has found it easy to monitor the programme through the site's QPulse HSE MS software, and he has used this to raised the profile of process safety. Although the HSEQ Department is the owner of the PSPI tool, the Site Manager has picked it up and is using it.

THE BOTTOM LINE. DOES IT WORK?

Since we defined the lagging indicators in 2003, we have recorded 80 process safety incidents. In the first year we noticed that we had had 10 fires. We had previously been aware that we did have fires, but had not realised just how many. These were all minor incidents, as were the other 70, but nevertheless it was mind focusing. Most of the fires were smouldering insulation, so we took some simple steps to reduce the leaks, and to remove the insulation from around the leak points. In the second year we had almost eliminated the insulation fires.

We have used the lagging indicator data to focus on other areas.

LEADING INDICATORS ARE MORE DIFFICULT

There is no question that this is the case. And some of the indicators we chose have not worked for us. The permit to work monitoring did not yield useful information, because the audit tool we were using was not sufficiently discerning. The process safety training indicator failed because we had not got buy in from the people required to collect the data. But we are now much better at communicating process safety information to shift technicians, are better at closing out inspection findings, and have a better control of our controlled inspection programme. While the lagging indicator programme is running exactly as it was set up in 2003, we have had to make more running adjustments to the leading indicators.

GOING GLOBAL

In 2006 it was decided to make the Carrington PSPI framework the basis for a corporate programme. We understood that for the programme to be a success, we needed to convince senior managers to pick up the tool and start using it. This would involve a change process. Prochaska's change model identifies 6 stages: Pre-contemplation, Contemplation, Preparation, Action, Maintenance and Relapse. We assessed that the company was at the Preparation stage, so the emphasis of our change process was initially on problem solving and establishing support systems. In case some people were still at the Contemplation stage, we promoted the expected positive outcomes. To help with the marketing we branded the programme with a satellite navigation logo, and the motto "PSPIs. Navigating to HSE success". Figure 2.

We made some adjustments to existing corporate guidelines, and wrote a new guideline on PSPIs. For the lagging indicators we used the Appendix 1 definitions. For the leading indicators, we had decided that there was no single set which would be appropriate across the company. Adopting a single set would have run the risk that sites be forced to collect data which was not useful, diverting efforts from indicators which might be more useful in their location. Also we felt it important that sites buy into their own set of indicators. The PSPI guideline therefore concentrated on giving the sites a simple methodology to follow to pick their own.

The corporate PSPI programme was officially implemented in April 2007. Each site was given two objectives:

- To collect and report to Corporate HSE the lagging indicators every month;
- By the end of the year to have selected their own set of five leading indicators.

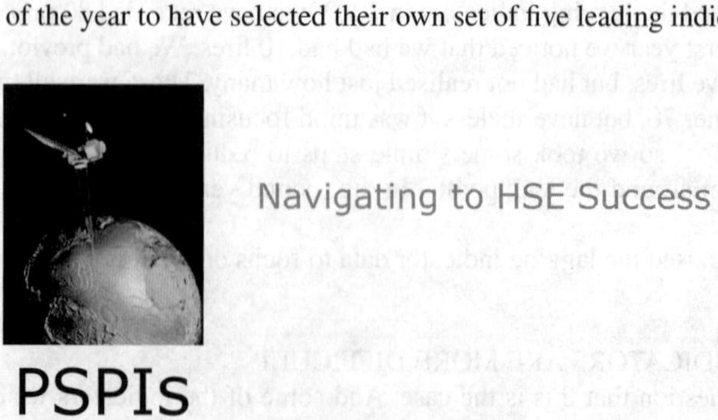

Figure 2. PSPI logo

SENIOR MANAGEMENT INTEREST AT THE CORPORATE LEVEL

The company Chief Financial Officer visited the site before a decision had been taken to roll it out globally. We took the opportunity to sell the concept to him. Although we suspect he is not an expert in process safety, he understands risk very well. He publishes a monthly letter on the company intranet, and we were very pleased to see the following words of support in his March 2007 letter, just before the launch in April:

> *"Our good start to 2007 continued in February, both in terms of safety and financial results. ... But as good as we are in personal safety – which is a shared value throughout Basell – we know that we must also maintain world-class process safety performance. ... Basell's HSE team will soon introduce Process Safety Performance Indicators (PSPIs) as a way to further improve our ability to control the risks associated with the processes and hazards in our plants. PSPIs will help us measure how well we have been controlling process risks and, more importantly, they will provide critical information we can analyze to determine what we have to do to improve."*

He had moved easily from Prochaska's Preparation stage to the Action stage.

AVOIDING MANAGEMENT BY "DRIVES", OR INITIATIVE OVERLOAD

In the absence of the PSPI tool, managers had little else available to them but to react to the latest incident. Managers then get only an incomplete picture, and steer the organisation in a direction which might not be helpful – "We need more process safety training", or "Our sites are not managing changes adequately". This management by "crisis" or by "drives" can be counter productive (Drucker 1968). Things always collapse back into the *status quo ante* three weeks after the drive is over. It puts emphasis on one aspect of safety, to the detriment of everything else. In the words of Peter Drucker:

> *"People either neglect their job to get on with the current drive, or silently organise for collective sabotage of the drive to get their work done. In either event they become deaf to the cry of "wolf". And when the real crisis comes, when all hands should drop everything and pitch in, they treat it as just another case of management-created hysteria.*
>
> *Management by drive, like management by "bellows and meat axe", is a sure sign of confusion."*

STATUS OF THE GLOBAL IMPLEMENTATION

LAGGING INDICATORS

It is early days, but from implementation in April until September 2007, there had been 357 process safety incidents recorded. Of these 55 were fires (all minor), of which 27 were associated with the same type of equipment. Each one of these has been investigated locally, but we are confident that we will be able to extract useful big picture learnings from this information, share it around the company, and reduce the number of fires.

LEADING INDICATORS

Many sites by now have chosen their leading indicators. Indicators related to integrity inspection seem to be a popular choice.

LACK OF AN INCIDENT DATABASE

We do not currently have a company wide corporate incident recording system or database. This means we have to work a little harder at the data collection and monthly reporting at the corporate level, to make sure that senior managers have the right information; they need to be able to understand what is going on on their patch, the sites which they are responsible for.

The company does have a knowledge exchange system where discussion topics can be posted and questions raised. This has been effective in sharing information about the PSPI system, and also sharing experiences and information about process safety incidents.

THE HSE GUIDANCE AND THE BAKER PANEL REPORT

In 2006, the HSE published guidance on PSPIs (HSE 2006). Unfortunately we did not have the benefit of it when we were developing our programme. It is an excellent document, and we would like to offer some comments.

As mentioned above, in some cases, having chosen the weak risk control system, we found it difficult to define a metric. We like the suggestion in the guidance to ask "What does success look like".

On how to choose which risk control systems to measure, we think the guidance could give more help – management team in smoke filled room or workforce involvement?

A significant difference between our two approaches is the HSE's recommendation to chose linked pairs of leading and lagging indicators. This is with the intention of giving "dual assurance". This is a good idea, but we wonder if it will make the process of finding suitable leading indicators more difficult, possibly too difficult for some. A set of disconnected but nevertheless individually relevant lagging and leading indicators might be good enough. In our experience, a major benefit of the PSPI programme has been that it has moved process safety in general onto the radar of senior managers; maybe it does not matter too much if the indicators are not exactly correct.

Hopkins (2007) examined the meaning of "leading" and "lagging" indicators in the HSE guidance and Baker (2007), and found that the terms were not used consistently. He identified three types of process safety indicator:

A. Measures of routine safety related activity, e.g. proportion of safety critical instruments tested on schedule.
B. Measures of failures discovered during routine safety activity, e.g. proportion of safety critical instruments which fail during testing.
C. Measures of failures revealed by an unexpected incident, e.g. numbers of safety critical instruments which fail in use.

Indicators of type A are consistently referred to as "leading", and type C as "lagging", but categorisation of type B is inconsistent within both documents. However, he concluded that this was of little consequence; it is more important to define indicators which are not only relevant but which also generate a statistically useful number of "hits".

WHERE NEXT?

DO WE WANT MORE OR FEWER INCIDENTS?!

We think that for the lagging indicators, at least initially, we want to encourage the reporting. Of course we do not want to have more incidents, we only want to have a better view on what incidents we are already having! For that to happen, senior managers must handle the tool with care; rather than a gun to the head of the site managers, it must be a vehicle for an informed discussion about process safety.

BENCHMARKING AND SETTING TARGETS

An interesting question which is already on the minds of senior managers is whether we should set corporate targets for the lagging PSPIs analogous to injury rates. For the time being, this would seem unnecessary as we are getting the benefit just from the profile PSPIs have brought to process safety, and from the big picture learnings. But at some point we will inevitably need to define targets. An interesting question is what the denominator should be for the PSPI rate – per plant, per ton of throughput, per ton of hazardous material inventory, per man hour?

For leading indicators we do not see a possibility to benchmark, since every site has a different set. The only objective is to have the system in place and working. Nevertheless, this objective and the expectation that sites demonstrate how they are using leading indicators to drive improvements in process safety performance has a powerful symbolic value, like the Sword of Damocles.

Hopkins (2007) advocates relating senior manager remuneration to PSPIs, but comments that this works because it is symbolic, affecting reputation and pride, rather than exercising any real financial leverage. Hopkins also cautions against perverse outcomes, which are a consequence of the measure itself being managed rather than safety. This author believes this risk can be controlled by ensuring that the PSPIs have the attributes listed in Table 2.

LAGGING INDICATOR DEFINITIONS

We do not have a category for dust hazards, and these are relevant for our business. We are thinking about what would be a suitable definition.

THE CHANGE PROCESS

The organisation is moving through the process of changing the safety focus towards process performance. We are moving through the Action stage of Prochaska's change model;

strategies should therefore emphasise how this new tool has put managers in a better position to control process safety, what the benefits are for them, and that the normal short term fluctuations in personal safety performance will not be blamed on efforts being diverted away. As we move through the maintenance stage we will need to ensure that adequate support is given, and that internal rewards are reinforced. We should also be prepared for relapses, for example a drop off in lagging indicator reporting, and what we will do about it.

CONCLUSIONS

There is no "silver bullet" for HSE performance. PSPIs are just one of the required elements of a properly functioning management system. And we know that there are many areas where we have to improve.

Before our PSPI programme was implemented, senior managers were already interested in process safety. But it had not been as easy for them to engage themselves as it had been in personal safety, since they had no process safety Sword of Damocles. Now they have a tool. Although with hindsight we should not have been surprised, we have discovered that the PSPI programme is driving improvements not only because of the useful information it produces but also, perhaps more importantly, because it is helping senior managers to demonstrate leadership.

Consequently we have concluded that while choosing the right indicators is important, it is more important to design a programme which is easy for senior managers to use, produces reliable data and which helps senior managers exercise their leadership accountabilities with regard to process safety.

REFERENCES

Baker J. (2007) *The Report of the BP US Refineries Independent Safety Review Panel.* Download from www.bp.com

British Standards Institution (1999) *BSI-OHSAS 18001Occupational health and safety management systems – Specification*. BSI, London.

Drucker P.F. (1968) *The Practice of Management*, Pan Books, London.

Groeneweg J. (2006), *The future of behaviour management*. Health and Safety Culture Management for the Oil and Gas Industry. Conference, Amsterdam, The Netherlands.

Health and Safety Executive (1995) *The Reporting of Injuries, Diseases and Dangerous Occurrences Regulations*. HSE Books, Sudbury.

Health and Safety Executive (2003) *BP Grangemouth – Major Incident Investigation Report*. Download from www.hse.gov.uk

Health and Safety Executive/Scottish Chemical Industries Association (HSE/SCIA) (2003). *Pilot project on voluntary reporting of major hazard performance indicators: Process safety and compliance measures – Guidance notes.*

Health & Safety Executive (2006) *Developing process safety performance indicators, A step by step guide for chemical and major hazard industries* HSG254, HSE Books, Sudbury.

Hopkins A. (2000) *Lessons from Longford, The Esso Gas Plant Explosion*. CCH Australia Ltd., Sydney.

Hopkins A. (2007) *Thinking about process safety indicators* Human & organisational factors in the oil, gas & chemical industries. Conference, Manchester.

Institute of Petroleum, (2005) *A framework for the use of key performance indicators of major hazards in petroleum refining*. Energy Institute, London.

Kletz T. (1993) *Accident data – the need for a new look at the sort of data that are collected and analysed*. Safety Science **16,** 407-415.

US Chemical Safety and Hazard Investigation Board (2007) *Investigation Report Refinery Explosion and Fire, BP Texas City*. Download from http://www.chemsafety.gov/

Appendix 1. Company Wide Lagging Process Safety Performance Indicators

Electrical Short Circuit

Electrical short circuit or overload which was either attended by fire or explosion or which resulted in the stoppage of a piece of process equipment.

Explosion or fire

An explosion or fire occurring in any plant or place, where such explosion or fire was due to the ignition of process materials (including lubricating or heat transfer oils), their by-products (including waste) or finished products.

Release of flammable material inside a building

The sudden, uncontrolled release inside a building of 3 kg or more of a flammable liquid, 1 kg or more a flammable liquid at or above its normal boiling point, or 1 kg or more of a flammable gas.

Release of flammable material in the open air

The sudden, uncontrolled release in the open air of 50 kg or more of any of a flammable liquid, a flammable liquid at or above its normal boiling point, or a flammable gas.

Release of a substance which could have an impact on the environment

The sudden, uncontrolled release of 50 kg or more of a substance which could have an impact on the environment.

Safety Related Protective System called into operation

Emergency depressurisation/flaring, operation of pressure relief devices, breaking of bursting discs, operation of final element instrumented protective functions, quenching of exothermic reactions, crash cooling etc. which are instigated either automatically or by operator intervention. This category is restricted to systems beyond warning devices. Excluded are protective systems designed solely to ensure the integrity of the process or product quality where there is no safety or environmental protective duty. Excluded are activations of these systems simply due to instrument malfunction, unless the malfunction caused process conditions to deviate away from safe operation.

Safety Related Unplanned Shutdowns

Occasions where a process or activity has to be shut down or halted because it would be otherwise unsafe to continue.

This covers emergency stoppage of an individual plant, either automatically or by operator intervention, to prevent a loss of containment or other dangerous situation irrespective of the time involved, or stoppage or suspension of a planned plant or process start up due to a failure or malfunction of part of the plant or equipment, if unchecked, it had the potential to give rise to an emergency.

Appendix 2. Carrington Site Leading Process Safety Performance Indicators

HSEMS element	Issue	Definition of metric	System of measurement	System of monitoring	Target
Inspection	Inspection findings are not being closed out	% of inspection recommendations progressed.	Inspection findings are recorded in an Access database. Report generated quarterly by Maintenance.	Quarterly Inspection Meeting, chaired by Site Mgr.	
Inspection	Inspections are overdue	% of controlled inspections which are due and complete at the end of the month. Reported by department and site overall.	Site Inspector generates a report from SAP at the end of the month.	Quarterly Inspection Meeting.	
Communication, Safe operation	Technicians are not receiving all relevant process safety information	% of Key Communications which are signed off by all of the intended recipients in 2 months.	Dept mgrs choose Key Communications, and report in standard format.	Site HSE Council, chaired by Site Mgr.	
Permit to work	Lack of ongoing information on how well it is working	% of PTWs inspected during weekly plant safety walks which are found to be all satisfactory according to the check list in Appendix 1 of CHSES.2	Operations managers to fill in table of results.	Monthly departmental HSE meeting, and monthly Site HSE Council.	
Training	People are overdue with important refresher or process safety related training	% of people who have attended defined safety training by due date.	Training Dept extract data from Training Records system, and report quarterly.	Data monitored at quarterly Site Training Steering Committee Meeting.	

SELECTING RISK CONTROL MEASURES – WHY ORGANISATIONS OFTEN DEMONSTRATE POOR RISK MANAGEMENT

Christopher J. Beale (FIChemE)
Ciba Expert Services, Charter Way, Macclesfield, Cheshire, SK10 2NX, UK

The selection of risk reduction measures is a fundamental aspect of risk management. Many different approaches are used for selecting measures including legal compliance, insurance/engineering code compliance, qualitative and quantitative risk analysis. Project experience in different countries involving chemicals with major accident hazard potential has been used to identify the range of methodologies which are used for specifying risk reduction measures and making judgements about the acceptability of process safety risks.

The different techniques are described, supported by examples. A range of large and small accidents, including Buncefield and Texas City, are then reviewed to identify why deficiencies in risk control measures occurred. The paper concludes by assessing the methods which are used for costing risk reduction measures and how these are combined with risk reduction engineering to specify risk reduction measures.

KEYWORDS: ALARP, COMAH, risk assessment, cost benefit analysis.

WHY SELECT MEASURES?

High levels of process safety are achieved by effectively managing people, processes and plant. Effective safety management systems address all of these elements of process safety. Plant issues pose particular challenges. History provides examples from a wide range of industries showing how plants have not been designed correctly. Mistakes have been made through ignorance, design error, poor decision making, short-sighted cost control and a poor analysis of uncertainty (Beale, 2006).

Modern approaches to process plant design seek to remove risks using inherent safety principles. When this cannot be done, a plant risk is created. Risk control measures then need to be used to reduce the risk to a residual level which is considered to be acceptable. Plant risk control measures are often expensive to implement and involve difficult decisions and judgements. Operating companies make these decisions using both their own processes and the relevant legal compliance frameworks.

A balance has to be struck between the need for speedy decision making, with the implied risk of poor analysis and incorrect conclusions, and over analysis, with the implied risk of avoidance of making decisions. Words spoken by the Archbishop of Canterbury in a different context summarise the views of many people regarding risk reduction engineering and option analysis – 'deep contemplation in the shadow of the question mark'.

Table 1. Drivers for selecting risk reduction measures

Country	Strategic risk reduction decisions	Detailed risk reduction decisions
Australia	Detailed QRA to predict risk levels at the site boundary. Measures selected to meet acceptable risk targets. Subject to a Safety Case regime.	Hazop completed by operating company and design contractors. Hazop reviewed by authorities.
Austria	Determined by authorities.	Risk analysis completed by operating company and reviewed by authorities. Strong emphasis on code compliance for key design areas.
China	Determined by authorities in compliance with national legal requirements.	Specified by local design institute. Designs subject to independent risk analysis at the discretion of the operating company.
Finland	Determined by operating company (NB – plant in a completely unpopulated area).	Compliance with detailed standards. Supported by risk analysis by operating company.
France	Determined by authorities to minimise offsite hazard ranges.	Compliance with detailed standards. Supported by risk analysis by operating company.
UK	Specified by operating company using a goal setting approach. Challenged by authorities under a Safety Case regime.	Hazop and risk analysis. Decisions made by project teams with peer review.
USA	Compliance with detailed standards.	Compliance with detailed standards. Supported by risk analysis by operating company.

Recent project experience involving major hazard chemicals such as butadiene, acrylonitrile, methanol and styrene, has shown that very different approaches are used for selecting technical risk reduction measures around the world. Table 1 summarises the main drivers for selecting strategic measures – those involving plant layout, inventories and significant investment – and the main drivers for selecting detailed risk reduction measures based on this experience.

TECHNIQUES FOR SELECTING RISK REDUCTION MEASURES

A wide range of different risk assessment and decision making processes are used in the process industries for identifying and specifying risk reduction measures. Most of the larger companies use a mixture of different techniques to suit the specific technical, cost and legal aspects of the project which is being undertaken. Commonly used techniques include:

FUNDAMENTAL ANALYSIS

Detailed laboratory safety tests are performed on the chemicals which are being handled to identify their hazardous properties. The tests are performed on actual samples to represent process conditions and worst case samples to represent possible deviations. This allows the chemical hazards to be identified and safe operating limits to be established for chemical storage and process activities. Opportunities for inherent safety are then assessed to remove hazards where possible. The process design team then design control and protection systems to ensure that the process cannot exceed it's measured safe operating limits. This approach is normally used for assessing chemical reaction runaway risks, fire/explosion risks inside process vessels and dust explosion risks. Well designed plants will be supported by accurate laboratory safety tests, the correct interpretation and application of the test results and the experience of the design team. This normally leaves limited room for meaningful option analysis for control and protection measures aimed at preventing major loss of containment or energetic releases. Risk reduction efforts are then focused on sitewide mitigation and emergency response measures, including fixed and mobile fire fighting systems. This approach is popular in many north European countries.

RISK MATRIX

Hazards are assessed qualitatively or semi-qualitatively to estimate their frequency of occurrence and hazard potential (Middleton & Franks, 2001). Each hazard is then positioned on a frequency/consequence matrix. The matrix typically includes risk tolerability criteria. Figure 1 shows the risk matrix which is used by the Ciba UK manufacturing sites. Importantly, this matrix includes three regions:

- **Intolerable risks**, where immediate risk reduction measures must be implemented.
- **ALARP region risks**, where option analysis is required above and beyond the requirements of relevant standards and good practice.
- **Broadly acceptable risks**, which are acceptable with compliance to relevant standards and good practice.

It is, however, important to realise that further analysis is required for any risks which are identified in the ALARP region (HSE, 1999). If the risk matrix is created as part of a hazop study (IChemE, 2000), ALARP region risks are often discussed by the team before making a final decision about which risk reduction measures should be employed. This uses the experience of team members but can lead to poor decisions as there is no objective assessment by an outside person, insufficient time may be devoted to risk reduction assessment and the team's decisions may be overruled at a later date in the project. To overcome these problems, some companies complete additional risk reduction assessments as a separate exercise during project safety reviews or when Safety Reports are created (COMAH, 1999).

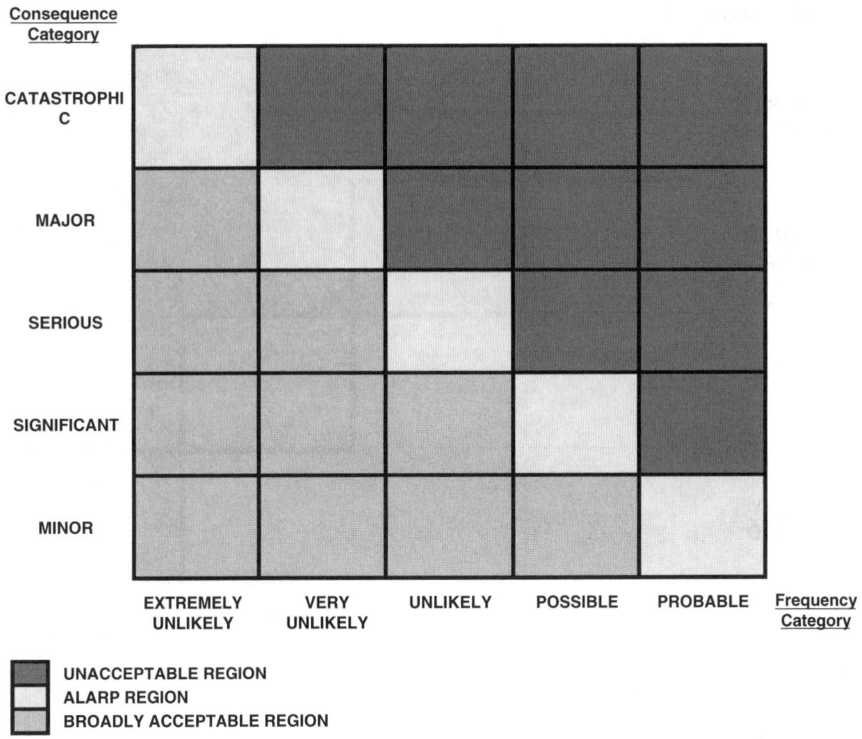

Figure 1. Ciba UK risk tolerability matrix

The risk matrix approach is very effective when there are a wide range of different hazards associated with a project and it has the benefit of being easily understood by wide groups of people, encouraging an inclusive approach to risk assessment. Technical staff can focus on the nature of each risk rather than on understanding an obscure and complex methodology and results can be communicated to senior management in an easily understandable format.

Practical problems occur in multinational organisations. They use management systems and decision making criteria which can be applied on a worldwide basis. Concepts which only apply to specific countries, such as ALARP, cannot easily be integrated into global EHS management systems. Indeed, risk is viewed in very different but equally valid ways around the world. As an example, corporate decisions about risk tolerability within Ciba are made using a risk matrix which measures consequence against the quality of risk control (see Figure 2). This allows risks to be ranked relative to one another as well as

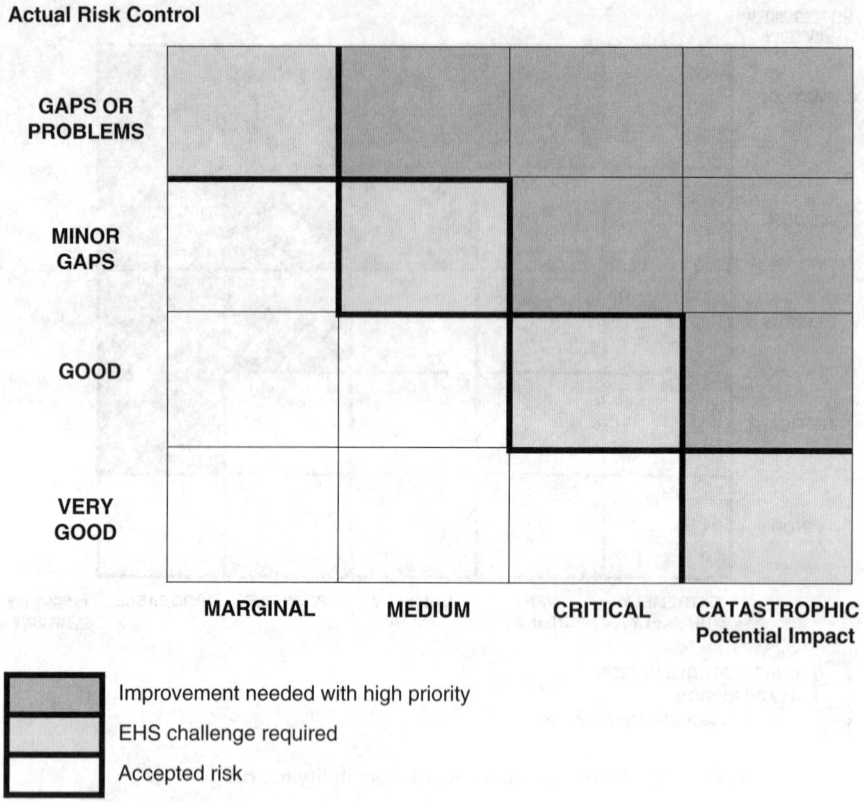

Figure 2. Ciba corporate risk tolerability matrix

identifying risks which are unacceptable and where immediate improvements are required. The senior management team can review risks globally, by business segment, by country or by site. Individual business segments and sites are then expected to drive risk reduction as part of a continuous improvement program.

PEER REVIEW

Most large organisations use peer review processes for challenging key business decisions. EHS risks and capital investment projects are key business decisions for responsible organisations, so it is easy for them to incorporate EHS 'challenging' into their general business decision making processes. 'Challenging' normally involves presenting an issue or a project to an independent review team. They then use their experience to check that the correct decisions have been made, including areas where EHS risks are not acceptably controlled and areas where excessive EHS expenditures are proposed.

Peer review will often work at different levels within an organisation. It may involve site staff for smaller projects and corporate staff for complex or large projects. Suppliers are often used to review design proposals as part of responsible care commitments. This can be very useful when the operating company has a limited knowledge about handling a hazardous chemical.

CHECKLISTS

Rather than relying purely on expert judgement, team decisions and risk assessment, it is possible to formalise a checklist of typical risk reduction measures. Priority risks can then be identified, for example, using a risk matrix (see above) and formally assessed using the checklist. This method provides a systematic framework for selecting risk reduction measures but has some important disadvantages. Firstly, it may stifle creativity and innovation because of the need to assess a long list of options formally for each identified risk. Secondly, it may be genuinely difficult to provide a concise list of risk reduction measures which are suitable for a diverse and complex industry such as specialty chemicals manufacturing. Thirdly, the selection of risk reduction measures is directly linked to the cost of each measure as well as the risk reduction benefits which would flow from adopting each measure. Cost estimation is difficult and time consuming and may divert specialist technical engineering design staff away from improving designs and towards costing designs which will not be built. Fourthly, a checklist approach could be used relatively easily for a new plant design. It is less useful for assessing existing plants and can lead to a blinkered view of actual risk and skew decisions and resources towards technical hardware improvements rather than operational control improvements such as human factors, plant maintenance and management of change. Table 2 summarises a checklist that was developed for a Ciba UK site. It should be noted that any cost estimates are unlikely to be applicable to other sites because of the problems associated with developing accurate cost estimates for risk reduction measures (see below).

COMPLIANCE WITH STANDARDS

Many problems in the process industry are similar and can be assessed using a common industry or corporate standard. The standard can be used as a solution to a generic problem and represents a risk assessment based on the experience and analysis which was used for creating the standard. This approach is in common use in many countries, such as the USA, China and northern Europe and is often embedded in the country's legal regime for safety.

Compliance with standards is not a foolproof guarantee of safety. Standards reflect recorded and accepted past experience but do not address unknown problems and uncertainty. They encourage a compliance driven rather than innovative and creative mindset and are often difficult to apply to novel situations. Grey areas, requiring interpretation,

Table 2. Example checklist for risk reduction options

Description of risk reduction measure	Installed cost (£)	N	Annual cost (£)	Running cost (£)	TAC
Pneumatic fire detection system	£2,000	15	£133	£100	£233
LHD cable fire detection system	£35,000	15	£2,333	£1,000	£3,333
Flammable gas detection system	£20,000	15	£1,333	£3,200	£4,533
Toxic gas detection system	£45,000	15	£3,000	£3,200	£6,200
Warehouse fire protection system	£200,000	35	£5,714	£500	£6,214
Tank farm fire protection system	£85,000	15	£5,667	£1,000	£6,667
Road tanker fire protection system	£15,000	15	£1,000	£1,000	£2,000
Plant fire protection system	£140,000	15	£9,333	£2,000	£11,333
Small manual fire protection system	£11,000	15	£733	£400	£1,133
Large manual fire protection system	£60,000	15	£4,000	£400	£4,400
Automated high level monitor	£106,000	15	£7,067	£1,000	£8,067
Manual high level monitor	£22,000	15	£1,467	£300	£1,767
Firewater supply pump	£23,700	15	£1,580	£2,400	£3,980
Firewater supply and run-off civils	£1.5M	35	£42,657	£20,000	£62,657
Large plant firewall	£83,000	35	£2,371	–	£2,371
Small area of Durasteel fire cladding	£1,200	35	£34	–	£34
Vessel passive fire protection per vessel	£7,000	15	£467	£200	£667
Steelwork passive fire protection per joist	£1,000	15	£67	–	£67
Small plant fire drains	£100,000	35	£2,857	£300	£3,157
Large plant fire drains	£170,000	35	£4,857	£500	£5,357
Fire system valvehouse	£20,000	35	£571	–	£571
Fire engine (second hand)	£20,000	6	£3,333	£3,000	£6,333
Acrylate inhibitor addition system	£20,000	15	£1,333	£300	£1,633
Dust explosion suppression system	£50,000	15	£3,333	£800	£4,133
One CCTV camera c/w monitor	£6,000	6	£1,000	£800	£1,800
Office block window filming	£22,000	35	£627	–	£627
DCS control system interlock	£1,000	10	£100	£100	£200
Hard wired control system interlock	£2,000	10	£200	£200	£400
Simple software driven SIL system	£5,000	10	£500	£500	£1,000
Complex software driven SIL system	£100,000	10	£10,000	£10,000	£20,000
Blowdown tank, piping and controls	£100,000	15	£6,667	£500	£7,167

NOTE

1. N is the estimated useful life for calculating annual depreciation costs.
2. Annual cost is calculated as installed cost/N.
3. Running cost is the annual inspection, maintenance, running cost.
4. TAC = Total annual cost.
5. 'Fire protection system' means an automated insurance compliant system.
6. DCS interlock costs assume that the plant already has a DCS control system.
7. Costings based on 2003 prices and need to be adjusted for inflation.

exist in most standards and flexibility normally exists to interpret phrases such as 'should' rather than 'must'.

Standards are used by most multinational companies to promote EHS consistency. This can produce standards which are extremely loose to cater for differences between individual countries and regions, destroying the original intent and control within the standard. This problem can be mitigated by having more general corporate standards with a requirement that there must also be compliance with locally applicable standards and laws.

If standards are overly tight, illogical EHS decisions often ensue. For example, a standard for fire protection of ambient flammable liquid storage areas is likely to require certain fire protection measures according to the material flashpoint. A liquid with a flash point of 40°C will pose a low fire risk in a cold north European country but will pose a significant fire risk in a hot country such as Mexico. Strict application of this requirement would result in overprotection of the European storage area compared to a risk assessment which was completed based on first principles.

QUANTITATIVE RISK ASSESSMENT (QRA)

These are the most complex and costly methods of risk assessment and are used for assessing complicated risks objectively. They are commonly used to assess human fatality risks and can also be used as a basis for plant design and layout decisions. The technique is most useful for assessing the relative risks associated with a number of possible options for a project. The following types of risk are most often calculated:

- **Societal risk**, how risk affects groups of people and can cause multi fatality events.
- **Individual risk**, how risk affects individual workers so that risks can be compared between different groups of workers in different industries and so that the risks to the most exposed workers can be assessed.
- **Location specific individual risk**, calculating individual risks at different locations assuming that a person is permanently present outdoors at each location. This can be used to optimise the location of buildings and other centers of population on site and control development beyond the site boundary.

This type of QRA is an accepted tool for risk assessment in major hazard industries where a relatively small number of chemicals are being handled but where accidents could have devastating consequences, such as chlorine storage, oil and gas, petrochemical processing and fuel storage sites. Risk is quantified and can be compared to quantitative cost estimates to form a numerical cost benefit analysis. This allows different risk reduction options to be compared easily against defined criteria or relatively.

QRA is much less useful for sites which handle a wide range of chemicals and have many different hazards. Some risks cannot easily be modeled for QRA purposes. These include reaction runaway and dust explosion risks. Constructing an accurate QRA model is time consuming and resource intensive and the QRA results will not reflect the totality of the site risks. For this reason, specialty chemical companies favour fundamental analysis and risk matrix approaches over QRA.

It should also be noted that there are considerable uncertainties associated with QRA modeling (Beale, 2006). For example, a QRA of the Buncefield fuel storage depot would have almost certainly failed to identify the potential for a large vapour cloud explosion following a cold release of petrol (Crawley, 2006). Any risk reduction measures would therefore have been allocated to known hazards such as pool and tank fires rather than devastating vapour cloud explosions.

MEASURES SPECIFIED BY CONTRACTORS

Most large engineering projects are completed by specialist engineering contract companies. They produce plant designs which are then approved by the operating company. The fundamental experience and creativity which is required to define risk reduction measures therefore exists within the contracting company. China provides an extreme example of how this system operates. The economy is centrally planned and the state and regional governments establish 'design institutes' (Zhang & Allen, 2007) which specialise in plant designs for different regions of China and for different industry sectors. The 'design institutes' serve as a centre of expertise and produce designs for individual sites. These designs are often not challenged critically, or there is limited scope for challenge within project timescales. Risk reduction measures are then very much specified by the 'design institute' rather than the operating company.

MEASURES REQUIRED BY THE AUTHORITIES

Most countries have a legal control regime for major hazard industries. Countries like the UK have a 'goal setting' legal regime where the operating company is required to identify hazards and demonstrate that the required safety management systems and risk control measures are in place. In other countries, such as France, the legal regime is much more closely based on the enforcement of standards supported by calculations of worst case hazard ranges. Additional risk reduction measures are then required by the authorities according to the location of these hazard ranges in relation to offsite areas. From an operating company perspective, this approach will often require hazardous chemicals to be stored in central site areas, thus minimising offsite hazard ranges.

This may result in plant layouts which increase the frequency of accidents as central site areas tend to be congested. It will often also mean that discussions about risk reduction measures are slanted towards actions required by the Regulatory Authorities to minimise offsite hazard ranges rather than requiring the operating company to thoroughly assess the risks that they are best placed to really understand.

EXAMPLES OF POOR SPECIFICATION

TEXAS CITY (USA), 23RD MARCH 2005

A large vapour cloud explosion occurred in the isomerisation unit at the BP Texas City refinery, causing 15 fatalities and more than 170 injuries. BP commissioned two reports

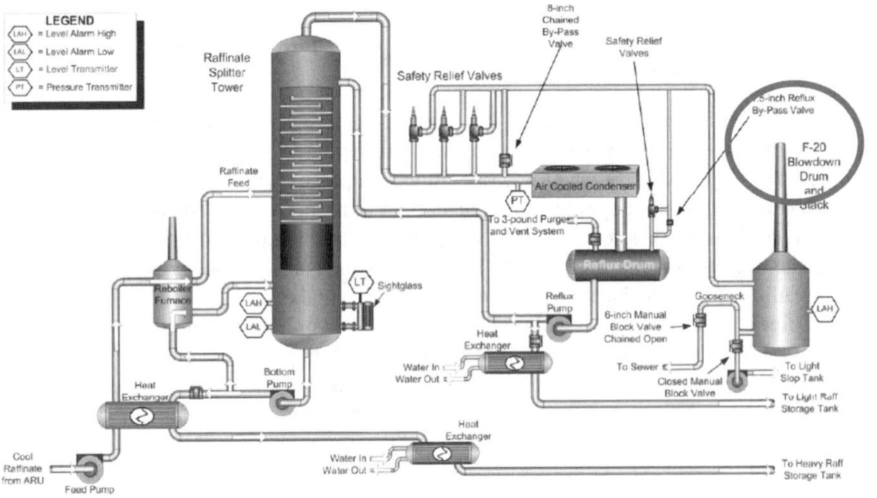

Figure 3. Design of raffinate splitter blowdown tank vent system

(BP, 2005a & BP, 2005b) and the Baker report was issued following an independent investigation (BPRISRP, 2007). A wide range of initiating causes and learning points were identified in these reports but one important issue related to the selection of technical risk reduction measures. The pressure relief system which protected the raffinate splitter tower vented into a blowdown drum which was connected to a tall stack. The stack vented to atmosphere rather than being connected to a flare system (see Figure 3). This meant that any large hydrocarbon vapour releases would be released into the refinery area, creating a large flammable cloud. If the stack had been routed to a properly designed flare system, these vapours would have been burnt off safely, removing the vapour cloud explosion risk.

A number of interesting points are identified in the Baker report:

1. The original blowdown tank stack was not connected to a flare when it was built in the 1950's, presumably in compliance with the relevant refinery design standards of the time.
2. OSHA, the Regulatory Authority cited the lack of connection of the blowdown drum stack to a flare in 1992 but subsequently withdrew the citation. This was presumably based on discussions with the operating company or a re-assessment of the safety risks.
3. The blowdown drum was replaced in 1997 by the then owners, Amoco. A like for like replacement was made even though this design feature did not comply with the latest corporate standards. The decision was presumably made on economic grounds and a view may have been taken that new design standards did not apply to old equipment.
4. BP acquired the Texas City refinery in 1998 but did not change the blowdown drum stack design. This type of issue would normally be identified in due diligence audits at

the time of the acquisition. BP would then have had to prioritise expenditure on the acquired company as part of a longer term program to bring the refinery up to the standards required by BP.

This highlights an issue faced by many sites. Plants may have been designed to comply with old standards and do not meet modern standards. In some cases, it is economic to upgrade the plant. In other cases, operating companies are faced with the option of either closing the plant or spending a large amount of money to upgrade the plant. Closing the plant could transfer fixed costs to other plants, prejudicing the long term future of the whole site. Funds may not be available for plant upgrades if adverse economic conditions are prevailing. These decisions can be very difficult to make.

After an accident of the scale of Texas City, BP is unlikely to build new plants with the design flaw of no vent connections to a flare. BP is lobbying for a change in US design codes to make other sites aware of this process safety issue. The world's newest refineries are being built in Asia and some still do not have the required vent connections to a flare system, even though they were constructed after the Texas City accident.

BUNCEFIELD (UK), 11TH DECEMBER 2005

One of the UK's most devastating peacetime accidents occurred at Buncefield when a petrol storage tank was overfilled. It is postulated that a large vapour/mist cloud then developed and ignited to produce a devastating explosion. A Major Incident Investigation Board was set up and a report was issued in 2007 (Buncefield MIIB, 2007). Initiating causes and learning points from the accident have been published. They cover a range of operational control, land use planning and emergency response issues as well as issues associated with not identifying that a vapour cloud explosion was a credible event at the site.

Logically, risk reduction measures would have been specified to address known and identified hazards. Protection measures against explosions would only have been specified if a precautionary approach had been taken. This might have involved measures such as flammable vapour detectors, high reliability inventory isolation systems and modifications to emergency plans and emergency resources.

The known and identified hazards were associated with fires and tank overfill should have been identified as a credible cause of fire. A robust option analysis would have identified the need for additional prevention and protection measures to prevent overfills. The IEC61508/61511 standard (IEC, 1998) addresses one key aspect of these systems – SIL (Safety Integrity Level) reliability requirements for critical interlocks. Most companies apply this standard by designing SIL levels to meet a required risk target using a LOPA (Layer of Protection Analysis), risk matrix, risk graph or fault tree analysis technique. Once an acceptable risk target has been met, option analysis is not normally carried out.

It is also interesting to note that it is suspected that the formation of the large flammable cloud was partly caused by liquid pouring over the tank and hitting deflector plates which had been installed to improve the performance of the tank fire sprinkler protection systems. The deflector plates may have increased the formulation of droplets, creating a flammable mist.

KAPRUN (AUSTRIA), 11TH NOVEMBER 2000

A fire occurred in a train on a steep funicular railway serving one of Europe's main ski areas. 170 people were killed (Beale, 2001). The transport system relied on unusual and specialist technology (funicular railways in mountain tunnels) with little or no provision for dealing with accidents. As such, there would have been very few standards which specifically addressed the design of funicular railway systems and any that did exist would have been based on a relatively limited amount of operational experience.

The railway system operated and was presumably considered to meet all required legal safety standards based on compliance with standards. If a more creative approach had been used based on risk assessment, some major design and safety issues would have been identified, namely:

- Identifying how a fire could start and spread in a train which was supposed to be fire resistant.
- The absence of fire fighting equipment (eg. fire extinguishers) inside the train or inside the tunnel, making it impossible to extinguish a fire.
- The absence of effective escape routes from the train and the tunnel as the train fitted tightly into a tunnel.
- Difficulties in access for emergency services. A long steep walk was required into the tunnel and there were no helicopter landing sites close to the tunnel for evacuating casualties.
- The reasons that the fire doors at the ends of the tunnel were open when they should have been closed to prevent fire and smoke spread.
- The apparent absence of an emergency plan and poor operator training for dealing with an emergency.

Additional risk reduction measures could have then been put in place to prevent and mitigate this accident.

SOLVENT RELEASE, CIBA SITE, 2003

Approximately 3te of cold hydrocarbon solvent was emitted from a bursting disc onto a works roof. The bursting disc was protecting a process vessel. The solvent temperature was well below it's flash point. This incident was minor but could have been far more serious.

This incident showed how important it is to identify risk reduction measures correctly at a detailed level of plant design as well as at a high sitewide/Safety Report level. A hazop study had been conducted for the vessel and the team failed to identify that it was possible to overpressurise the vessel if a solvent cleaning line to the vessel was left open. As such, risk reduction measures were not explored for this hazard and at least one important risk reduction measure was not specified for the plant. A simple and inexpensive measure, adding a high pressure feed isolation interlock, would have prevented the hazard.

The most likely reason for missing the hazard was fatigue at the end of a long hazop session. If each hazard had been option assessed in detail, this would have created even more fatigue. Experienced Ciba hazop leaders consider that the most effective way of specifying risk control measures at this level of detail is to rely on expert judgement and team experience rather than formal option analysis systems, but the creativity and alertness of the team must be maintained through what is often a very tedious process.

REACTOR VENT RELEASE, CIBA SITE, 2003

A vapour release of an odorous chemical occurred on a calm and cold winter's night. The release affected offsite areas some distance from the site boundary. The release occurred following a pressure relief event on a newly installed reactor.

The reactor design was based on a thorough process and engineering assessment of an old plant. A large capital investment was made to allow the old plant to be closed as it did not meet modern design standards. The new plant included an automated DCS control system, SIL rated interlocks, mechanical pressure relief systems and blowdown tanks, fire detection and fire protection systems. It was considered that all required risk control measures had been implemented. An oil and a water phase separated, creating a layer of accumulated monomer. The potential for layering had not been identified. This led to an uncontrolled polymerisation reaction. All of the safety systems worked as designed but a small vapour release was emitted from the blowdown tank vent. This was predicted to be a low frequency event.

Despite the fact that a wide range of risk reduction measures had been specified, all of which would have safely controlled a range of human, equipment and software failures, a relief event still occurred. After the incident, questions were raised about the need to install additional risk reduction measures. Emergency scrubbers, tall vent stacks or emergency incinerators could have been installed at very high operational and capex cost. Some of these measures would have caused additional safety risks. It was concluded that reasonable risk reduction measures had been installed and that the main cause of the accident was poor process control and inadequate process knowledge.

STYRENE STORAGE TANK, UK SITE, 2000

A fire risk assessment was being completed for an old tank farm which housed bulk styrene storage tanks. The tanks were housed in two adjacent containment bunds. Each bund was protected with an automated bund foam pouring fire protection system. The tanks in one bund had no sprinkler protection to provide cooling to the tank walls. The tanks in the other bund had sprinkler protection. This appeared to be completely illogical as fires in one bund would impact the other bund.

It was subsequently discovered that the tanks were installed at different times. When the first tanks were installed, the insurance company required a foam pouring system to be installed. Some years later, the tank farm was extended. The site had a different insurance company and they required sprinkler protection to the new tanks to comply with their requirements. The original tanks were not upgraded as they were existing tanks.

Design to standards had therefore created an inconsistent fire protection philosophy. Experienced fire fighters also recommended that the original design without sprinklers was safer. The sprinkler design allowed water to splash into the bund, breaking the protective foam layer and allowing fire burn back.

ECONOMIC CONSIDERATIONS

As well as identifying and assessing the risk reduction benefits of potential risk reduction measures, operating companies must also estimate the costs of each measure so a judgement can be made about the cost effectiveness of each option.

COMMONLY USED METHODS OF COST ESTIMATION

Companies use a range of techniques for this type of cost estimate:

- Expert judgement based on the knowledge and experience of design engineers and production managers. This technique is popular as it is practical and can be completed quickly. It does not, however, provide a formal demonstration that a cost benefit analysis has been completed.
- Analysis of historic costs for similar projects. This can be derived from real site project experience, experience at other worldwide sites and discussions with suppliers. The analysis can be completed quickly and some evidence can be provided that a structured approach has been used. It does, however have to be used carefully as it is easy to miss important practical issues which can have a major cost impact (see below).
- Broad brush engineering cost estimates. Most companies use capital investment decision making processes which involve progressive project screening and increases in accuracy. This allows unviable projects to be stopped quickly before large amounts of resource are devoted to the project. For example, in Ciba, this involves a Terms Of Reference (TOR) using a cost accuracy of +/– 50%, a Project Proposal (PP) using a cost accuracy of +/– 30% and a Final Project (FP) using a cost accuracy of +/– 10%. Once a project has been approved at these three stages, it passes to the implementation phase. Even the highest band of TOR cost estimate requires specialist engineering resources. These resources are limited and there is concern that engineers will be used to work on theoretical designs which will not be built rather than spending more time on real design work on projects which will be implemented.
- Detailed cost estimates. This provides an accurate cost estimate based on the practical project issues which affect costs. This is a time intensive activity.

HISTORIC COST DATA TRENDING

Cost estimates have to be used with great caution as experience has shown that there are often plant specific issues which have a major impact on cost. These issues include:

- Cost of plant downtime if modifications have to be made outside planned shutdowns.
- Cost of false trips for new safety systems on continuous plants.

- Instruments may be easy to install or the vessels may have inadequate mechanical stabbings. Major mechanical vessel modifications may be required.
- Equipment may be boxed in by other plant and equipment. Lifting and installation costs could be prohibitive.
- Newly installed equipment has to be compatible with existing equipment, which is sometimes old. It may sometimes be necessary to replace a whole control system rather than a single interlock.
- Costs are very dependent on the existence of existing infrastructure. A fire protection system will be much cheaper if a source of pressurised water is available locally, a valve house exists and feed pipe gantries are available. Firewater containment drains are much cheaper if they are close to existing drains. Some drains are very expensive to install because of the local site topography and the presence of other underground services locally.
- Economies of scale. Unit costs will reduce for larger projects as overheads can be spread across a wider range of activities.

The cost of many basic raw materials, including steel, has increased disproportionately faster than the general rate of inflation. This has had a major impact on the cost of some risk reduction measures in recent years. These cost increases must be factored into any historic cost analysis calculations.

CALCULATING LIFECYCLE COSTS

It is important to consider all relevant costs when carrying out a cost benefit analysis. The cost will typically involve a relatively large upfront cost which will be depreciated over the lifetime of the asset, impacting the company's balance sheet, profit and loss account and cash flow statement. There will also be ongoing costs such as staff training, equipment maintenance, inspection and testing and plant downtime. Downtime will be caused by planned maintenance and unscheduled equipment breakdown and maloperation. These are operational costs which affect the profit and loss account.

UNINTENDED CONSEQUENCES

It is easy to fall into the trap of 'silo' thinking when completing cost benefit analysis studies. The team may subconsciously focus on measures to reduce risk rather than inherent safety to remove risk. Deployed resources may be skewed towards major hazard safety improvements rather than occupational safety, environmental and asset protection investments.

Experience has shown that adding risk reduction measures to older plants actually causes a short term increase in plant risks in order to reach the longer term goal of lower risks (Beale, 2004). This is because stable systems which have been in use for a long period of time are changed. Risks are generated when plant changes are made. These often involve human factors. Systems and documentation are not always updated correctly, staff take time to understand the new systems and procedures can become confusing.

DECISION MAKING – LINKING COST AND RISK REDUCTION

Experience has shown that companies use five main techniques for selecting risk reduction measures:

1. ***Acceptance of externally driven requirements.*** If projects are being completed to tight timescales or if technical resources are in short supply, companies will sometimes accept decisions driven by other organisations due to the disruptive effect of challenging decisions, even though they are not considered to be correct by the operating company. Examples include: plant layout requirements driven by the Regulatory Authorities which take little account of practical onsite issues to minimise offsite hazard ranges: planting trees next to tank farm storage areas to provide visual environmental screening so as to secure planning permission from Local Planning Authorities; and installing fixed fire protection systems in low risk areas of the site to satisfy insurance requirements.
2. ***Acceptance of expert opinion.*** Some companies recognise that aspects of process safety can be complex and choose to rely on the advice of internal experts or specialist consultants when making certain decisions. This approach is often used in north European organisations when risk reduction measures are specified using a fundamental analysis technique. Examples include designing a reactor basis of safety when reactor runaway risks exist, designing powder handling systems which are subject to powder fires and dust explosions and specifying fire protection system requirements. The disadvantage of this technique is that it relies heavily on the analysis and experience of a small number of specialists. This approach is often also used when packaged equipment, such as power and nitrogen generation plant, is purchased from a specialist supplier.
3. ***Peer review/challenging.*** Most companies use peer review to challenge important decisions. A project team is assembled to produce a recommended design solution. This is based on their collective analysis and judgement. They then have to present their plans to an independent review team. The review team examines commercial, practical, financial and EHS issues and use their broad experience to check and improve the project design.
4. ***Simple cost benefit analysis (CBA).*** Operators of UK 'Top Tier' COMAH sites (COMAH, 1999) required to use a systematic approach for selecting risk reduction measures. If detailed CBA is not practical, it is still possible to list the additional measures which could be employed, discuss each measure and consider a broad cost estimate for implementation. A summary of why each measure was accepted or rejected is then provided to produce an audit trail. This provides an audit trail for decision making but is still subject to a large degree of expert opinion.
5. ***Detailed quantitative cost benefit analysis.*** When accepted techniques exist in an industry for numerically calculating risk levels, it is possible to construct QRA models. Sensitivity analysis can then be completed to calculate the predicted impact on risk levels if different risk reduction measures were implemented. Detailed costs can then be assigned to each measure to rank the cost effectiveness of each of the

identified risk reduction measures. An ICAF (Implied Cost of Averting a Fatality) is often used.

$$ICAF = C / (R_0 - R_i)$$

C = cost of risk reduction measure (£).
R_0 = baseline predicted average number of fatalities per year.
R_i = reduced predicted average number of fatalities per year after upgrade i.

This allows decisions to be made using relative risk ranking or absolute criteria. Examples of relative risk ranking criteria would be 'implement the ten measures with the lowest ICAF'. Examples of absolute criteria would be 'implement all measures with an ICAF < £400,000. Detailed CBAs are used in industries such as oil and gas and rail transport. They are rarely used in the specialty chemicals industry as process safety risks cannot easily or accurately be modelled using QRA.

CONCLUSIONS

A wide range of techniques can be used for specifying risk reduction measures. Some rely heavily on detailed standards; some are driven by regulatory requirements; some rely on expert judgement; others rely on detailed quantitative calculations. None of these methods are guaranteed to specify the correct range of measures. Different techniques are more effective in some situations than in others.

A balance has to be made between bureaucratic analysis and the need for operating companies to make timely investments in measures that will actually be effective. Operating companies also have to balance their choice of risk control measures across the different types of hardware and software risk reduction measures which are practical. A focus on technical measures is sensible for a new plant. For older plants, better risk reduction improvements may be obtained from people and system related measures, improving operational control. The search for measures must be balanced and should not focus entirely on technical hardware requirements.

Large organisations, such as multinational companies, have to manage a wide range of risks. They will need to use different techniques for specifying risk control measures according to the risks that they face.

REFERENCES

(Beale, 2001) 'Recent railway industry accidents: learning points for the process industry', IChemE Hazards XVI Symposium Series No. 148, 6-8 November 2001.

(Beale, 2004) 'Developing a major hazards learning culture - interpreting information from the Ciba Specialty Chemicals Bradford near miss reporting system up to 2003', IChemE Hazards XVIII Symposium Series No. 150, 23-25 November 2004.

(Beale, 2006)	'Uncertainty in the risk assessment process – the challenge of making reasonable business decisions within the framework of the precautionary principle', IChemE Hazards XIX Symposium Series No. 151 Burgoyne Memorial Lecture, 28-30 March 2006.
(BP, 2005a)	'Process and operational audit report, BP Texas City', BP Plc, James W. Stanley, 15th June 2005.
(BP, 2005b)	'Fatal accident investigation report, isomerisation unit explosion final report,' BP Plc, John Mogford, 9th December 2005.
(BPRISRP, 2007)	'The report of the BP U.S. refineries independent review panel', chaired by James A. Baker III, January 2007.
(Buncefield MIIB, 2007)	'Initial report', Buncefield Major Incident Investigation Board, 13th July 2007.
(COMAH, 1999)	The Control of Major Accident Hazards Regulations, 1999.
(Crawley, 2006)	'Buncefield was not unique', Frank Crawley, The Chemical Engineer, April 2006.
(HSE, 1999)	'Preparing Safety Reports: Control of Major Accident Hazards Regulations 1999', HSG190, HSE Books, 1999. ISBN 0 7176 1687 8.
(IChemE, 2000)	'Hazop guide to best practice', IChemE, 2000.
(IEC, 1998)	Functional Safety Of Electrical/Electronic/Programmable Electronic Safety Related Systems, Parts 1 to 7, 1998 (also published as BS EN 61508).
(Middleton & Franks, 2001)	'Using risk matrices', Mark Middleton and Andrew Franks, The Chemical Engineer, September 2001.
(Zhang & Allen, 2007)	'Investing in new facilities in China: critical success factors', Jimmy Zhang and Andy Allen, The Chemical Engineer, March 2007.

SOME OBSERVATIONS ABOUT MAJOR CHEMICAL ACCIDENTS FROM RECENT CSB INVESTIGATIONS

Hon. Gary Visscher
U.S. Chemical Safety Board

The U.S. Chemical Safety Board (CSB) conducts independent investigations of major chemical accidents that occur in the United States. Since 1998 the CSB has investigated nearly 50 chemical releases, fires, and explosions that resulted in death, injuries, community evacuations, and/or significant property damage.

While each accident is unique and presents its own lessons, taken together CSB-investigated accidents shed light on where major chemical accidents have been occurring, and some of the recurring characteristics of these accidents. The paper presents summaries of investigations, and identifies recurring features of recent CSB-investigated accidents.

This paper is prepared for a presentation to the IChemE XX Hazards symposium in April, 2008. It reflects the views of the author and does not represent an official finding, conclusion, or position of the U.S. Chemical Safety Board.

INTRODUCTION

"Having paid the price of an accident… we should use the opportunity to learn from it. Failures should be seen as educational experiences. … Having paid the tuition fee we should learn the lessons." (Trevor Kletz, *Still Going Wrong)*

After several major accidents at chemical plants in the 1980's brought worldwide attention to the issue, in 1990 the U.S. Congress passed the Clean Air Act Amendments, which included a three pronged approach to improve chemical accident prevention and increase public accountability for safety at companies and worksites involved in chemical production, processing, handling and storage in the United States.

Both the U.S. Environmental Protection Agency and the U.S. Occupational Safety and Health Administration were given new regulatory responsibilities.[1] The law also created a separate agency, the U.S. Chemical Safety Board (CSB), to conduct independent investigations of major chemical accidents ("any accidental release resulting in a fatality, serious injury or substantial property damages"), and to report "to the public" on the facts, conditions, and circumstances and the cause or probable cause of the accident.[2]

Since 1998 the CSB has completed 40 investigations and three safety studies, and currently has 7 additional investigations underway. The reports on all of the completed investigations are available on the agency website, www.csb.gov. In addition, short videos

[1]EPA's Risk Management Plan (RMP) rule and OSHA's Process Safety Management (PSM) standard.
[2]42 USC 7412(c)(6)(i)

are available on several recent investigations. The videos include animation that visually depict the accident and describe the accident's causes and the CSB's recommendations. The videos are available, for free, either by download or on DVD which can be requested on the website.

Each accident that the CSB investigates is unique and presents its own lessons for chemical safety and preventing similar accidents. There are also aspects and features in the accidents CSB investigates that seem to be particularly prominent and recurring. This paper surveys recent CSB investigations, and identifies three recurring features in this small, but growing, set of chemical accidents.[3]

I. WHERE MAJOR ACCIDENTS ARE OCCURRING

A news article from 1991, around the time that the CSB was created, described 14 major chemical accidents in the United States that occurred between 1987 and 1991, which together resulted in 79 fatalities, nearly 1000 injuries, and over $2 billion in damages. All of the incidents occurred at large chemical or oil and gas companies.[4]

If one compares this five year list of chemical accidents with the list of accidents that the CSB investigated during the five years from 2002–2007, there is a rather remarkable change. While two of the largest accidents that the CSB has investigated in recent years occurred at large companies (discussed in part III, below), most of chemical accidents that the CSB has investigated over the past five years occurred at small chemical companies or at companies that were not primarily involved in chemical processing, but used chemicals as a part of their operation.[5] A recurring feature in many of these accidents is a lack of awareness and prevention against chemical hazards that are well known, and, in most cases, are addressed by local, state, or national codes, standards, or regulations. Some recent examples follow.

[3]Using CSB-investigated accidents as a "database" of sorts requires an explanation of how such accidents are selected. As resources permit, CSB attempts to investigate the most serious – based on the consequences – chemical accidents that occur in the U.S. In order to do that, the staff monitors a wide variety (over 6000) news sources for reports on chemical accidents, and also receive daily reports of incidents from two government sources, the National Transportation Safety Board and the National Response Center. More serious accidents are evaluated against a simple incident selection tool that considers the number of fatalities, injuries, community impacts (evacuation, shelter in place), property damage, public interest, and potential lessons. In recent years 15 to 20 incidents have been considered for investigation using the criteria listed above, and of those the CSB initiates investigations at about 10 of the most serious.

[4]Schneider, K. (1991, June 19). Petrochemical disasters raise alarm in industry. *New York Times*. Entered in the Congressional Record, June 20, 1991, S 8235.

[5]The list of Completed Investigations and Current Investigations is available on the agency website, www.csb.gov

1. CONCRETE PRODUCTS MANUFACTURER

A concrete products manufacturing plant near Chicago, Illinois, Universal Form Clamp, Inc.(UFC), had converted a small portion of its plant into a chemical mixing area, where it produced chemicals used to treat concrete for certain purposes. The area had a 2200 gallon open top tank with steam coils to heat mixtures of chemicals. On the day of the date of the accident, the tank contained approximately 6000 pounds of heptane and 3000 pounds of mineral spirits. The tank had a temperature controller, consisting of a liquid filled temperature sensing bulb and pneumatic control unit. Investigation after the accident found that it likely malfunctioned due to not being installed or maintained in accordance with manufacturer specifications. The only additional safeguard against overheating the mixture was that at some point during the mixing operation an operator was supposed to climb to the top level of the tank and hand check the temperature of the mixture. There were no alarms and no temperature displays that would indicate a rising temperature. Contrary to existing building and fire codes (NFPA 30), exhaust fans for that area of the facility were at ceiling level; there were no floor level exhaust registers to remove vapors that accumulated near the floor. The local exhaust system on the mixing tank itself was broken and not working at the time of the accident. Design and construction of the chemical mixing area had taken place under the direction of a chemist and contract construction engineers. Apparently neither they nor the local government which approved the construction permit, recognized the hazard and the discrepancy from fire code requirements.

The accident occurred while a mixture was being heated. The temperature controller malfunctioned, causing the steam valve to remain open and the mixture to heat to the boiling point. The boiling mixture produced a heavy, flammable vapor, which spread to the adjacent areas where it was ignited by one of several possible ignition sources. Workers in the immediate vicinity of the tank saw the vapor cloud and were able to evacuate the facility. However a delivery man who happened to be coming into the facility at about the same moment that it was rocked by a large explosion was killed.

2. WASTEWATER TREATMENT PLANT

Another CSB-investigated accident that occurred at a worksite that was not primarily a chemical operation but maintained and used chemicals, occurred at a wastewater treatment plant in Florida, the Bethune Wastewater Treatment Plant. Workers were dismantling a metal roof that had sheltered a 10,000 gallon methanol storage tank but had become damaged during a recent storm. At the time of the accident the tank contained between 2000 and 3000 gallons of methanol. Although the workers were using a cutting torch to dismantle the roof, they did not assure that the methanol tank had been emptied before beginning their work. The employer (the city) also did not have a safety program for hot work, and workers had had infrequent training in chemical hazards. Sparks from the torch ignited fumes from the storage tank. A defective flame arrestor on the top of the tank allowed the flame to flash back into the tank, causing it to explode. Leaking fuel from the tank engulfed the ground operator of the lift truck from which the workers dismantling the roof were working, and both he and worker using the torch were killed. A third worker survived but was severely injured.

3. COMBUSTIBLE DUST EXPLOSIONS

In 2003 and 2004, the CSB investigated three large combustible dust explosions, each of which occurred in a manufacturing facility that used chemicals. A factor in all three of the incidents was a lack of awareness of the hazard of dust explosion from the materials used and the dust being generated in the production process. The largest of the three accidents (West Pharmaceutical) occurred at a manufacturer of rubber stoppers and similar products for medical devices and pharmaceuticals. The manufacturing process involved running rubber strips through a slurry of polyethylene and water. As the rubber was dried by running it in front of a fan, some of the polyethylene particles became airborne. The work areas of the plant were kept clean, but dust accumulated above a suspended ceiling; those who were aware of the accumulated dust did not know of its explosive properties, and the Material Safety Data Sheet for the slurry did not include information or warning about the combustibility of dust when the material dried. An unknown source ignited the dust and a secondary explosion largely destroyed the plant, killing 6 workers and injuring 38 others.

The lack of hazard awareness around combustible dusts in manufacturing settings in the three dust explosions investigated by the CSB led the agency to undertake a study of combustible dust incidents and explosions in the United States, and to take other steps (including a forthcoming video on dust explosions) to bring more awareness of the hazard of dust explosion.

Although several of the accidents that CSB investigated involved a general lack of chemical hazard awareness by personnel at facilities that are not primarily involved in chemical processing, other accidents occurred at small chemical processing plants that also did not consider or apply chemical process safety or hazard information to their process operations.

4. PAINT ADDITIVE MANUFACTURER/CHEMICAL PROCESSOR

One such incident investigated by the CSB occurred at a small chemical plant in North Carolina in January, 2006. The company manufactured a variety of powder coating and paint additives by polymerizing acrylic monomers in a 1500 gallon reactor. The company had received an order for slightly more of the additive than a normal recipe would make, so plant managers scaled up the recipe to produce the larger amount, and also changed the process by adding all of the monomer in the initial charge to the reactor. The impact of the changes, which plant managers did not fully calculate before beginning the reaction, was to increase the rate of heat release in the reaction to at least 2.3 times that of the standard recipe. The heat release rate exceeded the capacity of the reactor's cooling system, and the result was loss of control of the reaction. The reactor lacked additional safeguards to quench the reaction, and the reactor manway had been secured with only 4 of 18 clamps. It began to leak as the pressure in the reactor increased, forming a flammable cloud inside the building. The vapors found an ignition source, and the resulting explosion killed one worker and injured 14 others.

5. POLYETHYLENE WAX PROCESSOR

The CSB investigated a large fire at a polyethylene wax processing facility (Marcus Oil and Chemical) that caused considerable offsite damage. The process involved heating the wax to approximately 300 degrees (F). The system was designed to use nitrogen to move the molten material through the process system, but over time workers and supervisors found that the nitrogen generator on site did not always produce enough nitrogen, so they added an air compressor to the system. In addition, the company had altered the pressure tanks in which the wax was heated by installing steam lines, and the welding to repair the patch was poorly done and not in accordance with good industry practice. On the day of the accident the welded section of the tank gave way, creating a spark as it hit the concrete. The spark ignited the liquid and vapor hydrocarbons that poured out of the opening caused by the weld failure, and the fire backtracked into the tank. The oxygen introduced by the air compressor created an explosive environment in the tank; the 50 foot long tank was lifted about 250 feet, and the resulting fire ignited other parts of the facility, and burned for nearly 7 hours.

6. CHEMICAL "TOLLING" OPERATION

A small chemical plant located in Georgia (MFG) was contracted to manufacture triallyl cyanurate (TAC). In the course of the first production-size batch the reaction went out of control and overpressurized the 4000 gallon reactor. The overpressure activated the emergency vent, and released highly toxic and flammable allyl alcohol into the surrounding residential community. The release caused the evacuation of about 200 families, and 154 people received decontamination and treatment at the local hospital, including 15 police officers and emergency responders who assisting with the evacuation.

The company had conducted laboratory-scale testing of the reaction, and had also run smaller batches prior to conducting a production size reaction. The investigation found that in addition to a larger amount, the company had also changed the recipe, and did not account for the larger reactor (and reduced surface to volume ratio) in calculating the temperature of the reaction and the ability of the reactor's cooling system to contain it.

7. OIL FIELD CONSTRUCTION WORK

Another accident which the CSB investigated which resulted from a lack of basic safety precautions required by codes and standards occurred at an oil field in Mississippi (Partridge Raleigh). Contractors were connecting several adjacent storage tanks, and were welding an overflow pipe to the side of one of the tanks. While the workers had checked the tank on which they were welding for flammable vapors, they did not assure that the adjacent tanks were empty. Vapors from the two connected adjacent tanks vented near to where the welding was taking place; the vapors ignited and flames backtracked into the storage tanks, causing an explosion in the tanks. Three workers, standing on top of the tanks and without fall protection, were killed.

II. RELIANCE ON ADMINISTRATIVE CONTROLS

Lack of awareness of or precautions against well recognized chemical hazards, generally in smaller operations, has been one recurring feature in CSB-investigated accidents in recent years. Another has been the vulnerability of chemical plants and processes which rely on human reliability and judgment and/or administrative controls to safeguard against potentially catastrophic accidental release. The accidents underline the importance of companies realistically taking into consideration "the human factor" in their process safety. They also encourage companies to pay particular attention to this point in their process operations, especially of examining their training and hazard communication programs for operators and supervisors, and of utilizing automated controls when the risk or severity of an accidental release is high.

1. MEDICAL EQUIPMENT STERILIZATION PLANT

One such incident occurred at a medical equipment sterilization facility (Sterigenics). The facility used ethylene oxide to soak the medical equipment. The plant's regular process involved loading the medical equipment into a chamber, soaking it, then purging the ethylene oxide to an acid scrubber, using several nitrogen gas washes. Low amounts of ethylene oxide remaining in the chamber were then vented to an oxidizer when the chamber door was opened. If too much ethylene oxide remained in the chamber when the vent was opened, the heat from the oxidizer would cause it to explode. On the day of the accident, the chamber was being tested. There was no medical equipment in the chamber, and operators surmised that the ethylene oxide would thus be purged more quickly and not require as many gas washes. There was no monitor to indicate the level of ethylene oxide in the chamber, operators relied on their intuition. The additional "safeguard" to premature opening of the vent was that only a supervisor had the combination to open the chamber door before the end of the full cycle. In this case the supervisor agreed with the operators, and opened the door before the full cycle of gas washes had been completed. A significant amount of ethylene oxide remained in the chamber, which vented to the oxidizer. The result was a violent explosion that blew apart the chamber. No one was injured, though the facility received considerable damage.

2. CHLORINE FACILITY

The CSB also investigated a chlorine gas release at a chlorine repackaging facility (DPC) in Arizona. The release resulted in the evacuation of 1.5 square miles of the surrounding community, and medical attention for 16 persons, including 11 police officers who responded to the incident. The facility used a system for transferring the chlorine which captured chlorine vapors and sent them to a scrubber, where the chlorine combined with caustic soda to produce bleach. The reaction depleted the caustic, so it was critical that before the caustic was fully depleted, that the system be turned off, the bleach unloaded, and reactor recharged with caustic. The facility relied on operators to take regular samples, and to shut off the flow of chlorine to the scrubber while the sample was tested to prevent an accidental depletion of caustic. In practice, however, operators continued the flow of chlorine to the scrubber until the target concentration was reached, while periodically sampling the solution. The accident occurred while an operator was preparing to take a sample for laboratory analysis. The system did not have an automatic shut off, and reactor overchlorinated, causing the release.

3. ACETYLENE PRODUCER

The CSB investigated an acetylene explosion that killed 4 workers (ASCO) in New Jersey. The explosion took place when acetylene from the production generator flowed through water pipes to an outdoor shed where a propane heater was being used to keep recycled water tanks and piping from freezing during the winter. Ordinarily water in the pipes would be flowing into the generator when it was operating, and so would prevent acetylene from backflowing to the shed. The accident occurred when the operators switched off the public water supply that was used to start the production cycle, but forgot or was delayed in opening the recycled water line. The empty water line allowed the acetylene to backflow to the outdoor shed. A check valve that would have prevented the acetylene from backflowing failed to operate properly.

III. CORPORATE OVERSIGHT

While many of the accidents that the CSB has investigated in recent years have been at smaller companies which were unaware of or unprepared for well recognized or recognizable chemical hazards, incidents at larger companies can reflect the same lack of hazard awareness or prevention, though the causes may be different. Two of the largest accidents that CSB has investigated in the recent past have occurred at facilities of large chemical and oil and gas companies.

1. REFINERY EXPLOSION

Most notable was the March, 2005 explosion and fire at BP's Texas City, Texas refinery that killed 15 workers and injured about 180 others. The basic and immediate sequence of the accident is quite well known. Operators were restarting the raffinate splitter tower of the refinery's isomerization unit. For reasons not known but for which there were several possible explanations, they for several hours the operators allowed hydrocarbons to load into the tower without opening valves to allow liquid to flow out of the tower. The result was a liquid level 14-15 times the normal operating level. As the liquid in the tower was heated, it expanded and overflowed the tower, bursting pressure relief valves and overflowing the blowdown drum that was the only means of containment of an overflow from the splitter tower. Hot, flammable liquid and vapor erupted from a stack attached to the blowdown drum, causing a vapor cloud to form around ground level, which quickly found a source of ignition and exploded. Near the blowdown drum and in the path of the explosion were portable trailers where workers were meeting or using for office space. All of the fatalities and many of the serious injuries occurred in the destruction of the trailers.

The CSB's report on the accident found many, many factors that contributed to the accident – from operator and supervisor errors and miscommunications on the day of the accident, to laxness about safety procedures for unit startups and trailer siting, to long-standing failures to address hazards in the design and operation of the pressure relief system of the unit in which the release of flammable liquid and vapor into the refinery took place, to corporate management and corporate culture that undervalued process safety at the refinery.

2. VINYL CHLORIDE PLANT

CSB also investigated a major accident at a vinyl chloride plant belonging to a large chemical company, Formosa Plastics, which occurred in 2004. An operator was cleaning one of several reactors at the facility. After cleaning the reactor from the top level, he went to the lower level to drain the reactor. However in the process he turned the wrong direction and went to the wrong reactor, one that was on cycle. When he attempted to open the drain valve the interlock that prevented the reactor from being opened while on cycle prevented him from opening it. However the facility had a readily accessible override to the interlock, which was supposed to be used in case of emergency, but apparently was not always so limited. He used the override, as he opened the valve on the on cycle reactor, vinyl chloride poured out of the reactor. It found a source of ignition and exploded, killing 5 workers and injuring 3 others, and destroying the plant.

Both of these incidents have numerous important lessons and reminders about process safety and process safety management for chemical and petrochemical operations. While a full description of these is well beyond the scope of this article, two specific lessons for corporate management of process operations are mentioned here.

One is that most large companies monitor the performance of individual units, facilities, operations, and managers by numerical measures. These numerical indicators are used not only to monitor performance, but to improve performance, by basing pay and bonuses, budgets, or other types of recognition, on them. Thus, "what gets measured gets managed," and what gets reported to the corporate leaders is what gets attention and emphasis in operations, especially if pay and performance rewards are involved. BP's internal performance measurement and incentive systems for safety performance focused almost exclusively on injury rates, and did not include measurement of process safety performance. As a result, safety programs at the refinery focused on personal safety initiatives, and company officials received reports of improving safety performance at the refinery, based on lower injury rates, even as process safety deteriorated and the risk of major accident remained high. A principal lesson coming out of the Texas City refinery disaster is the importance of process safety indicators, in addition to personal safety measures, to monitor performance at any facility at which hazardous chemicals are used or processed.

The BP and Formosa accidents also highlight the important role that corporate oversight has in assuring that individual plants and facilities have pertinent information on process hazards, and act upon it: that the "information flow is open" and the "action loop is closed." For both BP and Formosa, previous incidents at other facilities of the company could have provided valuable lessons in identifying and correcting the hazards which ultimately led to a catastrophic accident, if the information had been used. BP's previous accident at the Grangemouth refinery highlighted some of the same issues as later were highlighted in understanding what went wrong at Texas City. At Formosa Plastics, other facilities of the company had experienced operators overriding interlocks and had put in place additional safeguards. But those had not been incorporated at the Illinois facility.

Further, the specific vulnerability in both cases – the pressure relief system/ blowdown drum in BP and operator access to open an on-cycle reactor in Formosa – had been identified by the plant prior to the accident, but was not corrected before the accident occurred. At BP, the vulnerability of atmospheric discharge from the blowdown drum had been identified years before the accident, but was not changed to send overflowing hydrocarbons to a flare system. At Formosa, another operator's override of the interlock two months prior to the accident led the plant supervisors to appoint a team to come up with alternatives, but no changes were made before the accident occurred.

CONCLUSION

The chemical accidents that the CSB investigates are each unique, and the number of major accidents is relatively small. The "low probability, high impact" events that CSB investigates do not necessarily predict where the next accident may occur or where the greatest risk of future accidents lies. The BP refinery explosion described above was a reminder of the tragic consequences that can attend any large chemical release. Certainly the above discussion of recent CSB-investigated accidents does not give reason for <u>any</u> plant or facility to feel safe, become complacent, or reduce its own vigilance against chemical accident!

Nonetheless, taken together, recent CSB investigations highlight some common or recurring characteristics of major accidents that have occurred in the United States in recent years. They suggest that many of the hazards of major chemical accidents in the United States are occurring at companies that are not primarily chemical companies, but keep or use large quantities of chemicals for other purposes. They also suggest that a particular area of vulnerability for facilities that use or handle chemicals are the controls and safeguards that rely on human judgment and reliability in order to prevent accidental release, and thus emphasize the importance that management should give to these points in their operations. Finally, recent major accidents that occurred at large companies highlight the critical and essential role of that corporate leadership and oversight have in chemical process safety, particularly in assuring that process safety is included, measured and valued in individual and business unit performance standards, and in assuring that plants and facilities not only have systems in place to assure that they are aware of problems and deficiencies, including particularly legal deficiencies, but that there is also assurance of timely correction of deficiencies.

Figure 1. Universal Form Clamp, Bellwood, IL (June 14, 2006)

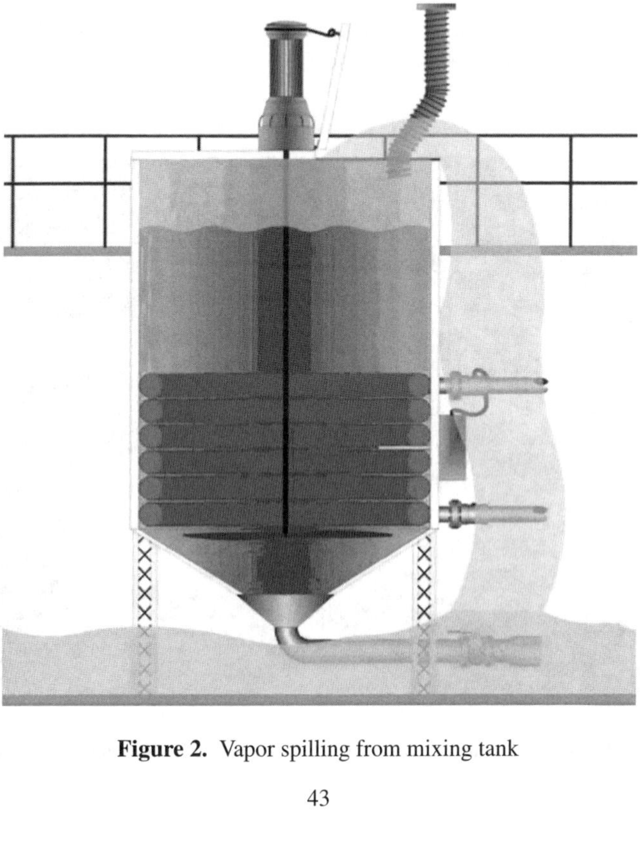

Figure 2. Vapor spilling from mixing tank

Figure 3. Bethune Point Wastewater Treatment Center: Daytona Beach, FL (January 11, 2006) Location of man-lift basket and 4-inch vent pipe

Figure 4. West Pharmaceutical Services, Inc.: Kinston, NC (January 29, 2003)

Figure 5. Synthron, LLC: Morganton, NC (January 31, 2006)

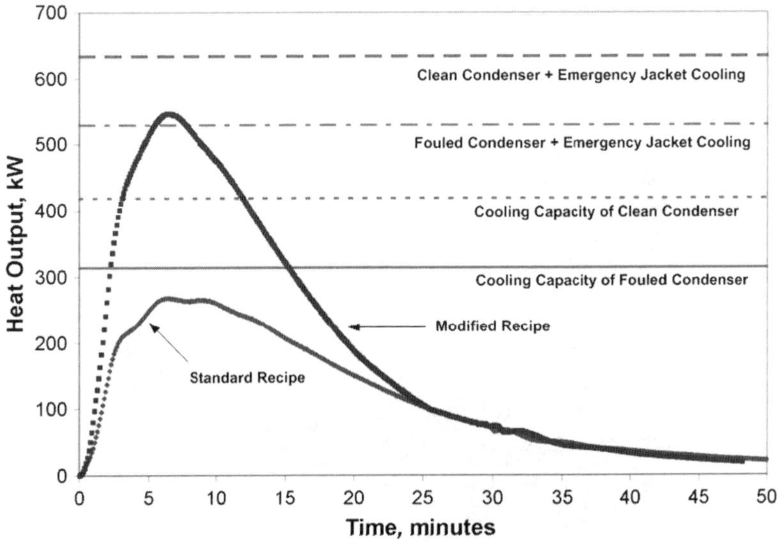

Figure 6. Reaction calorimetry heating curves for standard (lower) and modified (upper) recipes

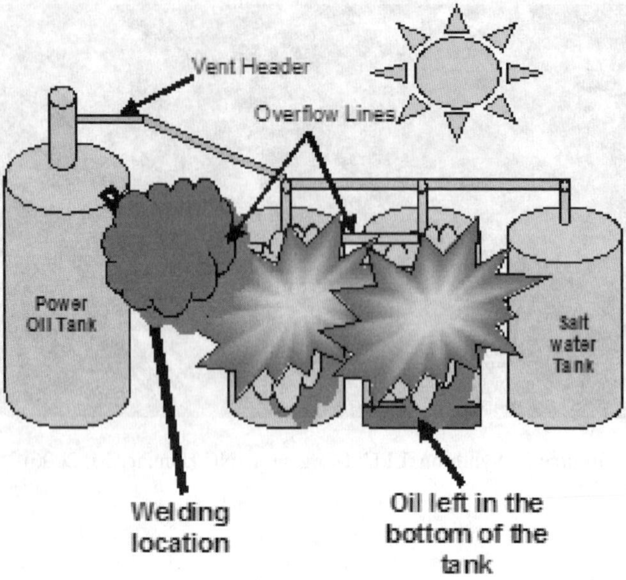

Figure 7. Oil Field Accident

Figure 8. ASCO: damage and debris around the decant water tanks

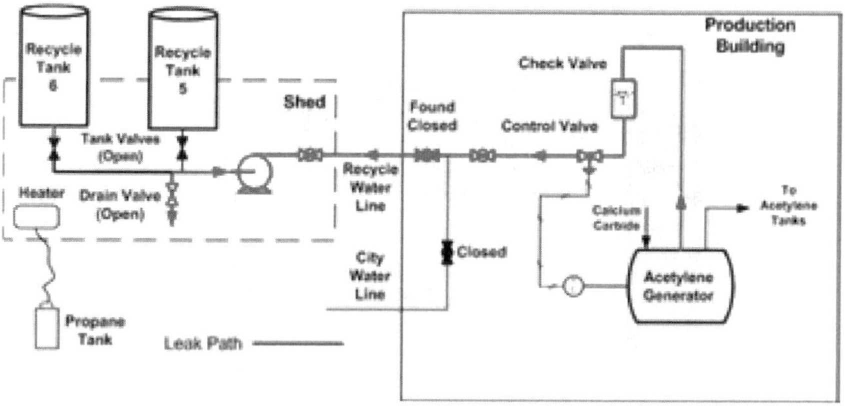

Figure 9. Acetylene system flow diagram

Figure 10. BP: Texas City, TX (March 23, 2005)

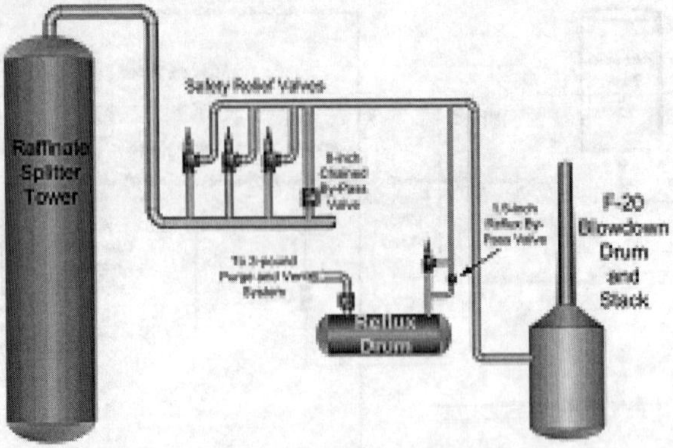

Figure 11. Disposal collection header system

BUNCEFIELD – LEGAL IMPACTS

Paul Bratt and Catherine Henney
Hammonds, Trinity Court, Manchester, UK

1. INTRODUCTION

1.1 THE INCIDENT

On 11 December 2005, an extensive fire broke out at Buncefield Oil Storage Depot in Hemel Hempstead. The fire resulted from a series of vapour cloud explosions – some very severe – as a result of which a huge cloud of vapour, smoke and smog engulfed the majority of the site. Remarkably, no-one was killed in the blast, although in excess of 40 people were injured. A large scale evacuation of the surrounding area was ordered, although nothing could save the large number of nearby properties – both commercial and residential – that suffered extreme damage.

1.2 THE FALL OUT

Two years on, the Environment Agency is continuing to monitor the extent of contamination surrounding the Buncefield site. Results obtained in February 2007 indicated that groundwater under and up to 2 kilometres to the North, East and South East of the site has been contaminated with hydrocarbons and fire fighting foam. Major treatment and disposal work has been carried out at 2 sewage treatment plants in the area, with up to 5 million litres of effluent uplifted from one of the sites.

1.3 THE INVESTIGATION AND REGULATORY RESPONSE

The Health and Safety Executive ("HSE") and the Environment Agency ("EA") commenced a joint investigation immediately following the blast, which is still ongoing. The HSE has issued Safety Alerts to the industry, and over 100 fuel storage sites were targeted for inspection. In addition, the independent Buncefield Major Incident Investigation Board ("MIIB") was also established to oversee and report upon the investigation.

A Task Group was also set up following the incident, comprising industry, trade association and employee representatives, and together the Group developed a number of recommendations for how the industry can enhance safety standards.

The HSE has published 2 separate consultations as a result of the incident, both of which are discussed in Section 4 below.

1.4 AIM OF THIS PAPER

This paper aims to discuss some of the issues, both legal and environmental, raised by the Buncefield incident in the light of the HSE consultations, the Control of Major Accident Hazards Regulations 1999 (the "COMAH Regulations") and the associated duties and obligations of site operators. The incident at Buncefield has massive implications not only for the future storgage of fuel and chemicals, but also for risk assessment regimes, and land development controls.

Unless there are major surprises still to be unearthed, it is very difficult to say that any of the causes of the incident were particularly novel to the operation of facilities such as Buncefield. What is presently unclear is how such proximate development was permitted near the site, and how management systems had failed to ensure that best practice had not been observed on site.

2 PROBLEMS HIGHLIGHTED BY BUNCEFIELD

Following the explosion on 11 December 2005, the MIIB published a series of progress reports detailing its findings in relation to the cause of the incident and highlighting problems or issues which may have contributed to it. This section briefly discusses some of those key issues.

2.1 BUND INTEGRITY

Whilst the bunds substantially remained standing throughout the incident, they did not fully contain the fuel and firewaters as a result of the explosion. Pools of fuel were burning in the bunds as a result of loss of fuel from the tanks, along with fires from the tanks themselves.

The MIIB took the view that the most urgent focus of attention in preventing future similar incidents should be on preventing loss of primary containment as a first port of call and should that fail, inhibiting rapid large-scale vaporisation, and thus any subsequent dangerous migration of flammable vapours[1].

The MIIB recommended[2] that the industry, together with the HSE, should review the purpose, specifications, construction and maintenance of secondary containment, in particular bunds around tanks. This work should lead to revised guidance, with the necessary standards capable of being insisted upon by law. The existing standards for secondary containment should also be reviewed.

The MIIB stated that revised standards should be applied in full to new build sites and to partial installations. Where not practicable to fully upgrade bunding on existing sites, operators should develop and agree with the HSE risk-based plans for phased upgrading as close to new plant standards as is reasonably practicable.

The authors find it remarkable that such steps were thought to be widely necessary in this day and age.

[1]p19, paragraph 66 of Initial Report to Health and Safety Commission and the Environment Agency (published 13 July 2006).

[2]p21, paragraph 73 of Initial Report to Health and Safety Commission and the Environment Agency (published 13 July 2006).

2.2 ALARM PROVISION

The MIIB found that the tanks on the site were fitted with an automatic tank gauging (ATG) system which was in turn connected to a control room from which tank levels were monitored. The ATG system was also integrated with an alarm system. Evidence from the ATG database temperature records and examination of valve positions, coupled with analysis of what took place on the day of the incident, showed that the protection system which should have shut off the supply of petrol to the tank to prevent overfilling did not operate[3].

The investigation of the MIIB revealed that the recording of monitoring, detection and alarm systems must be considered by the industry for improvement[4].

It was recommended that where a need for additional systems is identified, the Health and Safety Commission (the "HSC") and the HSE should satisfy themselves that current legal requirements (i.e. the COMAH Regulations 1999) are robust enough, and supported with sufficient resources, to ensure that these systems are provided and maintained at every fuel storage site where risks require them, without relying upon voluntary compliance. It has not been established whether changes in the law or in the resources available to the HSE are required to achieve this end. The legal requirements in the form of the COMAH Regulations 1999 are discussed in more detail at Section 3.

Nevertheless, it cannot be said that any of these recommendations represent new learning for operators of such facilities.

2.3 EMERGENCY PROCEDURES/ TRAINING

Operators of top tier sites, such as Buncefield, are regulated by the COMAH Regulations 1999, and are required by law to prepare adequate emergency plans to deal with both the on-site and off-site consequences of possible incidents[5].

In light of the emerging findings, the MIIB recommended that:

(a) Operators of oil storage depots should review their on-site emergency plans and the adequacy of information they supply to local authorities to ensure they take full account of the potential for vapour cloud explosion, as well as fires.
(b) The public health implications of potential vapour cloud explosions must be considered in both on-site and off-site emergency plans.
(c) The HSC and HSE should satisfy themselves that legal requirements are robust enough to ensure that any necessary changes to emergency plans are duly made.

[3]p7 of Initial Report to Health and Safety Commission and the Environment Agency (published 13 July 2006)

[4]p20 of Initial Report to Health and Safety Commission and the Environment Agency (published 13 July 2006)

[5]p21 of Initial Report to Health and Safety Commission and the Environment Agency (published 13 July 2006)

In terms of training, the MIIB recommends[6] that the sector should work with the HSE to prepare guidance and/or standards on how to achieve reliable industry practice through placing emphasis on the assurance of human and organisational factors in design, operation, maintenance and testing. In particular, appropriate training should be provided to staff for safety and environmental protection activities.

It is unclear why insufficient vapour cloud risk assessment had been carried out in this instance. Again, there is nothing novel about ensuring that the human and organisational factors of operating systems work in unison. What is surprising is that the MIIB appears to consider that such systems and processes were far from widespread.

3 THE LEGAL FRAMEWORK – THE COMAH REGULATIONS

3.1 AIMS AND OBJECTIVES

The main aim of the COMAH Regulations is to prevent and mitigate the effects of major accidents involving dangerous substances, which pose a serious risk of harm to the public and the environment.

The COMAH Regulations are enforced by the COMAH Competent Authority ("CA"), consisting of the HSE and the EA/SEPA (Scottish Environmental Protection Agency).

The CA inspects activities subject to COMAH and has the power to prohibit an operator from continuing to operate if they do not have in place suitable and sufficient control measures for the prevention and mitigation of major accidents.

Operators holding larger quantities of dangerous substances (referred to as 'top tier' sites) are subject to more onerous requirements than those who operate with lower quantities.

3.2 OBLIGATIONS/ DUTIES ON COMAH OPERATORS

The general duty on all operators is to take *"all measures necessary to prevent major accidents and limit their consequences to people and the environment"* (Regulation 4). It is a high standard and applies to all establishments within the scope of the Regulations. By requiring measures both for prevention and mitigation there is a recognition that all risks cannot be completely eliminated; proportionality is therefore a key element in the enforcement policy of the CA. Thus, the phrase "all measures necessary" will be interpreted on this basis.

Where hazards are high, appropriate high standards will be required to ensure risks are reduced to an acceptably low level, in line with the policy that enforcement should be proportionate. Prevention should be based on the principle of reducing risk to a level "as low as is reasonably practicable" in the case of risk to humans, and using the "best available technology not entailing excessive cost" principle for environmental risks, although it

[6]p18, recommendation 19 of Recommendations on the design and operation of fuel storage sites (published 29 March 2007)

indeed goes without saying that the ideal should always be, wherever possible, to avoid a hazard altogether.

Operators of COMAH sites have different obligations depending upon whether they are classified as a 'lower-tier' or 'top-tier' site. Lower-tier operators must prepare a document setting out their major accident prevention policy, or "MAPP"[7]. The MAPP not only sets out what is to be achieved in terms of prevention policy, but it should also include a summary of, and cross-refer to the operator's safety management system that will be used implement the MAPP. Top-tier operators must still comply with the requirement of a MAPP, but this is normally included in the COMAH Safety Report, which they are obliged to prepare and submit to the CA.

3.3 RISK ASSESSMENTS

All COMAH operators are under a statutory duty to undertake a suitable and sufficient risk assessment to determine the measures necessary to ensure that risks to health, safety and the environment are adequately controlled.

When the COMAH Regulations were implemented, the CA issued Guidance on the Environmental Risk Assessment Aspects of COMAH Safety Reports[8]. In such Guidance, the CA recognises that some hazards are "less readily identifiable than others", and that, for that reason, operators should consider ancillary risks posed by dangerous substances – for example, in circumstances where hazards may be posed by other substances formed/released during the course of any accident involving the primary dangerous substance with which its operations are concerned. Indeed, the Assessment Criteria in Appendix 2 of the Guidance suggests that the types of risks to be covered include *"immediate and delayed effects from uncontrolled releases arising from both normal and abnormal conditions on the installation"*, which includes *"commissioning, maintenance, operation and modification phases"*[9].

Interestingly, in relation to off-site emergency planning, the Guidance states that "*operators are not expected to prepare the off-site emergency plan*", although the majority of the information gathered during the risk assessment process will contribute to such a plan. However, it does recommend that *"the on-site and off-site emergency plans should complement each other and the requirements placed on operators include the provision of sufficient information to local authorities to allow this to happen"*[10].

This Guidance, although now somewhat outdated, is intended as a reference for operators in carrying out risk assessments for their COMAH Safety Report (in the case of top-tier operators), and to try to ensure that the control measures adopted are of a recognised standard of good practice for the industry. Nevertheless, the Guidance does not

[7]Regulation 5, COMAH regulations 1999

[8]Guidance on the Environmental Risk Assessment Aspects of COMAH Safety Reports, COMAH Competent Authority, December 1999

[9]Aspect 6, "Study Scope", page 48.

[10]page 34.

generally contain prescriptive requirements; this is because the regulatory approach of the CA is to leave the decision-making powers with the operator and merely ensure that the operator can justify the choice of control measures it adopts. It will only intervene to take enforcement action where the standard of compliance falls significantly below accepted good practice.

The Authors would query whether, (a) in the light of the incident at Buncefield, further, up to date guidance is required in this area, and (b) whether increased supervisory control should be given to the CA to ensure that suitable and sufficient risk assessments are carried out and appropriate control measures implemented from the outset.

3.4 EFFECT OF NEIGHBOURING TANKS

The Guidance also places importance on identifying the scope of the assessment in terms of which items of plant/site are included and their status. Therefore, installations should be divided into a series of technical units so as to make the risk assessment process manageable. This should include taking into account any effects from neighbouring units (including those owned by different companies). The assessment should ultimately cover all units in which dangerous substances are produced, used, handled or stored.

However, whilst it may be necessary for practical reasons and risk assessment purposes to consider the installation as consisting of a series of individual technical units, it is important to ensure that these are not viewed as closed systems which do not have any influence on each other. The scope of the risk assessment for each unit should be sufficiently flexibly defined to allow for cross-unit interactions.

3.5 RISKS OF MOVING VAPOUR CLOUDS

According to the Guidance, the hazard identification process should cover *"all the various different types of major accident hazards"*, including:

- Loss of containment accidents due to vessel or pipework failures;
- Explosions (batch reactors, tank explosion due to operator error – for example wrong contents and boiling liquid expanding vapour explosion);
- Condensed phase explosions relating to explosives;
- Large fires – for example warehouses and pool fires;
- Pressure relief valves lifting and venting to atmosphere;
- Events influenced by emergency action or adverse operating conditions – for example allowing a fire to burn rather than apply water, dump reactor contents to drain to avoid explosion, and abnormal discharge to the environment; and
- Other types of major accident hazard or abnormal discharge.[11]

Therefore, as part of the risk assessment process for sites such as Buncefield, the list of dangerous substances to be considered should be formulated based on the scope of the

[11]Section 2.3 Hazard Identification Process, page 22

operations and should take into account all possible substances occurring under both normal and abnormal conditions. At Buncefield therefore, the risks associated with moving vapour clouds – and the content of such vapour clouds – should have been identified by reference to what would, or could, occur under explosive or abnormal conditions.

4 LAND USE PLANNING AND SOCIETAL RISK: THE HSE'S PROPOSALS

4.1 INTRODUCTION

The risks from top-tier sites or major hazard installations, ("MHIs") as they are commonly referred to, are primarily managed through on-site accident prevention and mitigation measures required under COMAH, as discussed above.

The off-site risks of MHIs, however, are managed through the planning and development system. Local Authorities must consult the HSE on planning applications around these sites and the HSE will then advise planning authorities on whether the development should go ahead based on an assessment of the off-site risks from the site. The decision to grant planning permission remains with the planning authority, but experience shows that the HSE's advice is followed in the vast majority of cases.

4.2 LAND USE PLANNING

Sites storing hazardous substances above a certain level must have consent from the Hazardous Substances Authority ("HSA"), usually a part of the Local Planning Authority ("LPA"). In considering whether to give consent, the HSA is statutorily obliged to consult the HSE, who will then draw up a Consultation Distance around the site in question, identifying by way of "zones" the level of likely risk or harm to individuals within those zones. Proposed developments are also considered according to sensitivity levels, depending on the intended use; the number of people at the development; the intensity of the development; and whether the development is intended for vulnerable persons such as school children.

Following the Buncefield incident, the HSE launched a Consultation on Development Control around Large-scale Petrol Storage Sites[12]. The Consultation suggested some possible options for, and sought views upon, specific changes to the HSE's advice to LPAs on Land Use Planning ("LUP") around large-scale petrol storage facilities.

The HSE has based its current advice on the findings of the Advisory Committee on Major Hazards ("ACMH") and the principles and objectives that it has developed as a result of the ACMH's recommendations.

(a) Current LUP advice and HSE Principles

The current LUP advice given to LPAs consists of a "protection-based" approach, rather than by way of Quantified Risk Assessments ("QRAs"); this approach focuses on

[12]CD211 HSE Consultation: Proposals for Revised Policies for HSE Advice on Development Control around Large Scale Petrol Storage Sites

determining "risk boundaries" around the site, with risk being in relation to harm to individuals at particular locations around the petrol site, rather than taking into account societal risk as a whole (whereby the entire population around a site would be considered). Societal risk is only currently taken into account to a limited extent when considering each proposed development, and is not therefore considered in the Consultation; a further HSE consultation document[13], discussed below, examines whether societal risk should be brought into the LUP process for areas around on-shore non-nuclear MHIs.

(b) HSE Proposals in relation to its LUP advice

As a result of the Buncefield incident, the HSE's principles in relation to separation distances and vulnerable people (which the HSE considers to include children, the sick, the elderly, those with mobility difficulties, or those unable to recognise physical danger) may need to be re-assessed, particularly as regards developments within what is known as the "Inner Zone" (i.e. 120m from site).

It was clear from the Buncefield incident that, had the surrounding buildings in the Inner Zone been occupied, significant numbers of people could have been killed or injured. Further, given the nature of the incident and the fact that the vapour cloud spread without detection, questions were raised about the ability to carry out an organised evacuation in the event of a similar incident. The HSE therefore felt that the Consultation Distances may need to be reassessed.

The Consultation put forward 4 alternative options for changes to its policies:

(i) Option 1 – No change to HSE's LUP advice

(ii) Option 2 – Change size of the Consultation Distances and zones, based on hazard

The current Consultation Distances and planning zones would be altered so that these would cover approximately four times the area of land currently covered and a new "Inner Zone" of 250 m from the site will subsume all four areas currently within the Consultation Distance. Sensitivity levels would remain the same.

Whilst this would mean that risks to individuals within Consultation Distances would be better controlled, and there would arguably be greater reassurance to the public, development would be more restricted than at present.

(iii) Option 3 – Change size of Consultation Distances (as Option 2) and development sensitivity levels

As well as Consultation Distance areas and zones being enlarged, the type of development within the Inner Zone would be restricted to buildings that are "not normally occupied" (e.g. warehouses, outdoor storage, farm buildings and parking areas).

With this option, risk to individuals would be better controlled in line with the extent of damage seen at Buncefield, and economic disruption would be minimised in the event

[13]CD212 HSE Consultation: Proposals for Revised Policies to Address Societal Risk around On-shore Non-nuclear Major Hazard Installations

of an incident occurring. However, the limitation on development would be significantly greater and could be considered onerous.

(iv) Option 4 – Change size of Consultation Distances informed by risk, and adopt a new "Development Proximity Zone" ("DPZ") to give more restrictive advice

Consultation Distances and planning zones would be extended as in Options 2 and 3, as well as introducing a new DPZ with increased sensitivity levels. The DPZ would be at a radius of 130 m from the site, and any new development would be unadvisable unless it involved "not normally occupied" structures.

On 5 December 2007, the HSE released its finalised proposals for revised policies for its advice on development control around large-scale petrol storage sites such as Buncefield, which concluded Option 4 would best achieve the HSE's aim of maintaining a "sensible and practicable balance between risk and development". It found that, from the responses to its Consultation, 83% of respondents (which included a number of councils, industry members and regulators) were in favour of extending the Consultation Distance, and 79% agreed that the HSE should change its assumptions about the vulnerability of individuals living and working in the vicinity of large-scale petrol storage depots.

The HSE will continue to use its existing objectives and principles towards LUP advice, but these will be kept under review. Importantly, the role of the HSE will not change; it will remain as a consultee to planning authorities, both in its statutory capacity in relation to specific types of development which must be referred, as well as in a general advisory role with regard to consultation on policies adopted by the authorities for their development framework. The HSE does not, however, see the need for it to act as a statutory consultee for development plans as well – the onus remains on the planning authority to consult the HSE in the appropriate circumstances.

The revised policy will come into effect in summer 2008, but will be regarded as an interim policy whilst investigations into the Buncefield incident and vapour cloud explosions are continuing.

4.3 SOCIETAL RISK

Following the initial consultation issued by the HSE and discussed above, a further consultation document was released specifically addressing the question of "Societal Risk around Onshore Non-nuclear Major Hazard Installations", again as a result of lessons learned from Buncefield. This second Consultation is based on studies by a Government Task Group[14] that have been ongoing for some time since the introduction of the COMAH Regulations.

(a) HSE Consultation

The Consultation has a wider scope than the LUP Consultation discussed above, in that it considers all operators under the COMAH Regulations, or, more specifically, non-nuclear

[14]comprising representatives from HSE, DTI and DCLG (Department for Communities and Local Government)

on-shore MHIs, rather than addressing solely petrol storage facilities. The Consultation does not, however, deal with installations with off-shore risk, such as pipelines or nuclear facilities.

This Consultation considered the same process of HSE LUP advice as outlined in the first Consultation, discussed above. Whilst the HSE indicated that it believes that the advice and arrangements currently in place for controlling MHIs have been highly successful, it is felt the system could be further improved by taking into account societal risk in relation to both on-site measures adopted by the MHI itself, and the advice it gives to planning authorities as to development in the surrounding area.

(b) "Societal Risk" – what is it?

In terms of the current LUP system and the assessment of risks posed by nearby MHIs, "risk" is currently only assessed in relation to harm to individuals at particular locations around the MHIs, rather than taking into account "societal risk" in that area as whole. The Consultation differentiates societal risk from individual risk as follows:

> *'Individual risk' is the chance that a particular individual at a particular location ill be harmed… but [it] does not take account of the total number of people at risk from a particular event. 'Societal risk' is a way to estimate the chances of numbers of people being harmed from an incident [and] the consequences are assessed in terms of level of harm and numbers affected, to provide an idea of the scale of an accident in terms of numbers killed or harmed".*

The HSE therefore proposes to shift its risk focus to an analysis of the entire population distribution around the site in question, and the potential effects a major accident may have on that population as a whole, rather than on individuals within the area. The Consultation proposed that societal risk would therefore take into account, and ultimately depend upon *"what processes and substances are at the sites, and on the size, location and density of the population in the surrounding areas"*.

Examples of sites which the HSE considers could pose high societal risk are:

- Chemical plants that manufacture or use toxic substances;
- Large LPG and LNG[15] storage facilities; and
- Treatment plants that store chemicals for use in e.g. water purification.

(c) HSE proposals for dealing with societal risk

The Consultation suggests that, taking societal risk into account at the development plan stage may obviate the need for the HSE to provide separate guidance on the issue of societal risk for individual planning applications. Nonetheless, the HSE recognises that this may not capture all sites that should be subject to a societal risk assessment, for example where developments outside the Consultation Distance were not initially identified at development plan stage as posing a societal risk.

[15]Liquified Petroleum Gas and Liquified Natural Gas

For larger developments, such as housing estates, which are proposed close to certain MHIs but outside the current extent of the HSE's Consultation Distance[16], it is likely that societal risk may be increased to such a level where there is a chance that planning permission would be turned down. In these instances, the HSE considers that it would agree an extension to the Consultation Distance to ensure that the planning authority would have to consult the HSE in the case of such large developments. However, the HSE insists that this would not mean a full-scale extension of the Consultation Distance for all other cases. One may question, then, how this squares with the new 'Development Proximity Zone' and alterations to the Consultation Distance as proposed in the previous LUP Consultation.

4.4 CONCLUSION

The HSE issued an initial Regulatory Impact Assessment in relation to societal risk and land use planning in April 2007. It considered that using societal rather then individual risk as the concept for assessing the risks posed by MHIs would mean that the level of risk posed by a particular site could ***increase*** over time as the population in an area rises, while the level of individual risk would stay the same[17]. The result may be that the site operator would be required to put in place even more stringent on-site risk reduction measures that could be very costly to implement. Furthermore, the HSE stated that there could be negative effects on the 'wider society and the economy' if land use planning restricted the construction of new buildings around a MHI due to concerns over societal risk.

However, the benefits of taking account of societal risk would be that the off-site consequences of an incident should be smaller; there would be lower numbers of fatalities and injuries and reduced medical costs; and less damage to surrounding property and economic infrastructure.

Nevertheless, as discussed by the CIA in its statement (see section 4.5 below) there is potential for allocating the increased costs brought about by using the concept of societal risk in place of individual risk, between developers and site operators.

The HSE's work[18] indicates that there are only a limited number of existing sites where the introduction of societal risk calculations may result in future advice on planning applications, or future advice to site operators, being different from that which might otherwise have been the case. This is because:

- Only a relatively small proportion of all major hazard sites give rise to a level of societal risk that makes such considerations appropriate;
- The existing basis on which advice is given to planning authorities by the HSE would already prevent many of the developments within the existing consultation zone that could increase societal risk; and

[16]and therefore do not require HSE consultation

[17]p5 of Health and Safety Executive initial regulatory impact assessment (*published around April 2007*)

[18]In Health and Safety Executive initial regulatory impact assessment (*published around April 2007*)

- Around many of the sites being considered, there are limited opportunities to develop due to their location (e.g. the land around them is already built up).

The HSE does not foresee its proposals as bringing about a radical change to its LUP advice; the HSE will not have any power to refuse planning permission – its role will remain purely advisory. The only change to the current system will be one of improvement, by incorporating societal risk considerations into its LUP advice and extending this to include revised planning zones and increased sensitivity levels. Nevertheless, questions have been raised as to whether, as a result of the proposed changes to land use planning, there will be a negative effect upon MHI owners/operators – referred to as the "reverse COMAH" effect. This is discussed in more detail below.

4.5 "REVERSE COMAH"

The term "Reverse COMAH" refers to the effect brought upon COMAH operators as a result of changes in local land use; i.e. where development of land surrounding the COMAH site is such that the operator is forced to alter or increase its safety control measures in order to respond to the increased risk posed to the community as a result of the development.

The Chemical Industries Association (CIA) has recently issued a statement expressing concern over the effect of the HSE's proposals as they currently stand, in that COMAH operators look set to face huge financial burdens in order to reduce the societal risk created by a proposed development in the vicinity of its site.

Whilst the CIA recognises that development is needed to address issues of socio-economic importance such as unemployment and land blight, it is of the opinion that, in many cases, the associated costs of requiring an operator's COMAH Safety Report to be reviewed and further risk reduction measures implemented will be inequitable and should be borne by the developer who is proposing to build on the surrounding land.

The CIA fundamentally agrees that the principle of reducing risks "as low as is reasonably practicable" should still apply; however it feels that the criteria upon which judgments of risk are made could be improved.

(a) CIA Recommendations

The objectives and proposals put forward by the CIA can be summarised as follows:

(b) The HSE should make available to the public its individual risk methodology currently used for calculating the Consultation Distances. This would assist in engaging stakeholders in the societal risk methodology and ensure a greater understanding of the land use planning system.

(c) The UK should not be put at a competitive disadvantage in terms of implementing the Seveso II Directive, which means any measures adopted should be consistent with those adopted by other member states.

(d) LPAs should <u>not</u> be given increased responsibility in terms of assessing the risks of development plans and making planning decisions accordingly. If left to their own devices, LPAs could approve a number of developments that the HSE would advise against, and hence there is less control exerted over developers. At the very least, the current advisory system should remain in force until superseded by a more comprehensive societal risk policy is introduced. Indeed, in cases where LPAs do not follow the HSE's advice against a certain development, it should be legally required to inform the HSE, and the costs of risk reduction should be weighted against the developer.

(e) LPAs should lead the planning process, working in partnership with the industry and the HSE. Each LPA should have in place a structural plan for MHIs, which balances societal benefit against societal risk and sets out criteria and guidance in relation to proposed planning. The plan should attain central Government approval.
(f) Operators should be legally required to be involved in the consultation process for any proposed development within the Consultation Distance/ societal risk zones. Where the HSE is of the opinion that a risk exists, this should be communicated to the operator, who should be involved in a discussion about the consequences of allowing the proposed development to go ahead.
(g) In conjunction with (e) above, the cost of the burden on the operator to introduce risk reduction measures should be shared appropriately between the developer and the operator – this is particularly appropriate where the development threatens a "reverse COMAH" effect.
(h) Operators should have the right to object to any proposed development.
(i) In cases where development is proposed and the required compliance with the COMAH Regulations in terms of measures to be introduced to counter the increased risk cannot be achieved, there must be a procedure in place to determine what else can be done.

The HSE's Consultation on societal risk closed in July 2007. It therefore remains to be seen whether the response issued by the CIA, along with all other responses, will have an effect on the proposals initially put forward by the HSE.

5 CONCLUSION

The MIIB has conducted some sterling work in the aftermath of this incident, and has made a series of eminently sensible responses and recommendations. However, the Authors find it very difficult to escape the conclusion that this incident could have been avoided, and that contributory factors were identifiable. Off-site damage to buildings could have been avoided by thorough and proportionate assessment of the real risk profile of the installation. Had the explosion happened during conventional business hours (and not 06:00) quite a horrific loss of life could have ensued.

On the basis of the MIIB's findings thus far, the Authors are surprised by the apparently poor condition of the asset (in terms of safety systems, bunds etc) and the manner in which too proximate development was permitted.

It is unclear whether the CIA will get its way with regard to "Reverse COMAH" but much can certainly be done to ensure that suitable and sufficient risk assessment, safety procedures and management systems, all of which must be second nature to operational and safety professionals in the chemicals industry, are properly conducted, implemented and maintained.

CHALLENGING THE 'SHE' CULTURE IN THE RUSSIAN FEDERATION

Chris Brandon[1] and David W. Edwards[2]
[1]HSE Consultant, e-mail: christopher.brandon1@ntlworld.com
[2]Lead Engineer – HSE, Granherne – Moscow Office, e-mail: David.Edwards@KBR.com.

There is a gulf between the mature safety, health and environmental (SHE) culture of 'Western' companies and their Russian partner companies in the oil and gas sector. Comprehensive and highly prescriptive SHE regulations exist and many of these mirror western standards but they have been developed without consultation with the industries that they regulate. The corporate response, at least amongst the older generation of management, is to either merely comply with or pay lip-service to the law as cheaply as possible. In many cases, they pay the fines for non-compliance. SHE measures are seen as a cost to the business, rather than as an investment to achieve better performance and higher profitability. There is a 'blame culture' that inhibits managers taking any personal responsibility for SHE issues. Accident investigations generally seek to lay the blame on the unfortunate victim, rather than to discover the root-causes and prevent recurrence. Internal shareholder and public pressure for better performance has been, and still is, lacking.

This paper gives an anecdotal account of the differing approaches to SHE in the oil and gas industries in the UK and Russian Federation (RF). Accident statistics are compared and cultural differences described. The approach to SHE regulation is examined. Some suggestions for constructive change are made that might help the emerging generation of native young engineers and managers, who are well-trained and cognizant of the need for better SHE performance.

INTRODUCTION

Emerging economic powers, such as Russia, are striving to develop their oil and gas resources. Much of this development is in collaboration with 'Western' companies through joint ventures or partnerships. One of the bigger challenges faced by the western partner has been to reconcile the gulf between its inherent and mature Safety, Health and Environmental (SHE) culture and adherence to good SHE practices and those of its Russian partners. While it has been relatively simple to introduce corporate standards and procedures, transposing these to implementation at operational levels has been, and still is, a major hurdle. In Russia, owned subsidiaries are in themselves legal and autonomous entities and each General Director of such subsidiaries has the legal power to veto corporate efforts to impose SHE standards that he/she views as 'non-compliant' with the regulations or which are not in his/her best interests. In many instances the resistance stems from the necessary changes to the way SHE is managed and controlled in a way that places a direct and accountable responsibility with senior management. As the minority shareholder the consequent limited influence exerted by the western partner on improving

SHE performance gives rise to a serious concern that, in the event of a major accident, it will be accused of operating to double standards.

A SHE culture founded upon the risk-based approach and trying to do better than prescribed standards, promotion of a good reputation and care for the individual and the environment has matured in western partner companies over decades of burgeoning public concern and scrutiny and the increased expectations of the 'green' investor. An additional western 'driver' that clearly gets management attention is the relatively recent legislation making the most senior managers responsible and with the possibility of being found criminally negligent, if it can be shown that their failure to implement SHE measures contributed in any way to an accident.

The formerly state-owned industries of the Soviet Union did not have external visibility and, although Russian and other former Soviet Union republics have a long tradition of oil and gas activity, there is a different culture around SHE issues. Comprehensive and highly prescriptive SHE regulations exist and many of these mirror western standards but they have been developed without consultation with the industries that they regulate. The inflexibility of the prescriptive approach and its generic application often means that appropriate SHE measures are not developed on a site specific basis and in any case the corporate response, at least amongst the older generation of management, is to either merely comply with or pay lip-service to the law as cheaply as possible. In many cases, they pay the fines for non-compliance. SHE measures are seen as a cost to the business, rather than as an investment to achieve better performance and higher profitability. Enforcement of standards is correspondingly poor. There is a 'blame culture' that inhibits managers taking any personal responsibility for SHE issues. Accident investigations generally seek to lay the blame on the unfortunate victim (his own stupidity caused the accident) rather than to discover the root-causes and prevent recurrence. Internal shareholder and public pressure for better performance has been, and still is, lacking.

There is an emerging generation of native young engineers and managers, who are well-trained and cognizant of the need for better SHE performance in the new profit-focused privatized industries. However, they are frustrated by their inability to make progress against the entrenched attitudes and lack of personal commitment to SHE of the 'old-guard' management hierarchies. Monetary bottom lines and profit margins were much less important than production targets in the Soviet period and there is a strong resistance to change.

The situation is somewhat better in some of the former Soviet Union republics in that they are trying harder to change their SHE culture in a break from their Soviet past. However it may also be said that at the time of the collapse of the Soviet Union, these republics were left largely bankrupt. Industries were virtually defunct because of lack of investment and maintenance and much of the technological expertise returned to Russia. Alliances with western energy companies were made to provide the necessary investment, expertise and technology to develop natural resources and provide access to lucrative western markets. Consequently, western influence on the required and expected SHE standards was commensurately much greater.

This paper gives an anecdotal account of the differing approaches to SHE in the oil and gas industries in the UK and Russian Federation (RF). The paper seeks to explain differences in SHE performance between the RF and the UK and searches for ways to transfer some of the best international SHE practice, having regard to the sometimes large cultural differences and entrenched attitudes.

The authors are both engineers, working in SHE in the Russian Federation Oil and Gas industry. This paper is based on their personal experience, mainly in the oil and gas sector. No attempt has been made to produce a literature review of similar or related work in this field, although this would be useful in the future.

Both authors have a deep love of Russia, its people and culture. Any criticisms in the paper, either explicit or implicit, are made out of a desire to provoke debate and interchange of ideas and practice, in order to further the cause of improving SHE performance in all countries.

ACCIDENT RATES

International Labour Organization (ILO)[1] fatal injury rates are presented in Figure 1. Table 1 presents the rates for men and women in selected years.

The latest year that the ILO has statistics for fatal injuries for both the RF and the UK is 2005. In this year and over all employment, the number of fatal injuries per 100000 employees was 12.4 in the RF and 0.6 in the UK. The fatal accident rate in the RF was 21 times that of the UK in 2005. Moreover, the relative rates have been getting steadily worse.

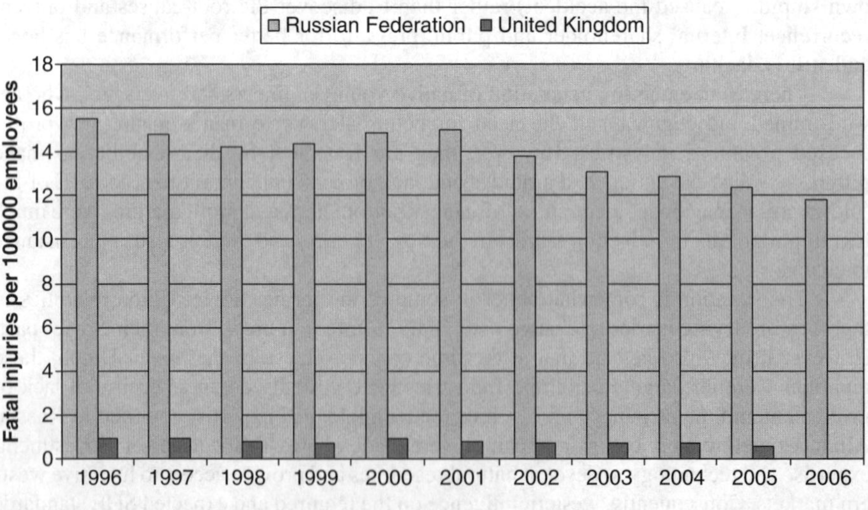

Figure 1. Fatal injury rates

Table 1. FATAL INJURIES reported per 100,000 employees

Year	1996		2000		2003		2005	
	RF	UK	RF	UK	RF	UK	RF	UK
Men	26.4	1.8	25.0	1.6	22.3	1.3	21.1	1.2
Women	1.8	-	2.0	-	1.7	0.1	1.9	0.1

The fatal accident rates in the RF and UK have fallen from 15.5 and 0.9 respectively in 1996, when the ratio was 17. These statistics indicate that the RF is far behind the best international performance on preventing fatal accidents and that the gap is widening. This is cause for concern.

ILO[1] non-fatal injury rates are presented in Figure 2 and differentiated by gender in Table 2.

The non-fatal injury rates for men reported by the ILO were 389 in the RF and 798 in the UK in 2005; the RF rate is less than half that of the UK. Again the rates in both countries have fallen since 1996, from 845 in the RF and 1008 in the UK, but the RF reporting rate has decreased significantly in relative to that of the UK.

A basic rule-of-thumb and an accepted ratio for approximate calculations is that there is 1 fatality for every 600 non fatal accidents; this is supported by the UK figures. If we apply this approximation to the RF fatality figures then the non-fatal injuries rate for men in Russia should be approximately 12,700, equating to 1 in 8 male employees. An extraordinarily high figure but based on field experience, knowledge of work ethic and

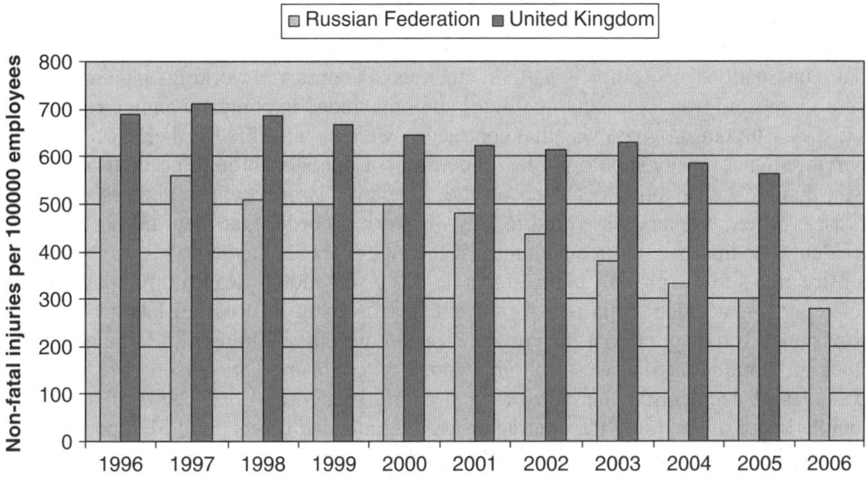

Figure 2. Non-fatal injury rates

Table 2. NON-FATAL INJURIES reported per 100,000 employees

Year	1996		2000		2003		2005	
	RF	UK	RF	UK	RF	UK	RF	UK
Men	845	1008	679	946	508	906	399	798
Women	285	364	269	334	225	347	180	324

management attitude it is not unreasonable to believe that at present least 1 in 10 persons in the RF will suffer a non-fatal injury every year.

In both countries it is clearly much safer to be a woman or perhaps women everywhere are more adept at letting men take the risk!

Such a high fatality rate is a considerable concern and it is possible that one of the reasons may lie in the extremely low level of non-fatal accident reports. From practically experience the authors are uncomfortably aware that many injury accidents are unreported and at times are quite deliberately covered up to avoid investigation and associated paperwork. Such action is commensurate with management being unwilling to take ownership of SHE issues. The low level of reporting suggests a very low frequency of investigation and commensurate recommendations for remedial actions to be taken to help prevent a potentially fatal recurrence.

ACCIDENT AND NEAR-MISS REPORTING

In the experience of the authors, accident statistics are hidden or suppressed even among companies with a high level of western influence. In one large Russian oil company with a major international oil company partner, fatalities of contractors working on company sites were segregated from company fatalities and not included in company annual reports and statistics. The reason given was that contractors were responsible for their own SHE and therefore the company should not be seen as a poor operator through no fault of its own, even though the fatalities occurred on the company's property. The authors are aware of cases where workers were paid to stay off work, in order that they did not report an accident. For instance, an acquaintance of the authors broke his leg while working on a construction site. He proudly claimed that he was give 3 months leave on full pay if he did not report his accident (this is a significant improvement on Russian labour sick-leave provisions). It transpired that the construction site was the building of a large new office block for a major Russian/western joint venture oil company. This person gave four other examples of accidents in his work area similarly treated and had heard of others. On completion of the building the contractor was commended for its good SHE performance and very low LTI frequency by the western partner.

Reporting of 'near misses' is nearly non-existent and such events are usually viewed as a 'lucky escape'. The near miss is rarely acknowledged as a significant event and almost never investigated to the same extent as a 'real' accident. Consequently many more SHE lessons remain unlearned and the learning process remains static.

Accident investigation is at best cursory and usually blames the victim for stupidity. Root cause analysis is not an accepted practice and its introduction by western partners is often resented; particularly where poor SHE management is cited with the implicit blame that bestows. So, not only are many accidents not reported and therefore not investigated but for those that are investigated the investigation usually fails to seek out and establish root causes, identify lessons learned and make recommendations to avoid recurrence. As a result lessons are not learned, changes are not implemented, hazards remain, risks are not mitigated and the circumstances contributing to the death or injury remain in place to recur.

EXPERIENCE OF CULTURAL DIFFERENCES

FIRE EXITS

A fire drill was held in Granherne's initial office in a Moscow business centre. After going down many flights of stairs, no-one could exit the building because the external doors were locked. People had to wait for 'security' to unlock the doors.

Fire Exits are usually locked in Russia! Why? "If they were not, we would not have security". This statement was made by our office manager, who was explaining why the fire exits were locked in Granherne's current office.

Russia seems to be obsessed with security. Go to any shop or office in Russia and, whatever else, there will be security guards. Security seems to take precedence over safety. The security guards unlock the fire exits. Key boxes, with the keys behind breakable glass panels, have now been installed next to the office fire exits.

FIRE ALARM

The alarm call points in the current Granherne offices do not sound a general alarm. They sound an alarm in a 'dispatchers' room, which is said to be continuously staffed. The dispatcher receives a visual alarm and it is up to him or her to start a tape with an alarm message for broadcast to selected parts of the building. Security will then unlock the fire exits.

ENVID AND HAZID 'WORKSHOPS'

ENVID and HAZID workshops are events where the project team identifies significant environmental and safety hazards and then assesses the associated risks. It is interesting to compare two such workshops, which were organised by Granherne for a Russian client company, in terms of the different approaches and the participants' responses.

The operating company participants in an ENVID workshop were required to attend by a director. They all turned up and took part. The ENVID was done by first compiling an environmental hazards register, listing the activity and environmental impact, etc. The register was provided to the workshop participants and they were led through the ENVID process by examining each hazard in the register, making changes and additions where necessary. There were no problems with this approach.

The participants in a HAZID were invited by a peer in the organisation. Many of them did not turn up, saying that they were too busy. Would they have found time, if required to attend by a senior person? The main difference to the ENVID was that the HAZID workshop started with a blank hazards register. There was much explanation of the HAZID process and there were two sessions where the participants were led through this process by identifying some hazards, assessing the risk and determining avoidance, control and mitigation measures. However, the response was very different to that at the ENVID. There was resistance to the HAZID process, because "it was not understood". Moreover, there were complaints that they (the operating company and head office people associated with the project) should not be doing the consultant's work for them. There was no understanding of the need for 'brainstorming' and that, in order for the project team to own the hazards, they must identify and assess them.

These experiences highlight two important cultural differences:

1. There is control over people's response to situations, even when this is contrary to safety – and an acceptance of this control.
2. There is a reluctance to take ownership of SHE issues, or even contribute to any work in this area, unless specifically required by superiors to do so.

Contrast this with the culture in the UK, where people would be aghast that fire exits should be controlled by security and where everyone is encouraged, even required, to take ownership of SHE matters.

SHE REGULATIONS IN THE RF

OVER-PRESCRIPTION

Good regulations exist in the RF and these mirror most western regulations. For example, the requirements for Environmental and Social Impact Assessments on new projects are very comprehensive and differ only in that they are more technically focused and have much less emphasis on the softer more discursive issues. Almost all the SHE regulations are prescriptive and, where overlap occurs, such as between the oil spill response regulations produced by the Ministry of Ecology and the Ministry for Emergency Situations, they become frequently ambiguous in their interpretation and application. Although there is a supposed consultation process between the Government and industry in drawing up the regulations this is practically non-existent and industry response to proposed legislation is mostly ignored. The regulations are often not enforced in a way designed to ensure that they are effective. For example, company emergency response procedures are written to meet the style prescribed by the regulations and for management this becomes more relevant than ensuring that the document content is appropriate to meet the best interests of the company.

Some of the prescriptive environmental standards, such as the allowable oil-in-water content for offshore discharges in the Caspian Sea, are technically impossible to achieve except under laboratory conditions and the fines for non-compliance are standardised. It is perhaps hard to understand why an unachievable standard was set. However, the problem is that if it is easier to pay the fine and be allowed to continue operating than to comply with the regulations, then the regulations become ineffective. This means that the quality of discharged water is much worse than would be the case if achievable standards were set and properly policed. This problem also highlights the absence of any meaningful process for consultation between Government ministries and industry to make sure that any proposed legislation is effective and sets achievable standards.

The problems that arise with very prescriptive legislation extend beyond mere compliance at the front end. A particular example of this is in the laws covering oil spill response. The legislation is comprehensive and even prescribes the length of time that can be taken to clean up an oil spill[2].

For example, in August 2003 the tanker "Viktoria", of 5000 tonnes deadweight, was loading at a port in the upper Volga River. The vessel exploded and spilled some 4000 tonnes of oil into the river. This accident did not receive much publicity but it is well documented outside Russia. The "Viktoria was registered as a foreign going vessel and was covered under international fund conventions for oil spill clean-up costs. (In fact the vessel was the first to be covered under the fund in the upper reaches of a river). Consequently the insurers sent a representative from the International Tanker Owners Pollution Federation (ITOPF) to witness the clean-up operations and to ensure that the accident was fully recorded and that costs and claims were justifiable. A summary of the ITOPF report was included in the Annual Summary of the IOPC Fund[3]. This report stated that some 50 km of the Volga shorelines were polluted and that clean up operations were still taking place in late November, until the encroaching ice prevented further work. Locals reported that oil remained visible in localized patches after the thaw in the following Spring.

Official reports in the Russian system stated that all the oil was fully contained locally and that all the oil was removed from the surface by the end of September, within the time required by the regulations. To have not reported this would have led to censure of the chief of the clean-up operations for failing in his duties. The fact that to have achieved such a feat was technically impossible becomes irrelevant. Sadly then, all details of the clean-up operations and arrangements subsequent to this are omitted from the official report and, again, valuable lessons are lost.

This example highlights prescriptive regulation inhibiting objective recording; whereby many valuable lessons regarding oil spill response on the Volga river have been lost. When the IOPC Fund annual report was shown to a member of the Ministry responsible for the clean-up operations, it was condemned as a lie. This further demonstrates the attitudes that prevail at ministry level and supports the view that there is reluctance to accept the need for change.

SHE VIEWED AS A COST, NOT AS AN INVESTMENT

A good example of this is the RF regulatory requirement that an oil or chemical industrial site should have a person in charge of emergency response and civil defence. The intent of the regulation is clear but compliance is minimal; persons recruited for this task have little or no specialised training and they have no authority and no control over resources. They are in place merely to provide the 'tick in the box'. Most western companies view such a position as an important component part of the process for protecting the company's assets and as an investment. The person will be well trained, be given good resources and the necessary authority to do her/his job.

Paying fines for non-compliance, rather than complying, because the alternative would be too expensive, is often the platform used to make payments to local inspection authorities. These payments have become, over time, an expected form of 'good-will' payment that also ensures a favourable scrutiny at the next inspection. An area where non-compliance is frequently found is in the failure to meet the full extent of fire prevention and fire fighting regulations. Insistence on compliance would mean that these fines would disappear together with the good-will of the inspectorates. This is a further and important secondary factor in determining the reason for resistance to change.

In summary the authors propose that Western companies see regulations as the minimum standards to be achieved when determining what SHE resources need to be in place for the proper protection of the company and its assets. In the RF the highest standard to be achieved is meeting the regulations and if this is not possible or cost-effective, 'we will pay the fines'.

THE CHALLENGE

It is perhaps too easy to be critical of a highly prescriptive approach that is difficult in the extreme to police and even more difficult to ensure compliance. A deeper look is required to see why this approach originated and the existence of the enduring post-Soviet 'hangover' that is still acting to prevent SHE improvements in the RF.

During the Soviet Union period all industry was state-owned and managers were tasked to meet production targets. There was no consultation on SHE matters, the state was the industry and therefore its own regulator. The Soviet Union was massive with many autonomous regions under central control exercised by the ministries in Moscow. The controllers, as regulators, realised that unless SHE requirements were to a common standard, strictly prescribed and highly regulated with financial penalties for non-compliance, they would not be implemented. Indeed the foresight of these regulators in seeking to establish a reasonable SHE culture is laudable and in this light it is easier to understand the process they chose. The 'hangover' stems from:

- The difficulty that western companies have in understanding and working with a prescriptive SHE regime.
- The strong resistance by the regulators to take a more constructive and consultative approach with industries. This is viewed by many of the old-guard in the Ministries as a prelude to losing the necessary levels of central control.
- The reluctance of the regulators to review and revisit legislation to ensure it remains appropriate.

- The lack of any formal requirement for senior management to take personal ownership and responsibility for SHE issues.

The inhibitors to change may be summarised as follows:

- A 'blame' culture that prevents objective accident reporting; investigations seek to lay blame rather than discover the root-cause.
- Absence of formal requirements for risk assessment and no task-based risk assessments.
- No personal responsibility or ownership of SHE at any level:
  - managers don't want the responsibility,
  - managers are not made to take responsibility,
  - this attitude 'cascades' through the organization.
- There is neither shareholder nor public pressure for better performance.
- Acceptance of control, with customary violation of over-prescriptive regulation.

HOW TO ENCOURAGE CHANGE

At almost all levels (except at senior middle management where most of the costs and responsibility for implementing change will be most felt) there is an increasing acceptance and support of the need for change and an increased awareness of the benefits that those changes would confer.

The concepts of cost benefit analysis are well understood in the RF but have not been applied to SHE. Applying this to SHE would in most instances provide the necessary justification for investment and training.

The authors would encourage the formation of Industry groups, for example representing Oil and Gas, Mining, etc, to create a cohesive lobby to take forward common points of interest. The priority of this group would be to encourage and build trust between the Regulators and industry with the aims of improving the consultation process and creating a good working relationship.

Compare the situation in Azerbaijan where, although a substantial amount of prescriptive legislation remains from its former Soviet Union days, the Government has conducted positive consultation with the oil companies and has agreed to changes to some procedures, whereby it agrees and sets achievable standards and targets, with penalties for non-compliance. Effectively there is the development of an embryonic self-regulatory system allowing a company to meet set targets in whichever way is best and without the restraints imposed by prescriptive methodology. The Government has acknowledged that this is much easier to monitor and control; any non-compliance is immediately apparent and penalties can be invoked where necessary. Persuading the Russian Ministries to adopt this approach is perhaps the most significant challenge and until this process begins it presents the ultimate hurdle to sustainable SHE improvement.

On a highly positive note, the Russian Government has recently tabled new legislation to promote the use of renewable energy and which describes the process for the remuneration of energy producers who introduce the provision of energy from renewable resources. This is encouraging for the environmental lobby but in a country where there is an abundance of steam coal and oil, produced much more cheaply than the investment required to sustain a renewable energy system, support for such change is perhaps harder to understand.

REFERENCES

1. International Labour Organization 'Laborsta' statistics, http://laborsta.ilo.org, accessed 25/9/7.
2. Russian Federation Regulation 613.
3. International Oil Pollution Compensation Funds – Annual Report 2003, http://www.iopcfund.org/publications.htm, accessed February 2004.

ACCIDENTS OF THE NEXT 15 YEARS?

Trevor A Kletz
Visiting professor, Loughborough University

From 1968 to 1983, the Heavy Organic Chemicals (later Petrochemicals) Division of ICI published a monthly *Safety Newsletter*, 171 issues in total. The circulation was at first small but grew rapidly and by the mid-seventies reached about 2500, as copies were sent to all parts of ICI, other oil and chemical companies, academics and regulators. It was an early example of open access.

Most of the incidents described are still recurring today so the IChemE is posting the *Newsletters* on the Internet. This paper describes their contents, illustrates them by examples and shows how they can be used to reduce accidents. This old information is still relevant as although designs have changed a more important factor – human nature – has not.

The *Newsletters* were not intended primarily for safety experts but for all those involved in design, operations, maintenance and construction, at all levels but especially at the professional level

KEYWORDS: Accidents, Experience, Information retrieval, Process safety, Safety

In 1968 I was appointed safety adviser to ICI Heavy Organic Chemicals (later renamed Petrochemicals) Division with responsibility for what we now call process safety. Among the many actions I took, described in (Kletz 2006), was the preparation of a monthly *Safety Newsletter*, usually 8 pages long. I sent copies of No 1 to the about 30 colleagues. Gradually, over the next fourteen years, the circulation and contents grew spontaneously. I did not advertise it, but added people to the circulation list at their request. By the mid-1970s the circulation was about 2500 and, as well as other ICI Divisions, included many outside companies in the UK and elsewhere, universities and the Health and Safety Executive. The *Newsletters* were not intended primarily for safety experts but for all those involved in design, operations, maintenance and construction, at all levels but especially at the professional level. I made it clear to the outsiders who received the *Newsletters* that they could be copied for circulation within their organisations but not offered for sale. Some companies circulated them widely.

Within ICI the *Newsletters* were seen by division directors, managers, foremen and, in some works, operators. The contents consisted mainly of reports on accidents of general and technical interest from ICI and also from other companies, which they supplied in exchange for the *Newsletters*. I did not copy the original reports, but rewrote them to bring out the essential messages. Many of the later *Newsletters* were devoted to specific themes, such as accidents due to plant modifications, preparation for maintenance, static electricity and human error. After I retired from ICI, I edited many items from old *Newsletters* and published them in a book called *What Went Wrong?* (Kletz 1998). Now in its 4th edition it is twice as long as the first edition and is my best-selling book. I have added many later reports and also written a supplementary volume, *Still Going Wrong?* (Kletz 2003) Both books are available from IChemE (see www.icheme.org/shop).

Many people were surprised that ICI allowed me to distribute reports of our errors all over the world but if we have information which may prevent accidents there is a moral duty to pass it on to other people. In addition it was to our advantage in several ways:

1. **Economic:** ICI spent a lot of money on safety. By telling our competitors what we did we encouraged them to spend as much.
2. **Pragmatic:** we got useful information from other companies in return.
3. **In the eyes of the public, the chemical industry is one**. The whole industry suffers if one company performs badly. To misquote the well-known words of John Donne:

> No plant is an Island, entire of itself; every plant is a piece of the Continent, a part of the main. Any plant's loss diminishes us, because we are involved in the Industry: and therefore never send to know for whom the inquiry sitteth; it sitteth for thee.

Colleagues and other companies were willing to let me describe their accidents and so-called "near misses" (actually near accidents), which usually reflected no credit on them, because I did not say where they occurred (except when the location was stated in the title of a published report). The *Newsletters* were thus an early example of "open access" though the phrase was not then used. When I retired from ICI the company gave me permission to reproduce or quote from them as much as I wanted, provided I did not say where they occurred or in which company. If anyone asked me - only a few did - where an accident had occurred I apologised for my poor memory. Now, as a further step in open access, IChemE are making all 171 *Newsletters* available on the Internet. Other companies' reports may be added later.

The information in the Newsletters is given in good faith but without warranty. Much of the advice is decades old and better methods of prevention may be available today. There are many possible solutions to most problems. However, the accidents happened, many are still being repeated today, and readers should therefore ask themselves, "Could this occur where I work and, if so, how do I or should I prevent it?"

In the period covered by the *Newsletters* (1968–1983) the Factory Inspectorate, and after 1974 the Health and Safety Executive, had a lighter touch than today. For this reason there are fewer references to the law than there would be if I was writing today.

In rent years ICI has been is very different from the ICI I knew. Except for the paint factories, with which I had little contact, all the plants owned in 1982 when I retired have been closed or sold to one of a large number of different companies. None of the incidents described n the *Newsletters* occurred on plants operated or owned by ICI in recent years.

I wrote everything in the *Newsletters* myself except for the engineering articles in the later issues most of which were written by Harland Frank, an outstanding engineer.

After I retired from ICI in 1982 the *Newsletters* continued for 18 months, written by my successor, Alan Rimmer, and were then abandoned when he retired early.

Many companies' monthly safety reports look like all the other memoranda we get. The *Newsletters* stood out as they were printed in fairly large print on good quality paper with clear diagrams, essential features if we want busy readers to recognise them as something worth reading and as something they can read and absorb in odd moments. For the layout of your safety reports take advice from those who design the leaflets that are sent to your customers. The *Newsletters* have been retyped for the Internet so that there is a unity of font and style but the wording is unchanged except for a few extra cross-references.

Many readers may wonder if information from 1968–1983 is still relevant. When I retired in 1982 and started working as a consultant as well as a visiting professor I thought that my life as a consultant would be no more than five years as after that I would be out of date. It has not happened. Many of the accidents described in the *Newsletters* are still recurring and many of the problems discussed are still puzzling people, as shown by the examples below. (See also the Afterthought on the last page.) It is also important to remember that while equipment has changed a more important factor, human nature, remains the same. Are you any more reliable than your parents or grandparents? Perhaps less, as when the *Newsletters* were written there were more people in design and operations and industry had not adopted the extraordinary practice of retiring people when their knowledge and experience were at their highest

US readers should note that some engineering and management terms have different meanings in the two countries. There are glossaries of them in the two books mentioned above but the following can be particularly confusing:

- In the UK a plant manager is usually someone at the lowest level of professional management, equivalent to a supervisor in the US. The UK equivalent of a US plant manager is called a works manager or factory manager. In the UK supervisor is usually another name for a foreman but can be anyone to whom other employees report.
- A chargehand is a rather old-fashioned UK name for a lead operator.
- Lagging is a UK name for insulation; flex is a UK name for a hose.

SOME *NEWSLETTER* ITEMS THAT ARE STILL RELEVANT

You can search the *Newsletters* for accident reports or information on particular equipment, substances and operations. For example, if you are thinking of fitting a sight-glass a search for that term will take you to *Newsletter* 35 (November 1971) where you will see that:

> Level glasses are always liable to break and it is, therefore, the policy of the Division to install ball check cocks in the lines connecting a level glass to the parent vessel. If the level glass breaks the pressure of the liquid in the vessel pushes a ball against a seat and stops the leak.

The ball check cocks form part of an isolation valve. *They will operate correctly only if the isolation valve is fully open or almost fully open.* They will not work correctly if the isolation valve is half-shut.

When a level glass connection broke recently there was a large escape of gas which caught fire and injured a man. The ball check cock did not operate because the isolation valve was nearly closed.

Please make sure that your operators know that these valves must be fully opened and then left just cracked off the back seat position.

On some plants the balls have been found to be missing. You may like to check that on your plant they are all present.

For comparison, the American Institute of Chemical Engineers publishes every month on the Internet and reprints in *Chemical Engineering Progress,* a Beacon, an illustrated one-page summary of an accident report. The July 2007 Beacon reports that while dealing with a leak and fire an operator tried to kick a valve closed and accidentally broke a sight glass, thus making the leak and fire worse. The recommendations do not mention ball check cocks.

The original leak came from a hose which had been repaired with tape. There are many references to hoses in the *Newsletters*, for example, *Newsletter* 44/4 (September 1972) reports that:

Before removing a hose a man tried to drain it by loosening the coupling nut. Hot water came out of the coupling and scalded him. In the past, men have been burnt by corrosive chemicals in this way.

Whenever hoses are used at pressure, a valve should be provided for blowing off the pressure, as shown below:-

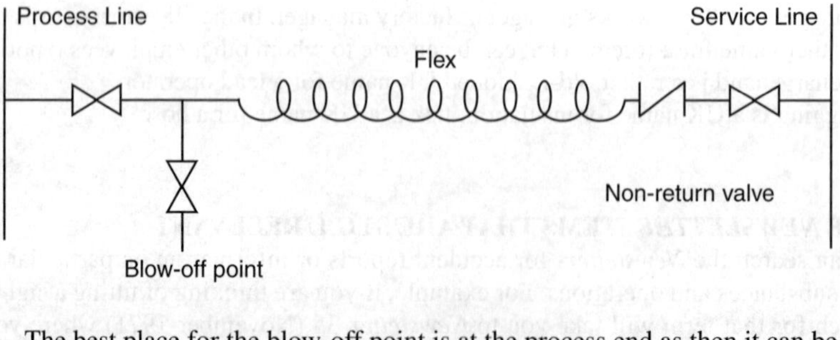

The best place for the blow-off point is at the process end as then it can be used to prove that the hose is clear, by opening the service and blow-off valves before the process vale is opened.

Note that in the diagram the hose is called a flex, at the time a common term in the UK.

There are many references to flexes in the *Newsletters*. The following one is from *Newsletter* 7/7 (January 1969):

> An accident involving oxygen is described in the October 1968 issue of "Accidents", published by the Factory Inspectorate. Welding was taking place inside a tank. The cylinders were outside and flexible hoses led to the welding set. One of the men lit a cigarette and noticed that it burned away more quickly than normal and that his lighter-flame was longer than usual. He did not, however, realise what this meant. When the welder started to weld a spark fell onto another man's pullover; it immediately caught fire and spread to his entire clothing. He later died from his injuries.

A search for "flex" brought up the following report, containing the word "flexible", from *Newsletter* 47/4 (December 1972):

> Corrosion was suspected on a distillation column. Ultrasonic thickness measurements were therefore made on the outside of the shell. These showed that although some corrosion had occurred, the thickness was still well above the design minimum.
>
> Some months later, when it was possible to take the column out of use, the lagging (insulation) was removed and it was discovered that part of the column was so thin that it could be flexed by hand.
>
> The thin spot was immediately opposite the vapour return line from the reboiler. The thickness measurements had been made on the other side of the column where the staging and ladders made access more convenient.

The lessons to be learned are:

1. Thickness measurements in distillation columns should be made at the points at which corrosion is most likely to occur. In the case described above, this was opposite the vapour return line. Often there is a baffle near the return line and corrosion is then most likely near the edges of the baffle. The geometry of the column must be studied.

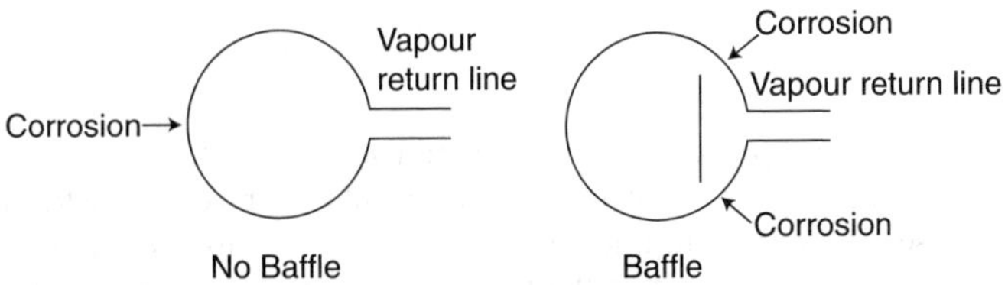

2. During design, access ladders should be positioned to facilitate thickness measurements at the points where corrosion is likely to be heaviest.

This report shows how a search for one term can lead to a voyage of discovery where all sorts of interesting and valuable information are brought to light. I hope I have convinced you that the information in the *Newsletters* is still relevant. Here are a few more items from more recent *Newsletters*:

Newsletter 96/4 (February 1977) described two incidents caused by reverse flow:

Reverse flow of catalyst

Some gases reacted in the inlet line to a convertor. The pipeline got so hot that it swelled and burst. At the previous shutdown the reactor had been swept out with nitrogen in the opposite direction to the normal flow and some catalyst dust had been deposited in the inlet pipe.

Reverse flow through a pump

Failure of a non-return valve caused gas at 25 bar to flow back up a liquid line when a pump stopped. This caused the pump and motor to rotate in the reverse direction at high speed. The motor was damaged beyond repair.

When failure of a non-return valve can have serious consequences, it should be registered for regular inspection. The use of two in series should be considered, preferably different types to avoid common mode failures.

There were also references to other incidents of reverse flow in an earlier Newsletter and to an article on the subject.

Newsletter 97/4 (March 1977) described a trip test that could not improve reliability:

On one Works, drums are filled with liquid product by an automatic device which weighs the drum and closes a valve when a pre-set weight is reached.

One day the valve failed to close. The drum was overfilled and the liquid splashed the filling operator's legs and feet. He tripped the supply manually and went to the locker room to change his overalls. While doing so he slipped and twisted his knee.

The investigation revealed that this was not the first time that the valve had failed to operate. The operators said that it had happened "once or twice before in the last year or two".

Amongst the actions proposed to "try to eliminate" the incident was the institution of trip testing once every two weeks. This will have no effect on the failure rate because 700 drums are filled every week and the trip is therefore tested 700 times per week! Failure of the trip mechanism will almost certainly be followed by a drum overflowing so a test is very unlikely to detect the fault.

The investigation report also suggested a much more effective way of reducing the failure rate: check the mechanism for adjustment and look for signs of wear and for this a lower frequency would seem appropriate, say once per month.

Newsletter 83 (January 1976) was devoted to accidents caused by the unforeseen effects of plant modifications:

Minor modifications are those so cheap that they do not require financial sanction. Often the only documentation associated with them has, in the past That is, before the mid-1970s), been a workshop chit or just a permit-to-work (clearance certificate).

A small leak of liquefied petroleum gas (LPG) from a passing drain valve on a pipeline produced a visible cloud of vapour about 5 feet across. The leak was soon stopped by closing a valve but the investigation brought the following to light:

1. The company's standards required two valves in series or a single valve and a blank.
2. The valve was made of brass, was of a type which was stocked for use on central heating and domestic water systems and was not of the correct pressure rating for LPG.
3. The valve was screwed onto the pipeline, although the company' standards di not allow screwed fittings on new installations except for domestic water lines and certain small bore instrument lines.
4. Since the LPG fire at Feyzin in 1966, which killed 18 people and injured 81, the company had drawn up standards for LPG, tried to publicise them and carried out numerous inspections to see if the equipment was up to standard. £30,000 (at 1970 prices) had been spent on the plant concerned on improving the safety of the LPG handling equipment. Nevertheless, subsequently someone installed a sub-standard branch. Presumably the detailed design of the branch was not specified correctly, if at all, on the chit placed on the plumbers. In addition, neither the man who installed the branch, nor the supervisor who handed the job back, nor the man who accepted the clearance back, none of the men who used the branch and none of the men who passed by, noticed anything wrong.

Like the plants in our gardens, our plants grow unwanted (and often unhealthy) branches.

Other minor modifications which have had serious affects on plant safety are:

Removing a restriction plate which limits the flow into a vessel and which has been taken into account when sizing the vessel's relief valve. A length of narrow diameter pipe is less likely to be removed than a restriction plate.

Fitting a larger trim into a control valve when the size of the trim limits the flow into a vessel and has been taken into account when sizing the vessel's relief valves.

The *Newsletter* also discusses other sorts of modification such as those made during start-ups, temporary ones, modifications made during maintenance and sanctioned ones, that is modifications for which money has to be approved and which are usually – though not always – considered more thoroughly than others.

None of the incidents described occurred because of a lack of knowledge of methods of prevention; they occurred because no-one foresaw the hazards and no-one asked the right questions. The *Newsletter* described ways to prevent similar accidents in the future.

Finally, here is a short item from *Newsletter* 68/5 on **how to stop ball valves or cocks on vertical lines vibratingen.**

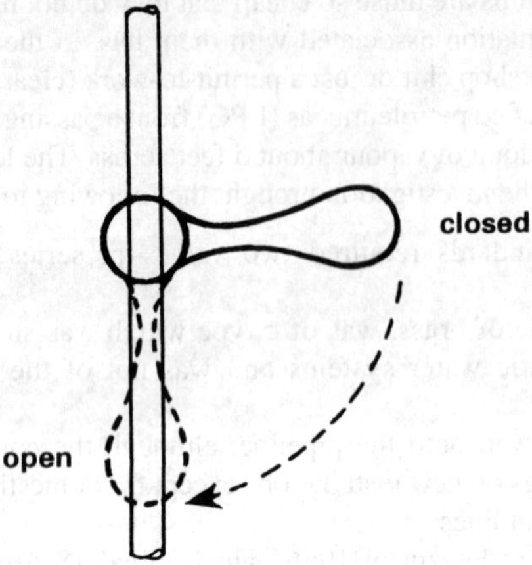

To prevent this happening the valves should be installed so that when they are open the valve handle points upwards.

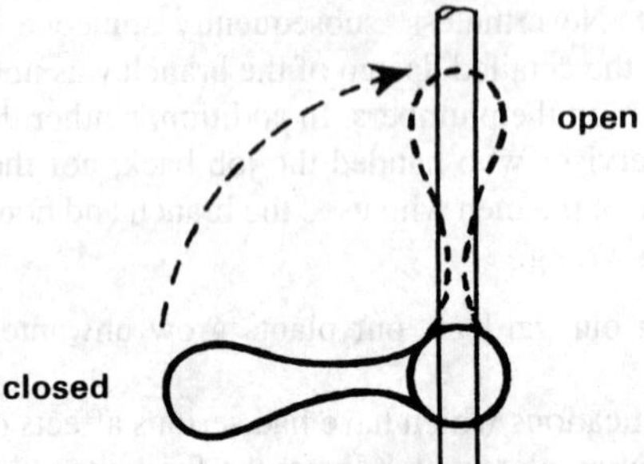

HOW TO GET THE BEST OUT OF THE *NEWSLETTERS*

I do not expect anyone to read right through the *Newsletters* as if they were a book but you may like to browse them, as I have done above, to see their scope.

At a safety meeting you can describe or distribute an accident report from the *Newsletters* and then ask those present why the incident occurred and if it could occur on the plant they operate or are designing. If it could, what have they done or should they do to prevent it happening. Remember that the advice given in the *Newsletters* may not be the best available today or the best for your company. "Believe in the motto: 'If it ain't broke …' And even if it is broke, you don't need to mend it the same way as everyone else." (Kellaway 2007). Also remember that discussions are a more effective method of learning than listening to a lecture or reading (Kletz 2006).

Alternatively, you can give a different accident report to everyone present and ask them to answer the same questions at the next meeting.

Another way of using the *Newsletters* is, when an accident occurs, to look in them to see if anything similar is reported in them. This will encourage your colleagues to use the *Newsletters* in the future.

Whichever way you use the *Newsletters* they could help you prevent the accidents described in them, most of which occurred between 1968 and 1983, happening again during the coming 15 years.

AFTERTHOUGHTS

Only that shall happen
Which has happened,
Only that shall occur
Which has occurred;
There is nothing new
Beneath the sun. – *Ecclesiastes 1:9*

"The reality is that mission statements have done little to change the corporate world for the better… People do not change by dint of a statement, no matter how carefully drawn up it might be" (Kellaway 2007). Telling them what has happened and will happen again unless they learn from it is more effective. Better, let them tell you what they think is the bet method of prevention.

REFERENCES

Kellaway, L., 2007, Systems: the good, the bad and the ugly, *Daily Telegraph*, 22 August, p. B4.

Kletz, T.A., 2006, How we changed the safety culture, *Hazards XIX, Process Safety and Environmental Protection, What Do We Know? Where are We Going*? Institution of Chemical Engineers, Rugby UK, pp. 83–94.

Kletz, T.A., 1998, What Went Wrong? *Case Histories of Process Plant Disasters*, 4th edition, Gulf Publishing, Houston, Texas, 1998.

Kletz, T.A., 2003, *Still Going Wrong: Case Histories of Process Plant Disasters and How they Could Have Been Avoided*, Butterworth-Heinemann, Burlington, MA.

Kletz, T.A., 2006, Training by discussion, *Education for Chemical Engineers*, 1(1):55–59.

LESSONS LEARNED EQUALS IMPROVED SAFETY CULTURE

F K Crawley
Department of Process and Chemical Engineering, University of Strathclyde,
Morrison Street, Glasgow G1 1XJ, UK

This paper is a challenge to all processing companies to adopt a new approach to the use of knowledge gained inside or outside its area of expertise to change the safety culture of the company. It argues that first there has to be a corporate culture from the top down which is open and willing to learn lessons and then the determination to put those lessons into place. It also argues that there has to be a more open approach within the process industry and from the Regulator, with the willingness to exchange news, good or bad for the benefit of all.

The collation of lessons learned, their distribution and teaching of those lessons will require a senior role who will command the respect of the senior and junior managers and who will have to develop new skills, which will involve "networking" in and outside the company, interpersonal skills, investigative skills and finally teaching and projection skills.

Finally, the body corporate has to take on the ownership of the lessons learned to achieve an improvement in then safety culture.

KEYWORDS: Culture, Lessons Learned, Databases, Cultural Change

INTRODUCTION

The definition of the word *culture* in Oxford English Dictionary is *the arts, ideas of a nation, people or group.* This means that any change in culture will be slow due to the inherent resistance to that change, as, by definition the culture will have evolved over many years and the group will feel happy with it, perceiving that there is no need for change. This is to be found in many organisations particularly following take-overs by other companies where there is a resistance to a change in the culture. However the culture may be one of complaisance and flaws may have become incorporated within it over the many years of its evolution. A point noted by Robens [1972] and still valid today.

Culture is an engineering paradox; it has no mass but high inertia!

There has been recognition in recent years that the general safety performance is not falling at the rate that the Government and the Regulator would wish. This has been discussed in many papers and need not be elaborated upon here. Recent press reports suggest that the trend is actually rising for the first time for some years. However there is a clear loss of analysis and a lack of detailed examination of the lessons that should be taken out of any event be it inside or outside the industry of interest. The reasons for this have been discussed in various papers [Crawley 2006, 2007] and have been put down to a number of causes including, lack of mentoring and training, a parochial approach to ones own industry, poor review of and improvement in corporate design standards due to the lack of in-house engineering resources, and the inability to see the problems, or to challenge a design feature, under the pressures of keeping projects "on time and on budget". There is also a potential flaw in the use of "standards" as the historic reason for various requirements or features has become lost in the mists of time with the inevitable reaction "I do not see the need for this, so, I will not apply it!" (Is this the corporate half life or is it more fundamental?) The analysis of the Buncefield Fire [Martin 2007] gives many pointers or lessons which do not apply uniquely to fuel storage (such as mass balancing) but also have application in other storage and processing industries.

More recently there has been a lot of attention given to databases and their usage [OECD 2005]. Undoubtedly databases do contain much information but they are not always totally accurate for reasons of confidentiality, they dwell on the lead up to the incident and how it was handled and it is not easy to extract the "lessons learned" from them as they pertain to a specific company. It may well be that the culture of Company X has those lesson to be learned under strict control but in Company Y they may not be so well controlled and it may be difficult for Company Y to recognise this failing and to take the appropriate actions. The inevitable question is this – "How can we find the lessons to be learned, how do we incorporate them into the company and how do we influence or change the culture?" Particularly where there is a resistance to "lessons learned" from incidents outside a single industry. It is suggested that there may be more significant cultural issues which require to be changed.

However, when there are changes they have to be managed or incorporated properly to ensure that they do not incorporate worse problems!

The resistance to lessons learned is exemplified by the analysis of the issues associated with Offshore Relief and Blow-down Systems [RABS 2001]. First, there was some reluctance to accepting that the problems were not company specific but were industry specific and that each company had found its own solution and failed to see the better industry solution. In other words there was a resistance to *cultural change*. Second, there was initial reluctance to the exchange of information on incidents. This was overcome by the simple expediency of eliciting information by showing that the problems were not company but were industry specific. This could not and would not have occurred in total isolation and required facilitation and a dialogue which discussed incidents or examples outside the industry and allowing members in the team to recognise the parallels within their industry. This required a very "open approach" from participating members and the willingness to listen.

ARE THERE SOLUTIONS?

Before this question can be answered it is important to understand what *culture*, and more particularly what *safety culture*, actually means. While the OED definition might be applicable to a Nation or possibly a company it does not necessarily apply totally to *safety culture* and requires further expansion. There are many forms or sub-sets of safety culture which are given by Hudson [2001]. These are

- ***1 Pathological*** – Safety is regarded as a problem caused by worker, the main driver is the business and a desire not to be caught by the Regulator. The Organisation cares less about safety than about being caught.
- ***2 Reactive*** – The organisation starts to take safety seriously but there is action only after an incident has taken place.
- ***3 Calculative*** – Safety is driven by management systems, with much collection of data, rather than learning from it.
- ***4 Proactive*** – There is a realisation that with improved performance the unexpected is a challenge; workforce involvement starts to move the initiative away from a purely top-down approach.
- ***5 Generative*** – The safety behaviour is fully integrated into everything the organisation does, there is active participation in safety at all levels, and safety is an integral part of the business. Organisations at this level are characterised by the term "*Chronic Unease*".

1 is clearly unacceptable, 2 is probably more representative of the past, 3 is probably representative of the present National Health Service and was part of the BP culture in Texas City [Baker 2006], 4 is more the present approach but 5 is the *Holy Grail*.

Bond [2007] discusses the Just Safety Culture which is given the definition; "*A way of thinking that promotes a questioning attitude, is resistant to complacency, is committed to excellence and fosters both personal accountability and corporate self-regulation in safety matters*". He also introduces *Operational Monitoring* which is a new term for what was called condition monitoring and key process parameter trending. One trend which should be monitored is the drift in a materials balance across a section of plant – see later.

The closing point made by Bond is as follows "*Adoption of an Operations Monitoring system combined with a Just Safety Culture approach would give a new emphasis to improving safety by sharing accident information*".

The only relevant issue not discussed but is probably included implicitly within these definitions, is the *open* and *no-blame culture*. The conclusion to this analysis is that to achieve a significant change in safety culture there has to be the correct organisational culture, open and no-blame, followed by a top-down and bottom-up approach to safety. This prerequisite is due to the fact that the safety culture is a sub-set of the corporate culture and if the corporate culture is not appropriate the safety culture will suffer. Hence there must be a corporate culture which may have to be a change at all levels to ensure that the safety culture is appropriate.

PROPOSED WAY FORWARD

Clearly there have to be solutions but before these can be put in place there have to be so prerequisites. First there has to be an *open attitude* to the analysis of incidents inside and outside the specific company, that is, there can not be a parochial approach. Second, incidents must be examined by the use of the "***WHY?***" questioning approach which should be targeted at the identification of the *Root Cause* of the incident. Third, there has to be the recognition that cultural changes may be painful. Fourth, there has to be an acceptance that in any vital and alive company changes in culture are a necessity of its own development and there has to be the correct culture within the company which is willing to accept the changes that will result – the *generative* approach [Hudson LP2001]. The body corporate then has to take on the ownership of the lessons learned so as to improve its safety culture.

Before reinventing the wheel it is worth looking at the events of the past discover if these problems have already been addressed, in part, elsewhere. Someone, somewhere must have had to face up to them and may have found forgotten solutions. In other words "There is nothing new under the Sun!" This can be illustrated the simple "lesson" of mass balances, (a lesson taught very early in any Chemical Engineering Course but then forgotten in real life), which was relevant in two major incidents Texas City [Mogford 2007], where a Distillation Column was overfilled due to instrument errors and also Milford Haven [HSE 1997], where there was a major recycle of fluids in the blow-down system. Nearly 40 years ago during the Initial Start up of a Plant there was the unexpected rise in the pressure drop across a Distillation Column. The pressure differential stabilised after about one hour but was significantly higher than expected. A simple mass balance around the Distillation Column was carried out over the previous two hours and an imbalance of over 10% was identified, representing a major hold up of process fluids within the column. The immediate reaction was that the "instruments were in error" but the error was too high to be real and a further mass balance over the previous day showed no faults. This represented use of a simple form of Operations Monitoring [Bond 2007]. Feed into the column was stopped and the pressure differential remained elevated even without reflux and re-boil so clearly there was a choke within the column – or was it instrument error? Within an hour the pressure differential suddenly fell to zero and the base level rose rapidly [PE]. Clearly there was a significant hold-up of process fluids within the column and to proceed would have lead to a major upset. The lesson that was learned (or was not forgotten after leaving University) and is relevant to all industries and was relevant in Texas City and Milford Haven and Buncefield is that during non-steady state operation which will include start up, shut down and upset conditions, Operations Monitoring, or monitoring/trending the inventory, on a continuous basis in each section of the plant MUST be in place. Why was it not done in the three cases quoted? Is it a lack of training or a lack of appreciation of what is actually happening? (Follow the book and do not think!) It is not "rocket science"; it was used on other situations nearly 40 years ago but was forgotten, Operations Monitoring is just good Chemical Engineering. There are many key parameters that can be monitored other than mass balances, including:

Heat transfer coefficients – to detect fouling
Pressure drop parameters in distillation columns – to detect fouling or tray damage
Thermal imaging – to detect faulty electrical equipment or even to scan the internals of a unit
γ-Ray scans – to detect build ups of solids in equipment
Compressor or pump efficiency – to detect fouling or wear

Vibration monitoring and analysis – to detect mechanical wear or faults
Acoustic monitoring – to detect leaks in piping or heat exchangers
Oil monitoring – to detect wear and predict the life of the unit

All of these tools have been used by the author and found to be of great use in condition monitoring.

It follows that the design of the Plant must incorporate the instruments required to mass balance each section and to monitor the key parameters. It follows that there has to be a cultural change in the design house that requires the analysis of the design to ensure that all of the diagnostic features (which may or may not be flow meters) are incorporated into the final design. One of the findings of [RABS 2001] was that the instrumentation of the Flare Knock-out Drum was insufficient to carry out simple diagnostics, particularly the source of major inflows of fluids and a mass balance across the Drum with the resultant likelihood of over-filling leading to liquid carry-over. (This occurs about once every two years in the North Sea.)

By the same argument the HAZOP process must adapt such that it is able to analysis the diagnostics – the parameter could be *inventory* or *performance* and the deviation could be *change*. Another lesson that should be learned. Then, if it is not integrated into a control or data recording system, the operations team must be given the basics training of carrying out mass balances or performance parameters and a comprehension of why they are necessary and the significance of any change, a lesson learned in the Longford fire [Hopkins 2000].

It is now worth examining the lessons to be learned from Chernobyl [Franklin 1986] and the Ramsgate walkway collapse [Crossland 1999]. While these occurred in two different industries the lessons are relevant to the process industry. In both incidents there was a plan for the execution of the work but equally the plan was changed without being subjected to an adequate review. At Chernobyl there were a number of delays which lead to the poisoning of the reactor resulting in poor reactor control. More particularly the longer the plan deviated from the objective the greater was the likelihood of the final event. At Ramsgate the bridge was not the correct size (measure twice and cut once) and required an on-site modification. In both cases the universal lesson to be learned is that once a plan can not be carried out fully and as intended there is every possibility for the incorporation of hazards and the work must be stopped immediately the plant or work revert to the previous state and the situation analysed in detail.

USE OF DATABASES

Can the information within the databases be used to identify the lessons? The answer to this is a guarded "Yes!" There is a slow change within databases from the traditional record of the sequence the events to lessons learned, so, extracting the lessons may be easier, however it is likely that the lessons may be more difficult to extract from older databases if the data is not accurate and is insufficient detail. Too often the cause of an incident is put down to the catch-all "human error" and then the detail stops. Human error is actually *management error* or a pitfall put in the way of the operations group. If management do not recognise this pit-fall it is a management error and the lesson to be learned will most certainly not be "improved training". This is particularly important on the more complex an integrated process plants. The fault in a valve design mentioned by Kletz [1999 (1)] was repeated elsewhere in the same company with a different valve design [PK] when a fitter broke the low pressure flange on a recently fitted ball valve, with the resultant a loss of confinement. The valve had been put in back-to-front and as installed the seat and ball were only retained against the pressure source by a grub screw, used only for assembly. The correct configuration was for the seat to be self activating and retained, on the pressure side of the valve by the upstream flanges. The fitter had assumed that the valve as a typical ball valve ball valve where the internals are fitted through a split joint in the valve and so the ball and seat were self retained, however there had been new design feature which had not been recognised by the engineer and explained to the fitter. This was clearly *management error*.

The HSE also issue Safety Flashes which report the incident and then require action by the recipients. These are potential sources of lessons learned but the request from HSE is usually in the form of "carry out an inspection" and the Flashes do not necessarily report on the detail in the flaw in the design or operations process.

It is also most unlikely that the Lessons can be extracted by one person working in isolation; it will require a small team to discuss the issues using the *WHY?* approach to identify the *Root Cause(s)* and then to draw out the correct lessons. Then the lesson must be formulated such that it is relevant to the industry, this may require some thought and may take time. Just as with RABS it will be necessary to elicit the relevant information using technical skills and experience. It is possible that the finer details, from which the lessons learned will be drawn, may not be available and then it may be necessary to obtain that detail by direct contact. There may be pit-falls here as discussed in the Hazards Forum [2006]. Some of the impediments to learning and information transfer come from the fear of litigation, a situation which has happened in the past, over regulation, increased media and pressure groups demanding accountability. This may require a change in the approach of the Regulator from the whip to the carrot!

The lessons may be well disguised and require to be extraction with care. The work will have to be carried out only in areas of concern and not reviewing trivia as the rewards must be proportionate to the effort. This in turn will require a depth of experience to separate trivia (or not critical lessons) from the underlying problems. There is an opportunity for an Academic or Learned Institute or the HSE to take a lead in this work utilising the skills and knowledge of retired professionals. Some of the lessons learned can still be found inside the Hazards Training Packages produced by the IChemE; however it is for the individual company to draw out the lessons learned and this may require new training skills and experienced personnel. (Is this not just part of the ***generative*** safety culture?)

As part of the change in safety culture it is essential that there is a change in the approach to incidents from recording the historic detail to the recording this and also the lessons that should be passed on. They should also include the "near misses" where the organic lessons to be learned will be found. This will mean that there must be new investigative skills. There will likely be more investigations and reports and these reports may have to be in more detail, may take longer to write and require more analysis (in conflict with lean and mean). This is also a symptom of the ***generative*** safety culture.

The next point that must be considered is what is to be done with the information and lessons learned. There is little point in each company inventing, or worse still re-inventing, the same solution. This clearly points to the need for a collection and distribution point [Hazards Forum 2006]. One such point could be the IChemE via the EPSC and another could be the HSE but in the latter case there must be a form of immunity from prosecution, this is unlikely. There are two particular industries where there should be an "in-house" collection system, the Nuclear Industry via UKAEA and the Offshore Oil and Gas Industry via UKOOA. An informal distribution, as was witnessed by the ICI Safety Newsletters, did achieve a cohesion in the Loss Prevention fraternity but it tended to be biased towards one company and suffered when the company was sued for not reporting an incident later reported in the newsletter in good faith [PC].

ROLE OF THE KNOWLEDGE MANAGER

The Knowledge Manager must have a number of attributes. He/she must be proactive and not reactive and have the authority and support of all levels and all disciplines in the company from the highest level and may have to carry out the role. The Manager must be accountable and responsible for the role. This profile is not easily matched!

The Knowledge Manager may have to develop a number of new skills. One skill will be the ability to elicit information from persons inside and outside the company; this in turn will require "*networking*" either officially or unofficially. Another skill will be the ability to find the *Root Cause* of the event, be it a real event or a near miss and then to convert this into a meaningful lesson. Finally, the lesson must be taught and corrective actions put in place, be they procedural or hardware. The harsh reality is that there is probably more to be learned from "*near misses*" that the less frequent incidents, this will require not only an open culture but also man-management skills and credibility such that the near misses are first of all reported promptly and then are investigated urgently. The problem may be that the Supervisors may not recognise that a near miss had occurred. The lessons learned can be both corporate and industry wide. The lessons must be distributed both upwards and downwards. The lessons from other industries must be analysed for relevance and for potential lessons for another company. Too often there is the expression "This does not apply to us!" Yes it does! The art or skill is to put the lesson learned it into context for the second company. This is not always easy but must be part of the role of the Knowledge Manager.

Finally the lesson must be taught and corrective actions put in place, be they procedural or hardware. The harsh reality is that there is probably more to be learned from "*near misses*" that the less frequent incidents, this will require not only an open culture but also man-management skills and credibility such that the near misses are first of all reported promptly and then are investigated urgently. The problem may be that the Supervisors may not recognise that a near miss had occurred.

Knowledge Management is a tool being discussed in many conferences. It does not occur spontaneously and must be managed actively bearing in mind the tendency for corporate memory fade, or is it memory fatigue? Knowledge must also be recycled not only to the operations group but also to the engineers. The latter is not evident in many industries with the trend to *Engineer, Procure and Install*. In these the corporate operations and engineering lessons learned, which are often to be found in the Corporate Codes and Standards, are not used in favour of some other codes. A simple review of any Corporate Code will usually show those features that arose from an incident some years ago. These can not be captured so readily by Design Houses unless there is an attempt to treat each piece of equipment as a standardised set of lessons, some of which may be irrelevant to some industries. In addition the objectives of the design houses are not necessarily in harmony with those of the operator. One lesson that must not be forgotten, or re-learned, is that codes and standards represent part of the corporate culture; they must not be forgotten and must be reviewed for accuracy on routine.

It is an unfortunate fact that those Scientists and Engineer who should be reading the *lessons learned* do not do so for many reasons. The Knowledge Manager must keep the lessons learned vital and there must be means of publicising them on a routine so encouraging the learning of the lessons and the assimilation into the working culture. This may require a significant budget. The bland use of "time since the last LTA" is not appropriate and campaigns are more likely to produce results. These can be viewed as re-enforcing campaigns designed to overcome memory fade. One other means of re-enforcing is the use of incidents outside the company and using the lessons learned from these to illustrate where the present company had the correct systems in place or where new systems must be devised.

Finally, and this may be at variance with the constitution of the regulator, the role of the regulator should be more open and supportive. This will involve giving advice, support or guidance and not instructions.

CAN WE USE THE LESSONS LEARNED TO CHANGE THE SAFETY CULTURE?

The answer to this question posed in the title is certainly – YES. However it will be necessary for the corporate culture to be receptive to changes and for the management of the lessons learned to take a much higher profile within the company with a leading figure, known as the Knowledge Manager having a high profile within the organisation. The framework of the role or job description of this person has been given. Actions must be spontaneous and prompt the solutions identified and implemented while the history of the incident is still real in peoples minds. Equally important is that the "lesson-to-be-learned" is recorded in the corporate archives for future reference and education. It has occurred in the past and with the will and direction it will happen for the benefit of industry as a whole and the Company in particular.

It is then for the body corporate to take on the ownership of the lessons learned so as to improve the safety culture. This has to be an iterative process to overcome the corporate memory fade.

CHANGE IN SAFETY CULTURE – THE EXPERIENCE OF ONE COMPANY

It was proposed earlier in this a paper that someone, somewhere had found solutions and that it was not necessary to re-invent the wheel. Kletz [2006] discusses the organisation within ICI in the 1960s and how the whole *safety culture* changed. The lessons learned and their influence on the company safety culture can be illustrated by just 4 examples taken at random. These are now part of the process industry safety culture.

1 In the 1960s there was a release of chlorine in an ICI factory in Merseyside. The cloud drifted across a school playground and some of the children were kept in hospital under observation over night. It would have been easy to examine the process and to propose engineering and procedural modifications but the *Lesson Learned* from this incident was that the root cause was a flaw in the design reviews which had not identified the design faults/errors during the design process. From this the traditional *6 Stage Hazard Studies* were developed (now 8 stage). The design culture of the company had changed.

2 Likewise in the 1960s the evolution of novel and large process plants resulted in some unexpected process problems. Once again it would have been easy to propose engineering and procedural modifications but the *Lesson Learned* was that root cause of the operational problems had not been identified and corrected "on the drawing board". The solution was to be found in an evolution of method study technique it then was given the name of *The Hazard and Operability Study* (HAZOP). Once again the design culture had changed and a new name (HAZOP) had been introduced into the industry.

3 A fatal accident [Kletz 1999 (2)] resulted from the poor operation of the permit to work (PtW) system. The PtW had been prepared by the previous night shift but signed on by the following morning shift, one of the root causes. Another of the root causes was that the oncoming shift was in work over-load as they had been on rotation and the Supervisor was trying to follow up all that had happened over the previous 3 days while trying to issue numerous PtWs. The *Lesson Learned* was not that the *Permit to Work System* was flawed but that root cause of the accident was the implementation of that system. It would have been easy to examine the permit itself and to propose changes to the permit but this would not have improved the implementation. This resulted in all Plant Managers talking through the incident with EACH shift pointing up where the implementation had failed. Further each month the Plant Manager was required to audit the Permits for that month, to identify weaknesses in the drawing up of the permit and to put into place improvements in the implementation of the PtW. This was written up as a report which was sent to the Works Manager. The operating culture had changed and the approach to all work under permit was subjected to a more critical analysis.

4 Finally the follow up to the explosion at Flixborough resulted in the realisation that to root cause of the incident was that of change, small or large, had to be *managed* [Henderson 1976]. All changes were subject to a detailed and recorded examination before they were implemented. The operational culture and approach to "changes" had changed. This process is now known as "*management of change*" but it is not recognised that a change in management must be managed with equal vigour.

POST SCRIPT

ICI was the source of many innovative systems in the fields of Loss Prevention. Many have been adopted by other companies and their origins have become lost in time. As of writing it might appear that ICI will cease to exist.

Let us hope that the words of W Shakespeare in Julius Caesar are not applicable to ICI,

> *"The evil that men do lives after them,*
> *The good is oft interred with their bones."*
> Act II Scene II lines 67–68

ICI taught us all many lessons – now, it would be a compliment to those pioneers who devised them if we incorporate them into the *corporate safety culture*.

Note: PC Private Communication; PE Personal Experience; PK Personal Knowledge

REFERENCES

Baker. JA. [2006] The Report of the BP U.S. Refineries Independent Safety Review Panel.

Bond. J. [2007]. A Safety Culture with Justice: a Way to Improve Safety Performance. 12th International Symposium on Loss Prevention and Safety in the Process Industries. Paper 64. IChemE Series 153, IChemE Rugby.

Crawley. FK. [2006]. How Can We Drive Down Incident Rates by use of Incident Records and Databases? Hazards XIX pp 16 – 29, IChemE Series 151, IChemE Rugby.

Crawley. FK. [2007]. Using Experience and Knowledge to Rejuvenate the SHE Learning Process. 12th International Symposium on Loss Prevention and Safety in the Process Industries. Paper 64. IChemE Series 153, IChemE Rugby.

Crossfield. Sir B. [1999] Port of Ramsgate Walkway Collapse Disaster. 71st Thomas Lowe Grey Lecture. Jan 1999, I Civil Engineers, London.

Franklin. N [1986]. The Accident at Chernobyl. The Chemical Engineer November 1986 pp17 – 22. IChemE, Rugby.

Hazards Forum [2006]. Notes of Brainstorming Session Held on 13th March 2006 at the Institute of Civil Engineers. Hazards Forum, I Civil Eng London 2006.

Henderson. JM, Kletz. TA. [1976]. Must Plant Modifications Lead to Accidents Process Industry Hazards – Accidental Release, Assessment and Containment. IChemE Rugby 1976

HSE [1997]. The Explosion and Fires at Texaco Refinery, Milford Haven, 24th July 1994. HSE Books 1997.

Hudson PTW, [2001] Safety Culture: Theory and Practice in the Human Factor in Safety Reliability. NATO Series RTO-MP-032 NATO, Brussels, 2001.

Hopkins. A. [2000]. Lessons Learned from Longford, CCH Australia, Sydney. Australia.

Kletz TA. [1999 (1)]. "What Went Wrong?" 4th Edition p 38. Elsevier MA. 1999.

Kletz TA. [1999 (2)]. "What Went Wrong?" 4th Edition p 1. Elsevier MA. 1999.

Kletz [2006] How We Changed the Safety Culture, pp 73–82, IChemE Series 151, IChemE Rugby.

Martin. C. [2007]. Worst Case Scenario. The Chemical Engineer pp 23–25 July 2007. IChemE Rugby.

Mogford. J. [2005]. Fatal Accident Investigation Report Isomerization Unit Explosion, Interim Report, Texas City Texas USA. March 2005 BP Web site.

OECD [2005] Report on the OECD Workshop on Lessons Learned from Chemical Accidents and Incidents, Karlskoga, Sweden, Sept 2005 ENV/JM/MONO (2005) 6 OECD Paris 2005

RABS [2001]. Guidelines for the Safe and Optimum Design of Hydrocarbon Pressure Relief and Blowdown Systems. ISBN 0 85293 281 1 Inst of Petroleum London 2001.

Robens, Lord. [972]. Safety and Health at Work ("The Robens Report"). HMSO, London.

UNLOCKING SAFETY CULTURE EXCELLENCE: OUR BEHAVIOUR IS THE KEY

John Hunter[1] and Ronny Lardner[2]
[1]EHS Leader, GlaxoSmithKline, Irvine, UK
[2]Chartered Psychologist, The Keil Centre Ltd, Edinburgh, UK

INTRODUCTION

THE INDIVIDUAL'S ROLE IN DEVELOPING A STRONG SAFETY CULTURE

This paper describes the development and deployment of methods to promote the behaviours which support a strong and sustainable safety culture. Most organizations in hazardous industries have embraced the need for a strong safety culture, and recognize that excellent safety leadership, effective supervision, and high levels of workforce involvement are essential safety culture ingredients (Flin, R et al, 2000; HSE, 1999; HSE, 2001). To support the development of a strong safety culture, both site and topic-based approaches have been adopted. This project used an alternative approach.

SITE-BASED APPROACHES

Site-based approaches typically involve some form of safety culture diagnosis, and a plan to address areas for improvement. The unit of analysis is the site or organization. The improvement plan typically includes a need for change in behaviours and practices at different levels of the organization.

TOPIC-BASED APPROACHES

Topic-based approaches involve interventions to address specific aspects of safety culture – for example supervisor or safety leadership development programmes, practices designed to encourage and promote workforce involvement such as appointing workforce safety representatives, or implementing a behavioural safety programme.

AN ALTERNATIVE APPROACH

Whilst the site or topic-based approaches are appropriate in some circumstances, they do not describe all the individual behaviours required to develop and support a strong safety culture, or specify how these behaviours relate to each other and are mutually supportive across different levels of the organization. Furthermore, site or topic-based approaches do not always lend themselves to integration into the organisation's existing safety management system or human resources systems.

GLAXOSMITHKLINE'S (GSK) IRVINE SITE, AND THE BACKGROUND TO THIS PROJECT

The site was established in 1973, covers 135 acres, and is located on the West coast of Scotland, approximately 30 miles from Glasgow. Irvine is one of GSK's largest primary manufacturing sites, with 650 staff involved in the production of both penicillin-based antibiotics and active pharmaceutical intermediates.

The approach to improving the safety culture at GSK Irvine, and the specific project described here are milestones on a journey for the site that commenced back in 2005. During that year the Factory Safety Committee had been discussing safety culture at several meetings, and ultimately agreed to make the first step in the journey by attempting to measure the existing culture. In October 2005 as part of the European Health and Safety Week, a site survey was launched using the existing HSE "Climate Survey Tool" (HSE, 1997). Over 400 employees (60% of the workforce) completed the 72-question survey. In early 2006 the Factory Safety Committee then analysed the survey, with the aim of identifying some key interventions and developing supporting action plans.

In March 2006, just after celebrating success at reaching 1 million hours Lost Time Injury-Free for the first time in the site's 33-year history, a process safety incident occurred when a 4500 litre reactor vessel exploded, badly injuring 2 operators. The site's top priority ever since has been an effective response to prevent any similar incident recurring, This project has been fundamental to achieving that goal.

After the explosion, the subsequent GSK internal investigation focused largely on the technical aspects of the root cause analysis of what happened within the reactor vessel. The investigation also examined some of the EHS behavioural issues concerning staff at all levels involved directly or indirectly with the incident. This led to a clear recommendation to review the behavioural findings from the incident in conjunction with the recently-completed site safety survey. To lead this project, the company appointed an experienced EHS manager (the first author).

At this stage it was decided to bring in some independent, external expertise and the second author was engaged to assist with the analysis of both the 2005 HSE climate survey and the behavioural aspects of the March 2006 incident investigation. The second author has been involved in the subsequent stages of the design and development of the EHS Behaviour Standard and associated interventions, whilst GSK staff have presented all training and communication sessions.

It was decided to develop a competency model which described (a) the specific managerial, supervisory and workforce behaviours which supported excellent EHS performance and (b) those which detracted from excellent performance.

DEVELOPING THE COMPETENCY MODEL

Five main sources of data were used to develop this competency model:-

1. Existing academic research, which identifyed leadership behaviours which support workplace safety outcomes (HSE, 2003)

2. Existing industry research conducted by the UK offshore sector's cross-industry Step-Change in Safety group, who developed a set of safety behaviours following a review of 11 offshore fatalities (Step-Change in Safety, 2004)
3. Previous work (Hayes et al, 2007) generously shared with GSK by Wood Group Engineering North Sea Ltd
4. Key EHS behaviours relevant to becoming a high-reliability organization (Weick, 1999) were integrated into the model
5. In-company research, to identify specific positive and negative HSE behaviours which had particular relevance to recent explosion, and the findings of the incident investigation.

It was decided to base the competency model on the behaviours which differentiate those who are more effective at managing health and safety, from those who are less effective.

There is an important distinction between the technical competences necessary to do a job (i.e. the ability to drive a fork-lift truck), and the personal competencies which differentiate between those who are more or less effective in a job. Although a group of fork-lift drivers may all possess the same technical competence, individual differences will exist in how effective they are in achieving their overall job objectives (e.g. safety, housekeeping, efficiency). Table 1 below summarizes the differences between behavioural competencies and technical competences, and how to analyse jobs to derive the behavioural competencies which support superior job performance.

In this project, the first job analysis method used was critical incident interviewing, (Flanagan, 1954) which asks interviewees to identify "critical incidents" they have personal knowledge of, which led to a good or poor result. In this case the result referred to HSE performance. Incident does not mean accident or loss, it could simply be someone's behaviour in a meeting which supported or undermined health and safety.

The second method was repertory grid technique (Kelly, 1955) which elicits the constructs or attributes which experienced people use to differentiate between good and

Table 1. Differences between behavioural competencies and technical competences

	Behavioural competencies	Technical competences
Focus	People who do the job	Jobs or tasks which people do
Level of Performance	Superior performance	Minimum standard
Outputs	Behaviours which contribute to superior performance	Key roles and tasks. Minimum knowledge, skills and abilities required
Appropriate job analysis methods	• Critical incident technique • Repertory grids • Studying documentation • Structured job analysis questionnaire • Observation	• Studying documentation Observation • Functional job analysis

poor job performers. The technique asked experienced people to think about managers, supervisors or others they know well, and who are (a) effective in managing health and safety or (b) less effective or ineffective. By comparing those in groups (a) and (b), it is possible to define the behaviour(s) which differentiate the effective and less effective performers.

These two job analysis techniques were used to extract the specific positive and negative EHS behaviours from interviews and focus groups held with managers, supervisors and technicians.

The next steps in model development were to examine the critical incident and repertory grid data, sort the positive and negative behaviours into related groups, and differentiate whether they were behaviours required of everyone on the workforce, only supervisors, or only managers. The wording of behaviours was refined, to aid clarity.

Four sets of positive and negative behaviours were identified for each level in the organization: everyone, supervisors and managers. Each set of behaviours was given a short descriptive label. The resulting overall scheme is shown in Figure 1 below. The total number of behaviours generated across all four sets and three levels totaled 100.

Figure 1 illustrates how it is only when the appropriate behaviours are displayed by ***all*** people in the organization that an excellent EHS result can be achieved. This approach

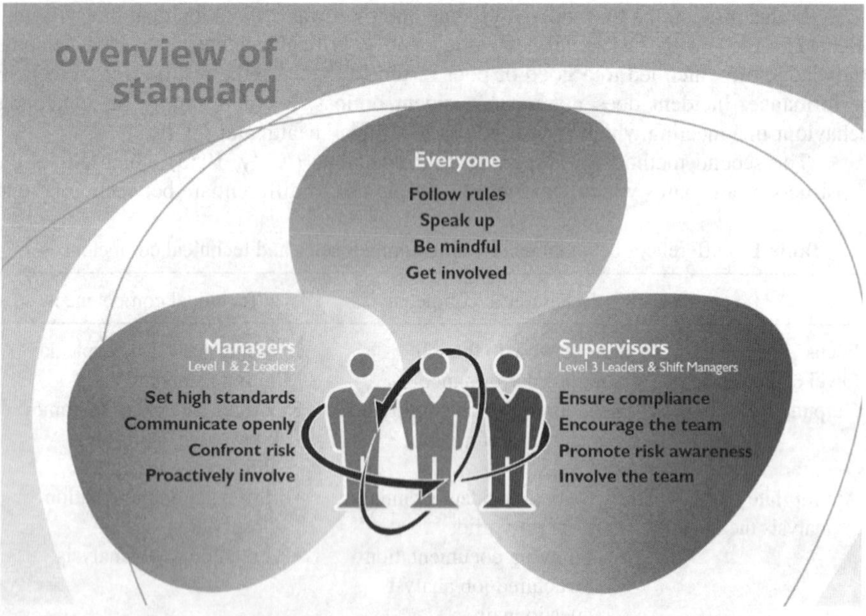

Figure 1. Overall scheme of EHS Behaviour Standard

can be contrasted with many "behavioural safety" programmes, which focus largely on ***workforce*** behaviours.

The 12 sets of behaviours were further examined, and it became apparent that four common themes ran through the sets of behaviours: standards; communication; risk management and involvement. These themes emerged from the data gathered, and were not pre-determined. Figure 2 illustrates how the sets of behaviours relate to the four themes.

By reading across these themes, it is possible to see the mutually-supportive inter-relationships between the sets of behaviours for each group. Similarly, it is possible to identify how a lack of the correct behaviours at any level can undermine the overall result. For example, management efforts to set standards, and workforce efforts to comply can be undermined by the wrong supervisor behaviours.

Figure 3 below is an example of the content of one of the twelve sets of behaviours, and shows the positive and negative behavioural indicators which were derived from the job analysis.

The resulting competency model, known in this company as the EHS Behaviour Standard, has the following important features:

- research-based
- simple to understand

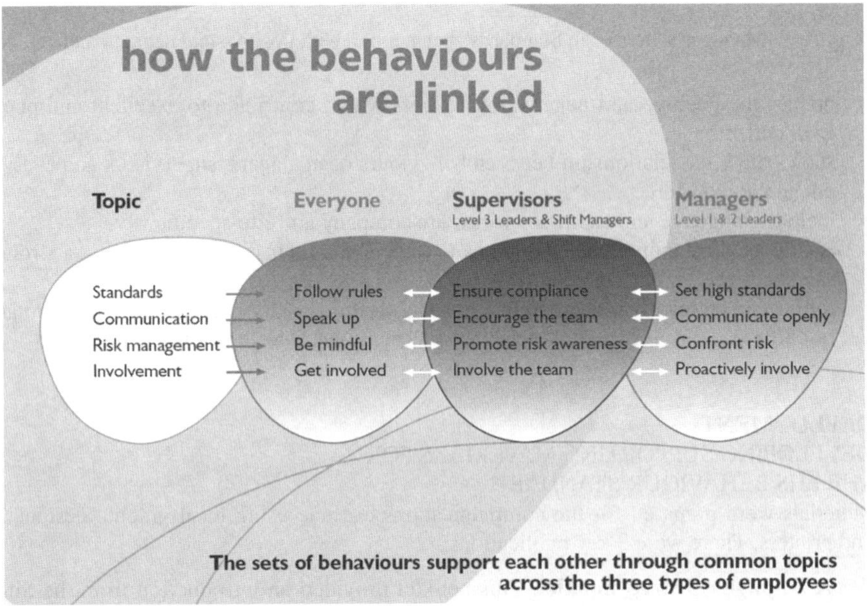

Figure 2. How EHS Behaviours relate to topics

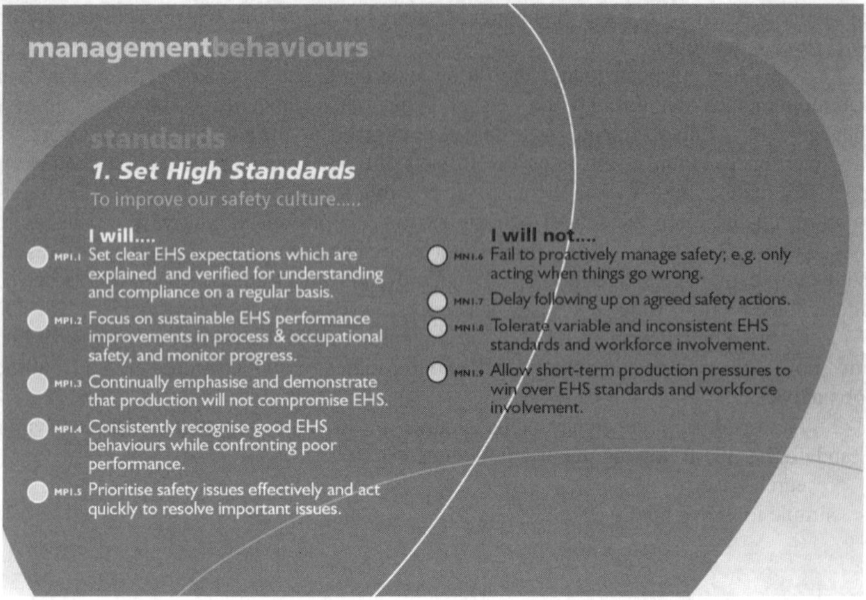
managementbehaviours
standards
1. Set High Standards
To improve our safety culture.....
I will....
MP1.1 Set clear EHS expectations which are explained and verified for understanding and compliance on a regular basis.
MP1.2 Focus on sustainable EHS performance improvements in process & occupational safety, and monitor progress.
MP1.3 Continually emphasise and demonstrate that production will not compromise EHS.
MP1.4 Consistently recognise good EHS behaviours while confronting poor performance.
MP1.5 Prioritise safety issues effectively and act quickly to resolve important issues.
I will not....
MN1.6 Fail to proactively manage safety; e.g. only acting when things go wrong.
MN1.7 Delay following up on agreed safety actions.
MN1.8 Tolerate variable and inconsistent EHS standards and workforce involvement.
MN1.9 Allow short-term production pressures to win over EHS standards and workforce involvement.

Figure 3. Manager's "Set High Standards" behaviours, with positive and negative indicators

- defines the positive and negative behaviours which contribute to excellent and poor EHS performance
- shows the inter-relationship between behaviours of managers, supervisors and everyone in the workforce
- includes language and examples which are company and site-specific
- can be used by individuals and teams to understand their role in developing a strong safety culture
- format can be readily integrated into the organisation's safety management and human resource systems, e.g. induction, selection, training & development, and appraisal

DEPLOYMENT

DEVELOPING SUPPORTING MATERIALS FOR THE EHS BEHAVIOUR STANDARD

Materials were prepared for the communications cascade using local graphic designers and printers. There were 3 key publications:-

1. **A 24-page A5 sized booklet**. This booklet provided an introduction from the Site Director, an overview of the approach to the standards and then sections on each of the 4 sets of behaviours for each of the 3 groupings of staff. The booklet also

included 2 pages at the back providing a process, with worked examples, as to how to assess oneself against the behaviour standard and to create an individual improvement plan.
2. **A small pocket-sized folding Zip-Card** containing summary information on the approach. This publication still however contained all 100 behaviours
3. **Postcards** The intention of the postcard was to encourage staff to indicate one strength and one area for improvement from the EHS Behaviours. These were recorded on the postcard at the communication sessions and "posted" to the Site Director. This enabled some analysis to be carried out on common themes and, more importantly, provided some follow-up as the postcards were passed back to the individuals directly via their team leaders 3 months after the training session they attended.

IMPLEMENTING THE EHS BEHAVIOUR STANDARD: RE-INDUCTION

The design of the implementation was informed by (a) the principles of ABC analysis, a behavioural change technique which emphasises the important of providing consequences to change behaviour (see HSE, 2002 for details), and (b) the knowledge that it was managers' and supervisors' behaviour which had to change first, followed by the workforce. It was decided to "re-induct" the entire workforce about the EHS Behaviour Standard.

For such a significant re-induction programme to be successful, securing management and supervisory ownership and involvement was critical. To that end, the first EHS Behaviour Standard re-induction sessions were carried out with that target audience. In addition, because of the subject matter and due to the site being unionised with two unions at shop-floor level, union Safety Representatives were also invited to the initial sessions. Two half-day re-induction sessions were delivered on consecutive days in January 2007 with a total attendance of 83 staff. The content was the same as used for subsequent workforce re-induction sessions' with some additional emphasis on the key responsibilities that management, supervisors and safety reps held with respect to safety performance and safety culture. Furthermore, the following individual objectives were communicated to this group of staff:-

Managers objectives

- Personally set an excellent example of the EHS behaviours at all times
- Ensure all team members participate in the EHS behaviours "re-induction"
- Active monitoring and sponsorship of team action plans arising from "re-induction"
- Ensure during Q3 2007 all managers, supervisors and safety reps complete 360-degree feedback on EHS behaviours, and that any identified development needs are built into their own PDP

Supervisors objectives

- Personally set an excellent example of the EHS behaviours at all times
- Take an active role in delivering the EHS behaviours "re-induction"

- Develop and support a team action plan on improving EHS behaviours arising from "re-induction"
- During Q3 2007 personally obtain 360 degree feedback on EHS behaviours, identify any development needs and build into personal PDP

To further support this group of key staff a set of Questions and Answers were prepared in advance by the project team and these were handed out at the end of the re-induction sessions.

Thereafter half-day re-induction sessions for the whole workforce then commenced some two weeks later. A series of ten sessions were set up over an intensive two-week period. Total attendance was excellent, with 657 attendees covering more than 95% of site staff. One key target set by the Site Director was to complete the re-induction programme before the first anniversary of the explosion – this was successfully completed with the last of the ten sessions held on the 28 February 2007.

Each re-induction session covered the following content, most of which was delivered as team-based activities, led by local managers and supervisors

- An introduction to the subject matter of "safety culture"
- An emphasis on the need for improvement at GSK, Irvine. This involved some commentary on the two injured operators from 2006 including their support for the approach to the EHS Behaviour Standard. Also the Site Director or a deputy recalled his personal account of the incident from 2nd March 2006. These personal messages were very powerful in what became an emotionally-charged atmosphere.
- An overview of the CAPA (Corrective And Preventative Action) Plans that were being progressed following the incident
- A simple worked based scenario to get staff thinking about their EHS behaviours
- An explanation of how the team created the Irvine EHS Behaviour Standard. This included reminding staff of their input via the 2005 climate survey
- An overview of the EHS Behaviour Standard and a walk-through the booklet
- The creation of Team Action Plans using a template supplied
- Discussions about follow-up supporting activities
- Individual commitments recorded using the booklet and the postcards.

These re-induction sessions worked extremely well because the team leaders were primed to facilitate discussion, and also because attendees were grouped into work teams to create their own team action plan. Furthermore, at the end of each session, time was allowed for staff to review their own levels of performance against the appropriate set of behaviours and commit to an individual area for improvement. These were captured in the booklets that they took away with them and on postcards which were sent to the Site Director.

FOLLOW-UP ACTIVITIES TO ENSURE SUSTAINED BEHAVIOUR CHANGE

Several follow-up activities were announced at the re-induction sessions to emphasise that the plan was to ensure this new approach to improving our management of safety was not going to be a one-off event. The commitment was there from the leadership team to ensure this became fully embedded in ways of working i.e. strengthening our safety culture. Follow-up activities included:-

- **Individual Commitment** – captured on postcards at the re-induction sessions, individually reviewed by Site Manager, who added a personal message, and returned 3 months later to the individuals via their Team Leader with a discussion about how they are progressing.
- **Team Action Plans** – developed during the workforce re-induction sessions, collated site-wide to check for consistency, and monitored for progress via departmental and personal objective tracking systems.
- **Manager and Team Leader Objectives** – as described above each manager and team leader has a set of specific objectives in relation to the EHS Behaviour Standards included in their 2007 PDP forms.
- **Checking understanding and soliciting feedback** from the workforce re-induction sessions. This was managed via a web-based survey form that was e-mailed to all attendees within 2 weeks of their training session. The questionnaire asked for responses to 7 questions to check understanding, 2 questions to assess their level of support for the programme and then optional extras to provide general feedback comments and to submit an idea for consideration. 575 staff had responded with 223 ideas submitted and 149 comments provided. A high level of understanding of the roll-out sessions, and support for the site's EHS Behaviour Standard was verified via the survey.
- **Winning prize for the best idea** helped encourage submissions and embed and sustain the programme.
- **Induction Programme** – a 30 minute interactive introduction to the EHS Behaviour Standard has been introduced into the site induction programme to ensure all new employees are covered as soon as they join.
- **Contractors** – it was always the intention to ensure that all contractors based on site were treated in the same way as GSK staff. All contractors (over 300) attended similar re-induction sessions in Spring 2007.
- **Publicity** – a number of banners and posters carrying the overview of the Behaviour Standards have been created and are on prominent display across the site.
- **Recognition** – use of the company reward programme has been encouraged to recognise any particular excellent examples of the correct display of EHS Behaviours. A number of these have already been awarded on site since the roll-out programme commenced.
- **360 Degree Feedback** – this programme for all managers and supervisors commenced with the site leadership team. Each person selects several colleagues/peers and several subordinates who together with their line manager assess the individual against each of the EHS Behaviour Standard "Manager" behaviours, using a web-based survey tool. Supporting consultants then analyse the returns and create a report that is fed back directly to the individual who is then assisted or coached into creating a more detailed individual improvement plan. The target is that all Managers, Supervisors and Safety Reps will complete this process in 2007.

- **Human Factors Training** – As a COMAH site there is an expectation by the regulator that systems are in place to manage human factors in relation to our major accident hazard installations. To date two "Human Factors Awareness" courses have been held on site for a cross section of staff from the Factory Safety Committee. Also a more detailed two day "Human Factors Analysis Tools" course (Lardner and Scaife, 2006) was run on site to enable experienced incident investigators to (a) better understand why people involved in incidents behaved as they did, and (b) write better behavioural recommendations which will influence behaviour of those immediately involved, and others, in the future.

RESULTS TO DATE

There is supporting evidence of benefits starting to be realised on a number of EHS KPIs. At the end of September 2007 the following highlights were noted as the best levels of EHS performance in the site's 33 year history:-

- Lowest ever recorded 12-month rolling LTIIR at 0.08.
- Highest-ever recorded number of hours worked since the last lost time accident – 1.45 million.
- Passing 1 million hours LTI free for only second time ever and in successive years (2006 and 2007).
- Lowest ever cumulative "Spill Index" – a key environmental measure.

A cross-section of experienced managers, supervisors, EHS professionals and safety representatives used the Safety Culture Maturity[1] Model (Keil Centre, 2007) to assess whether and how the site's existing safety culture had changed since the explosion in 2006. In their opinion, the site had improved by one level, and this was aided by the development and deployment of the EHS Behaviour Standard.

At an early stage in the project, the site HR Director expressed significant reservations about the ability of the site, aided by the second author, to achieve a cultural change. In July 2007 he commented "I was sceptical about the possibility of "culture change" – I thought it was expensive consultant mumbo-jumbo but I was wrong – I take my hat off to you – there has been a real change for the better over the past few months".

This project has been recognized internally, as it won the GSK Chief Executive Officer's EHS Award for 2007 in the "EHS Initiative" category. A total of 86 projects from GSK sites across 30 countries were submitted and reviewed by a panel of external judges.

The GSK Vice-President and Corporate Head of EHS commented:

"The efforts represented by this award application are not only critical to EHS performance at Irvine, but I believe that Irvine has set the mark for other sites to follow. It is a great effort and I'm very pleased that the external panel judged it so highly. Congratulations and well done to everyone at Irvine.

[1]Safety Culture Maturity is a registered trademark of The Keil Centre Ltd.

The approach described is now being adopted for use by GSK globally, as its preferred method of assessing and enhancing safety culture.

DISCUSSION AND CONCLUSIONS

The question has been asked – "If the EHS Behaviour Standard and methodology had been in place a year earlier, is it likely this would have prevented the reactor explosion?". Following the explosion, the incident investigation concluded that a holistic response was required, involving improvements to engineering, the safety management system and the existing safety culture. Using Reason's Swiss Cheese analogy, there were holes in multiple layers of system defences. The site identified 11 improvement areas from technical issues (Emergency Vent Design, Mechanical Integrity) to management systems (Permit to Work, Change Management) and safety culture and behaviours. Since then the approach has been to close all gaps in these system defences (or layers of cheese). In the opinion of the site management, without the focus on safety culture and behaviours, the chances of sustaining the benefits across the other improvement areas would be limited. It is impossible to say with any certainty whether the EHS Behaviour Standard and methodology would have prevented the explosion, without the engineering and safety management system improvements. The likelihood would certainly have been reduced, via management behaviours which place safety before production, supervisory behaviours which encourage wariness amongst their team members, and everyone being more involved in EHS, and speaking up about any concerns. An additional issue addressed very effectively by the EHS Behaviour Standard and methodology was the belief amongst some site employees that the deficiencies leading to the explosion were only relevant to the building where the explosion occurred. During development of the EHS Behaviour Standard it became apparent this was not the case, and this misconception was firmly rebutted during the workforce briefings.

This novel project helped a site develop and implement a safety culture improvement project, in the aftermath of a serious explosion. The methods used proved flexible, and acceptable to the target audience. A comprehensive implementation plan was developed and executed, with a strong emphasis on shaping manager and supervisor behaviours, as they in turn can strongly influence the wider workforce.

The flexible nature of the EHS Behaviour standard allowed to be built into the safety management system, so that its influence can endure

It is likely that this approach will be of interest to other organisations looking for practical methods to develop and embed the safety culture they desire.

REFERENCES

Flanagan, J. C. (1954) "The Critical Incident Technique" in *Psychological Bulletin*, 51, 327–58

Flin, R., Mearns, K., O'Connor, P. & Bryden, R. (2000). "Safety climate: Identifying the common features" in *Safety Science*, 34, 177–192

Hayes, A et al (2007) *Personalising Safety Culture: What does it mean for me?* Paper presented at Loss Prevention 2007, Edinburgh, UK 22-24 May 2007

HSE (1997) *Climate Survey Tool*: HSE Books

HSE (1999) *Effective supervisory safety leadership behaviours in offshore oil and gas industry* OTO 0065/1999 HSE Books

HSE (2001) *Involving employees in health and safety: Forming partnerships in the chemical industry* HSG 217 ISBN 0717620530

Health and Safety Executive (2002). *Strategies to Promote Safe Behaviour as Part of a Health and Safety Management System.* CRR 430/2002, HSE Books: Norwich UK.

HSE (2003) *The role of managerial leadership in determining workplace safety outcomes* http://www.hse.gov.uk/research/rrpdf/rr044.pdf

Kelly, G (1955) *The psychology of personal constructs* New York: Norton

Keil Centre (2007) http://www.keilcentre.co.uk/html/human_factors/safety_culture%20maturity%20model.htm

Lardner, R and Scaife, R (2006) Helping engineers to analyse and influence the human factors in accidents at work in *Process Safety and Environmental Protection*, 84(B3): 179–183

Step-Change in Safety (2004) *Fatality Report*

Weick, K., Sutcliffe, K & Obstfeld, D. (1999) "Organizing for High Reliability: processes of collective mindfulness", in *Research in Organizational Behaviour*, 21, 81–123

MANAGING RISK COMPETENCE†

Chris Urwin
Det Norske Veritas, Highbank House, Exchange Street, Stockport,
Cheshire, SK3 0ET. United Kingdom

This paper will explore what constitutes Risk Competence, and how organisations can manage it to ensure the appropriate risk behaviour of its people and improve process safety and environmental performance. Risk Competence includes the knowledge and skills required to identify and control hazards in the workplace, however it is important to look beyond the technical aspects of competence to include the psychological aspects which effect behaviour. How can organisation's ensure people perceive risks correctly, have appropriate values, beliefs and attitudes concerning risk, and commit to relevant norms and rules?

1. INTRODUCTION

Process industry today is under increased scrutiny concerning its process safety and environmental performance. Process industry managers know the dangers of getting process safety management wrong and that the bigger their companies are the more exposed they are. A major challenge facing process companies today is how to improve and demonstrate process safety and environmental performance through more effective risk management.

The process safety and environmental performance of an organisation is only as good as its employees. This paper will explore the following questions:

- What is Risk Competence?
- How can Risk Competence of employees and contractors be evaluated?
- How can Risk Competence of employees and contractors be developed to improve process safety and environmental protection?

We will address these questions by reference to a new model of Risk Competence and explore how it can be applied systematically to improve process safety and environmental protection.

2. WHAT IS RISK COMPETENCE?

The behaviour of young children, driven by curiosity and a need for excitement, yet curbed by their sense of danger, is risk management in action. Learning to walk or ride a bicycle

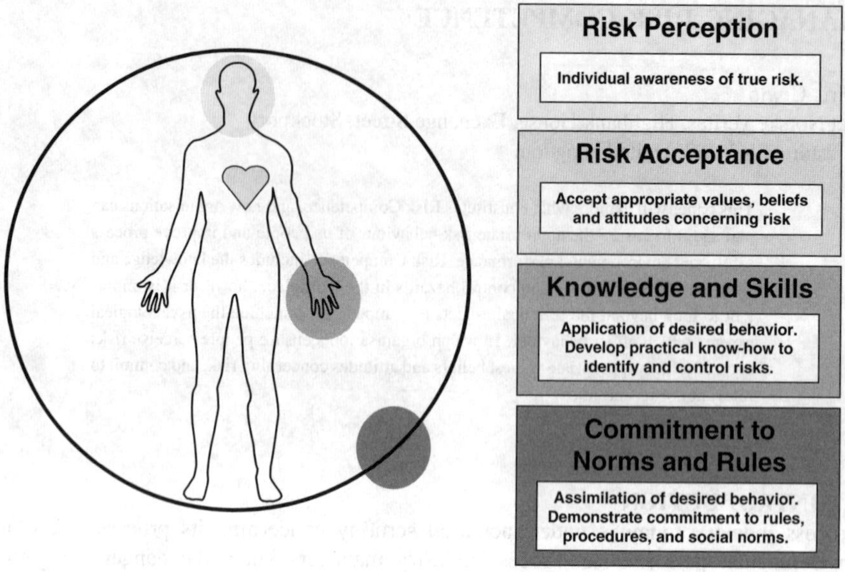

Figure 1. DNV Risk Competence model

cannot be done without accidents. In mastering such skills children are not seeking a zero-risk life; they are balancing the expected rewards of their actions against the perceived costs of failure.

We are all risk managers. Driving a car involves a high degree of risk evaluation and risk control. Crossing a busy road, competing in sports, carrying out home improvements, bringing up children, buying a house, travelling overseas, choosing a restaurant and making career moves are all activities involving degrees of risk we must evaluate and control. How do you go about managing complex risk issues such as these? Do you make the right risk decisions? Are you Risk Competent?

Drawing together psychological research on the risk behaviour of individuals, DNV has developed a new model of Risk Competence. Risk Competence may be defined as "an individual's risk perception, risk acceptance, knowledge and commitment to norms to be able to correctly identify and control the risks they are exposed to."

The Risk Competence model is shown in Figure 1 and its four elements are described in greater detail below.

RISK PERCEPTION

Risk may be defined as "the combination of the probability of an event and its consequence' (ISO/IEC Guide 73). This definition suggests that there is a true or objective risk

Table 1. Characteristics influencing whether events are perceived as "Safe" or "Risky"

"Safe"		"Risky"
Voluntary	vs	Coerced
Natural	vs	Industrial
Familiar	vs	Exotic
Not memorable	vs	Memorable
Not dreaded	vs	Dreaded
Chronic	vs	Catastrophic
Knowable	vs	Unknowable
Individually controlled	vs	Controlled by others
Fair	vs	Unfair
Morally irrelevant	vs	Morally relevant
Trustworthy sources	vs	Untrustworthy source
Responsive process	vs	Unresponsive process

associated with all events which can be calculated if one knows the probability of an event and its consequences. However, as individuals we are rarely in a position to calculate the true risk of an event. Rather we rely on our perception of risk to determine our behaviour.

As individuals we perceive the risks associated with activities differently and these perceptions are often very different to the true or objective risks. Often there is a inverse relation between our risk perception and the true risk. For example; people are afraid of flying and are happy to go by car, when the true risks indicate the opposite. People are most worried about their safety in the workplace and feel safer at home when the true risks are often the opposite – when did you last have an accident – was it at work or away from work? Most people's accidents occur doing sports or home improvement projects. In other words the things people are worried about and the things that actually hurt them are often inverted. This is a problem of Risk Perception.

Peter Sandman (1993) identifies 12 characteristics of events in Table 1 that influence whether an individual perceives an event is "safe" or "risky". Top of the list is whether the event is voluntary or coerced. For example the safety risk of a person parachuting from a plane voluntarily will be considered lower, compared to if they are pushed! If the event is natural or industrial is also important e.g. the objective health risk of radioactivity from a modern nuclear power station is far lower than for radon gas rising naturally in Cornwall in South West United Kingdom. However the public and media risk perception of getting cancer from nuclear power stations is far higher than for living in Cornwall[1].

[1]Radon is estimated to be responsible for 2500 deaths from lung cancer in Britain per year according to the UK Department of the Environment, British Medical Journal, May 1996.

Risk perception is important because it strongly influences risk decision making. Poor risk perception leads to poor risk decision making. In addition, employees, contractors, managers, regulators, the public, pressure groups and the media often perceive the risks of the same event very differently, which creates daily conflicts we read about in our newspapers.

Risk perceptions are deeply held by individuals and are difficult to change. Changing risk perception is best achieved through direct involvement of stakeholders and effective two way risk communication.

RISK ACCEPTANCE

Once we have perceived a risk, we then make conscious or unconscious decisions about its acceptance. This decision process means evaluating the risk against our personal risk acceptance levels. We may be aware for example that drinking cola increases the risk of putting on weight and diabetes and that smoking increases the risk of lung cancer. But do we accept these risks and change our behaviour? We may decide to accept the risk of drinking cola but not the risk of smoking indicating different levels of risk acceptance for these two risks.

Risk acceptance then refers to the behaviour of a person in a situation of uncertainty to engage in a certain behaviour or not to engage in it, after weighing up the estimated benefits against the costs.

Risk Acceptance is represented in the Risk Competence model as the "heart" because these are largely unconscious decisions guided by our values, beliefs and attitudes concerning risk. For example an individual may believe they are an excellent car driver which may then influence their decision to drive faster than the speed limit. Young people often have the belief they are invulnerable and accidents only happen to other people and so are more likely to accept the risks of drinking and driving. Behaviours that we do habitually can change our attitudes to risk and alter our risk acceptance, for example regularly driving at speed in traffic. If a person is a "sensations seeker" they value taking risks and will accept higher levels of risk than a person who is risk averse. Such people will engage in higher risk activities like skiing and motorbike riding. In a working context such people may not follow procedures and rules because they believe the rules are intended for other less competent people.

In the workplace, we would like individuals to adopt appropriate values, beliefs and attitudes concerning the risk they are exposed to. Only then will they make the correct risk decisions and desired risk behaviour.

KNOWLEDGE AND SKILLS

Competence in the workplace may be defined as "the knowledge, skills and attitudes to correctly perform a specific role." To be risk competent, individuals must have the knowledge and skills to correctly identify, evaluate, control and monitor the risks they are exposed to. These skills may include technical knowledge of the risks they are exposed to e.g. knowledge of how to handle hazardous chemicals. Necessary knowledge and skills may include the know how to conduct an effective hazard identification process, use a risk assessment matrix or knowledge of site rules and procedures.

Knowledge and skills are developed through education and training programs, coaching and experience on-the-job. The application of desired risk behaviour demonstrates and reinforces these knowledge and skills.

COMMITMENT TO NORMS AND RULES

Laws, rules, procedures, practices, conventions and norms exist at all levels of society. These norms and rules exist to safeguard against risks. For example, regulations protect the health and safety of employees in the workplace and stop company's damaging the environment. Company rules and procedures ensure quality product is delivered to the customer.

By complying with norms and rules individuals are protected from risk without always having to recognise the risks they are faced with. For example one can drive safely on unfamiliar roads by relying on the road traffic signs and knowledge of the highway code.

The extent to which individuals are committed to consistently apply rules and norms is the final component of risk competence. When compliance with rules and norms becomes habitual we can say that individuals have assimilated the desired behaviour.

3. HOW CAN RISK COMPETENCE BE EVALUATED?

The Risk Competence model may be used to evaluate the Risk Competence of individuals with respect to a particular task. For example is an employee Risk Competent to operate particular process plant. DNV have developed facilitated workshop methodology whereby the work team for a specific process area are guided through the Risk Competence model and evaluate as "go" or "no go" the Risk Perception, Risk Acceptance, Knowledge and Skills and Commitment to Norms and Rules of the team to complete critical tasks.

Any "no go" issues are captured and improvement actions identified.

4. HOW CAN RISK COMPETENCE BE DEVELOPED TO IMPROVE PROCESS SAFETY AND ENVIRONMENTAL PROTECTION?

Based on the research by leading psychologist Bandura (2000), we propose that the Risk Behaviour of individuals is function of two variables; Risk Competence and the External Environment. Risk Competence is made up of the four characteristics reviewed above. External Environment is made up of many factors including the management system and working environment. Risk Behaviour in turn influences the Risk Competence and the External Environment. These interrelations are represented in Figure 2.

This model may be used to improve the process safety and environmental performance of process plants. The model proposes that to improve process safety behaviour of individuals in process areas we must do two things. Firstly we should improve the External

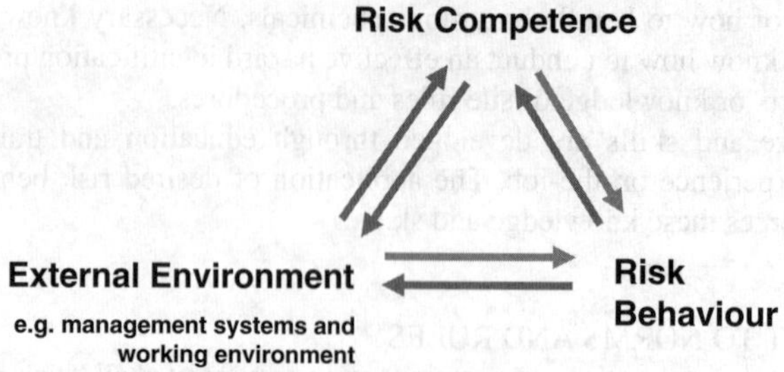

Figure 2. Influences on Risk Behaviour

Environment in which the individuals operate. The design of process plant should be correct to support the correct process safety behaviour, the procedures, practices and norms should also direct the correct behaviour. Secondly individuals should have the correct Risk Competence. They should have a good awareness of the true risks of the process hazards they face. They should accept the appropriate values, beliefs and attitudes necessary to manage those hazards and make the correct risk based decisions. They should have the

Management System Activity	Risk Perception	Risk acceptance	Knowledge & Skills	Commitment to norms and rules
Leadership	Stakeholder engagement, communication	Lead by example, management tours, recognition, motivation	Communication, coaching	Set expectations/ vision/values
Risk Recognition and Risk Evaluation	Point of work risk assessment	Group risk assessment, Team working	Training, Learning by doing	Learning by doing
Human Resources	Work-life-balance	Effective recruitment	Effective recruitment, set responsibilities	Effective recruitment, performance review
Training and Competence	Knowledge and skill training	Induction	Knowledge and skill training	Refresher training
Communication	Promotion campaigns	Group meetings, teambuilding	Risk communication	Coaching
Risk Control	Participation in work place design	Participation in developing rules	Learning by doing	Learning by doing
Maintenance and Inspections	Tours	Inspections	Coaching	Review meetings
Contractor Management	Participation	Contractor Forums	Contracting	Feedback loops
Emergency Preparedness	Emergency scenario analysis	Participation in Emergency teams	Training	Exercises, drills
Risk Monitoring	Participation	Perceptions surveys	Learning by doing	Task and behaviour observation, audits
Results and Review	Performance review	Business review	Reporting	Management review

Figure 3. System activities to develop Risk Competence examples

necessary technical knowledge and skills to evaluate and control process risks. Finally, they should demonstrate their commitment to the rules, procedures and social norms the company has put in place for process safety.

A systems approach is the only effective approach to manage the complexity associated with improving process safety Risk Competence. Figure 3 shows a menu of example management system activities which can be applied as required to improve the process safety risk competence of individuals leading to improved process safety and environmental protection.

5. CONCLUSIONS

This paper has introduced a model of Risk Competence, suggested a method to evaluate it and proposed how it may be used to improve process safety and environmental performance.

We suggest that achieving the necessary levels of Risk Competence among employees and contractors working in process areas will result in improved risk decision making and risk behaviour. In combination with other approaches, this can be used to improve process safety performance and reduce the risk of major accidents.

6. REFERENCES

Bandura A., 2000, Self Efficacy: The Exercise of Control, WH Freeman and Co. NY

Sandman P., 1993, Responding to Community Outrage: Strategies for Effective Risk Communication, American Industrial Hygiene Association

INTEGRITY MANAGEMENT FOR THE 21ST CENTURY WITH 20TH CENTURY EQUIPMENT†

Laza Krstin
ABB Global Consulting

Many of our assets in the process industries were built over 20 years ago and are still operating today – well beyond their anticipated design lives. By understanding the underlying causes of deterioration and thereby loss of integrity we can build a robust integrity management system and develop a practical action plan to operate these ageing assets in a safe, productive and cost effective manner.

Using a case study, the paper shows how some existing approaches can fail to provide the required degree of assurance. It also highlights how weak systems can combine with organisational and human factors to undermine years of apparently satisfactory performance.

The paper goes on to describe the key elements of an integrity management system, applicable to process facilities which are being operated beyond their design life. This approach is also potentially applicable to plants that have not yet reached their design life. It shows how consideration of asset life underpins process safety, maintenance and renewal policies to give a robust long term asset strategy.

INTRODUCTION

The process industries face challenging times. Increasing stakeholders' expectations and relentless economic pressures are compounded by the difficulties of managing an aged asset base. Many of our assets were built and commissioned through the 60s, 70s, 80s and are still operating today.

Many operating companies have introduced initiatives to maintain integrity or improve reliability such as Risk Based Inspection (RBI) and Reliability Centred Maintenance (RCM). However, they may still suffer from significant "unexpected" failures and losses of containment. Such incidents include leakage from storage tanks and failures of "non-critical" pipework which causes disruption to production, despite an apparently good maintenance and inspection history.

It is not just mechanical systems and equipment that needs to be addressed. Deterioration of instrument/electrical equipment and structural elements of plant can lead to an unacceptable risk to plant and to personnel. Even if the condition of the equipment is preserved, obsolescence can affect the useful life of certain types of asset, notably control equipment and machines.

Operating companies are increasingly "sweating the assets" – operating existing assets beyond their original design life, rather than building new plant, and tying up valuable capital. Indeed, many existing plants are now so far beyond their original anticipated life, that design margins have been used-up. For example, at the design stage, the selection of materials of construction for pressure equipment is based on the process fluid, the intended operating conditions and the expected rate of corrosion. A corrosion allowance is determined on the basis of the rate of corrosion over the desired operating lifetime. After many years in operation the corrosion allowance may be used up. The integrity of the equipment may potentially be further compromised by plant modifications, and a history of operating excursions outside of the operating envelope. Other life-limiting deterioration can be caused by operating cycles and stresses exceeding the fatigue design life, and the cumulative effect of operating for longer periods than assumed in the determination of the creep design life.

In the UK further challenges arise from economic factors – there has been little significant investment in new plant and reduced investment in refurbishment and maintenance of existing plant has been the norm since the late 80's. Pressures have been placed on maintenance budgets, resources, etc. to cut operating costs.

Yet, the industry is facing increased expectations from the public, employees and other stakeholders in such areas as environmental protection, continuous safety improvement, and enhancing the company's reputation as a "socially responsible" enterprise, whilst at the same time remaining profitable in a competitive global market. This is particularly acute in the wake of a serious incident and there have been a number of serious incidents around the world in recent years which have drawn attention to the importance of continued management of major hazards.

Companies' abilities to deal with these challenges may be restricted by an ageing workforce, an undoubted reduction in the recruitment and training of young people through the 90's, and a trend for reducing in-house core competence in favour of out-sourcing.

Operators are increasingly realising that achieving safety, reliability and plant integrity targets requires a holistic approach to integrity management. They are beginning to realise that safety, integrity and reliability are all linked and are all manifestations of a risk management system that is operating effectively.

There are many benefits that flow from effective integrity management, including:

1. Increased equipment availability/reliability
2. Increased output
3. Improved safety and environmental performance
4. Optimised maintenance costs
5. Statutory and regulatory compliance

INTEGRITY MANAGEMENT – A CASE STUDY

If integrity management is so important and worthwhile an objective, how can it be achieved? Who is responsible for it?

Integrity management is not just about assessing the condition of plant equipment. The elements of an effective integrity management system are best illustrated by a looking a specific case study. Effective learning from incidents provides a powerful means to improve integrity management of ageing plant, as long as the true root causes and contributory factors are determined.

On a chemical manufacturing site, an above ground piping system was used to transfer a hydrocarbon liquid product from a storage tank to a unit in another part of the site. The piping was NPS 4 stainless steel and approximately 1km long. Although the product was toxic to the environment, the duty was not arduous – ambient temperature and pump transfer pressure less than 10barg. After the pipework had been in service for many years, part of it needed to be rerouted to accommodate the demolition of a building, and this presented an opportunity to make the new section as an all-welded construction, removing a number of flanged joints that had had a history of leakage.

Some time after the modification had been commissioned a significant leak of the hydrocarbon product was detected coming from a filter at one end of the piping system. As the filter was in an out-of-the-way location, the leak was not detected immediately – by which time an estimated quantity of 150 te of the liquid had been released. It was concluded that approximately 1 te entered the nearby canal, 20 te evaporated, and 3 te were recovered, with the rest remaining in the ground with minimal prospect of recovery.

The direct cause of the incident was the failure of the filter due to over-pressure. The piping system was used on an intermittent duty, and during a shut-in condition, the liquid had been warmed by sunshine and ambient air, generating a pressure that eventually caused the filter lid retaining clamp to fail.

As with many incidents, the underlying causes of the failure arose from the cumulative effect of several factors across the life cycle of the pipework:

1. The original design specification did not appear to have considered thermal relief.
2. The flanged joints "sprung" in reaction to over-pressure and so acted as impromptu thermal relief devices.
3. The flange leaks were not adequately investigated, giving rise to the view that these were "troublesome flanges" and therefore their removal was seen as only having a beneficial effect from a maintenance point of view.
4. With no significant deterioration mechanisms to threaten the integrity of the piping (except possibly deterioration of pipe supports), the integrity of the pipework could be regarded as a "maintenance issue" to do with the flanged joints, rather than a focus for "inspection".
5. After the modification, with no flanges to relieve any over-pressure, the filter was the next "weak link" in the system.
6. The modification to reroute the pipework did not appear to consider the pressure relief requirements for the whole pressure system, apparently focussing only on the implications of the modification on the affected part of the pipework.
7. Leak detection systems and operating procedures were inadequate, to mitigate the consequences of a leak.

In practice, perhaps the most effective point of prevention for this incident might have been in the control of the modification, through improvements in the management of change procedure and increased competence of the technical review team who were responsible for the modification.

So, we can learn a lot from real incidents. But simply studying incident databases does not move us forward. We need to distil the key messages and turn them into useful guidance and action.

THE ASSET LIFE CYCLE

The asset life cycle has a number of stages from scope definition (the business case) through to demolition and disposal. Each stage is interlinked. Safe, reliable and cost effective operation into the future is dependent on all stages of the life cycle. In particular, modification to the plant, re-rating of equipment, and assessments of current condition to effect life extension beyond the original intended design life, all need to be addressed by going back to the first stages of the asset life cycle, scope definition and revisiting the original design basis.

To illustrate these issues, it's worth considering an example – pipework. Four years ago there was a major focus on pipework in the UK. Pipework accounts for the most serious and largest number of loss of containment incidents. Pipework does not generally receive the attention that main plant items receive and is often neglected. Maintenance and inspection policies often do not adequately reflect the importance of ensuring integrity against the consequences of loss of containment.

Furthermore, the life cycle for pipework is more fragmented than for any other functional area. Numerous groups, personnel, teams, suppliers, contractors etc. have a part to play at each stage. A robust management system is required to ensure coordination between each of these stages and that all aspects relating to integrity of the plant have been addressed.

Not only are many of these stages often outsourced, or implemented by a different organisation, but within each stage there are often further specialisations of resource, leaving to further fragmentation.

A study carried out by ABB Global Consulting for the UK's Health and Safety Executive (Ref 1) found that most incidents arose from the cumulative effect of a range of errors and vulnerabilities introduced throughout the life cycle of the asset through design, construction, operation and maintenance.

Setting up an effective integrity management system requires a structured approach relating to the identification and implementation of improvement initiatives, sharing of experiences within the process industry and learning from past incidents (not just your own company), taking a fresh perspective on significant issues and going through a process of highlighting the vulnerabilities relative to the plant. It is vital that priorities are defined and investment made available to address those priorities.

The issues of fragmentation of the pipework life cycle discussed above, and illustrated in the case study, are structural ones to do with the way the industry handles the subject of pipework. Such issues can only be effectively addressed by concerted management effort. It's not surprising therefore that many pipework integrity programmes fail to deliver sustained benefits.

Many "integrity improvement projects" regard pipework integrity as purely an in-service inspection exercise – integrity is the responsibility of the Inspection Department (or contractor) and if only they could identify the "magic" inspection technique all would be well. In these cases the typical outcome is a mass of inspection data that fails to provide the degree of assurance or a practicable improvement plan. In some other cases, the operating company sets up an ambitious project which attempts to tackle issues on all fronts, and ends up diluting its efforts and running out of steam.

Clearly it is not practical for every company with high hazard pipework to achieve a "100%" target against each benchmark factor in design, construction, maintenance etc. What is important, however, is to identify the key factors – those that are likely to make the largest impact. For example:

a) For existing plants, re-validating or re-engineering the assets to modern standards may not be practical. But, what is the real impact of the original design and construction standards on on-going integrity? And how do such standards affect the engineering of modifications and maintenance activities?
b) To what extent do operational practices affect the integrity of pipework?
c) What is the real affect of "maintenance cost reduction" projects on pipework integrity?

The principles underlined in the above example of pipework apply to all the asset types, and point to the need for a consistent approach.

THE ELEMENTS OF AN INTEGRITY MANAGEMENT STRATEGY

To answer the questions posed in the previous section, and to develop a pragmatic integrity programme, it is necessary to take a holistic look at the relevant factors, and to identify those where the site can set clear and realistic targets. Needless to say, this is likely to be different for each company, taking into account such factors as the health, safety and environmental impact of losses of containment; production consequences of credible failure scenarios; design and construction pedigree of the assets; plant upgrade plans etc.

Human factors become increasingly important in such a scenario – ranging from management understanding and support, and communications across the life cycle stages and organisations involved, through to the establishment of effective information systems, and sufficient understanding of the design and construction features and deterioration mechanisms by all the relevant groups (plant teams and external specialist resources).

What does the integrity management system actually consist of and what does it look like? Figure 1 shows how an integrity management system can be developed. The approach hinges on a coherent Asset Strategy that defines the requirements to be placed on the assets to support the long term business strategy. The Asset Strategy is implemented by a range of policies covering process safety, maintenance and inspection, renewal, and competence.

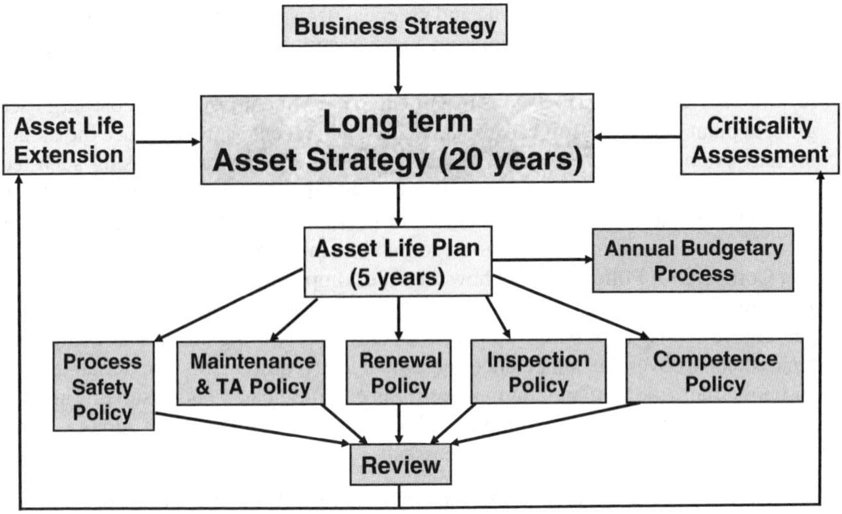

Figure 1. Asset management strategy

The main aspects of these policy areas include the following:

1. Process Safety policy should cover identification of the major hazards, and measures to eliminate, reduce and mitigate those hazards. This would include a documented design basis for the process, identification of residual risks and definition of risk reduction measures.
2. Maintenance policy should set out how safety, health, environmental risks, and the risks to production are to be monitored and controlled by engineering maintenance activities. This includes defining the optimum balance between on-stream and off-line maintenance; and how preventative maintenance is complemented by turnarounds and overhauls to ensure continued fitness for purpose and integrity for operation. It should also cover the policy for critical spares.
3. The Inspection policy should address the WHAT (what is to be inspected, types of equipment, specific areas of the equipment), HOW (how it is to be inspected, on-line versus offline, invasive versus non-invasive techniques), and WHEN (when should it be inspected, what is the period between inspections relative to known deterioration mechanisms). Gathering of data during inspections and storage of that data in a history file is key to addressing the issues of ageing and the ability for continued service into the future.
4. The Renewal policy would address the question of when does the equipment come to the end of its life? The key to this question is not about how old the equipment is, but about knowing what condition that equipment is in at the present time and how that condition changes over its operating life. In many cases, the decision to extend the life

of equipment is as much economic and practical decision as it is a technical one. How well are the life-limiting deterioration mechanisms known? What is the impact of equipment obsolescence? Which factors trigger the decision to carry out significant repairs and replacements? This is a particularly pertinent issue for most large scale process plant as much effort has been undertaken in recent years to reduce the duration of, and extend the period between, major planned shutdowns. By using such techniques as Risk Based Inspection (RBI) and Reliability Centred Maintenance (RCM), operators have focussed maintenance and inspection activities on minimising the need for maintenance and inspection work, so maximising production cycles.

5. The Competence Policy declares how the core competence of the organisation is to be maintained: what are the knowledge, skills and experience required of the key personnel, and how that is to be developed to ensure effective organisational competence. It should also cover the competence and availability requirements of external resource, and how their role in maintaining integrity is communicated and assured. The policy should also describe how learning from incidents and feedback from audits is used to strengthen integrity management processes.

Each policy area needs to be robust, supported by Procedures, Practices and Standards. It is further supported by competent resources, effective communication between groups, auditing and management reporting.

ASSET LIFE PLAN

We need to take a wider view – not just focus on known "critical assets" or problem areas. The diagram shows the key issues and aspects that need to be addressed to resolve those issues (Fig. 2).

Items of equipment can operate for many years, well beyond their original design life providing condition is determined and history is known and a plan is defined to ensure the item is maintained in an operable state, focussing on its vulnerabilities.

This requires a thorough understanding of the design basis, the design features and vulnerabilities, deterioration modes and operating and maintenance histories. In many cases, such information may not be readily available, placing more emphasis on the experience and expertise of the review team.

As part of an Asset Life Study, a re-validation of the design and of operation beyond the nominal design life is required. Using a multi-discipline team including external specialists to give an independent view and experience and good practice from other companies/industries, it is necessary to determine the current condition, deterioration modes, opportunities for improvement in asset care practices, costs and expenditure profiles for the projected life extension.

When it comes to revisiting the original design assumptions, it is necessary to ask: what was the original basis for the defined life of the equipment? As part of the asset life strategy for the plant, analysis should be carried out to identify items of equipment where life will be limited, and hence where the equipment will need replacing, within the

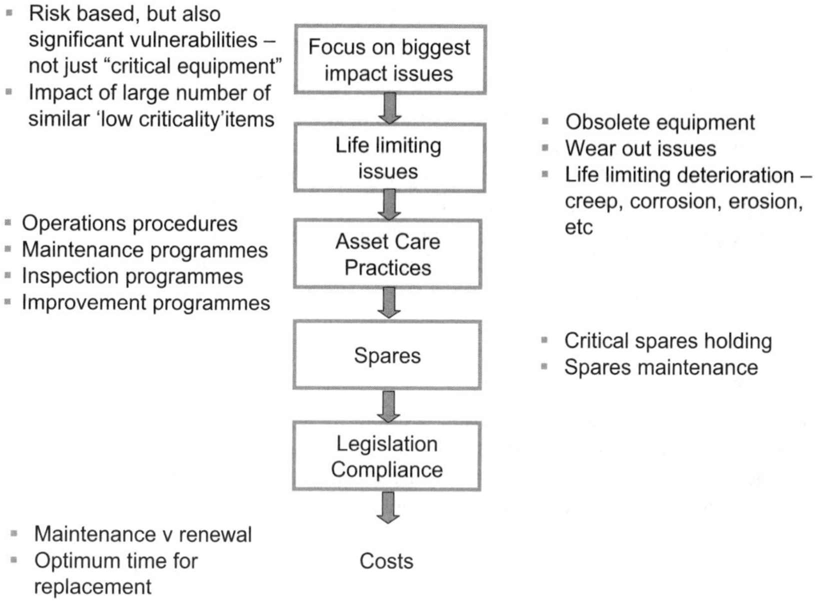
Risk based, but also significant vulnerabilities – not just "critical equipment"
Impact of large number of similar 'low criticality'items
Focus on biggest impact issues
Life limiting issues
Obsolete equipment
Wear out issues
Life limiting deterioration – creep, corrosion, erosion, etc
Operations procedures
Maintenance programmes
Inspection programmes
Improvement programmes
Asset Care Practices
Spares
Critical spares holding
Spares maintenance
Legislation Compliance
Maintenance v renewal
Optimum time for replacement
Costs

Figure 2. Asset life key issues

operating life or required extended operating life. This analysis should be based on safety grounds relative to the point where it is no longer economical or practical (e.g. spares availability) to keep the piece of equipment in operation.

The outcomes from such a study provide a technical justification for continued operation, but also can be used to generate cost information to help define maintenance budgets and rejuvenation investment plans.

ORGANISATIONAL COMPETENCE

To develop and deliver effective integrity management across the life cycle requires competence, communications and commitment. If any of these three areas are deficient in any way, integrity management will be compromised.

It is not sufficient just to collect data on plant condition, though that is itself a major task. It is also important to use the data effectively for decision-making. For example, measured wall thicknesses should be analysed:

1. to identify trends in deterioration and patterns of failure
2. to challenge the accuracy and validity of the data, depending on how critical the consequences of failure may be

3. to make the appropriate decisions about frequency and extent of future inspections and nature and timing of repair or refurbishment work

The various groups involved therefore need the necessary competence to carry out their tasks and to understand the need for communication across the organisational divides. Do your organisational and contract structures enable such communication?

In addition, this defines the data management requirements. Controlling of information through all stages of the life cycle and maintaining a history file is vital if we are to make robust decisions relating to the continued safe operation of plant and to the extension of operation of that plant in to the future. The information needs to be relevant, clear and concise and be readily accessible.

CONCLUSIONS

To summarise: if we are to safely operate, without incident, 20th century equipment well into the 21st century, robust systems relating to integrity management need to be in place. These include management and information systems that support data collection and management decision-making across the asset life cycle; methods and procedures that define the key integrity activities; competence development and training so that personnel (both in-house and contractor) are clear what their roles are in preserving integrity; and monitoring and auditing to reinforce the requirements and recognise and share good practice.

REFERENCE

1. D Jones and D Stanier, Piping Systems Integrity: Management Review, 2001 HSE RR253 available on www.hse.gov.uk.

BUNDING AT BUNCEFIELD: SUCCESSES, FAILURES AND LESSONS LEARNED

Aidan Whitfield[1] and Dr Mike Nicholas[2]
[1]Environment Agency, Kingfisher House, Goldhay Way, Orton Goldhay, Peterborough. PE2 5ZR, e-mail: aidan.whitfield@environment-agency.gov.uk
[2]Environment Agency, Frimley Business Park, Camberley, Surrey. GU16 7SQ, e-mail: mike.nicholas@environment-agency.gov.uk

A large fire that burned for several days and consumed over 40 million litres of fuel followed the explosion at the Buncefield fuel storage depot in December 2005. Fortunately nobody was killed but over 40 people were injured, there was extensive damage to property and pollution of the soil and groundwater. The performance of the tank bunds had a significant effect on fire fighting operations and the extent of the pollution. Some bunds remained intact but others suffered loss of containment during the fire, releasing fuel and firewater containing perfluorooctane sulphonates (PFOS) used in the fire fighting foam. The subsequent pollution of groundwater exceeded the threshold for reporting the environmental impact to the European Union under the Control of Major Accident Hazards (COMAH) Regulations 1999. This paper examines the secondary and tertiary containment systems, focussing on bund designs and the effect they had on the loss of containment.

The paper also describes several other recent incidents that involved loss of secondary and tertiary containment at COMAH sites and how these, together with Buncefield, led to the Competent Authority adopting a containment policy to raise standards across the fuel storage sector.

KEYWORDS: Buncefield, Control of Major Accident Hazards Regulations 1999, COMAH, perfluorooctane sulphonate, PFOS, secondary containment, tank bunds, tertiary containment.

THE EXPLOSION AND FIRE AT BUNCEFIELD

Buncefield is a major fuel storage and distribution depot located just outside Hemel Hempstead, some 40 kilometres north-west of central London. It was opened in 1968 and is the fifth largest oil depot in the United Kingdom (UK). Fuels are received from the coastal oil refineries via 3 separate pipelines. There are over 35 tanks on the site, ranging from 400 to 19,000 cubic metres capacity, located in 9 concrete and 2 earth bunds. All the tanks are built of steel, above ground, to conventional oil industry standards with fixed and floating roofs.

The site supplies aviation fuel by pipeline to Heathrow and Gatwick airports and distributes other fuels by road tanker to local customers. There are 3 operators on the site: Hertfordshire Oil Storage Ltd (HOSL); the British Pipelines Agency (BPA) and British Petroleum Oil (UK) Ltd (BP). Each of the 3 establishments is classified as top tier under the Control of Major Accident Hazards (COMAH) Regulations 1999. A joint Competent Authority (CA) comprising the Health and Safety Executive (HSE), the Environment Agency and the Scottish Environment Agency (SEPA), implements the COMAH regulations in Great Britain.

Just after 6 am on Sunday 11 December 2005 there was a massive explosion at the site, followed by a fire that involved over 20 tanks in 7 separate bunds. A large office building 100 metres beyond the site boundary was also set alight. The explosion injured several members of staff, destroyed the fire-water pumps and severely damaged the site offices so there was no effective fire fighting capability available from within the site. By the end of the day the fire in the office block had been extinguished and the fire on-site had been prevented from spreading to adjacent tanks.

On Monday 12 December 2005 the fire service mounted a plan to extinguish the fire. Large quantities of foam were applied to each tank and each bund in turn to extinguish the fire, then a foam blanket was maintained to prevent re-ignition. By Wednesday afternoon tank 12 at the north end of the site was the only one still alight. "Fire out" was declared on Thursday 15 December.

The subsequent investigation by the CA revealed that the initial loss of primary containment was the overfilling of a petrol storage tank T912, one of 3 tanks in a concrete bund, operated by HOSL. This continued for approximately half an hour and several hundred tonnes of petrol cascaded down the outside of the tank producing a large vapour cloud that exploded as a result of one or more sources of ignition.

The explosion produced overpressures that were much higher than expected and research is being carried out to understand why. The overpressure caused considerable damage to tanks and surrounding buildings.

During the initial stages of the fire, many of the bunds performed well. They contained both leaking fuel and fire-water, which allowed the fire service to operate close to the burning tanks and reduced the escalation of the fire. However loss of integrity of some bund walls developed over the following days, which allowed fuel and contaminated fire-water to flow out over the site. The fire services minimised the impact of this by re-circulating firewater and pumping it into other non-damaged bunds. Despite their efforts, a significant volume of liquid escaped into Cherry Tree Lane, a public highway that runs through the site. It then flowed under a bridge under the M1 motorway, where it soaked into the ground.

Cherry Tree Lane has a number of road drains connected to deep chambers, one of which contains a borehole at least 40 metres deep that penetrates the chalk aquifer. The on-site drains and road drains provided pathways for fuel and contaminated firewater to pollute the groundwater.

The fuels stored at the Buncefield terminal are classified under the Chemical (Hazard Information and Packaging for supply) (CHIP) Regulations 2002 as R51/53 "Toxic to aquatic organisms, may cause long-term adverse effects in the aquatic environment". In addition, the fire service applied 68 million litres of water and 786,000 litres of foam concentrate to fight the fire. Some of this foam contained perfluorooctane sulphonate (PFOS) which is persistent, bio-accumulative and toxic. About 33 million litres of fire water run-off were recovered from the terminal and taken off-site in road tankers for treatment and disposal. The operation to remove fuel and firewater from the site and its surroundings continued for many weeks after the fire was extinguished.

CONTAINMENT SYSTEMS

The safe storage of liquid dangerous substances is achieved by a combination of primary, secondary and tertiary containment systems:

- Primary containment is the most important means of preventing major accidents involving liquid dangerous substances. It is achieved by the equipment that has direct contact with the substances being stored or transported such as storage vessels, pipe-work, valves, pumps and associated management and control systems. It also includes equipment that prevents the loss of primary containment, such as high level alarms linked to shutdown systems.
- Secondary containment minimises the consequences of a failure in the primary containment system by preventing the uncontrolled spread of the liquid dangerous substance. Secondary containment is achieved by equipment that is external to and independent of the primary containment system, such as concrete or clay bunds around storage tanks. Secondary containment will also provide limited storage capacity for firewater management.
- Tertiary containment minimises the consequences of a failure in the primary and secondary containment systems by providing an additional barrier preventing the uncontrolled spread of the liquid dangerous substance. Tertiary containment is achieved by means external to and independent of the primary and secondary containment systems, such as site drainage and sumps, diversion tanks, impervious liners and/or flexible booms. Tertiary containment will be utilised when there is a small scale loss of primary containment from an area without secondary containment (e.g. a pipe flange leak or an overturned road tanker), and when there is a major incident that causes the failure of the secondary containment e.g. bund joint failure or firewater overflowing from a bund during a prolonged tank fire.

INCIDENT INVESTIGATION

A joint investigation team from the HSE and Environment Agency spent almost a year on the site. The most significant feature of the Buncefield incident was the vapour cloud explosion and it was this that caused the vast majority of the off-site damage to property. The incident investigation therefore focussed most of its effort on the initial loss of primary containment and the mechanism of vapour cloud formation. The key findings relating to the loss of secondary and tertiary containment systems were:

- The bunds substantially remained standing throughout the incident, but their ability to fully contain the fuel and firewater was lost.
- The bund design did not provide for effective fire-water management because there was no means of safely removing fire-water from below the layer of fuel and foam. The fire service reported that one of the bunds eventually filled and overflowed. When this happens it is difficult to predict where the fuel overflow will occur because the top of bund walls are practically level. As the bund reaches its maximum capacity, maintaining the foam blanket integrity can be increasingly difficult because the liquid surface becomes exposed to the wind and the draft of air being drawn in by the fire. Once fires are extinguished, it is important to maintain a foam blanket to reduce the risk of re-ignition and this introduces further quantities of firewater that will require containment.
- Many of the bund walls suffered loss of containment at the pipework penetrations and there was leakage due to the loss of seal between pipes and the bund wall.
- Damaged product pipework provided pathways for liquids to escape from bunds or to flow from one bund into another.
- The performance of the expansion and construction joints between concrete slabs varied between bunds. The joints that performed best had metal waterstops cast into them. The joints that had been modified by the installation of a metal plate in the inside face were provided with some degree of protection. In a number of bunds the joint materials were badly damaged resulting in loss of integrity.
- In at least two locations, the concrete bund floor buckled and broke.
- The tertiary containment systems did not contain fuel and fire-water within the boundaries of the terminal. Loss of power to pumps and inadequate drainage and lagoon integrity and capacity were key contributors to this.
- Although the terminal was located on a layer of clay, which helped reduce the impact of the incident on the underlying chalk aquifer, there were several on and off-site pathways that transported polluting materials directly into the groundwater.

THE DAMAGE TO PEOPLE, PROPERTY AND BUSINESSES

Fortunately nobody was killed but over 40 people were injured. The explosion caused significant off-site damage to industrial and domestic properties and the fire destroyed one major off-site office block. Residents and businesses had to find alternative accommodation while their properties were repaired. Civil damages claims in excess of £660 million have been lodged.

The tanks and bunds involved in the fire were completely destroyed, along with the offices and road tanker loading bays. They were all demolished and the site cleared down to ground level. The BP Oil tanks in the south-east corner of the site were virtually undamaged and the fuel in them was taken off-site in road tankers for re-refining several months after the fire. BP are planning to restart operations in their part of the depot in 2008.

THE DAMAGE TO THE ENVIRONMENT

The site is covered with a layer of clay soil approximately 5 metres deep, below which is a chalk aquifer used to supply potable water for the surrounding region. A number of boreholes have been sunk around the site and many groundwater samples have been taken in an attempt to establish the extent and severity of the pollution. The results suggest that groundwater under the site and up to 2 km to the north, east and south-east has been contaminated with hydrocarbons and PFOS. The accident has been reported to the European Union because the area of contaminated groundwater exceeds 1 hectare. The nearest public supply borehole is about 3 kilometres from the site and it has been shut down since the accident. A groundwater remediation plan is being developed. The clean-up is likely to be expensive and will take many years to complete.

ACTION TAKEN TO IMPLEMENT THE LESSONS LEARNED

Within days of the incident, the government set up the Buncefield Major Incident Investigation Board (MIIB). Their reports have explained the cause of the accident and made recommendations on issues such as design and operation of fuel storage sites, major incident emergency preparedness and work concerning the explosion mechanism.

The investigation board has recommended that the Competent Authority and the sector should jointly review existing standards for secondary and tertiary containment with a view to the Competent Authority producing revised guidance. The review should include, but not be limited to the following:

- developing a minimum level of performance specification of secondary containment (typically this will be bunding);
- developing suitable means for assessing risk so as to prioritise the programme of engineering work in response to the new specification;
- formally specifying standards to be achieved so that they may be insisted upon in the event of lack of progress with improvements;
- improving firewater management and the installed capability to transfer contaminated liquids to a place where they present no environmental risk in the event of loss of secondary containment and fires;
- providing greater assurance of tertiary containment measures to prevent escape of liquids from site and threatening a major accident to the environment.

They also recommended that revised standards should be applied in full to new build sites and to any major modification work at existing sites. They recognised that it may not be practicable to fully upgrade bunding and site drainage on existing sites. In such cases the operators should agree with the Competent Authority a risk-based plan for phased upgrading to achieve as close to new plant standards as is reasonably practicable.

The HSE, the Environment Agency and the oil industry set up the Buncefield Standards Task Group (BSTG) to co-ordinate implementation of the lessons learned. This was particularly effective at identifying and implementing the "quick wins" – a series of relatively simple measures to reduce the likelihood and severity of a Buncefield type accident. Their final report recommends engineering measures to prevent the loss of secondary and tertiary containment, including bund integrity, fire-resistant bund joints, firewater management and risk assessment. The BSTG has now been superseded by the Process Safety Leadership Group (PSLG).

The Competent Authority carried out a review of more than 100 fuel storage sites around the country where a Buncefield type accident could occur. The review was published in March 2007 and identified a significant number of sites that will need to carry out work to bring their primary, secondary and tertiary containment systems up to modern standards. Some of this work has already been completed.

In June 2007 the Competent Authority consulted on a containment policy that sets out the broad principles for determining what the standards should be. For above ground storage tanks containing substances that are flammable, highly flammable or extremely flammable the policy proposed that, in addition to provide secondary containment of the dangerous substance, the bund should have:

- adequate capacity and design to allow fire prevention and control measures to be taken;
- fire resistant structural integrity, joints and pipework penetrations; and
- a means of removing fire-water from below the surface of the liquid in the bund (for dangerous substances which are not miscible with water and have a lower density than water).

OTHER INCIDENTS

Whilst Buncefield has been the most dramatic and high profile process industry accident in Britain for many years, there have been several other recent accidents involving the loss of primary and secondary containment at fuel storage depots.

PETROL LEAK AT STRATH SERVICES (OPERATED FOR CONOCOPHILLIPS), MAYFLOWER TERMINAL, PLYMOUTH. JULY 2007

The terminal is a COMAH top tier establishment supplied by ship and distributing unleaded petrol and other fuels to south-west England.

An investigation by the Competent Authority is in progress as at December 2007, so the description given below should be regarded as provisional.

In July 2007, approximately 60 tonnes of unleaded petrol leaked from a hole in the base of a tank over a period of one to two weeks. When the leak was discovered by inventory discrepancy a water bottom layer was placed in the tank while the remaining petrol was transferred to other storage. The petrol leaked directly into the earth base of the bund below the tank and permeated the ground, emerging at a lower level nearby. The tank was due for a full inspection in September 2007.

A Prohibition Notice was issued under the Health and Safety at Work etc Act 1974 stopping operation of the terminal because flammable vapour was detected in the control room. The notice was lifted once the operator had installed preventive measures. A COMAH

Improvement Notice was issued requiring the operator to submit bund improvement proposals. The operator has complied so the notice has been lifted.

The reason for the leak is believed to be external corrosion leading to a small hole in the tank sump. The petrol is being recovered from the ground and does not appear to have moved beyond the site boundary.

DIESEL OIL LEAK AT CHEVRON, POOLE HARBOUR. OCTOBER 2006

Chevron operates an oil/fuel storage facility on Poole harbour, supplied by ship and distributing by road to local businesses. It is a COMAH lower tier establishment. There are 6 tanks each of 1,000 tonnes capacity, 3 in one bund and 3 in another. The nearby Poole Harbour is an SSSI and Ramsar site with high amenity value.

An investigation by the COMAH Competent Authority is still in progress as at December 2007, so the description given below should be regarded as provisional.

Approximately 25 tonnes of diesel fuel leaked from a hole in a tank base over several days at the end of October 2006. The fuel escaped into the ground through a defective joint in the concrete bund floor. When the leak was detected, water was added to the tank to float the fuel away from the hole and the remaining diesel pumped to another tank. The majority of the diesel entered a disused trade effluent sewer below the bund and was collected at the sewage treatment works. A recovery sump was excavated and further oil recovered from the sump and onsite boreholes. No oil entered the estuary.

The tank had been inspected 18 months previously and given a certificate for 5 years. Two other tanks in the same bund contained unleaded petrol. All 3 tanks were emptied and two COMAH prohibition notice were served to prevent them being used until the bund has been repaired. The notices are still in effect as at December 2007.

KEROSENE LEAK AT PETROPLUS, MILFORD HAVEN. AUGUST 2005

Petroplus, Milford Haven is a fuel storage and distribution facility mostly using ships and the national pipeline system. It is a top tier COMAH establishment. There are about 80 tanks that can store up to about 1.5 million cubic metres of fuels and were built in the 1960s as part of the Gulf refinery that closed in 1996. Milford Haven is a Special Area of Conservation.

On 2nd August 2005 Petroplus reported a leak of 653 tonnes of kerosene from a 40,000 cubic metres capacity storage tank. A water heel was established in the tank and the contents were transferred to another tank on-site. Kerosene was discovered in gardens, farmland, the foul water sewers, a stream and along the shoreline several hundred metres outside the site boundary. The incident resulted in substantial local concern and media interest. 550 tonnes of the oil was recovered from on-site boreholes and a skimmer was installed to remove oil from the local stream.

The leak was caused by pipework rubbing against the sump in the base plate floor, that had been replaced following a leak in 2001. The kerosene leaked out of the tank and down through the permeable floor of the bund into the ground. It went undetected for several days because it was not visible in the bund and the operators thought there were gauging errors.

Petroplus were prosecuted at Haverford West Magistrates court on 10 August 2006 and pleaded guilty to 3 charges; under the Water Resources Act 91 for causing kerosene to be released into controlled waters, under the PPC Act 2000 for breaching a permit condition and under the COMAH regulations 1999 for failure to take all measures necessary to ensure the mechanical integrity of the tank and prevent the accidental release of a hazardous substance into the environment. They were fined a total of £30,000 with £40,000 costs. They have also paid an estimated £3 million in clean-up costs. The pollution of groundwater exceeded the 1 hectare threshold defined in the COMAH regulations for reporting the environmental impact to the European Union.

The site was purchased by SemLogistics in February 2006. They tried to repair the tank that leaked at a cost in excess of £1.4 million, by lifting the tank walls and installing a bentonite mat below the tank. However they could not replace the metal base plate because the walls went out of shape, so they now intend to replace that tank completely. They are developing a programme to improve the tank base containment across the site.

ENVIRONMENTAL RISK ASSESSMENT (ERA)

The best way to achieve inherent safety is to remove the hazards, but this is not possible at a fuel depot whose prime purpose is the bulk storage of hazardous substances. In such cases multiple layers of protection are required to ensure the risks to people and the environment are reduced to an acceptable level. Detailed guidance on how to carry out an Environmental Risk Assessment (ERA) at COMAH establishments was published by the Competent Authority in 1999.

One of the important tools that can be used in an ERA is the Source-Pathway - Receptor model.

At 3 of the sites described in this paper, underground pathways were a significant contributor to the pollution of the environment:

- At Buncefield there were road drains leading to soakaways and an uncapped borehole on Cherry Tree lane
- At Milford Haven there was a geological fault below the tank through which the fuel leaked downwards rather than spreading out and becoming visible close to the tank.
- At Poole Harbour there was a redundant foul water sewer pipe underneath the tank bund which was still connected to the local sewage treatment works.

On all 4 of the sites there was loss of integrity in the secondary and tertiary containment systems at the time of the incident:

- At Buncefield the integrity and capacity of the tertiary containment (bunds, drains, lagoons etc) was insufficient to prevent releases off-site
- At Milford Haven and at Plymouth the tank bunds had permeable bases that extended under the tanks.
- At Poole Harbour there were defective joints in the concrete bund floor such that the bund was incapable of holding liquid.

It is essential that pathways and receptors are properly identified. The secondary and tertiary containment systems should be effective at blocking those pathways, but it must be recognised that loss of containment from these systems may occur during an incident.

CONCLUSION

Many of the fuel storage depots in the UK were built in the 1950s and 1960s. They were designed to the safety and environmental standards of the day, but since then containment standards have risen. All sectors of industry where bulk liquid hazardous materials are stored need to consider the incidents at Buncefield, Plymouth, Poole Harbour and Milford Haven and recognise the potential for loss of secondary and tertiary containment systems during an incident. It is essential to identify the pathways where liquids may be released and take mitigation measures to reduce the subsequent impact. Whilst primary containment is of the utmost importance, history has repeatedly shown that secondary and tertiary containment systems will be called upon as the last lines of defence. It is vital that containment systems at such establishments are carefully reviewed and the necessary actions are taken to ensure that they meet the standards that pertain in 2008. The actions may be expensive but the costs of off-site environmental clean-up often make them cost effective.

The COMAH Competent Authority is producing a containment policy that will set out the standards to be achieved for newly built establishments and the extent to which existing establishments must be upgraded.

Industry must learn the lessons from these incidents and implement measures to prevent similar incidents in the future - there is no excuse for complacency. It is time for operators to re-assess their secondary and tertiary containment to ensure that it is fit for purpose. Those who do not are putting people, property the environment and their own jobs at unnecessary risk. Those who act now may avoid becoming the subject of the next "lessons learned" paper.

Figure 1. Buncefield. Aerial picture of terminal fire & smoke plume (Photographed by Chiltern Air Support Unit)

Figure 2. Buncefield. Firefighters applying foam to tanks and bunds (Photographed by Chiltern Air Support Unit)

Figure 3. Buncefield. Catastrophic bund wall failure at pipe penetrations

Figure 4. Buncefield. Contaminated fire water flowing off-site into Cherry Tree Lane

Figure 5. Buncefield. Loss of integrity at a bund expansion joint

Figure 6. Buncefield. Loss of integrity at pipe penetrations

Figure 7. Buncefield. Damage to a lagoon liner revealing ground behind

REFERENCES

The Buncefield Major Incident Investigation Board (BMIIB) Reports. www.buncefieldinvestigation.gov.uk

COMAH Competent Authority. Report on the findings of the Oil/Fuel Depot Safety and Environment Reviews. www.environment-agency.gov.uk

Buncefield Standards Task Group (BSTG) final report "Safety and environmental standards for fuel storage sites", July 2007. www.buncefieldinvestigation.gov.uk

Process Safety Leadership Group website. www.ukpia.com/news_press/petrochemical_process_safety_leadership_group_formed.aspx?referrertabid=1863&linktext=Petrochemical+Process+Safety+Leadership+Group+formed

COMAH Competent Authority Policy on Containment of Bulk Hazardous Liquids at COMAH Establishments, consultation draft version 1.4, 20 June 2007. www.environment-agency.gov.uk

Guidance on the Environmental Risk Assessment Aspects of COMAH Safety Reports", COMAH Competent Authority, December 1999. www.environment-agency.gov.uk

PRACTICAL IMPLEMENTATION OF A PPC SITE PROTECTION AND MONITORING PLAN

Andrew Buchanan[1] and Julie Holden[2]
[1]Senior Engineer, Aker Kvaerner, Ashmore House, Richardson Road, Stockton on Tees, TS18 3RE, UK. E-mail: andrew.buchanan@akerkvaerner.com
[2]EHSS Manager, Elementis Chromium, Eaglescliffe, Stockton on Tees, TS16 0QG, UK. E-mail: Julie.holden@elementis-eu.com
Web: www.akerkvaerner.com/AKEHSandRisk

Elementis Chromium operates a chromium chemical production installation under a Pollution Prevention and Control (PPC) Permit. A condition of the permit required submission of a Site Protection and Monitoring Programme (SPMP) to the Environment Agency within six months of the permit issue.

The SPMP requires the baseline soil and groundwater quality for the site to be established at the start of the permit life, along with the implementation of an ongoing programme to demonstrate the effectiveness of pollution prevention measures. The objective of the on-going SPMP is to ensure that there is no deterioration in land quality under the permit and generate evidence that can be used to confirm this at any would be site closure. This is achieved through an effective process of planned inspection, testing and maintenance of the site infrastructure, supported by periodic environmental monitoring to confirm the site condition.

This paper will discuss practical experiences gained from the perspective of the Permitted Installation, who had to implement the on-going SPMP requirements, and the Environmental Consultant, who carried out the baseline ground investigation.

INTRODUCTION

Elementis Chromium operates a unique production facility in the UK producing a range of chromium chemicals. The process required permitting under the Pollution Prevention and Control (PPC) Regulations in 2005, this was a natural extension of its existing IPC Permits held since 1994. Among the new requirements under PPC, this site, in common with other PPC controlled sites had to implement a "Site Protection and Monitoring Programme" (SPMP) to prevent pollution to land and groundwater while operating under a PPC permit.

The SPMP requires the baseline soil and groundwater quality for the site to be established at the start of the permit life, together with the implementation of an ongoing programme to demonstrate the effectiveness of pollution prevention measures.

The objective of the on-going SPMP is to ensure that there is no deterioration in land quality under the permit and generate evidence that can be used to confirm this at any future site closure. This objective is achieved through an effective process of planned inspection, testing and maintenance of the site infrastructure, supported by periodic environmental monitoring to confirm the site condition.

The SPMP process involves the following steps:

- the collection of baseline "reference" data on existing land quality (including soil, groundwater and surface water),
- validation of protective infrastructure, e.g. bund and drain integrity,
- an on-going monitoring programme to ensure the continued functioning of protective infrastructure,
- an on-going environmental monitoring programme.

This paper will discuss practical experiences gained from the perspective of the Permitted Installation, who had to implement the on-going SPMP requirements, and the Environmental Consultant, who designed and carried out the baseline ground investigation.

SITE HISTORY

The Elementis Chromium site is located near the town of Eaglescliffe, Stockton-on-Tees in North East England and occupies an area of about 13.8 ha. A general location plan is shown as Figure 1.

The site was originally occupied by the Eaglescliffe Chemical Company Ltd. which was established in 1833 for the manufacture of sulphuric acid and fertilisers. Later production operations were added for animal foodstuffs, oleum, purple ore briquetting, copper extraction, zinc oxide, tin oxide and superphosphate production. Fertiliser production ceased in 1967 and a 1960 sulphuric acid plant was closed in 1984.

Production of chromium chemicals at the site started in 1927, with solid mineral process residues deposited in an above ground ("landhill") facility to the southwest of the current production site. The chromium operations were gradually expanded over the late 20th century to eventually occupy the majority of the site.

CLIENTS FACTORS FOR CONSIDERATION

With the long and complex history of the site as described above, there were concerns within Elementis Chromium that high concentrations of a variety of chromium and non-chromium pollutants might be encountered during the investigation. Contaminant mobility, potential risk to receptors both on and off site and liability implications facilitated the need for a thorough assessment of ground quality.

One of the main outcomes of the process, which will be discussed in greater depth throughout this paper, was that the concerns held prior to the investigation were unfounded. The client viewed the process as extremely positive, again views that will be discussed in detail throughout the paper.

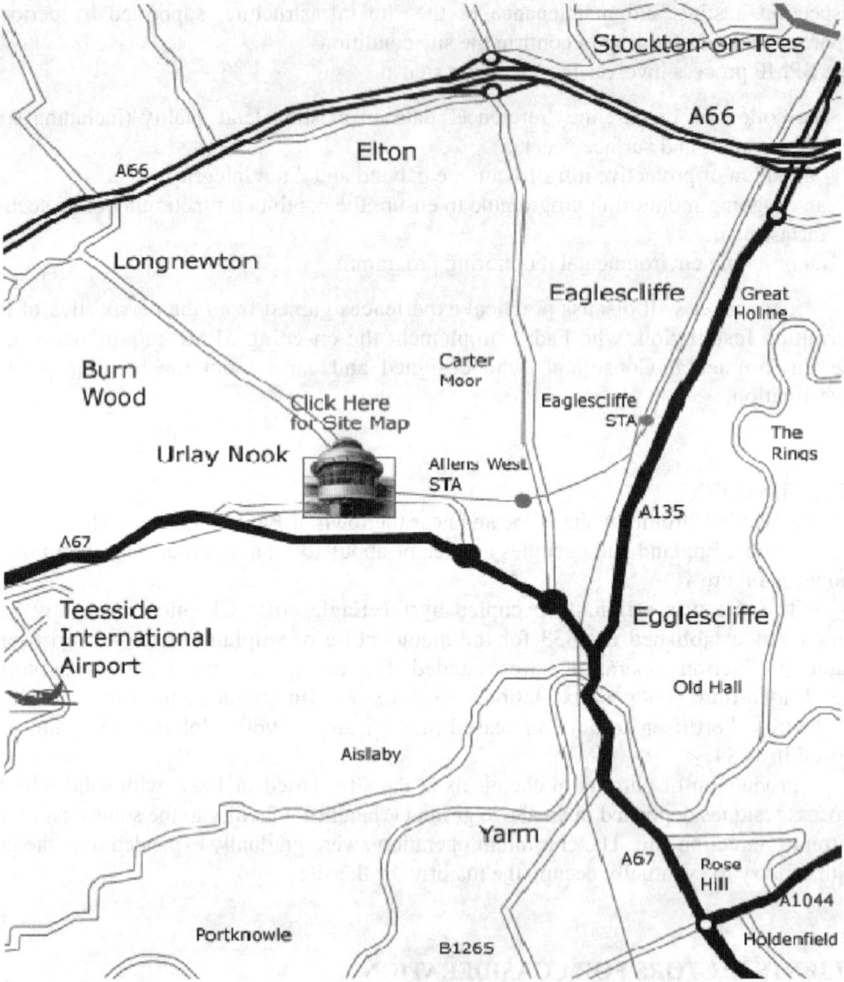

Figure 1. Site location plan

ENVIRONMENTAL CONTEXT OF THE SITE

As can be seen in Figure 2, the Elementis site is roughly triangular in shape and lies on the upper slopes of a gentle valley. A small beck forms the northwest perimeter of the site which is more or less surrounded by arable farmland. There is an ecology wildlife area established by the company on land formerly occupied by the Royal Navy to the East of the site.

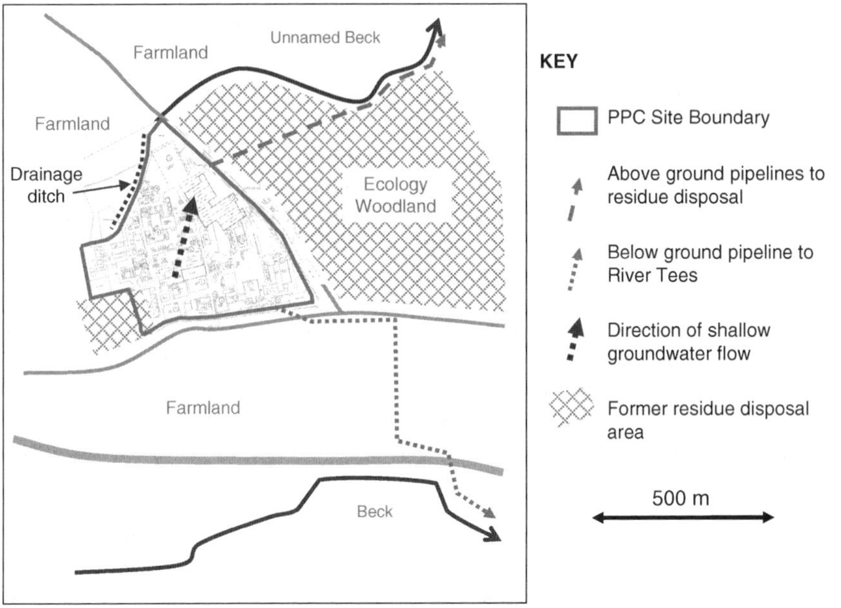

Figure 2. Environmental context of the site

Adjacent to the site to the southwest is a former above ground mineral residue disposal area operated by the company until 1949.

The Cenargo World Depot (a former COMAH site) is located about 600 m to the east and residential housing is located at Eaglescliffe about 1 km to the east.

GEOLOGY

The geological sequence beneath the site comprises made ground up to 2.0 m thick overlying Flandrian Alluvial Deposits consisting of alternating laminated clay and sands and Devensian Glacial Till of sufficient combined thickness to provide protection to the underlying Triassic Sherwood Sandstone.

Groundwater Vulnerability Map (Sheet 8) for the area indicates that the soils are classified as negligibly permeable over non-aquifer.

BASELINE INVESTIGATION OF LAND QUALITY

OBJECTIVES

The objectives of an SPMP baseline investigation of land quality are: to obtain information to allow the refinement of the conceptual model of the site and its surroundings; as well as to obtain baseline data in areas that are vulnerable to existing and future potential contamination. These data will serve as a reference against which the site condition will be assessed at closure.

DESIGNING THE INVESTIGATION

Sampling points which are shown on Figure 3 were selected on the basis of the pollution potential of the on-going PPC activities, together with historical information derived from site incident logs and previous ground investigations. Areas where historic contaminants are expected to be present are defined using this information and are shown in the shaded areas on Figure 3.

A technique that is frequently used when designing SPMP baseline investigations and one that was used at Elementis Chromium was to divide the site into zones depending on current site processes. The zones are also shown on Figure 3. The sample locations within each zone can then be selected based on this process information and the historical information. The cost benefit associated with the number of samples is also considered at this stage bearing in mind that the process is an iterative one and further sampling locations can be added in the future if necessary.

While historical site information combined with the findings of previous ground investigations affords a certain level of confidence in selecting the sampling locations, the uncertain nature of ground investigations means that the designer of the investigation must be prepared to modify any assumptions whenever further information becomes available. This also emphasises the need for the requisite time to be taken in order to compile a pre-investigation conceptual site model with the aim of attempting to understand as much as is possible about the environmental context of the site.

Locations where high concentrations of contaminants are encountered should be investigated further to identify the source and delineate the affected area. Under PPC, operators will be required to remove any contamination that was not identified at the start of the permit, therefore identifying areas of previous impact is important to limit future liability. Beyond the PPC requirements, the operator should be keen to identify cases of existing pollution as it allows them to evaluate the associated risks to the environment and to plan remediation strategies if required. This is one of the areas where Elementis Chromium welcomed the findings of the SPMP as it confirmed to them that the action they were taking, and continued to take, was protecting the local environment. Additionally, from a commercial point of view, it allowed the operator to identify and reduce any potential liability that may exist in relation to ground contamination.

A final action within the design of the investigation should be to assess the impact of the investigation itself and any potential impact to the environment that it may have.

RESULTS AND ISSUES ARISING FROM FIRST PHASE OF INVESTIGATION

Following the intrusive ground investigation and sampling collection process, selected samples were analysed for a range of determinants by a suitably accredited analytical laboratory.

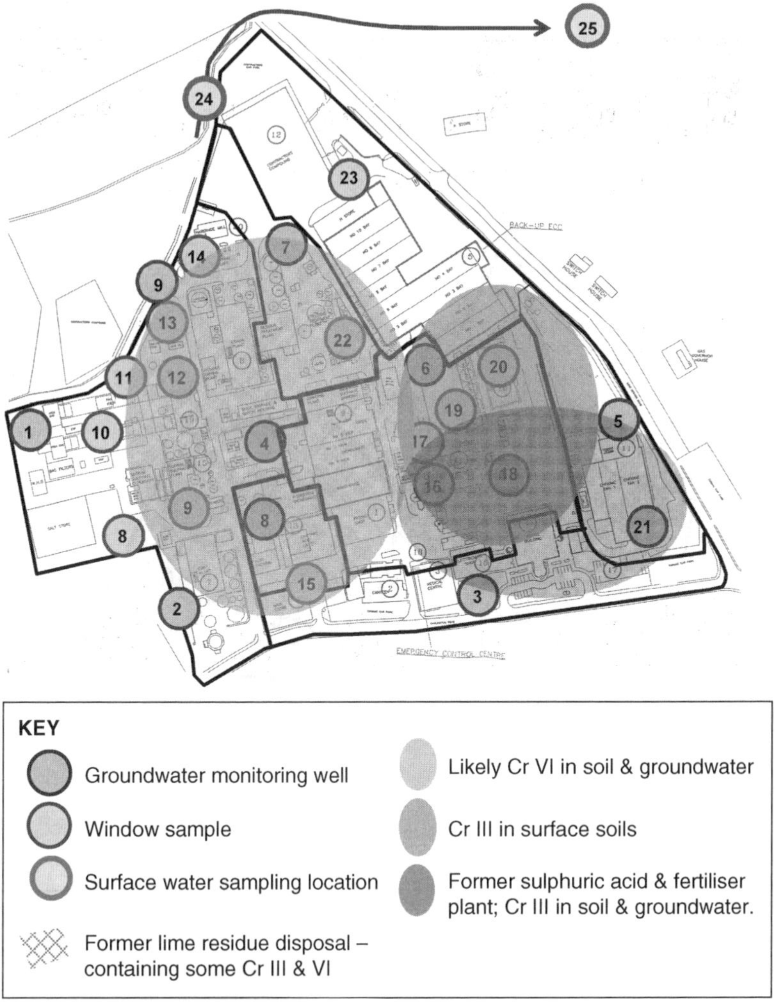

Figure 3. Sampling locations and areas of expected contamination

Data sets containing the analytical results from each zone were compiled enabling an accurate and robust baseline of land quality for each zone to be established. A conceptual site model, in plan and section view, was constructed based on the findings of the initial investigation and historical site data. The plan and section views of the initial conceptual site model are shown in Figures 4 and 5 respectively.

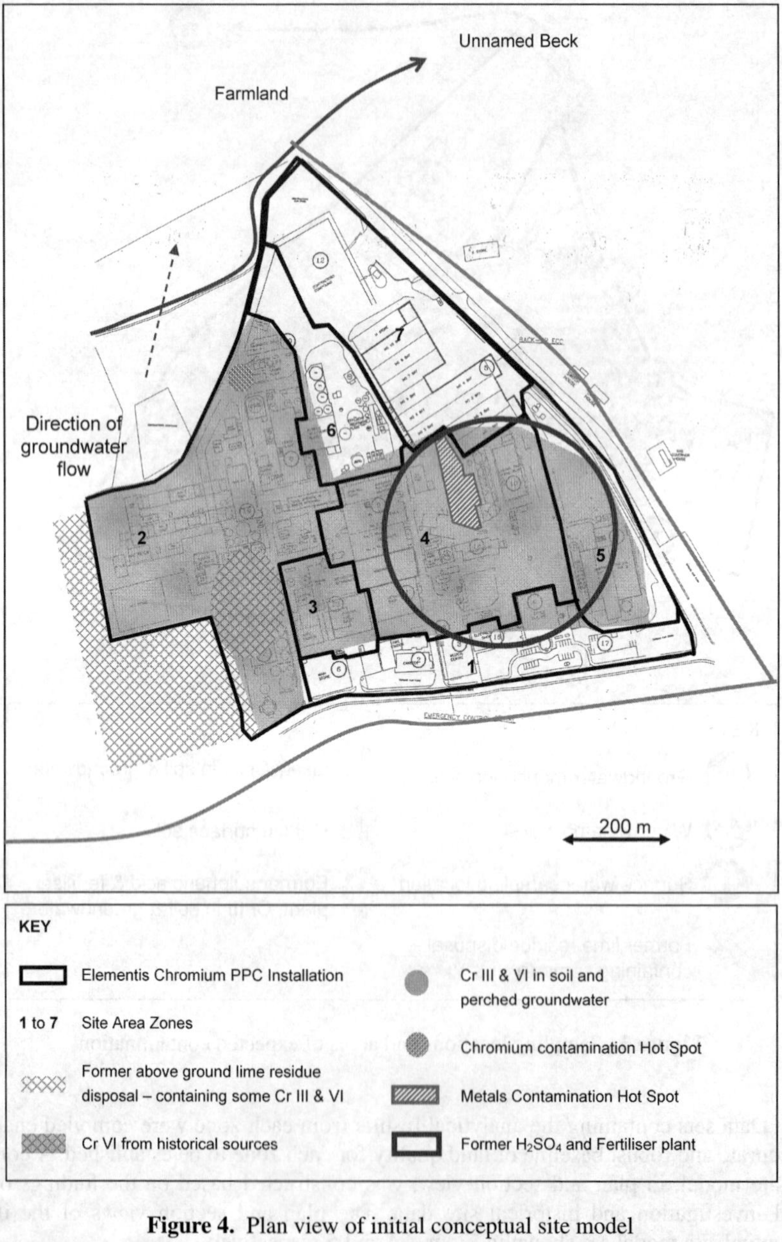

Figure 4. Plan view of initial conceptual site model

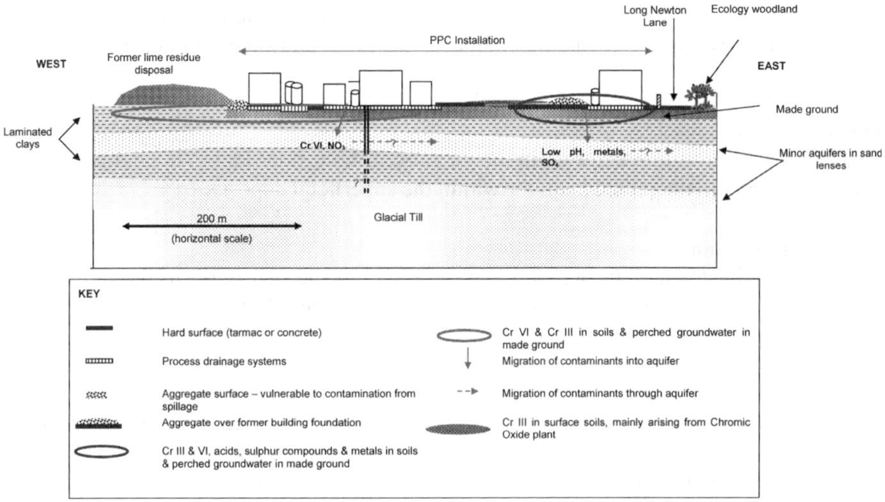

Figure 5. Section view of initial conceptual site model

Where data gained from the initial investigation were either of insufficient quality to compile a robust defendable baseline or left unanswered, questions relating to the potential risk to the environment, further intrusive investigation was required.

PLANNING AND EXECUTION OF FURTHER INVESTIGATION

To complete the PPC baseline evaluation, a further investigation of the following was required:

POINT 1

An area of elevated hexavalent chromium in surface soils which was not recognised as having a high pollution potential from the initial study

Second study action:

Locations were sited radiating out from is area "A" which was identified as the likely source of the elevated levels of hexavalent chromium, with samples taken at various depths at each location. The aim of this was to accurately map the horizontal and vertical extent of the contamination.

Outcome and Findings from further investigation:

The elevated chromium levels that were detected in the vicinity of Area "A" were confirmed to cover an area of approximately 700m^2. No particular distribution pattern was revealed by the further investigation however the limit of the vertical extent of the contamination was confirmed by a lack of contamination detected in groundwater monitoring wells.

POINT 2
High levels of chromium detected in the vicinity of the former Plant "B" and potentially penetrating into an unconfined aquifer within the superficial deposits

Second study action:
A new groundwater monitoring well was sunk within the plant "B" in the most likely source area for the contamination, where the highest concentrations of ground and groundwater contamination was likely to exist.

Outcome and Findings from further investigation:
The concentrations of chromium detected in the borehole within the former plant "B" were confirmed to represent the highest to be found anywhere on the site. As chromium production is no longer undertaken in the plant and the building has been decommissioned, there is no potential for additional chromium to be released in this area under the currently permitted PPC activities.

POINT 3
Groundwater flow regime for the site and surrounding area needed to be fully understood.

Second study action:
Groundwater was monitored for a further period of two months at each of the existing monitoring wells and two new wells including the new well in the plant "B".

Outcome and Findings from further investigation:
Additional rounds of groundwater monitoring led to the conclusion that there is no significant flow within the aquifer; the aquifer was not under pressure and groundwater height did not vary significantly across the site. The groundwater flow direction arrow from the initial conceptual site model has therefore been removed.

While this location represents the highest concentrations of contamination on the site, the fact that groundwater in the confined aquifer is not moving means that the risk to the environment from the contamination migrating off site is minimal. This inference can be made as the contamination that exists does not contain all three facets, i.e. source, pathway and receptor, required to represent a significant risk to the environment. In this case the source and receptors both exist but as there is no pathway connecting the two, the risk is perceived as minimal. This hypothesis was confirmed by the monitoring wells around the boundary of the site, where groundwater quality was found to be good.

Revised versions of the conceptual site models in plan and section view can be viewed in Figures 6 and 7 respectively.

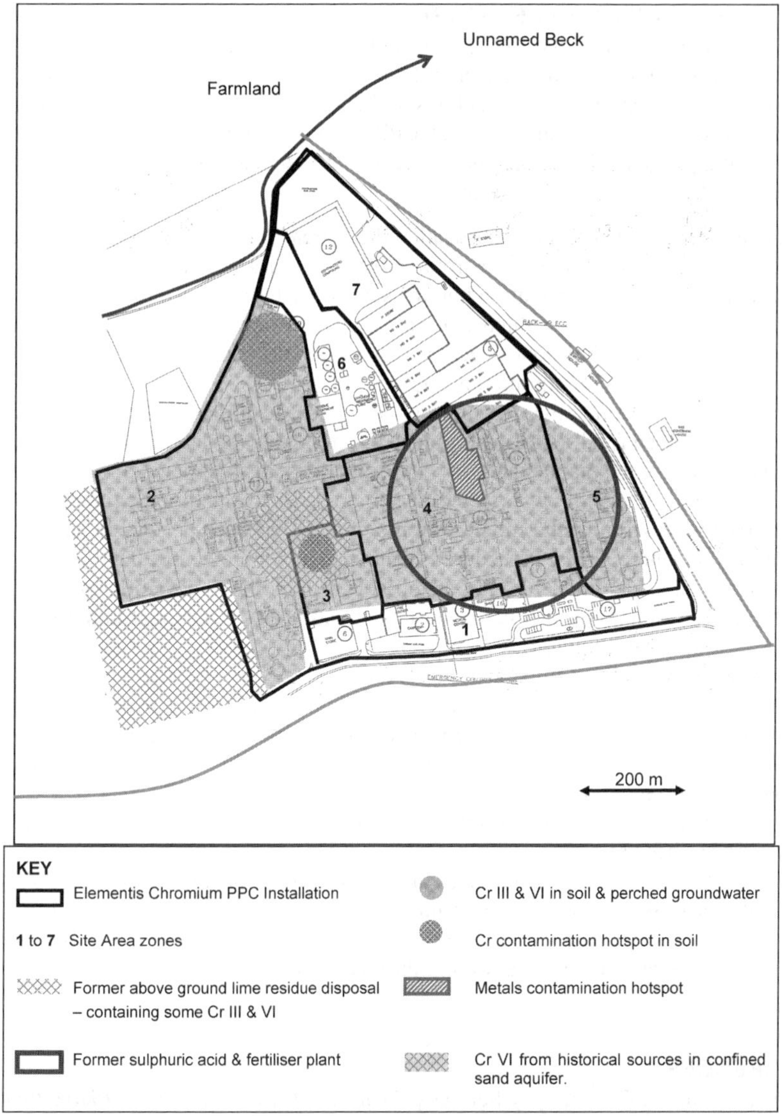

Figure 6. Revised conceptual site model plan view

CONCLUSION/LESSONS LEARNED

For Elementis Chromium, the PPC permit improvement condition to carry out a SPMP baseline investigation held some uncertainty in relation to the potential impact that the site was having on the environment. This links to the potential liability to which the company may be exposed, both within and beyond the PPC regulatory regime. The view the company takes is a positive one in terms of the impact the investigation has had to the business as can be described by the following examples:

- The investigation led to the identification of an area "A" that required action to ensure that it did not migrate into the environment. The area has subsequently been addressed; this potential source of pollution into the subsurface has been removed.
- The investigation demonstrated that contamination impacts are confined within the site boundaries, and indeed to specific areas within that boundary. It also demonstrated that the risk of off-site migration is minimal.
- Existing ground contamination impacts have been quantified and a good conceptual model developed for the site, underpinned by a robust set of data. This information will be beneficial for the maintenance of a fair valuation and assessment of risk in the event of a potential change of ownership of the company or the land in the future.
- An ongoing environmental monitoring programme has been developed and the existing site incident reporting procedure expanded to incorporate more detail of any potential environmental incidents. As a consequence of these developments Elementis Chromium can both protect the environment and mitigate potential liability to which that they would be exposed.
- All site personnel have attended training on the SPMP process, the reasons behind the programme and their individual role in the continuing monitoring programme. This has given the entire workforce continuing ownership of the new system of reporting incidents as well as responsibility for environmental protection, mitigating risk to the environment and reducing any exposure to potential risk and liability that may exist. By involving all site personnel in this way their proactive involvement in environmental matters has been secured.
- The bund inspection regime has been aligned to existing tank inspections. By simplifying the process of inspection and not enforcing a duel inspection regime we have ensured that the inspections fall into an existing, planned, preventative maintenance programme where actions are tracked through to completion.

REFERENCES

1. Aker Kvaerner Consultancy Services; *Elementis Chromium – First Phase Report of the PPC Site Protection and Monitoring Programme Including Reference Data*; Report No. 61016022/02/181/REP001; April 2007.
2. Aker Kvaerner Consultancy Services; *Elementis Chromium – Addendum to First Phase Report of the PPC Site Protection and Monitoring Programme*; Report No. 61016022/02/181/REP001; June 2007.

3. Aker Kvaerner Consultancy Services; *Elementis Chromium – Completion of PPC Baseline Ground Investigation*; Report No. 61016022/02/181/REP003; July 2007.
4. Environment Agency; *Horizontal Guidance Note H7: Guidance on the protection of land under the PPC regime*; August 2003.

A SUCCESSFUL REGULATORY INTERVENTION AT WILLIAM BLYTHE FROM DESPAIR TO DELIVERY

L Murray[1] and R. Costello[2]
[1]Health and Safety Executive
[2]William Blythe

This paper sets out the key ingredients of a successful regulatory intervention by the COMAH competent authority [CCA] at a chemical company, William Blythe, which had previously struggled to ensure good health and safety and environmental standards.

After a history of accidents, incidents and environmental releases at William Blythes, the CCA decided to take an alternative approach to routine inspection and enforcement. The approach required the company to accept there was a problem and demonstrate they were serious about bringing about positive change. For its part, the Regulator withdrew from inspection and HSE and EA staff only visited at the request of the company. Employee involvement was crucial. Finally, the HSE and EA demonstrated considerable flexibility abandoning a rigid interpretation of their internal rules and procedures.

The paper explains the steps taken by the CCA and the company.

Two years down the road, the number of environmental releases and Lost Time Accidents has reduced to virtually zero. The company has realised substantial business benefits (for example, reduced waste, reduced insurance claims, enhanced reputation and improved morale). The company's intiatives are now used as a model within their parent company Yule Catto. Finally in 2007 the company were nominated for a CIA award.

BACKGROUND

William Blythe Ltd, founded in 1845 in Accrington, Lancashire is one of the longest surviving chemical businesses in the UK. Blythes started as a manufacturer of inorganic chemicals for the locally based textile industry, producing sodium arsenate for dyeing processes, Hydrochloric acid by the salt cake process and sulphuric acid by the lead chamber process. It also produced various salts including zinc sulphate for use in the production of Rayon and zinc chloride for use in batteries and for the dissolution of cellulose. Chemicals for the war effort including lead nitrate were also produced.

In 1919 the company acquired a further business in Hapton a few miles away, manufacturing similar inorganic chemicals. A public company from 1928, William

A joint paper by Robert Costello, Operations Director, William Blythe and Linda Murray, H M Principal Inspector of Health of Safety, Hazardous Installations Directorate, HSE. September 2007.

Blythe was acquired by the Hickson group in 1969, in order to take advantage of the Arsenic chemistry with which Hickson were already involved. Holliday Chemicals acquired the business in 1991, as part of their strategy for growth, and latterly Yule Catto in 1998 took over Holliday Chemicals as part of their strategy for diversification from their traditional businesses. The two William Blythe sites rationalised to one in 2006 following growing competitive pressures from Europe for the sulphur dioxide based chemical business.

The business was founded on the manufacture of inorganic chemicals, and 160 years on still manufactures exclusively inorganic chemicals based on arsenic, copper, iodine, tin and zinc, with product applications as diverse as catalysis, electronic chemicals, glass-making, fire retardants, electroplating and printing.

The company is Top Tier COMAH and handles chlorine, arsenic, chromium, iodine and a range of oxidising materials. It employs about 110 people with a turnover of £30 million.

SETTING THE SCENE – A COMPANY PERSPECTIVE

CULTURE

The company have a loyal, traditional long serving Unionised local workforce. Employees are reserved and more comfortable in expressing opinions and concerns through shop stewards and union officials. The frequency of accidents and strength of the unions produced a strong claims culture. There was little or no focus or involvement in skills development, SHE improvements or occupational safety. There were no clear Trade Union objectives to improve conditions or to engage with the management to make progress in anything other than enhancement of terms and conditions.

There was generally a long serving and mature management who felt heavily challenged by change, cost reduction and rationalisation in recent years, where business survival challenges outweighed all others. A management team that had enjoyed many years of product introduction, diversification, innovation and profitability had had to adjust to the relentless loss of customers due to offshoring. Teams were devoted to output and cost targets, and generally managed in a strict rules based regime focussed on new customers, new products and new process routes. The team was very flexible and adaptive, capable of solving problems, both operational and commercial, with vigour and enthusiasm and was accustomed to adaptation of resources to meet urgent customer demands.

Prior to 1969 William Blythe managed its own affairs driven by shareholder results and expectations. Senior management operated from the original owners houses, on both sites. It was a respected company worthy of acquisition. The periods of ownership under first Hickson and then Holliday Chemicals were both generally beneficial, with regular investments taking place in production and product development facilities. Considerable autonomy was permitted to the business, which was generally self-sufficient. Following its acquisition by Yule Catto in 1998, central corporate influence was predominantly financial, although a Group SHE team carried out infrequent audits.

CURRENT BUSINESS CLIMATE

The current business climate is such that flexibility, speed of response, innovation and specialisation are even more important than in the past. Customers and suppliers are now global, with a greater proportion of raw materials imported from developing countries, and a greater proportion of products exported. Cost pressures for William Blythe have always been present, but in a climate of many decades of excellent profitability, it was easy to manage Group expectations, compared with the current position where increasing competition and reducing margins obviously mean that cost management is critical. Managers and staff have been aware of the recent business pressures through internal communications and briefings, and know what they need to do in terms of quality cost and output.

ACCIDENT AND INCIDENT HISTORY

Over the decades there have been several serious accidents, including fatalities, at both of the William Blythe sites. The most recent fatality was over 20 years ago, but some serious injuries have occurred since. The incidence of fires is low, as would be expected in inorganic chemical sites not handling flammable solvents. Serious chemical reaction hazards have been few, although one occurred at each site about 5 years ago, in which there were no injuries.

The Lost Time Accident frequency rate until 2005 was very high, relative to the Chemical Industry peer group. Most of the accidents were related to slips, trips and manual handling, but there were two in 2005 that caught the attention of the HSE which were chemical reaction hazards in which injuries ensued. One was related to the escape of HBr and one to an uncontrolled release from a reaction vessel.

The environmental performance in previous decades led to pollution in the local river, but this now operates as a grade 2 river following work by William Blythe and the closure of other businesses previously discharging into it. Previous breaches of discharge consent under IPC are not known to have caused any immediate harm to the environment. Airborne pollution was common 30 years ago, mainly due to the fog caused by traces of Hydrochloric Acid, but processes have since been adapted and no longer are there any visible atmospheric signs of chemical operations taking place on the site.

PRIOR REGULATORY INTERVENTION

The weakening indicators of performance for the changing face of the business were obvious;

- Reducing profitability due to UK customer base failing or transferring elsewhere.
- Declining employee morale brought about by a series of rationalisations and closures.
- Predominant Senior Management focus on business issues and not SHE.
- Failure to address the Change Management of the organisation following downsizing, resulting in lack of continuity and procedural gaps.
- An accident frequency rate higher than the Chemical Industry average.
- Regular presence of the Regulator following either breaches of consent or accident.

- The failure of the company to produce an acceptable Safety Report after two attempts. (Interestingly, three attempts were needed in the 1990s to produce an acceptable CIMAH Safety Case).
- The request by the Regulator to produce a Safety Report exemplar to reflect the Safety management of one process only. (There were 25 processes operating on the site).
- Poor engagement and involvement of the shop floor by management.
- Evidence of loss of containment in many areas.

SETTING THE SCENE – A REGULATORY PERSPECTIVE

The relationship with the regulator prior to the intervention was friendly, but it is fair to say there had been a less than frank and open dialogue with the CCA about health, safety and environmental issues.

Two attempts to produce a Safety Report had failed. In the five year period leading up to the intervention there had been repeated accidents, incidents and environmental releases. HSE had inspected on a regular basis (including team inspection); the EA were frequent visitors. Improvement Notices had been served (including a COMAH Prohibition at the company's sister site at Burnley, now closed). Concerns had been expressed about process control, human factors and engineering standards. The more effort the Regulator put in via inspection, the more the company were blown off course in trying to respond to an ever increasing agenda of concerns flagged up.

The company had a poor understanding of chemical reaction hazards and there were indicators of a lack of technical, managerial and operational competence. There was a poor attitude to health and safety; some senior managers viewing health and safety as a burden. Employees were happy to operate in a compensation culture.

Finally, the company tended to adopt an insular approach and did not look outside to benchmark good practice.

Two serious accidents and one major environmental relase in 2005 were the final straw for the CCA – a fresh approach was needed.

THE REGULATORY INTIATIVE – ESSENTIAL INGREDIENTS

It could be argued that the health, safety and environmental position the company found itself in at the end of 2005, was not just the responsibility of William Blythes; perhaps the CCA should have taken a more strategic approach. There was a need to move away from the company "having things done to them" eg visits, Improvement Notices, to a position where they took ownership of the problem.

It is possible that the regulator's well intentioned efforts to maintain a cordial relationship with the company (rather than head on confrontation) had led to misunderstanding about the severity of the problem. The starting point was to tell the company of our concerns.

A short letter to the company and their Head Office, stated that the CCA had lost confidence in the ability of the company to manage a COMAH site. They asked for an early meeting with senior managers to consider the way forward.

HSE made a presentation setting out the reasons for their concern (text and pictures). An intially shocked and tense group of senior managers quickly moved on and accepted there was a problem and committed itself to bring about change. The production of a substantive Action Plan by the company, and the drafting in of extra health and safety resource was a crucial step in convincing the CCA that the company meant business and was an essential step in building trust.

Almost in parallel, the employees were involved in the same process; they were shown the HSE presentation and the company explained its proposals to respond. The company could not make good progress unless the employees became part of the solution.

Good communication was essential, particularly in the early days. Regular keeping in touch meetings were arranged (monthly in the early part of 2006). Every effort was made to include employee representatives in this process.

For its part the CCA committed itself to "withdrawing" from site. No visits would be paid unless the company asked for them. During the first year of the initiative the company sought some help on Human Factors and Safety Report Predictive advice. Inspection did not recommence until year 2. The EA and HSE agreed to act in unison and have one common agenda. The CCA also promised to respond promptly to any requests/contact from the company (not always easy when HSE faced huge resource demands following the Buncefield incident).

The early period of the initiative did not always run completely smoothly; there were hiccups along the way. But the CCA kept faith with the company who were clearly trying very hard to improve. The CCA continued to demonstrate flexibility and did not adhere rigidly to the HSE/EA rules books, recognising that major change placed huge burdens on the company. For example HSE did not prosecute the company for the accident and incident at the end of 2005; nor did the EA take enforcement action for the environmental releases that continued to occur in the early days. A very late submission of the Safety Report has been accepted, allowing the company to concentrate on making important changes on site.

Slowly over the two years of this initiative, trust grew between the two parties. Keeping in Touch meetings moved from tense occasions to relaxed meetings involving a lot of laughter. The company have come a very long way in improving their management of health, safety and environmental issues.

WILLIAM BLYTHE'S RESPONSE

Yule Catto and William Blythe immediately responded to the request by the HSE to commit to a time bound improvement programme based on submitting a Safety Report at an agreed period in the future and agreed to provide resources to deliver the programme in the time frame suggested.

The Regulator proposed early deadlines for some critical steps in the process, such as carrying out a Baseline Competency Assessment on site, ranking the processes on site with respect to inherent risk, development of a formal Process Safety Management procedure and making an assessment of the Safety Climate. Reasonable timescales were agreed for completion of these elements of the programme. All other milestones were set by William Blythe and agreed with the Regulator.

Dates were agreed with HSE for review of progress, and representatives from Yule Catto were present at all these 'Keep In Touch' meetings. Relations with the regulator had become strained, mainly as a result of the repeated accident investigation visits, and the perception by the HSE that the Company were paying lip service to the improvement of SHE performance. The early meetings were therefore quite formal and somewhat strained. As soon as the Regulator realised that William Blythe was making progress, and committed to continuing the programme, the mood changed to being supportive.

SUPPORT PROVIDED BY THE REGULATOR

At all stages the approach of the Regulator was to help and support, and to give advice and direction. Examples include:

- Desk top review of Process Safety Management System by Regulator and feedback on progress.
- Inspection by Construction Inspector and follow up training of William Blythe staff on CDM by a HSE specialist.
- Inspection by the Mechanical specialist, and feedback and support on progress.
- Inspection by the Chemical reaction hazard specialist and feedback on progress.
- Inspection by the Human Factors specialist and feedback on progress.
- Inspection by the Control and Instrumentation specialist and support in developing an acceptable LOPA methodology for SIL assessment.
- The local EA inspector participated in all inspections.
- Organisation of a benchmark site visit by the Regulator to compare SHE communications and involvement.
- Attendance at employee meetings by the Regulator to give direct feedback on SHE progress from a regulatory perspective.

The period covered was in excess of 2 years, and in that period there were several incidents, where Regulatory intervention was required. One was related to asbestos disposal, one a waste disposal issue, one a release caused by a systems bypass. In all cases the Regulator investigated the incidents but decided against enforcement action, choosing not to be diverted from the overall programme of support.

HOW DID WE GET TO BE IN THIS POSITION?

Factors were:

- Complacency due to the maturity of the chemical processes, which were all inorganic. Fires were very rare, and hazards were generally from corrosive burns from corrosive materials.
- Pressure from the changing business dynamics.

- Poor change management prior to headcount reductions.
- Poor engineering standards. The effluent plant was built and developed to catch all realisable incidents. Local leaks could be caught in plant sumps or precipitated at the effluent plant.
- Poor operational standards. The operators changed jobs without being proved to be competent. They were rarely involved in process improvements.
- Most processes were carried out in reactors with limited process control and frequent escapes from open manways.
- The management style had developed in a very 'telling' manner, and employees were reluctant to raise issues.
- The middle management had little experience of standards in other companies.
- The trade unions were disaffected, not because of the SHE performance with which they were familiar through regular safety meetings and initiatives, but through the relentless cuts in the workforce.

WHERE ARE WE NOW?

We have:

- Reduced our serious accident and incident occurrence significantly.
- Reduced our excursions to sewer significantly.
- Reported four times the incidents and near misses than we did before.
- Engaged our employees in competency assessment, procedure development, human factors.
- Ensured everyone knows and understands that all change must be controlled and risk assessed.
- Used our Process Safety Management system to change the way that we have made chemicals for over 50 years. The yield is better, the process repeatability is better, the right first time is better. The process is safer. Everyone knows the hazards of the process.
- Fundamentally changed several processes. One process used to frequently discharge NOx into the workplace. This is no longer possible due to process redesign. Another is now being operated at a lower, safer temperature.
- Assessed internal competency and engaged specialists for SIL, for writing test and inspection protocols, loop testing interval, design etc.
- Employed some new members of the management team, with some fresh external experience. This does not reflect on those that have left, but allows us to refresh our ideas and challenge our beliefs.
- Measured the safety climate, and believe that it is improving.
- A comprehensive training and competency assessment process in place, including literacy and numeracy assessment and coaching, for existing and new employees.
- A comprehensive internal audit programme in place, including active participation by the Company's Directors.

WHAT NEXT?

We now recognise the advantages of working with the Regulator to help and support our busines with advice and specialist support to help us operate the business safely.

We recognise that the only way to run a business is safely, and to that end we will continue with the rigour of reviewing in detail all chemical processes to find safer ways of operation which should also repay financially.

We will continue to develop proactive methods of ensuring that we minimise the risk of loss during our operations.

Some of the enhanced systems developed over the last two years e.g. engineering procedures, training and competence assessment, are under consideration by Yule Catto for adoption as corporate standards.

LESSONS LEARNT

Ensure that the full impact of downsizing on maintaining safe operations is reviewed. Take account of not only the tasks but the competencies of those remaining

Don't underestimate the value of engaging the staff in change programmes.

Where systems and procedures exist, use them.

Ask the Regulators for help if needed.

Don't be afraid of admitting that you don't have the competency for some tasks.

WHAT DO WE WANT TO SUSTAIN AND HOW DO WE DECIDE?

M. Sam Mannan, Divya Narayanan and Yuyan Guo
Mary Kay O'Connor Process Safety Center, Artie McFerrin Department of Chemical Engineering, Texas A&M University System, College Station, Texas 77843-3122, USA
E-mail: mannan@tamu.edu

For bio-fuel to be a suitable substitute for fossil fuels, its sustainability as a fuel has to be established. This translates to establishing that bio-fuels have superior environmental benefits while being economically competitive with fossil fuel, and that they can be produced in sufficient quantities to satisfy the existing energy demand while providing a net energy gain over the energy sources used in producing them. As an integrated systems approach, sustainable development (SD) framework methodology has been proposed and employed to address these issues. Most sustainability studies have been concerned only with the environmental impact, while this methodology ensures SD of a process not with respect to just environmental impact, resource consumption but also with respect to societal and economic impacts. Moreover, it allows the user to decide what should be sustained and how to make such a decision. This methodology starts with the prioritization of the sustainability metrics (health and safety, economic, ecological and social components). Then the alternatives are subjected to a pair-wise comparison with respect to each SD indicator and prioritized depending on their performance. The SD indicator priority score and each individual alternative's performance score together are used to determine the most sustainable alternative. The feasibility and effectiveness of this methodology has been demonstrated by identifying the most sustainable bio-diesel process system from a set of alternatives.

1 INTRODUCTION

Most sustainability studies have been concerned only with the environment. In fact, sustainability is an integration of three issues which are the economic, environmental and social implications, thus it is a property of the entire system. For an engineering process, an optimal balance has to be achieved between all these implications. Hence in order to ensure a high degree of effectiveness in reducing the negative implications of a process, a broader concept of designing the process is necessary and the most suitable methodology for addressing this issue is sustainable development (SD), which is well defined by the Brundtland Commission of 1987 as the "*Development that meets the needs of the present generation without compromising the ability of the future generations to meet their own needs*". It aims at striking a balance between various impacts the process has on the environment, economy, society and safety while satisfying the requirements of all the generations of decision makers. Hence chemical companies have begun to assign strategic importance to SD by incorporating them into their decision making. In order to ensure a complete SD, appropriate tools and techniques are required for evaluating available choices and identifying the most sustainable alternative.

The definition of SD makes clear that the development of new technologies has to take into account economic and social issues (present generations) and long-term and large-scale environmental issues (future generations). Bio-fuels, an alternative fuel made from renewable biological sources, have gained more and more interest recently due to high energy costs, increasing demands, concerns about petroleum reserves and greater realization of the environmental impacts of fossil fuels. For bio-fuel to be a suitable substitute for fossil fuels, its sustainability as a fuel has to be established. This translates to establishing that bio-fuels have superior environmental benefits while being economically competitive with fossil fuels, and that they can be produced in sufficient quantities to satisfy the existing energy demand while providing a net energy gain over the energy sources used in producing them.

Currently, bio-diesel studies have been focusing on raw materials (Freedman et al., 1986; Ma and Hanna, 1999, Canakci and Gerpen, 2001), catalysts (Ma et al., 1998; Zhang et al., 2003), alcohols used (Zhang, 2002), and chemical reactions, such as transesterification (Wimmer, 1992; Ali, 1995; Ma et al., 1998) and thermal cracking or pyrolysis (Sonntag, 1979; Weisz et al., 1979). However, its sustainability as a fuel has not been studied widely.

In this paper, SD is performed on the life cycle of a bio-fuel system. The bio-fuel considered here for the SD is bio-diesel. Using proposed methodology the system boundaries for the bio-diesel system is defined, which includes a complete cradle-to-grave analysis of bio-diesel inclusive of the raw materials, the chemical reactants, the process conditions, the by-products, the waste treatment options as well as the disposal of the wastes, excess reactants and the used end product. Then SD indicators are used to quantify the impact bio-diesel has on the environment, economy, society and safety of the surroundings over its lifetime. The analytical comparison tool, Analytical Hierarchical Process (AHP) is used to prioritize the available alternatives depending on their degree of sustainability. Finally the end result of the analysis is a complete bio-diesel system that is sustainable from its cradle to its grave.

2 PROPOSED METHODOLOGY

A SD framework methodology has been proposed to identify the most sustainable design for a given chemical process from a set of alternatives. The proposed methodology has already been reported in detail (Narayanan, 2007). Thus, the principle and procedure of this methodology are briefly introduced in this section.

The SD framework consists of three major steps. Each of these three steps consists of a number of intermediate stages where certain analytical calculations and quantifications are performed. The three steps of the SD decision making framework are as follows

1. System definition and alternatives identification

System definition is the first step in the proposed methodology, which aims at defining the boundaries of the system and identifying the subsystems within the existing system. To enable a complete cradle-to-grave SD, a Life Cycle Analysis (LCA) of the system under consideration is performed. Slight modifications have been done to the process in order to customize it for the problem of SD. Once the system boundary has been established, the succeeding step is to divide the system into a number of subsystems to make the decision making process more robust. The main criterion for identifying the subsystems is to determine the decisions that need to be taken at each stage within the chemical process under development. This step is process dependent and has to be performed for each process for which the decision framework is being used to do sustainability development.

Once the system boundary and the subsystems have been identified, it is the time to recognize all the decisions that need to be made regarding the most sustainable process method or design for each subsystem. In order to proceed with this step, all practicable alternatives must be identified for each process or object under consideration. These alternatives were identified by performing literature survey on various studies performed on bio-diesel production. (Zhang et al., 2003; Tapasvi et al., 2005; Xun and High, 2004; Rudolph and He, 2004; Roszkowski, 2003; Demirbas and Karslioglu, 2007; Besnainou and Sheehan, 1997). The alternatives identified have been proven to be practicable though not sustainable, hence the main objective of the framework is to identify the most sustainable option from a list of practicable alternatives within each subsystem.

2. SD indicators/impact assessment

Once the alternatives have been identified, in order to do the comparison to identify the most sustainable option, the implication of each alternative on the economy, environment, society and safety must be quantified. This quantification is done by the calculation of certain SD indicators, safety indices and by performing a cost-benefit analysis. In this research work, four indicators are analyzed and applied in the SD of bio-diesel production, including the environmental indicators such as emissions, resource depletion, land and water usage, economic indicators such as expenditure, tax incentives, profit margins, safety indicators such as inherent safety of the entire process, hazards associated with the materials used and certain societal implications such as impact on the local economy and employment generation. For each subsystem, a set of indicators are identified. Most of the indicators are common to all subsystems, except for a few subsystem specific indicators.

3. Alternatives comparison and SD decision making

The final step is the decision making, where the most sustainable alternative is identified for each subsystem based on previously defined SD criteria which includes economic, environmental and social feasibility and other performance and safety criteria. Analytical Hierarchical Process (AHP), which is a multi-criteria decision making method was chosen for decision making. This method was subjected to minor modification to customize it to meet the requirements of SD.

3 CASE STUDY – BIO-DIESEL PROCESS

The proposed methodology for SD of a process has been applied to the bio-diesel system. The final product of the proposed methodology is a completely sustainable bio-diesel system. A step-by-step description of the entire SD method is illustrated with supporting tables and visualizations for the bio-diesel system in this section, but the detailed calculations for this process are not provided.

The entire life cycle of the bio-diesel plant has been taken into account in this work, which is inclusive of raw material manufacturing, transportation and storage, the actual process energy requirements, storage and transportation of the final product, treatment of effluents, disposal or reuse of excess reactants and raw materials and finally disposal of the final product after its usage. Within this system, subsystems are identified and suitable alternatives for each of these subsystems are subsequently identified and listed in Table 1.

3.1 RAW MATERIAL SUBSYSTEM

The raw material subsystem within the bio-diesel lifecycle system; is the first subsystem subjected to SD using the developed framework. In this case study the raw materials considered are soybean, rape seed, sunflower and beef tallow. Since these are widely cultivated, available and economically viable they are the most commonly used feedstock for bio-diesel production. The SD indicators used are environmental indicator including Environment Performance Indicators (EPI), land usage, and water usage; economic indicators including total capital costs, manufacturing costs, after tax rate of return, bio-diesel break even price; safety indicators, such as risk assessment matrix (RAM) index, and certain system specific indicators including fuel cetane number and fuel carbon %.

Table 1. The subsystems and the corresponding alternatives

Subsystem	Alternatives
Bio mass	Soybean
	Rape Seed Oil
	Sunflower Oil
	Beef Tallow
Catalyst	Basic
	Acidic
	Enzymatic
Alcohol	Methanol
	Ethanol
Production Process	Thermal Cracking
	Transesterification
Glycerol Extraction	Gravitational Settling
	Centrifuging
Bio-diesel Purification	Hexane Extraction
	Water Washing
Bio-diesel Mix Ratio	Direct Use
	Blending

Table 2. Priority scoring for SD indicators

Level	Score	AHP score	Definition-diff in level of priority
HIGH	3	1	Same
MEDIUM	2	2 or 0.5	1 Level
LOW	1	3 or o.33	2 Levels

The prioritization of these indicators was based on their degree of importance with respect to that particular subsystem. The scale is defined in Table 2 and used in AHP comparison to obtain the priority scores for the different SD indicators. The number scores are allotted to the SD indicators depending on their degree of importance to be used while performing the pair-wise comparison in AHP. The scoring scale varies from 1 to 3, with 1 representing equal importance or performance, 2 representing moderate difference and 3 signifies well marked difference between the two alternatives with one being strongly preferred over the other. The indicator with the higher level of priority is given the higher score and the other indicator is given the reciprocal of the score. The scaling used is qualitative for all the three SD indicators and based on historic data and expert opinion.

Table 3(a) indicates the priority levels assigned to the SD indicators for the raw material subsystem. As raw material is the highest contributor to the bio-diesel price, economic indicators are given the highest priority. Since feed-stock is used in the largest quantity among all the raw materials for bio-diesel production, its impact on the environment must be given high priority when considering the life cycle environmental impact of bio-diesel. There are no major safety-issues associated with raw materials manufacturing or use, thus safety indicators are given medium priority. Certain fuel properties such as cetane number and percentage of carbon depend largely on the raw material used and are hence used as indicators which are given high priority like the environmental and economic indicators. Table 3(a) also lists the numerical scores corresponding to the priority level for each of the SD indicators.

The calculated final priority score for each SD indicator with respect to the raw materials subsystem is illustrated in Table 3(b), which involves the neutralization of the pair-wise comparison scores.

An AHP template is used to compare the alternatives with respect to each of the SD indicators and prioritize them based on their performance using the pre-defined AHP

Table 3 (a). SD Indicator priority level assignment for raw materials subsystem

Indicator	Priority level	Number score
Environmental	HIGH	3
Economic	HIGH	3
Safety	MEDIUM	2
Fuel Performance	HIGH	3

Table 3 (b). Final priority score evaluation for raw materials subsystem

	Environmental	Economic	Safety	Fuel performance	Priority score
Environmental	0.286	0.286	0.286	0.286	0.286
Economic	0.286	0.286	0.286	0.286	0.286
Safety	0.143	0.143	0.143	0.143	0.143
Fuel Performance	0.286	0.286	0.286	0.286	0.286

scoring scales. The first SD indicator used for the comparison of the raw-material alternatives is the environmental indicator, EPI, which directly depends on the environmental impact of the pesticides and other chemicals used in the cultivation and other raw material related processes. The values for the EPI for each raw-material are given in terms of CO_2 weight equivalent emission, shown in Table 3(c). These values are based on the amount of green house gases emitted during fertilizer manufacturing, cultivation, harvesting and oil recovery as well as the amount of N_2O released during cultivation of the feedstock which is converted into CO_2 weight equivalents (Jungmeier, Hausberger et al., 2003). It was observed that soybean required much less fertilizer than both rape seed and sunflower. Rape seed cultivation requires large amounts of nitrogen fertilizers and hence its impact on the environment is higher in comparison to sunflower and soybean. Beef tallow was given the highest EPI score since more energy is input into the pre-processing of this raw material to be used as a feedstock for bio-diesel production.

Other than EPI, land and water usage are also used as environmental impact indicators. The land and water usage for the alternatives are qualitatively assessed as high, medium or low and are assigned corresponding numerical scores. Table 3(c) lists all three environmental indicators for all the raw material alternatives.

Once the environmental indicators are quantified, the next step is to perform the pair wise comparison of the alternatives using an AHP template, with respect to each of these

Table 3 (c). Environmental Indicators for raw materials subsystem

	Environmental				
		Land usage		Water usage	
Alternatives	EPI	Usage level	Number score	Usage Level	Number score
Soybean	40	MEDIUM	2	MEDIUM	2
Rape Seed	110	LOW	1	MEDIUM	2
Sunflower	70	HIGH	3	HIGH	3
Beef Tallow	140	MEDIUM	2	MEDIUM	2

Table 3 (d). Net Environmental Impact score for each raw material alternative

	Soybean	Rape seed	Sunflower	Beef tallow	EPI score	Land usage score	Water usage score	Environmental indicator score
Soybean	0.50	0.47	0.62	0.30	0.47	0.26	0.23	0.289
Rape Seed	0.17	0.16	0.10	0.30	0.18	0.14	0.42	0.224
Sunflower	0.17	0.32	0.21	0.30	0.25	0.14	0.12	0.153
Beef Tallow	0.17	0.05	0.07	0.10	0.10	0.45	0.23	0.234

indicators and obtain individual performance scores. The final indicator scores for all three environmental indicators for each of the raw material alternative are shown in Table 3(d).

Using the scores obtained with respect to EPI, land usage and water usage, the final environmental indicator score for each alternative is calculated by the following formula, and the results are listed in Table 3(d).

$$\text{Final Score} = \sum(0.33 * A_i)$$

Where, A_i = AHP score allotted to alternative with respect to environmental indicator i (i can be EPI, land usage or water usage); *0.33* = Score of importance given to environmental indicator i with respect to the other indicators (all indicators are given equal importance hence the score of 0.33).

The calculation results have shown that the most sustainable option with respect to environmental impact within the raw materials would be soybean as it has the highest environmental indicator score. AHP templates are also developed to quantify the other SD indicators and prioritize the raw materials with respect to economic (Zhang, Dube et al., 2003), safety and system specific indicators (NREL, 1994), shown in Table 3(e).

Table 3 (e). SD indicator quantification for raw materials

	Economic		Safety		Fuel performance	
Alternatives	Total costs($/kg)	Total manufacturing cost of biodiesel $/L	RAM index	Oxidation stability (rancimat induction period h)	Cetane number	Carbon %
Soybean	0.52	0.3	LOW	5.9	51.34	0.94
Rape Seed	0.67	0.69	LOW	9.1	54.4	0.044
Sunflower	0.48	0.56	LOW	3.4	49	
Beef Tallow	0.3	0.85	MEDIUM	1.2	58	0.92

Table 3 (f). SD score for raw materials subsystem

	Environmental	Economic	Safety	Fuel performance	SD score
Soybean	0.289	0.368	0.28	0.33	0.32
Rape Seed	0.224	0.145	0.37	0.23	0.22
Sunflower	0.153	0.24	0.24	0.31	0.23
Beef Tallow	0.234	0.247	0.12	0.13	0.19

The net SD score is determined for each alternative by taking an aggregate of the product of the alternative's indicator score and the corresponding indicator's prioritization score for each SD indicator. Table 3(f) lists all the indicator scores for the raw material alternatives with respect to each SD indicator and the final SD score which is used to determine the most sustainable option. It is found that soybean is the most sustainable raw material for bio-diesel production as it has an overall good performance in all the fields of SD, which is evident from its high SD score.

3.2 CATALYST SELECTION

Transesterification, which is treatment of triglycerides (present in the feedstock) with an alcohol in the presence of a catalyst to produce fatty acid alkyl ester (bio-diesel) and glycerine, is the most common method of producing bio-diesel. The transesterification process can be catalyzed by homogenous catalysts which can be alkalis, acids or enzymes (Vicente, Martinez et al., 2004). The first two types have received the greatest attention as they are more economically viable than enzyme catalyzed transesterification. For this subsystem, the SD indicators include certain system specific indicators such as reaction time in minutes and percentage of yield, besides environmental, economic indicators and safety indicators. The quantification of these indicators is shown in Table 4(a).

For the catalyst subsystem, environmental and safety indicators are given high priority and the economic and system specific indicators are given medium priority. The prioritization scores obtained for each of SD indicators by AHP is displayed in Table 4(b).

The catalyst alternatives are compared with respect to each of the SD indicators. For environmental indicators the EPI values are determined for sodium hydroxide (NaOH) for alkaline catalyst (Vicente, Martinez et al., 2004) and sulfuric acid (H_2SO_4) for acidic catalysts (Canakci and Van Gerpen, 1999). For economic indicators the percentage of difference in total manufacturing cost of bio-diesel and the bio-diesel break-even price are used as comparison parameters (Zhang, Dube et al., 2003). Certain system-specific indicators such as reaction time (min) and percentage of yield are used to compare the alternatives.

The score for each SD indicator with respect to each catalyst alternative is listed in Table 4(c). The final SD score is calculated for each alternative and from these scores basic catalyst was identified to be the most sustainable as it had the highest SD score. Basic catalysts have the best performance in both the environmental and safety implications and

Table 4 (a). SD indicators quantification for catalysts

Alternatives	Environmental	Economic			Safety		Specific indicators	
	EPI (for 100 units of release)	Total costs ($\$ \times 10^{-6}$)	Total manufacturing cost of bio-diesel ($\$ \times 10^{-6}$)	Break even price ($/ton)	RAM index	Number score	Reaction time(min) for 90% conv	Yield %
Base (NaOH)	70	0.32	6.86	857	MEDIUM	2	90	95
Acidic(H_2SO_4)	114.4	1.41	7.08	884	HIGH	3	4140	97
Enzyme	20	3.5	10.5	900	LOW	1	480	71

Table 4 (b). SD indicators prioritization score

Indicator	Prioritization score
Environmental	0.333
Economic	0.167
Safety	0.333
System specific	0.167

at the same time have favorable economic performance and also result in lesser reaction time and higher yield (Vicente et al., 2004). Acidic catalysts such as sulfuric acid have more impact on the environment as they result in acid rain; have higher human- and eco-toxicity levels than the alkaline catalyst. The Environmental Protection Agency (EPA) of USA gives sodium hydroxide a score of 3 and sulfuric acid a score of 7 on a scale of 10 for environmental impact and toxicity levels. Since basic catalysts have an overall good performance in comparison to acidic and enzymatic catalysts they are considered to be the most sustainable. This is evident from the highest SD score of 0.51.

3.3 REACTANT ALCOHOL SELECTION

The alcohols that can be used in the transesterification process are methanol, ethanol, propanol, butanol and amyl alcohol. Methanol and ethanol are used most frequently and hence are considered as the alternatives that are subjected to comparison for the identification of the more sustainable alcohol reactant. The SD indicators used are quantified and listed in Table 5(a) for methanol and ethanol.

The prioritization of the SD indicators for the alcohol subsystem is shown in Table 5(b). In the transesterification reaction, the alcohol to triglyceride ratio is 6 : 1 for alkali catalyzed reaction and 30 : 1 for acid catalyzed. Due to the large amount of alcohol required, it affects the price of bio-diesel; hence economic indicator is given high priority. Environmental, safety and system specific indicators are given medium priority.

Bio-diesel produced from ethanol and methanol have comparable chemical and physical fuel properties and engine performances (Peterson et al., 1995), but for economic reasons, only methanol is currently used for producing bio-diesel on an industrial scale due to the much lower price compared to ethanol. Methanol, however, is currently mainly

Table 4 (c). SD scores for the catalyst alternatives

	Environmental	Economic	Safety	System specific	SD score
Base	0.478	0.549	0.54	0.48	0.51
Acidic	0.172	0.310	0.16	0.17	0.19
Enzyme	0.350	0.141	0.30	0.35	0.30

Table 5 (a). SD indicator quantification for alcohol reactants

	Environmental	Economic		Safety		Fuel performance
Alternatives	EPI (for 100 units of release)	Total manufacturing cost of biodiesel ($\$ \times 10^{-6}$)	Break even price (\$/ton)	RAM index	Number score	Cetane number
Ethanol	70	10	900	MEDIUM	2	48.12
Methanol	14	6.86	857	HIGH	3	51.34

produced from natural gas. Thus, methanol-based bio-diesel is not a truly renewable product since the alcohol component is of fossil origin. Furthermore, methanol is highly toxic and hazardous, and its use requires special precautions. Use of ethanol for production of bio-diesel would result in a fully sustainable fuel, but only at the expense of much higher production costs. Table 5(c) illustrates the AHP scores obtained for the alcohol alternatives with respect to each of the SD indicators as well as the net SD score for each alternative. As can be seen in the table both the alternatives have the same SD score, but due to the above stated reasons, it is environmentally favorable and safer to use ethanol in the place of methanol though it is not a very economically favorable option.

3.4 BIO-DIESEL PRODUCTION PROCESS SELECTION

There are three most widely technologies to produce bio-diesel from plant oils or animal fats and they are pyrolysis, transesterification and microemulsification. Pyrolysis is the conversion of one substance into another by means of heat or by heat with the aid of a catalyst. It involves heating in the absence of air or oxygen and cleavage of chemical bonds to yield small molecules. The pyrolysis of vegetable oils, animal fats and natural fatty acids can result in the production of bio-diesel. Transesterification (also called alcoholysis) is the reaction of a fat or oil with an alcohol in the presence of a catalyst to form esters (bio-diesel) and glycerol. Micro-emulsion is the formation of thermodynamically stable

Table 5 (b). Prioritization of SD indicators for alcohol reactant

Indicator	Prioritization score
Environmental	0.333
Economic	0.333
Safety	0.167
System Specific	0.167

Table 5 (c). SD score for alcohol alternatives

	Environmental	Economic	Safety	System specific	SD score
Ethanol	0.750	0.250	0.667	0.333	0.50
Methanol	0.250	0.750	0.333	0.667	0.50

dispersions of two usually immiscible liquids, brought about by one or more surfactants. But micro-emulsions of vegetable oils and alcohols cannot be recommended for long-term use in engines as they are prone to incomplete combustion, formation of carbon deposits and an increase in the viscosity of the lubricating oil. Due to these drawbacks micro-emulsions are not usually used in large-scale production of bio-diesel. In this study, only pyrolysis and transesterification processes are compared for the production of bio-diesel. The SD indicators quantified for the production processes are the environmental (Impact degree), economic (total capital cost), safety (RAM index) and fuel performance (yield %) indicators, shown in Table 6(a).

As the system under study is a chemical process, economic and safety indicators are given high priority. As the environmental impact of the reactants involved in the process has already been included while selecting the sustainable alternatives, environmental indicators are given only medium priority. System specific indicator (yield %) is given the least priority while comparing the different bio-diesel production techniques. Table 6(b) illustrates the AHP prioritization score for the SD indicators for the production process subsystem.

Transesterification has much better environmental and safety performance than thermal cracking as thermal cracking requires bio-diesel to be produced in an oxygen-free environment and this requires more complex systems which increases the environmental impact as well as making the process more hazardous (Ma and Hanna, 1999). Moreover the bio-diesel obtained from transesterification has better emission performance than the bio-diesel obtained by thermal cracking. Transesterification is more economically favorable than thermal cracking due to lesser number of complex equipments. Due to all these favorable factors, transesterification is considered to be more sustainable than thermal cracking for producing bio-diesel. The AHP scores for each of the SD indicators as well as the final SD score for each alternative is shown in Table 6(c).

Table 6 (a). SD indicators for the production process alternatives

	Environmental	Economic	Safety	Fuel performance
Alternatives	Impact degree	Total capital costs	RAM index	Yield %
Thermal Cracking	HIGH	HIGH	HIGH	84
Transesterification	LOW	LOW	MEDIUM	98

Table 6 (b). Prioritization of SD indicators for production process

Indicator	Prioritization score
Environmental	0.189
Economic	0.351
Safety	0.351
System Specific	0.109

3.5 BIO-DIESEL PURIFICATION PROCESS SELECTION

Bio-diesel purification method is the final subsystem considered in this case study. Water washing and hexane extraction are considered as alternatives. Due to the evident impact of this subsystem on the total cost of bio-diesel, the economic implications are given the highest priority followed by safety issues. The reasoning for the priority scores allotted for environmental and system-specific indicators is similar to that offered for the bio-diesel production process subsystem. Table 7(a) and (b) show the SD indicator prioritization scores and the final SD scores respectively. It is found that the water washing has a much higher environmental score than hexane extraction, which is due to the avoidance of use of hexane thereby making the process inherently safer (Zhang, Dube et al., 2003). Water washing is also more economically favorable than hexane extraction due to simpler equipment and more readily available materials (water is cheaper and readily available than hexane). Due to these favorable features, water washing is usually preferred to hexane extraction and this was the result obtained from the decision framework developed.

In summary, using proposed SD methodology to analyze bio-diesel process, it is found that soybean is the most sustainable alternative over other raw materials, such as rape seed oil, sunflower oil and beef tallow. Basic catalyst is more sustainable than acidic and enzymatic catalysts. The optimal production process for bio-diesel is transesterification and the sustainable option for alcohol used in this process is ethanol. Water washing is considered to be the sustainable purification method used in the bio-diesel process, due to its good environmental and economic performance. All these identified sustainable alternatives for each subsystem are illustrated in Table 8.

4 CONCLUSION

The proposed methodology in this paper is an integrated systematic approach to sustainable engineering decision making, which has so far been treated only qualitatively. It elucidates

Table 6 (c). SD scores for the production process alternatives

	Environmental	Economic	Safety	System specific	SD score
Thermal Cracking	0.33	0.33	0.33	0.18	0.32
Transesterification	0.67	0.67	0.67	0.83	0.68

Table 7 (a). Prioritization of SD indicators for bio-diesel purification

Indicator	Prioritization score
Environmental	0.189
Economic	0.351
Safety	0.351
System Specific	0.109

Table 7 (b). SD scores for the bio-diesel purification process alternatives

	Environmental	Economic	Safety	System specific	SD score
Water washing	0.75	0.75	0.33	0.24	0.55
Hexane Extraction	0.25	0.25	0.67	0.76	0.45

Table 8. Sustainable bio-diesel process

Subsystem	Sustainable alternative
Bio mass	Soybean
Catalyst	Basic
Alcohol	Ethanol
Production process	Transesterification
Bio-diesel Purification	Water Washing

not only what we want to sustain but also how we do so based on environmental, economic and safety impacts. The feasibility and effectiveness of this methodology has been demonstrated by applying to identify the most sustainable bio-diesel process system from a set of alternatives, since the analysis results of the proposed methodology are in excellent agreement with the generic system accepted to be the most optimal and environmentally favorable by most researchers and commercial bio-diesel plant designers (Zhang et al., 2002; Haas et al., 2005; NREL). Therefore, the proposed methodology is useful in identifying sustainable options from a given set of alternatives and assessing new technologies in term of current generation and future generation.

In summary, this methodology is simple, flexible and user friendly, since the scoring scales for the SD indicators and alternatives comparison are not very system-specific. Thus, this methodology can also be customized to be applied to other engineering processes, such as, SD of fuel cell technology, solar cells, wind power, and other alternative energy sources.

ABBREVIATIONS

AHP	Analytical Hierarchical Process
EPI	Environment Performance Indicators
EPA	Environmental Protection Agency
LCA	Life Cycle Analysis
RAM	Risk Assessment Matrix
SD	Sustainable Development

REFERENCES

Ali, Y. (1995). Beef tallow as a bio-diesel fuel, PhD dissertation, Biological Systems Engineering, University of Nebraska-Lincoln.

Azapagic, A. (1999). "Life cycle assessment and its application to process selection, design and optimization." Chemical Engineering Journal **73**: 1–21.

Besnainou, J. and J. Sheehan (1997). Life cycle assessment of biodiesel, Toronto, Can, Air & Waste Management Assoc, Pittsburgh, PA, USA.

Canakci, M. and J. Van Gerpen (1999). Biodiesel production via acid catalysis. Transactions of the ASAE **42**(5): 1203–1210.

Demirbas, A. and S. Karslioglu (2007). Biodiesel production facilities from vegetable oils and animal fats. Energy Sources, Part A: Recovery, Utilization and Environmental Effects **29**(2): 133–141.

Freedman, B., Butterfield, R.O. and Pryde, E.H. (1986). Transesterfication kinetics of soybean oil, Journal of the American Oil Chemistry Society, **63**(10): 1375–1380.

Jungmeier, G., S. Hausberger, et al. (2003). Greenhouse gas emission and cost for transportation system. Comparison of bio-fuels and fossil fuels. Josnneum Research, Graz Technical University.

Ma, F., Clements, L.D. and Hanna, M.A. (1998). Biodiesel fuel from animal fat. Ancillary studies on transesterification of beef fallow, Industrial & Engineering Chemistry Research, **37**(9): 3768–3771.

Ma, F., Clements, L.D. and Hanna, M.A. (1998). The effect of catalyst, free fatty acids, and water on transesterification of beef tallow, transactions of the ASAE, **41**(5): 1261–1264.

Ma, F. and M. A. Hanna (1999). Bio-diesel production: a review. Bioresource Technology **70**: 1–15.

Narayanan, D, Zhang, Y. and Mannan, M.S. (2007). Engineering for Sustainable Development (ESD) in Bio-diesel Production. Trans IChemE, Part B, **85**(B5), 349–359.

Roszkowski, A. (2003). Perspectives of plant biomass as raw material to production of liquid fuels, Warsaw, Poland, Stowarzyszenie Elektrykow Polskich.

Rudolph, V. and Y. He (2004). Research and development trends in biodiesel. Developments in Chemical Engineering and Mineral Processing **12**(5–6): 461–474.

Sonntag, N.O.V. 1979. Reactions of Fats and Fatty Acids, Bailey's Industrial Oil and Fat Products, Vol. 1, 4th edition, 99 (Wiley-Interscience, MA, USA).

Tapasvi, D., D. Wiesenborn, et al. (2005). "Process model for biodiesel production from various feedstocks." Transactions of the American Society of Agricultural Engineers **48**(6): 2215–2221.

Taylor, J.R. (1994). Risk Analysis for Process Plant, Pipelines and Transport.

Vicente, G., M. Martinez, et al. (2004). "Integrated bio-diesel production: a comparison of different homogenous catalysts systems." Bioresource Technology **92**: 297–305.

Weisz, P.B., Haag, W.O. and Rodeweld, P.G. (1979). Catalytic production of high-grade fuel from biomass compounds by shape-selective catalysis, Science, **206**: 57–58.

Wimmer, T. (1992). Transesterification process for the preparation of C1-5-alkyl fatty esters from fatty glycerides and monovalent lower alcohols, PCT Int Appl WO 9200–9268.

Xun, J. and High, K.A. (2004). A new conceptual hierarchy for identifying environmental sustainability metrics, Environmental Progress, **23**(4): 291–301.

Zhang, Y. (2002). Design and economic assessment of bio-diesel production from waste cooking oil, MASc thesis, Department of Chemical Engineering, University of Ottawa.

Zhang, Y., Dube, M.A. et al. (2003). "Biodiesel production from waste cooking oil: 1. Process design and technological assessment." Bioresource Technology **89**(1): 1–16.

THE IMPLEMENTATION OF IPPC UNDER SCHEDULE 1, SECTION 4.5 PART A (1) A), FOR A SMALL MANUFACTURING ENTERPRISE, PRODUCING CONTRACT CHEMICALS UNDER A MULTIPRODUCT PROTOCOL

Sandra Hollingworth
NPIL Pharma (UK) Limited, Huddersfield Works, UK

NPIL Pharma (UK) Ltd at Huddersfield manufactures contract pharmaceutical intermediates and products in three plants ranging from kilo lab to 6.3 m^3 scale. Our business relies on rapid response to customers' demands with a constantly changing product portfolio. The site we are based on is owned by Signet, and five companies are present on the site. This paper describes how we managed our IPPC application, and the resulting maintenance activities and Improvement Conditions of the Permit.

1. BACKGROUND TO THE SITE AND APPLICATION

Chemicals have been manufactured at the Leeds Road, Huddersfield site for almost one hundred years. Initially British Dyestuffs, it was part of ICI for about sixty years before demerger to Zeneca in 1993. Further sales of parts of Zeneca (from1999 until 2005) have now caused the site to be split into five legal entities running plants. These are Syngenta, Arch, Noveon, NPIL and Dalkia who own and operate a power house. The site is owned by Syngenta with other companies owning their plants but leasing the land from Syngenta. All the companies use the Syngenta on- site effluent treatment plant, and services (electricity, steam, nitrogen) are supplied by Syngenta. There are also a number of other contracts between Syngenta and the tenant companies by which services such as site security, emergency response, domestic waste disposal etc are provided. There are no internal barriers within the site, although green fencing to designate the GMP areas surrounds the NPIL GMP plants.

An Application was prepared under the requirements of Schedule 1 of the Regulations, Section 4.5 Part A (1) a).

Preliminary discussions with the EA about how the companies were going to implement IPPC were held, and it was decided that the site would be one Installation, due to the one effluent treatment plant, but with five separate Permits; one for each legal entity manufacturing on site.

NPIL Pharmaceuticals Limited (formerly Avecia) manufactures contract pharmaceutical intermediates and products on three plants at Huddersfield, and occupies approximately 3 hectares of the total 100 hectare site. The plants are relatively small scale, ranging from kilo laboratory to a reactor size of 6.3 m^3. Our business relies on a rapid response to customers' demands, and while we have some long-term processes, we have a constantly changing product portfolio with the products and processes at different stages of development.

The site was already regulated under IPC with the three NPIL plants gathered together under one fairly generic authorisation. Every time a new process was introduced we wrote and applied for a Variation and appended it to our authorisation, which resulted in more than thirty variations in the last five years. The frequently changing portfolio, relatively low emission levels and batchwise nature of manufacture meant that all the emissions were calculated. Only one programme of measurement ever took place, measuring certain VOCs at the site boundary, which proved to be barely detectable.

When IPPC started to be introduced for other business sectors, it contained the concept of the Multi-product Protocol, which seemed a much better fit for our manufacture. It was agreed between ourselves and our local Inspector that this would be a slicker way of managing our new sitings into existing assets, and so we operated an IPPC Multi-product Protocol for about two years before applying for an IPPC Permit.

Our Application for an IPPC Permit was Duly Made in February 2006, (number UP 3738 LY) and a Permit was granted in December 2006.

2. PERMIT APPLICATION PREPARATION

Historically on Site, the chemical engineers who supported the plants also performed any necessary calculations to support the emissions also wrote IPC authorisations and variations. The Authorisations were "owned" by the plant managers, with the site SHE function acting in an advisory capacity, and operating as Site contact with the regulatory authorities. The splitting of the site into (in our case) much smaller identities meant that there were three process engineers on site, and one part time SHE manager. Nevertheless, it was decided to use in-house effort to prepare the permit due to the complex site situation and fluctuating manufacturing programme. A process engineer was appointed full time to this role to prepare the documentation and gather information from other sources as necessary.

Meetings began between the site partners around a year before the deadline for applications. The meetings were held monthly to discuss the approaches of the companies to various parts of the legislation, and feed back any comments received from the EA. One area debated was how the users of the effluent treatment plant were going to approach the section requiring reporting on emissions to sewer. After discussions with the EA and between ourselves it was decided that Syngenta would report on behalf of all the companies, as they held the license to discharge to sewer. As the Site was being taken as one installation, the site partners were keen to adopt similar standards and principles across all the permits.

Items prepared for the permit application were:

- Application Form
- Application Form Appendices (contained our multiproduct protocol)
- H1 Assessment
- Application Site Report and Appendices
- EPOPRA

The applications for the whole site were delivered in person (in approximately 20 boxes!) on schedule in February 2006.
The costs to NPIL up to the point of application were approximately

- Application fee of £30,000
- Cost of staff time to prepare application, approximately £50,000
- Purchase of additional material for application, approximately £3,000

Giving a total cost to NPIL of £83,000.

An estimated cost for the whole Huddersfield site is £500,000.

3. THE APPLICATION FORM

NPIL decided to use the electronic form provided and recommended by the EA. The application was written as generically as possible, assuming the maximum possible production rates. As we operate batch processes, with different cycle times, chemistries, unit operations and number of stages the figures used in the application were approximately double the annual production capacity achieved in the previous year. The activity was divided into two aggregations across the three plants.

The chemistry listed fifty different types of reaction, which we have either operated, or might possible wish to operate in the future. The equipment and plant operations were also described generically. Emissions were not listed but comments made referring to the multi-product protocol, and generic categories used for VOCs. The raw material list was three pages long, which represented all those we had used in significant quantities in the last year. We stated that we had no routine monitoring and that all figures given were by calculation.

In general we found that using the electronic templates in the application were helpful, as they led us through the application process. However, they were not written with plants operating under a multi-product protocol in mind, so we had occasionally to extend boxes, or write lengthy comments either on the form, or on separate sheets and append to the application.

4. THE MULTI-PRODUCT PROTOCOL (MPP)

This was written in line with guidance from the EA, and in operation for two years before the Application was filed. Our MPP is a Company Procedure that described the environmental assessments undertaken by project teams (mainly by the chemical engineer) during the siting project. There is also another internal company document called the User Guide that gives more detailed information to project teams, and a calculation spreadsheet to aid in the estimation of emission quantities and concentrations. We then decide if the change is minor, or requires a variation. So far, we have sited new processes within existing assets and required no new abatement equipment, so changes have been minor. To date in 2007 we have submitted around 20 assessments, mainly for low quantity short campaign manufactures in our two small development plants. The main issues we face in operating under the MPP are:

- Time pressures
  In our smallest plants, often we have about four weeks between a customer order and starting manufacture. It can take two for the plant to decide how they are going to run the process in the assets, and then they write the MPP document and perform the calculations. The EA have been very co-operative in turning round these minor changes in a few days, and even by return of e- mail on occasion, and have not held us up from manufacturing.
- It's not our process!
  Some customers allow or require us to develop their processes, in which case we have scope and time to replace more environmentally damaging solvents, remove some operations, reduce waste etc. However, some contracts are "take and make" with the customer specifying no development and a very short timescale. Some processes are from outside the EU, and have been developed in places with different legislative requirements. (For example a lot of processes developed in USA contain DCM.)

5. THE H1 ASSESSMENT

Again, we used the software provided by the EA, adapting our answers to our multi-product manufacturing. We chose eight processes as example processes, across the three plants. The processes included long term manufactures and some worst case examples of short term manufactures we had seen in the recent past. These used a wide variety of raw materials, chemistry and operating conditions. We assumed all the processes ran simultaneously, in effect doubling the capacity of our plants. The figures we used for emissions were those we had previously generated by calculation for our Environmental Impact Assessments using H1 (and E1) in previous years. All air impacts were insignificant with three exceptions and further detailed modelling was performed on these, which then proved to be insignificant.

6. THE APPLICATION SITE REPORT

The land for the whole site is owned by Syngenta, and NPIL holds a long term lease for certain areas. As the site had made chemicals for such a long period, before the site was split into more than one legal entity, a significant study was undertaken by Zeneca in 1999 to assess the ground contamination. It was agreed at this point that as the land would not be sold, the potential liability for any historical contamination would not be passed on to the buying company. The results of this study were made available to NPIL for the preparation of our Site Report. This report, conducted by an external specialist consulting company, contained information about the hydrogeology of the site, and a conceptual site model as well as various other items relevant to a site report. This has served as our baseline, as the enclosed nature of our operations and incident reporting system made us

confident we had not contaminated the land between 1999 and the time of application. Our Site Report was again written by our in-house staff.
Two factors specific to our location are

- The River Colne runs through our site, and is approximately 20 metres away from our closest plant.
- The site is adjacent to the A62, Leeds Road, and our plants are the closest on site to the road and human habitation.

Fortunately, our plants are relatively modern (20, 8 and 5 years old), self contained and enclosed, with high environmental integrity, and use of the D2A assessment table identified only two potential sources of pollution of the land or river, for which we proposed improvements. These were then included in our Improvement Conditions by the EA.

7. EPOPRA

This is the spreadsheet produced by the EA to calculate both the application fee, and the ongoing maintenance costs for IPPC, based on the principle that "the polluter pays". On the positive side, the plants are relatively small and low polluting, have good records, and the site has a history of good procedures and practices handed down from ICI / Zeneca/Avecia / NPIL. On the negative side, we are close to a river and human habitation, and operate complex and potentially hazardous chemistry, with a wide range of raw materials.
Our application fee came to £30,000, and our subsistence fee is £15,000, which is slightly higher than we were paying under IPC.

8. THE PERMIT

This was received in late December 2006, after two meetings to discuss the application with the EA, and NPIL providing some additional clarification and information as requested. No "Schedule 4" notices were issued. The permit contains a number of clauses common to all operators, such as a training plan, producing the Site Closure Plan and Site Protection and Monitoring Programme, reporting requirements etc with schedules of requirements specific to each company at the back. There is a schedule of emission concentration limits and annual emission limits to air, and a schedule of Improvement Conditions. The emission concentration limits were not set in the permit in our case. As we had not monitored what our concentrations actually were, the limits were subject to us undertaking a monitoring programme. This was recorded as one Improvement Condition.

There were ten Improvement Conditions, of which five were fairly minor, and common to all the companies on the Huddersfield Site. The most costly improvements identified specific to NPIL were:

- Improvements identified by us in the Site Report, to a roadway drain, and an area used for drum storage. The drain has been routed to an effluent drain from a Clearwater drain. The modifications to the drum storage area are significant, and so far a scope has been agreed, designed, and has been estimated at around £100,000.

- A monitoring programme of all registered release points to air, to demonstrate that the calculations performed are correct. This will then also confirm the annual losses to air, and demonstrate the abatement equipment is functioning as predicted. This programme was also required to demonstrate that the conditions of the Solvents Emissions Directive (SED) are being met, with regard to the emission limits of certain toxic solvents. Following the programme, we were then required to run the H1 software again, and confirm our releases are insignificant.

These improvement conditions will take at least a full year to complete, and costs in 2007 are estimated to be:

- NPIL staff time approximately £30,000
- VOC monitoring (External contractor) approximately £20,000
- Roadway drains improvements approximately £15,000
- Subsistence fee for 2007 of £15,000

Giving a total cost of IPPC work in 2007 of approximately £80,000.

A further expenditure in 2008 of around £100,000 is expected for upgrading the drum storage compound area.

9. THE SITE PROTECTION AND MONITORING PROGRAMME (SPMP)

This was written in house and sent to the EA as required, in February 2007. A response accepting our proposals was received in November 2007. Our proposals were based primarily on preventative measures, with only one baseline groundwater sample proposed. Our preventative measures consisted of existing procedures and policies as follows :

- Maintenance policies include the statutory preventative maintenance required of pressure vessels, registered pipelines, etc. plus a list of other work specified by the engineering manager which includes inspections to confirm bund, sump and drain integrity.
- Plant Environmental compliance checklists, which have daily/weekly checks for such things as sumps, leaks from vessels, pipes and storage drums and tanks, scrubber operation, condenser glycol system operation etc.

The one proposed groundwater sample utilised an existing borehole, originally dug for the 1999 study. This was inspected and judged to still be usable. It is located in the hydraulic gradient between our plants and the river, not far from the drum compound described above as requiring improvement. Our plan was this would be a one- off sample to establish a baseline, and not repeated unless we had a loss of containment onto unmade ground. An external specialist company has been appointed to undertake this work, as we do not have the appropriate UKAS accreditation in house.

It is proposed that sampling will take place in January, with the results sent to the Environment Agency by the end of February 2008.

10. MONITORING OF EMISSIONS TO AIR

One of the conditions imposed by the EA was that monitoring was conducted to MCERTs standards. Our small analytical department, which while being highly skilled and able to test pharmaceutical products to GMP standards, has no MCERTs accreditation. Hence, an external contracting company was appointed to undertake the measurement. Sampling took place between July and October 2007.

Our plants have 17 registered vents, and the EA initially required all of them being sampled. After discussion, our sampling plan involves actually sampling from only seven of these points. Four vents are used specifically for hydrogen, and it was argued that it was not safe to attempt sampling from these as they were nitrogen swept. They were highly inaccessible, with no sampling points already present, and it was agreed that drilling holes (and hence allowing air ingress) and taking electrical equipment to the vicinity would be too hazardous. Three vents were eliminated on the grounds of triviality, and two vents are not currently in use. One vent, from a dust scrubber, could not physically be sampled. It was situated on the top floor of a plant with the vent going straight through the roof, with no space after the fan to locate a sample point. The roof is not accessible, except under a special permit, using either a "cherry picker", scaffold, or roof boards. At a plant shutdown in April 2006, the scrubber vent was inspected, as well as the area of the roof around the vent. There was no evidence of any solids being emitted, and this was accepted as proof of triviality by the EA.

The three plants operated are at different scales, and also operate at different levels of occupation. The largest plant (called Pharms. Intermediates Plant, or PIP) operates mainly established processes with long term contracts for large quantities, with campaigns measured in months, and a planning horizon of several months. The plants operate at high levels of occupation and the batch processes are highly optimised for efficiency. Around 90 to 95% of the tonnage capacity of the company is from these plants, and an estimated 85% of the emissions. The plant has two manufacturing units, and can have as many as four batches in progress in one unit, and three in the other. For this plant, it was planned to sample from the aqueous packed scrubbers associated with each unit, for a continuous period lasting a full batch cycle, at a time when the plant was fully loaded. This meant 24 hour sampling for four days on each unit.

The two smaller plants, called Pharms. Development Plant (PDP) and the Early Phase Development Team (EPDT) manufacture products on a smaller scale, with the products generally being in clinical trial phase. The processes operated in PDP are generally being optimised, in clinical trial phase 3, with around 50% of one-off campaigns, often with less than ten batches. EPDT operates at even smaller scale, at the earliest clinical phases, and with even more one-off campaigns. In both plants, there is often only one batch in progress at any one time in each unit, with the overall batch time being long, with

7 to 10 days being common. Hence it was decided not to monitor these continuously, but the monitor when certain operations known to cause emissions such as vacuum distillations, product drying etc occurred on the plants.

The aim of the programme was to measure emission concentrations and quantities, and compare these with the calculated values, for generic unit operations. This was to demonstrate to the Agency that our calculation methods were reliable, and hence allow the plants to continue to operate under the multi-product protocol using these calculated values, and avoid installing on-line sampling, or having frequent routine monitoring.

As expected, the monitoring showed that our emission levels ranged widely depending on the operations performed by the plant, with a great many low duration peak concentrations relating to plant operations such as batch charges, blown transfers, and vessel purges with nitrogen. The interpretation of the data gathered on PIP was complicated by the fact that four batches were performing many such operations at the same time. However, by use of our GMP Process Instruction sheets, which have times of operations listed, and spending significant amounts of time on the plant, it was possible to attribute peaks to operations.

The data analysis showed reassuring agreement between our measured values of emission and our mass balance calculations for total emissions per batch. Our system appears to give more smoothing of peaks than we had allowed for, as in general the peak concentrations measured were lower than those calculated.

A report of the findings, including revised H1 assessment, was submitted to the Agency on 30th November 2007, as required. These figures will become the baseline for our Emission Limit Values in future.

11. AQUEOUS EFFLUENT MONITORING

One joint Improvement Condition was placed upon the site as a whole. This was to monitor the efficiency of the Syngenta Effluent Treatment Plant at removal of specific impurities. The plant already has on-line analysis at the outlet, and meets certain standards set by Yorkshire Water. Syngenta has an environmental group who test all process effluent for suitability of discharge into their system. All the site partners were to take composite samples from their plant effluent drains over the same three month period, with Syngenta also taking samples of what came out of the effluent plant. For continuous plants, or those making a set product portfolio, and discharging relatively large amounts into the plant, this is reasonable.

NPIL plants discharge relatively low volumes, in a batchwise manner, from a variety of processes using many different raw materials. The two smaller plants do not always run fully loaded, and for new, one-off and small tonnage manufactures, aqueous effluent is packed off and disposed of off-site. Also, the plants have large drum storage areas, and these plus the plant gutters and internal roadways are routed to the effluent plant. It is quite possible that any sampling could be misleading in wet weather, with the rainwater greatly in excess of process effluent. Sampling a composite sample out of the drain points would mean in our case attempting to analyse for extremely small amounts of pharmaceutical intermediates and solvents. It was decided to propose an alternate method for this Improvement Condition for NPIL as follows.

SAMPLING PHILOSOPHY

The streams that enter the Syngenta ETP were listed, and divided into categories as described below.

It is believed that Syngenta intends to run the sampling plan in early 2008. Therefore NPIL decided that for manufactures which are long term and ongoing, (i.e. those in PIP) to sample in July and August 2007. This sampling was **at source** rather than from the effluent drains/ sumps etc. As the plants operate to GMP, it will be possible to track which batches and discharges were made during the sampling period selected by Syngenta. This smoothes the analytical load on NPIL staff.

Other manufactures in PIPDEV and EPDT will be sampled as they occur during the sampling period set by Syngenta.

"Strong" streams

These streams are the process effluent streams, generally from aqueous wash and splits done during the work up of processes, or screening of salts and then dissolving them before disposal to Syngenta ETP. Syngenta Environment Group has tested all these as suitable for discharge. All the processes, which are considered to be long term manufactures and have "strong" effluent, were sampled. These streams were all from the PIP plant. They were sampled in August, and analysed by our internal NPIL Analytical Development Group for solvents and pharmaceutical products.

"Weak" streams

There are two sources of these streams; gas scrubbing containing traces of solvents, and plant cleaning containing traces of solvents and products. (During product changeovers, the plants are cleaned with solvents which are not sent to Syngenta ETP. However, before maintenance work, the plants are cleaned first with solvent, and then with water which is disposed of to Syngenta ETP.) Streams from established manufactures on PIP were sampled in July and August 2007. They were, as expected, weak, containing less than one per cent of solvents. Sources from PDP and EPDT will be sampled from manufactures that occur during the Syngenta monitoring period.

Streams which were not sampled or analysed

The plant gutters, and floors and roads close to the plants also drain to the Syngenta ETP either directly, or via the plant sumps. The sumps are overpumped daily to Syngenta, if there have been no losses of containment on the plants or storage areas. The normal sources of these streams are rainwater, or water from mopping floors, which is an operation that is performed daily for GMP plants. It is not proposed to sample these streams. In an emergency situation such as a loss of containment, the plant sumps are not pumped to Syngenta ETP.

This plan was accepted by the EA, and will be completed in 2008, concurrently with all the other companies on site.

12. THE SOLVENTS EMISSIONS DIRECTIVE (SED)

The aim of SED is to prevent or reduce the direct and indirect effects of emissions of volatile organic compounds (VOCs) into the environment, primarily air. The manufacture of pharmaceutical products is an IPPC Part (A)1 activity, that is also a SED activity, if the amount of solvents input in a year in >50 tonnes.

Initially, it was not clear whether NPIL at Huddersfield was subject to this legislation, as we do not ferment, formulate, package or sell any finished drug products. The majority of our output is pharmaceutical intermediates (not subject to SED). However, we do perform the chemical synthesis of crude API's in small quantities for clinical trials in our two smaller plants, and wished to retain this capability. This meant that all the activities on site, including intermediates, are subject to SED.

We were classed as an existing installation, and hence demonstration of compliance was required by the end of October 2007, using the emission limit values (ELVs) as below:

- VOCs with the risk phrase R40 (in our case, dichloromethane, DCM) – if emissions are >100 g/hr, the ELV is 20 mg/m3.
- VOCs with the risk phrases R45,46,49,60,61 – if emissions are >10 g/hr, the ELV is 2 mg/hr. Our current product portfolio did not use any such materials, but in the past we have used dimethylformamide (DMF), which is R61.

In general, we have adopted a policy of removal of DCM (and DMF) during process development. We are seeing less customer inquiries from Europe that use DCM, but in other parts of the world, for example, USA, where DCM is not classed as a carcinogen, it is more common. During the early clinical trial phases of a drug life, when relatively small amounts of material are required, with short lead times and high probability the drug will not actually reach the market, there may not be time or customer willingness to spend time (and money!) developing out the use of DCM. In this case, we use hired carbon absorber units to remove DCM on our smaller plants.

One such process was manufactured in PDP in September 2007, in parallel with laboratory work to replace DCM with a less toxic solvent. This campaign consisted of twenty batches of 60 kg each. This campaign was monitored to demonstrate compliance generically for the plant and technology of using carbon absorption. We gained prior acceptance from the EA that measuring emissions during a series of generic batch operations (charging from drums, vessel transfer, distillation etc) on one product would validate our equipment and the technology, and hence no variations would be required in the future for different processes using DCM on any of the three plants. However, monitoring during manufacture on the other plants could be required.

During the trial manufacture, process engineers performed mass balancing for every batch from the weights of solvent charged and discharged to calculate how much DCM was being lost into the carbon beds, and compare this with their previous assumptions. Monitoring both before and after the beds demonstrated that the beds could achieve the 20mg/m3 SED limit, and confirmed the reduction in concentration achievable across the beds. This was found to be greater than 99%, unless the bed had become exhausted.

The mass balancing work showed that on the initial batches, the plant was losing more DCM into the beds than we had thought. A review of exactly how the plant was operating was reviewed, and some re-training of the operators reduced this to close to the design figures. Some operational improvements which reduced loss of DCM into the absorber were slowing down the rate of distillation, and reducing or eliminating line and vessel nitrogen purge volumes or rates.

The results were shared with the local Inspector in mid October 2007, and the report was sent to the Environment Agency on 31st October.

In December, 2007, the company received a number of inquiries for manufactures using either dimethyl formamide (DMF), or dimethyl acetamide, DMAc, which are both R61 compounds. It was necessary for NPIL to apply for a variation to handle these compounds since we had not had the opportunity to demonstrate compliance with the SED limits before the 31st October deadline. The Variation Application has been submitted, and will be determined as a Simple Standard Variation, by our local Inspector. We expect to start manufacture in March 2008, with a monitoring survey similar to that conducted for DCM.

13. IPPC MAJOR AUDIT – 30TH OCTOBER 2007

This was the first major audit NPIL had had under IPPC. The audit lasted one day, and was conducted by two Inspectors. The areas audited in depth were maintenance, emergency procedures, our use and application of the multi-product protocol, operation of abatement plant, and general management procedures. There was also a Site Inspection. The audit report was issued within two weeks by the Agency.

The conclusions were very positive overall, with the Agency reporting, "NPIL illustrate sound overall compliance against their PPC authorisation. No breaches of the permit were noted".

Four actions were identified by the EA, which are being progressed by NPIL.

14. TOTAL COSTS, AND FUTURE WORK (BEYOND APRIL 08)

It is estimated that the costs to NPIL so far are £83,000 up to the point of application, and £80,000 between application and the end of 2007. Costs (not including the routine subsistence fee) in 2008 are expected to be around £120,000.

The EPOPRA was reviewed by the Inspector in December 2007 following the major audit and the return of VOC monitoring results from NPIL. The Inspector concluded that our score was unchanged from that which we had calculated ourselves at the time of application.

15. CONCLUSIONS

Compliance with the IPPC Regulations has been a considerable exercise for our company, requiring significant technical time, effort, and expenditure. The nature of our batchwise, non steady state, and multi-product manufacturing has made the process complicated, as the legislation was clearly not written with this in mind. We have been able to demonstrate to the Agency that we are operating BAT, and that our procedures for manufacture and calculation of emissions to atmosphere are good. Hence for normal emissions there has been no net benefit either to the company, or the environment.

However, the modifications around drainage should reduce the likelihood of pollution of the ground in the unlikely event of a spillage in the future.

REFERENCES

The following guidance produced by the Environment Agency was used.

1. Sector Guidance Note IPPC S4.02, Guidance for the Speciality Organic Chemicals Sector.
2. Guidance Note from the Department for Environment, Food and Rural Affairs and the National Assembly of Wales on the Implementation of the Solvents Emissions Directive.
3. Introductory Guidance on the new 2004 Solvents Emissions Regulations.
4. Solvents Emissions Directive – Overview of requirements.
5. Technical Guidance Note M1, sampling for stack – emission monitoring, version 4, July 2006.
6. Technical Guidance Note M2, monitoring of stack emissions to air, version 4.1, January 2007.
7. Technical Guidance Note H7, Guidance on the Protection of Land under the PPC regime: Application Site Report and Site Protection and Monitoring Programme.
8. Technical Guidance Note H8, Guidance on the Protection of Land under the PPC regime: Surrender Site Report.
9. Guidance on the use of a multi-product protocol (MPP) at chemical installations, version 2.1, March 2005.
10. Technical Guidance Note H1, Environmental Assessment and Appraisal of BAT.

INVESTIGATION INTO A FATAL FIRE AT CARNAULD METALBOX LTD

M Iqbal Essa[1] and Dr. Aubrey Thyer[2]
[1]Health and Safety Executive, Grove House, Skerton Road, Manchester, M16 0RB UK
[2]Health and Safety Laboratory, Harpur Hill, Buxton, Derbyshire, SK17 9JN UK

This paper covers a joint Health and Safety Executive/Health and Safety Laboratory (HSE/HSL) investigation of a fatal fire inside a 60 m high redundant steel chimney at the Carnauld Metalbox (CMB) Factory, Westhoughton, Nr. Bolton. The incident occurred at around 5.00 pm on 23/5/2002, whilst two contractors were working inside preparing the chimney for demolition. It is hoped that publishing the findings of this investigation will raise awareness of the issues and prevent further occurrences of this type of avoidable incident.

BACKGROUND

Prior to becoming redundant, the chimney in which the fire occurred had been used to vent fumes from a number of drying ovens/lines associated with the application of paint/lacquer to steel sheet used in the food canning industry. Over many years, these fumes coated the inside of the chimney and extract ductwork, giving a tarry layer several millimetres in thickness, which, it was thought, rose for a considerable height inside the chimney. The appearance of the deposit within the ductwork can be seen in Figure 1.

A specialist contractor was appointed to demolish the ductwork and chimney, using a hot cutting method from the interior of the chimney. Prior to this, a limited number of ad-hoc tests were undertaken by the contractor and CMB to establish the properties of the deposit. These concluded that it was not a fire risk, even though it burned while a flame was applied and created large amounts of smoke.

At the time of the incident two men were cutting holes through the wall of the chimney with an oxy-propane torch and were some 30-40 m above ground level. These holes were being made to allow the two men to form working platforms using scaffolding poles and wooden planks at various heights.

Having successfully cut a number of holes from the top in the upper portions of the chimney, they continued to work their way downwards towards the ground. Near the end of their working day a serious fire took hold whilst they were inside the chimney. This fire caused the ropes supporting their gantry to fail, allowing it to free-fall to the ground with the men in it. When the fire was extinguished the bodies of the two men were found in the wreckage of the gantry.

Figure 1. Appearance of tarry deposit in extract ductwork

THE INVESTIGATION

In common with many incident investigations, the exact cause of the fire could not be established, as the only two direct witnesses had been killed. The joint HSE/HSL investigation therefore concentrated on the four following areas.

1) Study of the accounts of eyewitnesses and supplementary information.
2) Examination of the oxy-propane cutting equipment being used by the victims.
3) Studies of the tarry waste including:
   a. chemical composition;
   b. properties of the chemicals liberated;
      thermal stability using thermo-gravimetry (TG) and differential scanning calorimetry (DSC); and,
   c. flash/fire point.
4) Studies of the potential combustibility of the tarry waste on samples of steel plate taken from the chimney.

EYEWITNESS ACCOUNTS AND SUPPORTING INFORMATION

EYEWITNESS STATEMENTS

Apart from a small number of points, little useful information could be gained from the direct observations of people who witnessed the event. Pertinent features were:

1) Material safety data sheets indicated that the paints/lacquers in use were all flammable;
2) It was plant policy to issue a periodic contract to chip the tarry deposit from the inside of ductwork and the lower reaches of the chimney;

3) Pre-demolition tests on the tarry deposit wrongly concluded that:
   a. the deposit was non-combustible, even though flames were formed on applying a cutting torch as well as significant quantities of smoke and it had historically been regarded as a fire hazard; and,
   b. for various reasons the contractor chosen was allowed to work on the inside of the chimney using hot cutting, even though other tenders suggested this was unsafe.
4) Hole cutting started at the top of the chimney and the workers progressed towards the ground, smoke seen to be emitted intermittently from the top of the chimney;
5) The thickness of the deposit in the chimney was reported to increase closer to the ground, with the centre portion containing a flaky deposit, rather than one which tightly adhered to the walls;
6) It was reported that the deposit glowed during cutting and produced black smoke;
7) No external indication of a fire was seen or heard until flames were seen coming from the latest holes cut through the wall, followed shortly after by flames shooting out of holes progressively higher up the chimney until they emerged from the top; and,
8) A small powder fire extinguisher taken into the chimney by the deceased was found to be discharged, it was not known if this had been used, or whether it was empty as a result of it being punctured when it fell to the ground.

No indication was given by any party of faults with the equipment, complaints that the deposit actually burned, (other than that it glowed and created black smoke), or early signs of panic from the two workers inside the chimney which would suggest the start of a fire. The major supporting evidence as to the potentially combustible nature of the deposit comes from the material safety data sheets (MSDSs) for the products used at CMB.

SUPPORTING INFORMATION ON PAINTS/LACQUERS USED AT CMB

All six MSDSs for paints/lacquers used in the processes at CMB were listed as being either flammable or highly flammable. Flash points listed range from less than 21°C to 32–62°C, and lower flammable limits range from 0.8–1.2%, when quoted.
Taking one lacquer as an example, the following compounds were present. (Table 1)

Table 1. Major components present in a typical lacquer formulation

Compound	Concentration range
n-Butanol	Above 20%
2-Butoxyethanol	10–25%
Methoxy propanol	1–5%
Methyl iso-butyl ketone	1–5%
Shellsol A	10–25%
Phenol	Below 0.3%

With the exception of Shellsol A, for which no data on flammability is available to the authors, all other materials present are known to be flammable.

In view of this, and information given in all six of the MSDSs studied, it is concluded that there was a potential for fires occurring in plant associated with the painting/drying process, as all these products would carry over into the ventilation system to varying degrees.

EXAMINATION OF THE CUTTING EQUIPMENT

The cutting equipment used by the contractors was examined at HSL and few deficiencies found. The cylinders and regulators being used appeared adequate, were of the correct types and were fitted with flashback arrestors at the regulator, although none were fitted near the torch. No damage was found to the equipment remaining after the incident, other than that which could be attributed to fire damage, or to the fall. It was evident, however, that there were a number of connections in the hoses as, by necessity, they had to be long enough to allow them to reach to the top of the chimney (at least 60 m). As the relevant portions of these hoses within the chimney were destroyed in the fire, it was impossible to determine their state and provide comment on whether they were leaking or not. It is, however, understood that the oxygen and propane hoses were checked for leaks by the contractors and they found that all was in order. Furthermore, examination of the undamaged lengths of the hoses indicated that they were in good condition and as such it is unlikely that they were leaking at the time of the incident.

EXAMINATION OF THE TARRY DEPOSIT

As the cutting operations undertaken in the chimney would lead to a large area of the tarry deposit inside the chimney being affected by heat, it was necessary to understand what products were liberated, in what quantities, and whether the material showed any energetic decomposition. The results of these studies are as follows:

CHEMICAL ANALYSIS OF DEPOSITS COLLECTED AT INCIDENT SITE

Two types of test were performed. Firstly, the determination of products liberated at room temperature and secondly, the analysis of compounds formed on heating.

Table 2 contains a list of compounds detected along with approximate relative proportions. The most prevalent species are those listed on the MSDS, or their thermal breakdown products.

EXAMINATION BY TG AND DSC

Results obtained using TG and DSC are given in Table 3.

As would be expected, these results show increasing mass loss with temperature, with 95 % of the materials volatilising at temperatures up to 520°C.

Table 2. Relative chromatographic peak areas

Compound name	Retention time (min)	Headspace in sample tin	Relative peak areas 100°C	200°C	400°C
n-Butanol	3.5	2400	10	1200	3030
Methanol/butene	2.15	180	—	—	320
Ethanol	2.25	470	—	40	230
Acetone	2.35	130	—	—	120
t-Butanol	2.45	90	—	—	—
Butyraldehyde	2.8	50	—	—	—
Isobutanol	3.1	30	—	—	—
Amine	4.4	—	—	230	400
Butyl formate	4.5	110	—	—	—
Butyl acetate	7.6	20	—	—	—
Xylenes	9.5–11.5	40	—	—	—
C_3-Benzenes	15–17	110	—	—	—
Phenol	16.7	—	—	120	—
Isophorone	21.3	—	20	170	—
Phthalic anhydride	24.5	—	—	1300	2500
Other aromatics	25–31	—	—	530	3300
Tributyl aconitate	31.5	—	—	—	440
Tributyl acetylcitrate	32.4	—	—	920	6600

Table 3. Mass loss data for samples of the tarry deposit examined using thermo-gravimetry

Sample A		Sample B	
Temperature 0°C	Percentage mass loss	Temperature 0°C	Percentage mass loss
25–100	1.0	25–100	1.0
100–139	5.0	99–140	4.8
139–451	55.2	141–451	49.9
451–599	33.4	452–599	39.0
599–800	0.4	599–800	0.5
Total % loss	95.0	Total % loss	95.2

The DSC studies on the samples showed slight decomposition and energy release on heating, but at levels which would not be deemed to have contributed significantly to the incident. This can therefore be ruled out as a possible cause for the fire.

PROPERTIES OF THE CHEMICALS LIBERATED ON HEATING

Table 4 lists the physical properties for compounds detected on heating. Included is information on flash point, autoignition temperature, flammable limits and vapour density. It is evident that all vapours for which data could be obtained are denser than air and would tend to sink once initial buoyancy due to heating was lost.

It can be seen that the lower flammable limits (LFL) for the compounds present vary from 1.4% for butanol to 3.3% for ethanol. Not unsurprisingly, no specific data is available on the LFL of the mixture of components found during the chemical analysis, so the actual LFL of the mixture of compounds cannot be determined with any certainty. It will, however, be between these limits.

Table 4. Physical properties of materials detected in the tarry deposit

Temperature °C	Compound	LFL % volume	UFL % volume	Flash point °C	Autoignition temperature °C	Boiling point °C	Density g.cm^{-3}
Headspace in sample tin at room temperature	n-Butanol	1.4	11.2	35–38	365	117.4	2.55
	Methanol	6	36.5	12	470	64.8	1.11
	Butene	1.6	9.3	–62	384	–6.3	1.93
	Ethanol	3.3	19	13	423	78.32	1.59
	Acetone	2.6	12.8	–18	465	56.2	2.0
	t-Butanol	2.4	8.0	10	480	82.8	2.55
	Butyl formate	1.7	8.0	18	322	106.0	3.52
100	n-Butanol	1.4	11.2	35–38	365	117.4	2.55
200	n-Butanol	1.4	11.2	35–38	365	117.4	2.55
	Phthalic anhydride	1.7	10.4	152	570	284	5.10
	Tributyl acetylcitrate	No data available		204	Not available	172–174	Not available
	Ethanol	3.3	19	13	423	78.32	1.59
	Triethylamine	1.2	8.0	–7	—	89.5	3.48
400	Tributyl acetylcitrate	No data available		204	Not available	172–174	Not available
	n-Butanol	1.4	11.2	35–38	365	117.4	2.55
	Phthalic anhydride	1.7	10.4	152	570	284	5.1

Examination of the boiling points listed in Table 4 shows that, of the major components present, n-butanol has the lowest boiling point at 117°C (and was also present in the highest proportion). It is therefore likely that the onset temperature for mass loss from the samples tested using TG was due to boil-off of n-butanol.

FLASH/FIRE POINT OF DEPOSIT

The information presented in Table 4 may lead to the conclusion that the flashpoint of the deposit should lie somewhere around 40°C, taking the flashpoint for butanol as a major component. However, the actual situation is more complex as the deposit comprised of a mixture of compounds, each of which would contribute towards the development of a flammable atmosphere; and it was also a thick, viscous liquid which inhibited normal vapour evolution.

As a result of the non-standard nature of the material, it was decided to undertake both a standard flash/fire point determination, as well as performing a number of ad-hoc tests on large quantities.

The ad-hoc tests involved heating the material in a 2 litre sample tin fitted with a lid with a central 29 mm diameter hole. A small pilot flame was applied periodically to the hole in the lid and the occurrence of any burning noted. Some evidence of transient burning of vapours was observed from 260°C and persistent burning at 285°C.

Following these ad-hoc tests, a further more accurate flash/fire point determination was undertaken using the Cleveland Open Cup flash point apparatus. Briefly, the behaviour of the sample was as follows.

70°C	begins melting
112°C	sample bubbles and fumes – similar to a boiling liquid
145°C	flash of flame if pilot flame is present when gas bubble in sample bursts
195°C	burning on liquid surface for over 5 s
200°C	burning for 20 s

These results equate to a flash point of 145 ± 5°C and a fire point of 195 ± 5°C. It should be noted, however, that these results could be subject to some variation as the concept of a flash point is normally applied to a mobile liquid, rather than the viscous tarry liquid formed when the deposit from the chimney melted.

HOT CUTTING TESTS ON SAMPLES OF COATED STEEL PLATE

A number of tests were undertaken where sections of coated and uncoated steel plates were cut with an oxy-propane torch. The objectives of these tests were to:

1) determine the behaviour of the deposit when heated using a cutting torch and to compare the appearance of any flames with those seen with 'normal' cutting operations; and,
2) to establish the pattern of heat distribution through the metal using thermal imaging.

A direct comparison between the behaviour of coated and un-coated steel plate is evident in Figure 2.

(a) (b)

Figure 2. Behaviour of coated and un-coated steel plate during oxy-propane cutting

It is immediately apparent from Figure 2 that large flames are produced on applying the cutting torch to the coated plate, along with significant quantities of smoke. This should have been taken as a strong indicator that, given the wrong circumstances, the deposit was able to continue to burn. This behaviour also would have been seen during initial screening tests to establish the properties of the material prior to commencing work on demolition.

The appearance of flames such as these led to CMB undertaking further screening tests, which concluded incorrectly, that the deposit did not present a fire risk. CMB and the contractors wrongly assumed that because the material did not continue to burn after the removal of the cutting torch it would not pose a risk. A working procedure was therefore adopted of pre-heating the deposit with the cutting torch and then scraping the softened material away from the area to be cut. This would have limited the potential for clogging the nozzle of the cutting torch with tarry material, but instead introduced the alternative hazard of allowing significant quantities of flammable vapours to form.

CONCLUSIONS

The circumstances surrounding this incident are complex, with many mitigating factors requiring consideration. However, our investigations have demonstrated that the deposit should have been regarded as a fire risk and that hot cutting should not have been allowed. If trained fire scientists had been consulted, this incident could have been avoided.

There was definite evidence before the fire that the paints/lacquers were listed as being flammable, and that the tarry residue was regarded as a fire hazard by the factory's insurers. Despite these factors, plant personnel, as non-fire specialists, were misled by the outcome of the ad-hoc flammability tests they conducted and wrongly concluded that the deposit did not burn.

Our investigations established that many compounds present in the original paints/ lacquers used at CMB were carried over into the extract system and were present in the tarry deposit. When heated, these compounds were evolved giving a flammable vapour mixture. The flashpoint of this mixture was determined as 145 ± 5°C and the fire point as 195 ± 5°C using the Cleveland Open Cup method. In large scale ad-hoc tests using several hundred grams of material, the flash and fire points were around 260°C and 285°C, respectively.

As well as the above direct evidence that the deposit represented a fire risk, further evidence of combustibility comes from tests where samples of steel plates taken from the chimney were cut using an oxy-propane cutting torch. During these tests it was confirmed that:

a) Large flames up to 60 cm high were formed while the torch was applied to the deposit and it continued to glow for a short time after removal of the flame; and,
b) Large areas of the deposit would have been subjected to heating, either by direct flame contact or thermal conduction, leading to the liberation of a large amount of flammable vapour in the chimney – if ventilation were poor, these vapours could have accumulated, possibly leading to an explosion risk.

Search for possible ignition mechanisms for the deposit pointed to four possible causes, these were:

1) Direct ignition of the deposit coating the chimney by the cutting torch, with the fire being fuelled by fresh material running from above;
2) Propane leak or rupture of the propane hose following a flashback, giving either an explosion in the chimney, or a large permanent flame from a ruptured/cut propane hose;
3) Oxygen leak, or accumulation of excess oxygen used in the cutting process, leading to enhanced combustion in the deposit by oxygen enrichment; or,
4) Accumulation of flammable vapours liberated by the tarry deposit when heated.

A major finding of our work was that whilst the above mechanisms could have ignited the material, it would not continue to burn when spread as a thin layer on a steel plate. It was therefore proposed that its combustion probably required the presence of a certain 'critical mass' in order to liberate sufficient heat to maintain combustion. In order for this to take place, either: the deposit in the area being cut must have been thicker; or, a large area of the chimney was suddenly heated - for instance by a gas/vapour explosion, or a major hose failure.

Experience gained during our experimental tests and observations inside the chimney after the fire, indicted that once a fire had become established it would rapidly heat material in the vicinity. The heated material would liberate flammable vapours and also melt. This flow of molten material into the burning zone would both feed the burning fire and give rise to increased fire spread by the direct flow of burning liquid.

As the only two direct witnesses to the event were tragically killed, a definitive answer as to the actual cause of the fire will never be known. What is certain is that any process involving flammable materials which can be deposited inside extract ductwork should be regarded as a fire hazard, and any restrictions on zoning or permits to work should also be extended into the extract system.

Subsequent to this investigation, the authors of this paper have found that a similar fire has occurred elsewhere in the UK where a process oven was used to drive off volatiles from a product. These volatiles accumulated in ductwork and a large chimney, and were also periodically removed to limit the fire risk - a direct comparison with the situation at CMB. In this second incident, changes to the oven's operating temperature led to a fire occurring in the ductwork which subsequently spread to the chimney, almost causing its collapse.

THE ONLY GOOD WASTE IS 'DEAD' WASTE – WASOP, A METHODOLOGY FOR WASTE MINIMISATION WITHIN COMPLEX SYSTEMS

Neil Blundell[1] and Duncan Shaw[2]
[1]Nuclear Installations Inspectorate, Health and Safety Executive, Bootle, Merseyside, L20 7HS, UK
[2]Aston Business School, Birmingham, B4 7ET, UK

Purpose: The international nuclear community continues to face the challenge of managing both the legacy waste and the new wastes that emerge from ongoing energy production. The UK is in the early stages of proposing a new convention for its nuclear industry, that is: waste minimisation through closely managing the radioactive source which creates the waste. This paper proposes a new technique (called Waste And Source material OPerability Study (WASOP)) for critically analysing a complex, waste-producing system to minimise avoidable waste and thus increase the protection to the public and the environment.
Design/methodology/approach: WASOP critically considers the systemic impact of up and downstream facilities on the minimisation of nuclear waste in a facility. Based on the principles of HAZOP, the technique structures managers' thinking on the impact of mal-operations in interlinking facilities in order to identify preventative actions to reduce the impact on waste of those mal-operations' on other facilities.
Finding: WASOP was tested with a small group of experienced nuclear regulators and was found to support their examination of waste minimisation and help them to work towards developing a plan of action.
Originality/value: Given the newness of this convention, the wider methodology in which WASOP sits is still in development. However, WASOP is believed to have widespread potential application to the minimisation of many other forms of waste, including household and general waste.

Keywords: HAZOP; nuclear; systems thinking; waste management.

INTRODUCTION

The UK's growing concerns over global warming and the limitation of landfill sites are driving a national agenda of recycling and reducing the manufacturing/consumption of items which produce unnecessary waste. This is over and above the business desire to reduce costs by plant integration as shown here for a toluene diisocyanate plant.

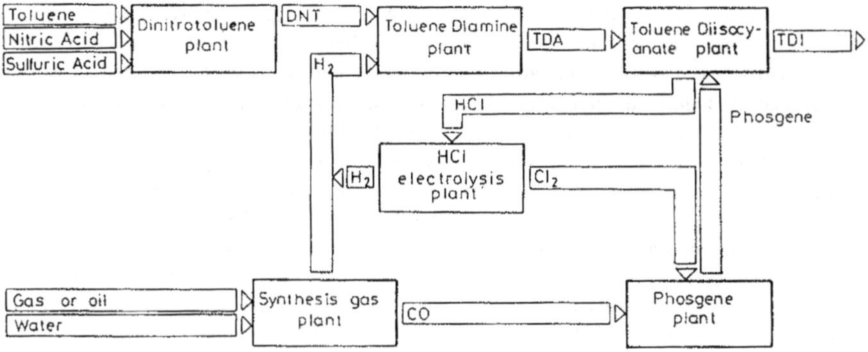

Radioactive material has the property of being easily detectable to extremely low levels and, similar to the effects now being exploited in nanotechnology, transfer of radioactive material from a source occurs at a molecular level by direct contact in a process called 'contamination'. This contamination thus generates further radioactive material which may become waste. This process is in addition to the generation of radioactive material caused directly from the fission process. It is an important principle that radioactive material is only truly waste or 'dead waste' when it can only be send to disposal.

The nuclear industry in the UK has legal requirements placed on it with a similar agenda which, put very simply, is to minimise the quantity of nuclear waste that cannot practicably be avoided.

In Practice, the agenda is to reduce the ability of a source material to smear its radioactivity around a nuclear site as it is transported to, and managed in, various facilities which operate to process that source material. The aim is to dispose of, a much reduced quantity of radioactive material through thoughtful approaches to waste minimisation which may include decontamination.

In this context, source material is radioactive material which will contaminate other material in its vicinity with its radioactivity. A waste is material which has become polluted by a source material and needs to be treated or disposed of. Waste, which is transported to other facilities, is itself a source material as it can smear its own radioactivity in downstream facilities.

When contrasting with other types of waste, an unusual feature of nuclear waste is the sensitive handling requirements and often the degradation that may occur due to the storage conditions and length of time it has to be stored, which can span decades. Also unusual is the requirement to manage some source material that is transported across a number of nuclear facilities, constantly smearing radioactivity and so producing wastes in all of: the production of the fuel; the nuclear reactor; short-term storage (around 3–10 years); downstream handling facilities; re-processing plants; long-term storage (over 50 years).

The management of a source material will cross several specialist facilities, potentially across geographically spread sites. Operations in general and the existing stock of radioactive materials in particular, may affect a facility's ability to receive additional stock from up-stream facilities, or deliver stock to downstream facilities. Delays in the sending or receiving of stock can cause the source material to behave in a way that creates more waste than it should under optimal operating conditions. Therefore, the management of source materials across a national nuclear capability is a complex task when problems in one facility can have knock-on effects up and down the supply chain of facilities (and across sites).

Again in simple terms, analysis of waste management must consider the effects of deviations in a system away from the normal operations and the design expectations.

The technique presented in this paper assists operational supervisors/managers of nuclear facilities in systematically analyzing the effects of (mal-)operations in up/downstream facilities on the production of waste in a facility-in-focus. [The technique examines a facility and its interactions with other facilities. From this point on, this facility under focused examination is called the *facility-in-focus*. This distinguishes it from other up/downstream facilities which interact with the facility-in-focus.]

We focus on mal-operations because when a facility is operating within-design it is assumed to have carried out an earlier analysis, perhaps based on use of the waste management hierarchy, and to be producing unavoidable waste (and the technique focuses on avoidable waste).

Conceivably the technique could be applied during design to new processes and facilities builds that are trying to avoid waste production, but that is not the focus of this paper which is to examine potential improvements to an existing system.

A systems thinking approach was taken to develop a technique which could first understand, and then plan to reduce, the effect of mal-operations in interlinking facilities for each significant waste deriving from a source material. What was developed was a structured approach that examines interactions with up/downstream facilities and identifies additional supervisory processes and engineering safeguards.

This paper outlines the technique, called Waste And Source material OPerability study (WASOP).

First WASOP's theoretical underpinnings are introduced as lying in systems thinking and in the *HAZard OPerability Study* (HAZOP) technique. Next the case study is introduced and used to show the application of WASOP to a hypothetical nuclear case. Discussion of the use of WASOP with nuclear inspectors and future developments conclude the paper.

This paper communicates the latest thinking from nuclear regulators on decision making methodology for supporting waste minimisation and will form part of future regulatory guidance.

UNDERPINNINGS OF WASOP

PHILOSOPHICAL UNDERPINNINGS

The philosophy of WASOP is to identify the best portfolio of actions that will severely limit the source materials' generation of avoidable waste resulting from disruptions in the facility-in-focus that are caused by (mal-)operations in its up/downstream facilities. This is a significant departure from the convention of managing nuclear waste that starts by discussing the existing radioactivity and how to deal with that (an approach reported by Hastings et al. (2007), exemplifying the innovation of this new convention).

The new convention proposed here is not waste management – which emphasizes the management of a material once it becomes waste. The convention is to manage the source material to limit it generating avoidable waste in the first place. This departure from an old mindset accommodates the new desires of site managers, politicians and stakeholder groups to deliver environmentally aware solutions through avoiding unnecessary waste production.

Behind the technique also sits the philosophy that we can prevent waste generation by understanding the behaviour of the source material in the facility-in focus, and then design actions to optimally manage that source material. However, understanding the behaviour of the source material in the facility-in focus is done separately to the WASOP, and so is not included in this paper. Though, it is important to note that, much of the information required for a WASOP would be already gathered for a HAZOP.

SYSTEMS THINKING UNDERPINNINGS

Designing actions for effective source material management in a facility, we must consider the up/downstream facilities which affect that management.

WASOP is informed by the principles of System Dynamics (SD) modelling (Forrester 1961). System Dynamics is "a perspective and a set of conceptual tools that enable us to understand the dynamics of complex systems" (pvii) (Sterman 2000).

SD models can represent the subtle relationships between issues through causal loop models which show the issues relevant to a problem (as nodes), and the positive/negative relationships between related issues (in arcs that link related nodes) (Sterman (2000) provides an accessible review). Causal loop models can inform the building of quantitative models of the dynamic relationships between factors in the model which can examine the stock (in our case, the stock of waste), and how that stock varies when other stock items (in our case, the stock of source materials) flow around the system.

Causal loop modelling is appropriate for exploring the concepts which underpin the management of source materials. Such models could be focused at operational levels and the issues they face when managing a source material in a facility (or equally they could be focused on the policy level models of site-wide, or national, strategy).

For WASOP, these models could explore the engineering issues associated with, for example, transporting waste between facilities and quantitative models could predict the additional wastes that are potentially generated.

However, in practice, problem structuring methods (Rosenhead and Mingers 2001), of which SD can be regarded as being, are criticized for being very difficult for novices to independently conduct (Westcombe et al. 2006). Part of the difficulty is in turning expert knowledge about a situation into a structured model which is theoretically and contextually valid.

Thinking about the situation in terms of nodes, arcs, causes, consequences, stocks or flows is not natural for many managers. Granted, SD facilitators could support managers who wish to use SD modelling, allowing modelling novices to benefit from the approach.

However, suitable SD facilitation support is not always as available as general meeting facilitation and so, to have widespread impact on operations in this context, a decision making methodology needs to be not intimidating as well as be easily usable. Also, the simplicity of the decision making process will ensure that the decision outcome is more reliant on the expert knowledge available than it is on the competence of the modellers to structure that knowledge in a certain way.

Despite this, the development of WASOP began with causal loop modelling and the initial development of quantitative models. Through this development it became clear than the highly technical nuclear knowledge involved in this application was not easily able to be modelled in SD terms, and certainly not by a novice modeller working only from an instruction sheet (i.e. without facilitator support).

Furthermore, although they could be developed, the required quantitative inputs to the model were not readily available, resulting in initial models being based solely on expert intuition. As such, it was decided to employ a systems perspective (in adopting the principle of understanding dynamic systems and material stocks and flows) and learn that the models WASOP would rely on had to be as simple as possible. Our approaches employ Sterman's notion (above) of SD being a perspective.

WASOP BUILDING FROM HAZOP

HAZOP is a technique for proactively managing a situation to avoid hazards being realised (Kletz 1997). It was developed in 1960 by researchers at ICI (Kletz 1997, Elliott and Owen 1968) who sought a methodology to rigorously investigate the hazardous effects of deviations from design of operations in processing plants. HAZOP requires experts to pool their knowledge to investigate potential operational weaknesses to allow those weaknesses to be suitably guarded against. Operational weaknesses are explored partly through considering the risk of failure or mal-operation. The technique can consider the smallest detail of piece of equipment to identify the cause and consequence of human, process or material failure (see Kletz (2006) or Redmill (1999) for reviews).

Keywords are used to structure the HAZOP study (for a range of keywords see Tyler et al. (2000)). For example, experts might consider the effects of a flow-pipe carrying NONE, MORE THAN, LESS THAN, HOTTER or COLDER (of) the material for which it was originally designed. Through considering the effects of deviations from design using the keywords, these experts can consider the effects on the integrity of the pipe and the material it carries. A key feature of HAZOP is the systematic approach to evaluating the effect of mal-operations.

The fundamental principles of HAZOP have remained relatively constant since early publications (Kletz 1997) but there have been a number of innovations using the fundamentals of the approach. For example: expert systems can provide analytical support to the process of HAZOP (e.g. Khan and Abbasi (2000) and Chae (1994)); mathematical simulation models have been used to help in training to explore issues around the magnitude of deviation of operating conditions (e.g. Eizenberg et al. (2006)). Also applications of HAZOPs have been extended beyond their original roots, for example: hazards for software and electronic systems that control production operations (Schubach 1997); hazards in business management processes e.g. financial accounts (Pitt 1994); assessing "human reliability and error analysis" (p306) in healthcare (Dhilon 2003).

The innovation presented in this paper is not in strengthening HAZOP, but it is in addressing the gap of waste minimisation through the structured approach which HAZOP brings to hazard minimisation.

Other techniques have utilised the HAZOP approach in a similar fashion e.g. ENVOP (Isalsk). However the focus has always been end result/objective driven. As indicated earlier the approach presented here aims, by analysing the system using the understanding of the fundamentals, to retain simplicity and allow the outcome to be unconstrained.

HAZOP structures the analyses of the system to take the experts through a decomposition of the complexity of the system. Through decomposing the complexity, the method aims to allow the complexity of source material behaviour and waste production to be better understood. The importance of providing useful techniques for understanding complex systems is illustrated by Lawley (1973) (quoted in Schubach (1997) p303) who asserts that "[HAZOP] is based on the supposition that most problems are missed because the system is complex rather than because of a lack of knowledge on the part of the design team". In alignment with this approach of decomposing complexity to understand it, WASOP in our nuclear context decomposes two aspects of facilities:

1) The system of facilities which collectively process a source material.
   System decomposition aims to identify transportations of source materials and materials that may become wastes or carry waste between facilities in order to appreciate the potential for smearing radioactivity across the site, as well as to understand which facilities depend on each other and may trigger operational consequences if there are mal-operations in the system. Importantly, only one level up/downstream is considered for the facility-in-focus – further up/downstream is considered when those other facilities are the facility-in-focus. System decomposition might be accessible from a site wiring diagram and so WASOP can effectively use existing documentation.
2) The interrogation of each source material transportation between facilities using keywords (see Table 1 for a summary of potential WASOP keywords).
   Keyword decomposition aims to take a structured approach to thinking through the widest range of possible consequences for a facility-in-focus, if mal-operations occur in up/downstream facilities. Although different applications may require a careful selection of alternative keywords, an illustrative list of keywords include considering the effects of:

   - *Nothing* being transported between the facility-in-focus and the up/downstream facility.
   - *More than* normal being transported between the facility-in-focus and the up/downstream facility.
   - *Less than* normal being transported between the facility-in-focus and the up/downstream facility.

Table 1. Potential keywords for the WASOP

– Nothing
– More than
– Less than
– Part of
– Other material (as well as the designed material)
– Other material (instead of the designed material)
– Reversing

- *Part of* the material (e.g. half-empty containers) being transported between the facility-in-focus and the up/downstream facility.
- *Other material (as well as the designed material)* being transported between the facility-in-focus and the up/downstream facility.
- *Other material (instead of the designed material)* being transported between the facility-in-focus and the up/downstream facility.
- Material *reversing* through this transportation route between the facility-in-focus and the up/downstream facility.

Some important issues which assist in the smooth running of the WASOP (many of which resonate with HAZOP) include:

- Before the WASOP, select a source material and facility-in-focus in which the source material is managed.
- Select a group of experts who can best inform the WASOP. These experts should bring the required depth and breadth of knowledge when considering the source material being managed in the facility-in-focus. In particular, they will bring sufficient knowledge of the chemical behaviour of the source material in various conditions of management and the operating conditions in the facility-in-focus and up/downstream facilities.
- A chair/facilitator should lead the group through the WASOP. They are responsible for providing the group with content support, that is encouraging the group to rigorously consider issues (and actions) as well as accurately capturing the detail of the discussion. They will also provide process support, that is ensuring the group understand the process, make progress through keywords, and attending to the social process of group working (Schwarz 2002).
- The group will need to identify the major wastes which emerge during the management of this source material, the system of up/downstream facilities which serve the facility-in-focus, and the keywords to be used for their application.
- Brainstorming and other techniques can be used to encourage the group to think laterally about the range of concerning issues using these keywords for this source material in this facility-in-focus.

- For each concerning issue, identify actions which can alleviate the concern or reduce the likelihood of the system producing avoidable waste.
- Document all issues and actions as appropriate to form a suitable audit trail.

The paper now moves to explain the context for the development of WASOP before discussing its application to the management of a source material in the nuclear industry.

CASE STUDY

THE CONTEXT

The production of waste is an inevitable feature of many production processes, often resulting in the need to dispose of that waste. Recognising this, the joint regulators of the nuclear industry (HSE's Nuclear Installations Inspectorate (NII), Environment Agency (EA) and Scottish Environmental Protection Agency (SEPA)) have delivered guidance on how the licensee can discharge its responsibilities for the management of radioactive waste (HSE, 2007). This has also reinforced the importance of the Nuclear Site License in this management as it is a legal contract that includes requirements on the licensee to manage all nuclear matter on its site and to minimise its radioactive waste. The disposal of this radioactive waste is governed by the Radiological Substances Act 1993 (RSA 1993). The extent of the waste management involved in the industry is highlighted in the Nuclear Decommissioning Authority's recent Strategy (NDA, 2006) which shows that the repository for low level radioactive waste is becoming full, thus requiring alternative approaches for that band of radioactive waste (in which waste minimisation is central).

A feature of this accumulated guidance is the clarification that radioactive waste management is a fundamental part of the safety case for a facility. This means that, when managing radioactive waste, there is a legal requirement on a licensee to demonstrate how they would manage radioactive waste (HSE, 2007). Best practice would indicate that the safety case for our context of the management of radioactive waste should include the adoption of the following principles:

- Auditable – so the waste, and its handling, can be tracked.
- Transparent – to avoid accusations of hiding nuclear material and to improve understanding.
- Clear – to avoid confusion and misinterpretation of the characteristics of the waste.
- Strategic and Planned – to demonstrate that the production of waste from the management of nuclear matter has been considered both locally and as part of an integrated site, or national, strategy.
- Managed – taking an operational view of facility conditions including deterioration.
- Optimised and minimised – to challenge management to deliver good practice.
- Integrated – to identify interdependencies with other nuclear facilities and matter.
- Delivered – a practical demonstration of operational compliance with that is the safety case.

Continuing the regulatory focus on using decision making methodology to further support operations, the HSE commissioned a project to help design a methodology for delivering waste minimisation at the UK's nuclear sites. At the time of writing, that project is in the process of developing, testing and evaluating a wider methodology, part of which is the WASOP technique reported here. Integral to the development of WASOP, in mid-2007 a day-long workshop was run involving four HSE/NII inspectors and two EA nuclear regulator inspectors with the aim of testing the technique and gaining feedback on its continual development. The case study material presented below is taken from the preparation for that workshop. Feedback from the inspectors on the utility of the technique is presented.

The paper now moves to introduce the waste that is being managed in a hypothetical 'system' and then use WASOP to show how waste minimisation can be analysed.

THE SOURCE MATERIAL AND ITS WASTE

In this example, the source material is spent nuclear fuel as fuel pins. Associated with this source material is a can which holds the fuel pins, a skip which holds several cans. Sludge which is inside the skip is associated with primarily the fuel but can spread to the can and skip and the water which surrounds all of the other components.

Put simply for the sake of illustration of the WASOP technique, wastes which potentially arise from association with the fuel pin include: the can which is disposed of downstream and is a consequence of the original process design; the skip which can be reused following decontamination; the sludge which contains radioactive particulates and the water surrounding the system. Other wastes are produced e.g. buildings and skip handling equipment, but these are unavoidable in this type of operation and so are outwith the scope of WASOP.

THE SYSTEM

Figure 1 shows the hypothetical system which is used to illustrate the WASOP technique. The facility-in-focus is the Fuel Pond and the upstream facility is PF1 while the downstream is PF2. The other boxes in the diagram represent extended parts of the system, but these are not immediately up/downstream from the facility-in-focus, so would be analysed when PF1 is the facility-in-focus. The WASOP keywords are be applied to each transportation route identified in Figure 1 i.e. 1–3b.

To explain Figure 1, several fuel pins are housed in a can which is housed in a skip. These are transported from PF1 to the Fuel Pond along transportation Route 1. When the skip (containing the can and fuel pins) arrives into the Fuel Pond it is flushed of water to remove the radioactive water which would contaminate the pond. The skip (still containing the can and fuel pins) is placed under fresh water in a pond container, and then moved to a location in the pond where it is cooled for several years in short-term storage. The pond container holding the skip is then removed from the pond and the skip (and its contents) is removed from the pond container and transported along Route 2 to PF2. In PF2 the fuel pin and the can are removed from the skip and processed. The skip is washed in PF2 and

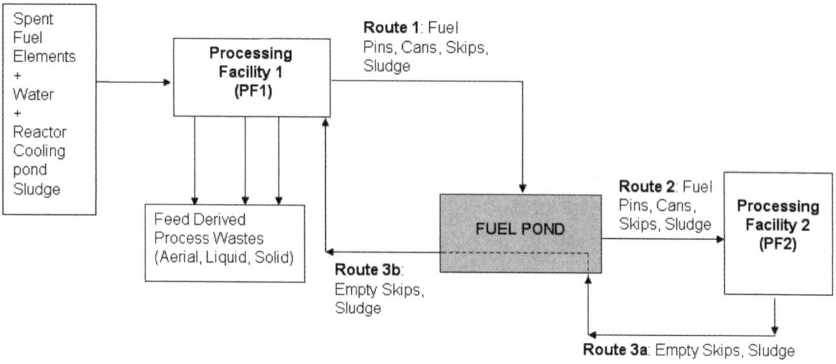

Figure 1. Nuclear material transportations between facilities (with the Fuel Pond as our facility-in-focus)

returned to PF1 along transportation Route 3 (a&b) where it is filled with another can that contains fuel pins. The breaking of Route 3 signifies that the only route for the skip to be transported back to PF1 is physically through the Fuel Pond where it is sometimes stored temporarily if Route 3b or PF1 is blocked. We add this break in route as it nicely complicates the system's dynamics for illustration.

Also able to be stored in the Fuel Pond is non-skipped fuel, albeit temporary storage. This can enter from PF2 through Route 3a. After storage it leaves through Route 2 to PF2. One 'unit' of non-skipped fuel is larger than one skip and so takes up considerably more space in the Fuel Pond.

APPLICATION OF WASOP

Table 2 illustrates a range of possible concerning effects of mal-operations in PF1 and PF2 and the transportation Routes 1-3b on the operation of the Fuel Pond with respect to the skips alone. Empty cells indicate that there are no concerning effects, either because the scenario is impossible, or beneficial.

It is assumed that the maloperation indicated by the WASOP word only occurs within the route being examined. All other routes are expected to be operating to normal flowsheet/capacity, Table 2, and importantly the discussion which informed its content, raised the following main issues:

Beyond the production of unavoidable waste, this system produces avoidable waste, excess empty skips, a main cause of which is a stagnation of full skips (in pond containers, but for simplicity this will be referred to only as skips) in the Fuel Pond. In part, this stagnation can result from either:

- empty skips blocking transportation Route 2 to PF2. This is partly caused by PF1 being unable to accept more empty skips and so the empty skips being temporarily stored in the Fuel Pond.

Table 2. Using the WASOP keywords for the Fuel Pond

	Route 1	Route 2	Route 3a and 3b
	The concerning effects of ... PF1 to Fuel Pond on the management of Fuel Pins in Fuel Pond	The concerning effects of ... Fuel Pond to FP2 on the management of Fuel Pins in Fuel Pond	The concerning effects of ... PF2 to Fuel Pond to FP1 on the management of Fuel Pins in Fuel Pond
No skips being transported along the transport route	• PF1 becomes blocked and so a bulid up of empty skips in Fuel Pond, eventually meaning that Fuel Pond is unable to receive from PF2, and from PF1 when it restarts	• Build up of full skips in Fuel Pond, eventually means the Fuel Pond is unable to receive from PF1, and from PF2 when it restarts • Stagnation and problems on restart if Fuel Pond feed-out is at capacity • Longer residence time in Fuel Pond, and so more contamination and degradation of materials	• Build up of empty skips in PF2 or Fuel Pond, eventually meaning that no empty skips in PF1 to be filled
More than normal skips being transported along the transport route	• Starve PF1 of empty skips (as they are not available to return to PF1), so temptation to order more skips causing more skips to clog the system and to eventually dispose of • Build up of skips in Fuel Pond, unless sending more to PF2 • Potentially longer residence time in Fuel Pond due to FP2 not processing at 'more than' rate (and so more contamination and		

Less than normal skips being transported along the transport route	• Build up of full skips in PF1 and build up of empty skips in Fuel Pond (which are unable to move along Route 3b due to space restrictions in PF1) – gridlock will result	• Build up of full skips in Fuel Pond, eventually meaning that Fuel Pond is unable to receive from PF1 • Longer residence time in Fuel Pond, and so more contamination and degradation of materials	• PF2 becomes blocked • Fuel Pond cannot send full skips to PF2, so soon will not be able to receive from PF1 • This block the entire system- but easier to move things in PF2, if turn off PF1 supply
Part filled skips being transported along the transport route	• Lots of part-filled skips will take up valuable space in Fuel Pond – inefficient use of space	• Build up of full skips in Fuel Pond as focus on processing part-filled skips	• Build up of empty skips in Fuel Pond as quicker to process part-filled skip in PF2, so faster turnaround of empty skip to send back to Fuel Pond
Other materials (as well as skips) being transported along the transport route		• For non-skips being stored in Fuel Pond (from PF2) – possible blockage of Route 3a and 2 if non-skips become dislodged during transportation when being returned to PF2	• For non-skips being stored in Fuel Pond (from PF2) – possible blockage of Route 3a if non-skips become dislodged during transportation on being sent to Fuel Pond • Takes up valuable space in Fuel Pond thereby reducing number of skips able to be stored

- PF2 being unable to accept full skips due to being off-line (which conceivably could happen for a period of several years in the nuclear industry).
- the Fuel Pond being congested with skips which are delaying and preventing the safe movement of skips inside the pond.

If, for one of these reasons, skips have a longer residence time in the Fuel Pond then they degrade by becoming more contaminated, more difficult to handle and decontaminate. This degradation can result in the loss of a skip from the reuse cycle which must be disposed of thus creating additional unforeseen waste.

It is possible that this unforeseen consequence may not have been recognised in the provision of waste disposal routings leading to the empty skip and pond container being left in the pond causing the loss of pond capacity or loss of shielding if containers were to be stacked vertically. Increased residence time in the pond may then lead to further contamination of that skip/pond container combination creating an even greater waste disposal problem.

For waste management in the Fuel Pond there is the issue of decontaminating, or disposing of: pond water; HEPA filters which catch aerial releases; degraded skips; damaged pond containers; degraded physical infrastructure from higher levels of pond water contamination, which is outwith the focus of this discussion. This waste is in addition to the processing of the source material when it arrives at PF2, and the complications which arise if the skip's structure is compromised.

From the analysis, there are two findings which we will use to illustrate the technique (there are more findings, but space prohibits an exhaustive discussion which would only repeat causes of the found consequences). First, there is a strong tendency for the Fuel Pond to become congested with skips. In part, the systems characteristics which result in congestion include:

- Sending 'more than normal' number of full skips from PF1 along Route 1.
- Sending 'normal' or 'more than normal' number of full skips from PF1 along Route 1 when there is a blockage from the Fuel Pond along Route 2.
- Sending 'no' or 'less than normal' number of full skips from PF1 along Route 1, meaning that PF1 can receive less empty skips along Route 3b (and so must be stored in the Fuel Pond). This is due to space restrictions in PF1 which can only accommodate a few skips at any one time.
- Sending 'no' or 'less than normal' number of full skips to PF2 along Route 2, eventually resulting in it causing restrictions on receiving full skips from PF1.
- Sending 'no' or 'less than normal' number of empty skips to PF1 along Route 3b, as this will congest the number of empty skips in the Fuel Pond.
- Sending 'part filled' skips from PF1 will lead to inefficiencies in the occupation of space in the Fuel Pond, requiring more skips to take the material from PF1 and be stored in Fuel Pond.
- 'Other wastes, as well as skips' being accepted from PF2 which take up space in Fuel Pond.

These issues that one may expect to discover in a tightly coupled system of this sort. The extensions of these points are three interesting findings which would become WASOP actions:

- Transporting full skips from the Fuel Pond to PF2 should be highly prioritized; else material will degrade in the pond. Degradation has many consequences for: pond contamination levels; processing the degraded material once it has been retrieved; reusing skips which have become too badly degraded; reusing pond containers which are too contaminated.
- If there is any hindrance in this system concerning the movement of full skips out the Fuel Pond, then operational supervisors in the Fuel Pond and PF1 and even PF2 should be prepared to stop all flows of skips into the Fuel Pond until the situation is rectified. This would aim to not further complicate the moving of skips in the Fuel Pond and ensure that materials are not degrading in the pond and creating waste any longer than absolutely necessary. However there may be circumstances where such an action would have consequences outside the facility-in-focus and its local network which highlights the importance of carrying out the WALARP step separately from WASOP.
- Operational supervisors should explore the possibility of decoupling the Fuel Pond from PF1 using a buffer store, perhaps by building a more penetrating decontamination facility to wash empty skips so that they can be safely stored outside (or at another facility) during times of congestion in the Fuel Pond. This will reduce the reliance on the Fuel Pond being the store for empty skips and containers, allowing it to store full skips from PF1, especially useful if PF2 is off-line for a considerable period.

Second, in times when PF1 lacks empty skips to fill, there may be a temptation to acquire more skips to transport material into the Fuel Pond. In part, the system's characteristics which drive this temptation include:

- Sending 'more than normal' number of full skips from PF1, resulting in a lack of empty skips available in PF1 due to there being insufficient storage of empty skips in PF1.
- Sending 'part filled' skips from PF1. This reduces the amount of material being sent from PF1 per skip, thus requiring additional skips.
- Sending 'no' or 'less than normal' number of empty skips to PF1 from the Fuel Pond. This may arise if the Fuel Pond is congested and unable to move the empty skips in a timely manner, or if the Fuel Pond receives 'less than normal' skips from PF2.
- When 'less than normal' empty skips are being sent from FP2 to Fuel Pond (i.e. a store of empty skips are building up in FP2), FP1 may view that the skips are far down the system and will take a long time to be transported to them.

The extensions of these points are two interesting findings which involve the Fuel Pond and which, again, would become WASOP actions:

- Do not purchase more empty skips when PF1 requires them. There are enough skips in the system and buying more will only further clog up the system with

additional (perhaps, half-full) skips. Also, for waste minimisation, the negative consequence of acquiring more skips is that more will need to be disposed of eventually. Instead of purchasing more, operational supervisors should locate the skips that are in the process and all should work on facilitating their movement to PF1. Thus enhanced communication channels between the facilities need to be encouraged.
– Build a buffer into PF1 so that the transportation of skips from PF1 is disconnected from the delivery of empty skips from the Fuel Pond.

DISCUSSION

This discussion centres on three issues. First, we discuss how WASOP sits within a wider methodology. Then, we discuss lessons learned from using WASOP during the workshop with six nuclear inspectors. Finally, we discuss opportunities for future research.

ANALYSING THE WIDER SYSTEM

By analysing the systemic effects of mal-operations in other facilities, WASOP ignores the effect of mal-operations within the facility-in-focus on the propensity to generate avoidable waste internally. Obviously a systems approach to analysing waste minimisation needs to consider all sources of waste generation in the system and so there is a requirement to consider that waste produced inside the facility. WASOP cannot help here, but WASOP does sit comfortably inside a broader method which examines these issues.

Just as HAZOP sits within the wider methodology of HAZAN (Hazard Analysis) (Kletz, 2006), WASOP sits within a wider methodology of WASAN (Waste And Source material ANalysis) (see Table 3). WASAN is a methodology which aims to derive a set of optimal conditions for managing source materials to reduce the generation of waste. Other components of WASAN are designed to interrogate waste minimisation within the facility-in-focus as well as eventually discriminate between potential actions.

Very briefly, to show context but not to fully explain the broader methodology, WASAN begins with the 'Waste And Source material Identification' which defines the scope of the system being analysed and identifies key components which are central to the analyses, for example, the nature of the source material and how it reacts in different storage conditions. Based on a shared group understanding of the source material and the facility, WASAN then examines operations inside the facility-in-focus. A Waste and Source material Management Hierarchy is used to explore ways in which waste inside the facility-in-focus can be, for example, minimised, reused or treated to ensure that which is eventually disposed of is of a reduced quantity and radioactivity than might otherwise be the case. This is the opportunity to analyse waste minimisation inside the facility-in-focus. WASOP follows next. Then the rank and sentence phase begins where the actions are considered as a portfolio of interacting actions that jointly contribute to the minimisation of waste in the system. The selection of actions is informed by making Waste As Low As Reasonably Practicable (WALARP) – knowing that some waste is inevitable, but identifying where effort is best placed to avoid

Table 3. The WASAN Methodology

'WASAN' for Fuel Pond 1
Waste And Source Material Analysis

A methodology for deriving a set of optimal conditions for managing source materials to reduce the generation of wastes

	Scope of analysis	'Internal to' facility management	'External to' facility management	Rank and sentence consolidation phase	Programme and delivery phase
What do we want to do?	To identify and understand source materials, their behaviour and the wastes which derive from them	To minimise waste generation by identifying management and engineering safeguards for the source material	To reduce the effect of (mal-)operations in interlinking facilities, for each of the significant wastes deriving from a source material	To highlight significant issues in the management of the source material by consolidating issues across the WASMAH and WASOP	To create a work programme to deliver the actions identified through the WALARP
How do we use this?	By identifying how to manage the source material both within, and external to, the faclity either by engineering in safeguards or setting up management processes	By using a structured approach to identify all reasonably forseeable methods for source material management within the facility	By using a structured approach that examines interactions with up/downstream facilities and identifies additional management processes and engineering safeguards	By ranking and sentencing the consolidated issues using relevant criteria (e.g. waste reduction significance, cost, timescale) to produce a set of source material management actions	The programme and its underpinning evidence from the WASAN is the deliverable for waste management under the safety case and is the input to a discharge authorisation
The name of the step	WASID Waste And Source material Identification	WASMAH Waste And Source material Management Hierarchy	WASOP Waste And Source material Operability Study	WALARP Waste As Low As Reasonably Practicable	WASP Waste And Source material Programme

unnecesary waste. The actions are then structured into a Waste And Source material Programme (WASP) to ensure they are prioritised in work programmes.

We present the details of the wider WASAN methodology to reinforce that a systems approach should not only examine one part of the system at the exclusion of other, critical parts (Forrester, 1961). A wider methodology beyond WASOP does exist, but space and its continuing development prevent a detailed presentation of the entire methodology at this stage although these will be described in future publications.

REFLECTIONS FROM USING WASOP

At the end of the workshop with the inspectors, we encouraged them to critique WASOP to reflect on potential improvements and future applications. From this, the importance of a range of issues emerged. First, there was strong agreement in the group that the WASOP process was successful in taking them systematically through the issues. Other comments are presented below:

Process of the WASOP:

- Have a chair/facilitator who is effective. Examples of the importance of this are that many times the group tried to explore the upstream when we were discussing downstream, or external issues when we were focusing on internal issues. While those issues can be logged, discussion of them should be held at the appropriate time in the methodology, so that a structured discussion can cover all issues in sufficient detail. Failure to chair effectively may result in whole issues being overlooked because of the complexity of the system being analysed and the lack of structure to the discussion.
- The experts should be able to share their knowledge to the development of the issues and have that knowledge accurately logged. A facilitator can help here in providing a process which supports the sharing of knowledge, partly through ensuring the sharing of airtime across participants. In our workshop a public projection screen was used so that every member of the group could see the record being made of key points, thus allowing them to feel ownership of the record. A rough copy can be printed out for them before they leave the room.
- When the group become familiar with the process then they should be able to move through it rapidly. It is significant that total novices to the method were able to complete two transportation routes to a facility-in-focus to an appropriate level of detail in about 60 minutes (a similar depth as contained in Table 2). If there is a need to streamline the process further then one could decide on which were the significant waste streams and focus mainly on those. An important issues is that WASOP needs to be appropriate for the user, and thus the inputs should be appropriate to the outcome (i.e. that it is not too much of a drain for resources given the outcome/benefit). In this, there is a need to ensure that the analysis does not expand out of control, providing marginal additional benefit.
- Participants should be reassured if they find commonalities across the WASOP discussion (e.g. the same issues arising from upstream and downstream plants). Commonalities show concentration – serving to validate the importance of issues and reinforce the need for action.

The source material:

- Ensure that everyone understands what is the source material, the nature of its management in the facility-in-focus, and general knowledge about up/down stream facilities. This is a fundamental issue which should not be assumed just because the experience of the participants. There may subtle, but important, differences in the perception of the source material, and so a common view needs to be built at the beginning (in the WASID).
- There may be situations when the group are unclear on what is the source material e.g. when the source material is smeared all around every part of the system. The advice here is to tightly define the facility-in-focus and the source material, systemically focusing on major wastes one at a time.

Actions:

- Discussion needs to examine the issues in enough rigor and appropriate breadth to ensure that important issues are not marginalized, but that irrelevant issues do not take prominence. To accomplish this the facilitator may need to remind the group that their task is to identify actions for the WASP and ask them to ground conversation in pursuit of that aim.
- While the WASP is important, the actions that emerge do not need to be restricted to what is currently feasible. Actions which may appear infeasible to some, could be actionable by others. Hence, the WASP action may be a feasibility study.
- Resulting from the workshop could be a list of actions which is impractical because it contains too many issues to simultaneously pursue. Hence, evaluation of the actions is necessary. WALARP aims to bring the set down to a manageable number of complementary actions which can be implemented in this system for maximum effect. The important point here is that WALARP is done near the very end, advocating that evaluation is not done during the WASOP. The process should be action generation (stage A) and then evaluation (stage B), not constantly iterating between these stages at will. Actions should be evaluated as a portfolio of systemic actions which can only be done near the end of WASAN.
- In our workshop the participants felt it important to steer away from identifying new bandages for old problems. Bandages do not solve the problem – they only cover the problem up with a temporary fix. Instead, they aimed to discover genuine actions which will have a positive lasting impact on waste minimisation and improving the fundamentals of the waste producing system.
- What is right for one facility might not be right for the wider system. For example, deciding to change operational relations between an upstream plant and a facility-in-focus might appear sensible for that relationship, but it might be problematic for facilities further upstream. This is not to say that the WASOP is not useful, but that actions need to be considered in a wider context and operating environment. This is a major issue in the site-wide utility of WASOP and WASAN.

The inspectors also fed back their belief in a strong potential for applying WASOP beyond nuclear, for example, new building of operational plants, environmental waste management, household waste. The topic of 'polluter pays' has resonance far beyond nuclear, and WASOP might contribute in the minimisation of waste and pollution.

The strength of WASOP for nuclear is partly in its affinity to HAZOP, which is widely respected and understood. Hence, these inspectors were familiar with HAZOP and were already comfortable with the notion of structuring analyses through keywords and systematically analysing a system through structured decomposition of interactions. Due to its simplicity in modelling approach, even for those who are not familiar with HAZOP (for example academic colleagues we have discussed this with) are able to understand the method of analyses, allowing them to devote their attention to the content of what is discussed and the conclusions which are being reached. The same might not be said for other methods which take a more complex diagrammatic modelling approach, requiring a steep learning curve on the process of analysis, as well as the content being discussed.

FUTURE WORK

There is an extensive programme of future work planned, including: further testing of WASOP and WASAN in nuclear and other context; continual reflection on the methodology and development in alignment with good decision making practice.

A key area of future work is to more firmly ground and evaluate WASOP and WASAN in the wider family of existing decision making approaches. For example, subtle lessons from other methods (e.g. failure mode and effects analysis (Stamatis, 1995) or fault tree analysis (Vesely et al., 1981; Toola, 1992)) might help to further strengthen the method.

Beyond WASOP, we will strengthen WALARP as currently this is an underdeveloped part of the methodology. We have considered taking a multi-attribute decision analysis approach to evaluating actions (French et al., 2007; Belton and Stewart, 2002) which is very popular in nuclear (Bertsch, 2007). However, this could require a considerable amount of analysis during a workshop which might over-engineer the methodology at this stage. Also for the same reason we have steered away from analytical hierarchy process (AHP) (Brent et al., 2007), but we might be able to select relevant parts of AHP, if not the entire method. Another method which is popular in nuclear is best practicable environmental option (EA/SEPA, 2004) which may offer lessons to option evaluation.

Perhaps a more radical development would be to reconsider the methodology in the light of a life cycle assessment approach (Alexander et al., 2000), by focusing on the lifetime smearing of the source material from conception to disposal.

The range of potential areas for future work is vast because this methodology is brand new and we do not discount anything regarding it continual improvement. Testing and evaluation with users will help us to ensure that all improvements align with the users' needs, and not simply development for the sake of it.

CONCLUSIONS

We have presented a new method for analysing Waste and Source material Operability, that is, the proactive avoidance or minimisation of waste resulting from mal-operations elsewhere in the system having impact on waste production in a facility-in-focus.

The philosophy behind the method is that a closer management of up/down stream facilities which interact with a facility-in-focus should aim to prevent fluctuations in their operations having a negative (waste producing) effect on the facility-in-focus. However, to accomplish this requires detailed knowledge of the type of mal-operations which might happen up/down stream, and their systemic effects on the facility-in-focus operations. WASOP structures such an analysis through encouraging experts to collaborate on dissecting the relationships between a facility-in-focus and its interacting facilities.

We, and the inspectors who tested the method, believe that this method is applicable to a wide range of waste producing systems. One feature of our present application is that waste can only be derived from operations associated with the handling and storage of nuclear matter – and this might not be the case in all waste-producing contexts. However, WASOP has potential applicability where the aim is to manage the source of waste and move towards a philosophy of the minimisation of waste generation. Future work will seek to test our belief, and we constantly search for alternative contexts (in energy and beyond) in which to apply this methodology.

APPENDIX – INSTRUCTIONS FOR CONDUCTING A WASOP

The facility-in-focus does not operate in isolation of other facilities. A WASOP looks at up/downstream facilities and explores their impact on the facility-in-focus.

1. Select: a source material; a facility-in-focus; a group of experts to conduct the analysis; a facilitator to manage the process and content.
2. Identify the major wastes which emerge from the source material.
3. Define the material transportations between the facility-in-focus and its up/downstream facilities i.e. facility interactions. Above these are defmed in Figure 1. It is important to note:
   - Only consider 'first-level' interaction (facilities interacting with the facility-in-focus), not 'second-level' interactions (facilities which interact with facilities which interact with the facility-in-focus).
   - Identify separately any wastes which have independent transportations to facilities or interactions between facilities. For example, in Figure 1, 'Fuel pins, cans, skips, sludge' represents all the major wastes contained in a skip. We would not treat these as separate wastes because they are dependent on one another, in that the skip always contain fuel pins, a can and sludge when moving from PFl to Fuel Pond. 'Feed derived process waste' however, is independent from 'Fuel pins, cans, skips, sludge' as it can leave the PFl without considering the movement of 'Fuel pins, cans, skips, sludge'. This analysis should separately consider the wastes which independently move between facilities.
4. Using a selection of appropriate WASOP keywords, systematically analyse each material transportation to explore the effect of mal-operations in up/downstream facilities on the operational performance of the facility-in-focus, for example:
   a. *Nothing* being transported between the facility-in-focus and the up/downstream facility.

b. *More than* normal being transported between the facility-in-focus and the up/downstream facility.
c. *Less than* normal being transported between the facility-in-focus and the up/downstream facility.
d. *Part of* the material being transported between the facility-in-focus and the up/downstream facility.
e. *Other material (as well as normal material)* being transported between the facility-in-focus and the up/downstream facility.
f. *Other material (instead of normal material)* being transported between the facility-in-focus and the up/downstream facility.
g. Material reversing through this transportation route between the facility-in-focus and the up/downstream facility.

It is important to note:

- Some keywords will not be appropriate for certain transportations.
- If an up/downstream facility has more than one interaction with the facility-in-focus (e.g. two types of source material independently being moved between facilities) then each interaction should have its own keyword analysis.

5. Review and validate the issues which have emerged through the WASOP. This review should: consolidate the discussion; identify issues which may require special highlighting and consideration; add issues which may be missing. The result from this activity will be confidence that the list exhaustively represents the breadth issues which require attention.
6. Identify at least one potential action to address each issue identified through the WASOP.
   a. if there is more than one action, then record all actions.
   b. if the group cannot identify any solution to an issue, then the action would be to find potential solution(s).
   c. some may be implemented in the facility-in-focus to allow it to reduce the effect of disruptions in the feed from upstream, or to downstream, facilities.
   d. some may be implemented in an upstream facility to allow it to reduce disruptions in its feed to the facility-in-focus.
   e. some may be implemented in a downstream facility to allow it to reduce disruptions in its acceptance of feed from the facility-in-focus.
7. Rank and sentence the actions into a plan of deliverable actions which will ensure waste is as low as reasonably practicable (WALARP). This can be achieved through a number of approaches, for example:
   a. Split actions into two sorts (to encourage discussion of the relative merits of the actions, not it is to identify a work programme).
      i. lesser actions, so-called because they require less resources (not because they necessarily have less effect). Lesser actions may be: less contentious; require little resource; require little preparatory work; implemented quickly.

   ii. major actions because they require major amounts of resources and/or preparation. Major actions may be: contentious; require large amounts of resource; require substantial preparatory work; those which involve length implementation.

b. conduct a full-scale multi-attribute decision analysis of the actions. Use the rating of each action against a range of measures to discuss the actions and build understanding of which are emerging as being 'good'.

c. Place all the actions on an effort/impact grid to identify how actions compare on these measures. Actions which require little effort compared to their impact are ones which have a good return and might be considered closely. Actions which have large effort compared to impact might have lower priority.

REFERENCES

Alexander, B., Barton, G., Petrie, J. and Romagnoli, J. (2000) Process synthesis and optimisation tools for environmental design: methodology and structure. *Computers & Chemical Engineering*, 24 (2–7), 1195–1200.

Belton, V. and Stewart, T. J. (2002) *Multiple Criteria Decision Analysis: An Integrated Approach*. Kluwer Academic Publishers, London.

Bertsch, V., Treitz, M., Geldermann, J. and Rentz, O. (2007) Sensitivity Analyses in Multi-Attribute Decision Support for Off-Site Nuclear Emergency and Recovery Management. *International Journal of Energy Sector Management*. Forthcoming.

Brent, A. C., Rogers, D. E. C., Ramabitsa-Siimane, T. S. M. and Rohwer M. B. (2007) Application of the analytical hierarchy process to establish health care waste management systems that minimise infection risks in developing countries. *European Journal of Operational Research*, 181(1), 403–424.

Chae, H., Yoon, E. P. and Yoon, E. S. (1994) Safety analysis using an expert system in chemical processes. *Korean Journal of Chemical Engineering*, 11, 153–161.

Dhillon, B. S. (2003) Methods for performing human reliability and error analysis in health care. *International Journal of Health Care Quality Assurance*, 16, 306–317.

EA/SEPA (2004) Guidance for the Environment Agencies's Assessment of Best Practicable Environmental Accessed from: www.environment-agency.gov.uk on 20th July 2007.

Eizenberg, S., Shacham, M. and Brauner, N. (2006) Combining HAZOP with dynamic simulation - Applications for safety education. *Journal of Loss Prevention in the Process Industries,* 19, 754–761.

Elliott, D. M. and Owen, J. M. (1968) Critical examination in process design. *The Chemical Engineer*, 233, 377–383.

Forrester, J. W. (1961) *Industrial Dynamics*, MIT Press, Cambridge.

French, S., Bedford, T., and Atherton, E. (2007) Supporting ALARD decision-making by Cost Benefit Analysis and Multi-Attribute Utility Theory. *Journal of Risk Research*. Forthcoming.

Hastings, J. J., Rhodes, D., Fellerman, A. S., Mckendrick, D. and Dixon, C. (2007) New approaches for sludge management in the nuclear industry. *Power Technology,* 174, 18–24.

HSE/EA/SEPA (2007) The Management of Radioactive Waste on Nuclear Licensed Sites, Accessed from www.hse.gov.uk/nuclear/wastemanage.htm on the 28th September 2007.

Khan, F. I. and Abbasi, S. A. (2000) Towards automation of HAZOP with a new tool EXPERTOP. *Environmental Modelling & Software,* 15, 67–77.

Kletz, T. A. (1997) Hazop – Past and future. *Reliability Engineering and System Safety,* 55, 263–266.

Lawley, H. G. (1973) Operability Studies and Hazard Analysis. AIChE Symposium Loss Prevention in the Chemical Industry. 105–116.

NDA, 2006, NDA Strategy, accessed from website www.nda.gov.uk on 28th September 2007.

Pitt, M. J. (1994) Hazard and operability studies: A tool for management analysis. *Facilities,* 12, 5–9.

Redmill, F. (1999) *System Safety: HAZOP and Software HAZOP*, John Wiley and Sons Ltd.

Rosenhead, J. and Mingers, J. (2001) *Rational Analysis for a Problematic World Revisited,* John Wiley & Sons, Chichester.

Schubach, S. (1997) A modified computer hazard and operability study procedure. *Journal of Loss Prevention in Process Industry*, 10, 303–307.

Schwarz, R. (2002) *The Skilled Facilitator*, Jossey-Bass, San Francisco, CA.

Stamatis, D. H. (1995) *Failure Modes and Effects Analysis: FMEA from Theory to Execution.* American Society for Quality.

Sterman, J. D. (2000) *Business Dynamics: Systems Thinking and Modelling for a Complex World*, Irwin McGraw-Hill, Boston.

Toda, A. (1992) Plant level safety analysis. *Journal of Loss Prevention in the Process Industries*. 5(2), 119–124.

Tyler, B. J., Crawley, F. and Preston, M. L. (2000) *HAZOP: Guide to Best Practice*, The Institution of Chemical Engineers, Rugby.

Vesely, W. E., Goldberg, F. F., Roberts, N. H. and Hassl, D. F. (1981) *Fault Tree Handbook (NUREG-0492).* U.S. Nuclear Regulatory Commission, Washington, DC.

Westcombe, M., Franco, L. A. and Shaw, D. (2006) New directions for PSMs – A grass-roots revolution? *Journal of the Operational Research Society*, 57(7), 776–778.

Isalski, W. H., ENVOP for waste Minimisation, IChemE Environmental Protection Bulletin 034, 16–21.

IMPACT OF EMERGENCY SHUTDOWN DEVICES ON RELIEF SYSTEM SIZING AND DESIGN

R. K. Goyal and E. G. Al-Ansari
Bahrain Petroleum Company, Bahrain Refinery, The Kingdom of Bahrain

In the sizing of individual relief valves protecting equipment or process or system, it is a common practice not to take cognizance of any immediate operator action or the action of any mitigating devices. However, an increasing number of consultants and practitioners are recommending not applying the same philosophy when it comes to designing an overall refinery flare system to cope with common mode failures (e.g., loss of power, cooling water supply failure). They propose taking credit for the action of devices such as unit emergency shutdown (ESD) systems, trips (for example, fired heater fuel supply cut-offs), or auto-starts of pumps whose actions reduce the potential load on the overall refinery flare system. Savings can thus be realized in the sizing of flare headers and other ancillary equipment. While there is no objection, in principle, to taking credit for ESDs in the design of relief systems, its application in practice deserves careful scrutiny. There are still many related issues that have not been adequately addressed by the proponents of the credit-taking approach. This paper highlights these concerns and offers practical advice to those facing relief system design decisions.

1. TAKING CREDIT FOR SHUTDOWN DEVICES

In a modern refinery, the practice of atmospheric discharge of gaseous hydrocarbons from pressure relief valve (PRV) tail pipes, irrespective of whether on-plot or off-plot, is neither permissible under environmental guidelines nor desirable from a safety standpoint. The common approach, therefore, is to tie all (or most) pressure relief discharges from a unit into a manifold or unit header, which is then routed to a refinery relief header connected to a suitably sized flare system. Two systems are sometimes preferred – a low-pressure system and a high-pressure system.

The key parameters in the design and sizing of such a relief/flare header or manifold are the flow rate, the driving pressure and the type of material expected to enter the header from the discharge pipes of various relief valves connected to it. This in turn depends upon assumptions made as to the concurrence of relieving from several sources.

If it is assumed the header is required to handle the numerical sum of the *rated* capacities of all the relief devices in all the units discharging to it, then its calculated design size will truly be of enormous proportions – and require an equally enormous flare stack to match! Clearly, such an approach is wasteful and unjustifiable, especially where it can be demonstrated that an event culminating in simultaneous relief from all the valves at their

Contents of this paper are based upon views and opinions of the authors, and do not necessarily represent BAPCO's present or past policy, codes, standards and/or practices.

respective rated capacities is impossible to occur (except, perhaps, as an extremely elaborate act of sabotage).

A certain degree of realism can be injected into the header design process by assuming that the maximum relief load will be equal to the sum of the actual expected maximum relief flows from those valves which could lift under a given emergency situation. For example, consider utility failure (power, cooling water, instrument air, steam, fuel oil/fuel gas, inert gas, or a combination based upon inter-relationship or common cause) or unit/plant fire. The header size derived will be smaller than that resulting from the total rated relieving-capacity assumption discussed previously. It will, however, be large enough to handle the relief load from all foreseeable emergency situations.

Hence, in sizing a header/flare system, there can really be no serious objection to utilizing a conservative time-line analysis approach or a dynamic analysis based on process parameter levels expected under "upset" conditions to calculate the required relief load, provided individual peak relieving rates get adequately addressed in the analysis.

Further economy in the header and flare system size can be realized by assuming that, in practice, several of the relief valves will not be required to lift in an emergency. Pressure in the vessels or equipment protected by them will not rise above the PRV set pressures due to the action of any "automatic instrumentation" installed that tends to pacify the source of pressure build-up. Automatic instrumentation here does not refer to the normally operating control systems and instruments used to operate the refinery [sometimes referred to as the Basic Process Control systems (BPCS) – *see* CCPS (1993) automation guidelines]. It refers to non-normal instrumentation such as emergency shutdown devices (ESDs), trips, safety interlock systems, auto-lockouts or auto-starts (all termed "ESD" for the purpose of this paper).

Size reduction sought on the basis of ESDs (i.e., taking credit for ESDs in relief and flare system design) – though it appears to have a "prima facie" justification – is nonetheless fraught with controversy and a source of genuine concern, especially among operations managements. The key question, therefore, is: **should we or should we not take credit for ESDs in the relief/flare system design?**

Before delving into the pros and cons of the practice of taking credit for ESDs, some clarifications and comments regarding the applicable standards and other related topics need to be made in order to better define the scope of the concerns and the real, underlying issues.

Take process vessels designed in accordance with ASME "Boiler and Pressure Vessel Code" Section VIII, Division 1. The need for pressure relief devices is included in Parts UG-125 to UG-136 of the code. Similarly, British standard BS-5500 "Unfired fusion welded pressure vessels," specifies the need for PRVs. In terms of relief header sizing, no distinction is made between PRVs installed for code compliance purposes and those installed for other reasons.

Once a decision is made to install a PRV at a given location in a refinery unit, its inlet piping needs to be designed per API RP-520, Part II, Section 4. Similarly, design of other parts of the relief system – such as PRV sizing, individual discharge piping and the header piping – can be carried out on the basis of the various API recommended practices. Applicable sections of the API RPs are illustrated in Figure 1.

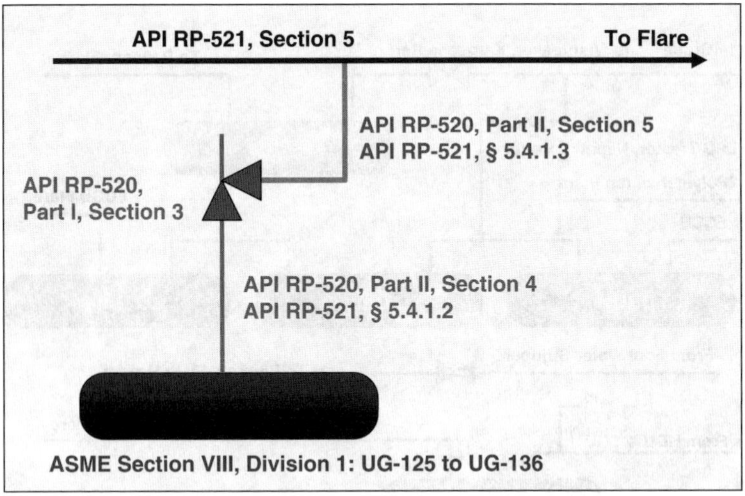

Figure 1. Applicable standards

Some designers in this field will argue that the API RP-520 Part I (January 2000), Part II (August 2003) and RP-521 (March 1997) are merely "recommended practices," implying that these need not be adhered to as meticulously as warranted by codes of practice or standards. It should be noted that these two RPs are extensively used by designers worldwide in order to identify the minimum requirements necessary for an acceptable design. For all practical purposes, the status of these two RPs is on a par with that of any other internationally recognized standard or code.

In addition to being connected to various PRV discharges, the unit manifold may also be connected to piping carrying excess gas which needs to be directed to the flare header from time to time as part of the normal operation in the refinery or as part of a controlled flaring activity following a minor plant upset. A utility failure scenario at a time when such flaring is taking place has not been considered in this paper.

2. ADVANTAGES OF TAKING CREDIT

Clearly, the biggest advantage of taking credit for ESDs is minimizing the size of the relief system required to handle the PRV discharges from a unit or the entire refinery. Relief and flare headers are typical of other safety-related equipment in a refinery – they cost a great deal of money to design and install to begin with, and then take up a significant portion of the regular maintenance effort.

Consider the flare system shown in Figure 2. The main flare is designed to take discharges from four crude distillation units, a crude gas recovery unit, a visbreaker, a kerosene rerun unit, a hydrodesulfurization unit, a LPG treater, a naphtha rerun complex,

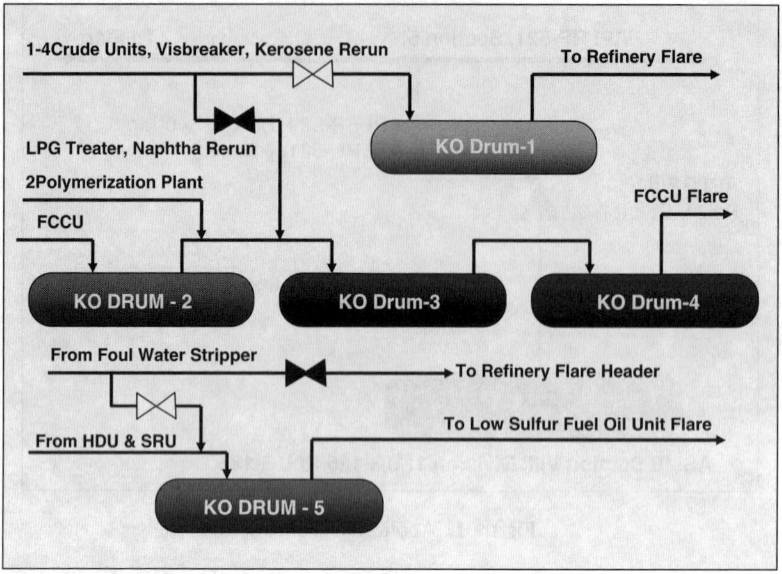

Figure 2. A typical refinery flare system

and several drip drums in the refinery gas circuit. The second flare (the FCCU Flare) is connected to the FCC unit, a crude unit associated with the FCCU, a polymerization plant, and a gas compression unit. During FCCU shutdowns for maintenance, the main refinery flare can also be taken out of service for maintenance by diverting its flare load to the FCCU flare.

The third, independent flare system – the LSFO (low sulfur fuel oil) Flare – serves the requirements of a hydrogen plant, a hydrodesulfurization unit, and a sulfur recovery unit. The foul water stripping unit is normally routed to the LSFO, but can be directed to the main refinery flare header if need be (this is to enable other units in the refinery to remain on-stream when the LSFO complex is down for maintenance). The presence of a Platformer/Unifiner unit brings into question the requirement of a fourth flare.

Header sizes of 36″, 42″, and larger are needed to handle the maximum possible flows from the units. In a large, well-spaced refinery requiring flare stacks to be located at a safe distance away, costs of installing large headers and associated equipment can be substantial. Furthermore, if some sections of the header system call for special metallurgy, then the costs escalate further.

If credit is taken for the unit/equipment ESDs in the belief that these will tend to reduce the expected flare load, then considerable savings in the investment costs can be realized by installing smaller size headers and ancillary equipment (valves, knockout drums, seal drums, etc.). Space required for the system would be smaller as would the civil/structural work. In some cases, a smaller system will result in lower regular maintenance costs (cleaning, inspection, etc.).

A major advantage of ESD credit taking is the use of existing relief/flare system for the purpose of permitting additional discharges into it. In the example illustrated in Figure 2, if ESD credit taking is not allowed then a fourth, new flare system will have to be engineered and installed to accommodate the relief load from the Platformer/Unifiner Unit.

If, on the other hand, credit is taken for the existing ESDs in various units of the refinery, then the existing flare headers can be modified to take in the relief load from the Platformer/Unifiner Unit at a fraction of the costs associated with a new flare system. The need to accommodate additional relief load is not just a hypothetical case. Many refineries have faced this problem – the need arising from a variety of reasons such as:

- Changes in product slate requiring alterations in process parameters
- Revamping or debottlenecking of existing units
- Technology upgrade
- Addition of new units
- Seeking compliance with more stringent environmental regulations
- Capacity increment.

Invariably, economic considerations must, and do enter into decision making on issues such as those illustrated in the example above. Consider a scenario in which ESDs are installed in the Platformer/Unifiner Unit and credit is taken for these ESDs in terms of reduction in the expected relief load, then it may be possible to accommodate this reduced load into one of the existing flare systems. If the life-cycle cost estimated for the installation and maintenance of the ESDs turns out to be greater than that associated with a new flare system, then the question of ESD credit taking is only of academic interest to the decision at hand.

It can be argued that there may be other reasons for the installation of ESDs to be considered. It could be due to the need to meet existing (or foreseeable, future) environmental regulations or part of an overall safety enhancement recommended by a HAZOP (hazard and operability) study team. Under these conditions, it might not be possible to maintain independence between these reasons and that related to the flare system.

Reduction in relief load means reduced flare stack diameter and length, reduced header and sub-header sizes, and hence lower investment. In addition to the effect on installation costs, and perhaps of greater significance, is the impact of relief load reduction on the following key parameters associated with the performance and siting of a flare stack:

- In-plant thermal radiation at grade
- Radiation received at adjacent equipment
- Radiation level at refinery fence-line
- Combined radiation from more than one flare
- Dispersion of combustion products
- Dispersion on flame failure
- Compliance with environmental regulations

- Health impact on immediate area
- Health impact on surrounding communities
- Quantity of product sent to flare.

3. TYPES OF ESDS

As mentioned earlier, the term "ESD" has been used in a generic mode in this paper. However, before discussing various ways in which credit could be taken for ESDs, there is a need to briefly describe the type of ESDs under question and the different terms being used in literature to refer to these. The emphasis is on *brief* descriptions rather than providing an elaborate set of definitions. A few noteworthy efforts in clearing up some of the confusion from safety system performance terms have been Gruhn (1993) and Beckman (1992 & 1993).

Safety Interlock System (SIS) is a term favored by the CCPS Safe Automation Guideline (1993). It consists of a dedicated controller (PLC) taking input from instrumentation installed for normal operating process control and/or sensors installed exclusively for the SIS. The output is in the form of dedicated alarms, event logger, and field actuation (automatic valve, motor starter or motor trip, etc.).

Three integrity levels can be considered for SIS design:

- Level 1 (low level) is a single path design with no redundant components
- Level 2 (medium level) consists of some redundancy (especially of components with known low reliability)
- Level 3 (high level) is a fully redundant system in which a high degree of reliability is achieved by means of redundant components, enhanced self-diagnostics, and avoidance of common mode failures (by selecting different types of sensors, etc.).

In addition to these, the term "**triple-modular-redundant** (TMR)" has been used to describe systems in which the objective is to achieve both high reliability and high availability (these are more popularly known as "two out of three voting" systems or simply as "2oo3").

Auto-lockout device refers to non-normal automatic instrumentation that trips or shuts a power or heat source. It is actuated by an abnormal condition and results in the stoppage of a process stream, a utility stream, and/or a piece of equipment that adds to a relief load. Examples are:

- Automatic steam supply shut off (valve closed) to a tower reboiler on high tower pressure
- Fuel gas supply shut off to the burners in a fired heater on high pass flow temperature, etc.

Auto-start devices, on the other hand, are those that attempt to reduce the flare load by *starting* some equipment. Examples are:

- An automatic start-up of a spare reflux pump (steam turbine driven) on electric power failure
- Automatic start-up of a cooling water circulation pump.

Shutdown systems for fired-heaters can consist of several levels; for example, individual main fuel trips, total "heat-off" and emergency shut down of the entire unit or a complex within the refinery. Process parameters that need to be brought into the ESD design logic can be determined by carrying out a quantitative risk analysis (QRA) of the costs in relation to the degree of desired availability and/or reliability of the installation. For further information on QRA methodology, *see* Goyal (1986 & 1993) and CCPS Guidelines (1989). Typical process parameters commonly considered in a QRA study for fired-heater shutdown systems are shown in Table 1.

The number of parameters from this list, which can be cost-effectively brought into the ESD design logic, depends entirely on the particular circumstances of a furnace installation. Hence, results from a QRA study identifying these parameters for one furnace installation cannot be directly used for a different furnace.

It should be noted that not all ESDs necessarily reduce the expected flare load. There can be situations in which an automatic trip can actually *increase* the expected flare load. An example of this is a steam-turbine-driven reflux pump which is expected to continue to work in the event of an electric power failure but gets cut out by a steam-load-shedding system acting to prevent failure of the overall plant steam supply system.

4. VARIOUS METHODS OF CREDIT TAKING

Either a time-line analysis or a dynamic analysis (sometimes referred to as "transient analysis") is generally performed to determine relief volumes under various emergency situations and on the basis of assumptions made about the impact of ESDs. It should be

Table 1. Parameters for furnace shutdown systems

#	Process parameter
1	High tube skin temperature
2	High individual pass outlet temperature
3	Low total heater pass flow
4	Low fuel gas pressure
5	Low pilot gas pressure
6	Low fuel oil pressure
7	Low atomizing steam pressure
8	High and low combustion air pressure (for forced draft)
9	Low combustion air flow (for forced draft)
10	High pressure in firebox
11	Low percentage of oxygen in flue gas
12	High percentage of combustibles in flue gas
13	High smoke density in flue gas
14	Low flue gas temperature (for air preheaters)

noted that the adequacy and applicability of methods currently available for sizing relief valves and a relief header to handle a *given*, predetermined relief load are beyond the scope of this paper. For more information on these topics, reference is made to several excellent articles by Cassata et al (1993), Coker (1992), Hall (1993), Leung (1992), Niemeyer and Livingston (1993), and Papa (1991).

The most popular method of taking credit for ESDs appears to be the "largest-load-failure" method. This can be illustrated through the example shown in Figure 3. The relief lateral for a given processing unit can be sized for the largest single relief valve within that unit under this method. For example, in Figure 3, if the Relief Valve "A" represents the largest load (i.e. it is greater than either "B" or "C"), then the Unit relief header could be sized to match the requirement of "A".

Consider a processing complex (say, part of a refinery) consisting of three units. Refer to Figure 4. The equipment in Unit-1 is protected by PRVs *A*, *B*, and *C*; in Unit-2 by *D*, *E*, and *F*; in Unit-3 by *G*, *H*, and *I*. The PRVs discharge into their respective unit relief headers, which in turn are connected to a common header for the whole complex. Additionally, assume that all equipment is protected with ESDs of reasonably high integrity, which act to prevent lifting of the PRVs under a specific contingency.

The sizing of inlet and discharge piping associated with each PRV is governed by rules given in API RP-520. If no credit is taken for the presence of ESDs, then Unit-1 relief header needs to be sized to accommodate the combined load from $A + B + C$. In the

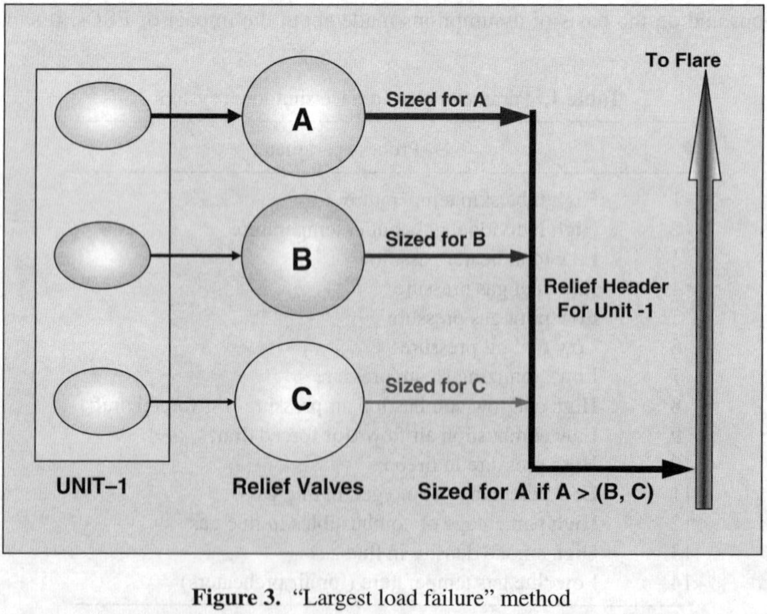

Figure 3. "Largest load failure" method

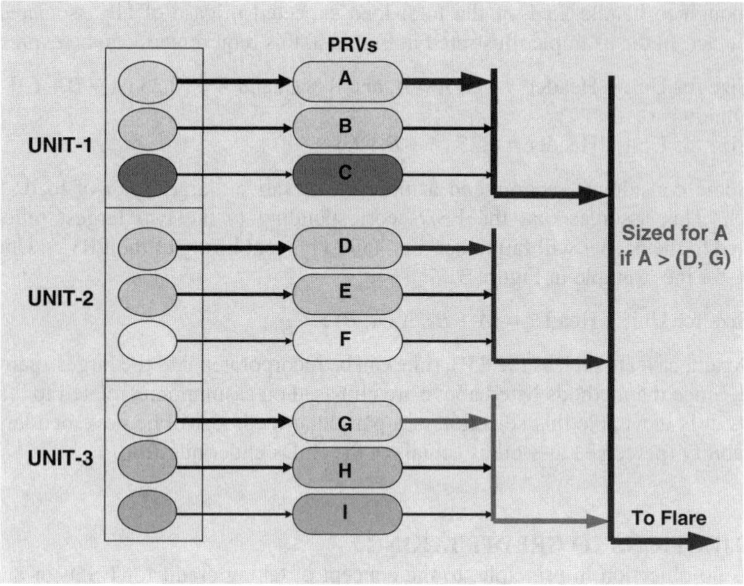

Figure 4. Further illustration of the largest-load-failure method

"largest-load-failure" method, it is assumed that the ESD corresponding to the largest individual load could fail under the stipulated emergency situation (i.e., fail to prevent the lifting of the PRV).

It should be noted that for the purposes of this discussion, the largest or maximum load is not necessarily the largest number of pounds per hour; it is the flow that results in the greatest friction loss through the header or pipe segment in question. Thus Unit-1 relief header need only be sized to take load from relief valve *A*, if *A* is largest among *A*, *B*, and *C*. Similarly, Unit-2 header can be sized for the load from *D* and Unit-3 for load from *G*; *D* and *G* being the largest among their respective unit loads.

This "largest-load-failure" concept is further utilized in sizing the common header for the complex. If the load corresponding to *A* is the largest among *A*, *D*, and *G*, then the common header can also be sized to accommodate load from *A* alone. The method can be extended and repeated to cover the entire refinery. Several variations on the basic theme of the largest-load-failure method have been proposed by engineering consultants and corporate engineering departments of operating companies.

An example of such a variation is inclusion of a caveat to ensure that the size of any relief manifold is not smaller than that corresponding to at least 25% of the total rated capacities of all relief valves connected to it. Another variation, closely related to the 25%-rated-capacity type, incorporates the requirement that the manifold size should be at least large enough to handle 25% of the total load expected in case of all associated ESDs failing to act. In the example illustrated in Figure 3, this requirement can be expressed as:

Size for Unit-1 Header = A, if $A > B$, and $A > C$, and $A > [0.25\,(A + B + C)]$
Otherwise,
Size for Unit-1 Header = $0.25\,(A + B + C)$

Some consultants recommend a more conservative "largest-pair-of-loads-failure" approach. This assumes that the ESDs corresponding to the two largest relief loads connected to the header will fail to act (i.e., fail to prevent lifting of the PRVs). Under this method, for the example in Figure 3:

Size for Unit-1 Header = $(A + B)$, if $(A, B) > C$.

Again, caveats such as the 25% rule, can be incorporated into the largest-pair-failure method. Since the methods listed above are all based on assumptions related to "failures" of ESDs, it is inevitable that selection of a particular method will be governed largely by the reliability (perceived as well as actual) of the ESDs under question.

5. OBJECTIONS TO CREDIT TAKING

There is no objection in principle, to the concept of taking credit for ESDs or any other shutdown devices/trips in evaluating relief system capacities. It is no different from any other cost versus risk-reduction benefit decisions faced by managements every day. In the highly competitive environment, which currently prevails in the refining business, the potential for savings associated with a smaller flare system cannot be dismissed lightly.

Nonetheless, before lending unequivocal support to the concept, a few concerns need to be aired and resolved. From the standpoint of operations and engineering managements these are considered to be extremely significant – in fact so much so as to disfavor the practice of ESD credit taking. Past incidents on record involving flare systems further add to a plant owner's anxiety in what is perceived as "cutting corners" in the system design. One example is the Grangemouth (U.K.) Refinery incident, *see* HSE (1989). Although not related to flare line sizing, it was, nonetheless, a major incident involving a flare system.

5.1 WHAT DO THE CODES RECOMMEND?

API RP-521 (March 1997) is considered the most widely used "guideline" in the design of relief and depressuring systems. An extract from paragraph 5.4.1.3.1 is worth reproducing here:

> *"... The discharge piping system should be designed so that the built-up back pressure caused by the flow through the valve under consideration does not reduce the capacity below that required of any pressure relief valve that may be relieving simultaneously."*

The above-mentioned statement is extremely clear and specific in terms of its content and guiding intent. It can be argued that ESD credit-taking violates the requirement quoted above in that if a smaller header size is selected it may permit the build-up of back pressure to such a level as to reduce the capacity of another PRV connected to the system *if the ESDs fail to act in the assumed manner*.

Nonetheless, the same source (i.e., API RP-521, paragraph 5.4.1.3.1) then goes on to state:

> *"...Common header systems and manifolds in multiple-device installations are generally sized based on the worst-case cumulative* required *capacities of all devices that may reasonably be expected to discharge simultaneously in a single overpressure event."*

The inclusion of "reasonably" in the above paragraph can be interpreted as providing justified support for the practice of ESD credit taking!

Apart from references of the type mentioned above, the latest publications of API RP-520 and RP-521 do not specifically sanction it nor do they oppose it. Furthermore, to our knowledge, there are no other internationally recognized standards, codes of practice or guidelines, which specifically permit or deny ESD credit-taking in relief system design.

5.2 LEGAL CONCERNS

The hydrocarbon processing and the chemical industries are sometimes portrayed in the media as being those causing many major incidents resulting in loss of life and property. Setting aside the validity of such claims, there is no denying that most reputable companies have been acutely aware of their responsibilities in terms of safety of the communities and the environmental issues since well before the onset of recent legislation on clean air and process safety management.

In the U.K., many companies embarked on a systematic search and evaluation of hazards in their plants in the mid 1970s. The driving force behind this effort was mostly self-imposed criteria by the industry rather than the force of law, *see* Al-Ansari (1990).

In Canada, the Canadian Chemical Producers Association (CCPA) published a policy on "Responsible Care" in 1983 and promulgated the "responsible care code of practice," CCPA (1989). In the U.S.A., the Center for Chemical Process Safety (CCPS) of the American Institute of Chemical Engineers picked up the challenge, and has admirably served the industry through the "guidelines" series of books. This "voluntary" effort by the industry culminated in API RP-750 (1990), which subsequently formed the basis for OSHA's proposed rule-making (July 1990) and regulation (February 1992).

A significant characteristic of the pre-OSHA era was that targets for risk reduction and/or risk acceptability began to appear in numerical, quantitative terms. CPQRA (Chemical Process Quantitative Risk Assessment) thus became an effective tool in the armory of the decision-makers – *see* CCPS (1989). Owner or operations management were readily willing to back decisions based on calculated risk. If a CPQRA analyst could demonstrate that the risk associated was negligibly small, then operating managements were willing to support a relief header sized by taking credit for ESDs.

In the post-OSHA period, the situation seems to have changed markedly. The punitive element, invariably associated with the law, has forced a major modification in the outlook of many operations managers. CPQRA no longer rules supreme. The first question management wants answered is: "Does this decision conform to existing international standards, codes of practice, or guidelines or best-known/approved practices?" Or, conversely: "Will we be in violation of, or interpreted to be in violation of any international code?" In the past, the fact that the API has been silent on the subject of ESD credit taking would have been just one factor in the overall decision-making process. Nowadays, this silence will get noticed with added alarm.

API must revise RP-521 to specifically permit ESD credit taking. Only then can operations management be expected to consider this a viable option in relief system design.

Acceptance of the ESD credit-taking practice despite the absence of a recognized standard backing the concept can lead to potential violation of the intent of OSHA 1910.119 (U.S. Federal Register, February 1992) "Process Safety Management ...." paragraph (d)(3)(H)(ii) which states that the employer shall document that equipment complies with *recognized* and generally accepted good engineering practices; the statement being applicable to relief system design and design basis as per paragraph (d)(3)(D) of the OSHA regulation.

Lack of a recognized standard leaves engineers and managers, who permit the design and installation of a relief system taking credit for ESDs, vulnerable to the possibility of unfavorable comment from official investigations of any loss or injury incidents involving relief system sizing. This concern should not be considered a mere speculation. Past experience of refinery management on incidents elsewhere, in which established industry practices were set aside in favor of calculated low-risk options, forces us to a closer scrutiny of this issue.

OSHA should be presented with the current situation related to the two paragraphs referenced earlier as a test case for interpretation of their mandate. They should be requested to state specifically that ESD credit taking does not constitute a violation of their intent. Only then will the practice be considered legally acceptable.

5.3 COMPROMISING A KEY SAFETY FEATURE

Even if the law permits taking credit for ESDs, a carte blanche approval can not be granted for this practice. Each application must be thoroughly analyzed on the basis of its particular situation.

In the field of loss prevention in the process industry, there are a few key features related to layout and design, which tend to enhance the intrinsic safety of a plant. For example:

proper ***spacing*** *(between equipment/units)*
proper ***size*** *(pipe/vessel size/wall thickness, etc.)*
proper ***steel*** *(correct metallurgy).*

These features, when incorporated into the layout and design of a refinery, provide a significant degree of safety by mitigating the consequences of process deviations and other incidents. Furthermore, they are, by and large, immune from the adverse effects of human error or other uncalled-for human intervention.

In well laid-out refineries, risk exposures will be limited because of the generous inter-unit distances. The EML (Estimated Maximum Loss) calculations carried out by the insurers in such cases reflect this lower risk, which, in turn, translates into lower premiums.

It can be argued that the ESD credit-taking practice compromises this safety margin. An "undersized" flare header receiving load from several units makes it possible for an equipment over-pressure event (which might lead to an explosion or fire) to occur simultaneously in more than one or all the units connected to the single flare system following a common mode initiating event such as power failure or cooling water failure.

In addition to the possible effect on insurance premiums, another area of concern is the fire-fighting and control capabilities, which need to be provided on site in a refinery. In well laid-out refineries, fire water systems and all other fire-fighting capabilities are based on the general assumption that emergencies will be limited to a single unit or area at a given time. Under-sizing a flare header raises a serious question as to the validity of that assumption.

Over years of disuse and potential neglect, some segments of a flare header system might get partially clogged by sludge deposits or liquid dropping out at low points or pockets in the system (present due to errors in design or construction). If such a system was originally sized to take the full load from all the sources feeding it (i.e., by not taking credit for the ESDs), then it will be more forgiving in the event of a partial blockage than a smaller system based on ESD credit-taking. Note this comment is not to be misconstrued in any way to mean condonation of design flaws (pocketed flare lines, etc.) and/or poor operating and maintenance practices.

5.4 INCOMPATIBILITY WITH SOME TYPES OF ESDS

The primary design basis and objective of some "ESDs" might be to provide furnace protection (i.e., minimize chances of heater explosions). As a result, the process parameters selected for input to such ESD systems may or may not be compatible with ESDs for which credit could be taken in relief system design.

Correct actuation of an ESD does not necessarily mean the relief load gets reduced to zero at the same instant. Residual heat in the fluid contained in a tower will often be sufficient to maintain flow through the relief valve for some time. Also, the time taken to discharge the vapor inventory from the PRV opening pressure down to the reseat pressure is not negligible.

In some cases, a heater ESD may be designed to close a valve in the burner fuel gas supply fitted with a minimum firing restriction orifice around the valve (note that use of these is discouraged nowadays). Furthermore, the heat capacity of the furnace, which depends on the type of refractory, will be another contributing factor to continuation of the relief discharge. It is imperative that all such factors are satisfactorily taken into account in time-line or dynamic analyses carried out to determine the maximum relief load expected from a given installation.

5.5 MULTIPLE-DEVICE UNITS: ONE OUT OF HOW MANY?

Some proponents of ESD credit taking have stressed that there exists a "very large margin of safety" under the method based on the assumption of the largest device failing. An example of this method is shown in Figures 3 and 4. Very large margins of safety would exist only when the largest device represents a large proportion of the total load. Admittedly, it is most likely to be the case in reality too, when the total number of devices or units connected to the common header is small (say, up to 5).

However, for the overall refinery, the assumption of failure of the mitigating device on the largest *single* individual load in the refinery **regardless of the total number of units attached to the combined header** needs to be investigated further. The "largest-load-failure" method theoretically allows an unlimited number of additional process units to be added to the system provided none of the individual relief sources is larger than the governing load.

Since the basic question is to determine if more than one mitigating device will fail concurrently when an initiating event occurs which causes the maximum combined header loading, the answer depends not only on the probability of failure of the individual devices but also on the total number of such devices. The binomial probability distribution function can be used to describe this case. Let "***p***" represent the on-demand failure probability of a single device and "***q***" the probability of the device acting successfully (therefore, $p = 1 - q$). For the sake of simplifying the analysis, further assume that failure probabilities of all the devices are equal. Then the probability of "***r***" or more concurrent failures from a total of "***n***" devices is given by:

$$P_r^n = \sum_{j=r}^{j=n} \frac{n!}{j! \bullet (n-j)!} \bullet p^j \bullet q^{(n-j)}$$

An example set of calculations derived from some assumed values of variables $\boldsymbol{p}$, $\boldsymbol{n}$ and $\boldsymbol{r}$ is given in Table 2.

From the data in Table 2 it can be seen that the probability of $\boldsymbol{r}$ or more failures from a given number of total devices decreases with increments in $\boldsymbol{r}$. Further, if a certain level of probability can be regarded as negligibly small (for example, say, 10^{-6}), then the number of devices which must be assumed to fail to achieve this negligibly small probability of system failure can also be determined. This is illustrated in Figure 5.

Results from an analysis of probability distribution can also be summarized in the form of data given in Table 3.

It is, therefore, evident that any general rule-of-thumb such as "largest load failure," "largest two loads failure," or "minimum 25% capacity" is, by itself, insufficient to ensure an acceptable level of safety under all situations. The over-riding criterion, therefore, needs to be on the basis of risk estimate derived from quantitative risk analysis.

Table 2. Failure probability distribution

	Probability of r or more failures from a total of n, given $p = 0.05$ and $q = 0.95$, for				
r	$n = 5$	$n = 10$	$n = 15$	$n = 20$	$n = 30$
0	1.00	1.00	1.00	1.00	1
1	2.262e-01	4.013e-01	5.367e-01	6.415e-01	7.854E-01
2	2.259e-02	8.614e-02	1.710e-01	2.642e-01	4.465E-01
3	1.158e-03	1.150e-02	3.620e-02	7.548e-02	1.878E-01
4	3.000e-05	1.028e-03	5.467e-03	1.590e-02	6.077E-02
5	3.125e-07	6.369e-05	6.147e-04	2.574e-03	1.564E-02
6		2.755e-06	5.281e-05	3.293e-04	3.282E-03
7		8.198e-08	3.518e-06	3.395e-05	5.735E-04
8		1.605e-09	1.830e-07	2.857e-06	8.465E-05
9		1.865e-11	7.418e-09	1.979e-07	1.068E-05
10		9.766e-14	2.324e-10	1.134e-08	1.162E-06

6. RELIABILITY OF ESDS

From the discussion included in the preceding sections, it follows that the question of taking credit for ESDs cannot be resolved without taking into account their reliability. Some guidance, therefore, must be given on the desired reliability characteristics of an ESD system. One way to accomplish this would be to provide some information on the minimum acceptable levels of reliability, availability, and maintainability associated with such an ESD system; i.e., the level of integrity of the ESD system.

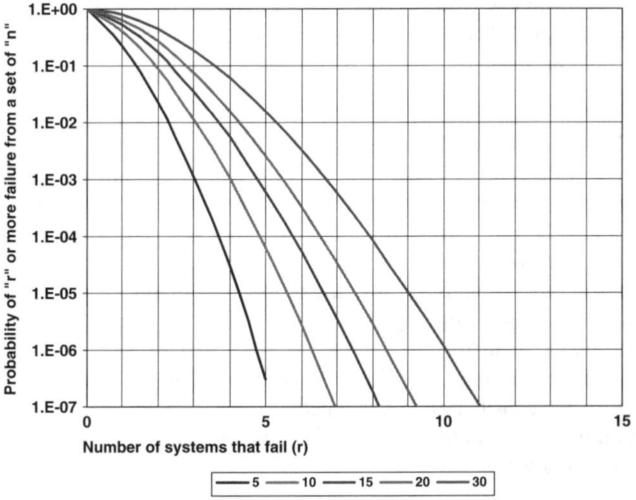

Figure 5. Failure probability distribution "r" failures from a a total of "n" devices

Table 3. Devices assumed to fail

Total number of devices (ESDs) that redice flare load	Number of devices that should be assumed to fail to achieve and extremely low level of overall system failure (10^{-5})
1–4	all
5	4
6–8	5
9–12	6
13–17	7
18–23	8
24–29	9
30–37	10
38–44	11
45–51	12

An easy, workable option from the standpoint of operations management would be to recommend that in each case a detailed reliability analysis be carried out of all instrumentation associated with ESDs for which credit-taking needs to be considered. Such an analysis should be conducted by an approved, qualified consultant. Nonetheless, this is easier said than done.

There are several key issues related to the reliability of ESD systems which need to be addressed by the reliability analyst to the satisfaction of operations management before credit-taking could be accepted.

6.1 HIGH-INTEGRITY ESDS VERSUS RELIABILITY OF PRVS

One claim often made is that in high-integrity systems (i.e., multiple redundant systems) the reliability of the ESD is usually greater than the reliability of pressure relief valves. A PRV tested annually has a *hazard rate* of 4×10^{-3} per year, and a duplicated trip system with weekly testing a *hazard rate* of 5.6×10^{-3} per year as mentioned in Lawley and Kletz (1975) and Lees (1980). From the above, the proponents of ESD credit-taking imply that since ESDs can be designed with high reliability, higher than that achievable with a PRV, therefore credit-taking is justified.

The hazard rate argument for the flare header design is, on the whole, irrelevant. If the decision under consideration was whether a PRV should be installed at a certain process location or not, then the hazard rates can be used as a criterion. The situation can be summed up as follows:

> *If you feel **extremely** confident that the ESD will work and will not permit an overpressure situation to arise, then don't install a PRV.*

If you feel a PRV needs to be installed, then don't undersize it because you feel the probability of it lifting is low due to ESD action. Provide a full-size PRV. See *Kletz (1984).*

If you decide to connect the tail-pipe to the flare header, then don't undersize the flare header because you feel the probability of PRV lifting is low. Provide an adequately-sized flare header.

6.2 RELIABILITY HAS NOT BEEN REALIZED IN PRACTICE

The experience with reliability of ESD systems (even the high- integrity systems) has been varied. With some of the earlier systems, Stewart (1971) reported that: "Sometimes it is shown by the safety assessment that the H.I.P.S. (High-Integrity Protective System) has not achieved its target design specification."

Undoubtedly, reliability of modern ESD systems is higher than that of systems designed in the past. This enhancement is the result of introduction of programmable digital devices, higher reliability of individual components and built-in redundancy and voting arrangements such as "2oo3 (2 out of 3)" and "1oo3." *See* Gruhn (1993) and Beckman (1992, 1993) for safety system reliability terms and calculation methods.

Nonetheless, the promise of high reliability has not materialized in practice. Even those systems that were selected and approved after rigorous FATs (Factory Acceptance Tests) have under-performed upon installation on-site. One major contributing factor is suspected to be poor software reliability.

Beckman (1993) states: "The reliability of the application software is a user responsibility, and it must be tested and validated by the user to ensure that it is free of faults, particularly latent faults (bugs)." Such a statement may find support among the vendors but is not likely to be favored by operating companies, i.e. the users. A significant portion of the responsibility must rest with the vendors. They must ensure the application software is not only free of bugs, but is also robust enough to withstand a certain degree of "rough treatment" in the field.

The operating companies, on their part, must provide accurate information on process variables, generate all deviation scenarios demanding intervention by the ESD system, participate fully in the application software development by the vendor and implement company procedures to carry out regular maintenance and testing of the installed system.

6.3 SPURIOUS TRIPS VIS-À-VIS HUMAN NATURE

Lack of reliable software sometimes results in a significant increase in the rate of spurious trips initiated by an ESD system. The consequential loss of revenue due to these unplanned stoppages can occasionally force an operations manager into taking the rather drastic measure of running the plant with the ESD bypassed. ESDs which are likely to be bypassed or switched over to "inhibit" mode should not be considered for credit-taking in relief header sizing.

One question often asked when designing ESD systems is: "What is an acceptable rate of spurious trips?" Lawley and Kletz (1975) considered a spurious-trip rate of 1.27 trips per year acceptable for the 1oo2 cross-connected high-pressure trip system they had studied. Is this a reasonably acceptable rate for other systems too?

From an end-user standpoint, the answer is given not in terms of spurious-trip rate alone, but in terms of its relationship to the ESD demand rate (i.e., frequency of process parameter deviations requiring ESD intervention). This is because the "perceived" reliability of the ESD system is equally important to the user. When next time the ESD shuts your plant down, are most of your operators convinced that it is a spurious trip? If yes, then they are likely to have little respect for the system installed. Such an ESD will be bypassed at the first opportunity.

A spurious-trip rate of 1 or 2 trips per year is acceptable only if the demand rate is also of the same order of magnitude. In this situation, operators will not regard the ESD system as a mere nuisance because each trip will be perceived as either being spurious or genuine, with equal probability.

If the demand rate is low, say 1 in 10 years (0.1 trips per year), then a spurious-trip rate of 1 per year may turn out to be unacceptably high. Additional cost of building a higher level of redundancy in the system to achieve a lower spurious-trip rate to match the demand rate would be justified here.

6.4 HIGH MTTR (MEAN TIME TO REPAIR)

Another problem created by poor software reliability is that the on-site times taken to diagnose and eliminate software faults have been longer than those predicted at the design stage. Operating companies do not possess software expertise of a level necessary to carry out quick and effective maintenance on these systems. They have to rely on vendor specialists to provide this service on call. The higher MTTRs mean that ESDs remain in a bypassed mode for significant durations of time. Once again, taking credit for such ESDs in relief system design is not advisable.

7. CONCLUSIONS AND PATH FORWARD

While there is no objection to the concept of taking credit for ESDs in the relief and flare header sizing and design, each application needs to be individually scrutinized to ensure plant safety is not compromised. Special attention needs to be given to potential impact on other units sharing the relief header.

The current API recommended practices (RP-520 and RP-521) appear to be silent on this issue. There are no other internationally recognized standards, codes of practice or guidelines which specifically permit taking credit for ESDs in relief system design. There is a need to initiate a dialog with the API and/or hold further discussions under the aegis of some other recognized body, such as the NPRA (National Petroleum Refiners Association), for guidelines to be established and placed on record.

Confirmation should be sought from OSHA that taking credit for ESDs in relief system design does not constitute any violation of the intent of OSHA 1910.119 Paragraph (d)(3)(H)(ii) which states that the employer shall document that equipment complies with recognized and generally accepted good engineering practices; the statement being applicable to relief system design and design basis per Paragraph (d)(3)(D) of the OSHA regulation.

In addition to giving approval to the concept, any future internationally recognized standards or codes must incorporate detailed guidelines on the types of ESDs for which credit-taking would be permissible. These should include reliability targets for high-integrity ESDs or a directive to conduct detailed reliability analyses of such systems.

From an operating company management standpoint, ESD credit-taking is not advisable before this practice is clearly recognized and/or approved in an international standard or code. Lack of such a standard leaves engineers and managers, who permit the design and installation of relief systems taking credit for ESDs, vulnerable to the possibility of adverse comment from official investigations of any loss or injury incidents involving relief system sizing. In a court of law, it would place them in a weak defensive situation.

Even if ESD credit-taking becomes an "approved" practice, operating company management are advised to exercise caution. An extremely risk-aversed inter-unit spacing in a well laid-out refinery is a valuable asset. It presents a natural barrier to the insurer's EML calculations. Do not erode this barrier by opting for "savings" in the relief header and flare system costs.

Designers and suppliers of ESD systems need to prove that the on-stream availability and reliability of their systems, so readily demonstrable on paper or in FATs, can be reproduced on-site, and are practically immune to environmental factors arising from geographical location or the work ethos of the client company. In a market place of ever-shrinking refining margins, the incessant pursuit of cost effectiveness in all decision-making is not merely a desirable activity, but the key to survival. However, cost effectiveness must never be misconstrued to mean indiscriminate cost-cutting.

LITERATURE CITED

Al-Ansari, Isa G. A., *BAPCO's Risk Assessment Programme*, Bahrain Society of Engineers, 1990.

American Petroleum Institute (API), *Management of Process Hazards*, Recommended Practice 750, Washington D.C., 1990.

American Petroleum Institute (API), *Guide for Pressure-Relieving and Depressuring Systems*, Recommended Practice 521, Fourth Edition, Washington D.C., March 1997.

American Petroleum Institute (API), *Sizing, Selection, and Installation of Pressure-Relieving Devices in Refineries, Part I. Sizing and Selection*, Recommended Practice 520, Seventh Edition, Washington D.C., January 2000.

American Petroleum Institute (API), *Sizing, Selection, and Installation of Pressure-Relieving Devices in Refineries, Part II. Installation*, Recommended Practice 520, Fifth Edition, Washington D.C., August 2003.

American Society of Mechanical Engineers (ASME), *Boiler and Pressure Vessel Code, Section VIII, Division 1, Parts UG-125 to 136*, Washington D.C., 1992.

Beckman, Lawrence V., *How reliable is your safety system?*, Chemical Engineering, issue January 1992 (pp 108–114).

Beckman, Lawrence V., *More on safety systems, Letters to the editor*, Hydrocarbon Processing, issue December 1993 (p 35).

British Standards Institution (BSI), *Unfired fusion welded pressure vessels, BS-5500,* London, U.K., 1976.

Cassata J.R., Dasgupta S., Gandhi S.L., *Modeling of tower relief dynamics*, Hydrocarbon Processing, Gulf Publishing, Houston, Texas, U.S.A., issue October 1993 (pp 71–76).

CCPA, *Responsible Care Codes of Practice*, The Canadian Chemical Producers' Association, Ottawa, Ontario, 1989.

Center for Chemical Process Safety (CCPS), *Guidelines for Hazard Evaluation Procedures (Second Edition)*, American Institute of Chemical Engineers, New York, N.Y., 1992.

Center for Chemical Process Safety (CCPS), *Guidelines for Chemical Process Quantitative Risk Analysis*, American Institute of Chemical Engineers, New York, 1989.

Center for Chemical Process Safety (CCPS), *Guidelines for Process Equipment Reliability Data*, American Institute of Chemical Engineers, New York, 1989.

Center for Chemical Process Safety (CCPS), *Guidelines for Engineering Design for Process Safety*, American Institute of Chemical Engineers, New York, 1993.

Center for Chemical Process Safety (CCPS), *Guidelines for Safe Automation of Chemical Processes*, American Institute of Chemical Engineers, New York, 1993.

Coker, A.K., *Size Relief Valves Sensibly – Parts 1 and 2*, Chemical Engineering Progress, August 1992 (pp 20–27), November 1992 (pp 94–102).

Goyal R.K., *Probabilistic Risk Analysis – Two Case Studies from the Oil Industry*, Professional Safety, July 1986, American Society of Safety Engineers, Des Plaines, Illinois, 1986.

Goyal R.K., *Practical examples of CPQRA from the petrochemical industries*, The Institution of Chemical Engineers (IChemE), Process Safety and Environmental Protection, Transactions of the IChemE, Vol 71, Part B, Rugby, England, U.K., May 1993.

Gruhn, P., *Safety system performance terms: clearing up the confusion*, Hydrocarbon Processing, February 1993 (pp 63–66).

Hall, Stephen M., *Size and Design Relief Headers*, Chemical Engineering Progress, March 1993 (pp 117–122).

Health and Safety Executive (HSE), *The fires and explosion at BP Oil (Grangemouth) Refinery Ltd.* —A report of the investigations by the HSE into the fires and explosion at Grangemouth and Dalmeny, Scotland, 13 March, 22 March and 11 June 1987, Her Majesty's Stationery Office, London, U.K., 1989.

Kletz, Trevor A., *Myths of the Chemical Industry*, The Institution of Chemical Engineers, Rugby, U.K., 1984.

Lawley, Herbert G. and Kletz, Trevor A., *High-Pressure-Trip Systems For Vessel Protection*, Chemical Engineering, May 1975 (pp 81–88).

Lees, Frank P., *Loss Prevention in the Process Industries*, Volumes 1, 2, 3, Butterworth, London, U.K., Second Edition 1996.

Leung, Joseph C., *Size Safety Relief Valves for Flashing Liquids*, Chemical Engineering Progress, February 1992 (pp 70–75).

Niemeyer C.E. and Livingston G.N., *Choose the Right Flare System Design*, American Institute of Chemical Engineers, Chemical Engineering Progress, New York, N.Y., December 1993 (pp 39–44).

NFPA, *Codes and Standards*, The National Fire Protection Association, Quincy, Massachusetts.

OSHA 29CFR Part 1910.119, *Process Safety Management of Highly Hazardous Chemicals*, Federal Register Vol. 57, No. 36, Washington D.C., February 24, 1992.

OSHA 29CFR (Code of Federal Regulations) Part 1910.119, *Notice of Proposed Rulemaking: Process Safety Management of Highly Hazardous Chemicals*, Federal Register Vol. 55, No. 137, Washington D.C., July 17, 1990.

Papa, Donald M., *Clear Up Pressure Relief Valve Sizing Methods*, Chemical Engineering Progress, August 1991 (pp 81–83).

Stewart, R.M., *High integrity protective systems*, The Institution of Chemical Engineers (IChemE), IChemE Symposium Series No. 34 (pp 99–104), Rugby, England, U.K., 1971.

BACKGROUND TO AND EXPERIENCE USING THE NII'S NEW SAFETY ASSESSMENT PRINCIPLES – LEARNING FOR THE HIGH HAZARD SECTOR?

Dr Andy Trimble
HM Superintending Inspector (Nuclear Installations), HM Nuclear Installations Inspectorate, Redgrave Court, Merton Road, BOOTLE L20 7HS

In January 2007 HSE's Nuclear Installations Inspectorate (NII) had redrafted and reissued its safety assessment principles (SAPs) following more than 10 years of use for the previous version. This paper reviews the drivers and outcomes from this exercise and also reviews experience in the first year of use in the nuclear chemical plant sector. We also review the way forward with our subsidiary technical assessment guides (TAGs) which complement the SAPs and further assist safety assessment in the nuclear sector.

The inspectorate now even better placed to carry out its work in a consistent and targeted manner. The clarity in regulatory expectation brought by the new SAPs has been welcomed in many quarters and our inspectors have also welcomed the increased clarity they bring. Overall, we believe that, once complete, the package of SAPs and subsidiary technical guides are as good as any in the world and form a sound basis for the inspectorate to move forwardand meet the challenges that face it. The lessons learned will also apply, selectively, in many other parts of the high hazard industries.

BACKGROUND AND INTRODUCTION

In the nuclear regulatory regime, the Health and Safety Executive's (HSE) Nuclear Installations Inspectorate (NII) does not specify what should and should not be in a safety case [12]. However, the regulatory goals are set out in our Safety Assessment Principles (SAPs) [e.g. 1]. These Principles were originally written for nuclear plant in design and they were also used to inform periodic safety case reviews required under licence conditions.

We decided, in the light of the experience gained over the last decade or so, to review our Principles and to make them more relevant to the environment in which we now regulate. It is important to note that the initial reviews showed that most of the original Principles are still relevant but could be made clearer in their application to the wide variety of plant we now regulate. We had already addressed some omissions in our subsidiary guidance [e.g. 9,10].

Whilst these new SAPs [5] (SAPs 06) were written explicitly for the nuclear sector, there are many parts that take on board thinking from the non nuclear high hazards sector and other parts that complement the high hazard sector thinking. Therefore, companies are encouraged to review their corporate safety processes against SAPs 06 (bearing in mind they are set as regulatory guidance) and, where appropriate, incorporate the applicable thinking into their own guidance and standards.

DRIVERS

The SAPs have evolved over time:

- 1979: first produced for nuclear reactors
- 1983: first produced for nuclear chemical plant
- 1988: the 1979 version modified following Sizewell B inquiry
- 1992: combined version produced, taking account of the Tolerability of Risk [2] framework

However, the 1992 SAPs have remained unchanged until 2006. Although they have needed expert interpretation from time to time, they have served us well in setting regulatory expectation for nuclear facility safety cases.

Increasingly, the 1992 SAPs were being used for assessing safety cases that had design elements that were constrained by what existed – for example, decommissioning safety cases [3]. In addition experience showed that safety thinking was developing and the SAPs were not giving best advice to our inspectors. Therefore, the NII has devoted scarce resource to this work to reap the long term benefit.

The prime drivers for change and sources of reference were:

a) IAEA standards: NII's policy is that our guidance will be consistent with the international atomic agency standards which provide an international benchmark. These IAEA documents have been evolving and continue to do so. This has been a driver to try and make the latest SAPs version easier to change by making them web based to avoid republishing a paper document. This does not mean that our SAPs are an attempt to clone the IAEA guidance – rather they reflect the safety thinking in the IAEA standards.
b) Increased emphasis on decommissioning: With the advent of the Nuclear Decommissioning Authority the level of decommissioning activity has risen significantly and we considered that there was sufficient experience to make it worthwhile incorporating our latest thinking into the new SAPs.
c) Aspects that had been part of regulatory good practice but not yet incorporated into SAPs:
   Leadership and Management of Safety
   Regulatory Assessment of Safety Cases
   Radiation Protection
   Accident Management and Emergency Preparedness
   Radioactive Waste Management

Decommissioning
Control and Remediation of Radioactively Contaminated Land

d) Improved presentational consistency: The basis of any safe facility is sound engineering (reflecting the Good Practice thinking in Reducing Risks, Protecting People [4] - R2P2) and so the engineering has been brought forward in the layout of the new SAPs. Similarly, people are key to any safe operation and so the people and organisational aspects, embedded in the leadership and safety case SAPs have also been brought forward in the layout. Finally there were minor internal inconsistencies between sections, for example, there were concepts in one part of the 1992 SAPs that were principles in one section and comparable concepts were supporting guidance in another. Whilst this did not detract from the value of the concepts, they had the potential to send inconsistent messages. The overall editing in the new SAPs has been a difficult and demanding task but we are convinced the outcome is now consistent enough and sends a much more coherent message than before.

Also the following occurred during the process of revision and were incorporated into the thinking as far as possible:

e) WENRA reference levels: NII is also committed to being consistent with the reference levels set by the Western European Nuclear Regulators Association. Although these apply only to existing power reactors and waste there are elements that represent international regulatory consensus on good practice and so, where appropriate, they have been incorporated into SAPs 06.
f) Potential new power reactor build: With the government's intention that the UK should build new nuclear power reactors there was a need to make our SAPs more transparent to an international audience of power reactor vendors who may not be familiar with the UK regulatory regime.

It is important to note that there has been no significant change in the underpinning law and, in particular, the so far as is reasonably practicable (better known as ALARP) obligation still remains. Also, in line with R2P2, ALARP is more than cost/benefit analysis. In most cases relevant good practice will be more important. Because SAPs06 reflect the most up to date thinking, they also reflect good practice. Therefore, part of good safety practice will be found in SAPs.

Similarly, because this work was primarily about incorporating present practice and consolidation, SAPs 06 did not automatically make current safety cases out of date. However, what we do expect is that licensees and other relevant duty holders will review their own criteria – bearing in mind the intent of SAPs – to see if there are improvements that can reasonably be put in place. We do not expect to see wholesale changes to licensees' safety documentation, either in the underpinning guidance or the safety cases which they support.

It is also important to stress that SAPs are assessment guidance for NII's assessors (as well as being adopted or recognised by other regulatory stakeholders in their spheres of responsibility). Therefore, they should not be used as design guides or to underpin operations. Duty holders are expected to develop their own criteria.

REVISED SAPS STRUCTURE

The Document structure is:

Introduction
Fundamental Principles
Leadership and Management for Safety
The Regulatory Assessment of Safety Cases
The Regulatory Assessment of Siting
Engineering Principles
Radiation Protection
Fault Analysis
Numerical Targets and Legal Limits
Accident Management and Emergency Preparedness
Radioactive Waste Management
Decommissioning
Control and Remediation of Radioactively Contaminated Land
Glossary, Annex etc

Following a period of extensive public engagement SAPs 06 have been published on the web (now the definitive version). Also published were:

- Resolution of comments from public engagement [11]
- Table linking 2006 and 1992 SAPs [8]
- Explanatory Note on Numerical Targets and Legal Limits [7]
- NSD Guidance on Demonstration of ALARP (T/AST/005 revised [6])

There were, among other things, commitments to carry forward some comments and some of the principles in the 1992 SAPs into supporting technical assessment guides (TAGs). There are processes in place to sentence these and the outcome will be published on the web.

It is important to recognise that ALARP is driven by HSE corporate thinking and T/AST/005 interprets this for then nuclear sector where there have been very specific challenges to interpretation over the years. So this document incorporates such thinking reflected in SAPs.

What is more significant from a usability point of view is the more consistent high level definition of a Principle. These are highlighted in the text and numbered with the relevant good practice below. This lends a clarity and consistency to the document that would be difficult to achieve in any other way. This clarity will be continued into TAGs to form a coherent and consistent suite of documentation covering the regulatory assessment technicalities either in greater detail or to give a different perspective and logic on how SAPs relate to each other and how they might be applied.

Also given the new way of presenting the Principles at a reasonably high level, then the amount of detail that can be included in the good practice and leave a document of manageable size was limited. Thus a comparison of the new SAPs with the 1992 version will show a number of omissions. For example, principle P45 which dealt with plant damage will be covered in TAGs where it can be more fully expounded for the range of facilities NII regulates. If necessary, there will be separate parts of TAGs covering certain generic plant, typically, power reactors, defence facilities and chemical plant.

The 1992 SAPs established the link between Tolerability of Risk and SAPs. This sent the unfortunate message in some quarters that the principal need in a safety case was to show compliance with the numerical analysis even though the introduction stated clearly this was not the intention. Therefore, in SAPs 06, we adopted a structure that better reflects the importance of the various aspects of safety management. However, it is important not to over compensate and neglect the rest of the Principles. In all cases a balance appropriate to operations and facilities under consideration needs to be struck. Assessment inspectors are always encouraged to take this holistic view when carrying out their work.

Although the new SAPs have expanded to 139 pages from 47, most of this expansion is due to the inclusion of material formerly elsewhere (e.g. [9,10,12]) and to greater clarity because we better have shown the underpinning thinking and defined many of the terms used. This is of particular importance to those who may be entering the UK nuclear industry for the first time as most of our established licensees have an understanding of our regulatory expectations for their operations. This is relevant not only to potential new build but also to companies entering the decommissioning field engendered by the Nuclear Decommissioning Authority's bidding process. To further aid understanding we have explained the underpinning thinking for the principles. Most often this is at the *"dialogue"* at the beginning of each relevant section e.g. the introduction to fault analysis at paragraphs 496 to 503. However, it is important not to overlook the underpinning philosophy in the introduction. There is much of value here including a closing statement:

> *"The principles are written bearing in mind the content of safety cases likely to be submitted to the NII. However, dutyholders may wish to put forward a safety case that differs from this expectation and, as in the past, the inspector will consider such an approach. In these cases the dutyholder is advised to discuss the method of demonstration with NII beforehand. Such cases will need to demonstrate equivalence to the outcomes associated with the use of the principles here, and such a demonstration may need to be examined in greater depth to gain such an assurance. An example of such a situation is the greater use of passive safe concepts."*

However, there is compelling guidance that, in essence, says [13] – bearing in mind the different purposes for which they are intended, there should be the minimum of differences between licensees' safety criteria and the NII SAPs. Plainly, it is each parties' interest to avoid extra work and the resulting delays if the outcome can be achieved in a more productive, mutually understood way. Thus NII often has such understandings with existing licensees to ensure that, when there are differences, then these are mutually understood. Establishing such understandings can be a long and challenging process.

Also in the introduction is a section on proportionality – an HSC policy imperative. Unlike the IAEA which publishes different documents for different facilities, HSE publishes general documents such as SAPs and applies proportionality. Thus the extent and rigour expected in any safety case will be in broad proportion to the underpinning hazard, among other things. This in turn is related to the harm potential of the materials being handled and the conditions under which they are handled – which is a way of defining hazard. In other words, the hazard is a function of the radio toxicity, mobility and driving force(s) under the plant conditions being considered and is consistent with legal interpretation of the Health and Safety at Work Act [17]. This policy has driven NII assessment for many years [14].

Although we use the word safety, this term includes regulating waste management on licensed sites and the final sections of SAPs demonstrate this. To ensure minimal regulatory overlap we have involved the Environment Agency in our development process and continue to do so as we develop the supporting TAGs. This should mean that the expectation from regulators should be consistent and minimise the regulatory burdens on industry.

Finally, we have clarified links to the law. This can be seen, for example, in references to nuclear site licence conditions and to dose limits from the Ionising radiations regulations [15]. In particular, some of the basic safety levels (BSLs – the levels our policy say should not normally be exceeded and equate broadly to the limit of tolerability from R2P2/TOR [4]) in the targets section have a designation BSL(LL) indicating their link to the law. This clarity helps inspectors to know when to insist on further improvements and how stringently they should pursue these. It also makes a more consistent approach to safety that is more transparent to all stakeholders.

SAPS AND TAGS

Although we have already published our TAGs (e.g. [6]) they are now inconsistent with the new SAPs (although that does not detract from the guidance in them). NII has now embarked on a process of revision. The content, approach and timetable for bringing these into the public domain is being developed and this is planned to produce a coherent and more comprehensive suite of documents that will also help our stakeholders understand our expectations.

The outline plan is:

By mid 2008 – first tranche of TAGs to be on the web for comment. These will include most of those relevant to new build as this is seen as the area of most pressing need.

By 2009 – all TAGs to be drafted and ongoing maintenance work to be undertaken based on experience and feedback.

This programme will be driven by the priorities at the time. Although this appears an easily achieved timescale, such tasks are not simple and our experience with the SAPs shows that being consistent across a suite of documents is not simple – especially when different TAGs may be drafted by several different authors working in separate technical disciplines. Pragmatically, we may need to tolerate minor inconsistencies to get the benefit of having a set of published documents. In line with SAPs 06 we would expect to revise this suite on the web and the definitive versions will be the electronic web documents.

In practice, some of the more significant TAGs (e.g. [14]) are in a late stage of drafting and may well be put out for comment earlier. As with the SAPs we are involving other regulatory agencies including the environment agency and the defence nuclear safety regulator (DNSR). As before, this should optimise regulatory demands on duty holders and thereby minimise regulatory compliance costs because these regulatory expectations should be more consistent.

The NII is prepared to put scarce resource to this work to gain the longer term benefits. However, this in itself will almost certainly generate further work with licensees to ensure their criteria are consistent enough even though minimal change may be required.

EXPERIENCE

It will have become obvious that since SAPs 06 have not significantly changed our guidance and thinking, that the effect on our work should be minimal. What has happened is that new inspectors, who usually start in assessment roles, now have much clearer and consistent guidance which assists their rapid development. This is crucial as the inspector age profile is skewed towards retirement and the need for knowledge management becomes ever more acute. As a result, NII is recruiting to address this and new inspectors can look towards a career in regulation with greater certainty of what is expected from them technically.

Our licensees, who were consulted about the SAPs 06, have also been reviewing their criteria to help meet the compelling advice to avoid inconsistencies between that and our SAPs. At least one is taking the opportunity to revisit their understandings with us. This work will underpin the way we regulate. If licensees present safety cases that meet our expectations or are to a comparable standard, then the assessment work is made a great deal easier and safety is better assured (provided the safety cases are implemented as intended). There is no safety benefit for either party to have long drawn out debate about methods and processes that do not deliver safety in operations. Thus, we are piloting an assessment approach that considers not only the paperwork but involves assessment inspectors taking a much higher role in safety case implementation. This is not something new but a change of emphasis to promulgate good practice more widely.

DISCUSSION

It is important to summarise the role of SAPs:

SAPs are	SAPs are NOT
Regulatory safety goals	Design criteria
Regulatory assessment guides	HSE guidance to dutyholders
Guides for regulatory judgments	Mandatory standards
Assistance in judging ALARP	To be met unconditionally
To be considered holistically	To be considered separate good practices
For regulatory use	To be adopted wholesale

Our expectation (as in the 1992 SAPs) is that modern facilities should have little difficulty in meeting these regulatory expectations. However, older facilities built to earlier standards will be judged against SAPs 06 and our expectation is that licensees should be able to demonstrate they have done (or will do) all that is reasonably practicable to reduce potential harm to people. SAPs 06 will be used as a benchmark as has always been the case. The clarity of this set of SAPs should help licensees and potential licensees understand our regulatory expectation.

There is one area where there may seem to be a contradiction. This is where risks increase, usually temporarily, in order to gain an overall benefit. This is seen most often in decommissioning [3] where the hazard needs to be managed by removing the facility's radioactive contents. Such actions can represent a reduction in safety on such ageing facilities but are a necessary part of achieving a long term stable state for their contents. Such situations are now better understood and where activities to reduce long-term risks mean risks rise in the short term, efforts should be made to carry out the activities such that the risks are minimised both in magnitude and time. It is important to stress (SAPs para 637):

> *"High risks that would exceed BSLs if evaluated as continuous risks should be avoided except in special circumstances. These circumstances should be justified in advance. They may include situations not originally foreseen in the design of the facility, or which are unavoidable because of the need to increase risks for a short time to reach a safer state in the long term."*

One theme that runs through this paper has been the holistic view that inspectors take in doing their assessment. This comes out very strongly in SAPs:

> *"Priority should be given to achieving an overall balance of safety rather than satisfying each principle or making an ALARP judgment against each principle. The principles themselves should be applied in a reasonably practicable manner. The judgment using the principles in the SAPs is always subject to consideration of ALARP."*

This simply acknowledges that engineering and management of safety are a matter of expert judgement to optimise the compromises that must be made to achieve a safe, workable, affordable environmentally friendly operation that delivers for the dutyholder. Inspectors realise that they need to judge whether or not such a balance has been struck.

Nuclear installations are in the high risk category in the Management of Health and Safety Regulations Approved Code of Practice (ACoP) [16]. However, they are not alone. Therefore, it is a logical deduction that other facilities may benefit from the nuclear experience now embedded in these new SAPs. Therefore, comparable high hazard industries are invited to review their corporate guidance and standards as a learning exercise so that they may be better able to demonstrate ALARP. Plainly, as with nuclear installations, not everything applies to every facility, discrimination should prevail.

POTENTIAL APPLICATIONS

As an example of how some of the concepts might be applied both in the nuclear and non nuclear sectors, consider the "technical" principles. These can be broken down into a number of broad categories [14]:

a. Design Basis accident analysis (DBAA)
b. Probabilistic safety analysis (PSA sometimes known as QRA)
c. Severe accident analysis (SAA)
d. Good Engineering Practice (GEP)
e. Waste Management

The first three are complementary forms of fault analysis. Dealing with each of these broad areas in turn:

DBAA: is a robust demonstration of fault tolerance. It links directly to the engineering principles which call for a preferred series of responses to faults. These vary from designs that are inherently safe to those that may require operator intervention in the fault sequence. The important feature of DBAA is that any uncertainty is allowed for by conservatism. Often this conservatism is in the input data and requires expert judgments about the degree of conservatism appropriate to any particular case. DBAA is concerned with faults with larger harm potential and not normally with more minor events. This methodology has much in common with Layers of Protection Analysis (LOPA) (although the output is different) which is already a well accepted methodology in the high hazard industries. In the nuclear sector DBAA is the bedrock of fault analysis and is used to derive the operating parameters for plant control.

PSA: The main purpose of PSA is to demonstrate a balanced design and it may also show that risks are minimised. The great strength of PSA is this overview. It is not covered by DBAA which deals with faults on a fault by fault basis. Undue reliance should not be placed on the numbers produced by PSA. These numbers are usually rather uncertain and so, while they are very useful in comparative terms, they must be used with caution as a definitive quantification of the overall risks from the operation considered. PSA is usually carried out using best estimate data.

SAA: A severe accident is one which is not necessarily expected in a plant lifetime but has the potential for high doses or environmental damage. It is not necessary for these doses or environmental damage to be realised. The prime difference between DBAA and SAA is in the way that data is used. SAA is carried out on a best estimate basis and may well be bounded by the DBAA if the level of conservatism is high. However, a sound understanding of the underlying phenomena during such accidents avoids the need for introducing unnecessary conservatism and hence unfruitful expenditure. The main aim of SAA is to provide an input to emergency planning and to identify reasonably practical improvements that can be implemented at reasonable cost. There may be cases where this may influence the options at the design stage of a project.

GEP: In every industry there are both pressures to reduce costs and increase cost effectiveness. However, most companies and most industries set basic standards below which any design should not fall. This ensures that for harm potentials smaller than would be covered by DBAA, the learning experience of the company and/or the industry are taken into account. Often GEP is embodied in design manuals or company standards. Quality engineering should not stray outside this standard. Nuclear safety cases have a significant section on engineering substantiation to demonstrate GEP has been met and that then engineering will continue to deliver the appropriate safety function for the foreseeable future (usually taken by convention at 10 years minimum). This has similarities with vessel inspections and lifting inspections carried out in the non nuclear sector - which are often not possible on nuclear process plant as access can be highly restricted due to radiation.

Waste Management: There are major additional external constraints as well as those required for safety. Much regulation is concerned with implementing government policy and good practice. Plainly, this also reflects public opposition to ill considered waste accumulation and storage (disposal is dealt with under Environmental Legislation administered by the Environment Agencies). This is becoming and increasingly large part of the work in the nuclear sector.

These aspects are often found in high hazard industry although the "labels" may be different. However, it is good practice to critically examine corporate standards against a range of good practice and such concepts are one input.

CONCLUSIONS

The inspectorate now even better placed to carry out its work nationally and internationally in a consistent and targeted manner. The clarity in regulatory expectation brought by the new SAPs has been welcomed in many quarters and our inspectors have also welcomed the increased clarity they bring. Overall, we believe that, once complete, the package of SAPs and subsidiary technical guides are as good as any in the world and form a sound basis for the inspectorate to move forward and meet the challenges that face it. The lessons learned will also apply, selectively, in many other parts of the high hazard industries.

REFERENCES

1. Safety Assessment Principles for Nuclear Plants HSE 1992 ISBN 0 11 882043 5
2. The tolerability of risk from Nuclear Power stations HMSO 1988 ISBN 0118839289
3. G A Trimble, SUMMARY OF THE CURRENT UK POSITION ON DECOMMISSIONING SAFETY CASES AND CONTROL OF OPERATIONS *Proc HAZARDS XIX* 792ff
4. Reducing Risks, Protecting People (R2P2) HSE 2001 http://www.hse.gov.uk/dst/r2p2.pdf
5. Safety Assessment Principles for Nuclear Facilities 2006 Edition HSE http://www.hse.gov.uk/nuclear/saps/saps2006.pdf
6. T/AST 005 Issue 3, NSD GUIDANCE ON THE DEMONSTRATION OF ALARP http://www.hse.gov.uk/foi/internalops/nsd/tech_asst_guides/tast005.pdf

7. Numerical targets and legal limits in Safety Assessment Principles for Nuclear Facilities, An explanatory note HSE December 2006 http://www.hse.gov.uk/nuclear/saps/explanation.pdf
8. SAPS 206 Edition – Stakeholder comments resolution http://www.hse.gov.uk/nuclear/saps/publiccomments.pdf
9. Management of radioactive materials and radioactive waste on nuclear sites http://www.hse.gov.uk/nsd/waste1.pdf
10. Decommissioning on nuclear licensed sites http://www.hse.gov.uk/nsd/decomm1.pdf
11. 1992 to 2006 SAPs Cross-reference Table http://www.hse.gov.uk/nuclear/saps/cross-reference.pdf
12. TECHNICAL ASSESSMENT GUIDE T/AST/051, GUIDANCE ON THE PURPOSE, SCOPE AND CONTENT OF NUCLEAR SAFETY CASES http://www.hse.gov.uk/nsd/tast051.htm
13. Sizewell B Public Inquiry Report by Sir Frank Layfield, Dept of Energy
14. T/AST/006 Issue 03, DETERMINISTIC SAFETY ANALYSIS AND THE USE OF ENGINEERING PRINCIPLES IN SAFETY ASSESSMENT http://www.hse.gov.uk/foi/internalops/nsd/tech_asst_guides/tast006.pdf
15. Work with ionising radiation. Ionising Radiations Regulations 1999. Approved Code of Practice and guidance L121 HSE Books 2000 ISBN 0 7176 1746 7
16. Management of health and safety at work. Management of Health and Safety at Work Regulations 1999. Approved Code of Practice and guidance L21 (Second edition) HSE Books 2000 ISBN 0 7176 2488 9
17. R v Board of Trustees of the Science Museum All England Law Reports. 10 Sep. 1993, part 3, 853–861.

ACKNOWLEDGMENT AND DISCLAIMER

Thanks go to many in HSE's Nuclear Installations Inspectorate for help and advice in developing this paper, in particular Mr G J Vaughan. The opinions here are those of the author. No part of this paper should be taken as definitive interpretation of HSE or NII policy, the law, or their application.

EXPLOSION PROCESSES AND DDT OF VARIOUS FLAMMABLE GAS/AIR MIXTURES IN LONG CLOSED PIPES CONTAINING OBSTACLES

Christian Lohrer[1*], Christian Drame[1], Detlef Arndt[1], Rainer Grätz[1], Axel Schönbucher[2]
[1] Federal Institute for Materials Research and Testing (BAM), Berlin
*Corresponding author: tel.: +49-30-8104-3249; fax: +49-30-8104-1217,
E-mail address: christian.lohrer@bam.de
[2] University of Duisburg-Essen

In this work results of experimental investigations towards explosions processes in long pipe systems are presented. The experiments were carried out in pipes with diameters of 0.159 m as well as 0.200 m, and overall lengths of up to 23 m. Three flammable gases mixed in air were used, including propane, ethene, and ethyne. All mixtures were set up to a similar maximum experimental safe gap (MESG) of about 0.94 mm – 0.97 mm (explosion group IIA). It was observed that the maximum pressures as well as the flame speeds differed significantly. The more the mole fraction of the investigated gases in air had to be reduced to achieve a similar MESG (compared to propane/air in stoichiometric composition), the less severe was the explosion.

Turbulence inducing elements enhance the heat and mass transfer in reactive flows. Therefore, the influence of various baffles (blockage ratios of 36%, 51%, 64%, 77%, and 91%) installed into the pipe system on the explosion characteristics was investigated for a stoichiometric propane/air mixture and an industrial mixture consisting of hydrogen, carbon monoxide, carbon dioxide, nitrogen, and air. As a result of the use of baffles, deflagration to detonation transitions (DDT) occurred in the pipe system. Absolute pressures of more than 100 bar joined by supersonic flame speeds of >2000 m/s were determined by piezoelectric sensors and photodiodes.

1 INTRODUCTION

Explosions and fires still cause serious losses in the process industries, followed by accidental releases, [Mannan, 2005], and [Kleiber, 2006]. The industrial principles of protection against gas explosions contain in a first step the avoidance of an explosive atmosphere and effective ignition sources. If neither of these methods nor a combination of both ensures safe handling and processing, constructional explosion protection measures in a second step reduce the impacts of explosions to an acceptable level, [Council Directive 1999/92/EC, 1999], [Bartknecht, 1993], and [Grätz and Förster, 2001].

Within the concept of explosion isolation, flame arresters, if wisely installed into the system, separate the part of the explosion from the tract, which needs to be protected. Therefore, flame arresters must be tested against flame transmission to the specific explosion group of the explosive mixture in accordance with IEC 60079-0 [2004]. The explosion group is determined by the maximum experimental safe gap for an explosive mixture (MESG). It represents the maximum gap of a joint which prevents any transmission of an explosion during multiple tests due to the effects of quenching, [IEC 60079-1, 2002]. The experimental safe gap of a gas mixture is not a constant value, but changes amongst others with the mole fraction of the flammable gas in air. Additionally, the efficiency of flame arresters is influenced by the local transition point from deflagrative to detonative reaction mechanism (DDT). The occurrence of the transition point depends on the length-to-diameter ratio (L/D) and the orientation of the pipe (vertical/horizontal), initial pressure p_0 and temperature T_0, flammable gas concentration X, ignition energy, wall roughness, and occurrence of turbulence inducing elements. According to EN 12874 [2001], EN 13237 [2003], and Grätz and Förster [2001] a deflagration is defined as an explosion propagating at subsonic velocity. The flame speed is smaller than the speed of the shock waves, which propagate at sonic speed. The reaction front at that time is constantly accelerating, and the time to the maximum pressure as well as the time of the pressure drop to the end value is comparatively long. In contrast, a stable detonation is an explosion propagating at supersonic velocity and characterised by a shock wave. Both the flame front and the shock wave are coupled with similar propagation speeds, which remain at a constant value. A stable detonation is furthermore characterised by an almost vertical pressure rise (up to 20 times of the initial pressure), followed by a fast pressure drop to the end value. The transition zone lies between the deflagration and the stable detonation regimes. It includes fast deflagrations, DDT, and overdriven detonations. Latter show pressures of >100 bar (abs.), but the pressure drop to the end value occurs slower than during a stable detonation.

In recent years, many investigations towards explosion processes of flammable gas/air mixtures have been carried out and reported in literature. Bartknecht [1993] reports of experiments with methane/air mixtures in a 1 m^3 vessel with central ignition. He found maximum explosion pressures p_{max} and maximum rates of pressure rise at stoichiometric compositions of the mixtures, with increasing impacts at higher initial turbulence intensities (Tu; ratio of the root-mean-square of the velocity fluctuations and the mean velocity). In addition, it was shown that the p_{max} decreased with an increasing surface-to-volume ratio of the investigated vessels at central ignition. Razus et al. [2007] investigated the influence of the mixture composition and the ignition point on the explosion characteristics of propylene/air mixtures in small spherical and cylindrical vessels. Maximum rates of pressure rise were determined at the stoichiometric composition. Due to the effects of heat loss to the cold vessel walls, the maximum rates of pressure rise at central ignition in the cylindrical vessel exceeded those with ignition at the top of the cylindrical vessel. Lohrer et al. [2007] observed similar dependencies of the p_{max} on the mixture composition for propane/air mixtures in a horizontal pipe system (D = 0.159 m, L = 23 m). Furthermore, a decreasing p_{max} with an increasing pipe length was measured due to the heat transfer of the hot combustion products to the pipe walls.

It has been known for a long time, that turbulence inducing elements enhance the heat and mass transfer in reactive flows, leading to a reduced start-up way of the ongoing detonation. These elements contain amongst others valves and devices for the measurements of mass flows. Pioneering works of Starke and Roth [1989], Andrews et al. [1990] and Phylaktou and Andrews [1991], showed that obstacles induce an acceleration of the combustion process. The deflagration experiments were carried out in comparatively small pipes (maximum lengths of 1.64 m and maximum diameters of 0.1 m) with the mixtures methane/air, propane/air, ethene/air, ethyne/air, and hydrogen/air. Furthermore, experiments in an explosion vessel (20 L; maximum dimension 0.545 m) with Liquefied Petroleum Gas (LPG) have been carried out by Masri et al. [2000] and Ibrahim and Masri [2001]. It was shown, that the flame speeds increased with an increasing blockage ratio BR from 10% to 78% (area percentage; defined as the blocked area divided by the cross-section area of the vessel). Lohrer et al. [2007] quantified the influences of two baffles on the flow velocities and the Tu of air flows through steel pipes (D = 0.159 m) with flange assemblies and baffles (BR = 36% and BR = 91%). The measurements were performed with the 3D constant temperature anemometry (CTA) technique. It was found, that flange assemblies (including gaskets) had no significant effect on the Tu. In contrast, both investigated baffles increased the Tu of the air flows significantly.

This work presents the results of explosion experiments in pipe systems with maximum diameters of 0.2 m and lengths of about 23 m. The investigations were carried out for various flammable gas/air mixtures of a similar MESG and an industrial mixture. In some cases the pipe system was set up with obstacles (BR of 36% to 91%) in order to reduce the start-up way of ongoing detonations.

2 EXPERIMENTS

2.1 MEASUREMENT TECHNIQUE

In the explosion experiments, flame speeds and dynamic explosion pressures were measured. To determine the flame front velocity in a reactive pipe flow, up to 6 photodiodes were mounted along the steel pipe. The pressure was measured with up to 4 piezoelectric sensors (PCB M113A22, 1.5 mV/kPa; PCB-J113A24/061M145, 0.73 mV/kPa). The measuring frequency f was set between $0.18\ \text{MHz} < f \leq 1\ \text{MHz}$.

2.2 EXPERIMENTAL SETUP

Two horizontal steel pipe systems with different flammable gas/air mixtures were used in the explosion tests:

- A: D = 0.159 m, L = 23 m, p_0 = 1.0 bar (abs.), different flammable gas/air mixtures of a similar MESG (0.94–0.97 mm; explosion group IIA): stoichiometric propane/air (4.2 Vol%), lean ethene/air (4.7 Vol%), and lean ethyne/air (3.6 Vol%),
- B: D = 0.2 m, L = 16 m, p_0 = 1.1 bar (abs.), industrial mixture: air (53.93 Vol%), carbon monoxide (20.33 Vol%), hydrogen (15.68 Vol%), carbon dioxide (9.53 Vol%), and nitrogen (0.53 Vol%).

In preliminary experiments the MESG of the gas/air mixtures for the test setup A had to be determined according to IEC 60079-1 [2002]. The tests were performed with propane/air, ethene/air and ethyne/air at an initial pressure of p_0 = 1.0 bar (abs.). With the assumption of a stoichiometric composition in air, ethene and ethyne belong to the explosion groups IIB and IIC, with a respective MESG of 0.65 mm and 0.37 mm. To achieve an almost

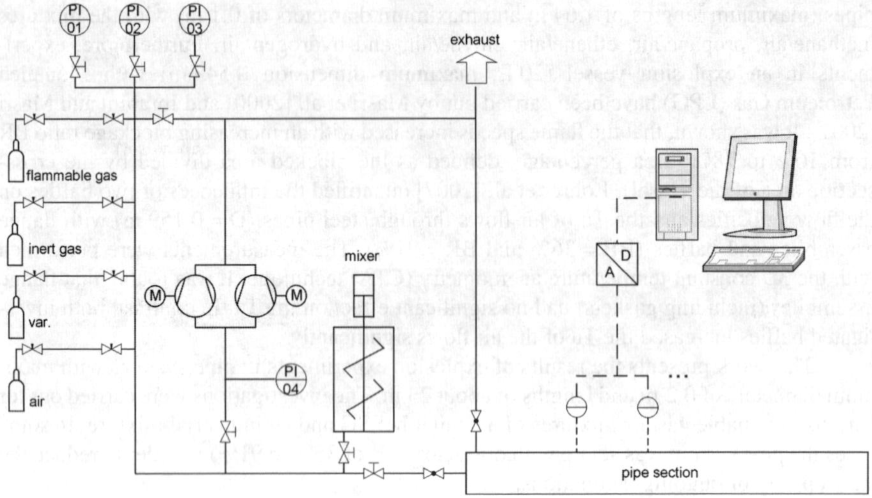

Figure 1. Principle of the experimental setup A, B and signal handling

equal MESG of about 0.94 mm–0.97 mm (compared to stoichiometric propane/air; explosion group IIA), the mole fractions of ethene and ethyne in air had to be reduced from their stoichiometric composition in air (6.5 Vol% and 7.7 Vol%).

The principle setup and signal handling can be seen exemplarily in figure 1, the setup A for the explosion experiments is presented in figure 2. For setup A, the pipe segments had at least 2 drillings every 0.5 m in axial direction for the adjustment of the photodiodes and pressure sensors. Since not all drillings were filled with sensors, plugs were used to close the leftover holes. In order to avoid extra turbulence generation, sensors and plugs closed up flush with the inner wall surface. Before the experiments were carried out, the pipe section was checked for leak tightness. Afterwards, the gas mixer and the pipe section were evacuated. Each flammable gas and air have been mixed with the partial pressure method and the pipe system was filled to the desired initial pressure. The ignition was started at the flange of one side of the pipe by a melting metal wire (ignition energy about 6 J to 12 J). Next to the ignition point, a photodiode triggered the measurement. The pressure signals have been "smoothed" by a low-pass filter to eliminate ground noise.

For setup A, the influence of the parameters flammable gas (without baffles; stoichiometric propane/air (4.2 Vol%), lean ethene/air (4.7 Vol%), and lean ethyne/air (3.6 Vol%) and BR of single hole baffles (for stoichiometric propane/air mixtures; 36%, 64%, and 91%) on the explosion characteristics was investigated. The baffles were installed in 0.6 m axial distance to the ignition point. As long as possible, at least 3 measurements for each parameter variation have been carried out. For the interpretation of the experiments, the values of p_{max} have been averaged over max. 4 sensors and 3 measurements.

Figure 2. Photograph of setup A to determine the explosion characteristics of various flammable gas/air mixtures in a closed pipe with and without obstacles (L = 23 m, D = 0.159 m)

Experiments in setup B were carried out with the above described industrial mixture, first without and than with two different baffles (BR = 51% [single hole] and BR = 77% [five holes]). Due to the different setup, the 77% BR baffle in these experiments was installed in 6 m axial distance to the ignition point. In addition, one experiment was performed with a pre-volume of 22 L from which the ignition was initiated, followed by a baffle with a BR of 51%. Both were installed in front of the pipe system.

3 RESULTS AND DISCUSSION

3.1 SETUP A

In first test series without baffles, the explosion characteristics of different flammable gas/air mixtures with a similar MESG of about 0.94 mm–0.97 mm were investigated and compared to each other. The results of the explosion experiments are presented in figure 3, figure 4, and figure 5.

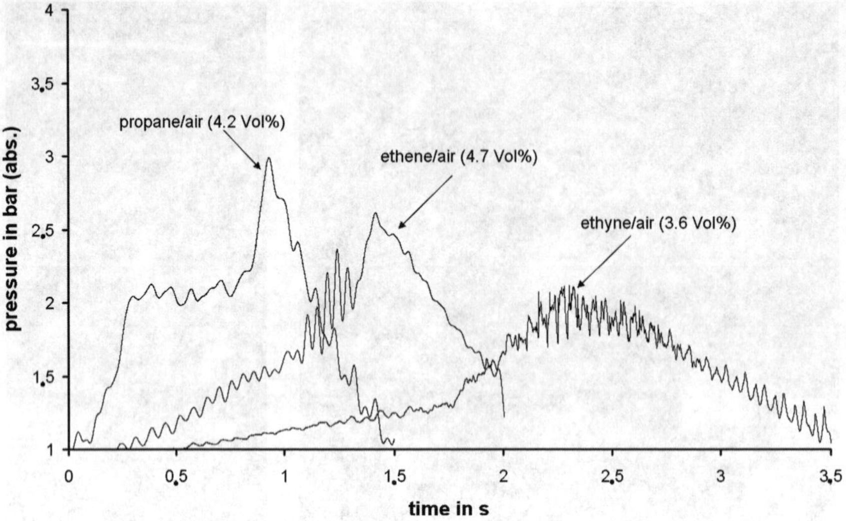

Figure 3. Explosion pressure versus time for three flammable gas/air mixtures of a similar MESG (setup A)

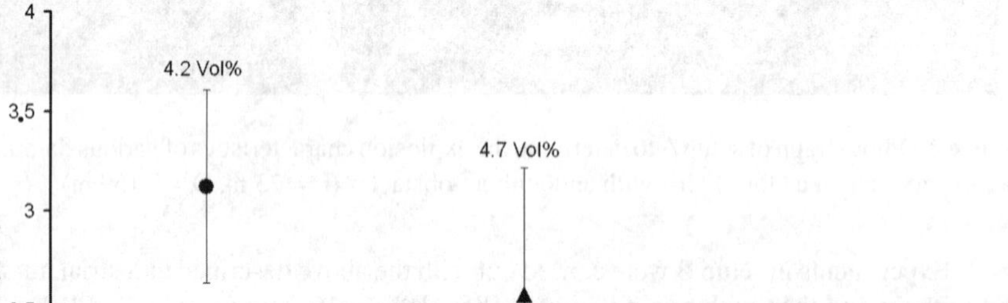

Figure 4. Maximum explosion pressure for three flammable gas/air mixtures of a similar MESG (setup A)

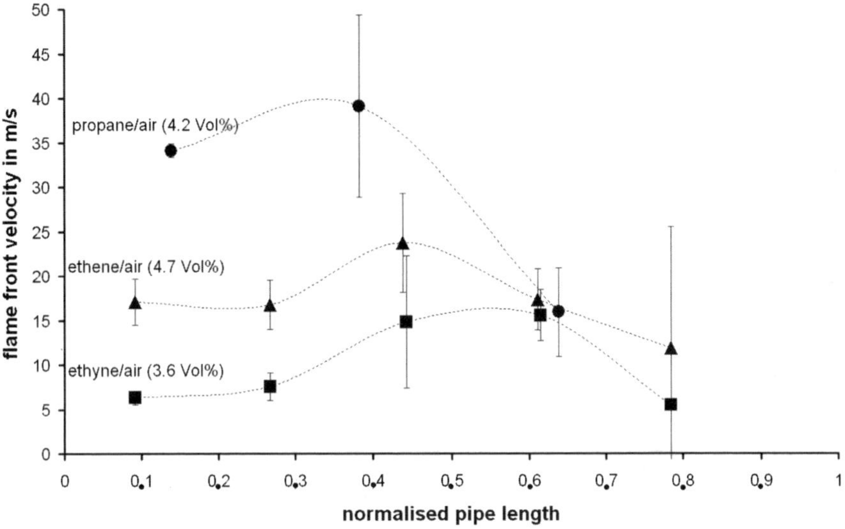

Figure 5. Flame front velocity versus normalised pipe length for three flammable gas/air mixtures of a similar MESG (setup A)

It was observed that the maximum pressures as well as the flame speeds differed significantly. For stoichiometric propane/air mixtures the highest explosion pressures of about 3.2 bar (abs.) were determined, followed by ethene/air and ethyne/air with respective p_{max} of about 2.6 bar (abs.) and 1.9 bar (abs.), figure 3 and figure 4. For ethene/air the mole fraction was reduced from the stoichiometric composition of 6.5 Vol% down to 4.7 Vol% (difference 1.8 Vol%) and for ethyne/air from the stoichiometric composition of 7.7 Vol% down to 3.6 Vol% (difference 4.1 Vol%). As can be seen, the more the mole fraction of the investigated gases in air had to be reduced to achieve an almost equal MESG (compared to propane/air in stoichiometric composition), the less severe was the explosion. Furthermore, the flame speeds of the stoichiometric propane/air mixtures (maximum values of 39 m/s) exceeded the values of the lean ethene/air and ethyne/air mixtures, figure 5. In all cases, flame acceleration was found up to the middle of the 23 m long pipe and deceleration towards the pipe end. Since the pipe system was closed at both sides, the axial propagation was detained, leading to higher radial spread and the flame fronts reached the cold pipe wall earlier and transferred heat to the pipe wall.

These results show that for flammable gas/air mixtures of a similar MESG the most conservative estimations towards the impacts of possible explosions will be achieved for stoichiometric compositions.

The experiments with stoichiometric propane/air mixtures were extended to the use of three different single hole baffles with BR of 36%, 64%, and 91% in the setup A. Figure 6

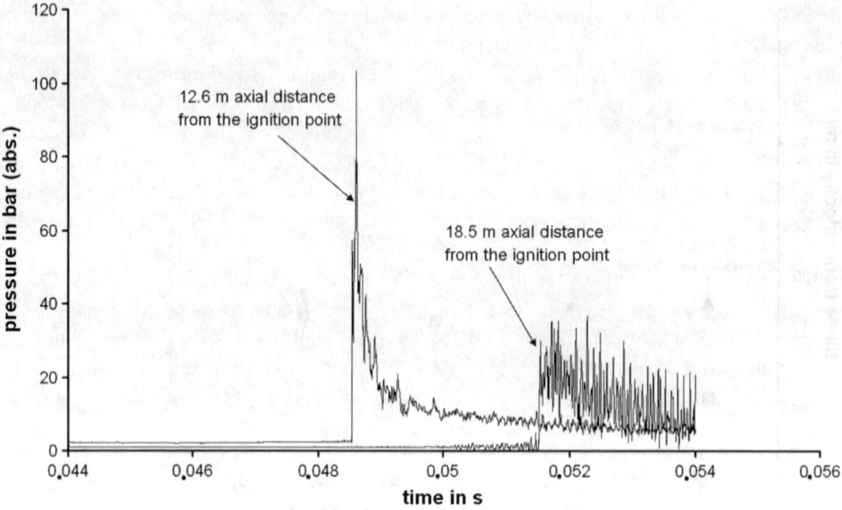

Figure 6. Pressure-time histories in 12.6 m and 18.5 m axial distance from the ignition point during a stoichiometric propane/air explosion with the use of a baffle (BR = 91%, setup A)

illustrates two pressure-time histories in 12.6 m and 18.5 m axial distance from the ignition point during a stoichiometric propane/air explosion with the use of a baffle (BR = 91%). Both pressure peaks describe different explosion regimes. In the first part of the pipe, the explosion starts as a deflagration (not visible with the chosen setup of the sensors) and finally accelerates to a detonation. The first peak in 12.6 m distance represents an overdriven detonation, characterised by a high pressure of 103 bar (abs.). Towards the end of the pipe at 18.5 m the local pressures decreased to 35 bar (abs.).

The influence of the BR on the p_{max} for explosions of stoichiometric propane/air mixtures in the setup A is presented in figure 7. The blank circles represent the pressure measurements in 12.5 m axial distance and the black points represent the pressure measurements in 18.5 m axial distance to the ignition point. Figure 8 gives the flame speeds versus the normalised pipe length (middle of the distance between the measuring points divided by the total pipe length) for the three investigated baffles with different blockage ratios. The flame speeds in these experiments were detected with 6 optical sensors along the pipe length.

For the baffle with a BR of 36% 5 out of 6 explosions occurred as deflagrations. Only in one experiment a DDT was observed, which is presented in figure 7 and figure 8. As can be seen for this case, the pressure in the rear part of the pipe system (53 bar (abs.)) exceeded the one in 12.6 m distance (5.2 bar (abs.)). Furthermore, the flame front accelerated to a maximum value of 480 m/s towards the end of the pipe. Both the pressure distribution in the pipe and the flame speed point to an ongoing detonation in the experiment

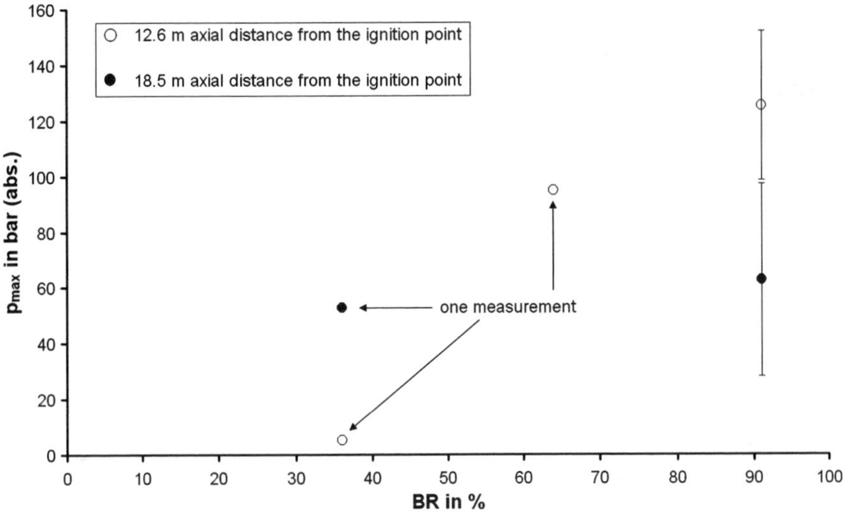

Figure 7. Maximum pressure versus blockage ratio of the baffles for stoichiometric propane/air explosions (setup A)

with the 36% BR baffle. Though, the overdriven detonation regime was not achieved in 18.5 m distance to the ignition point due to an insufficient turbulence generation in the main body of the flow. Lohrer et al. [2007] measured the turbulence intensities in 0.2 m distance behind a similar baffle across the pipe diameter for different air flows with the CTA technique. Due to the influence of the baffle on the flow characteristics, the Tu, in regions close to the wall, increased to a 5 times higher value (about 50%) compared to an empty pipe (about 10%). On the pipe axis the Tu dropped to a minimum value of about 3%.

As expected, the maximum pressures increased with an increasing BR due to an enhanced turbulence generation in the reactive flows. For the baffle with a BR of 91% the maximum Tu (55%) of various air flows was determined somewhere in the middle between the wall and the baffle opening. In the pipe axis Tu of about 10% were measured, [Lohrer et al., 2007]. The use of a baffle with a BR of 91% led to detonation processes in all three experiments. Figure 8 illustrates the acceleration process along the pipe. In the rear part of the pipe, flame speeds of >1500 m/s were determined. As can be seen in figure 7, the pressure values in 12.6 m distance (average p_{max} of about 125 bar (abs.)) exceeded the ones in 18.5 m distance (average p_{max} of about 62 bar (abs.)) to the ignition point. This observation confirms the assumption, that the deflagration was accelerated behind the baffle to an overdriven detonation.

An experiment with the medium baffle (BR = 64%) led to serious damages to the last two pipe segments and was therefore not repeated. The damages included a tear off of the pipe at the weld seam of the flange connecting the last two pipe segments, a crack in

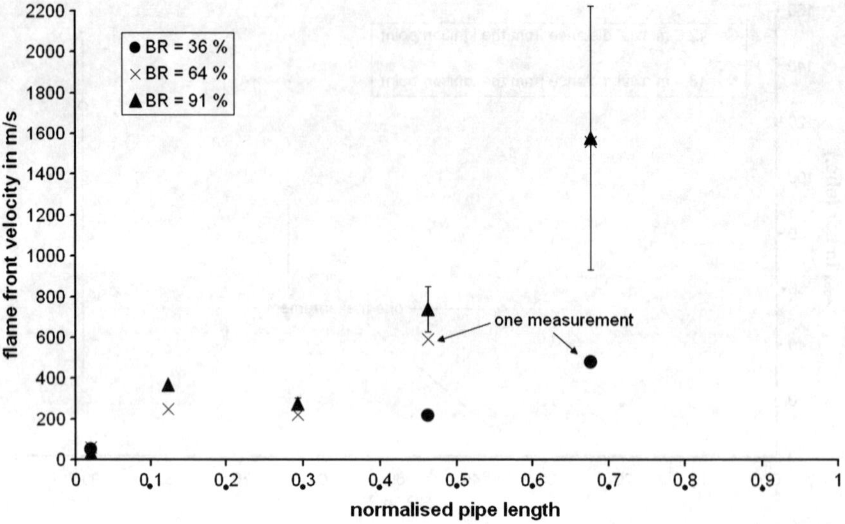

Figure 8. Flame front velocity versus the normalised pipe length during stoichiometric propane/air explosions with three baffles (setup A)

the weld seam at the end flange, and a 2 mm deformation in the middle of the blind flange (22 mm of thickness) which closed the pipe. Only one piezoelectric sensor (in 12.6 m distance) was able to measure the pressure time history. At this point a p_{max} of 95 bar (abs.) was determined (figure 7). Nevertheless, all photodiodes were able to measure the flame speeds along the pipe. The values are slightly smaller than those of the experiments with the 91% BR baffle (figure 8).

3.2 SETUP B

In contrast to setup A, setup B consisted of connected pipe segments with a diameter of 0.2 m and a total length of 16 m filled with a flammable industrial mixture. Both pressures and flame speeds were detected each with 4 piezoelectric sensors and photodiodes along the pipe axis.

Five preliminary experiments without turbulence inducing elements led to deflagrative reaction mechanism in the setup B. An exemplary pressure time history for this case is presented in figure 9. For these experiments a mean value of p_{max} = 3.56 bar (abs.) and maximum flame speeds of 72 m/s in the middle of the pipe were determined. As expected, the flame fronts decelerated towards the end of the closed pipe, which is characteristic for deflagration processes.

A totally different behaviour occurred with the use of baffles and a pre-volume. Figure 10 and figure 11 give pressure-time histories at 4 axial distances to the ignition

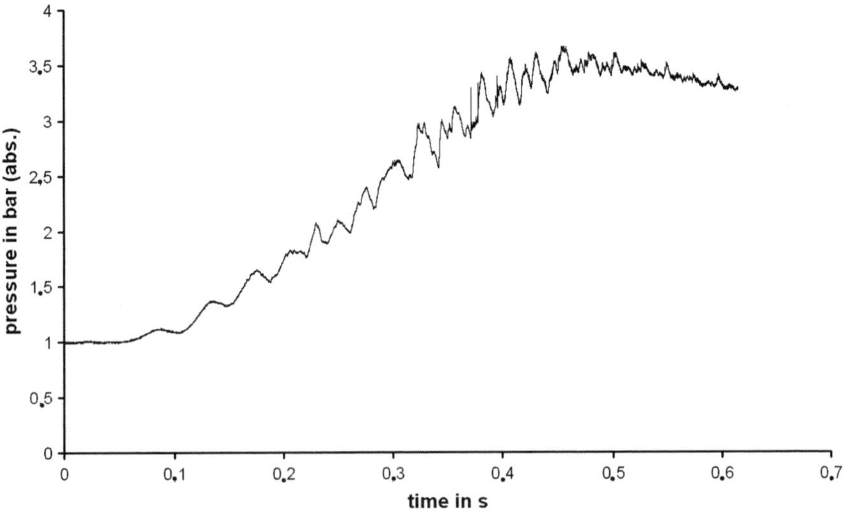

Figure 9. Deflagration pressure versus time for an industrial mixture (setup B; see "Experimental setup")

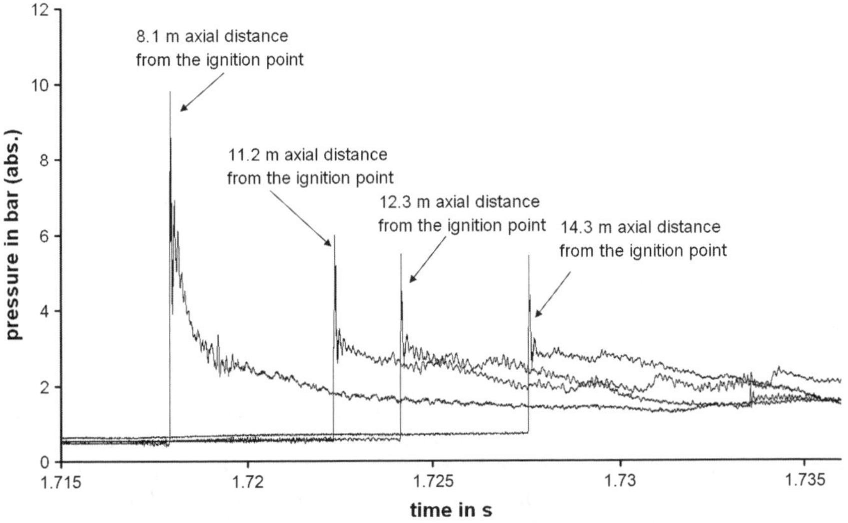

Figure 10. Pressure-time histories at 4 axial distances to the ignition point during an explosion of an industrial mixture with the use of a baffle (BR = 77%, setup B)

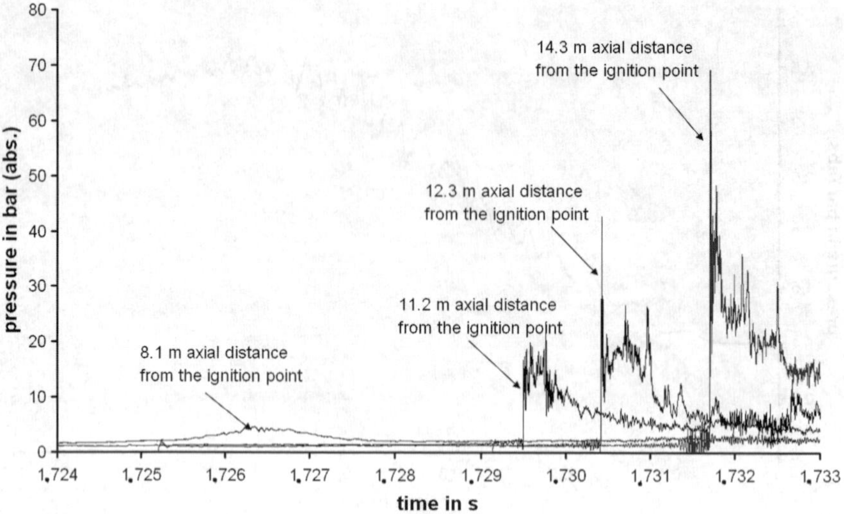

Figure 11. Pressure-time histories at 4 axial distances to the ignition point during an explosion of an industrial mixture with the use of a baffle (BR = 51%) and a 22 L pre-volume (setup B)

point with the use of baffles (BR = 77% as well as BR = 51% and a pre-volume). The velocities of the shock waves and the flame fronts in both experiments are merged in figure 12.

The almost vertical pressure rises in figure 10 (baffle; BR = 77%) suggest an overdriven detonation at the first sensor and a stable detonation at the following points since the pressure dropped from 9.8 bar (abs.) and remained constant at about 5.4 bar (abs.). Though, the pressure values seem too low for a fully developed detonation and the velocities of the shock wave and the flame front differed significantly from another (figure 12). The shock wave propagated at considerably higher velocities (688 m/s to 580 m/s) through the pipe than the flame front (507 m/s to 177 m/s). With regards to the comparably low pressures and the different velocities of shock wave and flame front, a fast deflagration seems more likely in this case. The steep pressure rise in this case could be explained by a jet ignition and vigorous combustion behind the baffle due to an instantaneously enhanced combustion area.

An ongoing overdriven detonation was observed with the use of a baffle (BR = 51%) and ignition from a pre-volume. As can be seen in figure 11, the maximum pressures along the pipe increased from 4 bar (abs.) at 8.1 m distance to 70 bar (abs.) at 14.3 m distance to the ignition point. The combustion process started as a deflagration and developed to an overdriven detonation. This process is illustrated in figure 12. The coupling of the shock wave and the flame front (which is characteristic for a detonation) occurred at the end of the pipe system. In this case, velocities of >1600 m/s were detected.

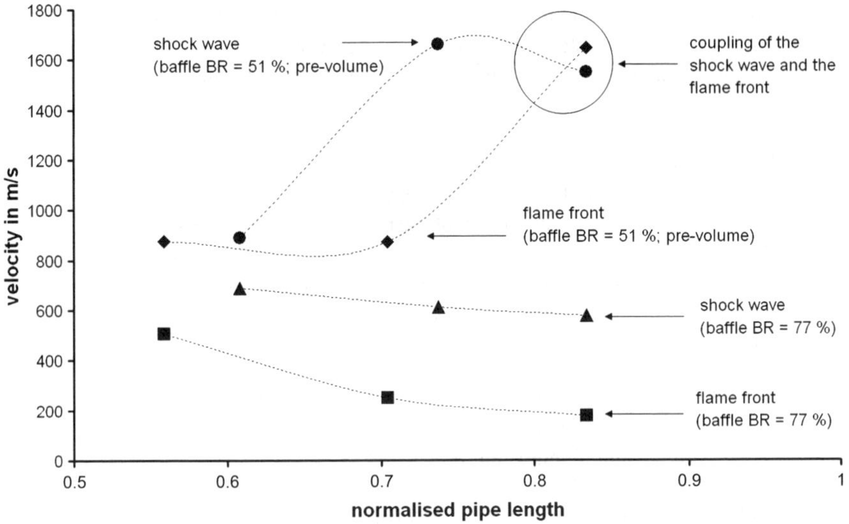

Figure 12. Shock wave and flame front velocities at 4 axial distances to the ignition point during explosions of an industrial mixture with the use of a baffle (BR = 51% and 77%) and a 22 L pre-volume (setup B)

4 CONCLUSIONS

Explosion experiments were performed in pipe systems with diameters of 0.159 m and 0.2 m and lengths of about 23 m and 16 m. The investigations were carried out for various flammable gas/air mixtures of a similar MESG and an industrial mixture. In some cases the pipe system was set up with obstacles (blockage ratios of 36% to 91%) in order to reduce the start-up way of ongoing detonations. It was found that the explosion characteristics of different flammable gas/air mixtures with a similar MESG differed significantly from another. The more the mole fraction of the investigated gases in air had to be reduced to achieve a similar MESG (compared to propane/air in stoichiometric composition), the less severe was the explosion. These results show that for classification of flame arresters (EN 12874, 2001) to be used in systems with various flammable gas mixtures of the same MESG, the most conservative estimations will be achieved for stoichiometric mixture compositions.

Furthermore, the experiments showed that baffles with blockage ratios of 36% – 91% increased the turbulence intensity of reactive flows tremendously, leading to DDT in the steel pipe systems. Absolute pressures of more than 100 bar (abs.) joined by supersonic flame speeds of >2000 m/s were determined by piezoelectric sensors and photodiodes. In consequence, technical equipments with comparable blockage ratios such as valves and devices for the measurements of mass flows, pipe elbows, bifurcations, in- and outlets as well as pipe cross section changes will reduce the start-up way of detonations. These conclusions have to be considered when designing constructional explosion protection measures in pipe systems in which flammable gas mixtures are transported.

5 NOMENCLATURE

A_d	m^2	area of the baffle opening
A_D	m^2	cross-section area of the pipe
BR	%	blockage ratio of the baffle; $BR = (1 - A_d/A_D) \cdot 100$
d	m	inner diameter of the baffle opening
D	m	inner diameter of the pipe
f	Hz	frequency
L	m	length of the pipe
p_0	bar (abs.)	absolute initial pressure
p_{max}	bar (abs.)	absolute maximum explosion pressure
T_0	K	initial temperature
Tu	%	turbulence intensity
X	Vol%	propane mole fraction in air

REFERENCES

Andrews, G.E., Herath, P. and Phylaktou, H.N. 1990. The influence of flow blockage on the rate of pressure rise in large L/D cylindrical closed vessel explosions, *Journal of Loss Prevention in the Process Industries*, 3: 291–302.

Bartknecht, W., 1993, *Explosionsschutz - Grundlagen und Anwendung* (in German). Berlin Heidelberg New York: Springer.

Council Directive 1999/92/EC, 1999, Directive 1999/92/EC on minimum requirements for improving the safety and health protection of workers potentially at risk from explosive atmospheres. Official Journal of the European Communities, Luxembourg.

EN 12874, 2001, Flame arresters - Performance requirements, test methods and limits for use, European Committee for Standardization (CEN).

EN 13237, 2003, Potentially explosive atmospheres - Terms and definitions for equipment and protective systems intended for use in potentially explosive atmospheres, European Committee for Standardization (CEN).

Grätz, R. and Förster, H. 2001. Deflagration, stabile/instabile Detonation - eine Flammend urchschlagsicherung für alle Fälle?, *proceedings of DECHEMA-Fachtreffen und 9. BAM/PTB-Kolloquium zur chemischen und physikalischen Sicherheitstechnik*, 48–56, (in German).

Ibrahim, S. S. and Masri, A. R. 2001. The effects of obstructions on overpressure resulting from premixed flame deflagration, *Journal of Loss Prevention in the Process Industries*, 14: 213–221.

IEC 60079 "Electrical apparatus for explosive gas atmospheres", Part 0: General requirements, 2004, International Electrotechnical Commission (IEC).

IEC 60079 "Electrical apparatus for explosive gas atmospheres", Part 1-1: Flameproof enclosures "d" - Method of test for ascertainment of maximum experimental safe gap, 2002, International Electrotechnical Commission (IEC).

Kleiber, M., Uth, H.-J. and Watorowski, J. 2006. Zentrale Melde- und Auswertestelle für Störfälle und Störungen in verfahrenstechnischen Anlagen (ZEMA) - Jahresbericht 2004 (in German). Federal Environmental Agency (UBA), Dessau, Germany. ⟨http://www.umweltbundesamt.de/zema⟩.

Lohrer, C., Drame, C., Schalau, B. and Grätz, R. 2007. Propane/Air Deflagrations and CTA Measurements of Turbulence Inducing Elements in Closed Pipes, *Journal of Loss Prevention in the Process Industries (2007), doi:10.1016/j.jlp.2007.06.003*

Mannan, S. 2005. *Lee's Loss Prevention in the Process Industries - Hazard Identification, Assessment and Control*, Volume 1 (3nd ed.), Elsevier Butterworth-Heinemann, 2: 20.

Masri, A. R., Ibrahim, S. S., Nehzat, N. and Green, A. R. 2000. Experimental study of premixed flame propagation over various solid obstructions, *Experimental Thermal and Fluid Science*, 21: 109–116.

Phylaktou, H. and Andrews, G. E. 1991. The Acceleration of Flame Propagation in a Tube by an Obstacle, *Combustion and Flame*, 85: 363–379.

Razus, D., Movileanua, C. and Oancea, D. 2007. The rate of pressure rise of gaseous propylene-air explosions in spherical and cylindrical enclosures, *Journal of Hazardous Materials*, 139: 1–8.

Starke, R. and Roth, P. 1989. An Experimental Investigation of Flame Behavior During Explosions in Cylindrical Enclosures with Obstacles, *Combustion and Flame*, 75: 111–121.

ACCEPTANCE CRITERIA FOR DAMAGED AND REPAIRED PASSIVE FIRE PROTECTION

Diane Kerr[1,*], Deborah Willoughby[1], Simon Thurlbeck[2] and Stephen Connolly[3]
[1]Health and Safety Laboratory, Process Safety Section, UK
[2]MMI Engineering Ltd
[3]Health and Safety Executive, Offshore Safety Division, UK
[*]Contact details: Diane Kerr, Process Safety Section, Health and Safety Laboratory, BUXTON, SK17 9JN, United Kingdom; e-mail: diane.kerr@hsl.gov.uk

KEYWORDS: PFP, Damage, Repair, Cementitious, Intumescent, Acceptance Criteria.

INTRODUCTION

The performance of ageing, weathered, damaged or repaired passive fire protection (PFP) is a major concern, particularly offshore where it is often a safety-critical factor in protecting structures and plant from the effects of severe fires. The two most commonly used types of PFP materials are cementitious and intumescent. Cementitious PFP works initially by holding the temperature of the substrate to around 100°C until all the bound water is vaporised and then acts as a passive insulator. Intumescent PFP has an organic base which, when subjected to fire, expands producing a stable char with good thermal insulation properties.

Whilst the performance of these materials when in good condition is well understood, little or no information is available which provides an understanding of their performance in a damaged or repaired state. In the absence of such information, verifying performance standards where PFP is employed as a risk-reduction measure cannot be undertaken reliably.

To address this shortfall, a two-phase Joint Industry Project was established to develop data on the performance of damaged PFP and to suggest an inspection methodology with acceptance criteria. This paper will describe specially produced damaged and repaired PFP test pieces (both cementitious and intumescent) and will summarise their performance in jet fire resistance tests.

The paper will also discuss the acceptance criteria developed for damaged and repaired PFP. The project's findings are providing important input into a new procedure for identifying when repair or replacement is necessary, and which forms of repair are most effective. The findings also highlight valuable lessons in quality assurance of the PFP application process.

The work reported here was undertaken in a collaborative project developed and managed by MMI Engineering Ltd. The work was sponsored by HSE, BP, Centrica

Energy, Canadian Natural Resources, Marathon, Total, and Shell. PFP materials were provided by International Coatings, Cafco International, and PPG Indistries, and specimen coating was undertaken by Salamis and RBG Ltd. Testing was undertaken on behalf of the Sponsors by the HSL.

TEST FACILITY AND INSTRUMENTATION

The tests were carried out using HSL's jet fire resistance test facility. The facility is designed to give an indication of how well passive fire protection materials will perform in a jet fire. It involves impinging a propane vapour jet fire on a target covered with PFP material. The test facilities and details comply with OTI 95 634 and Draft ISO standard ISO/CD 22899-1.

The test specimen was bolted between an open-fronted steel box (lined with ceramic boarding) and a protective rear chamber. A wall of lightweight building blocks was erected around the target assembly to prevent flames impinging on the sides of the target. A photograph showing the installed assembly is given in Figure 1.

The test specimen was instrumented with mineral insulated, stainless steel sheathed, 1.5 mm, type K thermocouples supplied and fitted by HSL. Thermocouples were installed in the back surface of the panels by means of drilled, interference fit, holes and were held in position in the web by bulkhead fittings. Measurements were logged using DT3003 32 channel Data Translation data acquisition boards fitted directly into a PC. A FLIR SC 2000 thermal imaging camera was used to measure the temperature distribution over the back of the test specimen.

SUBSTRATE

The substrate for a normal structural steel test consists of an open-fronted steel box, nominal internal dimensions 1500 mm × 1500 mm × 500 mm, made from 10 mm thick steel.

Figure 1. Installed Assembly (test 10 specimen, before test)

It has a 20 mm thick central web, 250 mm deep, to simulate corner or edge features such as stiffening webs or edges of "I" beams. The web is made up of two 10 mm thick steel plates, which are slotted before being welded together, to allow the insertion of thermocouples The box is flanged at the rear to allow attachment (by bolting or clamping) of a 1500 mm × 1500 mm × 1000 mm protective chamber when required.

In order to be able to test as many types of damage as possible, whilst keeping the number of jet fire tests to a minimum, it was decided to replace the back of the box with a panel comprising two half panels and a web. In this way, three different features could be assessed in each test. A piece of L-section was welded along the back of one edge of each of the half panels and holes drilled through these and the back of the web to allow the two half panels and the web to be bolted together (see Figure 2).

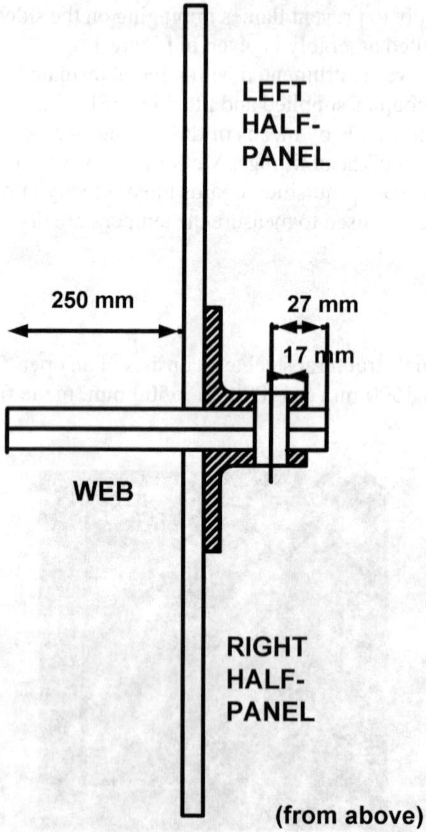

Figure 2. Half-Panel and Web Assembly

For the repair trials, it was considered that the number of joints should be minimised by welding the half panels and the web together to give a single panel incorporating the web. A second type of substrate was also used. This consisted of a simple plate panel, without a web, fitted to form the back of the box.

COATING

Test specimens for the damage trials were coated with cementitious or intumescent PFP materials supplied by Cafco International (Mandolite 550), International Coatings (Chartek IV and VII) and PPG Industries (Pitt-Char XP).

Specimens were coated with the required thickness of each material (as specified by the manufacturer) to achieve the following protection criteria, when subjected to a jet fire test:

Panels – at least 60 minutes for a temperature rise of 140°C.
Web – at least 60 minutes for a temperature rise of 400°C.
(temperature rise above initial substrate temperature)

Chartek IV is a relatively hard epoxy intumescent PFP material and Pitt-Char XP is a softer, more flexible material. A thickness of 12 mm of the respective intumescent materials was applied to the panels and webs, with reinforcement located in the middle third of the coating.

The cementitious test specimens were all prepared with a nominal 38 mm deep coating of Mandolite 550. Reinforcement was with plastic coated hexagonal wire mesh retained by helical pins in the middle third of the coating.

The repair test specimens were coated with a combination of cementitious material (Mandolite 550) and intumescent material (Chartek VII or Pitt-Char XP). These particular intumescent materials were chosen as both can be used with reinforcement that does not require pinning.

Coating of the majority of the substrates was witnessed and quality controlled by MMI Ltd. A photograph illustrating coating of a repair test specimen is provided in Figure 3.

TEST PROGRAMME

The first test series involved carrying out jet fire resistance tests on PFP specimens with the following types of induced damage:

- Gouges and nicks;
- Cracks;
- Loss of bonding and retention; and
- Water saturation (cementitious specimen)

The second test series was designed to test several typical 'in-situ' repairs found offshore. In practice, where damage has occurred to a small area of cementitious material, it is typically

Figure 3. Coating Application – Butt joint

removed and replaced with new intumescent material. Problems can arise at the interface between the materials, as the two types of PFP work by completely different mechanisms.

Three types of repair joint were investigated:

- Simple butt joints between the cementitious and intumescent PFP;
- Two step cementitious/intumescent joints; and
- A smooth chamfer joint between the materials

The induced damage and repair joints are described in more detail in Tables 1, 2 and 3. The joints are also illustrated in Figure 4.

RESULTS AND CONCLUSIONS

The most significant results and overall conclusions from the damage and repair trials are detailed below. The results from this collaborative project will also be accessible, via the HSE website, in due course.

DAMAGED INTUMESCENT TRIALS

RESULTS

Figures 1 and 5 show the specimen for trial 10 before and after testing. The purpose of this test was to determine the effect of gouging, cracking and disbondment on Chartek IV.

Left Panel

The left panel was physically damaged (gouged), but was bonded to the steel substrate. After the test, slight enlargement of the gouges was observed. Thermocouple temperatures and thermal images confirm that the main hot spot was recorded behind the horizontal

Table 1. Intumescent PFP tests

Test number	Purpose	Anomalies
01	Chartek IV control specimen	None. Panels and web all with same thickness (12 mm) of sound material and reinforcement
02	Effect of loss of material, to depth of reinforcement layer, and beyond on Chartek IV	Left panel – Sound material with 350 mm × 40 mm vertical gouge to depth of reinforcement. Right panel – Sound material with 350 mm × 40 mm vertical gouge beyond depth of reinforcement. Web – 350 mm long vertical 45° nick from edge of web beyond depth of reinforcement.
04	Chartek IV disbonding	Left panel – Disbonded with partially retained mesh. Right panel – Bonded with no mesh Web – Disbonded with partially retained mesh
10	Effect of gouging, cracking and disbondment on Chartek IV	Left panel – Sound material with 200 mm × 25 mm horizontal gouge in top half, 200 mm × 12 mm horizontal gouge in bottom half and 350 mm × 12 mm vertical gouge running downwards from centre. All gouges through mesh. Right panel – Disbonded material with 2, 3 and 5 mm cracks to substrate running full length. Web – Disbonded material with 5 mm crack to substrate running full length.
13	Effect of gouging, cracking and disbondment on Pitt-Char XP	Left panel – Sound material with 200 mm × 25 mm horizontal gouge in top half, 200 mm × 12 mm horizontal gouge in bottom half and 350 mm × 12 mm vertical gouge running downwards from centre. All gouges through mesh. Right panel – Disbonded material with 2, 3 and 5 mm cracks to substrate running full length. Web – Disbonded material with 5 mm crack to substrate running full length.

(25 mm × 200 mm) gouge above the centre line. The time until a 140°C temperature rise was seen behind this gouge was 4 minutes. The highest temperature rise after 60 minutes was 325°C.

Right Panel

Prior to the test, the right panel was cracked and the coating was deliberately disbonded from the steel substrate. After the test this panel was much more damaged than the left

Table 2. Cementitious PFP tests

Test number	Purpose	Anomalies
05	Control specimen	None. Panels and web all with same thickness (38 mm) of sound material, and reinforcement
06	Effect of loss of material, to depth of reinforcement layer, and beyond	Left panel – Sound material with 350 mm × 40 mm vertical gouge to depth of reinforcement. Right panel – Sound material with 350 mm × 40 mm vertical gouge beyond depth of reinforcement. Web – 350 mm vertical nick from edge of web beyond depth of reinforcement.
07	Effect of cracking	Left panel – Sound material with 1, 3 and 5 mm cracks running full length. Right panel – Disbonded material with 1, 3 and 5 mm cracks running full length. Web – Disbonded material with 5 mm crack running full length.
08	Effect of moisture content	Left panel – Sound material with panel fully saturated. Right panel – Sound material with panel 50% saturated. Web – Sound material with web fully saturated.
09	Effect of loss of bonding and loss of retention	Left panel – Disbonded material & steel reinforcement fixed around edges of panel. Right panel – Bonded material with no steel reinforcement. Web – Disbonded material and steel reinforcement partially retained.

panel (above), the material was bowed out from the substrate and cracked. There was considerable enlargement of the induced cracks, particularly the 2 mm crack. The highest temperatures were recorded at the lower ends of the cracks, in the high erosion zone. Thermal images confirm that there is a hot area around all three cracks in the high erosion zone. The time until a 140°C temperature rise was seen behind the cracks was 9 minutes. The highest temperature rise after 60 minutes was 262°C.

Web

The web was physically damaged (cracked) and the coating was deliberately disbonded from the steel substrate. Temperatures rose very rapidly behind the (5 mm) crack on the web. Major damage was caused, with bare steel being exposed for the bottom two thirds of the web. A 400°C temperature rise was achieved after 5 minutes. The maximum temperature rise of 705°C was achieved after 25 minutes, with the temperature rise at 60 minutes being 656°C. These temperatures are consistent with the Chartek IV being stripped off leaving bare steel exposed to the jet fire.

Table 3. Repair tests

Test number	PFP	Repair
03	Mandolite 550 with hexagonal wire mesh reinforcement repaired with Chartek VII with HK-1 glass/carbon fibre reinforcing scrim	Simple butt joint vertically and horizontally in a panel without a web
11	Mandolite 550 with hexagonal wire mesh reinforcement repaired with Chartek VII with HK-1 glass/carbon fibre reinforcing scrim	Left panel – Stepped cementitious joint across centre line. Right panel – Chamfered intumescent joint across centre line. Web – Stepped cementitious joint on left side and chamfered intumescent joint on front and right side of centreline.
12	Mandolite 550 with hexagonal wire mesh reinforcement repaired with Pitt-Char XP with FM glass fibre reinforcing scrim	Left panel – Stepped cementitious joint across centre line. Right panel – Chamfered intumescent joint across centre line. Web – Stepped cementitious joint on left side and chamfered intumescent joint on front and right side of centreline.

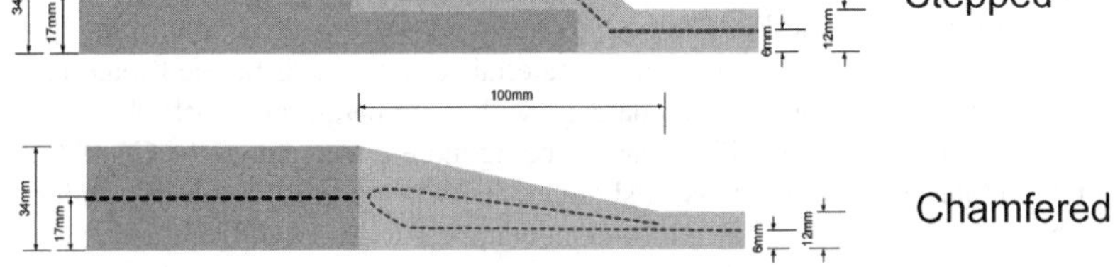

Figure 4. Repair joints

Figure 5. Specimen for test 10 (after test)

CONCLUSIONS

The main conclusions from the damaged intumescent trials were as follows:

(a) If there is a gouge in PFP on a panel, a hot spot is formed behind it in a jet fire. If the gouge is down to the substrate, 12 mm and 25 mm wide gouges give a 140°C temperature rise within 3 minutes and a 75 mm wide gouge gives a 400°C temperature rise within the same time. The temperature rise is slower if the gouge leaves PFP attached to the substrate (22 minutes until 140°C temperature rise on 75 mm gouge with material remaining.

(b) 45° nicks in PFP on webs lead to bare metal being exposed in a jet fire. Where the nick just exposed the substrate, a 400°C temperature rise was reached in 30 to 36 minutes and where the nick just exposed the mesh, in 46 minutes. The evidence suggests that the erosive forces shear off the char if it expands at right angles to them.

(c) A 5 mm crack in disbonded material has a much more severe effect on material on an edge feature (web) than on a flat surface (panel). A crack in the PFP on an edge leads to rapid failure of the PFP with the PFP being stripped off leaving bare metal and giving a temperature rise of 400°C within 4 minutes. This occurs with both Chartek IV (a relatively hard material) and Pitt-Char XP (a relatively soft material) and hence it is likely to occur with any epoxy intumescent material. Cracks in disbonded material on flat surfaces still lead to considerable damage with a temperature rise of 140°C occurring within 10 minutes, with the Pitt-Char XP being more effected than the Chartek IV.

(d) For the material on the panels, disbondment (with partially retained reinforcement mesh) and a complete lack of reinforcement appear to have only a minor effect on fire resistance, with lack of reinforcement having a marginally greater effect (58 minutes for a temperature rise of 140°C compared to greater than 60 minutes). However, as indicated above, if there are flaws in the surface of the material, this can lead to rapid failure. There appeared to be a slightly greater effect on the disbonded material with partially retained reinforcement on the web with a temperature rise of 400°C in 55 minutes rather than the required 60 minutes.

(e) Precise specification and quality control of intumescent PFP application (material thickness, uniformity, reinforcement position) is required in order to achieve the desired level of fire protection from the PFP.

SUMMARY

The overall conclusions are that the most severe effects are from damage to PFP on edges. Cracks in disbonded material lead very rapidly to the PFP being stripped off and nicks can give the same effect although it takes longer. For material on flat surfaces, gouges or cracks down to the substrate lead to a significance loss in fire resistance, particularly if the material is also disbonded. Disbondment or loss of reinforcement alone, (no surface damage) has only a minor effect on performance provided that the material is at least partially attached to the surface (by the reinforcement system if disbonded or by the bonding if the reinforcement system is corroded).

DAMAGED CEMENTITIOUS TRIALS

RESULTS

Figures 6 and 7 show the specimen for trial 07 before and after testing. The purpose of this test was to determine the effect of cracking and disbondment on Mandolite 550.

Figure 6. Specimen for test 07 (before test)

Figure 7. Specimen for test 07 (after test)

Left Panel
The left panel was physically damaged (5, 3 and 1 mm induced cracks), but was bonded to the steel substrate. On completion of the test, the dimensions of the cracks remained unchanged. The highest temperature rise of 166°C after 60 minutes was recorded at the top of the 5 mm crack.

Right Panel
The right panel was also physically damaged (5, 3 and 1 mm induced cracks). In addition, the coating was deliberately disbonded from the steel substrate. On completion of the test, the dimensions of the cracks remained unchanged but, in this case, the coating bowed out from the substrate in the bottom two-thirds of the panel. The highest temperature rise of 222°C after 60 minutes was recorded at the bottom of the 5 mm crack.

The thermal image at 60 minutes suggested that the whole of the right hand panel (with cracked, disbonded material) was hotter than the left panel (with cracked, bonded material). This may account for the coating bowing out from the substrate.

Web
The web was physically damaged (cracked) and the coating was deliberately disbonded from the steel substrate. Initially the web temperatures around the high erosion positions were higher than the temperatures in the high heat flux positions. Towards the end of the test the temperature at the centre of the web in the high heat flux zone exceeded the others. Bare metal was visible behind mesh on 95% of front face of the

web and on the sides of the web; the PFP material was (mainly) intact but coming away from the substrate, due to the erosion forces from the jet. If the reinforcement had not been intact, it is likely that the PFP would have fallen off. A full-depth crack was observed along the width of the web at the bottom right hand side. The highest temperature rise after 60 minutes was 473°C behind the middle of the 5 mm crack on the web.

CONCLUSIONS

The main conclusions from the damaged cementitious trials were as follows:

(a) 350 mm × 40 mm gouges on flat surfaces resulted in slightly higher temperatures if the gouge was through the retention mesh rather than down to the mesh although the damaged panels still both easily met the 140°C temperature rise in greater than 60 minute fire resistance requirement. The 350 mm long 45° nick in the web gave much higher temperatures than an undamaged specimen but still met the 400°C fire resistance criterion. Gouges and nicks of the size used would not result in failure, although the safety margin is lower for a nick on an edge feature.

(b) For the specimens with 1, 3 and 5 mm cracks, cracks in sound material were less serious than those in disbonded material. None of the cracks increased in size, probably because the retention mesh was intact. It is likely that the PFP on the web would have fallen off if the retention mesh had been corroded by water ingress. Cracks in sound material reduced the fire resistance time by 5 minutes whereas cracks in disbonded material reduced the time by 15 to 17 minutes.

(c) 100% saturation and 50% saturation of sound material made no significant difference to the fire resistance. There were no signs of water saturation causing spalling.

(d) Disbondment with full or partial retention appeared to have little effect on the PFP performance if there were no openings to allow jet the jet fire to get between the material and the substrate. Disbondment with cracking, especially on an edge feature, leads to significant loss of fire resistance.

(e) If the material is fully bonded to a flat surface and there are no jet fire ingress points, lack of retention appears to have no effect. However, loss of retention from the bottom of a gouge has a slight effect and from a nick in an edge feature has a significant effect.

SUMMARY

The largest temperature differences and consequent earliest times for water to be driven off all occur on a damaged web suggesting that damage to edge features is again most critical. The maximum substrate-heating rate after the water was driven off was 5.7°C min^{-1}. Although the test with water-saturated specimens had to be prematurely terminated, there was no evidence to suggest that these specimens would not have met their ratings. The only specimens not to meet their fire resistance specification were those with cracks.

REPAIR TRIALS

RESULTS

Figures 8, 9 and 10 show the specimen for trial 11 before and after testing. The purpose of this test was to determine the effect of chamferred and stepped repair joints.

Left Panel

The left panel incorporated a stepped repair joint. On this joint, fully expanded firm char was formed that was intact except for a crack near the edge of the panel on the left hand side where a small gap was visible where the Chartek had come away from Mandolite (see Figure 9).

Right Panel

The right panel incorporated a chamfered repair joint. At the interface between the Mandolite 550 and chamfered Chartek VII (see Figure 10), the Chartek had not come away from the Mandolite. However, there was a slight crack along the base of the Mandolite and along the Chartek. On the right side of the joint the char was more eroded and the mesh was exposed over a small area.

Web

The left hand side of the web incorporated a stepped repair joint. On the front of the left (stepped) side of the web, there was some damage to the Mandolite 550 and char had been eroded away near this zone.

Figure 8. Specimen for test 11 (before test)

Figure 9. Left panel after test (11)

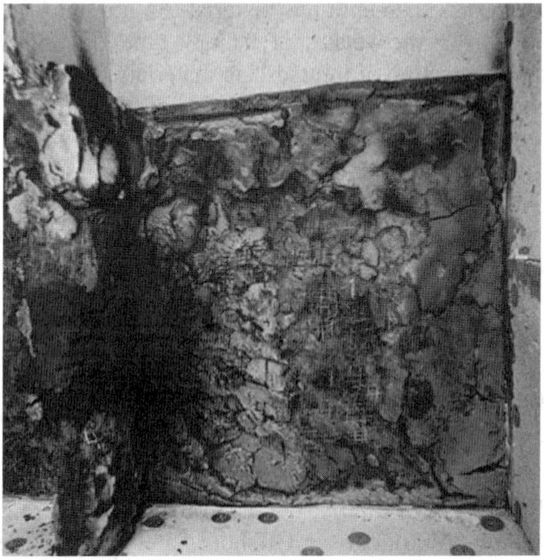

Figure 10. Right panel after test (11)

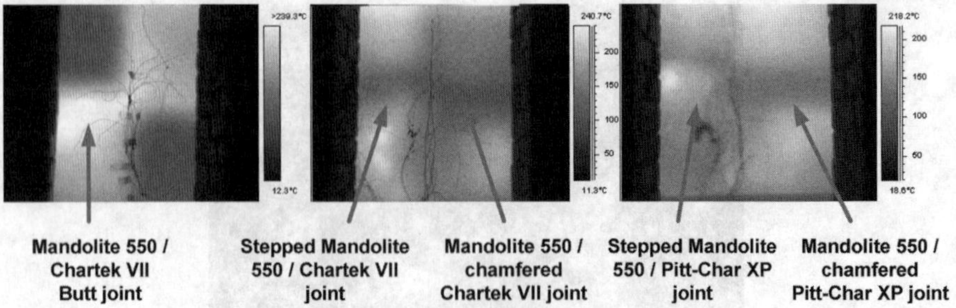

Figure 11. Thermal images (at 60 minutes) of repair joints

The right hand side of the web had a chamfered (Chartek VII) repair joint. After the test, the char on this side of the web was not firm and the mesh was exposed in the jet impact zone.

Overall, the temperature plots and thermal images indicated that, although both types of joint appeared acceptable, the chamfered joint performed better than the stepped joint. Thermal images comparing all three repairs are illustrated in Figure 11.

CONCLUSIONS

The following conclusions were made in regard to the three types of repair assessed:

(a) Butt joints between sound cementitious material (34 mm thick) and new intumescent material (12 mm thick) are the weakest of the three types of joint assessed. When the jet impacts on the thinner coating of intumescent material, it is directed across the surface until it hits the edge formed by the thicker coating of cementitious material, where it is then directed away from the surface. The combination of high erosive forces and high heat flux appears to result in the char eroding away faster in this configuration.
(b) Stepping the sound cementitious material and overcoating this with new intumescent material gives an adequate join but there are still erosive effects and, because the thickness of intumescent material is not much greater than normal, the repair is not as good as chamfering the intumescent material.
(c) Chamfering the intumescent material with an extra loop of reinforcement appeared to be the best form of repair. It worked well with both a relatively hard intumescent material and with a relatively soft one. In addition, it worked as a repair on an edge feature.

ACCEPTANCE CRITERIA FOR DAMAGED PFP

As reported by Thurlbeck (2006), current Offshore Safety Case submissions assume that PFP material is fit-for-purpose, performs as specified, and therefore performance standards

Table 4. Acceptance criteria for anomalies in epoxy intumescent PFP

Anomaly	Criteria	Suggested action
Loss of topcoat	Loss of topcoat is **ACCEPTABLE**.	Action should comprise removal of loose topcoat material to a clean and sound material, a check that adequate thickness remains, and then re-topcoating with a recommended topcoat. A check should also be made to identify the presence of corrosion at the substrate surface.
Cracking	Fully bonded, fully or partially reinforced, material which contains cracks of less than 2 mm width is **ACCEPTABLE**.	The crack should be sealed and made watertight to prevent water ingress into the material which may leach out active agents associated with the intumescent process.
	Material which is disbonded or has damaged reinforcement, and has cracks of any width or depth is **UNACCEPTABLE**.	Action should comprise removal of the material down to the substrate, and back to soundly bonded material, and the implementation of the Manufacturers recommended repair.
Disbonded	Disbonded material of an area of less than 1 square metre, and which shows no signs of cracking or surface anomalies, is **ACCEPTABLE**.	No repair is necessary but the anomaly should be monitored.
	Disbonded material of an area of greater than 1 square metre, or a disbonded area of less than 1 square metre which has surface anomalies, is **UNACCEPTABLE**.	Repair should comprise removal of the material down to the substrate, and back to soundly bonded material, and the implementation of the Manufacturers recommended repair.
Loss/removal of material on structural steelwork	Loss of bonded, reinforced material over an area which is less than the acceptable area, and to any depth of remaining material which may include damage to the reinforcement, is **ACCEPTABLE**.	Action should comprise removal of any loose material and the application of an appropriate topcoat system.

are met. Whilst this may be true, little or no information is available which provides an understanding of PFP performance in a damaged or repaired state. Verification of the performance standards could not, therefore, be reliably undertaken.

Required fire performance is generally specified in terms of time and temperature. In simple terms, the fire performance criteria specify survival times and are generally associated with ensuring an appropriate level of integrity is maintained to prevent an event from escalating, or for ensuring that personnel can be evacuated safely.

The fire performance requirements for a critical item should be detailed in the performance standards, along with the availability and reliability of the element, and the verification tasks to ensure that the performance is ensured.

The criteria are described by defining:

- Whether the protected item is a barrier or a loadbearing element;
- The type of fire/heat flux loading to which it is subjected;
- The critical temperature which should not be exceeded to ensure integrity and/or insulation requirements are met; and
- The time over which the temperature should remain below this critical temperature.

The data from the damage and repair trials described above, along with practical considerations, have been reviewed and a set of acceptance criteria produced, including a decision flowchart. This can be used by asset integrity personnel to assess the acceptability, or otherwise, of observed anomalies in PFP coatings and to determine whether the performance standard of a particular Safety Critical Element is maintained. Examples of the type of anomalies that (without corrective action) could be expected to lead to a failure (or reduced performance) include:

- Disbonded material of an area of greater than 1 m^2.
- Material which is disbonded or has damaged reinforcement, and has cracks of any width or depth.
- Loss of material, regardless of area, which is down to the substrate or has insufficient thickness to provide the required fire resistance performance.

Acceptance criteria for cementitious PFP and epoxy intumescent PFP material containing anomalies will be accessible, via the HSE website, in due course. An extract from the criteria, including suggested action where anomalies are observed, is shown in Table 4.

REFERENCES

Health & Safety Executive, 1996, Offshore Technology Report OTI 95 634, Jet-fire resistance test of PFP materials.

ISO standard ISO/CD 22899-Parts 1 & 2, (draft).

COMPETENCE ASSURANCE IN THE MAJOR HAZARD INDUSTRIES

M. Bromby and C. Shea
ESR Technology

INTRODUCTION

Competence Assurance as a management concept and practice is gaining acceptance and increasing use though it remains relatively novel.

This paper addresses Competence Assurance, and is intended to reflect our practical experience of the topic as safety and risk management consultants. We discuss our approach to the review and development of clients' Competence Management Systems (CMS), including how the reviews were undertaken, the key issues, findings and potential problems in implementing and using a CMS. In addition we discuss how failure to ensure a robust Competence Management System may have contributed to a number of serious accidents.

WHAT DO WE MEAN BY COMPETENCE AND ASSURANCE?

All organisations have a series of management processes that require staff to conduct a range of tasks. Each task must be done within specific parameters to a required standard. This implies that staff assigned to carry out particular tasks should have an acceptable level of competence to ensure the task is conducted and completed correctly i.e. in line with the relevant standard. In high hazard industries in particular this assumes tasks are conducted *safely*.

An idealised CMS cycle is illustrated in Figure 1 below. Once in place the CMS should become part of a wider management loop.

Our preferred definition of competence is selected from a publication regarded as "best-practice" guidance by the Health and Safety Executive (HSE), namely "Developing and Maintaining Staff Competence" (Railway Safety Publication No. 1, 2007), issued by the Office of the Rail Regulator and found on their website for public consumption. This was previously published by the HSE as Railway Safety Principles and Guidance Vol. 3A RSPG-3A). The rail industry was among the first to adopt and develop Competence Assurance among their stakeholders, and the guidance is suitable for implementation in any industry. The motivation for initially publishing this guidance was the perceived contribution of poor competence management as a significant contributory factor in several major UK railway accidents.

We define individual competence as "the ability to undertake responsibilities and to perform activities to a recognised standard on a regular basis. Competence is a combination of practical and mental skills, experience and knowledge." The awareness, knowledge, and competence should be appropriate for the job level and process safety responsibilities of the particular individual. It is important to note that competence is fluid, intimately linked to a

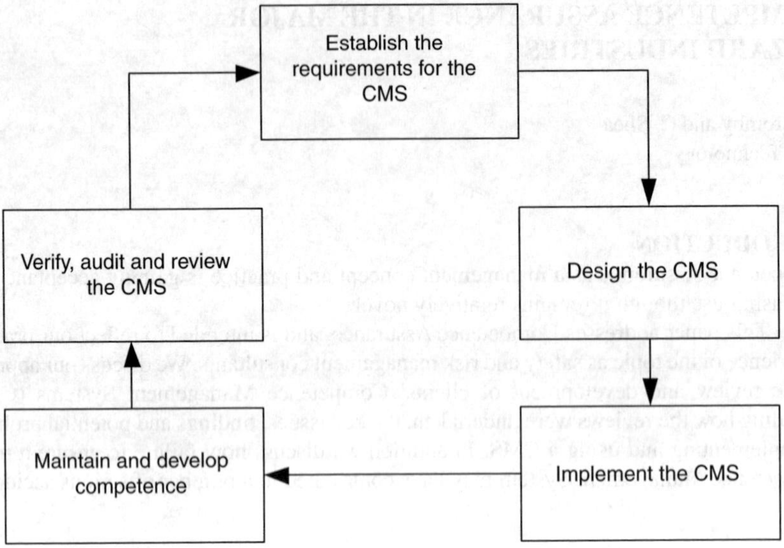

Figure 1. The Competence Management System Cycle (Railway Safety Publication No. 1 2007)

variety of factors including circumstance and individual characteristics such as age, personality and health/well being. In light of this Figure 2 summarises how individuals' competence may become inappropriate and not always increase with experience.

Competence Assurance is the process of ensuring and demonstrating that all staff's competence is properly managed at all times. The CMS should in theory and practice link all existing competence related elements of the management system. These elements include:

- policy;
- links to organisational strategy;
- job definition;
- selection and recruitment of staff;
- risk assessments;
- training, including induction, vocational (task) and refresher training;
- staff supervision;
- development, implementation and management of procedures and standards;
- performance monitoring, review and change;
- management of sub-standard competence.

Across an organisation, adoption of a CMS will result in new and/or amended management procedures. Some procedures will be generic with applicability across the

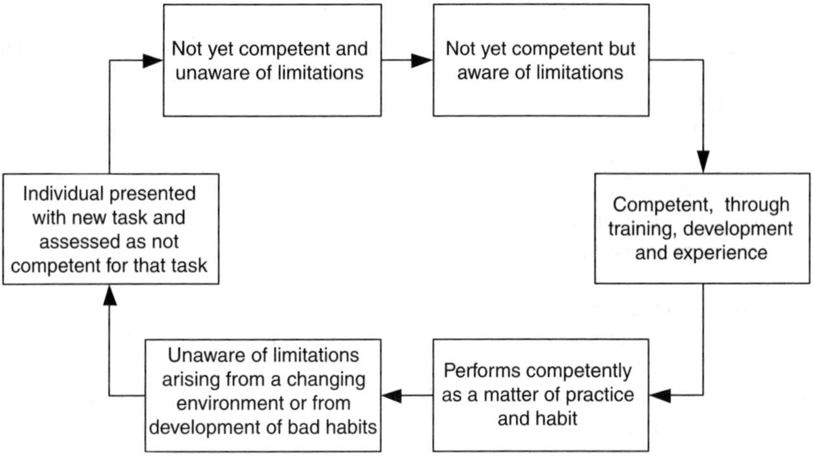

Figure 2. Competence stages for the individual. (Railway Safety Publication No. 1, 2007)

organisation e.g. competence monitoring and assessment while others will be department or area-specific e.g. HR procedures for the preparation of job descriptions. In parallel with this there should be appropriate staff briefings, for example describing and explaining the entire CMS, as well as training and other forms of communications to introduce new CMS related strategies and procedures.

WHY BOTHER?

Perhaps the most useful recent example of a large, successful organisation suffering catastrophic loss partly due to the absence of an effective and coherent CMS is the BP Texas City refinery accident. The accident on March 23, 2005 was one of the most serious US workplace disasters in the last 20 years. The Baker report (2007) emphasises the role that a lack of competence assurance played in the Texas City incident noting that 'BP has not effectively defined the level of process safety knowledge or competency required of executive management, line management above the refinery level, and refinery managers'. Further 'BP should develop and implement a system to ensure that its executive management, its refining line management above the refinery level, and all U.S. refining personnel, including managers, supervisors, workers, and contractors, possess an appropriate level of process safety knowledge and expertise'. (Baker report, 2007)

During follow on research during the process safety technical review at Texas City investigators found more failures in other refineries. A case study now referred to as the 'Whiting Rupture Disk Case Study' (Baker Report, Appendix D 2007), reported 'a breakdown in management oversight of training or assigning personnel to jobs with adequate technical knowledge relating to rupture disks.' The Baker report identifies a key root cause of this failure as 'a fundamental lack of knowledge about the safety implications of pressure between a rupture disk and a relief valve. Given this lack of knowledge, all the other actions BP personnel took (or failed to take) appeared to be reasonable and logical. However, this lack of knowledge raises the question of why personnel at every level—hourly staff, supervisors and managers—and in every work group—operations, maintenance, engineering, and management—lacked that knowledge. This lack of knowledge points to a breakdown in that portion of the management system that is responsible for ensuring workers have adequate technical knowledge' the CMS.

A 2001 incident investigation at a Toledo refinery on the release of reformate and water to the atmosphere notes training program deficiencies at Toledo. The investigation report concluded that "[the] current training system does not fully assess if a person has mastered the material being taught. . . . [The] operator had passed the written training test with a 100% and passed the field test with good ratings on the first try in 2000." Following these tests, the operator had been taken off shift for retraining due to an earlier incident. The retraining consisted of reviewing procedures and walking through systems. A formal training plan or written field test was not given. (Baker Report, 2007).

Clearly an important motivation for ensuring competence is the contribution of competence to safety. Competent staff play a fundamental role in achieving safe and effective performance.

An effective CMS should bring together the competence assurance aspects of various management systems that are often distributed throughout an organisation e.g. selection and recruitment may be facilitated by the Human Resources department while the actual recruitment and selection is undertaken by the relevant operational line managers. The often isolated elements in a management system dealing with competency must be developed or modified to support each other so that efforts in one area complement and facilitate the time and effort spent in another. This contributes to a coherent and comprehensive approach that can bring significant benefits, including the increasingly important ability to demonstrate to regulators, auditors and, occasionally, customers that staff competence is being effectively managed.

Not least, and as shown in Appendix 1, there is a significant list of UK legislation which requires competence to be formally managed.

THE CMS DEVELOPMENT OR REVIEW PROJECT

This section addresses what we consider to be an effective method of carrying out a CMS development or review project. We have selected a refinery as representative of a medium to large size organisation or business unit including the overarching complexities of corporate requirements and regulatory expectations.

The project may be led by either a competent senior/middle manager (better if they are from another site for independence) or an external consultant. For our purposes we will assume the client is the Site Manager.

FRAMING THE REQUIREMENTS

The first stage of the process as always is to decide on the scope of the project. An early strategy meeting with the client is useful to explore their expectations and determine which CMS model may be helpful and appropriate as a benchmark. This enables the project leader to ensure that the scope of the exercise is understood, agreed by all parties and therefore 'fit for purpose'.

It may be that instead of addressing the whole site organisation, a pilot study of one department or one plant area is chosen. This approach is recommended as it allows time to refine the methodology before addressing the entire organisation. It is important to emphasise that the effort required must never be underestimated. Although the initial project may only address one part of the organisation other parts of the organisation are inextricably linked and will invariably be drawn into the project to some degree.

To facilitate the project and foster positive and productive relationships it is always useful to have a local manager or individual with sufficient authority nominated to assist for example in organising interviews, providing local site information, and generally coordinating project requirements such as assigning meeting rooms and occasionally getting 'hand on' i.e. collecting individuals for interview as necessary.

PROJECT KICK OFF

The outline approach described here assumes that there is no formal CMS in place. If the existing site safety case and other arrangements such as risk assessments have provided adequate task identification and complementary/supporting risk assessments, it may be possible to start at step 5.

This list itemises the steps to include in a CMS project. It is at the discretion of the project team to amend and adapt the steps as appropriate for example conducting steps in parallel to increase efficiency.

For the organisational unit(s) to be addressed:

1. Inform all staff before and throughout the process about the project. The project may result in a significant organisational change while "competence" or rather "incompetence" is an emotive word. To support a healthy safety culture it is important to ensure staff are aware of what is happening, why it is happening, and feel their role and potential contribution to the project is recognised and valued.
2. Arrange the unit into functional groups, including any term contractors. These groupings should be vertical (e.g. between differing plants, administration staff, cleaners, etc.) and horizontal (e.g. plant operator, Control Room operator, Plant Supervisor, Shift Supervisor, etc.) The objective is to arrange the groupings so that those with the same roles, tasks and risks are identified.
3. In the analysis process include one or more representatives per group.

ANALYSIS

4. Conduct a Task Analysis and Risk Assessment on normal, degraded, and emergency operations by group. These will ensure that all tasks are identified and then prioritised

to address the most important first in terms of both cost and consequence. Ensure major hazards are addressed as well as Occupational Health and Safety hazards in all activities.

5. For each task, define all competencies which are required to carry out the task correctly:
   - Skills (e.g. welding);
   - Qualifications;
   - Knowledge;
   - Experience;
   - Training.
6. Identify and/or develop standards that define the correct method of carrying out the task and can be used to assess competence. (These may be standards, procedures, specifications, etc.)
7. Define monitoring and assessment requirements. Task observation is the simplest approach to ensure that task analyses are accurate and usually suitable for operator and some supervisory roles. Managers have more abstract competencies particularly in emergency roles. The use of simulators or more frequently interviews may be more appropriate.
8. Develop a procedure for monitoring of competence and ensure that contractors and sub contractors are included as appropriate.
9. Develop a procedure for managing sub-standard competence.
10. Develop a supplier assessment CMS and monitoring approach for visiting contractors.

HUMAN RESOURCES

11. Develop Job Descriptions.
12. Review induction training.
13. Review recruitment methodology.
14. Develop procedure for competence assessment.

OUTPUT

15. Train competence assessors and managers to manage the CMS.
16. Monitor, develop and assess competence.
17. Audit the CMS.
18. Address CMS changes.

The typical deliverable for this type of project is a report addressing:

- The current situation regarding the CMS;
- The strengths and weaknesses of the existing system when compared with a best-practice CMS model;
- A detailed development strategy, which will include:
  - What must be done;
  - Who must do it;

- When it must be done by;
- How changes will be managed;
- Identification of post-change responsibilities related to the CMS.

ISSUES

Different organisations will encounter different issues to address and resolve during the project. We discuss some of the most significant issues that we have experienced during CMS projects.

ORGANISATIONAL CULTURE

A hot topic at the moment is safety culture, arguably just one aspect of a wider organisational culture. A significant aspect of culture in this context is that awareness and understanding of staff attitudes and motivation will help projects efficiently and effectively target both organisational and staff requirements. With the best will in the world if staff are not engaged in a project or process success will be limited.

During one project we produced a CMS development strategy and participated in the Steering Group as advisors and to provide oversight of the implementation. We had an excellent CMS Project Manager who was very keen to communicate the project to all staff. We also had a very engaged Operations Manager who ensured that the project was presented to all shifts during a weekly session of 'tool-box' talks. As part of our oversight role, we carried out an audit of the organisation to investigate how staff were learning about the introduction of the CMS. We were disappointed to learn that ~20% of the operations staff knew nothing about the project and said they were not interested. More worrying was the ease with which they had avoided the tool-box talks simply by finding an excuse to be out on the plant. The lack of awareness by management regarding the reality of staff perceptions could quite simply undermine plant safety. If competence assurance is Safety Culture needs constant development, and people need to know that any change is an advantage (or at least no loss) for them.

BUY-IN FROM SENIOR MANAGEMENT

During the audit mentioned above, we also found an entire Directorate that had not been informed about the introduction of a CMS. A phased approach was being used to introduce the CMS. The Operations directorate was leading the project while the other Directorate was in a related but non-operational area. The latter Directorate did have a significant, if smaller, supporting role but was not included in the initial wave of communications and associated enthusiasm. The senior manager thus largely ignored the project and had not been encouraged to participate or support the work. As a result, other issues felt to be more pressing were given priority, and the manager did not attend key project meetings. The knock on effects included a significant and unexpected requirement to deliver the bulk of

CMS related communications a second time to Directorate staff and re-doubled efforts to ensure the Directorate felt involved and valued in the CMS project simply to avoid undermining much of the previous work.

PROJECT HIJACKERS

It is possible that some individuals may see change as an opportunity to enhance their role and/or status within the organisation. This is entirely natural and even necessary, as management should be motivated to improve. However, we have revisited organisations after delivering a CMS implementation strategy and found minimal progress with the resources allocated for the CMS project being used to promote an entirely different development plan. The best approach to manage this scenario is ensuring there is a strong, experienced and if possible senior project manager capable of coping with potential hijackers. Key to the project manager's success is the breadth and depth of their oversight. In this case they had no oversight and were thus vulnerable to manipulation as their lack of information provided through 'the usual channels' was severely limited.

PROCEDURES

It is not uncommon during a CMS review to find that procedures are not well managed. This may occur when they are not reviewed and updated appropriately, or may be missing, having never been developed. It is often the case that individual shift teams will develop their own ways of carrying out tasks, with the priority on ease of implementation rather than safety.

During one CMS review at a petrol refinery, operators reported that each of the four shifts on their plant had different ways of conducting plant operations. At the time of the review there were no formal procedures in place. The lack of procedures increased the risk of an individual doing something unsafe when working on a different shift that operated the plant in its own manner. An unambiguous, safe way of performing critical tasks that is accepted by all staff is clearly necessary. Formal procedures provide this framework. A substantial supervisory effort coupled with significant safety culture development work are also required to support implementation and embedding of the use of procedures. Often in a "no procedures" environment there is a prevailing negative attitude to the use of procedures that must be addressed in parallel with the introduction of formal procedures.

CONTRACTOR COMPETENCE

All organisations will use contractors, and many organisations have teams of contractors who are embedded into the site organisation as a long-term presence to carry out specific duties. These may range from supporting tasks such as cleaning, security and catering to the potentially higher risk activities such as plant maintenance and/or rail terminal operations. Other contractors will attend the site on a short-term basis to maintain or repair specialist equipment on breakdown or during a planned plant outage.

Major plants are commonly shut down for extended periods for a comprehensive maintenance outage. During this period there may be several hundred visiting contractors on the site, either to provide a general service (e.g. scaffolding, lagging, or painting) or specialist contractual support for specific plant maintenance and repair. The competence of every individual on site contributes to overall plant safety thus contractors' competence must be as well managed as permanent site staff.

However, at times, the ability to simply find, employ and manage such large numbers of contractors may be aggravated by the fact that major hazard sites are often found grouped together in relatively small and remote areas where there is a finite resource of contractors. In the absence of alternative suppliers and the potential severe cost and safety implications, significant effort may be justified to ensure they achieve acceptable levels of competence.

The negative consequences of poorly managed contractor competence were evidenced in the Phillips Petroleum accident in Pasadena, Texas on the 23rd of October, 1989. An explosion and ensuing fire occurred that resulted in 23 known dead and one missing. In addition, more than 100 people were injured to varying degrees. The accident report notes that 'metal and concrete debris was found as far as six miles away following the explosion.' (FEMA, 1989)

Post-accident enquiries revealed that, during maintenance, a pneumatically operated valve used for isolation had opened when the intention had been to close it. This occurred because the contractor had connected an air hose to the wrong control port which opened, rather than closed, the valve. Company procedures however required that air hoses should not be connected to valves during maintenance.

The investigation found that contractors had not been provided with company standards and procedures covering their work, and there was no system in place to evaluate contractors' competence in implementing the standards and procedures even if they had received the documents. An audit of the arrangements for the control of contractors could have identified this assurance gap and allowed time to manage the issue.

UK legislation continues to emphasise proper management of such issues. Under sections 2 and 3 of the Health and Safety at Work Act (HSW) (2002) the client company remains responsible for occupational health and safety and operational safety on the site, no matter where its resources come from. The self-employed and companies supplying the contractor labour also have a duty of care under the HSW while there are specific legislative requirements for competence to be formally managed in a variety of UK regulations as detailed in Appendix 1. The client company who operates the site must therefore verify that all contractors, sub-contractors, agency or self-employed are competent and managed within a CMS.

There are three ways this can be achieved:

1. Ensure that the contractors' management has a CMS which is appropriate and equivalent to the client site's CMS arrangements. This is probably the better option for those contractors who will be on site for a relatively short period, such as non-term contractors. The validity and equivalence of the contractor's CMS should be assessed initially as part of the supplier assessment process. Competence monitoring by the client site organisation can then be defined as part of the contractual arrangements.

2. Arrangement by the contractors' management to have their staff included within the client site CMS, and subject to the site monitoring, training and record-keeping regime. This is probably the best option for the term contractors who are on site for extended periods effectively becoming site staff.
3. Reliance on national or local industry training and competence schemes. In some areas industry is further addressing competence assurance through a 'passport' style scheme. To receive a passport a contractor must have a certain level of awareness of site risks and the mitigating safety requirements. Such an approach is useful though limited as each site has its own unique risk spectrum restricting the scheme to basic specifications of knowledge.

CHANGE MANAGEMENT

In our refinery example the introduction of a formal CMS is considered a significant change. Selection of an appropriate method to introduce and manage changes necessitated by the development or modification of a CMS will have a great influence on the success of the system. This is a delicate job and to embed a CMS in an organisation is quite simply best handled by experienced, motivated staff and change management champions with the support of all senior management.

CONCLUSIONS

The incidents cited throughout this paper highlight how competence assurance plays a vital part in continuously encouraging and improving the safe operation of major hazard installations which are heavily dependent on the competence of the people operating, maintaining and managing them, including both site staff and contractors.

As noted in the Baker report (2007) 'The passing of time without a process accident is not necessarily an indication that all is well and may contribute to a dangerous and growing sense of complacency. When people lose an appreciation of how their safety systems were intended to work, safety systems and controls can deteriorate, lessons can be forgotten, and hazards and deviations from safe operating procedures can be accepted. Workers and supervisors can increasingly rely on how things were done before, rather than rely on sound engineering principles and other controls. People can forget to be afraid.'

We believe that developing and embedding a robust CMS within an organisation will help avoid such dire circumstances. Far from requiring organisations to re-invent the wheel most organisations will have many elements of a CMS in place that can be brought together in a formal, co-ordinated system as discussed here.

ACKNOWLEDGEMENTS

We would like to acknowledge the contribution of ESR Technology in supporting our paper and in particular our colleague Maurice Bromby who provided years of experience and invaluable insights both on the job and while preparing this paper.

REFERENCES

1. BP US Refineries Independent Safety Review Panel Report, 2007, (Baker Panel Report)
2. Hoddinott D. 2002, "Developing and Maintaining Staff Competence" Health and Safety Executive (HMRI)
3. Lees, F.P., 'Loss Prevention in the Process Industries – Hazard Identification, Assessment and Control', Volume 3, Appendix 1, Butterworth Heinemann, ISBN 0 7506 1547 8, 1996
4. Phillips Petroleum Chemical Plant Explosion and Fire, Pasadena, Texas, October 23, 1989. Report 035, Major Fires Investigation Project, Federal Emergency Management Agency, United States Fire Administration National Fire Data Center.
5. Railway Safety Publication No. 1, 2007 - http://www.rail-reg.gov.uk/upload/pdf/sf-dev-staff.pdf

APPENDIX 1 – RELEVANT LEGISLATION

Competence Assurance is a legislative requirement in a variety of regulations including:

- Health and Safety at Work Act 1974
- Provision and use of Work Equipment Regulations 1998;
- The Control of Major Accident Hazards (Amendment) Regulations 2005;
- Offshore Installations (Safety Case) Regulations 2005;
- The Offshore Installations (Prevention of Fire and Explosion, and Emergency Response) Regulations 1995;
- Quarries Regulations 1999;
- Railways and Other Guided Transport Systems (Safety) Regulations 2006;
- The Control of Substances Hazardous to Health (COSHH) (as amended) 2002;
- The Health and Safety (Display Screen Equipment) Regulations 1992 (as amended 2002);
- The Ionising Radiation Regulations 1999
- The Control of Noise at Work Regulations 2005
- The Lifting Operations and Lifting Equipment Regulations 1998
- The Electricity at Work Regulations 1989
- Offshore Installations and Wells (Design and Construction, etc) Regulations 1996;
- The Construction (Design & Management) Regulations 2007 (CDM 2007);
- The Diving at Work Regulations 1997.

MOODLE E-LEARNING ENVIRONMENT – AN EFFECTIVE TOOL FOR A DEVELOPMENT OF A LEARNING CULTURE

Virve Siirak
Tallinn University of Technology, Chair of Working Environment and Safety, Estonia.
E-mail: vsiirak@staff.ttu.ee

The 21st century, structural changes in our industries and economics, globalization of our world need responsible new engineers and scientists. New competence requirements in the ICT (information and communication technology) sector and in information and knowledge work mean new challenges for national educational systems. In particular, traditional technical and higher engineering education needs critical evaluation and a broadening of curricula with knowledge traditionally included in social sciences. Understanding of human and organizational behaviour, cultural understanding, communication and language skills, and the capacity for conceptual thinking are important competencies needed in the future.

In this paper it is argued that blended learning with computer based learning in the Moodle e-learning environment based on social constructivist learning theory is an effective tool for teaching and learning of Occupational Health and Safety discipline (OHS) – including chemical risks and human factor issues, for future engineers and managers. The author has six years experience of computer based teaching. The author's own teaching experience of the Moodle e-learning environment for creating and providing different courses in the Tallinn University of Technology and in the Virumaa College of Tallinn University of Technology (located in the Ida-Viru County of Estonia and the Tartu College of Tallinn University of Technology, will be presented.

According to the questionnaires given to students at the end of each course, the teaching and learning in the Moodle e-learning environment as blended learning is very useful for development of a learning culture. The effectiveness and motivation for learning are higher than providing traditional methods of learning. New possibilities and dimensions for teaching and learning are opened which will develop the learning culture.

INTRODUCTION

Increasing access to more and better information is available by rapidly development of technology. The impact of rapidl development of Infocommunication Technology (ICT) to all aspects of the society is described by Bradley 2001, Järvenpää 2001. To help students turn information into knowledge, teachers need to know the new teaching strategies. Psychosocial risks, musculosceletal disorders and dangerous substances are among the priority areas for future Occupational Health and Safety research in European Union. The report *Priorities for Occupational Safety and Health Research in the EU-25*. According to this report a vast and increasing number of chemicals are present in workplaces, with about 100 000 different substances currently registered in the EU market. The chemical industry is Europe's third largest manufacturing industry, employing 1.7 million people directly, with up to three million jobs dependent on it. Exposure to dangerous chemicals occurs at many workplaces outside the chemical industry: for example agricultural workers use pesticides, detergents and microbiological dusts, and construction workers commonly use solvents and paints. According to the third European survey on working conditions (2000), 16% of employees in the EU handle or are in contact with dangerous substances for at least on quarter of their working time. There are three main research priorities in this field: the validation and improvement of models for workers exposure assessment, the exposure to nanoparticles and ultrafine particles and assessment and measurement methods for workplace exposure to biological agents. Chemical Engineers now have a rising capability of computing resources available to them.

In this situation, new challenges for the higher education are continued. The growing interest of blended learning (combination of traditional teaching methods of face to face and online media) in higher education is indicated by the increasing number of studies in this area (Poole 2006, Irons et al. 2002; MacDonald and McAteer, 2003; O'Toole and Absalon, 2003; Stubbs and Martin, 2003). For example at the University of Central England in Burmingham (UCE), academic staff are encouraged to incorporate both traditional and web-based ICT (information and communication technology) modes of teaching and learning in the courses they deliver, using Moodle software. Preliminary quantitave evaluations at UCE have releaved that over 70 percent of the 388 students in the sample from across all faculties claimed to have enjoyed using the web-based aspects of blended courses. Over 75 percent of a sample of 329 students felt that Moodle had helped them learn the subject and nearly 80 percent of the sample reported that they would like future modules be blended in this way (Poole, 2006; Staley 2005).

COMPUTER BASED TEACHING AND LEARNING EXPERIENCE IN TALLINN UNIVERSITY OF TECHNOLOGY

Tallinn University of Technology is a national technology university of international repute, an active cooperation network partner at the forefront of Estonia's knowledge-based economic development. (Strategy 2006)

In Tallinn University of Technology (TUT) (former Tallinn Technical University) The discipline of Risk and Safety Sciences based on ergonomics knowledge has been taught for more than 20 years (Kristjuhan 1994).

In September 1999 a new ergonomics laboratory with computers was installed in Tallinn University of Technology (TUT). These facilities help to inspire for searching Health and Safety information in Cyberspace for their studies based on Problem Based Learning (Siirak, 1999, 2000; 2001, 2002 Tint, 2000) The course of Risk and Safety Sciences for students of the Faculty of Chemical Engineering and Gene Technology (Bachelor level) was provided autumn 2001 by V. Siirak. During practices students were inspired to use the internet and find the modern Occupational Health and safety information from internet databases and to use the databases of the European Agency for Safety and Health and Work, the Health&Safety Executive, UK, the U.S. National Institute of

Occupational Safety and Health (NIOSH) etc. The special web-site for the students was prepared. For different courses.

MOODLE E-LEARNING ENVIRONMENT

Since autumn 2006 the courses of Risk and Safety Sciences and Working Environment and Ergonomics were created in Moodle e-learning environment. All courses are provided as blended learning: the traditional method face-to-face is blended of web-based support in Moodlle e-learning environment. Autumn 2006 the first course of discipline of Working Environment and Ergonomics was provided for full-time students of the Faculty of Economics and Business Administration in the Moodle e-learning environment. In this course participated 166 students (Bachelor level). Spring 2007 this course was provided with 72 participants for distance learning students of the Faculty of Economics and Business Administration. Spring 2007 also the course of the Risk and Safety Sciences for students of Production Engineering and Entrepreneurship in the Virumaa College of TUT was created and provided for 14 students and the Course of Risk and Safety Sciences for 72 students of Landscape Architecure and Construction in the Tartu College of TUT was created and provided. The course were provided as blended learning where the classroom activities were supported by Moodle e-learning environment. Autumn 2007 the Master level Course of Chemical Risks for future Industry Hygienists was provided as blended learning with Moodle e-learning environment. for 7 students. The assignments in Moodle learning environment were related with working with international professional databases of chemical risks (first year in Moodle e-learning environment).

AIM OF THE STUDY

The aim of the study is to find out how students appreciate the courses provided in Moodle e-learning environment.

MATERIAL AND METHOD

At the end of the courses a questionnaire was given to all groups of students. The questionnaires were given for 57 students in the course of Working Environment and Ergonomics (Bachelor level), (100% were filled) and for 14 students of the course of Risk and Safety Sciences in the Virumaa College of TUT (Diploma level) (100% were filled), for 42 students of the course of Risk and Safety Sciences in the Tartu College of TUT (Diploma level) and for 14 students of the course of Risks of Social Environment (Master level). (100% were filled). After the Master level course of the Chemical Risks for future Industry Hygienists the questionnaire was given for 7 students (100% filled).

Students had to answer to 5 questions:

1. How do you appreciate the Moodle e-learning environment?
2. Which part of the course was most interesting for you?

3. Which part of the course was unclear for you?
4. How do you will to use obtained knowledge in practice?
5. What do you like to learn more?

RESULTS

Of respondents 100% answered that Moodle e-learning environment is very effective learning tool. Students wrote that they are encouraged and motivated to learn more in Moodle e-learning environment and they do not like to learn courses which are not in Moodle e-learning environment. The materials in Moodle e-learning environment are clear. Of respondents 80% answered that their participation in forums and othet activities available in Moodle e-learning environment is very useful for learning from each other obtaining new knowledge. Some students wrote that availability of course materials and activities in Moodle e-learning environment encourage their interest for claasroom activities and the face to face contact with teacher is now in a new level. Students appreciate high that they can learn in Moodle e-learning environment the time and place suitable for each student. Some students wrote that the experience of learning in Moodle e-learning environment is helpful for development of their self- discipline.

Very interesting is that student who havedsuccessfully finished the course continued their participation in the course in Moodle e-learning environment. For example autumn 2006 the first course of the discipline of Working Environment and Ergonomics was provided. From 166 students finihed the course successfully, 130 students today continued their participation in the course in Moodle e-learning environment. E-learning community is established.

The Master Level Course of Chemical risks for industrial hygienists was the first year course in Moodle e-learning environment. The comparison of this courses before Moodle introduced is impossible.

DISCUSSION

According to my experience with computer based learning since the year 2001, providing of courses in Moodle e-learning environment are more effective than providing the courses where course materials are available on the website. Very effective is that students activities from participation in learning forums and learning from each other are encouraged. The problem is that sometimes students hesitate to participate in learning forums, they are sometimes afraid that other students can read their letters to learning forums. The efficiency of learning process is depending on the style of creation of the course. The course had to be designed simply and clearly encouraging students own activity in the learning process. The assignments encouraging students activity have to be provided. In comparison of previous results before Moodle was introduced (only materials were available on the website), students are more satisfied and motivated to learn the discipline. Students appreciate high that can fill all the assignments in Moodle e-learning environment, what is more suitable for students. Before Moodle was introduced the filling the different assignments was more complicated. There is no statistical differences of the satisfaction of the courses before Moodle was used. Before Moodle was introduced all the solutions of web based learning were highly appreciated by the students in comparison of traditional courses without using web tools. After Moodle was used, the motivation of students and their own activity in the learning process, the students interpersonal contacts (learning from each other) were successfully increased. After Moodle was introduced, for the teacher the online contact with students and monitoring their activities is successfully improved.

CONCLUSIONS

In the 21st century, where structural changes in our industries and economics, globalization of our world need responsible new engineers and scientists, new challenges for the higher education and learning culture are continued. and new possibilities, dimensions and solutions are opened.

One of the new possibilities and solutions is blended learning using the Moodle e-learning environment.

Moodle e-learning environment is a very effective learning tool supporting blended learning which encourages the students motivation for learning activity and interest to the course, developing the learning culture. We have to be open for future development of Moodle e-learning environment according to rapid development of ICT (information and communication technology) dimensions and new possibilities and solutions for developing the learning culture.

ACKNOWLEDGEMENTS

The author would like to express the gratitude to the European Commission and Estonian National ICT program for Higher Education for supporting the creation of the Course of Working Environment and Ergonomics in Moodle e-learning environment autumn 2006 (EITSA 2 ESF project REDEL).

REFERENCES

Bradley, G. *Information and communication technology (ICT) and humans – how we will live, learn and work.* In: Bradley, G. Editor. Humans on the Net. Stockholm:Sweden, Prevent, 2001.pp. 22-44. ISBN 91-7522-701-0.

Irons, L.R., Keel, R., and Bielema;C. L. *Blended Learning and Learner Satisfaction. Keys to user acceptance*, USDLA Journal 16(12), 2002

Järvenpää, E., Eloranta, E., *Information and communication technologies and quality of working life: implications for competencies and well-being*. In: Bradley, G. Editor. Humans on the Net. Stockholm:Sweden, Prevent, 2001.pp. 109-118. ISBN 91-7522-701-0.

Kristjuhan, Ü., *Increasing the efficiency of thinking in ergonomics research.* In: Nordiska Ergonomisällskapets Årskonferens NES'94, Stenungsund, Sverige, 1994.

O'Tools, J.M., and Absalom, D.J., *The impact of Blended Learning on Student Outcomes: Is there Room on the Horse for Two* ? Journal of Educational Media 28(2/3), 179-91.2003

Poole, J., *E-learning and e-learning styles: students' reactions to web-based Language and Style at Blackpool and the Fylde College.* Language and Literature 2006; 15; 307. The online version: http://lal.sagepub.com/cgi/content/abstract/15/3/307

Siirak, V., *Didactic experience of risk and safety education in Estonia.* In: Pacholski, L.M., Marcinkowski, J.S., (Eds): Certification and Accreditation of Ergonomics, Labour Protection and Work Safety Education. Proceedings of the 16th International Seminar of Ergonomics Teachers. Poznan University of Technology, Poland, 1999. Pp, 199-203. ISBN: 83-906191-2-5.

Siirak, V., Kristjuhan, Ü., *Experience of problem-based learning in ergonomics and safety education in Estonia.* In: D.de Waard, C. Weikert, J. Hoonhout, J. Ramaekers (Eds) Human-System Interaction: Research and Application in the 21st Century, The Netherlands: Shaker Publishing, 2000. Pp. 99-106. ISBN 90-423-0126-0.

Siirak, V., *Influencing behaviour through learning of ergonomics knowledge in Cyberspace: a new millennium strategy to the reduction of health risks and accidents at working environment in Estonia.* In: K.E. Fostervold, T. Endestad (Eds): At the gateway to Cyberspace-ergonomic thinking in a new millennium. Oslo: Nordiska Ergonomisällskapet. 2000. Pp. 225-228.

Siirak, V., Kristjuhan, Ü. *Changing Paradigms for Ergonomics and Safety Educational Technology in Estonia*, Proceedings of the Second International Conference ERGON-AXIA 2000 - Ergonomics and Safety for Global Business Quality and Productivity, Warsaw, Poland, 19-21 May, 2000, pp. 293-296, ISBN: 83-87354-54-6.

Siirak, V. *Some Possible Solutions for Improving the Estonian Working Environment Proposed by Students of Tallinn Technical University.* In: People in Control: An International Conference on Human Interfaces in Control Rooms, Cockpits and Command Centres. Conference Publication,. 19-21 June 2001, UMIST, Manchester, United Kingdom. IEE 2001. Pp. 340-344.

Siirak, V. *New challenge for ergonomics and human factors education in technical universities.* In: Promotion of Health through Ergonomic Working and Living Conditions. Outcomes and methods of research and practice. Proceedings of NES 2001 Nordic Ergonomics Society 33rd Annual Congress, 2-5 September 2001, University of Tampere, Finland. pp. 210-212. ISBN: 951-44-5168-6.

Siirak, V. *New Challenges for Human Factors and Ergonomics Education in Technical Universities-* CD ROM, WorkCongress6: 6th International Congress on Work Injuries, Prevention, Rehabilitation and Compensation, 30 November-3 December 2004, Rome. Italy. INAIL. Italian Workers Compensation Authority. Italian Workcover. Directorate of Communication. External Communication and International Relations Unit. Piazzale Giulio Pastore 6. I-0014 Rome RM, EUROPEAN UNION.

Siirak, V. *Multi-media and the internet as educational tools for solving the problems of ergonomics and safety*, In: Human Factors in Transportation, Communication, Health and the Workplace. Shaker Publishing, The Netherlands, 2002.pp. 471-472. ISBN 90-423-0206-2.

Siirak, V. *Experience of new teaching strategies of occupational health and ergonomics at Tallinn Technical University.* In: Best Practices in Occupational Safety and Health, Education, Training and Communication: Ideas That Sizzle. Proceedings of 6th International Conference Scientific Committee on Education and Training in Occupational Health, ICOH, October 28-30, 2002, Baltimore, Maryland, USA, pp. 224-226.

Siirak. V. *Computer Based Learning as an Effective Tool for Prevention of Chemical Risks* - CD ROM: 8th International Symposium of ISSA Research Section, Athens (Greece) 19-23 May 2003. E.L.I.N.Y.A.E. (Hellenic Institute for Occupational Health and Safety) ISSA Research Section Symposium 2003.

Siirak, V. *Some experience of computer based learning in OHS education for future engineers* In: ICIE 2005 1st International Conference on Interdisciplinarity in Education 17-19 April 2005, Athens.Greece Book of Abstracts. p.31.

Stubbs, M., and Martin, I., *Blended Learning One Small Step*, Learning and Teaching in Action 2(3), 2003.

Staley, A., *Students' Perspectives of Moodle*, Digital Future: The Newsletter of the Learning Technology Development Unit 2; 2-3, 2005.

Strategic plan of Tallinn University of Technology 2006-2010.

Tint, P., Siirak, V., *Computer-based learning in Occupational Safety and Health and health problems with computer use.* In: D.de Waard, C. Weikert, J. Hoonhout, J. Ramaekers (Eds) Human-System Interaction: Research and Application in the 21st Century, The Netherlands: Shaker Publishing, 2000. Pp. 107-109.ISBN 90-423-0126-0.

USING ARX APPROACH FOR MODELLING AND PREDICTION OF THE DYNAMICS OF A REACTOR-EXCHANGER

Yahya Chetouani
Université de Rouen, Département Génie Chimique, Rue Lavoisier, 76821, Mont Saint Aignan Cedex, France. E-Mail: Yahya.Chetouani@univ-rouen.fr

The main aim of this paper is to establish a reliable model of a process under its normal operating conditions. The use of this model should reflect the true behaviour of the process and allow distinguishing a normal mode from an abnormal one. In order to obtain this reliable model for the process dynamics, the black-box identification by means of an ARX (Auto-Regressive with eXogenous input) model based on the least squares criterion has been chosen. This study shows the choice and the performance of this modelling approach. An analysis of the inputs number, time delay and their influence on the behaviour of the prediction is carried out. A reactor-exchanger is used to illustrate the proposed ideas concerning the dynamics modelling. Satisfactory agreement between identified and experimental data is found and results show that the identified model successfully predicts the evolution of the outlet temperature of the process.

KEYWORDS: reliability; safety; modelling; ARX; reactor-exchanger

1. INTRODUCTION

Process development and continuous request for productivity led to an increasing complexity of industrial units. In chemical industries, it is absolutely necessary to control the process and any drift or anomaly must be detected as soon as possible in order to prevent risks and accidents. Moreover, detecting a fault appearance on-line is justified by the need to solve effectively the problems within a short time (Chetouani, 2006).

We are interested in the anomaly detection module intended to supervise the functioning state of the system (Chetouani, 2007). The former has to generate on-line information concerning the state of the automated system. This state is characterized not only by control and measurement variables (temperature, reaction rate, etc.), but also by the general behaviour of the process and its history, showing in time whether the behaviour of the system is normal or presents drifts. In the context of numerical control, fault detection and isolation (FDI) proves a vital complement to the adaptive means of dealing with instabilities in nonlinear highly unsteady systems. Under normal conditions, the fault detection module allows all information to be processed and managed in direct liaison with its general behaviour. In other case, it detects any anomaly and alerts the operator by setting on the appropriate alarms.

The intrinsic highly nonlinear behaviour in the industrial process, especially when a chemical reaction is used, poses a major problem for the formulation of good predictions and the design of reliable control systems (Cammarata et al., 2002). Due to the relevant number of degrees of freedom, to the nonlinear coupling of different phenomena and to the processes complexity, the mathematical modelling of the process is computationally heavy and may produce an unsatisfactory correspondence between experimental and simulated data. Similar problems arise also from the uncertainty for the parameters of the process, such as the reaction rate, activation energy, reaction enthalpy, heat transfer coefficient, and their unpredictable variations. In fact, most of the chemical and thermo-physical variables both strongly depend and influence instantaneously the temperature of the reaction mass (Chetouani, 2007). One way of addressing this problem is the use of a reliable model for the on-line prediction of the system dynamic evolution (Leontaritis et al., 1985).

The main aim of this study is to obtain a powerful model of reference allowing to reproducing the process dynamics in normal mode. The present study focuses on the development, and implementation of an ARX model for the one-step ahead forecasting of the reactor-exchanger dynamics. The performance of this stochastic model was then evaluated using the performance criteria. Results show that the ARX model is representative for the dynamic behaviour of the nonlinear process. Experiments were performed in a reactor-exchanger and experimental data were used both to define and to validate the model. The identification procedure, the experimental set-up and prediction results are described in the following sections.

2. INPUT-OUTPUT MODELLING APPROACH: ARX IDENTIFICATION

Modelling is an essential precursor in the parameter estimation process. Identification strategies of various kinds by means of input–output measurements are commonly used in many situations in which it is not necessary to achieve a deep mathematical knowledge of the system under study, but it is sufficient to predict the system evolution (Fung et al., 2003; Mu et al., 2005). This is often the case in control applications, where satisfactory predictions of the system that are to be controlled and sufficient robustness to parameter uncertainty are the only requirements. In chemical systems, parameter variations and uncertainty play a fundamental role on the system dynamics and are very difficult to be accurately modelled (Cammarata et al., 2002). Therefore, the identification approach based on input-output measurements can be applied.

In this study, the chosen method adopted for process modelling is based on a parametric identification of an ARX model. The choice of this strategy is justified by the fact that it is simple to implement it. The evolution of the estimated output allows to follow the dynamics evolution of the process and to reflect the fault presence by the variation of the estimated parameters of the identified model (Iserman, 1993).

ARX modelling was the subject of studies in several fields such as chemical engineering (Rivera et al., 1995; Rohani et al., 1999), agriculture and biological science (Fravolini et al., 2003; Frausto et al., 2003), medicine (Liu et al., 2003), energy and the power (Yoshida et al., 2001), Energy economics (Ringwood et al., 1993).

In this paper, we propose the ARX identification for modelling the dynamic behaviour of a reactor-exchanger. The aim is to analyze the model orders, the time delay and the validation of the identified model.

The ARX structure describes the input effects $u(t)$ on the process output $y(t)$. The ARX model is represented by the following expression:

$$y(t) = -a_1 y(t-1) - \cdots - a_{na} y(t-n_a) + b_1 u(t-1-n_k) + \cdots + b_{n_b} u(t-n_b-n_k) + e(t) \quad (1)$$

where $e(t)$ refers to the noise supposed to be Gaussian. a_{n_a} and b_{n_b} are the model parameters. n_a and n_b indicate respectively, the order of the polynomials of the output $A(q)$ and the input $B(q)$. The parameter n_k is the time delay between $y(t)$ and $u(t)$.

The polynomial representation of the equation (1) is given as follows:

$$A(q)y(t) = B(q)u(t - n_k) + e(t) \quad (2)$$

where $A(q)$ and $B(q)$ are given by:

$$A(q) = 1 + a_1 q^{-1} + \cdots + a_{n_a} q^{-n_a} \quad (3)$$

$$B(q) = b_1 q^{-1-n_k} + \cdots + b_{n_b} q^{-n_b-n_k} \quad (4)$$

q^{-1} is the delay operator such as:

$$u(t = 1) = q^{-1}\, u(t) \quad (5)$$

$A(q)$ and $B(q)$ are estimated by the least squares identification (Ljung, 1987, Ljung, 2000).

3. EXPERIMENTAL RESULTS

3.1. EXPERIMENTAL DEVICE

The reactor-exchanger is a glass-jacketed reactor with a tangential input for heat transfer fluid. It is equipped with an electrical calibration heating and an input system. It is equipped with Pt100 temperature probes. The heating-cooling system, which uses a single heat transfer fluid, works within the temperature range between −15 and +200 C. Supervision software allows the fitting of the parameters and their instruction value. It displays and stores data during the experiment as well as for its further exploitation. The input of the reactor-exchanger $u(t)$ represents the heat transfer fluid temperature allowing the heating-cooling of the water. $y(t)$ represents the outlet temperature of the reactor-exchanger. The process is excited by an input signal which is very rich in frequencies and amplitudes in order to have a data set suitable for the estimation procedure. The sampling time is fixed at 2 seconds. Before starting the estimation of parameters, the database is divided into two separated sets. The first set is used for the estimation of parameters and the second one for the model validation. The first set is sufficiently informative and covering the whole spectrum. The second set contains sufficient elements to make the validation as credible as possible.

3.2. ESTABLISHMENT OF ARX MODELS

A set of models is built by fixing $n_a = [1, \ldots, 5]$, $n_b = [1, \ldots, 5]$ and $n_k = [1, \ldots, 10]$. The models having n_a lower than n_b are rejected in order to respect the physical aspect of

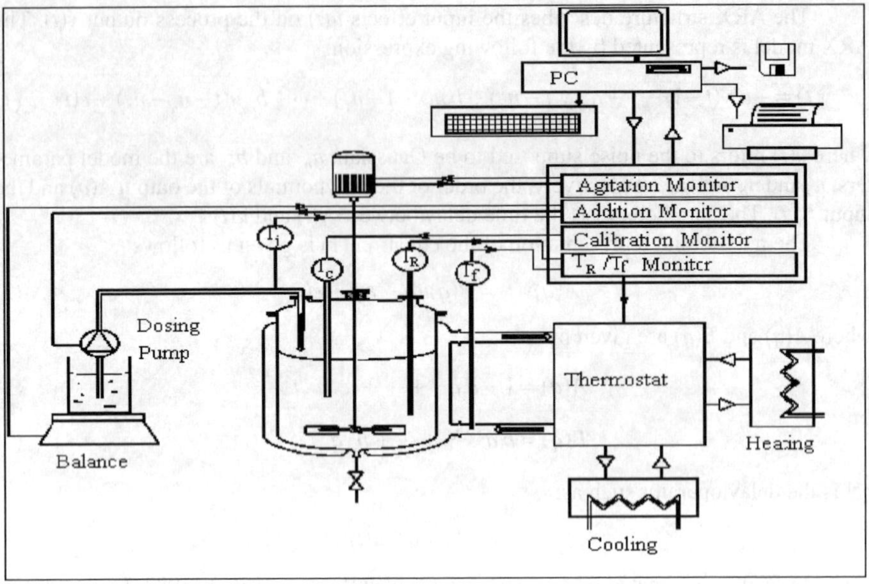

Figure 1. Experimental device: A reactor-exchanger

the process. Consequently, a set of 150 models is worked out and estimated while examining the stability of each model by the Lyapunov criterion (Ljung, 1987).

3.2.1. Estimation of the time delay

There are several methods for estimating the time delay (Ljung, 1987; Chen et al., 1989; Ljung, 2000). In this paper, the adopted approach is based on the evaluation of the quadratic criterion (Ljung, 2000). This criterion is as follows:

$$V(\theta) = \frac{1}{N}\sum_{t=1}^{N}\varepsilon(t,\theta)^2 \tag{6}$$

$\varepsilon(t,\theta) = y(t) - \hat{y}(t)$ and $\hat{y}(t)$ represent respectively the prediction error and the associated predictor. The quadratic criterion value is calculated in function of the time delay value $n_k = [1, \ldots, 10]$. This method is applied to two simple ARX models ($n_a = n_b = 1$) and ($n_a = n_b = 2$). The choice of these simple models allows observing the criterion evolution according to the time delay but without compensating it (time delay) by a high complexity model. The criterion evolution according to the time delay for those simple models is shown in figs. 2 and 3.

By examining fig. 3, it is easy to observe the presence of the minimal value of the criterion for $n_k = [5,6,7,8]$. But, in fig. 2, this presence is supported clearly for $n_k = [6,7]$.

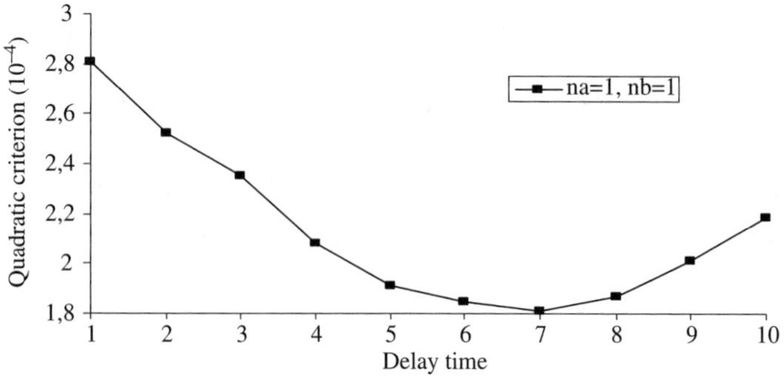

Figure 2. Criterion evolution according to the time delay n_k for $n_a = n_b = 1$

Therefore, it is better to consider, first, that the time delay values are both $n_k = 6$ and $n_k = 7$. Then, each model having a different time delay ($n_k = [6,7]$) will be rejected.

3.2.2. Quality of fit

The quality of fit criterion allows a judicious selection of models. This criterion proposed by Hagenblad et al. (1998) is based on the analysis of the prediction error and of the output variance. It is given by the following expression:

$$Q = 100 \times \left(1 - \sqrt{\sum_{k=1}^{N} \left(\hat{y}(k) - y(k) \right)^2} \Bigg/ \sqrt{\sum_{k=1}^{N} \left(y(k) - \frac{1}{n} \sum_{i=1}^{n} y(i) \right)^2} \right) \tag{7}$$

Fig. 4 shows the criterion evolution according to the different models $M_{n_a\,.\,n_b}$. The models $M3.2$, $M4.2$ and $M5.5$ have a good quality of adjustment compared to the other models

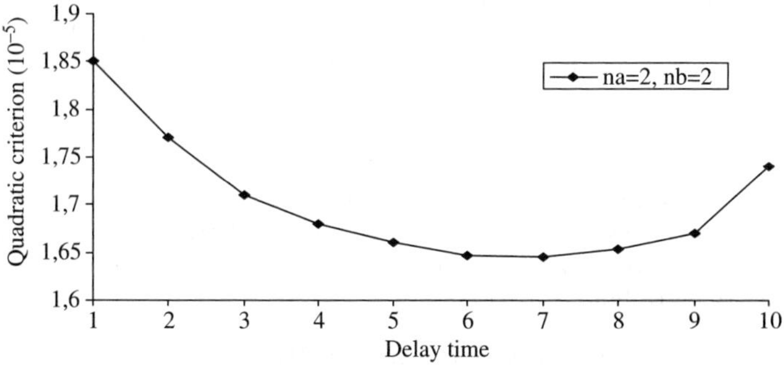

Figure 3. Criterion evolution according to the time delay n_k for $n_a = n_b = 2$

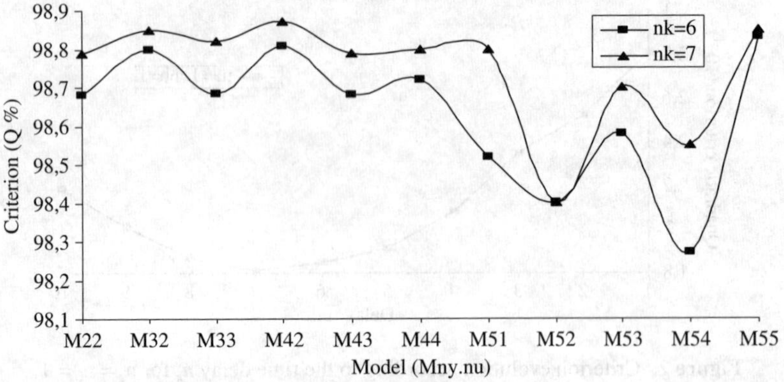

Figure 4. Criterion evolution according to the different models $M_{n_a \cdot n_b}$

(important peaks). The model $M5.5$ is not being chosen because it is too large. The peak of the model $M4.2$ is more important than that of the model $M3.2$. Consequently, the model $M4.2$ is more representative for the dynamic behaviour than the model $M3.2$ and thus for the two time delay values ($n_k = 6$ and $n_k = 7$). In conclusion, the model ($M4.2.7$) having $n_k = 7$ is the most suitable one for reproducing the process dynamics.

3.3. RESIDUAL ANALYSIS

Once the training and the test of the ARX model has been completed, it should be ready to simulate the system dynamics. Model validation tests should be performed to validate the identified model. Billings et al. (1986) proposed some correlations based model validity tests. In order to validate the identified model, it is necessary to evaluate the properties of the errors that affect the prediction of the outputs of the model, which can be defined as the differences between experimental and simulated time series. In general, the characteristics of the error are considered satisfactory when the error behaves as white noise, i.e. it has a zero mean and is not correlated (Cammarata et al., 2002; Billings et al., 1986). In fact, if both these conditions are satisfied, it means that the identified model has captured the deterministic part of the system dynamics, which is therefore accurately modelled. To this aim, it is necessary to verify that the auto-correlation function of the normalized error $\varepsilon(t)$, namely $\phi\varepsilon\varepsilon(\tau)$, assumes the values 1 for $t = 0$ and 0 elsewhere; in other words, it is required that the function behaves as an impulse. This auto-correlation is defined as follows (Zhang et al., 1996; Billings et al., 1986):

$$\phi\varepsilon\varepsilon(\tau) = E(\varepsilon(t - \tau)\varepsilon(t)] = \delta(\tau) \qquad \forall\tau, \tag{8}$$

where ε is the model residual. $E(X)$ is the expected value of X, τ is the lag.

This condition is, of course, ideal and in practice it is sufficient to verify that $\phi\varepsilon\varepsilon(\tau)$, remains in a confidence band usually fixed at the 95%, which means that $\phi\varepsilon\varepsilon(\tau)$ must remain inside the range $\pm\frac{1.96}{\sqrt{N}}$, with N the number of testing data on which $\phi\varepsilon\varepsilon(\tau)$ is calculated.

Billings et al. (1986) proposed also tests for looking into the cross-correlation among model residuals and inputs. This cross-correlation is defined by the following equation:

$$\phi u\varepsilon(\tau) = E(u(t-\tau)\varepsilon(t)) = 0 \qquad \forall\tau \tag{9}$$

To implement these tests (8, 9), u and ε are normalized to give zero mean sequences of unit variance. The sampled cross-validation function between two such data sequences $u(t)$ and $\varepsilon(t)$ is then calculated as:

$$\phi u\varepsilon(\tau) = \frac{\sum_{t=1}^{N-\tau} u(t)\varepsilon(t+\tau)}{\left(\sum_{t=1}^{N} u^2(t)\sum_{t=1}^{N} \varepsilon^2(t)\right)^{1/2}} \tag{10}$$

If the equations (8, 9) are satisfied then the model residuals are a random sequence and are not predictable from inputs and, hence, the model will be considered as adequate. These correlations based tests are used here to validate the neural network model. The results are presented in fig. 5.

In these plots, the dash dot lines are the 95% confidence bands. Fig. 5 shows that the evolution of the cross-correlation of the ARX model is inside the 95% confidence bands. The auto-correlation of the ARX model exceeds the threshold (95%) for few points. This explains the non-dependence of the residual signal from the input one. Therefore, this model is considered a reliable one for describing the dynamic behaviour of the process. Fig. 6 represents the prediction error between the real output temperature and the estimated one.

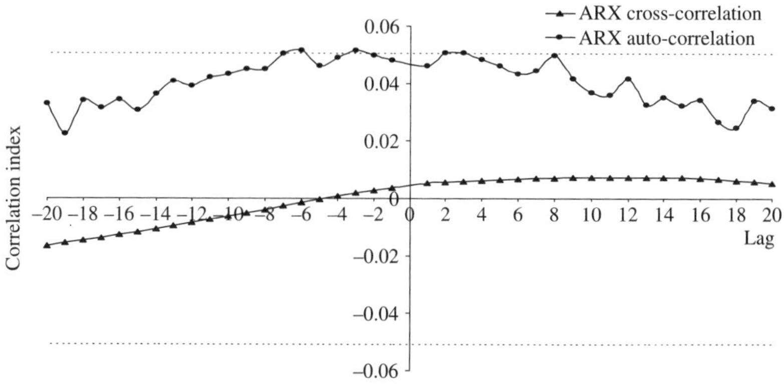

Figure 5. Results of model validation tests

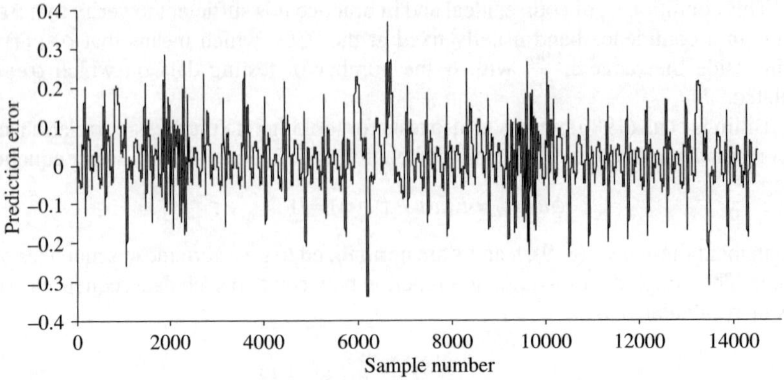

Figure 6. Prediction error of the output temperature

The main advantage of the proposed approach consists in the natural ability of the ARX approach in modelling nonlinear dynamics in a fast and simple way and in the possibility to address the process to be modelled as an input-output black-box, with little or no mathematical information on the system.

4. CONCLUSION

This work aims to identify process dynamics by means of an ARX model in order to provide reliable predictions. This study shows that the identification of the reactor-exchanger dynamics by means of input-output experimental measurements provides a useful solution for the formulation of a reliable model. In this case, the results showed that the model is able to give satisfactory descriptions of the experimental data. Finally, the identified model will be useful as a reference one for the fault detection and the isolation (FDI) which can occur through the process dynamics.

REFERENCES

Billings, S.A., Voon, W.S.F., 1986, Correlation based model validity tests for nonlinear models, *International Journal of Control*, 44: 235–244.

Cammarata, L., Fichera, A., Pagano, A., 2002, Neural prediction of combustion instability, *Applied Energy*, 72: 513-528.

Chen, S., Billings, S.A., 1989, Representation of nonlinear systems-The NARMAX model, *International Journal of Control*, 49: 1013–1032.

Chetouani, Y., 2006, Fault detection in a chemical reactor by using the standardized innovation, *Process Safety and Environmental Protection*, 84: 27-32.

Chetouani, Y., 2007, Use of Cumulative Sum (CUSUM) test for detecting abrupt changes in the proccess dynamics, *International Journal of Reliability, Quality and Safety Engineering*, 14: 65-80.

Fravolini, M. L., Ficola, A., La Cava, M., 2003, Optimal operation of the leavening process for a bread-making industrial plant, *Journal of Food Engineering*, 60: 289-299.

Frausto, H. U., Pieters, J. G., Deltour, J. M., 2003, Modelling Greenhouse Temperature by means of Auto Regressive Models, *Biosystems Engineering*, 84: 147-157.

Fung, E.H.K., Wong, Y.K., Ho, H.F., Mignolet M.P., 2003, Modelling and prediction of machining errors using ARMAX and NARMAX structures, *Applied Mathematical Modelling*, 27: 611-627.

Hagenblad, A., Ljung, L., 1998, Maximum likelihood identification of Wiener models with a linear regression initialization, *Proc. 37th IEEE Conference on Decision and Control*, USA.

Iserman, R., 1993, Diagnosis of machines via parameter estimation and knowledge processing, *Automatica*, 29: 815-835.

Leontaritis, I.J., Billings, S.A., 1985, Input–output parametric models for nonlinear systems, part I: deterministic nonlinear systems, *Int. J. Control*, 41: 303-328.

Liu, Y., Birch, A.A., Allen, R., 2003, Dynamic cerebral autoregulation assessment using an ARX model: comparative study using step response and phase shift analysis, *Medical Engineering & Physics*, 25: 647-653.

Ljung, L., 1987, *System identification, theory for the use*, Prentice-Hall, New Jersey.

Ljung, L., 2000, *System identification toolbox user's guides*, The Math Works, Natick.

Mu, J., Rees, D., Liu, G.P., 2005, Advanced controller design for aircraft gas turbine engines, *Control Engineering Practice*, 13: 1001-1015.

Ringwood, J. V., Austin, P. C., Monteith, W., 1993, Forecasting weekly electricity consumption: A case study, *Energy Economics*, 15: 285-296.

Rivera, D.E., Gaikwad, S.V., 1995, Systematic techniques for determining modeling requirements for SISO and MIMO feedback control problems, *J. Process Control*, 5: 213-224.

Rohani, S., Haeri, M., Wood, H. C., 1999, Modeling and control of a continuous crystallization process, *Computers & Chemical Engineering*, 23: 279-286.

Yoshida, H., Kumar, S., 2001, Development of ARX model based off-line FDD technique for energy efficient buildings, *Renewable Energy*, 22: 53-59.

Zhang, J., Morris, J., 1996, Process modelling and fault diagnosis using fuzzy neural networks, *Fuzzy Sets and Systems*, 79: 127-140.

INTEGRATING RISK INTO YOUR PLANT LIFECYCLE – A NEXT GENERATION SOFTWARE ARCHITECTURE FOR RISK BASED OPERATIONS

Dr Nic Cavanagh[1], Dr Jeremy Linn[2] and Colin Hickey[3]
[1]Head of Safeti Product Management, DNV Software, London, UK
[2]Regional Manager, DNV Software, London, UK
[3]Safeti Product Management, DNV Software, London, UK

Over the last three decades, technology for assessing the risks associated with operating major accident hazard facilities has been continuously developed. Over this period the accuracy and speed of the modelling on which this technology is based has improved enormously. The tools for modelling the effects of hazardous releases in terms of emergency response, safety management and Quantitative Risk Analysis (QRA), for example, are now well validated and used extensively by industry. Also, the processing power necessary for using these tools is now routinely available on a typical desktop computer. Other quantitative tools are beginning to appear which use related technology to assist in improving operational performance, particularly for inspection and maintenance planning activities. These tools are also progressively integrating more directly with operational management systems like SAP, ERP and ERM.

Commercial analysis tools like Phast, FRED, Trace and Canary for effects modelling, Safeti, Shepherd and RiskCurves for QRA and Orbit and RBMI for RBI are becoming more and more widely used. These tools are generally used standalone and independently of one another and other design and operational systems, even though they share much common data with the latter. Applying risk technology more directly into the plant life-cycle through integration with design and operational management systems has not kept pace with improvements in other areas.

As developers of Phast, Safeti and Orbit, we are committed to our technology being used throughout the plant life-cycle and that it is as closely integrated with our customers' value chain as possible. This paper describes our vision for a next generation architecture supporting this integration, the development of which is ongoing. A prototype of this architecture, "The Safeti™ Risk Framework", will be presented along with a longer term vision for a fully integrated risk based operations system linking risk technology with mainstream design applications and operational management systems through application of other risk management techniques

CURRENT SITUATION

The maturity of risk management in the process industry has seen much advancement in the areas of hazard analysis, risk analysis and risk assessment. Global recognition of the need for professional risk management has driven the creation of demanding legislative requirements and successful commercial products and services. This has occurred in parallel with greater need for transparency to the public, increased scrutiny of process plant activities and greater demand for better business performance.

The wide range of advanced tools and methodologies used in risk management have been enhanced greatly by the IT revolution. Advanced consequence modelling tools like Phast continue to be developed to meet the evolving and more stringent needs of hazard and risk analysis in the process industries.

The process industry has in parallel started to take advantage of advanced products and services for other areas of business management. Advanced systems are used for process control, asset management, management systems and financial management.

COMMONALITY

Process facilities have a range of attributes which are drawn upon, measured and controlled as part of the business value chain. Attributes such as people, materials, plant, buildings, transport, utilities, governing legislation, processes, weather conditions and market economics are all part of the dynamic environment process plant operators conduct business within. Specialist tools have been developed to support business optimisation through measuring, monitoring and/or controlling each of these attributes. For this reason many tools and services used across a process plant and within process industry organisations handle and use the same information. (Cavanagh and Linn 2005)

Worthington and Cavanagh 2003, introduced the concept of a data asset as illustrated in Figure 1. Data is contributed throughout the lifecycle of a plant from many sources. Plant design and CAD applications contribute during the design phase, process simulation during design and operational phases, GIS and safety management tools during design and operation, and so on. These data sources add value in relation to how well they can be kept up-to-date, shared and re-used. The Risk Framework provides a means of accessing, maintaining and sharing this data asset.

One example of a system which uses the latest technology and methodologies to measure and manage attributes of a processing facility is a fully integrated process control system. Such systems are used to measure the state of equipment and materials throughout the site. The control system tracks process conditions and, with operator control, makes adjustments to keep process conditions within predetermined limits. This approach ensures design conditions are met and sustained, process efficiency is optimised and also that non-design and potentially hazardous conditions do not arise. In addition to process control, control systems often contain hazard detection devices such as hazardous gas and fire detectors. These devices feed back to decision logic in the control system enabling hazard mitigating response to be carried out. The continuous live feed of data surrounding process plant conditions are common with the generic data used in hazard and risk analysis. Yet in contrast to live control system control and response, a risk management study is off-line and merely a snapshot of design or identified potential non-design conditions.

A second example of a system which uses the latest technology and methodologies to measure and manage attributes of a processing facility is an enterprise wide asset

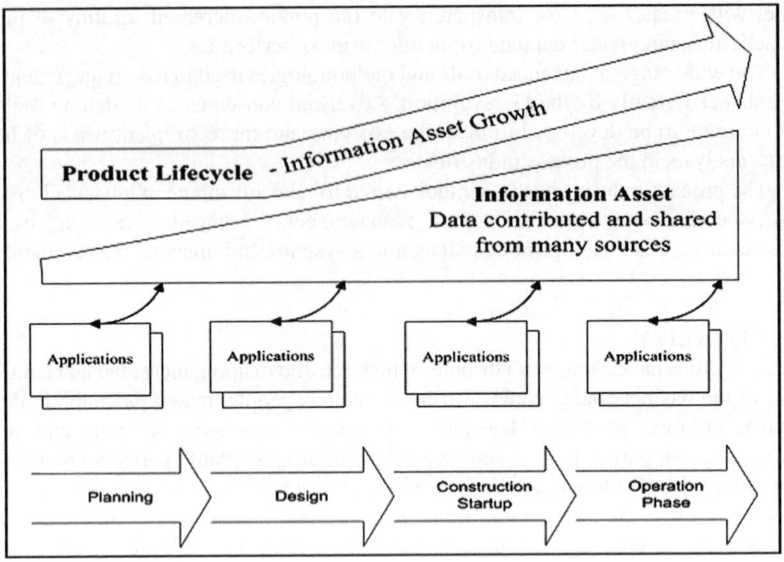

Figure 1. The data asset concept

management database tool. Such systems are used to store and track the wide range of data associated with plant equipment. Modern extensions of these tools provide support for maintenance and inspection management. Risk analysis is already incorporated into asset management tools enabling risks to be quantified for Risk Based Inspection (RBI) using software such as Orbit.

RISK AS A DECISION MAKING TOOL

Today risk analyses tend to be performed as an offline activity by risk specialists either to meet the needs of legislative requirements or as part of a plant modification to reduce risk. These tend to be snapshot studies that are filed once the relevant decision has been made or legislative requirement satisfied.

In recent years techniques have been developed to extend the applicability of risk technology beyond basic assessment of the severity of an incident or its likelihood of being realised. Techniques like the bow-tie approach and Layer Of Protection Analysis (LOPA) take account of the barriers and mitigators put in place to prevent an incident from occurring or escalating or to mitigate its effects if it does occur. All these related activities can be monitored and high risk operations avoided or extra safeguards put in place.

Used effectively in operational decision making these techniques can reduce operational risk and the likelihood of an incident that may result in a loss of life or to the profitability of a plant. So called Risk Based Operations or RBO enables decisions to be made based on knowledge and understanding of risks attributable to certain operations or processes, both before and after any operational changes are implemented.

As has been mentioned earlier QRA tends to provide a snapshot of the risks associated with a plant under a particular set of conditions. If extended to cover a multitude of scenarios or to take account of changing operational conditions, these quantifications of risk can, when combined with operational risk management techniques, provide a real time measure of the risk to which a plant is exposed. These kinds of systems are able to provide operational managers with quantitative real-time data, rather than static assessments, to support their decision making in an ongoing basis.

This kind of risk-based decision support offers increased benefit from risk analysis, bringing traditional QRA technology and methodology into the operational phase of the plant life-cycle. By using this information in a more dynamic and holistic manner, QRA is brought from the back-office into the daily operational management of your facility.

THE SAFETI™ RISK FRAMEWORK

It is our vision that risk management tools will be used throughout the lifecycle of the plant from design to operation and beyond. Risk management should support engineering design and day to day operation of the plant through live measures of activities and circumstances and use these to model the implications on and potential changes to a business's overall risk exposure. This will provide instant decision support and accurate perception of real time risks leading to continuous risk optimisation and reduction.

We believe best risk management practice is now achievable through the parallel evolution of software, data management and internet technologies already proven and in use by businesses globally.

The vision for the Safeti™ Risk Framework (Figure 2) is to help make risk based decision support a reality at all stages of a plant's lifecycle. By integrating the wide range of existing risk management tools and currently non risk-based process plant management tools into one complete system, risk becomes a key input to the decision making process.

Data of relevance to the risks associated with a process facility will feed in and out of all of the existing tools in a flexible manner. The diagram above illustrates how, for example, asset data relevant to a QRA can be reused for an RBI. It also demonstrates how, for example, a Matrix of Permitted Operations (MOPO) as part of a risk assessment system can use information from the asset database and risk measures to feed the control system or management system for risk based decision support.

The Risk Framework concept creates benefits at a number of levels. Reuse of data, live risk based decision support and integrated business management are some of the many benefits derived from the approach suggested.

The following scenarios illustrate how the Safeti™ Risk Framework can help to avoid undesirable situations arising from typical process plant circumstances.

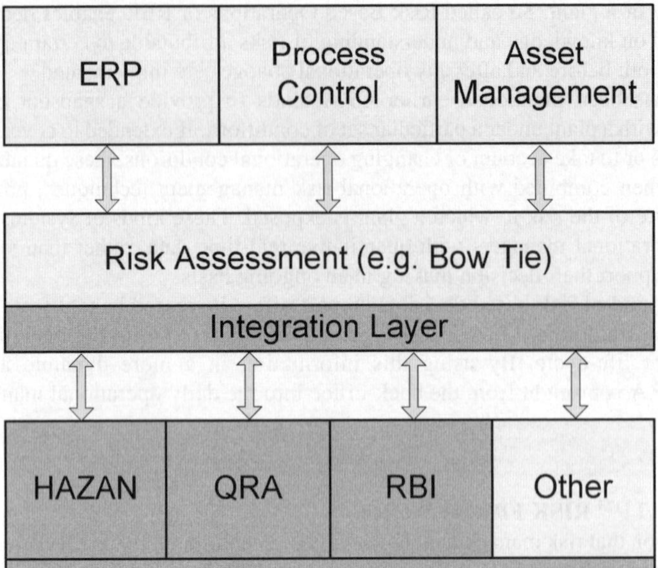

Figure 2. The Safeti™ Risk Framework concept

SCENARIO 1. MY PUMP HAS A FAULT AND THE PARALLEL RESERVE PUMP IS SCHEDULED TO BE DOWN FOR MAINTENANCE WHICH HAS NOT STARTED YET. CAN I USE THE RESERVE PUMP?

This is a commonly occurring situation in the process industry. It involves interaction between a control system for switching the pump on or off, a permit to work system which is part of the management system and a number of assets; pumps, valves and pipework.

Many approaches have been developed to deal with the safe interaction between these business processes. For example, electrical isolation of the pump can be applied to override the control system preventing it from starting a device when it is in an unsafe state that the control system cannot measure. In addition, a permit to work system is a document based activity used in management systems to control and monitor the status of process equipment for clarity and safety during manual operation and maintenance.

In scenario 1 the following are a selection of potentially undesirable events arising from failure of interaction between the controlling and monitoring business processes on the site:

1. The paper based permit to work document indicating that the pump was unsuitable for operation had been logged incorrectly giving the impression that the pump was fit for service.

2. The control system is independent of the permit to work system and maintenance scheduling tools. It may detect that the state of the process equipment is normal and allow for remote commands to be sent to the process equipment for start up leading to undesirable circumstances.

Information involved in this scenario which will be linked via the Safeti™ Risk Framework:

- Asset database
- Equipment design conditions
- Live process conditions, supplied from control system devices
- Maintenance Schedule
- Live status of risk picture on site based on current circumstances
- Rules and responsibilities within the management system governing maintenance schedule and permit to work

The undesirable outcomes of this scenario could be avoided by having one hub with access to all interrelated risk relevant information. An electronic permit to work system tied to high risk equipment provides transparency of plant status for safe operation. Also, the live intelligent permit to work system, integrated with the control system provides a software back-up for the hardware electrical isolation. This creates an extra layer of protection in cases of management system processes failure. Ultimately the status of each piece of process equipment could be found by drilling down through details on an enterprise wide risk dashboard.

SCENARIO 2: DUE TO INCREASED PRODUCTION AND SIMULTANEOUS CONSTRUCTION AND ENGINEERING PROJECTS, MORE STAFF ARE REQUIRED ON MY SITE. CAN I TEMPORARILY LOCATE THEM AT THIS LOCATION?

The spatial layout of population can change the risk picture of a process facility. Locating temporary buildings, for example, can become a critical part of a facility's risk exposure. This is the type of decision where risk based support can be critical in the safe operation of a site.

It is often the case that the offline, periodic reports created to comply with some European countries' local implementation of the EU Seveso Directive or the consequence based US Risk Management Programme (RMP) form the foundation for the overall risk management of a process plant. These reports are therefore usually offline – taking idealised standard operating conditions or hypothetical non-ideal operating conditions.

They are also usually out-of-date soon after they have been created due to the dynamic operating environment of modern process plants.

A site being managed with mature risk management processes will undertake what-if studies using tools such as Phast and Safeti. These effectively predict what new risk or hazard levels may be posed due to required changes to a site. An arbitrary line is often drawn under activities for which potential hazards and overall risks are assumed to be tolerable. The overhead of performing risk assessment is often a disincentive to it being carried out.

The undesirable events posed by selecting different siting locations for temporary staff facilities can be described using firstly individual risk contours overlaid on population siting options giving a qualitative assessment of risks. In this example based on one simple failure case individual risk contours are generated in the form illustrated in Figure 3.

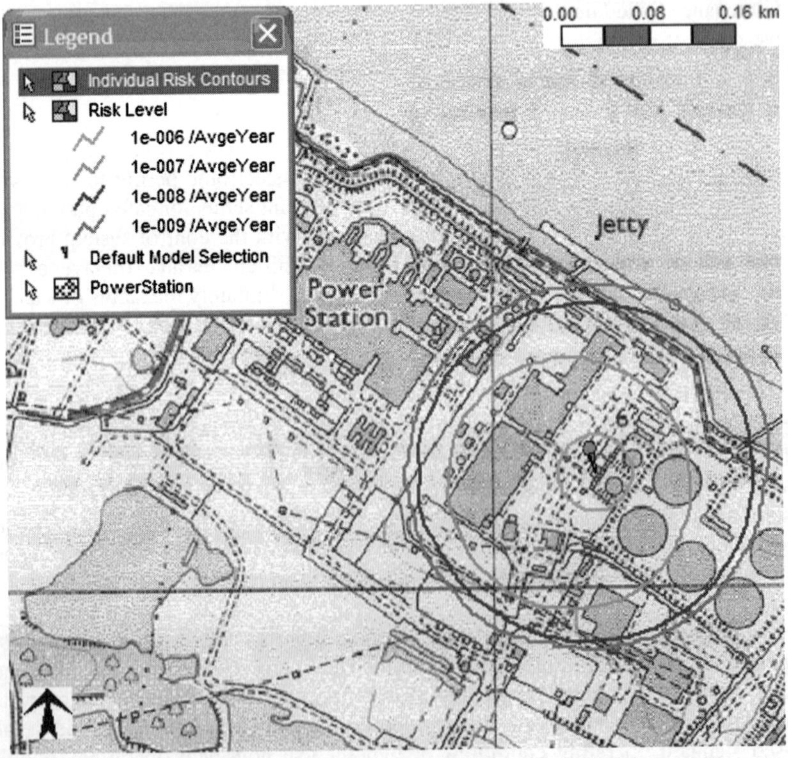

Figure 3. Individual risk contours of Scenario 2

For this particular facility we have two temporary office siting options as illustrated in Figure 4. Both options require temporary accommodation to be sited close to the source of hazard also illustrated in Figure 4. But each option will pose different levels of societal risk. This is traditionally enumerated using F-N curves as illustrated in Figure 5. As can be seen in the societal risk comparison from the F-N curves in Figure 5, option 2 is the preferred location in the context of risk.

The Safeti™ Risk Framework concept will help to make such decision easier by having all necessary input data continuously linked to the organisation's risk management console as illustrated in Figure 6. What-if studies can be performed quickly and thoroughly.

In addition to what-if analysis the same functionality within the Risk Framework can be used as a continuous monitor of real time risks, providing a live risk dashboard for the site.

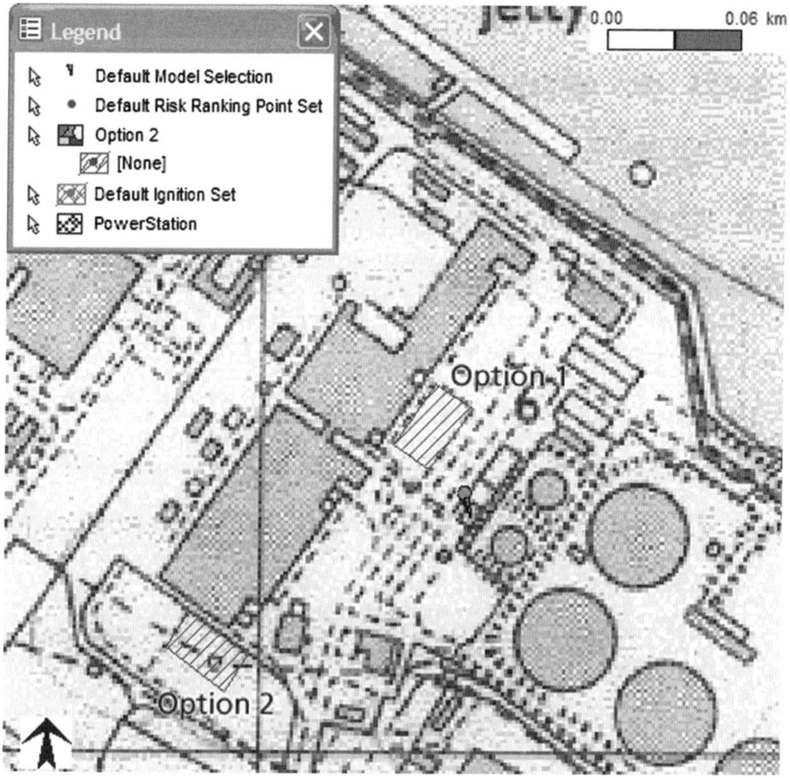

Figure 4. Two temporary office siting options

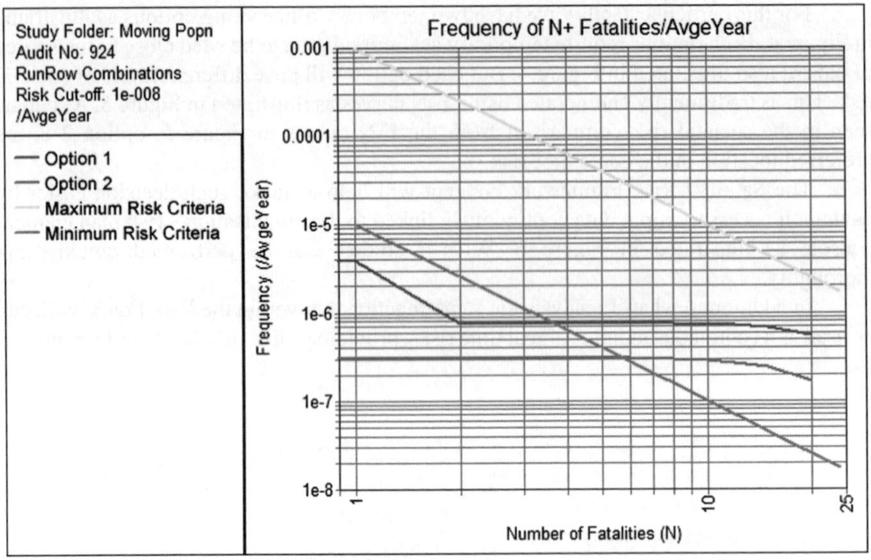

Figure 5. Societal risk comparison of two office location options

This requires the Risk Framework to be continuously processing the risk relevant data on the site to calculate the live risk picture. The typical measures of risk – societal and individual – would be calculated continuously as various aspects of site conditions change. With individual and societal (e.g. risk integral) risks continuously displayed with drill-down capabilities for easy identification of main contributors in terms of both equipment and personnel.

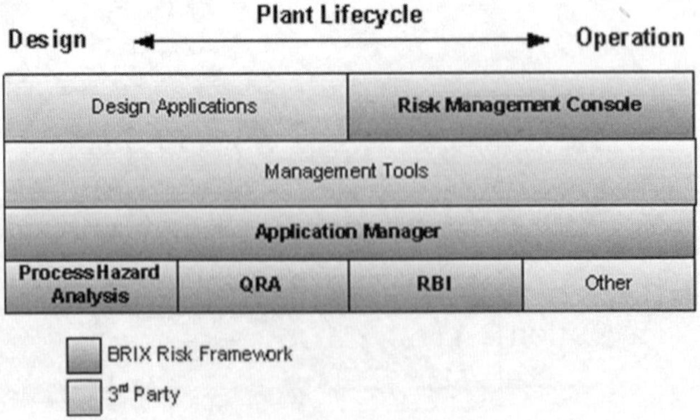

Figure 6. The Safeti™ Risk Framework architecture

THE SAFETI™ RISK FRAMEWORK ARCHITECTURE

DNV Software already manages an array of world leading risk management software tools. The Risk Framework concept is an evolution of these tools providing, amongst other things, the next generation architecture for these tools. This is the enabler facilitating integration of existing technology with new technology and 3rd party systems. The resulting risk framework service platform supports full lifecycle management. This ensures cost effective reuse of an organisation's data asset and delivers real time risk based decision support for design and operation.

At the core of the Risk Framework is the plant's asset database. It is from here that all activities begin. From hazard analysis through frequency analysis to control and management system decision support through the Risk Management Console/Dashboard all data pivots on the underlying asset database. At this stage the Safeti™ Risk Framework exists in prototype form and will be the basis for the next generation of the Phast application which is well underway. A key development is the evolution from a scenario based to an equipment based model.

This facilitates integration of the underlying data models for QRA, RBI and PHA in Safeti, Orbit and Phast respectively. This is also the first step in integration with 3rd party applications, CAD and other databases and provides the enabler for integration with other operational systems. A further development is the ability to map data from one model or data source to another using a configuration utility rather than hard coded into the software. Again, this development has previously been prototyped and is now well under way.

REFERENCES

Cavanagh, NJ. and Linn J., April 2005, Beyond Compliance – The Future Role of Risk Tools, AIChE Global Safety Symposium, Annual Conference of Centre for Chemical Process Safety, Atlanta, Georgia, USA

Worthington, D.R.E. and Cavanagh, N.J., June 2003, The development of software tools for chemical process quantitative risk assessment over two decades, ESREL 2003 Conference, Safety and Reliability, Maastricht, The Netherlands.

ASSESSMENT OF HIGH INTEGRITY INSTRUMENTED PROTECTIVE ARRANGEMENTS†

Alan G King
Hazard & Reliability Specialist, ABB Engineering Services, Billingham, Cleveland UK. TS23 4YS

To comply with the requirements of Functional Safety standards such as IEC 61508 and IEC 61511, it is a requirement that end users undertake calculation of the probability of failure: "The probability of failure on demand of each safety instrumented function shall be equal to, or less than, the target failure measure as specified in the safety requirement specifications. This shall be verified by calculation".

In many instances, this sort of calculation is straightforward and presents relatively few challenges. However, there are other situations where the design of the safety function aims for high reliability and its consequent complexity demands a different approach.

This paper describes as a case study an assessment of the probability of failure of a typical high integrity high pressure trip system for a "top tier" COMAH site in UK. The site had decided to replace the original pressure switches with new pressure transmitters and also replace some of the trip logic with a new safety PLC. The reminder of the trip system remained the same. They required a calculation to demonstrate that the probability of failure was still acceptable – below the target value for this function.

The calculation method for this high integrity safety function required the identification of string sets – success paths through a block diagram of the safety function, and then the minimum cut sets – failure groupings. Once the minimum cuts sets have been listed, the independent failure probability for each of these groupings can be calculated. Additionally, the dependent common cause failure probabilities are calculated. These can be summed to give the overall failure probability for the whole safety function. The practical method discussed in this paper works for the sort of more complex of arrangements often found in many SIL 2 and SIL 3 safety functions and which cannot be assessed with the simple approaches used for single channel SIL 1 loops.

KEYWORDS: IEC 61511, IEC 61508, Functional Safety, Risk Reduction, Minimum Cut Sets

INTRODUCTION

It has always been the case that the management of high hazard plants has needed to demonstrate the application of suitable means to manage the risks associated with their operations. In many countries across the world, this has become a legal requirement for operating in those regions. This is the case across Europe, in North America, and in many industrial countries across the world. However, this generally implies that there is not only a need to demonstrate appropriate risk management but also a demonstration of the use of current industry good practice. This in turn involves the use of and compliance with relevant national and international standards.

The use and management of instrumented protective functions such as trips, alarms and interlocks are of key importance in the effective management of risks on many sites. These fall into the category of functional safety[1]. The standards representing current good practice for functional safety using electrical, electronic or programmable electronic means are IEC 61508 [1] and IEC 61511 [2][2], together with the other sector standards that have been generated from IEC 61508. These standards are generally agreed across the world to represent current good practice in this field. These standards have been in the public domain for several years now, and the regulators in many countries are looking for compliance now with these standards, or at the least a programme of action leading towards compliance, as a means of demonstrating the use of current good practice.

These standards cover the whole of the safety lifecycle – from the initial concept through to operation and maintenance[3]. Within the requirements relating to design and operation of instrumented safety functions, there is the demonstration that each safety instrumented function achieves a necessary target performance. This is the performance needed for effective management of the level of risk. The focus for this paper is the specific requirement in the standards that end users undertake a calculation of the probability of failure: "The probability of failure on demand of each safety instrumented function shall be equal to, or less than, the target failure measure as specified in the safety requirement specifications. This shall be verified by calculation"[4]. Calculation is therefore a mandatory requirement for compliance with the standard. In many instances, such calculations of failure probability are simple and straightforward. This is true for single channel safety instrumented functions with probabilities of failure in the range for Safety Integrity Level 1 (SIL 1)[5]. However, for safety functions designed for higher reliability, for example those aiming to achieve SIL 2 or SIL 3, the consequent complexity can demand a different approach.

[1]Functional Safety here refers to those systems that rely on the correct functioning of electrical or other systems to achieve the required level of safety.

[2]IEC 61511 is the Process Sector standard derived from the generic standard IEC 61508 on instrumented Functional Safety.

[3]And through to the eventual decommissioning of the systems.

[4]IEC 61511-1 Clause 11.9.1

[5]See References [1] and [2] for definitions of SIL 1 through to SIL 4.

BACKGROUND

This paper describes as a case study an assessment of the calculation of probability of failure of a typical high integrity high pressure trip system for a "top tier" COMAH site in UK. The site had decided to replace their original pressure switches with new pressure transmitters and also replace some of the trip logic with a new safety PLC. The remainder of the trip system remained the same. They required a calculation to demonstrate that the probability of failure was still acceptable – below the target value for this function.

The original configuration of the safety instrumented function is shown in Figure 1.

The safety instrumented function is initiated by high pressure and acts to stop the flow of both reactants (A & B). For high pressure protection, the critical requirement is to stop the flow of Reactant B. Thus, the safety function can be seen as limited to that part which senses high pressure and stops the flow of Reactant B. The sensors are both pressure switches and for the action to stop the flow of Reactant B the pressure switches operate on a 1 out of 2 basis – either pressure switch sensing high pressure is sufficient to trigger successful operation of the function.

The plant wished to improve the system by replacing the pressure switches with pressure transmitters, so that the analogue value of pressure from each could be made available to the operators. Additionally, they decided that they would purchase a safety PLC for the plant trip system as a whole. This safety PLC would then provide the interface to the new pressure transmitters and also provide a means of relaying the analogue values to the plant control system for display to the operators. The new arrangement is shown in Figure 2.

From this it can be seen that the changes only affect the part of the trip system upstream of the two original relays R5 and R6. The remainder of the system is unchanged. There are two new relays R1 and R2. These are part of the output arrangement for the safety PLC.

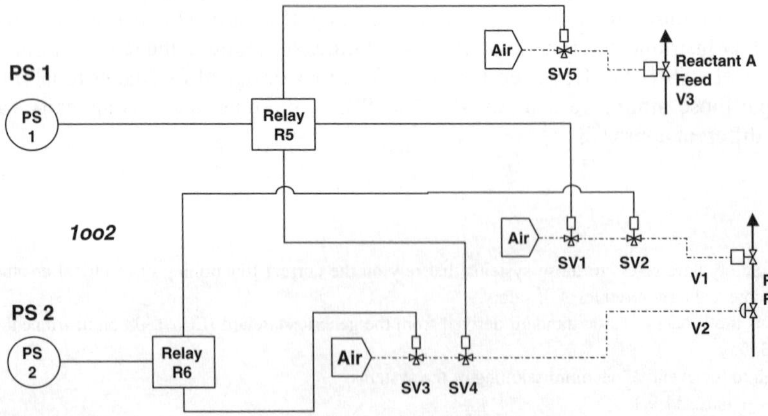

Figure 1. Original trip system

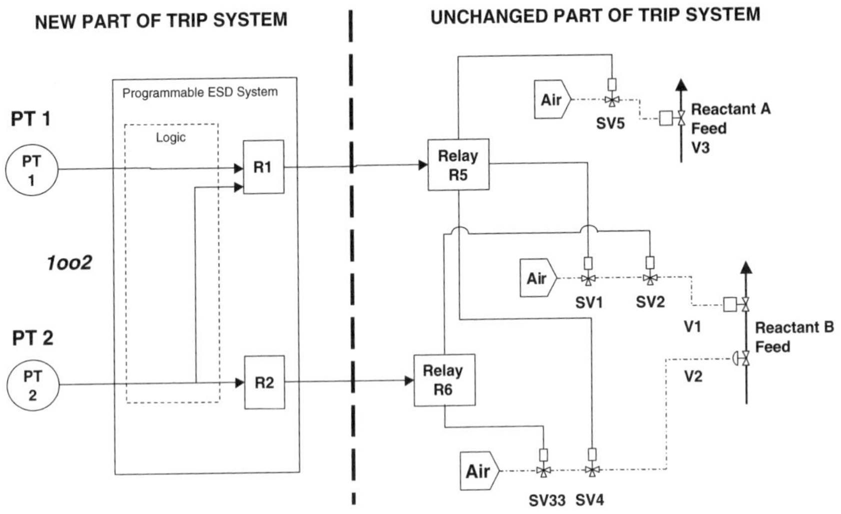

Figure 2. Proposed modification

DESCRIPTION OF THE METHOD

BLOCK DIAGRAM

The diagram in Figure 2 is too detailed as a basis for calculation. The first stage is therefore to simplify the function into a block diagram, showing only the parts of the safety function that are essential for high pressure protection. This is illustrated in Figure 3 and shows the sensors and logic block (representing the safety PLC). The two new relays R1 and R2 have been combined with their corresponding relays from the original system R5 and R6 respectively. This is to simplify calculations. Furthermore, the output side of the safety function only shows the routes through to the trip valves V1 and V2, which are the two valves that can block the flow of Reactant B. Closure of either of these valves represents success for the safety function.

From this diagram, we can identify the success paths through the function. There are six of these success paths:

1. PT1 – Logic – R1 & R5 – S1 – V1
2. PT1 – Logic – R1 & R5 – S3 – V2
3. PT2 – Logic – R1 & R5 – S1 – V1
4. PT2 – Logic – R1 & R5 – S3 – V2
5. PT2 – Logic – R2 & R6 – S2 – V1
6. PT2 – Logic – R2 & R6 – S4 – V2

These are known as string sets. However, what we are interested in for the probability calculations are not the success paths but the failure groups known as minimum cut sets.

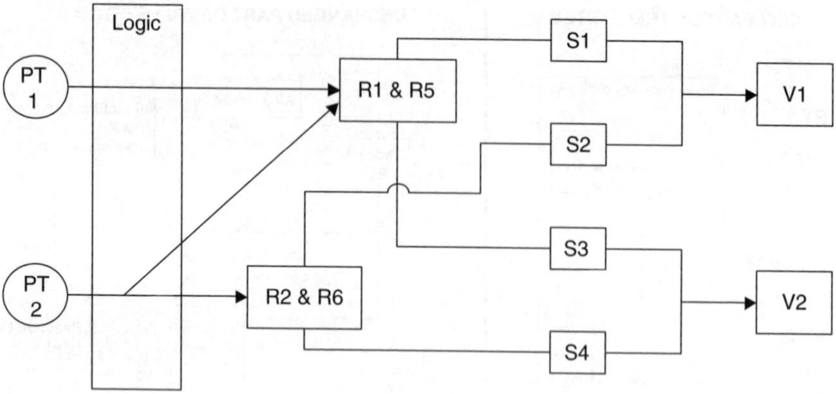

Figure 3. Block diagram of trip function

It is possible to generate the minimum cut sets for the function from the string sets but it is not straightforward. In practice, it is easier to identify the minimum cut set groups for functions of limited complexity, such as Figure 3, by inspection.

MINIMUM CUT SETS

A cut set is a group of items in the safety function whose failure will cause the function to fail. A minimum cut set is a group of items whose failure is just sufficient to cause failure of the overall function. Restoration of any one of the items from failed to working will cause the overall function to work successfully. For the function shown in Figure 3 there are 17 identifiable minimum cut sets. These are shown in Table 1 below.

It can be seen from Table 1 that some minimum cut sets have one member, some have 2 members, some 3 members and some 4 members. For calculation of independent failure probability, the minimum cut sets may be thought of as voting groups on a basis of 1 out of 1, 1 out of 2, 1 out of 3 and 1 out of 4 respectively. If we assume that the duration of any proof testing is short compared with the interval between tests then we can use the simplified formulae for the average probability of failure on demand (PFDavg), where θ is the dangerous failure rate and T is the test interval:

$$\text{PFDavg (1oo1)} = 0.5 \times \theta T$$
$$\text{PFDavg (1oo2)} = \frac{4}{3} (0.5 \times \theta T)^2$$
$$\text{PFDavg (1oo3)} = 2 (0.5 \times \theta T)^3$$
$$\text{PFDavg (1oo4)} = 16/5 (0.5 \times \theta T)^4$$

These formulae for the independent failure probability are shown for minimum cut sets with identical types of members. For example, Minimum Cut Set No 2 with four solenoid valves (S1, S2, S3, S4), or Minimum Cut Set No 5 with two trip valves (V1, V2). Where a minimum cut set has members of different types, for example, No 6 with a pressure

Table 1. Minimum cut sets

No	Minimum cut set	No	Minimum cut set	No	Minimum cut set
1	PT1, PT2	7	PT2, S1, V2	13	R2&R6, S1, V2
2	S1, S2, S3, S4	8	PT2, S3, V1	14	R2&R6, S3, V1
3	PT2, R1&R5	9	R1&R5, S2, S4	15	S1, S2, V2
4	R1&R5, R2&R6	10	R1&R5, S2, V2	16	S3, S4, V1
5	V1, V2	11	R1&R5, S4, V1	17	PLC Logic
6	PT2, S1, S3	12	R2&R6, S1, S3		

transmitter and two solenoid valves (PT2, S1, S3) the formulae have to be modified – the exponent is removed and the single bracketed part is replaced with the product of the PFDavg for each member of the cut set. Additionally, calculation of dependent failure probability is required, where a cut set contains a number of similar or identical items[6].

CALCULATIONS

For the safety instrumented function described above, calculation was carried out for each minimum cut set. The site had their own preferred failure rates for use in PFDavg calculations and the proof test interval was set at 3 months. The calculation summary is shown in Table 2. It shows the calculation spreadsheet for independent failure. The figures in brackets in the "Formula" column are the PFDavg results for the cut set items. These were based on the site preferred failure rates and test interval using the formula:

$$\text{PFDavg (item)} = 0.5 \times \theta \times T$$

where θ is the dangerous failure rate and T is the test interval

It should be noted that whilst the spreadsheet calculates to many decimal places the input data is only good to two significant figures (if that). Thus, the total Independent PFDavg comes to 5.0×10^{-5}. However, many of the cut sets have potential for dependent failure. The PFDavg for this must be calculated and added to the value for independent failure. Dependent failure probability has been calculated using the beta factor method and the formula:

$$\text{PFDavg (Dependent)} = \beta \times 0.5 \times \theta \times T$$

Where θ is the dangerous failure rate for the item in question, T is the proof test interval and β is the Beta Factor. A value of 15% has been used for the Beta Factor as a conservative figure for identical items in close proximity.

[6]The process valves were designed to fail to the safe position on loss of instrument air. Consequently, failure of instrument air has not been included in the consideration of common cause failure. Common cause failure of the valves due to contaminated instrument air adversely affecting their performance is covered by the dependent failure assessment.

Table 2. Independent PFDavg calculations

	Minimum cut set	Formula	Independent PFDavg
1	PT1, PT2	= 4/3 (0.001663)^2	= 3.69E-06
2	S1, S2, S3, S4	= 16/5 (0.004167)^4	= 9.64506E-10
3	PT2, R1&R5	= 4/3 (0.001663) × (0.000448)	= 9.93344E-07
4	R1&R5, R2&R6	= 4/3 (0.000448)^2	= 2.67755E-07
5	V1, V2	= 4/3 (0.004167)^2	= 3.47222E-05
6	PT2, S1, S3	= 2 (0.001663) × (0.004167)^2	= 5.77257E-08
7	PT2, S1, V2	= 2 (0.001663) × (0.004167) × (0.004167)	= 5.77257E-08
8	PT2, S3, V1	= 2 (0.001663) × (0.004167) × (0.004167)	= 5.77257E-08
9	R1&R5, S2, S4	= 2 (0.000448) × (0.004167)^2	= 1.55599E-08
10	R1&R5, S2, V2	= 2 (0.000448) × (0.004167) × (0.004167)	= 1.55599E-08
11	R1&R5, S4, V1	= 2 (0.000448) × (0.004167) × (0.004167)	= 1.55599E-08
12	R2&R6, S1, S3	= 2 (0.000448) × (0.004167)^2	= 1.55599E-08
13	R2&R6, S1, V2	= 2 (0.000448) × (0.004167) × (0.004167)	= 1.55599E-08
14	R2&R6, S3, V1	= 2 (0.000448) × (0.004167) × (0.004167)	= 1.55599E-08
15	S1, S2, V2	= 2 (0.004167) × (0.004167)^2	= 1.44676E-07
16	S3, S4, V1	= 2 (0.004167) × (0.004167)^2	= 1.44676E-07
17	Safety PLC[7]		= 1.0000E-05
		Total Independent PFDavg	= 5.0225E-05

The calculation in Table 3 has only been done for those cut sets with all items identical, as these will be the ones most susceptible to dependent failure. It is possible to calculate and include contributions for those cut sets where some items are the same but one is different. However, the size of the contribution from this would be much smaller and may be neglected.

The overall PFDavg for the system is therefore:

$$\begin{aligned} \text{PFDavg (System)} &= \text{Independent PFDavg total} + \text{Dependent PFDavg total} \\ &= 0.00005 + 0.00157 \\ &= 0.00162 \end{aligned}$$

It is worth noting that the PFDavg in this example is dominated by the dependent failure contribution and were it to have been omitted the result would have been around 1.5 orders of magnitude out.

[7]A notional illustrative probability has been used in these calculations for a Safety PLC suitable for SIL 3 safety functions. In any actual calculations, the probability used should be based on assessment of the actual Safety PLC architecture and its predicted performance.

Table 3. Principal dependent failure PFDavg

	Minimum cut sets	Principal dependent PFDavg
1	PT1, PT2	0.000249
2	S1, S2, S3, S4	0.000625
3	PT2, R1&R5	
4	R1&R5, R2&R6	0.000067
5	V1, V2	0.000625
6	PT2, S1, S3	
7	PT2, S1, V2	
8	PT2, S3, V1	
9	R1&R5, S2, S4	
10	R1&R5, S2, V2	
11	R1&R5, S4, V1	
12	R2&R6, S1, S3	
13	R2&R6, S1, V2	
14	R2&R6, S3, V1	
15	S1, S2, V2	
16	S3, S4, V1	
17	Safety PLC	
	Total dependent failure PFDavg	0.001567

RESULTS

The above calculation shows that the overall PFDavg is in the range for SIL 2. It is at the higher performance end of the range for SIL 2. The operating site was looking for an overall PFDavg that would be at least as good as the previous system and therefore allow them to meet their target. The previous arrangement had a PFDavg of 0.00268 and so the site could demonstrate that the changes would be an improvement and the new system would meet the requirements for the plant.

CONCLUSIONS

Whilst the calculation of the failure probability for a single channel safety function is relatively simple to do, safety functions with the type of architecture described in this paper present more of a challenge. This is often the case with safety functions aiming to achieve performance in the range for SIL 2 or SIL 3. This paper has demonstrated that there is a systematic way to approach calculation of the PFDavg of these more complex arrangements and to show that in this respect the requirements of IEC 61508 and IEC 61511 to show by calculation can be met.

REFERENCES

IEC 61508: "Functional safety of electrical/electronic/programmable electronic safety-related systems", International Electrotechnical Commission, Geneva, 1998 & 2000

IEC 61511: "Functional safety – Safety instrumented systems for the process industry sector", International Electrotechnical Commission, Geneva, 2003

CONSIDERATIONS FOR LAYER OF PROTECTION ANALYSIS FOR LICENSED PLANT

Jo Fearnley
Senior Consultant, Aker Kvaerner Consultancy Services, Aker Kvaerner, Ashmore House, Stockton on Tees, TS18 3RE, UK
E-mail: jo.fearnley@akerkvaerner.com

Chemical plants are routinely built around the world using a standard process design package supplied by the licensor of technology. Typically included in the package is the identification of the instrumentation to be included within the emergency shut down system, and the required safety integrity level identified for each safety instrumented function. These are presented as necessary requirements to meet the licensor's internal minimum safety standards. Other requirements to meet minimum safety standards are also often included, such as mandatory procedures. The input to, and detail of, the assessment on which these requirements are based is not provided within this licensed process design package.

This raises the question of what additional safety studies are required as part of the engineering design package. This will depend on a number and variety of factors; these will be discussed, with examples, and the implications for SIL assessment debated.

BACKGROUND

A licensed process design package (PDP) is rapidly becoming the most common basis for chemical plant expansion across the world. Tried and tested designs for producing a wide range of chemicals are licensed by technology owners, and the client pays a premium for the benefit of investing in a design which is known to deliver a quality product. The basic process will remain unchanged, although progressive improvements to equipment design or catalyst may be part of the evolution of the design over the years.

The actual scope of the licensed PDP will differ from one chemical to another as typically the PDP is targeted at those aspects of the design where a change to the base design could affect the end product. Hence, although the reaction and purification unit operations are fundamental to the process, other activities, such as raw material handling, raw material purification, catalyst charging, separation, recycle and product handling, may or may not be within the scope of the licensed process. It is therefore not possible to talk about the standard content of a PDP, and hence what will not be included.

When companies developed, designed, constructed, commissioned and operated their own process plants they were in control of the associated safety studies, even if a contractor was used for the detailed design and construction aspects. In such cases the progression of successively detailed hazard studies, such as the six stages of hazard study originally developed by ICI plc in the 1960s, was part of the standard design process for many larger companies. As a result these studies were well controlled, documented, monitored, and comprehensive.

However, with licensed plants the structured progression through a series of studies often gets lost in the development from front-end design to detailed design to construction to commissioning and eventually operation. The confusion is increased by the frequent switch between contract companies at the various stages, leading to a lack of continuity in both personnel and communication/information flow.

Underlying the requirements for safety studies at different stages in the design process is the information provided by the licensor and the standards required by the client. Between the two is the contractor, who needs to take great care to understand the discrepancy between the licensor supply and the client demand such that it is adequately covered in the bid and the programme. The risk is that in the commercial drive to reduce costs and win the contract this could be overlooked.

Dependent on the scope of the contract, consideration will need to be given to those aspects of the design within the PDP and those outside the licensor's remit, but within the contractor's remit; i.e. the off-plots. The situation is further complicated by elements of the design which are outside battery limits (OBL), i.e. to be completed by others, or already existing. There will frequently be differing safety, health and environmental (SHE) assessment requirements for the PDP, the off-plots and the OBL interactions. The usual situation is that the PDP will contain some integral hazard studies, whilst none of the off-plots will be covered, and information regarding the OBL interactions will depend on their build status compared to the licensed plant.

As part of my role as a risk consultant for Aker Kvaerner Consultancy Services I am routinely involved with the range of risk assessments required for licensed plants. One aspect of this consultation is the fundamental question:

'What safety, health and environmental assessments do we need to do?'

Unfortunately there is no single answer to this question as the combination of the licensor, the client, the contractor and the contract itself mean that every project is totally individual. This paper discusses the issues raised by this uncertainty when considering the requirement for safety instrumented functions (SIF) to comply with EN-61511.

In order to determine the safety instrumented functions for a plant, and the other design and operating aspects which provide prevention, control or mitigation of hazardous scenarios, a detailed hazard identification for the whole plant is necessary, which covers the areas within the PDP and those within the off-plots. It is not possible to cover the areas outside battery limits, but an awareness of impact on. and from. these is needed to be comprehensive.Once the hazardous scenarios have been identified, typically at an early stage in the project, the design should, where possible, provide inherent design features which eliminate the risk, or, where this is not possible, prevent, control and mitigate that risk. The assessment of the identified hazardous scenarios, considering the design of the plant, is frequently done using the layer of protection analysis (LOPA) methodology from EN-61511 for process plants, to identify the relevant protective features of the design and ultimately determine the instrumented systems which are required to have a safety integrity level (SIL) rating.

As a simplistic starting point it is normally the case that for a licensed plant design the section within the PDP will have the SIL ratings of the SIFs identified, whereas the off-plots will not.

PROCESS DESIGN PACKAGE – PDP

The licensor will have completed safety studies as part of the development of the PDP, and the design will also typically have been developed over the years to incorporate operational experience. However the PDP seldom, if ever, includes the SHE assessments in a form which is comprehensive and easily transferable between licensor and client. It is this lack of definitive documentation which leads to the question 'What safety, health and environmental assessments do we need to do?'

At any stage in the design process, the basic questions to ask are:

- What has already been done?
- What is available?
- What does the contract require to be done?
- What does good/best practice require to be done?
- What safety targets/risk tolerabilities are required/appropriate?

Unfortunately, just because a risk assessment has already been completed does not mean that it is available for direct use. For example the licensor will have a wide range of prevention, control and mitigation features built into the design of the PDP supplied. However what is often lacking is the full detail of the completed hazard assessments which lead to the design as it stands. A document commonly provided as part of the PDP is the list of alarms and trips for the plant, which need to be part of the independent safety system rather than part of the basic process control system (BPCS). For these instrumented safety systems the required safety integrity level (SIL) will also be provided within the PDP. The process trips within the design which are not subject to a SIL rating are those within the design of the BPCS. The detailed assessment which determined which are SIL rated and which are not is not normally available; only the output detailing those hazards determined by the licensor as requiring specific design or operational features will be available. This assessment is often referred to as the PDP layer of protection analysis (LOPA) document. The document is, however, not calibrated or quantified, but simply contains a list of scenarios and the requirements for associated procedures, alarms and (SIL rated) trips.

This is the fundamental cause of differences between projects, as for some clients it is enough to accept the standards set by the licensor, whilst others insist that their own company standards are applied to all new build projects. For the former the supplied design can be used as the basis for the project, whereas for the latter it may be necessary to complete all risk assessments to the client standards and hence alter the PDP if there is a discrepancy in integrity levels determined. For licensed technology it is part of the license that it is only acceptable to increase the safety functions, not decrease them.

The underlying factor which affects this is that the risk criteria on which the licensor has based the risk assessments are not typically available for comparison with the standards of the client. Such risk criteria are often considered confidential and not something which a company wishes to be shared, as they are perceived as a reflection on the company concerned if they are not as restrictive as someone else's criteria. However to complete a safety integrity level assessment, either using a risk matrix or layer of protection analysis requires that the criteria for the boundaries between intolerable, tolerable and broadly acceptable levels of risk are quantified. The interaction between the frequency of an event occurring and the severity of that event will determine the risk arising from the event, and hence its acceptability. To reduce the risk the frequency and / or the severity needs to be reduced.

To compound the problem, it is normally obvious from the output data numbering which is supplied within the PDP LOPA report that there are a large number of scenarios which were in the original licensor assessment for which the output has not been supplied; i.e. only those scenarios deemed significantly hazardous to require instrumented systems, or mandatory procedures to address them are included. The problem with this approach is that it is therefore not obvious which other design or operational features have been assumed by the licensor as the protection against the various scenarios. The licensor's argument is that the requirements are built into the PDP, and hence provided it is designed to the stated requirements the other scenarios will be adequately covered. This is fine where the resolution of the risk is covered by an inherent protection feature, e.g. that the equipment design temperature is high enough such that it cannot fail, or the relief device is sized to relief the pressure safely, or a vent is routed to a flare to address environmental issues. However some of the basis of the risk assessment may be a function of the tolerable risk accepted by the licensor, or may be based on assumptions about the environs of the client's site which are not valid.

Included within the PDP LOPA are physical, mechanical and inherent layers of protection, as well as control and operational considerations, which may not be specifically highlighted as layers of protection. This means that if there are any changes to the standard design it is not immediately apparent whether the SIL requirements are affected, as the change could affect a scenario not identified as having a SIL implication in the licensor package. It is therefore very important that any change in the design from the basic licensor package is specifically assessed for related hazardous scenarios and hence for potential SIL requirements, which can only be done in conjunction with the licensor. Examples of these layers of protection include relief devices (pressure and vacuum), bunds, flare systems, vent headers, design conditions of equipment and pipework, restrictive devices and equipment types. Care needs to be taken with any change as in some cases what may appear a positive change, such as increasing the ground area of a bund, could have a negative effect as it could increase the size of a pool fire and hence the radiation effect.

LOCATION-RELATED FACTORS AFFECTING THE PDP SIL ASSESSMENTS

There are other general factors which may affect the licensor LOPA which has resulted in the identified SIL requirements. These include layout, geographical considerations, operating philosophy or population density. The risk criteria used as the basis for the SIL assessments for the PDP will be based on a standard plant environment, and is very unlikely to have been adjusted for the specific environment in which the client plant will operate. Due to the lack of underlying information regarding those hazards which have been screened out, it is difficult to determine whether the client specific environment will change the risk assessment.

An easy location specific factor may the vulnerability of the client site to certain geographical effects, which will not be part of the basic design package. An example is if the client site will be in a region prone to earthquakes, in which case there will possibly need to be vibration/motion trips that initiate the shutdown of a hazardous installation that would not be in the standard licensor package. Alternatively if it is in a desert then sand may invalidate a particular protective measure such as a bund, as sand build-up may restrict the capacity available unless additional precautions are taken. A further example is if the site is in a region prone to flooding, which may block drainage routes leading to an unexpected pool fire.

Changes to the layout may affect the SIL classification identified for a particular risk, as the consequential effects may be different. For example, if the occupied building locations are changed compared to potential flammable release sources then this could change the number of people who could be at risk if an event were to occur. Consideration is needed not only of the plant to be licensed, but also any other production units which are close enough to have an inter-plant effect. Higher numbers of people could also be at risk if the plant is to be located near to a site boundary with a population living close to the external boundary.

Operating philosophy changes may change a SIL classification, as the exposure/vulnerability of the operating personnel may be changed. Modern plants are typically intended to be remote-operated, and as such have a low time at risk factor for certain events with a localised effect as a low proportion of time will be spent on site. However if a manual operating regime is planned then this layer of protection factor will not be valid.

It will be necessary to consider and discuss all these potential factors between the licensor, client and contractor to ensure that the final design appropriately considers the risks, and relevant SIL assessments may need to be reviewed to assess the validity of the data and hence the design basis of the PDP. Part of this consideration is the awareness that SIL classification is an order of magnitude technique, so a significant change is one which has an order of magnitude effect on the layer of protection analysis. This is complicated by the fact that the original LOPA may have identified, for example, a SIL 1 requirement for a SIF, but this analysis was right on the boundary between the SIL 1 and the SIL 2 result. Hence it is possible that a small change could move an assessment into the next level of required protection, and without the values used in the original PDP LOPA it is not possible to identify the scenarios where this is the case. This is further complicated where a scenario had not been SIL classified initially, and yet a change may take it to SIL 1.

Underlying all these considerations if the principle of whether the client/local directives require that the 'as low as reasonable practicable' (ALARP) or 'so far as reasonably practicable' (SFARP) principles are applied, and whether the licensor package utilises these principles. This could add a significant level of complication to the discussions about acceptable risk criteria, and is highlighted here, but is not further discussed.

CONSIDERATION OF OFF-PLOTS AND SIF REQUIREMENTS – UTILITIES

The utility requirements of the site will be different dependant on whether the new plant is on a brown-field or green-field site; what facilities already exist; what is being built in the same timescale; and what the plans for the future are. As a result the utility requirements for a licensed plant are seldom, if ever, included in the basic PDP as the demands are endlessly variable. The effect of the local environment on the utility requirements also changes the specification significantly.

As such the only standard statement for all licensed plants is that the utilities will need to have a series of appropriate SHE assessments completed during the lifetime of the project, including a hazard identification process which will enable appropriate assessment of the risks identified to enable the SIL assessments to be completed. These will need to consider not just the licensed plant which is the subject of the project, but also the interactions between the other plants on the complex. Therefore even the scope of such studies will not be clear cut, as the utilities may be part of the off-plots or provided by the client or a third party as a series of OBL connections. In the latter case the safety studies will need to focus on what the project effect could be on the OBL plant, and what the potential effects of the OBL facilities could be on the project. Hence the critical aspect in this case is clear communication and two-way transfer of information to ensure that all mutual concerns have been addressed.

The utilities include not just steam, water, nitrogen, air etc, but also flare systems, waste treatment systems and pre-treatment systems as necessary. There can therefore be a significant scope in the utilities aspect of a licensed plant that needs to be considered.

For different projects the variability of licensor, client, contractor and stage in the process design will mean that it will not be certain that the SIL assessments will have been completed, so an initial check for the SIL assessment status at the transfer of a licensor project into a new stage is good practice.

CONSIDERATION OF OFF-PLOTS AND SIF REQUIREMENTS – RAW MATERIAL AND FINISHED PRODUCT HANDLING

The other areas which may fall into the off-plots scope are raw material and/or finished product handling facilities. Again this is because the scope can vary so widely between facilities, as it depends on what is already present. The supply of raw materials, for example could range from a supply pipeline from an upstream, integrated source, through to transport offloading facilities and significant storage, handling and purification systems. Catalyst and additive handling may be additional or inclusive in the PDP. The finished product handling is seldom included in the PDP as the requirements for storage or packaging will depend on the local infrastructure and transport situation.

As for the off-plots utilities these aspects of the off-plots will need appropriate SHE studies to be completed, including a hazard identification process leading to SIL assessments. The complexity and extent will depend on the hazards of the materials to be handled. These areas can be quite extensive for some process plants, and hence a significant amount of resource may need to be committed to the SIL assessments, and as for utilities the comprehensiveness of the SIL assessment data available needs to be reviewed when a project moves from one stage in the design process to another, and/or from one contractor to another.

PDP SIL REQUIREMENTS – SAFETY, ENVIRONMENTAL AND FINANCIAL

The SIL requirements indicated within the PDP LOPA report are based on the requirement to protect against human harm, and usually to protect against environmental harm as well. However, they will seldom, if ever, include protection against financial / asset loss. Therefore, in addition to considerations of the differences between client and licensor risk criteria for human and environmental harm, the contract needs to be very clear on whether there are risk criteria to be considered for financial and / or asset loss. The level of acceptable risk for financial loss is even more variable between companies than that for human harm or environmental damage. Consideration of financial loss can lead to significant changes in instrumented systems, typically around equipment which on catastrophic failure could lead to significant outage, loss of revenue and / or replacement costs whilst not necessarily having a high level of safety or environmental risk. This issue becomes even more complicated if there are reliability and availability clauses within the contract, as installing additional instrumented functions to protect against equipment damage could lead to spurious trips, leading to additional downtime. The cost of such instrumented functions can therefore escalate significantly if building in duplication and voting systems to get the correct balance between protection against failure to operate and protection against spurious operation.

ALARM RESPONSE AND CONTROL ACTIONS AFFECTING SIL REQUIREMENTS

There is one area which can have a significant impact on the PDP LOPA acceptability to a client. This is when the client does not accept response to an alarm as a valid layer of protection. This is not common, as the IEC 61508/61511 standards allow alarms and control actions to be considered as layers of protection where they are independent from each other, although with a limited probability of success, but there are companies where their internal standards are stricter than the IEC standards.

In this case all the relevant SIL assessments will need to be reviewed and alternative protection identified, or SIL requirements changed appropriately. However it is only for the scenarios identified in the PDP LOPA report where this can be readily achieved, and there may be other scenarios where the use of an alarm response has reduced the risk sufficiently for the scenario not to be listed. It is therefore necessary to have direct discussion between the interested parties to address the issue. Where direct response to an alarm is not typically considered a layer of protection by a client this may be overcome by including a written procedure for the response to the alarm, to increase the integrity of response. However this is client dependent, and in reality the response time available between alarm and incident needs to be sufficiently long to enable a procedure to be accessed, read and acted upon for this to be a real layer of protection.

Another difference specific to certain clients is the number of layers of protection within the control system which it is acceptable to consider in a SIL assessment. Normal practice limits control systems to providing a maximum of two independent layers of protection, and care must be taken to assure true independence of these within the control system. Detailed consideration of the potential for common mode failure of the control system, which would affect the independence, is needed. Common mode failure mechanisms could occur due to a variety of causes, such as input/output cards, parallel cable routing, physical location, common equipment supplier or duplicated equipment types. It the client only accepts a single layer of protection relating to the control system, taking a pessimistic view the one failure could affect the whole control system, then a similar review to that detailed for the alarm response is required.

ACHIEVING SIL REQUIREMENTS FOR THE FINAL DESIGN

As the data is not available, it is often a subjective view as to whether the licensor package identified SIL requirements and other specified layers of protection are equivalent to those which would have resulted from a SIL assessment based on a client calibrated risk graph. One method to overcome this concern is to reassess a representative selection from the supplied PDP LOPA report, using the client risk criteria. If the assigned values seem to be comparable, then the assessment is usually accepted, but where it appears to be underestimated then a decision needs to be taken as to whether to complete a comprehensive, or partial, process hazard and SIL assessment for the design package, in order to satisfy the client internal standards.

To achieve the specification of the SIL requirements for the final design all the above aspects need to be consolidated into a single package. The use of LOPA to determine that a SIL rated safety instrumented function is required is just the start of the process. The actual achievement of the layers of protection identified is part of the detailed design package, considering the complete SIF design, equipment specification, installation, maintenance, proof testing, and auditing.

REFERENCE

Fearnley, J., 2007, Safety Integrity Levels – Considerations for New and Existing Assessments, 12th International Symposium Loss Prevention and Safety Promotion in the Process Industries, 2007, Edinburgh

HARMFULNESS AND HAZARD CATEGORISATION – IMPACT OF EMERGING TECHNOLOGIES ON EQUIPMENT DESIGN IN THE MINING INDUSTRY

Dr David S Dolan[1] and John Frangos[2]
[1]Fluor Australia Pty Ltd
[2]Toxikos Pty Ltd

For the design, fabrication and installation of pressure equipment, knowledge of the harmfulness of the contents and the hazard allocation for the operating conditions is required.

For some processes where the substances in question are pure or common this is a straightforward procedure. However, when the process contains mixtures of substances the classification of the harmfulness of the substance and the hazard allocation becomes complicated, especially when the components of the mixture present toxicological interactions. This categorisation is further complicated by the differences between the Australian, American and European standards.

As technology advances, especially in the rapidly progressing mining industry, it is becoming increasingly important that this procedure is established and understood to ensure that plant is classified correctly and all appropriate standards and codes are complied with. Here, with a mixture of substances, a toxicologist may be required to develop a profile of the toxicological properties of the substances, the effects of concentration and the interactions with other components in the contents.

This paper details the multidisciplinary approach to the classification of the harmfulness of the contents and to the hazard allocation of a system, using examples taken from projects in the mining industry and also explores the variations between different standards used worldwide.

INTRODUCTION

The design of pressure equipment in most countries is governed by regulation, codes and standards. The purpose of these is to provide the framework for the safe, economic and equitable design, manufacture and use of the equipment. This paper discusses the use of the codes and standards particularly in Australia and with reference to the EU and USA with regards to hazard level and harmfulness of the contents.

In the 1970's, German pressure equipment law introduced a term with [Pressure times Volume] (Druckbehalter, 1974) and had considered the problem of hazard quantification. This has evolved into the current European Union Pressure Equipment Directives as enunciated in PED 97/23/EC of the European Parliament and of the Council of 29 May, 1997. This Directive considers pressure, volume or volume equivalent (i.e. pipe diameter) and fluid service. The Australian standards evolved to consider several factors that impact on the quantification of risk. These include pressure, volume or volume equivalent (i.e. pipe diameter), location, service or duty and the equipment contents (AS4343:2005).

Designers have the responsibility to ensure that the design meets the appropriate standards and the identification of hazards, risk assessment and control of risk to health or safety forms part of this process.

In Australia, the standards for the design of pressure equipment requires the designer to evaluate the harmfulness or toxicity of the contents of the equipment then use this information along with the pressure, volume and service conditions or location of the equipment to determine the quality assurance requirements in design, manufacture and operation. The codes and standards set this out and are readily interpreted for common substances where the codes provide adequate data on the substance properties. However, when the codes do not contain data on the contents, or the contents are mixtures that are uncommon the designer needs to work with toxicologists to determine the characteristics of the contents.

This paper discusses the classification of harmfulness and hazard levels of the contents in different standards for pressure equipment. Examples of how the different standards arrive at different results for the levels of quality assurance are given for several cases in the mineral processing industry where new processing technologies have resulted in the use of large high pressure equipment. The harmfulness level of the contents of some of this equipment could not be classified using the relevant standards or codes so the services of a toxicologist were used to find the relevant information. These differences in results may also occur in other industries, but will not be considered here.

These requirements have evolved with the development of high pressure and temperature processes which require the mineral processing engineer to investigate the harmfulness of contents and to coordinate the services of the toxicologist with the pressure equipment designers.

BASIS OF DESIGN FOR PRESSURE EQUIPMENT

The design, manufacture, installation, commissioning, operation, inspection, testing and decommissioning of pressure equipment in Australia and New Zealand is governed by regulations and a series of standards. The 'parent' Standard is AS/NZS 1200 - Pressure equipment. (AS/NZS 1200:2000). This standard lists the standards and codes applicable to pressure equipment for all periods of the equipment life cycle. The Standard AS 3920.1 Assurance of Product Quality - Part 1: Pressure equipment manufacture, (AS 3920.1:1993) describes the methods of selecting the degree of external design verification and fabrication inspection that are required. This selection is based on the hazard level of the equipment which is determined in accordance with the Standard AS 4343 – Pressure Equipment – Hazard Levels (AS 4343:2005).

Other Australian standards apply for advanced design and construction and to specialised services such as serially produced pressure vessels, sterilizers and LP gas vessels for automotive use.

The standard AS 4343 defines five hazard levels from Level A (high hazard) to Level E (negligible hazard). The hazard level depends on the design pressure, the volume of the equipment or the diameter of the pipe, the situation where the equipment is to be located or used (service and site factors), the degree of harmfulness of the contents and whether the fluid contents are gaseous or liquid. Thus the larger the volume, the higher the pressure, the more dangerous the fluid contents and the more likely the location could result in further damage, the higher the hazard level. There are four harmfulness categories for contents ranging from a non-harmful fluid, harmful fluid, very harmful fluid through to a lethal fluid. Each of these categories is further divided as to whether the contents are considered to be a liquid or a gas.

The standard lists the harmfulness for over 900 fluids as pure substances. The harmfulness parameters are quite readily determined by the designer or the engineering design team when the contents are one of the pure fluids listed in the standard. The difficulty arises when the contents are not listed, are substances that are diluted or are mixtures of various ingredients.

This commonly occurs when the pressure equipment is processing intermediate streams in the middle of a major processing facility. Thus the materials are "in process" or in a state of manufacture and the composition of the contents are under going change.

CATEGORIES OF FLUID SERVICE OR CATEGORIES OF FLUID CONTENTS AS USED FOR PRESSURE EQUIPMENT

The standards and codes used in countries such as Australia, USA and Europe use different definitions of fluid service or categories of fluid contents.

Firstly we will outline the similarities and differences in the descriptions of fluid contents or service conditions between the codes and standards in Australia, the USA and the European Union.

In Australia, Standard AS 4343 divides the fluid contents into the four categories: lethal, very harmful, harmful and non-harmful.

Lethal contents are classified as "containing a very toxic substance or highly radioactive substance which, under the expected concentration and operating conditions, is capable, on leakage, of producing death or serious irreversible harm to persons from a single short-term exposure to a very small amount of the substance by inhalation or contact, even when prompt restorative measures are taken". Guidance is given by examples. Contents are classified as lethal if the exposure limit is less than or equal to 0.1 ppm or equivalent.

Very harmful contents are "containing a substance, which, under expected concentration and operating conditions, are classified as extremely or highly flammable, very toxic, toxic, harmful, oxidizing, explosive, self-reactive, corrosive, or harmful to human tissue, but excluding lethal contents." This class includes carcinogenic, mutagenic and tetatogenic substances.

Harmful contents are "containing a substance which, under the expected concentration and operating conditions, is classified as a combustible liquid or fluid irritant to humans, or is harmful to the environment, above 90 °C, or below –30 °C, but excluding lethal or very harmful fluid."

Non-harmful contents are contents not covered by the above categories except for concentration effects such as oxygen depletion and pressure.

The Standard AS 4343 refers to several standards by the Australian National Occupational Health and Safety Commission (NOHSC) that can be used to determine the level of harmfulness. These standards are NOHSC 1003 – National Exposure Standards for Atmospheric Contaminants in the Occupational Environment (NOHSC1003:1995), NOHSC 1008 – National Standard for Approved Criteria for Classifying Hazardous Substances (NOHSC 1008:1999) and the Australian Safety and Compensation Council (ASCC) Hazardous Substance Information System (HSIS) The latter of these is a database accessible through the internet. In addition, the Australian Code for the Transport of Dangerous Goods by Road and Rail (Dangerous Goods Code ADG Code) (ADG: 2007) lists contents that are dangerous.

The first thing to note when comparing the codes and standards is that the use of the words "harmful" and "harmfulness" are not synonymous in various disciplines. This is discussed further below.

In the USA, two codes primarily cover the design of pressure equipment in chemical and mineral processing facilities. These are the ASME Boiler and Pressure Vessel Code Section VIII Division 1 (BPV-VIII-1:2007) and the ASME Code for Process Piping, B31.3 (ASME B31.3:2006). The ASME BPV code has two service conditions; vessels that are to contain lethal substances (lethal service applications) and vessels for any other contents.

The ASME Code for Process Piping, B31.3 (ASME B31.3:2006) notionally uses four classes for fluid contents. These are High Pressure Fluid Service, toxic (Category M), flammable and damaging to human tissue (Normal Fluid Service) or contents not included in any of the preceding classes (Category D). The Category M fluid service is defined in a very similar way to the definition of lethal contents in Australia. The Normal Fluid Service is very similar to the combination of the Australian Very Harmful and Harmful contents grouping. The Category D fluid service is very similar to the Australian contents group of Non-Harmful. Thus the ASME B31.3 range of classifications is reasonably close to the Australian classification.

In the EU, fluid contents are classified into two groups: Group 1 and Group 2 in accordance with Pressure Equipment Directive 97/23/EC (PED 97/23/EC: 1997).

The Group 1 fluids comprise dangerous fluids which are defined as: explosive, extremely flammable, highly flammable, flammable (where the maximum allowable temperature is above flashpoint), very toxic, toxic and oxidising. Guideline 2/7 of the PED lists the specific "risk" phrases for the classification of Group 1 fluids. Risk phrases are used to describe a hazard, and are applied to individual substances at defined cut-off based on concentration. The risk phrases are:

- R2, R3 for explosive
- R12 for extremely flammable
- R11, R15, R17 for highly flammable
- R26, R 27, R28, R39 for very toxic
- R23, R24, R25, R39, R48 for toxic
- R7, R8, R9 for oxidising

Group 2 fluids comprise all other fluids not covered by group 1.

Also in the EU, a dangerous fluid is a substance or preparation covered by the definitions in Article 2 (2) of the Directive on Dangerous Substances, Directive 67/548/EEC (Directive 67/548/EEC). According to the notes to Guideline 2/7, not all fluids that defined as a dangerous substance in accordance with this directive are a Group 1 fluid. Thus care is needed in the correct classification to Group 1.

It can be seen that each jurisdiction has a different way of describing and grouping the fluid contents of pressure equipment. Table 1 – "Comparison of fluid classifications between Australian, European and USA standards and regulations" provides an interpretation of the various categories or groupings and attempts to draw parallels between the various classifications. There are some general similarities but no direct relationships across the practices in Australia, the EU and the USA.

THE ENGINEERS DILEMMA – HOW TO MAKE SENSE OF ALL OF THIS

In Australia, for substances that are not mixtures and that are listed in the Standard AS 4343 the engineer can readily arrive at the appropriate harmfulness for the equipment contents.

If the designer does not find the contents listed in AS4343, it cannot be assumed that the substances to be contained in the equipment are not harmful. Firstly one needs to go to the National Exposure Standards in the Adopted National Exposure Standards (NOHSC 1003 1995). This standard applies to harmful gaseous substances. One can deduce the harmfulness of some gaseous substances from there. However if the contents are not listed in this standard the designer must go to the lists in the Australian Hazardous Substances Information System (HSIS Internet Database) in which there are many thousands of substances listed. There the designer will find the substance name, the CAS Number, the UN Number and the classification using the risk and safety phrases for the pure substance. The database includes the cut-off concentrations at and above which the risk and safety phrases apply. This is when the designer needs the advice and assistance of the toxicologist because the terminology used in the risk and safety phrases is not expressed in the terms of harmful, very harmful or lethal as used in the standard AS4343.

APPLICATION TO AN EXTRACTIVE METALLURGY PROJECT

In recent years the technologies used in the extraction of nickel and cobalt from lateritic nickel ores by the mineral processing industry have developed significantly with the availability of advanced materials that enable the economic fabrication of pressure equipment with superior corrosion resistance and strengths.

An example of one of these processes is the "high pressure acid leach" or HPAL process with subsequent recovery of the metals from the process slurries and solutions. This process utilises the reaction of the finely ground laterite ore with sulphuric acid at operating temperatures in the range of 250 to 270 °C (design temperatures of 260 to 280 °C) and design pressures of 4800 to 7000 kPag. Products produced in this process include nickel sulphate and cobalt sulphate and excess acid is used. Depending on the process route, nickel sulphide, cobalt sulphide, nickel ammonium sulphate and cobalt ammonium

Table 1. Comparison of fluid categories between Australian, European and USA standards and regulations

AUSTRALIA AS4343	DEFINITION		EU PED 97/23/EC: 1997	USA ASME BPV-VIII-1:2006	USA ASME B31.3:2006
Lethal	Highly Radioactive^	where a single short-term exposure to a very small amount by **inhalation or contact** can result in death or serious irreversible harm even when prompt restorative measure are taken	^Radioactive excluded from 97/23/EC	Lethal (no mention of contact, only inhalation)	Category M
	Toxic				
Very Harmful	Extremely Flammable		Group 1	Other	Nominal Fluid Service
	Highly Flammable				
	Very Toxic				
	Toxic				
	Oxidising				
	Explosive				
	Harmful		Group 2 excluding radioactive		
	Self-reactive				
	Corrosive				
	Harmful to human tissue*				
	Carcinogenic				
	Mutagenic				
	Teratogenic				
	Radioactive^				
Harmful	Combustible liquid				
	Fluid Irritant to humans				
	Harmful to the environment, above 90oC or below -30oC				
Non-Harmful	All other contents normally not harmful				None (Category D)

sulphate may be produced as further intermediate products before the final production of nickel and cobalt as powders. Typical concentrations for some of the major components for some typical process streams in the processing of the laterite ores are shown in Table 2. Depending on the ore being processed, other components in these streams can include compounds of iron, copper, chromium, manganese, aluminium, zinc, magnesium and/or silica in solution or as finely dispersed solids in slurries. For simplicity these other constituents are not considered here.

The risk phrases for some of these substances are shown in Table 3. Of course, rarely do these substances exist in pure from, they are usually in aqueous solution and in most cases they are present as mixtures in aqueous solution. This is when the designer requires the services of the toxicologist to determine the level of harmfulness of the contents so the correct fluid contents classification is used.

The determination of the harmfulness of these streams requires the interpretation of the combined impact of the mixture. This requires careful consideration of the characteristics and composition of the mixture. The health assessment of hazardous substances is complicated by the reality that most toxicological testing is performed on single chemicals, but human exposures are rarely limited to single chemicals. Potential exposures resulting from pressure vessels generally involve a complex mixture of substances. A particular issue is whether a mixture of components, may be hazardous due to additivity, interactions, or both. For mixtures that are made up of relatively heterogeneous components, it is also important to consider that the toxicity may be due to a small proportion of the mixtures constituents, for example, immediately following a release of petroleum hydrocarbons, inhalation exposure to the more volatile components, especially the low molecular weight alkanes, may be the immediate concern.

In the absence of data and health criteria for the mixture of concern or of data for a sufficiently similar mixture, the standard toxicological approach recommended by practically all regulatory guidance including the NOHSC and EU Dangerous Preparations Directive 1999/45/EEC has been to use the exposure and health criteria for the individual components of the mixture. The process involves evaluation of whether the exposures or risks for the components can reasonably be considered as additive based on the nature of the health effects. However it is the responsibility of the toxicologist to evaluate whether toxicological interactions among the components are likely to result in greater (or lesser) hazard or risk than would be expected on the basis of additivity alone. The concern for the toxicologist is that in terms of occupational health following exposure, toxicological interactions may increase the health hazard above what would be expected from an assessment of each component singly, or all components additively.

Toxicological interactions can either increase or decrease the apparent toxicity of a mixture relative to that expected on the basis of dose-response relationships for the components of the mixture. Table 4 provides definitions of terms used in describing interactions.

The toxicity of the constituents of a mixture therefore needs to be considered carefully to assess whether there is evidence that constituents in combination may interact in a different manner than additively, if not, additivity is assumed for the purposes of health hazard classification.

Table 2. Typical stream compositions for the significant components in the leach and other process streams for a nickel and cobalt laterite processing facility

Stream--->	(1) Leach feed laterite slurry	(2) Leach product	(3) "Purified" solution	(4) Sulphide precipitation	(5) Sulphide slurry product	(6) Cobalt reduction feed	(7) Nickel reduction feed
Stream components							
temperature °C	200	250	99	90	90	100	100
Gas/Vapour							
H_2S, % vol./vol.	–	–	–	30	–	–	–
Solution							
Ni as $NiSO_4$, g/L	–	6 to 10	5 to 6	5 to 6	0.04 to 0.08	–	40 to 60
Co as $CoSO_4$, g/L	–	0.3 to 0.7	0.15 to 0.5	0.15 to 0.5	0.01 to 0.02	40 to 60	–
H_2SO_4, g/L	0.05	50 to 70	0.5 to 1	0.5 to 1	5 to 10	–	–
$(NH_4)_2SO_4$, g/L	–	–	–	–	–	300 to 500	300 to 500
NH_3, g/L	–	–	–	–	–	20 to 40	20 to 40
Solids							
Wt % solids in Slurry	40	–	–	2 to 4	2 to 4	–	–
Ni as NiS, wt%	0	–	–	50 to 55	50 to 55	–	–
Co as CoS, wt%	0	–	–	5 to 7	5 to 7	–	–
Risk Phrases							
	Classification according to AS4343						
Stream harmfulness	Harmful	Very harmful	Very harmful	Lethal	Very harmful	Very harmful	Very harmful
	Classification according to Directive 97/23/EC – Annex II						
Fluid group	2	2	2	1	2	2	2

Table 3. Risk phrases for the significant components in the leach and other process streams

Substance name	Classification	Cutoffs	Source
Cobalt sulphate	Carc. Cat. 2; R49 Xn; R22; R42/R43 N; R50-53	Conc >= 25%: T; R49; R22; R42/43 >= 1%Conc < 25%: T; R49; R42/43 >= 0.01%Conc < 1%: T; R49	Eu
Aqueous ammonia	Corrosive; R34 N; R50	Conc >= 10%: C; R34 >= 5%Conc < 10%: Xi; R36/37/38	Eu
$CoSO_4$(0.5%) – NH_4OH(40%) complex	Corrosive; R34 Carc. Cat. 2; R49	Conc >= 10%: C; R34 >= 5%Conc < 10%: Xi; R36/37/38 >= 0.01%Conc < 1%: T; R49	Derived
Cobalt sulphide	Xi; R43; N; R50-53	Conc >= 1%: Xi; R43	Eu
Nickel sulphate	Carc. Cat.3; R40 Xn; R22; R42/43 N; R50-53	Conc >= 25%: Xn; R40; R22; R42/43 >= 1%Conc < 25%: Xn; R40; R42/43	Eu
Nickel sulphide	Carc. Cat. 1; R49; R43 N; R50-53	Conc >= 1%: T; R49; R43 >= 0.1%Conc <1%: T; R49	Eu; A
Hydrogen sulphide	F+; R12 T+; R26 N; R50	Conc >= 10%: T+; R26 >= 5%Conc < 10%: T; R23 >= 1%Conc < 5%: Xn; R20	Eu; A
Sulphuric acid	C; R35	Conc >= 15%: C; R35 >= 5%Conc < 15%: Xi; R36/38	Eu; A

Table 4. Interactions terminology (ATSDR 2004)

Term	Description
Interaction	When the effect of a mixture is different from additivity based on the dose-response relationships of the individual components.
Additivity	When the effect of the mixture can be estimated from the sum of the exposure levels weighted for potency or the effects of the individual components.
Influence	When a component which is not toxic to a particular organ system does not influence the toxicity of a second component on that organ system.
Synergism	When the effect of the mixture is greater than that estimated for additivity on the basis of the toxicities of the components.
Potentiation	When a component that does not have a toxic effect on an organ system increases the effect of a second chemical on that organ system.
Antagonism	When the effect of the mixture is less than that estimated for additivity on the basis of the toxicities of the components.
Inhibition	When a component that does not have a toxic effect on a certain organ system decreases the apparent effect of a second chemical on that organ system.
Masking	When the components produce opposite or functionally competing effects on the same organ system, and diminish the effects of each other, or one overrides the effect of the other.

In Australia, a substance is considered hazardous in pressure equipment if it classifiable based on health related criteria or if it is considered dangerous under the Australian Dangerous Goods Code or harmful to the environment. The process of classification based on health related criteria involves the placing of a chemical substance into a particular hazard category by identifying the hazard based on criteria stipulated in the Approved Criteria for Classifying Hazardous Substances (NOHSC:1008). The output at this point is a set of risk phrases. In Australia, the Hazardous Substance Information System (HSIS[1]) database provides a list of chemical substances for which the classification has been conducted and thus risk phrases are available.

Mixtures are classified by first determining the risk phrases for each ingredient and the concentration cut-offs that apply to each risk phrase. The interactions between ingredients are then considered to produce a hazard classification and set of risk phrases to describe the mixture. For instance a pressure vessel containing heated aqueous process stream (Stream 6 in Table 2) containing cobalt sulphate at 0.5% and ammonia at 40% would be classified in the following manner using HSIS data:

Cobalt Sulphate Classification

Substance Name	Classification[2]	CutOffs
Cobalt Sulphate	Carc. Cat. 2; R49 Xn; R22; R42/R43	Conc >= 25%: T; R49; R22; R42/43 >= 1%Conc<25%: T; R49; R42/43 >= 0.01%Conc<1%: Toxic; R49

Ammonia Classification

Substance Name	Classification[2]	CutOffs
Aqueous ammonia	Corrosive; R34	Conc >= 10%: C; R34 >= 5%Conc<10%: Xi; R36/37/38

Process Stream classification

Mixture	Classification	CutOffs
Process stream Cobalt sulphate 0.5% Ammonia 40%	Corrosive; R34 Toxic R49 (carcinogenic by inhalation.	Conc >= 10%: C; R34 >= 5%Conc<10%: Xi; R36/37/38 >= 1%Conc<25%: T; R49; R42/43

[1]Most of the classifications with the HSIS - Australian List of Designated Hazardous Substances have been taken from Annex I of the European Dangerous Substances Directive (DSD) 67/548/EEC. The Australian classification system is essentially the same as the European DSD.

[2]Human health classification only.

Although occupational hazard experts are conversant in the above terminology the application of the hazard classification to pressure equipment using AS4343 is not an intuitive process.

Standard AS4343 is intended to protect workers and the environment from accidental or short term release from a pressure vessel. There are four hazard levels described partly based on health hazard. The dilemma is how to relate these four hazard levels to the risk phrases identified within NOHSC 1008(2004). As the terminology between the two standards is different, expert judgement is required to bridge the gap. Table 5 provides one possible translation between AS4343 and NOHSC:1008. There is a further complication in the translation between AS4343 and health hazard classification. The standard provides guidance that categorises individual substances into a hazard level. The guidance provided within AS4343 refers the reader to the Australian Dangerous Goods Code (ADG). This code classifies substances based on the United Nations harmonised rules for classifying dangerous goods. The ADG is predominantly based on physical hazards but to a small extent it also classifies substances according to their acute toxicity and ability to cause corrosion. Thus there is an overlap between the hazard classification criteria of NOHSC:1008 and the ADG. Unfortunately the definitions and classification cut-offs differ between these two codes further confusing the hapless non-expert. Fortunately there is hope for the non-expert as a global harmonised classification scheme has been agreed at an international level and over the next five years will be implemented by various nations around the world including Australia and the European Union. Using a single system will help standardise interpretation of hazard criteria for human health and will thus simplify downstream applications of these hazard classifications such as that applied within AS4343.

Based on the above analysis and from inspection of Table 5 for interpretation of the Risk Phrases R34-Causes burns, R42/43- May cause sensitisation by inhalation and skin contact and R49-May cause cancer by inhalation, it is deduced that the cobalt ammonium sulphate stream is "Very Harmful".

A similar process to that described above is used by the toxicologist to determine the harmfulness level of the remaining streams in Table 2.

DETERMINATION OF THE HAZARD LEVEL AND CONFORMITY ASSESSMENT CATEGORY FOR TYPICAL VESSELS AND PIPING IN A MODERN HYDROMETALLURGICAL PROCESSING PLANT

For an example, the harmfulness levels for the contents shown in Table 2 will be used to determine the hazard level for typical pressure vessels and process piping in a modern nickel and cobalt hydrometallurgical processing plant.

Typical dimensions, design pressure and temperature of each vessel and pipe are given in Tables 6 and 7. In several of the examples, the fluids are liquids above their boiling point at atmospheric pressure and are considered to be a gas according to AS 4343. Table 8, which is Table 1 from AS4343, is used to derive the Hazard Level for each of the vessels and piping using the harmfulness, fluid state, values of pressure times volume (pV)

Table 5. Rough translation of the hazard levels of AS4343 to human health and non-health risk phrases

AS4343 Harmfulness level	AS4343 Terms for classification of contents	Hazard classification[3] (hazard category and risk phrases according to NOHSC:1008)	Risk phrases for health and non-health effects and ADG Code Class
(1) Highest harmfulness – "lethal contents"	Very toxic substance or highly radioactive substance	Very toxic (T+) , a harmful substance which can cause irreversible effects after acute exposure	R26, R27, R28, R32, R39 ADG Code Cl. 7
(2) High hazard – "very harmful contents"	Very toxic, toxic, harmful, very corrosive, corrosive or harmful to human tissue. Extremely or highly flammable, oxidizing, explosive, self reactive	Very toxic (T+), toxic (T), very corrosive, corrosive (C), carcinogenic (Carc.), mutagenic (Muta), teratogenic (Repr.), a skin or respiratory sensitiser (Xn) and classifications based on chronic health effects but where the evidence is not sufficient to classify the compound as a probable hazard to humans. Extremely flammable (F+), flammable (F), Oxidizing (O), explosive (E)	R23, R24, R25, R29, R31, R32, R34, R35, R40, R41, R42, R43, R45, R46, R48, R49, R60, R61, R62, R63. R1, R2, R3, R4, R5, R6, R7, R8, R9, R10, R11, R12, R14, R15, R16, R17, R18, R19, R30, R44. ADG Code Cl. 2.1, 3, 5, 8
(3) Moderate hazard – "harmful contents"	Fluid irritant to humans or combustible liquid or harmful to the environment, above 90 C or below – 30 C	Harmful (Xn), irritants (Xi), dangerous to the environment (N)	R20, R21, R22, R33, R36, R37, R38, R64, R65. R50, R51, R52, R53, R54, R55, R56, R57, R58, R59.
(4) Extra low/no hazard – "non-harmful contents"		Non hazardous but mild irritants	None

[3]The hazard classification for the mixture contained within the pressure vessel. This requires evaluation of each individual constituent and the lowest relevant concentration cut-off level for each constituent specified for the hazard classification in the NOHSC:1008.

Table 6. Determination of hazard levels and conformity assessment categories for various pressure vessels

Pressure Equipment	Vessel 1 Preheater	Vessel 2 Leach Autoclave 1	Vessel 3 Leach Autoclave 2	Vessel 4 Pressure Filter	Vessel 5 Pptn Feed Vessel	Vessel 6 Sulphide Precipitator
Fluid Contents	Feed Slurry	Leach Product	Leach Product	Purified Solution	Sulphide Slurry	Sulphide Precipitate
Tan to Tan Length, m	7	35	20	3.5	5	7.5
Vessel Inside Dia, m	1.8	5	4	2	5	7.5
Vessel Volume, L	20,866	752,673	284,838	15,184	163,625	552,233
Vessel Design Pressure, p, MPa	3	4.5	7	0.7	0.5	0.5
Vessel Design Pressure, P, Bar	30	45	70	7	5	5
Vessel Design Temperture, C	200	270	270	90	90	90
	Classification according to AS4343 - Modification to value of pV is Nil					
Fluid State	Gas	Gas	Gas	Liquid	Liquid	Gas
Fluid Harmfulness	Harmful	Very Harmful	Very Harmful	Very Harmful	Very Harmful	Lethal
Value pV, MPa.L	6.3E+04	3.4E+06	2.0E+06	1.1E+04	8.2E+04	2.8E+05
Mod's to pV for special conditions	Zero	Zero	Zero	Zero	Zero	Zero
Hazard Level	B	A	B	B	B	A
	Classification according to AS4343 - Modification to value of pV is 3 for Location as Major Hazard Facility					
Fluid State	Gas	Gas	Gas	Liquid	Liquid	Gas
Fluid Harmfulness	Harmful	Very Harmful	Very Harmful	Very Harmful	Very Harmful	Lethal
Value pV, MPa.L	6.3E+04	3.4E+06	2.0E+06	1.1E+04	8.2E+04	2.8E+05
Mod's to pV for special conditions	3	3	3	3	3	3
Modified value pV, MPa.L	1.9E+05	1.0E+07	6.0E+06	3.2E+04	2.5E+05	8.3E+05
Hazard Level	B	A	A	B	B	A
	Classification according to Directive 97/23/EC - Annex II					
Fluid State	Gas	Gas	Gas	Liquid	Liquid	Gas
Fluid Group	2	2	2	2	2	1
PD, Bar.L	625,994	33,870,296	19,938,641	106,291	818,123	2,761,165
Conformity Assessment Category	IV	IV	IV	IV	IV	IV

Table 7. Determination of hazard levels and conformity assessment categories for various pipe duties

	Pipe 1	**Pipe 2**	**Pipe 3**	**Pipe 4**	**Pipe 5**	**Pipe 6**
Pressure Equipment	**Preheater**	**Leach Autoclave**	**Leach Autoclave**	**Pressure Filter**	**Pttn Feed vessel**	**Precipitator**
Pipe duty	**Preheater Feed Pipe**	**Autoclave Feed Pipe**	**Leach Disch. Pipe**	**Filter Feed Pipe**	**Feed Pipe**	**Precipitator Piping**
Fluid Contents	**Feed Slurry**	**Leach Feed**	**Leach Product**	**Purified Solution**	**Sulphide Slurry**	**Sulphide Precipitate**
Pipe Inside Dia, mm	250	250	250	500	600	600
Pipe Design Pressure, p, MPa	3	7	7	0.7	0.5	0.5
Pipe Design Pressure, P, Bar	30	70	70	7	5	5
Pipe Design Temperture, C	270	200	270	90	90	90
	Classification according to AS4343 - Modification to value of pD is Nil					
Fluid State	Gas	Gas	Gas	Liquid	Liquid	Gas
Fluid Harmfulness	Harmful	Harmful	Very Harmful	Very Harmful	Very Harmful	Lethal
pD, MPa.mm	750	1750	1750	350	300	300
Mod's to pD for special conditions	Zero	Zero	Zero	Zero	Zero	Zero
Hazard Level	B	B	B	C	C	B
	Classification according to AS4343 - Modification to value of pD is 1.5 for Location as Major Hazard Facility					
Fluid State	Gas	Gas	Gas	Liquid	Liquid	Gas
Fluid Harmfulness	Harmful	Harmful	Very Harmful	Very Harmful	Very Harmful	Lethal
Value pD, MPa.mm	750	1750	1750	350	300	300
Mod's to pD for special conditions	1.5	1.5	1.5	1.5	1.5	1.5
Modified value pD, MPa.mm	1125	2625	2625	525	450	450
Hazard Level	B	B	B	B	C	B
	Classification according to Directive 97/23/EC - Annex II					
Fluid State	Gas	Gas	Gas	Liquid	Liquid	Gas
Fluid Group	2	2	2	2	2	1
PD, Bar.mm	7500	17500	17500	3500	3000	3000
Conformity Assessment Category	III	III	III	Art. 3, Para. 3	Art. 3, Para. 3	III

Table 8. Hazard levels of pressure equipment (Part of Table 1 of AS4343 published with permission of SAI Global Ltd)

Equipment – Type and conditions (see Notes 6 & 9)					Hazard level (see Notes 5, 7 & 8)																			
1 PRESSURE VESSELS (except vacuum vessels and boilers) – includes unfired, fired, static & transportable vessels																								
Fluid Type of Contents (see Notes 1 & 2)			**Volume** (V) L	**Pressure** (p) MPa (see Notes 10 & 4(a)(v))	**Value of pV, (as modified by Notes 4 & 10) MPa.L (see Note 3)**																			
					0.1–0.3	0.3–1	1–3	3–10	10–30	30–10^2	10^2–3×10^2	3×10^2–10^3	10^3–3×10^3	3×10^3–10^4	10^4–3×10^4	3×10^4–10^5	10^5–3×10^5	3×10^5–10^6	10^6–3×10^6	3×10^6–10^7	10^7–3×10^7	3×10^7–10^8	$>10^8$
1.1	Lethal (see Note 11)	Gas	>0.05	>0.05		C							B							A			
		Liquid	>0.2		E	D		C					B							A			
1.2	Very harmful	Gas	>0.2				E	D		C			B							A			
		Liquid	>1.0				E			D	C		B									A	
1.3	Harmful	Gas	>0.2				E		D		C		B									A	
		Liquid	>1.0				E				D		C			B							A
1.4	Non-harmful (see Note 5)	Gas	>0.2				E			D	C					B						A	
		Liquid	>10				E						D	C		B							

(or pressure times diameter (pD) for piping) and modification factor for special conditions. Two modification factors are considered in these examples. The first group of examples uses a factor of zero for no location or service modifier while the second group of examples uses a factor of three for the pressure vessels and 1.5 for piping on the basis that the vessels and piping are located in a major hazard facility.

Two of the vessels in the example have hazard levels of A and the other four are B. When the vessels are considered to be located in a major hazard facility, Vessel 3's hazard level increases from B to A while the other remain the same. Four of the pipes have hazard levels of B and two have C. When the modifier is increased, pipe 4 hazard level increases from C to B while the remainder are unchanged. These hazard levels determine the required degree of external design verification and fabrication inspection.

In Australia, the design and fabrication verification requirements are specified in AS 3920.1 – 1993 Assurance of product quality – Part 1: Pressure equipment manufacture (AS 3920.1 – 1993). Table 2.1 of the standard, titled "Relationship between hazard level of equipment and required degree of external design verification and fabrication inspection with and without a manufacturer's certified quality system" is used to select the level of verification.

The extent of verification required for the manufacture of the vessels and piping in the examples is described below for the case where the designer and fabricator have certified quality systems to ISO 9001 and ISO 9002.

Table 9. AS 3920.1 – 1993 Table 2.1 – Relationship between hazard level of equipment and required degree of external design verification and fabrication inspection with and without a manufacturer's certified quality system (Part of Table 2.1 of AS3920.1 published with permission of SAI Global Ltd)

	Design			Fabrication	
Hazard level of equipment (see Appendix B)	Certified quality system status	Design verifying body (see Note 1)		Quality system status	Fabrication inspection body required (see Notes 2 & 3)
A	AS/NZS ISO 9001	Yes		AS/NZS ISO 9002	Yes
B	AS/NZS ISO 9001	Yes	or	AS/NZS ISO 9002	No
	No CQS	Yes		No CQS	Yes
C	AS/NZS ISO 9001	No (Note 4)	or	AS/NZS ISO 9002	No
	No CQS	Yes		No CQS	Yes
D	AS/NZS ISO 9001	No		No CQS	No
	No CQS	Yes (Note 5)			
E	No CQS	No		No CQS	No

If the pV or pD modifier is zero, Vessels 2 and 6 require full independent verification of design and full independent fabrication inspection, while Vessels 1, 3, 4 and 5 only require full design verification. Pipe no's 1, 2, 3 and 6 require full independent design verification while pipe no's 4 and 5 do not need any independent verification. If the equipment is to be located in a major hazard facility where the pV or pD modifiers are 3 and 1.5 respectively Vessel 3 and Pipe no. 4 change up one level of verification.

The same pressure vessels and piping are also evaluated using the criteria in PED/97/23/EC to determine firstly the fluid contents grouping and then the categories of modules in accordance with Annex II. The modules define the conformity assessment procedures required for design verification and fabrication inspection. All of the pressure vessels are designated as Category IV, the highest level of conformity assessment. The conformity assessment categories for two of the pipes are the lowest level; Article 3, Paragraph 3. The remainder of the pipes are category III; the second highest level.

The hazard levels E, D, C, B and A in AS 3920.1 and AS4343 are similar but not the same as the Groups in the PED of Article 3, Paragraph 3, I, II, II and IV. Thus it can be seen that, for the examples given, the use of the Australian standards and the PED arrive at different quality assurance or conformity assessment requirements for the equipment in the same service. The PED is stricter for the pressure vessels while the Australian standard is stricter for the piping.

CONCLUSIONS

The quality assurance procedures for pressure equipment in Australia, the EU and the USA use classifications for the fluid contents as part of the procedure for arriving at the level of conformity assessment to be applied to the design and fabrication.

In Australia, the classifications are derived from the Australian Dangerous Goods Code and Australian List of Designated Hazardous Substances, taken from Annex I of the European Dangerous Substances Directive –DSD, or if not included there, the classification is to be derived using the procedures in the National Standard for Approved Criteria for Classifying Hazardous Substances. Often there is no clear path for arriving at one of the four harmfulness classes from these sources and it is recommended for the designer to use the services of a toxicologist for guidance in the classification. In the EU the derivation is deduced from the DSD using the Risk Phrases specified in the Pressure Equipment Directive Guideline 2/7 though care is needed when using the classifications in directive 67/548/EEC.

In the USA, the ASME Code for process piping has four fluid classes which are reasonably well defined in the code. The ASME BPV code only specifies substances that are lethal by inhalation with little guidance as to their classification.

Once the fluid contents have been classified the selection of the hazard level or category for conformity assessment is readily determined in AS3920.1 or PED Annex II.

It would appear that there are significant differences between the Australian and the EU PED conformity assessment or quality assurance requirements for pressure vessels or piping in the same duty. The PED requirements for pressure vessels are generally stricter while the reverse is the case for piping.

ADDRESS
Correspondence concerning this paper should be addressed to Dr David S. Dolan, Fluor Australia Pty Ltd, GPO Box 1320, Melbourne, Victoria, 3001, Australia *or email* david.dolan@fluor.com

REFERENCES

ADG: 2007, Australian Code for the Transport of Dangerous Goods by Road and Rail (Dangerous Goods Code ADG Code – 2007).

AS/NZS 1200:2000, Australian Standard/New Zealand Standard: Pressure equipment. AS/NZS1200 – 2000. Standard available at *www.saiglobal.com.*

AS 3920.1:1993, Australian Standard: Assurance of product quality – Part 1: Pressure equipment manufacture. AS3920.1 – 1993. Standard available at *www.saiglobal.com.*

AS 4343:2005, Australian Standard: Pressure equipment – Hazard levels. AS4343 – 2005. Standard available at *www.saiglobal.com.*

ATSDR 2004, Guidance Manual for the Assessment of Joint Toxic Action of Chemical Mixtures. US Department of Health and Human Services, Agency for Toxic substances and Disease Registry. May 2004.

Directive 67/548/EEC, The Directive on Dangerous Substances. 67/548/EEC.

Druckbehalter, 1974, Unfallverhutungsvorschrift, Carl Heymanns Verlag KG, Gereonstrasse 18-32, 5 Koln 1, Germany, Serial VBG 17.

HSIS, Australian Safety and Compensation Council (ASCC) Hazardous Substance Information System. Internet based database available at http://hsis.ascc.gov.au.

NOHSC1003:1995, National Exposure Standards for Atmospheric Contaminants in the Occupational Environment. NOHSC 1003 – 1995.

NOHSC 1008:1999, National Standard for Approved Criteria for Classifying Hazardous Substances. NOHSC 1008 – 1999.

PED 97/23/EC: 1997. Directive 97/23/EC of the European Parliament and of the Council of 29 May 1997 on the approximation of the laws of the Member States concerning pressure equipment.

DESIGNING FOR SAFETY – HOW TO DESIGN BETTER WATER TREATMENT WORKS

Peter Bradley
BScHons MScTech (Process Safety & Loss Prevention) CMIOSH RMaPS AIEMA
United Utilities, Warington, UK.

The paper describes the initiatives that have taken place in a large water utility Company – United Utilities plc, to improve the safety of new water and wastewater treatment works. These initiatives were designed to improve the standards of safety for personnel during the operation and maintenance of the assets. The focus of the initiative was to ensure that the entire project team understood the hazards and difficulties created by not considering operational and maintenance safety throughout the development of the project. This was a quality improvement initiative that produced safer treatment works.

KEY TERMS: Designing for Safety, Safer by Design, Operation, Maintenance, Quality, Construction.

1. INTRODUCTION

The water industry is set challenging targets by its regulators, and achieving those targets stretches the delivery teams. Historically, the delivery of the capital programmes was fragmented, with none of the client, design, construction and operation (client) teams being actively involved throughout the process. Consequently, the views of parts of the team may have been overlooked, and hazards built into new projects. Because of the fragmented nature of the delivery, the whole project team did not always learn the lessons, and the same hazards were repeatedly identified on a number of projects. In pre-handover inspections of newly constructed works occupational health and safety specialists identified commonly occurring faults that could have been identified earlier in the design and construction process. These faults were expensive and difficult to resolve retrospectively. In order to remedy this, it was necessary to close the feedback loop, and allow the whole project team to see for themselves, and to learn from these hazards.

A programme was developed to raise the awareness of the project teams. This started as a series of presentations, showing photographs of the commonly occurring hazards. This developed into 'designing for safety workshop tours' of the best water and wastewater treatment works, and then finally a series of "discovery tours" were held. The Company, designers and the project teams adopted improved practices that resulted in safer treatment works. Treatment works constructed presently now have far fewer built-in problems that make the works easier to operate and maintain and are consequently safer.

The project management triangle comprises of the elements of cost, quality and time. In the majority of projects emphasis is given to cost and time in getting the project completed to budget and on time due to the commitments given to the regulator in satisfying regulatory requirements. The quality focus in this case is to meet regulatory standards for water. Frequently the quality element of the projects itself is more difficult to measure and is often relegated to the 'post-mortem' operational stage of the project when the key participants have moved onto their next project.

2. BODY

2.1 THE PROBLEM

When new water or waste water treatment are constructed a formal safety inspection is performed by an occupational health and safety specialist as part of the commissioning process to identify any remaining hazards that need to be addressed and resolved prior to handover to the Company. It was identified by the occupational health and safety specialists that the problems identified were repetitive.

Examples of these problems were as follows.

a. changing lights bulbs. (Picture 1 – problem, Pictures 2–3 solutions)
b. the cleaning and replacement of ultrasonic level sensors. (Picture 4 – problem, Picture – 5 solution)
c. fixed lifting equipment unable to lift and move equipment to a set down point or lorry.
d. unsafe manual handling tasks with heavy weight and difficult journeys. (Picture 6)
e. safe access to under floor equipment.
f. accessing high level workplaces with a tool kit.

2.2 THE SOLUTION

It was decided that it would be beneficial to gather the client representatives together with the designers and constructors of works together and make them aware of the commonly occurring problems. Three approaches were used:-

2.2.1 Approach one

The meetings were of a two-hour duration and were attended on each occasion by thirty participants comprising Directors, Construction Managers and Design Engineers in the Company and Partner Organisations. A total of sixty key players attended these events. The presentation lead by a health and safety specialist took the following format

- Introduction – to establish the attendees engineering background knowledge and their experience of designing and constructing water and wastewater treatment works.
- Outline of legislation – to understand the legal requirements.
- Presentation showing photographs of good and bad design and the safety implications.
- Discussion, and Summary

Many of the attendees expressed surprise that such hazards had not been identified and designed out prior to construction. One expressed amazement at the "schoolboy howlers".

Picture 1. How easy is it to change this light bulb?

Senior management agreed that steps should be taken to ensure that improvements were made, and agreed to sponsor a series of site tours for the project teams that lead to Approach 2.

2.2.2 Approach two

After these initial 'awareness raising' events it was decided to take the project teams to sites to see water and wastewater sites that had recently been constructed. (Picture 7) These 'see for yourself' tours for the project teams took place on newly completed treatment works. Parties of twenty to thirty designers were shown around the works showing them the difficulties in operation and maintenance that can result from poor design. Experienced Process and Maintenance personnel were present on the tour, and they were able to handle detailed questions. This enabled the message to be given to a much larger audience which enabled greater numbers to be numbers to be briefed (around 150).

Picture 2. This light bulb can be easily replaced

2.2.3 Approach three

The third step was to hold 'discovery' tours. These were held for between eight and fifteen attendees from one organisation or project team. The structure differed from previous tours with the emphasis changing from the attendees being told what was wrong on new works to more of a 'voyage of guided discovery' where they were split into two groups and taken to twelve pre-identified locations on the works. The attendees were challenged to identify good

Picture 3. These light fittings descend from the ceiling so that the light bulb is easily replaced

Picture 4. These ultrasonic sensors accumulate spiders and cobwebs and are difficult to clean and replace

Picture 5. Ultrasonic sensor with hinge

Picture 6. This strainer lid weighs in excess of 20 kg and is in a constricted location – how can it be lifted?

Picture 7. Designers 'seeing for themselves' the works that 'we made earlier'

and bad aspects of the design and construction. The tours were lead by safety, operational and maintenance experts also acting as facilitators to ensure that no pertinent points were missed. The delegates returned to the meeting room to discuss what they had found. There was 'a buzz' – after all what was being examined and scrutinised was our best water treatment works. The delegates had not constructed the works and consequently were not responsible for the

Picture 8. Lostock Water Treatment Works – our best water treatment works! Ultrasonic level sensor – how is it cleaned and replaced?

design faults. This allowed free discussion of the good and bad points that had been identified. Each team produced a list of frequently occurring faults and this was structured and prioritised into a 'hit list' of problems to avoid during the construction of new treatment works. These lists were used as 'lessons learnt' prompt sheets for future projects and subsequently incorporated into the Access, Lifting and Maintenance Review Procedure.

This was the most successful of the three approaches. Due to the group size the groups formed much more of a team mindset in trying to spot the strength and weaknesses. Elements of game-playing and competition were introduced which brought out the 'fun' element which is not normally associated with a safety training event. It was turned into a game – people are relaxed when playing a game and are open to new ideas. It was fun. This event seemed to spark the team into becoming better designers.

Case Study – Lostock Water Treatment Works

The project team for Lostock Water Treatment Works (Picture 8) attended a "discovery tour" of 'Approach Three' type sessions. The priority problems were identified and meetings were held to debate and agree the way forward. As a result of these tours the following events happended:-

The team revisited the site to 'ask questions' and 'tap into the knowledge' of the process controllers. They visited several other similar sites to learn from others – they were determined to improve on previous sites. As a result of these initial activities the Team was awarded the first 'Designing for Safety Award' from the Company in recognition of their active commitment to the subject. The works, supplying safe drinking for 500,000 people in Greater Manchester, was built and due to the exceptionally high standard of the design and construction the construction team received a further 'Designing for Safety Award'.

2.2.4 The 'Designing for Safety' seminars were repeated in following the approach three format and the following outputs occurred:-

a. A CD-ROM self-learning package was produced with the key learning points. This took the form of several PowerPoint Presentations on Designing for Safety based on the most important learning points of the tours. This was issued to the partner organisations and their design teams.
b. A lifting, access and maintenance review was brought into the design process to consider the common problems in a 'HAZOP' type' approach. This also requires that any lifting equipment must be demonstrated to be proved 'fit for purpose' prior to handover.
c. Designing for Safety Working Group re-convened and new edition of the Company Designing for Safety Document was issued.
d. The Company became more conscious and demanding in its needs for quality on new works. Improved 'signature' or model designs were developed and used as 'building blocks' for project specific designs. There was also greater involvement of the

operators and maintainers in the safety inspection at the pre-handover inspections and earlier in the design process.

e. Designing for Safety Awards were issued for the Company programme of work to Companies that had demonstrated commitment and competence in designing and building better works.
f. Designers used 3D – design software with operators prior to construction on a project.

3. RESULTS

The new water treatment works produced by the focus team was by far the safest that had been produced to date. Improvements have been seen on the majority of projects. Those projects where hazards are identified at the pre-handover stage are now flagged for investigation to the Engineering Department. They can then take any necessary steps to ensure that the processes are modified and to ensure learning. Overall, the amount of snagging and corrective work has reduced – reducing costs, and the works that are built are safer to operate and maintain.

4. DISCUSSION

Delivering a safe new treatment works, needs the involvement of the whole team – from good specification, and standards, good outline and then detailed design, to diligent construction and commissioning and then provision of feedback to the whole team about the safety of operation and maintenance. If the construction team are instructed about hazards that should have been designed out in the design phase, or the snagging team are informed of hazards that have not clearly been specified by the client, then appropriate learning cannot take place.

Just as no one wishes to have an accident – everyone would like to do a good job. Good engineers are problems solvers and solution providers – give them a challenge and they can rise to it. After all it was engineers who put a man on the moon.

The Institution of Civil Engineers website currently states in a promotion of engineers "When you understand civil engineering, you see the world differently" The truth is we all see the world differently, sometimes, safer design and construction comes through seeing the world through another persons eyes – the operators!

5. CONCLUSIONS

Unless an effective feedback loop is established throughout the delivery process, lessons will not be learned, and the same mistakes may be repeated. In a delivery process that engages a number of different organisations, it is more difficult to ensure that feedback is given to the appropriate team members throughout the design and construction process.

On all treatment works the standard of completed treatment works that we now have built is much higher than when we commenced this approach in 2000.

Awareness has increased with all the approaches used and the construction team is now aware of the need for designing for safety.

The works are easier to maintain and operate.

'Designing for Safety Awards' act as a useful incentive that can be issued by the Client and can be used to advertise the success of the Project Team in promotional material on their Company websites (see Internet links).

6. RECOMMENDATIONS

This entire initiative promoted a learning approach with the project team. It was vital to raise awareness of senior management with Approach One. Approach Three was more successful than Approach Two because 'Team Learning' and peer group pressures to perform were present.

Use Company documents to support needs and preferences – signature designs, asset and company standards. Some preferences are not obvious to designers e.g. – we don't like 'ships ladders'.

Designers and constructors should always visit previous works to understand from the operators and maintainers the lessons learnt from the operation and maintenance of similar projects. This can form part of their continuing professional development.

Recognition and reward are useful incentives – good designers should be recognised and they should receive Designing for Safety Awards' for good practices. After all everyone remembers the 'Diane Memorial Fountain in Hyde Park and the wobbly Millennium Bridge but when was the last time a designer received praise and recognition?

REFERENCES

The Health and Safety at Work Act 1974

Management of health and safety at work: Management of Health and Safety at Work Regulations 1999, Approved Code of Practice and Guidance, L21 Health and Safety Commission, ISBN 0 7176 2488 9

Managing health and safety in construction, Construction (Design and Management) Regulations 2007, Approved Code of Practice, Health and Safety Commission, ISBN 978 0 7176 6223 4

Managing construction for health and safety, Construction (Design and Management) Regulations 1994, Approved Code of Practice, L54, Health and Safety Commission, ISBN 0 7176 0792 5

Manual handling: Manual Handling Operations Regulations 1992, Guidance on Regulations L23, Health and Safety Executive ISBN 0 11 886335 5

Workplace Health, Safety and Welfare: Workplace (Health, Safety and Welfare) Regulations 1992 – Approved Code of Practice and Guidance L24 (Health & Safety Commission) ISBN 0 11 886333 9

AWARDS GIVEN BY UNITED UTILITIES

Two awards for Safety in Design at Lostock Water Treatment Works to a team comprising Costain Ltd – Principal Contractor, MWH & Haswell Consulting Engineers – Designers.

GUIDANCE ON EFFECTIVE WORKFORCE INVOLVEMENT IN HEALTH AND SAFETY

David Pennie[1], Michael Wright[1], Paul Leach[1] and Mark Scanlon[2]
[1]Greenstreet Berman, 10 Fitzroy Square, London W1T 5HP
[2]Energy Institute, 61 New Cavendish Street, London W1G 7AR, UK

Workforce involvement (WFI) can be a complex and sensitive issue and it can be difficult to engage with workers or make involvement more effective. To help organisations within the petroleum and allied industries, practical guidance, for WFI, has been developed on behalf of the Energy Institute by Greenstreet Berman Ltd.

This paper introduces the issues surrounding WFI and describes the 3-step approach, case studies and assessment exercises detailed in the guidance. It also discusses how such an approach might be beneficial in improving health and safety, as well as bringing other benefits.

INTRODUCTION

BACKGROUND

According to the Health and Safety Executive (HSE): ***"involving workers in health and safety leads to healthier and safer workplaces and produces a range of benefits for workers and managers"***[1].

This view is based on evidence from a considerable body of research and is shared by many other similarly respected organisations.

Guidance on workforce involvement (WFI) was developed on behalf of the Energy Institute (EI) by Greenstreet Berman Ltd (GSB). The key objective of the EI was to develop for publication:

"Simple petroleum and allied industries guidance on how to achieve effective workforce involvement. It should navigate users through the necessary steps on a pathway to effective workforce involvement, identifying available resources and inputs to assist attainment of each step."

WFI can bring significant benefits. This is i llustrated by a number of examples, identified during the development of the guidance:

- One organisation, in the high hazard industry, reported that increasing WFI was associated with a 50% reduction in reportable accidents[2];
- Organisations with formal safety committees have reported 40–50% lower injury rates[3];
- Occupational illness was lower when employees were involved in safety[3];
- Around 50% of high performing companies use worker suggestions, ideas and feedback[4];

- An organisation reported that when workers were involved in equipment review and design there was a reduction in material and overtime costs[2].

Involving workers in making decisions about health and safety makes sense because the people who carry out work are well placed to say how work might be improved. Workers have more direct experience of unsafe conditions and how it affects their job.

WFI can lead to these types of benefits for a number of reasons, for example:

- **Compliance is improved** – workers involved in the development and review of policies and procedures have an interest in maintaining the rules, they have helped to develop, and are more likely to support and comply with them;
- **Concern for safety is increased** – involvement highlights that everyone is responsible for safety, which in turn can mean individuals start to take greater responsibility for the health and safety of themselves and colleagues;
- **Morale and trust are improved** – working together can increase understanding and trust across an organisation;
- **Decision making is better** – decision making is better because, by involving workers, managers become more informed about the issues affecting their business.

The evidence for the benefits of WFI have led respected organisations, for example, the Health and Safety Executive (HSE), to make WFI central to their philosophy on tackling health and safety:

There is also evidence that serious incidents have occurred, in the petroleum and allied industries, where lack of WFI in process safety may have been a contributory cause.

In March 2005, a catastrophic process accident at Texas City refinery in the US resulted in 15 deaths and more than 170 injuries. Investigations into this event and the subsequent report by Baker J et al[5], concluded that: ***"... Texas City, has not established a positive, trusting, and open environment with effective lines of communication between management and the workforce."***

The report also stated that the organisations involved in these incidents were managing occupational safety particularly well, with very low personal injury rates. There was, however, an apparent imbalance between the effort put into personal injury versus major accident prevention. As illustrated by the following quote:

The organisation ... ***"in recent years, has achieved significant improvement in personal safety performance, but did not emphasize process safety, mistakenly interpreting improving personal injury rates as an indication of acceptable process safety performance at its U.S. refineries"***

Interestingly this does not seem to be be unique, with similar conclusions made by the investigators[6] into a major incident at the Grangemouth complex in the UK during 2000.

"Commendable success in managing personal injury rates down to a very low level, together with a failure to adequately distinguish these successes from process safety management. This imbalance between the effort put into personal injury versus major accident prevention was by no means unique ... the Competent Authority and others have found similar tendencies in other comparable businesses."

This focus on occupational safety may reflect that in the past organisations within the petroleum and allied industries may have concentrated efforts on behavioural modification. This may have been at the expense of involving workers in other critical aspects of safety management, such as process safety.

ISSUES CONCERNING WFI

Many organisations are aware of the benefits of WFI, however, research by ECOTEC Ltd[7] and anecdotal evidence suggests that attempts to improve WFI are not always effective.

This may be because increasing involvement may not be straightforward and there are barriers preventing successful engagement of workers. Interviews with stakeholders, as part of our work to develop the guidance, helped to identify, possible reasons why WFI may not always be successful, for example:

- **Lack of senior manager commitment** – senior managers do not demonstrate commitment to WFI, sending the message to workers that their views are not valued;
- **Poor problem diagnoses** – underlying causes to problems are not identified and the wrong solutions and methods to improve WFI are applied;
- **Poor planning** – involvement programmes are not planned properly and hence people become unclear about roles and responsibilities;
- **Poor resourcing** – individuals involved in the involvement programme do not have time to become actively involved;
- **Lack of feedback** – the effect and results of involvement programmes are not communicated so individuals are unaware of the benefits and how they can help improve safety;
- **Initiatives are not sustained** – initiatives come to an end and are not continued so long lasting benefits are lost.

ACTIVITIES TO DEVELOP THE GUIDANCE

The work to develop the guidance was directed by the EI Human and Organisational Factors Working Group (HOFWG) using funding from EI Technical Partners. This group represented a range of stakeholders, for example:

- Regulators;
- Onshore oil and gas industry;
- Offshore oil and gas industry;
- Psychologists;
- Ergonomists;
- Designers.

The project comprised a number of tasks to develop the guidance:

- Consideration of research and existing guidance to determine the issues concerning workforce involvement and identify good practice;

- Conducting interviews, with a range of professionals from different petroleum and allied industries, to establish user requirements;
- Conducting interviews with others who have expertise in workforce involvement;
- Developing a specification document that details guidance structure and content;
- Engaging with stakeholders to review and comment on this framework document;
- Finalising the guidance in a form that best meets the industry's needs.

THE SCOPE OF THE GUIDANCE

The design of the guidance was initially based on information provided by EI's HOFWG, who stated that the guidance should:

- Identify the pertinent legislative requirements for both managers and workforces;
- Describe a step-by-step process illustrated with issues, case studies, and hints and tips;
- Identify and develop tools and resources for managers and workforces to help stimulate or structure workforce involvement.

Feedback from stakeholders and interviews also recommended that the guidance should explain why it was important to be fully committed when trying to increase WFI. This was because although potentially very beneficial trying to tackle WFI, without a clearly thought out and planned approach, could actually make things worse rather than better.

The rationale for the guidance was therefore similar to other existing environmental, safety and quality management models. A strong emphasis being placed on planning and taking a step-by-step structured approach with practical tools provided to help application. In addition because of the perceived complexity of WFI advice was also provided on the issues that can impact on the success of WFI.

GUIDANCE OUTLINE

The guidance provides an introduction which aims to explain and define WFI. This is because it is important to have a clear understanding of the concept before trying to improve WFI.

Understanding WFI is made more difficult, than might first appear, because there are many different terms that can be used to describe it. For example: worker involvement; workforce involvement; worker consultation; worker participation; worker engagement; partnership working and participative ergonomics.

Generally, however, participation, engagement and involvement appear synonymous and therefore for the purposes of this report are viewed as serving the same purpose. They are, however distinct from consultation. This distinction is explained by the HSE workforce involvement model.

This model shows, that at its most basic level, communication between management and workers is simply about keeping workers informed so they can do their jobs.

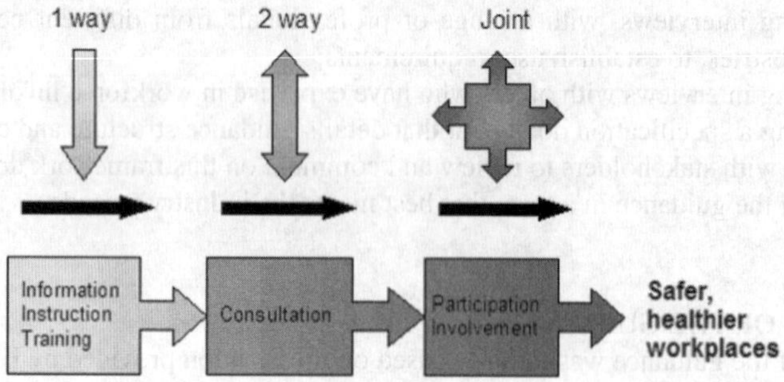

Figure 1 HSE workforce involvement model

The next more developed level, and what is required as a minimum in UK Health and Safety (H&S) legislation, is about consulting workers to find out their views and opinions. Decisions at this level are made by managers based on their understanding of an issue and their interpretation of feedback from a well designed consultation process.

The last level, shown in Figure 1, represents the active involvement of workers in the decision making process. This means managers and workers seek agreement together on health and safety and how, as partners, they will achieve commonly shared objectives. The final decision, and responsibility for health and safety, still ultimately resides with management.

This guidance focuses on improving this last level, providing help and advice on how to involve workers more effectively in health and safety.

To acknowledge the extensive role of contractors in the petroleum and allied industry WFI has been defined, in the guidance, by GSB as:

"The ways in which employees, including contractors, are encouraged to take part in making decisions about managing health and safety at work".

GUIDANCE STYLE

This guidance is designed for everyone who wants to find out more about Workforce Involvement (WFI) and how it can be improved and made more effective. It is aimed at employers, managers, safety representatives, trade union officials, contractors and all workers within the petroleum and allied industries. This is because WFI is viewed as collaboration and not managers imposing WFI initiatives on workers. Encouraging workers to read the guidance will also mean they are more likely to understand and buy-in to the benefits of WFI.

To make it more accessible, simple language has been used and the size of the guidance has been limited to thirty pages. Images, like the cartoons below, are also used to break up the text (Figure 2).

Figure 2. Cartoons to illustrate attitudes to WFI

The guidance also uses case studies to demonstrate the practical ways that different organisations have involved workers in different aspects of health and safety management. Case studies are also used to illustrate how to overcome the potential barriers to WFI. Genuine quotes, taken from industry, are also used to help convey key messages, for example:

"Once you have involved employees, you have to take on board what they are saying and use their ideas – you can not partially involve them. Employees have to see that their involvement has influenced its outcome. If an idea is wrong, fine, but go back and explain to that person why it is not a good idea"[8]

GUIDANCE STRUCTURE

This guidance provides a simple three step approach to ensure that efforts to improve WFI are more likely to succeed.

1. Assess current levels of WFI and identify the enablers and barriers to involvement;
2. Consider examples of activities to help improve WFI and then implement these in the workplace;
3. Ensure that workforce involvement is sustained and continues to lead to business improvements.

Using a methodical and thorough approach, that follows each of these key stages, is more likely to ensure WFI is effective. Being committed and taking the time and trouble of getting it right is more likely to lead to successful outcomes in the long run.

These 3 key stages and the elements that make up these stages are also presented in Figure 3 below and then discussed in more depth.

WHERE ARE WE NOW

Prior to trying to improve WFI it is important to determine the current level and quality of WFI. It is also important to identify what could be preventing WFI from being

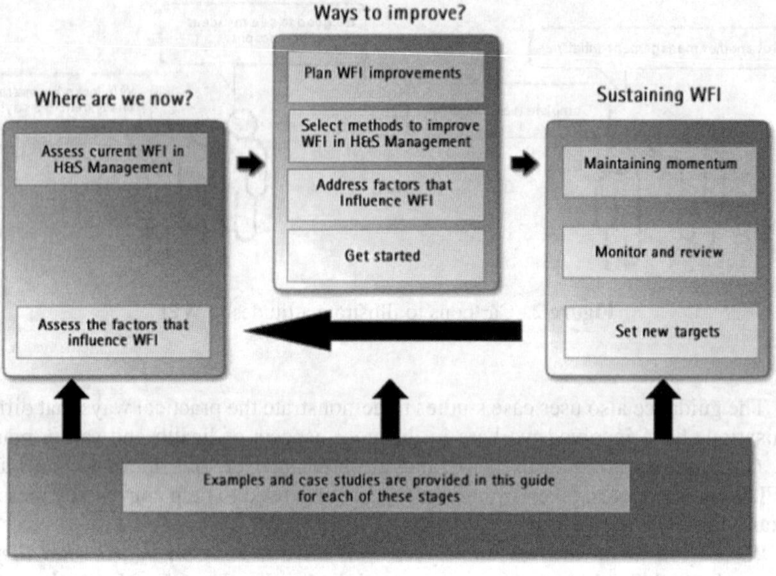

Figure 3. Three step approach for improving workforce involvement

effective. The guidance therefore provides advice and a series of questions to help the reader assess:

1. The current level of WFI in managing health and safety;
2. The factors that can influence the success and effectiveness of WFI.

The assessments are not aimed at providing a thorough, detailed analysis of current workforce involvement. They are designed to help generate discussion between workers and managers about where WFI might be improved. Information is provided in the guidance to help answer these questions. It also provides advice on how to complete, rate and review the findings from the assessment. Examples of the assessment questions are provided next:

WAYS TO IMPROVE

The purpose of this section is to help workers and managers plan and implement improvements to WFI. It does this by providing a series of case studies that illustrate how organisations, in the past, have involved workers in different aspects of health and safety management. For example how to involve workers in process safety:

Example 1: Process Hazard Analysis (PHA)
In order to enhance process safety within a facility an organisation decided to carry out Process Hazard Analysis (PHA) involving workers across the plant.

A team was selected comprising those with experience of operations and/or PHA. The manager of the area under review produced a charter that detailed objectives, scope, roles and responsibilities and formal communications for the PHA.

The PHA team (operators, engineer, trainer, technical safety specialist) identified potential hazards and issues. These were then discussed within the PHA meeting.

Recommendations were made to help combat the facility's significant areas of risk, for example recommending the sprinkler system for fire protection was put on a routine check.

When this check was conducted, two burst water pipes were found and fixed.

All recommendations were formally communicated through team briefings and fully supported by management.

This section also provides case studies detailing the different ways organisations have overcome the barriers to WFI, for example, poor levels of trust between managers and workers:

Example 2: Developing Trust
Specially arranged onshore meetings were used to develop trust between managers and workers based on an offshore platform.

The meetings, held at a hotel, helped to uncover some underlying problems because those involved reported they felt more freedom to express concerns.

There was also less distraction and less opposition from individuals who may not have taken WFI initiatives seriously in the past. The events also helped to demonstrate to workers that management were committed and took involvement seriously.

The guidance also suggests how to review the results from the assessment using a neutral forum like a workshop and provides advice on 'getting started', for example, how to set up an effective team to run a WFI initiative.

SUSTAINING WFI

Feedback from the interviews and HOFWG suggested that following initial improvements, impetus in increasing involvement can falter. This section of the guidance, therefore, provides advice on ways to help sustain workforce involvement:

- Maintaining momentum through the life-cycle of a programme;
- Reviewing and monitoring performance;
- Setting new targets once a programme is complete or coming to an end.

Practical advice is also provided on, for example, how feedback might be provided to workers:

- Staff meetings, committee meetings;
- Posters, newsletters, company magazines;
- Tool box talks and team briefing;
- Informal discussions;
- Forums and workshops.

CONCLUSION

The guidance, outlined in this paper, aims to provide simple WFI guidance, for the petroleum and allied industries, on how to achieve effective workforce involvement.

The need for such guidance is strongly made because:

- The guidance was commissioned by the Energy Institute – the leading professional body for the energy industries;
- The Health and Safety Executive has a made a declaration of support for WFI;

Table 1. Assessing the current level of WFI in managing health and safety

Aspect of H&S	How much are workers involved in the different aspects of H&S Management	Not sure	Seldom if ever	Some times	Yes, often
Equipment	Have workers been involved in reviewing existing safety **equipment** (PPE, lifting equipment etc)?				
	Have workers been involved in decisions to purchase new safety **equipment** (PPE, lifting equipment etc)?				
Workplace design	Have workers been involved in decisions to redesign the layout of the work environment when **facilities are being rebuilt or refurbished?**				
Task & procedure design	Have workers been involved in reviewing and/or developing safe working **procedures?**				
	Have workers ever been involved in improving the safe **design of tasks** or ways of working?				
Process safety	Have workers been involved in analysing or reviewing **process safety arrangements?**				

- Research has recommended the need for ***"a guide to worker engagement"*** [4];
- Reports investigating several major accidents in the high hazard industries have made recommendations to improve WFI.

The guidance states the necessity of understanding issues concerning WFI, carefully planning new initiatives and considering how they will be sustained into the future. The guidance helps to ensure that effective WFI is more likely to happen, by providing the appropriate rationale, strategy and tools.

Table 2. Assessing the factors that can influence the success and effectiveness of WFI

The questions are focused on health and safety but th ey can also be used to assess WFI in other areas of work.

Attitude

1. How do you rate the level of trust between the management and workers?					
Very poor		**Adequate**		**Very good**	**Do not know**
1	2	3	4	5	0
Can you give a reason for your answer?					
2. Do you think workers are suspicious about the motivations for workforce involvement?					
Suspicious		**Not sure**		**Not suspicious**	**Do not know**
1	2	3	4	5	0
Can you give a reason for your answer?					
3. Do you think workers want to be more involved in health and safety?					
No		**Not sure**		**Yes**	**Do not know**
1	2	3	4	5	0
Can you give a reason for your answer?					
4. Do you think things will change as a result of more workforce involvement?					
No		**Not sure**		**Yes**	**Do not know**
1	2	3	4	5	0
Can you give a reason for your answer?					

The importance of getting it right is that the failure of one WFI initiative can make it much more difficult to get employees involved again in the future. The guidance highlights the danger of imposing WFI initiatives on workers. It also explains how they can become cynical, about WFI schemes, if nothing every changes in their workplace.

Investigations of past incidents in the petroleum and allied industries, would seem to indicate, a focus on occupational safety. This may have been at the expense of not involving workers in other critical aspects of safety management, such as process safety.

The benefit of this guidance is that it aims to be different by encouraging a more holistic approach to tackling WFI, recognising that workers can be involved in many aspects of safety management. This may help organisations utilise workers more effectively, tapping into a very significant resource.

REFERENCES

[1] Health and Safety Executive, 2007, Topic pack – worker consultation and involvement, version 2 *HSE Books*.

[2] Entec UK Ltd, 2000, examples of effective workforce involvement in health and safety in the chemical industry, CRR291, *HSE Books ISBN 0 7176 1847 1.*

[3] University of Aberdeen – Industrial Psychology Research Centre, 2005, review of research findings on workforce involvement *RR IPRC 2005/04.*

[4] Health and Safety Executive, 2006, an investigation of approaches to worker engagement RR516 *HSE Books.*

[5] Baker, J et al, Independent Safety Review Panel, 2007, the report of the BP U.S. refineries.

[6] Health and Safety Executive, 2003, major investigation report BP Grangemouth Scotland, a public report prepared by HSE on behalf of the Competent Authority *HSE Books.*

[7] ECOTEC Ltd, 2005, Obstacles preventing worker involvement in health and safety RR 296, *HSE Books.*

[8] Bell et al, 2003, Employee involvement in health and safety: some examples of good practice, *Health and Safety Laboratory*.

Please note: Guidance on Effective Workforce Involvement in Health and Safety can be obtained from the Energy Institute: 61 New Cavendish Street London W1G 7AR (Telephone: 0207 467 7100) Website: http://www.energyinst.org.uk/.

DEVELOPMENT OF AN EFFICIENT SAFETY AND LEARNING CULTURE IN ROMANIAN SMALL AND MEDIUM ENTERPRISES (SME'S) THROUGH VIRTUAL REALITY SAFETY TOOLS

Dr. Stefan Kovacs
INCDPM "Alexandru Darabont" Romania

Small and Medium Enterprises (SME's) are the most dynamic in the Romanian economy. Adapted to the rapid market changes, Romanian SME's are a booster for the whole industry. Unfortunately, SME's are not so lucky regarding incidents and occupational accidents. More than 75% of the accidents recorded in 2006 were located here. Taking into account these aspects, our research was oriented constantly towards the development of new and efficient safety tools for SME's. After various studies we focused on two specific problems: the design of an efficient safety and learning culture developer together with the development of a multi-role assessment system. The two instruments are built around virtual environments which could realistically model almost every work situation- allowing the SME work teams to develop and exercise specific safety skills and also to define reference models for assessments. These two instruments were included in the Integrated Safety Management Unit – a complex safety management structure for SME's. This paper presents the most important aspects regarding these efficient solutions for improving safety in SME's.

THE INTEGRATED SAFETY MANAGEMENT UNIT (ISMU)

ISMU is the management centre of the safety solutions for a SME[1]. In this respect it stores and manages the developed knowledge and optimises the informational flows assuring an efficient usage of all these tools in order to improve safety into SME's[2]. ISMU schema is presented below:

THE VIRTUAL ENVIRONMENT BASED SAFETY AND LEARNING CULTURE DEVELOPER

Safety Culture (SC) could be defined as "A general term for the degree to which the culture of an organisation promotes and cooperates with safe and healthy work practices[3]".

[1]Zhang H., Wiegmann D.A., von Thaden T., 2002, SAFETY CULTURE: A CONCEPT IN CHAOS?, Proceedings of the 46th Annual Meeting of Human Factors and Ergonomics Society, Santa Monica Human Factors and Ergonomics Society 2002

[2]Stefan Kovacs 2007, SOME RESEARCHES REGARDING CBT USAGE IN THE TREATMENT OF WORK DISORDERS, in Abstracts of the WCBT (World Congress on Behavioural and Cognitive Therapies), Barcelona 2007

[3]www.edp-uk.com/glossaries/terms.htm

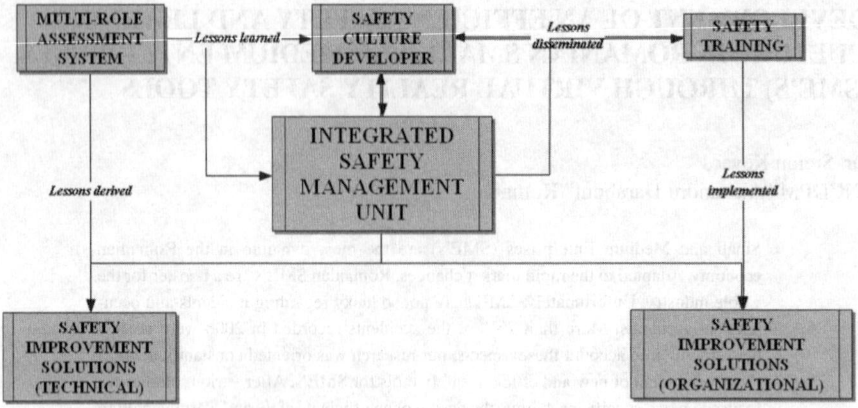

Figure 1. Integrated safety management unit

A difference must be made between Safety Culture – which is developed in time – and Safety Climate – which captures the immediate safety state[4]. The main idea of the Virtual Environment Based Safety and Learning Culture Developer (VE) is to develop and improve safety skills and attitudes in a virtual, controlled environment. After these skills are imprinted correctly the worker or the work team could use them into the workplace. The worker and the work team have at their disposal the virtual environment in order to **learn, try, test and eventually implement the germs of an effective safety culture**.

MECHANISM OF THE VIRTUAL ENVIRONMENT (VE)

The mechanism of the VE is shown in the next image.

This mechanism is very straightforward. For example, Safety Culture Framework (SCF) 1 manages safety culture aspects regarding dangerous behaviours. The framework is based on role playing, the person with the dangerous behaviour seeing directly the most negative consequences of his/hers behaviour; the same person could also play the role of the team leader which must take immediate measures to prevent an accident and so, understood the responsibilities of his/hers supervisors.

DEVELOPMENT AND USAGE OF THE SAFETY CULTURE VIRTUAL ENVIRONMENT

The safety culture development instrument is seeking continuously in the workspace for safety problems. Once identified, these problems are transposed in the virtual environment

[4]Cooper M.D. 1997, Evidence from Safety Culture that Risk Perception is Culturally Determined, The International Journal of Project & Business Risk Management, Vol 1.(2), pg. 185–202

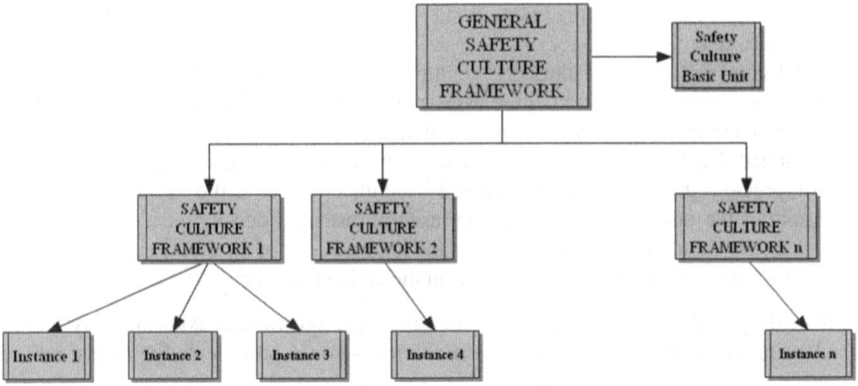

Figure 2. Mechanism of the virtual environment

in order to use them as interactive lessons for the development of a better safety culture. The development of the virtual environment includes the following main steps:

- analyse – identify problems, needs arising from these problems, facilitators – people who could make the difference between safety and non-safety;
- develop Safety Culture Frameworks(SCF);
- build instances – instances are built using the SCF on specific unexpected events that occur in the workplace and could degenerate into incidents or accidents- instances are saved as interactive use-cases which workers could exercise with;
- use – use the virtual environment already defined by the frameworks and instances in order to improve safety culture through the cycle:
  - learn about:
    - effective and efficient safety rules and procedures;
    - own commitment to safety;
    - team commitment to safety;
    - good organisational skills;
    - management involvement;
  - interact and try the learned aspects on the use – cases existing in the virtual environment;
  - develop the necessary safety culture skills and attitudes;
  - became confident in using these skills;
- discard or save- discard use cases when they are no more relevant (for example at the radical change of a technology) or save them in a case-base

VE is based on a "frame of reference"[5]

[5]Marek J., Tangenes B 1985, Experience of risk and safety, Work Environment: Stratfjord Field, Universtetforlaget, Oslo

The following picture shows the interaction between the real world inside the SME and the virtual environment.

All starts with SME existing safety culture. At the beginning, in order to establish how valuable it is, a complete safety culture assessment[6] is performed at the global level. In this process, based on the identification of the problems, assessment of needs, etc. the main SCF's are developed. Next time, the starting point could be a post-mortem assessment of actions in the past (previous day, month, year). The SCF's are then inserted in the VE and together with the results of the post-mortem analysis are used at instances development – as interactive use – cases that would be effectively used in the improvement of the safety culture.

As seen in Figure 3 the VE is based on the following pillars:

- **identify probls** – problems could be identified through various assessment systems[7];
- **identify the needs** arising from the problems – could be knowledge needs, attitude needs, organisational needs, etc.;
- **identify the facilitators** – facilitator could be anyone at the workplace- but generally they are people that through their own qualities could act as developers of the safety culture inside SME's. An ideal goal of the system is to transform every work team member into a facilitator.
- **identify the needed actions** – actions that would be performed in order to solve the problems; these actions could be like:
  - train;
  - re-organise;
  - allocate more resources to the facilitators;
  - interact more frequently the work teams with line management or upper management;

Now is the time to see if the existing safety culture could be improved or must be radically changed Radical changes are using the Safety Culture Basic Unit which could even give a optimal safety culture embryo if there is no viable safety culture in place.

THE VIRTUAL ENVIRONMENT AND THE INDIVIDUAL USER

At the individual level the improvements/changes regarding SC are based on three main subjects of SC development:

a) risk understanding: if causes, actions and effects of risks are not well understood by the worker he will be not able to identify, prevent/mitigate them;
b) training – SC training in the VE is focused on acquiring risk related skills, mainly through interactive simulation using best – worst case scenarios. Navigating through these scenarios will motivate the worker to be risk efficient in order to protect him and the others;

[6]Reason J. 1997, Managing the risks of organisational accidents, Alsweshot, Ashgate

[7]Transports Canada 2007, TP 13844 - Score Your Safety Culture, http://www.tc.gc.ca/CivilAviation/systemSafety/Brochures/Tp13844/menu.htm

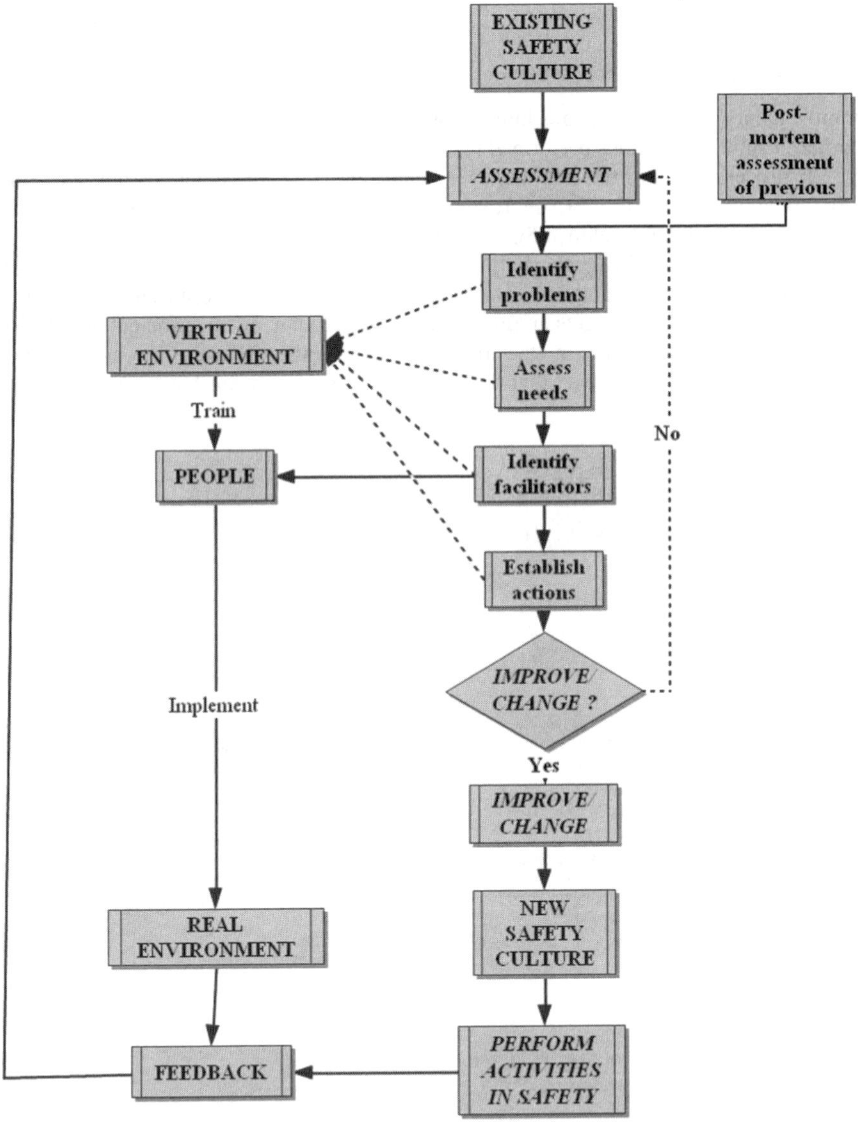

Figure 3. The interaction between the real world and the virtual environment

c) improvement of team work; if there is no cooperation inside the team or if the cooperation is not efficiently enough then VE will focus on role playing in order to optimize this cooperation.

Improvements or radical changes into the SC must reflect into the improved safety of the worker in simulated actions performed inside VE in interactive case studies. The leader of the VE process (usually the line manager) shall observe if these actions are performed correctly and in the spirit of SC improvement; if so, the worker has the OK to proceed to the real world and the developed SCF's and instances are stored into the knowledge base for reuse. If not, the feedback shall provide corrective changes in order to improve SCF's and instances; for example, a workplace is not well enough described by the instance – because this, the worker forgets an important safety step having no sufficient details. The corrective change consists in the development of the instance by adding these details. Our experience showed that these are the most significant failures of SCF's and

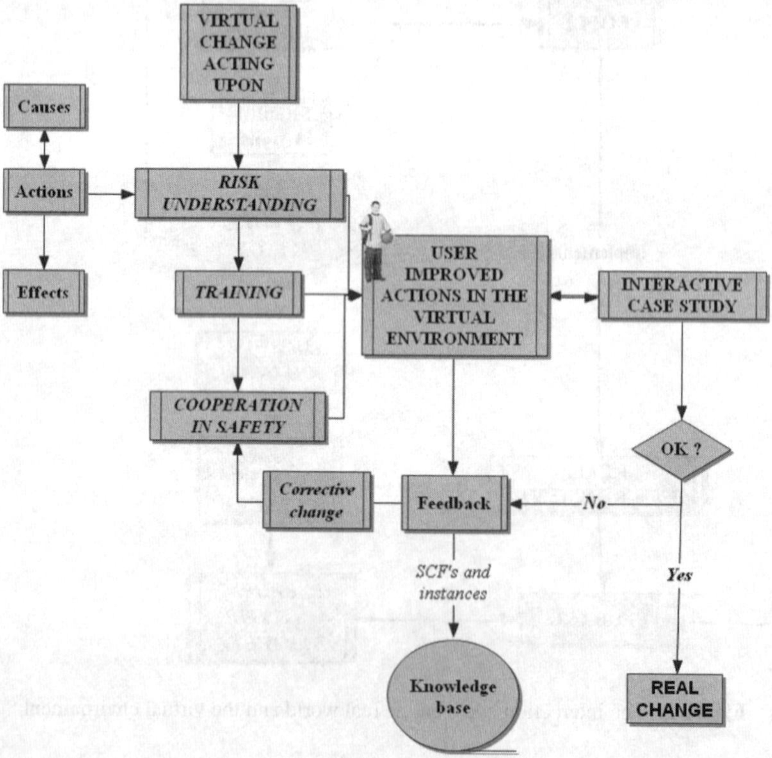

Figure 4. The individual user and VE

instances- the absence of the necessary details or quite contrary, the presence of too many details that could confuse the user. After the training in VE the user (worker) should be able to exercise his newly acquired skills and perform similarly in the real life. Figure 4 shows the interaction between the individual user and VE.

Inside the VE there could be also planned SC improvements/changes on a large scale – in the event of radical changes at the workplace.

THE MULTI-ROLE ASSESSMENT SYSTEM (MRAS)

MRAS was developed so that SME's which usually are not well resourced in order to sustain an extensive safety assessment – could benefit as much as possible from this operation. So, MRAS performs quality, safety and environment assessments.

MRAS is starting with an ideal image given by the VE for the main activities performed inside SME – using as a reference base the main safety, quality and environment documents that the SME must comply with.

MRAS is conceived as a three levelled structure-like in the following image.

Table 1. Summary example of a MRAS assessment

	Action	Attributes					
MRAS	Waste tanker load	Worker or work team	Task	Machine	Environment	Interaction	Score
Strategic points	Pre-action (action preparing)-fixing the hose coupling to the exhaust pipe	4	5	5	5	3 (Chief of the work team failed to explain the correct pipe manoeuvring)	4.4
	Action-emptying the exhaust pipe into the waste tanker	5	5	5	4	4	4.6
	Post action-de-coupling the hose from the pipe	3 (worker was careless in the de-coupling of the pipe)	4	4	4	5	4

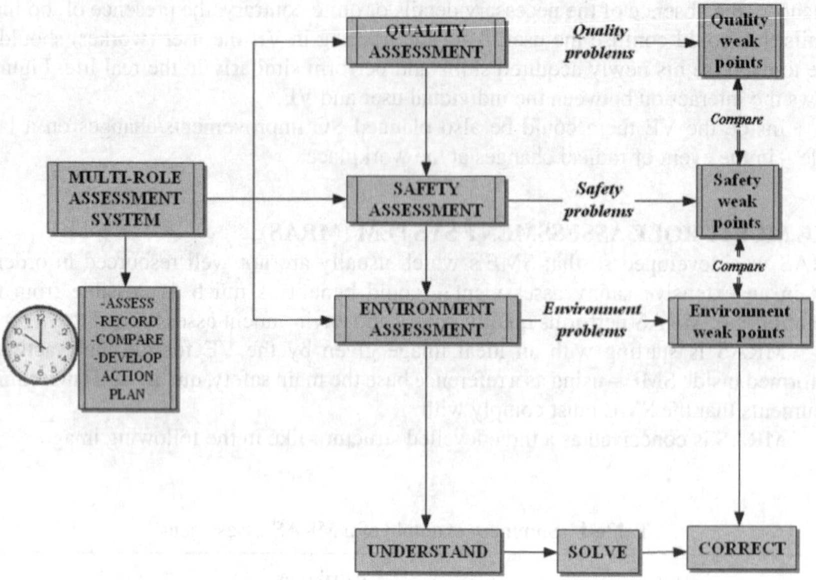

Figure 5. MRAS structure

MRAS is assessing the main activities performed by the SME- so this is a dynamic assessment. MRAS is performing its assessment taking into account attributes and strategic points which describe, together, work situations. A number of checklist items are defined for each situation. Every item could be assessed on a 0 (most unfavourable) ... 5(most favourable) scale. A summary example of such an assessment is presented below.

All the items equal or below 3 are considered as weak points and must be corrected. If the item is assessed equal or below 2 points an immediate correction must be made.

Table 2. MRAS benchmarking

	Quality			Safety			Environment		
MRAS BENCHMARKING	L1	L2	L3	L1	L2	L3	L1	L2	L3
Preparation of ingredients for the chemical process	5	5	4	4	4	4	4	4	3
Performance of the process	5	4	5	5	5	4	4	4	3
Transport of the main resulting products	4	4	4	4	4	4	4	5	5
Waste removal	4	4	5	4	4	4	3	5	5

The score is computed as an arithmetic mean of the attributes. All weak safety points must be documented. A cumulative example for quality, safety and environment benchmark is given in Table 2.

L1..L3 were three similar locations of petrochemical units (Bucharest, Piteşti and Craiova). It is very easy to see that L3 is a little behind L1 and L2 and must be helped in order to overcome its deficiencies.

MRAS MODULARITY

MRAS could be adapted accordingly with the size, activity and objectives followed by the SME in the assessment. A general image of this aspect is presented below.

A template of a MRAS report is presented in Table 3.

Table 3. MRAS report template

	QUALITY, SAFETY AND ENVIRONMENT AUDIT REPORT	
Auditor:	Date:<DD/MM/YYYY>	Place of audit:
MRAS level: B M E	I agree with the results <Manager>	I disagree with the results <Manager>
A.VULNERABILITY ANALYSIS		
B.QUALITY ASSESSMENT B.1.Quality weak points; B.2. Operational plan to optimise quality B.3. Connections between quality and safety *(for example Internal Document 23A is common)* B.4.Compliance with ISO 9001; B.5.Compliance with other quality documents (please name the documents);		
C.SAFETY ASSESSMENT C.1.Safety weak points C.2.Immediate operational safety plan *(for weak points below 2)* C.3.Operational safety plan *(for weak points equal 3)* C.4. Common quality and safety weak points C.5.Conexions between safety and environment *(example: Operation X is dangerous for safety and environment)*		
D.ENVIRONMENT ASSESSMENT		
E.COMPLIANCE ASSESSMENT E.1.Compliance with quality documents E.2.Compliance with safety documents E.3.Compliance with environmental documents E.4.Operational plan to solve compliance problems		
F.QRA *(if needed)*		
G.GENERAL BENCHMARKS		
H.CONCLUSIONS		

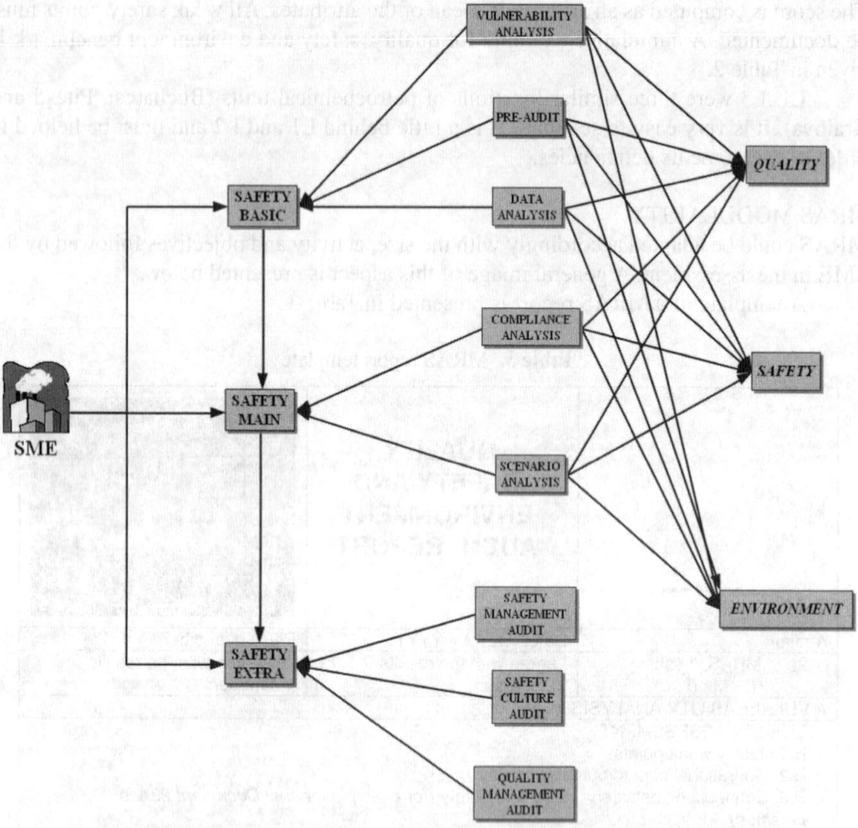

Figure 6. MRAS modules

IMPLEMENTATION ASPECTS

Implementation was performed till 2000 through two channels:

– The Romanian Centre for SME's; it implemented the system in more than 1000 Romanian SME's from various industrial activities;
– The Romanian Institute for SME's implemented the system in 200 SME's from the research and learning domain. The system was implemented gradually, starting with the safety culture developer. Benchmark criteria were developed in order to assess the efficiency of the system. The main benchmarking criteria are given in Table 4.

Table 4. Main benchmarking criteria

Safety culture development inside SME
Previous (5 year media if possible) registered work related complains
Test period work related complains
Previous (5 year media if possible) registered safety related sanctions
Test period safety related sanctions
Test period actions regarding safety
Previous safety culture global index
Test period safety culture global index
Multi-role assessments inside SME
Previous (5 year media if possible) registered quality problems
Test period identified quality problems
Previous (5 year media if possible) registered safety incidents
Previous (5 year media if possible) registered safety accidents
Previous (5 year media if possible) registered safety problems
Test period identified safety problems
Previous (5 year media if possible) registered environmental problems
Test period identified environmental problems

The set of established criteria[8] gives the possibility to have an objective feedback regarding the efficiency of the developed system not just globally but also for a specific SME or group of related SME's. The benchmarking set includes all the significant data that could be used in order to assess the interesting aspects.

RESULTS

The process industry SME's that were included in our test lot had given some interesting results. These results were computed for a reference process industry SME and are presented in Table 5 as median values

It is possible to see an evident improvement of the criteria that are defining safety culture through the implementation of the system.

CONCLUSIONS

The developed solutions had, as the main advantage their operational efficiency, together with a special adaptability regardless the specific type of economic activity. They are affordable even for Romanian SME's and could facilitate-through VE – the modelling of

[8]Stefan Kovacs, Apostol George 2006, Meme Based Cognitive Models Regarding Risk And Loss In Small And Medium Enterprises, in Proceedings of the Seventh International Conference on Cognitive Modelling, Trieste 2006, iccm 06 pg. 377

Table 5. Obtained results

Criteria	Value
Safety culture development inside SME	
Previous (5 year media if possible) registered work related complains	30
Test period work related complains	11
Previous (5 year media if possible) registered safety related sanctions	22
Test period safety related sanctions	6
Previous (5 year media if possible) number of persons trained in safety matters	10
Test period number of persons trained in safety matters	30
Previous (5 year media if possible) registered actions regarding safety	5

every major SME activity. The multi-role assessment system contributes not just at the safety improvement in the SME but allows also a general economic improvement through quality and environment. The Integrated Safety Management Unit (ISMU) is conceived in order to integrate not just these tools but every new computer-based tool developed for SME safety. As the solutions were implemented gradually into the SME safety structures from 2000, their performance shows us that they are viable. This integration could be performed easily, with the final objective the preservation and development of safety knowledge inside SME. As SME are the most risk prone enterprises in Romania their safety protection is of vital concern for our safety research.

We could estimate that our research was succesfull, leading to a decrease of incidents and accidents produced in the Romanian SME's by 20% in 2 years of reference (2004–2006).

KNOWLEDGE TRANSFER – CRITICAL COMPONENTS IN OCCUPATIONAL HEALTH AND SAFETY – AN ESTONIAN APPROACH

Marina Järvis and Piia Tint
Tallinn Technical University, Kopli 101, 11712 Tallinn, Estonia
E-mail: marina@staff.ttu.ee, tint@staff.ttu.ee

Defining and understanding the knowledge-sharing process facilitates the application of knowledge management to problems, systems, and situations in individual organizations and in the field of occupational health and safety in general. In this paper the process of knowledge transfer in occupational health and safety at the state level and the main barriers are described. The authors offer a possible tool for knowledge management – a Sectoral Profile on Occupational Health and Safety for knowledge creation and transfer and underscore the need to focus on the extent to which decision-makers and others receive and use such information and knowledge.

KEYWORDS: knowledge management, knowledge transfer, networking, training, Sectoral Profile, occupational health and safety

INTRODUCTION

Occupational health and safety (OH&S) is one of the main concerns of today's business. Due to complexities of the products, process or equipments used to create the products or services, sudden accidents or accidental events could happen at any time. The impact of these disasters may be too much costly for the enterprises [1]. The requirements of effective management of OH&S often constitute a big challenge for many contemporary enterprises, which operate under conditions of increasing competitions of the global market and continuously rising requirements for products and services [2]. Information and knowledge are the central resources in the achievement of the goals of OH&S management. Active interest in OH&S requires that the workers and employers have the right information at the right time to make decision affecting health and safety. Knowledge and information is a precondition for action [3] and providing useful information to decision-makers (including employers, government officials, practitioners, unions, and workers) is essential in addressing OH&S issues. Well-informed decisions are needed at the political and administrative levels, as well as at the organizational level and in practical actions. The challenge is to provide OH&S information in such a form that each workplace can utilize it for its own purposes in a cost-effective manner [4]. Businesses of all sizes invest time and resources dealing with OH&S issues. Over time, corporations gain a significant amount of knowledge. Knowledge management (KM) has become an important process in knowledge intensive companies over the past few years [5]. It has been widely recognized that knowledge sharing is an effective approach to maintaining organizations' sustainable competitive advantages [6]. The knowledge of individuals, through the process of knowledge sharing, could gradually accumulate and convert to the overall knowledge for an organization. The KM literature yields several articles that describe knowledge sharing as it occurs in sample organizations [7].

Much material, information and intellectual capacities on OH&S are dispersed among different ministries and government agencies, employers' and workers' organizations, universities, and other institutions. This information from OH&S authorities and national institutions is a valuable asset, and like any asset, it works best if it is well managed in order to develop of working conditions and promotion of workers' health accordingly.

Generally, KM research has focused on identifying, storing, and sharing the transaction-related knowledge and has described efforts within and between companies to consider knowledge as a manageable asset [2, 7–9]. Few systematic attempts have been made broadly outline the requirements and flow of information [9], but an analysis of the OH&S knowledge creation, transfer and utilization at the state level has not been prioritized. Defining and understanding the roles of knowledge cycle elements facilitate the application of KM to problems, systems, and situations in individual companies and in the field of OH&S in general. This paper is a first step in the process of applying KM principles to the field of OH&S in Estonia and it has four objectives:

- to discuss the importance of KM as an effective business practice and to assess how this practice is performed at the state level in the Estonian system of OH&S;
- to analyse the process of knowledge transfer at the state level, that need to be captured, refined and aggregated, and brought to the right places at the right time;
- to assess the possible knowledge transfer barriers;
- to discuss a possible tools of KM like network and a Sectoral Profile on Occupational Health and Safety for the evaluation of knowledge creation and transfer and underscore the need to focus on the extent to which decision-makers and others receive and use such knowledge. The principle of the Sectoral profile on OH&S in Estonian agriculture was used as an example in order to describe the knowledge transfer in the field of OH&S.

KNOWLEDGE MANAGEMENT IN OCCUPATIONAL HEALTH AND SAFETY

Knowledge has been recognized as a new resource in gaining organizational competitiveness. A variety of definitions of KM exists. For Lomax [5] KM is "The process of capture, refinement, aggregation and sharing of data and information between employees, departments, subsidiaries and partner organizations to achieve a position of knowledge-base competitive advantage". Sherehiy and Karwowski [2] also suggested, that the principles and tools of KM should be used to facilitate the management of the existing individual (personal) knowledge, structural knowledge (i.e. knowledge codified into manuals, reports, databases, and data warehouses), and organizational knowledge (activity of learning within the organization) in the fast domain of practical application [2, 10]. Knowledge management (KM) and knowledge management systems (KMS) have been positioned as strategies and tools that enable organizations to create and transfer knowledge in order to sustain competitive advantage. While KM as a strategy gained legitimacy, KMS have struggled to show a causal relationship to knowledge creation and knowledge transfer. The Knowledge Management Systems process in the field of OH&S is shown in Figure 1. [11]

One possible definition of KM was proposed by Davenport and Prusak in 1998 [12]: "Knowledge management draws from existing resources that your organization may already have in place – good information systems management, organizational change management, and human resources management practices". From the definition of KM it is clear that any advancement in this field need to adopt an integrated [12], interdisciplinary and strategic perspective. KMS are able to accumulate social capital and showing its effect on the creation and transfer of knowledge. A number of studies discuss the impact of so-called social capital on productivity, innovation and sustainability [13]. Social capital describes the ability of the organization, team, group, community or nation to work together. There are many characterizations of social capital. According to Robert Putnam [14] social capital has the following features: active participation in social networks, reciprocity, trust, and respect for social norms, communality, and initiative. The characterization of social capital, when applied to working life describes the situation that in Europe used to describe with the words: good job–good workplace.

KM in OH&S within the enterprises as well as at the state or regional level is needed so that: i) society's scarce resources are not wasted by duplicating work; ii) OH&S information is easily available and accessible to all users of that information; and so that the information is kept as a structured entity instead of as fragmentary, unorganized bits of information [4]. In practice, KM combines various concepts from different disciplines,

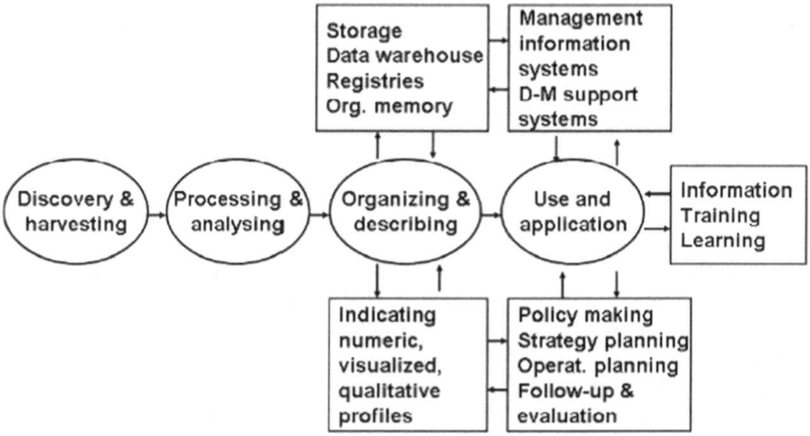

Figure 1. Knowledge management systems [11]

such as organizational theories, human resource management, artificial intelligence, ergonomics, and informational technologies [15]. The concept of KM can also be used to describe the collection of technique, methods, process, structure, and cultures developed to improve the creation, sharing and utilization of knowledge [2].

EXPLICIT AND TACIT KNOWLEDGE

Organizational KM in OH&S treats mixture of two kinds of knowledge: tacit knowledge and explicit knowledge. *Explicit knowledge*, sometimes referred to a codified knowledge, is objective knowledge that can be transmitted in formal, systematic language [4]. An example of explicit knowledge on OH&S consists of governmental and local regulations, standards, norms, and safety requirements, which are stored as written documents or procedures. According to Sherehiy and Karwowski [2], explicit knowledge in the area of OH&S and ergonomics are accident records, safety regulations, safety guidelines, theories and axioms, company records [2]. In the context of the management of OH&S, special attention should be given to tacit knowledge, because the research topics are often identified through direct human experience in the workplace, and the results of the research are often immediately applicable to the solution of a problem. When people solve complex problems in the field of OH&S, they bring knowledge and experience to the situation, and as they engage in problem solving they create, use, and share tacit knowledge. Zeleznikow [16] stated that *tacit knowledge* is highly personal, context specific as well as deeply rooted in an individual's actions and experience, which could be technical (i.e. know-how of an expert) or cognitive (i.e. based on values, beliefs and perceptions), and it hard to formalize, making it difficult to communicate or share with others. Examples of tacit knowledge are: safety engineer's experience, safety hazard recognition, perceptual and cognitive skills, physical experiences, rules of thumb and synthesis of facts [2]. In addition, in the context of OH&S management system, examples of tacit knowledge include individual knowledge of experienced worker and specialist as well as estimating and tendering skills acquired over time through hands-on experience, understanding the technological process, interaction with clients/ customers, awareness of occupational hazards and possible health effects, prevention measures, their responsibilities and right.

BACKGROUND AND PREVIOUS WORK

Despite the growing interest in KM studies, little research was carried out in the field of OH&S. Sherehiy and Karwowski [2] proposed model of KM for occupational safety, health, and ergonomics (OSHE). Schulte et al. [10] described the examples of current and effective KM practices within occupational hygiene in the USA. In 2006, Butler & Murphy [17] tried to exam the relationship between knowledge and work going forward. Schutle et al. [9] identified the special areas of dissemination of occupational and environmental safety and health information: the information needs of the changing workforce, new and young workers; small business [9]. Many researchers have described the process through which knowledge is created, developed, retained, and transferred in firms [18, 19], and the role played by leadership [20–22] and decision-making styles in influencing these processes [23]. In most discussions of KM, the focus has been the organization as whole (e.g. Kotter & Heskett, 1992 [24]), organizational knowledge management [8] and on the right technology, proper organizational culture. There have been several systematic research studies in knowledge sharing within organizations and the 124 barriers from the KM literature were identified [7–8]. Such barriers concerned with source reliability, motivation to share, ability to learn and apply new knowledge and so on [6–8]. OH&S knowledge sharing is a field that has not received a great deal of researcher attention. In addition, the focus has not been on knowledge transfer at the state level and on knowledge transfer barriers. No research or other systematic studies have addressed knowledge transfer and possible barriers to knowledge sharing in the field of OH&S in Estonia. In order to understanding the success and failure of·KM efforts in the field of OH&S, there is need to understand the process of knowledge creation, sharing and utilization at the state level. A profile on OH&S in Estonian agriculture was compiled with a rapid assessment approach and was used as an example of possible KM tool that helped to describe and explain the current situation in the field of OH&S, the process of knowledge cycle at the state level and to assess the possible barriers to knowledge transfer in Estonia.

SECTORAL PROFILE ON OCCUPATIONAL HEALTH AND SAFETY

A Sectoral profile is valuable in its own right, being a contextual summary of issues of importance with specific focus. Profile (a situation summary) is a tool that is used for policy formulation and monitoring purposes, and for informing stakeholders about the state of affairs such as OH&S in Estonia. It is a document that also includes statistical indicators, which are interpreted and qualified in a profile, because it is more flexible and more informative than a collection of indicators. A profile is more than a set of indicators because it provides an understanding and context that cannot be communicated by numbers only. Profile and indicators of OH&S are used to describe state of affairs, provide early signals for problems in the work life, prioritizing activities, monitor trends, assess the effectiveness of programs, identify the information sources, as well as present a baseline against which progress is measured [25–28]. A profile aims at being understood also by decision-makers who deal with aspects of social dimensions other than OH&S, and who might see it useful to link elements of OH&S into their field of responsibility. The target groups of the profile are administrators, decision makers, politicians, labour inspectors, Trade Unions, employers' organizations, academic institutions, planners and managers, company management, OH&S specialists and OH&S expert institutions, local authorities and stakeholders [26, 29]. A Sectoral profile on OH&S in Estonian Agriculture was compiled under the umbrella of the Sub-network on OH&S in Agriculture as a part of the Estonian-Finnish Twinning Project on Occupational Health Services (2003–2004). The purpose of the Sectoral profile in OH&S in agriculture was to understand OH&S system in the local context and from the perspective of stakeholders, to increase the awareness about OH&S situation of national and local decision makers, companies, OH&S specialists, labour inspectors as well as stakeholders by promoting the compilation of profiles at national and sub-national level as well as to provide a written summary that documents the state of affairs [26, 29]. In order to identify the information and knowledge sources, facilitate information sharing and dissemination, and raise awareness about OH&S in general, the Sectoral profile included headings as Overview of production; Labour Force demography; OHS legislation; OHS infrastructure and system; Information Strategy; Occupational hazards and risks; Occupational Health Services; Occupational and work-related diseases, work injuries; Sectoral OH&S Network in Agriculture; Inventory of educational and training materials (OH&S); Results from SWOT analysis by stakeholders; Main organizations involved in agriculture etc. Sectoral profile could be compiled by using several qualitative techniques such as desk-reviews of documents, conversations, group discussions, observations, walk-trough assessments. On addition, the Sectoral profile can also be used, within reason, for making comparisons with other economic sectors within a country and between similar sectors in other countries [26, 28–29].

KNOWLEDGE TRANSFER IN OCCUPATIONAL HEALTH AND SAFETY

Information and knowledge are critical components of OH&S decision-making, policy development, regulation, compliance, training, education, enforcement and risk management in general. In order for that knowledge to create value, it must be shared. The state may play an important role in process of knowledge generation and transfer, dissemination by establishing necessary legal infrastructure to support research development, network, and collaboration between authorities and enterprises [29]. Although knowledge sharing and knowledge transfer are often used interchangeably. *Knowledge sharing* refers to an exchange of knowledge between two individuals: one who communicates knowledge and one who assimilates it. Knowledge sharing focuses on human capital and the interaction of individuals. *Knowledge transfer* focuses on structural capital and the transformation of individual knowledge to group or organizational knowledge, which becomes built into processes, products, and services [8]. Knowledge transfer may occur between and among individuals, within and among teams, among organizational units, and among organizations. A major focus of knowledge transfer is on the individual who can explicate, encode, and communicate knowledge to other individuals, groups, and organizations. It is essential that enough information material on various risk and hazards of the work environment are available for the workplaces and workers. The dissemination of the OH&S information and transforming information into knowledge, i.e. into human knowledge capital, which can be used in many different ways to solve problems, to learn more in the field of OH&S, etc.

Many Estonian legislations (like Estonian Act on Occupational Health and Safety) and regulations contain stipulations about disseminating and applying information concerning OH&S. From a legislative perspective, improved dissemination of information and knowledge should encourage awareness, urge precaution, and lead to a reduction in occupational morbidity and mortality. In addition, to laws and regulations, voluntary consensus standards (e.g. OHSAS 18001) and corporate policies stipulates large role for information dissemination and knowledge transfer. However, there is little known how employers, workers and OHS specialists receive, analyze and use this information. The analysis of the information needs should be based on a situation analysis and assessment of the status of the work life, occurrence and trends of OH&S problems, ongoing activities and operations, resource available, scientific and professional information available for experts, academia, social partners and enterprises [11]. According to Aaltonen [30], especially supportive safety information is needed in small companies because of the lack of safety knowledge. There is also need to define the target groups for OH&S information and knowledge because the contents of the information and knowledge need to be modified according to the needs of the information receivers. This requires an analysis on which are the groups of persons in need of information about OH&S, good solutions and practices at the workplace level. Knowledge, information and guidance materials are needed at every level of organization, but the type and content may be different. Small companies have different safety information and knowledge needs than large one.

There are some good and continuous channel for OH&S information and knowledge dissemination in Estonia: publication in journals, books, magazines, documents, brochures, CDs, lectures, posting on the web. According to Lagerlöf (2000) [31] research transfer is the process by which relevant research information is made available in a strategic manner for practice, planning and policy making. Technology transfer in OH&S is the application of new technologies or ideas to address OH&S problems [9]. Media is a very important factor when planning the information dissemination for the general public and for raising general awareness. One of the most effective factors is a regularly published newsletters and journals where OH&S issues are widely and extensively dealt with. The Estonian Newsletter on Occupational Health and Safety is the main channel for regular dissemination of information and knowledge Estonia [Figure 2]. In addition, the labour inspectors can disseminate regular information to the workers and employers during their visits to the workplaces.

The main ways of knowledge transfer in the field of OH&S are via communities of practice, the internet and training [10].

Communities of practice/Networks: People also share knowledge in different network: the organisation network, they belong to a team, a project, but they also form communities of practice (CoP) with people in other parts of the company or with experts outside the company. Knowledge sharing requires networks of specialists, but these are difficult to organise because of the increasing mobility and turnover the personnel and lack of time to share knowledge [5]. The field of OH&S could be considered as a CoP and it is important to know how knowledge, information, practice and values are shared, conserved, transformed within community.

The OH&S community of practice in Estonia is linked by Estonian Occupational Health Physician Association, non-governmental organizations, Occupational Health Services (OHS), consultants and researchers. In Estonia, the services provided by an OH physician, an OH nurse, a hygienist, a psychologist or a specialist in ergonomics are considered to be OHS. These service providers are all called 'occupational health specialists'. According to the Estonian legislation, only entrepreneurs or private medical companies

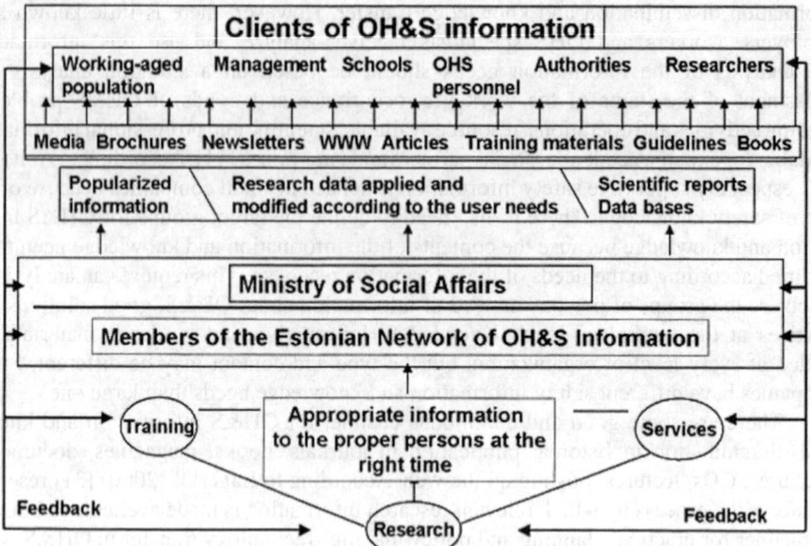

Figure 2. Process of information dissemination on OH&S in Estonia [11]

may provide the OHS. Another key function in this knowledge stewardship involves the accreditation of laboratories by the Estonian Accreditation Centre. According to data from Estonian Accreditation Centre, there are about 23 certificated laboratories in Estonia who may perform occupational hygiene measurements. Often the OH&S knowledge that a company applies to problems comes not from within, but from consultants retained for such purposes. In this case, the knowledge and expertise of the consultant is a marketable asset [29]. Utilization of consultants' services is one method of KM. Estonia has several national networks, for instance, the national Network on OH&S was built up during the Estonian-Finnish Twinning Project on Occupational Health executed in the years 2000–2002. The main functions of a networks include: sharing information and knowledge between organizations; increasing OH&S awareness in society; developing strategies, methods and instruments, and supporting training and education in OH&S [27–28].

The internet is becoming a primary source of OH&S information and knowledge. As Estonia has put a lot of emphasis on the development of Internet access, it can be expected that the workplaces use Internet actively. Therefore, training as well as information and knowledge transfer are performed via Internet. This electronic technology uses the Internet to train workers and employers as well as provide self-teaching courses and training modules. This is an additional challenge but there are already a lot of information available through Internet among the various organization involved in the National

Network on Occupational Health and Safety in Est onia. Information and knowledge dissemination on the Internet is generally considered to be passive, but list servers, video-conferencing, training, and other interactive formats are also available. Easy availability of and accessibility to well-managed information and knowledge can empower future workers and encourage life-long learning. Despite the potential of the Internet in Estonia, however, a systematic assessment has not made to evaluate to how this technology is used in the field of OH&S and by whom. Comprehensive and multi-disciplinary approaches, including an understanding of the needs and behaviours of online information users, will be required to improve the health, safety, and competence of employees and managers.

Training and education are focused forms of dissemination of information and knowledge in the area of OH&S. A wide range of groups, including employers, labour unions, academia, private training companies, Estonian Accreditation Centre conduct training in the field of OH&S. Training and education are generally conceived in the OH&S field as worker and employer oriented or used OHS experts. The Estonian Act on Occupational Health and Safety and many other regulations contain requirements for worker training, which involves instruction in recognizing known hazards and using available methods of protection. Worker education in contrast prepares one to deal with potential hazards or unforeseen problems in order to find the possibility to eliminate the hazards at the workplaces. The training and education of OH&S experts is a part of graduate degree programs to obtain competency and certification in a particular field. The OH&S specialist and working environment specialists need a lot of information, part of which is transferred through training but another part by information and knowledge dissemination. Various channels are needed: informative web-pages that provide the information in an easy to access format, databases, textbooks, guidelines, etc.

KNOWLEDGE TRANSFER BARRIERS

There are long-standing barriers to sharing knowledge in OH&S in Estonia. One of these is a lack of commitment of the government and social partners to be able to draw up policies and strategies for further development of Estonian OH&S system, by knowledge transfer. In addition, there is very little motivation from the legislation for employers to deal with OH&S issues [25, 32]. The compilation of the Sectoral profile on OH&S [26] defined, that the OH&S infrastructure in Estonia is still weak and that there is lack of the research activities in this the field [29]. The number and density of experts in the area of occupational psychology, toxicology, ergonomics and occupational hygiene in relation to the total workforce is still low and only a minority of employees has access to the OH&S specialist [25–26, 32]. One possible reason is that there is impossible to get degree education in some of these fields in Estonia and legislation reforms on OHS are needed [29, 32]. Barriers related to information and knowledge transfer via Internet include an information overload, shifting customer (people from many backgrounds) base for OH&S information. In addition, transferring knowledge and information via the web sometimes is not free. Knowledge and information on the web requires resources and effort, because the material must be constantly maintained and updated. Another barrier to knowledge sharing and disseminating is the interdisciplinary nature of OH&S. Different OH&S specialists should be able to communicate across boundaries of component disciplines [10].

According to literature, the barriers to information and knowledge dissemination include constraints in the will to disseminate, inadequate resources, motivation to share, ability to learn and apply the new knowledge, and the lack of knowledge of what to disseminate or how to do it [9–10]. Often, OH&S information from state authorities may not get to small business employers because they are not the focus of the information as well as may not know how to reach the small business employers, and even if the employer is reached, the information may not be what is needed to make a decision. In the term of information seeking behaviours, the following categories of barriers have been defined: personal characteristics of the seeker; social and interpersonal characteristics; environmental or situational characteristics; are sources credibility [6, 9–10]. Next barrier is the increasingly global nature of knowledge creation, transfer and use. International standards development and the global harmonization of hazard classification and labelling systems are examples of initiatives to facilitate consistent and universal exchange of knowledge and information resources in occupational health and safety [10].

DISCUSSION AND CONCLUSIONS

The process of OH&S knowledge transfer at the state level together with possible knowledge transfer barriers were identified from literature and results of the Sectoral profile on OH&S in Estonia. The authors used the principles of the Sectoral profile on OH&S in agriculture as an example. Sectoral profile on OH&S could be utilized by other sectors of the economy industries, not only in agriculture. It should be used as a possible tool for managing the safety knowledge, which provides situational understanding and clarity of the current OH&S system in the local context and from the perspective of stakeholders. The compilation of the Sectoral profiles on OH&S would strengthen the sectoral approach, information and knowledge dissemination and use. The target audience of the profile could be administrators, professionals, and others who deal with OH&S in the local context as well as specialists from other sectors who benefit from understanding the OH&S situation.

There is potential for organizations to learn, adopt and apply best practice, knowledge and information in the area of OH&S from other companies and various state authorities. Further research is needed in order to understand the factors involved in OH&S knowledge transfer and translated into practice, especially in focusing to knowledge management for young workers, non-Estonian speaking, for employers and employees in small business.

In order to overcome the knowledge transfer barriers, there is need to strengthening of national OH&S system in Estonia as well as awareness of the public through tripartite collaboration, and this includes legal provisions, enforcement, compliance and labour inspection capacity and capability, knowledge management strategy, information exchange, research and support services.

REFERENCES

1. Keles, R., 2006. E-safety for distributed occupational environments. Proceedings of 5th International Symposium on Intelligent Manufacturing Systems, May 29–31: 112–117.
2. Sherehiy, B., and Karwowski, W., 2006. Knowledge Management for Occupational Safety, Health, and Ergonomics. Human Factors and Ergonomics in Manufacturing, 16 (3), 309–319.
3. Takala J., 1998. Information: A precondition for action. In: Stellmann J., editor. Encyclopedia of occupational health and safety. Geneva, Switzerland: International Labour Organisation. P 22.2–22.4.
4. Lehtinen, S., Tammaru, E., Korpen, P., and Rünkla, E., 2004. Information in occupational health and safety – brining about impact in practice in Estonia. In Occupational health services in Estonia. Estonian-Finnish Tinning Project on Occupational Health Services 2003–2004. Finnish Institute of Occupational Health, Helsinki. ISBN: 951-802-616-5, 59–64.
5. Barnard, Y.F., 2005. Developing Industrial Knowledge management: Knowledge Sharing over Boundaries. In: Proceedings of the International Conference on Advances in the Internet, Processing, systems, and Interdisciplinary research (IPSI-2005 France). Carassonne, France. IPSI Belgrade, Serbia, Serbia and Montenegro (www.internet-conferences.net).
6. Kwok, S.H., and Gao, S., 2005/2006. Attitude towards knowledge sharing behaviour. The Journal of Computer Information Systems, 46, 2; 45–51.
7. Lindsey, K.L., 2006. Knowledge Sharing Barriers, in Encyclopedia of Knowledge Management. ISBN: 1-59140-573-4, 499–506.
8. Jacobson, C.M., 2006. Knowledge Sharing Between Individuals. In Encyclopedia of Knowledge Management. ISBN: 1-59140-573-4, 507–514.
9. Schulte, P.A., Okun, A., Stephensen, C.M., Colligan, M., Ahlers, H., Gjessing, C., Loos, G., Niemeier, R.W., and Sweeney, M.H., 2003. Information Dissemination and Use: Critical Components in Occupational Safety and Health. American Journal of Industrial Medicine 44:515–531.
10. Schulte, A., Lentz, T.J., Anderson, V.P., and Lamborg, A.D., 2004. Knowledge Management in Occupational Hygiene: The United States Example, Annals of Occupational Hygiene, 48(7):583–594.
11. Lehtinen, S., 2002. Mission of information dissemination. In Information Dissemination Strategy on Occupational health and safety in Estonia. ISBN 951-802-481-2.
12. Davenport, T.H., and Prusak, L., 1998. Working knowledge: How organizations manage what they know. Boston: Harvard Business School Press.
13. Rantanen, J., 2004. Occupational health and safety as a resource for social development. In Occupational health services in Estonia. Estonian-Finnish Tinning Project on Occupational Health Services 2003–2004. Finnish Institute of Occupational Health, Helsinki. ISBN: 951-802-616-5, 18–20.
14. Putnam, R.D. 1993. Making Democracy Work. Civic traditions in modern Italy, Princeton, NJ: Princeton University Press, 258+ xv pages.

15. Liebowitz, J., (Ed.) 1999. Knowledge management handbook. Boca raton, FL: CRC press.
16. Zelenkow, J., 2006. Legal Knowledge Management in Encyclopedia of Knowledge Management. ISBN: 1-59140-573-4, 578–582.
17. Butler, T., and Murphy, C., 2006. Work and Knowledge in Encyclopedia of Knowledge Management. ISBN: 1-59140-573-4, 884–889.
18. Argote, L., McEvily, B., and Reagans, R., 2005. Managing knowledge in organizations: An integrative framework and review of emerging themes. *Management Science,* 49(4), 571–582.
19. Nonaka, I. and Takeuchi, H. (1995). The knowledge creating company. Oxford University Press.
20. Bryant, S.E., 2003. The role of transformational and transactional leadership in creating, sharing and exploiting organizational knowledge. Journal of Leadership & Organizational Studies, 9, 32–42.
21. Vera, D., and Crossan, M., 2004. Strategic leadership and organizational learning. Academy of Management Review, 29, 222–240.
22. Zyngier, S., 2006. Knowledge Management Governance, in Encyclopedia of Knowledge Management. ISBN: 1-59140-573-4, 373–379.
23. Cepeda-Cariòn, G., 2006, Competitive Advantage of Knowledge Management, In Encyclopedia of Knowledge Management. ISBN: 1-59140-573-4, 34–38.
24. Kotter, J.P. and Heskett, J.L., 1992. Corparate culture and performance. New York: Free Press.
25. Kempinen M., and Kurppa, K., 2004. Ülevaade Töötervishoiust ja Tööohutusest Eesti Põllumajanduses. [Sectoral Profile on Occupational Health and Safety in Estonian Agriculture. English summary.] Estonian Occupational Health Centre, Tallinn, Estonia, ISBN: 9949-10-841-1, 112 p.
26. Kempinen, M., and Kurppa, K., 2004. Sectoral Profile on occupational health and safety in Estonian agriculture. Within the EST-FIN Twinning on Occupational Health Services. In Occupational health services in Estonia. Edit. Lehtinen, S. Finnish Institute of Occupational Health, Helsinki: 53–58. ISBN 951-802-616-5.
27. Kurppa, K., 2002. Building networks on occupational health and safety in Estonia. In The Finest bridge. Finnish-Estonian Collaboration in Occupational Health. Eds. Ylikoski M, Lehtinen S, Kaadu T, Rantanen J, Finnish Institute of Occupational Health, Helsinki, 76–82.
28. Kurppa, K., Tammaru, E., Kempinen, M., Rünkla, E., Sõrra, J., and Lehtinen, S., 2006. Sectoral Network on Occupational Health and Safety in Agriculture to Support Enteprises and Family Farms in Estonia. *Industrial Health,* 44, 3–5.
29. Järvis, M., and Tint, P., 2007. Knowledge management in occupational health and safety: Estonia example. In Ergonomics in Contemporary Enterprise. Proceeding of the 11th International Conference on Human Aspects of Advanced manufacturing: Agility and Hybrid Automation. 4th Intern.Conf.ERGON-AXIA. Edit.by Pacholski & Trzeielinski. ISBN 978-09796435-0-7, 517–534.

30. Aaltonen, M., 1996. A Consequence and Cost Analysis of Occupational Accidents in the Furniture Industry. Finnish Institute of Occupational Health. ISBN 951-802-130-9, 13–35.
31. Lagerlöf, E., 2000. Research dissemination. Arvete Och Hälsa 16: 1–6.
32. Martimo, K-P., 2004. Strengthening of the service provision of occupational health in Estonia. In Occupational health services in Estonia. Estonian-Finnish Tinning Project on Occupational Health Services 2003–2004. Finnish Institute of Occupational Health, Helsinki. ISBN: 951-802-616-5, 26-32. ISBN 951-802-616-5.

INDUSTRIAL PROCESSING SITES – COMPLIANCE WITH THE NEW REGULATORY REFORM (FIRE SAFETY) ORDER 2005

Steven J Manchester
BRE Fire and Security
E-mail: manchesters@bre.co.uk

The aim of this paper is to inform Managers of industrial processing sites about the fire legislation that came into force on 1st October 2006. The paper will go through the main parts of the legislation, highlighting key areas and requirements. The main body of the paper will then discuss the responsibilities and the tasks required by Senior Managers in order to ensure all parts of the sites, including the administration, process and storage areas, will comply with the new legislation. A key area that will be discussed will be the potential conflicts with the existing DSEAR legislation that applies to many processing sites.

KEYWORDS: Process, Industrial, Fire, RRFSO, DSEAR, Compliance, Responsible Person, Competent Person, Risk Assessment, Emergency Plan.

INTRODUCTION

As part of a commitment to reduce death, injury and damage caused by fire, the Government has reviewed the current fire safety law and has made a number of changes through the Regulatory Reform (Fire Safety) Order 2005 (RRFSO) which became law in England and Wales on 1st October 2006. Similar requirements have also been introduced in Scotland under the Fire (Scotland) Act 2005. Two thirds of all fire deaths occur in the home and the government is directing the efforts of the fire brigades in prevention strategies in this area and consequently less direct attention on the commercial sector.

The main effect of the changes is a move towards greater emphasis on fire prevention in all non-domestic premises – for offices, shops, factories, leisure and other buildings. Hence, industrial processing sites come under these new regulations and it includes all the offices, and process buildings on the premises as well as storage areas and any other places where personnel can get access. Fire certificates are abolished and cease to have legal status. Responsibility for complying with the law now rests with the 'Responsible Person'. Non-compliance could ultimately lead to prosecution, with the maximum penalty being two years in gaol. The role of the fire and rescue service has also changed as they become enforcers of the new regulations, in a similar manner to the HSE with regard to DSEAR and COMAH. The new law means a much greater emphasis on fire prevention and businesses need to be taking steps to identify and deal with fire risks.

On industrial processing sites the demarcation lines between what is enforced by HSE and what comes under the Fire & Rescue Service is not particularly clear to many people. The HSE are responsible for 'Process Fire Precautions' (PFP) which are specific process related fire safety requirements. Whereas 'General Fire Precautions' (GFP) as defined in the Regulatory Reform (Fire Safety) Order are enforced by the Fire Service.

OVERVIEW OF THE FIRE LEGISLATION

The new fire legislation covers all non-domestic premises and communal areas of domestic premises, e.g. blocks of flats, care homes, houses of multiple occupancy. For processing sites this will cover all manned operational sites and offices. The main requirements are:

- Appoint a 'Responsible Person'
- Appoint at least one 'Competent Person'
- Undertake a Fire Risk Assessment
- Take general fire precautions
- Formulate an Emergency plan

The RRFSO also contains six other legal duties covering the safety of employees, consultation with employees, informing other employers in buildings under your control, issues of other buildings that are in your control, means of contacting emergency services and employee co-operation. Once again the onus is the 'Responsible Person' to ensure that their business meets the requirements of the new law.

RESPONSIBLE PERSON

The collection of fire safety duties which form the core of the order are placed upon the person called the 'Responsible Person'. As part of the requirements to comply with the Order, the 'Responsible Person' is required to carry out a fire risk assessment of the premises, or instruct a 'Competent Person' to undertake this task on their behalf. The responsible person is:

a) The person who is the employer or has a workplace which is to any extent under his 'control'.
b) in relation to premises not falling within a) the person who has control of the premises in connection with carrying out a trade, business or other undertaking; or the owner, where the person in control of the premises does not have control in connection with carrying out a trade, business or other undertaking.

For small to medium sized companies this would be the Managing Director or Chief Executive. For large organisations such as utility companies that have very large numbers of sites, the Responsible Person will need to be appointed at a senior executive management level, who will be responsible for ensuring compliance at a corporate level. There will also undoubtedly be a Responsible Person to ensure compliance at individual sites or groups of sites, who will be required to ensure fire safety measures are implemented locally. For example, if the enforcing authority finds that there is a problem when inspecting a site with the management systems or a company policy relating to fire then the corporate Responsible Person would be informed of the non-compliance. If the problem found was a local issue, e.g. fire exit blocked by someone placing a filing cabinet in front, then the local Responsible Person would be informed. The Responsible Person must:

- identify the significant findings of the risk assessment and the details of anyone who might be especially at risk in case of fire
- record the significant findings of the risk assessment

- provide and maintain such general fire precautions as are necessary to safeguard those who use your workplace
- provide information, instruction and training to employees about the fire precautions in the workplace.

It should be emphasized that it is the Responsible Person who is liable in law for ensuring their workplace is in compliance with the RRFSO.

COMPETENT PERSON

The 'Competent Person' may be appointed in-house from available staff with sufficient training, and/or experience, in fire risk assessment, or companies may choose to appoint an external consultant to undertake the risk assessment on their behalf. If internal staff are tasked with undertaking these assessments then they must have sufficient training and experience to fulfill this role to the satisfaction of the enforcing Fire & Rescue Service.

In appointing an external consultant there are a number of key factors that should be borne in mind:

Have they sufficient experience in undertaking fire risk assessments?

Since the introduction of the new legislation it seems everyone is now an expert in fire risk assessment! Contractors offering this service should be asked to provide examples of their previous work. For example, companies whose main work has been in asbestos surveying or servicing fire extinguishers may now also be offering fire risk assessments, but have they the technical knowledge and experience in risk assessment techniques to undertake this role?

Are they truely independent?

Many companies are now offering this service and a 'bolt-on' to their usual services for little of no cost. However, the client should be aware that many of these companies may be biased towards directing the assessment to focus on a certain area to the detriment of other more important areas. If a company whose main business is providing fire extinguishers or smoke detectors undertakes your fire risk assessment don't be surprised if the report recommends more extinguishers or detectors!

Have they had sufficient training?

Contractors should be able to show that they have been through a suitable training course and preferably have a Certificate of Competence. There are a number of reputable companies that are providing training on how to undertake fire risk assessments, with some also providing an assessment of competency. Unfortunately, there is no Government requirement or scheme to benchmark against. So it is left to the person being trained to ensure that the training provided is by a reputable company.

Does the fire risk assessor have sufficient insurance cover?

In undertaking a fire risk assessment, the assessor is taking on a certain amount of liability for the report produced and therefore must have sufficient Professional Indemnity and

Public Liability Insurance cover. This will be of particular importance if, as an operator of a large number of sites, you decide to place all your fire risk assessments for all your sites with a single service provider.

Quality of report

The key output from the fire risk assessment exercise is the production of the report. This must stand up to scrutiny by the local Fire & Rescue Service. If, in their opinion, the report is not suitable and sufficient to meet the requirements of the RRFSO, then they could issue an Enforcement Notice that will require the assessment to be undertaken again.

FIRE RISK ASSESSMENT

A properly carried out fire risk assessment will help to decide the nature and extent of the fire precautions that need to be taken. There is a general requirement to reduce the risk to a level which is as low as reasonably practicable.

Fire Risk Assessments must also consider all employees and all other people including contractors, visitors and members of the public, who may be affected by a fire in the workplace. There is also a requirement to make adequate provision for any disabled people with special needs who use, or may be present, at the premises. A key point to note is that they must not be seen as 'one off' exercises but as 'living documents', and should be kept under review and revised where necessary. Changes may need to be made, for example, when the fire risk or hazard may have changed due to alterations to the building, the nature of the work, the number of employees, or changes to the fire safety management processes.

The assessment must include as a minimum the following:

- Management systems
- Active fire protection systems
- Passive fire protection systems
- Detection
- Staff Training records
- Maintenance records
- Evacuation/emergency routes
- Significant findings
- Actions and timescales
- Plans/drawings of the buildings
- Photographs to illustrate potential hazards or non-compliances

GENERAL FIRE PRECAUTIONS

There is a requirement in the regulations to undertake what is known as 'general fire precautions'. In relation to the premises, this means;

1. measures to reduce the risk of fire on the premises and the risk of fire spread;
2. measures in relation to the means of escape from premises;

3. measures for securing that the means of escape can be safely and effectively used;
4. measures in relation to the means for fighting fires,
4. measures in relation to the means for detecting fires and giving warning in case of fire on the premises; and
6. measures in relation to the arrangements for action to be taken in the event of fire, including: instruction and training of employees and measures to mitigate the effects of fire.

The above does not include special, technical or organisational measures required to be taken in any workplace in carrying out work processes, e.g. explosion protection systems on process equipment which come under DSEAR.

GUIDANCE

Twelve guidance documents have been produced to assist those people tasked with undertaking fire risk assessments. Working for the Department for Communities and Local Government (CLG), BRE has written eight of the thirteen guides, that have been produced to date. The guides cover the following categories of buildings and work places:

- Offices and Shops
- Premises providing sleeping accommodation
- Residential care
- Small and medium places of assembly
- Large places of assembly
- Factories and warehouses
- Theatres and cinemas
- Educational premises
- Healthcare premises (responsibility of the Department of Health)
- Transport premises and facilities
- Open air events
- Means of escape for disabled people
- Animal premises and stables

For industrial and processing sites the factories and warehouses guide[1] is probably most applicable.

ENFORCEMENT

For the vast majority of sites the enforcing authority is the local Fire & Rescue Service. The inspecting officer has the power to do anything necessary for the purposes of carrying out his duties to enforce the Order. There are three Notices that can be issued to the Responsible Person.

ALTERATIONS NOTICE

If the enforcing authority is of the opinion that something on the premises constitutes a serious risk, they may serve an Alterations Notice to the responsible person in order that they notify the authority of any alterations to the premises that might result in a significant increase of the risk.

ENFORCEMENT NOTICE

If the Responsible Person has failed to comply with any provisions of the order, the enforcing authority may issue an Enforcement Notice that requires the responsible person to take steps to remedy the failure.

PROHIBITION NOTICE

A Prohibition Notice can be served on the Responsible Person if, in the opinion of the inspector, the use of the premises involves or is likely to involve a risk to persons that is so serious that the use of the premises should be prohibited or restricted. In effect they have the power to stop work at the site.

DANGEROUS SUBSTANCES

Throughout the RRFSO there is reference made to 'dangerous substances'. These are substances that can be classed as explosive, oxidising, extremely flammable, highly flammable or flammable; a preparation that because of its physio-chemical or chemical properties creates a risk, or any gas, vapour, mist or dust that can form an explosive mixture in air.

If dangerous substances are present in the workplace, their use must be eliminated or reduced as is reasonable and have to be taken into account in the risk assessment. In addition, the use of dangerous substances comes under the Dangerous Substances and Explosive Atmospheres Regulations (DSEAR). Under these regulations a written risk assessment is also required and, depending on the volumes and the nature of the substance and usage, could involve hazardous area zoning.

For industrial processing sites typical substances that can be deemed a 'dangerous substance' would include:

- Flammable gases, e.g. methane, hydrogen, propane
- Flammable vapours, e.g. solvents, paints
- Flammable dusts, e.g. chemicals, pharmaceuticals, foodstuffs, paints, rubber

A risk assessment report compiled to satisfy DSEAR would not be sufficient in itself to meet the requirements of the RRFSO but could inform or form part of the fire risk assessment. For those companies yet to undertake their DSEAR risk assessments on their premises, it can be useful to have it undertaken together with the fire risk assessment, as there are areas which may overlap and substantial cost savings may be made by having them done at the same time.

COMPLIANCE OF INDUSTRIAL PROCESSING SITES

Industrial processing sites can vary in size and complexity enormously from simple small operations that may have half a dozen employees in a single building with an office area, to large sites that can comprise of numerous process buildings, office blocks, power plants, storage tanks, warehouses and complex process equipment.

UNMANNED AREAS

For parts of sites which are usually unmanned but will have personnel present for short periods, these sites will still be required under the RRFSO to have a fire risk assessment, although the assessment would be expected to be quite brief. It could take the form of a simple checklist that may be compiled in a generic form to cover these types of areas or buildings.

PROCESS BUILDINGS

In the main works there can be numerous buildings in which personnel will be working and which come under the RRFSO. As well as the obvious need to assess the office accommodation and meeting rooms, other buildings that contain process plant would also need to be assessed as staff can be expected to undertake work activities in them. In addition, these types of buildings may contain dangerous substances which would also need to be assessed under DSEAR as well as the RRFSO.

Buildings containing processing plant that use or produce flammable materials could pose particular risks to persons working on the process plant. Some plants are large complex processes that encompass numerous multi-storey levels and can virtually fill a building. There are a number of concerns with these plants:

- Fires from self-heating
- Dust or gas explosions
- Means of escape
- Building structure comprising of flammable materials

Fires from self-heating

This phenomenon occurs with certain types of material, for example dried sewage sludge, due to exothermic chemical/and/or biological action. It can occur at numerous stages of a process either with relatively small quantities of the material under hot fully oxygenated conditions for short periods or when stored in bulk in silos or hoppers in cooler temperatures but for longer periods. In respect of the substance being liable to undergo spontaneous combustion, it can be classed as a 'dangerous substance' and thus comes under the requirements of both DSEAR and also the RRFSO. In either case it would be necessary to show to the enforcement authorities that an isothermal self-heating assessment has been undertaken to the European Standard prEN 15188, which includes testing of the material and use of Thermal Ignition Theory to calculate the safe operating parameters, in terms of critical ignition temperature, safe storage volumes and time to ignition for the full scale process.

Dust and gas explosions

Organic materials in the form of a dust cloud suspended in the atmosphere can form an explosible cloud if present in sufficient concentration, typically above 40 g/m^3. Similarly, flammable gases and vapours can also form explosible mixtures in air, when present a concentration above their lower explosive limit. As such, these processes will come under the requirements of DSEAR and thus requires a risk assessment and hazardous area zoning exercise, in addition to the other requirements of these regulations. Explosions, either gas or dust, do not explicitly fall under the RRFSO. However, a consequence of an explosion in many cases leads to a subsequent fire in the building that contains the process. Thus, it may be argued that for processes that are inside buildings, the possibility of a dust or gas explosion leading to a fire **must** be taken into account and assessed as a risk when undertaking a fire risk assessment for the building. Hence, some knowledge of dust and gas explosions must be required by the person undertaking the fire risk assessment in order for this risk to be adequately assessed.

Means of escape

Some of the larger types of process plant have complex arrangements for access to parts of the process, which can pose problems for means of escape in case of fire. Of particular concern are those plants that virtually fill the building in which they are housed, with very little space between the top level of the process and the roof and/or the sides of the building. To make matters worse, a single exit point from the process to ground level could mean excessive travel distances from certain areas of the plant to the place of safety outside the building. This issue would again need to be addressed when undertaking a fire risk assessment and in forming an emergency plan under the RRFSO.

Building structure

As part of the fire risk assessment an important area that should not be neglected should be the building itself. As an example, some buildings particularly in the food processing industry, may be constructed from insulated core panels. It is important that the nature of the insulation used is determined as typical materials used vary greatly in their behaviour. Some materials, particularly expanded polystyrene (EPS), can be particularly hazardous if ignited in a fire both from the rate of heat release (size of the fire) and the toxic combustion products. The risk is increased significantly if the wall panels are cut through to allow the passage of equipment or services (see Figure 1).

Vehicles

There are particular hazards associated with the use of vehicles such as fork lift trucks which must also be considered. For example, battery charging can give rise to hydrogen

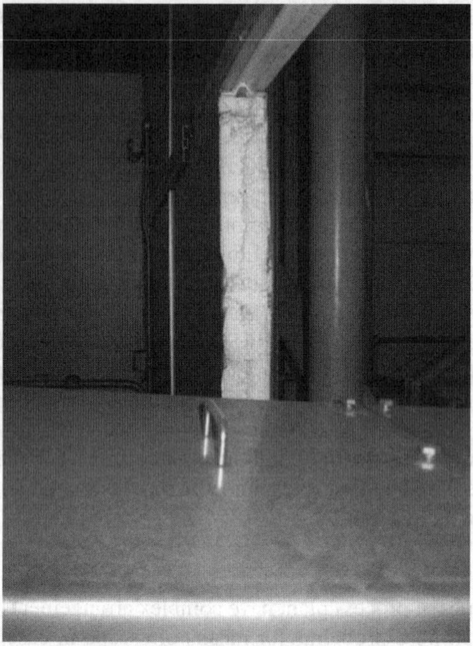

Figure 1. Expanded polystyrene insulation panel in thermal dryer building cut to allow conveyor access through side wall

which is highly flammable and could cause an explosion if it is present in sufficient concentration. Charging points should therefore be located in well ventilated areas and, if inside a building, sited against a 30 minute rated fire-resisting wall.

STORAGE AREAS

Large storage buildings and warehouses have their own particular hazards, mainly arising due to the very large volumes of combustible materials present and the lack of compartmentation. Fires in these buildings can result in very rapid spread leading to the loss of the whole building and contents. For this reason extensive use of sprinklers is normally found throughout the building and on the storage racking in many cases. Although, this fire protection provision is not primarily for life safety, usually being required by the insurance company for property protection reasons, it needs to be assessed and taken into account in the fire risk assessment. Experiments[2] have shown that fires started in boxes stored on a 10m high racking system frequently reached to the top of the racking within two minutes.

Areas which aren't buildings as such but would also come under the DSEAR and possibly the RRFSO are items such as flammable material storage tanks. This would be due to the presence of a flammable material which could form an explosive atmosphere or fire hazard in case of leakage or spillages. In most instances this would not lead to a subsequent fire but this would still need to be taken into consideration when assessing the risk to life safety of personnel working on top of or around these tanks. If fire is a potential hazard then a fire risk assessment of the tanks would need to be undertaken to comply with the RRFSO, perhaps as part of an overall site fire risk assessment.

MANAGEMENT SYSTEMS

A key element of the RRFSO is the requirement to have in place suitable management systems. This encompasses keeping records and information in a number of areas, including having a written record of the Responsible and Competent person(s). Specific documentation that the Fire and Rescue Service may wish to see during a site visit could include:

- Fire Safety Policy
- Emergency Plan
- Fire Risk Assessment
- Permit to work system for hot working
- Staff training
- Evacuation drills
- Maintenance records

FIRE SAFETY POLICY

Most responsible companies will have a written statement outlining their corporate policy on fire safety, either specifically or perhaps as part of a general Health & Safety Policy. Many large companies will also have a Health & Safety Manual which would be expected to cover fire issues.

EMERGENCY PLAN

This will detail the expected actions on personnel in the event of a fire occurring. It would be expected to be specific to a building or site, although it may also contain corporate safety instructions as well. For example, an internal emergency phone number for personnel to use in the event of a fire that alerts a receptionist to call the fire service may be a corporate action, but specifics assembly points for staff to evacuate to would be specific to a building.

FIRE RISK ASSESSMENT

As described in detail in this paper, a fire risk assessment for each building on a site would be expected to be made available to the Fire and Rescue Service when requested. These documents should also be available for staff.

PERMIT TO WORK

Where hot working may be undertaken, particularly in high risk areas such process buildings, it is good practice to use a permit to work scheme. These can help to reduce the risks from fire started accidentally by repair and maintenance works.

STAFF TRAINING

A requirement of the legislation is that staff know what to do in the event of a fire and where they are expected to muster on evacuation of the building. For process areas other specific tasks may also be required of the staff. If staff are expected to use a fire extinguisher they should be trained in their use and must know what type of extinguisher to use for the different classes for fire. A higher level of training would obviously be required for those staff given specific fire fighting roles. Records of the staff training can be expected to be requested by the enforcing authority.

TRAINING

It is important that personnel undertaking duties under the RRFSO have training to undertake these tasks. In particular, depending on the nature of the buildings being assessed, the fire risk assessment may be quite a complex document requiring knowledge of a wide range of fire safety systems such as detection, passive and active fire protection, means of escape and travel distances, dangerous substances, human behaviour and the ability to interpret building plans and drawings. There is guidance available, as described earlier, to assist in this process, but specific training may be required if a suitable and sufficient assessment is to be made that satisfies the enforcing authority.

The role of the Responsible Person is fundamental to compliance with the RRFSO and they need to be clear what their duties entail and the possible penalties they could face by not undertaking them to the requirements of the legislation. Training in the duties of this role is another area for consideration.

EVACUATION DRILLS

These are a key part of ensuring everyone knows what to do in the event of the fire alarm sounding and where they need to evacuate to so that everyone can be accounted for when the Fire Service arrive. These drills should be undertaken at least annually and records kept detailing when they were undertaken and any outcomes or actions to be taken from them.

MAINTENANCE RECORDS

It is important that records are kept showing that the fire safety systems used are properly maintained. This includes portable fire fighting equipment, automatic fire detection systems and fixed fire fighting installations (e.g. sprinklers). Passive systems, such as fire doors, should not be overlooked. It is also important when purchasing fire protection systems, that they have been independently certified, by a body such as LPCB, which will assess the performance of the system or component against a standard.

CONCLUSIONS

The new fire legislation is aimed at cutting deaths from fires in non-domestic premises and indirectly this will reduce damage to buildings and reduce business interruption. The responsibility is laid squarely at the door of the Responsible Person to ensure the regulations are met. A key part of interpreting the RRFSO is recognising who the Responsible Person is for different types of business. For large businesses with many sites there will need to be an overall corporate Responsible Person but with specific sites also having a Responsible Person to ensure local compliance.

Many industrial processing sites will come under DSEAR as well as the RRFSO. A DSEAR risk assessment alone will not cover all the requirements of the RRFSO. There will be some overlap between DSEAR and the RRFSO but with someone who is competent in both these areas, conflicting risk assessments should be avoided.

A key aspect of the new fire legislation is the need to maintain good records, have suitable management procedures in place and to ensure adequate staff training and awareness for fire. The importance of good Management Systems should not be underestimated.

REFERENCES

1. Fire Safety Risk Assessment: Factories and Warehouses. Department for Communities and Local Government, 2006.
2. 'Sprinklers: High piled and rack storage'. Fire Surveyor, Vol. 9 No. 1, 21–26. February 1990.

THE REVISED EN 13463-1 STANDARD FOR NON-ELECTRICAL EQUIPMENT FOR USE IN POTENTIALLY EXPLOSIVE ATMOSPHERES

Dipl.-Phys. Konrad Brehm
Bayer Technology Services GmbH, D-51368 Leverkusen
E-mail: konrad.brehm@bayertechnology.com
Dr. Richard L. Rogers†

As a consequence of the Directive 94/9/EC which defines the requirements for electrical and non-electrical "mechanical") equipment intended for use in explosive atmospheres there was a demand for harmonized standards. For electrical equipment those standards have a long history but not for mechanical equipment. On this field it was a quite new topic and the task of the European working group CEN/TC 305 "Potentially explosive atmospheres – Explosion prevention and protection". Within WG2, "Equipment for use in potentially explosive atmospheres", established in 1994, the mechanical standard series EN 13463 were produced. The first version of the EN 13463-1 (2001), "Basic methods and requirements", was revised after the first experiences with this standard. The revision ("revised EN 13463-1") was focused on the mandatory risk analysis, i.e. ignition assessment procedure carried out by the manufacturer. Furthermore, it now contains assessment tools for all possible 13 ignition sources, requirements for mechanical equipment, the ignition assessment procedure, including the evaluation of malfunctions.

Still discussed is the case were the mechanical equipment is an integral part of a process (i.e. pumps, dryers, mixers etc.) and the user has to integrate this equipment into the safety concept of the process.

1. INTRODUCTION

Starting July 1st, 2003, electrical and non-electrical (i.e. "mechanical") equipment has to fulfill the requirements of Directive 94/9/EC if the equipment is intended for use in explosive atmospheres and supposed to be placed on the European Market. Up to this time, only requirements for explosion protected electrical equipment were standardized and accompanied by a great number of product standards. Mechanical equipment was covered only by the less specific Machinery Directive 98/37/EC.

In line with the EU-Directives following the "New Approach", the Directive 94/9/EC sets out the essential safety- and health-requirements for the products, whereas harmonized standards linked to this Directive describe a way to comply with those requirements. Therefore, it was necessary to prepare specific standards for mechanical equipment. This is the task of the introduced European Working Group CEN/TC 305 "Explosive Atmospheres–Explosion Protection". Within CEN/TC 305/WG 2 is engaged in producing standards with requirements for mechanical equipment and started its work in 1994.

The seven different ignition protection concepts for mechanical equipment are described within the EN 13463-series. The first version of the EN 13463-1, "Basic Methods and Requirements" was published 2001 and was now revised.

2. THE STRUCTURE OF EN 13463-SERIES

EN 13463ff series comprises of a basic part, referred to as part 1 and a number of specific standards dealing with the individual ignition protection concepts. In Table 1 those standards are listed including there current status.

Technically, EN 13463-series follows as closely as possible the structure of the comparable standards for electrical equipment, especially with respect to the ignition protection classes, which are very close to those defined in EN 60079. The last of the standards were published in July 2005 and since then, no part of the EN 13463-series has been revised. EN 13463-1 (2001) has been in use by manufacturers and users of equipment since its publication in 2001 and feedback from users is incorporated in a revised version (revised EN 13463-1).

The original EN 13463-1 (2001) is structured such that any equipment designed in accordance with this standard will fulfill the requirements for equipment group II, category 3 according to Directive 94/9/EC. On one side, the other standards treating the different ignition protection classes should be aware of those ignition sources which cannot be avoided by the measures dealt with in EN 13463-1. On the other side, they enable a lower frequency of occurrence of the ignition source by their application. However, it must be noted that the type of protection "Protection by flow restricting enclosure 'fr' " is limited to equipment category 3.

One goal of assessing ignition sources – which must be facilitated by the manufacturer - is to identify whether or not the ignition sources, as listed in EN 1127-1 /3/, are potential ignition sources for the specific equipment being considered. As part of this

Table 1. Standards of EN 13463-series non-electrical equipment for potentially explosive atmospheres

EN 13463-1	Part 1: Basic method and requirements
EN 13463-2	Part 2: Protection by flow restricting enclosure (fr)
EN 13463-3	Part 3: Protection by flameproof enclosure (d)
EN 13463-4	Part 4: "Intrinsic safety" – cancelled – Note: Content to be taken into account by revising part 1
EN 13463-5	Part 5: Protection by constructional safety (c)
EN 13463-6	Part 6: Protection by control of ignition source (b)
EN 13463-7	Part 7: Protection by pressurized enclosures (p) (Protection by pressurization described in EN 60079-2 can also be used for non-electrical equipment.)
EN 13463-8	Part 8: Protection by liquid immersion "k"

assessment, it is distinguished, whether or not an ignition source becomes effective either during normal operation, in case of frequent (foreseeable) malfunction or in case of rare malfunction. Additionally, protective measures have to be taken to avoid the ignition sources becoming effective.

Depending on which of the described operation conditions might result in an ignition source becoming effective, the manufacturer can determine the relevant category for the equipment. The categories of equipment group II according to Directive 94/9/EC are given as follows:

Category 3: Ignition sources will not become effective under normal operating conditions

Category 2: Ignition sources will not become effective under normal operating conditions and in case of frequent malfunction

Category 1: Ignition sources will not become effective under normal operating conditions and in case of frequent malfunction and rare malfunction

Especially by implementing the assessment of the ignition sources by the manufacturer, it appears that the currently available standard EN 13463-1 (2001) does not provide sufficient support. Although this standard includes a great number of technical requirements, it turns out to be too complex for manufacturers to applyan assessment of ignition sources systematically. Moreover, without the knowledge of an experienced user, who pursues the specific applications of a graded safety concept, an assessment on non-electrical equipment is often done inappropriately.

Due to this reason, CEN/TC305/WG 2 initiated the revision of the EN 13463-series beginning with EN 13463-1.The core intention of the revision aims at accomplishing the risk assessment required by Directive 94/9/EC. It should be noted that according to 94/9/EC the risk assessment is an assessment of the ignition sources. The following sections discuss the revision of EN 13463-1.

3. REVISED EN 13463-1

3.1. OBJECTIVE OF REVISION

The main objective of the revised EN 13463-1 is to aid the manufacturer in carrying out the ignition hazard assessment by clearly outlining the requirements of the standard in chronological order. This was achieved by:

- extending the definition of ignition sources
- outlining a step-wise ignition hazard assessment procedure
- including statements about all (possible) ignition sources defined in EN 1127-1 (1997)
- providing details on the requirements for the prevention of ignition sources from becoming effective.

3.2. EXTENDED DEFINITIONS OF IGNITION SOURCES

Due to support the manufacturer during the identification of ignition sources which are related to the equipment and evaluating whether those ignition sources can become potential ignition sources of the equipment was one of the main objectives in the revised EN 13463-1. Especially during the investigation under which circumstances an ignition source is classified as potential and evaluated as effective are the basic points of the ignition hazard assessment according to Directive 94/9/EC. During his assessment procedure the manufacturer has to answer the following questions:

a) Which of the ignition-sources mentioned in EN 1127-1 is a possible ignition source related to the equipment?
b) Which ignition source related to the equipment has the potential to ignite the explosive atmosphere the equipment is intended to be used in?
c) What is the frequency of the occurrence of effective ignition sources, i.e. turning those potential ignition sources into ignition sources leading to the ignition of the explosive atmosphere and thus determining the equipment category?

To arrive at a scheme for the ignition hazard assessment resulting from these questions, it was necessary to have explicit terms for these different types of ignition sources mentioned above.

As a consequence the CEN/TC305/WG2 agreed on to the following definitions:

- **Equipment related ignition source** (question a) Possible ignition source (remark of the authors: i.e. mentioned in EN 1127-1, (1997)), which is caused by the equipment under consideration regardless of its ignition capability.
- **Potential ignition source** (question b) Equipment related ignition source which has the capability to ignite an explosive atmosphere (i.e. to become effective) (remark of the authors: the equipment becomes subject to Directive 94/9/EC, if it has at least one own potential ignition source).
- **Effective ignition source** (question c) Potential ignition source which is capable to ignite an explosive atmosphere. The likelihood of presence of the ignition source (during normal operation, expected or rare malfunction) determines the equipment category.

3.3. IGNITION HAZARD ASSESSMENT

By using these definitions, it is possible to perform the ignition hazard assessment in the form of a table. An example of such a table for equipment related and potential ignition sources is shown in Tables 2 and 3 for a compressor for natural gas:

3.4. HOT SURFACES

This chapter now bundles all information and requirements relating to hot surfaces as an ignition-source. This includes the establishment of the maximum surface temperature, especially the determination of the temperature class, evaluation of small parts, the correction factors for process and design temperatures and thus the correct marking. For example surfaces with temperatures <200 °C an area <1000 mm² can be tolerated even for T4-equipment.

Table 2. Example of an initial assessment of equipment related ignition sources for a compressor conveying flammable gas (IIA, T2)

Possible ignition sources (List from EN 1127-1)	Equipment related Yes/No	Reason
Hot surfaces	Yes	Gas compression, moving parts of the coupling
Flames, hot gases	No	Not present
Mechanically generated sparks	Yes	Internal moving parts may come into contact housing due to male functions and create sparks
Electrical ignition sources	Yes	Used control devices
Stray electric currents and cathodic corrosion protection	No	Not present
Static electricity	Yes	Isolated metal parts (e.g. housing, rotor)
Lightning	No	Not present
Electromagnetic waves	No	Not present
Ionizing radiation	No	Not present
High frequency radiation	No	Not present
Ultrasonic	No	Not present
Adiabatic compression	No	But user has to consider the internal parts of the compressor (mention in the instruction for use)
Chemical reaction	No	User has to consider the suitability materials (e.g. corrosion) of the conveyed gases

3.5. MECHANICALLY GENERATED SPARKS

The definition of limits for mechanically generated sparks is important for the decision whether or not mechanical sparks are a potential or even effective ignition source. For a lot of manufacturers the assessment of this ignition source turns out to be problematic. To decide if this possible ignition source is an equipment related ignition source causes no problems for most manufacturers, but to decide if it is a potential ignition source leads to a lot of problems. In most cases, the design documentation allows to extract which equipment parts can impact with which relative velocity and thus what energy level can be reached. But it is very difficult for manufacturers to decide if these circumstances lead to mechanically generated sparks. Therefore, the revised standard includes chapters with energy and speed limits for different material combinations.

The assessment of mechanically generated impact sparks results in one of the following two situations:
Situation 1: When does the ignition source mechanically generated spark -solely by single impacts- need not be considered a potential ignition source independent from the category?
These values are important in two ways. If the values defined in the standard are not exceeded by the operation of the respective equipment, the ignition source "mechanically generated impact sparks" is classified "not a potential ignition source". If there is any other potential ignition related to the equipment, the equipment is under the scope of Directive 94/9/EC, otherwise it is not. The following requirements will be defined in the revised standard:

If the following conditions are met, a mechanically generated spark is classified as not a potential ignition source.

Either

a) the impact velocity is less than 1 m/s and the maximum potential impact energy is less than 500 J and
   1) Aluminum, titanium and magnesium in combination with ferritic steel is not used, or
   2) Aluminum is only to be used in combination with stainless steel (≥16,5 % Cr), if the steel cannot corrode and no iron oxide and/or rusty particles can be deposited on the surface (appropriate reference to the properties of the stainless steel shall be given in the technical documentation and instructions for use), or
   3) Hard steel shall not be used in combination with hard steel, or
   4) Hard steel shall not used where it can impact in granite, or
   5) Aluminum is only to be used in combination with aluminum if no iron oxide and/or rusty particles can be deposited on the surface.

or

b) where a combination of non-sparking materials is used the impact velocity is less than or equal to 15 m/s and the maximum potential energy is less than 60 J for gas/vapor-atmospheres or less than 125 J for dust atmospheres.

Situation 2: When do ignition sources generated solely by single impacts, depending on the category, exceeding these values need not to be considered as an effective ignition sources?

The energy values documented in this standard are dependant on the equipment category and explosion group. They are provided for manufacturers who want to reach a certain equipment category for an intended explosive atmosphere. If these values are not exceeded the equipment fulfills the requirements for the intended equipment category according to the equipment related ignition source mechanical generated sparks.

The limits of mechanically generated sparks are based on published values as well as on extensive tests executed by BAM (2005)/8/.

Additionally, the chapter about mechanically generated sparks will also include statements about grinding sparks resulting from the recently finished MechEx-Project /4/.

Table 3. Example of an ignition hazard report for a compressor conveying flamable gas (IIA, T2)

Ignition hazard assessment report according to the revised EN 13463-1: Compressor							
	1		2				
	Ignition hazard		Frequency of occurrence without application of an additional measure				
	a	b	a	b	c	d	e
No.	Potential ignition source	Description of the basic cause (which conditions originate which ignition hazard?)	during normal operation	during foreseeable malfunction	during rare malfunction	not relevant	Reasons for assessment
1	Hot Surfaces (6.2)	Running of comcressor gas compression		#			Increased temperature due to compression/running against closed valve
3	Mechanically generated sparks (6.4)	Contact between rotating and static parts		#			Internally: mixture above upper explosive limit no risk of explosive mixtures Externally: - Pressure side: externally no moving parts - Power side/coupling: moving parts in enclosure with oil filling. Gear is adequately designed for expected torque and load
4	Electrical ignition sources (6.5)	Electrical control units and instrumentation			#		Equipment is in accordance with II 2G IIA T3 or better
5	Static electricity (6.7)	Spark discharges due to non earthed parts		#			Metal parts are connected and earthed (resistance to ground < 1 MO)
		Brush Discharges due to the use non-dissipative coatings (surface resistance exceeding 1 GO			#		There are no non -dissipative coatings in use
6	Adiabatic compression and shock waves (6.13)	Running against/ inbetween closed valves		#			Monitoring of the valve position within PC-System

3			4					
Measures applied to prevent the ignition source becoming effective			Frequency of occurrence incl. measures applied					
a	b	c	a	b	c	d	e	f
Description of the measure	References (standards, technical rules, experimental results)	Technical documentation (evidence including relevant features listed in column 3a)	during normal operation	during foreseeable malfunction	during rare malfunction	not relevant	resulting equipment category inrespect of this ignition hazard	necessary restrictions
Control of exhaust-gas-temperature and shut down at 230 °C within PC-System (quality: IPL1)	EN 13463-1 Basic methods and requirements EN 13463-6 Control of ignition sources	test-report xxx about thermal test instructions for use			#		2G	T2
Leakages of grease will be detected during routine inspections. Shut down of the compressor via vibration control at the gear/compressor (Gas supply "OFF"). Start-up/shut-down via control-system.	EN 13463-1 Basic methods and requirement EN 13463-8 Liquid immersion	Maintenance according to instructions for use			#		2G	T3
	EN 13463-1 Basic methods and requirements	Equipment list xyz			#		2G	IIA T3
	EN 13463-1 Basic methods and requirements	Earthing according to instructions for use			#		2G	IIB
largest area less than 25 mm²	EN 13463-1 Basic methods and requirements	Parts list yyy			#			
Monitoring of the valve position within PC-System (quality: IPL1)	EN 13463-1 Basic methods and requirements EN 13463-6 Control of ignition sources	test-report xxx about thermal test instructions for use			#		2G	T3
Resulting equipment category including all existing ignition hazards:							2G	IIA T2

3.6. ELECTRICAL IGNITION SOURCES

The electrical ignition sources were included in this standard with their links to the relevant IEC-standards of IEC EN 60079ff /5/. Thus, the interfaces to the standards on the field of electrical equipment were defined as the basis for unified standards.

3.7. REVISED STRUCTURE AND DOCUMENTATION

In addition to the fixed chapters 1 to 3 the revised structure of EN 13463-1, consists of the following main parts:

- Equipment groups and categories, explosion groups (Chapter 4)
- Description of ignition hazard assessment procedure (Chapter 5)
- Assessment of possible ignition sources according to EN 1127-1 including their relevance for the specific equipment (Chapter 6)
- Additional aspects (e.g. dust deposits and other material in the gaps of moving parts, openings of enclosures, non metallic parts of the equipment, removable parts, materials used for cementing, light transmitting parts (Chapter 7)
- Tests Procedures (determination of the maximum surface temperature, mechanical (impact) tests, non-metallic parts) (Chapter 8)
- Documentation (Chapter 9)
- Annexes, Bibliography

The chapter 9, "Documentation", was extended. Now it differentiates in between the technical documentation which has to be archived by the manufacturer and the one which is given to the user. Furthermore it gives guidance to the manufacturers of non-electrical equipment of equipment group II category 2 regarding the documentation given and stored at the Notified Body.

4. SPECIAL CASE: MECHANICAL EQUIPMENT WITHIN PROCESS PLANTS

4.1. IGNITION HAZARD ASSESSMENT FOR MECHANICAL EQUIPMENT WITHIN PROCESS PLANTS ("PROCESS EQUIPMENT")

EN 13463-1 (2001) states that the manufacturer of equipment has to perform an ignition hazard assessment for all possible applications and define measures to prevent ignition sources from becoming effective. The mechanical equipment within the interior of process plants ("process-equipment") is often not used permanently but only during certain well defined phases of the process. During those phases the occurrence of explosive mixtures can deviate from the overall occurrence of explosive mixtures of the complete process, which often would require category-1-equipment. If the occurrence of the explosive mixture depends on the process phase, an optimized/tailor-made ignition hazard assessment requires the knowledge of detailed process information, which is only available from the user's side.

Often, the manufacturer of fast running process equipment (e.g. mixers with rotor-stators or choppers) is asked to design process-equipment for interiors in which explosive dust/air mixtures are either present for long periods or on a frequent basis. In that case (i.e. "category-1-equipment"), the manufacturer has to evaluate all possible applications in a potentially explosive atmosphere and rare malfunctions leading to effective ignition source and this without knowing the process. On the other hand, the user has detailed knowledge of the process and can perform a customized risk assessment for the phases of the process when the equipment is actually in use. This is done by choosing standard-, category-3 or -2-equipment (no ignition source in normal operation or even in the case of malfunction), assessing the presence of explosive mixtures during the phase of operation and defining the necessary measures for prevention of ignition sources during that phase. A basis for the user's risk assessment can be the ignition hazard assessment of the manufacturer of category-3 or -2-equipment which needs to be reassessed. On top of this, if necessary, the user defines special measures for explosion protection in his risk assessment. The documentation of this additional (i.e. the user's) ignition hazard assessment is part of the explosion protection document (-> Directive 1999/92/EC, Annex II B /7/).The procedure is shown in Table 4.

5. SUMMARY/NEXT STEPS

The revised EN 13463-1 is strongly influenced and improved by the feedback of users since publication. The standard now more adequately describes the step-wise performance of a risk assessment, makes reference to all possible ignition sources, provides additional information on the evaluation of mechanical sparks and gives a revised guideline for accompanying documentation. Despite these informative changes, there are still applications in which the standard does not provide adequate direction. These include mechanical equipment used in the interior of process plants ("process-equipment") where the risk assessment often cannot be performed without more advanced knowledge of the user.

Table 4. Procurement of equipment intended for use in explosive atmospheres

	Electrical equipment	Non-electrical-("mechanical") equipment-process-equipment
Interior of process-plants	Category according to the presence of explosive mixtures "in terms of a zone"	Category independent of process- and operation (i.e. category 2 or 3) + risk assessment by the user (acc. to Directive 1999/92/EC, Annex II B /7/)
Exterior of process-plants	Category according to zone	Category according to zone

Remark: The use of equipment which deviates from the manufacturer's instruction for use must be reassessed in the risk assessment of the user. (acc. to Directive 1999/92/EC, Annex II B or its implementation into national regulations)/7/

The revised EN 13463-1 provides an excellent starting point for the revision of all other standards in the series.

6. LITERATURE

/1/ Directive 94/9/EC of the European Parliament and the Council of 23. March 1994 on the on the approximation of the laws of the Member States concerning equipment and protective systems intended for use in potentially explosive atmospheres

/2/ EN 13463-1 : 2001: Non-electrical equipment for potentially explosive atmospheres – Part 1: Basic method and requirements

/3/ EN 1127-1:1997: Explosive Atmospheres Explosion Prevention and Protection – Part 1: Basic concepts and Methodology"

/4/ Hawksworth, S., Rogers, R., Proust, C., Beyer, M., Schenk, S., Gummer, J. and Raveau, D.: Ignition of explosive atmospheres by mechanical equipment, Symposium Series No. 150, Crown Copyright 2004

/5/ IEC EN 60079-0 : 2006: Explosive Atmosphere – Part 0: Equipment – General Requirements (Final Draft)

/6/ Brehm, K.: Alles neu im Explosionsschutz? (i.e. "Everything New in Explosion Protection", Technische Überwachung Band 45 (2004) Nr. 11/12 – Nov./Dez.

/7/ Directive 1999/92/EC of the European Parliament and Council of 16 December 1999 on minimum requirements for improving the safety and health protection of workers potentially at risk from ex-plosive atmospheres (15th individual Directive within the meaning of Article 16(1) of Directive 89/391/EEC)

/8/ BAM Research Report on mechanical generated sparks (2007)

IMPLEMENTATION EXPERIENCE OF ATEX 137 FOR A PETROCHEMICAL SITE

Jo Fearnley[1] and Roald Perbal[2]
[1]Senior Consultant, Aker Kvaerner Consultancy Services, Aker Kvaerner, Ashmore House, Stockton on Tees, TS18 3RE, UK
[2]Manager Industrial Safety, SABIC Europe B.V., P.O. Box 475, 6160 AL, Geleen, The Netherlands
E-mail: jo.fearnley@akerkvaerner.com and Roald.Perbal@SABIC-europe.com

Directive 99/92/EC (ATEX 137) deals with the safety and health protection of workers potentially at risk from explosive atmospheres (ATEX). Assessments of all existing plants to the new standards were required to be completed by 30 June 2006. This presentation shows the practical experience from a large petrochemical site with the implementation of the Directive and reflects the balance of work completed across a range of areas including:

- Update of existing hazardous area classifications to new standards
- Explosion risk assessment of existing mechanical equipment safety
- Explosion risk assessment of existing electrical equipment
- Explosion risk assessment of temporary work places
- Inspection and maintenance of equipment used in potentially explosive atmospheres
- ATEX training of operators, supervisors and contractors.

The added value of the implementation of the Directive with respect to actual risk reduction and increasing awareness within the organization will be discussed.

The Directive 99/92/EC is better known as ATEX 137 and its aim is to protect the health and safety of workers potentially at risk from explosive atmospheres. The Directive requires an overall assessment of explosion risks and provision of measures to eliminate, prevent or protect against explosions. When completing such a risk assessment the Directive requires that the following considerations are taken into account:

- The likelihood that the explosive atmosphere will occur,
- The likely persistence of the explosive atmosphere once formed,
- The likelihood that an ignition source will be present.
- The likelihood that the ignition source will be active and effective.

By considering the above the risk of ignition actually occurring can be assessed. When this risk is combined with the installation concerned, the substances present and the processes in use, and the possible interactions between them, then the scale of the potential consequential effects can be assessed.

This paper reviews the experience to date of implementing ATEX 137 on the SABIC Europe petrochemical site at Geleen in The Netherlands.

The site has been in existence for many years, with up to 30 years old plants. The site decided to implement ATEX 137 to continuously improve on process safety performance and to maintain regulatory compliance, taking into account cost effectiveness. The key requirement for the site is to assess systematically the explosion risk from the workplaces and the equipment in use, as indicated in Figure 1.

One of the basic principles for successful ATEX 137 implementation is that it should not be perceived as a paper exercise, but as a practical means of raising the explosion safety standards at the site. To assess the possible added value a brief review of recent incidents was completed to identify how the implementation of ATEX 137 could improve site safety. During a 14 month period from 2004 – 2005 there were 97 incidents recorded in the site incident reporting system. Of these it was estimated that 11, or 11%, could probably have been prevented if ATEX standards had been in place, see Table 1.

In order to manage the cost effective implementation of ATEX 137 it was necessary to clearly define the work scope for the project. It was decided to carry out the implementation as a project such that it was completed consistently across the large and complex petrochemical site, and also to optimise resource requirements.

The work scope for the project was as detailed below:

- Establish the best practice for ATEX implementation
- Review Hazardous Area Classification
- Complete an electrical gap analysis for each plant
- Complete a mechanical gap analysis for each plant

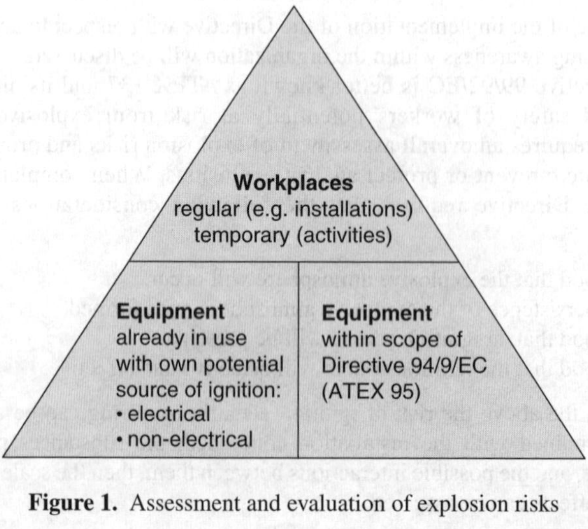

Figure 1. Assessment and evaluation of explosion risks

Table 1. Incidents related to ATEX

Cause of incident	No. of incidents
Temporary workplace	6
Not conscious of working in zoned area	2
Procurement	1
Conformity of design	1
Process failure (in regular workplace)	1

- Develop action plan to address gaps identified
- Complete generic ignition risk assessments for mechanical equipment
- Complete specific mechanical ignition risk assessments
- Carry out an expert survey for other process ignition risks
- Develop training packages
- Deliver ATEX training for selected personnel
- Create explosion protection documents (EPD)
- Include ATEX in job safety assessments (JSA) for temporary workplaces
- Ensure safety, health and environment (SHE) management system includes all ATEX requirements.

This is schematically shown as a flowchart in Figure 2.

As the first step a multidisciplinary team of experts developed the risk assessment methodology for the site. The intention is to align the work processes developed with the SHE management system such that it becomes aligned with the site risk evaluation procedures. This has proved to be more difficult than expected, as ATEX 137 didn't specify residual risk tolerance criteria especially for normal operation.

The starting point was to review the best practice for ATEX, using internal and external resources to define what the site viewed as best practice. This required a variety of experts, and a managed workshop to develop the basic principles for the project team.

SABIC Europe's next action was to agree a way forward with the Dutch Labour Inspectorate. The proposal was to agree on the use of the best practice methodology and to seek agreement for using a mixture of generic and specific ignition risk assessments.

The Labour Inspectorate agreed with the philosophy that for electrical equipment ignition risk assessments (IRA) would only be undertaken if the hazardous area classification (HAC) from the past had changed. Further for mechanical equipment it was agreed that generic IRAs could be used for equipment in zone 2 areas, whereas equipment in zone 0 or 1 would have a specific IRA completed. Other ignition sources would be identified through the use of an expert survey, e.g. to determine hot surfaces and potential electrostatic ignition sources. The focus would initially be on regular work places, including process installations, followed by temporary work places.

The first practical exercise was to review all of the existing HAC assessments to verify that they matched the new HAC standards. The areas of concern which arose were

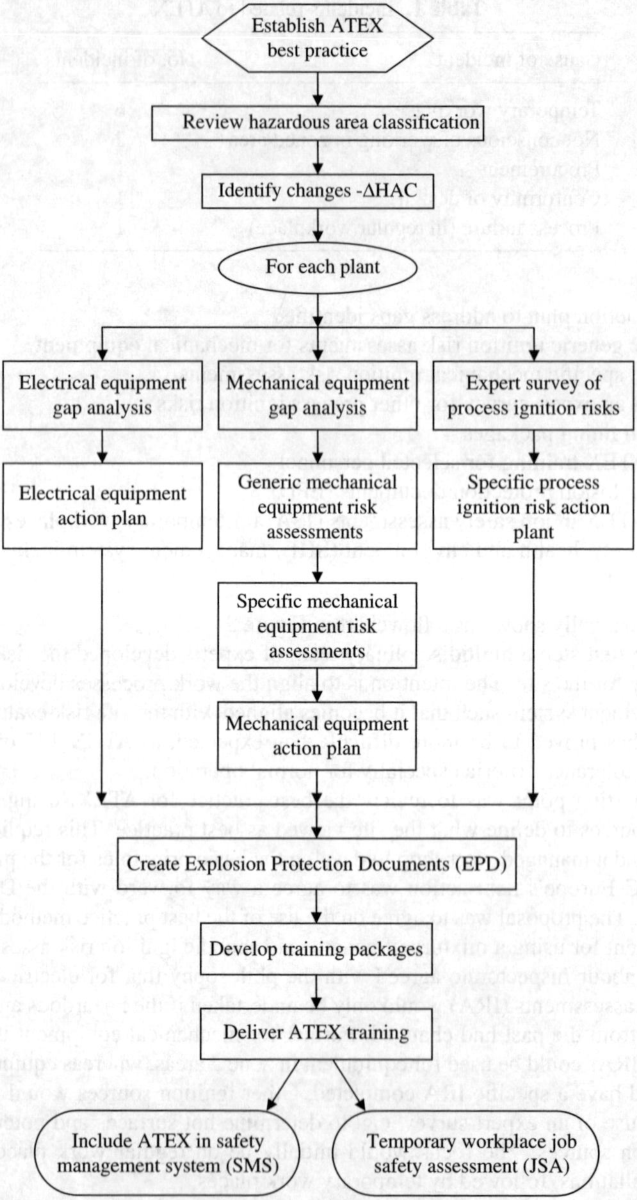
Establish ATEX best practice
Review hazardous area classification
Identify changes -ΔHAC
For each plant
Electrical equipment gap analysis
Mechanical equipment gap analysis
Expert survey of process ignition risks
Electrical equipment action plan
Generic mechanical equipment risk assessments
Specific process ignition risk action plant
Specific mechanical equipment risk assessments
Mechanical equipment action plan
Create Explosion Protection Documents (EPD)
Develop training packages
Deliver ATEX training
Include ATEX in safety management system (SMS)
Temporary workplace job safety assessment (JSA)

Figure 2. Flowchart for ATEX implementation

dust explosion hazards, vessel internals, ventilation and openings between buildings or into adjacent hazardous areas.

The exercise highlighted how limited the available information is on dust explosion risks, such as minimum ignition energy values, and lead to additional testing on certain materials handled to obtain the correct data for the explosion risk assessments.

The outcome of this changed hazardous area classification (ΔHAC) was a set of pre- and post-project HAC drawings for the site.

Vessels had typically not been internally zoned previously, so for each vessel it was necessary to verify whether there was any electrical or mechanical equipment internal to the vessel which should be subject to a risk assessment. For vessels which were inerted the principle of fault tolerance was used to determine the required reliability of the inerting system to ensure that the vessel remained inerted under all foreseeable process conditions and so was not subject to HAC zoning, in particular where there was a potential internal source of ignition.

The fault tolerance (FT) principle is based on the number of independent faults or system failures that need to occur before the potential ignition source becomes active in the unprotected situation. If an ignition source is not caused by a fault or failure but is inherently present during normal operation, or if the occurrence of the ignition source and the formation of the explosive atmosphere have a common cause, then the FT is −1. If a single fault already leads to an ignition source then the FT is 0. If two independent, simultaneous faults need to occur to give an ignition source then the FT is 1, and so on.

ATEX 137 does not specify acceptable risk tolerance criteria for use in the explosion risk assessments, but a target value for tolerable ignition risk can be derived from the requirements and standards under the ATEX 95 Directive. This Directive indicates that the sum of the fault tolerance (including all protection measures) and zone for the intended use of the equipment shall always have at least the value of two in order to achieve the requisite overall level of protection [Perbal et al., 2006]. To derive the required reliability of the inerting system for a vessel, the sum of the internal grade of release, the fault tolerance of the equipment within the vessel, and the reliability of the independent protection layer(s) (IPL) (expressed as a safety integrity level (SIL)) must equal at least two. An IPL is defined as a device, system or action that is capable of preventing the scenario from proceeding to its undesired consequence, independent of the initiating event or the action of any other layer of protection associated with the scenario. Hence, an IPL shall be effective, independent and auditable. This principle is expressed in Figure 3 and Table 2, and can be used generally for determining either the integrity of the safety instrumented system to be used for the process under consideration (PUC), or the required fault tolerance of the equipment under control (EUC).

In general the ΔHAC did not create too many changes to the existing zones, but due to the considerations of ventilation and openings some areas extended further than previously, in particular through openings or into adjacent hazardous areas. One example was where drive belts and motors for a compressor were located outside a building to separate them in the past from the zoned area. However, now the wall opening is treated as a potential source of release, creating a zone 2 area where the motor and drive belt system

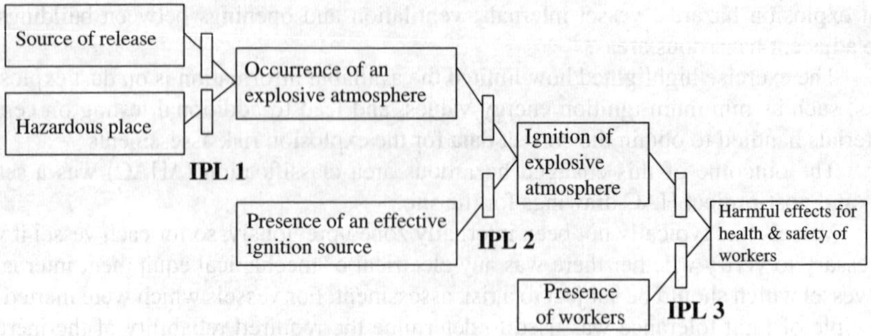

Figure 3. Explosion scenario with independent layers of protection

are located. Another example is within the skirt of a furnace, where the lack of ventilation now requires the area to be zoned as an enclosed space with no natural ventilation, with a risk of release of flammable gas. Again the existing equipment was not specified for the zone 2 now required. Hence, these situations needed a more close explosion risk assessment.

The detailed gap analysis for all changes found by the ΔHAC was split into electrical and mechanical reviews.

For electrical equipment a detailed inspection of areas where the zone had changed was completed. This included areas now zoned which had not previously been, and areas where the zone had increased, e.g. from zone 2 to zone 1. Further the gas group and temperature class ratings of equipment were checked generally, using a random auditing process across all zoned areas as it was not practical to check every piece of equipment. It was generally found that the selection of electrical equipment in the past was very

Table 2. Explosion risk assessment using fault tolerance

Explosion risk assessment	Probability/Frequency	Factor
Formation of explosive atmosphere (PUC[1])	Grade of release: 0/1/2	
IPL 1: preventing formation of explosive atmosphere	SIL: 1/2/3	
Fault Tolerance of equipment present (EUC[2])	FT: –1/0/1/2	
IPL 2: preventing ignition of explosive atmosphere	SIL: 1/2/3	
IPL 3: mitigating harmful effects of an explosion	SIL: 1/2/3	
Sum of factors	Target value ≥ 2	

[1]Process under control (PUC) is the intended operation of the process within the design parameters, in absence of any protection measure associated with the explosion scenario.

[2]Equipment under control (EUC) is the intended use of equipment within the design parameters, in absence of any protection measure associated with the explosion scenario.

good, and due to the previous philosophy of using zone 1 specified electrical equipment as standard, thus even previously less hazardous classified areas were found to be generally compliant.

As expected, some areas for improvement were found, however considering the huge numbers of items of electrical equipment this was not excessive. Only a number of pump motors across the whole site were found to be non compliant and need corrective action. Also cables for intrinsically safe systems were identified as a gap analysis issue, as the cable lengths in use were found to be generally acceptable for gas group IIB based on a typical maximum cable length, but needed further assessment in case of IIC.

An area of concern which was identified was that some installation and maintenance standards for electrical equipment were not effectively implemented with respect to ATEX, which needed to be improved by means of not only theoretical but also practical refresher training. All items of equipment will be checked through the regular maintenance and inspection programmes in which ATEX will be included. The ATEX 137 Directive does not specify requirements for maintenance and inspection, so these have been developed as part of the best practice review.

The mechanical equipment gap analysis was completed to identify what equipment was found in each zone, and hence to assess the extent of the work to complete relevant ignition risk assessments. It was estimated that 95% of mechanical equipment within zoned areas was in a zone 2. The types of mechanical equipment in each zone were listed to determine how to optimise the risk assessments. Most equipment in zone 2 areas could be classified into a generic type. A list of the equipment suitable for a generic ignition risk assessment was drawn up, and the assessments completed for each plant, based on a common standard. The generic ignition risk assessments are listed in Table 3.

For zones 2 generic ignition risk assessments were completed for mechanical equipment operating in gas group IIB and temperature class T3. Where the gas group or the temperature class was higher then a specific risk assessment was deemed necessary. A further 260 specific ignition risk assessments for the whole site are in the process of being completed for mechanical equipment mostly being present in zones 1.

An expert survey of the site revealed various problems with hot surfaces which could act as a potential source of ignition. Typically these were on the polymer plants where high temperature steam is used. Un-insulated high pressure steam pipework could easily have a surface temperature above the auto-ignition temperature (AIT) of the flammable materials in the vicinity. A general policy was implemented to require that for lines in zones 1 the piping is clearly labelled on the plant. In zones 1 these lines are not permitted to be de-insulated during operation. Lines in zones 0 where the process temperature may exceed 80% of the AIT are not allowed without temperature control.

Another identified risk area were oil reservoirs internally heated by steam coils. There is a risk of exposed steam coils igniting oil vapours, so low level switches have been installed to control the ignition risk. A further example is where gas venting via seal pots could release flammable gas. It was decided that such seal pots should have level controls to ensure that there is always a water seal.

Table 3. Generic mechanical ignition risk assessments

Ref.	Generic mechanical ignition risk assessments
01	Pump including stuffing box
02	Top entry mixers
03	Splined drive shaft couplings
04	Centrifugal fans
05	Chain transmission
06	Clutch with friction plates
07	General purpose gearbox with anti-friction bearings
08	Plunger and diaphragm pumps
09	Reciprocating compressors crosshead type with oil circuit
10	Rotary feeder
11	Oil flooded rotary screw compressors
12	Shaft couplings
13	Canned motor pumps
14	Magnetic driven centrifugal pumps
15	Bearings
16	Centrifugal compressors
17	Special purpose steam turbines
18	Dry running screw compressor and roots blowers
19	Hoists
20	Dry running rotary screw vacuum pump
21	Liquid ring type vacuum pump compressor with single double tandem seal and all metal coupling
22	Shaft coupling Eupex generic
23	Rotary positive displacement pumps
24	Side channel pumps
25	V-belt transmissions
26	Pelletizers
27	Catalyst pumps
28	Drum sieves
29	Screw conveyors

Temporary work place risk assessments have proved problematic to resolve adequately. The use of the fault tolerance principle was initially intended to be rolled out across the site, with the permit issuers using the method as part of the job safety assessment (JSA) for the planned work. However this has proved to be too complicated and the technicians and supervisors are not confident in using the technique. As a result the assessment of temporary workplaces in compliance with ATEX 137 is still done on a qualitative basis by means of JSA.

An example of how ATEX 137 has not been fully understood when preparing a JSA is the use of tenting for weather protection. The use of a tent changes the ventilation from good natural ventilation to an enclosed space with poor ventilation. This changes the hazardous area zone within the tent, which therefore changes the risk; but it has been found that this has not been identified through management of change by those involved in the job.

The training for ATEX 137 has been very successful within SABIC Europe, and has been a key focus of the Dutch Labour Inspectorate when auditing the site. The training requirements have been carefully identified for defined groups within the company, especially the ATEX Experts are actively involved in management of change (MOC) and pre-start-up safety requirements (PSSR). These people are identified as key personnel across all plant areas that will have a day to day involvement in ensuring compliance in ATEX implementation. They are the local plant experts who other plant personnel can ask for advice, and are key communicators regarding ATEX. Also the ATEX Expert has a role in ignition risk assessments and job safety assessments for temporary workplace to ensure ATEX compliance.

Linked with the ATEX Experts is the ATEX Committee, who makes overall policy and decisions for the site regarding ATEX implementation, and who have been approving the work done by the ATEX project team. The ATEX Committee's role is to ensure consistency of standards across the site and to approve the revised hazardous area classification assessments. Their role is also to advise the site management of their duties and liaise with the Dutch Labour Inspectorate, external committees and other companies to monitor developments.

The collation of explosion protection documents (EPD) for the site has mainly been completed by external contract personnel, due to the high workload involved, based on a template developed by the ATEX Committee. The need for the EPDs is not viewed positively, as collating the information into one document has not provided any tangible benefit or added value for the operations. The EPD is of use as a compliance auditing document only, and relies on the correct hyperlinks through the electronic record systems.

There has been a significant resource input to the ATEX project on the Geleen site by SABIC Europe. A large number of people have been involved; employees, consultants and contractors; and the project has been running for over two years, with the scope evolving over time. As the urgent or short term aspects are completed the longer term issues are being addressed, especially how to integrate everything robustly into the recently restructured SHE management system.

The benefits of the project have been the raising of the explosion safety standards of the site, as non-compliant risks have been identified and addressed. The training has been very beneficial with a high awareness across the teams, indicated by proactive questioning and feedback.

The Dutch Labour Inspectorate has started auditing on the implementation of ATEX 137 at major hazard sites during the second half of 2007, after following special ATEX training during the first half of the year. They are firstly focussing on roles, responsibilities and training of personnel as well as the actual hazardous area classifications, and have until now been satisfied with SABIC Europe's approach regarding ATEX 137 implementation as no major non-conformances were identified during the two audits completed to date.

The requirements for compliance with the ATEX 137 Directive in Great Britain were covered by the Dangerous Substances and Explosive Atmospheres Regulations (DSEAR), 2002. There were differences in compliance dates for some aspects of the Directive, but in general many of the issues identified in this paper are similar to those found by companies seeking to comply with DSEAR in the UK. Discussions with various contacts has indicated that meeting the compliance dates has often not been fully achieved, and some aspects of compliance are still ongoing within many companies, both in the UK and across Europe.

REFERENCE

Perbal, R., Fernie, L., 2006, Implementation of the ATEX Directive 99/92/EC and a practical methodology for explosion risk assessment in existing plants, 2nd International Conference on Safety & Environment in Process Industry, Naples.

PRACTICAL APPLICATION OF STATIC HAZARD ASSESSMENT FOR DSEAR COMPLIANCE†

Graeme R Ellis, Senior Safety Consultant
ABB Engineering Services, Daresbury Park, Daresbury, Warrington, Cheshire WA4 4BT, UK

KEYWORDS: Static hazard assessment, DSEAR, risk based

INTRODUCTION

This paper describes the methodology and practical experiences of carrying out a Static Hazard Assessment for Chemtura Manufacturing (UK) Ltd at the Trafford Park site, in compliance with the Dangerous Substance and Explosive Atmosphere Regulations 2002 (DSEAR).

Companies handling substances capable of creating explosive atmospheres are required to carry out a formal risk assessment [Ref 1]. This must consider the extent of foreseeable explosive atmospheres within and external to the process, and ensure that suitable equipment is installed to control all potential ignition sources.

Within the Process Industry there has been a focus on classifying hazardous areas caused by leaks of flammable substances from the process and assessing the electrical and mechanical ignition sources within these areas. Less attention has been given to potential explosive atmospheres within equipment and the need to apply measures to control all ignition sources including electrostatic discharges.

Many substances handled in the Process Industry can create explosive atmospheres in air due to flammable vapours or combustible dusts. Low conductivity liquids and highly resistive powders are capable of generating electrostatic charges during processing resulting in the potential for a static discharge with sufficient energy to cause ignition. It is thought that fires and explosions caused by static discharges occur weekly in the UK.

The latest standard for the control of static hazards [Ref 2] provides detailed guidance on the control measures that need to be applied. Many companies are unclear about the level of risk posed by static hazards and tend to apply a range of generic control measures without assessing their criticality or reliability.

A risk based methodology was developed for the Static Hazard Assessment described in this paper. The initial step involved collection of physical property data on the conductivity of liquids and resistivity of powders, with low hazard substances screened from further assessment. The operations involving medium and high static risk substances were then reviewed to identify credible explosive atmospheres within equipment and possible charging mechanisms.

A risk grid approach was used to determine the required reliability of the 'anti-static' control measures with a Layer of Protection Analysis (LOPA) for high risk cases. For events where the 'anti-static' measures were judged 'critical', a review was carried out to ensure full compliance with relevant standards, and follow the more onerous design options where applicable.

BACKGROUND

The Trafford Park site first started operations as 'The Geigy Colour Company Limited' on Christmas Eve 1939. In the last 30 years the Site has been owned by Ciba Geigy and then as FMC before being bought by Great Lakes Chemical Corporation in 1999. In July 2005 Crompton Corporation merged with Great Lakes to form Chemtura Corporation. A range of speciality chemicals including phosphorus flame retardants and fluids as well as industrial water treatment additives are manufactured on the site in a number of batch processing plants.

The Chemtura site is regulated by the Health and Safety Executive (HSE) and Environmental Agency (EA) under the Control of Major Accident Hazard (COMAH) regulations as a 'top tier' site. As required by the COMAH regulations, a safety report was submitted to the HSE in 2000 and updated in 2005. A general risk assessment was used to identify major accident hazards, some of which were related to explosive atmospheres and static ignition. A number of actions were raised during the COMAH risk assessment to provide further demonstrations that relevant good practice is being followed to control static hazards.

It was decided to carry out a structured review of static related hazards covering all operations on site, in compliance with DSEAR and demonstrating relevant good practice. The scope of this assessment needed to include dust explosion hazards that were excluded from the COMAH risk assessment as dusts are not classified as 'dangerous substances' under this regulation.

This work would complete the various activities being carried out on the site for compliance with DSEAR, including re-assessments of hazardous areas on the site to ensure that suitable electrical and mechanical equipment is installed in zoned areas.

REQUIREMENTS OF STANDARDS

The Dangerous Substance and Explosive Atmosphere Regulations 2002 (DSEAR) are intended to protect people from the harmful physical effects caused by dangerous substances with the potential to cause fires and explosions, including flammable vapours, combustible dusts and reactive materials. This includes the creation of an 'explosive atmosphere' of the substance in air and ignition leading to harmful effects from thermal burns, blast overpressure or oxygen depletion. A risk assessment is required to identify fire and explosion hazards during processing and to agree appropriate controls to eliminate the hazard or if this is not possible then to reduce the risk as far as reasonably practicable.

Further guidance is provided on control and mitigation measures [Ref 3]. Hazardous area classification should be used to identify where explosive atmospheres may form, specifying zones dependent on the frequency and duration of the explosive atmosphere. This must include areas within processing equipment where the potential for an explosive atmosphere exists, in addition to external areas where leaks can lead to explosive atmospheres outside of the process.

The risk assessment must identify all possible ignition sources, assess whether these could cause a fire or explosion causing harmful effects, and introduce control measures to prevent the ignition sources occurring. Ignition sources include heat energy, electrical energy (including discharges of static electricity), mechanical energy and chemical energy.

Reference is made to the relevant British Standard [Ref 4] for advice on the control of electrostatic discharges. This code of practice provides information on the generation of static electricity in solids, liquids, gases and on persons, with recommendations for suitable control measures.

STATIC HAZARD ASSESSMENT METHOD

A structured methodology was developed for the assessment to meet the requirements of the standards and building on existing risk assessment techniques in use on the site. The assessment involved a Process Safety consultant from ABB and a team of experienced technical and operations staff from Chemtura.

The methodology involved a 7-step process as described below.

STEP 1 – DATA ON DANGEROUS SUBSTANCES

Relevant data was initially gathered on all flammable and combustible substances handled on the site with the potential to cause an explosive atmosphere in plant equipment or generate a significant electrostatic charge. This data included flash point and electrical conductivity for liquids, and minimum explosion concentration, minimum ignition energy and volume resistivity for dusts and powders. Where available for raw materials this data was obtained from published sources. For some intermediate materials, such as recovered xylene, tests were carried out by Chemtura on samples taken from the plant.

STEP 2 – IDENTIFY EXPLOSIVE ATMOSPHERES

All activities on the Trafford Park site involving flammable liquids and combustible dusts were assessed to identify the potential for explosive atmospheres within equipment. This was based on major accident scenarios from the COMAH Safety Report with an additional assessment of the powder handling operations.

The assessment considered the potential for explosive atmospheres either during normal processing or foreseeable upset conditions such as failure of nitrogen blanketing systems and/or heating of substances above their flash point due to temperature control failure.

STEP 3 – ASSESS CONSEQUENCES

For all credible explosive atmospheres identified in step 2, the consequences of ignition were assessed. Scenarios were screened from further assessment if the explosive atmosphere is judged to be too small to cause significant harmful effects or if the explosion is contained within the process. The latter case involves consideration of the maximum credible explosion pressure in comparison with the equipment design and test pressure rating.

Scenarios involving powders were screened from further assessment if the minimum ignition energy for the dust is >10J. Above this limit no measures are required to avoid static hazards [Ref 2], due to there being no credible operations in the Process Industry that can generate electrostatic discharges with such high energy levels. This screening was not permitted if other materials that could form an explosive atmosphere were present in addition to the powder, such as a flammable vapour.

In cases where an explosion could cause significant harm to people, the severity of the event was assessed using the following word models based on the Chemtura risk matrix, as used for COMAH risk assessments.

- Level 1 Catastrophic – multiple fatalities on-site
- Level 2 Disastrous – fatality or multiple major injuries on-site
- Level 3 Serious – major injury or multiple severe injuries on-site
- Level 4 Significant – Reportable accident/LTA
- Level 5 Minor – first aid cases.

STEP 4 – IDENTIFY CHARGING MECHANISMS

Where the ignition of an explosive atmosphere can cause significant harm, potential charging mechanisms during processing activities were identified that have the potential to lead to a static discharge. Consideration was given to normal charging mechanisms for liquids such as flow in pipes, splash filling into tanks or stirring in vessels. For powders the mechanisms included charging from bags to vessels and pneumatic conveying. Consideration was also given to the potential for static discharges caused by charged operating personnel approaching an explosive atmosphere.

Scenarios were screened from further assessment on the basis that the charging mechanisms involve low static risk liquids with conductivity >1,000 pS/m or low static risk powders with resistivity <10^6 ohm/m [Ref 2], if processed within well earthed and conductive equipment. It is very unlikely for liquids or powders outside these limits to generate significant charges during normal processing activities found in the Process Industry.

STEP 5 – ASSESS CRITICALITY OF 'ANTI-STATIC' CONTROL MEASURES

A criticality assessment was carried out for 'anti-static' control measures to determine those judged to be 'critical'. This used a risk based methodology similar to the method used on the site for Safety Integrity Level (SIL) determinations for safety instrumented systems. The method considers a number of factors to determine if the 'anti-static' control measures need to be 'Standard' or 'Critical', with the latter requiring the highest standard of control measures.

The first stage involves assess the likelihood for an explosive atmosphere within equipment taking account of whether an explosive atmosphere is a normal occurrence or whether one or more failures need to occur before an explosive atmosphere can be formed. From this assessment the following hazardous area zones are determined:

- Zone 0 – explosive atmosphere continuously, for long periods or frequently, normally assumed >1,000 hr/yr, probability assumed =<1.0
- Zone 1 – explosive atmosphere in normal operation occasionally, normally assumed 10–1,000 hr/yr, probability assumed <0.1
- Zone 2 – explosive atmosphere not likely in normal operation and only for a short period, normally assumed <10 hr/yr, probability assumed <0.001

An occupancy factor is selected as F_B (100%) for heavily occupied areas or where the operator causes the explosion to occur, or F_A (10%) for areas that are not usually occupied.

An avoidance factor is selected as P_B (100%) where avoidance is not likely or P_A (10%) to account for avoidance factors such as explosion relief, explosion suppression or where it is judged unlikely for the operator to suffer the stated level of harm.

Using these factors and the consequence level in step 3, the criticality of 'anti-static' control measures is determined from the risk graph in Table 1. This risk graph has been calibrated to achieve a risk below the target of below 3×10^{-5} per year for operator fatality. The static ignition frequency used in the calibration was estimated as 3/yr for 'Standard' measures and 0.03/yr for 'Critical' measures. These values are considered to be conservative estimates although there is a lack of published sources for ignition frequencies.

The risk graph determines 'anti-static' control measures as at one of the following levels; S (Standard) or C (Critical).

STEP 6 – LAYER OF PROTECTION ANALYSIS

For cases where the risk graph gives a Q (Quantify) result, a more detailed risk assessment technique is required. In this case it involved a Layer of Protection Analysis (LOPA) as used by Chemtura for SIL determination for high hazard safety instrumented systems and

Table 1. Risk graph for determination of criticality for 'anti-static measures'

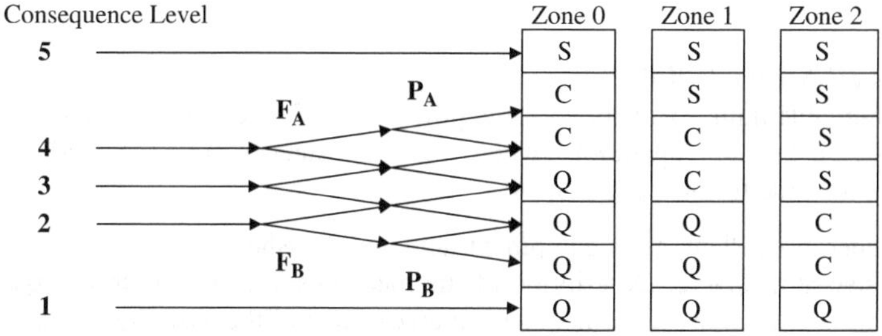

for aspects of the COMAH Safety Report. The same approach was used for this study based on an Excel spreadsheet.

STEP 7 – VERIFICATION OF 'ANTI-STATIC' CONTROL MEASURES

All 'anti-static' control measures in place at Trafford Park for the events in step 4 were reviewed against relevant good practice [Ref 2]. If the control measures were assessed as 'Critical', a proportionately more in-depth review was completed looking for compliance with the most stringent requirements from the standards. For example, where earth cables need to be attached to mobile equipment, there should be an earth proving interlock to prevent the charging mechanism.

The verification process involved checking the system design on the P&ID's, discussions with Chemtura staff and a site audit to inspect the equipment systems. This stage of the assessment resulted in a number of recommendations for improvements to the 'anti-static' control measures.

PRACTICAL EXPERIENCE

PHYSICAL PROPERTIES DATA

Gathering conductivity and resistivity data on the wide range of liquids and powders handled on the site was made difficult by the limited number of sources. This type of information is generally unavailable on Material Safety Data Sheets (MSDS) but can be found for common substances in specialist publications [Ref 5]. In some cases it was necessary to take samples of substances from the process for analysis, especially for process intermediates. An example was waste xylene that had previously been assumed as high conductivity and low risk due to the presence of impurities including 2–3% acetone. Analysis of samples showed the conductivity for waste xylene to be typically around 400 pS/m, and therefore requiring further assessment as a medium risk substance.

Of nineteen flammable liquids handled on the site, ten were screened out with high conductivity and therefore low static risk, with the highest risks posed by an organic peroxide, xylene, diesel and therminol heat transfer fluid.

All three of the organic powders handled on the site were known to be high resistivity from previous specialist analysis. One of these materials was screened from further assessment as its' minimum ignition energy was above 500J.

EXPLOSIVE ATMOSPHERES

For all flammable liquids and combustible powders handled on the site an assessment was carried out on the unit operations where these substances are processed, and a table completed with the following headings:

- Unit Operation: Plant, main equipment item and flowsheet reference
- Explosive atmosphere: Description of the mechanism by which an explosive atmosphere may form in the equipment and an assessment of the hazardous area as zone 0, 1

or 2. This made reference to the flash point for vapours, inert blanket systems on the equipment, heating systems and any trips or alarms. For powder handling operations it also considered the potential for dust clouds above the LEL to be formed. In some cases involving vapours it was concluded that an 'explosive atmosphere' was not credible and the unit operation was screened from further assessment.

- Consequence of Ignition: The consequences of ignition of the explosive atmosphere in terms of harmful effects to people were assessed and the worst credible severity rating set as level 1 to 5. This assessment considered the maximum explosion pressure and the likelihood for vessel rupture, plus the scale of the credible explosion effects. Reference was made to consequence assessment of scenarios in the COMAH risk assessment to ensure consistency. Some events were screened from further assessment due to the explosion being contained in the equipment or of such a small scale that significant harm was not credible.
- Charging Mechanism: The credible mechanisms for static charge generation were identified for each unit operation, such as pumped offloading into a storage tank, powder flow from a 'big bag' during emptying, or charged operators approaching an explosive atmosphere. These mechanisms were screened out as not credible for substances with low static risk, i.e. high conductivity liquids or low resistivity powders processed in well earthed and conductive equipment.
- Criticality: For credible charging mechanisms, the criticality of the control measures was assessed as 'Standard' or 'Critical' using the risk graph or LOPA method described earlier. This column gives the results of the assessment and any assumptions made for occupancy and avoidance factors.

ANALYSIS OF RESULTS

Thirty four unit operations were assessed in total with five screened out as no credible mechanism for an explosive atmosphere. One event was screened out for minor consequences due to the small volume of explosive dust/air present. Sixteen of the events were screened out for no charging mechanism, mostly due to the low static risk substances involved. Nine events were assessed using the risk graph approach with seven rated 'Standard' and two rated 'Critical'. The remaining three events were rated high risk on the risk graph method and were assessed using LOPA, all resulting in a rating as 'Standard'.

VERIFICATION OF CONTROL MEASURES

The twelve events with static control measures rated as 'Standard' or 'Critical' were verified against the requirements of the standards, and a number of improvement recommendations made, with examples as follows:

- Provide 'anti-static footwear for operators and re-instate equipment and procedure for checking operator earth resistance at start of shift.
- Earth proving system at peroxide drum offloading to be interlocked with offloading pump.

- Provide nitrogen inerting system on peroxide and xylene storage tanks with low pressure trip of transfer pump.
- Check that air diaphragm pump cannot deliver at velocity above 7 m/s even with air regulator adjusted in error.
- Confirm routine inspection and testing of fixed earthing on equipment to achieve earth resistance below 10 ohm.
- Confirm that powder is supplied in 'type C' big bags with provision for attaching an earthing clamp for interlocked earth proving system.
- Provide 'anti-static' flexible hoses on dust extraction system to ensure earth continuity of system.
- Provide earth clamp system for solids charge chute to reactor.
- Provide dip-pipe on head tank filling line to prevent splash filling.

EXAMPLE 1: WASTE XYLENE ROAD TANKER LOADING

Waste solvent is top loaded from a storage tank via a pump to a road tanker barrel. The flash point of waste xylene is at or above 21 degC, meaning that an explosive atmosphere could occur regularly in the tanker. It is not feasible to nitrogen blanket the tanker prior to loading. The atmosphere inside the tanker is therefore assessed as zone 0, as it will regularly contain an explosive xylene/air mixture.

An explosion in the tanker would cause the manlid to be blown open (this is loose during tanker loading) causing a flash fire from the opening and possible missile hazards. This was assessed as likely to cause a serious injury to a person in the immediate vicinity of the road tanker, therefore consequence level 4. It was assumed that occupancy in the area is 100% as the tanker driver stays during loading, and the avoidance of a serious injury was judged as 10% based on the position of the driver and the flash fire expected to be directed vertically from the opening. Using the risk graph approach with consequence level 4, zone 0, occupancy 100% and avoidance 10%, the 'anti-static' control measures were rated as 'Critical'.

Electrostatic charge could be generated by flow through the 3″ charging line but the flowrate of 12 m^3 per hour limits the line velocity to 0.7 m/s. This is well within the limit for single phase flow from the standard [Ref 2] of 7 m/s, and below the limit of 1 m/s for 2-phase flow. Charge could also be generated by splash filling of the road tanker from the top entry fill line as there is no dip-pipe on the tanker. A recommendation was raised for modifications to the filling line to prevent splash filling. As part of the tanker loading procedure the operator connects an earth clip to the tanker to ensure that charge cannot accumulate on the tanker. It is possible for the clip to be left off in error. It was recommended that an earth proving unit is provided with an automatic interlock to the loading pump to prevent loading if the earth clip is not effectively connected.

EXAMPLE 2: DUST FILTER EXPLOSION

A dust filter handling explosive dust was judged to be capable of causing a major injury to operators in the area if an explosion occurred. Using the risk graph method the 'anti-static' measures were assessed as level Q, therefore requiring a more detailed quantified assessment.

Using the LOPA method, the frequency of static ignitions capable of causing an ignition were initially estimated as 3/yr, based on 'Standard' control measures. The atmosphere within the filter was classified as zone 1 due to the potential for frequent explosive atmospheres. These figures were considered very conservative as they suggest an explosion would occur 3 times per year. In practice there have been no explosions in the filter over several years of operation. The risk of major injury is then reduced by 0.01 for the explosion relief panel on the filter, by 0.1 for the low occupancy in the area, and by 0.1 for the chance of suffering a major injury. The LOPA assessment concludes that the 'anti-static' control measures should be at the 'Standard' level.

CONCLUSIONS

This paper has described a Static Hazard Assessment carried out on the Chemtura COMAH 'top tier' site at Trafford Park, in accordance with the requirements of DSEAR. A structured methodology was developed that could be applied to other sites that handle a range of flammable or combustible materials. The approach was to screen out low risk scenarios and focus attention on cases where the risk of static ignition causing an explosion is judged to be significant. Cases were screened out when there was no potential for an explosive atmosphere, no significant harmful effects from an explosion or a low risk of electrostatic discharge due to the high conductivity of the substance.

For cases where the static hazard was judged to present a significant risk, a risk graph approach was used to assess the criticality of the 'anti-static' control measures. This identified situations where the control of static is judged to be 'Critical', and the most stringent measures need to be applied. This structured approach provides a justification for taking extra measures for some situations whilst applying generic measures in others. A number of specific improvements were identified during the assessment that provide a targeted reduction in static risks and a demonstration that the risk of static related major accidents has been reduced to 'as low as reasonably practicable'.

The limited number of static hazards assessed as requiring 'Critical' control measures reflects the good design principles that have been applied at the Trafford Park site. The general design approach has been to avoid explosive atmospheres in equipment by operating below the material flash point or using a nitrogen inert atmosphere where this is not possible. For situations where an explosive atmosphere cannot be prevented, mitigation measures have normally been provided in the form of explosion relief systems to reduce the effects of an explosion.

REFERENCES

1. HSE, L138, 2003, Dangerous Substances and Explosive Atmospheres Regulations 2002, Approved Code of Practice and Guidance, HSE Books

2. BSI, 2003, Electrostatics – Code of practice for the avoidance of hazards due to static electricity, PD CLC/TR 50404:2003, British Standards Institute
3. HSE, L136, 2003, Control and Mitigation Measures, Approved Code of Practice and Guidance, HSE Books
4. BSI, 1991, Code of practice for control of undesirable static electricity – Parts 1 & 2, BS5958-1 & BS5958-2:1991, British Standards Institute
5. Britton, L., 1999, Avoiding static ignition hazards in chemical operations, CCPS, American Institute of Chemical Engineers

EXPERIENCES WITH THE APPLICATION OF IEC 61511 IN THE PROCESS INDUSTRY IN AUSTRIA

R. Preiss, K. Findenig, and M. Doktor
TÜV Austria

Within this paper different approaches of handling the requirements of IEC 61511 (starting from risk analysis up to the Safety Integrity Level (SIL) verification) are discussed.

If pressure equipment is protected against exceeding the design limits (pressure, temperature) by means of a Safety Instrumented Function (SIF), this SIF has to undergo a conformity assessment procedure according to the PED. The corresponding requirements of TÜV Austria, acting as a notified body for pressure equipment, on the risk analysis resulting in SIL classification of such SIF are given. Due to a lack of acceptance of semi-quantitative methods in risk analysis of some Austrian authorities, the risk graph as given in IEC 61511-3 Annex E is sometimes required, whereby other approaches seem to be much more comprehensible.

Regarding explosion protection issues, an overview of the SIF requirements for ignition protection of non-electrical equipment and safety devices used for the control of ignition hazards is given. Furthermore, a concept for handling primary explosion protection in SIL classification is proposed.

Many vendors of equipment used in SIF do not have SIL certificates, PFD/PFH-values, SFF values, etc. available since often the components are unique, specially designed or purely mechanical devices. Thus, alternatives proving an equivalent level of safety of a SIF, or using B10-values for pneumatic and electro-mechanical components may be used in such cases.

Finally some practical recommendations on SIF component selection and signal transmission are presented.

1. INTRODUCTION

EN 61511 [1, 2] which defines the requirements on functional safety of safety instrumented systems for the process industry sector, has been in force several years. The big difference to previous standards, like DIN V 19250 [4], is that some requirements are now given on a probabilistic basis, i.e. the probability of failure on demand of a safety system has to be determined and compared with allowable values. The standard also allows different approaches for the determination of the SIL of a safety function, e.g. risk graphs and Layer-Of-Protection-Analysis (LOPA), but does not specify mandatory requirements on the calibration of these methodologies.

Besides others, these two facts concerning determination and validation of safety instrumented functions lead to a lot of uncertainties and different approaches on how to handle the standard's requirements in practise. In the following some of them are discussed and corresponding recommendation given.

2. SIL DETERMINATION

2.1. RISK ANALYSIS – DIFFERENT METHODOLOGIES TO DETERMINE A SIL

2.1.1. Risk graphs

Usage of risk graphs for SIL determination is the common approach for Austrian facilities, especially in cases which require an approval by the authorities. The main reasons are that this approach is well-known from the application of the old standard DIN V 19250 [4] – the equivalent risk graph is now given in EN 61511-3 [2], Annex E – and that the quantitative or semi-quantitative approach as it is used for the LOPA approach [5] is not in line with the common safety culture in Austria. This is mainly based on the fact that no official numbers for risk tolerance criteria, as for example can be concluded from the HSE document R2P2 [6], exist and thus the authorities resist to accept such numbers. Moreover, some local authorities do not accept use of the semi-quantitative risk graph given in EN 61511-3 [2], Annex D and the qualitative one according to Annex E has to be used.

TÜV Austria, which acts as a consultant in hazard and risk analysis as well as a notified body according to the Pressure Equipment Directive [7] generally supports risk graph approaches as well as LOPA for SIL determination. If pressure equipment is protected against exceeding the design limits (pressure, temperature) by means of a SIF, this SIF is considered to be a safety accessory in the sense of the PED. Thus, a conformity assessment procedure for this SIF is required, and usually this is done within the conformity assessment procedure of the overall assembly. Such an assembly can be a rather simple system, e.g. a pressure vessel with a safety valve against overpressure due to thermal expansion and a SIF acting as a level-high-shutdown-system against overpressure due to a liquid inlet flow, or it can be a very complex system as a whole plant within a refinery. Since the effectiveness and the adequateness of the SIF is confirmed by a successful conformity assessment procedure, the risk analysis and the procedure for SIL determination, including calibration of the risk graph, has to be acceptable for the notified body. If the quantitative risk graph is used the calibrations shown in Tables 1 and 2 are the common ones for (personnel) safety and environmental issues, i.e. a calibration leading to a higher tolerated risk is not accepted.

In the safety SIL table the occupancy is defined as "FA" for normally unmanned operation of the relevant part of the plant (exposure is less than 10%) and "FB" when the relevant part of the plant is attended locally on regular basis with more than 10% exposure time or during the specific time on demand (e.g. start-up or shutdown of equipment) or the relevant part is located near a continuously occupied road. The occupancy "FA" shall not be used in cases where the consequence is given by public fatalities.

The possibility of a person (operator) avoiding the hazard ("PA") is defined as follows (all conditions have to be fulfilled):

- the operator is alerted by an independent device;
- the operator has independent facilities to avoid the hazard or which enables all persons to escape the hazard;
- the time between the operator being alerted and a hazardous event occurring is definitely sufficient for the necessary actions.

Table 1. Saftey SIL classification (risk graph Annex D)

Safety SIL table									
							Demand rate [1/yr]		
							W3	W2	W1
Consequence		Occupancy		Exposure avoidance			$W \geq 1$	$1 > W \geq 0.1$	$0.1 > W \geq 0.01$
C_A	Minor injury						a	—	—
C_B	Severe injury	FA	<10%	PA	Yes		1	a	—
				PB	No		2	1	a
		FB	>10%	PA	Yes		2	1	a
				PB	No		3	2	1
C_C	Up to two fatalities (workforce)	FA	<10%	PA	Yes		2	1	a
				PB	No		3	2	1
		FB	>10%	PA	Yes		3	2	1
				PB	No		4	3	2
C_D	More than two fatalities (workforce) or fatalities in public	FA*)	<10%	PA	Yes		3	2	1
				PB	No		4	3	2
		FB	>10%	PA	Yes		4	3	2
				PB	No		b	4	3

*) Not to be used in case of risk of fatalities in public

Nevertheless, some company internal requirements use a more stringent calibration of the risk graph, which is – of course – acceptable for us when acting as a notified body.

If the qualitative risk graph according to EN 61511-3, Annex E, is used it is recommended that limiting values are also used for calibration, especially for the W parameter. Otherwise, the SIL classification can lead to very subjective results, depending on the interpretation of the terms "very slight probability" (W1), "slight probability" (W2) and "relatively high probability" (W3).

Nevertheless, the different consequence calibration of the risk graphs according to EN 61511-3 Annex D and Annex E, as suggested in the standard, may lead to considerable different SIL requirements. The consequence calibration as used in Table 1 is, generally, located in between.

2.1.2. Layer of protection analysis (LOPA)

For the LOPA methodology as described in [2, 5] the risk tolerance criteria for personnel safety given in Table 3 can be used and is considered to be rather conservative. The basis for these numbers, especially concerning the one for a single fatality, are several company internal standards of big oil companies and the HSE document R2P2 [6]: the maximum tolerable risk (in units of probability of death per year for an individual) for workforce

Table 2. Environmental SIL classification (risk graph Annex D)

Enviroment SIL table							
					Demand rate [1/yr]		
					W3	W2	W1
Consequence		Exposure avoidance			$W \geq 1$	$1 > W \geq 0.1$	$0.1 > W \geq 0.01$
C_0	No release or negligible damage to environment				a	—	—
C_A	Minor damage to environment, to be reported to authorities	PA	Yes		1	a	—
		PB	No		2	1	a
C_B	Release inside fence with environmental damage	PA	Yes		2	1	a
		PB	No		3	2	1
C_C	Release outside fence, temporary (short/medium term)	PA	Yes		3	2	1
	environmental damage outside fence	PB	No		4	3	2
C_D	Release outside fence, permanent (or long term) environmental damage outside fence	PA	Yes		4	3	2
		PB	No		b	4	3

from all scenarios is given by 10^{-3}, the negligible risk for workforce from all scenarios is given by 10^{-6}. Since it is most useful for LOPA to use a criteria for any one scenario affecting an individual, the above number of the maximum tolerable risk for workforce from all scenarios is multiplied by 10^{-1} up to 10^{-2} (depending on the number of potential risks in a plant and the degree of the applied conservatism in the risk analysis). Hence, the final maximum tolerable risk for workforce in one LOPA Scenario is given by 10^{-4} to 10^{-5}, as the more conservative value is shown in Table 3.

A possible LOPA procedure which differs from the one described in [5] is such, that the consequence is not expressed in terms of injury or fatality, but in terms of loss of containment (LOC). The LOC category depends on the media released, the size of release, and the location of release. The LOPA goal is given as risk tolerance criteria for the release frequency of such LOC categories. An example for this LOPA procedure is shown in Tables 4 and 5.

Table 3. LOPA risk tolerance criteria

Consequence	Minor injury	Severe injury	Single fatality (workforce)	Multiple fatality (workforce)
Risk tolerance criteria [1/yr]	10^{-3}	10^{-4}	10^{-5}	10^{-6}

Table 4. LOPA categories for LPG release (example)

Media LPG			
LOC category	Size of release	Type of release	Location of release
1	Large	Catastrophic failure of a vessel (storage tank, reactor, etc.)	—
2	Medium	Catastrophic failure of smaller equipment Leakage point or opening of atmospheric safety valve	In process installations, near buildings, near public areas
3		Catastrophic failure of smaller equipment Leakage point or opening of atmospheric safety valve	Not in process installations, nor near buildings, public areas
4	Small	Limited release in quantity and time (thermal expansion valve, sample point, etc.)	—

Usage of this methodology may avoid considerable differences in risk analyses performed by different teams for similar installations, since parameters like occupancy, avoiding the exposure, number of people injured, number of fatalities, etc. are not used anymore. But, on the other hand this methodology may lead to less conservative results compared to the conventional LOPA or risk graph approaches in cases with possible severe injury/fatality due to a small release of substance, or cases with high occupancy.

A principle generally valid for all methodologies of risk analysis and SIL determination is that measures against a hazardous scenario, based upon action of the BPCS can only form one single layer of protection, independent from the real numbers of such measures. This is in line with the statement given in EN 61511-1 [1], subsection 9.4.2, that the risk reduction factor for a BPCS (which does not conform to EN 61511 or EN 61508) used as a protection layer shall be below 10. Otherwise it would be possible to install several measures via the BPCS and claim an overall risk reduction higher than 10.

2.2. SIL DETERMINATION FOR OVERPRESSURE SCENARIOS

Regarding overpressure of pressure equipment the required measures/safeguards against a scenario are often discussed in the context of the extent of the overpressure. Frequently

Table 5. LOPA LOC risk tolerance criteria (example)

LOC category	1	2	3	4
Risk tolerance criteria (LOC frequency 1/yr)	10^{-5}	10^{-4}	10^{-3}	10^{-2}

Table 6. LOC of pressure equipment depending on the overpressure and present damage mechanisms

LOC	Corrosion and other damage mechanisms might be present	It is guaranteed that corrosion and other damage mechanisms are not present
No LOC	P ≤ PS	PS < P ≤ 1.1 PS
Small LOC due to leakages possible	Not applicable	1.1 PS < P ≤ PT
LOC and rupture/fragmentation of the containment	P > PS	P > PT

it is claimed that an overpressure below the proof test pressure of the equipment would not lead to rupture and thus, would not lead to loss of containment or fragmentation of the containment. A possible way how to handle such situations is given in Table 6 (PS denotes the maximum allowable pressure, PT the proof test pressure, and P the overpressure under consideration).

The further consequences of LOC depend – of course – on the media released (liquid, boiling liquid, gas, flammability, toxicity, formation of explosive mixtures, etc.).

Company-internal HSE standards covering this topic may be more or less conservative: every pressure above the design pressure is considered to lead to LOC, or a pressure below the bursting pressure (which is often assumed to be twice the design pressure) is considered not to lead to containment fragmentation. Of course, the latter statement is not used for fast pressure built-up.

A principle generally applied shall be that if a SIF is used as the only overpressure protection device, this SIF must at least hold a SIL of 1, independent of the possible harmlessness of the resulting consequence and the considerations given in Table 6. Hence, it follows from the requirements of EN 61511-1 that such a SIF cannot be implemented in the BPCS. This is also in line with the requirements given in [15].

Another principle with respect to overpressure scenarios which result in SIL 3 or SIL 4 requirement is the following: if a safety valve is present and it can be proved that its design, maintenance and testing, as well as the service conditions assure proper operation, a risk reduction equivalent to SIL 2 can be claimed for it. Thus, the additional required risk reduction equivalent to SIL 1 or SIL 2 has to be performed by a SIF (and this may be more rigorous as required from authorities).

2.3. SIL IN THE FIELD OF EXPLOSION PROTECTION

2.3.1. Non-electrical equipment

In case of ignition protection of non-electrical equipment for potentially explosive atmospheres (in context of the ATEX 94/9 directive [11]) by means of ignition control the relevant standard EN 13463-6 [8] uses the term "IPL" (ignition prevention level) in the levels 1

Table 7. IPL requirements for ignition protection of non-electrical equipment group II

Risk of ignition	Category 3 equipment (to be used in zone 2, 22)	Category 2 equipment (to be used in zone 1, 21)	Category 1 equipment (to be used in zone 0, 20)
Normal operation	IPL 1	IPL 2	Not applicable
Operation including malfunctions normally to be taken into account	Not relevant	IPL 1	IPL 2
Operation including rare malfunctions	Not relevant	Not relevant	IPL 1

and 2 (see Table 7). In EN 13463-6 the IPLs are related to categories according to EN 954-1 (safety of machinery – safety related parts of control systems) [9]. TÜV Austria performed a comparison of the categories according to EN 954-1 with the SIL according to EN 61508 / 61511 in the context of an equivalent level of safety, the result is given in the equivalence Table 8.

A problem that often arises in practise is, that manufacturer of non-electrical equipment, e.g. pumps, state in the operating manual, that e.g. dry-running and running-against-closed-valve is not permitted, but no information concerning the required level of ignition protection is described. Thus, it is within the responsibility of the user to ensure a safe use of the equipment via an adequate safety function.

2.3.2. Safety devices used for the control of ignition hazards

Safety devices required for the safe functioning of equipment with respect to explosion risks (in context of the ATEX 94/9 directive [11]) are covered by the standard prEN 50495 [10], which uses the term SIL. Examples of such devices are level detectors for the control of submersible pumps and gas detectors when used to control potential ignition sources.

Within this standard, the requirements given for equipment group II are summarised in Table 9.

Table 8. Comparison of IPL acc. to EN 13463-6, category acc. to EN 954-1 and SIL acc. to EN 61508/6151

IPL acc. to EN 13463-6	Category acc. to EN 954-1	SIL according to EN 61508/61511
—	B	—
—	1	1
IPL 1	2	1
IPL 2	3	2
—	4	3

Table 9. SIL and fault tolerance requirements for safety devices controlling potential ignition sources in equipment group II

Hazardous areas	Zone 0 Zone 20			Zone 1 Zone 21			Zone 2 Zone 22	
Fault tolerance of EUC	2	1	0	1	0	−1[a]	0[a]	−1[a]
Fault tolerance of safety device	—	0[b]	1[b]	—	0	1	—	0
SIL of the safety device	—	SIL1	SIL2		SIL1	SIL2	—	SIL1
Category of the combined equipment	1			2			3	

[a]"−1" indicates that the EUC (equipment under control) performs a source of ignition in normal mode of operation (e.g. general purpose equipment)
"0" indicates that the EUC is assessed as safe in normal operation in zone 2/22 without a safety device
[b]"0" indicates that one single fault may cause the safety device to fail
"1" indicates that two faults may cause the safety device to fail, the safety device is seen as redundant

HFT requirements (according to EN 61508/EN 61511) of the SIF are given by the "Fault tolerance of safety device" values in Table 9. These requirements may differ from the requirements according EN 61508-2 [3] for type A or type B equipment as well as from the EN 61511 requirements, but have to be fulfilled.

2.3.3. Primary explosion protection

Primary explosion protection is in principle used to avoid explosive atmospheres or to reduce the probability of the creation of such an atmosphere. The corresponding measures and activities do not fall under the regulation of the ATEX 94/9 directive [11] (in the broad sense that no ignition protection is performed), but fall under the ATEX 1999/92 directive [12] and are considered in the whole explosion protection policy of a facility.

Thus, the controls of some primary explosion protection measures, like e.g. inerting with nitrogen, have to be classified with the usual process risk approaches like the risk graph or LOPA. The big difference of this approach in comparison the ATEX 94/9 requirements is, that in the latter case the question of the consequence of an explosion is never raised. E.g. considering a zone 0 and usage of equipment with a fault tolerance of 0 (i.e. it is safe for normal operation, see Table 9), two faults of the SIL 2 safety device are necessary to create an explosion. Thus, in the worst case, the simultaneous occurrence of 3 independent failures with a probability of 10^{-3} yr^{-1} (assuming of probability of failure of the EUC itself without the safety device being 10^{-1} yr^{-1}) would lead to an explosion (and in the worst case to multiple fatalities). This theoretical worst case consideration seems not to be reflected in practise since the single failure probabilities are usually much lower.

Nevertheless, two consequences can be concluded:

- the protection against an explosion requires at least 3 quasi-independent layers, considering the concurrence of an infrequent equipment fault creating an ignition source and an infrequent process fault creating a potential explosive atmosphere as a layer.

- the risk tolerance criteria depend from case to case on the consequences, but shall never be less than 10^{-4} (considering the above given probabilities not as low as theoretically possible).

A simple example is the nitrogen inerting system of a chemical reactor. Let's assume that in the case of loss of nitrogen an explosive atmosphere is present, inside the reactor a zone 2 is defined, the equipment inside the reactor is ATEX 94/9 category 3 equipment (i.e. it is adequate for service within zone 2) and no additional sources of ignition are present. Hence, one layer of protection is given by ignition safety of the category 3 equipment. In case of loss of nitrogen (loss of one protection layer) an independent system has to shutdown the reactor. Using the semi-quantitative number above renders that (the probability of loss of nitrogen) × (the probability of failure on demand of the reactor shutdown system) × (the probability of category 3 equipment to become an ignition source) has to be less than 10^{-4} yr^{-1}.

3. DESIGN AND INSTALLATION OF SAFETY INTEGRITY FUNCTIONS

3.1. INTERPRETATION OF ARCHITECTURAL CONSTRAINS

According to EN 61511-1, subsection 11.4.3, for all subsystems (e.g. sensors, final elements and non-PE logic solvers) except PE logic solvers the minimum hardware fault tolerance shall be as shown in Table 10, provided that the dominant failure mode is to the safe state or dangerous failures are detected, otherwise the fault tolerance shall be increased by one.

The standard states, that to establish whether the dominant failure mode is to the safe state it is necessary to consider each of the following:

- the process connection of the device;
- use of diagnostic information of the device to validate the process signal;
- use of inherent fail safe behaviour of the device (for example, live zero signal, loss of power results in a safe state).

Since it might be difficult in practise to identify the dominant failure mode, one can use the SFF (Save Failure Fraction) value of the device if it's available: if the SFF is larger than 50% it can be assumed the above condition is fulfilled.

Table 10. HFT requirements on subsystems except PE logic solvers acc. to EN 61511-1

SIL	Minimum hardware fault tolerance (HFT)
1	0
2	1
3	2
4	Special requirements apply (see EN 61508)

Nevertheless, if there is no inherent fail safe behaviour of the device (e.g. a motor driven valve which shall close in case of demand), it is strongly recommended to increase the required HFT by 1. Furthermore redundant power supply is required, but to obtain serious estimations of the system integrity it is necessary to consider the reliability of the energy supply, and this seems to be a quite challenging task, especially in case of electrical power supply.

According to EN 61511-3, the minimum fault tolerance specified in Table 10 may be reduced by one if the devices used comply with all of the following:

- the hardware of the device is selected on the basis of prior use ("proven-in-use");
- the device allows adjustment of process-related parameters only, for example, measuring range, upscale or downscale failure direction;
- the adjustment of the process-related parameters of the device is protected, for example, jumper, password;
- the SIF has an SIL requirement of less than 4.

To decide if the device can be selected on the basis of prior use, the standard states that appropriate evidence should be available. This includes demonstration of the performance of the components or subsystems in similar operating profiles and physical environments as well as the volume of the operating experience.

The usage of proven-in-use components is often claimed, but no corresponding certification or at least a data base which renders sufficient data are available. Therefore two possibilities can be used in practise:

- strict application of the standard, i.e. acceptance of proven-in-use statements only in case of sufficient evidence presented by the supplier, or
- credit to the experience of the users (operating companies) if corresponding field reliability data is available. This corresponds to a possible approach as given in VDI/VDE 2180-3 [16].

In any cases, the required PFD data for the numerical SIL conformation is required for proven-in-use components, too.

Alternative fault tolerance requirements may be used providing an assessment is made in accordance to the requirements of EN 61508-2 [3], Tables 2 and 3. These alternative requirements differentiate between so-called type A and type B equipment (basically depending on the knowledge about failure modes) and use the SSF (Save Failure Fraction). Using this alternative it is possible to create a SIL 3 safety function with zero HFT. In this case, corresponding documents from the device manufacturers are only accepted together with certificates from independent bodies.

3.2. USAGE OF COMPONENTS WITHOUT PFD DATA

3.2.1. Equivalent level of safety approach

Fore some applications, the availability of components with sufficient functional safety data (SIL, PFD, PFH, SFF, etc.) is limited, especially for actuators, e.g. valves, electromechanical components (high voltage switches, contactors). In such cases, a full SIL assessment of a SIF is not possible, i.e. is restricted to the sensor elements, the logic solver, and the I/O units. Hence, the challenge in practise is to find an approach to determine an "equivalent level of safety" such that the risk analysis with regard to the required risk reduction by a SIF can be fulfilled.

In the field of safety of machinery, EN 954-1 [8] is used for many years and the corresponding SIFs are classified in categories. Due to a lack of alternative applicable standards to EN 61511 for the process industry, a comparison of these categories according to EN 954-1 with the SIL according to EN 61511 in the context of an equivalent level of safety was performed, the result is given in the equivalence table 8 (2nd and 3rd row).

Using this comparison table the requirements for determination of an equivalent level of safety can be summarised as follows:

- The sum of the PFD values of the sensor and logic solver part must be less than 51% of the total allowed PFD (based on the usual PFD distribution of a SIF)
- The equivalent category to the SIL according to the comparison table must be proved. Within this approach the HFT requirements according to EN 61511 are automatically fulfilled. It must be noted, some categories higher than 1 require frequent diagnostic functions. This requirement might be in problem in cases of ESD valves, which cannot be operated several times a year; partial stroke tests might by the only answer for this problem.

3.2.2. Usage of B10 values

For many pneumatic and electro-mechanical components so-called B10-values are available from the manufacturers. The B_{10d}-value is the average number of cycles until 10% of the components fail dangerously. If only a B_{10} value is known (i.e. the average number of cycles until 10% of the components fail, without consideration of the type of failure), the B_{10d}-value can be assumed to be 50% of the B_{10} value.

According to EN 13849-1 [13] (which will replace EN 954-1), the failure rate λ can now be calculated by usage of the number of operations per hour n_{op} to

$$\lambda = 0{,}1 \cdot n_{op}/B_{10d} \qquad \text{(Eq. 1)}$$

and thus, the probability of undetected dangerous failure λ_{DU} is given by

$$\lambda_{DU} = \lambda \cdot (1 - SFF) \qquad \text{(Eq. 2)}$$

If the safe failure fraction of the component is unknown, it follows that

$$\lambda_{DU} = \lambda, \qquad \text{(Eq. 3)}$$

and thus, for a 1oo1 voting, with

$$PFH = \lambda_{DU}; \qquad \text{(Eq. 4)}$$

$$PFD_{avg} \approx \lambda_{DU} \cdot \frac{T_1}{2} \qquad \text{(Eq. 5)}$$

If the components are operated in the low demand mode, additional information of the manufacturer like minimum life time and required proof test period T_1 are required.

3.3. SIS COMPONENT AND SIGNAL RECOMMENDATIONS

Application of analogue transmitters without moving mechanical parts shall be preferred contrary to simple digital sensor elements (e.g. simple switches) or sensors based on mechanical measuring principles because:

- PFD values are usually relatively low (high diagnostic coverage within the device possible)
- By usage of 4–20 mA signal (HART – standard) wiring diagnostic is simple and straightforward (SFF = 1 for wiring)
- Usually very robust and highly reliable

In case of application of simple digital switches sensor according to NAMUR standard should be preferred. By using this Standard, the digital signal is converted to a analogue 4–20 mA signal, and thus, short circuit and open circuit are diagnosed.

Important issues when analogue sensors are used are correct calibration (at or near the required trip point) and setting under consideration of the error of measurement.

Signal transfer between logic solver (CPU) and field I/O units via modern bus systems shall only be used if the Bus system and its protocol for safety application is certified according to EN 61508. Otherwise conventional wiring is required (as it is – at the time being – for field I/O units for EExi applications).

To reduce faults and disturbance of the SIS equipment, the power supply shall be according to EN 60204-1 [13].

The quality of lightning protection of the SIS should be such, that the availability of the system is not influenced considerably.

3.4. SOFTWARE ISSUES

To ensure a correct implementation of the SIFs according to risk analysis, it is required to use continuous function charts (e.g. including cause-effect-matrix, set-points, delay times, override functions, etc.) and requirements specifications which establish qualifications, validation, manage of change, etc.

The software has to be certified according to EN 61508.

Changes of a SIF shall not be done until a corresponding risk analysis shows admissibility; documentation of the changed software (program signature) is required.

4. CONCLUSIONS

To obtain consistency with regard to safety of process installations in Austria by usage of EN 61511 for functional safety issues, it is essential to introduce some interpretations and recommendations which exceed the standard's normative content. The ones presented in the paper concerning calibration of risk analysis procedures, SIL in the context of explosion protection, interpretation of architectural constraints of SIF, usage of components without PFD data, etc., are a step in this direction, helping the users to clarify obscurities and to implement proper safety functions even if not all requirements of the standard can be followed strictly.

REFERENCES

[1] EN 61511-1, Functional safety – Safety instrumented systems for the process industry sector, Part 1: Framework, definitions, system, hardware and software requirements, CEN, 2003 + Corrigendum 2004.
[2] EN 61511-3, Functional safety – Safety instrumented systems for the process industry sector, Part 1: Guidance for the determination of the required safety integrity levels, CEN, 2003 + Corrigendum 2004.
[3] EN 61508-2, Functional safety of electrical/electronic/programmable electronic safety-related systems, Part 2: Requirements for electrical/electronic/programmable electronic safety-related systems, CEN, 2001.
[4] DIN V 19250, Control technology – fundamental safety aspects to be considered for measurement and control equipment, DIN, 1994.
[5] Layer of Protection Analysis, CCPS, 2001, ISBN 0-8169-0811-7.
[6] HSE document: Reducing Risks, Protecting People (R2P2), 2001, ISBN 0-7176-2151-0.
[7] Directive 97/23/EC on the approximation of the laws of the Member States concerning pressure equipment.
[8] EN 13463-6, Non-electrical equipment for potentially explosive atmospheres, CEN, 2007.
[9] EN 954-1, Safety of machinery – safety related parts of control systems, Part 1 – general principles for design, CEN, 1996.
[10] prEN 50495, Safety devices required for the save functioning of equipment with respect to explosion risks, CEN, 2006.
[11] Directive 1994/9/EC on the approximation of the laws of the Member States conerning equipment and protective systems intended for use in potential explosive atmospheres.
[12] Directive 1999/92/EC on minimum requirements for improving the safety and health protection of workers potentially at risk from explosive atmospheres.
[13] EN 13849-1, Safety of machinery – safety-related parts of control system, Part 1 – general principles for design, CEN, 2006.
[14] EN 60204-1, Safety of machinery, Electrical equipment of machines, Part 1 – general requirements, 2006.
[15] VDI/VDE 2180-1, Safeguarding of industrial process plants by means of process control engineering, Part 3 – Introduction, terms, comments, 2007.
[16] VDI/VDE 2180-3, Safeguarding of industrial process plants by means of process control engineering, Part 3 – Plant engineering, realisation and operation, 2007.

INDEPENDENT REVIEW OF SOME ASPECTS OF IP15 AREA CLASSIFICATION CODE FOR INSTALLATIONS HANDLING FLAMMABLE FLUIDS

Philip Nalpanis[1], Stavros Yiannoukas[1], Jeff Daycock[1], Phil Crossthwaite[2], and Dr Mark Scanlon[3]
[1]DNV, Palace House, 3 Cathedral Street, LONDON, SE1 9DE.
E-mail: philip.nalpanis@dnv.com, www.dnv.com
[2]DNV, Highbank House, Exchange Street, STOCKPORT, SK3 0ET.
[3]Technical Manager – Safety, Energy Institute, 61 New Cavendish Street, London, W1G 7AR.
E-mail: mscanlon@energyinst.org.uk, www.energyinst.org.uk

1. INTRODUCTION

The third edition of IP *Area classification code for installations handling flammable fluids* (IP15) was published in July 2005 and incorporated both technical clarifications and editorial amendments. It is widely used in both the upstream and downstream sectors of the petroleum industry. In addition, IP15 is regarded as a key methodology for addressing the area classification requirements of the Dangerous Substances and Explosive Atmospheres Regulations (DSEAR) 2002.

In developing the second edition of IP15, some research studies were commissioned to strengthen its evidence base; this included the research published as IP *Calculations in support of IP15: The area classification code for petroleum installations*. Further issues were identified with a weak evidence base whilst developing the third edition of IP15. These were held over to form the subject of further research conducted by DNV for the Energy Institute's Area Classification Working Group, as part of their technical work programme. The research is reported in Energy Institute (2007).

2. OBJECTIVE AND SCOPE OF WORK

The purpose of the research was to provide an independent evidence based review of some aspects of the IP15 methodology by researching the following technical issues:

1. Sensitivity effects of input parameters on dispersion characteristics.
2. Area classification for liquid pools.
3. Application of area classification methodology to LNG.

It is intended that the Energy Institute's Area Classification Working Group will restructure and update IP15 (to be rebranded EI15) during 2008, informed by the results from this study.

3. APPROACH

3.1 FLUID COMPOSITIONS

The fluid compositions as set out in IP15 Annex C were used.
The categories refer to:

Liquids	Gases
A: LPG	G(i): Natural gas
B: Hot process intermediate (e.g. crude oil flashing or gasoline above its boiling point)	G(ii): 80% Hydrogen (typical petroleum refinery hydrogen stream)
C: Straight run gasoline	

3.2 SENSITIVITY EFFECTS OF INPUT PARAMETERS ON DISPERSION CHARACTERISTICS

The previous dispersion analysis, as published in IP15 Annex C, was carried out essentially for a single set of parameters as set out in Table 1, which also sets out the parameter sensitivities analysed here.

Table 1. Parameter values used for dispersion values

Parameter	Value modelled for IP15	Range of values modelled here
Ambient temperature	30°C	–20°C, 0°C, 20°C, 40°C
Storage/process temperature	20°C	Same as ambient + Fluid Category A (LPG) at –40°
Relative humidity	70%	50%, 90%
Wind speed	2 m/s	1.5 m/s 2 m/s 5 m/s 9 m/s
Stability class	D	F D D D
Surface roughness length	0.03 m	0.1 m, 0.3 m, 1.0 m
Release direction	Horizontal	Horizontal
Release height	For R1: 5 m For R2: 1 m	For R1: 5 m For R2: 1 m
Release angle	For R_1: horizontal For R_2: unknown	For R_1: horizontal For R_2: –30°, –45°, –60°
Sample time	18.75 s	No variation
Reference height	10 m	No variation
Hazard distances	To LFL	To LFL and 0.5 LFL

Most of the parameters were initially varied singly in turn, however all four stability-windspeed combinations were modelled for the other parameter variations. The investigations were carried out for all the fluid categories listed in Section 3.1.

Dispersion was characterised by the distance to LFL measured by two different radii: R_1 represents the hazard distance for a release that does not interact with the ground; R_2 represents the hazard distance for a release that does interact with the ground. These are illustrated in IP15 Figure 5.6.

3.3 AREA CLASSIFICATION FOR LIQUID POOLS

For the liquids listed in Section 3.1 and the ambient temperatures, surface roughness lengths and weathers given in Table 1, the following parameter variations were analysed:

- Pool sizes and depths: as set out in Table 2
- Surface types: concrete, dry soil
- Fluid temperature: ambient, except fluid category A: refrigerated at –40°C
  Fluid category C also modelled at 100°C

The spill volume has been modelled as being released instantaneously onto the ground and allowed to spread until it reaches the pool diameter shown in Table 2. Evaporation and dispersion will take place as soon as the pool starts spreading outwards.

3.4 APPLICATION OF AREA CLASSIFICATION METHODOLOGY TO LNG

LNG is largely methane but also typically includes small quantities of ethane, propane and CO_2 and sometimes other component. However, based on DNV's experience in several LNG studies, it has been modelled it as pure methane for the present study.

Typical rundown, storage and loading temperatures for LNG are in the range –170° to –160°C; therefore releases from a storage temperature of –165°C have been modelled.

Typical pressures in LNG systems range from 1.5 bar(a) to 10 bar(a); therefore these two pressures have been modelled, and also an intermediate pressure of 5 bar(a).

Table 2. Pool spill volumes modelled

	Spill volume (m^3) for pool depth			
Pool diameter (m)	Concrete (0.005 m)	Dry soil (0.02 m)	0.1 m	1 m
1	0.0039	0.015	(Not modelled)	(Not modelled)
3	0.035	0.14	0.71	(Not modelled)
10	0.39	1.6	7.9	79
30	(Not modelled)	(Not modelled)	70	707
100	(Not modelled)	(Not modelled)	(Not modelled)	7854

Two ambient temperatures have been modelled, approximately bracketing the range of values given in Table 1: −20°C and 30°C. Other parameters have been varied as in Table 1.

3.5 SOFTWARE

DNV's proprietary, commercial software PHAST (**P**rocess **H**azard **A**nalysis **S**oftware **T**ool) has been used for this study. PHAST has been licensed to over 500 organisations worldwide. PHAST's modelling has undergone extensive validation, as described in Witlox & Oke (2008) being presented at this conference.

The key model within PHAST for this study is the Unified Dispersion Model, described in Witlox & Holt (1999). The atmospheric dispersion modelling takes account, at every time-step, of whether the plume spreading and dilution is driven by initial momentum, plume density, or atmospheric turbulence. It also includes liquid pool spreading and evaporation, for which different models are adopted depending whether the spill is on land or water, and whether it is an instantaneous or a continuous release (Witlox & Holt 1999). The pool spreads until it reaches a bund or a minimum pool thickness. The pool may either boil or evaporate while simultaneously spreading. For spills on land, the model takes into account heat conduction from the ground, ambient convection form the air, radiation and vapour diffusion. These effects are modelled numerically, maintaining mass and heat balances for both boiling and evaporating pools. This allows the pool temperature to vary as heat is either absorbed by the liquid or lost during evaporation.

4. INVESTIGATION OF SENSITIVITY EFFECTS OF INPUT PARAMETERS ON DISPERSION CHARACTERISTICS

4.1 VARIATION WITH WEATHER (WINDSPEED AND STABILITY)

4.1.1 R_1 Distances to LFL (releases at 5 m height above ground)

For lower pressures, the weather has no discernible impact on the hazard distances for the two gases G(i) and G(ii). This is because the releases are momentum driven jets, with the velocity difference between the released fluid and its surroundings driving the entrainment of air and resulting lowering of concentration. For higher pressures, higher windspeeds do lead to more rapid dilution and hence shorter hazard distances for the two gases, compared with lower windspeeds. The higher windspeeds also lead, at all pressures, to shorter hazard distances for the three liquids A, B and C released at ambient temperature and for fluid A (LPG-like), refrigerated and released under 5 bar(a) head (e.g. pump pressure).

However, for fluid A (LPG-like), refrigerated and stored at atmospheric pressure (modelled as a discharge under 10 m tank head), a case not considered in IP15, higher wind speeds give longer dispersion distances. The refrigerated fluid forms an evaporating pool on the ground, with evaporative mass transfer therefore increasing with wind speed.

Contrary to what is often believed regarding atmospheric dispersion, stability F does not always give the longest hazard distances (except for fluid B released from a 10 mm hole, and then the distance is only marginally longer than for weather 2D). Stability F does occur sufficiently frequently that it cannot be discounted. The received wisdom only applies when passive (sometimes called "gaussian") dispersion is the dominant mechanism of dispersion. In the case of flammable materials discharged under pressure, the dominant dispersion mechanisms in the early stages are turbulence induced by the momentum of the release and then dense gas dispersion (i.e. dispersion of a cloud substantially denser than air). Usually the LFL is reached before the cloud is moving with a velocity close to the ambient windspeed and the cloud's density has approached neutral with respect to air, the conditions for passive dispersion to dominate. Hence atmospheric stability tends to have less influence on dispersion to LFL. However, it does influence dispersion to lower concentrations.

4.1.2 R_2 Distances to LFL (releases at 1 m height above ground)

Based on preliminary analysis, releases at 1 m height above ground have been modelled at an angle of 60° below horizontal in order to examine the influence of ground effects on the hazard distances.

At 10 bar(a) and for all hole sizes, fluid C shows much longer hazard distances for weathers 2D and 1.5F, i.e. low windspeeds, than for weathers 5D and 9D, i.e. moderate to high windspeeds. For fluids A and B the same behaviour is exhibited at 10 bar(a) for the 2 mm hole size. The behaviour can be understood by looking at side views of the plumes. For fluid A at low windspeeds, the plume centreline hits the grounds whilst the concentration is still well above LFL. This results in the plume losing much of its momentum and being directed horizontally (so long as it is not buoyant), with the concentration decreasing to the LFL. At the higher windspeeds, the centreline concentration is close to or below LFL when the centreline hits the ground. By contrast, for fluid C similar behaviour is shown for all weathers, with the plume centreline having hit the ground with the concentration well above LFL.

For fluid G(i), distances to LFL vary reasonably consistently with windspeed. By contrast, for releases of fluid G(ii) at 100 bar(a), for a 2 mm hole there is sharp decrease in hazard distance from low windspeeds to a windspeed of 5 m/s, and for a 10 mm hole there is a sharp increase in hazard distance from low windspeeds to a windspeed of 5 m/s. This can be understood as follows. For the smaller hole size (2 mm), the behaviour is similar to that noted above for fluid C. For the larger hole size (10 mm), at the lower windspeeds the plume becomes buoyant and lifts off, whereas at the higher windspeeds it is "knocked down" to remain at ground level.

For refrigerated fluid A (LPG-like), both under tank head (10 m) and under pressure (5 bar(a)), the hazard distances are much shorter at higher windspeeds as compared with low windspeeds, with the longest hazard distances being for weather 1.5F. The dispersion is similar to that for fluid A at ambient temperature and under 10 bar(a) pressure.

The modelling results indicate hazard distances in most cases much smaller for leaks up to 5 mm that might be expected from scaling down of the hazard distances for 10 mm leaks. Closer examination of the dispersion modelling, and in particular the pool evaporation, indicates that the pool area decreases much more quickly with hole size than the release rate and that in many cases there is no significant pool evaporation. The fraction of the initial release that rains out (and forms a pool) decreases slightly with hole size (74% to 81% for 10 mm leaks vs. 65% to 72% for 2 mm leaks, depending on wind speed).

Overall, for all fluids except G(ii), weather F1.5 gives the longest hazard distances for releases where ground effects influence dispersion. For fluid G(ii), the low molecular weight renders the fluid buoyant even when the concentration approaches LFL, so at high windspeeds the plume resulting from larger releases is knocked down and higher windspeeds give the longest hazard distances.

4.2 VARIATION WITH AMBIENT TEMPERATURE

For the gases G(i) and G(ii), variations with ambient temperature are mostly small and not systematic. The exception is for G(i) at –20°C, which has a hazard distance 10% to 20% higher than under the IP15 base case conditions.

For fluids B and C, there is a small systematic increase in hazard distance with ambient temperature but this is less than 10%. Compared with the base case results, the variation is only a few percent. Hence the variation with ambient temperature for these fluids is not significant.

For fluid A under pressure there is an apparent decrease in hazard distance with increasing ambient temperature, especially for the smaller leak sizes. This results from the thermodynamics: as the ambient temperature increases, the liquid fraction in the discharge and the liquid droplet diameter decrease, and the velocity increases. The velocity increase enhances entrainment of air and hence promotes more rapid dispersion, reducing the hazard distance. The larger liquid fraction and droplet size at lower ambient temperatures maintain the centreline concentration well above LFL to a greater distance, also increasing the hazard distance at lower temperatures. Hence in colder climates it may be necessary to increase the classified area dimensions compared with the values given in IP15 around equipment containing a fluid similar to A (i.e. LPG-like) under pressure, by up to 70%.

For fluid A (LPG-like), refrigerated at –40°C, the hazard distances increase slightly with ambient temperature both for atmospheric storage with 10 m tank head and for pressure of 5 bar(a) (Figure 1). For the case of a release under pressure, the increase is less than 10% over the ambient temperature range modelled except for 10 mm holes, for which the increase is 11%. However, compared with releases under IP15 conditions (i.e. fluid A under pressure of 6.8 bar(a) and at 20°C), these releases give hazard distances up to 80% higher. For the case of a release under tank head, a 1 mm hole gives hazard distances up to 50% longer than a release under IP15 conditions, a 2 mm hole hazard distances only slightly longer, and a 10 mm hole hazard distances up to 40% shorter. Based on these results, hazard distances for refrigerated LPG significantly exceed those given in IP15 for fluid category A stored under pressure.

4.3 VARIATION WITH HUMIDITY

The modelling results show that there are no significant variations in hazard distance with humidity for any of the fluids, hole sizes or pressures, including fluid A (LPG-like) refrigerated.

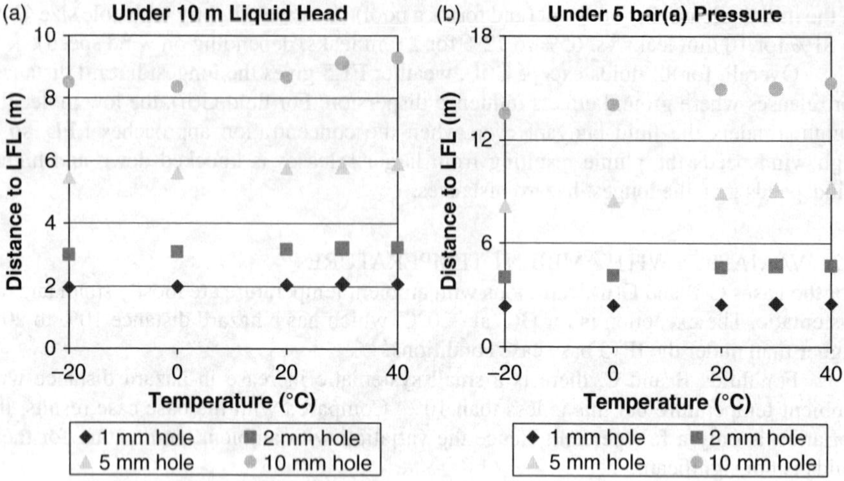

Figure 1. Comparisons of R1 hazard distances to LFL for refrigerated LPG between different ambient temperatures

4.4 VARIATION WITH SURFACE ROUGHNESS

For all fluids except G(i) the hazard distance falls off with increasing surface roughness. This is as expected, since increasing surface roughness promotes turbulence, enhancing dilution of the vapour cloud and hence reducing the hazard distances.

For fluid G(i) the hazard distance is more-or-less independent of surface roughness. This is because the dispersion is driven by the initial momentum of the release and atmospheric turbulence does not influence the dispersion before the cloud centreline concentration has reached LFL.

5 INVESTIGATION OF AREA CLASSIFICATION FOR LIQUID POOLS

5.1 HAZARD DISTANCES: BASE CASE

There is no distinction between R1 and R2 for the results presented in this section, since all releases are at ground level.

Several general observations can be made on the modelling results:

- Pools of fluid A generally give larger hazard distances than those for other fluids studied for various pool diameters/depths, surface types and weathers.
- Going from the smallest spills to the largest, there is a trend in the maximum distance with weather from 9D to 1.5F. This means that, for very small spills (such as might result from the breaking of a coupling), the hazard distances are longest for high winds; for larger spills, the hazard distances are longest for low windspeeds and stable conditions.

- For spills of 0.1 m and 1 m depth, the hazard distance is independent of surface type for fluids B and C whereas the variation of hazard distances with surface roughness for fluid A is slight and shows no systematic variation.
- For fluid C released at 100°C, the hazard distances are longer than for the corresponding releases at 20°C at low windspeeds.

5.2 HAZARD DISTANCES: SENSITIVITIES

As the results presented in Section 4.3 showed no significant variation of hazard distances with humidity, this sensitivity has not been examined. The sensitivities with temperature and surface roughness are presented in Sections 5.2.1 and 5.2.2 respectively.

5.2.1 Variation with ambient temperature

Figure 2 shows distances to LFL for a range of ambient temperatures (releases of fluid A are refrigerated; releases of fluids B and C are at ambient temperature), for weather conditions 1.5F and 9D (i.e. the lowest and highest windspeeds).

For spills of fluid A onto concrete (Figure 2(a)) there is a general increase with temperature in the distance to LFL for low windspeed; for high windspeed the smallest spill gives a much longer hazard distance for higher temperatures whereas the temperature has little influence on the hazard distance for larger spills. For spills of fluid A onto dry soil (Figure 2(b)), the hazard distances are generally shorter than for spills onto concrete but the trends with temperature are stronger.

Spills of fluid B onto concrete (Figure 2(c)) generally show a similar influence of temperature to fluid A. For low windspeed, at temperatures of −20°C and 0°C there is minimal pool evaporation and hence the hazard distance under these conditions is zero or negligible. Fluid B shows little dependence on surface type (compare Figure 2(c),(d)).

Spills of fluid C show very little variation with temperature other than for high windspeed and a temperature of 40°C, which results in much longer distances to LFL for a 3 m diameter, 0.1 m deep pool. Fluid C shows little dependence on surface type.

5.2.2 Variation with surface roughness

Figure 3 shows distances to LFL for a range of surface roughnesses, for weather conditions 1.5F and 9D (i.e. the lowest and highest windspeeds).

In general these show the expected decrease in hazard distance with increasing surface roughness, by a factor of at least 2 over the range modelled for both low and high windspeeds. However, there were some anomalous results which should be disregarded as they are artefacts of the modelling.

6 INVESTIGATION OF APPLICATION OF AREA CLASSIFICATION METHODOLOGY TO LNG

6.1 LIQUID RAINOUT

When a liquid is released, the droplets will evaporate as they travel through the air but will also fall towards the ground under the influence of gravity. Whether or not they reach the

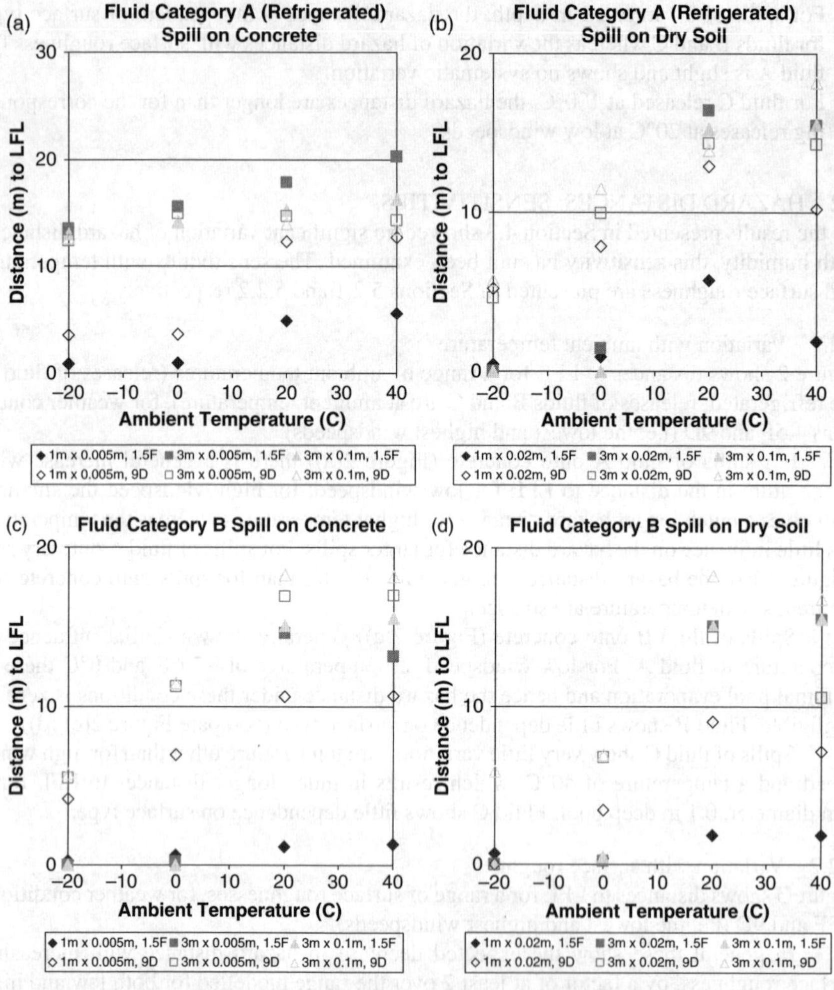

Figure 2. Example comparisons of R_1 hazard distances between different temperatures for spills of fluid categories A and B

ground depends on the release height, rate and velocity as well as the material properties. If they do reach the ground, they form a spreading and evaporating liquid pool.

The amount of rain-out was calculated for releases initially horizontal at three heights: 5 m, 1 m, and 0.1 m. Rain-out only exceeds 50% for releases at 0.1 m height and 1.5 bar(a). For higher pressures there is no rain-out at all; for releases at 0.1 m, the larger

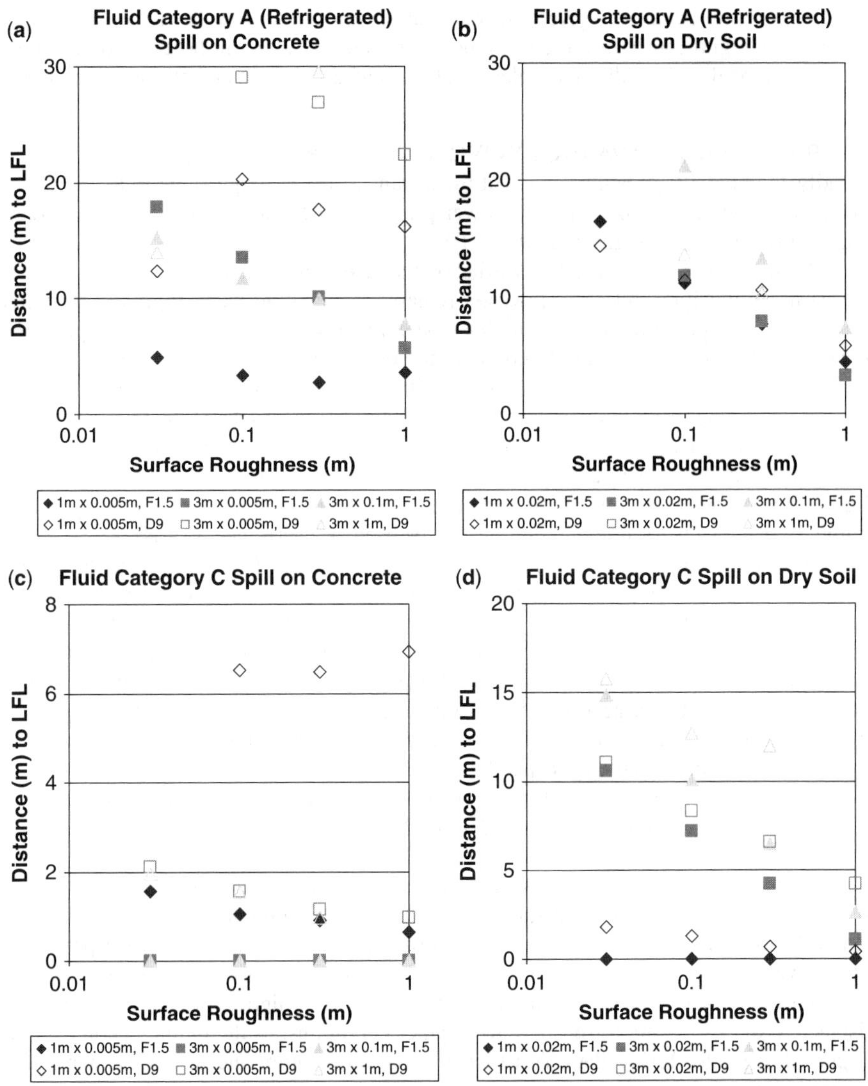

Figure 3. Example comparisons of R_1 hazard distances to LFL between different surface roughnesses for liquid spills

hole sizes (5 mm, 10 mm) do give some rain-out for a pressure of 1.5 bar(a). In order to see the effect of significant rain-out and evaporation, releases at 0.1 m height and 1.5 bar(a) have been modelled in addition to releases at 1 m and 5 m height and all three pressures.

6.2 BASE CASE AND VARIATION WITH WEATHER

Broadly, the modelling results show that the distances to LFL (equivalent to R_1 in Section 4.0) increase with windspeed for the lower pressures (1.5 bar(a), 5 bar(a)) but decrease with windspeed for the highest pressure (10 bar(a)).

Of the materials for which hazard distances are presented in IP15, fluid G(i) is the closest in composition to LNG. However, releases of this fluid were originally modelled for a storage temperature of 20°C and have been modelled in this study for temperatures down to −20°C. It is interesting to compare the results for LNG with these other results. The dispersion behaviour of fluid G(i) shows little variation with temperature.

It was found that LNG gives much longer hazard distances than fluid G(i). This is not surprising as the release rates are much higher, LNG being liquid and fluid G(i) gas; also the modelled release velocity is about 43 m/s for LNG but sonic for fluid G(i), which means that fluid G(i) will entrain air much more rapidly through induced turbulence at the edge of the plume. Hence results for fluid G(i) should not be used as a surrogate for LNG.

The same trends with windspeed for releases at 1 m and 0.1 m height above ground are seen as for releases at 5 m height. Considering first the releases at 1 m height, the distances to LFL are shorter than for releases at 5 m height for the lower pressures and smaller hole sizes, increasing to exceed them for the higher pressures and larger hole sizes. These variations can be explained by a combination of plume interaction with the ground and rain-out. Only the largest releases 5 m interact with the ground, and then only at the lowest windspeed, but the larger releases at 1 m do interact with the ground, even for the highest pressure and highest windspeed.

For releases at 0.1 m height, the distances to LFL are smaller than from releases at 1 m height for the smaller hole sizes (1 mm, 2 mm) but larger for the larger hole sizes (5 mm, 10 mm).

Figure 4 shows the variation with weather conditions graphically. The effect referred to above of weather 1.5F on the distance to 0.5 LFL for 10 mm holes is clearly seen. More generally, the distances are more sensitive to weather conditions for the lower pressures; alternatively, the windspeed is less important for the higher pressure releases as dispersion is driven more by the turbulence induced by the resulting higher release velocities. The distance to 0.5 LFL is more sensitive to weather conditions: as the plume becomes more dispersed, it is also losing the effect of the initial momentum and hence the windspeed becomes more significant.

6.3 VARIATION WITH AMBIENT TEMPERATURE

For releases at both 1 m and 5 m height above ground at ambient temperatures, the distances to LFL are up to 20% shorter for releases at −20°C compared with 30°C.

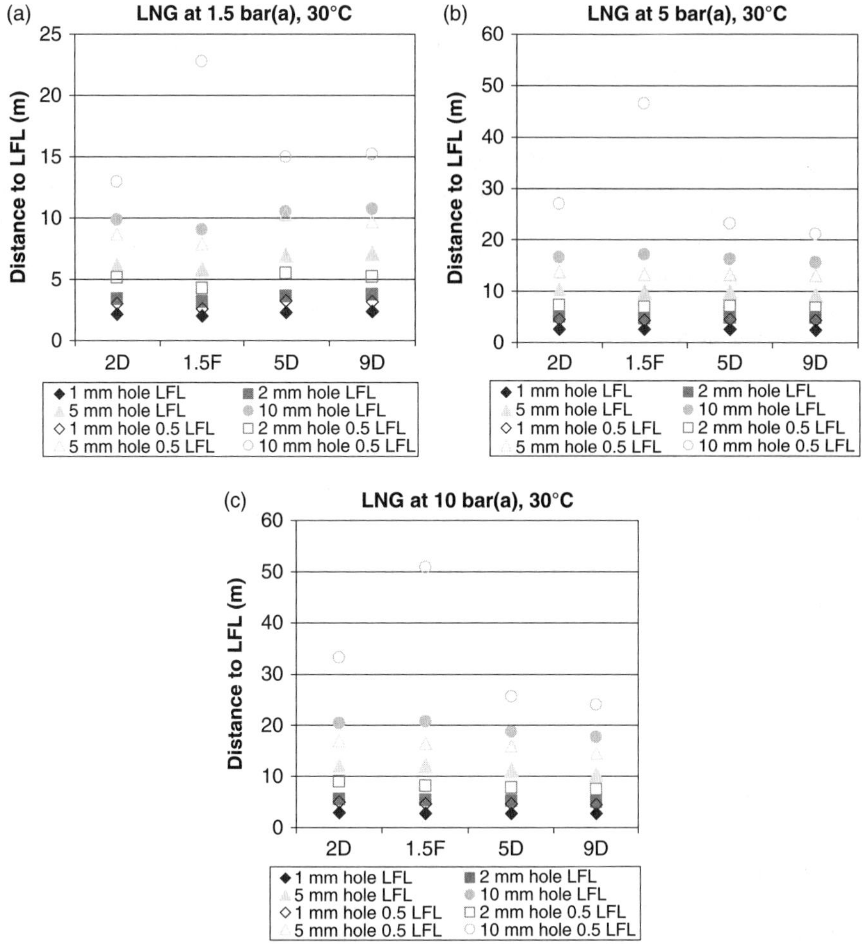

Figure 4. Variation of distances to LFL with weather for LNG spills (Release height 5 m)

6.4 VARIATION WITH HUMIDITY

The effect of varying the relative humidity (RH) is no more than about 10%, and no clear trend is apparent.

This contrasts with the results in major hazard releases of LNG, where the effect of relative humidity is marked: as RH increases, hazard distances decrease. These results are seen in modelling and have been observed in experiments (e.g. Maplin Sands). However, the size of the releases in both trials and major hazard modelling is much greater than in the present study and so the absence of any identifiable trend with RH for such small releases does not conflict with the results for large releases.

6.5 VARIATION WITH SURFACE ROUGHNESS

For releases at 5 m height above ground there is a consistent decrease in hazard distances from the Base Case surface roughness of 0.03 m to 1 m, the largest roughness modelled, of around 25%. This trend is as expected. For releases at 1 m height, most cases exhibit the same trend but the largest releases (10 mm) under weather 2D do not show such a clear trend. This is because the plume centre grounds and hence dispersion behaviour is modified by the interaction with the ground.

6.6 VARIATION WITH SURFACE TYPE

All releases over water will result in longer hazard distances than the equivalent releases over land because the surface roughness is lower, following the trend demonstrated in the previous section (6.5). The results show modest increases in distance to LFL: up to about 7% for weather 2D, 25% for weather 9D.

An additional effect can be expected when the release results in a spill onto the water surface due to the different heat transfer properties of land and water. Only the releases at 0.1 m height and 1.5 bar(a) result in significant rain-out. There are larger increases in hazard distance (up to 38% longer) for weather 2D, showing the enhanced vapour generation on water compared with land. For weather 9D the increase is generally less marked than for releases at 5 m.

For spills onto water, the phenomenon of "rapid phase transition" (RPT) has not been considered.

7 CONCLUSIONS AND RECOMMENDATIONS

7.1 SENSITIVITY TO PARAMETER VARIATIONS

In the following summary, variations of ±20% in hazard distance from base cases are discounted given the uncertainties in dispersion modelling.

7.1.1 Fluid category A

- For fluid category A under pressure:
  - the variation in hazard distance with weather category compared with the base case (2D) is mostly less than ±20%, whereas for higher windspeeds and larger hole sizes the reduction in hazard distance compared with base case reaches 27%.
  - the variation of hazard distance with relative humidity is negligible.
  - the reduction in hazard distance with increasing surface roughness (up to 1 m) is less than 20%.

- For fluid category A, refrigerated at –40°C and stored at atmospheric pressure:
  - the hazard distances are up to 80% higher than for the same fluid stored under pressure at ambient temperature. Figure 1 indicates suitable hazard distances for refrigerated LPG.
  - higher wind speeds give longer dispersion distances than the same fluid stored under pressure at ambient temperature. This needs to be taken account of for installations storing refrigerated LPG.
  - the variation of hazard distance with relative humidity is negligible.
  - the reduction in hazard distance with increasing surface roughness (up to 1 m) is up to 31%. However, in most cases the surface roughness will be significantly less than this hence the distances shown in Figure 1 should be used.

7.1.2 Other fluid categories

For fluid categories B, C, G(i) and G(ii):

- For lower storage/process pressures, the weather has no discernible impact on the hazard distances for the two gases G(i) and G(ii). For fluid G(ii), high windspeeds give the longest hazard distances. Higher windspeeds also lead, at all pressures, to shorter hazard distances for the three fluid categories A, B and C.
- The variation of hazard distance with ambient temperature is less than 20% over the range modelled.
- The variation of hazard distance with relative humidity is negligible.
- The reduction in hazard distance with increasing surface roughness (up to 1 m) is less than 20%.

7.2 LIQUID POOLS

7.2.1 Base case

A comprehensive approach to area classification for liquid pools due to spillage has been developed for various fluid categories, pool diameters/depths and weathers for two surface types. The following specific points may be noted:

- For very small spills, the hazard distances are longest for high windspeeds; for larger spills, the hazard distances are longest for low windspeeds and stable conditions.
- In the majority of cases, the surface type (concrete or dry soil) makes no significant or systematic difference to hazard distances.

7.2.2 Sensitivity to parameter variations

- For fluid category A there is significant variation with increasing ambient temperature compared with the base case results. Figure 2(a),(b) indicates the range of variation.
- For fluid category B the variation for a higher ambient temperature than base case (20°C) is less than 10%. For lower temperatures, it is recommended that the hazard distances in Figure 2(c),(d) are used as they are significantly shorter than the base case.

- For fluid category C the variations of hazard distance with ambient temperature are mostly small.
- For all of fluid categories A, B and C the variation of hazard distance with relative humidity is negligible .
- For all of fluid categories A, B and C the variation of hazard distance with surface roughness is significant as illustrated in Figure 3.

7.3 LNG

- Figure 4 provides a suitable basis for guidance on hazard distances for IP15.
- Except for very small releases (i.e. 1 mm hole size), the hazard distances for releases initially horizontal at 0.1 m height are mostly longer than for the corresponding releases at 1 m height due to rain-out, liquid pool formation and re-evaporation occurring.
- The variation in hazard distance between ambient temperatures of –20°C and (+)30°C is less than 20%.
- The variation in hazard distance with relative humidity is less than 10%.
- The reduction in hazard distance with increasing surface roughness (up to 1 m) is less than 25%. However, in most cases the surface roughness will be significantly less than this.
- For releases resulting in significant rain-out (those at 0.1 m height and 1.5 bar(a) pressure), the hazard distances are significantly increased at low windspeed compared with releases onto land.

8 REFERENCES

Energy Institute 2007. *Research Report: Dispersion Modelling and Calculations in Support of the Area Classification Code for Installations Handling Flammable Fluids*, 2nd edition, ISBN 0 978 85293 489 0.

IP 2005. *Area Classification Code for Installations Handling Flammable Fluids ('IP15')*, Model Code of Safe Practice in the Petroleum Industry Part 15, 3rd edition, ISBN 978 0 85293 418 0, London: Energy Institute.

Pasquill, F, and Smith, F B, 1983. *Atmospheric Diffusion*, 3rd ed., Chichester: Ellis Horwood.

Witlox, H W M and Holt, A, 1999. A unified model for jet, heavy and passive dispersion including droplet rainout and re-evaporation, *International Conference and Workshop on Modelling the Consequences of Accidental Releases of Hazardous Materials*, CCPS, San Francisco, California, 28 September – 1 October, 315-344.

Witlox, H W M, and Oke, A, 2008. Verification and Validation of Consequence Models for Accidental Releases of Hazardous Chemicals to the Atmosphere, *Hazards XX – Process Safety and Environmental Protection, Harnessing Knowledge – Challenging Complacency*, IChemE Symposium, University of Manchester, Manchester, UK, 15 – 17 April.

POTENTIAL FOR FLASHBACK THROUGH PRESSURE/ VACUUM VALVES ON LOW-PRESSURE STORAGE TANKS SYNOPSIS

A Ennis[1] and D Long[2]
[1]Haztech Consultants Ltd, Unit 13, Meridian House Business Centre, Road one, Winsford Industrial Estate, Winsford, Cheshire, CW7 3QG, UK
E-mail: www.haztechconsultants.com
[2]Protego UK

Low pressure or "atmospheric" storage tanks are commonly used for the storage of flammable hydrocarbons. In order to minimise losses of volatile organic compounds to the environment, these tanks are often fitted with pressure / vacuum valves (PV valves). These enable the tank to operate at pressures typically between +56mbar and -6mbar before the valve opens.

The atmosphere within the tank is often in the flammable region and hence during outbreathing, the vapours vented from the PV valve are also flammable. These may be ignited by an external ignition source such as static discharge or hot work occurring near the vent. The common assumption has always been that the tank would be protected from internal explosion by the presence of the PV valve on the basis that the flame could not pass back through the valve.[1]

This paper presents the opinion that ignition can, in fact, flash back through a PV valve based on consideration of flame speed, minimum experimental safe gap (MESG) and flow rates. Experimental work done by Protego will also be presented demonstrating that flashback can occur, especially under conditions of continuous venting. It is therefore recommended that consideration should be given to installing a flame arrester on tanks protected by a PV valve.

INTRODUCTION

Low pressure or "atmospheric" storage tanks are often used for the storage of flammable materials i.e. those with flashpoints of less than about 50°C. Thus, under normal operating conditions it is normal for there to be a flammable atmosphere inside the tank. In order to minimise vapour losses from the tank,

Taking into consideration the normal operation of LP storage tanks, it can clearly be seen that vapour will be periodically vented from the tank to atmosphere. The volume and flowrate of the emission will be dependent upon a number of factors such as:

- Pumping in (filling) rate
- Diurnal temperature changes
- Internal heating

Under certain conditions, flammable vapour may be vented from the tank for considerable periods, especially for larger tank sizes. Thus, there will be a cloud of flammable material vented, generally at high level, over the roof space of the tank.

In particular, during filling of a tank with a flammable liquid e.g. Hexane or similar, it is possible that a flammable mixture could be vented for a period of more than 30 minutes.

LOW PRESSURE STORAGE TANK DESIGN

One of the key features of these tanks is that they have extremely limited resistance to pressure with typical design limits of +56 and −6 mbar. Although these are the nominal design limits, tanks will, in practice, often withstand approximately +100mbar and −10 mbar. API 650 gives 100 mbar internal design pressure and more under fire relief cases. EN 14015 gives a test pressure of 100 mbar, a pressure often used as the full ope ning pressure for PV Valves. API 620 also allows 100 mbar Maximum Allowable Working Pressure and up to 120 mbar for fire relief. Ultimate protection for these tanks is provided by frangible roof seams. Note that each of these standards uses slightly different terminology.

Thus, compared to traditional pressure vessels, it can be seen that low-pressure tanks are very fragile. This causes certain design problems, especially in the design of pressure relief systems. Pressure and vacuum relief devices need to function with a minimum of pressure drop if required flowrates are to be achieved with reasonably sized devices.

It is often required to minimise emissions from tanks by minimising the amount of venting that takes place. This is achieved by maintaining the tank between the maximum and minimum design pressures. Thus, a relief device is required. Typical pressure relief devices are either lute pots or PV Valves. Lutes have generally fallen out of favour because of the inherent problems of maintaining the liquid seal and the attendant pollution problems. See diagram 1. The other method for maintaining a limited pressure in the tank is the Pressure/Vacuum Valve. These are now commonly used throughout the chemical and associated industries and are the standard method of minimising vapour emissions.

An internal explosion of a stoichiometric hydrocarbon mixture will generate pressures in the region of 7–8 bar (if totally confined in a strong vessel). A low-pressure storage tank will explode at a much lower pressure. It is not feasible to design storage tanks to withstand this level of internal pressure, especially in larger sizes. It is, therefore, necessary to provide some form of protection against an external ignition entering the tank via the vent lines. External ignition can occur due to several reasons:

- Electrostatic discharge e.g. lightning
- External hot work e.g. maintenance
- Failure of external electrical systems e.g. lights, instruments
- Portable electrical equipment e.g. misuse of mobile phones
- Impact spark e.g. due to dropping tools

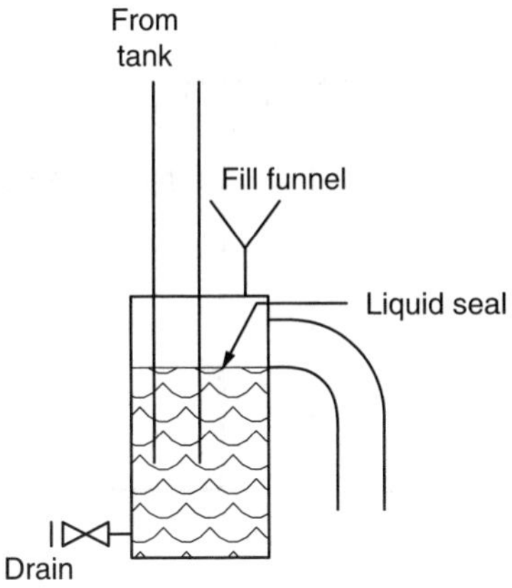

Diagram 1. Typical lute pot installation

FLASHBACK PROTECTION

It has always been assumed, at least within the UK, that protection against the flashback of an ignition into the tank is provided by the provision of a Pressure / Vacuum Valve (PVV) on the vent line. This has generally been endorsed by the HSE[5] with the result that the vast majority of tanks containing flammable materials which are equipped with PV valves do not have associated flame arresters in the vent line. A typical PV Valve configuration is shown in Diagram 2 below. Operation of the valve under pressure and vacuum are shown in diagrams 3 & 4.

Theoretically, in order for the protection to be effective, one of two conditions must be met as follows:

Either:

The velocity through the valve must be greater than the turbulent flame velocity

Or:

The gap between the valve pallet and the seat must be less than the Maximum Experimental Safe Gap (MESG)

Some typical MESG values are:

Gas	MESG (mm)
Propane	0.965
Methane	1.14
Ethylene	0.65
Hydrogen	0.50
Hexane	0.95
Cyclohexane	0.94
Ethyl acetate	1.04

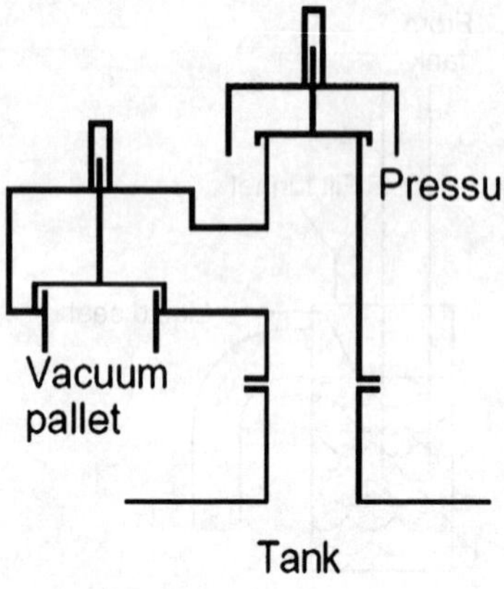

Diagram 2. Typical PV valve configuration

It can be seen that these values are a maximum of just over 1mm.

Note: The MESG is the largest gap through which a flame will not pass. This parameter is used in the specification of flame arresters to ensure that a flame will not pass through the arrester. MESG is measured for stoichiometric mixtures (the worst case).

Taking a typical 3″ (75 mm) diameter PV valve, the opening for the pallet is in the order of 2″ (50 mm) at a flow in the order of 600 m^3/h^6. It can be seen that the opening is far greater than the MSEG and thus it must be assumed that, if the flame has sufficient velocity, it may pass back through the PV Valve.

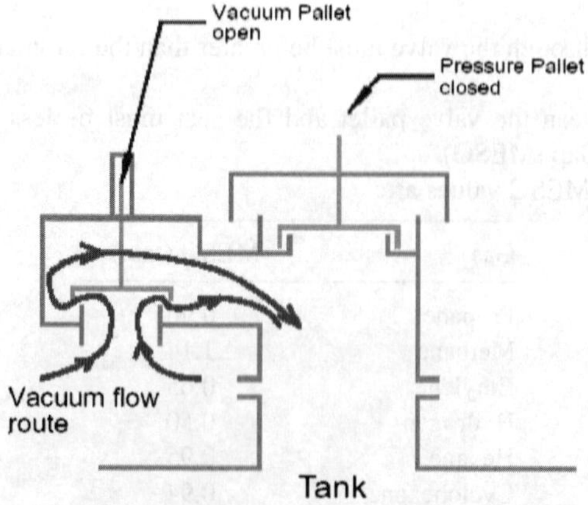

Diagram 3. PV valve under vacuum

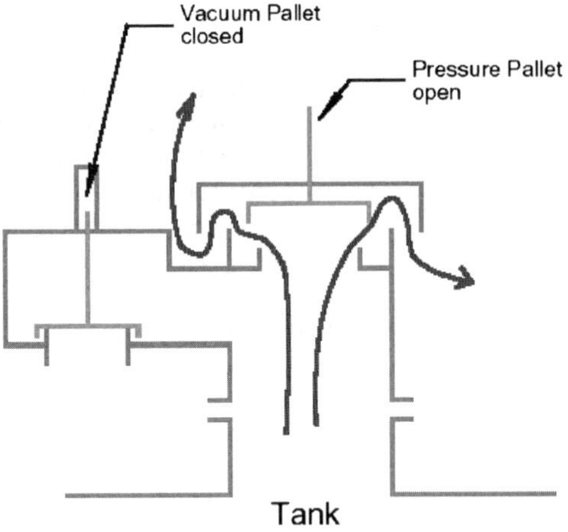

Diagram 4. PV valve under pressure

A 2″ (50 mm) pallet lift gives a superficial gas velocity in the order of 14 m/s through the valve based of Protego flow capacity charts for the DN80/3" valve[6], although this could be significantly less for lower tank design pressures where higher lift is required for to achieve the desired flow under normal operating conditions. It can be seen that this is a relatively low velocity for turbulent flame speed. In fact, the superficial gas velocity through the valve could be as low as 2-3 m/s with a lift of 5mm or more, considerably more than the MESG.

BURNING VELOCITY & FLAME SPEED

A key consideration in the transmission of flame through the valve is the velocity at which the flame propagates. There are two factors involved in this, these being the fundamental burning velocity and turbulent burning velocity. The fundamental burning velocity (sometimes known as the laminar flame speed) is a constant of a given flammable mixture and is the speed at which a flame will travel in a laminar flowing mixture. Typically, these are in the region of a few cm/s for most common hydrocarbons and somewhat higher for hydrogen. Some examples for common materials are[7,8]:

Methane/air = 35 cm/s
Propane/air = 45 cm/s
Hydrogen/air = 350 cm/s
Acetylene/air = 158 cm/s

Fundamental burning velocities are measured under strictly controlled conditions where the flow of the gas is strictly controlled to ensure a laminar and linear flow through the measuring apparatus, normally a long straight tube. Measures are taken to ensure that the gas flow is as laminar as possible, including a very smooth tube wall and a long, straight upstream tube.

Laminar flame speed, is, however, of little use in a real-world situation such as lowpressure tank vents because the nature of the flows and equipment that are in use are far more likely to result in turbulence within the gas mixture. The key factor in the transmission of the flame is the turbulent flame speed. The turbulence of the gas mixture has a significant effect on the flame speed as does the degree of confinement. Turbulent mixtures in confined spaces (such as pipelines) can reach speeds of 300 m/s.

Turbulence in the gas will increase the flame speed since the size of the flame front will be increased by eddies resulting in increased reaction rate. Increased reaction rate and flame speed form a positive feedback loop by causing an increase in turbulence and hence the flame can accelerate. In extreme cases, flame speeds of greater than 1000 m/s can be reached in long pipelines where deflagration can transition to detonation.

In practice, under real-world conditions, there is always a degree of turbulence associated with venting from tanks. Considering a typical LP tank vent system configuration it is clear that under many of the flow conditions that may commonly occur the flow through the system will be turbulent. The turbulence will be caused partially by the changes in section and direction within the system and partly related to the velocity (as the Reynolds number increases).

Note that these maximum velocities are achieved by near-stoichiometric mixtures and flame velocities do fall off away from stoichiometric mixture. However, flame velocities are still high enough to be greater than the flow velocity through the vent system (including the PV valve).

PROTEGO TESTS

Tests done on conventional PVVs by Protego in Germany indicate that PVVs may, in some circumstances allow an ignition to pass through the valve into the tank. Obviously, this would result in an internal explosion and destruction of the tank. Three tests are shown with this paper, these are as follows:

IGNITION IN GAS MIXTURE OUTSIDE CONVENTIONAL PV VALVE

In this test, a flammable gas mixture was held inside a polythene bag secured outside the test valve. The gas mixture was then ignited with both pallets of the valve closed. The flame accelerates through the mixture towards the valve and flashes back through into the vessel. In this case, the test vessel is a short section of open ended pipe. This simulates the remote ignition of a cloud of flammable gas that has been vented from the PV Valve. In this case, the cloud was confined in a polythene bag simply to prevent the dispersal of the flammable cloud. The bag does not have a significant effect on the flame propagation or behaviour.

The most likely source of the passage back into the vessel is via the vacuum pallet. Since the vacuum pallet is designed to open at pressures less of than −10 mbar, it is possible that the pressure outside the valve caused by the burning gas can lift the vacuum pallet against its' weight thus allowing passage of the flame, pressure outside the valve being equivalent to a vacuum inside.

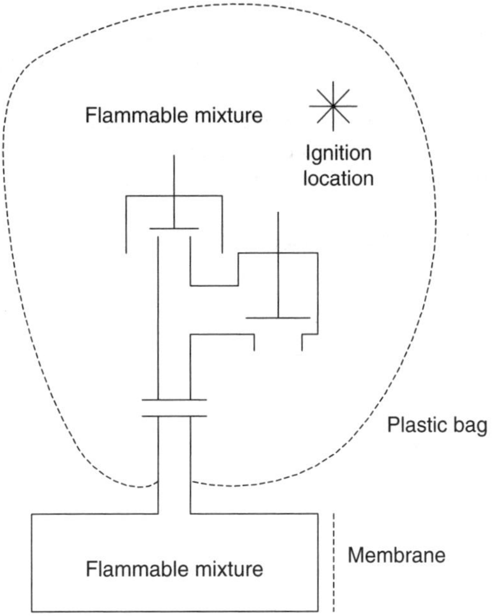

It is known that vapour cloud explosion do not produce a great deal of pressure unless in congested or confined regions, it is, however, foreseeable that an energetic (near stoichiometric) gas mixture might foreseeably generate more than 10 mbar overpressure in the region of the valve, especially if the region around the valve is congested by pipework etc. The issue here is that ignition of a reasonably sized vapour cloud outside of and surrounding a PV Valve may foreseeably result in the generation of sufficient pressure to lift the vacuum pallet. This will then provide a clear route for the flame back into the vessel.

ENDURANCE BURNING

If vented material from the PV Valve continues to burn for an extended period then it is foreseeable that the valve cap, pallet etc will become hot. If this continues to the point at which the temperature of the metal reaches the autoignition temperature of the vapour then a flashback may occur. This test shows flashback through the valve occurring after approximately six minutes. Note that during this test the flame was not played directly onto the valve but was directed to one side and had no significant effect on the heating of the valve.

Thus, heating of the valve occurred solely as a result of the burning of the gas coming out of the PV Valve.

The reason for the ignition of the gas so quickly is that the pressure pallet is of light construction, as required for the duty, and thus will heat up comparatively rapidly. In the video it can clearly be seen that the steel valve bonnet reaches red heat very quickly. This item is generally constructed of light gauge steel since it is only there to provide weather protection. This kind of temperature will also rapidly destroy any polymeric seals on the pallet destroying the valve integrity.

It is noted that this circumstance does rely on the sustained ignition of the gas mixture leaving the valve and the resultant heating of the valve material. It is, however, foreseeable that there will be circumstances, such as during tank filling, where venting may occur for an extended period.

Thus, this is another event that may be considered credible, if a rare event.

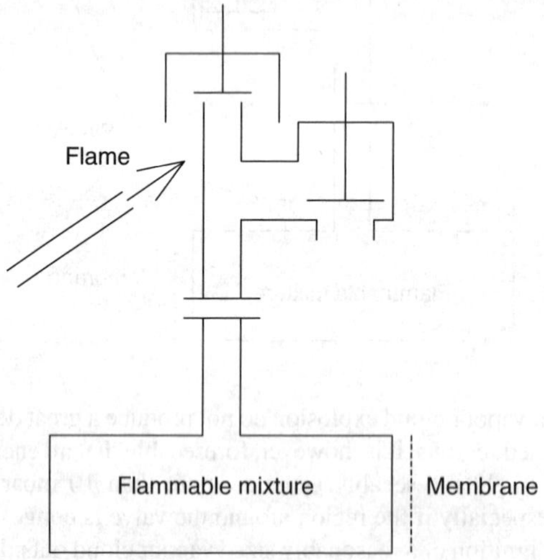

IGNITION OF VENTED GAS MIXTURE

In this test, a flammable gas mixture was vented from the PV valve with an ignition source located outside the valve and not impinging on the valve body. It can be clearly seen on the video that as the valve pallet lifts and flammable gas it vented, it ignites from the pilot flame. The flow of gas results in pallet lifts of approximately 25 – 30 mm. The flame resulting from the ignited gas is seen surrounding the valve. The pallet then drops back until another pulse of gas is emitted. This cycle occurs several times until the flame flashes back through the valve. The resultant pressure results in the destruction of the valve pallet and bonnet.

Since the pallet lift is clearly greater than the MESG of the mixture, the mechanism for transmission must be the turbulent burning velocity of the gas being greater than the superficial gas velocity leaving the valve.

This is considered to be a credible event since the combination of pallet lift and an external ignition source are a foreseeable combination.

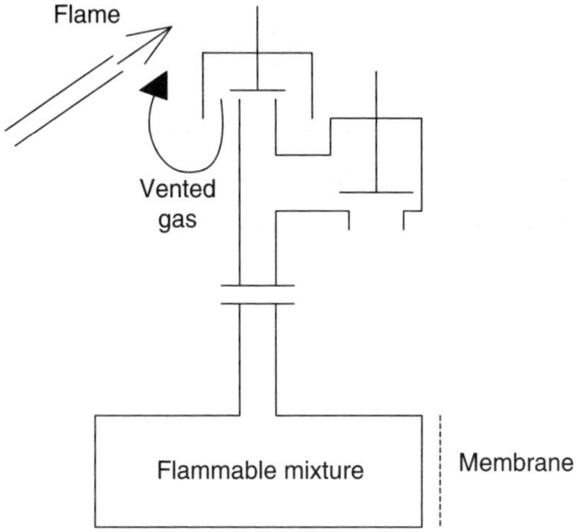

CONCLUSIONS

It can be seen that the installation of a PV Valve on a low-pressure storage tank containing a flammable vapour does not provide total protection against the flashback of an external ignition into the tank. The combination of the lift of the PV Valve pallet (which will, in the vast majority of cases be greater than the MESG of the vapour), the turbulent flame speed and the superficial gas velocity coming out of the valve can result in circumstances where it is theoretically possible for the flame to pass into the vessel.

Based on the tests conducted by Protego, is is clear that this theoretical consideration is, in fact, borne out in reality under test conditions. The test conditions used, whilst they might be considered to be arduous, are, in fact credible, especially in the third case where the ignition of the gases venting from the valve results in flashback into the vessel.

Equally of concern is the flashback through the valve occurring when the valve is closed (the first case) where ignition of an external flammable cloud flashes back through the vacuum pallet.

Thus, it is concluded that there is a small but significant risk of flashback through a PV Valve under conditions of external ignition. On this basis, consideration should be given to the provision of a flame arrester in the vent system in order to provide additional

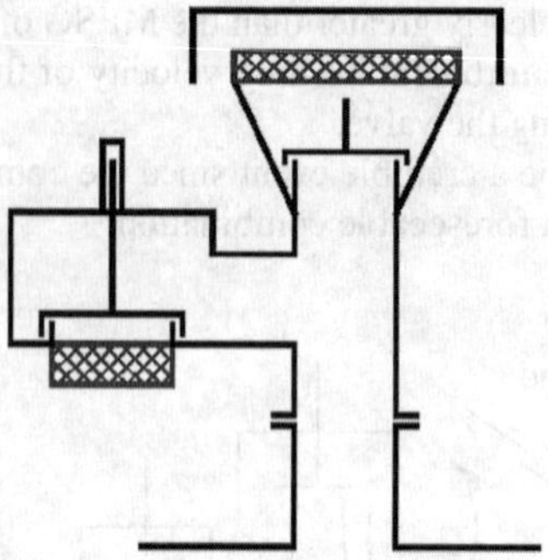

Flame arrester elements shown in blue. Note that other configurations are possible

Diagram 5. PV valve with integral flame arresters

flashback protection. The flame arrester could be located either at the inlet to the valve (between the valve and the tank, Diagram 5) or else on the outlet of the valve.

REFERENCES

1. BS EN 12874: 2001 Flame arresters
2. API 650 Welded Steel Tanks for Oil Storage, 10th Edition
3. EN 14015:2004 Specification for the design and manufacture of site built, vertical, cylindrical, flat-bottomed, above ground, welded, steel tanks for the storage of liquids at ambient temperature and above
4. API 620 Design and Construction of Large, Welded, Low-Pressure Storage Tanks, 10th Edition
5. HSE Contract Research Report CRR281 "Investigations into concerns about BS EN 12874:2001 flame arresters; HSE 2004
6. Protego Pressure/Vacuum Relief Valve Catalogue for type VD/SV-HR
7. Garstein M, Levine O & Wong EL; Fundamental flame velocities of hydrocarbons; Ind Eng Chem V43, pp2770-2772; 1951
8. Coward HF, Jones GW; Limits of flammability of gases and vapors; USBM Report No.503, 1952

METHODS OF AVOIDING TANK BUND OVERTOPPING USING COMPUTATIONAL FLUID DYNAMICS TOOL

SreeRaj R Nair
Senior Engineer, Aker Kvaerner Consultancy Services
Aker Kvaerner, Ashmore House, Richardson Road, Stockton on Tees, UK, TS18 3RE,
E-mail: Sreeraj.Nair@akerkvaerner.com
Web: www.akerkvaerner.com/AKEHSandRisk

Atmospheric storage tanks are one of the main containment methods used on major hazard sites to store feed, intermediate and finished products. The storage tanks can be of different sizes, shapes and positioned at various levels (below ground and aboveground).

It is good practise to have secondary and tertiary means of containment in order to contain and mitigate an event in case of a primary containment (tank) failure. There are many standards and codes specifying the containment philosophy of tank bunding and structural requirements.

Over the years, there have been many cases where the secondary means of containment was not successful / adequate in retaining the material in the event of tank (primary containment) failure. The recent incident at Buncefield [12] is an example. About tank bunds, the following questions arise:

- Are the existing design standards and codes for designing secondary containment adequate?
- Are there any better means of designing secondary containment to assist liquid retention following tank failure?
- How do we ensure that the existing tank bunds are safe secondary means of containment?

This paper tries to answer the above questions by reviewing the adequacy of an existing bund for a tank farm containing multiple storage tanks. Several tank failure types are evaluated using Computational Fluid Dynamics (CFD) and the impact of the failures on bund integrity are investigated. Consequently, the effectiveness of several consequence reduction techniques to reduce overtopping potential is presented.

KEYWORDS: Tank failure, Bund overtopping, secondary containment design, Computational fluid dynamics

INTRODUCTION

Storage of hazardous materials (flammables and toxics) has the potential of loss of containment hazard associated with it and such hazards could affect people and environment. Secondary containment is often used as a second line of defence to prevent, control or mitigate such hazardous events. A secondary containment system is defined as [1]: 'any item of equipment which may help to prevent the spread of an accidental release of a hazardous substance.'

In the case of storage of hazardous materials, the secondary containment could be in the form of bunds and dykes, double skinned tanks and vessels or concentric pipes.

Bunds are generally used around storage tanks where flammable or toxic liquids are held. Bunds are also used within plant areas as a layer of protection for bulk liquid vessels and reactors.

This paper looks into the following aspects of the bunds as secondary containment of a hazardous material storage area:

- Requirements of the bunds (statutory and design)
- Adequacy of the requirements in efficient secondary containment
- Adequacy of the existing bunds and options for improvement.

BACKGROUND

This paper is based on a study conducted on an existing secondary containment for multi-component fluid (mixture) storage. The fluid is both toxic and flammable and is stored in a fixed roof tank with a storage capacity of 4560 m^3 (4,000 tonne). The tank (hereafter referred as Tank A) is in a common bund with another intermediate product storage tank (Tank B). The general tank and bund arrangement is as given in Figure 1.

The study has been performed using computational fluid dynamics (CFD) tool *fluidyn*-NS in order to:

- Determine the liquid behaviour in the event of loss of containment
- Determine the pressure exerted on the bund walls
- Determine the best option to ensure an efficient secondary containment.

The primary emphasis of the study was to determine the degree of retention of materials within the exiting bund (Secondary containment) under several loss of containment scenarios in order to identify solutions which would reduce the major hazard risk and minimise the contamination of land and/or water course.

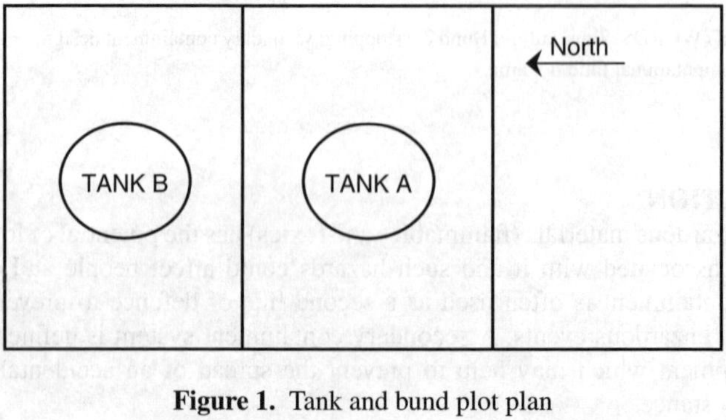

Figure 1. Tank and bund plot plan

EFFECTIVENESS AND ISSUES IN BUNDS AS SECONDARY CONTAINMENT

The secondary containment can be less effective if not adequately designed, constructed and maintained. In some cases inefficient bund design even could worsen the situation, e.g. high bund wall to reduce bund overtopping may provide sufficient confinement for explosion. The common issues identified in bunds as an effective means of containment in the event of failure of the primary containment (tank and allied system) are given below:

DESIGN AND CONSTRUCTION ISSUES

- Bund capacity: Adequacy to contain the volume of stored material in the tank/s
- Dimensions and layout of the bund: The distance between the tank and bund walls, wall design (height and width of the bund walls), the equipment and piping within the bund area
- Material of construction: Best suitable material based on the fluid within the tank (mechanical strength, the vaporisation rate and resistance to thermal shock) and ability to withstand the atmospheric deterioration
- Integrity: Bund wall strength against the static and dynamic loading from the fluid in the event of an incident, Construction / expansion joints and pipe penetrations can fail in the event of leak and fire if it is not adequately designed and constructed
- Bund floor and surface water drainage: If the bund floor is pervious or surface water drainage is inadequate, the material spill could penetrate and reach the soil as well as the water bodies
- Common bunding: Incompatibility of the materials stored, the spacing between and the cascading effects of an incident could be of concern in common bunding.

OPERATION AND MAINTENANCE ISSUES

Improper and inadequate maintenance could result in deterioration of the bund like:

- Growth of plant life within bund affecting the integrity
- Cracks on walls and floors could result in seepage in the event of liquid release
- Accessibility restrictions to the tank and allied facilities for routine activities
- Improper surface water drainage (valve left open or closed).

There are many incidents where the bunds were not efficient in containing the fluid in the event of failure of primary containment. Some of the issues identified are given below:

- The bund joints and manifolds failed and the fluid with firewater reached near by water bodies (Buncefield incident, 11 Dec 2005 [12]).
- Restriction of fire fighting effectiveness due to high bund wall
- Overspill of the fluid due to fire water resulting from insufficient bund capacity
- Overtopping of the bund and destruction of bund wall (LPG storage tank incident, Qatar [2,4], liquid fertilizer tank incident Ohio, 2000 [8]).

Some other issues of concern could be:

- Bund overtopping from a leak on tank shell and the release jet hitting the ground outside the bund.
- Stored liquid can vault an inclined side or pile up rapidly at the face of a bund wall and then flow over the top.
- A strong shock wave forming at the bund wall and then returns towards the storage tank.

WHAT DO THE STANDARDS AND STATUTES SAY ABOUT SECONDARY CONTAINMENT?

There are many standards and codes, research reports and guidance, which lists and guides through the requirements and design specification for various means of secondary containment. The documents that address the secondary containment are:

- SRD R 500, The Design of Bunds, Safety and Reliability Directorate, United Kingdom Atomic Energy Authority.
- Technical measures document for secondary containment, HSE, UK.
- NFPA 30, Flammable and Combustible Liquids Code, US.
- HS(G)176, The Storage of flammable liquids in tanks, UK.
- COMAH Guidance, HSE, UK.
- Contract Research Report, CRR 324/2001, HSE UK.
- CIRIA Report 163, Construction of bunds for oil storage tanks.

In general, for bund capacity and integrity, the following are the main requirements specified in the above documents.

Capacity: Codes differ in their recommendations on bund capacity, which vary between 75% and 110% of the normal capacity of the tank protected. The basis of the recommendation is that bund should have sufficient capacity to contain the largest predictable spillage [4, 6, 10, 11]. Data quoted by Barnes from the General Accounting Office (GAO) report identifies a capacity range from 50% to 139% [2]. Where two or more tanks are installed within the same bund, the recommended capacity of the bund is 110% of the largest tank or 25% of each tank within the bund [9].

Bund wall dimension: There are no general rules regarding the ratio between wall height and floor area. Codes vary greatly with respect to bund wall height recommendations [4]. A low bund wall facilitates fire fighting. In the US NFPA stipulate a minimum of 1.5 m for walls. In the case of flammable and combustible liquids both UK and US NFPA codes of practice restrict the bund wall heights to 5 feet (1.5 m) and 6 feet (1.8 m) respectively [2]. Many codes of practice do not state maximum height for bund walls. For high walled bunds, consideration will need to be given to the possibility of tanks floating as the bund fills, causing catastrophic failure [6]. It is recommended that a freeboard of 250 mm is provided to protect against dynamic effects [9]. It is also advised that the bund wall should be sloped to prevent liquid accumulation beneath the storage tank.

Mechanical strength: Care must be taken in the design of the bund wall to withstand the dynamic loads upon bund walls when a large liquid release occurs (Dynamic load at the base of the bund wall may be six times the hydrostatic pressure) [2, 7, 11]. The bund walls should also be impervious to liquid and the wall should be capable of withstanding full hydrostatic head [9]. The bund wall should have sufficient strength to contain any spillage or fire fighting water [2]. The secondary containment shall be designed to withstand the hydrostatic head resulting from a leak from the primary tank of the maximum amount of liquid that can be stored in the primary tank [10].

Materials of construction: Should be capable of withstanding the mechanical and thermal shock that occurs on catastrophic failure of the primary containment [6].

Integrity: The bund should be liquid tight (especially if pipes and other equipment penetrate through the wall) [2]. It is recommended to route any pipes over the wall of the bund to avoid the penetration together of the bund wall [9]. The floor of the bund should be concrete or other material impervious to the liquid being stored [2].

ADEQUACY OF THE GUIDANCES AND CONTAINMENT ISSUES OF CONCERN

The standards and codes mentioned in the above section give guidance on the design requirements of the bunds. Some of the requirements are material specific and holds good for that application whereas those may be irrelevant for a different material. It is noted that the topic of storage is dominated by flammables and detailed specifications are available for LPG, LNG, Hydrogen and Ammonia. However, for toxic materials, inherently safer design and high integrity design are normally specified.

With the situation under consideration (as given in the background), it is noticed that sufficient guidance could not be found for the following issues.

- Bund wall strength: Some of the codes address the need for bund to be able to withstand dynamic loading, but no stringent requirements are made. Bunds made without dynamic loading criteria may still hold good for small leaks and spills but may completely fail in case of an instantaneous release.
- Bund capacity: The general requirement is for 110% of the largest tank within the bund area, but the dimensions are not specified. The 110% volume/capacity could be achieved by large bund area with low bund walls or small bund area with high bund walls. Both options have its own merits and demerits depending on the material handled, the topography and the ease of handling.

Health and Safety Executive (HSE) commissioned Liverpool John Moores Univeristy (LJMU) to perform simulations of catastrophic failure of a storage tank and to measure the dynamic pressures on the bund wall and the quantity of liquid that overtops the bund. The results of the experiments have been published in the CRR 333 [8].

ESTIMATING TANK BUND OVERTOPPING

The design of a bund wall can have a significant influence on the stored fluid behaviour in the event of a tank failure. By acknowledging the fact that effectiveness of secondary containment depends upon an adequate design, which considers all possible failure modes of the primary containment, the following containment issues are dealt herewith.

- Dynamic effects from the wave generated by a catastrophic tank failure which could result in bund failure, overflow or both,
- A spigot flow (jetting), this occurs when tank is punctured resulting in liquid jet to hit the ground or beyond the bund.

This paper will address these issues based on bund overtopping and overpressure on bund walls.

Bund overtopping: Two methods have been used to estimate the bund overtopping:

- LJMU correlation given in CRR 333 [8], and
- Computational fluid dynamics modelling (CFD).

LJMU CORRELATION METHOD

The correlation derived from the experimental investigation by LJMU [8] is used in this method to estimate the bund overtopping following catastrophic failure of a storage vessel. The function below is recommended:

$$Q = A \times \exp(-B \times (h/H)) \quad (1)$$

where:

Q (Overtopping Fraction)
A, B = constants
h = height of the bund
H = height of material in tank

The range of validity is $0.66 \leq (r - R)/R \leq 5.32$ where r is the bund radius and R is the tank radius.

Input values

The equivalent bund radius (r) is 41 m and tank radius (R) is 11.5 m. The tank and bund dimension ratio falls within the range of validity of equation 1.

The height of the bund is 1.5 and the height of the material in tank is 13.22 m (12.1 m liquid height + 1.12 m plinth height). The tank and bund dimensions are given in Figure 2.

The constants A and B are 0.6359 and 2.4451 respectively for Middle category tank (R/H ~ 1) and 150% bund capacity [8].

The total tank volume is 4562 m^3.

Result

By substituting the values and solving equation 1, an overtopping of 48.1% (2194 m^3) is estimated.

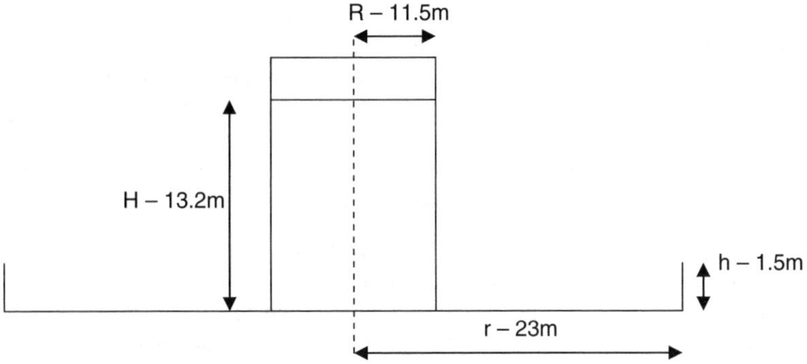

Figure 2. Tank and bund dimension – LJMU

Limitation of the method:

- This method can be used only for overtopping following catastrophic rupture (shell disappearing) failure cases,
- This method does not take into account obstructions like another tank sharing the bund (Figure 1)

COMPUTATIONAL FLUID DYNAMICS METHOD

A representative set of four different scenarios have been considered in order to determine the bund overtopping and pressure exerted on the bund walls:

1. Complete tank rupture,
2. Zip opening of tank bottom,
3. Big hole on tank shell,
4. Small hole on tank shell.

The fluid behaviour following the release in all the four scenarios has been modelled using computational fluid dynamics approach.

Modelling tool

The following modelling tools were used:

CADGEN – geometry and grid generation
Fluidyn-NS – digital model to solve wave effects

CFD tool *fluidyn*-NS is 3D software offered by Fluidyn and Transoft International.

Grid Generation (CADGEN)

The geometrical model has been considered in three dimensions. The software *fluidyn*-CAD has been used to create the geometry of the tanks and bund under consideration.

It was based on the site map as well as the geometrical description of the tanks and the walls/bunds (thickness, height, and distance to the tanks).

The volume within the boundary was divided into discrete cells (the mesh/grid). In this study structured grid has been considered and the grid is finer close to the surfaces (ground, walls) in order to ensure a good description of the pressure loads. For partial ruptures (zip) or the jet (small and large holes), the grid is finer at the opening to improve the precision. Following assumptions have been made for simulations:

- The flow is isotherm, incompressible and laminar. These choices are considered as the best to represent the cases in this study.
- Turbulence effects are neglected and surface tension is not taken into account. From simulations exercises done previously, it was noted that turbulence and surface tension has no / negligible effect on wave behaviour,
- The pressure values estimated on the monitor points (on bund wall) correspond to the dynamic pressure exerted by the fluid wave motion following release,
- The gravity is set up at 9.81 m/s^2.

Following assumptions have been made for boundary conditions:

- The walls (ground, base, tank envelope, retention walls) are considered rigid, adiabatic and smooth. Other boundaries are set up as pressure outflows,
- The liquid is considered motionless at the beginning in the tank. At the initial time (t = 0 s), an opening is created on the wall of the tank for the partial rupture scenarios. In the case of the catastrophic failure, the tank shell is deemed to have vanished. A wave is then formed which impacts the walls surrounding the retention.

Model input

Scenario Description:

Case 1 – Catastrophic rupture of the tank: This scenario considers the liquid behaviour in the event of an instantaneous removal of the tank shell and the subsequent collapse of the column of liquid.

Case 2 – Horizontal zip open: This scenario considers a tear at the bottom of the tank shell. The dimensions of the tear were 18 m lateral and 0.2 m wide. The orientation of the opening is as given in Figure 3.

Case 3 – Big hole in the tank shell: This scenario considers failure of one of the tank inlet pipeline, with a horizontal release of the fluid mixture. The hole is 2 m above the ground and its diameter is 0.2 m. The release direction was oriented towards east.

Case 4 – Small hole in the tank shell: This scenario considers failure of one of the tank inlet pipeline, with a horizontal release of the liquid. The hole is 1.5 m above the ground and its diameter is 0.076 m.

Geometry of the retention bund

Tanks A and B has the same dimensions: diameter 23 m, height 12.6 m. The height of the liquid inside the tank for simulations is 12.1 m. The height of the basis (plinth height) is

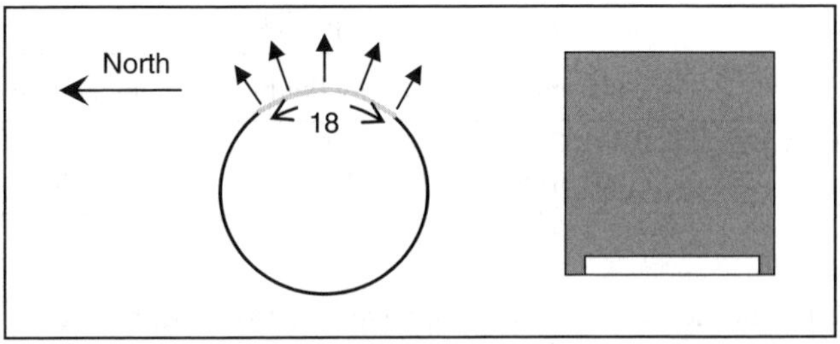

Figure 3. Schematic description of horizontal tear

taken at 1.12 m. The geometry of the bund is detailed in the sketches in Figure 4. The geometry has been simplified by assuming that:

- Every pipe and other equipment are not taken into account,
- The slope of the terrain in the retention bund is not taken into account.

The slope (0.274°) and piping within the bund is not expected to make any significant difference in the fluid behaviour in the current study and hence not taken account of.

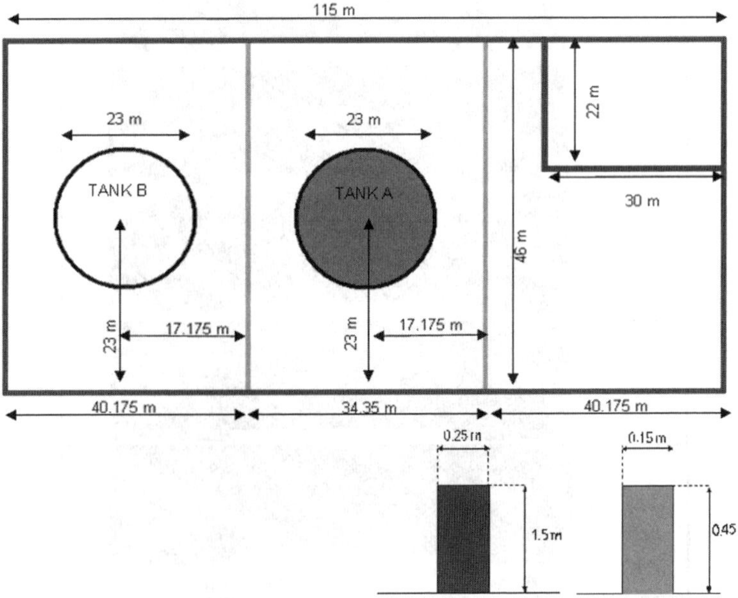

Figure 4. Layout of the retention bund

Table 1. Fluid properties

Product	Viscosity (mm^2/s)	Density (kg/m^3)	Temperature (°C)
Fluid mixture	0.441623	800	20
Air	1.895 10-5	1.29	20

Fluids

The liquid stored in the Tank A is an intermediate product with mixture of toxic and flammable materials with water. The operating temperature is set at 20°C. In this case, the thermo physical properties of the fluid will be set as given in Table 1.

RESULTS

Case 1: Complete rupture of the tank (Shell vanishing)

The flow of the fluid as a function of time is shown in Figure 5. It shows that the liquid reaches bund walls at 1.36 seconds after the tank failure and begins to overflow after 1.69 seconds.

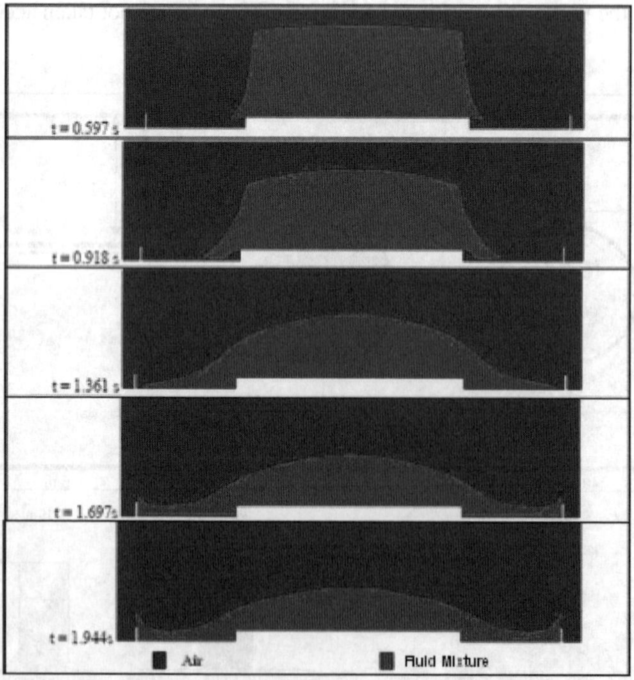

Figure 5. Case 1 – fluid flow from catastrophic rupture of the tank

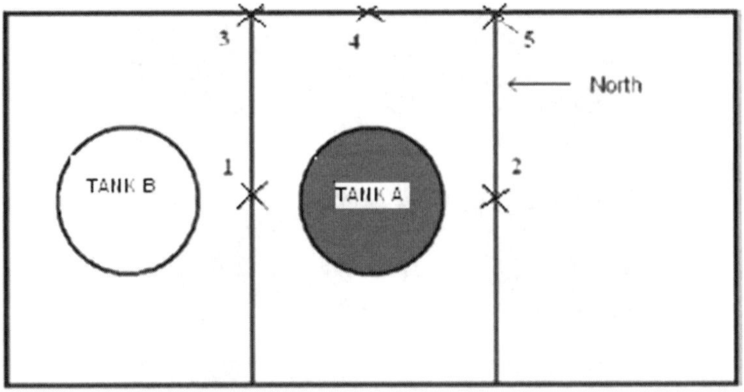

Figure 6. Case 1 – monitor point groups for reading dynamic pressure on bund wall

Overpressure on the walls

The calculations have been run until 31.6 s after the beginning of the tank collapse in order to estimate the maximum overpressure on the bund walls. Overpressure calculations were performed on the five monitor point groups on the bund wall as shown in the Figure 6.

Maximum overpressure estimated was 1.05 bar on the bund wall corners (monitor points 3 and 5) and this occurred after 1.86 s after instantaneous release. The maximum pressure on east wall (monitor point 4) was estimated to be 0.99 barg.

Bund overtopping following complete rupture of the tank

With a storage capacity of 4562 m^3, it is estimated that 72.8% of the volume of the fluid of Tank A spilling over the retention wall.

The result estimated using CFD method (72.8%) is 1.5 times higher compared to the result estimated using LJMU method (48.1%). The difference in estimation from two methods could be as the result of following:

- LJMU method is based on experiments using water as the fluid and the methodology need not be accurate for fluids with different physical properties (e.g. for fluid which is viscous than water). CFD is based on the fluid viscosity and hence the wave effect for another fluid could be different even with same release conditions (similar tank and bund).
- LJMU method is limited to a single tank surrounded by a single bund wall. For our present study, there is intermittent kerb walls and another tank in the same bund. CFD takes account of the actual dimensions of bund wall and also any obstructions or restricting structures on the wave path (like kerb walls and near by tank)

Case 2: Release from Bottom Zip Tear on the Tank

The flow of fluid as a function of time is shown in the Figure 7. It shows that the onset of liquid overflow occurs at 1.34 seconds.

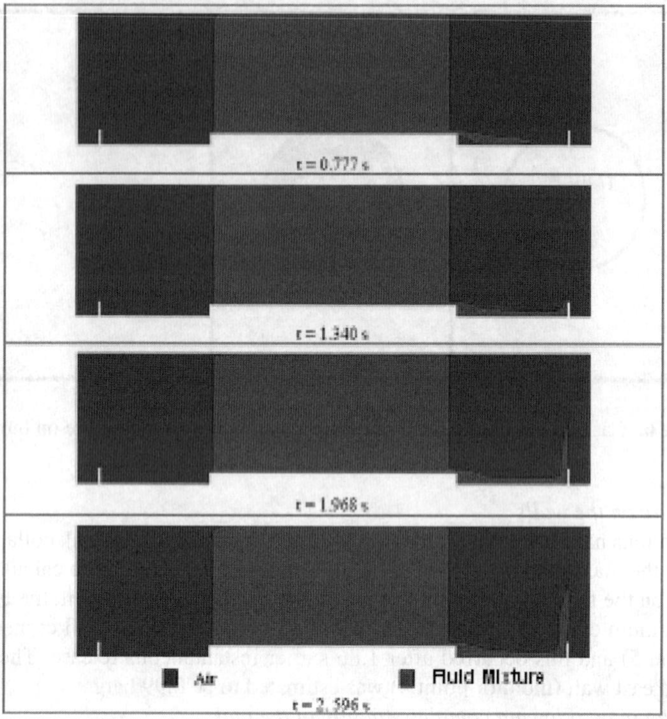

Figure 7. Fluid flow Case 2 – horizontal zip open

Overpressure on the walls

Overpressure calculations were performed at three monitor point groups on the bund wall as shown in the Figure 8. Maximum overpressure estimated was 0.106 bar on the bund wall (monitor group point 2) facing the release direction (in this case East wall). This was obtained after 4.11 s of release.

Bund overtopping following bottom zip tear on the tank

With a storage capacity of 4562 m³, it is estimated that 49.8% of the volume of the fluid of Tank A spilling over the retention wall.

Case 3 and Case 4: Release from Big Hole and Small Hole on the Tank

Overpressures on the walls

For Case 3, the maximum overpressure estimated was 0.0255 bar on the bund wall facing the release direction (in this case bund wall on east side). This was obtained after 1.3 s of release.

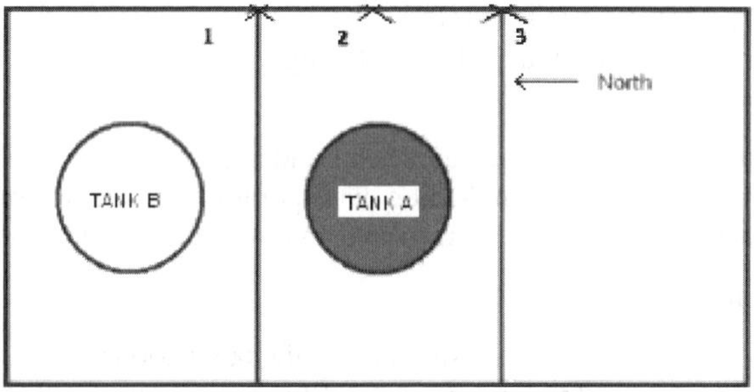

Figure 8. Case 2, 3 and 4 – monitor points for reading dynamic pressure on bund wall

For Case 4, the maximum overpressure estimated was 0.0024 bar on the bund wall facing the release direction (in this case bund wall on east side). This negligible pressure on the bund wall from the release was obtained after 1.63 s of release.

Bund overtopping

There is no overflow for both Case 3 and Case 4, as the liquid release falls and remains within the retention bund. The illustration of fluid flow is following Case 3 is given in Figure 9.

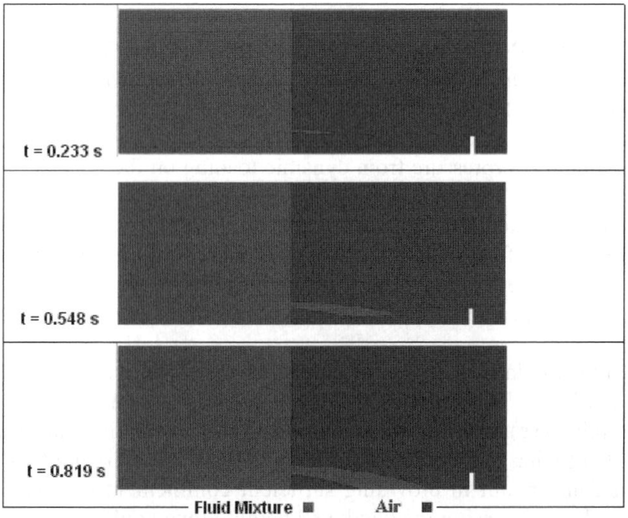

Figure 9. Case 3 – fluid flow from big hole

ADEQUACY OF THE BUND DESIGN:
The following are inferred from the CFD modelling exercise:

Bund Capacity:

- Adequate to contain small release events and inner kerb wall restricts the pool size and there by considerable reduction in evaporation and dispersion,
- Exiting bund capacity inadequate to contain instantaneous and major release events.

Pressure on Bund Walls:

- Maximum overpressure was estimated on the bund corners.
- Need to ensure the structural integrity of the bund wall to withstand the dynamic loading (maximum 1 barg) from wave effects.

DESIGNING EFFECTIVE SECONDARY CONTAINMENT:
In order to address the deficiencies, changes in bund design have been envisaged. The changes in design are to be made considering that, the new containment takes account of no overspill or minimal overspill. From the incident history [1], the key issues identified that could limit the overspill are:

- The height of the bund wall,
- The separation distance (between tank shell and bund wall/other tank), and
- The strength and integrity of the bund wall.

DESIGN OPTION: BUND WALL WITH INCREASED HEIGHT
CFD modelling has been performed for wave behaviour following catastrophic rupture (Case 1) with bund wall height of 3.5 m (1.5 m + increased height of 2 m). The schematic diagram of the fluid flow is shown in Figure 10.

The maximum overpressure from dynamic loading on the bund wall was 1.05 bar (no change in maximum overpressure due to increase in height). However, with the high bund wall, it is estimated that1642 m^3 of the fluid would overtop. Even though the overtopping is reduced from 72.8% to 32.6% (by increasing the bund wall height from 1.5 m to 3.5 m), the overtopping volume is significant. Results from similar studies [8] also indicate a serious problem if bunds are to be relied upon in the event of a catastrophic tank rupture.

From CFD simulations, it was revealed that the wave rises as high as the tank B shell height (~13 m). This implies that even by raising the bund wall as high as tank shell (in this case) will not be sufficient for 100% containment. Also increasing the bund wall height has other concerns as it could restrict the fire fighting effectiveness and access, it could result in providing sufficient confinement for explosion. Hence, it is concluded that, increasing the bund wall height alone could not solve the overtopping issue.

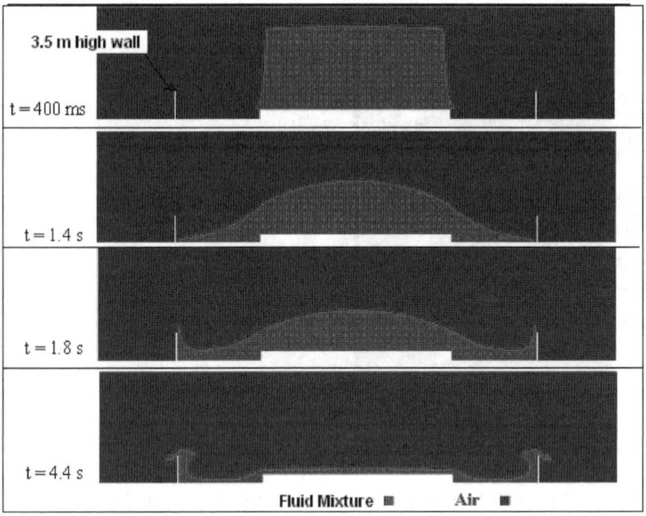

Figure 10. Fluid flow – bund wall with increased height

DESIGN OPTION: BUND WALL WITH DEFLECTOR ATTACHMENT

Bottom zip tear scenario (Case 2) is considered to be a credible scenario compared to catastrophic rupture (shell vanishing) and for this exercise release from bottom zip tear is considered.

CFD modelling has been performed for wave behaviour following bottom zip tear (Case 2) with bund wall height of 2 m (1.5 m + deflector at 45° angle). The schematic diagram of the deflector arrangement is shown in Figure 11.

The maximum overpressure from dynamic loading on the bund wall was 0.06 bar (0.106 bar without deflector attachment). With the deflector, the wave is deflected and fluid remains within the bund area. The overtopping estimated is negligible (11.4 m^3 or 0.25%) and hence, this is considered as a suitable design option for avoiding or minimising bund overtopping.

Some other design options, which could be considered to address the deficiencies in bund design, are:

- Wave action blocker (mesh/gate) in between tank shell and bund wall
- Additional bund outside the current bund on the tank farm perimeter
- Surge drop channel between tank shell and bund wall

WAVE ACTION BLOCKER IN BETWEEN TANK SHELL AND BUND WALL

In this option, a mesh/grill member or intermittent stopper walls to be installed in between the tank shell and the bund wall in order to block or interfere the wave action.

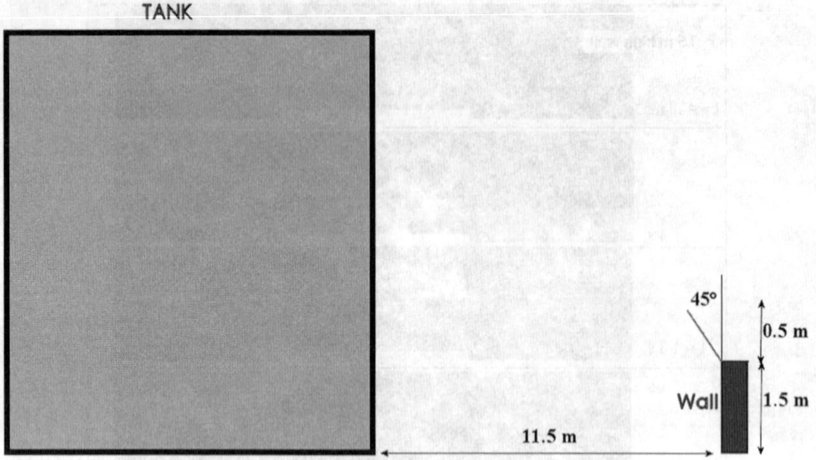

Figure 11. Design option – bund wall with deflector

The additional blocker installed will reduce the wave effects and thus limit the overspill. The disadvantages of this option could be the difficulty in maintaining the system and restriction in access.

ADDITIONAL BUND OUTSIDE THE CURRENT BUND ON THE TANK FARM PERIMETER
In this option, an additional bund to be constructed surrounding the existing bund to limit the spread of fluid spill. The size and position of the additional bund shall be defined based on the hazard management plan and the site emergency preparedness. The disadvantages of this option could be that more area is required for containment and increased rate of evaporation from liquid pool.

SURGE DROP CHANNEL BETWEEN TANK SHELL AND BUND WALL
In this option, a channel to be built in between tank and bund wall in order to reduce the surge resulting from the instantaneous release or bulky continuous release of fluid. This wave motion could be restricted by big single channel or a sequence of channels based on the fluid properties and release type of consideration.

CONCLUSION

This paper looked into the adequacy of ensuring bunds as an effective secondary containment and is based on a study performed on containing releases from a tank in a common bund. Computational fluid dynamics (CFD) modelling tool was used to simulate wave effects of the fluid motion and to estimate the two identified issues of concern; overspill and overpressure on bund walls. Liverpool John Moores University (LJMU) correlation [8] was also used to estimate the overspill volume and compared with CFD method.

A literature search on various standards, codes and statutory design requirements have been performed and generalised. It was noted that no stringent specifications are available to address tank bund design and maintenance against overflow and overpressure.

The study using two methods estimated the bund overflow as 48% (LJMU correlation) and 73% (CFD modelling) following a catastrophic failure of the tank. Using CFD, bund overflow estimations have been performed for releases from horizontal zip open (50% overtopping), big and small hole (no overtopping). The maximum pressure (1.05 barg) estimated on the bund wall was on the corners following the catastrophic failure of the tank.

The study further looked into various design options to avoid or minimise overtopping using CFD tool. Option to increase the bund wall height will reduce the overtopping, but not a preferred option for 100% containment. Option with deflector attachment on existing bund wall results in negligible overtopping and considered as a solution to avoid bund overtopping in the discussed case. For an efficient secondary containment in tank farms, any single option or combination of options discussed in this paper shall be used based on the requirements, material handled and site limitation.

REFERENCES

1. Clark, S.O., and Deaves, D.M., 2001, Effects of secondary containment on source term modelling, HSE: Contract Research Report – 324/2001
2. Barnes, D.S., 1990, The design of bunds, HSE: SRD/HSE R 500
3. HSE, 1998, HS(G) 176 – The storage of flammable liquids in tanks'
4. Less, F.P., 1996, Loss Prevention in the process industries – Hazard identification, assessment and control, Butterworth Heinemann: Second edition
5. Hunt, D., 2005, Computational Fluid Dynamics – Tessella Support Services plc: Issue V1.R1.M4
6. HSE, Technical measures document referring to secondary containment, http://www.hse.gov.uk/comah/sragtech/techmeascontain.htm
7. Wilkinson A, 1991, Bund overtopping – The consequences following catastrophic failure of large volume liquid storage vessels, HSE: SRD/HSE R 530
8. Liverpool John Moores University, 2005, An experimental investigation of bund wall overtopping and dynamic pressures on the bund wall following catastrophic failure of a storage vessel, HSE: Research Report 333
9. Mason, P.A., Amines, H.J., Sangarapillai, G., Rose, G., 1997, Construction of bunds for oil storage tanks, CIRIA: Report 163
10. NFPA, 2008, Flammable and Combustible Liquids Code, NFPA 30
11. HSE, 1995, Control of fire-water run-off from CIMAH sites to prevent environmental damage, EH 70, HSE Books
12. Buncefield Investigation Homepage, http://www.buncefieldinvestigation.gov.uk/

HAZARDS IN THE MARITIME TRANSPORT OF BULK MATERIALS AND CONTAINERISED PRODUCTS

J B Kelman, Head, Fire and Explosion Department,
CWA International, Balmoral House, 9 John St. WC1N 2ES London.

Risk management of the shipping of raw materials and of finished products and manufactured goods is of vital importance to the process industries. This paper focuses on the supply chain in maritime context and the incidence of hazardous events, such as vessel stranding, fire and breach of containment, which occur within international shipping. The risk factors involved require an in depth knowledge of the maritime industry, the materials and products in question and the mode of safe storage while in marine transit.

A targeted review of previous energy transport incidents and their analysis, and an analysis of the growth in transport in the alternative fuels sector is provided. The increased transport of raw materials for the biofuels industries and the distribution of alternative energy source products presents new challenges for the maritime industry.

Examples of developing risks associated with this growth in the transport of unstable raw materials, temperature and moisture sensitive products are presented. This includes the fumigation of grains, self-heating of raw vegetable oil products and the stability of developing energy storage mediums. Guidance for the safe transport of these raw materials and products is suggested.

INTRODUCTION

The marine transport industry tonnage accounts for 90% of world trade. This is borne by the worlds merchant fleet which comprises around 50,000 vessels, and some 650 million deadweight tonnage (dwt) as of January 2006. Whether it involves the delivery of raw product to a manufacturer, or a final product to an end user, marine transport is a critical part of the supply chain. These figures make it the most significant link in connecting the producer with consumer. The merchant fleet primarily comprises general cargo vessels (38%), tankers (25%), bulk carriers (14%), container carriers (7%) with passenger vessels and others comprising the remaining (19%). The container fleet made up some 160 million dwt of the total tonnage showing that container ships total around 4 times as much tonnage as the average. The significant change over the last 15 years is the growth of the container fleet and the corresponding decline in the general cargo fleet. The study of container carriers and their hazards is worthwhile in assessing the risk of incidents. General cargo vessels have generally had higher incidents rates than the rest of the world's fleet, in a large part due to the nature of the fleet operations, numerous, smaller, older ships operating in less developed area of the world. The growth of the container fleet, while taking trade from the general cargo fleet, has put additional stresses on crew and command.

Successful fleet operations require good ship and crew management, minimising risk and maintaining safety standards. Close monitoring of ship operating procedures and reacting to potential problems at the earliest possibility assists in minimising hazards in shipboard systems and cargo and reducing incident rates.

While the maritime transport industry continues to evolve, new products and new carriage methods introduce new and less well understood hazards. This is true of container shipping. The largest container ships within the New World Alliance are estimated to be capable of carrying around 14,000 twenty foot equivalent unit containers (TEU). The accumulated risk, and potential loss from a single incident has reached new proportions. It may be noted that high value cargoes are increasingly using container ships for transport. An example in the food sector include frozen shrimps, which may near US$ 1m for a single refrigerated container. Previously the cargo would have moved in dedicated refrigerated "reefer" vessels. A single incident that causes power loss to a vessel has the potential to create significant losses of these types of perishable cargoes. While these cargo changes take place, the crew of container ships do not have the same involvement with their cargo that a reefer crew may have had. This detachment of crew and cargo contains its own risks.

The changes in shipping transport are not only affected by the growth in the market, but also by changes environmental legislation, that have unforeseen consequences [Beale 2000]. The banning of halons has made dealing with shipboard fires more difficult, and had the unusual effect of impacting some fumigation processes, resulting in the use Aluminium Phosphide rather than Methyl Bromide for fumigating grain cargoes. Whereas Methyl Bromide was a fire suppressant, phosphine is spontaneously combustible at higher concentrations, leading to some shipboard fires in situations that were previously unheard of.

HAZARDS

There is a large variation in the hazards to cargo within the marine industry dependant on the vessel type and its cargo. Shipboard incidents may also be impacted by the perceived hazard that the cargo presents. In contrast to what may be expected, frequently the most outwardly hazardous goods have a lower frequency of serious incident than the more benign cargoes. Liquefied Natural Gas is a case in point with an unsurpassed safety record. Much of this results from a high level of training and new, well maintained vessels. The close link between the cargo and the crew maintains a constant awareness of the risk. In contrast, general cargo vessels have a poor safety record, with roughly nine time higher total losses than LNG carriers [DNV 2006]. Petroleum products, and liquefied natural gas are seen as dangerous, and handled carefully, food cargoes are perceived as "safe" and less attention may be paid to careful handling. The economic risk however, may not be reflected by the perceived safety of the cargo.

GROUNDINGS AND STRANDINGS

Groundings or strandings comprise the majority of serious or very serious incidents for cargo vessels resulting from adverse weather, navigational errors or machinery failures and in some rare cases, uncharted dangers. The impact on cargo operations may range from simple delay to total loss. Cargo damage may be direct mechanical damage, loss of containment or contamination by sea water or bunker products. Losses may not be limited to initial stranding events. Additional hazards may be encountered during the recovery and trans-shipping of cargo.

Groundings and strandings resulting from adverse weather tend to have more serious consequences. The recent events surrounding the *Pasha Bulker* in Australia show the danger in which a ship may find itself, even if, in this case there was no cargo onboard, see Figure 1. However, for the process industry relying, in this case, on coal, the extended stranding of the *Pasha Bulker* still resulted in costly delays. In the similar, earlier case of the *Sygna*, also waiting off Newcastle in adverse weather the result of the stranding was a total loss.

The *Pasha Bulker* was not laden, avoiding the complications involved in trans-shipping cargo. The transhipping of cargo from a stranded vessel increases the hazard exposure of the cargo through several mechanisms. In the case of oil and chemical products, accounting errors for the product can add up as the main cargo is removed in smaller consignments, even if there is no apparent loss of containment. The level of contamination also increases as the product is transferred, both from sea water and other product.

Major strandings of laden ships, such as the *MSC Napoli* and *APL Panama*, lead to significant cargo loss. This may be either through loss of the cargo overboard, as in the *MSC Napoli* and through indirect loss such as spoilage of frozen or refrigerated cargoes. The cost or the loss of power loss should not be underestimated. The case of shrimps has been mentioned above, but it is worth noting the delays in getting the large container carrier *Hyundai Fortune* (which suffered a major explosion and fire in March 2006) to port

Figure 1. Pasha Bulker aground off Newcastle, Australia, 2007

and under repair resulted, in no small part from the biohazard resulting from hundreds of tonnes of defrosted rotting fish in unpowered reefer containers.

The subsequent trans-shipping of containers also exposes container cargoes to additional hazards through repeated non standard handling, such as helicopter lifts in the case of the *APL Panama*, see Figure 2, and poor security of containers in regions not designed for container storage.

FIRE

Fires account for around 25% of all losses in the containership fleet, and around 10% of the fatalities. Although most fires start in engine rooms and are contained by engine room carbon dioxide systems, hold fires tend to spread and cause more widespread damage. In many situations it is the variety of cargoes being carried that make control of hold fire difficult. Hazards are increased due to the difficulty of access once a fire has initiated.

In all vessels the problem of cooling the cargo and the fire is paramount. While the widespread use of water is possible on land, it has serious effects on vessel stability and can only be used with care. While the use of CO_2 systems is reasonably effective in suppressing fires, it has limited impact on cooling down the seat of the fire, and is ineffective if the hold has been breached. Leakage of the CO_2 from the hold, or the opening of the space after the fire assumed to be extinguished frequently results in the fire rekindling. There is also a limited supply of CO_2 on vessels, limiting the ability to flood an onboard space repeated times. While fires in container cargoes are difficult to fight, they also tend to propagate through a hold relatively slowly, and generally only upwards. Other cargo in the hold of a container vessel is therefore unlikely to be affected by the fire unless in the immediate vicinity or particularly susceptible to heat or smoke damage.

Figure 2. Container removal from the APL Panama on Ensenada Beach, Mexico, January 2006

Examining the type of cargoes that are prone to fire initiation, they fall in two broad categories: those that are susceptible to external sources of ignition, such as hydrocarbons, and those that are capable of self heating to ignition, if held at slightly elevated temperatures or exposed to moisture or incompatible products. While the ignition hazards of hydrocarbons are well recognised, the self heating or auto-ignition of commodities is less clearly understood. Products ranging from organic seeds, and processed vegetable oils through to direct reduced iron and even rechargeable batteries, have resulted in unexpected fires through self heating.

While container ship operation tends to be highly professional with few foundering, strandings or collisions, the risks of fire damage to cargoes is reflected in the statistics. Groundings and strandings tend to impact the vessel more than the cargo, while fires result in serious cargo losses. Although fires are a direct hazard to cargoes, indirect side effects may have as much impact on the cargo loss as the fire itself. In many cases extensive water use for fire fighting results in more spoilage of the cargo than the fire. This is often the case in container vessels, where a fire in the upper hold damages relatively few containers, while the fire fighting water floods all containers in the lower tiers.

INCIDENTS

A study of some incidents within marine transport highlights the complexity of the problems that may be encountered and also the rapid progression of an apparently small error into a serious loss situation. Individual vessels are, in effect, a small self contained community, housing crew, power generation and cargo, while transiting a frequently hostile environment. Even when the sea is not overtly hostile through severe weather, it remains corrosive. In the vicinity of land, the vessel is subject to currents, tidal streams and on occasion, reduced visibility and congested traffic. All these factors require the utmost vigilance on the part of the vessel's crew to maintain safe passage for the vessel and its cargo. The following case studies highlight the consequences of lapses in concentration and the failure to ensure that safe practices are followed at all times. It may be difficult for crew to maintain a high level of alertness, either due to sleep deprivation in heavy weather conditions or on long uneventful voyages. In most events, a single lapse will not result in an incident, occasionally it may result in a small inconvenience or loss, and in extremis, it may result in the loss of the entire vessel. Here we give examples of a range of events from small cargo loss resulting from poor loading and water ingress, with relatively minor fire damage through inappropriate stowage of apparently non-hazardous goods, through to an apparently small stowage error resulting in total loss and major environmental impact.

HEAT SENSITIVE PRODUCTS

On November 11th 2002, the near new container carrier *Hanjin Pennsylvania* suffered an explosion and fire which spread to containers including fireworks. The initial explosion caused extensive damage, breaching the holds in front of the accommodation and initiating a widespread fire. To add to the hazard, some of the remaining containers contained fireworks, which later ignited and these too, exploded. While it has been difficult to resolve the exact cause of the fire, initial suggestions that the fireworks were responsible have to all intents and purposes, now been discounted. It is believed that a cargo of undeclared calcium hypochlorite was responsible.

This event highlights several hazards within the marine supply chain. Clearly products such as calcium hypochlorite which act as an oxidising agent and which are unstable at mildly elevated temperatures need careful handling. They are now highly regulated in the IMDG Code, and subject to strict conditions of carriage.

Since 2002 some shipping lines have banned the carrying of Calcium Hypochlorite but mis-declarations appear to remain. In June 2007 it was reported that the *Zim Haifa* suffered an explosion and fire from Calcium Hypochlorite which had been declared as Calcium Chloride and certified as safe by the shipper [Lloyds Casualty Reports, 2007]. If confirmed by investigations, this would be a clear breach of both the IMDG Code [Amendment 30, 2001] and of ZIM Shipping's own restrictions.

Finally, shipboard practices and operations make some parts of the ship less suitable for storage of some cargoes than others. Either side of the accommodation stack and engine room, holds tend to be subject to higher thermal loads than elsewhere on the vessel. While this is not an issue for the vast majority of cargoes, in some circumstances it may be critical. In the case of the *Hanjin Pennsylvania*, a combination of events had disastrous consequences. The mis-declared cargo, the positioning of it in a slightly warmer section of the ship, and the cargo of fireworks nearby all contributed to the initial explosion and fire, and the subsequently difficulty in bringing that fire under control.

An unlikely heat sensitive product is standard rechargeable Nickel Metal Hydride (Ni-MH) batteries. Warnings on the batteries are clear. Do not short circuit, may ignite, leak, explode or get hot, do not dispose of in fire. Less well known is the possibility of them suffering thermal runaway, getting to temperatures sufficient to ignite paper and releasing hydrogen. In normal operation the batteries are safe. Despite the fact that they may get hot, the heat is generally dissipated rapidly to the environment and thermal runaway does not occur. Thermal runaway is a result of the self discharge tendency of Ni-MH batteries. Ni-MH batteries slowly lose charge over time, releasing heat as they do so. This is a slow process at ambient conditions which will occur over a period of weeks. As temperatures are increased the metal hydride begins to decompose and the hydrogen pressure in the cell increases. At higher temperatures the cell will begin to vent hydrogen to prevent the cell case bursting.

In the case of transport in containers, where several hundred thousand cells may be close packed the opportunity for thermal runaway is increased, as the heat dissipation is restricted by the packaging. If the batteries are held at an elevated temperature they self heat as shown in Figure 3. It can be seen that even at relatively low temperatures around 60 degrees, the batteries are heating more rapidly than would be expected from their environment and continue to heat well beyond the ambient temperature. In the data presented here, the battery temperatures are monitored with an external thermocouple, and internal temperatures are expected to be higher.

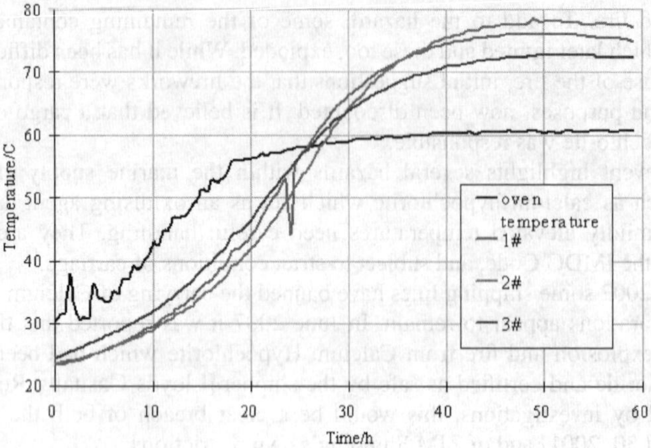

Figure 3. The self heating characteristics of Ni-MH batteries

Results from testing Ni-MH batteries when heated, indicate that at temperatures above 90 centigrade the batteries begin to vent hydrogen. This combination of thermal runaway and hydrogen venting provide a potentially disastrous combination of events in a confined space. Increasing temperatures and the release of hydrogen gas develop an explosive environment.

On May 28th 2005 the container carrier *Punjab Senator* departed Singapore on passage to Colombo, Sri Lanka. On the morning of May 30th, around 07:30, a container in Hold 6 exploded with a subsequent fire. The hold was injected with CO_2 and water applied for boundary cooling. The fire was brought under control by 15:00 that afternoon.

On investigation of the *Punjab Senator* incident, the container responsible for the explosion was found at the bottom of Bay 58, immediately aft of the bridge and situated over the machinery space beside the engine. [Federal Bureau of Maritime Casualty Investigation, 2006] The machinery space contained a settling tank that butted up hard under the hold space. Temperature traces from this show temperatures in the tank of around 80–85 centigrade in the day preceding the explosion and fire.

It appears most probable that the high fuel tank temperatures in the machinery space immediately below the container of batteries prompted them to undergo thermal runaway. The near sealed environment of the container prevented both the dissipation of the heat generated by the batteries and the diffusion of the hydrogen. Consequently the hydrogen build up occurred in close proximity to the heating batteries. Short circuit tests on Ni-MH batteries show that temperatures may be reached that will char and ignite paper. Figure 4 shows batteries close packed for shipping with two sample batteries placed on top. The batteries on top have been subject to a hard short circuit, which has caused to charring to the paper on the short circuit strip. The high discharge capacity of these types of batteries

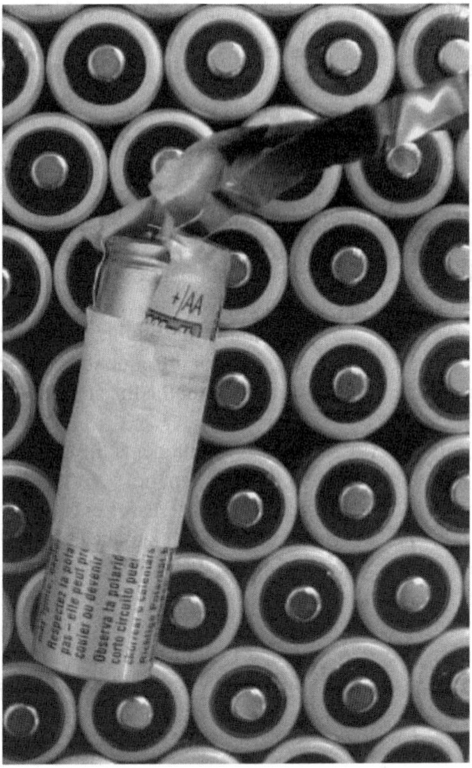

Figure 4. Packaged batteries with hard shorted cell showing paper charring

mean that a short circuit is capable of generating temperatures high enough to ignite cellulose products. It is likely that the progressive heating of the batteries resulted in short circuits that provided the mechanism to ignite the now explosive mix of hydrogen in the container.

While the hazards from large battery banks are well recognised, such as the "gassing" of lead acid batteries on high charge, smaller rechargeable batteries appear less dangerous. However, once large numbers are assembled for transport in a confined space, significant hazards exist. Shippers recommend that Ni-MH batteries are loaded away from heat sources in ships, but Ni-MH batteries are not currently listed as Dangerous Goods and load masters may not be aware of all the Bills of Lading for a container full of mixed products. In a ship holding thousands of containers, individual knowledge of the contents of every container is not feasible. The increasing requirement for rechargeable energy sources, both for small portable items such as laptops or power drills, and for larger energy storage mediums such as Ni-MH batteries for hybrid and electric vehicles means that the transport of batteries will only increase. The hazards of this must be recognised and suitable precautions put in place.

UNDECLARED CARGO

Mis-declaration of cargo increases the risks to the vessel's compliment of crew and all cargo on board. The investigation into the contents of the containers on board the *MSC Napoli* is likely to shed some light on the level of mis-declarations that occur in the shipping industry. The ability to check declared container contents against a physical examination enables investigators to have a snapshot of what may be assumed to be a typical container ship cargo. While the above example highlights the fire risks associated with mis-declarations, container collapse is another not infrequent occurrence. Shippers making conservative declarations of container weights threaten the stability of container stack both above and below deck.

The *M/V Xin Qing Dao*, a 5618 TEU container ship demonstrated the danger of heavy seas in losing deck containers. Even when not overweight 31 containers went overboard and another 29 were damaged when the vessel encountered heavy weather off Brittany in October 2004. Overweight container stacks make events like this more likely as the loads on the containers at the bottom of the stacks increase beyond the design limit.

In recent events, the *Annabella* in February 2007, suffered a container collapse in a forward hold when containers totalling 225 tonnes were loaded in a location limited to 150. The incident was further compounded by the lower containers having a maximum stacking load of 100 tonnes, and the upper most 3 containers containing hazardous cargo. (MAIB 2007)

RAW MATERIALS

Heat and moisture are two of the major hazards that threaten marine cargoes. The above examples of calcium hypochlorite and Ni-MH batteries highlight the temperature sensitive nature of chemical cargoes, but a combination of heat and moisture is also hazardous to organic cargoes. This is becoming more critical as the proportion of high organic oil content cargoes increases as a result of the growth in alternative fuels, such as, palm seed, copra, rapeseed and cottonseed, all examples of non-mineral raw product used for oil generation. There are numerous raw vegetable and grain products that are transported in bulk, that when subjected to high humidity levels and mild heat have a tendency to self heat or spontaneously combust. Cargoes with a high propensity to combust spontaneously are listed under the IMDG Code as Class 4.2 Substance Liable to Spontaneous Combustion.

The self heating process is accelerated in Class 4.2 products when temperatures are above 30 centigrade on loading, and when moisture levels are above the equilibrium moisture content of the product when the external relative humidity is over 75%. The excess moisture in the product coupled with the elevated temperature promotes microbial activity that exudes heat and moisture. When a product is in a hold with limited ventilation, limited capability for the dissipation of heat and the diffusion of water vapour, the conditions further promote self heating. Over 40 centigrade, the evolution of heat promotes the oxidation of the unsaturated oils in the products. As with any oxidation process there is a further evolution of heat and if left uncontrolled accelerated heating will occur, with spoilage of the product.

It should be recognised at this stage that the self heating process is a slow one with progressive change from biotic activity to oxidative fat cleavage of the unsaturated oils. As the temperatures increase from 55 to 75 degrees the microbial activity will cease, as the organisms responsible are killed off by the excessive heat, and the fat cleavage processes take over. However, the high inherent moisture levels in the product act as a heat sink during this stage and there is only a slow rise as moisture is ejected from the cargo. If unchecked the temperature may rise towards 90 degrees, and the product will emit clouds of water vapour. The vapour cloud over the cargo will act as a fire suppressant for some time, and only when this has fully dispersed is there an elevated risk of spontaneous combustion. The heating process remains slow and large volumes of smoke may be expected before the auto-ignition temperature of the product is reached.

In addition to the moisture implications for self heating products such as those with high levels of unsaturated fats, there are now fumigants for grains that are susceptible to auto ignition when mishandled and wetted. Such an example is that of Aluminium Phosphide (AlP). While it is designed to slowly release phosphine gas through the reaction of the AlP with the natural humidity in a hold, water ingress may rapidly accelerate the phosphine release. At concentrations greater than 2% by volume, phosphine will react with oxygen, even when oxygen levels are depleted, [Kondo S. 1995].

This has presented a relatively new hazard as the fumigant for grain transport by sea was previously methyl bromide, a fire suppressant. However methyl bromide is also an ozone depleting chemical, and its use has been banned. The replacement of methyl bromide by a spontaneously combustible gas has led to new and unexpected hazards for grain transport.

The dangers may be demonstrated by a vessel which had been loaded with mixed grain in Argentina and made passage to Chile. The holds were fumigated with Aluminium Phosphide. After passage through the Straits of Magellan, heavy weather was encountered that apparently caused water ingress into forward holds. The excess water on the Aluminium Phosphide led to rapid release of phosphine gas which appears to have spontaneously ignited in the forward hold. The incident was exacerbated by the poor distribution of the Aluminium Phosphide pellets in the hold. Although the hazard of using Aluminium Phosphide will always exist it is only a combination of events that turn the hazard into an incident. The poor distribution of fumigant bags, the heavy weather and the failure of the hold's water tight integrity were all required to cause the incident.

MINOR FAILING, MAJOR LOSS

On the 4th of January 1993, the *M/T Braer* suffered engine problems that resulted in vessel grounding in the Shetland Islands with loss of the vessel, its cargo of 85,000 tonnes of light crude oil and a potentially disastrous oil pollution problem. The study of the incident here highlights how a relatively small initial failing led to such a major loss. It is the nature of the marine industry and maritime transport, in the sometimes adverse environment of the sea, that allows a minor error to rapidly escalate into a catastrophic incident. In the case of the *Braer*, the tempest actually reduced the impact of the subsequent oil pollution, by assisting in the rapid dispersion of the light crude. This is in contrast to the Exxon Valdez disaster of 1989.

The *Braer* incident is worth closer analysis for both the circumstances which lead to the engine failure and the failure of the crew to recognise the implication of the events as they unfolded. The time line of events shown in Table 1 is highlighted with those critical points at which intervention might have avoided the final disaster.

Table 1. Timeline of m/t Braer events

Date	Time	Event	Critical point
January 3th	13:00:00	Braer departs Mongstad, Norway, forecast for southerly gales.	
	15:00:00	High and low level alarms on the auxiliary boiler are sounding, apparently due to weather.	
January 4th	10:00:00	Spare pipes noted to have broken free and deck air pipes appear to bent	Investigation of air pipes warranted but not undertaken
	12:00:00	After midday watch change, second assistant engineer adjusted water level on aux boiler to prevent tripping	
	19:30:00	Auxiliary boiler trips out and weather was suspected as cause	Reason for tripping not fully investigated.
	20:30:00	Boiler alarms sounding, believed to be the air transmitter controlling the alarms	
	21:00:00	Auxiliary boiler switched to diesel and shut down	Switch to contaminated fuel
	23:30:00	Difficulties in re-firing boiler Fuel oil temperature drop noticed Main engine changed to diesel oil	Switch to contaminated fuel
January 5th	00:30:00	Salt water contamination of auxiliary boiler diesel supply	Not recognised that the main engine was also on contaminated fuel
	02:00:00	Superintendent to boiler room	

(*Continued*)

Table 1. Continued

Date	Time	Event	Critical Point
	02:30:00	Chief Engineer called to boiler room.	Chief Engineer failed to recognise that the main engine is on contaminated fuel
	03:30:00	Water contamination of diesel oil settling and service tanks discovered. Attempted to drain off water	
	04:00:00	Engine speed reduced to conserve diesel fuel	
	04:10:00	Master advised and decided to proceed to anchorage in Moray Firth. Water still being drained off tanks	
	04:40:00	Main engine stopped. Braer 10 miles off Sumburgh Head	
	04:42:00	Generator stopped, all main power lost. Continuing attempts to drain off water.	
	05:15:00	Coast Guard advised but no assistance requested	
	05:26:00	Master requested tug as soon as possible	Could have requested tug some 2–4 hours earlier.
	06:00:00	Braer 6 miles south of Sumburgh Head	
	06:54:00	Evacuation of personnel commences	
	08:54:00	Evacuation of crew complete	
	11:19:00	Vessel Grounds	

The *M/T Braer* departed Mongstad in Norway on January 3rd at 13:00 for Canada. The forecast was for storm force southerlies and progress was slow. The vessel was rolling heavily and shipping heavy seas. The heavy weather appeared to be causing the auxiliary boiler high and low level water alarms to sound. The auxiliary boiler provided steam to preheat the fuel oil for the main engine.

Four spare pipes about 5 metres long and half a meter diameter were stowed on the port side of the upper deck against the port engine casing bulkhead. They had been secured in a temporary frame and spot welded in place. On the morning of January 4th after breakfast, the Chief Officer and the Chief Engineer viewed the pipes from a window of the mess room. They noticed that the pipes had broken free of their framing and were rolling around on the aft deck between the engine casing bulkhead and the ship's railings. Immediately inside the railings were the deck air pipes and the loose pipes were banging against these. One of them appeared to be bent. The situation was discussed with the Master, and the matter was to be left until the weather improved.

Clearly, this is the first critical event. The Chief Officer and Chief Engineer could have investigated the possible damage to the deck air pipes and re-secured the spare pipes that had broken free. While the Master felt that the weather needed to abate for the problem to be rectified, in hindsight he may have felt differently. It is clear that crew intervention at this stage may have prevented the situation escalating. The weather, mechanical failure and the crew inaction all combined to propagate the incident.

At 20:30, under the watch of the Third Assistant Engineer, the boiler alarms again started to sound, and the engineer thought there might be something wrong with them. He decided to shut down the boiler to install a spare transmitter. The boiler was switched to diesel fuel before shutting down to aid restarting later. By 21:30 the spare transmitter had been installed and the engineer started the firing sequence to restart the boiler but there was a flame failure.

The second and probably irreversible decision was to attempt to repair the alarms. While the alarms may have been causing problems, the boiler was apparently still functioning. Switching the boiler to contaminated fuel spelled the end to its operation. It would not have been possible for the Third Assistant Engineer to know that the fuel was contaminated, but consultation with the Chief Engineer who had observed the damage to the deck pipes, may have allowed a more circumspect approach. It would appear that the attempt to fix the alarms was not necessary as other explanations for the soundings had been put forward, that of the heavy rolling of the vessel.

At 23:30, the engineer noticed that the main engine fuel temperature had fallen from 120 to 95 centigrade. On noticing this, the Third Assistant Engineer called the Chief Engineer and told him he intended to change the main engine over to diesel oil. This was authorised before the cause of the boiler failure was established.

The contribution of the weather at this stage is clear. It was a major cause of the pipes breaking free, without the heavy water over the decks, the air pipes would not have been subject to water ingress, and without the heavy rolling of the ship, the water may have been drained from the contaminated tanks. The weather would also make extra physical demands on the crew, as the tanker was reported to be rolling as much as 30 degrees.

At 02:30, the Chief Engineer was called to the engine and he noted that the main engine was running on diesel. At 00:30 the fuel supply to the boiler had been discovered to be contaminated and at 03:30 the diesel oil settling tank was inspected and found to be contaminated, as was the diesel oil service tank. A reduction of speed was ordered to conserve fuel. The course was altered in an attempt to make an anchorage where the diesel could be drained of water. At 04:40 the main engine stopped, followed by the main generator, and the vessel reverted to emergency power.

The vessel was now in need of assistance and radioed Aberdeen Coast Guard with a request for a tow. Delays in organising towage resulted in the vessel grounding and being lost.

The failure to recognise that the main engine was also on contaminated fuel, lost between 2 and 4 hours in which to summons assistance. The cause of the failed auxiliary boiler was established at 00:30 on the 5th, but the Chief Engineer was only called two hours later. At this time the main engine was running on the same fuel supply. Without a change in fuel or the ability to filter the supply, the operation of the main engine must have been in doubt.

It can be seen from the above analysis, that while weather and machinery failure were outwardly the cause of the loss of the *Braer*, the crew decisions were complicit in the loss, despite their high qualifications and experience. It must be said that the conditions they encountered were extreme, and the unlikely combination of events hard to predict, but opportunities for prevention did present themselves, both for rectification of the problems and for the earlier summoning of assistance. In light of the current shipping boom and the deficit in experienced crew, hazards like those leading to the loss of the *Braer* may well go unnoticed. Training and vigilance remain of utmost importance in maintaining incident free shipping.

CONCLUSIONS

The hazards in the marine supply chain involve a complex interaction of natural effects, hardware serviceability and vigilance on the part of the crew. In most circumstance the degradation of any one of these will not lead to an incident but a combination of any two raises the risk of an incident significantly.

Maintaining crew competency is a challenge for all shipping lines in the current climate of growth. The separation of the crew from the cargo, as occurs in container ships, further reduces the crew's perception of the risks that relate to their cargoes. The inability of a crew to be able to monitor all containers on a vessel means that risks to vulnerable cargoes may result in incidents. This is especially true of heat sensitive products, whether it is perishable food items or heats sensitive products, such as batteries.

It is clear from the incidents highlighted in this paper that the changing format of shipping, the change in regulations and the growth in alternative energy transport is introducing new risks, and pressures on crews and shipping agents. While the IMDG Code assists in highlighting cargoes that require special care, new products and the carriage of products in new ways, are not necessarily covered. As the shipping trade continues to evolve, the change in risks, of both environmental and economic, need to be address at all levels. Officers and crews need to be educated on their changing cargoes, the IMDG Code needs continual upgrading and insurers need more information on the behaviour of the products for which they provide cover.

REFERENCES

Amendment 30 (2001) of the IMDG Code

Beale CJ., (2000) "Identifying areas in which environmental improvements can conflict with safety requirements for chemical plant design and operation", Hazards XV pp 25–39

Casualty Reports (2007), Lloyds List, September 13th
DNV(2006) "Level of Safety for Ships"
Federal Bureau of Maritime Casualty Investigation, (2006) "Explosion and Fire on CMS Punjab Senator", Report 187/05, 15th December
Kondo S., Tokuhashi K., Nagai H., Iwasaka M., and Kaise M., (1995) "Spontaneous Ignition Limits of Silane and Phosphine", Combust. Flame, 101:170–174.
MAIB Report (2007) "Report on the investigation of the collapse of cargo containers on *Annabella* Baltic Sea", Report No 21/2007, September

THE CAUSES OF IBC (INTERMEDIATE BULK CONTAINER) LEAKS AT CHEMICAL PLANTS – AN ANALYSIS OF OPERATING EXPERIENCE

Christopher J. Beale (FIChemE)
Ciba Expert Services, Charter Way, Macclesfield, Cheshire, SK10 2NX, UK

Intermediate bulk containers are in widespread use in industry for handling hazardous and non-hazardous materials. As the use of IBCs has increased, so the number of chemical leaks and fires from IBCs has increased. This paper analyses the causes of IBC leaks at a large Ciba UK manufacturing site. The analysis is based on incident reports, near miss reports, interviews with site staff and interviews with site emergency response staff. Common causes of leaks are identified. Generic leak frequencies are then calculated for IBC leaks. The analysis includes different types of operations involving IBCs including transport, temporary storage, warehouse storage, waste product storage and process applications.

KEYWORDS: Intermediate Bulk Container, Learning From Incidents.

INTRODUCTION – THE USE OF INTERMEDIATE BULK CONTAINERS IN INDUSTRY

IBCs are in common use in industry as they allow relatively small quantities of chemicals to be transported between suppliers, manufacturers and customers and around site areas efficiently. IBCs are produced in a wide range of different sizes, shapes, materials of construction and designs to suit specific user requirements. They can be purchased as standard designs as a commodity product or they can be custom built for specific uses.

Custom built designs are more expensive. They tend to be used for regular shipments of product within a site or between different sites.

Standard designs tend to be used for single or medium use duty and are often used for delivering products to customers. Suppliers will often operate a recycling scheme, picking up used IBCs when new deliveries are made. Recycled IBCs are then cleaned and checked prior to re-use.

This paper is based on a study of 1,000 litre IBCs, with a variety of different designs.

IBC DESIGN PRINCIPLES

The integrity of an IBC depends on three critical components (see Figure 1):

1. A pallet which allows the IBC to be moved easily by fork lift trucks. Pallets are commonly made of heat treated timber, plastic or steel. Timber pallets are more susceptible to mechanical damage. Plastic and metal pallets have a longer design life but have low surface friction resistance. This can cause slippage when handled by fork lift trucks.

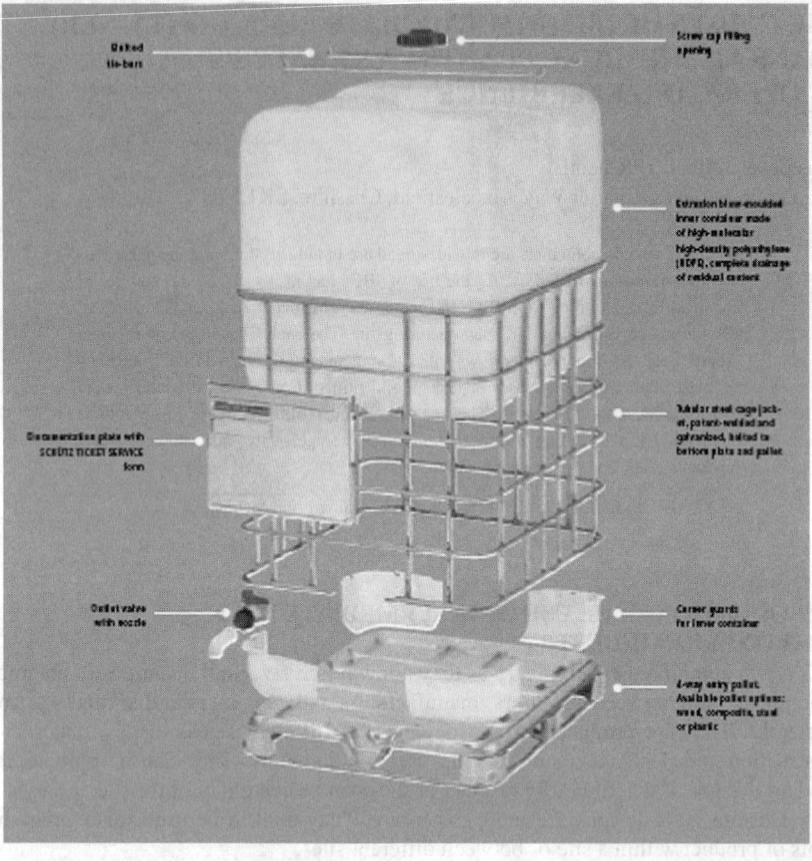

Figure 1. Typical IBC design features

Metal pallets can also cause sparks when they contact hard surfaces like concrete. The pallets are normally stamped, indicating the manufacturer and date of manufacture.

2. A container which holds the chemical. It is often made of translucent plastic, which is light, strong and easy to manufacture. Coloured plastics are also used for specific applications such as differentiating between different categories of chemical. Black plastic is used for reducing the risk of photo-initiation of monomers. White plastic is used for reducing heat input when containers are transported in hot countries. The container normally has a top screw cap which is used for filling operations. Small bore connections are normally provided, with larger bore connections for viscous liquids. An outlet valve is provided at the base of the container. This is normally recessed into the container to prevent the valve from being damaged when in transit. The valve is

normally fitted with a screw cap for additional leak integrity. Screw caps often have a tamper proof seal. Some IBCs are fitted with simple pressure relief devices which are attached at the fill point. IBCs are normally specified with an ullage space above the liquid level. If the ullage space is correctly calculated, a pressure relief device will not be required to meet normal transport requirements. For example, a 1,000 litre IBC will typically have a brimful capacity of 1.050 litres. Mistakes can be made by not calculating the ullage volume correctly or by failing to allow for the specific gravity of the product in the ullage calculation.

3. A metal cage which surrounds the container and is attached to the pallet. The cage can be manufactured with welded tubes, mesh rods or metal sides. Nameplates are attached to the cage so that the contents can be identified. The cage can be designed to provide static protection to the IBC.

The combination of pallet, container and cage provide mechanical integrity for the IBC. The mounting which holds the cage onto the pallet has to be strongly fixed. IBCs which are used for transporting hazardous chemicals often have additional and/or stronger fixings to improve integrity.

RISK DRIVERS

The following factors influence the risk of chemical leaks from IBCs:

- Safe operation is largely determined by the way that people handle IBCs.
- They are moved between different sites and reliance is placed on different companies in the supply chain.
- They are often re-used. Checks are required to ensure that the containers are clean and have not been damaged.
- Large numbers of IBCs are handled on chemical sites, often by relatively unskilled staff.
- There is a perception that they have limited hazard potential due to their relatively small size.
- They are often stored in groups in remote or unmanned site areas. Research has shown that this can allow fire incidents to spread rapidly, releasing large flammable/combustible inventories into uncontained site areas (Atkinson & Riley, 2006).

FRAMEWORK FOR ANALYSING IBC FAILURE MODES

Four consequence categories for IBC failures have been identified:

1. **Offsite chemical releases during transport** by truck, in port or on ships. These incidents are often limited to the release of relatively small inventories of chemical from one or a small number of IBCs. Releases tend to occur from poor packing/stacking, road traffic accidents or chemical reactions. These releases can cause significant nuisance and shipping delays because chemicals may be dispersed in small quantities over a wide geographic area. Particular problems occur when leaks involve toxic, sticky, odorous or environmentally sensitive chemicals.

2. **Chemical releases which do not ignite.** These releases occur frequently on sites which handle large volumes of IBCs. The consequences of release are often localised because of a combination of the relatively small inventories feeding the release, the fact that releases normally involve single or small numbers of containers and the relatively low hazard potential of many of the chemicals which are handled in IBCs. This type of release is typically seen as a low priority process safety issue at many sites.
3. **Chemical releases which ignite.** These are very rare events based on data from UK Ciba sites. IBCs are essentially small chemical storage tanks which can easily be manipulated by people and which have very limited hardware safety features. The containers may be handled where ignition sources exist; they may be stored in remote areas of the site which have limited fire detection and protection systems; and they may be stored close to other flammable or combustible containers, producing a scenario which allows rapid fire escalation.
4. **Fuel sources, causing fires in other areas of the site to escalate.** IBCs are often stored in relatively large groups in site areas such as waste storage areas, temporary container holding areas and warehouses. Research has shown that this can lead to rapid fire escalation, fed by large flammable/combustible inventories and limited pool containment systems (Atkinson & Riley, 2006).

Data about fires and fire escalations is rare. Data about unignited releases can, however, be found through staff interviews and an analysis of near miss records. The Ciba Bradford site has used a database near miss reporting system since 2001 (Beale, 2004). Database records from between 1/1/2005 and 30/9/2007 (a period of two and three quarter years) have therefore been analysed to identify reported chemical releases and near miss incidents involving IBCs. This provides a detailed profile of IBC releases over the period 2005–2007. Incidents which occurred before 2004 would only have been recorded if they were significant. These significant releases were identified from incident reports and from discussions with line managers and emergency response specialists.

All of these reported and recorded incidents have been analysed to identify IBC failure mechanisms leading to loss of containment and the relative frequency of occurrence of each type of failure mechanism.

FAILURE MODES

"IBC incidents are almost always caused by people mistreating or mishandling IBCs. They are rarely caused by mechanical or structural failure."

This is how an experienced line manager summarised his experience of failures and near misses at his site. Interviews with line managers and emergency response staff show that the following failure mechanisms have occurred:

Transport Incidents

1. IBCs are badly stacked inside trucks or the loads are badly secured. This allows movement in transit, causing IBCs to be damaged or topple.

2. Trucks are involved in road traffic accidents. IBCs are damaged and may leak inside the truck, onto roads and into drains.

Warehouse Incidents

3. Poor stacking causes IBCs at the top of a stack to fall onto other IBCs or onto the ground.
4. IBCs which are not stacked carefully into a warehouse compartment can protrude into the area where fork lift trucks operate. The next time that a fork lift truck accesses the area, it catches the protruding IBC, dislodging the IBC. This can cause the IBC to fall to the ground and can also cause the IBC to snag the warehouse racking, causing structural racking failure. Structural racking failure is most likely when tall and narrow warehouse aisles are used. Figure 2 shows an example of the aftermath of a racking failure.

Onsite Handling Incidents

5. IBCs fall off or unbalance a fork lift truck because they are not loaded carefully onto the forks, because the fork lift truck is not driven carefully or because the fork lift truck strikes an object or a pothole. The IBC is then dropped onto the ground.
6. An IBC is pierced with the fork lift truck lifting arms, puncturing the IBC and releasing it's contents. Figure 3 shows an example of a pierced IBC.
7. Fork lift trucks crash into static objects when moving IBCs. Static objects could be trucks, other fork lift trucks, stacks of IBCs, warehouse racking and warehouse walls.
8. IBCs jam or are misaligned on conveyor handling systems, causing deformation or toppling onto the ground.

Chemical Reactions

9. Containers are incorrectly labeled, often because they hold waste material, by-products or intermediate products. This can result in chemical storage in the wrong

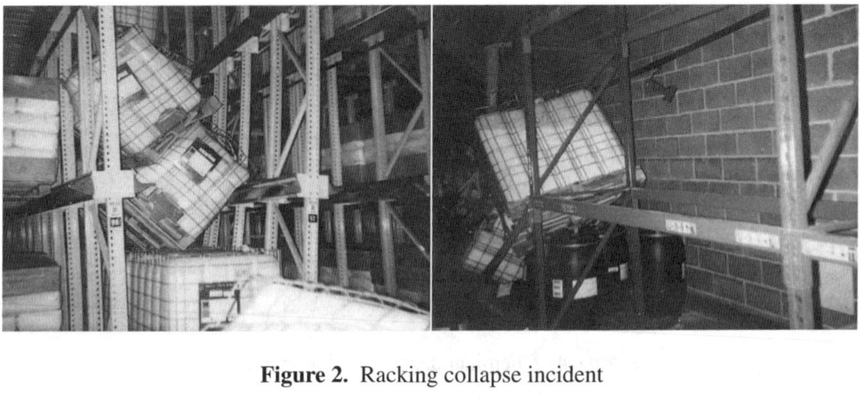

Figure 2. Racking collapse incident

Figure 3. IBC pierced by fork lift truck

location or chemicals may be left in storage accidentally for long time periods, increasing the risk of an undesired reaction inside the container.

10. Fork lift truck operators leave IBCs in the wrong area of site. If they are stored close to incompatible chemicals, this could lead to a chemical reaction.
11. Waste material is run-off into IBCs, where it is accidentally mixed with incompatible material or it generates an unstable mixture.
12. Reactive chemicals, such as monomers, are left in IBCs and a polymerisation reaction is initiated. Typical causes would be inadequate quantities of inhibitor, contact with an impurity, lack of circulation and aged stock. IBCs are generally not fitted with pressure relief devices, so these failures tend to cause container swelling or failure. Figure 4 shows the aftermath of an IBC chemical decomposition reaction.

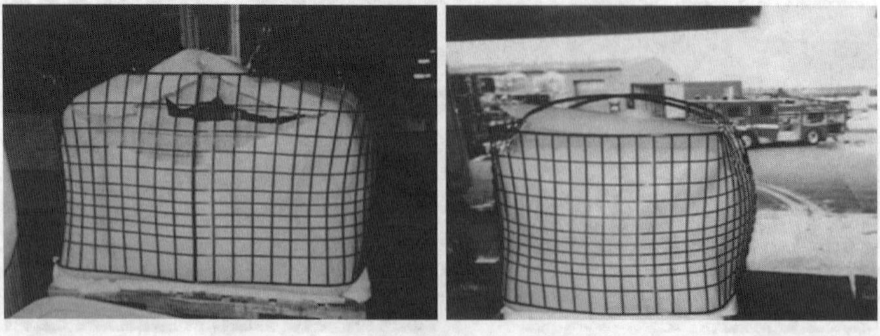

Figure 4. Chemical decomposition incident

Process Operations

13. IBCs are normally filled under operator control. They fill relatively quickly. If the operator is not concentrating or is distracted, containers can be overfilled.
14. The container is pumped out with the top cap in place, causing the IBC to be sucked in.
15. The contents of the IBC are charged to the wrong vessel or tank. This could cause an uncontrolled reaction in downstream process plant.

Mechanical Failures

16. The pallet at the base of the IBC is broken causing the container to slump. These failures are most likely with wooden pallets and rarely cause chemical leaks.
17. The outlet valve leaks. Most IBCs are fitted with external caps with tamper proof seals, thus providing an additional barrier against valve leaks.
18. Outlet valve connection leaks. This is normally caused by mechanical pressure on the top side of the valve mechanism. This causes the mechanism to bend with a failure at the connection point to the main IBC body. These failures can be prevented by fitting supports under the valve connection.
19. Deliberate tampering with the outlet cap and valve during transit or in a process area. Many IBCs have tamperproof seals to minimise this type of scenario.
20. Outlet valve left open in error causing the contents to leak to ground.

Damage To Safety Systems

21. In rack warehouse sprinkler systems are damaged when IBCs are not stored carefully. This should cause the system to operate, causing a revealed failure. This scenario is most likely when sprinkler heads are poorly located or when operators try to place multiple or large containers into a bay which is sized for a smaller container.
22. Fork lift trucks damage fire hydrants when manoeuvring around the site.

Table 1 summarises the reported causes of incidents and near misses involving IBCs for one large manufacturing site over the 33 month period. This data is based on employee generated reports using the site near miss reporting system. 107 reports for IBCs were raised in this time period. Table 2 summarises the type of site operation which was occurring when each report was made. Table 4 summarises the significant IBC incidents which were recorded in the period 1990–2003 based on incident reports and staff interviews.

FAILURE FREQUENCIES

Table 3 summarises the generic leak frequencies per year for the large manufacturing site based on incident records over the 33 month period where loss of containment was known to have occurred. It is not possible to determine the size of the leak from the reports and events could range from pinhole releases (1 mm equivalent hole diameter) to larger 75 mm hole diameter releases. Most of the releases are known to have been associated with low hazard products.

Table 1. IBC near miss and incident summary 2005–2007

Fork Lift Trucks (FLT)	**Operational Errors**
Pallet overturned in transit 9	Stored in too high a bay in warehouse 8
FLT collision with racking 5	Incorrect or no labeling 7
IBC punctured (spiered) by FLT 5	Stored in wrong area of site 5
FLT collision with IBC 4	Wrong chemical delivered to works 3
FLT collision with wall 4	Inappropriate container used 1
FLT collision with process equipment 2	
FLT collision with FLT 2	
Tried to stack 2 IBCs in 1 warehouse bay 2	
FLT damage to fire sprinkler system 1	
Container slipped off racking 1	
FLT mechanical failure 1	
FLT overbalanced 1	
Process	**Integrity**
Leak during filling 6	Container leak 11
Waste product polymerises 3	Pallet leak 2
Leak during emptying 2	Seal too big for container 2
Overfill 1	Loose outlet valve 1
Hose hit by operator 1	Warehouse support beam collapse 1
Leak during IBC switchover 1	
Supply Chain	**Automated Packing Machines**
Contamination in container 3	Snagged on packing machine 1
Lid not fastened tightly 3	Fell off packing machine 1
Load incorrectly packed in HGV 2	Set to reverse not forward 1
HGV collision with IBC 2	
Driver offloaded himself with pallet truck 1	
Filled hot, capped, cooled, imploded 1	

Note: Number of reported loss of containment events over a 33 month period, 2005–2007.

CONCLUSIONS

22 IBC failure and error mechanisms have been identified based on the experience of line managers and emergency response staff who work with IBCs. A detailed analysis of reported near misses and incidents over a 33 month period between 2005 and 2007 identified 107 records relating to IBCs. Fork lift truck movements inside warehouse areas, warehouse storage and fork lift truck movements in site areas each accounted for about 20% of the failure reports. Filling/emptying operations and process incidents each accounted for about 13% of the failure reports. 13% of the failure reports were caused by errors in the supply chain with site deliveries.

Table 2. IBC near miss and incident summary 2005–2007 by operation

Operation	Number of events	%
Receipt at goods inwards	14	13
Fork lift truck movement in warehouse	19	18
Storage in warehouse area	20	18
Fork lift truck movement around site	22	21
Filling/emptying	14	13
Process use	16	15
Packing in warehouse	2	2
TOTAL	**107**	**100**

Note: Based on reported loss of containment events over a 33 month period, 2005–2007.

Table 3. IBC leak frequency analysis

Cause of loss of containment	Number	%	Frequency/site/yr
Container leak	11	35	4.0
Leak during filling	6	20	2.2
Spiered by fork lift truck	5	17	1.8
Uncontrolled polymerisation	3	10	1.1
Leak during emptying	2	6	0.7
Container overfilled	1	3	0.4
Loose outlet valve	1	3	0.4
Spill during IBC switchover	1	3	0.4
Operator contact with hose	1	3	0.4
TOTAL	**28**	**100**	**11.3**

Note: Based on reported loss of containment events over a 33 month period, 2005–2007.

The reports suggest that the large site suffers about 11 IBC loss of containment incidents per year. These range from small pinhole leaks, such as nail penetration through base to catastrophic failures, such as polymerisation reactions. 35% of the leaks are from the container. This will include brand new containers and re-used containers. 20% are caused by leaks during filling. 17% are caused when fork lift trucks puncture the IBC with their forks. 10% are caused by uncontrolled reactions inside the IBC.

This analysis could be used to identify IBC failure modes for a risk analysis and as a source of generic frequency data for base events which could be used in a fault tree or Layer Of Protection Analysis (LOPA) study for a major accident hazard scenario. Although

Table 4. IBC incident summary 1990–2003

Date	Type	Cause	Consequence
1993	Warehouse – IBC explosion at Goods In	Truck container full of IBCs from the USA being offloaded. IBCs should have contained low hazard product but it was contaminated with hydrogen peroxide which then decomposed.	IBC exploded liberating 200 kg of product. Evidence of deformation and internal pressure build up in remaining IBCs in the container.
1994	Fork lift truck failure	Securing pin on fork lift truck failed when an IBC was being lifted.	IBC fell to floor. No loss of containment.
1996	Warehouse – fork lift truck failure	Hydraulic failure.	Damage to warehouse racking affecting low hazard product.
1996	Warehouse – racking collapse	Pallets not centered on racking support rails. Pallet fell dragging other containers with it.	16 IBCs fell, 12 from high level. Small spill of low hazard product.
1997	Fork lift truck failure	Bleed plug expelled. Loss of hydraulic pressure on fork lift truck used for accessing high levels.	Truck cab fell rapidly from height and then slowed. No injury.
1998	Warehouse – fork lift truck collided with racking.	Driver steering error. Fork lift truck crashed into racking.	Localised racking collapse affecting low hazard product.
1998	Warehouse – racking collapse	Pallet not stacked accurately in storage bay.	Localised racking collapse when pallet moved.
1998	Transport accident at major port	Tug driver pulling two containers which he believed were both empty. One was actually full. Load was unbalanced. Container toppled over at roundabout inside port area.	Container held 18 IBCs. IBC leaked low hazard liquid into container. Container had to be cleaned out.
1999	Fork lift truck failure	Lifting chain on fork lift truck collapsed.	IBC fell to floor. No loss of containment.

(*Continued*)

Table 4. Continued

Date	Type	Cause	Consequence
1999	Fork lift truck failure	Lifting chain on fork lift truck failed. Empty IBCs had just been moved.	No impact.
2000	Road traffic accident on major motorway	HGV passed too close to transit van and trailer causing trailer to jack knife. Trailer contained IBC.	Viscous material spilt on motorway. Motorway closed for long period.
2003	Warehouse – fork lift truck collided with racking.	Driver drove down narrow aisle with forks protruding about 300 mm. The forks hit the main structural frame of the warehouse racking, weakening the structure.	Localised racking collapse affecting 4 tiers of racking containing low hazard product.

the leak data from 2005–2007 is considered to be comprehensive for the site, care should be exercised because the data covers a relatively short 33 month analysis period and it is difficult to ascertain the size of the leaks which underpin the analysis.

REFERENCES

(Atkinson & Riley, 2006) 'Controlling the fire risks from composite IBCs', G. Atkinson and N. Riley, IChemE Hazards XIX Symposium Series No. 151, 28–30 March 2006.

(Beale, 2004) 'Developing a major hazards learning culture – interpreting information from the Ciba Specialty Chemicals, Bradford near miss reporting system', C.J. Beale, IChemE Hazards XVIII Symposium Series No. 150, 23–25 November 2004.

LIQUID DISPERSAL AND VAPOUR PRODUCTION DURING OVERFILLING INCIDENTS

Graham Atkinson[1], Simon Gant[1], David Painter[1], Les Shirvill[2] and Aziz Ungut[2]
[1]HSE
[2]Shell Global Solutions

There have been a number of major incidents involving the formation and ignition of extensive flammable clouds during the overfilling of atmospheric pressure tanks containing gasoline, crude oil and other volatile liquids [1–4]. These incidents are characterised by widespread fire and overpressure damage.

The purposes of this paper are threefold:

1. to discuss physical processes of liquid dispersal, vaporisation and air entrainment that lead to the formation of a flammable cloud.
2. to describe an approximate method of calculation that can be used to determine whether the formation of a flammable cloud is possible for a given filling operation – a scoping method.
3. to describe the implications for safety and environmental standards for fuel storage sites in the UK.

1. PHYSICAL PROCESSES

1.1 LIQUID FLOW

The nature of the liquid release from an overfilled tank depends primarily on the flow rate and on the tank design. Three categories of tank have been identified that differ significantly in the character of the liquid release in the event of overfilling.

Type A: Fixed roof tanks with open vents (typically with a internal floating deck)
Type B: Floating deck tanks with no fixed roof
Type C: Fixed roof tanks with pressure/vacuum valves and possibly other larger bore relief hatches.

1.1.1 Liquid release from Type A tanks

This is the type of tank that was involved in the Buncefield incident. This tank was typical of Type A tanks with a number of open breather vents close to the edge of the tank at a spacing of around 10 m around the perimeter.

Tanks of this sort may be provided with a fixed water deluge system, which delivers water to the apex of the conical top of the tank. In the event of a fire, injected water flows down over the tank roof. Typically there is a "deflector plate" at the edge of the tank, which redirects water draining from the top of the tank on to the vertical tank wall.

In the event of tank overfilling, liquid will flow out of the open vents, spreading a little before it reaches the tank edge. The flow rates during overfilling are typically much higher than cooling water flow for which the deflector is designed. A proportion of the liquid release is directed back on to the wall of the tank and a proportion simply flows over the edge of the plate. This is illustrated in Figure 1.

Some tanks, including the tank involved in the Buncefield incident, have wind girders part way down the tank wall to stiffen the structure. Any liquid falling close to the tank wall will hit this girder and be deflected outwards, away from the tank wall. This outward spray may intersect the cascade of liquid from the top of the tank. This is illustrated in Figure 2.

The lateral spread around the tank perimeter of the free cascade of liquid formed from each breather vent is slightly greater if a deflector plate or wind girder is present. With these features present, the spray typically extends approximately 3m around the tank perimeter. If the vents are spaced at 10 m intervals and the elevation of the vents is similar, the final result is a series of liquid cascades that cover approximately 30% of the total tank perimeter.

1.1.2 Liquid release from Type B tanks

Floating deck tanks with no fixed roof typically have a large wind girder close to the top of the tank wall. This is fully welded to the side of the tank (to avoid stress concentration) and may be used as an access way (Figure 3). Small bore holes drain the top girder shelf but in the event of an over fill almost all of liquid overtopping the wall of the tank will flow out over the edge of the top girder forming a cascade. Typically the top girder is wide enough that liquid will not subsequently contact the tank wall and will therefore form a free cascade.

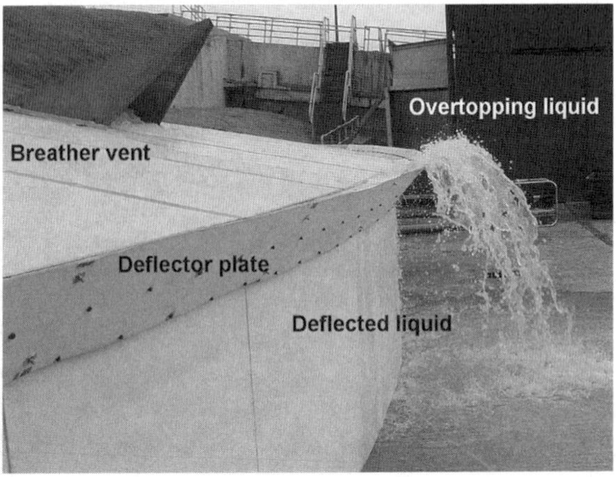

Figure 1. Liquid release from a vented fixed roof tank with a deflector plate

Figure 2. Intersection of free cascades from a Type A tank with a deflector plate

Figure 3. Top grider (walkway) in floating roof tank

The proportion of the tank perimeter over which this cascade extends is likely to depend on the construction of the tank. Any variations in the elevation of the tank wall will tend to concentrate the release on one side of the tank. Similarly any damage to the tank wall by the floating deck or access to this deck prior to the overflow may concentrate the release in an even smaller fraction of the tank perimeter. It is unlikely to extend round the full tank perimeter.

1.1.3 Liquid release from Type C tanks

Pressure/vacuum valves provided for pressure balancing during filling and emptying operations will generally not be adequate to relieve the liquid flow during overfilling. Liquid will come out of larger bore pressure relief hatches if these are fitted or from a split in the tank if they are not. Normally the tank construction should ensure that any split is at the junction between the tank top and wall.

In any case, it is likely that the release will be concentrated in a cascade covering a relatively small proportion of the total tank perimeter.

1.2 LIQUID DISPERSAL

There do not appear to have been any previous studies of high volume, low momentum liquid releases that accelerate and disperse under the action of gravity. Some large-scale tests on water and petrol undertaken in the aftermath of the Buncefield incident have provided some useful indicators but there is a pressing need for more data.

In the first few metres of fall the large scale liquid strings and lamellae formed in the release separate and accelerate, dividing into large droplets with a diameter of order 10 mm. The fate of these large fragments depends on the mass flux density of liquid in the cascade (i.e. the amount of liquid falling through each square metre per second). If the flux density is relatively low most of the initial liquid fragments rapid shatter to form a range of secondary droplets a few millimetres in diameter. The characteristic size is clearly a function of the liquid surface tension. Comparisons between 15 m high water and petrol cascades at similar mass densities showed that, at ground level, the droplets of water are variable in size in the range 2–5 mm whereas the characteristic size of petrol droplets are around 2 mm.

If the liquid flux density is very high, the aerodynamic drag forces on individual droplets in the core of the cascade will be lowered and some of the large fragment initially formed may persist for the full height of the drop.

All of the droplets then hit the ground. In cascades with high liquid mass flux densities the droplet impact speed may considerably exceed the terminal velocity for a single drop. Again the number and size of smaller secondary droplets formed on impact depends on the surface tension, impact speed and the nature of the impact surface i.e. wetted solid or deep liquid.

An initial estimate of the size range of secondary droplets produced by a petrol cascade impinging onto a bund floor can be made using the droplet splashing model of Bai *et al.* [4]. This predicts secondary droplets of diameter 130–200 microns for impingement on a dry floor and 100–180 microns diameter for a wetted floor. The total mass of splash products is very dependent of the depth of liquid on the impact surface and may even exceed the incident droplet mass in some circumstances.

In this paper, the phrase "vapour flow" is used to describe the air drawn into a liquid cascade and any gas produced from the liquid evaporating and mixing with the air. The fineness of droplets in the splash zone is very significant because the vapour flow driven by the cascade (described in Section 1.3) passes through the splash zone. There is an opportunity for very rapid exchange of mass, heat and momentum. Exchanges of heat and mass in the splash zone drive the liquid and vapour flows closer to thermodynamic equilibrium. Fine (100–200 micron diameter) droplets rapidly picked up by the vapour flow in the splash zone absorb momentum from the vapour flow and this may have a significant effect on its subsequent dispersion.

It is worth pointing out that the settling velocity for droplets in the size range 100–200 microns is 0.2 to 0.8 m/s. This means that droplets this size may remain airborne for a time of order 1–5 seconds during which they may be convected a distance of order 10 metres from the base of the tank. This means that some liquid droplets may remain suspended in the vapour flow as it impacts on the bund wall or other tanks within the bund.

1.3 AIR ENTRAINMENT

Jets of air or buoyant plumes entrain air through the action of shear driven vortices. A dense liquid cascade entrains air in a different, somewhat less complex way. Individual falling drops drag the air within the cascade downwards and air is drawn in through the sides to compensate. There are shear forces and induced vortices at the edge of the cascade but if the cross section is large these processes make little difference to the total volume flux of air – which is the quantity of primary interest.

A comparison has been made of detailed CFD predictions, which have included all the aerodynamic processes involved in falling sprays, and a simple momentum conservation model which ignores the induced shear flow on the spray periphery. This has shown that for the scenarios considered here it is adequate to use the latter, simpler treatment, which is described in Annex 1. Typical results obtained using the simple momentum conservation model are shown in Figure 4. In overfilling incidents the mass flux density is likely to be in the range 1 to 10 $kg/m^2/s$. This corresponds to maximum droplet velocities of 10–13 m/s and vapour velocities of 4–6 m/s.

CFD methods of the sort reported in Section 3 are capable of calculating droplet and vapour velocities both in the liquid cascade and in the vapour flow spreading out from the foot of the tank. These calculations fully encompass exchange of mass, heat and momentum between liquid and vapour phases.

1.4 VAPORISATION OF LIQUID

The fineness of liquid dispersal controls the extent to which liquid and vapour approach thermodynamic equilibrium. Example results from a CFD study of heat and mass transfer in the cascade are shown in Figure 5.

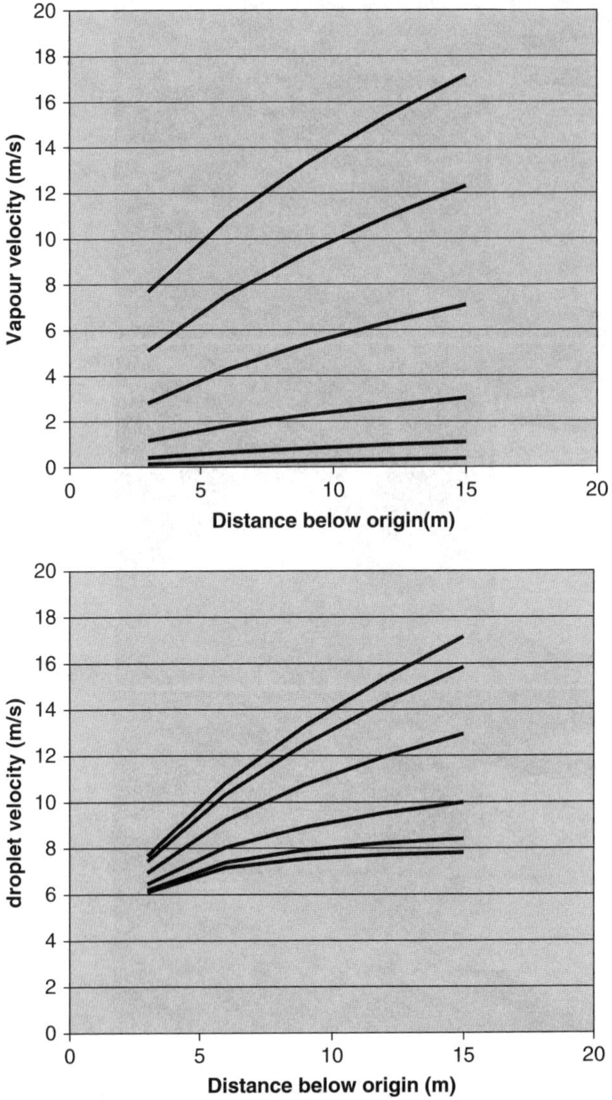

Figure 4. Vapour and droplet velocities induced by liquid cascades of different densities. The highest velocities shown in both plots (for comparison) correspond to free-fall with no air resistance. The lower velocities correspond respectively to liquid flux densities of 100, 10, 1, 0.1 and 0.01 kg/m^2/s

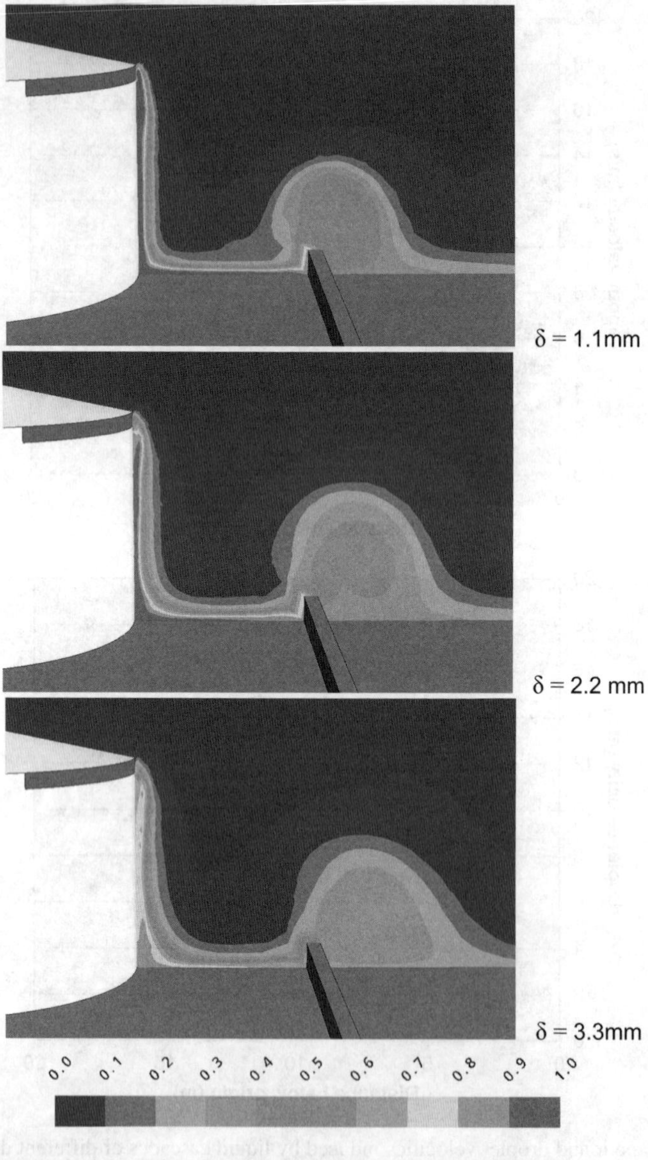

Figure 5. Contours of the ratio of predicted vapour volume fraction to the saturation volume fraction. A value of 1.0 indicates that the vapour is saturated. The three predictions are for different initial droplet size distributions using the Rosin-Rammler diameters shown

For droplets of a diameter of 2 mm or less, droplets and vapour in the core of the cascade (where the mass flux is concentrated) are very close to equilibrium. Areas on the fringes of the cascade where there is a greater proportion of fresh air are clearly further from equilibrium.

The CFD modelling shown in Figure 5 does not include droplet splashing – droplets in the model disappear on impact with the ground. The presence of the pool of liquid in the bund around the base of the tank is also ignored. It is likely that in most circumstances the splash zone at the base of the tank is an additional area where vapour and very finely divided liquid are vigorously mixed for a significant period of time, which pushes the whole of the flow closer to equilibrium.

In the scoping method described in Section 2 it is assumed that the liquid released and the gas flow that it entrains in the cascade and splash zone are in thermodynamic equilibrium. This is a conservative assumption in the assessment of vapour cloud production but available information on liquid dispersal and heat and mass transfer calculations suggest it is also reasonably close to the truth in most cases.

One important exception to this may be tanks where high volume releases are concentrated in very small sections of the tank perimeter. Releases from many Type C tanks could be of this sort. Very high liquid mass flux densities O(100 $kg/m^2/s$) could result. In this case liquid dispersal would be limited and the spray would be composed of very large droplets or streams of liquid. For the very large liquid fragments, the rate of vaporisation could be limited by the ability of lighter, more volatile fractions to diffuse to the surface of the liquid in contact with the air. This is significant in the analysis of the potential for Type C tanks to produce flammable clouds when overfilled with liquids composed of only a small volume fraction of volatile material e.g. light crude oils.

1.5 NEAR FIELD DISPERSION

Generally, dispersion of a release of flammable vapour cloud is treated separately from the source term (unless a full CFD treatment of the whole release is possible). To take this approach it is necessary to identify where the source term ends and the dispersion calculation should begin. The choice taken here for this point of separation is at the base of the tank or at the edge of the zone where the vapour flow is deflected into the horizontal.

Care has to be taken in joining source term and dispersion calculations in this way. High vapour velocities O(5m/s) are typically induced by the cascade at the foot of the tank. Even though the flow is denser than air, such a flow will entrain air as it flows out across the floor of the bund. This entrainment process occurs whether the flow impacts on a bund wall (as in Figure 5) or not. Any entrainment of fresh air after the bulk of the liquid has rained out will result in a reduction in vapour concentration. Contact between the vapour and liquid pool on the floor of the bund may on the other hand increase the concentrations, although this may be limited since the vapour close to the floor of the bund may be close to being saturated already.

There is a tendency for the entrained air to move through the cascade towards the tank wall (the Coanda effect). This means that the bulk of the vapour flow passes through

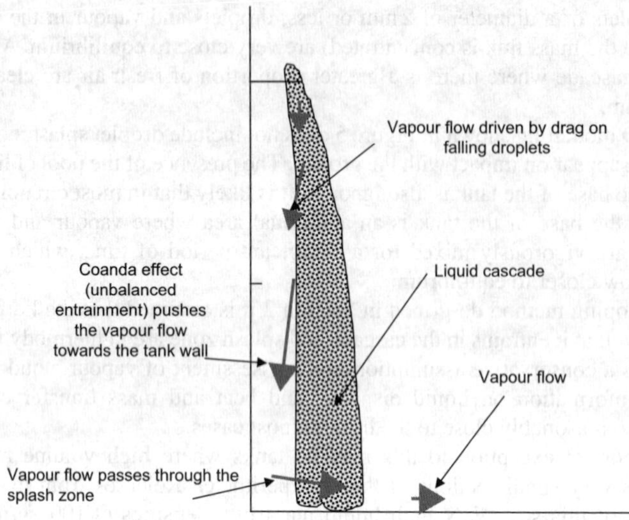

Figure 6. Schematic showing vapour flow driven by a free liquid cascade

the droplet splash zone at the base of the tank – see Figure 6. Droplet splash products are capable of absorbing part of the vapour jet momentum and consequently suppressing the tendency for entrainment – even in the near-field. This effect is still under investigation. Large-scale experimental releases of hydrocarbons are needed to obtain reliable data on the flow behaviour for this case.

2. SCOPING METHOD

2.1. APPROACH AND ASSUMPTIONS

The scoping method described here is based on principle that production of vapour concentrations within the flammable range at the base of the tank will bring liquids "in scope". This is a somewhat conservative, but reasonable, assumption that might be refined if more was known about the splashing process and its effects of the near-field dispersion.

The method provides a means of determining whether a given filling operation in a given tank can lead to the generation of a flammable cloud. Such a scoping method is clearly of interest in determining the appropriate level of protection against overfilling. The volume and concentration of flammable vapour close to the source are outputs but to predict the potential extent of the cloud would require a dispersion model.

Although it may appear initially counter-intuitive, the likelihood of producing flammable vapour for many substances increases as the amount of fresh air entrainment is reduced. Enhanced air entrainment leads overall to greater evaporation but the vapour produced is often below the lower flammability limit.

The scoping method is divided into a number of stages which are described below:

A. Proportion of tank perimeter covered by liquid release
It is assumed that in all cases the liquid released is distributed over 30% of the tank perimeter. In the case of Type C tanks this may be an overestimate. In principle this might lead to non-conservative overestimation of the induced vapour flow, however this is unlikely to lead to serious underestimates of risk because of the relatively low sensitivity of the induced flow to the liquid mass flux and the tendency for vapour concentrations to fall short of equilibrium at very high liquid mass fluxes.

B. Liquid mass flux in the cascade
The distance the spray extends away from the tank wall is assumed to be 1.5 m over the full height of the cascade. This is a reasonable minimum figure based on observations on water cascades. Wind girders part way down the tank can increase the width to in excess of 3 m but any broadening of the liquid cascade increases the total induced air flow and tends to reduce the maximum vapour concentration. Given the cross section of the cascade and the total liquid release rate the liquid mass density can be calculated.

C. Entrained air flow
Given the liquid mass density the volume flow of entrained air can be taken from a plot such as that shown in Figure 4. The height over which air is entrained is not the full height of the tank because it typically takes several metres for primary aerodynamic break up to be complete and there is likely to be re-entrainment of contaminated air from the splash zone in the last few metres of fall. It has therefore been assumed that air is entrained over a minimum height of 6 m. For very high tanks (>15 m) this may be an underestimate leading to minor underestimates of airflow and overestimation of risk.

Observations of petrol releases suggest that 2 mm is an appropriate droplet diameter for this calculation. The airflow is insensitive to this choice of diameter within a reasonable range.

D. Equilibrium calculations
The concentration of vapour at the foot of the tank is estimated by assuming thermodynamic equilibrium. Given total liquid flow rates and air entrainment rates (and the temperatures of both) the final temperature and vapour concentration can be calculated straight forwardly. Examples of results of such a calculation for a winter grade petrol are given in Annex 2. Water vapour condensation should be included in the enthalpy balance but only makes a substantial difference if the humidity and ambient temperatures are high.

E. Comparison with flammability limits
If the vapour concentration calculated in D exceeds the Lower Flammable Limit it is possible that overfilling of the tank will produce a flammable cloud.

The method described above accounts for the fact that the temperature drop due to evaporation of spray droplets may reduce the saturation vapour pressure sufficiently to avoid the production of flammable vapour. This means that in some cases a substance that is flammable at room temperature, such as toluene, may not produce flammable vapour in the cascade from a tank overfilling release. In reality, in such cases, the liquid from the tank overfill will accumulate within the bund and may eventually rise to ambient temperatures and start to produce flammable vapour. This hazard could be modelled using standard pool-evaporation models.

Results of such scoping analyses on typical high volume refinery liquids and crude oils are shown in Figures 7 and 8. Composition data for the mixtures analysed are shown in Annex 3. In all cases the temperature of the released fluid was 15 °C and the ambient temperature 15 °C. The independent variable is the total liquid release rate divided by the total tank diameter.

3. IMPLICATIONS FOR SAFETY AND ENVIRONMENTAL STANDARDS AT FUEL STORAGE SITES

The technical work described in this paper was carried out in support of the Buncefield Standards Task Group (BSTG). The BSTG was formed soon after the Buncefield incident and consisted of representatives from industry and the joint Competent Authority for the Control of Major Accident Hazards (COMAH). The aim of the task group was to translate the lessons from the incident into effective and practical guidance.

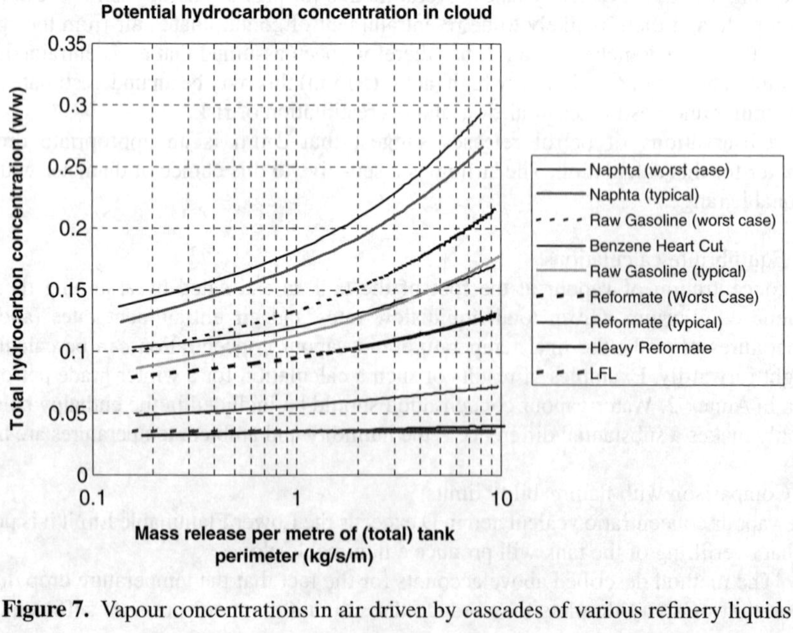

Figure 7. Vapour concentrations in air driven by cascades of various refinery liquids

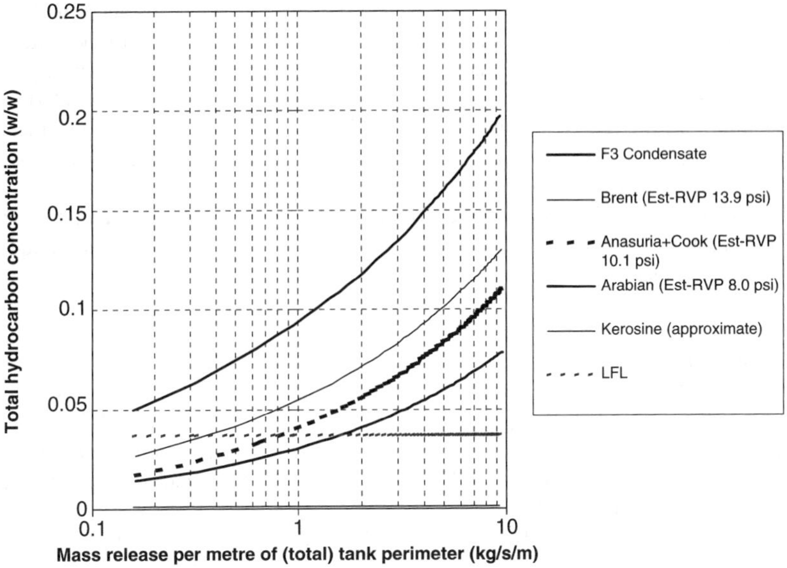

Figure 8. Vapour concentrations in air driven by cascades of various crude oils

To ensure focussed and timely responses to the issues arising from Buncefield the scope of application for the work of the task group was defined in the initial report by BSTG (5). This was confirmed in the final report of July 2007 (6) and is repeated here:

- COMAH top- and lower-tier sites, storing:
- gasoline (petrol) as defined in Directive 94/63/EC [European Parliament and Council Directive 94/63/EC of 20 December 1994 on the control of volatile organic compound (VOC) emissions resulting from the storage of petrol and its distribution from terminals to service stations], in:
- vertical, cylindrical, non-refrigerated, above-ground storage tanks typically designed to standards BS 2654, BS EN 1401:2004, API 620, API 6508 (or equivalent codes at the time of construction); with
- side walls greater than 5 metres in height; and at
- filling rates greater than 100 m^3/hour (this is approximately 75 tonnes/hour of gasoline).

The results of the work reported in this paper confirm the scope of application for the initial response to Buncefield. That is to say that all types of storage tank described in section 1.1 are believed to be capable of generating a cascade of liquid droplets in the event of overfilling with hydrocarbon liquid. If that liquid hydrocarbon is gasoline then there is the potential for the formation of a large flammable vapour cloud.

This work also indicates that there is the potential for other substances with similar physical properties to behave in a similar way in the event of a loss of primary containment following overfilling. Work continues in order to establish an agreed definition for the extension of scope to a limited number of other substances. This might also lead to a better understanding of the release conditions that might lead to this scenario. The further work continues under the Petroleum Process Standards Leadership Group which has been formed to take forward the work started by the BSTG.

In the meantime the results of the work of BSTG have been taken forward as a series of actions required of operators. The final report (6) details these actions and includes the supporting guidance.

REFERENCES

1. Maremonti M., Russo G., Slazano E. and V. Tufano *Post–accident analysis of vapour cloud explosions in fuel storage areas*. Trans. IChemE, 1999, **77:** p.360–365.
2. Yuill, J. *A discussion on losses in process industries and lessons learned*. in 51st Canadian Chemical Engineering Conference (see http://psm.chemeng.ca), Halifax, Nova Scotia, Canada, 2001.
3. Buncefield Investigation – Third Progress Report. 2006 Major Accident Investigation Board. (available from http://www.buncefieldinvestigation.gov.uk).
4. Chang, J.I. and Cheng-Chung, L. *A study of storage tank incident*, J. Loss Prevention, 2006 **19:** p.51–59.
5. Bai, C.X., Rusche, H. and Gosman, A.D., (2002) *Modelling of gasoline spray impingement,* Atomisation and sprays, **12:** p. 1–27.
6. Buncefield Standards Task Group *initial report – recommendations requiring immediate action* 12 October 2006 (available from http://www.hse.gov.uk/comah/buncefield/bstg1.htm).
7. Buncefield Standards Task Group *final report – safety and environmental standards at fuel storage sites* 24 July 2007 (http://www.hse.gov.uk/comah/buncefield/final.htm).

Annex 1: Gas flow driven by liquid cascade

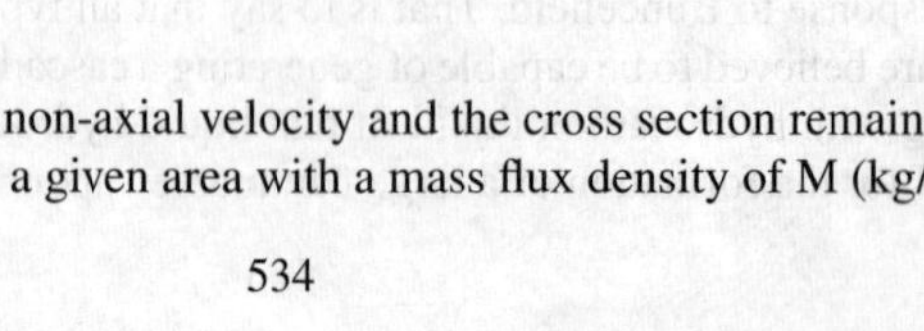

Assume

1. The spray has little initial non-axial velocity and the cross section remains constant.
2. The spray is uniform over a given area with a mass flux density of M ($kg/m^2/s$).

3. The induced gas phase velocity is constant across the section. The additional gas mass flow required is presumed to be entrained through the vertical boundary of the spray and rapidly mixed across the section.
4. The spray is monodisperse (i.e. all droplets are the same size).

Droplet dynamics

$$m_{droplet}\frac{du_{droplet}}{dt} = m_{droplet}\cdot g - \frac{1}{2}C_d\rho_{vap}A_{drop}(u_{droplet} - u_{vapour})^2$$

Vapour dynamics
Vapour velocity at a horizontal control surface below the origin of the spray

$$\rho_{vap}u_{vapour}{}^2 = \sum_{droplets}\frac{1}{2}C_d\rho_{vap}A_{drop}(u_{droplet} - u_{vapour})^2$$

The summation is carried out over droplets above the control surface
Additional relations used

$$N(x) = \frac{M}{m_{droplet}u_{droplet}(x)}$$

This relates the number density of droplets to M the mass flux density (kg/s/m^2) in the spray

$$\frac{A_{drop}}{m_{droplet}} = \frac{3}{4r_{drop}\rho_{drop}} \text{(characteristic of spherical droplet)}$$

These equations can easily be integrated (numerically) form the origin of the cascade to yield droplet and vapour velocities.

Annex 2: Characteristics of vapour produced by a cascade of winter petrol (Ambient temperature 0 °C). Liquid flow rate 550 m^3/hr
The conditions given below are calculated based on equilibrium between the liquid and vapour phases. A given flow rate of liquid is mixed with a given flow rate of fresh air and allowed to reach equilibrium in terms of both temperature and concentration.

Initial liquid composition (Liquid temperature 15 °C)

n-butane (as a surrogate for all C4 hydrocarbons)	9.6%	wt/wt
n-pentane (as a surrogate for all C5)	17.2%	wt/wt
n-hexane (as a surrogate for all C6)	16%	wt/wt
n-decane (as a surrogate for all low volatility materials)	57.2%	wt/wt

Rate at which air entrained into cascade 96 m³/s
Final vapour and liquid temperature −8.5 C.

Vapour composition

n-Butane (as a surrogate for all C4 hydrocarbons)	6.0 %	wt/wt
n-pentane (as a surrogate for all C5)	6.1 %	wt/wt
n-hexane (as a surrogate for all C6)	2.06%	wt/wt
Total hydrocarbons (in air)	14.17 %	wt/wt

Residual liquid composition

n-butane (as a surrogate for all C4 hydrocarbons)	2.4%	wt/wt
n-pentane (as a surrogate for all C5)	11.5 %	wt/wt
n-hexane (as a surrogate for all C6)	16.3 %	wt/wt
n-decane (as a surrogate for all low volatility materials)	69.6 %	wt/wt

Annex 3:

	Paraffins						Aromatics				Naphthenes		
Composition % (w/w)	C4	C5	C6	C7	C8	C9	C6	C7	C8	C9	C5	C6	C7
Naphta (worst case)	9	58	20				4				7	2	
Naphtha (typical)	2	56	21	6	1		3	1			2	5	3
Raw gasoline (worst)	2	20	20				35	15	8				
Raw gasl'ne (typical)	1	9	21				35	13	7	14			
Benzene heartcut			50				50						
Reformate (worst)			22	27	3		21	25	2				
Reformate (typical)			4	18	17	4	5	24	23	5			
Heavy reformate			4	5	3		1	31	34	22			

	Paraffins						Aromatics		Nap
Composition (w/w)	C2	C3	C4	C5	C6	C7	C6	C7	C5
F3 condensate		0.3	4.4	6.5	4.1	6.5	4.7	1.4	2.8
Anusa	0.02	0.4	1.78	2.72	2.3		1.42		0.28
Brent	0.07	0.74	1.75	2.65	2.27	2.84	2.53	1.25	1.5
Arabian		0.57	0.76	1.75	1.53	1.68	1.22	0.37	0.08

The balance of the crude oil mixture is modelled as a range of low volatility alkanes (not shown).

CONTINUOUS MONITORING OF RISKS – PEOPLE, PLANT AND PROCESS

Dr. John Bond

The aviation industry has demonstrated that the flight hazards associated with the operation of large public transport aircraft can be monitored and the recognised risks controlled to improve safety significantly. They have introduced this new approach with:-

- Flight data monitoring and analysis covering people, equipment and operating conditions.
- Just Culture.
- Sharing information.

This approach has had the active support of the Civil Aviation Authority and the Air Accident Investigation Branch and makes the approach less error prone and more error aware. Its objective is not to decrease the safety accountability of the operator but to increase the safety accountability of everyone who designs, constructs, manages, operates and maintains the system.

The paper will describe ways that this approach can be applied to computer controlled petrochemical plants to monitor all the identified hazards, to analyse the operation against the limits set in operating procedures and hence provide information to control risks on a continuous basis and to improve safety.

KEYWORDS: Monitoring risks; Just Culture; risk control; evolutionary management.

INTRODUCTION

Accidents usually occur when a series of errors in the equipment, in the process or in the method of operation coincide. Seldom does a single error result in an accident. If the people, the equipment and the process are monitored the data obtained can be used to identify errors that occur and hence establish the level of the risk involved in the whole operation. Once the frequency of errors is established then they can be addressed, the process altered and the new level of risk established. The production activity thus incorporates the provision of information for the improvement in safety. This evolutionary management system becomes a dynamic and progressive philosophy leading to reduced risk all round.

The recent report on the Texas Refinery Fire (BAKER 2007) produced a number of recommendations and under the section "*Measuring process safety performance*" it states "As a result, BP's corporate safety management system for its U.S. refineries does not effectively measure and monitor process safety performance" Thus Recommendation 2 titled "INTEGRATED AND COMPREHENSIVE PROCESS SAFETY MANAGEMENT SYSTEM" required:-

> " BP should establish and implement an integrated and comprehensive process safety management system that systematically and continuously identifies, reduces, and manages process safety risks at its U.S. refineries."

I suspect that these criticisms apply equally to many companies, but what is a comprehensive process safety management system? I am sure that there is much debate about what it comprises but I would expect it to start at the top of the company with a corporate:-

- statement on leadership and responsibility at all levels of the organisation;
- requirement to comply with all management systems; and
- development plan for a safety culture that was just, identified hazards and monitored the risk arising from the personnel, plant and process.

The Buncefield Standards Task Group Final Report (HSE 2007) requires an active monitoring operation of tank storage but makes no mention of direct monitoring of operator's work.

It is my view that a system of monitoring all the identified hazards in a process operation (equipment, process and personnel) to ensure that the risks are under control to an acceptable level is now required. Any variation of the operation from the finalised Standard Operating Procedures (SOP) would be noted and a prompt decision made whether to alter the SOP, carry out maintenance of the equipment or provide additional training as appropriate. This would provide the demonstrated control of the risks involved that is required by the Baker Report (BAKER 2007).

THE AVIATION INDUSTRY

The civil aviation industry has introduced over recent years a comprehensive method of monitoring the operational flying standards of large civil airliners. This includes the monitoring of identified hazards involved in each flight operation, the analysis of the data against the boundaries set in the Standard Operating Procedures (SOP) for the flight, a Just Culture approach and a sharing of information on accidents and near-miss events. As a result they have knowledge of and are in control of all the identified risks involved in the flight operations. Any modification of the SOPs can be studied in following flights to ensure that the risk has been reduced. This management system has been so successful that it is now being applied to helicopters (CAA 2002).

The Flight Data Monitoring system (CAA 2003) (see Figures 1, 2 and 3) records data on the crew's operational performance, the performance of the equipment and the flight conditions for each flight. The data is removed after the flight by a disc or by telemetry and analysed. In British Airways this is done automatically by the Special Event Search and Master Analysis (SESMA) software which can evaluate the whole flight against the

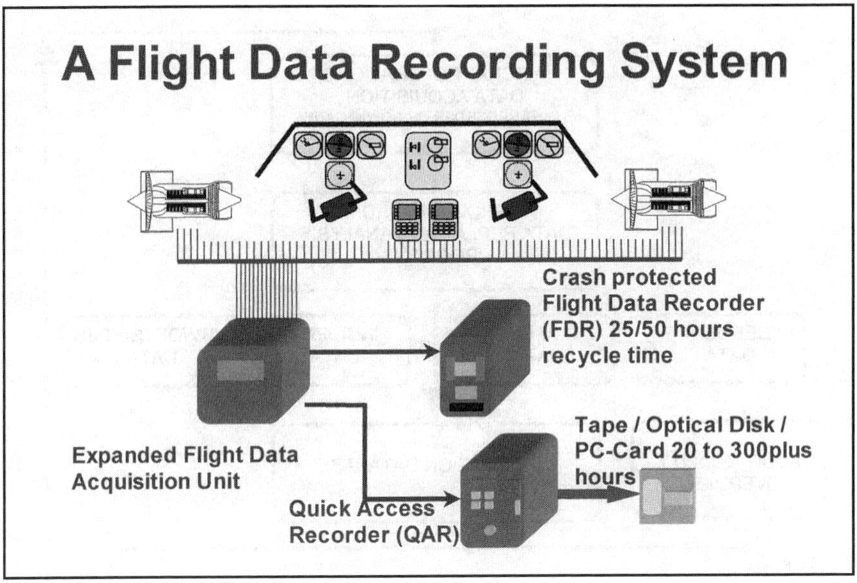

Figure 1. (CAA 2003)

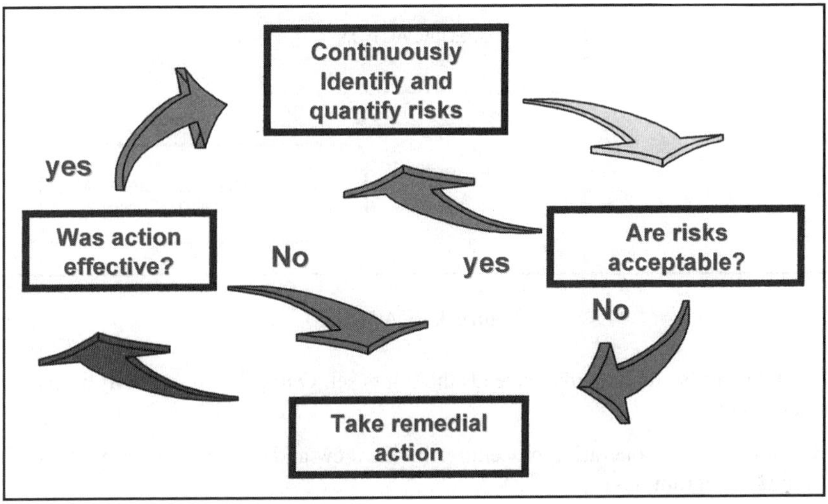

Figure 2. (CAA 2003)

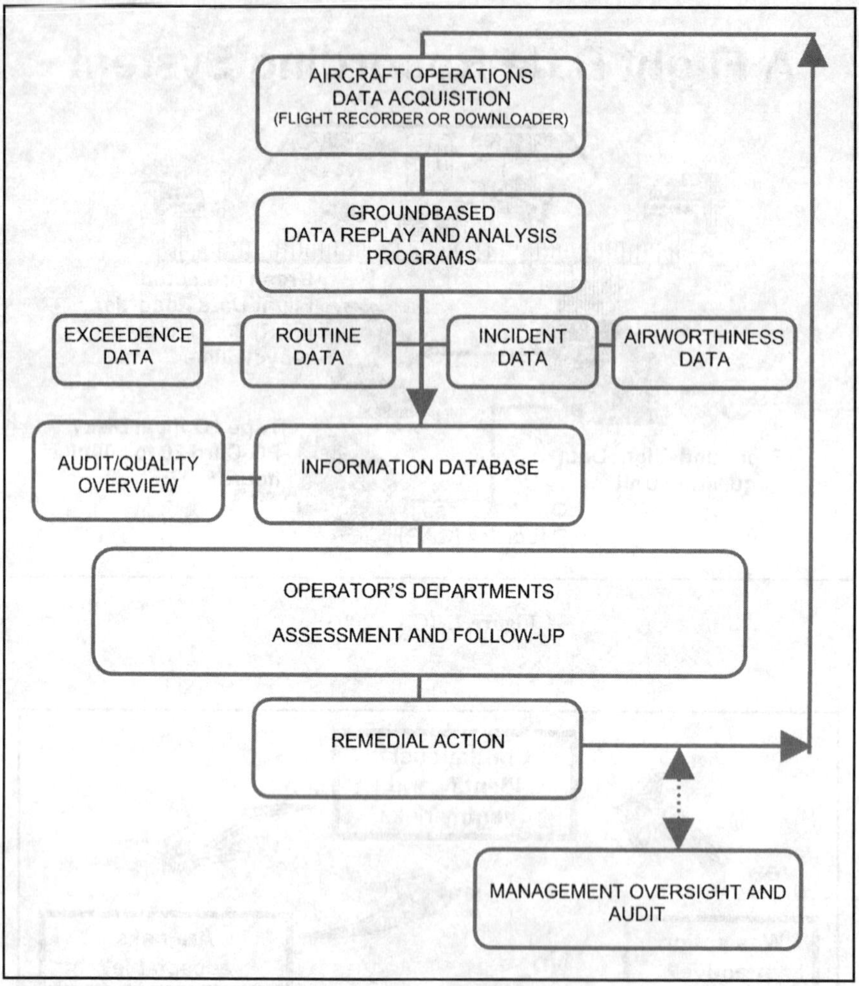

Figure 3. (CAA 2003)

SOPs and records any data that exceeds the limits set. Output from the analysis can automatically show:-

- Compliance with operating procedures by the crew and can feed this if necessary into training programmes
- The state of the equipment and can feed this information into maintenance programmes.

- Flight operating conditions for unusual flight conditions to be studied by appropriate personnel.

The Civil Aviation Authority defines (CAA 2003) the Flight Data Monitoring as a "*... systematic, pro-active and non-punitive use of digital flight data from routine operations to improve aviation safety.*"

The choice of data to be collected is based on the identified hazards involved in the operation and could be described as a study of hazards and operability similar to a HAZOP. Each operator configures his Flight Data Monitoring programme to reflect his SOPs and any exceedence of the boundaries set is identified by the SESMA system. These are then noted and appropriate action taken to ensure safety of operations. This ensures that all the risks are continuously monitored and their level evaluated and reduced where necessary. The whole process becomes an evolutionary management system which monitors and reduces the risk of flight operation.

The monitoring of the pilot operations on an airliner in the UK covers a wide variety of some 100 plus conditions such as climbing speed too low, flaps out speed exceeds limit set in SOP, too deep landing, tail scrape and many others. About 400 to 500 events each month are discovered, usually only minor infringements of the limits but some more serious resulting in 10 to 20 crews being contacted for an explanation. The pilot's union are involved as middlemen in these discussions and hence confidentiality maintained.

The Flight Data Monitoring system thus provides information on the level of risk of each identified hazard and provides information both within the aircraft and external e.g. weather, Air Traffic Control, Airport.

The Just Culture approach was found to be a necessary part of the Flight Data Monitoring system as it involved the work of the crew. In rejecting the blame culture, the International Civil Aviation Organisation (ICAO), states in Section 4.5.40 of Safety Management Manual (ICAO 2006):

> *"If an accident was the result of an error in judgement or technique, it is almost impossible to effectively punish for that error. ...... If punishment is selected in such cases, two outcomes are almost certain. Firstly, no further reports will be received of such errors. Secondly, since nothing has been done to change the situation, the same accident could be expected again."*

A Just Culture has been defined (GLOBAL 2004) as:

> *"A way of thinking that promotes a questioning attitude, is resistant to complacency, is committed to excellence, and fosters both personal accountability and corporate self-regulation in safety matters."*

The approach has been described by James Reason (REASON 1998 and 2005) and has been applied also in the Air Traffic Management area of the aviation industry to improve safety. He states that:

> *"A prerequisite for a just culture is that all members of an organisation should understand where the line must be drawn between unacceptable behaviour,*

deserving of disciplinary action, and the remainder, where punishment is neither appropriate nor helpful in furthering the cause of safety".

The civil aviation industry also has two major information systems to share information and lessons learnt comprising the Mandatory Occurrence Reporting Scheme (MORS) and the Confidential Human Incident Reporting Programme (CHIRP). It is for the adoption of these approaches to safety that the Just Culture becomes important.

The civil aviation industry in Australia, Canada and the UK has fully adopted the management system involving all three approaches:

- Flight Data Monitoring system with continuous analysis.
- The Just Culture approach.
- The sharing of accident and near-miss information

This has resulted in an increase in the reporting of incidents while keeping low the fatal accident rate. The fatal accident rate (fatalities per million hours flight) over a period of ten years for UK Registered/operated Large Public Transport Aeroplanes (3 year moving average) was only 5 fatal accidents causing 8 fatalities over the ten year period. Australia, Canada and the UK currently have a 3 year moving average fatal accident rate of zero compared with a figure of 13 for the whole world. These figures are based on Western designed aircraft with similar training courses but different management and regulatory systems

This approach to controlling the risks involved may not be directly applicable to other industries but the principles of:

- Monitoring the identified hazards associated with personnel, equipment and conditions on the flight, analysing the data against the limits set in the SOP and hence identifying the level of risk in the operation.
- Setting up a Just Culture approach to ensure the sharing of information
- Sharing information with other companies

are applicable to many other industries. The three stages are interdependent with one another to get the full value of the risk minimising process.

The difficulty of obtaining acceptance by the work force is much more of a problem if they are not brought into the process. As with civil aviation staff the consequences of errors by process operating staff can result in fatalities and therefore both staff and management have an interest in reducing risk. In the UK civil aviation industry the pilots union are an integral part of the safety culture.

This approach in the civil aviation industry was spearheaded by the ICAO, the CAA, the AAIB, the European Commission, Eurocontrol and the companies who have all taken an active part in its development. It is not surprising that Lord Broers, Past President of the Royal Academy of Engineering stated (BROERS 2005):

> *"One crucial recommendation emerges* (from the debate in the Royal Academy of Engineering). *That the investigation of accidents should concentrate on finding the cause of the accidents not the person or persons to blame. The latter*

only leads to defensiveness and cover up. The investigation should seek the cause of the accident so that it may be eliminated in the future. The airline industry's remarkable safety record is thought by some to be because the investigators seek the cause of accidents rather than hunt down the person to blame."

MONITORING AND ANALYSIS IN THE PROCESS INDUSTRIES

The design of plant is now always subjected to a HAZOP study where the hazards are identified and the risks assessed. Modifications to the design are then made to ensure an acceptable level of risk. Monitoring of the risks during operation is based on well recognised standards. Equipment is inspected on a schedule dependant upon experience and records are kept of all equipment. Instrumentation is tested on a regular basis and also recorded. However, corrosion of equipment causes a number of serious accidents and online monitoring of pipework for corrosion is needed.

The hazards of the process are also well established prior to design and are kept in mind at the HAZOP stage and suitably dealt with. The process is monitored continuously to ensure that the right product is produced and in an efficient manner.

The hazards associated with people making an operational error are not always dealt with sufficiently. It is assumed that the training and competency assessment would prevent most errors. Monitoring of process operators is carried out by the supervision provided.

Continuous monitoring by computer of risks associated with the process operators, the plant equipment and the process is possible on the larger plants and has been shown to be very effective in the aviation industry. It should be applied in the process industry and could begin with the start-up operations, extended through the whole of the process operation and with the shut down operations

Establishing the finalised SOP is an opportunity to ensure that all of the identified hazards are monitored on a continuing basis. Analysis of the data obtained during operation of the process will:-

- Identify and quantify operational risks associated with the people, the plant and the process on a continuous basis;
- Identify and quantify any changing risks in the operational work;
- Formally assess the risks to determine which are not at an acceptable level;
- Where the risks are not acceptable take remedial actions;
- Demonstrate that the risks are being monitored continuously;
- Provide leading safety performance measurements as indicators.

The whole monitoring programme and analysis becomes an evolutionary safety management system to reduce risk and provide improvements in the level of safety.

MONITORING OF PROCESS OPERATORS

Continuous monitoring of process operators was frowned upon by the unions because of the attitude to 'blame the operator' which would lead to disciplinary measures. If the blame culture was completely removed from the monitoring process, the advantage in safety could be demonstrated bearing in mind it is the operator that is the one usually injured in an accident. Such a monitoring operation was carried out (BOND 1975) when random sampling of operators and craftsmen was used to see if all of the operations were being carried out to the SOP. All process and maintenance operations were defined with two safeguards being required to protect the men against all reasonably foreseeable hazards. Activity Sampling was carried out on a random basis to establish what the true position was based on a 95% confidence limit. This work was carried out with the knowledge and support of the unions on the understanding that no names would be recorded on any sample taken. Hence there was no blame associated with the procedure. It was found that compliance with the SOP after the first training period was 78% for operators and 54% for craftsmen. After further training the compliance was raised to 79% and 73% respectively. The training of the men and the random sampling were unfortunately stopped not by the unions but by senior management! My experience of monitoring process operators and craftsmen was that there were relatively few problems if the reasons were made clear and if it was explained that you were honestly not seeking anyone to blame.

In the past the work of process operators running process plant was monitored by foremen, chargehands and senior operators. With the reduction of supervision people there has been a greater reliance on human factors including training and competency of the operators. This is important but, as with pilots, operators are human and can make occasional errors. When errors do occur investigations usually show that there has been an operation outside the limits set in the SOP. The training or competency process is then usually blamed.

With the computer control of process operations the monitoring of process operators would be possible particularly for start-up, normal operations, shut down and emergency procedures. The following operations would be critical operating parameters affecting safety and could be monitored:

Start-up Procedure where errors occur:

- Correct sequence of start-up
- Valves in correct position
- Flow conditions correct
- Alarms commissioned
- Compressors operating satisfactorily
- Thermal shock due to heating up too fast

Normal Operations where errors occur:

- Reliance on high level alarms to stop transfer operations to tanks
- Alarms failed to be back on line after testing
- Alarm isolations, response and reinstalling
- Relief valves lifting due to pressure resulting from the temperature of the LPG being pumped into the tank is too high.
- Warning of temperature excursions

Consider just two examples, thermal shock and transfer to a tank.

Heating up of pipes and flanges at a maximum rate of, say, 25°C per hour as specified in the SOP, allows equality in the heating up process and maintains integrity of the system. If the heating up rate is exceeded there is a greater possibility of losing containment of material. The heating up rate of items could be continuously monitored and any that exceeds the specified rate set in the SOP would be noted. A very short and small excess might be considered only a small risk but should be noted by the data analysis and the shift personnel warned that the specified rate must be adhered to. A small amount of overheating should be considered a near-miss.

Transferring material to a tank should be a fixed amount such that the Normal Fill Level is not exceeded. Reliance on the Level Alarm High (LAH) to cut off a transfer is not acceptable. I have experience of three cases where this was used and the tank overflowed because the alarm did not cut off the transfer. How many times the pump trip switch had been relied upon and operated to stop the transfer I do not know but I suspect many times. Monitoring the operations could identify this case as exceeding the SOP and stopped before an accident actually happened.

Monitoring the work of process operators and analysis of the data identifies the operational irregularities which could foreshadow accidents. A full knowledge of the risk level occurring in the operations ensures that the management is in full control of the process. It provides active safety indicators of all the hazards identified for the whole plant.

MONITORING THE OPERATION OF EQUIPMENT

Monitoring the condition of some equipment is carried out by detailed crack inspection but crack identification could be continuously monitored and then analysed by comparing the data with the limits set in the appropriate standards. Other areas which should be monitored are for example:-

- Compressor vibration,
- Operations out of sequence,
- Abnormal pump pressures,
- Reaction temperatures profiles out of normal pattern,
- Relief valves lifting below their set pressures,
- Alarms left isolated,
- Surge pressures in pipework.

MONITORING THE OPERATION OF THE PROCESS

With computer controlled plants many critical factors are monitored continuously and variation from the normal SOP is immediately high lighted by an alarm so that the process can be brought back in line. The frequency of some of these events could be part of the process data to be established and considered whether any alterations are necessary. Some aspects of the process could, with advantage, be monitoured and the data compared with the limits set in the SOP for example,

- Reactor temperature variation
- Reactor pressure variation
- Variations in impurities
- Quality and energy efficiency of the process

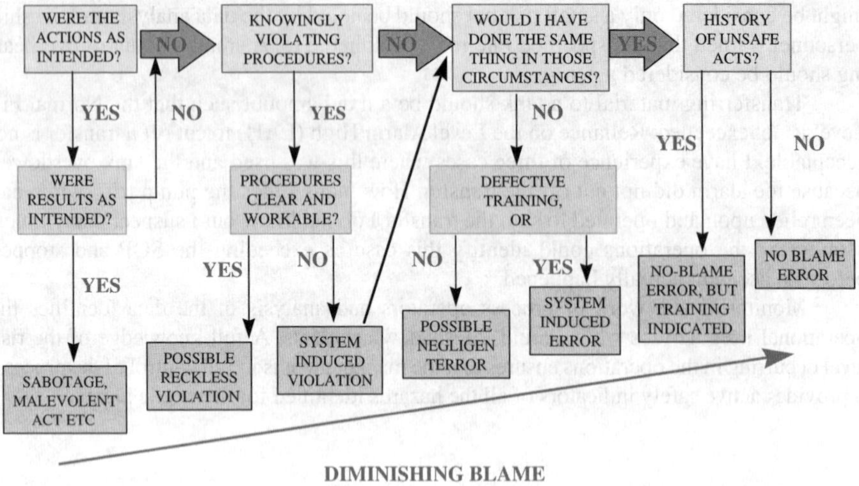

Figure 4. Disciplinary action

THE IMPORTANCE OF A JUST CULTURE

If the process staff are to be monitored it is imperative that the concept of the Just Culture system is adopted throughout the company as without it there will be no reporting of incident information and the monitoring system will quickly break down. Figure 4 shows the logic of the system adopted by one UK company.

The CAA recognises the Just Culture approach and encourages companies to take the disciplinary action. The CAA has stated (ALCOTT 2006):

> *"We promote a just culture. Since 1976 the CAA has run a mandatory Occurrence Reporting Scheme where we have asked industry to submit to us the quite low level incidents that are happening in the industry. We have given a guarantee that we shall not take punitive action against those people who report to us, except in cases of gross negligence. We expect industry to behave in the same way and to use that data with us for continuous improvement .... We can then work towards the 'let us not let this happen again' type of scenario."*

SHARING INFORMATION

Hazards can be recognised not only from the operations carried out by the operators but also by managers recognising the advantages of sharing lessons learnt from accidents. The Responsible Care Programme, the COMAH Regulations, the recommendations of the Texaco Refinery Fire of 1994 and the recent Buncefield Standards Task Group Final Report all require sharing of lessons learnt from accidents. All staff must be treated equally and therefore managers must also be monitored. The sharing has to be in the whole industry, not just the organisation.

REGULATORY AND INVESTIGATION BODIES

In the aviation industry accident investigations are carried out by the Air Accident Investigation Branch (AAIB) who is independent of the CAA regulatory body. The AAIB has the right of entry to investigate accidents involving aircraft. In the Regulations (AAIB 1996) they have an objective:

> *"The sole objective of the investigation of an accident or incident under these Regulations shall be the prevention of accidents and incidents. It shall not be the purpose of such an investigation to apportion blame or liability. The provider of any evidence given to the AAIB cannot be used in other court actions. This ensures that the full evidence can be given to the investigators. The results of the investigation and all recommendations are made available to the public."*

A similar situation exists with the other transport industries in the UK and with the Chemical Safety and Hazards Investigation Board in the USA.

CONCLUSIONS

The Health and Safety Executive (HSE) has a duty of investigating accidents as well as being the regulatory body. It might be thought advisable that these two duties be separated, as in the aviation, rail, marine industries and some European regulatory bodies. This would allow the investigation to be carried out to establish all the causes of the accident rather than to find a person to blame. This would leave the disciplinary side to consider whether a violation of a regulation had been established. Because a company with a Just Culture approach would take disciplinary action against its employee as necessary the HSE could then be relieved of much of its work.

The monitoring system of people, plant and process allows the critical operating parameters for all sections of a process to be under the control of the management. They will have knowledge of the level of risk of each identified hazard and any variation will be quickly identified and can be readily rectified.

This approach of an evolutionary management system should meet the requirements of the Baker Report to "… implement an integrated and comprehensive process safety management system that systematically and continuously identifies, reduces, and manages process safety risks …"

If hindsight is defined as wisdom after the event, learning the level of risks by the monitoring operation and analysing the cause of the errors becomes an evolutionary process of converting hindsight into foresight. Any modification to the operation in order to improve safety can then be readily monitored and an evolutionary management system is thus developed.

REFERENCES

AAIB 1996 The Civil Aviation (Investigation of Air Accidents and Incidents) Regulations 1996 Statutory Instruments 1996 No. 2798.

Alcott 2006 "The Economics and Morality of Safety – The Civil Aviation Industry". B. Alcott, Royal Academy of Engineering, ISBN 1-903496-26-8. April 2006.

Baker 2007 "*The Report of the BP US Refineries Independent Review Panel*" The Baker Panel Report 2007. www.safetyreviewpanel.com

Bond.1975 "The Two Safeguard Approach for Minimising Human Failure Injuries." J. Bond, The Chemical Engineer, April, 1975.

Broers 2005 "*Risk and Responsibility*" Lord Broers BBC Reith Lecture 2005.

CAA 2002 "Final Report on the Helicopter Operating Monitoring Programme (HOMP) Trial" CAA Paper 2002/02.

CAA 2003 "Flight Data Monitoring" Civil Aviation Authority, Safety Regulation Group CAP 739 2003 ISBN 0 86039 930 3 www.caa.co.uk

Global 2004 "A Road Map to a Just Culture: Enhancing the Safety Environment" Global Aviation Information Network First Edition September 2004.

HSE 2007 "Safety and environmental standards for fuel storage sites" Buncefield Standards Task Group Final Report 2007.

ICAO 2006 "ICAO Safety Management Manual" International Civil Aviation Organisation Doc. 9859 AN/460 First Edition 2006. Available on the internet.

Reason 1998 "Achieving a safe culture: theory and practice." James Reason *Work and Stress* 1998, Vol. 3 page 293–306.

Reason 2005 "Managing the Risks of Organisational Accidents" James Reason Ashgate Publishing Ltd. 2005, ISBN 1 84014 105 0.

THE NECESSITY OF TRUST AND 'CREATIVE MISTRUST' FOR DEVELOPING A SAFE CULTURE

Johnny Mitchell, MSc
Occupational Psychologist, The Keil Centre, Edinburgh, UK
E-mail: Johnny@keilcentre.co.uk

KEYWORDS: Trust, creative mistrust, safety culture, communication, transformational leadership

1. INTRODUCTION

Numerous studies have investigated the effects of organisational practices such as performance-based pay, team-working and safety training, on the number of injuries in the workplace. Most have not found a strong relationship. In fact, one study found that the effects of 10 high performing work practices had minimal effect (8% variance) on lost-time injuries after taking into account the nature, size and age of the organisations. Of course, this isn't always the case and some companies significantly reduce injuries by introducing new working practices. So, why do some organisations manage to reduce the number of accidents by introducing improved organisational practices while some don't?

Chmiel (2007), recently suggested that it is not the organisational practices per se that have a direct effect on safety behaviour and accident involvement, but the safety climate in which a company operates. The safety climate is reflected through perceived management values, commitment and attitudes. 'Perceived' is the key word here as many management teams are wholly committed to safety and yet the workforce would not perceive them to be. Few would argue against the importance of developing this perceived commitment and therefore it is vital to determine what aspects of the organisation lead employees to perceive that management is committed to safety.

This paper argues that trust is the key to leading employees to perceive management as committed to safety, and acts as a vital 'lubricant for the functioning of a safe culture'. In order to support this claim, this paper examines the results of research carried out in a large maintenance organisation to examine the relationship between trust on communication and safety behaviours. This paper also examines the concept of 'creative mistrust', makes the assertion that it is not the logical opposite of trust and that it needs to be developed alongside trust in order to create a safe culture. Finally, a practical model for developing trust and 'creative mistrust' is proposed and discussed.

2. A QUICK DEFINITION OF TRUST

A widely used definition of trust was proposed by Mayer, Davis & Schoorman (1995; p.712) as "the willingness of a party to be vulnerable to the actions of another party based on the expectation that the other will perform a particular action important to the trustor, irrespective of the ability to monitor or control that other party." Most theorists agree that making oneself vulnerable to another is a key part of trust (Rousseau et al, 1998).

However, typically in high trust relationships there is an absence of vulnerability even when the consequence of a trust violation is potentially high. For instance, if an employee admitted to having a near-miss to a trusted work-mate they might not feel vulnerable. However, this subjective feeling of vulnerability increases if they shared this information with a work-mate they did not trust. Therefore, trust can be viewed as the trustor's willingness to engage in behaviour that could make them *objectively* vulnerable if the trust was violated.

3. WHY IS TRUST IMPORTANT?

Many studies have hypothesized about the role that trust plays in developing a safe work culture. For instance, it is argued that trust facilitates an informed culture (open reporting of near misses and errors) as it has been consistently linked with open communication characterised by knowledge sharing between organisational members (Bonacich & Schneider, 1992) and has a positive relationship with employees challenging unsafe behaviour (Burns, 2004; cited by Flin & Burns, 2004). Furthermore, trust has been found to enhance co-operation, organizational commitment and the acceptance of organizational goals and decisions (Dirks & Ferrin, 2001). The influence trust has on these processes suggests that it is an essential aspect for the development of a safe culture.

While these assertions all ring true, this paper asserts that first and foremost trust influences safety culture by influencing the workforce's perception of management's commitment to safety. As safety culture is set by management and permeates down through the organization it is argued that management has the biggest impact on safety climate and associated safety behaviours (Conchie & Donald, 2006). Therefore, the importance of trust in gaining organisational commitment, cooperation and the acceptance of organizational decisions and goals (Dirks & Ferrin, 2001) is absolutely pivotal for this process. Management have the least opportunities to demonstrate trustworthiness as they have limited face-to-face contact with employees (considered to be an irreplaceable element for building trust) and have limited shared experiences, repeated interaction and shared social norms (all deemed important in building trustworthiness). Key safety messages are often communicated down through supervisors and other forms of indirect communication such as posters, videos etc.

In summary, management are in the unenviable position where their trustworthiness is pivotal if staff are to accept and act upon messages about safety, but they have limited opportunities to develop this trust.

Of course, trust is also crucial between colleagues and supervisors to facilitate the challenging of unsafe behaviours, the encouragement of the right behaviours and the reporting of near misses and errors. As these groups typically interact frequently and share many experiences there is plenty of opportunity for this trust to develop. However, as the 'way things are done' in an organisation is often determined by management, it is proposed that the trust of management has the biggest impact in developing a safe culture.

4. A STUDY INVESTIGATING THE RELATIONSHIP BETWEEN TRUST, COMMUNICATION AND SAFETY.

In order to identify the relationship between trust, communication and safety behaviours (both compliant safety behaviours that perform and maintain workplace safety and pro-active safety behaviours that help develop the environment that supports safety) a large transport maintenance organisation employed the author to survey a number of front-line workers in several locations around the UK. The study engaged 179 participants in completing an anonymous questionnaire that provided reliable measures of 'best practice' safety communication, levels of trust of various groups (colleagues, supervisors and managers) and measures of safety compliance and pro-active safety behaviour.

The research identified that:

- Trust of management had a significantly stronger effect on the safety behaviours of maintenance workers than their trust of supervisors or colleagues.
- Lower trust of management related to an increase in the number of times workers had behaved in a way that could have caused an accident in the previous year.
- The effects of management's safety communication on worker's safety behaviour was mediated by the amount they trusted management.
- Willingness to report near misses did not have a relationship with levels of trust (at any levels).

In short, this research highlighted that workers trust of management has a strong positive relationship with their safety behaviours. Furthermore, management communication is primarily associated with trust and this largely accounts for its relationship with safety. It is therefore vital to consider the ways in which communication builds and develops trust in order to improve safety behaviours.

Somewhat surprisingly the willingness to report near misses was not related to levels of trust. Further analysis revealed that the consequences of reporting near misses (e.g. the time it takes, paperwork, disciplinary action, the feeling that it makes no difference) were strongly related to whether people are willing to report near misses. In order for employees to be willing to report near misses it would require both the consequences to be addressed through effective processes (e.g. less paperwork, ease of reporting) and trust that reporting near misses to management will make a difference and they will be treated fairly.

5. HOW IS TRUST BUILT?

In order to understand how trust in management can be developed it is vital to understand how trust is developed between a trustor (the trusting party) and a trustee (the party to be trusted).

The model above, proposed by Mayer and colleagues (1995), highlights three characteristics responsible for trust: ability, benevolence and integrity (these are mediated by the trustors natural propensity to trust).

- **Ability** refers to the skills, competencies and characteristics that enable the trustee to have a positive influence on a specific domain.

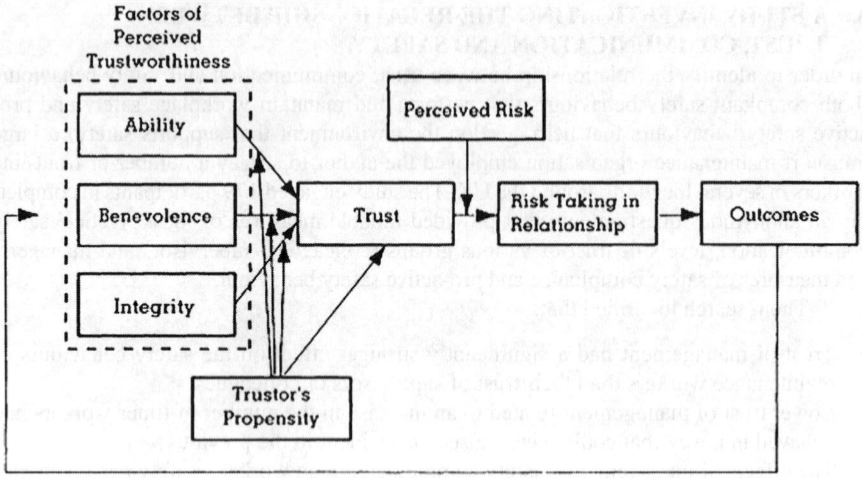

Figure 1. A proposed model of trust (Mayer et al., 1995, p.715)

- **Benevolence** is the degree to which the trustee is acting in an altruistic way, for the sole good of the trustor.
- **Integrity** refers to the degree to which the trustee is seen to adhere to a set of principles that the trustor finds acceptable (e.g. keeping promises).

These factors are seen as varying along a continuum and while they may vary independently of each other, Mayer et al., (1995) argue that when all three factors are high the trustee is deemed trustworthy. Further to this, Mayer et al., (1995) argue that integrity will play a key role at the beginning of the relationship and benevolence will grow in importance as parties develop a relationship and learn more about each other's intentions. Research supports Mayer et al's three factor model and the strong positive relationship it has with trust (Davis, Schoorman, Mayer & Tan, 2000).

6. THE IMPORTANCE OF COMMUNICATION IN DEMONSTRATING TRUSTWORTHINESS

Mayer et al., (1995) highlights the importance of communication in demonstrating the characteristics that develop trust (ability, benevolence and integrity). Various studies have researched the type of communication that builds trust. For instance, Brown (1999) studied the characteristics of organisational communication that impacted upon trust during a period of change at an aluminum facility. The results showed that openness, promptness and face to face methods of communication had a positive impact on trust while mass communication tended to diminish trust. It was found that communication throughout the period of change enhanced trust.

It is argued that communication methods associated with transformational leadership styles offer an appropriate model for enhancing both trust and occupational safety (Zacharatos., et al., 2005). Transformational leaders intellectually stimulate, inspire, and are individually considerate of employees (Bass & Avolio, 1994). They motivate employees to set aside personal gain and arrive at a mutual understanding and shared goals (Bass & Avolio, 1994), which logically supports the adoption of safety culture. It is proposed that this style of management influences safety through trust (Bass, 1990; Jung & Avolio, 2000), so that trust is repaid through increasing commitment to goals. The communication methods associated with transformational leadership are, amongst others, listening (consideration), encouragement, motivating and challenging (Barling, Loughlin & Kelloway, 2002). This would suggest that transformational leaders would challenge employees to improve safety, listen to their ideas, consider their circumstances and motivate them to improve safety.

Studies have supported the use of transformational leadership for safety. For instance, Yule (2003; cited by Flin & Yule, 2004) found that in the UK energy sector, leaders seen as transformational led business units with a significantly lower rate of injury. Yule identified a number of critical behaviours such as communicating an attainable picture of safety performance, engaging key staff in decision making and being clear and transparent when dealing with safety issues. Cohen and Cleveland (1983) compared 42 heavy industry sites and found that employees work more safely when they are involved in the decision making process, have specific responsibilities and authority and receive prompt feedback on their work.

These examples demonstrate some of the ways in which trust can be developed through communication. However, in order to develop trust there needs to be action as well as words and senior management have the opportunity to demonstrate trustworthiness by listening and addressing key worker issues, providing adequate resources, demonstrating concern, encouraging participation and by setting a good example. Whitener et al. (1998) proposed that five factors influence employee's perceptions of managerial trustworthiness. These include behavioural consistency, behavioural integrity, sharing of information and delegation of control, open communication and a demonstration of concern for the welfare of others. The model outlined in figure 2 highlights the various ways in which these factors can be considered.

7. WHAT IS 'CREATIVE MISTRUST'?

The danger with encouraging an atmosphere of trust is that if it is carried into every area of work employees will blindly follow and lose the ability to think objectively for themselves. The term 'creative mistrust' was coined by Hale (2000) who argued that employees need to adopt a more questioning attitude and avoid accidents and incidents that are a result of blindly trusting technologies, systems and processes. 'Creative mistrust' is similar in many ways to the concept of 'mindfulness' (developed through studying high reliability organisations) where emphasis is on constantly being aware, never being satisfied with safety performance and looking to anticipate new problems or old problems in different guises. However, as Joyner and Lardner (2007) remark, 'mindfulness' isn't

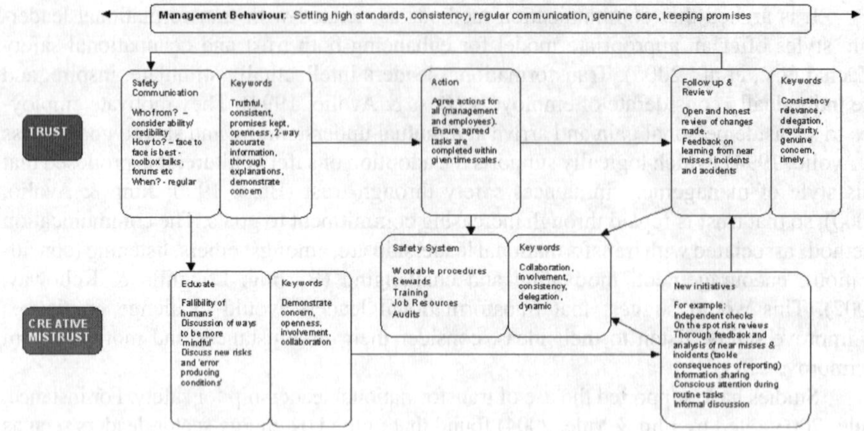

Figure 2. A model giving examples of how trust and 'creative mistrust' can be developed in organisations

just about what people notice, it's about what people do with what they notice. To truly reap the benefits of a 'questioning attitude' takes workers with personal responsibility and ownership who are able to overcome the, often subconscious, temptation to do nothing about valid concerns.

Extreme levels of trust in an organisation may encourage individuals to strive for agreement or 'groupthink' and ignore independent thinking and creativity. 'Groupthink' has been implicated in a number of major incidents where unsafe behaviours and actions have gone unchallenged (Reason, 1997). By developing a sense of 'creative mistrust' employees are encouraged to question each others practices in order to gain understanding of their intentions and methods. In this way, employees can collaborate to seek safer ways of working. Another area in which developing a sense of 'creative mistrust' may significantly benefit organisations is 'human error'. Most would agree that under certain conditions even the most competent employees can make a mistake. In an atmosphere of 'creative mistrust' this would be recognised and checks would be in place to pick up errors made. For instance, consider the process of isolating equipment. Errors can be made at any number of stages from design to de-isolation and under certain conditions (multi-tasking, tiredness, distractions) mistakes will be made. Independent checks can be put in place to ensure errors are managed more effectively.

'Creative mistrust' is not the polar opposite of trust and in order for workers to engage in the behaviours outlined above they must be able to trust the reactions of their colleagues. For instance, the process in which employees report a potential accident source is reliant on the trust that management will respond positively (e.g. listen, act) to this information. Or the process in which an employee challenges a colleague for behaving unsafely is based on the trust that the colleague will respond in the right way (listen, non-aggressively). Thus, the development of 'creative mistrust' should be created alongside a culture of 'trust'.

8. A MODEL FOR DEVELOPING TRUST AND 'CREATIVE MISTRUST'

The model outlined in Figure 2 brings together the research on trust and 'creative mistrust' in order to provide some guidance on how they can be developed to improve safety culture. The model is not meant to be an exhaustive or detailed list of how to evoke safety culture change but it is hoped that it will provide reminders and prompts to consider when communicating with and responding to employees. The expectations of managers need to be set early on as their behaviours override the whole process of safety culture change. One of the key aspects to remember is that trust is easy to break and very difficult to build. As management have less time to communicate face-to-face it's vital that the expectations of staff are continually met by the attitude and behaviour of management. Even if these expectations are met, the 'perception' of managers will only change if these trustworthiness attributes are made overtly visible.

9. CONCLUSIONS

It is evident from the literature and from the experience of the many organisations that have attempted to improve their safety culture, that it is not just what you do but how you go about doing it. This paper argues that the key to improving safety culture is to develop trust in management in order to build a strong positive perception of management's values, attitudes and commitment to safety. Trust is the lubricant with which perceptions can be changed, communications can be heard and change embraced. In other words, trust is a necessary precursor to an effective safety culture. Trust can be developed through consistently demonstrating ability, benevolence and integrity. In order to demonstrate this management will have to show face, set the example, get the workforce involved and be open in their communications. The likely return for this effort is a workforce highly engaged and committed to creating a safe culture.

Alongside creating an atmosphere of trust it is also necessary to develop a workforce who 'creatively mistrust' technology, processes and human nature. A workforce who constantly question and who are willing to take personal responsibility for acting on anything they find suspect. Trust and 'creative mistrust' are not mutually exclusive and should be developed together in order to create a safe culture.

10. REFERENCES

Barling, J., Loughlin, C. & Kelloway, E. K.(2002). Development and Test of a Model Linking Safety-Specific. Transformational Leadership and Occupational Safety. *Journal of Applied Psychology 2002, Vol. 87*, No. 3, 488–496

Bass, B., & Avolio, B. (1990). The implications of transactional and transformational leadership for individual, team and organizational development. *Research in Organizational Change and Development, 4,* 231–72.

Bass, B., & Avolio, B. (1994). *Improving organizational effectiveness through transformational leadership*. New York: Sage.

Bonacich, P., & Schneider, S. (1992). Communication networks and collective action. In Liebrand, W. B. G., Messick, D. M., & Wilke, H. A. M. (Eds). *Social dilemmas: Theoretical issues and research findings*, 225–245. New York: Pergammon Press.

Brown, M. (1999). Communication, trust, and organizational change at a manufacturing facility: A critical incident technique analysis. *Dissertation Abstracts International Section A: Humanities and Social-Sciences*. Vol 59(7-A).

Chmeil, N. (2007). Safety in the third age. *People and organisations at work, Summer edition*. Division of Occupational Psychology.

Cohen, H., & Cleveland, R. (1983). Safety program practices in record-holding plants. *Professional Safety,* March, 26–33.

Conchie, S. M., & Donald, I. J. (2006). The role of distrust in offshore safety performance. *Risk Analysis*, Vol. 26, No. 5.

Davis, J. H., Schoorman, F. D., Mayer, R. C., & Tan, H. H. (2000). The trusted general manager and business unit performance: Empirical evidence of a competitive advantage. *Strategic Management Journal, 21*(5), 563–576.

Dirks, K. T., & Ferrin, D. L. (2001). The role of trust in organizational settings. *Organization Science, 12*(4), 450–467.

Flin, R., & Burns, C. (2004). The role of trust in safety management. *Human Factors and Aerospace Safety,* Volume 4 (4), pp. 277–287

Flin, R., & Yule, S. (2004). Leadership for safety: industrial experience. *Quality & Safety in Health Care, 13*(supplement II), 45–51

Hale, A. (2000). Editorial: Cultures confusions. *Safety Science, 34*, 1–14.

Joyner, P., & Lardner, R. (2007). Mindfulness: realising the benefits. Paper presented at the *Loss Prevention Conference*, Edinburgh, Scotland.

Jung, D. I., & Avolio, B. J. (2000). Opening the black box: An experimental investigation of the mediating effects of trust and value congruence on transformational and transactional leadership. *Journal of Organizational Behavior, 21*, 949–964.

Mayer, R. C., Davis, J. H., & Schoorman, F. D. (1995). An integrative model of organizational trust. *Academy of Management Review, 20*(3), 709–734.

Reason, J. T. (1997). *Managing the Risks of Organizational Hazards*. Ashgate: Alsershot.

Rousseau, D. M., Sitkin, S. B., Burt, S. R., & Camerer, C. (1998). Not so different after all: A cross-discipline view of trust. *Academy of Management Review, 23*(3), 393–404.

Whitener, E., Brodt, S., Korsgaard, M., & Werner, J. (1998). Managers as initiators of trust: An exchange relationship framework for understanding managerial trustworthy behaviour. *Academy of management review, 23,* 513–530.

Zacharatos, A., Barling, J., & Iverson, R. (2005). High-performance work systems and occupational safety. *Journal of Applied Psychology*, Vol.90, No.1, 77–93.

USING THE BEST AVAILABLE TECHNIQUES TO CHANGE BEHAVIOUR IN THE CONSTRUCTION INDUSTRY

Martin Worthington[1], Samantha Hughes[2], and Arvinder Saimbi[3]
[1]Morgan Sindall
[2]Morgan Ashurst
3Morgan Professional Services

This paper will describe the approach that has been taken to begin a behavioural change process in a large multi disciplined construction company. We will describe the preparation for change, the realisation that we had to do more than we were doing. That despite having robust systems and procedures in place serious accidents where still occurring. How we have used the Health and Safety Executive's climate survey tool and cultural maturity matrixes to benchmark our culture. Our research into the approaches that are delivering change in other industries. We will explain how the leadership of the company was engaged and how champions and coaches where selected to help lead change for the front. How a programme of training and awareness raising was implemented. How we have realised that our solutions need to be tailored to each individual situation. We will look at some case studies of how we have used different approaches with success. How we have reviewed our progress in our journey and our next steps in our goal that is Looking to an Incident Free Environment.

1 INTRODUCTION

Over three years ago Morgan Ashurst (at that time AMEC Construction Services) undertook a benchmarking exercise to compare its own approach to SHE matters against industries outside the construction industry.

After a decade of steady improvement it was realised that even with mature processes in place there was a 'step change' required to move the business to a higher level of safety performance.

This benchmarking exercise and the need for change was further catalysed by investigations into a number of incidents with high potential for harm, that showed although the 'paperwork' was in place it was the underlying human behaviours at various levels within the management team that had led to the incidents.

To this end a major component of the exercise was to look at various human behavioural approaches that were being used in both the Oil and Gas and Nuclear industries and what subsequently emerged as the 'hearts and minds' programme.

In addition to the approach, two fundamental leadership decisions were made and these were:-

1. To make sure that we not only addressed the prevention of incidents in the workplace, but also led a cultural change that is based on both working safe and being safe at home.

To this end the LIFE campaign and the LIFE mission was developed to meet the specific needs of our business.
This also includes our LIFE vision being built into our community engagement programmes through school visits and promotional material.

LIFE Mission Statement

"To create a company that believes no injury or occupational illness is acceptable and all members of the team are committed to living within an incident free environment"

2. The approach was to be intrinsic to the way we did things with absolute ownership to be within our organisation. Although investment was made in the development, launch and roll out, the real investment came from people, commitment and determination to succeed.
This paper focuses on the application of the 'hearts and minds' in our construction led organisation.

2 DEVELOPING A LAUNCH PLATFORM

Once the benchmarking exercise had been completed and the intelligence analysed and assessed, the next major piece of work was for our organisation to ask a few hard questions of ourselves and ascertain what level of maturity we were at.

There were a number of tools and techniques used to provide an assessment of where we were in terms of maturity. These included:-

- Use of the Health and Safety Executive's (HSE) Health and Safety Climate Survey tool that was used to gauge the perception of both our own employees and those within our supply chain. The survey involved over 3,000 people across the business.
- Application of some of the tools available through the 'Hearts and Minds' programme. This included the use of the maturity matrices.
- A series of structured interviews were conducted throughout the business and this engaged a range of people from the Managing Director through to the construction operatives.

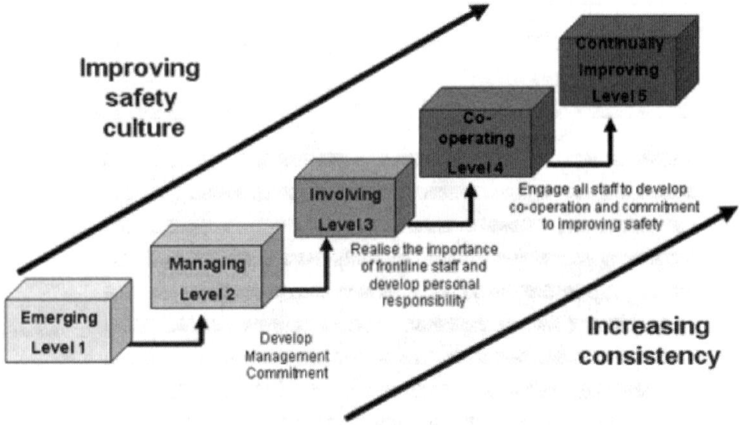

Figure 1. Maturity levels

On completion of the information gathering the data was analysed to identify the key issues and assess our perceived maturity. The key issues were highlighted as:-

- There was an inconsistency in our leadership and application of Health and Safety requirements across the UK business.
- Where performance was seen as 'lumpy' there was an apparent cycle of reactive management occurring i.e. quick intervention following an incident but insufficient resource in maintaining and improving standards.
- There were low levels of reporting of learning events (traditionally termed 'near misses') that were mainly driven by focussing on outcome (actual harm) rather that the potential.
- A distinct lack of understanding at various levels, in particular supervisory, as to the impact of behaviours and leadership styles on safety performance.

In considering these facts, by use of peer review forums etc, it was felt that by using the maturity level model (Figure 1) that the business was somewhere in between level 2 (managing) and level 3 (involving).

3 MOVING UP THE MATURITY LEVELS

On agreeing what level of maturity the business was at, the next major decision was to agree how we should improve our maturity and ultimately our level of SHE performance.

The main considerations being:-

1. The level of maturity we were at
2. The tools and techniques that were available to us

The Overall Picture

The chart below provides an overview and comparison of all Factors given in the key point summary on the left.

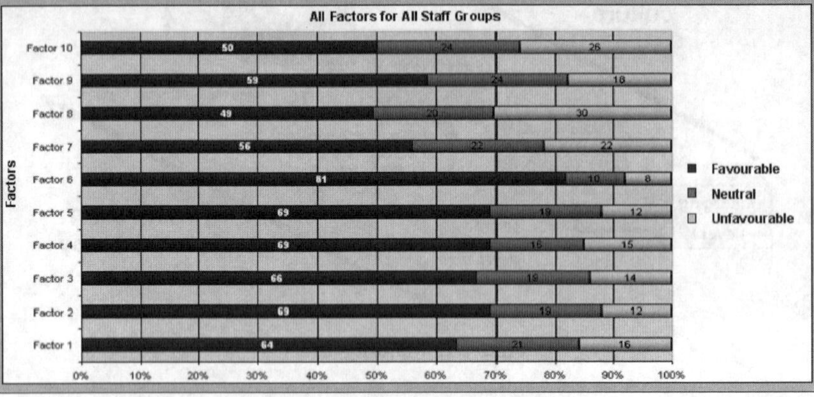

How are we going to move forward from the survey?

Factor	Survey Finding	Proposed Follow up Actions
1 & 8	Lack of involvement in developing processes. Not enough recognition for the contribution of H & S Committees etc.	Project Management/VOICE Consultation forum to be established to act as a consultation body. Guidance on best practice incentives to be developed.
2	Immediate bosses need to communicate H & S matters better.	Standardise the supervisors induction scheme across DPS. Hold ABC 'Safety Watch' workshops with supervisors.
3 & 7	Not enough checking by supervisors to ensure people are working safely.	Being progressed through the delivery of the LIFE/ABC programme.
5	There is a perception that workmates won't react strongly to rule breaking.	Operative ABC behavioural workshop in place and being rolled out across DPS through the introduction of LIFE Coaches etc.
10	Learning events (near misses).	Complete review of existing approach. Site visits to be carried out to discuss this issue, capture best practice, develop a framework for implementation.

A big thank you to nearly 4,000 people who were involved with the survey and in helping us to develop real improvement actions that will get us nearer to our LIFE goal.

Figure 2. H&S climate survey results

3. The organisation structure that needed to be in place
4. A vehicle for actively changing the behaviour and approach to health and safety matters.

Points 1 and 2 have already been discussed and 3 and 4 were being developed in parallel, in conjunction with the intelligence that was emerging.

In relation to the organisation structure there was a framework agreed that would support the **A**ctive **B**ehavioural **C**hange (ABC) programme that was being developed following the benchmarking exercise. The structure was based around: -

- A sponsor who would lead and commit to 'Looking to an Incident Free Environment' (LIFE) and the **A**ctive **B**ehavioural **C**hange Programme that was being rolled out, forming the core element. This sponsor was the Managing Director. (See figure 3b)
- LIFE Champions who were embedded within the organisation (e.g. regional operational directors) to ensure that the behavioural change programme was given the leadership commitment and co-ordination at a high level.
- LIFE Coaches, these were individuals who had volunteered to become coaches who could facilitate workshops and thereby accelerate our programme into all corners of the business. In the majority the coaches were not SHE professionals but project managers, directors etc.
- A LIFE support kit that included:-
  - Various 'training' modules for ABC that are being organically developed as we are going through our journey. Forums that involved a full cross section of the business and included stakeholders such as clients and supply chain partners.
  - Tools that include LIFE induction packs, best practice SharePoint website, active maturity matrices, observation checklists.

4 ORGANIC DEVELOPMENT

When we started on our LIFE journey over 3 years ago one thing was obvious, this was, that the journey would either grind to a halt or take us somewhere where we did not want to go if we did not display the right leadership or provide full commitment and maintain and continually improve our approach as we matured.

A **B**ehavioural **A**ssurance **M**odel was developed to identify the key values, checks, balances and measurement, arrangements that needed to be put in place to ensure continuous improvement (see below).

In relation to the measurement of our performance, a structured process was in place that includes:-

- Periodic reuse of the H&S Climate Survey.
- Application of the maturity matrix in each business unit to assess and gauge maturity levels, based on the outcome of LIFE assessments undertaken by the business improvement groups.
- Database of Learning Events and best practice that is shared across the business and industry.
- Engagement of others within and outside the construction industry. The main vehicle for this is the UK Behavioural Change and Worker Engagement (BCWE) forum that meets 3 times a year and is supported by a research project that aims to provide an evidence based approach and 'toolkit' to industry and in particular small to medium size enterprises.

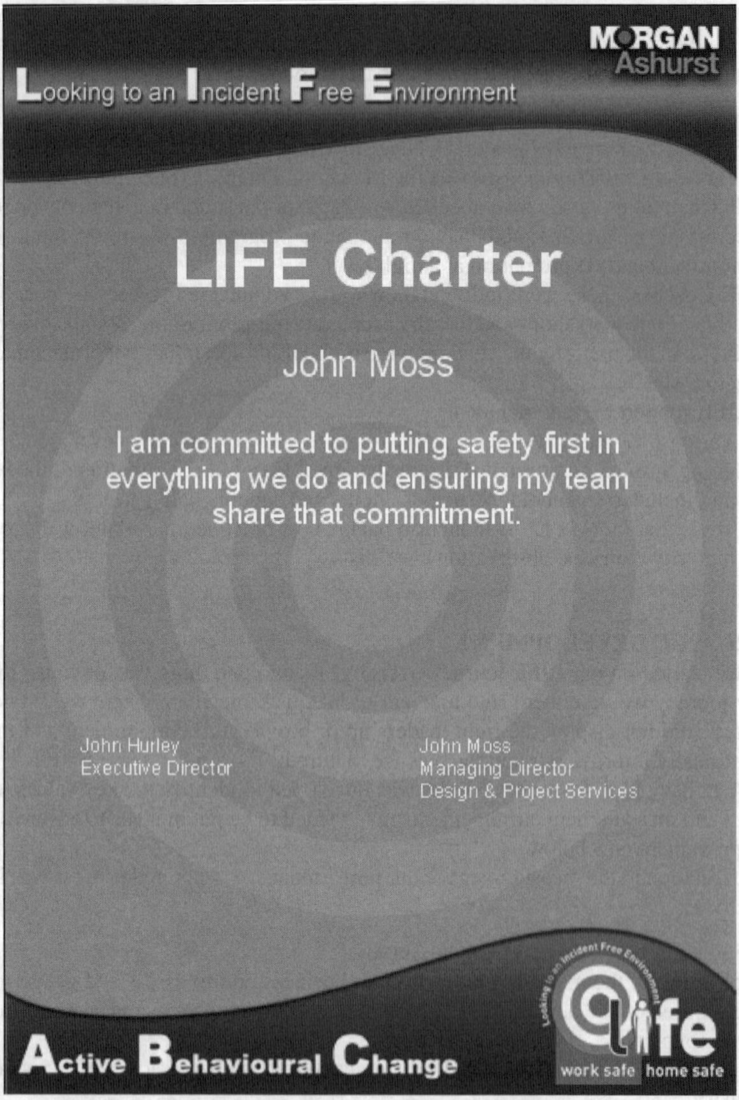

Figure 3a. Example of managing director's personal LIFE charter

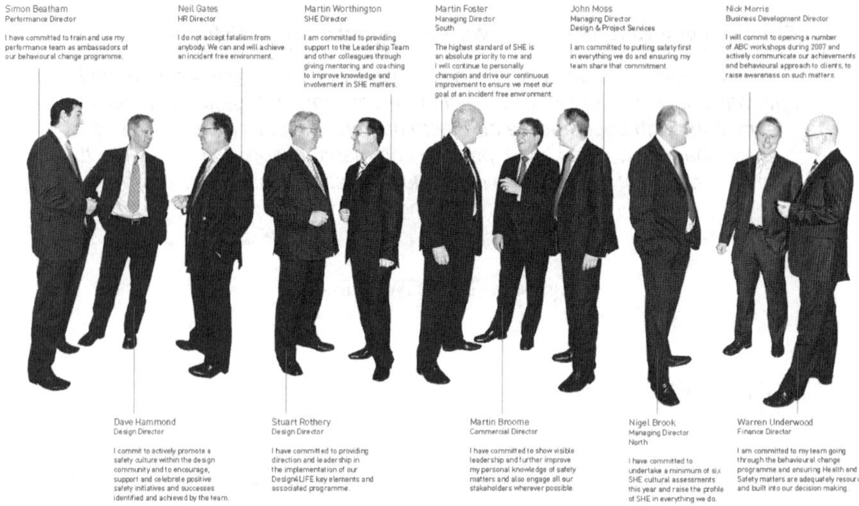

Figure 3b. Example of business leadership team LIFE charter.

- Tying into existing research projects and subsequent recommendations such as 'Engaging the workforce' undertaken through the Caledonian University.
- The use of LIFE improvement plans that were owned by the applicable business teams and used SMART principles to ensure they could be evaluated and used as a value adding performance tool.

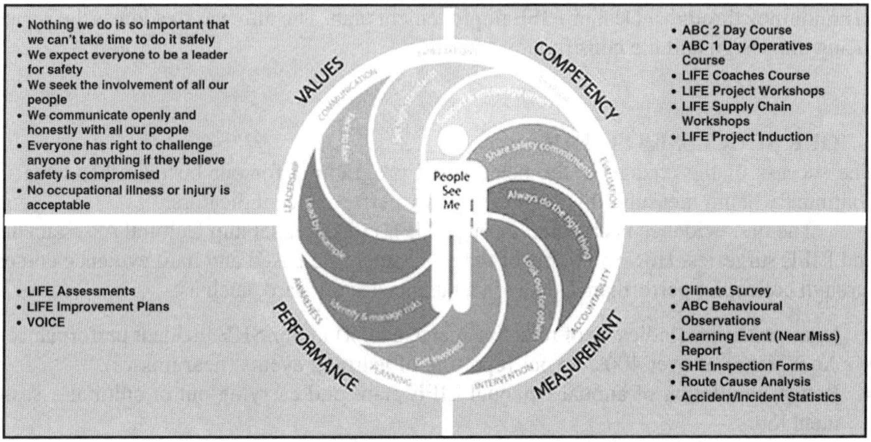

Figure 4. Behavioural assurance model (BAM)

The Behavioural Change and Worker Engagement Mission Statement

"To work together under a common purpose by developing a cohesive and pragmatic approach to behavioural change and worker engagement that will evolve through best practice and learning across industry with a view to changing the way we lead, plan, procure and manage work activities. With the collective goal of reducing the incidents that result in harm and personal suffering"

Figure 5. The BCWE mission

5 DESIGNING FOR LIFE

As part of the safety organic development, it was identified, that in addition to the operational focus we were applying across the business there would be a major opportunity to make safety considerations further up the decision making line to the design phase we engaged in our designers, process engineers etc in our LIFE scheme. This helped us to develop an intervention tool that would greatly enhance and ultimately add to our SHE improvement.

From this Design^{4}LIFE was developed and included the 'design community' in our LIFE journey through a Design^{4}LIFE improvement plan. The aim was to address the issues facing the design and the construction activities.

6 OUR SUCCESSES SO FAR

The success of the change in the behavioural programme for our business has been in continually being measured by a number of key performance indicators.

The first evidence is captured through LIFE talks, Leadership Cultural Assessments and LIFE surgeries. However some of the more scientific based and hard evidence comes through collecting performance data via a balanced score card, such as:-

- Less variability (indicator of re-active management) in our SHE incident performance.
- An increase of over 400% in our reporting of learning events (near misses).
- Better completion of annual Personal LIFE plans and carrying out of cultural assessment tours.
- Improvement in maturity levels based on the maturity matrix approach
  - Over 90% of projects achieving zero reportable accidents in 2006.

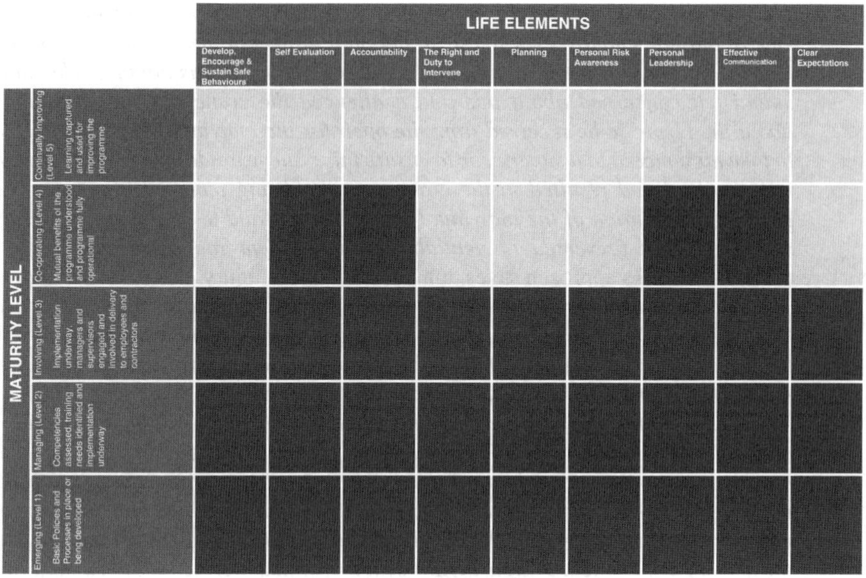

Figure 6. Example of maturity matrix

  - Reduction in severity of accidents and days lost due to accidents.
  - Improved margins of around 10% in all factors of the H&S Climate Survey. The biggest increase being in the factor for risk taking behaviours.
  - Active engagement and involvement of some of our key stakeholders.
  - Approaching 4000 people actively involved in our ABC programme.
  - Transfer of knowledge across the construction industry via the BCWE forum.
  - Recognition by our peer groups of the LIFE achievements through involvement, support and a number of industry awards.
- Real time case studies of improvements.
- Welsh Water Framework's outstanding achievement of a 3 year period with zero reportable accidents.
  - *Through the personal commitment of the project team and the application of the LIFE tools and techniques to monitor and underpin safe working on site and improve performance the Welsh Water Framework has achieved a 3 year period with zero reportable accidents. With multiple construction sites across the south east of Wales effective communication is key and the project team achieve this by engaging the workforce and benefit from approachable and committed workers who are actively involved in decisions regarding safety matters on the project, where conversations about safety have become part of their daily life.*

- Positive Intervention by site employee prevents a potentially serious incident.
  - *One of our site employees whilst working on a project in the North West noticed that as a crane was being self erected, a vital component was missing. The individual felt concerned about this and challenged the crane operator. Unhappy about the response he received from the operator the individual felt comfortable to take his concern to a supervisor to ensure this issue was addressed. His concern was resolved and resulted in the crane operator being removed from site. The positive intervention of the individual in not being afraid to challenge an unsafe situation clearly prevented a potentially serious incident from occurring.*
- Use of the LIFE Talks approach saves dumper driver from injury
  - *The benefits of using the LIFE Talks 9 step approach, which is a simple proactive technique for observing and reinforcing safe behaviour and correcting at-risk behaviour was highlighted on one of our construction projects in Scotland. A machine operator was loading excavated material onto a dumper, when he observed that the dumper driver had remained on the dumper waiting to be loaded and was not wearing his seatbelt. Concerned for the driver's safety, the machine operator stopped the activity and approached the driver, using the LIFE Talks approach to have a safety conversation with driver and challenge him about his at risk behaviour. At the end of the conversation the machine operator had secured the drivers commitment to wearing his seat belt whilst driving and to stand down and away from the dumper during loading operations.*

  A week later the dumper driver reported a learning event where the dumper he was driving went over uneven ground and the front wheel hit a soft spot. In this instance the driver was wearing his seatbelt, following the challenge on his behaviour the previous week which undoubtedly prevented him being thrown from the dumper and being seriously injured.

7 CONCLUSION

The construction industry, by its nature of a constantly changing work environment and the processes that are needed to recognise this, is a difficult industry to achieve a consistent high level of SHE performance. Most of the major organisations are on some form of journey with behavioural change programmes which support the procedures and processes that have been put in place.

Flagships such as the construction of T5 at Heathrow have enabled a range of behavioural tools and techniques to be developed and used in the industry. Many approaches are emerging, however, in our case, the focus of this paper has been 'Hearts and Minds' that are supporting and raising the bar in terms of SHE performance.

The true benefits come as we begin to share our learning, good and bad, across the industry and create an environment on projects that supports positive interventions and the correct behaviours that ensure a 24/7, and fully inclusive approach to health and safety matters.

IMPLEMENTING AND SUSTAINING HUMAN RELIABILITY PROGRAMMES OF WORK – A MANAGERS' GUIDE

Alison Hubbard and Jamie Henderson
Human Reliability

1 INTRODUCTION

Health & Safety Executive (HSE) guidance suggests that sites subject to the Control of Major Accident (COMAH) regulations should use predictive qualitative assessment techniques to identify potential human factors issues (HSE, 2007). The guidance acknowledges that 'this will be a relatively new area for many dutyholders' and explains that 'our expectation is that they conduct qualitative analyses of human performance – identifying what can go wrong and putting remedial measures in place'. Largely as a result of this regulatory interest, sites subject to these regulations are starting to undertake human reliability risk assessments.

Over the past few years, Human Reliability has had considerable experience of applying qualitative risk assessment techniques in a range of organisations. Our observation is that, once the commitment has been made to undertake this type of work, the ultimate success of the work depends, to a large degree, on how certain organisational issues are managed. Drawing on our practical experience, this paper is designed to provide some forewarning of these issues for organisations taking their first steps in this field.

To this end, an overview of one type of qualitative human reliability risk assessment is presented. Guidance is provided regarding planning, resources and timescale issues. Finally, some common pitfalls are discussed.

2 PROCESS DESCRIPTION

HSE guidance suggests that sites subject to the COMAH regulations should use predictive qualitative assessment techniques to identify and manage potential human factors issues (HSE, 2007). They provide an outline description of one such technique, describing a 7 stage process:

- Step 1: consider main site hazards;
- Step 2: identify manual activities that affect these hazards;
- Step 3: outline the key steps in these activities;
- Step 4: identify potential human failures in these steps;
- Step 5: identify factors that make these failures more likely;
- Step 6: manage the failures using hierarchy of control;
- Step 7: manage error recovery.

Over the last few years Human Reliability have had considerable experience of applying an analysis technique that is consistent with this process. Due to constraints of space, it is not possible to provide a fully detailed description of the analysis technique; however, Human Reliability is currently developing a handbook for the process. The following sections give a brief overview of the process.

A – IDENTIFY AND PRIORITISE ACTIVITIES WITH THE POTENTIAL TO AFFECT MAJOR SITE HAZARDS

Process plants typically have hundreds of activities to undertake on site. In order to get the most benefit from the analysis process, it is important that effort is directed at the most critical site tasks. The existing site COMAH site safety report, in conjunction with a list of site procedures, can be used as an input to a task prioritisation process. This activity should be undertaken in a workshop involving individuals with a good understanding of site processes and hazards. Tasks related to hazard areas where human failures have the potential to lead to significant consequences should be identified. HSE (2007) guidance suggests that the following types of activity, in particular, should be examined:

- tasks that have the potential to initiate a major accident sequence (e.g. inappropriate valve operation causing a loss of containment);
- tasks designed to mitigate the consequences of failures (such as activation of ESD systems);
- tasks designed to prevent an incident (e.g. maintenance of safety systems).

If a large number of tasks are identified then it may be useful to further prioritise tasks using a screening process. Human Reliability use a diagnostic tool to assist with this activity. For each task, a simple scoring system is used to assess the following factors:

- Opportunities for recovery from error,
- Task complexity,
- Task familiarity,
- Requirement to defeat safety systems,
- Quality and number of hardware defences and safeguards.

All of these factors influence the vulnerability to failure. These scores are combined with an assessment of the task hazard level to provide an overall Task Criticality Score (TCS) that is used to rank order the tasks.

B – ANALYSE TASKS

The next stage of the process involves developing a clear understanding of how these critical tasks are currently performed. This is an important stage of process, since an incomplete understanding will significantly reduce the accuracy of the subsequent analysis.

Existing procedures can be used as the basis for the analysis. However, written procedures typically vary in quality, and our experience is that different individuals and shifts often carry out even the most important tasks in different ways. To ensure the analysis reflects actual working practices, individuals with a good working knowledge of the task should participate in the analysis process. This sometimes leads to task review and a reassessment of best practice.

It is recommended that a formal analysis process, such as Hierarchical Task Analysis (HTA) be used to lend structure to the analysis (see, for example, Kirwan & Ainsworth, 1992). One of the advantages of this is that it groups tasks steps according to higher-level goals. Experienced operators often find it easier to describe *what is done* rather than the *purpose* of the task steps. Organising the task into higher-level goals helps participants to consider why they do things in a particular way. This often leads to suggestions for better task performance. HTA works well for many process industry tasks, it is particularly effective for analysing tasks such as preparation for maintenance. However, there will be some types of task, such as control room response scenarios with high decision-making content, where the approach may need to be adapted or a different analysis method used.

Task walkthroughs should also be used to ensure that the practical dimensions of the task are fully understood. In our experience, operators accept problems that have existed for a long time as part of the job. Having an independent facilitator can help operators to challenge these accepted conditions and practices.

C – IDENTIFY POTENTIAL HUMAN FAILURES

As for previous analysis stages, workshop sessions should be arranged. It is useful if the same people who attended the task analysis workshop can also attend these sessions. However, as a minimum, one individual should have an understanding of the practical aspects of the task and one should have an understanding of the potential consequences of actions.

Human Reliability use a process based on the SHERPA system (Embrey, 1986), a mature technique that has been used extensively over the past 25 years. It is similar to the process used in HAZard and OPerability Studies (HAZOP), using guidewords (e.g. action omitted, action too early) to analyse steps in the task analysis for deviations that may have serious safety consequences.

The process also identifies existing Risk Control Measures (RCMs) designed to prevent human failures from contributing to a Major Accident Hazard (MAH) event, these include, for example, alarms, automatic sequencing, deluge systems and relief valves.

D – ANALYSE FACTORS THAT MAKE THESE FAILURES MORE LIKELY

Performance Influencing Factors (PIFs) are the characteristics of people, tasks and organisations that influence human performance and therefore the likelihood of human failure. PIFs include time pressure, fatigue, design of controls/displays and the quality of procedures. Evaluating and improving PIFs is one approach for maximising human reliability. Some PIFs influence individual task steps, for example ease of access to an emergency valve. Other PIFs have a broader influence, such as quality of training or level of task experience. The potential impact of these factors for the task in question should be considered.

E – MANAGE OUTPUTS AND IMPLEMENT APPROPRIATE RISK MANAGEMENT STRATEGIES

The principal output of the risk assessment will be the specification of actions to reduce risks arising from the analysed activities. In many cases, the existing RCMs will be adequate. If this is the case, then they should be documented in the analysis output to demonstrate that the related risks are being managed. If the RCMs are not adequate, or some aspects of the process need to be altered to make the task safer, then these decisions should be made with reference to cost-effectiveness and hierarchy of control considerations.

It is vital that all proposed actions are managed using an action-tracking process. Most organisations have these systems. At a minimum, the system should prioritise the actions required, allocate responsibilities and provide a clear timeframe in which they should be completed. In all cases, the reasons for choosing a particular option, even where the decision has been to take no action, should be noted and fed back, along with the outputs, to participants in the process and the wider organisation.

In order to maximise the benefit from the analysis work the outputs can also be used to support activities other than risk assessment. For example, the structured description of the task is very useful for providing clear, action oriented, step-by-step procedures and job aids. The findings from the risk assessment can be used to annotate these procedures. For example, providing warnings and underlying reasons for actions, and can be used to produce detailed training standards for these critical tasks.

3 GUIDANCE FOR PLANNING HUMAN FACTORS RISK ASSESSMENTS

From our experience of a range of organisations, the success of the programme will depend to a large extent on how well it is supported and resourced. This section describes some of the important planning decisions that need to be made before the process begins.

PLANNING THE SCOPE AND DURATION OF THE INITIAL PROJECT

As this topic is likely to be new to the organisation, the first objectives should be to introduce the concepts to the relevant staff and to deliver results that illustrate the benefits of the process. Major accident critical tasks should be identified and prioritised at an early stage. This should give an idea of the scale of the work to be done, but it need not define it completely.

It is recommended that organisations set themselves achievable goals for the first year of application. This may be by the number of tasks to be analysed, for example, ten in the first year, or by the number of workshops organised. The site facilitator and the external consultants can then plan these sessions and identify the resources required to complete them. The sessions should be scheduled for when the largest pool of contributors is likely to be available. For example, the summer months, where many people take holidays, are best avoided. Remaining tasks can be scheduled for analysis in subsequent years. Tasks should be analysed in terms of their priority, but some flexibility can be exercised to suit the interests of the participants. For example, if a particular task is known to cause difficulty to operators, this can be given increased priority, in order to demonstrate the benefits of the process and gain the support of the participants.

SELECTING SITE FACILITATORS

If the organisation does not have internal human factors expertise, then external consultants can supply this. However, even if specialists are engaged, they will require the close support of at least one individual on site to act as a process *facilitator*. The role of this facilitator is varied, but they must be allowed to spend a significant percentage of their time on the project, for some larger organisations this may be as much as a day a week. The types of support that they should be able to provide includes the following:

- Organising dates for workshop sessions and securing the release of task experts (e.g. field operators, control room operators, maintenance technicians) to attend these meetings.
- Supplying appropriate materials for the analysis process (e.g. procedures, P & IDs)
- Reviewing analysis outputs and presenting issues to appropriate site authorities to be addressed.
- Providing feedback to project participants, and the wider organisation, about safety improvements that have arisen from the analysis work.

Ideally, the facilitator will also take an interest in the technical aspects of the process and attend some or all of the analysis sessions. The precise role of the person is not critical; however, they should have a good understanding of the plant and have contacts across the site. It is particularly important that they are respected by, and able to communicate effectively with, both engineering and operating level staff. In the past this role has been filled successfully by process engineers, however, these individuals often have significant demands on their time.

PROVIDING THE RIGHT PARTICIPANTS

This is a bottom-up process, in that the quality of the output relies heavily on having people with the right knowledge and skills participating in the workshops. For the task analysis workshops, this means including people with extensive practical task experience (i.e. operators). This is so they can describe real working practices and task constraints arising from local working conditions, as well as any differences that exist between shifts.

The process also relies on knowledge of the potential consequences of failures in task performance. This means that at least one person with good knowledge of the plant system is required. For example, if a maintenance task were being analysed, in addition to the maintenance staff that carry out the task, an engineer with process knowledge would also be necessary.

We have found it useful to hold training sessions during project start-up, to educate potential participants and their managers about human factors issues. This training is designed to explain how the risk analysis works, but also explain the importance of human factors, usually with reference to high profile accidents.

PROVIDING RESOURCES FOR ADDRESSING IDENTIFIED ISSUES

Once an organisation has committed to a piece of work, it is usually relatively straightforward to secure an undertaking to provide resources for the analysis work itself. However, budgetary support for actions that are likely to arise from the risk assessment may be more difficult to secure. This is for a number of reasons, but perhaps primarily because the scale of these actions is unknowable until the analysis is complete.

This is important because the credibility of any process, and particularly processes that are relatively new to an organisation, depends on the participants seeing genuine benefits arising from their inputs. Therefore, it must be appreciated at the beginning of the project, by the appropriate managers, that improvement work (e.g. to plant equipment) will need to be undertaken to address deficiencies identified by the analysis. One way of supporting this is by giving the process facilitator a small working budget to undertake simpler actions. For more substantial issues, they must have access to the appropriate decision makers, and have the ability to get more significant improvements onto site action plans. All of these decisions should be taken on a cost-benefit basis. However, where the outcome is that no action will be taken, this information should be fed back to the project participants, along with an explanation of why this decision has been made.

MANAGING PROCESS OUTPUTS

Along with the specific identified actions, the analysis process typically generates other outputs that require management. These can include:

- Procedures and job aids
- Training
- Workplace improvement schemes.

The successful management of these outputs depends heavily on existing site systems. Relevant systems can include, for example, procedures management, training, competence assurance, change management and action tracking systems. We have found that it is worth spending some time at the start of the analysis process to establish the likely compatibility of process outputs with these existing site systems. This ensures that the maximum benefit from the analysis effort can be obtained. For example, if one outcome of the risk assessment is a significantly different way of doing the task, then this will need to be disseminated to operators through training and updated procedures.

PLANNING FOR ONGOING WORK

One danger with this kind of work is that it is treated as a one-off activity. The temptation, whether this is explicitly stated or not, can be to fund an initial project, over say, the duration of a year, where many of the site safety critical tasks are examined, and then assume that all human factors issues have been fully addressed. A human factors risk assessment can never be a one-off project. Sites, for example, will always be developing new ways of working, using new equipment and changing levels of manning.

Ideally, sites should have their own expertise to manage human factors issues in an integrated manner. It is worth, therefore, considering at the start of this type of project how the site envisages managing these issues in the future. If the expected outcome is that it will be managed internally, then consideration must be given to training the staff who will be implementing the recommendations in order to effectively transfer the skills and knowledge. Moreover, time must be allocated to the individual(s) charged with undertaking these responsibilities to enable them to complete the work. Too often, the responsibility is added to a long list of other duties, with no recognition of the additional time it will take.

Some sites find it simplest to outsource the work to external consultants, thus enabling their key staff to remain focussed on their day-to-day site duties. In this case, it is important to maintain points of contact between the external consultants and site staff. Here, the role of facilitator (see previous discussion) is pivotal.

WIDENING THE SCOPE

This task-based approach to human reliability risk assessments has been implemented successfully in large and small COMAH sites. One of the reasons for its success is that it quickly draws attention to task specific, MAH safety issues (e.g. task design, valve labelling, access, unreliable instrumentation, poor interface design, alarm issues). These issues are easy to communicate, clearly linked to risk management and specific enough to address.

For a site new to the topic of human factors, this clear link between human factors issues, MAH and solutions can be the easiest introduction to human reliability and human factors. Other approaches, for example human factors audits of site systems (e.g. shift handover, procedures, training, change management), whilst important, can result in general recommendations that can be hard for a site to address. In addition, when first introduced to human factors, some people find it difficult to make the link between these general issues and process safety. Starting with a task-based risk assessment helps a site identify specific issues as well as providing an insight into strengths and weaknesses of more general site systems. Having established weaker areas in such systems, these can then be subjected to more detailed human factors reviews. Having Human Factors expertise within the organisation or from an outside source can help to capitalise on issues that are raised during the qualitative risk assessment process.

4 POTENTIAL PITFALLS

Every organisation is different in terms of their culture and the way work is managed. In our experience, different organisations face different obstacles when carrying out human reliability programmes. Some issues have been alluded to in the previous sections, for example, involving the right people in the analysis. This section includes an overview of some problems and issues that have arisen in previous projects.

KEEPING THE FOCUS ON MAJOR ACCIDENT HAZARDS

The primary focus of these types of analyses is major accident hazards. However, as tasks are scrutinised, other issues related to occupational process safety and process efficiency are uncovered. These are also important issues, and are useful to pursue if time is available. However, where time is limited this can result in the lengthy and complex analyses. Therefore, facilitators must be careful to keep the focus on major accident hazard issues during workshops.

MAXIMISING PROCESS EFFICIENCY

Qualitative Risk Assessment needs to be a systematic, rigorous and thorough. However, as with any similar process, there is a danger of the analysis itself becoming the goal, rather than improved process safety performance. Therefore, the analysis team needs to remain vigilant to ensure that the balance between effort put into the assessment and the resultant benefits is appropriate.

In order to achieve this we suggest the following strategies:

Transfer of lessons learnt between tasks

Tasks of similar types can be grouped together during the initial prioritisation stage. For example, the top four critical tasks identified by the screening process may all involve process isolations. Rather than repeat the complete analysis for four similar tasks, the most critical process isolation could be used to identify general issues regarding the management of isolations, as well as specific issues related to that particular isolation. It is likely that the general issues will be common to all isolation operations, saving time when analysing further tasks of this type.

Keep the level of detail proportionate to risk

Human Reliability advocate the use of structured task analysis, since we have found that reliance on walkthroughs and written procedures can lead to significant issues being missed. One of the strengths of HTA is that it allows the level of detail in one part of the analysis, where the hazard is greater, to be higher than in less hazardous parts of the task. With experience, the analyst can use this feature of the analysis technique to reduce the effort required.

Responding to identified issues

There can be a danger, when responding to issues identified by the analysis, to rely on training and procedures as solutions. These are an important part of a site's risk management systems. However, a proper review of hierarchy of control and cost-benefit principles must be undertaken for all identified issues before deciding on appropriate responses. This can be assisted by a employing a multidisciplinary team (e.g. operators, maintenance engineers, managers, human factors specialists) working together to identify the best response.

An additional point is to beware of the sticking plaster approach. As described previously, the nature of the process means that task specific issues are readily identified. The facilitator should consider whether the issues raised are indicative of wider problems (e.g. a problem with a stiff and difficult to operate valve may indicate problems with the valve maintenance regime). In particular, management factors such as allocation of resources, determining priorities, managing change should be considered as important potential issues.

OBTAINING A CONSENSUS BETWEEN PARTICIPANTS

Our experience is that even the most critical tasks are regularly performed in different ways by different shifts and individuals. Often these deviations are insignificant, but occasionally they have short or long term consequences for process safety. There is not space here to discuss fully the reasons for these deviations; however, the way that organisations manage these variations in critical task performance has a significant impact on process safety. We believe that it is important to use representatives from different shifts when conducting human factors risk assessments. Operators should feel able to explain why they might need to deviate from standard practice: if nothing else, this leads to suggestions of ways that tasks might be improved. Usually there are good reasons for these variations, and providing a forum for operators to discuss these issues will help in their resolution. However, once the analysis has been completed it is important that the outputs be transmitted to training and competence departments, to ensure that all operators understand why tasks should be done in a particular way.

LACK OF BUY-IN TO THE PROCESS

Management support, beyond the initial commissioning phase, is critical to the success of these types of analyses. This can be a particular problem on smaller sites, where there are fewer specialists, and people have many tasks competing for their attention. The success of the analyses depends on the inputs from site staff. If they are taken away, or remove themselves, from workshop sessions to undertake other duties, then it is difficult to get an adequate quality of output. Moreover, the status of the process suffers in comparison with other site activities. Proper planning can reduce the prospect of this happening. Participants should have sufficient work time allocated to the project. Moreover, it is important that managers keep the profile of the project at an appropriate level, and assign it comparable priority with other on-site activities and ongoing projects.

A related issue is the speed with which recommendations can be implemented. Proposed changes have to compete for site resources with outputs from other projects and, for less critical alterations, it may be several months, or even years, before a change will occur. During these periods, it is important to maintain feedback to participants regarding the priority and status of these actions. In the absence of feedback, and without immediately apparent outcomes from their inputs, then there is the danger that participants will become disillusioned with the process.

FAILING TO SUPPORT PARTICIPANTS

Often, the individuals involved in the analysis process are also the same people that will be called upon to effect the changes arising from the analysis. If participants lack the resources to address all the issues they raise, this can at best lead to disillusionment with the process, and, at worst, result in information regarding issues being withheld. Therefore, as previously discussed, process facilitators must be supplied with the time and resources to be able to capitalise on the insights gained from the analysis process.

5 CONCLUSION

In summary, the HSE provide a useful, straightforward guide to carrying out human reliability risk assessments. This paper provides additional useful insight into successful implementation of this seven-step process. Human Reliability have implemented these types of assessments at a number of organisations within the process industries, and have observed that the success of these types of project depends, to a large extent, on the way they are planned and resourced. This paper has identified some issues that, when addressed at the planning stage, substantially influence the benefits that arise from implementing human reliability assessments and increase the likelihood of them being successfully integrated with other site safety management systems. We have also identified some common pitfalls with these types of techniques that can be avoided throughout the programme.

REFERENCES

Embrey, D.E. (1986) SHERPA: A systematic human error reduction and prediction approach. *Paper presented at the International Topical Meeting on Advances in Human Factors in Nuclear Power Systems*, Knoxville, Tennessee.

Health & Safety Executive *Core Topic 3: Identifying Human Failures* http://www.hse.gov.uk/humanfactors/comah/core3.pdf

Kirwan, B., & Ainsworth, L. K. (1992) *A Guide to Task Analysis*. London: Taylor & Francis.

IMPROVING SHIFT HANDOVER AND MAXIMISING ITS VALUE TO THE BUSINESS

Andrew Brazier[1] and Brian Pacitti[2]
[1]Consultant, Llandudno, UK. E-mail: andy.brazier@gmail.com
[2]Infotechnics Ltd, The James Gregory Centre, Aberdeen AB22 8GU, UK.
E-mail: brian.pacitti@infotechnics.co.uk

Recent accidents at Buncefield and Texas City have illustrated how poor shift handover can contribute to major accidents. This is not a new discovery, but given the ever greater interest in human factors, it is one that is finally receiving attention.

Shift handover is a complex, high risk activity that is performed very frequently. Normally we would try to 'engineer out' high risk frequent tasks, or at least automate them to minimise the likelihood of error. However, this is not an option for shift handover.

There are two complimentary approaches that can be used to improve shift handover. The first is to improve the handover process by supporting the people involved with better systems, tools, and competencies. The second is to change perceptions by maximising the value of the information collected as part of the handover process and increasing its use. This creates additional stakeholders in the process and subsequently ensures a more effective feedback cycle regarding the quality of handover.

This paper will examine the human factors involved in shift handover. Also, it will illustrate that information about minor incidents, human errors, and reliability issues is often collected; and will demonstrate how this can be collected and disseminated effectively and efficiently.

INTRODUCTION

Continuous operation requires people to work shifts. Ensuring safety and efficiency requires critical information to be communicated between these shifts. All communication is prone to error. This makes shift handover a highly critical activity that is performed frequently with little opportunity to engineer out potential errors.

ACCIDENTS CAUSED BY POOR SHIFT HANDOVER

The role of shift handover has been highlighted in a number of recent high profile, major accidents. Following the 2005 Texas City refinery explosion, BP released their internal investigation report to the public [BP 2005]. This identified that poor shift handover was a contributor to the accident, citing the failure to communicate the failure of a hard-wired high level alarm between shifts as a contributing event. By way of an explanation the report stated that "there was no written expectations with explicit requirements for shift handover." The subsequent inquiry carried out by the Chemical Safety Board [CSB 2007] agreed with these finding, stating that "the condition of the unit – specifically, the degree to which the unit was filled with liquid raffinate – was not clearly communicated from night shift to day shift."

In the UK, whilst at the time of writing this paper the full facts of the case were still not know, it seems clear that shift handover had a role in the 2005 explosion at the Buncefield oil storage terminal. One of the recommendations from the Buncefield Standards Task Group (BSTG) [BSTG 2006] was that "effective shift/crew handover communication arrangements must be in place to ensure the safe continuation of operations." Further details on this subject were provided in the BSTG's final report [BSTG 2007].

But this is not a new discovery. The inquiry into the 1988 Piper Alpha disaster found that prior to the accident critical information about the status of the condensate pumps was not communicated at shift handover. This meant operators started a pump that was not in an operational state. And before that, following the discharge of highly radioactive material from the nuclear processing plant at Sellafield in 1983, it was found that failures of communication between shifts created confusion regarding the contents of a particular tank that was pumped to sea.

Whilst a number of accidents have identified shift handover as a contributory cause, it seems likely that this is an under-reported issue. Put simply, because there has not been much attention paid to shift handover and relatively little information published on the subject, people have not been looking for evidence of failures with shift handover when investigating incidents and accidents.

THE PROBLEMS WITH SHIFT HANDOVER

The Health and Safety Executive's guidance document HSG48 [HSE 1999] states that reliable communication is highly critical to safety, and that shift handover falls into this category. Failures of communication occur for a number of reasons. In general either the information being communicated is incomplete or inaccurate; or the person on the receiving end misunderstands the meaning of the information they are given. There are many reasons why this can occur, with information being presented poorly being the underlying factor.

Whilst all communication is error prone, the more complex the situation the more likely errors are to occur. For shift handover, situations such as during maintenance and deviations from normal working should are considered to be 'high-risk' because errors are likely and their consequences can be significant.

IMPROVING SHIFT HANDOVER

HSG48 provides advice for improving shift handover. This includes carefully specifying information that needs to be communicated, using aids such as log books during handover, using more than one communication medium (e.g. both written and verbal), allowing sufficient time and developing communication skills and behaviours.

The advice in HSG48 is good and has been reemphasised by the final report from the Buncefield Standards Task Group. However, despite its importance and the occurrence of numerous incidents influenced by poor shift handovers, it is quite a surprise that there is not more advice available. The reasons for this are not clear, but it is easy to surmise that it is a 'soft' and intangible subject that has probably fallen into the 'too hard' category for many years.

COMMUNICATION THEORY

In the absence of specific advice regarding shift handover, there is plenty for communication in general that should be applicable to shift handover.

One thing is clear, people tend to underestimate how complex the communication process is and consequently over estimate their ability to communicate effectively. The reality is that error is a natural and inevitable aspect of communication because language is inherently imprecise and ambiguous.

A successful communication is one where a person receiving a message achieves exactly the same understanding of that message as the person transmitting it intended. However, the following factors can interfere with this process [Lardner 1999]:

1. It is not possible to transfer meanings from one person to another directly. Rather, the receiver creates meaning in his or her mind;
2. Anything is a potential message, whether it is intended or not;
3. The message received is the only one that counts;
4. Taking the above together, unintentional meaning is likely and potential miscommunication is the norm.

Communication requires effort by both parties to avoid miscommunication. Although not infallible, face-to-face communication is generally the most reliable, not necessarily because it is a better way of transferring understanding, but because it allows immediate discussion. In contrast, written communication is generally less reliable because of this lack of immediate feedback.

BEHAVIOURAL ASPECTS

One of the reasons shift handover is a difficult topic to address is that individuals' behaviours have such a significant impact on its effectiveness. In this context the following are relevant:

- People need to be willing to say if they do not understand what they have been told;
- They need to be willing to challenge what they have been told;
- They need to be able predict what someone else needs to know;
- They need to show that they are interested in what they are being told;
- They need to make time for the handover.

No procedure or management system can address these issues directly. Whilst guidance can be provided to help people understand what is expected of them, there will be a requirement for continuous supervision and coaching to ensure bad habits are avoided and to drive continual improvement. Given the pressures of work, it is unlikely that this will happen automatically. Shift handover practices are likely to evolve over time. Sometimes this will result in improvement, but at other times short cuts and bad practice may be the result.

THE CHALLENGES TO IMPROVING HANDOVER

As well as shift handover being a particularly complex activity, there are other reasons why making improvements are difficult. In particular, we must recognise that the individuals involved may not always have incentives to put in the effort required.

The most important person in any handover is the person finishing their shift. The quality of the information they provide and their communication skills will have the greatest influence on how well informed the person starting their shift is. However, at the end of an 8 or 12 hour shift even the most conscientious person will be interested in getting home. Also, some may have the attitude that any problems they leave are going to be dealt with by someone else (i.e. the incoming shift).

It is true that the person starting their shift can influence the quality of the handover they receive. Asking questions and being interested will tend to improve the quality of the handover. However, they are not in a particularly powerful position because they do not know what questions to ask, especially if key data has not been logged.

Finally, we must recognise that the people typically responsible for improving performance are often not present (i.e. managers) or busy themselves (i.e. supervisors) when handovers are taking place. Most handovers will be carried out unsupervised.

AN ALTERNATIVE PERSPECTIVE

Changing behaviours is always difficult, especially when the individuals may not perceive a direct benefit. Therefore, it would be useful to have an alternative approach to improving shift handover.

Having studied the handover process it is clear that a lot of information communicated at handover could have a much wider use if it could be made available in an appropriate format. If other people were to start accessing that information it would increase the number of stakeholders in the process. More people would have a vested interest in improving handovers and would be more likely to intervene if the information they need was not forthcoming. Ultimately it seems likely that a better consensus would be reached over what needs to be communicated at shift handover.

The proposal here is that the log books and handover reports used at handover could be used to record information that can then be shared. The beauty of this information being that it reflects what actually happens in practice. Face to face communication will remain the most important part of the handover, but the log books and handover reports will add structure and detail.

A STUDY OF THE INFORMATION RECORDED FOR USE AT HANDOVER

In order to determine what type of data is recorded for use at handover, a study was conducted at an offshore oil production platform [Brazier 1996]. Copies of pages from log books and handover reports from across the platform were collected covering a seven day period. This information (a stack of paper weighing 3 ½ kg) was carefully examined to determine what information had actually been recorded about events occurring during the period of interest.

Analysis of the information showed that many of the events recorded in log books and handover reports could be particularly useful for safety and reliability studies, given that they were rarely reported in other systems and this information was a largely untapped resource. The study categorised the information as being related to human error, minor incidents, routine events, and solutions to problems.

HUMAN ERRORS

There is a general consensus that human error is a significant cause of accidents and incidents. Also, that many of those errors occur frequently with minimal consequence, and only on occasions combine with other events and conditions to cause an accident. Therefore, it is particularly useful to know about the errors that occur routinely, but these are rarely reported through formal channels. The following errors were found in log books examined in this study:

- Valve 'inadvertently' closed – delayed the return of equipment to service;
- Gas leak from a newly fitted gasket – system had to be shutdown and joint remade;
- Parts missing from replacement components (two events) – delays in critical repairs because components were supplied with parts missing;
- Part missed when assembling equipment – oil leak occurred, equipment had to be shutdown, dismantled and reassembled again;
- Incomplete modification – pipework modified but control system was not. An additional task was required to rectify;
- Error in job description – diving work was delayed because instructions referred to an incorrect valve location;
- Unable to find an up to date drawing – delays whilst a drawing showing all recent modifications was found;
- Data lost from computer disk – system temporarily unavailable whilst backup data was recovered.

None of these errors had significant consequences, which explains why they were not reported through more formal channels. However, any error indicates a problem that in other circumstances could have contributed to more serious events. Having information about the errors that occur more readily available would be useful for ensuring risk assessments are accurate and for prioritising human factors activities.

MINOR INCIDENTS

Most companies now have near miss reporting systems that means all incidents should be reported no matter how minor the consequences are. In practice the effort of reporting an incident is often perceived as outweighing the benefit. This study found 15 events recorded in log books and handover reports that could be considered as incidents, in addition to the human errors described above. Examples included:

- Unplanned hydrocarbon releases (three events) – small oil slicks observed on the sea following activities and a valve found to be leaking gas;
- Equipment failures (eight events) – chemical dosing pumps failing simultaneously, generator and compressor trips, compressors failed to start and an emergency shut-down valve did not close during a test;
- Equipment found to be inoperable (four events) – pressure override switch broken, pump operating at high temperature, pig receiver door damaged and pressure gauge pipework blocked.

The fact that incidents are being recorded suggests that log books and handover reports may provide an alternative mechanism for capturing these events. The advantage is that the reporter does not have to report the same information twice whilst the information is still immediately available for shift handover. However, the study showed that information about why these incidents occurred was often missing. This suggests that effort will be required if such an approach is taken to ensure the information about why events occur is recorded to allow further investigation.

ROUTINE TASKS

It is a bit of an anomaly, but the tasks people perform most frequently are often the ones we know least about. This is because they are often not covered by procedures and, when performed successfully, there is little indication to show they occurred. Because of this it is easy to think they are not important, but this is not the case. In fact, when considering plant reliability it is particularly useful to know what routine tasks are performed, their frequency, their duration and success rate. This study showed that this information can often be extracted from log books and handover reports. Of particular interest was that operators recorded 120 different routine tasks that had been performed in the seven day period. Whereas maintenance tasks are often captured in a management or recording system, operations tasks are usually not captured anywhere else.

The study concluded that the information about routine tasks recorded in log books and handover reports is probably more accurate than other sources of information because it is a record of what actually happens (i.e. rather than a pre-defined schedule which may not always be followed). Therefore, it was potentially particularly useful for reliability studies.

SOLUTIONS TO PROBLEMS

One reason we continue to employ people on facilities, despite advances in technology, is that they are good at dealing with unforeseen events and developing ad hoc solutions to problems. Understanding how people deal with events can give us a very useful insight into how they understand the systems they deal with. Where successful, solutions to problems can be shared so that others can use them in the future. However, it is important to know about any temporary or experimental solutions, as they may contribute to problems in the future.

This study found that solutions to problems were often recorded in log books and handover reports. For example:

- The need to release trapped pressure to reset an alarm;
- A production well that will only flow at low pressure;
- Another production well that would restart flowing if left for a while;
- The need to reduce gas pressure to start a turbine;
- Use of a 'similar' spare part where the correct one was not available;
- Manually manipulating a valve to stop it sticking;
- Using plastic sealing compound instead of a gasket to prevent a leak.

In each of these cases the operators were clearly solving the immediate problem. The successful solutions may be useful for others. In a number of cases it appears that the solution may not be fully approved, and hence it may be important in the future to know what has been done. A number of additional instances were recorded where people logged their suspicions about the cause of a problem being experienced, but where they had not been able to test them out. In these cases these assumptions may assist personnel tasked with solving the problem in the future.

USING DATA COLLECTED FROM LOG BOOKS AND HANDOVER REPORTS

The study described above identified the type of information communicated at handover. The following summarises three published papers that describe how this type of information has been used in practice.

COMPONENT RELIABILITY

A group of companies working in oil and gas production in the North Sea collaborated to collect equipment reliability data from operational experience to form a "Reliability Data Handbook" [Moss 1987]. Data was collected from maintenance and operating logbooks relating to hours of operation and stand-by, failure events and repair time. Overall, despite problems with extracting information from systems where this use was not envisaged (i.e. hand written logs etc.), the study was considered to be worthwhile and demonstrated that such an approach had potential, especially if the reliability, content and accessibility of records could be improved.

ECONOMIC OPERATION

In another study a change of operation was planned for a power station [Campbell 1987]. The company wanted to be sure that this could be achieved without increasing costs due to plant breakdown and other reliability problems. They did this by extracting data from shift log books that would indicate the main causes of problems. An assessment of this data, backed up with discussion with engineers, allowed an economic model to be developed. Sensitivity analysis allowed the identification of the items of equipment within the plant that were most critical to system reliability. From this information it was possible to identify the most appropriate operating regime. The conclusion from this study was that the data was available and did allow an accurate and useful economic model to be developed.

RELIABILITY

In the third study a power station had experienced a number of reliability problems for some time. A fault tree model of the system was developed using site specific data obtained from logbooks covering 29 years of operation [Galyean et al. 1989]. From this the major contributors to system unreliability were identified, allowing decisions to be made about future modifications. Extracting the data was difficult because the logbooks were all hand written, but once manually transferred to a computer database it was found to be particularly useful method for storage and manipulation. Also, there were some concerns about variations in the quality of records, especially as the events of most interest were generally associated by a high level of personal stress that meant those records were not always as detailed as would have been liked. However, despite these concerns it was felt that the results had made a significant contribution to the understanding of system reliability and that the cost and effort required to collect the data was worthwhile.

IMPROVING HANDOVERS AND MAKING DATA AVAILABLE

The studies described above show that the information used at handover can have a much wider application. However, extracting that information is not usually easy because much of it is in handwritten log books. Even though computers are now being used to record events and prepare handover reports, many use word processor or spreadsheet packages, or simple databases. Whilst these make the information more legible and hence may help people carrying out handovers, they do little to improve the availability of data and so do not fundamentally increase the number of stakeholders in the process with a direct incentive to drive improvement.

A MORE SOPHISTICATED DATABASE

It is one thing to capture information. It is another to make it readily available for people to use. A database can assist in this process, but to be effective the following needs to be understood:

- To get the full picture, it is usually necessary to have input from more than one area of the business;

- It is useful to be able to consider logged information alongside the relevant 'hard' process data;
- Information may be required in different formats for different purposes.

To address these issues it becomes apparent that the current approach of having lots of individual log books and handover reports will never allow the potential of the data stored to be realised. Instead a system is required that acts as a source of information that can be used during shift handover, but has a much wider use across the business. At the same time it must be easy for the people performing the handover to use, as without this it is possible that the quality of handover is reduced in an effort to make more information available to other potential users such as engineers and managers.

PRACTICAL REQUIREMENTS

History is littered with lots of examples where technology based products have failed to achieve their potential because they were not used as intended. In this case it is no use developing a database system that people do not record information in or extract the useful data. Therefore the following are the minimum requirements:

- Logging of information must be simple;
- To be really valuable the information must not only say what has happened, but explain why;
- Information must be highly visible so that people know what is happening;
- Analysis of historical information must be possible.

It is essential that end users are actively involved in developing the system to both embed their experience into the system and to ensure they understand the objectives of the new approach. In this case operator involvement is key to ensuring information logging will be practical and efficient. However, there will be other end users who have different needs, particularly in extracting and analysing data. A consensus is required about what constitutes the optimum solution for all end users.

It is important to recognise that this is not just an exercise of transferring a current log system to computer, as the benefits of this are quite limited. Therefore, as some of the concepts will be new to the end users, it is equally important to involve people who can input information about what can be achieved with such as system.

Finally, whilst technology can assist the handover process, it can have negative consequences if not managed correctly. Robust systems must be in place that addresses the requirements for good communication at shift handover and between different job functions (most significantly between and day workers).

THE PROPOSED SOLUTION

There is no intention to replace current arrangements with something completely new. Therefore, the proposed solution builds on the existing use of shift logs and handover reports. It assists by automating the process as far as possible and provides numerous functions and procedures that ensure handovers are as comprehensive and consistent as possible. Also, it makes information far more visible and this means there are more stakeholders with an interest in maximising the quality of shift handovers and who are more likely to intervene where these requirements are not achieved.

The old adage of 'rubbish in, rubbish out' certainly holds true in this case, especially since much of the information we are talking about is 'soft' in nature, being based on the observations of people rather than 'hard' plant data from instruments and control systems. The aim is to ensure all critical information relating to past, present, and future events are captured, will be visible and is supported by additional information including process data.

This will only be successful if the information that forms the basis of the handover is of high quality. This solution therefore encompasses the creation of operational logs and the subsequent use of that information for the benefit of managing the operation (of which handovers is a major use). In order to successfully underpin the handover process the solution must fulfil a number of key requirements:

- FACILITATE LOGGING OF INFORMATION: whatever method of capturing operational information is used, it will be, to some degree, an imposition on the operator. The proposed solution encourages quality logs by allowing most information associated with an entry to be made with a few mouse clicks. The operator is then required to type only value added information, which generally explains why an event occurred. The operator is guided through the logging process in a structured way, ensuring all essential information is captured. This has added benefits for less computer literate operators as it minimises the input required.
- PROVIDE A STRUCTURED LOGGING ENVIRONMENT: it is essential that any solution provides the flexibility to capture all the varied operational activities required in logs across the operation. At the same time, it is important to impose a level of structure on the logs to encourage consistency of input. Simply providing for free-format text entry provides flexibility but does not allow for a structured approach. The solution allows 'Event Hierarchies' to be pre-defined. Each log can have its own hierarchy tuned to the specific logging requirements. Each hierarchy point (an event that can be logged) can have its own template. This template defines the structure of the log entry and can accommodate any additional information to be captured, any rules on how the entry is shared or copied with other logs and any external documents to be attached or referenced. The level of structure imposed by the template is defined by the users. A template could in its simplest form be a simple free format text field. This template approach helps ensure that the same event logged over time will be logged in the same way and is of great benefit when reviewing and reporting on logs.
- ALLOW FOR EASY SHARING OF INFORMATION: the solution allows log entries to be very easily shared across multiple logs throughout the operation. This can be automated if required to ensure important information is highly visible to the appropriate people or issues are effectively escalated.
- ALLOW QUICK SEARCHING AND REPORTING OF LOGS: the solution allows easy access to logs, whether it is the current shift reviewing the previous shift logs,

engineers carrying out analysis of historic logs or management reporting across multiple logs. Providing the structured templates allows reports to be very easily built. This helps turn the logs from an operational record to be filed away into a live repository of valuable information. The logs become valuable assets of the business.

THE SOLUTION IN PRACTICE

The handover process can encompasses a wide range of information including the logs of past events, the current plant status, and issues for future shifts. This is true for shift handovers that occur on a daily basis and for 'trip handovers' that may occur weekly, monthly or event longer.

The solution has a number of functions specifically designed to facilitate the handover process, including; the ability to view logs across any timeframe, ability to flag important information, ability to add log entries to ToDo lists in order to action issues across shifts and the ability to create, assign and track logbook tasks across shift teams.

One of the most important aspects of the solution is the ability for each area to configure its own log structures. This is a user centred solution that achieves buy-in from the user base and subsequently leads to higher quality logs. This is much more desirable than imposing a rigid system on operators that does not meet their own individual or departmental requirements.

It must be emphasised that any computerised solution to managing shift handovers can only support (and not replace) a well thought out and well followed handover procedure. The wider considerations are those of company and operational culture and discipline. A culture of open communications, continuous learning and continuous monitoring of process quality can be underpinned by a well developed computerised system but cannot ultimately be controlled by such a system; the whole process starts and end with the organisations biggest asset – its people.

In implementing such a solution it is important to recognise that, as with any intervention in any system, there are always potential negative consequences. In this case there is the potential that making information more readily available over a computer network may mean that people talk to each other less often. Whilst it is felt the way the solution works means more people will become interested in what is going on and hence are actually more likely to ask questions and discuss events, this is something that needs to be monitored as part of its implementation. Also, it is recognised that this solution may not currently be appropriate in places where a significant proportion of the workforce either do not have access to a computer or lack the appropriate skills.

This solution has been in use with great success within several large operations with user bases in the 100s. One large site in the UK has used it for over three years. It replaced many paper based and individual computer based logs with a single, integrated solution that allows operational knowledge to be shared 24 hours a day whilst interfacing with other existing operational systems. In another case a large power generation and distribution company used the solution to develop an integrated logging system across its diverse range of sites and corporate level functions. Following a pilot project, the solution was implemented across the company in four months. As well as practical benefits such as more consistent logs and improved availability of information, cultural improvements have been experienced including operators having a better understanding of the value of high quality logs, shift handovers are much more efficient because the high visibility of information enables the oncoming team quickly get 'up to speed' and allows them to ask insightful questions to ensure they fully understand the issues.

CONCLUSIONS

There is no doubt that shift handover is a critical activity and poor handovers have contributed to major accidents. However, it has received relatively little attention and the guidance available is rather limited.

The goal of shift handover has been defined as "the accurate, reliable communication of task-relevant information across shift changes, thereby ensuring continuity of safe and effective working." To do this oncoming personnel have to gain an accurate understanding of plant status so that they are able to make correct decisions and initial appropriate actions as required.

Improving shift handover requires systems to be in place that include procedures, training and assessment, monitoring and audit. Also, it is necessary to address the behavioural aspects, which may be something that organisations have tended to shy away from. Structured log books and handover reports can assist the face-to-face aspect of a shift handover. Also, it is important to recognise that certain circumstances such as ongoing maintenance and deviations from normal operation create higher risk and need careful consideration during handover.

This paper proposes an approach to improving shift handover that is complementary to developing and improving the communication aspects. It aims to make information recorded at handover a more valuable resource. Studies show information about all aspects of operation are often recorded in shift logs and handover reports, including information about human errors, minor incidents, routine tasks and solutions to problems. This can be used across the business to improve safety, reliability, production, and environmental performance.

To make the information more available it is suggested that a computer based database solution is required. This goes beyond simply converting log books into computer form, and instead results in a comprehensive source of management information that has many uses, as well as supporting shift handover. The advantages of this approach include:

- Important information becomes more visible;
- Better information is available when making operational and strategic decisions;
- Time is saved in logging events, meaning more value-added information can be recorded;
- Information flows much better across the organization;
- A full audit trail is provided.

A computer based approach has many potential benefits, but it must be remembered that the behaviour of users will have the greatest influence on shift handover effectiveness. Any improvement must reflect the human factors involved. However, shift handover is a critical activity and should be a high priority for any organisation working in a hazardous industry. Key issues include [IP 2006]:

- Provision of clear procedures/written guidance describing the key information to be exchanged and how this should be done (e.g. word of mouth, in writing or both);
- Providing training and having systems to ensure employees are competent to use handover procedures;
- Carrying out regular and thorough monitoring and auditing;
- Involving employees in the examination and improvement of the practices;
- Updating systems in light of information from incidents and accidents due to shift handover problems and bringing this to the attention of employees.

REFERENCES

BP 2005, Fatal Accident Investigation report – Isomerization Unit Explosion Final Report – Texas City, Texas USA.

Brazier 1996, Sources of Data for Human Factors Studies in the Process Industry. PhD Thesis, Edinburgh University.

BSTG 2006, Initial report – recommendations requiring immediate action, published by Buncefield Standards Task Group on the HSE website.

BSTG 2007, Safety and environmental standards for fuel storage sites – Buncefield Standards Task Group final report.

Campbell 1987, Small Size Event Data Banks. Reliability data bases: proceedings of the Ispra course, ed. Amendola, A. Keller, Z. publ. D. Reidel Publishing Company.

CSB 2007, BP Texas City – Final Investigation Report.

Galyean, WJ. Fowler, RD. Close, JA. Donley, ME. 1989. Case Study: Reliability of the INEL - Site Power System. IEEE Transactions on Reliability. vol. 38, no. 3.

HSE 1999, Reducing error and influencing behaviour.

IP 2006, Human Factors Briefing Notes number 10 Communication, Institute of Petroleum human factors website.

Lardner 1999, Safe Communication at Shift Handover: Setting and Implementing Standards. European Process Safety Centre (EPSC) Human Factors conference 2006.f

Moss 1987, Reliability Data from Maintenance Records. Reliability Data Bases: Proceedings of the Ispra Course, ed. Amendola, A. Keller, Z. publ. D. Reidel Publishing Company.

AN INVESTIGATION INTO A 'WEEKEND (OR BANK HOLIDAY) EFFECT' ON MAJOR ACCIDENTS

Nicola C. Healey[1] and Andrew G. Rushton[2]
[1]Health and Safety Laboratory, Harpur Hill, Buxton, Derbyshire, SK17 9JN
[2]Hazardous Installations Directorate, Health and Safety Executive, Redgrave Court, Bootle, L20 7HS

A number of recent and high profile accidents in the process industries have occurred on bank holidays or at weekends. This has led to the suggestion that there was a potential for higher numbers of accidents, particularly those classed as major accidents, to occur on a bank holiday or a weekend.

If this perception should represent an actual effect, there could be serious implications for the work of the safety regulators of major hazard industries and the industries themselves.

The Health and Safety Laboratory (HSL), was commissioned by the Health and Safety Executive (HSE) to analyse a selection of data on major accidents. The aim of the analysis was to identify whether there was evidence of a weekend or bank holiday increase in the numbers of major accidents.

Attempts were made to normalise the data to account for the reduced operations and staffing levels sometimes present at process industry sites on weekends or bank holidays. Unfortunately, sufficient data could not be found to normalise the accident data in these respects.

Main findings are:

- Analysis of the aggregated data and sub-sets consistently failed to provide evidence for either a weekend or bank holiday increase in the number of major accidents.
- Fairly consistent evidence was seen of a mid-week peak and a weekend dip in the number of accidents reported.
- Chi-Square tests consistently showed either 'no significant difference', or 'significantly more accidents occurring on a weekday' in comparison with accidents occurring on a Saturday, Sunday or bank holiday.

In essence, the results show either no evidence of an effect, or that the effect is too weak to stand out over other factors within the data available.

INTRODUCTION

Unfortunately, major accidents involving dangerous substances in the process industries have occurred and continue to occur despite continued efforts aimed at their prevention, control and mitigation. The control and regulation of process industry major hazards is informed by analysis of those accidents (as well as by first principles approaches to hazard management).

Several high profile incidents have occurred at weekends. Examples include those at Flixborough, Seveso, Milford Haven and Buncefield (Parker, 1975, Orsini 1977, HSE 1997, Newton 2006). The publication, in 2005, of a report into a bank holiday incident at the Humber Refinery in the UK (HSE 2005) was the prompt for the present work.

Speculation based on anecdotal evidence, has led to the suggestion that there is a potential for higher than proportionate numbers of accidents, particularly those classed as 'major' accidents, to occur on a bank holiday or a weekend (Saturday or Sunday). Here, this postulated increase of accident incidence on a weekend day or bank holiday is termed a "weekend (or bank holiday) effect". The term bank holiday is not used in the strict sense, but is used loosely to refer to the kinds of 'national holiday' days when roles that are normally filled on week days will not be filled.

If this perception should represent an actual effect, there could be serious implications for the work of the regulators regarding major hazard activities and industries. The confirmation of any clear weekend (or bank holiday) effect might justify further research clarifying the causes of the effect and the targeting of resources on reduction of those causes or the introduction of additional barriers tailored to the identified effect paths.

This paper reports the findings of research carried out by the Health and Safety Laboratory, commissioned by the Health and Safety Executive. An analysis of data from a selection of major accident lists and databases was performed, to search for significant differences in the numbers of major accidents occurring on weekdays when compared with weekend days and bank holidays. Significant differences might provide evidence in support of a weekend or bank holiday effect on the occurrence of major accidents.

More specifically, the objectives of the exercise were:

- To review five collections (lists or databases) of major accident information (data sources);
- Identify whether these accidents occurred on a weekday, weekend or bank holiday;
- Analyse the data to determine significant (or non-significant) differences in accidents occurring on a weekend or bank holiday in comparison to non bank holiday weekdays.

Throughout this paper the term 'accident' is used to refer to accidents and incidents recorded across the five sources of data.

Full details of the analysis are reported by Healey 2007. This paper provides a summary of the work and some discussion of the context and implications.

PREVIOUS WORK

A literature search found no previous work specifically investigating the presence of a weekend or bank holiday effect on the occurrence of major accidents.

A study of fatal accidents at work associated with maintenance (HSE, 1985) found that accidents occurred throughout the working week with a peak observed mid-week. "Surprisingly, only 13% happened during weekend maintenance work, but this may be affected by the recession or reduced production pressures". If accidents were proportionately spread over the entire week (that is, if there were no reduction in activity or numbers of people at risk and accidents were evenly distributed) then of course about 30% of accidents would occur at weekends (2/7 ~0.3).

DATA SOURCES AND ANALYSIS

A selection of five data sources, with appropriate inclusion criteria, were identified and utilised throughout this work. Appendix A gives an outline description of the sources, an indication of the criteria for inclusion of data in each source and, where appropriate, additional criteria applied in the data selected for this study.

The data sources covered a varying range of the process industries, and varied in their inclusion criteria.

Each of the five data sources was interrogated and information extracted. Journal articles and websites (see references) were used to supplement the 'date' and 'country' information from the accident data sources to identify the day of the week an accident occurred on. Using this information most accidents could be tagged according to whether they occurred on weekdays, weekends or bank holidays.

There was apparent replication of some accident reports both within and between the various data sources. In some cases the replication was clear, whilst in other cases it was not so clear. Where possible, replicated accident reports were removed to reduce the likelihood of duplication in the analyses. HSL relied on its subjective judgement to remove this replicated information.

ATTEMPTED NORMALISATION OF THE DATA

There are potentially numerous and varied factors which might influence the occurrence of accidents on weekdays, weekends or bank holidays. One obvious factor is the extent to which operations may (or may not) differ: some sites will operate in similar mode regardless of the day of the week, others may operate very differently (perhaps not at all) on a bank holiday. Another factor is the presence of population (both on and around the major hazard site). The effects of these factors may be simple (e.g. fewer operations leading to fewer opportunities for accidents) or complex (e.g. fewer people on site may result in fewer casualties in the event of an accident, or, a greater number of people on site, by virtue of their appropriate actions, may inhibit the escalation from an initiating event to a reportable accident).

A fair comparison of the incidence of accidents (and their consequences) would take account of other factors which raise or lower the potential for such accidents at weekends or bank holidays. To some extent the occurrence of accidents may be inhibited at weekends or bank holidays if operations are reduced or consequences may be inhibited if fewer people are present. So, for example, a fair comparison would take account of reduced operations (or exposure to consequences) outside weekdays.

In the discussion above, a simple definition of "weekend" (i.e. a Saturday and Sunday) has been used. However if there is a "weekend effect" it may not be relevant to distinguish between midnight on Sunday and resumption of normal "office hours" on Monday. A weekend effect may also be somehow related to (or, at best, confounded with) diurnal effects (which have been studied separately, Fortson 2004).

It was, therefore, desirable to explore whether the analysis was sensitive to scaling of the data to factors such as 'degree of weekend working' and 'population present'. So it was envisaged that alternative analyses would be undertaken with the data "normalised" in respect of these factors.

Literature and abstract searches were carried out in order to identify a method (i.e. a relevant data source) for normalising the major accident data, in line with differences in the productivity levels and work patterns typically observed in the different industries reporting accidents. An appropriate source for this information was not identified.

Furthermore, the accident information held in the data sources was not detailed enough to identify information such as staff shift patterns or patterns in productivity levels, and normalisation of the accident data was not possible (neither for these factors nor for other factors).

ANALYSIS

Analysis was performed on the aggregated data from all sources, on data from each source separately (or a selection of that single-source data) and on subdivisions of the aggregated data (for example, the aggregated set of accidents with reported consequences including ten or more fatalities).

The Chi-square test was applied to the accident information collected from the five data sources to assess the significance (or non-significance) of differences between the proportions of major accidents occurring on weekdays, weekends and bank holidays.

A variety of t-test was used to determine the statistical significance of differences in the relative proportions of 'weekend' to 'weekday' accidents in subsets of the data, when compared with the same proportions in the overall aggregated data.

RESULTS

An 'overview analysis' was performed on the aggregated data from all sources. The raw aggregated dataset consisted of information relating to 4333 accidents, collected across the five data sources. Replicated accident reports (numbering 487) were removed from the dataset before the overview analysis.

The remaining sample of 3846 accidents were classified according to the day of the week on which they occurred.

Figure 1 shows the breakdown of the accidents, by day of the week. Information was 'not available' (N/A) for some accidents. The chart shows a mid-week peak and a decrease in accidents reported as occurring at the weekend (Saturday and Sunday).

Table 1 shows the breakdown of the sample, by bank holiday accidents. To determine a bank holiday both the country, and exact date were required. This additional data was not always available leading to a higher number of accidents classed as 'N/A'.

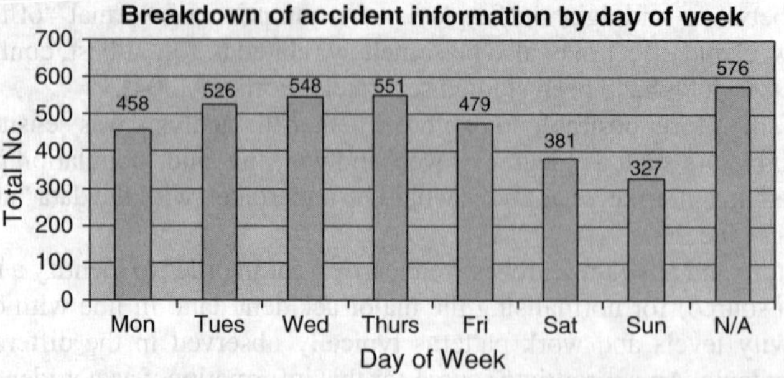

Figure 1. Breakdown of accident information by day of week (n = 3846)

Results of Chi-square calculations showed there were significantly more accidents on a weekday (Monday–Friday) than on a Saturday, Sunday, weekend (Saturday and Sunday) or bank holiday in this overall sample. A little over 20% of accidents occurred on a weekend (compared with ~30% for a flat distribution). Significantly more accidents occurred on a Saturday, in comparison to a Sunday.

ANALYSIS: BY DATA SOURCE AND BY CONSEQUENCE (NUMBERS OF FATALITIES)

Further analyses of the accident data were also performed to explore the sensitivity of features in the data to the data source and to a measure of accident consequence (number of fatalities).

The data from each data source in isolation was analysed. The results for each data source, comparable with Figure 1 for the aggregated data, are presented in Appendix A.

The aggregated data was subdivided by ranges of numbers of fatalities (a crude indication of accident severity). The results are detailed in Healey 2007.

In most cases, for either individual data sources or for subsets of the aggregated data based on bands of consequence (ranges of numbers of fatalities), the features of the "overview analysis" were confirmed.

Table 1. Breakdown of bank holiday accident information (n = 3846)

Bank holiday	Total no. accidents
Yes	46
No	2308
N/A	1492
Total	3846

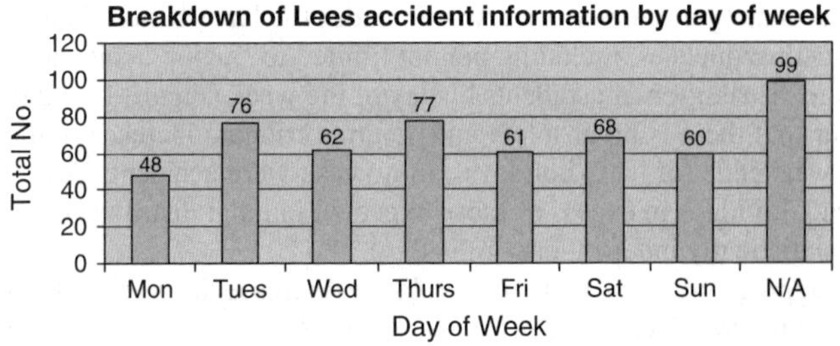

Figure 2. Breakdown of Lees accident information by day of week (n = 551)

There were some apparent distinctions between features of the "overview analysis" and the analysis of individual data sources or sub-sets of the aggregated data. For example, a relatively high proportion of 'weekend' to 'weekday' accidents was seen in both the Lees data (Figure 2, Appendix A) and a 'ten or more fatality accidents' subset of the aggregated data, when compared to the overall aggregated data. A Chi-square test showed 'no significant difference' between accidents on weekdays and weekends for these two cases, where the overall, aggregated data had shown significantly more accidents on weekdays.

Further statistical analysis was necessary to establish whether this change was significant or merely a consequence of the smaller amount of data in a single source or sub-set of the aggregated data.

A form of t-test was used to determine the statistical significance of differences in the relative proportions of 'weekend' to 'weekday' accidents in individual data sources or sub-sets of the aggregated data, in comparison with the overall, aggregated sample. A statistically significant higher proportion of 'weekend' to 'weekday' accidents was observed in both the Lees data and a 'ten or more fatality accidents' subset of the aggregated data, when compared to the overall aggregated data.

Therefore, the dip in weekend accidents noted earlier (HSE 1985) and here in the "overview analysis" is significantly less pronounced in some subsets of the data.

Although these findings do not show evidence in support of a weekend or bank holiday effect, the results of the t-test suggest that some subsets of the data (in particular the Lees data, and a 'ten or more fatality accidents' subset of the aggregated data) are different in nature to the overall, aggregated data. It is possible this is evidence of a weak, relative association of larger/more severe accidents with weekends.

In no case was there evidence of an excess in the incidence of accidents on weekends (or bank holidays). The full results are reported by Healey 2007.

DISCUSSION

The aggregated data studied here is principally distinguished from the data reviewed earlier by HSE (HSE 1985) by its focus on the process industries (i.e. excluding construction etc).

The aggregated data (particularly because it includes the large IChemE dataset) includes a wide range of consequences, including but not limited to 'major' accidents.

From the distribution of accidents by day of the week (Figure 1) and statistical analysis, it is clear that there is no absolute and disproportionate increase in the incidence of accidents on weekends (or bank holidays). In no case were accidents occurring on weekends or on bank holidays in excess of those expected in a flat distribution (where accident frequency is statistically independent of the day).

In the aggregated data, there is a relative increase in weekend accidents when compared to the broader-based study (from ~13% in HSE 1985 to ~20 % in the aggregated data here).

There were significantly more accidents on a weekday (Monday – Friday) than on a Saturday, Sunday, weekend (Saturday and Sunday) or bank holiday in this overall sample. A little over 20% of accidents occurred on a weekend (compared with ~30% for a flat distribution). Significantly more accidents occurred on a Saturday, in comparison to a Sunday.

Analysis of the individual data sources which are more clearly restricted to "major" accidents (for example Lees), or subsets of the data more restricted to large scale consequences (for example accidents associated with ten or more fatalities) showed a weak relative increase in the incidence of accidents on weekends. But these increases were of little or no statistical significance, or else were confounded by the reduced amount of data.

These analyses produced no general evidence of a disproportionate increase in accident numbers on weekends or bank holidays and no conclusive evidence that more consequential, influential or spectacular accidents correlate positively with weekends or bank holidays. There were some indications that more consequential, influential or spectacular accidents are relatively (but not absolutely) more likely at weekends, though this may merely be due to a confounding factor (i.e. industries capable of more consequential, influential or spectacular accidents may be more likely to be operating on weekends and bank holidays).

When viewing the data classified by consequence (simply represented by number of fatalities), there may be a confounding effect in that weekend (or bank holiday) incidents may be less consequential than corresponding weekday events. For example, had the Flixborough disaster occurred during a normal working day (and been in other respects unchanged) the fatalities in that incident might have been ten times greater (Rushton, 1998).

There are, undoubtedly, some safety professionals who believe there is a "weekend (or bank holiday) effect". Perhaps this belief is mistaken, or perhaps it is based on sound (if unsubstantiated) intuition.

It is easy to conjecture how a weekend effect might arise. There may be fewer staff to detect incipient causes (accident initiators or barrier failures), supervise interventions, respond to incipient causes or diagnose 'leading' indicators. For example there may be more situations put "on hold" awaiting the attention of weekday staff or contractors. It is easy to speculate on other features of bank holiday and weekend working which might weaken or remove barriers to the initiation of or escalation to an event.

It is possible there is a "weekend effect", that is too weak to stand out over other factors within the data available. Where common sense and/or engineering judgment indicates a greater propensity for initiation or escalation of events in some particular circumstances (including but not limited to weekends or bank holidays), then that propensity should of course be given proportionate attention in the management of hazards.

CONCLUSIONS

There are some safety professionals who believe there is a disproportionate incidence of accidents (particularly 'major' accidents) on weekend days or bank holidays (a "weekend (or bank holiday) effect").

Study of the day of occurrence of a large number of accidents in the process industries shows no statistical evidence of a weekend or bank holiday increase in the numbers of major accidents.

The main findings were:

- Analysis of the aggregated accident data from five sources and sub–sets of that data consistently failed to provide convincing evidence for either a weekend or bank holiday increase in the number of major accidents.
- Fairly consistent evidence was seen of a mid-week peak and a weekend dip in the number of accidents reported.
- Chi-Square tests consistently showed either 'no significant difference', or 'significantly more accidents occurring on a weekday' in comparison with accidents occurring on a Saturday, Sunday or bank holiday.

There is some weak evidence of a relative increase in incidence of accidents on weekends in the process industries compared to industry in general. There is weak evidence of a relative increase in incidence of 'major' accidents on weekends in the process industries compared to accidents in the process industries in general.

In essence, the work has shown either no evidence of an 'effect', or that the 'effect' is too weak to stand out over other factors within the data available.

Although evidence was not found in support of a weekend or bank holiday effect, at this juncture, such an effect has not been disproved. It is possible that a weekend or bank holiday effect may exist, but was not revealed in this study due to the masking effect of other factors.

A comprehensive understanding of the organisational factors underlying major accidents is vital in order to inform regulators and major hazard industries, regarding regulation and the prevention of major accidents. Where common sense and/or engineering judgment indicates a greater propensity for initiation or escalation of events in any particular circumstances, then that propensity should be given proportionate attention.

In recognition of the limitations associated with this research, it is recommended that these findings be considered as a foundation for further work, with a focus on addressing some of the limitations encountered in this research.

DISCLAIMER

The views expressed in this paper are those of the authors alone and are not a statement of HSE policy.

REFERENCES

EC, 1997, Council Directive 96/82/EC of 9 December 1996 on the control of major-accident hazards involving dangerous substances, Official Journal of the European Communities, Luxembourg (also known as "the Seveso II directive"), as amended by Directive 2003/105/EC.

Fortson, K., 2004, The Diurnal pattern of on-the-job injuries, *Monthly Labor Review,* September 18–25.

Healey, N., 2007, An investigation into a weekend or bank holiday effect on major accidents. HSL report. RSU/RM/07/04.

HSE, 1985, Deadly Maintenance – A Study of Fatal Accidents at Work, ISBN 0-11-883806-7 HMSO.

HSE 1997, The explosion and fires at the Texaco Refinery, Milford Haven, 24 July 1994, ISBN 0 7176 1413 1.

HSE, 2005, Public report of the fire and explosion at the Conoco Phillips Humber Refinery on 16 April 2001 at http://www.hse.gov.uk/comah/conocophillips.pdf

Kirchsteiger, C., 2001, MARS 4.0 – An Electronic Documentation & Analysis System for Industrial Accidents Data, European Commission, DG JRC, EUR 19766 EN, Ispra.

Newton, Lord, 2006, Buncefield: Initial Report to the Health and Safety Commission and the Environment Agency of the investigation into the explosions and fires at the Buncefield oil storage and transfer depot, Hemel Hempstead, on 11 December 2005 at http://www.buncefieldinvestigation.gov.uk/reports/initialreport.pdf

Orsini, B.,Parliamentary Commission of Inquiry on the Escape of Toxic Substances on 10 July 1976 at the ICMESA Establishment and the Consequent Potential Dangers to Health and the Environment due to Industrial Activity, Rome (English translation, HSE, 1980)

Parker, R.J., 1975, The Flixborough Disaster: Report of the Court of Inquiry, HMSO, ISBN 0113610750.

Rushton, A.G., 1998, Lessons Learned from Past Accidents, in Risk assessment and management in the context of the Seveso II directive (Kirchsteiger C, Christou MD and Papadakis GA Eds.), Industrial Safety Series Volume 6, Elsevier.

DATA SOURCES

Institution of Chemical Engineers Accident database. A CD of Chemical and Processindustry accidents, created and maintained by IChemE.

Lees, F. 1996 Loss Prevention in the Process Industries: hazard identification, assessment and control. Volume 3 Appendices. 2nd Edition.
Large Property Damage Losses in the Hydrocarbon-Chemical Industries: A Thirty-Year Review. 13th Edition, 1990.
Large Property Damage Losses in the Hydrocarbon-Chemical Industries – A Thirty-Year Review. 19th Edition, 2001.
Major Accident Reporting System, accessed September 2006 at:
http://mahbsrv.jrc.it/mars/Default.html
http://mahbsrv.jrc.it/mars/MARS-Technical-Guideline-February-2001.pdf
MHIDAS (Major Hazard Incident Data Service) database. Enquiries directed to the Health and Safety Executive, Hazardous Installations Directorate, Business Assurance and Operations Analysis Team, Bootle.

SUPPLEMENTARY INFORMATION

Day of the week information accessed October 2006 at: http://www.searchforancestors.com/utility/dayofweek.html
Day of the week and 'Bank holiday' information accessed October 2006 at: http://www.timeanddate.com/calendar/
'Bank holiday' information accessed October 2006 at: http://www.bank-holidays.com/
World Commercial holidays, 1996. Business America, Dec 95, Vol. 116 Issue 12, p21

APPENDIX A: SAMPLE OF DATA SOURCES AND APPROPRIATE RELEVANCE CRITERIA

- ***Loss Prevention in the Process Industries. Lees.*** This is a personal selection/collection by the author (F.P. Lees) of major accidents reported by the process industry. The collection is in the form of a list. So, the inclusion criterion was the subjective discretion of Lees. All accidents in the list were effectively classified as 'major' by the author due to their inclusion in the selection, and were all considered in this analysis.

The sample of 551 accidents recorded in the Lees database were organised according to the day of the week they occurred on.

Figure 2 shows the breakdown of the Lees accidents, by day of the week, 'N/A' was used where information was 'not available'.

- ***Large Property Damage Losses in the Hydrocarbon-Chemical Industries: A Thirty-Year Review. Marsh McLennan.*** This is a list reviewing 100 large property damage or losses occurring in the hydrocarbon processing and chemical industries over a thirty-year period. (Information relating to an additional 15 accidents, taken from the latest version of the review was also included). All accidents were included in this analysis. The inclusion criterion was economic loss (in the qualifying period and limited by number).

The sample of 115 accidents recorded in the Marsh McLennan database were organised according to the day of the week they occurred on.

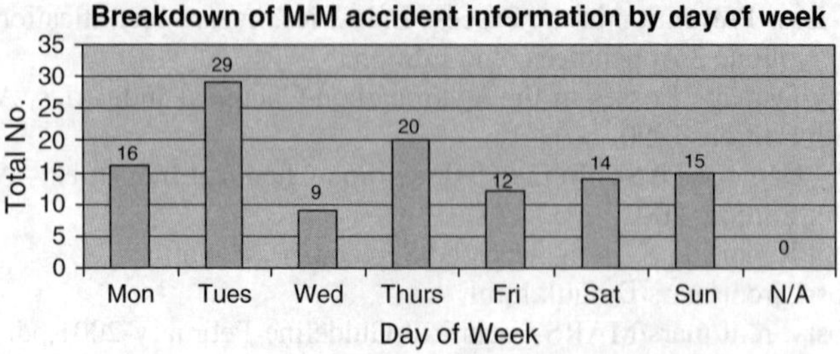

Figure 3. Breakdown of Marsh McLennan accident information by day of week (n = 115)

Figure 3 shows the breakdown of the Marsh McLennan accidents, by day of the week.

- ***MHIDAS. The Major Hazard Incidents Data Service.*** This is a collection of world-wide accidents recorded using information taken from the public domain. The collection is in the form of a database. The inclusion criterion is incidents involving hazardous materials that had an off-site impact, or had the potential to have an off-site impact. Such impacts include human casualties or damage to plant, property or the natural environment. All accidents in the database are effectively classified as major due to the offsite impact (or potential for offsite impact). The most recent 860 accidents recorded excluding those occurring in long standing members of the European Union, i.e. Belgium, France, (West) Germany, Italy, Luxembourg, Netherlands, Denmark, Ireland, United Kingdom, Greece, Portugal, Spain were considered for this analysis. This reduced set was used in order to minimise the number of accidents likely to be included within the remit of the other data sources, therefore minimising replicated incidents in the analysis. (The current work was limited to the inclusion of 860 accidents due to project resources.)

Figure 4 shows the breakdown of the MHIDAS accidents, by day of the week, 'N/A' was used where information was 'not available'. The graph shows a mid-week peak and a decrease in accidents reported as occurring at the weekend (Saturday and Sunday).

- ***EU MARS. The Major Accident Reporting System.*** This is a distributed information network compiling information from 15 databases in each member state of the European Union. All accidents in the compiled database are effectively classified as major accidents, as the reporting process relies on the major accident definitions in the Seveso directives (EC 1997). All accidents in the database were included in this analysis.

The sample of 603 accidents recorded in the EU MARS database were organised according to the day of the week they occurred on.

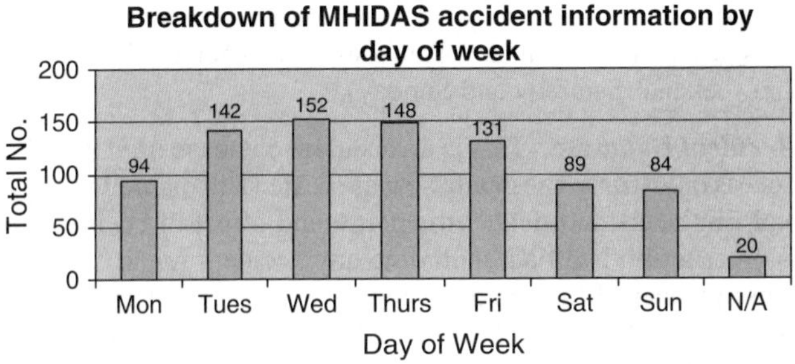

Figure 4. Breakdown of MHIDAS accident information by day of week (n = 860)

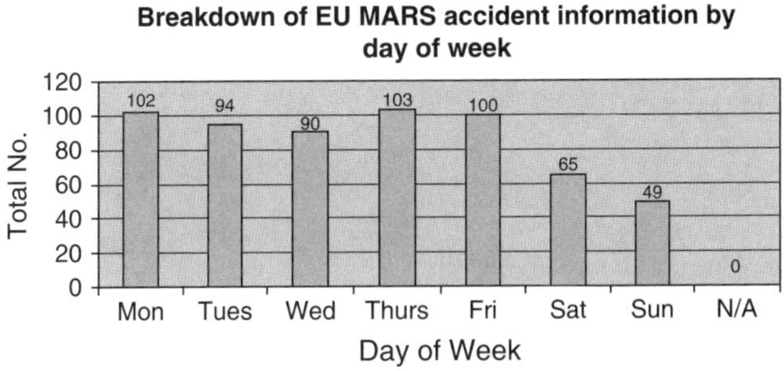

Figure 5. Breakdown of EU MARS accident information by day of week (n = 603)

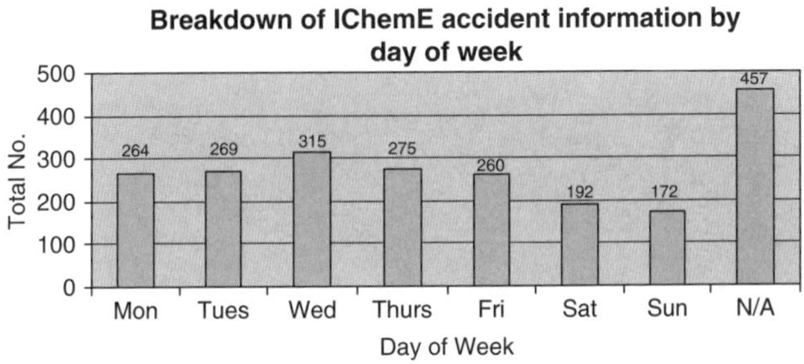

Figure 6. Breakdown of IChemE accident information by day of week (n = 2204)

Figure 5 shows the breakdown of the EU MARS accidents, by day of the week. The graph shows peaks on Monday and Thursday and a decrease in accidents reported as occurring at the weekend (Saturday and Sunday).

- ***IChemE Accident Database.*** This is an accident database used worldwide by chemical and process industries. The database was created and maintained by the Institution of Chemical Engineers and details the cause and effect of accidents and incidents. The inclusion criterion is broad, including any accident report in the process industry. The consequences of the accidents are many and varied and so it is unclear that all the accidents would meet a definition of "major" (such as in the Seveso Directives). All accidents in the database were including in this analysis.

The sample of 2204 accidents recorded in the IChemE database were organised according to the day of the week they occurred on.

Figure 6 shows the breakdown of the IChemE accidents, by day of the week, 'N/A' was used where information was 'not available'. The graph shows a mid-week peak and a decrease in accidents reported as occurring at the weekend (Saturday and Sunday).

IS HAZOP ALWAYS THE METHOD OF CHOICE FOR IDENTIFICATION OF MAJOR PROCESS PLANT HAZARDS?

Alfredo Verna[1] and Geoff Stevens[2]
Arthur D. Little Limited, Science Park, Milton Road, Cambridge, CB4 0XL, UK
[1]E-mail: verna.alfredo@adlittle.com
[2]E-mail: stevens.geoff@adlittle.com

Hazard and Operability Analysis (HAZOP) has long been established as one technique for risk identification in Process Plants and many firms incorporate HAZOP procedure into their safety management approach. HAZOP studies appear at various phases of project development including FEED (Front End Engineering and Design), as part of Detailed Design and for engineering modifications to existing plants.

An alternative approach to HAZOP is described suitable where potential hazards are well understood and where time is limited but independent assurance is desirable. The method was applied as a review of a design that had already been fully studied by HAZOP but still revealed a number of important new hazard issues. Apart from describing the new technique, the paper examines reasons for shortcomings of HAZOP conducted too early in the engineering development and conditions under which it can be cost effective.

INTRODUCTION

Hazard identification is one of the key steps in systematic Safety Risk Management which can be illustrated in diagrams such as Figure 1.

The process illustrated is iterative and can be applied at a number of stages in the project cycle. Because it is recognized as a primary methodology for conducting hazard analysis, some suppliers offer HAZOP at any stage of the design[1]. The approach in practice needs to be varied according to the documentation (especially the diagrammatic representation) available at each stage. Even in the operating phase, HAZOP can be used, for example, when modifications are planned or when a major turnaround is anticipated including safety related upgrading.[2]

HAZOP STUDIES CONDUCTED WITH LIMITED TECHNICAL DEFINITION

The contract stage at which design information emerges will vary with the contractor's approach and the degree of integration they have achieved[3]. At the early stages of the project (development of the design basis manual and to some extent the Front End Engineering) the technical definition includes limited (sometimes no) information on the proposed control instrumentation and Safety Interlocks either for machine protection or plant emergency shut-down. It is not uncommon for details of machinery protection functions to be delayed until vendor selection has been completed and this can be quite

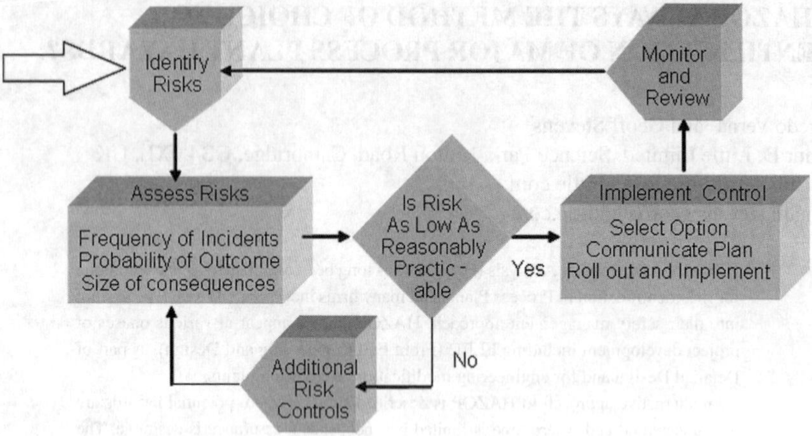

Figure 1. Block flow diagram of the process of Safety Risk Management. Source: Arthur D Little

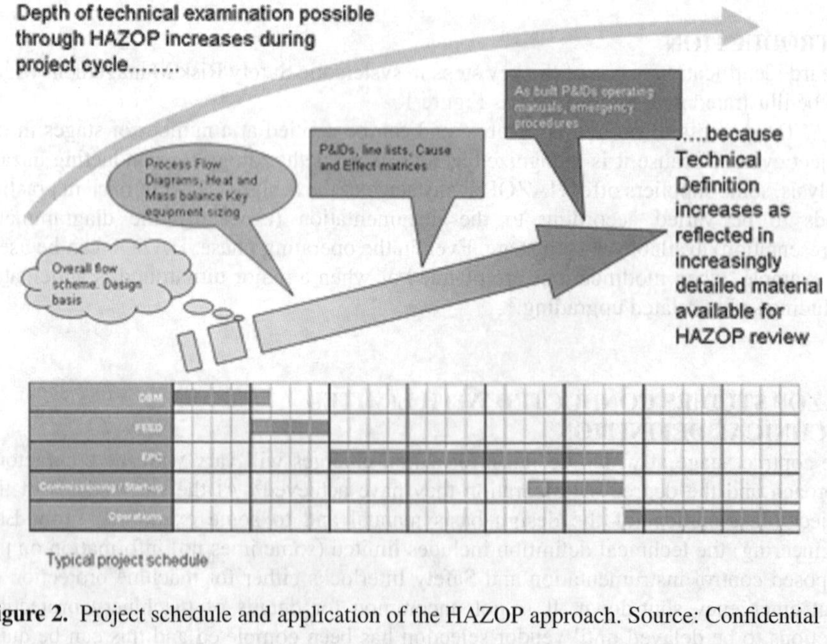

Figure 2. Project schedule and application of the HAZOP approach. Source: Confidential

advanced in the EPC (Engineer Procure and Construct) phase of the project. If a full recording HAZOP technique is employed at each of the key stages in the project the following outcomes can be expected:

- In the early stages there is sufficient information to identify the main hazardous inventories and the generic hazards associated with the type of equipment available. If a coarse HAZOP is undertaken, each deviation with hazard potential is likely either to recommend the issue is revisited when better definition is available or (in the case of major hazards posing intolerable risk) to suggest a review of the design.
- When Process Flow Diagrams are available, the HAZOP can deal with potential hazards associated with each deviation but there may not be sufficient definition of control systems or materials selection to determine if adequate levels of protection are provided.
- Once P&IDs are available, including control arrangements, main trip functions, equipment design conditions and piping material, a full recording HAZOP can be conducted. Typically this review works well for the engineering contractor's scope but there may be a number of "black boxes" where information is missing from equipment vendors. The HAZOP record is likely to note issues to be referred to the "vendor package" HAZOP.
- After vendor selection the P&IDs and the control system can be matched to the vendor's arrangements including machinery protection trips. This may be the earliest point at which Cause and Effect diagrams for vendor trips can be compared to proposed ESD and plant or plant section protections. It needs to be recognized that the reliability of the plant and its ability to meet the owner's service factor expectations can be materially affected by choices at this stage and these come late in the project schedule when changes imply a high cost and schedule penalty.
- Once in operation a full documentation including the as built arrangements is available. In practice operators may already have made some changes, for example removing or overriding some interlocks and trips which inhibited start-up. Some additional maintenance or start-up lines may also have been added either for convenience or to make good oversight in the earlier stages of the project. Two issues arise in the HAZOP at this stage, firstly diagrams which are not actually "as built" and need to be modified and secondly the operator's typical belief that a potential hazard which they themselves have not experienced is not likely to happen on their plant.

The main practical point from this discussion is that HAZOP study at any stage faces some difficulties. In the EPC/Vendor/Operations Phase these issues can be managed by effective project control, particularly scheduling. In the case of Conceptual Design and FEED the difficulties may be inherent if HAZOP is attempted with limited technical definition. In these circumstances other techniques may be more effective and this paper describes one such alternative approach.

INDEPENDENT HAZARD REVIEW

The independent review is based on the use of a workshop firstly to review the major hazards and then to assess if the proposed risk controls are appropriate.

Rather than piping and instrumentation diagrams (P&IDs), the review works using Block Flow Diagrams and Diagrams of Pressure, Temperature and Composition to highlight the main process hazards which must be managed. The PTC diagram identifies each of the major items of equipment and shows the temperature, pressure and material characteristics as shown in Figure 3.

For each item of equipment such as pumps, heat exchangers, furnaces, reactors, compressors and so forth typical failure modes can be readily identified based on previous HAZOP studies and on accident accounts. The block flow diagram can be annotated with these main potential hazards and their significance judged by reference to the P-T-C chart.

For any particular item of equipment for example Heat Exchangers, generic arrangements and options for level of protection can be drawn up using analyses from previous HAZOP studies.

During the review workshop we check whether the level of protection proposed for all main potential hazards in each section of the Block Flow Diagram is appropriate. Where the protections are found to be lacking or inadequate, changes will be recommended. These are recorded in the same systematic manner as used for HAZOP studies.

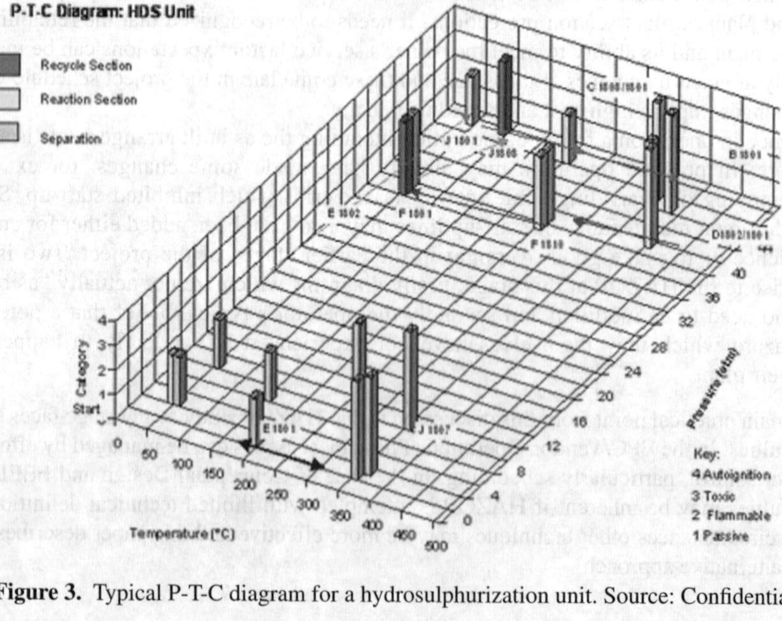

Figure 3. Typical P-T-C diagram for a hydrosulphurization unit. Source: Confidential

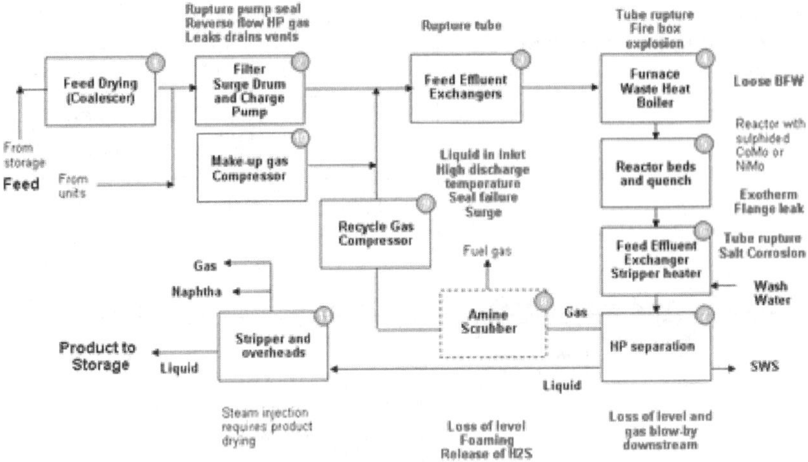

Figure 4. Main hazards in the plant. Source: Confidential

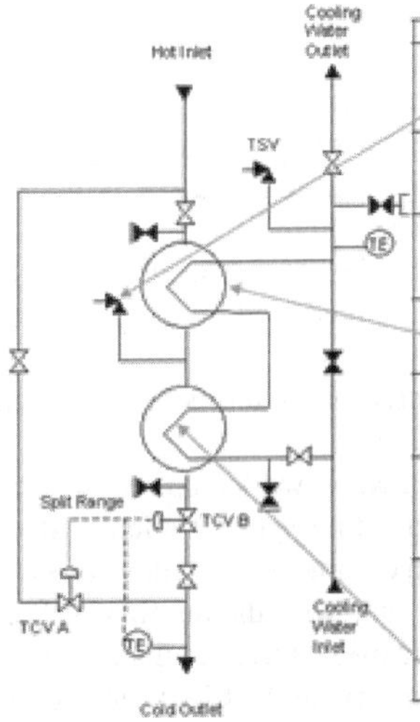

Consequence	Protection principle	Implementation
Shell failure, resulting in loss of containment and release of hydrocarbon	Don't leave exchanger full of liquid when isolated	Maint. Procedure requiring to empty exchanger immediately after isolation Provide PSV for large shells
Shell/tube failure	Design criteria	Design shell/tube for highest upstream pressure Provide PSV
Shell/tube failure	Design criteria	Design shell/tube for highest upstream temperatures TAH on inlet
Shell failure	Design criteria	Open vent during steam out Design shell for vacuum
Tube rupture, which can lead to shell rupture caused by overpressure	Ensure shell can withstand overpressure	Design shell at DP×2/3DP of tube side Provide PSV sized for tube rupture if above criterion not met
Thermal expansion, possibly leading to rupture and loss of containment	Provision of thermal relief if cold side can be intercepted	Provide TRV on cold side; connected to blowdown if HC can be released to atmosphere
Tue leak, leading to contamination of counter flow stream (oil service) or release of hydrocarbon at cooling tower (water service)	Analysis of fluid potentially contaminated Removal of ignition sources (explosion proof electrical equipment at cooling towers)	Analysis (e.g. HC detector at cooling tower) Classified electrical equipment at cooling towers Provide by pass to allow repair without shutting unit

Figure 5. Protection for heat exchanger. Source: Confidential

A main hazards tabulation is developed for each plant Section.....

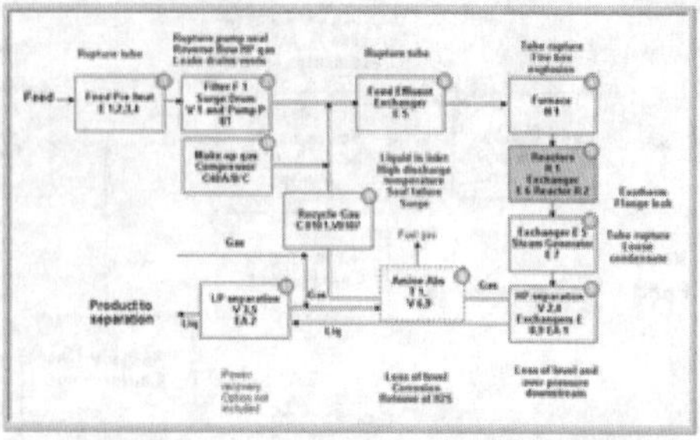

Topic	Specific Issue	Hazard Description	Existing control	Issues/Questions
Hydrogen rich gas leak and ignition	Failure of hydrogen quench line	Line rupture or leak followed by ignition may result in intense flame capable of severe damage	1. Shut C1 to reduce pressure and inventory in the hydrogen line. 2. Keep P1 running to maintain a cooling hydrocarbon flow through the reactors. 3. Cut H1 furnace to reduce the temperature of the hydrocarbon reactor inlet stream. 4. Depresssure	No specific shutdown sequence will proposed to deal with this situation, all leaks in HP section will be managed by initial depressurisation. Basic design requirements should be implemented (smallbore pipework removal, minimisation of flanges, re

Figure 6. Main hazards in the plant. Source: Confidential

Depending on the stage of the project at which the independent hazard review occurs, documentation like PFDs or P&IDs as well as any available hazard identification reports – including HAZOP – will be used as a basis for the review. However, the study will only be focused on protections and controls against the main potential hazards associated with the process being examined.

The hazards are based not only on findings from HAZOP studies conducted on similar plants but also on accident accounts. As for the selection of the most effective prevention and protection measures for specific hazards a useful input can be provided by what other operators have implemented in similar units (see Figure 7).

This review approach allows a Risk Register to be developed in which the Loss Aversion offered by additional actions over and above those proposed by the project team can be assessed [4]. If a demonstration of completeness is required, a Loss Profile simulation can be carried out to demonstrate that the assessment is credible in terms of losses which have occurred elsewhere in the industry[5].

Regarding the resources required, the independent review workshop requires much less time than a full HAZOP study. For example, for a refinery unit typically it takes between one and two days to carry out a review workshop compared to 10 days or so for a HAZOP study. These are indicative times and of course vary with the number of P&IDs requiring review (for HAZOP) and the complexity and intrinsic hazards in the process (for the Review Workshop). The hazards and recommendations arising from a review, embodied in a risk register, require a similar period of one or two days for risk assessment and ranking, if a Risk Assessment Matrix is used.

Location	Date	Loss	Units	Description
A	2000			[illegible]
[illegible]	[illegible]	[illegible]	[illegible]	[illegible]
C	[illegible]	[illegible]	[illegible]	[illegible]
D				[illegible]
D				[illegible]
E				[illegible]

	REF 1			REF 2		REF 3	REF 4	REF 5
	55 bar	74 bar	108 bar	54 Bar	50 Bar	67 Bar	36 Bar	40 Bar
PSV sized full-open valve + 50% BP (PSV placed on LP separator to protect downstream equipment, safety valve correctly sized)	?	?	yes	PSV exists but size unknown	PSV exists but size unknown	PSV exists but size unknown	PSV exists but size unknown	PSV exists but size unknown
BPCS LC (only control valve)	yes	yes	yes	yes	yes	yes	yes e/ C/FAH	yes
Independent LAL								
LSLL on LCV (independent cut-off valve driven by primary element)				Yes, controlled by PAHH on LP separator				yes
LSLL on XV (independent cut-off valve driven by independent instrument)	yes	yes				yes		
2 out of 3 LSLL on XV (as above with 2 out of 3 voting logic)			yes		Yes LCV and XCV both close.			

Figure 7. Accident records and operators survey tabulations. Source: Confidential

One benefit of the Review Workshop approach is that it delivers a balanced allocation of resources between hazard identification, assessment and the evaluation of improved controls. With HAZOP the potential is to spend a disproportionate time on the identification procedure (with the temptation to cut short the assessment of risk and the implementation of actions).

Before adopting a Review Workshop approach there are some important provisos.

1) EXPERTISE

The ability to conduct the Workshop depends on the data available to the facilitator, especially knowledge of the hazards identified in several previous HAZOP studies of similar plant as well as understanding of the root cause of accidents which may have occurred elsewhere on plants of the type being studied. HAZOP study also requires expertise but provided the HAZOP facilitator is experienced with the method, he can rely more on the knowledge of team members regarding process hazards.

2) LIMITATION TO MAJOR HAZARDS

It must be clearly understood that a Review Workshop only addresses major hazards and will not pick up detailed issues which require line by line analysis. Generic matters such as leaks from vents or drains and manual operation such as sampling can be addressed through typical arrangement drawings but there will not be time to check that these have been consistently applied across all plant P&IDs. Similarly, materials selection can be examined as a general topic but there is not the time to check line by line that materials of construction have been correctly selected. For this level of detail, a full recording HAZOP study is recommended for example when P&IDs are to be "Approved for Construction" in situations where no other independent design review is to be conducted.

REFERENCES

1. For example http://www.upop.com/services/2090.html
2. G C Stevens and M Marchi: (June 2001) *A Benefit/Cost approach for prioritising expenditure during plant turnaround* 10th International Symposium on Loss Prevention and Safety in the Process Industries
3. E Tipton, S Mullick and A McBrien (2007) *Focus on integrated FEED* Hydrocarbon Engineering
4. P Beall and G Stevens (2007) *Making HAZOP the method of choice for identification and assessment of process hazards* 9th Process Plant Safety Symposium. AIChE Houston
5. G C Stevens, A Verna and M Marchi, (October 2001) *Getting full value from HAZOP. A practical approach to the assessment of risks identified during studies on process plants.* IChemE Hazards XVI Manchester England

HAZOP FOR DUST HANDLING PLANTS: A USEFUL TOOL OR A SLEDGEHAMMER TO CRACK A NUT?

Alan Tyldesley
Haztech Consultants Ltd, Meridian House, Road One, Winsford, Cheshire

The HAZOP technique was developed by ICI during the 1960s, and has been refined and codified extensively since then, and applied well outside the chemical industry. In ICI, HAZOP was one of six stages of study - from initial design to beneficial operation. It was generally applied when the design was virtually complete and was intended to mop up outstanding safety issues and expose potential operability and maintainability issues. A top-down study usually preceded the HAZOP to expose the most significant potential hazards when it was still possible to modify the design before retrospective changes incurred extra costs. However, HAZOP has comparatively rarely been applied to the powder handling industries, and the intention of this paper is to explore the reasons, and the circumstances in which it perhaps ought to be more widely used.

Choosing the right guide words is important at the outset of a HAZOP study, and there are different lists commonly used for batch type processes, and continuous processes. The powder handling industries have plenty of examples of both types of operation. Examples of essentially continuous processes include flour milling; chipboard manufacture milk spray drying and sewage sludge drying. Batch type operations are commonly found in places which blend different components, such as animal feed mills, and bakeries, and also at the beginning of many processes where powders are brought to site in road tanker loads or introduced to the process from sacks and IBCs. Many sites will have activities of each type. The standard guideword lists might need to be refined at the outset of a study, but this should create no difficulties.

The origins of HAZOP go back to complex chemical plant, which may have chemical and physical processes going on simultaneously, recycle streams and phase changes, together with large amounts of heat being input or withdrawn from the process. In contrast plants handling dry powders are very rarely designed to carry out any form of chemical process, few will have recycle loops, and many run at or close to ambient temperature. Where there are no chemical reactions, serious risks such as exothermic runaways, release of gases, causing the wrong reaction by adding an incorrect component or in the wrong amount, boilovers caused by overheating or loss of mixing in a two phase system are not an issue.

If there is no recycle requirement, some complexities are avoided, and if there is no need to add or remove heat, another major source of problems is eliminated. Put bluntly, most powder handling plants in places which handle food or bulk polymers as powders look like very simple operations to engineers used to designing or running chemical processes.

Powder handling plants do however often have some common hazardous features that can often be avoided in the gas and liquid type of processes for which the HAZOP study was developed. In particular a key example is the extent to which flammable atmospheres are inevitably formed inside the process equipment. Most continuous processes using flammable liquids and gases are designed to exclude air as far as possible, e.g. continuous processes running above atmospheric pressure. Batch processes may be provided with an inert gas blanket to achieve the same aim, while large tanks for flammable liquids may have a floating roof for the same purpose.

The situation is different with powders. The commonest operation run above atmospheric pressure is pneumatic conveying, and this nearly always uses air. Somewhere in the system a flammable atmosphere is bound to form, not only during continuous running, but perhaps at different locations during start up or shut down. Furthermore, many processes are far from completely enclosed, making inerting an unattractive option, and this is not likely to change as long as low value materials are handled.

Adding to the problem, is difficulty of excluding ignition sources from many processes. Sometimes these arise directly from the product or powder itself, which may gradually self heat in the presence of air until at some point flaming combustion is possible. Milk powder and many vegetable products containing unsaturated oils are prone to this phenomenon.

Just as common is the presence of mechanical ignition sources. Static clumps of powder in contact with moving parts in a screw conveyor or mill are liable to get hotter through frictional processes. Better design of the process or equipment can reduce this problem, but it cannot be eliminated.

A third ignition source that can be hard to exclude is static where the powder is highly insulating, since any movement of the powder will generate charge. The methods of avoiding this are often simple, but lack of attention to detail of earthing has caused many incidents.

Faced with the widespread extent of flammable atmospheres inside the process, many dust handling plants are designed on the basis that sooner or later, an ignition source is likely to arise, and we must cater for the consequences of an ignition. The widely adopted techniques of explosion venting and suppression are in effect an admission that explosion prevention cannot be achieved. This is backed up by explosion isolation, which is a recognition that flammable atmospheres or at least the powder and air components that have the potential to create a flammable atmosphere can spread far through a system and make it very vulnerable to extensive damage far from the site of ignition. This approach is set out in the I Chem E book on dust explosion prevention and protection, which discusses the choice of a basis of safety, limiting these effectively to venting, suppression, inerting or control of ignition sources. There is no reference here to the use of the HAZOP tool.

From the point of view of the HAZOP study carried out to assist in designing a safe plant, it might be argued that there is no need to go looking for all the process deviations that might cause a hazard, if we have designed all the parts of the plant to cope with the worst case event, i.e. the ignition of a flammable atmosphere. It is a resource intensive process, and needs quite a range of expertise. The value will be easily lost unless issues identified are discussed until there is an agreed conclusion, and suitable recording of this.

I don't think that these sorts of considerations control how often HAZOP is used in the powder industries. Instead it is more likely a consequence of the training and mindset of those who design powder processes, but it does give us a reason to pause and wonder what this design tool has to offer.

I would argue that even if we intend to build in comprehensive explosion venting or suppression, the designer should aim to minimize the risk of ignition, and that is likely to require consideration of unintended events or conditions. Well designed explosion protection will prevent danger to operators, or limit plant damage to trivial levels, but there remains a downtime cost of cleaning out burnt material, and checking that all is safe to restart.

A HAZOP study is intended to stand back from looking at what is intended to happen in a process, and ask what could cause a deviation from this, and what would be the consequences – WHAT IF? If only a few deviations from the intended condition can be envisaged it is likely that these will be considered without the need for a special study. In the case of a very manual operation like adding sacks of product into a blender making a two or three component mix, and bagging off the blend, a HAZOP would surely be an unnecessary tool. Likewise a local dust collection system, with a single filter and modest number of collection hoods is unlikely to merit a study.

Many continuous powder handling processes have a whole series of linked operations. As an example, from the chipboard industry we would expect to see sequentially logs turned to chips, the chips being dried, the dried product being graded, separate streams for dust, and chips of different sizes, some intermediate storage, and perhaps dust going to a bagging unit, or combustion plant, while the chips go on to be mixed with binder and turned to board. Dust will be present to a different extent in different places through the process. Not all these steps have explosion risks, but all have operability issues if one step in the process or item of equipment goes wrong. Are we sure we know how to get the process running again if power is lost to one of the conveyors, or how some flow blockage will be detected before it causes such a back up of material that the whole plant has to be shut down?

Among the statistics of dust explosion incidents, drying plants feature quite prominently. The most serious incidents are associated with continuous processes, largely because these are typically much larger, but fires and explosions in batch driers are not uncommon.

The hazards associated with driers are no surprise, as soon as you start to heat material you take it closer to the temperature at which ignition occurs. Monitoring temperatures is often difficult, as large temperature gradients can exist within a small distance in the absence of any convectional mixing, and the low thermal conductivity of most bulked powders. What sort of deviations might we need to consider? The volatile content of the material being dried can vary, and this will affect the flow properties; the residence time in the heated zone may be affected by process upsets elsewhere; and there is always the probability that some material may be heated longer than intended. Leaks from or into the system may create flammable atmospheres where they were not intended. It is probable that a HAZOP for a continuous drying process will highlight issues like this needing to be resolved.

It has long been recognized that dust explosions have the ability to spread through a process plant from unit to unit, even though there is probably not an explosive atmosphere throughout the plant at the time of ignition. The shockwave from an initial small explosion disturbs dust, and creates new dust clouds. The solution to this is to provide explosion isolation, but that immediately begs the question of where, and how? The practical reality is that it is often not possible to isolate every item of plant, and moreover, that many of the isolation techniques are not 100% effective. Rapid acting isolation valves may react too slowly, rotary valves may have clearances worn until they allow flame to pass, or water spray systems linked to spark detectors may fail if a spray head is blocked.

What is needed is an overview of the process, looking at the plant as a whole, and an assessment of the probability of particular deviations which could result in a dust explosion which propagates with serious consequences. This type of HAZOP might well be most effectively carried out at the outline design stage. Suitable instrumentation to detect the deviation might be more cost effective than an explosion barrier whose reliability was debatable.

Start up and shut down of continuous plant with multiple unit operations always needs consideration at the design stage, whether the plant has fully automated controls, or is largely under the control of the operators. In particular, the sequence required for an unplanned or emergency shutdown must not be overlooked.

A large coal mill associated with a power generation or cement plant is likely to run for extended periods, and generate considerable heat. The heat will be carried away in the milled product, and may dissipate in the downstream process. One normal way of operating a controlled shutdown is to run the mill feed to empty, and remove residual product in the mill through an inspection port into a movable hopper. In the case of an unplanned stoppage, coal will remain in the mill, and may draw heat from the casing. If the stoppage is brief, restart may cause no problems, but if the restart is delayed, the coal may have started to smoulder, and burst into flame when fresh air is blown through the system. When the plant is first tripped, the operators may not know whether it can be restarted in a few minutes, or whether much longer will be needed to rectify the problem. Different actions may be needed in the two cases, and good operating instructions will be needed, rather than any change in plant design. A HAZOP is not the only way of identifying this issue, but done properly, it might well ensure that the potential problem is not overlooked.

One cause of hazards not associated with combustion is overpressure caused by a deliberate supply of compressed air, when not all the parts of the plant can withstand the full pressure of the air supply, or where the air pressure can produce unintended consequences.

This was the case with the very old incident at General Foods in Banbury, where the initial problem was caused by a pneumatic conveying system. A single transfer system was used to fill multiple bins, and a failure in a diverter valve caused one bin to be overfilled. The filter on the bin soon blocked, and then air pressure from the conveying system caused the filter unit to become detached from the bin. A large dust cloud quickly formed, and ignited, creating an explosion which caused substantial damage.

There have been similar more recent incidents where we have seen filters mounted on silos become detached but the consequences have not been as severe, because the silos were outside.

A different problem with the same type of origin came to light following incidents of ignition within pneumatic tanker discharging systems. Typically these use top pressure above the powder in a road tanker, and a venturi system at the discharge point to fluidise the powder and induce flow in the transfer line. It transpired that in the case of the fire incidents, the operator had turned the venturi off towards the end of a transfer, and left the top pressure in the tanker to empty the last of the load. This had the unintended consequence when the tanker barrel emptied of causing reverse flow in the venturi and powder entered the blower. Subsequently friction in the blower caused burning material to be blown down the transfer line.

A particular generic example comes from dense phase conveying where the air pressures used in the blow egg are much greater than those found in a lean phase system, typically up to 7 bar. There is a steady drop in pressure along the transfer line in normal operation. The amount of air to be disengaged is comparatively small, but it has to go somewhere. If air at 7 bar entered an empty transfer line, what would happen? If the body of the downstream filter equipment could not withstand anything like this pressure, could the filter elements pass the amount of air that would flow? Would this still be the case with a partly blocked filter? It is clear that robust arrangements to prevent weak plant being overpressurised are necessary. It seems to me that HAZOP is a useful tool to consider this type of operational problem, as it forces the designers of the system to consider what could happen when all does not operate as intended.

Many powder plants suffer from flow blockage from time to time, and often the first reaction from the operator is to get the hammer out, and vibrate the accumulation until it is displaced. A flow blockage may have no safety consequences, but it will have to be moved sooner or later. A HAZOP might help identify the causes of a reduction of flow, and from the discussion identify improvements to the instrumentation which would identify problems before flow stops completely.

Most powder handling systems are essentially enclosed to prevent major release of dust to the surroundings. Any major release is likely to create a dust cloud and the risk of an explosion. I have often asked operators of powder handling plant how they would detect a release from some fault such as a failed filter element, flexible coupling, or torn explosion vent panel. This may not be an issue if the area is constantly supervised, but if plant is highly automated, and the operator rarely leaves the control room, it is necessary to consider the adequacy of the instrumentation. A HAZOP study might be appropriate, and it could be done either at the design stage or on an existing plant.

CONCLUSIONS

The title of this paper set a question, and I'd like to draw together some conclusions.

There is no general reason why the HAZOP technique should be seen as unnecessary or unsuitable for application to powder handling plants.

Many powder handling processes are comparatively simple, and may not merit the effort of a HAZOP study. This is most likely to be the case where, there is only a single unit operation involved, where there are no particular problems associated with stopping or starting the process at any point, and where the process operates at essentially atmospheric temperature and pressure.

The traditional approach to mitigating the consequence of dust explosions by explosion venting or suppression together with explosion isolation between items of plant will enable the most serious potential risks to be controlled, but used alone will fail to identify potential ways of improving the design to reduce the risk of a process upset which has the potential to lead to a fire or explosion incident. HAZOP will be appropriate in some cases.

Selective HAZOP's, using appropriate guide-words (prompts), applied ideally at an early stage of design may be useful as a means of identifying hazards in powder handling plants other than those caused by fire and explosion events, specifically the implications of using compressed air together with some plant or equipment that is not built to pressure vessel standards.

HAZOP is particularly useful in considering multistage processes, where the wrong sequence of operations during start up or shut down can create problems.

HAZOP is likely to be useful when considering the potential for unintended releases from the process, and how these would be identified and controlled promptly.

Most particularly, a HAZOP study on a process or design that incorporates all the normal dust explosion precautions is likely to identify issues that are more operational than safety related, and a study could usefully be undertaken with this as the primary aim.

REFERENCES

Corn starch dust explosion at General Foods Ltd, Banbury, November 1981, HMSO ISBN 011 8836730

BS IEC 61882 : 2001 Hazard and Operability Studies, Application Guide

Fire and Explosion hazards during pneumatic discharge of road tankers handling explosible dusts HSE internal circular see www.hse.gov.uk/foi/internalops/fod/oc/200-299/283_14.pdf

Dust Explosion Prevention and Protection, a practical guide, ed John Barton, I Chem E, 2002 ISBN 085295-410-7

A RULE-BASED SYSTEM FOR AUTOMATED BATCH HAZOP STUDIES

C. Palmer[1], P.W.H. Chung[1], and J. Madden[2]
[1]Department of Computer Science, Loughborough University, Leicestershire, LE11 3TU, UK
[2]Hazid Technologies Ltd, Beeston, Nottingham, NG9 2ND, UK

The hazard and operability study technique (HAZOP) is widely-used for identifying potential hazards and operability issues in process plants. To overcome the repetitive and time-consuming nature of the technique, automated hazard identification systems that emulate HAZOPs for continuous plants have been developed. This work considers batch processes, in which material undergoes processing in distinct stages within the plant equipment items according to a set of operating procedures, rather than each equipment item remaining in a "steady state", as is normal for continuously operating plants.

In batch plants deviations which can lead to hazards can arise both from deviations from operating procedures and process variable deviations. Therefore, the effect of operator actions needs to be considered.

CHECKOP is an automated batch HAZOP identification system being developed as a joint project between HAZID Technologies Ltd and Loughborough University. To produce a product in a batch process, a plant operator follows a series of operating instructions. For an operating procedure to be analysed by a computerised system, such as an automated HAZOP system, it must be formally represented. Deviations may be applied to the operating instructions to simulate batch HAZOP. CHECKOP's simulation engine applies the operating instructions to the plant configuration model, which describes the plant connectivity and state Each operating instruction acts to change the state of the plant. The rule-based system tests for potential hazards which result from the operating instructions and their effect upon the plant model.

This paper will focus upon describing the rule-based system. The rules may be categorised according to whether incompatible equipment state or incorrect operation is being investigated. Generic rules may be derived. Examples of incorrect plant operation, which cause the rules to be activated, will be shown. Future development of the system will be described.

INTRODUCTION

A widely used hazard identification technique within the process industry is HAZOP (hazard and operability study). A study considers all possible deviations of a plant from its intended operation, by using deviation guidewords (No, More of, Less of, Part of, Other) applied to each of the process variables (flow, pressure, temperature, etc.) in the plant in turn (Lawley, 1974). The HAZOP of batch processes requires extra time-related or order-related guidewords (Early, Late, Before, After, Quicker, Slower) which are utilised by introducing a particular error into an operating procedure (Bickerton, 2003; Mushtaq and Chung, 2000). The HAZOP technique seeks to identify the hazard or operating consequences arising from the deviation. Postulating the interaction between deviation, consequence, hazard and performance demands the use of expert engineering knowledge.

To overcome the repetitive and time-consuming nature of the technique, automated hazard identification systems that emulate HAZOPs have been developed. A significant recent innovation in the automation of the continuous plant HAZOP procedure is the Hazid System from Hazid Technologies Ltd. Hazid combines a sophisticated knowledge base with a fault-propagation engine to apply deviations and propagate their effects throughout the complex plant networks.

Much of the research on automated HAZOP identification, based on signed-directed graphs, has concentrated on continuous plants (McCoy et al, 1999a and 1999b; Venkatasubramanian & Vaidhyanathan, 1994). However, very little work has been done in automated hazard identification of batch plants. The limited work done is based on the Petri-net representation (Kang et al, 2003; Srinivasan & Ventkatasubramanian, 1998a and 1998b). In batch processes the plant operation moves through a number of stages, rather than each equipment item remaining in a "steady state", as is normal for continuously operating plants. In batch plants deviations which lead to hazards can arise both from deviations from operating procedures and from process variable deviations. The effects of operator actions need to be considered.

The signed-directed graph approach used in continuous plant HAZOP emulators is found to be unable to capture the information needed to consider the deviations in operating instructions necessary for the HAZOP of batch plants. The main problem is that it does not keep state related information as the HAZOP analysis moves from one operating instruction to the next. A method of representing a sequence of actions (operating instructions) to achieve a state is needed.

This paper describes CHECKOP, a prototype automated batch HAZOP identification system. CHECKOP uses a state-based approach to HAZOP analysis. The paper commences by describing a simple batch plant case study. CHECKOP contains a rule-based system which identifies potential hazards or operability issues. This paper will focus upon discussing the rule-based system. Examples from the case study are used to illustrate the rules.

A SIMPLE EXAMPLE PLANT

In order to demonstrate how CHECKOP is used to aid batch HAZOP of a process, the simple batch processing plant illustrated in Figure 1 is considered. The plant consists of a reactor, two feed tanks and a product tank. The reactor shown produces a product *P* from two reactants *A* and *B*, using the simple chemical reaction $A + B \rightarrow P$. An excess of reactant *B* is used so that reactant *A* is completely consumed in the reactor.

OVERVIEW OF CHECKOP

CHECKOP is being developed as a joint project between Loughborough University and Hazid Technologies Ltd. McCoy et al (2006) describe an early version of the system.

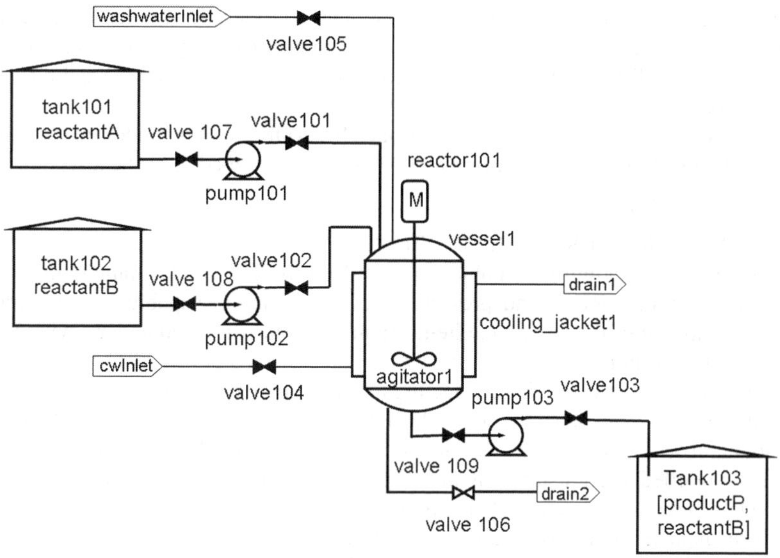

Figure 1. A simple batch processing plant example

To simulate a batch HAZOP CHECKOP systematically applies the HAZOP deviation guidewords to the operating procedure. CHECKOP infers the consequences if a certain instruction in the procedure is not executed, or if the instruction is carried out too early or too late, etc. A report is produced providing warnings against any undesirable situations that may result from the deviations.

CHECKOP's system model consists of four sections:

- an object-oriented plant configuration model
- the operating procedures
- a state-based simulation engine
- a rule-based system

The plant configuration model describes the plant connectivity and state. The topographical relationships of the plant model are derived from the plant piping and instrumentation diagram (P & ID). Initially, the plant is specified to be in its "idle" state with all valves closed and pumps stopped. Operating instructions act to update the states of the equipment items.

To produce a product in a batch process, a plant operator follows a series of operating instructions. For an operating procedure to be analysed by a computerised system, such as an automated HAZOP system, it must be formally represented. Deviations may be applied to the operating instructions to simulate batch HAZOP. The simulation engine applies the operating instructions to the plant configuration model. Each operating instruction acts to change the state of the plant. The rule-based system tests for potential hazards or operability issues which result from the operating instructions and their effect upon the plant model.

Brief details on the first three sections of CHECKOP's system model are given below. The main focus of this paper, the rule-based system, is described in the next section.

THE PLANT CONFIGURATION MODEL

CHECKOP employs a unit-based object-oriented approach to model plant items and their connectivities, temperatures and pressures. This approach is capable of predicting the dynamic behaviour of the equipment items, in normal operation and under deviation from normal plant operation.

Process plants are built by connecting together smaller sets of units to carry out the required functions. The behaviour of each of these types of units can be modelled generically so that it will apply to any plant in which the unit is used. Each item of equipment in a plant is modelled as an instance of an equipment model, taken from a library of process unit models, which forms a knowledge-base.

The plant is modelled as a unit model which is itself composed of unit models. These unit models are the equipment instances. Instances are assigned a unique identifier. The units can be composed of sub-units. The generic models of the model library may also contain instances. This enables information re-use within the model library. For example, the model library may contain a reactor model, composed of instances of a vessel, an agitator and a cooling jacket. The instances are identified, using object-oriented notation, as: reactor.vessel1, reactor.agitator1 and reactor.coolingjacket1 (see Figure 2).

The unit models contain connections, attributes and actions. Connections enable the ports of equipment instances to be linked together to form a plant model. Each plant equipment instance has a state determined by the current value of its attributes. Actions alter the state of equipment instances by changing attribute values. The connection information and some attributes (e.g. pressures and temperatures) for the plant model are derived from the plant P & ID. Further attribute information and the actions permitted for each equipment item are contained by the generic models of the model library.

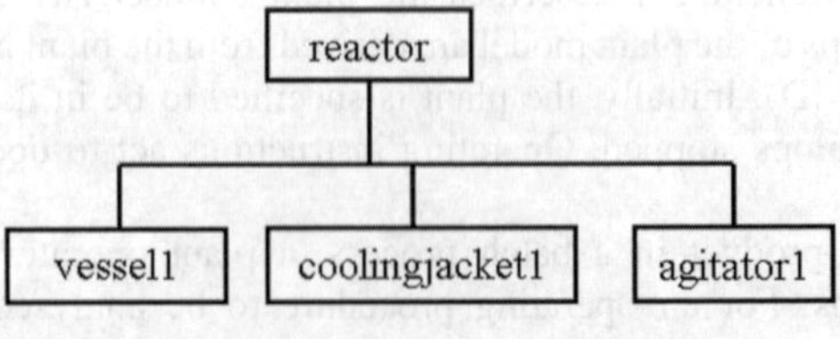

Figure 2. An example unit model

THE OPERATING PROCEDURES

To safely produce a product in a batch process, a plant operator follows a sequence of operating instructions. Each operating instruction directs one or more actions to change the state of the plant. As a plant operation moves through a number of stages, the states of the plant equipment instances are updated. Implicit in the operating instructions is the existence of a complete plant model representation.

An operating procedure is a sequence of operating instructions. For example, to produce productP the simple batch plant would require the following instruction sequence:

```
1) charge reactor101 with reactantA from tank101 until reac-
   tor101.liquid_amount = 30 %vol/vol,
2) operate reactor101.agitator1,
3) cool reactor101.cooling_jacket1 until reactor101.cooling_
   jacket1.liquid_temperature < 25 Celsius
4) charge reactor101 with reactantB from tank102 until reac-
   tor101.liquid_amount = 60 %vol/vol,
5) discharge reactor101 with reactantB, productP to tank103,
6) stop reactor101.agitator1,
7) shut_down reactor101.cooling_jacket1,
```

Each instruction has an implicit set of assumptions about the state of the plant. For example, the instruction "`Charge reactor101 with reactantA from tank101 until reactor101.liquid_amount = 30%vol/vol`" assumes that tank101 is a source of reactantA, is connected via a flow path to reactor101 and contains a sufficient quantity of reactantA to fill reactor101 to a level of 30%. The operating instructions are composed using a set formal template representation. For details of the formal template representation to describe operating procedures see Palmer et al (2006).

An instruction may be an action primitive or be composed of more instructions or action primitives. The following action primitives exist: open, close, operate, run, stop, check and wait. An example of a more complex instruction is "`charge reactor101 with water from washwaterinlet until reactor101.liquid_amount = 65 %vol/vol`" which is composed of the following action primitives:

```
open valve105,
wait until reactor101.liquid_amount = 65 %vol/vol,
close valve105
```

An action primitive operates on an equipment instance. For example, the action primitive "`open valve105`" acts on the equipment instance "valve105". This action primitive creates a flow path between the washwaterinlet and reactor101.

An instruction or action primitive may optionally contain a condition, such as "`until reactor101.level_amount = 30 %vol/vol`", which indicates when completion occurs. Identifiers (e.g. reactor101) link the operating instructions and their constituent action primitives to the instances within the plant configuration model.

THE STATE-BASED SIMULATION ENGINE

Relating the operating instructions to the plant configuration model allows the state of the plant configuration model to be changed at each stage of operation. An action primitive denotes which of the actions of the equipment instance will be executed by the simulation engine. For example, "`open valve105`" indicates that the action "open" contained by the equipment instance "valve105" will be executed.

An action updates the plant model state by changing the attribute values of equipment instances. The new attribute values may be contained by the action model or the condition of the action primitive.

The new plant state resulting from an action may necessitate further changes to the plant model. For example, applying the action "`open valve105`" to the simple batch plant will create a flow path between the washwaterinlet and reactor101 (see Figure 1). This flow path will allow water to transfer from the washwaterinlet to reactor101. After each action is executed the simulation engine tests for the presence of flow paths and updates the attributes describing the contents of connected equipment instances.

THE CHECKOP RULE-BASED SYSTEM

This section describes the CHECKOP rule-based system and explains describes how the rules are structured and categorised. Examples of the rules are given, listed within their categories. Instructions which relate to the simple batch plant are given to demonstrate how the rules are utilised. Examples of incorrect plant operation, which cause the rules to be activated, will be shown.

Given an operating procedure, which may or may not be complete or correct, the rule-based system verifies if it achieves its desired results and does not also lead to any additional, unexpected effects. Any potential problem identified will be reported. Formalising the operating procedure allows alternative orderings of the operating instructions to be considered. This allows the procedure to be modified. Simulation demonstrates the effect of the modified procedure on the plant model. Operating procedure deviations may be generated by automatically applying the HAZOP guidewords to the operating instructions. The system rules capture the important effects of the deviation for hazard reporting.

A rule-based system is well suited to represent the complex, unstructured knowledge required to test for potential hazards or operability issues which result from the operating instructions. The rule-based system is simple to understand and flexible. Existing rules may be changed or new rules may be easily added. This allows the system to be updated if new information becomes available as a plant design develops or to be adapted to specific plant configurations.

A rule consists of the structure "If ... Then". For example,

> If the instruction is to charge a reactor and the reactor does not contain space,
> Then indicate a hazard or an operation problem.

If the stipulations of a rule's "If" section are met, the rule is said to be "activated" or "fire".

Three types of rule exist within the CHECKOP rule-base:

1. Generic
2. Specific
3. Independent

A rule may be classified as "generic" if it relates to more than one action primitive or instruction. For example, a generic rule could relate to the action primitives "open" "close" and "operate". Another generic rule could relate to both a "charge" and a "discharge" instruction. Generic rules are not employed to test the operating procedure but form patterns from which rules specific to a certain instruction may be derived. Independent rules do not depend on the structure of another rule. Independent rules relate to the equipment instances referred to by an operating instruction. Specific and independent rules are used to test the operating procedure. Currently the rule-base contains approximately thirty rules.

The rules may be categorised according to whether they investigate:

- incorrect operation
- incompatible equipment state.

These categories may be further divided depending upon whether a rule relates to an action primitive or to a more complex instruction which contain further instructions or action primitives. Rules which test for incorrect operation are applied before the operating procedures are simulated as plant state information is not required. Rules which investigate incompatible equipment state examine the set of assumptions each instruction contains about the state of the plant. As the simulation engine updates the state of the plant model for each instruction encountered, rules which investigate incompatible equipment state must be applied with each instruction.

If one of CHECKOP's rules is activated then a low level warning or a more important error message is issued to indicate a hazard or operation problem depending on how serious its potential effects are. The warning or error message identifies the nature of the hazard or operation problem and whereabouts in the operating procedure the problem occurs. For example,

```
Operation number 2: Charge cannot proceed as reactor101 is
   full
```

The rules described in the following sub-sections issue error messages unless stated otherwise.

RULES INVESTIGATING INCORRECT OPERATION

Incorrect operation occurs when the sequence of operating instructions is performed in the wrong order, an operating instruction is omitted or an extra instruction is executed.

Incorrect operation results from oversights in the operating procedure or operator error. In addition to causing operability issues, incorrect operation may also lead to hazardous consequences.

Rules applicable to action primitives

Rule type 1:

If an action primitive is to be performed and an action primitive which reverses its state does not exist within the operating procedure Then indicate a hazard or an operation problem.

Example rules of this type:

1. If open an equipment instance and a later instruction to close it does not exist within the operating procedure Then indicate a hazard. For example, if the instruction "`open valve101`" is found within the operating procedure, a later instruction "`close valve101`" must also occur within the procedure otherwise this rule will fire. A low level warning is generated if the action primitive to close the equipment instance is not found within the same instruction as the one which contains the action primitive to open the equipment instance but is found later within operating procedure, as this instruction sequence may allow an unexpected or extraneous flow to occur.
2. If operate an equipment instance and a later instruction to stop it does not occur within the operating procedure Then indicate an operation problem. For example if the instruction "`operate pump101.pump_drive`" is found within the operating procedure, a later instruction "`stop pump101.pump_drive`" must also occur within the procedure or this rule will be activated.

Rule type 2:

If an action primitive is to be performed and it reverses the state of the action primitive which occurs immediately prior to it Then indicate an operation problem. Rules of this type issue low level warnings if activated as a problem may not be conclusively indicated.

Example rules of this type:

1. If close an equipment instance and the immediate previous instruction is to open the equipment instance Then indicate an operation problem. For example the instruction couplet,

```
open valve101,
close valve101,
```

would cause this rule to fire.

2. If stop an equipment instance and the immediate previous instruction is to operate the equipment instance Then indicate an operation problem. For example, this rule would be activated by:

```
operate reactor101.agitator1,
stop reactor101.agitator1,
```


Rule applicable to complex instructions
Rule type 3:
If an instruction X occurs within the operating procedure and a previous instruction Y does not occur Then indicate a hazard or an operation problem. For example, if an instruction "`discharge reactor101`" is present in the operating procedure and a previous instruction "`charge reactor101`" does not occur within the procedure. If Y does occur but an instruction occurs in the operating procedure between Y and X which negates the effect of Y Then indicate hazard or an operation problem. For example the following instruction sequence would cause a specific expression of this rule to fire,

```
charge reactor101 from tank101, (instruction Y)
discharge reactor101 to tank103,
shut_down reactor101.cooling_jacket1,
discharge reactor101 to drain2, (instruction X)
```

Example rules of this type:

1. If discharge an equipment instance and a previous instruction to charge the equipment instance does not occur Then indicate an operation problem. Implicit in the "discharge" instruction is the assumption that the equipment instance is discharged until it is empty.
2. If wash an equipment instance (X) and a previous instruction to discharge the equipment instance (Y) does not occur Then indicate a hazard. If a previous instruction to discharge the equipment instance does occur but an instruction to charge the equipment instance occurs in the operating procedure between Y and X Then indicate a hazard. The following sequence of instructions would cause this rule to fire,

```
discharge reactor101 to tank103, (instruction Y)
shut_down reactor101.cooling_jacket1,
charge reactor101 from tank101,
wash reactor101, (instruction X)
```

RULES INVESTIGATING INCOMPATIBLE EQUIPMENT STATE

An incompatible equipment state may occur when an equipment state does not match that given in the operating instruction or conflicts with plant safety. Rules investigating incompatible equipment state compare the state of the plant model with the state of the plant as implied or defined by an instruction. For example the instruction, "`Charge reactor101 with reactantA from tank101 until reactor101.liquid_amount = 30%vol/vol`" implies that tank101 must contain enough reactantA to fill reactor101 to a volume of 30%. This instruction defines that upon completion reactor101 should contain a volume of 30% reactantA. Rules investigating incompatible equipment state fall into two groups: those which are applied before an instruction is simulated; and those applied after its simulation.

RULES APPLIED BEFORE AN INSTRUCTION IS SIMULATED

Rules applicable to action primitives

Rule type 4:

If action primitive brings about an existing equipment state Then indicate an operation problem. This rule indicates an instruction included in the procedure is deemed unnecessary.

Example rules of this type:

1. If operate an equipment instance which is already operating Then indicate an operation problem. For example this rule will fire if the instruction "`operate pump101.pump_drive`" is applied to a pump_drive instance, pump101.pump_drive, which is already of state "operating".
2. If stop an equipment instance which is already stopped Then indicate an operation problem.
3. If open an equipment instance which is already open Then indicate an operation problem.
4. If close an equipment instance which is already closed Then indicate an operation problem.

Rule type 5:

If an action primitive occurs in the operating procedure where the plant is not in an appropriate state for the action to take place then indicate a hazard.

Example rules of this type:

1. If the instruction is to operate an instance of a pump drive, the adjoining suction valve instance is not of a state "open", or the adjoining discharge valve instance is not closed or the following instruction does not open the discharge valve Then indicate a hazard. For example, considering pump101 in the simple batch plant, if an action primitive "`operate pump101.pump_drive`" is found within the operating procedure, the suction valve, valve107, must be open, the discharge valve, valve101, must be closed and the next instruction in the procedure must be "`open valve101`", otherwise this rule will fire.
2. If the instruction is to stop an instance of a pump drive, the adjoining suction valve instance is not of a state "open", or the adjoining discharge valve instance is not closed or the following instruction does not close the suction valve Then indicate a hazard. For example this rule will be activated if an action primitive "`stop pump101.pump_drive`" is to be performed, unless the suction valve, valve107, is open, the discharge valve, valve101, is closed and the next instruction in the procedure is "`close valve101`".

Rules applicable to complex instructions

Rule type 6:

If the material defined in the instruction is not the same as that contained in the source equipment instance Then indicate an operation problem. For example, if the instruction "`Charge reactor101 with reactantA from tank101`" is to be performed and tank101 does not contain reactant A then a specific version of this rule will be activated. This rule applies to instructions which effect the transfer of material.

Example rules of this type:

1. If the equipment instance supplying the material for a charge instruction does not contain the same material as defined in the instruction Then indicate an operation problem. This rule needs to check the contents of the second equipment item described in the instruction. E.g. for "`charge reactor101 with reactantA from tank101`", checks that tank101 contains reactantA.
2. If the equipment instance supplying the material for a discharge instruction does not contain the same material as defined in the instruction Then indicate an operation problem. This rule verifies the contents of the first equipment instance described in the instruction. E.g. if the instruction occurs "`discharge reactor101 with productP to tank103`" in the operating procedure, checks that reactor101 contains productP.

Rule type 7:

If the source equipment instance does not contain a sufficient quantity of material to fulfil a condition defined in the instruction Then indicate an operation problem. E.g. for the instruction "`charge reactor101 with reactantA from tank101 until reactor101.liquid_amount = 30 %vol/vol`" a specific expression of this rule will test that tank101 contains enough material to fill reactor101 to a volume of 30%. This rule also applies to instructions which effect the transfer of material. Like the previous generic rule described above, two specific expressions of this rule exist: one which tests a charge instruction and one for a discharge instruction.

Rule type 8:

If the capacity of the sink equipment instance is less than the value of a condition defined in the instruction Then indicate a hazard. For the example, in the instruction "`Charge reactor101 with reactantA from tank101 until reactor101.liquid_amount = 15 tonnes`" the volume of reactor101 should be greater than or equal to 15 tonnes. Again, this rule applies to instructions which transfer material between two equipment items and two specific versions to the rule exist: one for a charge instruction and one for a discharge instruction.

Rule type 9:

If the condition of the instruction is not consistent with that of the equipment instance referred to by the condition Then indicate an operation problem. For example, a specific expression of this rule will fire if the instruction "`charge reactor101.vessel1 from tank101 until reactor101.liquid_amount = 30 %vol/vol`" occurs within the operating procedures and the volume of reactor101.liquid_amount is greater than 30%. This rule indicates that an instruction is unnecessary or that it is occurring out of sequence.

Example rules of this type:

1. If the instruction is "charge" and the value of the condition of the instruction is greater than that of the equipment instance referred to by the condition Then indicate an operation problem.
2. If the instruction is "cool" and the value of the condition of the instruction is less than that of the equipment instance referred to by the condition Then indicate an operation problem. For example, this rule will activate if the instruction "`cool reactor101.cooling_jacket1 until reactor101.cooling_jacket1.liquid_temperature < 25 C`" is to be applied to the batch plant and the value of reactor101.cooling_jacket1.liquid_temperature is already less than 25 C.

RULE APPLICABLE AFTER AN INSTRUCTION IS SIMULATED

Independent Rule Applicable to a Complex Instruction:

If a condition defined in an instruction is not achieved Then indicate an operation problem. For example when the instruction "`Charge reactor101 with reactantA from tank101 until reactor101.liquid_amount = 30 %vol/vol`" is complete, if reactor101 does not contain 30% by volume of reactantA this rule will be activated.

CONCLUSIONS AND FUTURE WORK

CHECKOP's plant model captures the state of the plant's equipment items. By simulating the effect of each operating instruction on the plant model the resulting plant state can be demonstrated and compared to the intended state of the procedure. Formalising the operating procedure allows alternative orderings of the operating instructions to be examined. This enables the procedure to be modified for batch HAZOP analysis.

The CHECKOP rule-base verifies that an operating procedure does not lead to unexpected consequences. Hazards or operability issues are detected which arise if the operating procedure is deviated from. Currently most of the rules detect hazards relating to flow. Further rules need to be added to detect hazards relating to pressure and temperature. The rule-base can be extended to identify hazards for specific plant set-ups.

The CHECKOP modelling system needs to be enhanced to capture chemical information in order that the plant model can be updated when a reaction occurs. This will allow the consequences to be detected of incompatible substances meeting or unforeseen chemical reaction occurring. More case studies are required to assess the capabilities of the rule-base, including more complex plants which utilise sequential or parallel processing.

REFERENCES

Bickerton, J., 2003, HAZOP applied to batch and semi-batch reactors, *Loss Prevention Bulletin*, 173: [1] 10–12.

Hazid. Available at <http://www.hazid.com> [Accessed 06/08/07].

Kang, B., Dongil S. and Yoon, E.N., 2003, Automation of the safety analysis of batch processes based on the multi-modeling approach, *Control Engineering Practice*, 11: 871–880.

Lawley, H.G., 1974, Operability Studies and Hazard Analysis, *Chemical Engineering Progress*, 70: [4] 45–56.

McCoy, S.A., Zhou, D. and Chung, P.W.H., 2006, State-based modelling in hazard identification, *Applied Intelligence*, 24: 263–279.

McCoy, S.A., Wakeman, S.J., Larkin, F.D., Jefferson, M., Chung, P.W., Rushton, A.G., Lees, F.P. and Heino, P.M., 1999a, HAZID, A Computer Aid for Hazard Identification 1. The STOPHAZ Package and the HAZID Code: An Overview, the Issues and the Structure, *Transactions of the Institution of Chemical Engineers*, 77: [B] 317–327.

McCoy, S.A., Wakeman, S.J., Larkin, F.D., Chung, P.W., Rushton, A.G. and Lees, F.P., 1999b, HAZID, A Computer Aid for Hazard Identification 2. Unit Model System, *Transactions of the Institution of Chemical Engineers*, 77: [B] 328–334.

Mushtaq, F. and Chung, P.W.H., 2000, A Systematic HAZOP procedure for batch processes, and its application to pipeless plants, *Journal of Loss Prevention in the Process Industries*, 13: 41–48.

Palmer, C., Chung, P.W.H., McCoy, S.A. and Madden, J., A Formal Method of Communicating Operating Procedures, 2006, *Hazards XIX*, IChemE Symposium Series 151: 448–457.

Srinivasan, R. and Ventkatasubramanian, V., 1998a, Automating HAZOP analysis of batch chemical plants: Part I The Knowledge representation framework, *Computers and Chemical Engineering*, 22: [9] 1345–1355.

Srinivasan, R. and Ventkatasubramanian, V., 1998b, Automating HAZOP analysis of batch chemical plants: Part II Algorithms and application, *Computers and Chemical Engineering*, 22: [9] 1357–1370.

Venkatasubramanian, V., & Vaidhyanathan, R., 1994, A knowledge based framework for automating HAZOP analysis, *AIChE Journal*, 40: [3] 496–505.

A CONSISTENT APPROACH TO THE ASSESSMENT AND MANAGEMENT OF ASPHYXIATION HAZARDS

K. A. Johnson
Sellafield Ltd, Risley

Asphyxiation by inert gases is a hazard throughout the chemical process industries and beyond. Best practice management of hazards associated with access to inerted vessels and confined spaces is well understood and well documented; the hazards associated with service supply lines and other process systems running through occupied buildings are not so well understood. This paper was inspired by the need to find simple, practical approaches to meet industry aspirations for best practice.

This paper presents the use of a zoning type methodology to:

- achieve a consistent approach to assessment and risk reduction;
- focus design such that the hazards are eliminated or minimized;
- assist operators in defining provisions & procedures to manage the hazard.

Risk assessment approaches, analogous to those used for flammability hazards, are proposed to assign zones that consistently identify the level of risk and enable appropriate management methods to be selected and deployed.

This is an application of existing data on leakage rates and dispersion models derived for flammable gases applied in the alternative scenario of oxygen deficient atmospheres. It is an approach that can be applied to any asphyxiation hazards from service and process systems pipework both inside and outside buildings although the focus is for inside buildings.

KEYWORDS: Asphyxiation, Hazards, Area Classification

INTRODUCTION

Asphyxia, the word is from the Greek *a* – meaning "without" and σφυγμός (sphygmos) meaning, "pulse or heartbeat". Asphyxiation is a condition in which the body becomes defiicient in oxygen due to an inability to breathe normally. Oxygen deficiency without remedial action can progress rapidly from diminished mental and physical capacitiy, to unconsciousness, to brain damage and ultimately to death.

Asphyxiation is a hazard encountered on many sites and many industries. Anyone using or generating gases or vapours that can displace oxygen has the potential hazard. When working with asphyxiant materials in confined spaces there are regulations and working practices to manage the situation. (SI 1713 1997) However, there are times when the work place is not apparently confined and the asphyxiation hazard is not immediately apparent. Such was the case when working on the planned review of the safety case for a plant at Sellafield. The plant used nitrogen to manage a flammability hazard within the process. Access to the inerted vessels and the precautions required were, apparently, well understood. However, the problem stemmed from nitrogen supply lines running through rooms within the building. These rooms were ventilated by a cascaded induced draught system designed for radiological containment and not intended to manage asphyxiation hazards. This raised questions about the identification, assessment and management of asphyxiation hazards both inside buildings and in outside locations across the whole site.

This paper reports progress on ongoing work to develop a consistent approach to the identification, assessment and management of asphyxiation hazards initiated by that safety review.

THE NATURE OF THE HAZARD

The key to a successful and elegant hazard management strategy is to understand the nature of the hazard and how it arises. In the case of asphyxiation hazards, this requires:

- An appreciation of the physiology of asphyxiation;
  and
- An understanding of how oxygen deficient atmospheres can arise.

PHYSIOLOGY

Humans need oxygen to survive; too little and humans suffer diminishing physical and psychological abilities with reducing oxygen ultimately leading to death; too much oxygen and humans can die from the accumulation of fluid in the lungs (oedema). A summary of these effects is provided in Table 1. (This is a compilation of data from a number of sources.) The life supporting oxygen range at sea level is 19% to 23%. The transport of oxygen into the blood is a combination of mass transfer and absorption chemistry; the mass transfer element means that partial pressure is the true driving force. This to the chemical engineer explains why people experience problems at altitude but can breathe pure oxygen at reduced pressure.

Curiously, the desire to breathe is triggered by rising carbon dioxide levels in the body detected in the carotid sinus rather than by reduced oxygen. This creates the sensation known as air hunger, which is the urge and desire to breathe. When there is not enough carbon dioxide to cause air hunger and trigger the breathing reflex victims suffer the symptoms of lack of oxygen without knowing it. Effective remedial action is needed otherwise the condition can very rapidly lead to unconsiouness, brain damage and death.

A typical human breathes between 12 and 20 times per minute at a rate primarily influenced by carbon dioxide concentration and thus pH in the blood. This obviously increases if the individual is carrying out heavy physical work, such as an operation or maintenance activities. With each normal breath, a volume of about 0.6 litres is changed from an active lung volume (tidal volume + functional residual capacity) of about 3 litres; breathing heavily from exertion can considerably increase the 0.6 litre figure.

Table 1. Effects of reduced oxygen atmospheres

Oxygen (vol%)	Effects and symptoms
23.5	Maximum "safe level" (23% is often the high level alarm set point for most oxygen detectors).
21	Typical O_2 concentration in air.
19.5	Minimum "safe level" (19% is often the low level alarm set point for most O_2 detectors).
15–19	First sign of hypoxia. Decreased ability to work strenuously. May induce early symptoms in persons with coronary, pulmonary or circulatory problems.
12–14	Respiration increases with exertion, pulse up, impaired muscular coordination, perception and judgement.
10–12	Respiration further increases in rate and depth, poor judgement, blue lips.
8–10	Mental failure, fainting, unconsciousness, ashen face, blueness of lips, nausea, vomiting, inability to move freely.
6–8	6 minutes – 50% probability of death 8 minutes – 100% probability of death.
4–6	Coma in 40 seconds, convulsions, respiration ceases, death.

When a person enters an oxygen deprived atmosphere, the oxygen level in the arterial blood drops to a low level within 5 to 7 seconds. Loss of consciousness follows in 10 to 12 seconds and if the person does nor receive any oxygen within 2 to 4 minutes, heart failure and death follow.

Moving affected and unconscious persons from a nitrogen atmosphere into fresh air is not enough to promote recovery. The patient has to be physically resuscitated in order to restore the oxygen supply to the brain.

(Data complied from: IGC Doc 44/00/E and US Chemical Safety Board Safety Bulletin 2003)

Unconsciousness in cases of accidental asphyxia can occur very rapidly, typically within one minute. Loss of consciousness results from critical hypoxia, when arterial oxygen saturation is less than 60%.

"At oxygen concentrations [in air] of 4 to 6%, there is loss of consciousness in 40 seconds and death within a few minutes". (DiMaio 2001) If the atmosphere were to be completely devoid of oxygen, the sequence of effects should be expected to occur even more quickly.

At an altitude greater than 43,000 ft (13,000 m), where the ambient oxygen concentration is equivalent to 3.6% at sea level, an average individual is able to perform flying duties efficiently for only 9 to 12 seconds without oxygen supplementation. (DiMaio 2001) The US Air Force trains aircrews to recognise their own and other individual's subjective signs of approaching hypoxia. No two people react the same; some individuals experience headache, dizziness, fatigue, nausea, or euphoria, but some become unconscious without warning. (DiMaio 2001)

Loss of consciousness may be accompanied by convulsions and is followed by cyanosis and cardiac arrest. (DiMaio 2001) The onset of irreversible brain damage occurs at about 4 minutes and about 7 minutes of oxygen deprivation causes death of the cerebral cortex and presumably the medulla oblongata, which controls breathing and heart action. Only two breaths of an oxygen deficient atmosphere can induce unconsciousness and death! (IGC Doc 44/00/E)

HOW DOES THE HAZARD ARISE?

Asphyxiant gases are used for a variety of purposes across many industries; general examples include the exclusion of oxygen for fire safety or reasons of chemistry, as a transport medium in ejectors of Reverse Flow Diverters (RFDs) and welding operations. Hence, we can think of the hazard as present:

- through deliberate action either continuously or for significant periods of time; or
- accidentally through leakage.

Identifying the need for a hazard management strategy for the former is relatively straightforward and one might argue covered by Confined Spaces Regulations. (SI 1713 1997)

The latter is more problematic. Service pipes run through occupied areas in both outdoor and indoor locations can leak through fixtures and fittings and can even break. This is the hidden hazard exposed by the periodic safety review of an operational facility. What is intriguing is that unlike flammable gases there is apparently little or no guidance on the management of this hazard. The guidance that exists examines bulk far field effects rather than near field effects. A large body of analysis and standards exists for managing the near field effects of releases of flammable gases these include: IP15, BSEN 600979-10:2003, IGE SR25 etc. This paper examines ways of evolving the approach used for managing flammable materials so that it can be used for managing asphyxiant materials.

HAZARD ASSESSMENT AND MANAGEMENT

There are two concepts that need addressing when managing and assessing a potential asphyxiation hazard:

1. The Far Field.
   This represents the general areas of the space surrounding the source;
2. The Near Field.
   This is the hazard close to the source of release where localised effects are significant.

THE FAR FIELD

The far field is the general area of a room; it represents the area first moved into upon entry into a room or location. The guidance to be found on managing asphyxiation hazards concentrates on this macro scale field often assuming perfect mixing.

The atmosphere in the far field area is normally monitored by static oxygen depletion monitors. The ventilation system should be designed to ensure that any leakage is diluted and well mixed so that the air movement presents the oxygen depletion instrument with a representative sample of the room environment. Thus, in the far field the safety of personnel is based upon a good ventilation system diluting any leakage, and promoting mixing in the room and a good monitoring regime (well placed instruments) checking for oxygen depletion. This is what the vast majority of people understand as the hazard of asphyxiation. Examining the far field in this way neglects the local effects in the near field and overlooks any 'dead zones' in the ventilation. Dead zones being areas of very little or no air movement.

The ventilation systems of most plants on nuclear sites are designed as radiological protection systems providing containment as opposed to an asphyxiation hazard protection system. Good mixing is not a success criterion. This also applies to systems designed to provide human comfort in occupied areas. These use only 5 to 6 air changes per hour, which is the level required for normal human occupancy. In such systems 'dead zones' probably exist. For normal containment ventilation or comfort ventilation, this is not a problem, but when the hazard is asphyxiation, the danger within these 'dead zones' is that nitrogen or any other asphyxiant gas can accumulate and develop localised oxygen depletion volumes.

Gases that are lighter than air will have some mixing through buoyant flow. Heavier than air gases do tend to accumulate at floor level. However, there is a myth that neutrally buoyant gases, like nitrogen, (having a similar density to air) mix readily with air. Having a similar density to air means there is no great driver for mixing and localised variations in oxygen content of air can and do exist. The 'dead zones' have to be identified and if possible eliminated.

THE NEAR FIELD

The near field represents the effects local to any gas release and the mixing zone around the release. It is a hazard that is not readily addressed by bulk ventilation. As the potential source of release is approached there is a possibility of a gas jet or a localised volume of inert atmosphere from a 'seep'. If the jet, or the localised oxygen depleted atmosphere is inhaled, there can be effects. If the amount of gas inhaled is enough to negate the detection of carbon dioxide by the body then the next breath will not follow and the individual will be in serious difficulties.

If the gas or vapour released is flammable, there is a considerable body of information on the assessment of the extent of the mixing zone. This is readily applied to any gas. Therefore, this body of information was used to try to define the nature of the problem.

There is a current trend to turn immediately to CFD (Computational Fluid Dynamics) modelling for such situations but we were looking for simple desktop approaches that could be applied. Source terms can be derived by looking at standard hole sizes, which can

be found for various types of fixtures and fittings in a number of sources notably Cox and Lees 1990 and IP15. The latter even contains some limited information on mixing extents for nitrogen. The strength of IP15 is that it is based on CFD modelling for the flammable materials. However, simple desk top jet mixing models are also used such as the modified Froude approach described by Burgoyne 1984, Brennan 1984 and Marshall 1977.
Where overall Jet Length J is given by:

$$J = 2.4Fr^{0.5}d \tag{1}$$

The Froude Number Fr is defined here as: $Fr = \dfrac{\rho_g^{1.5}u^2}{\rho_a^{0.5}\left(\rho_a - \rho_g\right)dg}$ (2)

Where u jet velocity given by: $u = \dfrac{G}{\rho_g A}$ (3)

and $A = \dfrac{\pi d_{ps}^2}{4}$ (4)

When the release is sonic the maximum pseudo diameter of the jet is used rather than the actual diameter of the release orifice used for subsonic emissions:

$$d_{ps} = d_o\sqrt{C_d\frac{P_1}{P_2}\left(\frac{2}{\gamma+1}\right)^{\frac{\gamma+1}{2(\gamma-1)}}} \tag{5}$$

d_o diameter of orifice
C_d discharge coefficient
P_1 line pressure
P_2 final pressure
γ ratio of specific heats

Now and the distance X to a given concentration E is given by

$$X = \frac{2050}{E}\left(\frac{G}{M^{1.5}T^{0.5}}\right)^{0.5} \tag{6}$$

where X distance to dilute to concentration E
G mass flow rate
M molecular mass
T absolute temperature

If the calculated distance to the required concentration E is less than the Jet Length J then the Equation (6) applies. If however the distance X exceeds the Jet Length J then the distance to the specified concentration lies outside the jet length, the momentum jet becomes wind controlled. The distance to the concentration (downwind) is calculated from:

$$X = 2.4Fr^{0.5}d + \left(\frac{920Q}{E}\right)^{0.55} - \left(\frac{920Q}{C}\right) \tag{7}$$

where
$$C = \frac{2050}{2.4Fr^{0.5}d}\left(\frac{G}{M^{1.5}T^{0.5}}\right)^{0.5} \tag{8}$$

Or a low momentum dispersion model such as

$$X = \frac{86.75}{E^{0.6}}\left(\frac{Q^2\rho_g}{(\rho_g - \rho_a)g}\right)^{0.2} \quad \text{(Burgoyne 1984)} \tag{9}$$

These are sufficient to define the extent of the hazard. Future extensions to this work are to look at further CFD validation of these methods and what refinements to these desktop models exist.

The jet mixing model has to be used with caution as it is appropriate for a sonic flow and significant sub sonic releases, however there are three weaknesses using this approach:

1. The model assumes that the jet flows axially to a 0.5 ms^{-1} wind, which will not be present in indoor locations;
2. The model assumes dilution by air; if the far field oxygen concentration in a room is reduced this will have the effect of increasing the dilution distance;
3. If the jet impinges on anything it can lose it's velocity and the dispersion becomes a low momentum plume type dispersion.

As a pragmatic consequence, the recommendation is to use the low momentum plume type model, which is the more pessimistic approach. However, refining the risk based approach (The Institute of Petroleum 1998) to area classification used for flammable systems that establishes outer boundaries for Zone 2 hazardous areas whilst taking account of the risks associated with contributory factors would be advantageous. This is a future area of work.

This work led to a practical rule of thumb to allow for a near field of about a 1 metre radius around a release source for a jet release, unless it is a line breach.

Localised ventilation systems can be used to reduce the size of the near field, but they cannot eradicate it altogether – there is always a near field. The issues within the near field are:

- Identification of hazard potential;
- Awareness of the hazard by personnel on the facility;
- The use of procedures as part of the hazard management strategy.

The following safety rules are proposed for near fields that may pose asphyxiation risks:

1. There should be no lone working where near fields may exist.
2. There needs to be a good understanding of the emergency response and actions to be taken if there is an incident.
3. There needs to be close monitoring of any individual working within a near field, by someone trained in what to look for.
4. PPE (breathing air hoods, etc.) can also be used but only as a last resort.
5. The logical alternative would be to identify 'dead zones' highlighting their extent (possibly using markings on the floor) and the provision of additional oxygen depletion monitoring in that area.

ZONING PROPOSAL

Having used the concepts and principles used in defining the extent of flammable gas zones it is a logical extension to consider a Zoning system for asphyxiant gases. The benefits of doing this are:

- achieving a consistent approach to risk assessment and risk reduction;
- focusing designs such that the hazards are eliminated or minimized;
- assisting operators in defining provisions and procedures to manage the hazard.

Risk assessment approaches, analogous to those used for flammability hazards, can be used to assign zones that consistently identify the level of risk and enable appropriate management methods to be selected and deployed.

This is an application of existing data on leakage rates and dispersion models derived for flammable gases applied in the alternative scenario of oxygen deficient atmospheres. It is an approach that can be applied to any asphyxiation hazards from service and process systems pipework both inside and outside buildings although the focus is for inside buildings. This approach concentrates on line fittings and joints, one way to minimize the hazard potential is to use all welded pipe lines.

The proposal is to implement a three zone system:

- Zone 30 where, during normal operations, an oxygen deficient atmosphere is present continuously or for long periods of time.
  These are most likely to be the conditions inside process equipment;
- Zone 31 where, an oxygen deficient atmosphere is likely to occur in normal operation occasionally.
  These are likely to be around engineered access points into equipment;
- Zone 32 where an oxygen deficient atmosphere is not likely to occur in normal operation but, if it does occur, will persist for a short period only.
  This is most likely to surround process and service pipework and equipment, especially when located inside buildings.

As a simple example derived from actual plant experience of how this can be applied: consider the room illustrated in Figure 1, it contains equipment supplied with nitrogen and is accessed through an antechamber. The ventilation inflow is induced through door louvres and extracted at high level through a duct that is part of the engineered building extract system. Figure 1 also illustrates a means of recording the information representing it as a map of the specific problem. The nitrogen supply lines with fixtures and fittings, which are the potential leak points, can be recorded on the map. A ventilation survey highlights vent

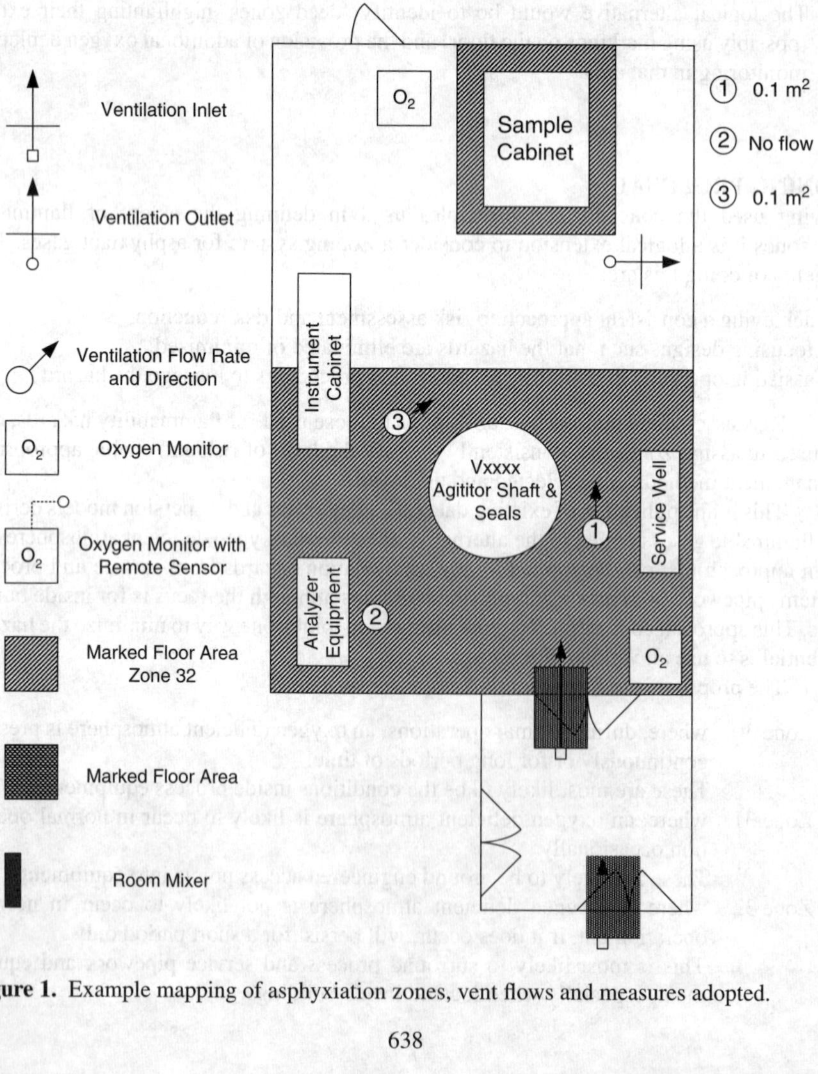

Figure 1. Example mapping of asphyxiation zones, vent flows and measures adopted.

flows and directions. In this case, it also highlights that any obstruction in front of the louvred doors has a major impact on vent flows and dead zones. This survey information is also recorded on the map. Marking out the boundaries of the near field around the potential leak points and the 'dead zones' shows the overlaps between sources leading to the identified Zone 32 area. Now the most frequent operation in the room is undertaken at the Sample Cabinet at the back of the room. Access to this requires an operator to pass through the Zone 32 area. In this way the deficiencies are highlighted and the measures to assist in the management and awareness of the hazard are recorded. The possible measures to manage the situation include:

- the use of additional oxygen monitors to monitor the atmosphere on the pathway to the sample cabinet. This demonstrates that the route is inhabitable before the room is entered;
- marking the floor to indicate to any operator that this is a potential hazard area and not to loiter;
- managing access to and occupancy of the room through Permit to Work procedures. This could include no lone working. The role of the 'buddy' would be to observe from the antechamber and be trained to recognise the behaviours that could indicate the onset of asphyxia and initiate a prepared recovery plan.
- Other measures to consider could be installed room mixers to reduce the dead zones or re-engineering the inlets to sweep stagnant areas and to remove the obstruction issue.

This example is for an existing facility but the mapping concept can also be applied as a design is developed, but it's use at the design stage can help show how to demonstrate how the potential hazard has been minimised.

CONCLUSIONS

Zoning itself is not a protection system but highlights problems to both designer and operator; it draws attention to managing the near field effects and 'dead zones' often overlooked by existing guides. A set of guidelines is to be developed for each zone that establishes the minimum standards and best practices in various situations. Key to these are the ventilation capability for purging and mixing. One simple provision effectively used as part of a suite of measures for Zone 32 is to mark the zoning around potential leak sources on the floor. This provides a powerful visual indication of an invisible potential hazard, which can be used in a number of ways to assist in the protection of people.

This project is ongoing and future work in planned to refine the approach. This will focus on:

- Preparing detailed codes of practice and guidelines for use within the company;
- Developing and demonstrating risk reduction measures in practical applications;
- Using CFD to validate the simple desk top source term models.
- Evaluating and extending risk based approaches to hazardous area classification to asphyxiation hazard management.

REFERENCES

Brennan, E.G., Brown, D.R., and Dodson, M.C., 1984 Dispersion of High Pressure Jets of Natural Gas in the Atmosphere IChemE Symposium Series No 85 1984

BS EN 60079-10:2003, Electrical apparatus for explosive gas atmospheres - Part 10: Classification of hazardous areas.

Burgoyne, J.H., 1984 IChemE Coarse Notes February 1984

Cox, A.W., Lees, F.P., and Ang, M.L., 1990 Classification of Hazardous Locations. IChemE 1990 ISBN 0852952589

DiMaio, V., and DiMaio, D., 2001 Forensic Pathology Second Edition Chapter 8 Asphyxia. ISBN 084930072

IGC Doc44/00/E, 2000 European Industrial Gas Association, Hazards of Inert Gases,

IGE/SR/25 Hazardous Area Classification of Natural Gas Installations, The Institution of Gas Engineers Communication 1665

IP15, Model Code of safe practice Part 15. Area classification code for installations handling flammable fluids, 3rd edition 2005, Energy Institute.

Marshall, J G.; 1977, The Size of Flammable Clouds Arising from Continuous Releases into the Atmosphere IChemE Symposium Series No 49, 1977

SI 1713: Statutory Instrument 1997 No. 1713 The Confined Spaces Regulations 1997

The Institute of Petroleum 1998 A risk based approach to hazardous area classification. ISBN 0852932383.

US Chemical Safety and Hazard Investigation Board 2003 Safety Bulletin. Hazards of Nitrogen Asphyxiation

ACKNOWLEDGEMENTS

This paper was prepared as part of Nuclear Decommissioning Agency (NDA) funded activities.

The author wishes to thank and acknowledge Dr Ray Doig for his encouragement and contribution to this paper.

Keith A Johnson
Nuclear Chemical Engineering Centre of Expertise Leader
Process Engineering Capability
Sellafield Ltd
H350 Hinton House
Warrington Road
Risley
Warrington
WA3 6AS

INTERPRETATION OF THE HCR FOR QRA – AND ITS APPLICATION BEYOND THE NORTH SEA

Dr S.A. Richardson
CEng CITP, Shell Global Solutions

The use of quantitative risk assessment (QRA) has become widely accepted in the petrochemical industry as a way of assessing plant safety, both when considering design alternatives for new plant and when evaluating the safety of existing plant.

One of the difficulties with quantitative risk assessment is finding a suitable source for the release frequencies and ignition probabilities. The UK Health and Safety Executive's (UK HSE's) hydrocarbon release database (the HCR) is one of the best sources [UK HSE, 2005]. It is a collection of information on releases in the UK sector of the North Sea since 1992. For a modest annual fee the records for all hydrocarbon releases can be downloaded. The database also holds information on the amount of installed equipment, and, although this cannot be downloaded, the web-based interface allows queries to be posted which return the number of events, and the number of equipment years relevant to the query. In this roundabout way it is possible to obtain information on the installed equipment base.

The HCR has been the basis of many analyses of frequency of releases and probabilities of ignition, and the results from these analyses are increasingly being incorporated in the rule sets of operating companies, thereby displacing rules based on a number of older sources including the "Hydrocarbon Leak and Ignition Database" report by the E&P Forum (which subsequently became the International Association of Oil and Gas Producers) [E&P Forum, 1992] and the book "Classification of Hazardous Locations" published by the Institute of Chemical Engineers [Cox, 1991].

Although the HCR data has been gathered since 1992, and currently comprises some 3500 entries, it is still statistically sparse on the larger releases that result in major safety hazards. This provides a particular challenge for those wishing to estimate the releases from equipment for which the installed base is relatively small (for example there are a very large number of flanges but relatively few compressors).

However, for the HSE consultant with global reach, one of the biggest challenges is whether the release and ignition frequencies deduced from the HCR can be applied in operations where the gas contains appreciable H_2S (sour gas), or where the winter conditions might require enclosure of the modules and forced ventilation, or for completely unrelated areas such as service stations handling compressed natural gas or even hydrogen.

In order to decide whether the HCR data can be used it has been necessary to delve slightly deeper into the data, and aspects of this are discussed.

WHAT DATA IS RECORDED IN THE HCR ?

The requirement to report hydrocarbon releases stems from RIDDOR (the Reporting of Injuries, Diseases and Dangerous Occurrences Regulations 1995). The UK HSE has clarified the requirements under RIDDOR in OTO 96 956 "Revised guidance on Reporting of

Offshore Hydrocarbon Releases"[UK HSE, 1996]. This indicates that: "The definitions aim to obtain reports of confirmed hydrocarbon releases at the lower end of the incidents scale with a potential for escalation, in addition to fires, explosion and other serious stoppages."

Further clarification was published by UKOOA (the UK Offshore Operators Association), in co-operation with the UK HSE, in 2002 [UKOOA, 2002]. This states that all ignited releases are reportable, as are all 2-phase or condensate releases. "For all other releases to be reportable … then the potential for ignition/escalation needs to be examined, particularly in the case of releases in the minor range". It then goes on to state that amongst other things, if an alarm or withdrawal of people from the area occurs then the release is reportable.

The details recorded in the database are extensive, and include information about the module type, the equipment, the hole size, pressure, density, quantity, duration, and a variety of cause codes.

WHAT ARE NOT RECORDED?

Small releases and fugitive emissions that are no hazard, and cause no alarms.

The details of the release mechanism are not recorded in the database. Releases are coded so that they can be categorised, but there is no free text description of the event.

HOW IS THE DATA USED TO PROVIDE QRA RULES?

The approach adopted is usually to statistically analyse the hole size distribution for different classes of equipment, and derive frequencies, relative to the installed equipment population, of those holes. This hole size distribution is then applied to the same classes of equipment in a different situation and the release rates calculated based on the fluid type, and the pressure within the equipment.

In screening QRA several conservative assumptions are made, and one of the main ones is to group all holes in a specific size range, and assume their size is at the top end of the range. If such an analysis gives rise to concerns, then refinement to more narrower hole size ranges would often be a next step.

As a relatively small release rate can occur under negligible pressure from an opened pipe, in some circumstances analysts screen such releases out of the hole size distribution.

The rules on hole size distributions arising from this form of analysis have been used in locations significantly different from offshore in the North Sea (for example in refineries) and the predictions for the number of major events are, in broad terms, in agreement with reported experience.

This hole-size based approach has a number of disadvantages when the statistics are applied to significantly different situations, for example:

- A small release that poses no hazard in a low pressure system need not be reported in the HCR. However in a very high pressure system, or in a sour gas system this 'unrecorded' hole could pose a hazard.

- Some releases grow over time, and will be reported in the HCR when they are detected and acted on. In a high pressure or sour gas system these may be detected much earlier, and so the hole size at which they are acted on would be smaller than on a sweet gas system.

These two factors act in opposite directions, and it is far from certain that the under-estimation of risk arising from the first point will be balanced by the over-estimation of hole size arising from the second point.

Figure 1 shows the size distribution for gas releases in the HCR for both low pressure systems (<30 bar) and high pressure systems (>70 bar). The data have been normalised on the basis of the total number of releases of over 3 mm hole size in each case. Although the distribution of hole sizes is very similar for holes of over 3 mm (which is releases of 0.1 kg/s and over in the high pressure systems), it can be seen that the recording of holes of below 3 mm differs strongly between high and low pressure systems.

Analysis of the relative frequency of releases in high and low pressure systems is not possible as the UK HSE does not permit the installed equipment data to be downloaded – and it is not available via the web-based interface as a query. It is not known whether the equipment database stores the operating pressure.

It is also the case that in management of gas releases in ventilated modules it is primarily the release rate that poses the risk and not the hole size. Furthermore, it is the release rate that is used to determine the ignition probability for a release and not the hole size. This encourages us to analyse the HCR by release rate.

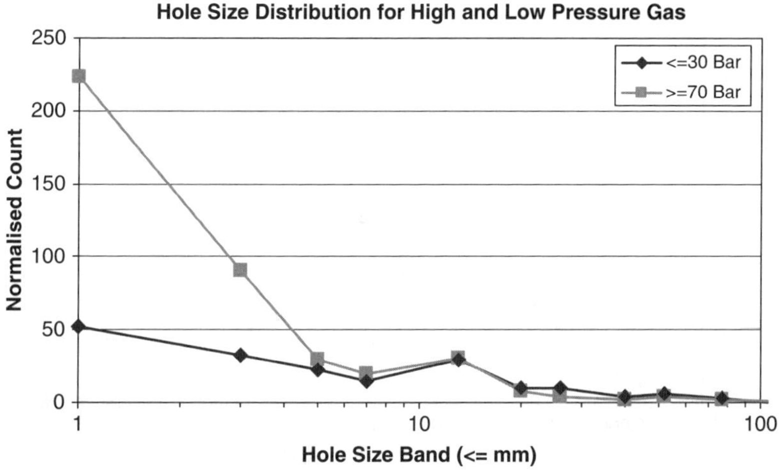

Figure 1. Gas releases from HCR for high and low pressure

ANALYSIS OF THE HCR BY RELEASE RATE

The HCR does not record the release rate, but it does record the actual pressure, hole size, and density of the fluid at that pressure. It also records the total quantity of the release and the duration of the release. Either or both of these can be used to estimate the release rate using standard relationships. The graph below is taken from all gas releases reported in the HCR database at the time of writing. In estimating the release rate the sonic flow relationship (with a discharge coefficient of 0.8) was used if sufficient data was available, otherwise the amount released divided by the duration was used (though this was in the minority of cases). The results show the distribution of gas releases according to release rate.

It can be seen that the recorded number of releases peaks between 0.01 kg/s and 0.1 kg/s. It is fairly evident that larger releases are less frequent, but the left hand side of the graph shows the effect caused by the fact that non-hazardous releases need not be reported under RIDDOR – and of course there is a size threshold below which a release might go undetected.

The UK HSE classify offshore gas releases as follows [UK HSE, 2005]:
Major (>1 kg/s for over 5 minutes OR over 300 kg released)
Minor (<0.1 kg/s for less than 2 minutes OR less than a kg released)
Significant (Anything lying between Major and Minor)

We could probably add
Insignificant (Minor releases not required to be reported under RIDDOR)

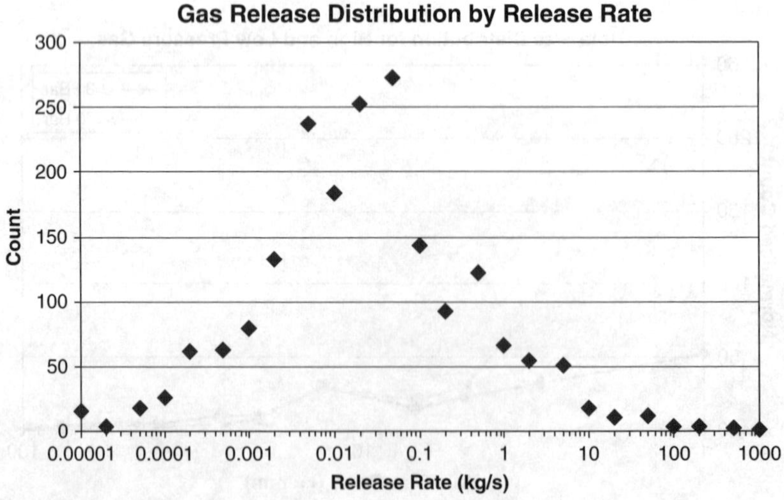

Figure 2. Gas releases in the HCR displayed by release rate

Let us simplify things by considering explosion hazards and using the following rules of thumb for naturally ventilated modules based on the above:

- Release over 1 kg/s can form a substantial flammable gas cloud in a naturally ventilated module.
- Releases between 0.1 kg/s and 1 kg/s could form a minor explosion hazard, and might trigger a gas detector.
- Releases of less than 0.1 kg/s are unlikely to pose a serious explosion hazard, and generally are of minor or no consequence.

These categories will be referred to later in the text. However it is clear that the transition from minor to insignificant occurs in the 0.01 to 0.1 kg/s region, and even for the releases reported in this area it must be appreciated that the estimation of hole sizes that may be a millimetre or less is extremely difficult.

When rules sets for calculating release frequencies are deduced from the HCR data a lower hole size or release rate threshold is normally applied so that the very large number of small releases are not all rounded up into the smallest release rate category (which will typically be in the **Significant** range).

STATISTICAL V MECHANISTIC UNDERSTANDING OF RELEASES

The UK HSE has, in the past, published analyses of the UK HCR data [UK HSE, 2002], but since making the data available for download in 2002 no further summaries have been published, although numerous analyses are carried out by consultants commissioned by petrochemical companies and other bodies such as UKOOA. These studies have in the main been statistical analyses of the data, and as such are quite good where the statistical sample is adequate. Where the HCR has a smaller population of a certain sort of equipment there has been a tendency to fit the same relationship as discovered for large populations to the sparse data.

The problem of applying a statistical relationship without regard for the mechanics of the equipment can lead to some odd conclusions – and there is the risk that this could lead to prediction of significant risk of major release events that cannot occur – something along the lines of a 6 inch hole in a 3 inch pipe – though slightly more subtle.

A further problem occurs when considering the failure rates for individual equipment types. There is very little scope for analysing improvements over time that might result from changes in equipment design or more sophisticated control systems. A mechanistic analysis of releases could lead to such improvements, and UKOOA have published some excellent guidelines on how to reduce the release frequencies for flanges, instruments and other equipment [OKOOA, 2004], but it could take years before the resulting improvements manifest themselves in the HCR, and as a result the benefits are not reflected in QRA studies.

Sometimes technology can make a step change, like computer-controlled compressors with dry gas seals, and it could have a significant impact. At present we have a problem in not being able reflect this in the rules, and the plant designer can make very little difference to the site QRA by selecting higher specification equipment.

A few examples of how a mechanical appreciation of equipment could help interpret release records are for instrument fittings and for compressors.

INSTRUMENT FITTINGS

In figure 3 it can be see that the data for releases from instrument fittings shows that small releases are more common, but also there is a second maximum in the size range of 7 to 13 mm. The reason for this peak is immediately evident if one considers the arrangement of a typical remote instrument (see Figure 4), where a small bore tube – typically of 3/8″ or 10 mm diameter connects the instrument to the valve and flange. It is this line becoming broken or pulled out of the screwed connector that gives rise to the peak in the recorded data.

The hole size arising from a full-bore break in a 3/8″ pipe is typically 7.6 mm (it may be smaller according to the pressure rating of the tube). Depending on the pressure this gives a release rate that is marginal in whether it results in a hazardous gas cloud. As it is bracketed by an interval of 7 to 13 mm in the analysis above, it will typically be rounded up to 10 or 13 mm even in a detailed study. This is quite conservative as release rates change as the hole area, not the hole diameter (i.e. the release rate from rounding a 7 mm hole up to 10 mm is TWICE what it should be). Due to the large number of instruments on

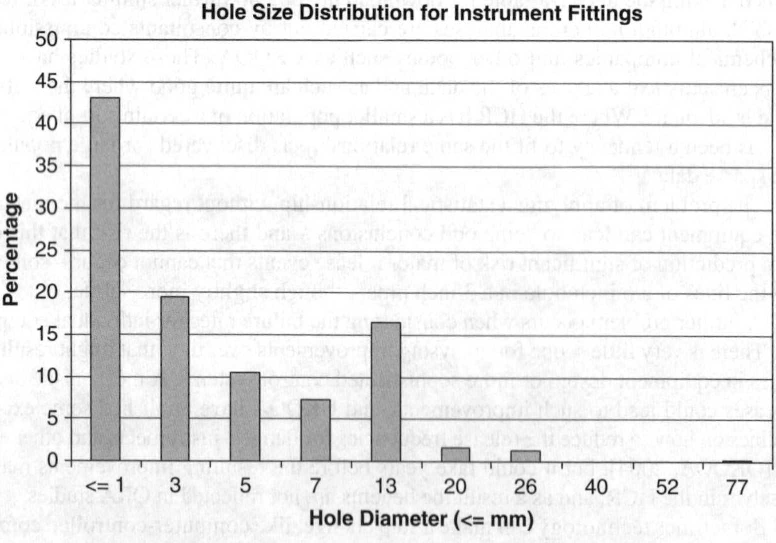

Figure 3. HCR data compared to statistical model for instrument releases

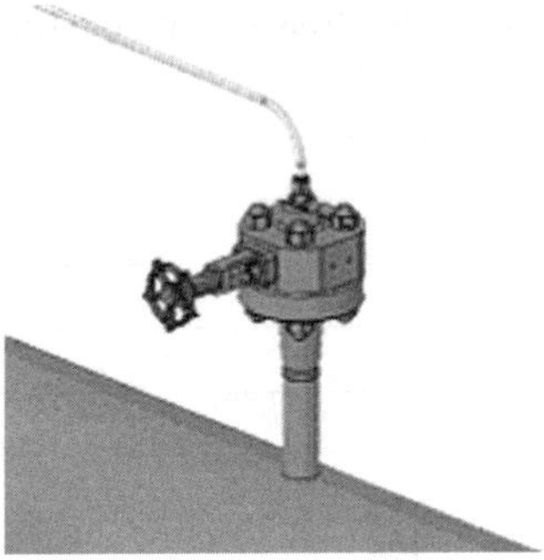

Figure 4. Schematic of an instrument fitting showing small bore pipe

some types of plant (there can be hundreds), and the frequency of small bore pipe breakage, this can have a significant effect on a QRA.

Understanding the mechanics of the release can prevent a smooth curve being fitted through this blip in the data, which would otherwise lead to over-estimation of releases in the larger hole sized bands – which could result in a lot of attention being focussed on managing a risk that doesn't exist – potentially masking some other hazard.

CENTRIFUGAL COMPRESSORS

Centrifugal compressors are a good example of sparse data. The HCR records only one gas release from a hole of over 13 mm diameter. This occurred in 1998, and the hole was 48.2 mm diameter.

Analysis of the release rates reported for centrifugal compressors (Figure 5) indicates that only the single large release was over 1 kg/s. The cause is recorded as "Opened" and the mode as "Routine Maintenance". The difficulty is in whether it is legitimate to extrapolate from the small releases to the larger ones – and it would help if we understood the mechanics of the smaller releases.

One source of releases is the compressor seals. Modern compressors typically have dry gas seals backed up by labyrinth seals, and they tend to have fairly sophisticated seal gas management systems that can detect problems and even take executive action. However, even in the event of sudden and complete failure of the dry gas seals, the labyrinth seals restrict the flow of gas from the compressor. These are substantial metal constructs and would appear to set an upper limit on all conceivable seal leaks of about 10 mm equivalent diameter. For this reason a statistical extrapolation of small seal leaks above 10 mm would be completely misleading. Ideally any extrapolation of probabilities of larger releases should be based on a mechanistic understanding of the smaller releases, for which some

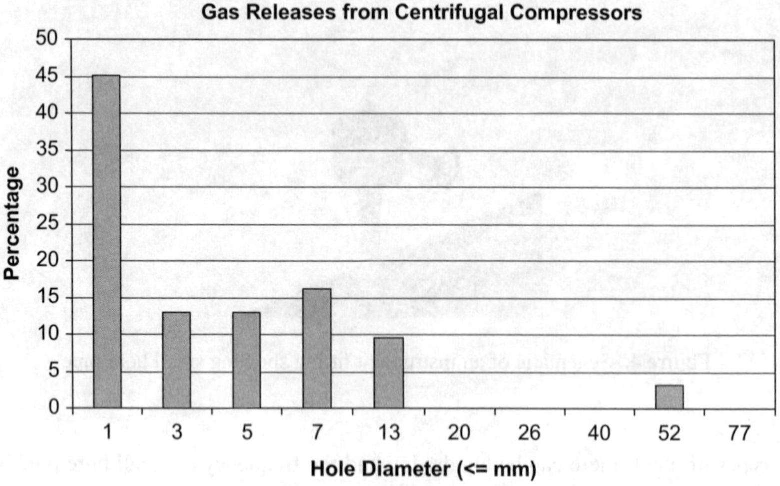

Figure 5. Gas releases from centrifugal compressors

text describing the nature of the release would be helpful (possible something that could be added to HCR recording).

APPLYING HCR DERIVED FREQUENCIES IN OTHER LOCATIONS

ARE RELEASES INSTANTANEOUS?

An assumption in the purely statistical analysis of reported leaks is that they occur instantly. Consider the instant forms of release:

- Dropped objects
- Overpressure burst
- Erosion burst
- Leaks on repressurisation
- Valve opened in error

And then there are releases that are more progressive

- Corrosion holes
- Flange leaks
- Fatigue (though this can be quite fast)
- Valve stem seals
- Seal leaks in pumps and compressors

No leak will ever be recorded in the HCR if it hasn't been detected – and in a noisy offshore environment with difficult access a small leak could be difficult to detect. Once a leak is detected, the action taken will depend on the hazard it poses and operational opportunities. A small leak forming a few bubbles on a valve stem might be standard fugitive emissions within specification for the valve. Such a leak will not be recorded in the HCR as it is not hazardous.

THE PROBLEM OF FORCED VENTILATION

In some environments, including the North Sea, the weather can be so cold that enclosure of the plant becomes important for the workers – and compressors too suffer from the cold and refuse to start unless at a comfortable temperature (e.g. 5C).

In cold conditions it is often the case that the ventilation rate achievable in a sizeable module is of the order of 12 air changes per hour (ACPH) – this being limited by the power required to heat the incoming air. This level of ventilation is roughly an order of magnitude lower than you would have on open process units in the North Sea – even on a fairly quiet day, and even allowing for the weather cladding around a North Sea process unit. As an example calculation - if you can get 1 m/s air flow through a 30 metre wide process unit, then the air is changed every 30 seconds, so that is 120 ACPH.

If the ventilation is an order of magnitude lower, then the release rate that can form a flammable cloud is also an order of magnitude lower. Referring back to figure 2 – releases above 0.1 kg/s that were classed as marginal become more serious in a forced ventilated module. Releases of below 0.1 kg/s that may not even have been included in the release statistics used to make the rules for release frequencies have become marginal and may need to be brought into the risk assessment – even though we can see that the data in the HCR for these small releases seems incomplete.

WHAT ABOUT SOUR GAS?

In a number of countries we are now seeing sour gas with considerable H_2S content, e.g. 10% or more. The LFL for methane is about 5%, which is 50,000 ppm. If our example raw gas contains 10% H_2S then at the LFL you have 5000 ppm H_2S. Sour gas alarms will likely be set at 10 and 20 ppm, and at 500 ppm (two orders of magnitude more dilution than for sweet gas flammability) you are at levels that can have serious consequences.

If a QRA study is going to determine the toxic risk it needs to predict release frequencies two orders of magnitude smaller than we need for North Sea QRA's.

- 0.01 kg/s will cause gas alarms
- 0.001 to 0.01 kg/s is marginal and may cause alarms
- <0.001 kg/s might be OK

It is clear from Figure 1 that the HCR doesn't record releases down to these low levels with any reliability. Many of these smaller releases might not even be detected with sweet gas.

However small releases of sour gas ARE detectable, thank to the H_2S which itself acts as a tracer. If the North Sea was mainly sour gas, then releases would be detected at much smaller sizes, and would be reported as such, and so the curve in Figure 2 would have far more reports at the lower release rates (although the difficulty in assigning hole sizes to such minor releases would remain).

Supporting evidence for this comes from Canada [CAPP, 2003] where a number of sour gas fields are in production. Tests of different plant using a variety of sensitive methods of leak detection and analysis have revealed that the emissions are an order of magnitude lower for the sour gas plants than for the sweet gas plants. Although this could be due to superior equipment, it is more likely that smaller leaks can be detected in the sour plant and are acted upon quickly.

If we could separate the progressive leaks from the instantaneous leaks in the HCR data then maybe we could offer a rule set more suitable for sour plant, as we could anticipate that the progressive leaks would be detected at much lower levels – but for now it is clear that using rules derived from the HCR and then applied to sour gas plants could lead to two problems:

- You could significantly over-estimate the toxic hazards by using hole sizes that would have been detected and repaired at a smaller size.
- You could significantly under-estimate the toxic hazards as the frequencies you are using could ignore all releases below 0.1 kg/s, which could still be significant as a toxic hazard.

Which of these two opposing factors dominates could depend on the individual plant. There is no reason to suppose that they cancel out.

SUMMARY

In this report we have focussed on two factors:

i) The nature of the events recorded in the HCR – statistically sparse for larger releases, and with a lower cut-off dictated by the RIDDOR reporting requirement
ii) The problem of interpreting the data in a statistical way without taking into account the mechanics of releases

It is hoped that some insight has been given into the suitability of using HCR data in areas where very high pressure or sour gas could make the statistics of small releases much more important.

Even with typical North Sea operations it can be seen that an understanding of the mechanics of releases can help in deciding whether release frequencies should be extrapolated to larger sizes where data is sparse or missing.

When attempting to apply HCR derived release frequencies in a predictive manner for very different circumstances (such as high pressure hydrogen) then it is doubtful the

HCR data can be used directly, and modifying HCR-based rules is difficult because it is not obvious which releases would have been detected earlier under other circumstances.

However the HCR provides a valuable insight into classes of failure, especially covering the mix of human, mechanical and design factors – though inevitably more detail would help further, even in current interpretations. Maybe the best route to deriving rules for very different operating conditions would be to decompose the HCR releases according to instantaneous or progressive releases, and causative factors, add in any new causative factors (e.g. different materials problems) and then use this to drive an FMEA approach for equipment types and thereby assemble a new rule set appropriate for plant operating under significantly different conditions.

Table 1. Emission factors for sour v sweet gas plant

Fitting	Emission factor (kg/hr/fitting)
Valves (sweet gas)	0.04351
Valves (sour gas)	0.00518
Flanges (sweet gas)	0.00253
Flanges (sour gas)	0.00031

REFERENCES

[1] Cox, A.W., Lees, F.P., Ang, M.L., May 1991, "Classification of Hazardous Locations", I Chem E. ISBN 0 85295 258 9.

[2] CAPP - Canadian Association of Petroleum Producers, 2003 "Calculating Greenhouse Gas Emissions" (http://www.capp.ca/raw.asp?x=1&dt=PDF&dn=55904).

[3] E&P Forum, "Hydrocarbon Leak and Ignition Database" Report No 11.4/180, May 1992 (note: E&P Forum subsequently became the International Association of Oil and Gas Producers).

[4] UK HSE, 1996, "Revised Guidance on Reporting of Offshore Hydrocarbon Releases", Offshore Technology Report, OTO 96 956.

[5] UK HSE, 2002, "Offshore Hydrocarbon Release Statistics", HID Statistics Report HSR 2002 02.

[6] UK HSE, 2005, "Hydrocarbon Releases System" (https://www.hse.gov.uk/hcr3/index.asp), v 1.1.3.

[7] UKOOA, 2002, "Supplementary Guidance for Reporting Hydrocarbon Releases". (http://www.ukooa.co.uk/issues/health/index.cfm)

[8] UKOOA, 2004, "Hydrocarbon Release Reduction Toolkit", UKOOA Publication EHS 19.

AN IMPROVED APPROACH TO OFFSHORE QRA

Brian Bain[1] and Andreas Falck[2]
[1]DNV Energy UK
[2]DNV Energy Norway

QRA is now an established method used worldwide for the evaluation of risks on offshore installations. The technique is increasingly used as a tool throughout the planning phase and more closely integrated with the design processes. The scope of QRAs may also now be extended to cover other types of loss such as asset and environmental damage. However, there are many issues in the use of QRA which may challenge the value which such studies provide. These include;

- integration with the design process
- appropriate involvement of the operators and decision makers
- lack of consistency
- complexity of the overall model structure
- uncertainties
- lack of functionality
- likelihood of errors
- knowledge of analysts
- ability to update existing studies, and
- incorporation of new data and methodologies

Each of these issues present challenges to an efficient and effective approach. Some models may have been used for a long time and users may have become complacent - no longer striving to improve the model's accuracy or functionality. This paper looks at these issues and suggests steps which can be taken to achieve an improved approach.

KEYWORDS: Offshore, QRA, Safety Case

INTRODUCTION

Risk analyses have been actively used by the offshore industry in the North Sea for more than 25 years and are now established tools which are used worldwide. However, the requirements for Quantitative Risk Assessment (QRA) differ between different geographical areas and they may also have been modified with time as the technique has matured and legislative requirements have changed.

QRA has developed from primarily being a verification activity to being a tool which can be used throughout the planning and design process, although the extent to which this latter function is applied may vary between operators and within different parts of the world.

While the main focus of QRA historically has been to evaluate personnel risk, other outcomes such as environmental damage, asset damage and loss of production have lately been calculated/estimated.

This paper outlines the issues and shows what steps can be taken to revitalise the QRA method and move to a more effective approach. The development of offshore risk models has been used as an example but many of the issues are equally applicable to onshore studies.

CURRENT CHALLENGES IN OFFSHORE QRA

There are many challenges in the continual development of QRA models as a decision support tool. Issues include;

- consistency in models
- handling of uncertainties
- implementing improvements in model functionality
- knowledge of analysts
- effective presentation of results
- ability to update existing studies

Some of these issues have been described in detail [Gadd et al, 2003] and [Bain, 2003] but are summarised below.

DIVERSIFICATION OF APPROACHES

The seeds of problems in the current QRA modelling in the UK were possibly sewn in the early 1990's with the need to rapidly develop models to support Safety Case submissions. There was a conflict between the need to produce assessments quickly and the need for adequate accuracy and consistency. This speed of development made it difficult for the offshore oil and gas industry to evolve standard ways of tackling the various technical problems. Inevitably, different operators and different consultants adopted different approaches to each of these aspects. This has led to a multitude of model types.

Although the basis of QRA is straightforward, its implementation in a model with the capability of providing results to support decision making in the design process is complex. It involves many aspects of physical modelling each of which can have large uncertainties and there might be little consensus on the most appropriate methodology and computational tools to use. There are many different aspects to consider, each of which has a number of different approaches which could be followed. It is little wonder, therefore, that models created independently of each other have evolved in very different ways.

There were some views that QRAs should be created or customised to deal with specific installations or the specific requirements of a client. While there is some merit in this, the number and variety of solutions offered went beyond the need to meet these requirements and similar situations have been dealt with in entirely different ways. For example, assessment of fire hazards can be addressed by a variety of phenomenological models, empirically derived equations or simple rule sets. The same can be said of ignition modelling, release modelling and the structure of event tree risk calculations.

Even where there is consensus on an approach or set of data to use there may still be differences in how this is implemented. In the UK and most other counties, the UK HSE's hydrocarbon release database (HCRD) [HSE, 2007] is acknowledged by most practitioners as the best source of data on process release frequencies. However, different QRA providers will still interpret the result in different ways leading to different calculated risk levels.

MODEL COMPLEXITY

In many parts of the world, the impression has formed that QRA studies are simple commodity services which any competent practitioner can undertake with a high level of accuracy. In fact, the process of providing a QRA which produces meaningful answers is quite involved. There is an expectation that a simple QRA with a reduced set of input parameters can identify the difference in risk associated with different design or operational options but this often isn't the case. Ignition control is an example of a key safety systems and it is therefore often important to evaluate the effect of different measures and configurations as part of a QRA. In order to do so, it is necessary to simulate their effect in the model itself. Adopting a simplified model with generic ignition probabilities will not be able to differentiate between alternative strategies. In many parts of the world clients may express a desire for a simplified approach and while this is well motivated and has some benefits there are also some obvious disadvantages. In this market there is little incentive for a QRA service provider to invest in developing the tools they use in their analysis. The model is likely to be less accurate and less capable of being inspected to understand the drivers which lead to a particular risk level for a given hazardous event.

In Norway the trend is going in a slightly different direction with a stronger focus on detailed and complex models and where the customer often is very active in the risk analysis process.

INVOLVEMENT OF OPERATOR IN THE PROCESS

Few operators carry out QRA studies using their own safety departments. Typically they are contracted to an external consultancy. On occasions they may be further removed from the detail through other intermediaries, e.g. where the operator employs an engineering consultancy who contract a consultant to prepare a safety case who in turn rely on a separate group of providers to perform the QRA. This distancing makes it harder, both to reflect the operational aspects of the installation in the analysis and for the persons who are in a position to implement improvements to benefit from information and insights uncovered in the model. Typically, analysts carry out the work in an office and base their assumptions on the information provided by the client in the form of drawings, tables, information in the existing Safety Case and other documentation. Some of this information may be out of date but this may not be apparent. An analyst is not always afforded the opportunity to visit the platform and assess at first hand the effect a given hazard might have.

USE OF RESULTS

In the UK the main purpose of QRA has been to demonstrate compliance with tolerable risk criteria in the Safety Case legislation [HMSO, 2005]. Consequently, there was a focus on demonstrating that the criteria for Temporary Refuge Impairment Frequency (TRIF) and Individual Risk per Annum (IRPA) were met. A typical approach was to initially carry out the analysis using conservative assumptions and progressively refine the model as necessary to meet the acceptance criteria. When compliance with these criteria had been reached, the incentive to develop the analysis further declined.

There is still a requirement to demonstrate ALARP (As Low As Reasonably Practicable) and this puts further requirements on the QRA model. Simple models and approaches as described above are less likely to be able to support such processes in a satisfactorily manner.

The approach in Norway has been different; here QRAs have been more integrated with the design process and there has been more of a drive towards accuracy through the use of CFD modelling, and probabilistic tools, particularly for explosion analysis [NORSOK, 2001]. In the past, risk analyses were often carried out in isolation from the main design process and the overall planning. The findings of the QRA were not always effectively implemented because the best solutions were often identified late in the design process resulting in costly variations and compromises which could have been achieved more simply if these issues had been identified earlier. Experience in the use of the technique, changes in legislation, and some very costly incidents such as Alexander Kielland, Piper Alpha and the P36 disasters have influenced the use of QRA. Today it is a tool that is actively used throughout a project's planning and design phase. It is used for decision support as well as to explore the safety implications of the choices being made. The QRA activities are closely integrated with the design processes and are in many respects considered as routine.

UPDATING OF EXISTING STUDIES

QRAs are normally intended to be "living" studies which can be updated to reflect changes in configuration and operation of the installation. Unless there are major changes in design or operation, there will normally be an expectation that risk levels will be within the tolerable criteria and therefore the motivation to carry out the work to a high degree of rigour and to develop more accurate approaches is lessened. The time available to update a QRA will normally be far less than for the original analysis and so the opportunity to modify it to include improved data and methodologies will be compromised.

Very often the analyst updating the study is different from the one who carried out the original work. The process of building up a detailed knowledge of the model may take days or even weeks. In such circumstances it may be necessary to accept that a complete understanding won't be achieved. This will diminish the ability of the analyst to fully understand the interaction of the various parameters and hence the ability to identify the key safety critical parameters and remedial measures which could provide risk reductions.

Changes to the model may be made in a way which addresses the aims of the present study but they may be done in a less than robust manner which will be difficult to understand by others at a later date. It also increases the risk of inadvertently introducing errors into the model.

It is also likely that there will be a different person commissioning the study on behalf of the operator. It is therefore important that risk assessment reports are written in a way which clearly describes the status of the installation and the assumptions made. This will assist in identifying changes that have taken place in the intervening years.

Updates to QRAs are normally driven by technical or operational changes. Changes in analytical models and our knowledge of physical phenomena may, however, also necessitate modifications. In such cases it is important to analyse the effect of improved modelling and the effect related to changes in configuration and operation in separate stages. This allows the actual risk change to be evaluated as opposed to changes in calculated values which are purely the result of different methodologies. For example, in 1998 fire and explosion experiments performed by the Steel Construction Institute gave new knowledge to the industry regarding potential explosion risk. This resulted in the re-analysis of most platforms in the North Sea and changed the design parameters for many installations.

It is easy to see why a "vicious circle" might be set up whereby the value the client places on the model and the lessening ability of the QRA provider to maintain their understanding and make improvements leads to a decline in the overall quality of the model and further reduction in its value to the client - quite the opposite of what should be aimed for.

UNCERTAINTY IN RISK ANALYSIS

It is well known that uncertainties in risk and consequence assessments can be considerable. Furthermore, the accuracy of such studies will vary extensively with detailed analyses providing more accurate estimates than coarse assessments. A systematic way of managing uncertainties in risk estimates is particularly important when the accuracy of the results is critical, e.g. if probabilistic explosion results are to be used as the basis of the design of a blast wall, or if the results are close to an acceptance criterion. In structural engineering it is normal practice to reflect uncertainties through safety factors. Use of safety factors has not been normal practice in the area of risk assessments; instead a "best estimate" approach is the norm, while the nature and effect of uncertainties may be discussed but not quantified. It has become apparent that, because of the uncertainties in the analysis and the wide variety of approaches which can be taken, the results obtained by different QRA providers will differ significantly.

A risk analysis may, for example, be compared to a weather forecast. Based upon models and available data, one tries to say something about what can be expected. The accuracy of the weather forecast is dependant on analytical skills, available tools, quality of data and the degree of detail required. The uncertainty of a quantitative risk analysis will be related to aspects such as relevance and degree of detail of:

- Analytical models.
- Failure data (scarce or no data on equipment representing new technology).
- Engineering judgement (response assessment, human reliability etc.).

The uncertainties in even the most detailed risk assessment may be significant and the user of the results needs to be aware of this in making decisions based on them.

In general, all evaluations of risk in the analysis are sought to be "best estimates", i.e. no systematic conservatism (or optimism) is included in the evaluation. However, the assessment of issues where the uncertainty is significant tends to be on the conservative side in order to account for the uncertainty. In addition, sensitivity analysis is often performed to investigate which input parameters influence the results to the greatest extent. Here, the effect of a small change in an input parameter on the result is quantified. The results from these sensitivity analyses can then be used to focus more attention on evaluating the key parameters which influence risk.

AN IMPROVED APPROACH

A risk analysis by itself is of no value unless it is used as input to real decisions. The risk analysis must therefore focus on issues where it can identify practical improvements in design or operation. The analysis must be suitable for its purpose and this will impose several requirements on the analysis itself and how it is integrated in the decision process. In order to get more benefit from QRAs and hence an improved ability to identify cost effective safety improvements, a number of steps are suggested. These are discussed in turn.

INTEGRATION OF ANALYSIS IN THE DESIGN PROCESS AND CLIENT INVOLVEMENT

In order to achieve the goal, the risk analysis must be initiated earlier and be more integrated in the design process;

- Effective communication between the risk analysts and the design team is essential.
- The risk analysis process should be synchronised with the engineering activity.
- Finally, the QRA results need to be "translated" into engineering terms.

Many QRAs suffer from lack of client involvement. As a result, the quality of data based on subjective judgements will not be as high as it should be. The client's staff may be in the best place to provide both the factual data and judgements on the likely consequences of initial and escalated events. Similarly, the results of the analysis need to be fully understood by the client in order to correctly interpret them as an input to decision making.

The active use of QRA for decision support for the planning and design phases of a platform lifecycle poses several challenges to the risk analysts, the engineering team and the decision makers. QRA models are very abstract representations of the actual scenarios being modelled. Effective communication is essential, both to ensure a proper understanding of the design problems so that these can be effectively addressed in the QRA, and also to ensure that the QRA results are understood by the design team and decision makers.

Another important communication aspect is that the risk analysis process is synchronised with the engineering activity. It needs to provide the right information at the right time. As the design progresses, the level of detail in the design increases and the uncertainties reduce. The risk analysis needs to reflect this in order to address the key required decisions as the design progresses. It is therefore necessary to aim for a living QRA, i.e. a risk model of the platform that is updated and refined as required. Assumptions being made at an early stage to compensate for missing information need to be followed up and eventually replaced by factual information when available.

Finally, the QRA results need to be translated into engineering terms. Risk is measured in terms dictated by the risk acceptance criteria; PLL (Potential Loss of Life), FAR (Fatal Accident rate), etc. and risk reduction will typically be measured in terms of these. This is not relevant information for the engineering team. The requirements must be specified in terms of, for example, design capacities for explosion barriers or location of critical equipment. The risk analysis needs to be sufficiently detailed to address the effects of safety critical elements so that alternatives can be assessed. Consequently, the risk analysis needs to be closely integrated with detailed engineering studies.

A graphical representation of how the QRA views the platform layout and the hydrocarbon equipment may aid client involvement by providing a more convenient means of presenting data and results.

STANDARDISATION

The single most important aspect which places a barrier in the way of improvements in QRA is the lack of standardisation. With numerous variations of model to maintain, possibly a different one for each installation analysed, there is unlikely to be the opportunity to appreciably enhance a given model's capability within a given project budget.

There is a desire for consolidation but this conflicts with a desire to minimise the spend on a given study. Some improvements can be made as the model moves towards compatibility with a perceived best practice approach but this is a long and inefficient process which will take many iterations.

Some operators have come to recognise this problem and how it impairs their ability to compare risk levels between different installations. Some have decided to contract the provision of QRA services, at least within a geographical region, to a single provider who has been tasked with developing a consistent approach. Other operators are documenting their requirements for a standard approach which they require their QRA providers to follow. The Dutch government have, for example, specified one particular tool to be used for onshore QRA's One major operator is developing a standard tool for high level QRA to be applied to its installations worldwide. However, they will still rely on different providers for detailed QRA. These approaches have some clear benefit but is likely to still leave room for interpretation leading to different results from two analysts complying with the same requirements.

Some countries have sought to introduce standards to improve consistency but most have put the onus on operators to apply suitable techniques and engineering practices.

Standardisation of approach is almost essential for the long term development of QRAs. It will have the following advantages.

- More efficient development because cost can be spread over a number of projects.
- Greater accuracy because the model will be scrutinised by more analysts who can feedback information on errors and possible enhancements.
- Consistency of results between installations.
- Easier to justify the writing of user manuals and training material.

STRUCTURE

It is clear that appropriate data and methodologies are important in creating a successful QRA model. Less obvious is the need for an appropriate structure and its importance is generally underestimated. The clarity of how the model is operated and how the component parts interact with each other is essential for developing a good understanding of it. A complex intertwined model with little or no documentation will be difficult to understand. It will take a new user longer to understand than would otherwise be the case and is more likely to contain errors. It also makes updating the model harder and errors are more likely to be introduced when making changes to it. The more complex a structure becomes the more effort is required to change it into a simpler one which is easier to follow. The temptation to make changes using a "quick fix" is greater but this makes the model structure more difficult to follow for future users.

The size and complexity of a QRA model can vary considerably. At one end of the spectrum there are highly integrated models with automatic transfer of data between the component parts. At the other end, the study may consist of independent pieces of analysis where the part referred to as the QRA model itself consists only of the analytical structure for combining the results of the separate frequency and consequence analyses. Some elements of the overall QRA may be external programmes or calculations whereas the part regarded as the model itself will have a greater degree of connectivity. This is illustrated in figure 1. The greater the proportion of elements within the QRA boundary, the faster it is likely to be to process changes in data through to final results.

Models which require the manual transfer of data between one component and another are time consuming to operate. When a change to an item of data is made and the model has to be re-run the user may be faced with the time consuming task of re-running large parts of the model and having the tedious task of carrying out data transfer operations. Data transfer operation can include;

- Viewing data in the output from one model and typing it to the input of another.
- Cutting and pasting operations between spreadsheets.

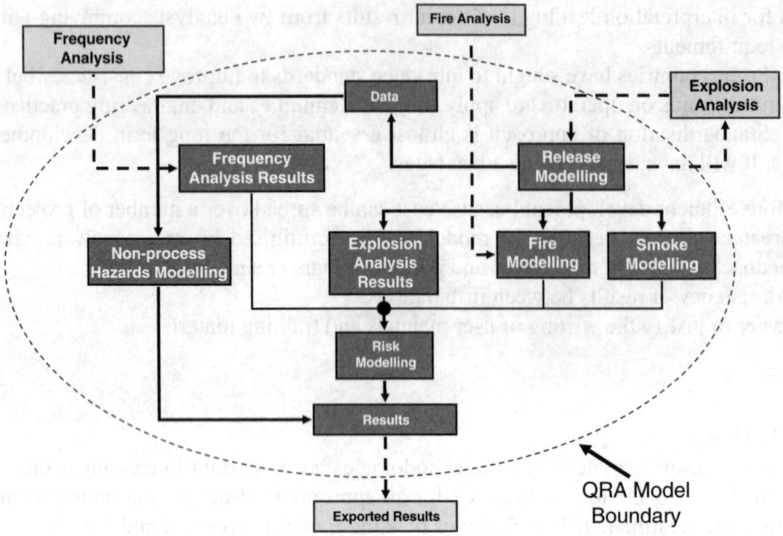

Figure 1. Internal and external components of a QRA model

- Linking of spreadsheet cells.
- Use of programming embedded in spreadsheets, e.g. visual basic macros, to manage the transfer between parts of the model.
- Automated transfer between models in toolkit type models.

One further type might be termed "operated adjusted data transfers" in which the analyst considers the output from one or more models and makes a judgement on the value to be entered into the next part. This might include situations where, for example, the size of a fire and the layout of the platform are considered together in order to make a subjective judgement on fatality rates in different areas. This has detrimental effects on the speed and consistency with which models can be run but this may be offset by allowing the analyst to take a fuller account of all the parameters which affect the results rather than being bound by prescriptive rule sets which may not always be appropriate. Figure 2 shows an example of a data transfer tool.

EVENT TREES

At the heart of the analysis of a given hazard in a QRA is some form of event tree. This may be represented in the classical format, in some variation presented graphically or as a series of calculations presented in tabular form. Generally, some form of diagrammatic form is preferable since it is easier to follow the logic when it is split into manageable steps

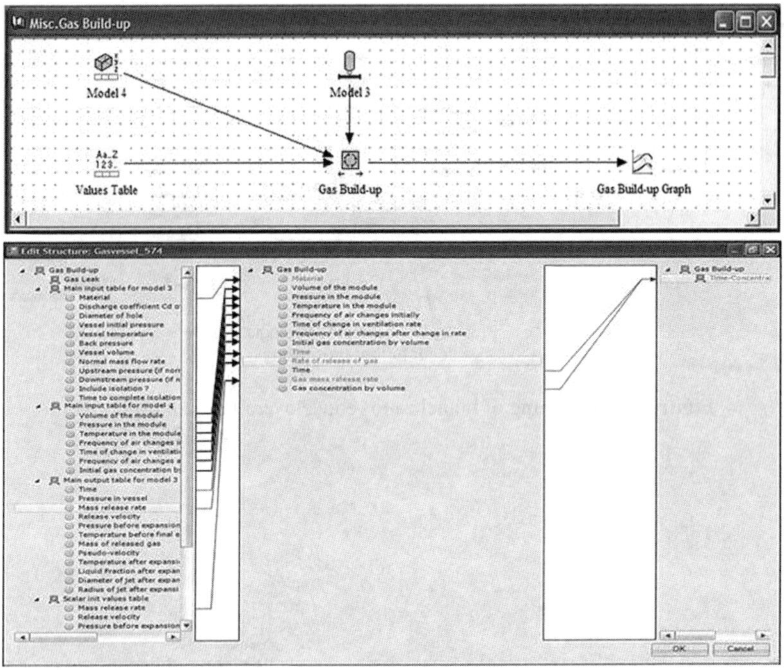

Figure 2. Example of data transfer tool in risk analysis software

rather than a complex equation which is difficult to understand. The complexity to which an event tree should be developed is a matter of opinion. The more parameters considered the more accurate and detailed the solution should be. This may be important in improving resolution for F-N curves, or in order to reflect the impact of different safety systems. However, the number of parameters which could be considered can be large leading to a very great number of end outcomes which have to be handled. However there are some techniques which can be employed to increase the number of parameters while still maintaining a relatively simple diagrammatic form.

One method is to combine branches at an intermediate point in the event tree where the remaining branch structures are the same. For example, consider the segment of an event tree shown in Figure 3.

A more advanced technique is the use of linked event trees where the end branch probabilities of the first become the top events for a second event tree which is then run iteratively to produce sets of end outcomes as illustrated in Figure 4. This approach reduces the overall event tree structure to a number of segments which are relatively easy to

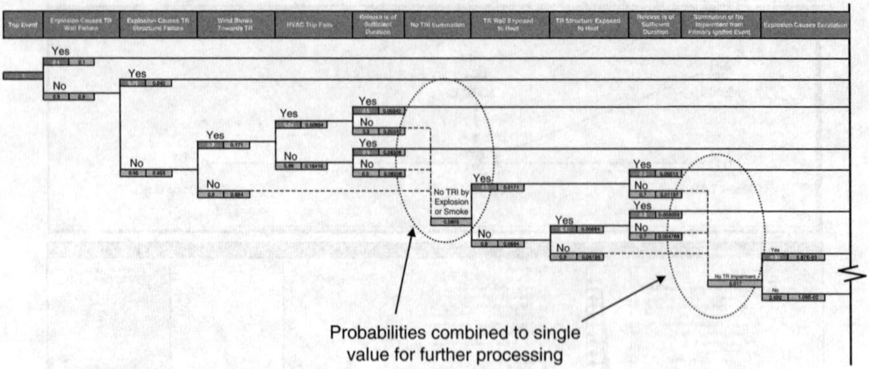

Figure 3. Combining of branches to reduce overall size of event tree

Figure 4. Linked primary and secondary event trees

understand but which combine to produce a very detailed structure addressing a large number of safety critical parameters.

LOCATION OF DATA

In a typical QRA, data is distributed throughout the structure and mixed with the analytical elements and results. Figure 5 shows this situation diagrammatically. This may not appear to be a problem but it creates difficulties when trying to continually develop a standardised approach.

In this situation it becomes almost impossible to maintain the consistency of the various models. Either the same methodology change has to be applied to all of the models or one model has to be upgraded, copied to provide a series of templates for the others and then data replaced.

The alternative approach which resolves this issue is to separate out the data from the analytical parts of the model. This situation is shown in Figure 6. This configuration allows for the same model to be used to analyse any number of data sets each representing a different platform or different variations of the same installation. It is now possible to modify the single model structure and in effect update all the QRA studies simultaneously. The user will still have to update the data as the configuration of the platform changes but this is a relatively simple task. Only if the methodology change requires the structure of the data file itself to be modified do the individual data files need to change and this will normally be a relatively short task.

METHODOLOGY

The use of an appropriate methodology is self evident, but what constitutes "appropriate" in this context may be a matter of opinion. "Appropriate" may not necessarily mean most accurate since this may involve the use of techniques which are expensive and time

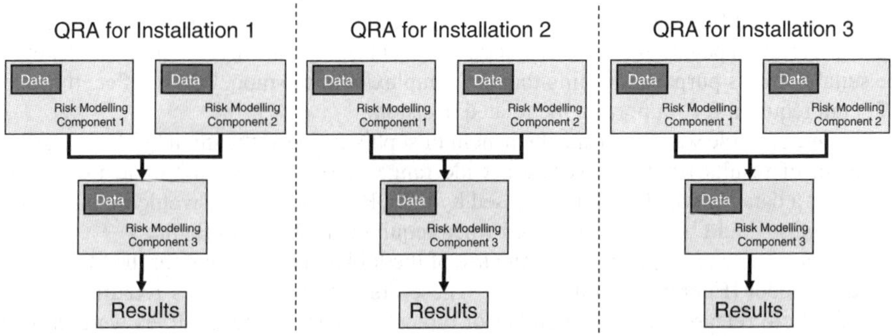

Figure 5. QRA models with embedded data in each component

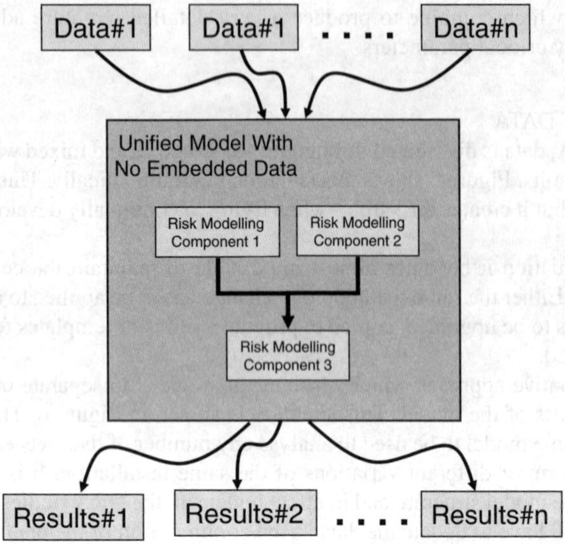

Figure 6. Installation specific data sets processed through a common model

consuming. Furthermore, the most detailed forms of dispersion, explosion and fire consequence analyses look in depth at a set of very specific scenarios relating to the location, rate, direction and composition of a hydrocarbon release, whereas a QRA has to consider the overall effects of all the possible combinations of these parameters. Detailed analyses will normally require specialised software outside the QRA model itself and so there will be issues of transferring the results in an appropriate manner into the overall analysis so that it can be combined with the frequency data in order to calculate risks. This is both time consuming and a potential source of errors. The main issue is however that the model must be suitable for its purpose meaning that the complexity of the model must reflect the level of detail required for supporting the related decision.

One possible way of getting the benefit of sophisticated software in QRA is to generate a set of results from it covering a wide range of typical scenarios and use this to construct a data set which can be accessed by the QRA. The data set would be part of the model and so could be readily accessed when required, as shown in Figure 7.

One simple example of this is the use of the lookup correlations for the "UKOOA" ignition model [Energy Institute, 2006]. These are a series of curves relating ignition probability to release rate for a number of typical scenarios. These curves were derived from a spreadsheet model which implemented the full methodology as described in the same report and which covers the development of the method. Whereas, the use of the full analysis requires the user to obtain and enter a significant amount of data relating to the

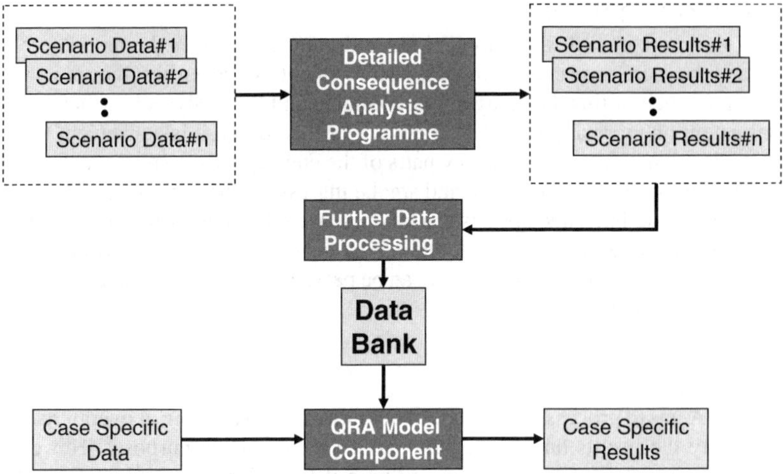

Figure 7. Use of detailed analysis to create data bank for subsequence input to QRA

installations configuration and ignition sources, the look-up version requires only the selection of the appropriate curve and the release rate, as shown in Figure 8 for typical examples.

A more complex application of this approach has been used by DNV Energy to make the detailed information available from the Computation Fluid Dynamic (CFD) modelling available in a simplified form within spreadsheet based QRA models.

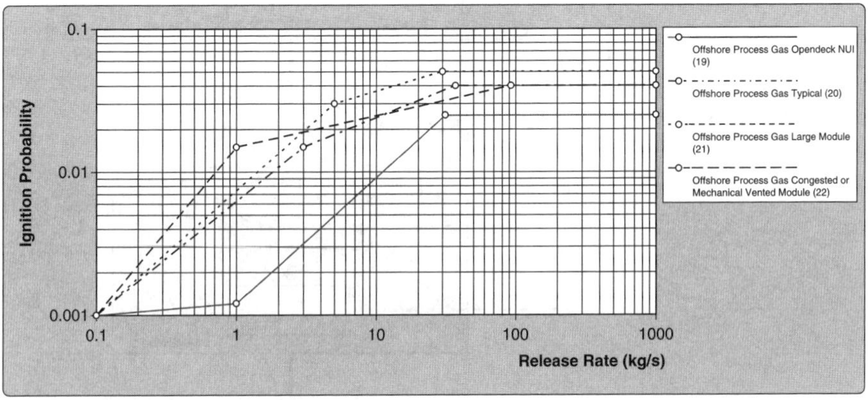

Figure 8. Look-up correlation curves from the UKOOA ignition model

METHODOLOGY OPTIONS

Generally the more complex the applied methodology the greater the amount of data it requires and the more time consuming it is to implement. In the early stages of an installation's design, much of this data may not be available and the analysts may have to rely to a greater extent on more approximate methods. For this reason it is convenient to provide a number of options for some of the key parts of the consequence analysis such as release, dispersion, explosion, fire, escalation and smoke ingress. In the early stages of the design the simpler less data intensive options can be chosen. In the later stages these are replaced with more detailed alternatives. The key here is to ensure that the outputs from the various alternatives are in the same form so that it can be passed on to the next stage of the analysis as indicated in Figure 9.

NSPECTABILITY

A typical QRA has to process a multitude of combinations in arriving at overall risk values. As a necessity the results have to be summarised for reporting purposes. However, this tends to hide the detail of the analysis and with it the information which describes the

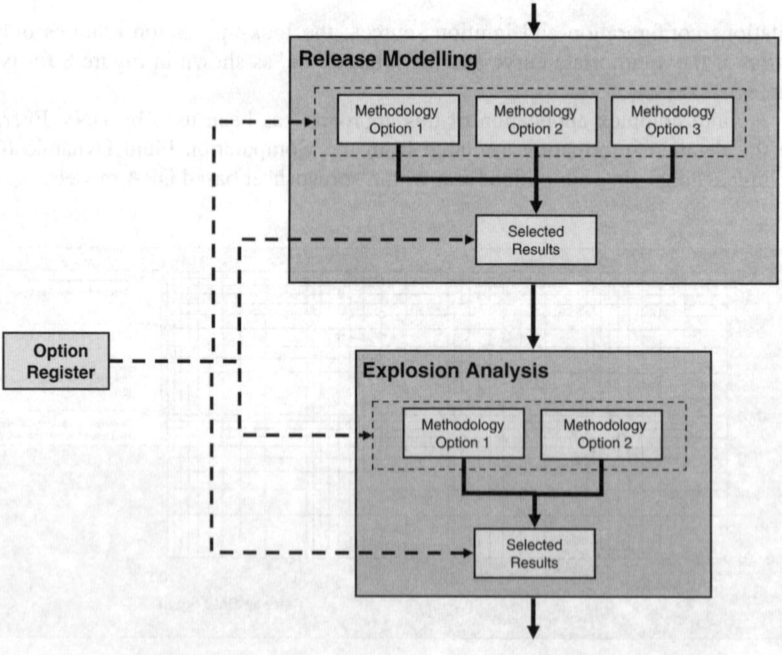

Figure 9. Consequence analysis with processing options

reasons for the contributions of the individual components. Without the ability to gain an understanding of these issues the QRA process loses much of its value since it is this detailed understanding of the safety critical parameters that may lead to the identification of risk reduction measures.

A good QRA will allow the analyst to access the detailed information in a convenient manner. One approach is to save all the intermediate data so that it can be referred to later. This, however, may lead to the model itself being excessively large. A more convenient approach is to arrange for the model to conduct the analysis in iterative loops in which only the results needed for later parts of the analysis are stored. When an event of interest is identified the user can cause the model to recalculate that particular scenario for inspection.

In one model used by DNV Energy the user is able to access a detailed event tree relating to the risks for a worker in;

- a given area when the release starts
- a given manning configuration
- a given isolation/blowdown failure combination
- a given ignition condition, and
- a given release size of
- a given hydrocarbon release scenario

This allows the analyst to view the progress of the calculation in detail. This may enable them to identify that some items of input data are inappropriate, and hence can be corrected, or to identify the significant safety critical parameters for the installation.

RESOLUTION/DEGREE OF DETAIL

The resolution to which a QRA model is constructed is important. In this context "resolution", sometimes referred to as "granularity", is the degree to which the various possible scenarios are treated independently. It is common, for example, to average the effect of release direction when determining the consequences for structures and workers in the various areas of a platform. One approach is to consider six broad directions (up, down, north, south, east and west), assign a failure probability or fatality rate for each direction but then to average these before moving on to the next stage in the analysis. The model does not calculate the number of fatalities which would be expected for a release in each of the directions but only the average. This does not affect the calculation of risk for the platform workers but it does make it more difficult to examine the model to gain an understanding of the impact of the hazard. Identifying why the fatality rate might be a certain value for a certain release direction may be apparent to the analyst but the origins of the average may be more obscure. Typically many more parameters such as manning density, explosion strength, HVAC (Heating, Ventilation, Air Conditioning) shutdown and the effect of weather conditions on evacuation fatality rates are all averaged. This means that when inspecting the flow of data through an event tree the analyst is looking at an average effect of many different parameters rather than a specific scenario which would be easier to evaluate.

Resolution is particularly important if the model is required to produce F-N curves. This is because while intermediate parameters can be averaged without making a difference to the overall risk measured by traditional parameters such as IRPA, FAR and PLL the distribution of f-N pairs is affected by the averaging process.

A fatality rate of 0.1 applied to an exposed population of 20 will result in an average of 2 fatalities. This might mean;

- that on each occasion 2 persons will be killed
- that on 90% of occasions everyone will survive but that everyone will be killed in the remaining 10%, or
- that 50% of the time 2 persons will be killed, 25% of the time 4 persons will be killed and 25% of the time that all will survive.

There is actually a spectrum of possibilities all covered by the same fatality rate. Each of these will result in the same contribution to personal risk but the f-N pairs generated are different and so the shape of the F-N curve will be changed. This means that the relatively few incidents which result in large numbers of fatalities can become diluted by the more numerous incidents where the number of fatalities are small or zero. This situation is illustrated in figure 10 which shows two curves from a hypothetical QRA. Both curves portray the risk distribution from a set of hazardous scenarios and have the same PLL. The difference is that some parameters have been averaged when arriving at the result depicted by the "Low Resolution" curve. This may make the difference between passing and failing an acceptance criterion especially if a risk aversion factor greater than unity is used.

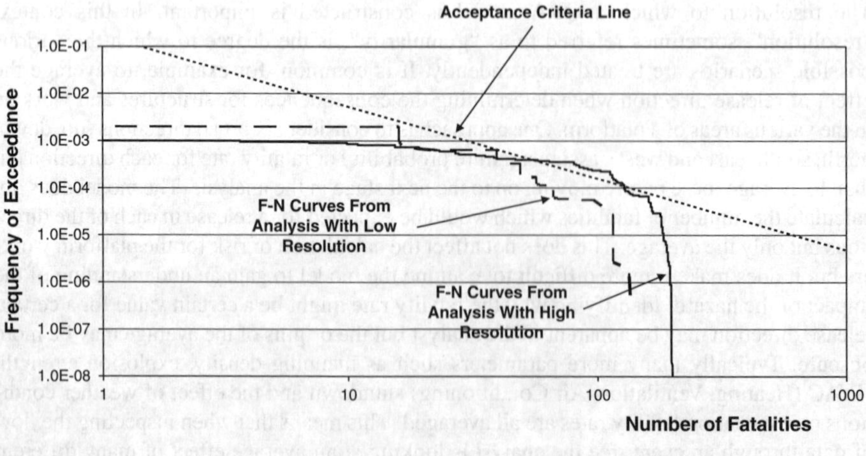

Figure 10. Variation in F-N curve output with resolution

Conversely, there is also a danger of making the QRA so complex that the user loses an overview of the situation. The complexity must be balanced with respect to the state of knowledge, availability of data etc. It may also be inappropriate to have detailed methodologies employed in one part of the model while other parts employ relatively simple techniques.

More complexity can be added but it then becomes increasingly important to implement it in a clear logical structure.

One way of dealing with this issue is to use a set of acceptance criteria measuring the risk for different groups of people in different ways. Achievement of a combination of criteria for maximum individual risk together with average risk for a group of people and F-N curves will give a better presentation of the risk than measuring only one of the criteria.

COMPETENCE

The above aspects focus on the structure and methodology of the model itself, but how it is used is also important. It may be possible to construct a model and make its operation so automatic that the role of the analyst is reduced to entering data, and initiating the run and transferring the results to a report. In these situations it can become easy to accept the results without spending the time to check that the model is adequately representing the frequency and consequences of the various risks or in deriving an understanding of the key drivers in the analysis.

The range of consequence analysis, techniques and mathematical operation in a QRA may be quite extensive and cover topics which the analyst will not have met in their education. In order to get the best value from a QRA the analysts have to be adequately trained in the various aspects of consequence analysis and the operation of the model being used. Although some university programmes on risk assessment exist, most analysts come from a more general engineering background so the obligation will tend to fall on QRA providers to educate their junior analysts through structured training and on the job experience combined with appropriate supervision.

ADDED VALUE

In most parts of the world, QRAs focus on the threat to human lives. While this should remain the primary focus it should be noted that many of the results can also form the basis of an analysis of other forms of loss, particularly environmental damage, asset damage and loss of production. Rather than studying these types of loss separately they can be combined into a single analysis which can evaluate the total risks resulting from the hazards present.

The QRA can provide results such as expected number of days of production loss per year and how this is distributed (many small stops or fewer larger ones) together with the expected amount and distribution of accidental oil spills. Results can be portrayed as an annual average or as frequency-cost curves. This can again be used as input to specific economic and environmental studies and to emergency planning.

CONCLUSIONS

Risk Analysis has been successfully used to support decisions relating to the safety of offshore installations. However, the benefits of using the technique may not always have utilised its full potential. This is partly because the creation and use of QRA models is a complex process which requires a great deal of effort to implement effectively and also because the diversity of methods has led to inconsistencies in output. Steps are therefore being taken to address these issues which should result in the technique delivering greater value as an input to engineering decision support with consequent benefits to cost effective safety, environmental and business risk management.

The key factors in creating an effective approach can be summarised as follows;

- Active involvement of the customer in the process and better synchronising between the analysis and the decision process.
- Standardisation of approach and methodologies
- Use of a clear structure which the analyst can understand
- Provision of a mechanism in the model to allow analysts to inspect the various scenarios to gain understanding of the key factors affecting risk.
- Provision of alternative methodologies appropriate to the stage in the design and the availability of data.
- Using appropriate higher degrees of resolution in the QRA.
- Striving to continually improve levels of competence among practitioners.

When implemented, the quality and consistency of results obtained should provide a better basis for decision making. The same approach should also deliver benefits in the field of onshore risk analysis.

REFERENCES

1) "Problems in the Maintenance and Development of QRA Models for Offshore Platforms", B. Bain, Proceedings of the 12th International Conference on Major Hazards Offshore, London, December 2003.
2) "Ignition Probability Review, Model Development and Look-Up Correlations", IP Research Report, Energy Institute, January 2006.
3) "Good practice and pitfalls in risk assessment", S. Gadd, D. Keeley and H.Balmforth, Health & Safety Laboratory, HSE Research Report No. 151, 2003.
4) "Risk and emergency preparedness analysis", NORSOK Standard Z-013, Rev 2, 2001.
5) "Hydrocarbon release system" http://www.hse.gov.uk/offshore/hydrocarbon.htm, HSE, 2007
6) "The Offshore Installations (Safety Case) Regulations 2005", Statutory Instrument 2005 No. 3117, ISBN 0110736109.

HANDLING OF REACTIVE CHEMICAL WASTES – A REVIEW

J C Etchells, H James, M Jones, and A J Summerfield
Health and Safety Executive

A study has been made of 142 incidents reported to HSE, caused by unintentional or inadequately planned mixing of incompatible waste chemicals, or the decomposition of thermally unstable wastes. 62% of the incidents occurred at waste producer sites, the remainder occurring during waste treatment and transit. The immediate effects of such incidents included fires, explosions, chemical releases and drums rocketing off-site. In some cases employees were killed or injured. Five common reaction types accounted for over 68% of incidents where the chemistry was known. In most cases these reactions could be linked to specific industry types, in particular the chemical industry and engineering/metal treatment.

This paper reviews the incidents and their causes, many of which were failures to take simple precautions, such as properly characterising, packaging and labelling the waste, particularly during "bulking up" into storage containers. The guidance available to prevent such incidents has been identified and, where gaps were found, suggestions to take matters forward with industry are made. A particular issue is the screening procedures required before waste chemicals are mixed, particularly in large tanks and reaction vessels, both at waste producer and waste-treatment sites. An ongoing research project on scale-up, being carried out for HSE by HSL is described. The Environment Agency (EA) is taking an active interest in this project.

INTRODUCTION

A number of incidents have been reported to HSE which occurred during the storage, handling and transport of chemical waste, caused by unintentional or inadequately planned mixing of incompatible chemicals, or in some cases by the decomposition of thermally unstable substances. The incidents occurred, in particular, at waste producer sites (such as the chemical industry), and also at waste-treatment sites and during transport. Concern amongst HSE inspectors led to the formation of a small HSE working group to review the causes of these incidents and identify where further action is needed, including the need for guidance. This paper discusses the main findings of the WG and how HSE is taking matters forward in collaboration with the Environment Agency (EA). The need to minimise the amount of waste produced in order to avoid/minimise the impact of such incidents is outside the scope of this paper, although it clearly should be considered.

Table 1. Number of incidents by location

Waste producer	100	62%
Waste treatment company	40	25%
Transport	14	8%
Information not available	9	5%

ENFORCEMENT AGENCIES

HSE and EA/SEPA (Scottish Environment Protection Agency) both have roles for regulating the supply, handling and transport of hazardous wastes. HSE take the lead on safety issues. The role of EA/SEPA includes enforcing pollution prevention and control legislation (PPC), regulating hazardous waste movements and licencing and inspecting waste storage/treatment sites, including those at the waste producer. There are memorandums of understanding between HSE and EA/SEPA defining the arrangements for liaison and co-operation between the agencies.

INCIDENT HISTORY

A review of incidents reported to HSE was carried out, to provide information on key technical areas and industries involved. 142 incidents involving reactive wastes were identified for the period 1989 to 2005. The average number of incidents per year (over the 11 years where data was most reliable) was approximately ten.

The incidents ranged from relatively minor explosions in drums to major releases of hazardous chemicals. The effects included fires, explosions, chemical releases and drums rocketing off-site. In some cases employees were killed, in others large numbers of people were evacuated from the surrounding area and the incidents led to environmental effects, such as contamination of the watercourse.

A breakdown of the number of incidents by location is given in Table 2. The majority of incidents (62%) occurred at waste producer sites, and typically involved unintended mixing/reaction in drums and containers, which occurred for example as a result of mixing incompatible chemicals together (often termed "bulking up"). The main industries involved (other than waste-treatment) were chemical manufacture and supply (41%), and engineering and metal treatment (22%). Smaller incidents at schools, colleges, universities, hospital teaching labs accounted for 20% of the incidents. Other industries involved included printing, ceramics, nuclear, electronics and leather processing. Poor control of intentional mixing of chemicals for treatment was involved in 15% of incidents at waste producer sites.

Failing to take simple precautions was the most common factor involved in incidents at waste-producer sites, for example checking the label for compatibility with other chemicals (particularly during bulking up), appreciating the composition of the waste and its hazardous properties, and contamination or the addition of an incompatible chemical.

Table 2. Business activities at waste-producer sites where incidents occurred

Of the incidents at waste producer companies, 85 involved companies whose main activity could be readily identified:

Main business activity of waste producer company	Number of incidents	Percentage
Chemical manufacture and supply (including solvents, lubricants, fertilisers, paints and varnishes, plastics, rubber chemicals)	35	41
Engineering, metal plating, metal finishing, motor vehicle repair	19	22
Schools, colleges, universities, hospital teaching labs	17	20
Printing, reprographics	3	4
Ceramics products and services	2	3
Nuclear	1	1
Contact lens manufacture	1	1
Electronics manufacture	1	1
Geologists	1	1
Leather processing	1	1
Textile coating	1	1
Packaging manufacture	1	1
Total	85	100

CHEMICALS INVOLVED

The types of chemical reaction involved in the incidents, and the associated industries involved, have been reviewed. It was found that 6 reaction types accounted for 68% of the incidents, and 87% of incidents where the chemistry was known. These are summarised in Table 3. 22% of the incident records had insufficient information to deduce what reaction had been involved, and it is a matter of concern that the contents of the waste does not appear to have been sufficiently known by the waste handlers.

A significant number of the incidents, perhaps not surprisingly, involved strong acids, particularly nitric and sulphuric acids, reactive metals, sodium hypochlorite and reactive monomers.

Although most industries appear to show a fairly uniform distribution of reaction scales from small containers to bulk tanks and vessels, in the plastics industry most incidents occurred in drums. There is evidence to suggest that the practice of "reacting off" some "waste" monomers in drums is occurring at some sites.

The main issues are discussed below.

Table 3. Chemical reactions involved in the accidents

Chemical reaction	% of Total
Reactions involving potentially reactive metals, n.b aluminium, magnesium, sodium, with water and acids to produce hydrogen.	19%
Nitric acid in combination with various solvents and acids to produce unexpected unstable nitrations.	14%
Reaction of hypohalites, e.g. sodium hypochlorite (often in bleach), with acids to produce chlorine	10%
Unstable monomers reacting together in runaway polymerisations	9%
Sulphuric acid and metals producing hydrogen	8%
Acid/base neutralisations	8%
Unknown Chemistry	22%
Other	10%
Total	68%

WASTE CHARACTERISATION

Waste streams need to be characterised sufficiently to ensure that they can be safely handled and stored, for example to prevent them being inadvertently mixed with incompatible chemicals, and so that they can be safely treated. A number of incidents occurred due to failures to adequately characterise the waste, both at waste producers and during subsequent handling and treatment.

Guidance on characterisation is given in reference 1, published by the EA. Although primarily aimed at waste-treatment operators, it also contains useful guidance for waste producers. Properties requiring characterisation include:

- Quantity;
- Chemical analysis (individual constituents and their percentages, as a minimum, are usually required);
- Form (solid, liquid, sludge, viscosity etc.);
- Hazardous properties, e.g. flammability, toxicity, corrosiveness, reactivity, thermal stability, etc;
- The specific process from which the waste is derived

The type of analysis required should be part of a risk-based approach. This can require the exercise of considerable professional judgment. Some guidance on this is given in references 2 and 3. However most of the factors that need to be considered are standard good chemical engineering practice, including:

- The need to identify the critical components of the waste stream, e.g. heavy metal concentrations in wash waters, acid concentration, etc. These play an important role in defining storage requirements, and ultimately any treatment routes;

- The physical nature and uniformity of the waste, for example the number of phases present;
- The variability (and potential variability) of the processes that have produced the waste. For example, waste from trial runs and batch processes are likely to be more variable than that from established and continuous processes;
- The treatment process or handling regime to be used, and its sensitivity to variations in waste characteristics. "Off spec" product is a particular risk, for example, if it contains unreacted or thermally unstable material, and procedures to deal with this should be in place where necessary.
- the need to record the results of the characterisation tests systematically.

SAMPLING

Where the contents of a waste container/stream are not already known, a suitable sampling regime is needed to ensure that the samples obtained are truly representative. The development of the sampling regime should form part of the technical assessment and take particular account of the variability of the waste, arising from:

- the normal process;
- foreseeable deviations;
- variability within the container such as settling and multiple phases.

The analysis required will depend on the nature of the waste, the treatment process to be used and what is already known about the waste. In particular, the following checks should be made:

- A check on the constituents;
- Determination of all relevant hazardous characteristics, for example pH, flashpoint;
- Further analysis relevant to the waste stream or treatment method, for example cyanide or chloride content.

It is important that the technique selected produces a representative sample, for example by taking a core sample to the base of the container, or by producing a composite sample from several individual samples taken at different points in the container. The most suitable sampling technique will depend on:

- the physical and chemical characteristics of the waste;
- the type of container to be sampled; and
- any special circumstances, such as highly corrosive or toxic properties, the presence of multiple phases or volatile constituents.

A wide variety of sampling techniques are available, both for on-line sampling, e.g. from continuous waste streams, and for sampling of containers, such as drums and tankers. Information on sampling of hazardous wastes is given in References 4 to 18. Until recently, most of these were not specific to waste materials. However CEN/TR 15310, parts 1–5[14–18] were published recently and give useful information on waste sampling criteria, techniques, field sampling, packaging and transport and developing sampling plans.

Open sampling tubes are often used by the industry. Reference 8 specifies that one end should be narrow, so that the thumb can easily close it (the standard is not specific to hazardous waste and consideration of personal exposure during sampling must be included in the risk assessment). The other end is drawn to an orifice diameter that is specified for the viscosity of the material to be sampled. Reference 8 also states that open sampling tubes are not suitable for sampling very viscous liquids because of the difficulty in getting a representative sample and retaining it in the tube.

BULKING OF BATCHES

Bulking is the placing of smaller quantities of what should be similar materials for storage and/or transport into larger containers. It does not include the intentional mixing of chemicals for treatment, which is discussed later. Bulking up of chemicals accounts for the approximately 70% of incidents at waste producer sites, due to:

- Inadvertent mixing of incompatible chemicals;
- The use of contaminated/inadequately cleaned or emptied storage containers;
- Contaminated waste, e.g. waste containing reactive materials such as peroxides;
- Self-reactive materials or mixtures (e.g. monomers) being placed in the container either too hot, or for too long, or with insufficient inhibitor;
- Water leaking into waste containers (mainly drums) and reacting.

EXAMPLES INCLUDED:

An operator was injured after being in the vicinity of an exploding IBC. It had contained alkaline solution and was wrongly identified as being suitable for disposal of dilute acid. It exploded six hours after the dilute acid was added.

An operator decanted sulphuric acid into a drum containing waste chlorinated solvents for disposal. The subsequent reaction resulted in chemical burns to his face, neck and arm.

Scrap resin in a number of 200 litre drums in a warehouse started to react. Fumes issued from one, and others showed signs of pressurisation. The drums were cooled and the bungs removed. The next day, one drum was found 9–12 m away.

In some cases incidents also occurred in road tankers, due to bulking up of incompatible chemicals.

Reference 1 gives guidance for bulking of batches at waste treatment sites. This requires such operations to take place under the "direct instruction and supervision of a suitable manager/chemist and should be under local exhaust ventilation in appropriate cases". There does not appear to be equivalent guidance for waste producer sites, many of which may not employ chemists (e.g. printers and metal treatment works) and HSE are planning to prepare a short information sheet which will be available on the web-site. This will stress the need for operators to consider in advance the likely range of operations to be undertaken, so that appropriate instructions can be given to their operating personnel.

USE OF COMPATIBILITY CHARTS AND TOOLS

Where chemicals are to be mixed or bulked a number of tools have been developed to assist in assessing their compatibility, such as matrices and computer tools. These are useful aids, but they should be interpreted by a competent person and not relied on in isolation. For example they may not take account of concentration effects and may only be able to consider binary mixtures, so the effects of impurities, reaction intermediates and catalysts may not be accounted for. Matrices may also assume a homogeneous reaction mixture, which is known not to be the case with many waste streams. In addition they do not always take account of temperature effects. One of the models failed to predict incidents from the HSE data where heating occurred, for example, when steam entered a tank of waste containing dinitrochlorobenzene.

One approach that has been adopted in the USA is based on work conducted by the US Environment Protection Agency[19]. This uses a matrix to define a generic compatibility look up table to identify potential incompatibilities. Comparison of the model with the HSE accident data showed a good performance in predicting most of the incidents that occurred. However it is important that the matrix is not relied upon in isolation for the reasons mentioned above.

A Chemical Reactivity Worksheet[20] is available free of charge from the US Environmental protection Association (EPA) and National Oceanic and Atmospheric Administration (NOAA). It includes a database of reactivity information and a way of virtually "mixing" chemicals to find out what dangers could arise from accidental mixing. The EPA/ NOAA website does warn that, whilst they believe it can accurately predict whether or not a reaction will occur between two chemicals, there are difficulties in predicting more complex mixtures.

STORAGE, PACKAGING & LABELLING

Poor standards of storage (particularly segregation), packaging and labelling can initiate and contribute greatly to the speed of escalation and severity of incidents involving chemical wastes.

Problems with labelling were found to include lack of labelling, incorrect or inadequate labelling and multi-labelling.

- *Chlorine evolved due to inadvertent mixing of hydrochloric acid and sodium hypochlorite residues from carboys being emptied for cleaning. Both carboys were labelled "hypochlorite waste".*

Guidance on labelling of waste for transport to transfer or treatment stations is given in reference 1. In cases where the waste is to be transferred for supply purposes, e.g. for blending into cement kiln fuel (cemfuel), then CHIP classifications[21] may be required.

Where waste is to be treated on site, no specific guidance was found, although the transport classifications form a good basis. For reactive wastes there is also a need to consider chemical compatibilities within the transport classes, for example using the EPA compatibility tables.

Packaging, considerations should include the integrity of the packaging and its compatibility with its contents, or any previous contents.

- *A wooden pallet containing three drums of chlorate waste caught fire whilst being moved prior to loading onto a vehicle, possibly due to the pallet reacting with spilt chlorate.*

Most container manufacturers should provide guidance on the compatibility of their products. However, manufacturers' guidance does not normally include any consideration of compatibilities for re-use.

With regard to storage, management issues include storage area design, segregation, separation, housekeeping, fire prevention and emergency response.

EA, in collaboration with HSE, has recently redrafted its guidance on the storage of hazardous wastes[22] and relevant findings from the HSE study have been incorporated.

WASTE TREATMENT (MIXING)

There are a number of options for the treatment of hazardous wastes, e.g. incineration, biological treatment. This paper discusses the mixing of incompatible wastes as part of their treatment, e.g. acid-base neutralizations, cyanide destruction by hypochlorite.

A search of the incidents occurring at waste producer sites found 15% to be due to intentional mixing over the 25-year period. Examples included:

Runaway reaction during de-odourisation of chemical waste containing hydrogen peroxide. Probably caused by over-addition of caustic soda. Explosion and resulting fire caused widespread damage. One person killed, two injured, more than 500 evacuated.

Chemist sprayed with corrosive mix when neutralising an acid copper solution using caustic soda pellets. Too much added at once.

EA Document S5.06[1] gives guidance on waste treatment. It specifies that any waste should be adequately characterised prior to treatment and the need for compatibility testing. Where chemical reaction hazards exist it refers to HSE's guidance "Designing and Operating Safe Chemical Reaction Processes"[23] for further guidance. However problems arise where companies may not subject such processes to the same rigour as they would for planned chemical reactions. HSE and EA are considering the way forward on this issue.

INCIDENTS AT WASTE-TREATMENT SITES

A number of serious incidents also occurred at waste treatment sites, particularly during bulking up and waste treatment. Several of these resulted in serious off-site consequences and damage to the environment, for example pollution of watercourses. In addition, several incidents caused substantial local disruption and public concern, and led to prosecutions by EA and/or HSE.

Following such incidents, the industry has been required to improve its screening procedures, including compatibility checks, sampling and laboratory tests. It has responded by using simple Dewar flasks to pre-screen their samples of material for compatibility before mixing them on the large scale, both for storage and treatment. Temperature rises of between 6–10°C in the Dewar vessel over a period of 10 minutes have been suggested by the industry to indicate an exothermic reaction of concern, with gas bubblers being used to measure any gas generation. Whilst this approach may be valid for smaller volumes, HSE has questioned the validity of these test methods for scale up to large volumes. Comparisons of the heat losses between Dewar flasks and large-scale vessels are given in the literature, for example references 23–26.

As a result of these concerns, HSE have asked HSL to review existing data/literature on cooling rates and heat losses data for various pieces of calorimetric equipment, including Dewar vessels, and examine further the scale up limitations with particular reference to the waste treatment industry. Whilst it would be expected that individual occupiers would justify this approach for their own individual installations, such information would provide useful information for inspectors and should be of wider interest to the industry. This Work is being reported separately[27].

ENSURING WASTE SAFE FOR TRANSPORT

Nine incidents involving transport were recorded. The most common cause was inadequate cleaning of tankers between loads. This led to unplanned reactions that resulted in explosions, loss of containment and fires. Examples included:

- *A runaway reaction occurred in a tanker collecting caustic waste, due to metallic aluminium in the tanker.*
- *Hypochlorite solution in a tanker reacted with methanol that had previously been transported.*
- *An exothermic reaction in a road tanker containing tallow oil. The tank had not been cleaned out following a previous load of nitric acid sludge.*

HSE are planning targeted guidance on the specific issue of tanker cleaning in the form of a short information sheet.

CONCLUSIONS

A large number of incidents are reported to HSE involving the storage, handling and transport of chemical waste, most of which are caused by unintentional or inadequately planned mixing of incompatible chemicals.

Of the incidents analysed, 62% occurred at waste producer sites. Of these, 70% occurred during bulking-up. EA Guidance on this is available for waste treatment sites, and this could usefully be applied at waste-producer sites. HSE are planning to produce a short information note on this topic.

An analysis of the chemical incompatibilities that have caused the incidents showed that six common reaction types accounted for 87% of incidents where the chemistry was known. In most cases these could be identified with particular industry types.

The incident analysis indicated that many companies had failed to adequately characterise their waste products, and hence establish a suitable disposal route. Guidance on the characterisation of such wastes has been identified. Most of the factors that should be considered are standard good chemical engineering practice. Poor standards of storage, packaging and labelling also contributed to the speed of escalation and the severity of incidents. Joint EA/HSE guidance on this topic has been prepared.

Poor control of intentional mixing was involved in 15% of the incidents at waste producer sites. Although the total number is not large, their potential consequences can be severe and it is important that these operations are subjected to rigorous risk assessments, as for any chemical operations.

25% of incidents occurred at waste treatment sites, including many of the more serious accidents, both during mixing and storage. EA and HSE are active in inspecting waste treatment sites and a research project on the issue of compatibility testing is underway. The incidents also illustrate the importance of adequately controlling and characterising the wastes, including upstream at waste-producer sites and during transfer.

REFERENCES

1. EA Sector Guidance Note IPPC S5.06 Guidance for the Recovery and Disposal of Hazardous and Non Hazardous Waste, available on the Environment Agency website
2. "Design and development of a hazardous waste reactivity testing protocol", Wolbach CD, Whitney RR and U Spannagel, report no. EPA-600/2-84-057, US Environmental Protection Agency, February 1984
3. Management of Process Industry Waste, Bahu, R., Crittenden, B., & O'Hara, J., IChemE, ISBN 0-85295-324-0
4. "Methods for sampling chemical products - Part 1: Introduction and general principles", BS 5309-1, 1976
5. "Sampling procedures for inspection by attributes – Part 0: Introduction to the BS 6001 attribute sampling system", BS 6001-0, 1996 (ISO 2859-0, 1995)
6. "Encyclopaedia of Chemical Technology", Kirk-Othmer, 4th edition, 1997, Vol. 21 pp.626–650
7. "Waste analysis at facilities that generate, treat, store and dispose of hazardous wastes", A Guidance Manual, 1994, available on the US EPA web-site
8. "Methods for sampling chemical products Part 3: Sampling of liquids", BS 5309-3, 1976
9. "Methods for sampling chemical products Part 4: Sampling of solids", BS 5309-4, 1976

10. "RCRA Waste sampling draft technical guidance – planning, implementation and assessmcnt", US Environmental Protection Agency, EPA530-D-02-002, August 2002, available from the US EPA web-site
11. "A practitioner's guide to testing waste for onward reuse, treatment or disposal acceptance", Environmental Services Association Research Trust (ESART), WRc Ref: UC6656, July 2004
12. "Guidance on sampling and testing of wastes to meet landfill waste acceptance procedures", Environment Agency, Version 1, April 2005: available from the EA web-site
13. "Characterization of waste - sampling of waste materials - framework for the preparation and application of a sampling plan", EN 14899, CEN, 2005
14. prCEN/TR 15310-1 "Characterization of waste - sampling of waste materials - Part 1: guidance on selection and application of criteria for sampling under various conditions"
15. prCEN/TR 15310-2 "Characterization of waste - sampling of waste materials - Part 2: guidance on sampling techniques"
16. prCEN/TR 15310-3 "Characterization of waste - sampling of waste materials - Part 3: guidance on procedures for sub-sampling in the field"
17. prCEN/TR 15310-4 "Characterization of waste - sampling of waste materials - Part 4: guidance on procedures for sample packaging, storage, preservation, transport and delivery"
18. prCEN/TR 15310-5 "Characterization of waste - sampling of waste materials - Part 5: guidance on the process of defining the sampling plan"
19. A Method for determining the compatibility of hazardous wastes, H.K.Hatayama, J.J.Chen, E.R. de Vera, R.D. Stephens, D.L.Storm, EPA Paper – 600/2-80-076. 1980
20. Chemical Reactivity Worksheet, available from the US Office of Response and Restoration web-site
21. CHIP for Everyone, HSE Guidance Booklet, HS(G)228, ISBN 0-7176-2370-X, HSE Books
22. Proposed EA, SEPA, NI EHS and HSE Joint Guidance on the Storage of Hazardous Wastes, downloadable from the EA web-site
23. *Designing and Operating Safe Chemical Reaction Processes*, HSE Guidance Booklet, HS(G) 143, ISBN 0-7176-1051, HSE Books
24. Wright, TK & Rogers, RL, Adiabatic Dewar Calorimeter, IChemE Symposium Series No. 97, 121–132, 1986
25. J A Barton & R L Rogers (editors), 1997, *Chemical Reaction Hazards*, 2nd edn., ISBN 0 85295 341 0, IChemE, Rugby
26. Singh, J, Safe Scale-up of Chemical Reactions, Chemical Engineering, May 1997
27. Dewar Scale Up for Reactive Chemical Waste Handling, L Vechot, J Hare, hazards XX, 2008

DEWAR SCALE-UP FOR REACTIVE CHEMICAL WASTE HANDLING

Luc Véchot* and John Hare
Health and Safety Laboratory, Buxton, Derbyshire, SK17 9JN, UK
*Corresponding author

The use of non-pressurised Dewar flasks has been proposed by some parts of the chemical waste treatment industry to determine the exothermic reaction incompatibility of mixtures. Temperature rises of between 6–10°C in the Dewar vessel over a period of 10 minutes have been suggested by the industry as criteria to indicate an exothermic reaction of concern. This paper reports work sponsored by the Health and Safety Executive (in discussion with the Environment Agency) to investigate the limits of this method for scale-up to vessels typically used in the waste treatment industry.

A literature review of the specific heat losses from Dewar flasks and large-scale vessels is compared to specific heat losses of Dewar flasks measured experimentally. Typical values of thermal characteristics of large-scale vessels used in the waste industries have also been assessed. The specific heat loss in the Dewar flask and large vessels are very different. Scale-up limits of four types of Dewar have been calculated for different values of overall heat transfer coefficients for large-scale vessel.

Thermal behaviour of exothermic reactions in a Dewar flask has been compared to that predicted in large vessels using reaction kinetics and heat transfer models. For fast and highly energetic reactions the reaction energy release rate can be significant compared to the heat losses and the Dewar flask can detect runaway reactions. However, for low energy reactions or reactions with long induction time, the heat losses can be significant compared to the heat release rate and the Dewar test can then miss exotherms or give non-conservative results.

It appears that the 6–10°C criterion proposed by the waste treatment industry might be observed when the heat losses do not have a significant importance compared to the reaction heat release rate. However, the reaction completion time at large scale would be shorter than at the Dewar scale. In some cases, 10 minutes might be sufficient to detect the exotherm but not the runaway reaction. The test should therefore be run to reaction completion in order to fully detect exotherms. Reliable conclusions about the scale-up of Dewar data can be obtained when the chemical reaction kinetics are well known. Unfortunately this is not generally the case in the waste-treatment industry. So, unless the specific heat loss of the Dewar has been shown to be less than large-scale vessels, this method in isolation is likely to be unreliable for scale up to large vessels.

KEYWORDS: Dewar flask, waste treatment, runaway reaction

1. INTRODUCTION

In some parts of the waste treatment industry, part of the procedures used by some companies to assess exothermic reaction incompatibility of mixtures is to use non-pressurised Dewar flasks. The Dewar calorimeter is a flask containing a vacuum jacket that minimises the heat losses from a reacting mass to the surroundings. The results of these tests have been applied directly to quite large vessels. Temperature rises of between 6–10°C in the Dewar vessel over a period of 10 minutes have been suggested by the industry to indicate an exothermic reaction of concern, with gas bubblers being attached to such flasks to detect any gas generation.

The objective of this research was to investigate the range of applicability of the Dewar flasks in assessing thermal hazards. This was undertaken in three steps:

a) Thermal characteristics of Dewar flasks were gathered from literature and measured experimentally.
b) The limits of the use of Dewar flasks were calculated for different values of overall heat transfer coefficients for large-scale vessel.
c) The thermal behaviour of two exothermic reactions in a Dewar flask were compared to that predicted in typical large-scale vessels used in the waste treatment industry using reaction kinetics and heat transfer models.

The use of Dewar flasks is recommended in the United Nations "Recommendations on the Transport of Dangerous Goods" (ST/SG/AC.10/11/ Rev.4)[1] to simulate transport packaging. The criterion used for the scale-up is the specific heat loss, noted as $\dot{q}_{loss}$ (W.kg^{-1}.K^{-1}):

$$\dot{q}_{loss} = \frac{UA}{m} \tag{1}$$

This criterion can be understood as the cooling potential of a vessel showing a Newtonian cooling behaviour. Several authors[2,3] have demonstrated that this method is suitable in the case of a well-stirred tank, showing a homogeneous temperature distribution. Homogeneous systems were therefore investigated as a first approximation. The scale-up theory in the case of pure solids, high viscosity liquids or slurries, is much more complex, as the heat transfer is governed by the bulk material and not by the package or the vessel[3]. This makes such heterogeneous systems even more difficult to scale up.

2. THERMAL CHARACTERISTICS OF DEWAR FLASKS

Thermal characteristics of Dewar flasks have been gathered from the literature sources (see Table 1). These data were compared to experimental measurement of thermal characteristic of two commercial Dewar flasks.

2.1 DATA FROM THE LITERATURE

Table 1 presents some thermal data (including the specific heat losses values) for different types and arrangements. The general orders of magnitude of the specific heat loss are:

- 500 ml Dewar flask (unknown material): between 0.04 and 0.077 W.kg^{-1}.K^{-1}
- 1000 ml glass Dewar flask: 0.018 W.kg^{-1}.K^{-1}
- 1000 ml stainless steel Dewar flask in adiabatic oven: 0.195 W.kg^{-1}.K^{-1}

Table 1. Dewar flask thermal characteristics

Type of Dewar	Volume	Thermal inertia factor ϕ	Specific heat loss $\dot{q}_{loss}$	Cooling rate	Heat transfer coefficient or conductivity	Source
Cylindrical (filled with 400g of asphalt salt mixture)	0.5 litre	1.89 (Cp = 840 J/K/kg)	0.03 W/l/K	–	λ = 1.46 W/K/m	4
Cylindrical	0.5 litre	–	0.04 W/kg/K	–	U = 0.6 W/m2/K	2
Cylindrical (filled with 400 ml of water)	0.5 litre	–	Between 0.062 to 0.0768 W/kg/K	–	–	5
Glass cylindrical + cork bung	0.5 litre	$1.05 < \phi < 1.5$	0.03 W/l/K	1.6 K/hr	–	6, 7
Glass cylindrical + rubber bung	0.5 litre		–	2 K/hr	–	
Cylindrical Sensitivity = 0.01 K/min P_{max} = 35–50 bar, T_{max} = 300°C	1 litre	$1.1 < \phi < 1.2$	–	–	–	8
Glass cylindrical	1 litre	–	0.018 W/kg/K	–	–	9, 10
Stainless steel cylindrical in adiabatic oven	1 litre	–	0.195 W/kg/K	–	–	
Spherical + PTFE bung	1 litre	1.34	–	0.8 K/day	λ = 0.18 W/K/m	11

2.2 EXPERIMENTAL MEASUREMENTS

As indicated in Table 1, the thermal characteristics of a Dewar flask will depend on its construction and experimental configuration. Literature data have been compared to experimental measurements of the specific heat losses for two unstirred commercial Dewar flasks (Table 2):

- "1 litre glass Dewar in a stainless steel container" which can be closed by a clipped on vacuum lid.
- "1 litre stainless steel Dewar" which can be closed by simple stainless steel lid (without vacuum). A layer of insulating polymer is stuck on the internal surface of the lid.

The flasks were placed on a laboratory table (not in a temperature controlled oven) and filled with 800g of hot water (70–80°C). Cooling of the water was recorded for both closed and open configurations. Ambient room temperature was also recorded.

The cooling curves obtained were then used to assess the specific heat loss coefficients for the Dewar flasks, using the following Newtonian cooling model:

$$T(t) = (T_{ini} - T_{ext\,ini})\exp\left(-t\left(\frac{\dot{q}_{loss}}{Cp_l}\right)\right) + T_{ext} \tag{2}$$

Figure 1 (a) and (b) show the results for both glass and stainless steel Dewar flasks in a closed configuration. The experimental cooling curves fit the Newtonian cooling model well, which allows the calculation of the corresponding $\dot{q}_{loss}$. As expected, the glass flask (0.086 $W.kg^{-1}.K^{-1}$) showed less heat losses than the stainless steel flask (0.103 $W.kg^{-1}.K^{-1}$) (Table 2). The value of $\dot{q}_{loss}$ for the closed glass Dewar measured experimentally was approximately five times higher than the value in the literature for a typical "similar" Dewar (0.018 $W.kg^{-1}.K^{-1}$, Table 1).

Table 2. Thermal characteristics of two 1 litre Dewar flasks in both closed and open configuration

	Closed Dewar		Open Dewar	
	Glass	Stainless steel	Glass	Stainless steel
Internal Volume (m^3)	1.04×10^{-3}	1.14×10^{-3}		
m water before cooling (kg)	0.8	0.8	0.8	0.8
m water after cooling (kg)	0.8	0.8	0.72	0.73
$\dot{q}_{loss}$ ($W.kg^{-1}.K^{-1}$)	0.086	0.103	0.5	0.5

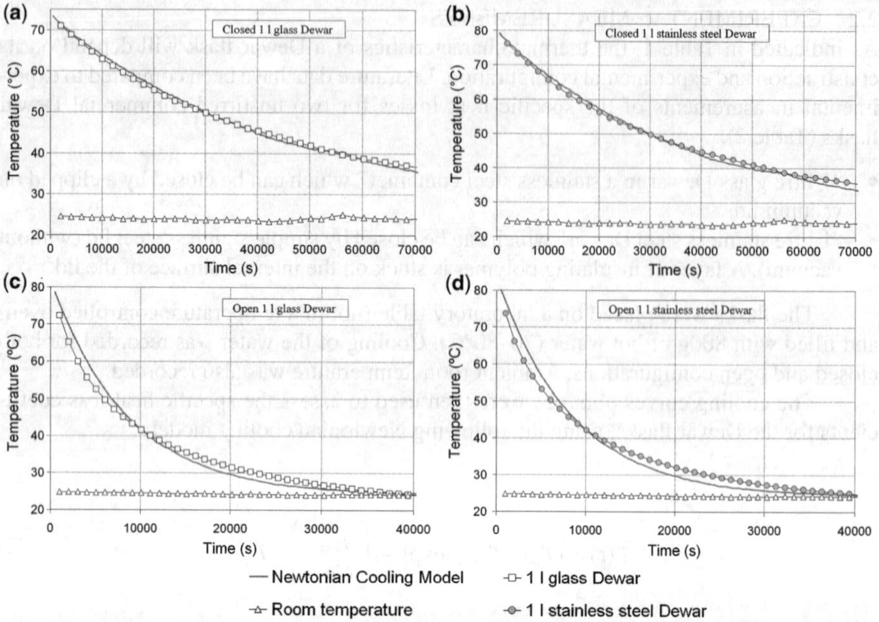

Figure 1. Both glass Dewar and stainless steel Dewar cooling curves in open and closed configurations

This experiment shows that for the same volume of Dewar (1 litre), significant differences can be found between literature values and those measured for a specific system. It is therefore very important to measure the value of the specific heat loss for the particular Dewar type instead of using a value from literature without any additional investigations. The cooling curves for the open configuration are presented in Figure 1 (c) and (d). The Newtonian cooling model does not fit the experimental data properly for the two Dewar flasks. This is due to the fact that heat exchange by free convection at the water surface and by evaporative cooling occurs. This is confirmed by the decrease of liquid mass in the open flasks (approximately 10%, see Table 2). The approximated values of $\dot{q}_{loss} = 0.5\ W.kg^{-1}.K^{-1}$ can, however, be stated for the open configuration assuming the behaviour approaches Newtonian behaviour. This value indicates that using an open Dewar instead of a closed Dewar can increase the heat losses by a factor of five.

3. LIMITS OF APPLICABILITY OF DEWAR FLASKS

The range of applicability of the Dewar flask method can be defined by comparing the specific heat loss of a Dewar flask and large vessels.

3.1 LITERATURE REVIEW

Several authors have undertaken the comparison of specific heat loss for large-scale vessels and Dewar flasks. Fierz[2] indicated that assuming a $\dot{q}_{loss}$ of 0.08 W.kg^{-1}.K^{-1}, a 500 ml Dewar flask will have the same cooling behaviour as 50 l package, whereas 0.04 W.kg^{-1}.K^{-1} makes it equivalent to a 500 l package.

Rogers[10] ran experiments to measure $\dot{q}_{loss}$ and half-life time (time taken for the temperature to fall to half its original value) for different size vessels, including Dewar flasks. He showed that:

- a 1 litre glass Dewar could simulate small plant reactors up to 12.7 m^3; and
- a 1 litre stainless steel Dewar in an adiabatic oven (with $\Delta T = -1K$) could simulate a 25 m^3 vessel.

Wright *et al*[11] showed that the cooling rates of 0.5 m^3 and 2.5 m^3 plant vessels are equivalent to those of 250 ml and 500 ml Dewar flask, respectively. This information seems, however, to be different to Fierz's work[2] described previously. The variation in $\dot{q}_{loss}$ for different Dewar flasks, as described in section 2.2, may provide some explanation for this.

The UN Recommendations on the Transport of Dangerous Goods[1] states that Dewar vessels, filled with 400 ml of substance, that have $\dot{q}_{loss}$ between 0.08 and 0.1 W.kg^{-1}.K^{-1} shall be representative of 50 kg packaging. No literature reference could be found for scale-up over 25 m^3. The volume of the vessels used in the waste treatment industry can be well above this limit (up to several hundred cubic meters).

3.2 SCALE-UP LIMIT: GENERAL APPROACH

A general approach has been developed to define some scale-up limits. The specific heat loss for different vessel volumes and overall heat transfer coefficients (Figure 2) have been calculated using equation 1. The following vessel features were used:

- Diameter to height ratio: D/H = 0.8
- Fill level = 80%

We assume that the thermal inertia (ϕ) is 1 for all the vessel volumes.

For a given value of the overall heat transfer coefficient, $\dot{q}_{loss}$ decreases with increasing the vessel size. This is due to the fact that the vessel heat exchange surface to mass ratio (*A/m*) decreases strongly when increasing the size of a vessel.

Four values of specific heat loss for Dewar flasks (from literature and our experimental determination) are indicated in Figure 2. Provided the reaction is allowed to run to completion, then for given values of U and vessel volume:

- if $\dot{q}_{loss\ \mathrm{Dewar}} < \dot{q}_{loss\ vessel}$: the use of Dewar vessel is conservative (heat losses are more important at the large-scale)
- if $\dot{q}_{loss\ \mathrm{Dewar}} < \dot{q}_{loss\ vessel}$: the use of Dewar vessel is non conservative (heat losses are more important at Dewar scale)

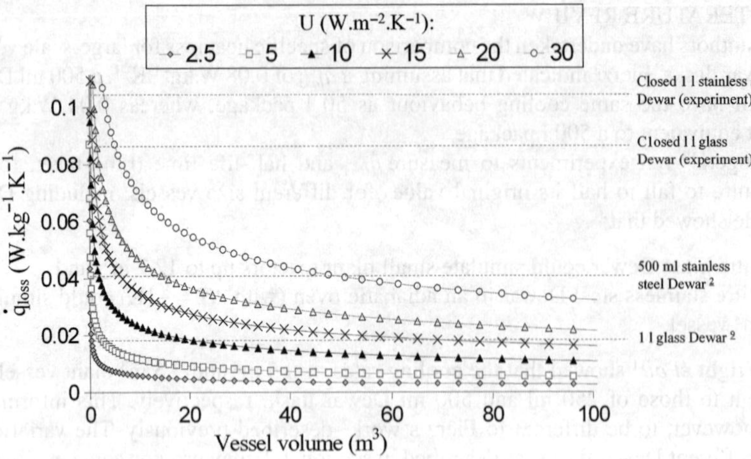

Figure 2. Calculation of $\dot{q}_{loss}$ as a function of the vessel volume and the overall heat transfer coefficient (U). Comparison with $\dot{q}_{loss}$ for four dewar flasks.

- if $\dot{q}_{loss\ \mathrm{Dewar}} = \dot{q}_{loss\ vessel}$: the volume of the vessel is the maximum vessel volume for which the use of the Dewar vessel is conservative.

Following the above rule, Table 3 summarises the maximum reactor volumes for which the use of each Dewar is conservative in the case of water (ρ = 1000 kg/m^3). The scale-up limits proposed in Table 3 are only suitable for the specific vessel features and contents chosen for this calculation. This, however, gives an order of magnitude of the range of applicability of Dewar flask for different values of the overall heat transfer coefficient of the large-scale vessels. This type of calculation could be applied to other substances. It is also recommended that a good quality Dewar is selected and that the tests are run in a closed Dewar (minimising heat losses), providing that the suitable safety measures are followed.

4. SCALE-UP OF DEWAR FLASK DATA FOR THE WASTE TREATMENT INDUSTRY

The previous part of this paper gave limits to the use of Dewar flasks by simply comparing the specific heat loss to that in large-scale vessels. We now consider the additional complication of the kinetic aspects of the scale-up and time dependence. The thermal behaviour of two exothermic reactions in a stainless steel Dewar flask are compared to that in two large-scale vessels from the waste treatment industry, using reaction kinetics and heat transfer models.

Table 3. Maximum vessel volume for which the use of dewar is conservative (calculation for water at 27°C, $\rho = 997$ kg/m^3)

	1 l Glass Dewar[9]	500 ml Stainless steel Dewar[2]	1 l Closed glass Dewar (exp)	1 l Closed stainless steel Dewar (exp)
$\dot{q}_{loss}$ Dewar (W.kg^{-1}.K^{-1})	0.018	0.04	0.083	0.09
U of the large scale vessel (W.m^{-2}.K^{-1})	Maximum reactor volume (m^3) for which the use of a dewar is conservative vessel features: D/H = 0.8, fill level = 80%			
2.5	0.33	0.03	0.003	0.002
5	2.64	0.23	0.028	0.015
10	21.17	1.9	0.23	0.11
15	71	6.5	0.77	0.4
20	169	15.5	1.8	0.9
30	573	52	6.2	3.1

The suitability of the criterion suggested by parts of waste treatment industry to determine the exothermic reaction incompatibility of mixtures (temperature rise between 6–10°C in the Dewar flask over a period of 10 minutes) is also evaluated.

4.1 OVERALL HEAT TRANSFER COEFFICIENT FOR A STAINLESS STEEL DEWAR FLASK

We chose to simulate the thermal behaviour of the 1 l stainless steel Dewar flask described in 2.2. The thermal balance corresponding to the Newtonian cooling of the water contained in this Dewar flask is given by:

$$\phi m C p_l \frac{dT}{dt} = -UA\left(T - T_{ext}\right) \tag{3}$$

With ϕ, the thermal inertia factor:

$$\phi = \frac{mCp_l + m_{vessel} Cp_{vessel}}{mCp_l} \tag{4}$$

Table 4. Overall heat transfer coefficient for the 1 l stainless steel Dewar flask

A (m^2)	0.0434
m water (kg)	0.8
m flask (kg)	0.911
Cp water (J.kg^{-1}.K^{-1})	4186
Cp stainless steel (J.kg^{-1}.K^{-1})	477
ϕ	1.13
$\dot{q}_{loss}$ (W.kg^{-1}.K^{-1})	0.103
U (W.m^{-2}.K^{-1})	2.147

The specific heat loss criterion is then:

$$\dot{q}_{loss} = \frac{UA}{\phi m} \tag{5}$$

The overall heat transfer coefficient can then be assessed from the value of the specific heat loss criterion measured experimentally (see Table 2). For this Dewar flask the value of the overall heat transfer coefficient is 2.147 W·m^{-2}·K^{-1} (see Table 4).

4.2 ASSESSMENT OF THE OVERALL HEAT TRANSFER COEFFICIENT FOR LARGE-SCALE VESSELS USED IN WASTE TREATMENT

Some information about the design of typical storage vessels and reaction vessels within waste treatment companies was canvassed by way of a questionnaire. No experimental measurements of the thermal characteristics were available. Assessment of the overall heat transfer coefficient using heat transfer models was therefore undertaken. The results are presented for two vessels (Table 5):

- A 112 m^3 stainless steel stirred reactor equipped with a flat blade turbine stirrer.
- A 40 m^3 stainless steel stirred reactor equipped with a blade stirrer and 3 baffles.

For the assessment of the overall heat transfer coefficient, it was assumed that the vessels are filled with water at 100°C and the temperature of the air surrounding the vessels was 27°C (Table 6). For these two reactors the heat transfer through the vessel wall is realised by three mechanisms (Figure 3):

i) forced convection in the liquid, the reactors being stirred;
ii) conduction in the stainless steel wall; and
iii) natural convection in the air outside the vessel

Table 5. Thermal characteristics of both 112 m^3 and 40 m^3 stainless steel stirred vessel

	Stainless steel stirred reactor	
Volume (m^3)	112	40
Height (m)	6.096	4
Diameter (m)	4.8768	3.6
Wall thickness (m)	0.0127	0.006
Agitator diameter (m)	1.21	3
Agitator speed (rpm)	20	36
Baffles	No	Yes: 3 of 3 m × 0.15 m
T_2–T_{ext} (°C) (assumption)*	75°C	75°C
λ_{SS} ($W.m^{-1}.K^{-1}$)	16.2	16.2
h_{liq} ($W.m^{-2}.K^{-1}$)	8.5×10^3	5.7×10^4
h_{ext} ($W.m^{-2}.K^{-1}$)	6.2	6.2
U ($W.m^{-2}.K^{-1}$)	6.2	6.2

The overall heat transfer coefficient (U) is then given by:

$$U = \left[r_1 \left(\frac{1}{r_1 h_{liq}} + \frac{\ln(r_2/r_1)}{\lambda_{SS}} + \frac{1}{r_2 h_{ext}} \right) \right]^{-1} \tag{6}$$

h_{int} and h_{ext} are the internal and external convective heat transfer coefficients, respectively. λ_{SS} is the thermal conductivity of stainless steel.

Table 6. Properties of water at 100°C and air at 27°C [Incropera *et al*, 2001][12]

	Water at 100°C	Air at 27°C
λ ($W \cdot m^{-1} \cdot K^{-1}$)	$680\ 10^{-3}$	$2.63\ 10^{-2}$
Cp ($J \cdot kg^{-1} \cdot K^{-1}$)	4217	1007
ρ ($kg \cdot m^{-3}$)	957.85	1.1614
μ (Pa·s)	$2.79\ 10^{-3}$	$1.85\ 10^{-5}$
β (K^{-1})	$7.5\ 10^{-4}$	$3.663\ 10^{-3}$

* For a liquid temperature of 100°C, it is realistic to assume that the vessel external surface temperature is at least 95°C because of the high efficiency of the agitation and the high conductivity of the stainless steel. Moreover, decreasing the vessel external surface temperature to 80ºC does not affect significantly the value of U.

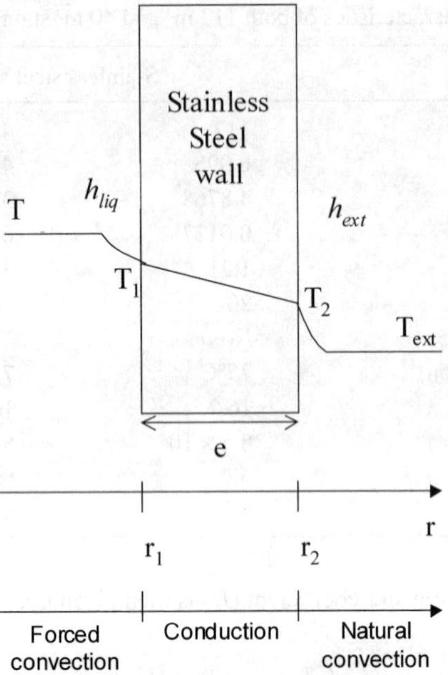

Figure 3. Heat transfer model for the two stirred stainless steel vessels (112 m^3 and 40 m^3)

The correlations used to assess these coefficients are given in the Appendix.

For the two reactors, the stainless steel wall is a good heat conductor and agitation is highly efficient. The greatest part of the resistance to heat transfer comes from external natural convection. The assessed overall heat transfer coefficients are 6.2 $W.m^{-2}.K^{-1}$ for the 112 m^3 and 40 m^3 vessels (Table 5).

Assuming that the vessels are filled at 80% with water at 27ºC, then Table 7 shows that the corresponding value of specific heat loss is higher in the stainless steel Dewar flask than in the large-scale vessels by an order of magnitude. This indicates that this Dewar flask will not be able to reproduce the thermal behaviour of the large-scale vessels.

4.3 SIMULATION OF CHEMICAL REACTIONS

The values of overall heat transfer coefficient calculated above can then be used to simulate the thermal behaviour of two exothermic reactions. The thermal balance for a closed vessel containing a reacting mixture is given by:

$$\phi m Cp_l \frac{dT}{dt} = m\dot{q}_R - UA(T - T_{ext}) \qquad (7)$$

Table 7. Assessment of $\dot{q}_{loss}$ for large scale vessels and Dewar flask filled at 80% with water at 27°C

	U ($W.m^{-2}.K^{-1}$)	A (m^2)	m (kg)	ϕ	$\dot{q}_{loss}$ ($W.kg^{-1}.K^{-1}$)
112 m^3 Reactor	6.23	92	8.95×10^4	1.016	6.3×10^{-3}
40 m^3 Reactor	6.25	46	3.196×10^4	1.023	8.8×10^{-3}
1 l Closed stainless steel Dewar	2.147	4.84×10^{-2}	9.09×10^{-1}	1.11	1.02×10^{-1}

The reaction energy release rate ($\dot{q}_R$) is linked to the reaction kinetics by the following expressions:

$$\dot{q}_R = \Delta H_r \frac{dX}{dt} \tag{8}$$

$$\frac{dX}{dt} = C \exp\left(\frac{-E}{RT}\right) f(X) \tag{9}$$

with ΔH_r the reaction energy, E_a the activation energy and X the conversion.

The temperature rise rate in the vessel containing a chemical mixture can be obtained by numerically integrating the above differential equations.

The reactions encountered in the waste-treatment industry vary widely, depending on the chemicals being treated, their concentrations and any contaminants present. Due to their unpredictable nature there is very little information available on the reaction kinetics. Two chemical reactions on which kinetic data are available were therefore used to investigate the Dewar for scale up. These were:

- the autocatalytic hydrolysis of acetic anhydride
- the first order decomposition reaction of 20% *tert*-butyl peroxy 2-ethylhexanoate (Trigonox 21) in a solvent (Shellsol T).

4.3.1 Hydrolysis of acetic anhydride

Snee *et al.*[13] proposed an autocatalytic kinetic equation for the hydrolysis of acetic anhydride as:

$$\frac{dX}{dt} = 24166 \exp\left(\frac{-52378}{RT}\right)(1 + 6.5X^{0.85})(1 - X)^{1.25} \tag{10}$$

The reaction energy (ΔH_r) is 446.4 kJ/kg and the initial temperature is 45°C. The Cp of the chemical mixture is evaluated to be 2400 $J.kg^{-1}.K^{-1}$.

Table 8. Specific heat losses for the vessels containing the acetic anhydride/water mixture

Vessel	ϕ	$\dot{q}_{loss}$ (W.kg^{-1}.K^{-1})
1 l SS Dewar (exp)	1.21	1.03×10^{-1}
112 m^3 Stainless steel reactor	1.03	6.8×10^{-3}
40 m^3 Stainless steel reactor	1.04	9.35×10^{-3}

Table 8 presents the calculation of the specific heat loss for the different vessels containing this chemical mixture. It appears that the heat losses will be much more important in the Dewar flask than in the large vessels. From a purely thermal point of view, i.e. only regarding the specific heat loss (as considered in sections 2 and 3 of this paper), the use of this Dewar flask would be inappropriate as a mean of simulating the chemical reaction runaway potential of these size vessels.

Figure 4 (a) shows the thermal behaviour in the Dewar flask in comparison to that modelled for the large-scale vessels when this chemical reaction occurs. A runaway reaction occurs at large scale and the 40 m^3 and 112 m^3 reactors give approximately the same results. Even with a significant larger value of specific heat loss from the Dewar, it still allows the detection of the runaway reaction. This is possible because the relative importance of the heat losses in the Dewar flask is low. Indeed, the ratio of the power produced by the reaction to the power lost to the surrounding (at least a factor of 10) for the stainless steel Dewar is important at the start of the reaction. This leads to a rate of temperature rise greater than 0.48°C/min (see "ratio power prod/loss" on Figure 4 (b)). This rate of temperature rise is close to the criterion proposed in the waste treatment industry. However, in the Dewar flask the reaction completion time is longer and the maximum temperature achieved is less than that seen in the large-scale vessels.

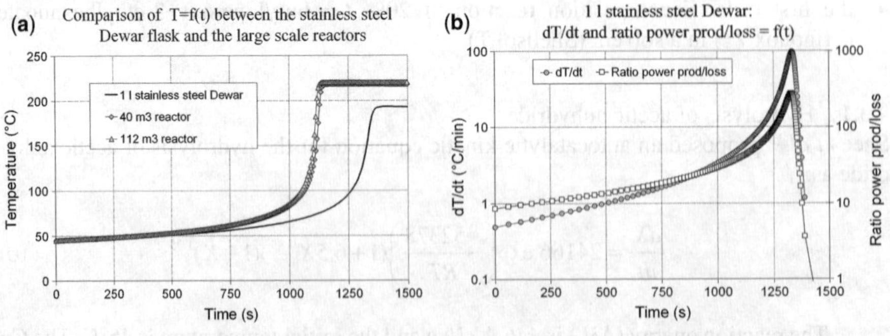

Figure 4. Modelling of the autocatalytic hydrolysis of acetic anhydride ($T_{ini} = 45°C$)

Table 9. Specific heat losses for the vessels containing 20% Trigonox 21 in Shellsol T

Vessel	ϕ	$\dot{q}_{loss}$ (W.kg^{-1}.K^{-1})
1 l SS Dewar (exp)	1.3	1.1×10^{-1}
112 m^3 stainless steel reactor	1.04	7.8×10^{-3}
40 m^3 stainless steel reactor	1.06	1.07×10^{-2}

4.3.2 Decomposition of 20% Trigonox 21 in Shellsol T

Snee *et al.*[14] investigated the kinetics of the decomposition of 20% Trigonox 21 (*tert*-butyl peroxy 2-ethylhexanoate) in Shellsol T. They proposed the following first order kinetic equation:

$$\frac{dX}{dt} = 8.5 \times 10^{13} \exp\left(\frac{-124462}{RT}\right)(1 - X) \tag{11}$$

The reaction energy is: ΔHr = 198 kJ/kg and the initial temperature for this experiment is 70°C. The *Cp* of the chemical mixture is 2000 J.kg^{-1}.K^{-1}. Table 9 shows that on considering only the specific heat losses then the use of this Dewar flask would be inappropriate. Figure 5 (a) shows that a runaway reaction would occur at large scale, with the 40 m^3 reactor exhibiting a longer completion time than the 112 m^3 reactor. However, the more important observation is that the reaction would not even be detected in the stainless steel Dewar flask test.

This can be explained by the fact that the induction time of this reaction is quite long. At low temperature and low conversion, the reaction heat release rate is lower than

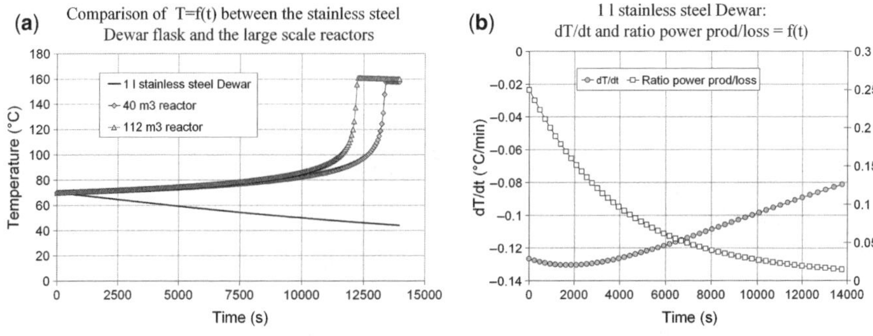

Figure 5. Modelling of the decomposition of 20% Trigonox 21 in Shellsol T (T_{ini} = 70°C)

the heat loss rate (Figure 5 (b)). This results in temperature decreasing in the Dewar flask. In this case, the relative importance of heat loss is clearly not negligible. The use of this Dewar flask for predicting large-scale behaviour for this particular chemical reaction is therefore inappropriate.

4.3.3 Sensitivity study with the first order reaction

For the decomposition reaction of 20% Trigonox 21 in Shellsol T the sensitivity to the initial temperature was investigated further. This allowed us to study the effect of different reaction rates. When increasing the starting temperature to 93°C the reaction rate is quite high from the beginning and the completion time at large scale is decreased (see Figure 6 (a)). The runaway reaction is also detected in the Dewar flask. Indeed, the reaction energy release rate is high enough to reduce the effect of the heat losses from the flask (ratio power prod/loss > 2.6 (Figure 6 (b)). The corresponding temperature rise in the Dewar flask is approximately 6ºC over 10 minutes (this corresponds to the criterion under evaluation).

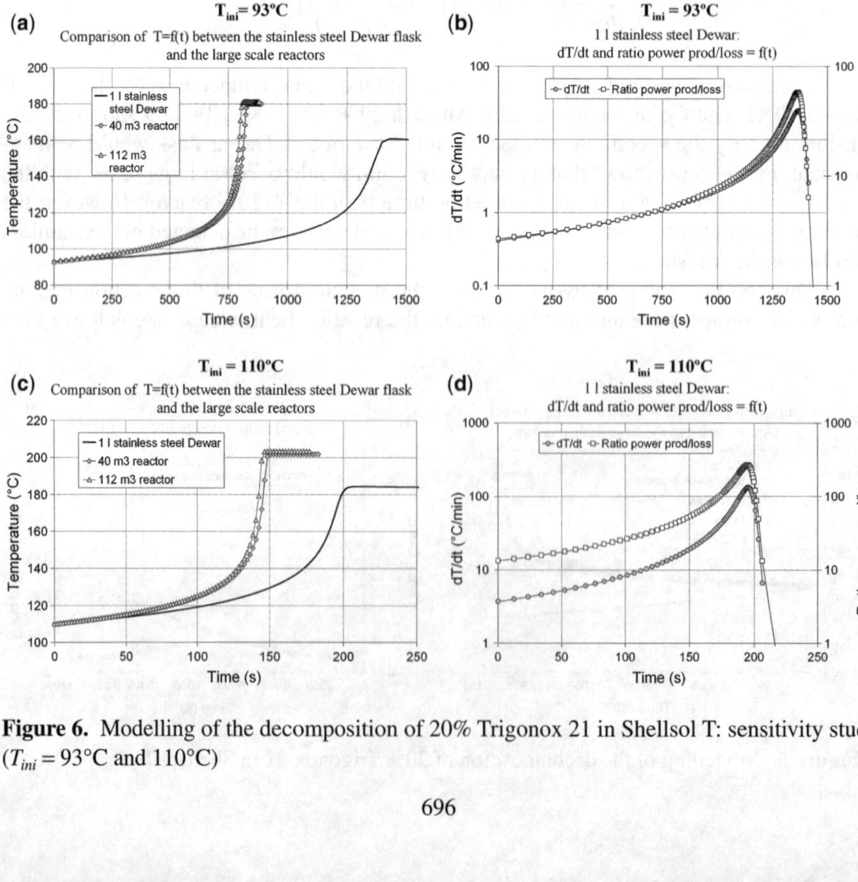

Figure 6. Modelling of the decomposition of 20% Trigonox 21 in Shellsol T: sensitivity study (T_{ini} = 93°C and 110°C)

Even if the runaway reaction is detected, the heat losses still have a significant influence on the actual results for the Dewar flask. Indeed, the time for completion is 1.7 times greater than that in the large-scale vessels. Hence, running the chemical reaction in this Dewar flask for 10 minutes would be sufficient to detect the exotherm but not the runaway reaction.

When increasing the starting temperature to 110°C, the reaction rate is again increased, resulting in a quicker completion time at large scale (150 s, Figure 6 (c)). The Dewar flask can detect the runaway reaction. The difference in the reaction completion time between the Dewar flask and the large-scale vessel is reduced. Running the chemical reaction in this Dewar flask for 10 minutes would be sufficient to detect the runaway reaction. This result has, however, been obtained when the heat losses are at least 13 times lower than the reaction energy release rate (Figure 6 (d), leading to a rate of temperature rise in the Dewar flask greater than 3.7°C/min). This rate is approximately 3 times greater than the criterion proposed by the waste treatment industry. The maximum temperature will, however, still be smaller within the Dewar experiment.

4.3.4 Reliability of the 6–10°C criterion

The simulation of the exothermic reactions showed that the 6–10°C criterion in a Dewar flask might be observed when the heat losses do not have a significant importance compared to the reaction heat release rate. In such a situation, the Dewar could detect a runaway reaction successfully. However, the reaction completion time at large scale would be shorter than at Dewar scale. In some cases, running the chemical reaction in a Dewar flask over 10 minutes could be sufficient to detect the exotherm but not the runaway reaction. The test should therefore be run until the reaction is complete to fully detect exotherms. In some cases, particularly for low energy reaction or long induction time reactions, exotherms may not be detected. This will depend strongly on the type of Dewar flask used and on the chemical reaction parameters.

5. CONCLUSIONS

A study of the specific heat loss criterion allowed the development of an approach to defining the range of applicability of Dewar flasks for different values of the overall heat transfer coefficient of large-scale vessels. The overall heat transfer coefficients, and therefore specific heat loss criteria, have been assessed for two typical vessels used in the waste treatment industry using heat transfer models. It appears that, *only* regarding the specific heat loss criterion, none of the tested Dewar flasks are suitable for direct simulation of these large-scale vessels.

The simulations of the thermal behaviour of two exothermic reactions in a 1 litre stainless steel Dewar flask and two different volume large-scale vessels show that:

- For fast and highly energetic reactions, the reaction energy release rate can be significant compared to the heat losses. The Dewar flask can therefore detect runaway reactions.

- For low energy or long induction time reactions, the heat losses may be significant compared to the heat release rate. The Dewar can, depending on the parameters, then miss exotherms and give non-conservative results.
- When an exothermic behaviour is detected in a Dewar flask, a 10 minute test, as proposed by the waste treatment industry, could be insufficient to detect a runaway reaction. The test should be run until the reaction is complete to fully detect exotherms.
- At large scale, the maximum temperature is likely to be higher and the reaction completion time shorter than at Dewar scale.

Reliable conclusions about the scale-up of Dewar data can be obtained when the chemical reaction kinetics are well known. Unfortunately this is not generally the case in the waste-treatment industry. So, unless the specific heat loss of the Dewar has been shown to be less than large-scale vessels, this method in isolation is likely to be unreliable for scale up to large vessels.

NOMENCLATURE

A	Exchange surface (m^2)
C	Pre-exponential factor (-)
Cp_l	Liquid specific heat capacity ($J.kg^{-1}.K^{-1}$)
Cp_{vessel}	Vessel specific heat capacity ($J.kg^{-1}.K^{-1}$)
D	Vessel external diameter (m)
E	Activation energy ($J.mole^{-1}$)
Gr	Grashof number (-)
h_{ext}	External convective heat transfer coefficient ($W.m^{-2}.K^{-1}$)
h_{int}	Internal convective heat transfer coefficient ($W.m^{-2}.K^{-1}$)
m	Liquid mass (kg)
m_{vessel}	Vessel mass (kg)
Nu	Nusselt number (-)
Pr	Prandtl number (-)
$\dot{q}_{loss}$	Specific heat loss ($W.kg^{-1}.K^{-1}$)
$\dot{q}_r$	Reaction energy release rate ($W.kg^{-1}$)
r	Vessel radius (m)
Re	Reynolds number (-)
t	Time (s)
T	Liquid temperature (K)
T_{ext}	Room temperature (K)
$T_{ext\ ini}$	Initial room temperature (K)
T_{ini}	Initial liquid temperature (K)
U	Overall heat transfer coefficient ($W.m^{-2}.K^{-1}$)
X	Chemical reaction conversion (-)

Γ	Viscosity factor (-)
β	Volumetric thermal expansion coefficient (K^{-1})
Φ	Geometric factor (-)
λ	Thermal conductivity ($W.m^{-1}.K^{-1}$)
ϕ	Thermal inertia or phi-factor (-)
ρ	Density ($kg.m^{-3}$)
μ	Dynamic viscosity (Pa.s)
ΔHr	Reaction energy ($J.kg^{-1}$)

APPENDIX: CORRELATION FOR CONVECTIVE HEAT TRANSFER COEFFICIENTS CALCULATION

NATURAL CONVECTION:

$Nu = a(GrPr)$

For a fluid at a vertical wall, the following correlation can be used [Althaus *et al*[15]]:

$Gr.Pr < 10^{-3}$:	$Nu = 0.5$
$10^{-3} < Gr.Pr < 10^{3}$:	$Nu = 1.18\ (Gr.Pr)^{0.125}$
$10^{3} < Gr.Pr < 2.10^{7}$:	$Nu = 0.54\ (Gr.Pr)^{0.25}$
$Gr.Pr > 2.10^{7}$:	$Nu = 0.135(Gr.Pr)^{0.33}$

FORCED CONVECTION:

For agitated jacketed vessels, the correlation is [Rogers *et al*[9]]: $N_u = \Phi\ Re^{0.33}\ Pr^{0.67}\ \Gamma^{0.14}$

For 300 = Re = 7.5 10^5 and 2.2 = Pr = 2500 [Althaus *et al*[15]] :

- *turbine mixer:* without baffles: $\Phi = 0.54$ / with baffles: $\Phi = 0.76$
- *blade mixer:* without baffles $\Phi = 0.38$ / with baffles $\Phi = 0.78$

REFERENCES

1. Recommendations on the Transport of Dangerous Goods, Manual of Tests and Criteria, 4 revised ed., 2003, United Nations, ST/SG/AC.10/11/Rev. 4, United Nations, New York and Geneva
2. Fierz H., 2003, "Influence of heat transport mechanisms on transport classification by SADT-measurement as measured by the Dewar-method", Journal of Hazardous Materials, Volume 96, Issues 2–3, 31, 121–126
3. Kossoy A.A., Sheinman I.Ya., 2007, "Comparative analysis of the methods for SADT determination", Journal of Hazardous Materials, Volume 142, Issue 3, Pages 626–638
4. Li X.-R., Sun J.-H., Koseki H. and Hasegawa K., 2005, "Experimental determination of the minimum onset temperature of runaway reaction from a radioactive salt disposal in asphalt", Journal of Hazardous Materials, Volume 120, Issues 1–3, 11, 51–56

5. Committee of experts on the transport of dangerous goods and on the globally harmonized system of classifation and labelling of chemicals, 2002, Amendments to calcium hypochlorite entries of UN Nos 1748, 2208 and 2880 of Class 5.1, ST/SG/AC.10/C. 3/2002/5
6. Rogers R.L., 1989, "The use of Dewar calorimetry in the assessment of chemical reaction hazards", Hazards X: Process safety in chemical plants, Symposium series No. 115, 97–102, IChemE, Rugby, UK
7. Rogers R.L., 1989, "The advantages and limitation of adiabatic Dewar calorimetry in chemical hazard testing", Plant/Operation progress, Volume 8, No 2, 109–112
8. Kersten R.J.A., Boers M.N., Stork M.M. and Visser C., 2005, "Results of a Round-Robin with di-tertiary-butyl peroxide in various adiabatic equipment for assessment of runaway reaction hazards", Journal of Loss Prevention in the Process Industries, Volume 18, Issue 3, 145–151
9. Barton J., Rogers R. (ed.), 1997, "Chemical Reaction Hazards – A guide to safety", 2nd Edition, IChemE
10. Rogers R., 1991, Fact finding and basic data part 1: hazardous properties of substances, IUAPC Conference Safety in Chemical Production, Basle (Blackwell, UK)
11. Li X.-R., Koseki H., 2005, "Study on the early stage of runaway reaction using Dewar vessels", Journal of Loss Prevention in the Process Industries, Volume 18, Issues 4–6, 455–459
12. Wright T.K., Rogers R. L., 1986, "Adiabatic Dewar calorimeter", Hazards in the process Industries: Hazards IX Symposium Series No. 68, 4/W:1, IChemE, Rugby, UK
13. Incropera F. P., DeWitt D. P, 2001, "Fundamentals of Heat and Mass Transfer", 5th Edition, John Wiley & Sons
14. Snee T., Butler C., Cusco L., Hare J., Kerr D., Royle M., Wilday J., 1999, "Venting studies of the hydrolysis of acetic anhydride with and without surfactant", Health and Safety Laboratory internal report No PS/99/13
15. Snee T., Ahmed S., Butler C., Carver F., Hare J., Royle M., Wilday J., 1997, "Venting studies of the decomposition of tert-butyl peroxy–2-ethylhexanoate in Shellsol T", Health and Safety Laboratory internal report No PS/97/10
16. Althaus E, Jakubith M., Memofix, 1993, "Chemistry and Chemical Engineering", VCH Publishing

WHAT KIND OF RELATIONSHIP DO YOU HAVE WITH YOUR TOLLERS?

Craig Williams, Markus Luginbuehl and Phil Brown
Syngenta

This paper describes a chemical process that had been operated in the laboratory and on pilot plant without incident. At this point the process was outsourced to a toller to operate on the large scale and an unforeseen incident occurred which resulted in an uncontrolled exothermic reaction. Fortunately, there were no serious consequences, but the incident raised the possibility that something had been missed in the original hazard/risk assessment.

The paper describes the procedure of hazard and risk assessment that was carried out prior to handing over the data package to the toller and discusses where the process may have failed. It gives details of the incident and how Syngenta safety specialists then worked with the toller to determine the likely causes of the temperature excursion, with a discussion around the actual chemistry of the process. It shows how a slight increase in concentration, dramatically affected the reaction kinetics.

The paper focuses on the relationship that manufacturers have with their tollers and questions whether concerns over future litigation issues can adversely affect the data transfer process, sometimes resulting in incomplete safety data packages being handed over, and increasing the risk of unsafe toll operations.

1. INTRODUCTION

Manufacturing once accounted for almost 40% of the UK's output. It now represents less than half that. It has declined steadily over the past 30 years, giving way to competition from abroad, particularly the Far East where labour is much cheaper. The minimum wage in this country is now around £6 an hour, whereas in China decent labour can be as little as 40p an hour. This has led some economists to argue that manufacturing in the UK is no longer a viable proposition. Manufacturing has also borne the brunt of the slowdown in world trade and is believed to be shedding about 10,000 jobs a month. Even die-hard manufacturing experts believe that British industry needs to adapt to the new conditions with many of the opinion that companies need to "change or die".

One way to do this is for the business to re-focus on what they do best and leave the rest to others. Contract manufacturing seems to fit neatly into this practice. In addition to allowing companies to focus on core competencies, contract manufacturers offer numerous other advantages over in-house manufacturing, including lower costs, flexibility, access to external expertise and reduced capital[1]. Recent research has shown that manufacturers are beginning to see the value of choosing to work with strategic outsourcing companies rather than those who can merely beat the competition on price. This is a reflection of the increasing closeness of outsourcing relationships, with more and more parties entering into longer-term partnerships rather than simply closing narrowly defined supply deals.

The dynamics of such relationships are changing accordingly, with factors such as the development of trust and close management contact becoming increasingly important[2].

This paper describes how such a relationship between the business and a contract (toll) manufacturer was essential in ensuring that a potentially hazardous chemical process was safely scaled up from the laboratory scale to full scale production.

2. PROCESS DETAILS

The initial process solution is material **A** in dimethyl sulphoxide (DMSO) with potassium hydroxide, potassium carbonate and a phase transfer catalyst. This is heated to 80–85°C under vacuum (10–20 mbar). The reactor is set to distil via a short column and still head condenser, routed to a binary separator. Dichloroethylether (DCEE) is charged over 2 hours at 80–85°C. Distillates contain water which is separated at the binary separator and the DMSO/DCEE lower phase returned to the reactor. On completion of the addition, the reaction is held for 4 hours to complete. The reaction mass is screened to remove inorganics and distilled at ca 140–150°C under reduced pressure to remove DMSO.

3. INITIAL SAFETY DATA

The process was examined in the safety laboratory, looking at the heat of reaction by heat flow calorimetry and adiabatic Dewar calorimetry and investigating the thermal stability of the raw materials, reaction mass and distillation products. The process assessed was quite dilute, involving the use of ca 12 mol:mol DMSO:**A**. The main concern at this stage was the fact that the solvent being used was DMSO.

3.1 DMSO

Dimethyl sulphoxide (DMSO) is a very good solvent that solubilizes many inorganic compounds and many other difficult-to-dissolve-materials. It is widely used in the chemical industry. It is a polar solvent that is high-boiling and thus is easy to dewater. DMSO is known to slowly decompose at temperatures at or above its boiling point (189°C). However, these properties lead to some of the specific hazards of this material:

- DMSO is not inert; the reactivity of solvents that play only a physical role in reactions is easily forgotten. DMSO decomposes with a significant release of heat and gas.

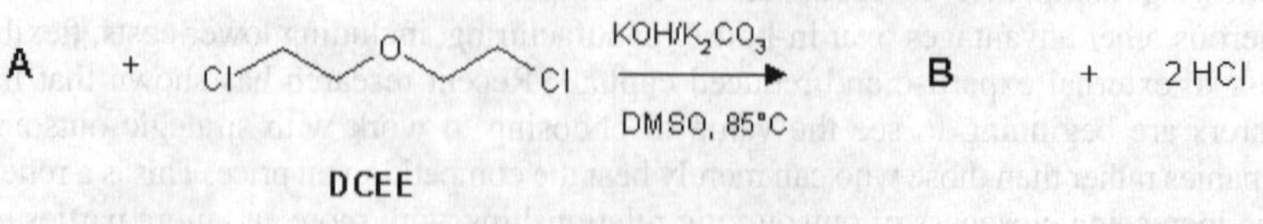

Figure 1. Reaction scheme

The decomposition reaction is already evident at the boiling point; DMSO cannot be distilled at ambient pressure without decomposition. This decomposition is self-accelerating and is accelerated by both acid and base, particularly with even trace amounts of organic or inorganic halogens. In some cases, the reaction is virtually instantaneous whilst others are characterized by the steady accumulation of heat and pressure with eventual runaway.

- The high boiling point hinders the dissipation of heat through vaporisation.
- Like many other solvents, DMSO reacts with many other reactive compounds, such as acid chlorides. However, in contrast to other solvents, the reactions with DMSO involve a large release of heat that cannot be dissipated through vaporization.

Many serious incidents involving the handling of DMSO have been reported.[1]

3.1.1 Thermal Stability Data for DMSO

Initial Differential Scanning Calorimetry (DSC) tests of pure DMSO (Fig. 2) indicated that it was stable up to its boiling point. DMSO recycled from the distilled reaction mass is likely to contain residual DCEE which could be expected to hydrolyse and reduce the pH of the DSMO, thereby having a detrimental effect on the thermal stability. Therefore recycled DMSO was generated in the lab and tested by DSC to determine the effect of the recycle. The dynamic DSC test (Fig. 3) indicated that the onset and peak temperatures of the decomposition were shifted to lower temperature and the decomposition itself was much sharper which indicated possible autocatalysis. This means that if held at any given temperature the DMSO would eventually decompose rapidly after an induction time, which would be dependent on the hold temperature. Isothermal DSC testing was therefore carried out to determine the likely induction times around the process temperature (Table 1).

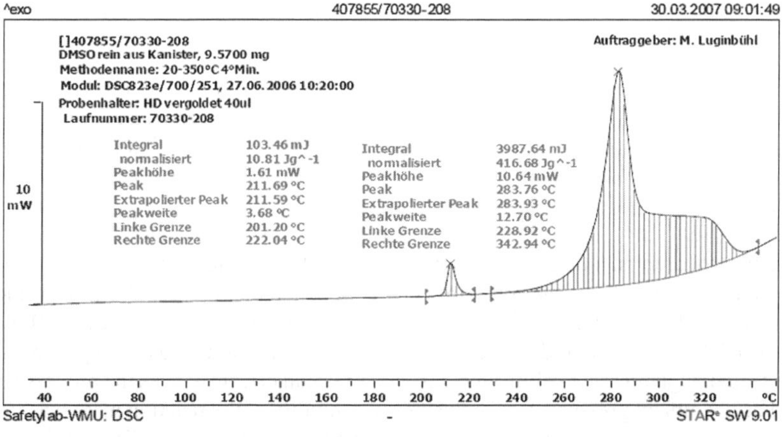

Figure 2. Fresh DMSO

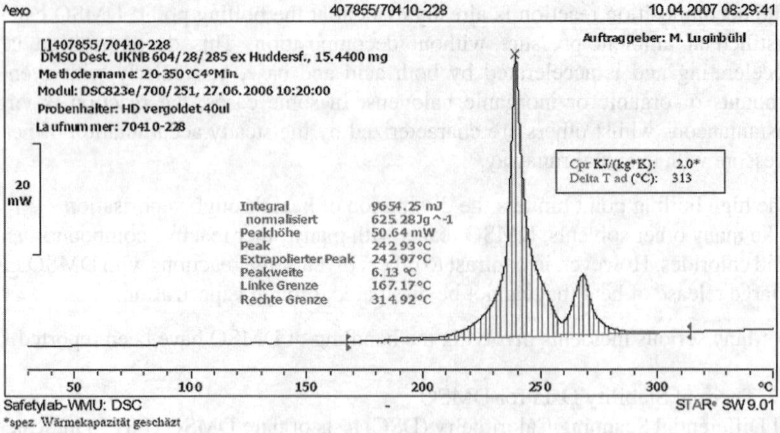

Figure 3. Recycled DMSO

Table 1. Recycled DMSO – induction time with temperature (from DSC)

Isothermal hold temperature (°C)	Induction time (mins)
170	25
160	50
150	116

The isothermal tests showed that the induction time at 150°C was only 116 min, to the start of decomposition, with the peak rate of the decomposition occurring after ca 290 mins. Whilst it is unlikely that DMSO alone would be subjected to such high temperatures, the reaction batch and crude product solution may be. During the normal reaction, the temperature of the batch should not exceed ca 85°C (unless cooling was lost and the reaction allowed to runaway, see 3.3.1). The main concern would be the DMSO recovery distillation stage which, if carried out in a batch distillation, could result in DMSO being heated to ca 150°C for prolonged periods. The maloperation scenario of agitator loss would also be a problem as the batch would be subjected to the service temperature (probably steam at ca 180°C).

The DSC tests were indicative of the effect in a sealed situation (as they were carried out in sealed, high pressure cells), whereas under normal circumstances the process vessels would be at ambient pressure. Hence the thermal stability of the reaction mass was investigated under ambient pressure conditions using Dewar calorimetry. The results of that experiment are discussed in section 3.3.2.

3.2 MEASUREMENT OF THE HEAT OF REACTION

The reaction was investigated by both isothermal heat flow calorimetry and adiabatic Dewar calorimetry to determine the heat of reaction and the expected adiabatic temperature rise.

3.2.1 Isothermal Heat Flow Calorimetry

The reaction was carried out by controlled addition of DCEE over 2 hours to the batch whilst maintaining the reaction temperature at 85°C. The reaction was allowed to work-off for 3 hours after completing the addition.

Addition of DCEE resulted in an exothermic activity with a fairly even power output (see Fig. 4). After addition was stopped, there was a slow work off of accumulated reactant.

Heat of reaction (total), $\Delta H = -238.65$ kJ/mol (-57 kcal/mol) **A**
Accumulated heat = -66.22 kJ/mol (-15.6 kcal/mol) (ca 28%))
Heat capacity, Cp batch = 89.47 cal/K (2.14 J/K.g)
Adiabatic temp rise, $\Delta T = 82.8$ K
ΔT (accumulation only) = 22.7 K
No significant gas evolution was detected.

For the reaction conditions described above, with a batch heat capacity of 2.14 J/K.g, an adiabatic temperature rise of ca 83 K would be anticipated for a reaction without cooling. Hence, from a starting temperature of 85°C, the worst case possible is that the batch temperature would reach 168°C if cooling were lost at the start of the reaction.

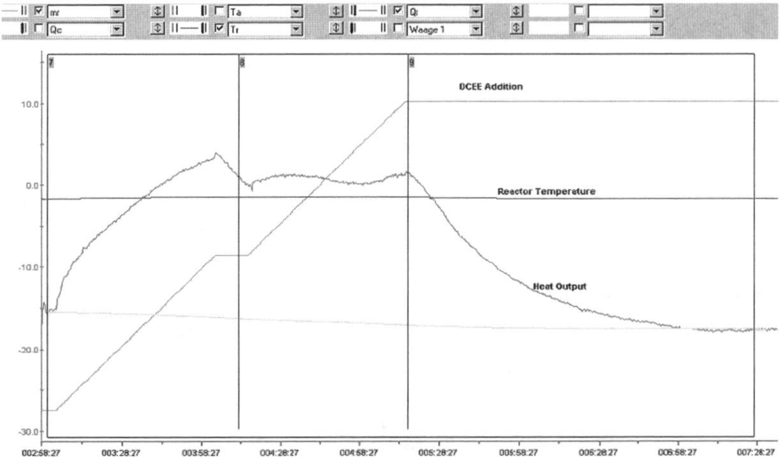

Figure 4. Dilute process (2hr DCEE addition) in RC1

3.3 ADIABATIC DEWAR CALORIMETRY

3.3.1 Reaction

The reagents were charged to a 250 ml narrow neck Dewar. The Dewar was then placed into a shield oven set on adiabatic control. The reagents were heated to 82°C with agitation and DCEE charged over 2 hours. The batch was then held adiabatically with agitation for 2 days (see Fig. 5).

The DCEE addition resulted in an immediate exothermic reaction and the batch temperature reaching a maximum of 162.3°C.

Adiabatic temp rise, $\Delta T = 80.3$ K.
Cp batch = 89.47 cal/K (2.14 J/K.g)
Heat of reaction, $\Delta H = -30.07$ kJ (−7.18 kcal)
= −233 KJ/mol (−55.3 kcal/mol) A

The reaction was accompanied by gas evolution at a maximum rate equivalent to 0.04 l/min/kg batch, which slowed down to a rate of ca 0.003 l/min/kg batch as the reaction completed. This gassing continued at a similar rate as the batch temperature dropped slowly to ca 152°C over the next 48 hours.

3.3.2 Reaction Mass Thermal Stability

3.3.2.1 Product B Solution in DMSO The product **B** solution (in DMSO) was charged to a narrow necked squat Dewar flask. The Dewar was then placed in a shield oven. The sample was agitated and heated to 180°C over ca 60 min before switching the heater off. The sample was held adiabatically with slow stirring for over 20 hours (see Fig. 6).

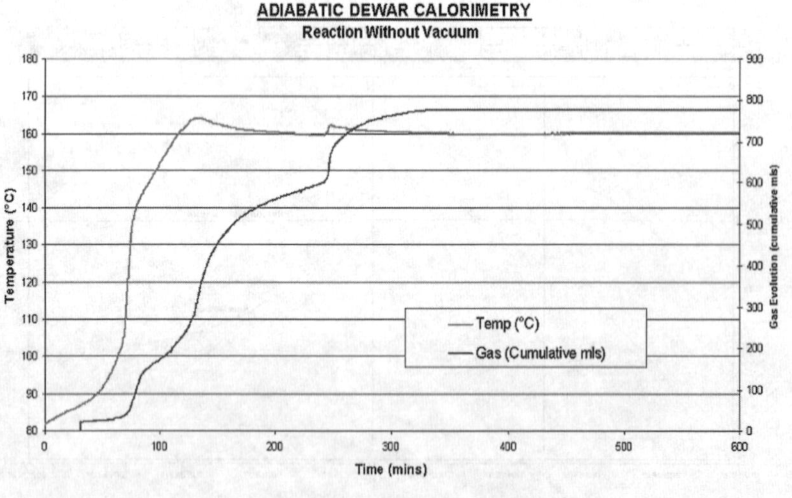

Figure 5. Reaction profile in adiabatic Dewar

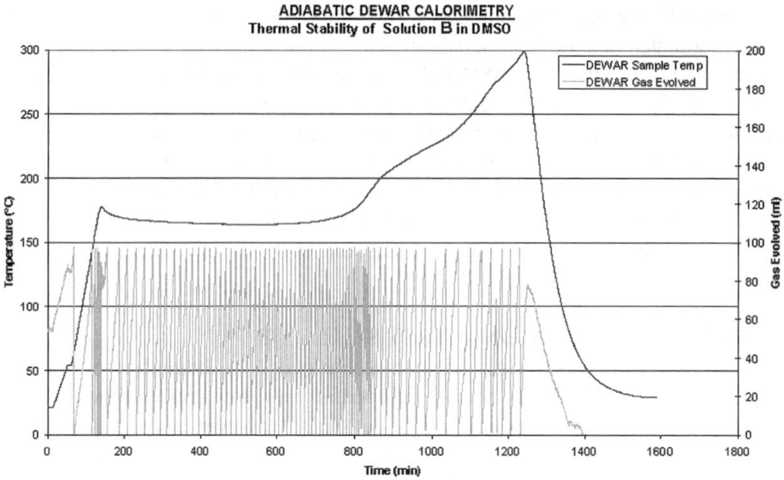

Figure 6. Reaction mass thermal stability

Heat-up to 180°C was accompanied by gas evolution which became rapid from ca 150°C. After the heater was switched off, the sample temperature fell to ca 164°C over the next 7 hours. The gas evolution rate initially dropped then slowly increased with time. The temperature and gas rate then started to increase again; the temperature rose to 300°C over the next 11 hours, at a maximum rate of 23 K/hr at ca 175°C which occurred about 4 hours into the exotherm. The gas rate reached a maximum at 175°C then tailed off. At this point the test was terminated by the automatic maximum temperature trip. During the experiment ca 71% of the batch was distilled out and the total amount of gas evolved was ca 36 l/kg.

This test showed that if the reaction mass is heated to ca 180°C (i.e. to the service temperature) it could go into thermal runaway after ca 8 hours. However, the decomposition may be tempered somewhat by distillation of the DMSO, and consequently a significant amount of distillate was collected during the experiment.

3.4 CONCLUSIONS FROM THE SAFETY DATA

The main safety concern in terms of chemical reaction hazards of the process was considered to be the thermal stability of the reaction mass, the bulk of which is DSMO solvent. The thermal hazards of DMSO are well documented, and the tests carried out in the safety lab also suggested that there were potential problems both in the reaction stage and the DMSO recovery distillation stage.

It was shown in the sealed tests (DSC) that recycled DMSO can decompose autocatalytically with quite a short induction time at the recovery distillation process temperature. Also, if a maloperation were to occur in the reaction stage, with loss of cooling, the batch temperature could possibly reach the DMSO decomposition temperature. Adiabatic Dewar calorimetry indicated that the decomposition under ambient pressure conditions is tempered somewhat by the distillation of DMSO and is therefore not considered a significant problem in an open vessel, but a blocked vent leading to a pressurised situation could result in a catastrophic event. Further tests (not detailed here) carried out on relief vent sizing for this scenario suggested that the decomposition was not relievable (i.e. the vent size would be prohibitively large). These findings suggest that a short-path, low residence time vacuum distillation would be preferable to a batch process for this recovery stage. The reaction stage was considered safe to operate with appropriate control measures to avoid loss of cooling.

4. PILOT PLANT CAMPAIGN – FIRST SCALE UP

While the safety testing was being carried out, the process was being further developed in the lab. The reaction had initially been developed at fairly low concentration (ca 12 mol: mol DMSO:**A**) which gave the optimum product yield. However, as the product became a viable option for scale-up, economic considerations came to the fore and a more concentrated process (ca 8.5 mol:mol DMSO:**A**) was developed to save processing costs associated with DMSO recovery.

It was decided to use the concentrated process for the first scale-up in the pilot plant, which was used to produce ca 500 kg of material for toxicity screening. A Process Risk Assessment (PRA) team (including safety specialists) evaluated the process and determined the basis of safety for the pilot plant campaign after consideration of the safety data generated for the dilute process. The team decided that this data could be used and applied to the concentrated process (by calculation) and did not consider that any further safety testing was necessary.

The pilot facility consisted of a general purpose reaction vessel, fitted with an all purpose distillation column. The campaign was carried out and monitored by specialist technical operators, as was standard practice for pilot plant operations. The process ran to plan, with reaction profiles similar to the laboratory experiments, product quality and cycle times were as predicted, and no safety concerns were noted.

5. CONTACT WITH TOLLERS

At this stage, the business decision for the product was to send out the process to contact manufacture. A number of (Swiss and European) tollers were approached, all of whom had been involved with the company on previous occasions with positive results. A series of face to face meetings took place at which the process was discussed in detail and a data package was handed over. The data package contained details of the process that had been investigated thus far, with protocols, batch data and engineering details from the pilot plant campaign. All the safety data that had been produced to date was shared, but no interpretation, conclusions or recommendations were included. This is common practice and is done to make sure the toller does not (unintentionally) misinterpret any conclusions made by the Syngenta safety specialist and requires the toller to use the raw data, apply it to the situation in their plant, and to determine the basis of safety for themselves. This is to try to avoid future litigation issues should there subsequently be an incident with the process.

PRA data for the pilot plant campaign was not part of the package but information on all known safety hazards was discussed. The tollers took the information away for consideration, some carrying out preliminary laboratory work to determine the feasibility of operation of the proposed process.

5.1 DECISION ON TOLL MANUFACTURER

After a number of weeks, the tollers were invited to discuss their proposals for operating the process. They had brought laboratory data for discussion and one toller also invited their own (contracted) safety consultants to discuss some queries they had around the safety data. The discussion on the safety data was mainly around clarifying the interpretation of the raw data available. No further safety advice was given on our part but the discussion raised a number of additional queries which one toller felt required further safety testing before they could make a final decision on the process operation. This testing was carried out by the toller's safety consultants but the results were not shared with us.

After a further few weeks, the tollers returned with their final proposals. At this point some of the tollers dropped out on the grounds that they felt they would be either not able or not willing to manufacture the product. The reasons given were either economic grounds or inadequate kit or through safety concerns. The toller that was awarded the contract gave reassurance that they had adequate kit and an economically viable process. They had no concerns over the safety of the operation.

Whilst open discussions had been carried out throughout the decision process to this point, it is always possible that a toller may have given an overly positive impression of their situation in order to win the contract. Whilst every effort is always made to ensure that all the relevant information is passed on by the toller, at this stage there also has to be a good degree of trust. The chosen toller was a relatively small operation, with a small technical team, able to and capable of making personal decisions on the ground, with a relaxed attitude to the process hazards and a feeling that they understood them and were able to manage them. They also had synergies with our company as they were already manufacturing some other products for us under contract (including DCEE). Therefore, the technical team were quite well known to us and some good working relationships had been developed.

5.2 FOLLOW-UP CONTACT WITH TOLLER

The toller carried out their own PRA (with their safety consultants) then went ahead to manufacture product **B** on a 5000 kg scale. After they had carried out a number of batches, they contacted us to say they had been observing unusual and unforeseen temperature excursions during the reaction stage. They had been seeing large sudden exotherms part-way through the reaction, with the batch temperature reaching up to 150°C on occasions.

No reason could be seen for this occurrence as this effect had never been seen in any of our laboratory tests. However, the information raised serious concerns about the safety of the process as the temperature excursions that were being reported were taking the batch close to the temperature at which DSMO may be expected to decompose autocatalytically. Due to the safety concerns, a Syngenta team visited the toller's site to see if there was any obvious reason for the temperature excursion. They were looking for particular engineering problems with the kit being used or significant process changes.

The team found that the process was being operated as expected with no obvious changes to the original process that had been handed over. In terms of the engineering, a check on temperatures and pressures around the distillation column during the reaction suggested that it was not functioning as it should. The column function was to condense the distillate, collect and separate the water and return the DMSO/DCEE phase to the vessel. The temperatures and pressures in the column indicated that the column was being flooded and was not removing water as desired. Engineering calculations quickly determined that the column was undersized for the required duty. However, this could still not immediately explain the temperature excursions that had been noted.

Further discussions were carried out to try to determine the source of the problem. From the information received from the site visit, it was suggested that the problem of the temperature excursions was probably linked to the column inadequacies. It was postulated that if the column was under-sized then water would not be removed properly, which may have an effect on the reaction rate. No such effect had been seen in the original safety tests, which had been carried out without vacuum (i.e. no water was actually removed during the calorimetry experiments) or in the pilot plant campaign. However, it was quickly realised that the original safety testing had been carried out in a dilute system, and the more concentrated process had not been re-assessed in the laboratory. The pilot plant campaign had been with the concentrated process, however the column present on the pilot plant reactor was adequately sized and therefore it was likely that all the water had been removed very efficiently from the reaction in this case.

At this stage it was agreed that as the safety concerns were so serious that further safety testing would be carried out by Syngenta to investigate the problem.

6. FURTHER SAFETY DATA

The safety testing carried out was on the reaction stage for the concentrated process, again this involved using both isothermal heat flow calorimetry and adiabatic Dewar calorimetry. The isothermal heat flow calorimetry would show the extent of the power output due to the expected and/or unexpected exothermic reaction and the adiabatic Dewar calorimetry would show the expected consequences of a loss of cooling scenario.

6.1 ISOTHERMAL HEAT FLOW CALORIMETRY

The reaction was carried out as previously described (3.2.1) with a 2 hour addition of DCEE, this time using the more concentrated process (see Fig. 7).

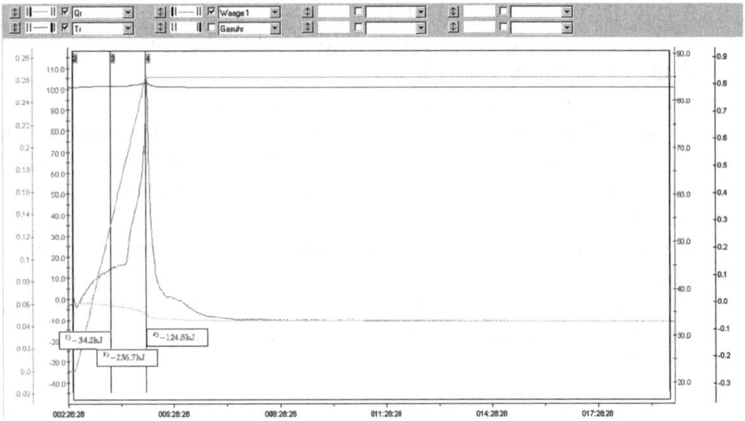

Figure 7. Concentrated process (2hr DCEE addition) in Rc1

The rate of heat output during DCEE addition was similar to that previously seen with the dilute process up to ca 1.5 hours into the addition. At this point a rapid increase in heat output was observed, indicative of a very rapid exothermic reaction, which then subsided just as rapidly. The total heat for the reaction was similar to that previously measured for the dilute process.

A further test was carried out using a 4 hour addition of DCEE (Fig. 8) which again showed a rapid exotherm, but this occurred later and had a lower instantaneous power output.

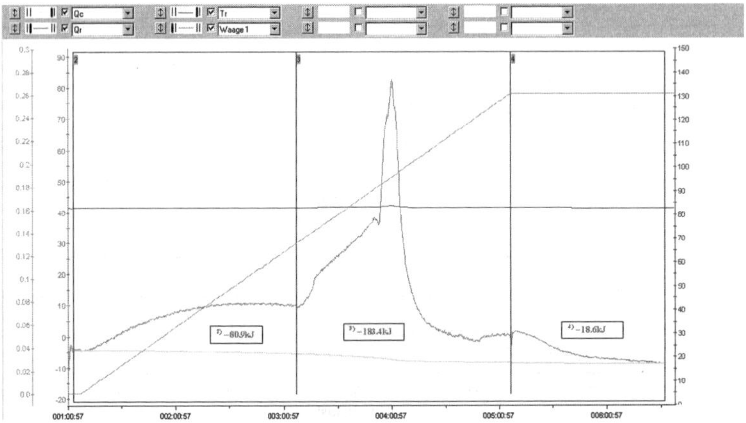

Figure 8. Concentrated process (2hr DCEE Addition) in RC1

6.2 ADIABATIC DEWAR CALORIMETRY

The same reaction was carried out in a Dewar to simulate the worst case, zero cooling, scenario, with the DCEE added over 2 hours.

Very little reaction was observed over the first hour and then a rapid exotherm occurred. The rates of temperature rise and gas evolution were so great that the batch primed and ca 60% of the batch was lost from the Dewar vessel. The experiment was repeated with 4 and 6 hour DCEE additions and the rate of temperature rise/gas evolution compared (see Fig. 9 & 10).

The rates of temperature rise and gas evolution for the three experiments are summarised in the table below:

These additional tests suggested that a slower addition rate of DCEE would allow greater control of the temperature excursion should it occur.

6.3 CONCLUSIONS FROM ADDITIONAL DATA

This additional safety data strongly suggested that the concentration of water in the batch played a critical role with respect to the reaction rate. The rate of reaction was thought to be dependent on the rate of deprotonation of the molecule **A**. This is therefore dependent on the availability of base, in this case KOH. At the start of the reaction, as very little water is present, the deprotonation of **A** is dependent on the phase transfer catalyst, which is required to help solid KOH come into contact with **A** in DMSO solution. As water is produced, KOH dissolves, and at a critical water concentration becomes miscible with

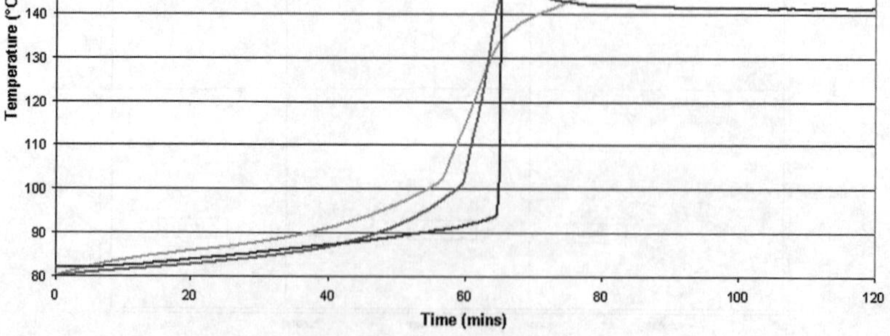

Figure 9. Concentrated process (Comparison 1 – temperature rise)

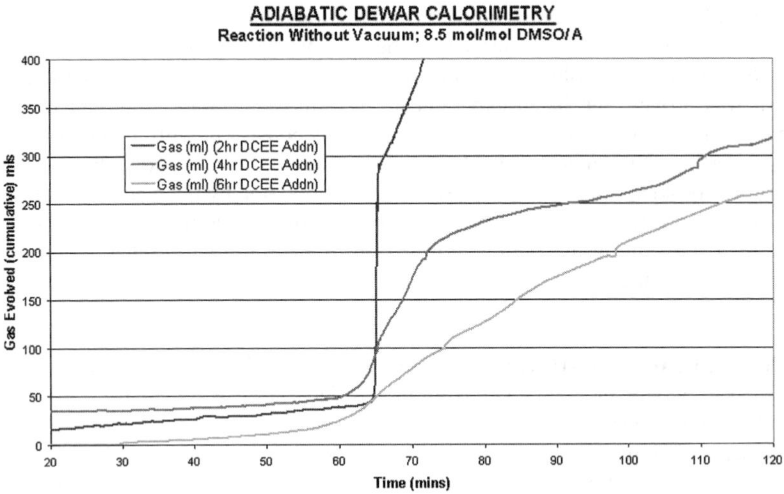

Figure 10. Concentrated process (Comparison 2 – gas evolution)

Table 2. DCEE addition rate – comparative data summary

DCEE addition time (hrs)	Maximum instantaneous rate of temperature rise (°C/min)	Maximum instantaneous rate of gas evolution (ml/min)
2	180	900
4	10	50
6	4	8

DMSO. This immediately makes significantly more KOH available for reaction with **A**. As the rate of reaction starts to increase, more water is produced, dissolving more KOH which increases the rate further. In addition, the exothermic reaction causes the batch temperature to rise, which also increases the rate, producing an overall rate of reaction that is much faster than the expected exponential reaction rate. The situation is also exacerbated by the relatively high accumulation of DCEE to this point.

6.3.1 Manufacturing Plant Observations

The rapid rate of reaction seen in the calorimetry work (6.1 & 6.2) was shown only to occur if water was not removed from the reaction mass, or if water entered the mass before significant reaction had occurred. On the production plant, it was found that the temperature excursions were occurring almost every other batch, even though the reaction was being carried out under vacuum and with a distillation column in place. It had been noted that the temperatures and pressures across the column were not in line with expectations and indicated that the column was becoming flooded early in the distillation. When vacuum was broken, the batch experienced an immediate rapid exotherm, raising the batch temperature to around 150°C. This effect was consistent with the laboratory findings, suggesting that water was held up in the column and returned to the batch when vacuum was broken. This then increased the rate of reaction by the mechanism described in section 6.3. Further monitored batches showed similar column problems and similar exotherms occurring during the reaction period, indicating that the distillation was not removing water effectively.

7. CONCLUSIONS AND RECOMMENDATIONS TO THE TOLLER

From the safety data and data taken from the production plant, calculations were carried out to determine the size of distillation column required to remove water effectively from the 5 te scale plant reaction. The calculations did indeed reveal that the column currently in use on the toller facility was very much under-sized and was not fit for the duty expected. Therefore, the persistent temperature excursions would certainly continue if this column were left in place, with the inevitable safety concerns therefore remaining.

To ensure safe operation of the process and to help with future process efficiency, a number of suggestions were made to the toller:

- For the immediate future, the process could be run with a 4 hour DCEE addition which would lessen the impact of any thermal excursion and minimising the MTSR (maximum temperature of the synthesis reaction).
- An additional element of process control could be adopted to stop addition of DCEE if the rate of temperature rise started to increase. This would minimise the amount of DCEE accumulation and would again reduce the MTSR.
- The distillation column could be replaced with a column fit for purpose i.e. one that was correctly sized for duty. This would allow the process to be carried out as planned, with reduced cycle time and increased yield, therefore raising efficiency and increasing commercial viability.

7.1 REMEDIAL WORK AND RESULTS

The cost of the replacement column was too much for the toller to bear. However, the strategic importance of the product and efficiencies that would be achieved together with the risk to the business should an incident occur, were sufficient for the Syngenta business to sanction the cost of the project ($M's). This required several months lead time, during which the toller continued operation having adopted the other suggested process changes to ensure it could be run more safely.

The process changes showed an immediate improvement and although temperature excursions still occurred, they were much less frequent, less severe and eminently manageable. The column was put in place with one more design modification to improve safety; a device was installed at the base of the column to prevent the contents of the column returning to the reactor should vacuum be lost. The process was then run as designed without incident and continues to run, incident free to this day.

8. SUMMARY AND CONCLUSIONS

This paper has detailed one technical example of the possible benefits of maintaining a good ongoing working relationship with a contract manufacturer. At the point of process handover, a comprehensive data package was provided to the toller. The data, particularly the safety data, did not include recommendations for scale up and did not include Process Risk Assessment (PRA) data. However, the exchange of information on handover included detailed discussions of every known process hazard and from the technical package, the toller was able to carry out their own PRA and arrived at a basis of safety comparable to that determined in-house.

As is often the case, scale up of the process highlighted problems which did not come to light in the laboratory or on the pilot plant scale. The problem was one that was not foreseen as it related to a maloperation, whereas the data handed to the toller related only to the desired process. Once the problem was highlighted, the good working relationship that had been built up with the toller, both during this campaign and from previous out-sourcing projects, helped both parties to work together to acquire all the necessary information to resolve the issue successfully.

Whilst this example is one of a successful working relationship between a business and a toller, in reality this is probably an unusual case and the outcome could have easily been much different. Questions one could ask:

- Would the relationship and the outcome have been the same if the toller had been in the Far East?
- If the suggested process changes had not been adopted, what action would the business have taken?
- If the remedial work had not been successful, how much more resource would the business give to solving the problem?
- If the advice given had resulted in an incident, could the business be liable?

In this case there were a number of factors which made it advantageous to have a closer relationship:

- The toller was the sole supplier of the product at that time
- The product was a "blockbuster" product in its first year of sales and therefore hugely important to the business to maintain security of supply

Having this close relationship provided the advantage that the business found out about problems at an early stage and were able to minimise the Business Interruption Risk (BIR). However, it is not always advantageous or desirable to have such a close relationship. The tendency with contract manufacturers is normally to leave them well alone.

After all, we have an only finite resource – which is why products are outsourced in the first place. It is ultimately the contractor's responsibility, it is their asset.

In the end it comes down to common sense and maintaining a balanced approach.

- When dealing with contract manufacturers, it is important to share as much information as possible at the start. This means obtaining as well as giving information – find out as much as possible about the manufacturers themselves, such as the standard and suitability of their manufacturing facility and particularly their level of understanding of the process risks and hazards.
- It can often be advantageous to have an ongoing relationship with the toller but the level of relationship will depend on the situation – when trying to avoid BIR and to maintain security of supply, a closer relationship will mean that problems are communicated earlier. However, it is important not to spoon-feed the toller and impose your way of thinking, it could lead them down the wrong path. Don't make the assumption that you know better – it is not always the case – and remember, they know their own kit better than you do. Closer relationships also require more resource, the one thing we wanted to free up in the first place.
- When dealing with manufacturers in developing countries, we need to accept that standards are not going to be as high as in the established manufacturing facilities in the Western World. We have a moral obligation to use the relationship during the contract to help them advance. However, there comes a point where we have to accept the standards and live/work with the remaining risk.

REFERENCES

1. Charles Davis, Per Hong, Contract Manufacturing: Realizing the Potential, Executive Agenda Vol 7 No 4.
2. Medical Device Network 1 March 2005.
3. J Hall, Ciba-Geigy Ltd., Loss Prevention Bulletin, December 1993.

THERMAL STABILITY AT ELEVATED PRESSURE – AN INVESTIGATION USING DIFFERENTIAL SCANNING CALORIMETRY

IJG Priestley[1], P Brown[1], Dr J Ledru[2], and Prof E Charsley[3]
[1]Hazards Group, Syngenta Huddersfield
[2]University of Aberdeen
[3]University of Huddersfield

During laboratory scale development of a new chemical process which is to be operated at elevated pressure a material was found to undergo an unexpected thermal decomposition.

Initial DSC testing had indicated that melting appeared to be a pre-requisite for decomposition and based upon this and the fact that melting points are elevated at increased pressures the material had been expected to be thermally stable under the proposed operating conditions.

The unexpected thermal decomposition resulted in a more extensive investigation into the thermal stability of the material being performed. Work was carried out primarily at ambient pressure in order to obtain an understanding of the mode of decomposition. The work was then extended to pressures of up to 30 bar and although this gave us a further insight into the decomposition it failed to simulate the process conditions which could reach 600 bar. A collaboration with The University of Aberdeen has enabled DSC measurements to be carried out at 500 bar confirming predictions about the melting point behaviour but also giving an unexpected view of the thermal decomposition. A second collaboration with The University of Huddersfield provided a further insight into the thermal decomposition of the material.

INTRODUCTION

The vast majority of solid materials handled within the Agrochemical industry have a generally predictable pattern for thermal decomposition. Typically they are stable up to the melting point and only start to decompose at elevated temperatures. The decomposition may or may not be accompanied by gas evolution. This has led to the erroneous belief from some quarters that if a melting point exists then the material will be thermally stable at all temperatures below this.

If the material is thermally stable up to the melting point then handling under normal conditions is relatively straight forward as, provided you ensure that the maximum heating medium temperature is limited to a temperature below the melting point, the risk of thermal decomposition is minimised. However, as chemicals become more complex we are starting to find that materials do not fit the general pattern and the erroneous view is being challenged.

Address for correspondence: IJG Priestley, Syngenta HMC, process Hazards Group, Leeds Road, Huddersfield, HD2 1FF, email: ian.priestley@syngenta.com

This paper highlights an investigation in to one such material in an unusual application.

EXPERIMENTAL

Initial investigations in to the thermal stability of the material were carried out using a Mettler DSC821 ambient pressure DSC at 5 K/min over the temperature range 25–450°C with a gold plated high pressure crucible. Several runs were carried out on the material and all showed that reproducible results can be obtained. A typical thermogram is shown in Figure 1. These test results showed an endotherm in the range 123–152°C with an indicated melting point of ca 139°C, followed by a series of exothermic events which could be separated into an initial decomposition with a heat output of ca 610 J/g, followed by a second sharper exotherm with a heat output of around 750 J/g and a third broader event having a heat output of ca 1040 J/g. The total heat output was ca 2400 J/g.

When using an aluminium crucible at 10 K/min the melting point was ca 138.4°C and the initial exotherm was around 640 J/g, after this point although thermal data was obtained the thermogram shows a clear indication that the containment in the crucible has been lost Figure 2.

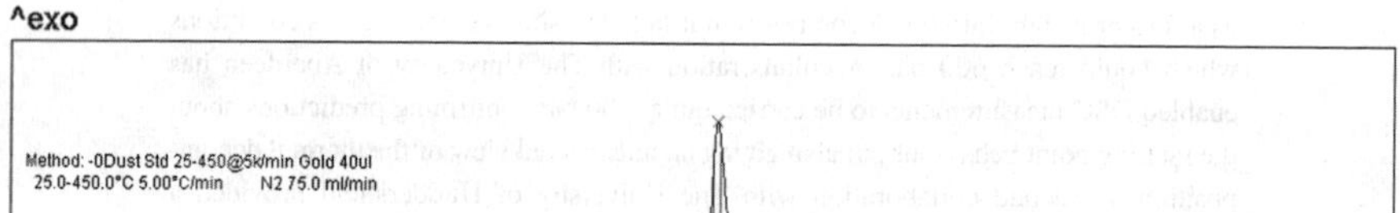

Figure 1. DSC scan from 25°C to 450°C at 5 K/min in a HP gold plated crucible

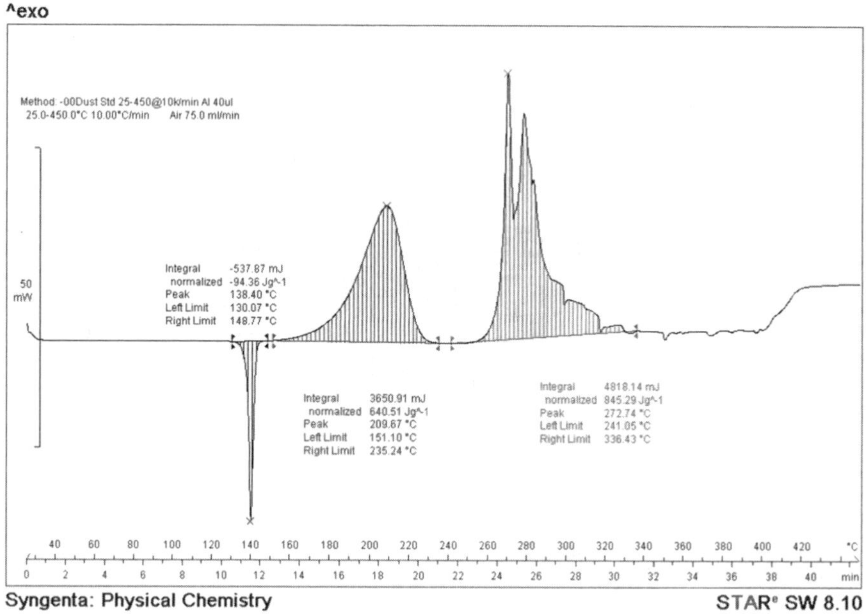

Figure 2. DSC scan from 25°C to 450°C at 10 K/min in an aluminium crucible ex Syngenta

Subsequent investigations were then carried out using a Mettler 27HP DSC Figure 3 and an open gold plated stainless steel crucible at 3-bar nitrogen pressure. This showed a melting endotherm around 140°C followed by two thermal events the first having a heat output of ca 290 J/g and the second around 270 J/g. When using an aluminium crucible similar results were obtained.

Further work was carried out at Aberdeen University using a Perkin Elmer Pyris Diamond DSC at ambient pressure from 20–400°C (Figure 4) then at 500 Bar over the range 20–200°C (Figure 5).

At ambient pressure a melting peak occurred at ca 139.3°C, two separate exothermic peaks were then apparent with respective heat outputs of ca 620 J/g and ca 180 J/g, with evidence that integrity of the crucible is lost sometime after the first exotherm.

Under 500 bar pressure two separate runs showed melting points in the range 152–153°C but no evidence of thermal activity up to 200°C (Figure 5).

TG/DTG work carried out at Huddersfield University showed that on heating at a rate of 10 K/min three separate thermal events were apparent. The first from 145°C to 226°C resulting in 17.3% mass loss, the second from 226°C to 309°C with 60% mass loss and the final event from 309°C to 445°C resulting in 7.3% mass loss. In parallel to this thermal microscopy showed that the first event resulted in significant gas evolution which appeared to stop on exceeding ca 220°C.

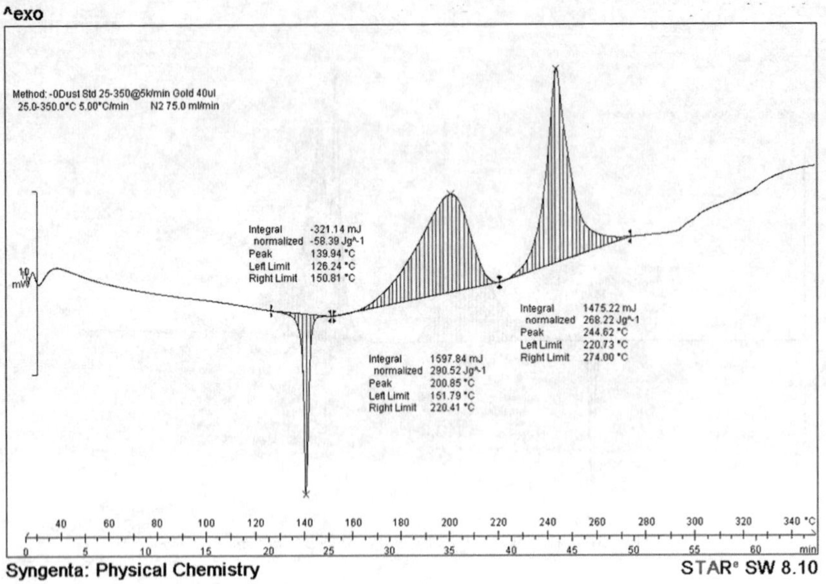

Figure 3. DSC scan from 25°C to 450°C at 5 K/min in an open HP gold plated crucible under 30 bar pressure

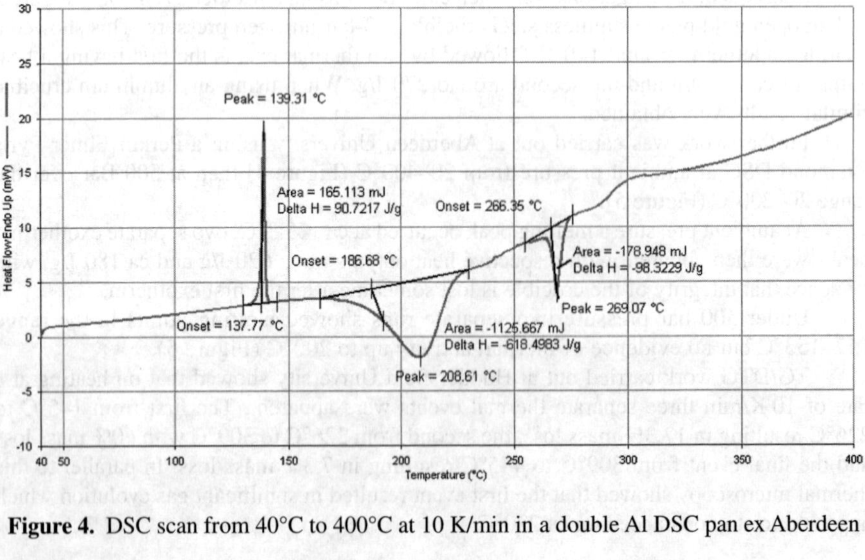

Figure 4. DSC scan from 40°C to 400°C at 10 K/min in a double Al DSC pan ex Aberdeen

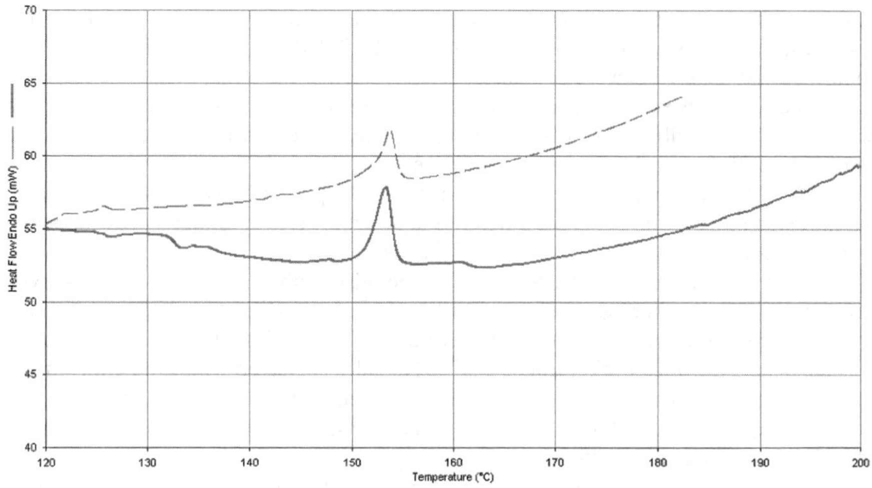

Figure 5. DSC scans from 120°C to 200°C at 10 K/min in an aluminium crucible

DISCUSSION

Development work highlighted the potential for a new application involving formulating the material in to a plastic which is extruded at elevated temperature and pressure. This was a new area for both the manufacturing procedure for this material and also in terms of the work required to define safe operating parameters for the formulation.

During the development of the AI only a limited amount of work was carried out to investigate the thermal stability of the material using DSC and as no exothermic effects were observed below the melting point it was concluded that it would be thermally stable up to this temperature (ca 139°C). Therefore if the 'rule' that the material will be stable up to the melting point is assumed no issues would be expected in operating up to say 120°C. However, if the material were to melt and decomposition occurred then the overall temperature rise under adiabatic conditions would be >1400 K. Experimental work using other techniques has shown that the decomposition would be accompanied by the evolution of copious quantities of gas and this would clearly be a problem on a plant scale, leading to the possibility of a catastrophic vessel failure.

The initial test work did not provide an accurate simulation of the process situation therefore further testing was carried out using a Mettler 27HP DSC which allowed a back pressure of nitrogen to be applied to the sample.

At first glance the overall profiles of the thermograms obtained with and without applied pressure are similar (Figures 1 & 3). However on closer inspection it is apparent that under pressure the melting point has increased by ca 0.8 K and the first exotherm has reduced from ca 610 J/g to around 290 J/g. Whilst the minor change in the endothermic peak could be due to sampling the significant difference in the first exothermic peak could not be explained in such a way. Re-calibration of the 27HP DSC followed by repeat runs confirmed that the effects were genuine.

There was also no appreciable difference between results obtained in aluminium or gold plated crucibles. This indicates that catalytic effects due to the material of construction of the crucibles were not an issue.

The applied back pressure was only ca 5% of the pressure that would be exerted in the extrusion process and given the difference between the two thermograms extrapolation to the process conditions was not thought to be advisable.

Further work was then carried out in collaboration with University of Aberdeen who have developed a method based upon a Perkin Elmer Pyris Diamond DSC which can be operated at pressures of up to 5000 Bar over the range 20–300°C. It should be noted that the data obtained from a Perkin Elmer DSC is typically displayed with exothermic effects shown as negative inflexions, ie, the inverse of the Mettler data representation.

The equipment developed at Aberdeen is described in detail in the reference paper[1] and a brief description is provided here for clarity.

The HP-DSC operates on the power compensation principle, using a Perkin Elmer Pyris Diamond DSC. This is equipped with an autoclave which can be pressurised by means of a hand operated spindle pump. The pump is filled with a silicone oil which acts as the pressurising medium. The HP-DSC head consists of two self-contained silver furnaces in ceramic housings which closely fit inside the autoclave.

A branched chain silicone oil is used as the pressurising medium, this particular oil having a usable temperature range of 20 to 300°C (cf. operating range of the DSC instruments used up to 50 bar of 25–450°C) over the pressure range 50–500 MPa (500–5000 bar).

In order to provide a comparison with the previous work a thermogram was produced under ambient pressure conditions using a Perkin Elmer DSC (Figure 4).

Consideration of these two thermograms show good agreement with the enthalpy of fusion being 90.7 J/g and 94.4 J/g respectively and the actual melting point peak differing by <1K. The initial exotherms are also in very close agreement, at 618 J/g and 640 J/g. Subsequent exothermic behaviour does however differ, with both thermograms showing evidence of loss of containment. This deviation is related to crucible type and mode of operation and was not felt to be significant in this application. It was therefore concluded that there was a very good comparison between data generated on different instruments in different locations. After some discussion it was agreed that work at a pressure more representative of the pressure that would be encountered in the extrusion process would be carried out.

Our particular area of interest is 30–60 MPa (300 to 600 Bar), this is at the lower end of the operating range for the high pressure differential scanning calorimeter. It was agreed that in order to validate the technique investigations would be carried out at a pressure of 50 MPa.

Two separate experiments carried out at Aberdeen are shown in Figure 5, both show endothermic activity with a melting point around 152–153°C, but neither shows any appreciable exothermic activity.

During safety reviews concern was expressed as to the affect that applying very high pressures to a material could have on its thermal stability. During the initial investigations it was not known if melting was a pre-requisite for decomposition to occur. If melting is a pre-requisite then elevation of melting point would not be a problem as the sample could tolerate higher temperatures before decomposition would occur. If melting is not a pre-requisite for decomposition then this is potentially more hazardous as decomposition could be apparent below the melting point of the material, ie, decomposition would occur during the formulation process.

It is well known that applying pressure to a material can increase its melting point. The phenomenon of melting point elevation under pressure was first reported in 1826 by Perkins[2], yet it was not until 1849 that a theoretical and experimental investigation was carried out[3,4], Clausius then put the work on a sound theoretical basis in 1850[5]. Numerous other investigators examined the phenomenon however the most successful work is by Simon and Glatzel[6] who derived what is now known as the Simon equation.

The degree of melting point elevation depends upon the structure of the material but typically this amounts to a 1–3 K rise for every 100 Bar increase in pressure. A more accurate determination can be obtained using the basic version of the Simon equation as follows:

$$\frac{P - P_{tp}}{a} = \left(\frac{T_m}{T_{m,1atm}}\right)^b - 1 \qquad (1)$$

Where P = pressure
P_{tp} = triple point pressure
T_m = equilibrium melting point at P
$T_{m,\,1\,atm}$ = melting point at 1 atm
a & b = constants related to structure

P_{tp} can normally be neglected as it is much smaller than P.

Although specific values for a & b for the material being studied were not available, values were available for compounds with some structural similarities[7,8]. Taking these values allowed a range of melting points to be calculated which are shown in Figure 6 and Table 1, the shaded area in the graph representing the expected melting point limits at any given pressure from 0–600 bar.

The calculations indicate that at 500 bar the melting point would be expected to be in the range 150.3°C to 156.9°C, the experimental values obtained were151.9°C and 153.1°C respectively. It can therefore be concluded that the melting point elevation seen experimentally is in good agreement with the theoretical prediction.

The DSC results indicate that increasing the pressure reduces the exotherm that occurs after melting and it could therefore be concluded that increasing the pressure is therefore making the process safer. It was unclear as to why the exotherm had been reduced by almost 50% when applying 30 bar and disappears completely when 500 bar pressure is

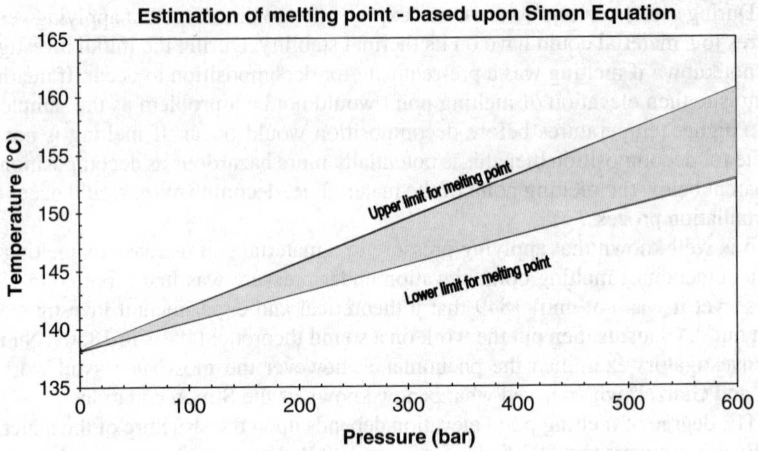

Figure 6. Estimation of melting point using Simon equation

applied. As a number of adiabatic calorimeters designed to operate under pressure are known to experience a loss in sensitivity as pressure increases it was suggested that the cause could be related to the equipment. This was discussed with the manufacturers (Mettler Toledo) who indicated that whilst the possibility of this occurring could not be completely discounted any error introduced from this source would be limited to a few %.

Table 1. Calculated values for melting point derived from the Simon Equation

Pressure(bar)	Melting point limit 1 (°C)	Melting point limit 2 (°C)
1	138.1	138.1
10	138.3	138.5
50	139.3	140.0
100	140.6	141.9
150	141.8	143.8
200	143.0	145.7
250	144.3	147.6
300	145.5	149.5
350	146.7	151.4
400	147.9	153.2
450	149.1	155.1
500	150.3	156.9
550	151.5	158.7
600	152.7	160.6

We are therefore dealing with a real issue rather than an artefact of the experimental method.

By this time a small scale laboratory trial had been carried out on the extrusion process and this yielded unexpected results. It was found that it was necessary to heat the carrier polymer to around 120°C in order to ensure that it was sufficiently free flowing and also ensure that the other components of the formulation were uniformly distributed throughout the final product. Given the thermal data no issues would be anticipated on the scale proposed nor would chemical interactions between the components be expected to be a problem. However, on exiting the extruder instead of the desired product – long spaghetti like strands – a foamy material of variable diameter and density was formed. It was suggested that this was as a result of poor mixing in the extruder but this did not seem to be a plausible explanation and even after pre-milling the components the foaming effect was still apparent.

It was still unclear why the exothermic peak in the DSC disappears under pressure. It is known from other experimental investigations that at elevated temperatures the test material will decompose with the evolution of copious quantities of gas. Work by Bogdanov et al[9] describes transformations of solid organic material under high pressure. Of particular note is work carried out on benzyl peroxide which can undergo up to 73% decomposition simply by applying pressure and altering the stress and shear deformation, however no thermal data is reported. Previous unpublished work by Priestley had observed that applying pressure to materials such as oxalates can under certain conditions result in the evolution of CO_2. In addition to this a literature search highlighted an application in which microcellular foam can be formed by the action of a gas on a thermoplastic in an extruder[10].

Taking this information into account, a scenario was postulated in which decomposition is occurring at elevated pressure however as the decomposition produces gas the application of pressure suppresses either the reaction itself or simply leads to any gases produced dissolving in the reaction mass. At elevated pressures the gases remain in solution and on extrusion through a die the dissolved gases are rapidly released due to the pressure drop leading to the foamy texture in the extruded material.

This scenario does not provide a complete explanation as there does not appear to be a valid thermochemical reason as to why no heat is evolved. It is unlikely that the heat of solution of decomposition gases in the reaction mass would be sufficiently endothermic to offset a heat of decomposition of the order of 600 J/g.

A more likely although as yet unproven explanation is that the application of pressure does in fact suppress the decomposition of the material up to a temperature of ca 200°C, however as the material exits the die head it is still hot (ca 120°C) and the rate of cooling is not rapid enough to prevent the sudden gas evolving decomposition of the material suspended in the semi-molten plastic. For this scenario to be valid it would mean that the material was decomposing below its melting point at ambient pressure.

In order to provide some supporting evidence for this latest hypothesis two parallel work programmes were carried out.

1) In house investigation in to the thermal stability of the material below the melting point at ambient pressures.

2) An investigation of the properties of the material using reflected light thermomicroscopy and TG/DTG carried out by Professor Charsley and The Centre for Thermal Studies at Huddersfield University.

Extensive in-house work using a variety of techniques has clearly shown that the material can undergo thermal decomposition at temperatures below the melting point with decomposition occurring from as low as 95°C after extended isothermal hold periods. The work has also highlighted that the overall decomposition mechanism is apparently autocatalytic and further investigation of this is currently being carried out in order to gain a more complete understanding of the mechanism for decomposition.

The work at Huddersfield University showed that on heating the material started to undergo some minor transformations from temperatures as low as 70°C. On melting gas evolution was apparent immediately and this continued up to ca 226°C at which point it stopped abruptly even though heat evolution was still occurring. Comparison with the thermoanalytical data (Figure 7) shows that this temperature is coincident with the completion of the first exotherm and the start of the second. This suggests a complex mechanism for decomposition with the first part involving the generation of both heat and gas and the second heat only. This evidence also provides a possible explanation of why there is a

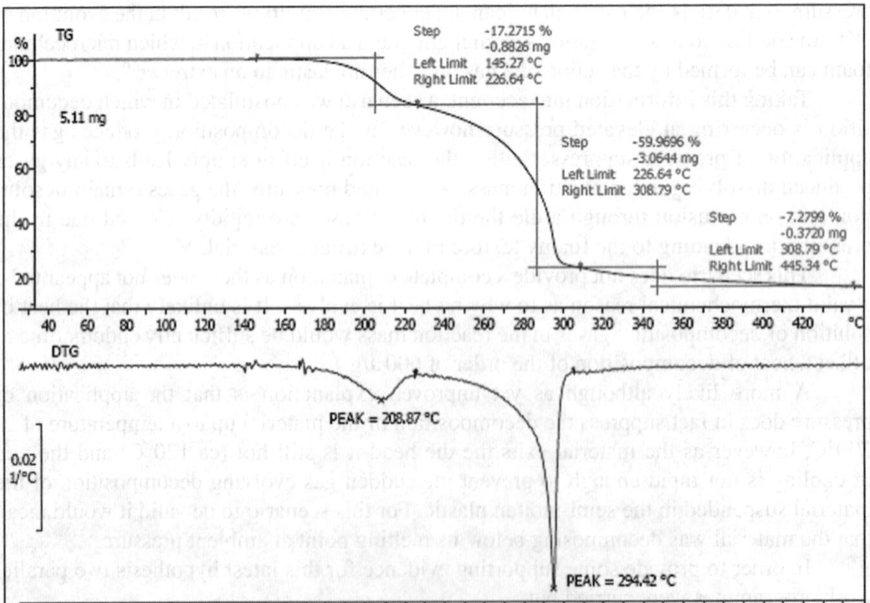

Figure 7. TG/DTG curve heating rate 10 K/min

reduction in heat output on applying pressure, as following Le Chateliers principle, if the evolution of gas is suppressed then the degree of conversion of the starting material will also be suppressed, hence the amount of heat produced will decrease.

Work by Miller et al[11,12] carried out on the decomposition of the explosive HMX and also nitromethane at elevated pressures showed that for a unimolecular mechanism the application of pressure will significantly decreased the rate of the decomposition. This suggests that the mode of decomposition of our material may in part be due to a unimolecular process, further investigation will however be necessary to confirm this.

Our work has clearly identified that the material can undergo thermal decomposition below its melting point.

From a safety viewpoint due to the construction of the extruder decomposition of the material whilst it is contained within the equipment would not be a problem. Discharging the material at elevated temperatures but ambient pressure has been shown to result in decomposition. The effect of pressure on the decomposition is not fully understood however from the insight obtained so far it has been possible to provide guidance to the Formulation development team to enable them to develop an extrusion process which does not result in decomposition of the material during production.

CONCLUSIONS

1. Under pressure the melting point of the material in question increases. The experimentally determined high pressure melting point is in good agreement with theoretical calculations.
2. Applying moderate pressure suppresses the initial decomposition.
3. At an applied pressure of 500 bar no exothermic effects are observed below 200°C.
4. At ambient pressure the material can undergo a gas evolving thermal decomposition below its melting point.
5. The mode of decomposition of the material is complex and requires further investigation in order to provide a complete understanding of its nature.
6. Notwithstanding the 'unusual' mode of decomposition a safe method for extruding the material can be defined.

REFERENCES

1. J Ledru, CT Imrie, JM Hutchinson, GWH Hoehne, Thermochimica Acta 446 (2006) 66–72
2. J Perkins, Phil Trans Roy Soc London 116, 541 (1826)
3. J Thomson, Trans Roy Soc Edinburgh, 16, 575 (1849)
4. W Thomson, Phil Mag. 37, 123 (1849)
5. R Clausius, Pogg. Ann, 79, 376, 500 (1850)
6. FE Simon and G Glatzel, Z Anorg U. Allgem. Chem, 1929 Vol 178
7. SE Babb, Reviews of Modern Physics, 1963, Vol 35 Part 2 April

8. SE Babb, Journal of Chemical Physics, 1963, Vol 38 p2743–49
9. AY Bogdanov, AA Zharov and VM Zhulin, Bulletin of the Academy of Sciences of the USSR, Division of Chemical Science, 1986, Vol 35, p 233
10. EF Kiczek, AI Dalton Jr, US patent 5034171, July 23 1991
11. PJ Miller, GJ Piermarini and S Block, J Phys Chem 1987, 91, 3872–3878
12. PJ Miller, GJ Piermarini and S Block, J Phys Chem 1989, 93, 457–462

CASE STUDIES IN HAZARDS DURING EARLY PROCESS DEVELOPMENT

Alistair Boyd, Paul Gillespie, Mark Hoyle and Ian McConvey[1]
Process Engineering Group, AstraZeneca, Silk Road Business Park, Macclesfield, SK10 2NA

A number of examples of potentially hazardous situations are presented and the outcome of the subsequent evaluation of each of the safety challenges is discussed. The paper considers examples in the area of: solvent stability under reaction conditions, high pressure amination reaction in a hydrogenator, unexpected crystallisation at reflux, uncovering exothermicity of a scaled up reaction and the preparation of distillation residues for further experimental evaluation. In each case a general conclusion is reached that is helpful for: safety specialists, process developers and engineers.

1. INTRODUCTION

The speed of process development to manufacture active pharmaceutical ingredients is increasing to ensure that the demands of patients for new medicines can be met. In fact speed to 'First Time in Man' (FTIM) is a critical benchmark for the industry. This benchmark needs to be met whilst maintaining Safety, Health and Environment requirements and Quality standards.

During the early phase of development the quantity of material required to carry out full hazard evaluation is not normally available. Therefore the avoidance and control of hazard can be improved by the vigilance of process technologists in identifying potentially unsafe or environmentally hazardous situations during early development.

Production of intermediate and active pharmaceutical compounds for toxicological testing is generally carried out in large-scale laboratories (LSL) or kilo-labs. Typically these facilities operate with glass reactors (up to 100 litres in volume), which have low design pressures. Though most laboratories are also equipped with metallic pressure-rated reactors for use in hydrogenation reactions. The speed of delivery from the LSLs is of paramount importance, since toxicology testing of the active is a critical path activity.

This paper will present a number of Case Studies where potentially hazardous conditions were identified in early process development and remedial actions taken to avoid scale-up issues. Examples will be given where the following underlying effects needed to be considered with Dimethyl Sulphoxide (DMSO) stability under reaction conditions, the use of ammonia in a high pressure batch reaction, crash crystallisation/ precipitation from a solvent at reflux, the hazards of a reported thermo-neutral reaction,

[1]Address for correspondence: Dr I F McConvey, AstraZeneca, Process Research and Development Silk Road Business Park, Charterway, Macclesfield, Cheshire, SK10 2NA, e-mail: ian.mcconvey@astrazeneca.com

and ensuring that the correct composition of distillation residues was generated when a sample was generated under low temperature and low pressure conditions. It is expected that the lessons in this paper will raise the awareness of process developers and other safety experts to these potential issues and help to avoid potentially dangerous conditions in similar situations elsewhere.

2. DIMETHYL SULPHOXIDE STABILITY UNDER REACTION CONDITIONS

In a proposed process, DMSO, amine intermediate and cesium carbonate were charged to a reactor and heated to 65°C. The results of Carius tube experiment indicated that this mixture would be free from self-heating and gas evolution up to at least 149°C on scale-up.

A bromo-ethoxyalkane was then added to this mixture and the batch held at 70°C overnight to complete reaction. A Carius tube experiment on this mixture resulted in a rapid exotherm and associated rapid pressure rise from 140°C, leading to bursting of the test container (Figure 1). Assuming normal kinetics one would expect this exotherm to be seen from as low as 80°C in bulk. However, the rapidity of the exotherm and pressure rise indicated that the reaction/decomposition was probably autocatalytic in nature. In this type of reaction the sudden decomposition observed could occur after an induction period at a lower temperature. From the rate of reaction/decomposition noted from the experimental test it would be unventable and any vessel containing the mixture would probably rupture violently if it were to occur.

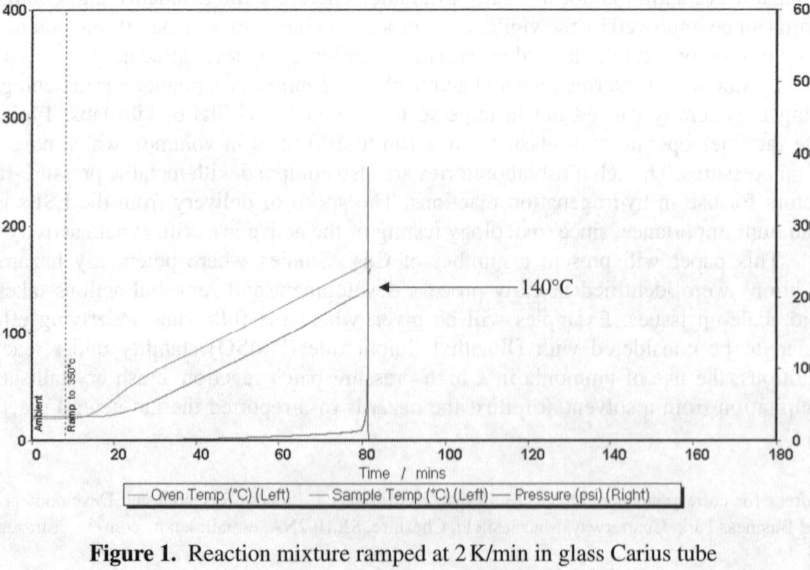

Figure 1. Reaction mixture ramped at 2 K/min in glass Carius tube

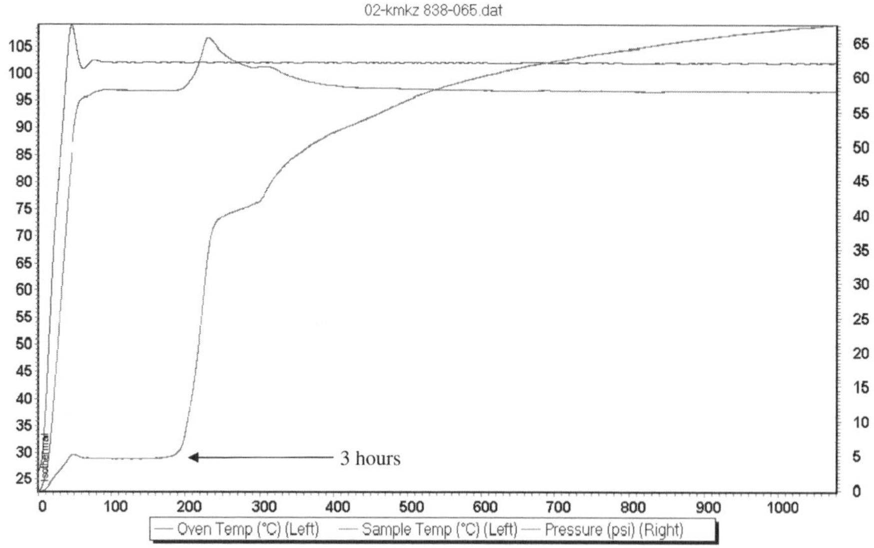

Figure 2. Isothermal at 95°C – Reaction mixture – exotherm noted after 3 hours

To determine if autocatalysis was occurring a Carius tube experiment was carried out with the oven set at ~102°C. In the experiment the sample temperature remained at 95°C for about 3 hours at which point a sudden pressure and temperature rise occurred (Figure 2). This confirmed that the reaction/decomposition was autocatalytic. It should be noted that heat losses from the Carius tube under isothermal conditions are very high and the fact that an exotherm was observed indicates rapid and very energetic reaction/decomposition. In the context of the proposed process the results showed that holding the batch at 70°C overnight (this was required to complete the slow reaction) would almost certainly result in the batch decomposing with potential for over-pressurisation and rupture of the reactor. If the reaction were to be carried out at a lower temperature then a longer hold period would be required to complete reaction and again accessing the exotherm would probably still occur due to its' autocatalytic nature.

A further ramped Carius tube experiment was carried out on the sample after the isothermal experiment was complete. In the experiment a large exotherm occurred from 123°C (Figure 3). This indicated that the decomposition exotherm was complex, i.e. possibly normal kinetics as well as an autocatalytic element was involved, and holding the sample at 95°C resulted in destabilisation of the reaction mixture – (as noted by the reduction in onset temperature from 140°C to 123°C).

From the small-scale experimental results it was concluded that the reaction stage could not be carried out safely on a pilot plant scale. The root of the problem was the

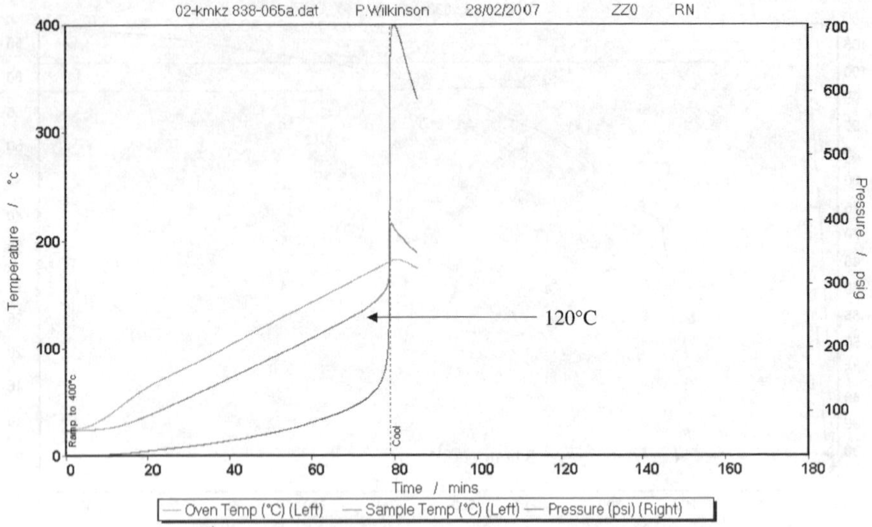

Figure 3. Ramped test at 2K/min after isothermal hold at 95°C for 17.5 hours

instability of DMSO in the presence of bromide, which was generated in this reaction as cesium bromide.

DMSO is far from being an inert solvent and its' decomposition can be catalysed by a range of reagents/impurities, including bromide. Many incidents have previously been reported[1].

After discussion of the experimental results with the Process Chemist an alternative solvent was suggested (N-methylpyrrolidone (NMP)). The results of Carius tube experiments showed that using this solvent would allow the batch reaction to be carried out up to at least 100°C without chemical reaction hazard (Figure 4). Using NMP was not without issues during the subsequent work-up and product isolation, however, these issues could be developed further without a major hazard potential.

3. HIGH PRESSURE BATCH REACTION IN A HYDROGENATOR USING AMMONIA

A high pressure amination reaction needed to be accommodated in a large-scale laboratory. (max 100 litre vessel volume). Most kilo-labs or large-scale labs operate glass reactors and have low design pressures. Some are equipped with metal pressure-rated reactors, usually for hydrogenation reactions, and such a vessel was used in this case but with ammonia. The case study details the method by which it was ensured that the process could be accommodated safely in a hydrogenator pressure vessel.

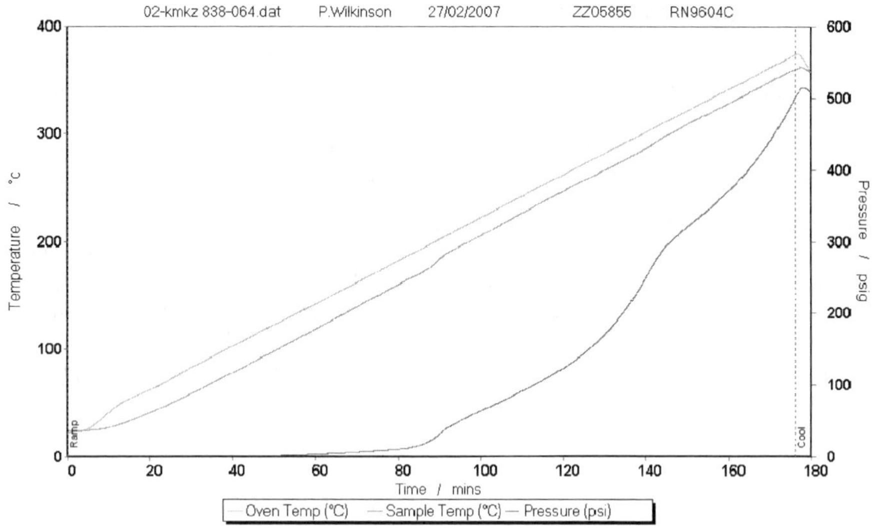

Figure 4. Same reaction mixture in NMP ramped at 2 K/min

The reaction to be accommodated was the amination of an organic chloride to form a primary amine by the following reaction:

$$R\text{-}Cl + NH_{3\,(aq)} \rightarrow R\text{-}NH_2 + HCl_{(aq)} \quad [a]$$

$$R\text{-}NH_2 + HCl_{(aq)} \rightarrow R\text{-}NH_3^{+}Cl^{-}{}_{(aq)} \quad [b]$$

$$NH_{3(aq)} + HCl_{(aq)} \rightarrow NH_4Cl_{(aq)} \quad [c]$$

The reaction was achieved using an overall stoichiometric mixture (two equivalents) of aqueous ammonia (10.76% w/w) and the organic chloride, as shown in reactions [a] to [c] and heating to 135°C. Thermal stability testing had not detected any thermal runaway or decomposition reactions in the reaction mixture. Examination of the reaction conditions indicated that the process would need to be performed in a closed system to prevent excessive ammonia (and water) loss, thus a pressure rated hydrogenation vessel was chosen for the accommodation.

The hydrogenator vessel had a bursting disc with a minimum rupture pressure of 10.45 barg (at 133°C), and a jacket fed with heat transfer fluid up to 200°C. The maximum batch temperature alarm and trip were configured at 140°C to initiate crash cooling. It was therefore necessary to determine that the maximum developed pressure would not cause inadvertent activation of this bursting disc.

The maximum developed pressure within the vessel was determined by summation of:-

1. The inert compression of the head space (as the reactor was nitrogen purged after charging);
2. The vapour pressure of 10.76% w/w ammonia at 140°C;
3. The vapour pressure of HCl (generated by reaction scheme).

The molecular weight of most pharmaceutical compounds is significantly large relative to solvents and other reagents, and it can be assumed that volatility and therefore vapour pressure due to the compound itself would be negligible.

In a closed reactor system any permanent gasses exert their partial pressure, and this will increase as both temperature increases and liquid density decreases. Knowing the liquid densities enables the vapour volume (*V*) to be determined, and then partial pressure can simply be calculated using ideal gas laws:-

$$P_2 = P_1\left(\frac{V_1}{V_2}\right)\left(\frac{T_2}{T_1}\right) \tag{1}$$

which in this case resulted in a pressure of

$$P_2 = 1.013\cdot\left(\frac{50}{36.38}\right)\cdot\left(\frac{413}{273}\right) = 2.106\,\text{bara} \tag{2}$$

The initial headspace volume was taken as being simply nitrogen, i.e. the partial pressure of aqueous ammonia was neglected, as errs to a safer result. In practice, nitrogen will inevitably dissolve in the solution to a low level.

The component and total vapour pressure of aqueous ammonia was readily obtained from literature sources, e.g. Perry[2]. The data in Perry was presented in table format and (for this example) was both interpolated and extrapolated, since the nearest available data was for 9.5 & 14.3% w/w ammonia up to 120°C.

It was assumed that the vapour pressure of aqueous ammonia followed the Antione equation, and the natural logarithm of the vapour pressure versus the reciprocal of the absolute temperature was plotted to yield straight lines. This confirmed the validity of the data. As the plot yielded straight lines, a linear interpolation to the required concentration was carried out and then extrapolated to the required temperature. The resultant vapour pressure of 10.76% w/w ammonia at 140°C was determined to be 7.283 bara.

The molar consumption of ammonia in the reaction is equal to the formation of hydrogen chloride. Therefore at the end of the reaction the hydrogen chloride concentration would be equal to that of the initial ammonia charge. The total vapour pressure of aqueous ammonia solutions was far greater than those of aqueous hydrogen chloride, thus the highest vapour pressure would occur during the initial heating of the reactor. It is noted that for this reaction system, any HCl will react and form aqueous ammonium chloride,

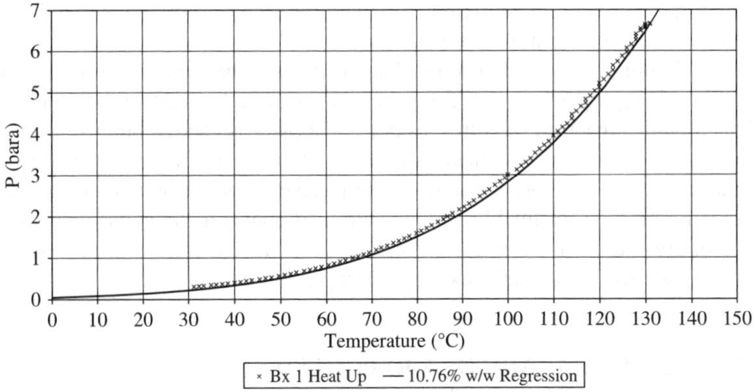

Figure 5. Theoretical total vapour pressure (including inert) versus batch 1 actual

which will have negligible vapour pressure contribution. For this reaction, the sublimation of ammonium chloride can be regarded as insignificant under the reaction conditions.

The calculated maximum developed pressure was therefore 7.283 + 2.106 = 9.392 bara (8.379 barg). This was below the minimum burst pressure of the relief device fitted on the proposed hydrogentator. Therefore it was concluded that a 10.79% w/w ammonia solution at 140°C would not cause the bursting disc to activate.

The calculated pressure was not exceeded in practice when the reaction was carried out (Figure 5).

4. CRASH CRYSTALLISATION/PRECIPITATION AT REFLUX

A semi-batch process was operated in the LSL with addition of the Bredereck's reagent at reflux (batch boiling point, 100°C). However, after the addition was complete exothermic crystallization/precipitation of the product, due to unexpected self-seeding, occurred during reflux. This resulted in rapid boiling of the batch. If the condenser had liquid logged this could have resulted in over-pressurisation, or if the condenser had failed or been overwhelmed then it could have resulted in loss of solvent and potential thermal decomposition of the residue.

An attempt to measure the heat of crystallization directly was unsuccessful due to significant thickening of the batch during seeding. This heat can be estimated using the heat of fusion data, assuming no solvent interaction, in this measured as 106 J/g by DSC. From this, an adiabatic temperature rise of 22 K was calculated for the current process concentration which explains the rapid boiling observed during the crash crystallisation at reflux. To avoid the potential rapid boiling hazard a higher boiling solvent (bpt 151°C) was incorporated whilst keeping the operating temperature at 100°C. This gave a 50 K temperature margin between operating temperature and the boiling point of the solvent, i.e. well above the maximum temperature the mixture could theoretically achieve (122°C). However, it was noted that some minor refluxing of a low boiling by-product generated during the reaction could still occur (but would be minor).

Although not a chemical reaction hazard per se, the potential for rapid boiling, condenser liquid logging, etc., exists for any process in which product supersaturation may occur. This includes reaction mixtures, as per the above case study, but also in operations where the reaction mixture is concentrated by distillation prior to a controlled cooling crystallization. Information/observations from the development chemist, as to the potential for self-seeding, is then invaluable. One should also consider the consequences of over-distillation of the batch, which could increase the potential for crash crystallization, and its possible consequence as the heat sink would also be reduced.

5. THE SCALE-UP HAZARD OF A REPORTED THERMO-NEUTRAL REACTION

Rapid process development and manufacture of bespoke molecules by Commercial Manufacturing Organisations (CMO) is often crucial to meeting product delivery times for toxicology trials during drug development. A requirement to develop and manufacture an intermediate on a relative small scale was recently undertaken by a CMO and they failed to manufacture the material. This required rapid development and manufacture of the intermediate within AstraZeneca to attempt to meet these stringent timelines.

Information provided by the CMO on their initial development of the process included a statement that *'Stage 2 is thermo-neutral and therefore does not pose a hazard on scale-up'*. However, the development chemist within AstraZeneca noted a nominal 'creep' in temperature and reported this to the Safety Group. The process developed by the CMO was a batch reaction, i.e. all the reagents were charged to the vessel and heated to the reaction temperature. Such processing methods can potentially be unsafe if loss of control occurs during operation. Hence, this was investigated further.

Adiabatic Calorimetry is ideal for investigating batch reaction scenarios as it can directly simulate minimal heat losses associated with large-scale plant vessels. In an initial experiment the reagents were charged to the calorimeter and warmed to the start temperature (100°C) (Note: little reaction occurred below this temperature due to solubility effects and catalyst activation). On attaining the start temperature the reaction appeared to initiate and generated considerable heat (Figure 6). The temperature of the reaction mixture reached the boiling point of the solvent in approximately 19 minutes and some solvent was lost (~11% of the total solvent content). Calculation equated this to a heat of reaction of –192.5 kJ mol^{-1} (limiting reagent), when taking into account the heat capacity of the calorimeter and the vapourised solvent. The heat of reaction could theoretically give an adiabatic temperature rise of 166 K, assuming all the heat simply heated the reactants and none were lost to the containing vessel or surroundings, i.e. the worse case scenario. Hence, from the start temperature (100°C) it could reach a theoretical maximum batch temperature of 266°C. This was well above the boiling point of solvent being used (bpt 202°C) and therefore solvent boil over could occur. Theoretically, vaporisation of ~40% of the total solvent charge could occur. In a practical case the reactor system will inevitably

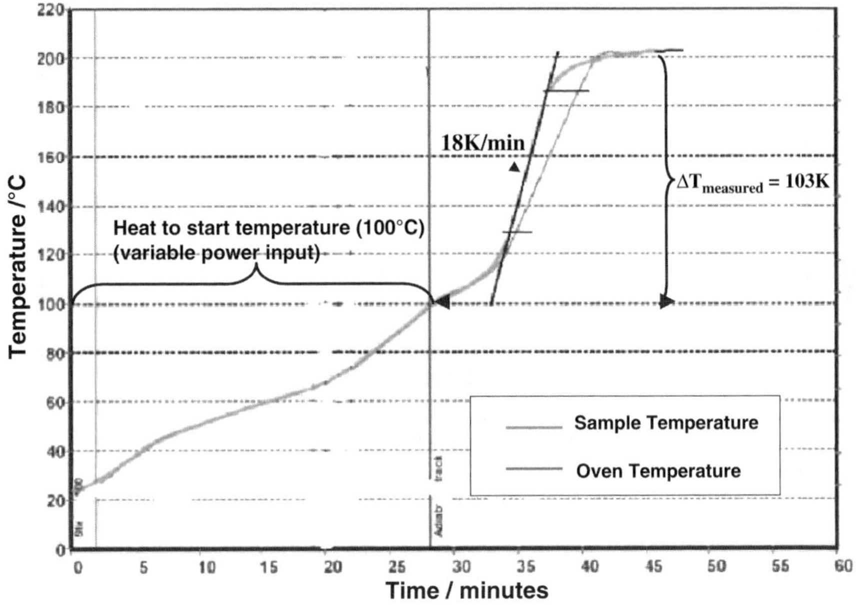

Figure 6. Initial experiment on 'Batch Process' in adiabatic calorimeter – heat to start at 100°C.

provide some heat sink but the contribution of this will become less with increase in scale. Therefore, the operation of the process developed by the CMO at increased scale would pose some considerable hazard.

An attempt was made to operate the process in a semi-batch mode by controlling the rate of addition of one of the reagents. Unfortunately, in the short amount of time available to develop this method it was chemically unsuccessful. Discussion between chemist, engineer and process safety assessor suggested that the best way forward for the chemistry to work would be to add the metal salt powder being used in a controlled manner. Operationally this was considered the worst option with respect to powder handling and toxicity issues, but methods to avoid these issues were being worked on for later manufacturing campaigns.

Hence, the batch processing option was revisited and a method of making this inherently safer considered. The heat capacity of the batch was increased by addition of more solvent so as to safely accommodate the potential adiabatic temperature rise. Calculation showed that increase from 2 to 5 relative volumes of solvent would result in a temperature rise of ~80 K in the adiabatic calorimeter. Repeating the experiment with the increased solvent level validated the calculation. The same temperature profile was noted as before with temperature rising rapidly from 100°C to 160°C and then a decrease in rate (Figure 7). The temperature drifted slowly upwards to 188°C during the 4.5 hours

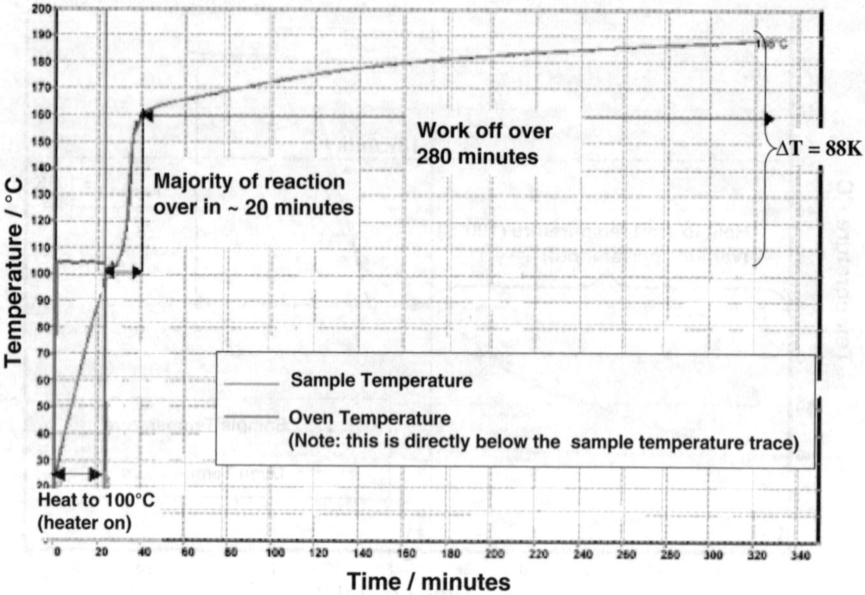

Figure 7. Time/Temperature profile from follow up experiment on batch reaction with increased solvent content

work-off. It was concluded that the calculated temperature rise was fairly accurate, albeit 8 K out over the whole experiment. This may be accounted for by some agitator power input, as the reaction mass was noted to be viscous, or possibly some minor variance in the heat capacity of the Calorimeter or reaction contents over this temperature range. Calculation using the measured temperature rise (88 K) suggested an adiabatic temperature rise of ~112 K would be the theoretical maximum. This would just take the batch temperature above the boiling point of the solvent (202°C). In reality the reactors in the plant have a significant heat capacity and it is considered that the Adiabatic Calorimeter essentially simulates a reactor vessel of at least $10\,m^3$, and this scale was not being exceeded for this manufacture. (Note: the heat capacity of the Calorimeter was measured and the phi factor calculated to be 1.3 for the reaction mass used). If application of cooling to the reaction vessel were to fail at the worst possible moment, i.e. when the batch had just reached the start temperature, then the maximum temperature noted in the Calorimeter experiment would not be achieved at the scale being proposed.

The reaction mass was thermally stable to this maximum temperature (188°C) when carried out in a separate test, with only nominal exothermic behaviour being noted at greater temperature, after allowance of an appropriate safety margin. Hence, the manufacture was carried out with confidence and indeed the maximum temperature rise was not achieved.

In summary, the operation of a difficult chemical process was made inherently safer at the desired scale of operation by simply increasing the heat capacity of the batch, i.e. addition of extra solvent.

This is a good reminder that absence of temperature rise in a laboratory scale experiment does not necessarily mean that the reaction is thermo-neutral or thermally non-hazardous. It may simply be that the rate of heat loss from the laboratory vessel is greater than the rate of heat produced in the reaction, hence giving no apparent temperature rise at that scale.

6. DISTILLATION RESIDUE COMPOSITION: FROM A 2METHF/ WATER/ METHANOL MIXTURE

Distillation processes can be potentially hazardous as over-distillation can leave residues prone to exothermic decomposition at the vessel service temperature. Therefore, an investigation of the thermal stability of the residues is generally undertaken. To perform this investigation a sample is generated at low pressure and temperature to eliminate exposure to elevated temperatures and thus reduce the risk of thermal decomposition prior to thermal stability testing. Unfortunately, in doing this the VLE boundaries/separatrices can change and it is possible that the residue could have a significantly different composition to that of an atmospheric distillation, in this particular instance that was the case.

A limited amount of physical properties data for this system is available in the literature and it is summarised below:

	Normal boiling point	Liq density (g/ml@20°C)
Water	100°C	1
Methanol	65°C	0.791
2MeTHF	79°C	0.855

Azeotropic composition 2MeTHF/water[3] = 89.4/10.6% wt
Azeotropic boiling point 2MeTHF/water[3] = 71 °C

2MeTHF solubility in water is 14%w/w at 20°C[3]
Water solubility in 2MeTHF is 4.4%w/w at 20[3]

The approximate composition of a reaction liquid phase, not including the reaction product(s), prior to the start of distillation was: 27.8%w/w methanol, 61.9%w/w 2-methyl-tetrahydrofuran (2MeTHF) and 10.3%w/w water.

The initial reaction sample was distilled down to a limited volume under reduced pressure (5 Torr) and resulted in a two-phase mixture (organic and aqueous phases). This was not the result of carrying out the distillation under atmospheric pressure whereby water was easily removed. It was obvious that the VLE boundary had changed under the reduced pressure conditions to favour concentration of water rather than removal.

SMSWin[4] and ProPred[5] were used to predictively model the ternary system using the UNIFAC method (Universal Quasichemical Functional Group Activity Coefficient).

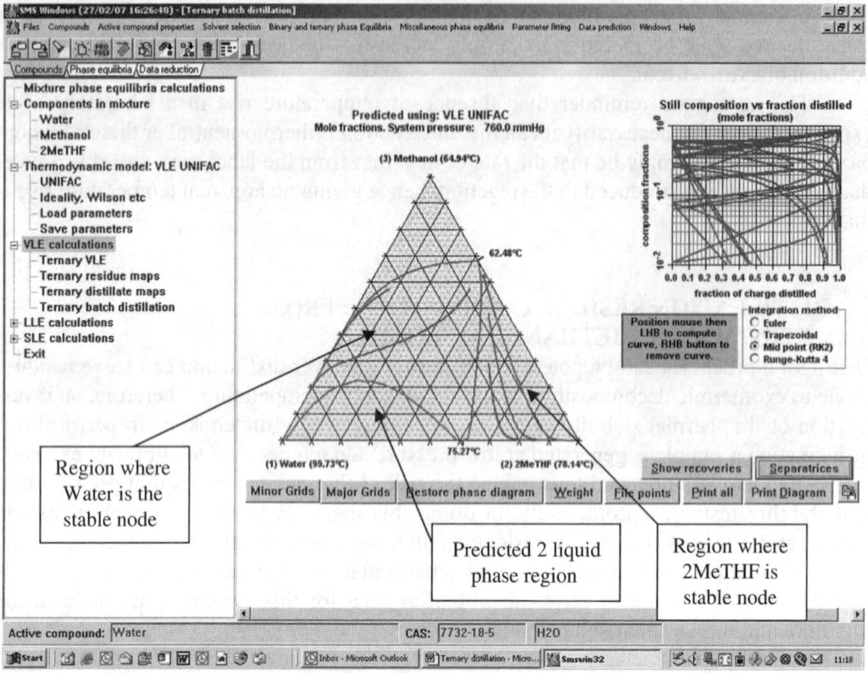

Figure 8. Predicted phase equilibria for water/2MeTHF/methanol

The ternary diagram for the system including predicted LLE (liquid-liquid equilibria) is shown (Figure 8). For successful removal of water under 5 Torr pressure, it was shown to be necessary to add an appropriate amount of anhydrous 2MeTHF to move the composition into the region of the diagram where the 2MeTHF was the stable node. This would ensure the effective removal of water from the system to below the specified target.

The two-phase LLE region is also shown on the ternary diagram (Figure 8). There was a lower level of confidence in the accuracy of this LLE region and if necessary a more detailed model would have had to be generated. However, the reduced distillation was attempted with the additional 2MeTHF added and was shown to work by analysis of water level in the residue.

In summary, due account of the change in VLE/LLE equilibria when carrying out low temperature, low pressure, distillations to simulate atmospheric pressure distillation residues needs to be understood. It may require appropriate changes to the initial composition prior to the distillation to ensure the correct final composition is achieved in the final sample.

The pink lines shown are the separatrices (a separatix is an equation to determine the borders of a system). Once within a region shown on the diagram it is not possible to cross these lines except by adding material from an external source.

8. GENERAL CONCLUSIONS FROM THESE CASE STUDIES

DMSO STABILITY

- DMSO is a potentially reactive chemical and thermal instability can be induced by a range of chemicals/impurities.
- Alternative solvents should be investigated before choosing DMSO as a reaction solvent. Similarly, with other aprotic solvents such as dimethyl formamide (DMF), dichloromethane (DCM), etc.
- Ideally, some form of thermal stability test of reaction mixtures using DMSO or other potentially reactive solvents should be carried out before use, particularly before heating.

ACCOMMODATING HIGH PRESSURE REACTIONS IN LARGE SCALE LABORATORY HYDROGENATORS

- The accommodation of high pressure reactions should take account of the any inert compression of the head space (if the reactor is purged after charging), vapour pressure of the reaction mixture at the maximum vessel temperature, the vapour pressure of any generated volatile in the reaction, taking into account any gas evolution which may take place, in specifying maximum pressure achievable.

CRYSTALLISATION OF MATERIAL AT REFLUX

- Crash crystallization at or close to reflux, or during any process in which product supersaturation may occur, e.g. distillation, can be potentially hazardous. Chemists observations from small scale experiments are useful in determining the need for investigation of such events, i.e. does it happen and, if so, then investigate heat of crystallization and potential consequence.

TEMPERATURE OBSERVATIONS IN LABORATORY EXPERIMENTS

- Observations from small-scale laboratory experiments can be very misleading. Be careful of taking verbatim the fact no temperature rise was noted in small-scale laboratory experiments. This may simply be due to the scale of operation.

GENERATION OF REPRESENTATIVE DISTILLATION RESIDUES AT LOW TEMPERATURE AND PRESSURE

The change in VLE/LLE equilibria when carrying out low temperature, low pressure, distillations to simulate atmospheric pressure distillation residues needs to be understood. It may require appropriate changes to the initial composition prior to the distillation to ensure the correct final composition is achieved in the final sample.

ACKNOWLEDGEMENTS

The help and expertise of the following people within AstraZeneca are acknowledged in producing this paper: Nigel Burke, Sue Burns, Steve Hallam, Matthew Harrison, Sue Jenkinson, Darren Maude, Julie McManus, Steve Raw, Paul Wilkinson and Stephen Whittingham.

REFERENCES

1. Bretherick's Handbook of Reactive Chemical Hazards (6th Edn. P336)
2. Perry, R.H., Green, D., "*Perry's Chemical Engineers' Handbook*", McGraw-Hill.
3. Methyltetrahydrofuran (2004) PENN Specialty Chemicals www.pschem.com
4. SMSWin is a software package which CAPEC (Computer Aided Process-Product Engineering Center) maintain and are further developing for integration with ICAS. SMSWin has a database of compounds and their properties, a collection of property models for phase equilibrium calculations, which are especially suitable for solution properties involving solids. See also User Guide to Solvents, Melts and Solutions by J W Morrison. www.capec.kt.dtu.dk
5. ProPred - a toolbox for estimation of pure component properties of organic compounds (part of ICAS (Integrated Computer Aided Systems) –see CAPEC) www.capec.kt.dtu.dk

ADAPTING THE EU SEVESO II DIRECTIVE FOR THE GLOBALLY HARMONISED SYSTEM OF CLASSIFICATION AND LABELLING OF CHEMICALS (GHS) IN TERMS OF ACUTE TOXICITY TO PEOPLE: INITIAL STUDY INTO POTENTIAL EFFECTS ON UK INDUSTRY

Mary Trainor[1], David Bosworth[2], Anna Rowbotham[1], Jill Wilday[1], Susan Fraser[1] and Ju Lynne Saw[1]
[1]Health and Safety Laboratory, Harpur Hill, Buxton, Derbyshire, SK17 9JN, UK
[2]Health and Safety Executive, Redgrave Court, Merton Road, Bootle, L20 7HS, UK

Within the EU, the risks of major accidents from chemical installations are regulated under the 'Seveso II' Directive. This paper describes an initial study into potential implications for regulation of UK installations arising from changes to the classification of acute toxicity to people when the EU adopts the Globally Harmonised System of Classification and Labelling of Chemicals (GHS). The study's aim was to identify a means of adapting the Seveso II Directive for GHS that: will not increase the Directive's scope and attention unless this increases safety from major accidents; will not increase the risk of a major accident by creating gaps in the regulation of installations; and will be transparent and straightforward for industry to apply. The outcome was to identify a possible option, the 'Simple Alignment', whereby references to the EU classifications Very Toxic and Toxic are replaced by GHS acute toxicity hazard Category 1 and Category 2 respectively for all exposure routes and physical states. To prevent regulatory gaps, the adapted Seveso II Directive would include further Named Substances, such as the lower molecular weight gases ammonia and sulphur dioxide, which have a less severe GHS acute toxicity category but are currently in the Seveso II regime and correspond to installations with major accident hazard potential that would not otherwise fall within the scope of the Directive. These substances would be identified using Technical Criteria that could, for example, be used by an EU Technical Committee to include further Named Substances in future. The other options considered were rejected either because of cost or because of the potential to significantly increase the scope of Seveso II. The outcome of this initial study, together with work by the German and Dutch Seveso II regulatory authorities, is being taken forward through an EU Technical Working Group.

KEYWORDS: GHS, Seveso II Directive, acute toxicity, major accident

INTRODUCTION: THE SEVESO II DIRECTIVE AND THE GLOBALLY HARMONISED SYSTEM OF CLASSIFICATION AND LABELLING OF CHEMICALS

In the EU, the risks of major accidents[1] from chemical installations are regulated through the 'Seveso II' Directive (96/82/EC as amended) for the Control of Major Accident Hazards Involving Dangerous Substances [ECC, 1997 & 2003]. The Directive covers accident prevention and mitigation.

Seveso II applies to establishments where dangerous substances may be present or generated in quantities in excess of specified threshold tonnages – the 'Qualifying Quantities'. The status of regulated establishments is either 'lower-tier' (Directive Articles 6 and 7 apply) or the more highly regulated 'top-tier' (Article 9 additionally applies) depending on whether lower or higher Qualifying Quantities apply. The Qualifying Quantities differ according to which of the Seveso II 'Dangerous Categories' the dangerous substances fall into *on the basis of their classification* and whether they are Seveso II 'Named Substances'. There are ten Dangerous Categories, they relate either to substances': physico-chemical properties such as flammability and explosivity, toxicity to people, or toxicity to the aqueous environment.

This simple threshold tonnage approach operates as an approximate screen to determine the appropriate degree of regulation of establishments under Seveso II. The screen is approximate since off-site risk in the vicinity of any specific installation depends on factors such as: a substance's packaging or containment and inherent physical properties such as vapour pressure; the process and storage conditions; and the geography of the local area. The approximate nature of the screen is explicitly recognised in Seveso II in so far as an installation may be granted a 'derogation' exempting the operator from preparing a full Seveso II 'safety report' if there is no major accident hazard potential.

At present, the basis of classification of substances and mixtures (preparations) is the EU's classification system according to the provisions of The Classification, Packaging and Labelling of Dangerous Substances Directive, CPL (67/548/EEC as amended) and The Classification, Packaging and Labelling of Dangerous Preparations Directive (99/45/EC as amended) [ECC, 1967 & 1999]. Approximately five thousand substances are listed in Annex 1 of CPL with a 'harmonised classification' that is legally binding in the EU. Other substances must be self-classified by the supplier or Seveso II installation operator.

The EU is replacing this classification system by the Globally Harmonised System of Classification and Labelling of Chemicals (GHS) [ECC 2007], [UN, 2005]. At an international level, it is anticipated that major benefits of adopting GHS will include:

[1]Major accident `shall mean an occurrence such as a major emission, fire, or explosion resulting from uncontrolled developments in the course of the operation of any establishment covered by this Directive, and leading to serious danger to human health and/or the environment, immediate or delayed, inside or outside the establishment, and involving one or more dangerous substances' [ECC, 1997].

reducing classification costs to industry by having a single system in use; increasing the consistency and transparency of those public protection levels that are based on classification of chemicals [ECC, 2006a]; and reducing animal testing [UN, 2005].

The adoption of GHS at EU level is a major endeavour because there is not a one-to-one correspondence between GHS and the current EU classification system. Seveso II is only one of over twenty regulations that will potentially be affected. A proposed regulation on `Classification and Labelling of Substances and Mixtures based on the Globally Harmonised System' was published in August 2006 [EEC, 2006a] and updated in June 2007 [EEC, 2007] following stakeholder consultation. The currently proposed EU timescales for the adoption of GHS are for classifications of substances to be mandatory from Dec 2010 and of mixtures from June 2015 [Bierman, 2007].

The timescales for GHS are being coordinated with those for the introduction of the new EU regulatory framework for chemicals called REACH (Registration, Evaluation and Authorisation of Chemicals) under which enterprises that manufacture or import more than one tonne of a substance per year will be required to register it in a central database [European Commission, EIDG & EGD, 2007]. The registration process will include regulatory scrutiny of the substance classifications submitted by enterprises.

An EU 'ad hoc Technical Working Group on Seveso II and GHS' (TWG) is considering the implications for Seveso II when GHS is adopted. Essentially, the differences between the EU and GHS classification systems mean that if GHS classifications are used there is a potential for changes to:

- the *scope* of Seveso II where establishments move between being regulated under Seveso II and not being regulated under Seveso II, or vice versa; and
- the *regulatory attention* of Seveso II where establishments move between lower-tier and the more highly regulated top-tier status or vice versa.

AIMS OF INITIAL STUDY INTO IMPLICATIONS FOR UK INDUSTRY OF THE OPTIONS FOR SEVESO II IN TERMS OF ACUTE TOXICITY TO PEOPLE WHEN GHS IS ADOPTED

This paper describes an initial study into the implications for UK industry of the options for Seveso II in adopting GHS substance classifications for acute toxicity to people. (The study's remit did not include classification of mixtures or classification for toxicity to the aquatic environment and physico-chemical properties.) The study was carried out by the Health and Safety Laboratory, HSL, working with, and on behalf of, the Health and Safety Executive, HSE, which is the lead UK Competent Authority (regulator) for those aspects of Seveso II which relate to harm to people. The objective was to inform the considerations of a group of interested EU Member State Competent Authorities drawn from the EU TWG.

The overall aim of this initial study was to identify a means of adopting GHS for the Seveso II Directive in terms of acute toxicity to people that:

- will not increase the Directive's scope and regulatory attention unless this increases safety from major accidents since this would pose a needless cost burden on the chemical industry[2] and dilute the UK regulatory effort;
- will not increase the risk of a major accident by creating gaps in the regulation of installations; and
- will be transparent and straightforward for industry to apply.

THE OPTIONS FOR SEVESO II WHEN GHS IS ADOPTED AND APPROACHES USED TO STUDY THEM

At EU level, two approaches are under consideration for Seveso II in terms of acute toxicity to people when GHS is adopted:

- The first approach, the 'Dual Classification Option', is to continue to use the current EU classification system to determine the Seveso II Dangerous Category of a substance, whilst requiring industry to classify by GHS under other EU legislation. This option was proposed in [ECC, 2006b]. It is of interest because there would be no change to the regulation of installations under Seveso II.
- The second approach is to replace the EU classification system by GHS using one of the possible 'Alignment Options' whereby references in Seveso II to EU classifications are replaced by references to specified GHS classifications.

HSE's view is that the above long-term Dual Classification Option is not acceptable for Seveso II because using two classification systems in parallel would present an additional cost to both industry and EU Member State regulators compared to using GHS classifications alone.

To assess the possible Alignment Options, a two-part approach was used. The primary approach was to analyse the implications for UK installations based on consideration of: the operation of the Seveso II Aggregation Rule and Qualifying Quantities; and the differences between the EU and GHS classification systems for acute toxicity to people. The second, supplementary, approach was an initial study into the implications at a substance-by-substance level; it is limited by two confounding factors:

1. lack of knowledge of the GHS classifications that will be in use for individual substances within the EU since these will be made, at a future date by industry[3]; and

[2]The costs to UK industry of complying with Seveso II (which is implemented in the UK through the 'COMAH' Regulations) are estimated in [Brazier, 2003]. For example: the cost of analysis and safety report writing starts at about £35k (approximately 50k euros) for storage and warehouse installations, rising to about £220k (approximately 300k euros) for petroleum refineries; and for a fifth of companies considered, Seveso II safety report preparation diverts resources away from other safety activities.

[3]Under the proposed GHS regulation [ECC, 2007] enterprises would classify substances on the market by the end of the transitional period for substances; this is currently proposed to be Dec 2010 [Bierman, 2007]. This GHS classification does not require substance testing: it would be *notified* to the EU Chemicals Agency

2. the extreme difficulty of identifying substances that are not currently within the Seveso II regime[4] in terms of acute toxicity to people, but may be brought in depending on how Seveso II is adapted for GHS, since at present they have no Seveso II regulatory significance and are therefore not listed in any Seveso II related database.

INSTALLATIONS FALLING IN THE SCOPE OF SEVESO II IN TERMS OF SUBSTANCE CLASSIFICATIONS FOR ACUTE TOXICITY TO PEOPLE

Two of the ten Seveso II Categories of Dangerous Substance relate to acute toxicity to people: the Toxic and Very Toxic Categories. A substance falls in these categories if its *overall* EU classification (the most severe of the classifications for the oral, dermal and inhalation exposure routes) is Toxic (T) or Very Toxic (T^+).

Table 1 shows the Seveso II Qualifying Quantities for these Categories of Dangerous Substances. Dangerous substances present at an establishment in quantities greater than 2% of the relevant Qualifying Quantity need to be considered - the 'Aggregation Rule'. The Aggregation Rule also applies to Seveso II Named Substances for the relevant Categories of Dangerous Substances. Table 1 also shows the Qualifying Quantities for two examples of Named Substances that are acutely toxic.

In practice, in the UK many of the Seveso II installations that meet the Qualifying Quantity conditions for the Toxic or Very Toxic Categories of Dangerous Substances, do so on the basis of the Aggregation Rule. For example, installations manufacturing pharmaceuticals or agrochemicals tend to produce a range of substances with overall classification as T or T^+. Similarly, some installations also fall within the scope of Seveso II because they meet the Qualifying Quantity conditions for one or more Categories of Dangerous Substances relating to physico-chemical properties such as flammability: examples include refineries, and some manufacturing plants using toxic substances as intermediates.

Hence, for many UK Seveso II installations, any potential for a reduction in regulatory status and attention arising from a reduction in the severity of the acute toxicity classification of some substances, will in practice be offset by the operation of the Aggregation Rule and their Seveso II status for Dangerous Categories relating to physico-chemical properties. However, the converse is not the case: increases in the regulatory status of installations may arise from an increase in the severity of the acute toxicity classification

unless a substance has already been *registered* through the REACH legislation. Thereafter, the classification may change, for example when a substance is newly registered through REACH (when the classification will be subject to regulatory scrutiny). 'It is anticipated that for some substances the classifications will vary. Over time, it is expected that notifiers and registrants will agree on a single entry [classification]' [ECC, 2007].

[4]For ease of reference, we refer to any substances that either fall into one of the Seveso II Dangerous Categories, or are named in Seveso II, as 'falling within the Seveso II regime' whether or not they lead to installations falling within the scope of Seveso II.

Table 1. Seveso II qualifying quantities for: the toxic and very toxic categories of dangerous substances not named in the directive, and two examples of named substances

Dangerous substances	Qualifying quantity in tonnes of dangerous substances: Lower-tier (Seveso II Articles 6 and 7 apply)	Top-tier (Seveso II Article 9 additionally applies)
Very Toxic Category	5	20
Toxic Category	50	200
Phosgene (a Very Toxic Named Substance)	0.3	0.75
Chlorine (a Toxic Named Substance)	10	25

of some substances, and in some instances this effect may be amplified by the operation of the Aggregation Rule.

OVERVIEW OF THE EU AND GHS CLASSIFICATION OF ACUTE TOXICITY TO PEOPLE

Conceptually, the EU and GHS classification systems are broadly similar in terms of acute toxicity to people *except* in the treatment of inhalation exposures to substances classified as gases under GHS.

Both systems assign substance classifications based on their acute toxicity following exposure via the oral, dermal and inhalation routes. For inhalation exposures, the physical state of a substance is taken into account: that is to say whether it is classified as a gas or vapour, or as an aerosol or particulate. Both systems rank the acute inhalation toxicity of substances classified as vapours, and as aerosols or particulates, in terms of the *mass* inhaled in a given volume. However, unlike the EU system, the GHS system makes a classification distinction between vapours and gases. For inhalation exposures to those substances classified as gases under GHS[5], the EU and GHS systems are fundamentally different:

- the EU system is set up to rank acute inhalation toxicity in terms of the *mass* inhaled in a given volume, whereas
- the GHS system is set up to rank acute inhalation toxicity in terms of the number of *molecules* inhaled in a given volume.

[5]Substances classified as gases under GHS are those for which the test atmosphere is a gas or a vapour near the gaseous state [UN, 2005]. A `vapour near the gaseous state' is not defined. No reason is given for classifying gases and vapours (gases in contact with the liquid or solid state) differently.

Table 2. 4hr LC_{50} and LD_{50} acute toxicity classification boundaries used under the GHS and EU classification systems for each exposure route/ physical state combination

Exposure route/physical state		Definitions of 4hr LC_{50} or LD_{50} (with units) used to set acute toxicity classification boundaries	Classification boundaries	EU classification	GHS classification
Oral		LD_{50} mass fraction (mg/kg)	<5	T+	1
			5 to 25		2
			25 to 50	T	
			50 to 200		3
			200 to 300	Xn	
			300 to 2000		4
Dermal		LD_{50} mass fraction (mg/kg)	<50	T+	1
			50–200	T	2
			200–400		3
			400–1000	Xn	
			1000–2000		4
Inhalation	Aerosols & Particulates (GHS terminology mists & dusts)	4hr LC_{50} mass fraction (mg/l)	<0.05	T+	1
			0.05–0.25		2
			0.25–0.5	T	
			0.5–1		3
			1–5	Xn	4
	Vapours	4hr LC_{50} mass fraction (mg/l)	<0.5	T+	1
			0.5–2	T	2
			2–10	Xn	3
			10–20		4
	Gases (EU as for vapours)	4hr LC_{50} mass fraction (mg/l) **EU only**	<0.5	T+	
			0.5–2	T	
			2–20	Xn	
		4hr LC_{50} *volume* fraction (ppmV)* **GHS only**	<100		1
			100–500		2
			500–2500		3
			2500–5000		4

* For individual substances the conversion factor is: 4hr LC_{50} mg/l = 4hr LC_{50} ppmV × Molecular Weight g/mol ÷ 24,450 .

Table 2 shows the LD_{50} and 4hr LC_{50}[6] boundaries used to classify a substance under the EU system as T+, T, or the less severe 'Harmful' (Xn), and under GHS as Category

[6]For a particular species, the LD_{50} is the dose that will kill 50% of the exposed population whilst the LC_{50} is the equivalent airborne concentration for a specified exposure period.

(Cat) 1, 2, 3 or 4 of which Cat 1 is the most severe. It can be seen that the correspondences between the boundaries fall into three groups:

1. For inhalation exposures to substances classified as vapours, the boundaries for the EU T and T^+ and GHS Cat 1 and Cat 2 classifications are identical.
2. There is a straightforward shift in some boundaries for dermal and oral exposures and for inhalation exposures to aerosols. Therefore, for these exposures, an alignment can either be chosen for which substances may move to a less severe classification but not a more severe one, or alternatively where substances may move to a more severe classification but not a less severe one.
3. There is a correspondence depending on molecular weight for inhalation exposures to substances classified under GHS as gases but as vapours/gases under the EU system; this is illustrated in Figure 1. It can be seen that some lower molecular weight substances that have inhalation classification as T or T^+ in the EU system, will not have a severe GHS classification (GHS Cat 1 or 2). Examples are the industrially important substances ammonia, sulphur dioxide, and ethylene oxide (a Seveso II Named Substance). Conversely, some higher molecular weight gases that are not classified as T or T^+ in the EU system will have a relatively severe GHS classification (GHS Cat 2).

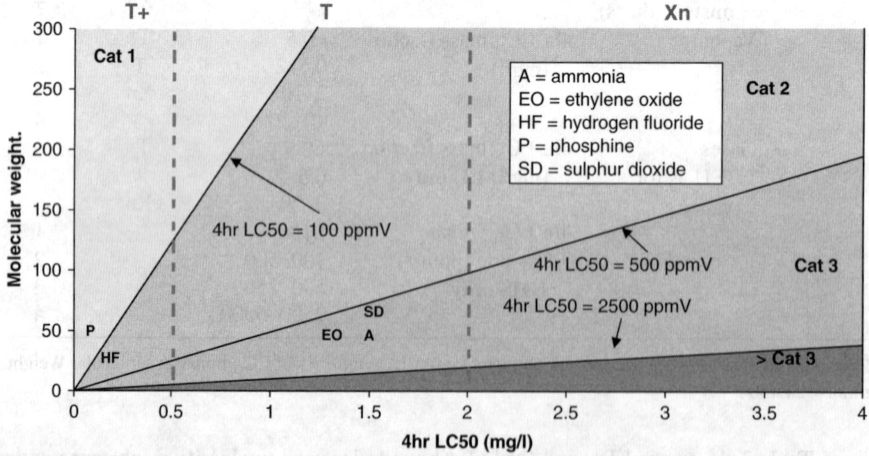

Figure 1. The 4hr LC_{50} boundaries used to define acute toxicity inhalation classifications for substances classified in the EU system as gases/ vapours but as gases under GHS: shown by dashed lines for EU T^+, T and Xn, and by full lines for GHS Cat 1, 2, 3 or 4. Also shown are the 4hr LC_{50} values of four example lower molecular weight substances

OUTCOME OF PRIMARY ANALYSIS: THE SIMPLE ALIGNMENT OPTION WITH TECHNICAL CRITERIA

Based on the above, we considered the suitability of various alignments including the 'CPL' and 'Precautionary' Alignments discussed below. We identified a possible option that meets the study's aims, the:

- 'Simple Alignment' where references to the EU T^+ and T classifications are replaced by GHS acute toxicity Cat 1 and Cat 2 for all exposure routes and physical states. We refer to this as GHS Cat 1 being GHS T^+-equivalent, and GHS Cat 2 being GHS T-equivalent. This is supplemented by
- Technical Criteria to be used to retain other substances with a less severe GHS category that are currently within the Seveso II regime and correspond to installations with major accident hazard potential that would otherwise fall outside the scope of the Directive. This could, for example, be implemented by the addition of extra Named Substances in the Directive, and the use of an EU Technical Committee to include further Named Substances thereafter.

Our reasoning for this option, in order to best meet the UK aims stated above for adapting Seveso II for GHS, and assuming the EU uses the GHS classification distinction between gases and vapours, is as follows:

1. This alignment is the same for all physical states for inhalation exposures. Therefore, knowledge of the physical state used for classification is not needed thus maximising both the ease of use and transparency of this alignment.
2. This alignment minimises the potential for substances to move to a more severe equivalent classification, hence minimising the potential for increases in scope and oversight of Seveso II. This potential arises only for some higher molecular weight substances classified under GHS as gases where changes from T to GHS T^+-equivalent, or Xn to GHS T-equivalent are possible. (We are considering whether further technical criteria could be used to address this.)
3. With the exception of these higher molecular weight gases, this alignment means that substances may move to a less severe equivalent classification but not to a more severe one. Hence, with this exception, the regulatory scope and attention of Seveso II cannot increase. The opportunity for gaps in regulation to arise is limited by the operation of the Aggregation Rule and Dangerous Categories of Substances relating to physico-chemical properties. The use of the Technical Criteria as described above would act as a safety net to ensure that such gaps cannot arise.

PRIMARY ANALYSIS OF THE CPL AND PRECAUTIONARY ALIGNMENTS

We also considered the suitability of two other Alignments Options that have been of interest at EU level: the 'CPL Alignment' and the 'Precautionary Alignment'.

The CPL Alignment Option was the initial EU proposal for classification and labelling purposes [ECC, 2006a]. It differs from the Simple Alignment in aligning T with

GHS Cat 2 and 3 for the oral route and inhalation exposures to aerosols. It has the drawback of requiring knowledge of the physical state used for inhalation classifications. It also has significant potential to increase the regulatory scope and oversight of Seveso without increasing safety from major accidents, by bringing Xn substances into the Seveso II regime through GHS classifications for the oral route. The proposal has now been dropped pending review [ECC, 2007]; we do not consider it further.

The Precautionary Alignment Option aligns Seveso T^+ with GHS Cat 1 and Cat 2, and Seveso T with GHS Cat 3 for all exposure routes and physical states. It is of interest because it is the most straightforward alignment that would not result in a reduction in the scope of Seveso II because substances could not, in practice, move out of the Seveso II regime. Like the Simple Alignment, it has the advantage that the alignment is independent of the physical state for inhalation exposures.

However, this alignment does not meet the UK's aims because it has significant potential to increase the regulatory scope and oversight of Seveso without increasing safety from major accidents. For example, from Table 2 and Figure 1, it can be seen that Xn substances may be newly brought within the Seveso II regime as GHS-T equivalent through GHS classifications for the oral and dermal routes and inhalation exposures to vapours and higher molecular weight gases, or as GHS T^+-equivalent through GHS classifications for higher molecular weight gases.

INITIAL SUPPLEMENTARY STUDY INTO IMPACT ON UK INDUSTRY: SUBSTANCES, CLASSIFICATIONS, AND ASSESSMENT OF IMPACT ON INSTALLATIONS FORMING BASIS OF STUDY

The initial supplementary study into potential impact on UK installations considered two groups of substances:

1) Substances currently classified as T or T^+ that are important in the UK in terms of Seveso II - the 'UK Seveso T and T^+ substances'.
2) Substances that are not currently classified as T or T^+ but could newly be brought into the Seveso II regime under the Simple or Precautionary Alignment Options. To attempt to identify candidates, we trawled the EU High Production Volume Chemicals, HPVCs. These are the approximately 2,500 chemicals that were on the European market before September 1981 and are produced or imported in quantities exceeding 1,000 te per year. Data provided by manufacturers and importers on HPVCs such as tonnage and toxicity is held on the IUCLID database [Hansen, 1999]. HPVCs do not include all the substances of interest (examples are low production substances such as intermediates and reagents, or substances that have only been high production volume since 1981). We trawled approximately 1,300 HPVCs: all those with an EU harmonised classification that means they may potentially be brought within the Seveso II regime; and the approximately 40% of highest production volume[7].

[7]We used a list provided by the European Chemicals Bureau, ISPRA, in October 2006 of HPVCs ordered in groups of descending tonnage volume using latest reported volumes from manufacturers.

A confounding factor in identifying the second group of substances, and in using both groups for the study, is lack of knowledge of the GHS classifications that will be made by EU industry subject. We assigned relatively rapid *informal* GHS substance classifications (see Appendix) which without doubt will differ from the detailed industry GHS classifications for some substances. The informal GHS classifications were assigned to:

- 71 (30%) of the 238[8] UK Seveso T, T^+ substances. This was done using readily available toxicological data. (See Appendix for the data sources used and the constraints on substances that could be included.)
- 29 candidate substances that may be newly brought into the Seveso regime under the Precautionary Alignment. This was done using industry toxicity data from the IUCLID database. Substances with an EU harmonised classification that is inconsistent with this data were excluded from the study. No candidates were found that would be brought in under the Simple Alignment.

The regulatory impact on UK installations arising from classification changes to these substances was considered in a series of three meetings with HSE specialists. These were informed by Internet information on industrial use of the substances, together with HSE information on tonnages at specific UK installations for some substances. The aim was to identify changes to the scope or regulatory attention of Seveso II. We did not aim to, and could not, identify borderline establishments such as those which are borderline top-tier and would become borderline lower-tier. We do not consider that such cases have regulatory significance given that the Seveso II Qualifying Quantities act as an approximate screen only.

INITIAL STUDY INTO IMPACT ON UK INDUSTRY OF PRECAUTIONARY ALIGNMENT OPTION: OUTCOME

Two of the candidate Xn substances that may be newly brought into the Seveso II regime under the Precautionary Alignment based on industry toxicity data in IUCLID were found to have UK regulatory significance:

- sodium dodecyl sulphate [CAS 131-21-3] which is used in detergents and foamy personal hygiene products like shampoo, shaving foam and bubble bath;
- calcium diproprionate [CAS 4075-81-4] which is used as a mould inhibitor in processed foods such as cheeses, non-alcoholic drinks, confectionaries and some meat products, as well as in livestock and poultry feeds.

As a result, some UK formulators of processed foods, animal feeds, detergents or frothy personal hygiene products might become Seveso II sites. (A formulator blends ingredients to make final products: therefore a site may have stock tanks or other relatively large storage of ingredients.)

[8]See list at http://www.hse.gov.uk/hid/haztox.htm. HSE uses this in connection with: the assessment of Seveso safety reports, and the provision of advice on land-use planning in the vicinity of installations.

Our trawl of candidate substances from IUCLID will only have identified a small fraction of the substances that might be newly brought into the scope of Seveso. Hence, whilst the GHS classification assigned by industry in future may differ for these two example substances, our view is that this initial study confirms the potential for the Precautionary Alignment to widen the scope of the Seveso II Directive without giving an increase in safety from major accidents.

INITIAL STUDY INTO IMPACT ON UK INDUSTRY OF SIMPLE ALIGNMENT OPTION: OUTCOME

For the 71 UK T, T^+ substances considered, the effect of the Simple Alignment on overall classification is that:

- between 29% and 43% of overall T^+ classifications drop to overall GHS T-equivalent (between 15 and 22 substances out of 51); and
- between 30% and 45% of overall T classifications drop to overall GHS-equivalent classification less than T – we refer to this as 'GHS sub-T equivalent' (between 6 and 9 out of 20 substances).

We quote a range because for some substances the overall classification is dependent on the physical state assumed for inhalation exposures. No examples were found of substances moving from T to GHS T^+-equivalent. Table 3 gives the classifications for example substances and their current proposed EU harmonised GHS classification[9].

The classification changes were only found to have regulatory significance for two substances: the lower molecular weight gases ammonia and sulphur dioxide. As a result of their overall classification change from T to GHS sub-T equivalent, UK installations with major accident hazard potential would fall outside the scope of Seveso. To address this, they would be made Named Substances.

No other adverse regulatory impact was found. For one substance, it was found that there would be a reduction in the scope of Seveso but that this would be beneficial as the reduction applies to installations with no major accident hazard potential to people[10]. For one substance, information on inventories at UK installations is limited but there is none to suggest that there would be an unacceptable regulatory impact. For the remaining

[9]The current proposal [ECC, 2007] is that EU harmonised acute toxicity classifications would be translated to *minimum* GHS classifications except for T^+ for dermal exposures which translates directly to GHS Cat 1. Industry would increase the classification over this minimum where appropriate. An alternative proposal of interest at EU level is *maximum* harmonised GHS classifications that industry would decrease where it can be demonstrated that this is appropriate.

[10]Potassium dichromate: some UK surface engineering industry establishments are brought into the scope of Seveso II solely on the basis of this substance's inventory. Under the Simple Alignment they would be out of scope. http://www.hse.gov.uk/surfaceengineering/comahguidance.pdf gives an HSE analysis showing that there is no off-site major accident hazard potential in terms of acute toxicity to people for the quantities of potassium dichromate typically stored.

Table 3. Effect of simple alignment on the overall classification for example substances with breakdown by Oral (O), Dermal (D), and Inhalation (I) routes. Listed are the: current EU classification; *informal* GHS classification followed by its EU equivalence under the simple alignment; and proposed harmonised (H) GHS classification marked * where a minimum

Substance name	Substance CAS No.	Classification	O	D	I	Over-all	Effect of simple alignment on overall classification
Anhydrous ammonia	7664-41-7	EU	—	—	T	T	Change: EU T to GHS Xn- equivalent based on *informal* GHS classification
		GHS	—	—	3	3	
		GHS-equiv	—	—	Sub-T	Sub-T	
		H-GHS	—	—	3*	3*	
Ammonium dichromate as CrVI	7789-09-5	EU	T	Xn	T^+	T^+	Change: EU T^+ to GHS T-equivalent based on *informal* GHS classification
		GHS	3	4	2	2	
		GHS-equiv	Sub-T	Sub-T	T	T	
		H-GHS	3*	4*	2*	2*	
Methylene dithio-cyanate	228-652-3	EU	T	—	T^+	T^+	No change: EU T^+ to GHS T^+-equivalent based on *informal* GHS classification
		GHS	3	—	1	1	
		GHS-equiv	Sub-T	—	T^+	T^+	
		H-GHS	3*	—	2*	2*	

substances, there is no change in regulatory attention or scope of corresponding installations due to the operation of the Aggregation Rule and the Dangerous Categories related to physico-chemical properties.

No substances were found that would be newly within the Seveso remit. As described above, the potential for this to occur is limited, arising only through GHS inhalation classifications for some higher molecular weight gases.

Our view is that this initial supplementary study supports the conclusion that the Simple Alignment with Technical Criteria is a suitable option and identifies ammonia and sulphur dioxide as examples of substances that would need to be Named Substances in the adapted Seveso II Directive. Further work is underway at HSL to develop the Technical Criteria.

STATUS OF STUDY AND NEXT STEPS

In March 2007, this initial UK study was disseminated to the group of interested Member States drawn from the EU TWG. Together with studies from the Netherlands and Germany it formed the basis of discussion on acute toxicity at the group's 2nd meeting in September 2007. It is anticipated that at the 3rd meeting in November 2007, an agenda will be mapped out for technical sub-groups to consider the issues for toxicity to people, the aqueous environment and physico-chemical properties in order to facilitate data sharing and pooling of expertise between Member States.

ACKNOWLEDGMENTS

The study described here was carried out working closely with HSE experts in the fields of process safety, toxicology, and Major Hazards policy. We are particularly grateful to Peter Ridgway for his valuable advice and insights into substance classification, and to Sandra Ashcroft, Tim Beals, Andrea Caitens, Dave Carter, Richard Cary, Steve Porter, Ralph Rowlands and Kirstin Wattie. We also thank Ole Nørage of the European Chemicals Unit, ISPRA, for helpful information from IUCLID on HPVCs ranked by tonnage.

REFERENCES

Bierman T., 2007, Seveso and GHS, Presentation on behalf of European Commission Environment Directorate General at Informal GHS and Seveso Meeting, HSL, Buxton, UK, 14.9.07.

Brazier A. and Waite. P, 2003, Safety Report Regime – Evaluating the Impact on New Entrants to COMAH, HSE Research Report 092, ISBN 07176 2173 1, HSE Books.

ECC (European Communities. Commission), 1967, Council Directive 67/548/EEC on the Approximation of Laws, Regulations and Administrative Provisions Relating to the

Classification, Packaging and Labelling of Dangerous Substances, *O.J. European Communities* 16.08.67, **L196**: 1–98.

ECC (European Communities. Commission), 1991, Council Directive 91/414/EEC of 15 July 1991 Concerning the Placing of Plant Protection Products on the Market, *O.J. European Communities* 19.8.1991, **L230**: 1–32.

ECC (European Communities. Commission), 1993, Council Directive (EEC) 793/93 of 23 March 1993 on the Evaluation and Control of the Risks of Existing Substances, *O.J. European Communities* 5.4.93, **L84**: 1–8.

ECC (European Communities. Commission), 14.1.1997, Council Directive 96/82/EC of 9 December 1996 on the Control of Major Accident Hazards Involving Dangerous Substances, *O.J. European Communities*, **L10**: 13–33.

ECC (European Communities. Commission), 1999, Directive 99/45/EC of the European Parliament and of the Council of 31 May 1999 Concerning the Approximation of the Laws, Regulations and Administrative Provisions of the Member States Relating to the Classification, Packaging and Labelling of Dangerous Preparations, *O.J. European Communities* 30.7.999 **L200**:1–68.

ECC (European Communities. Commission), 31.12.2003, Directive 2003/105/EC of the European Parliament and of the Council of 16 December 2003 Amending Council Directive 96/82/EC on the Control of Major-Accident Hazards Involving Dangerous Substances, *O.J. European Union*, **L345**:97–105.

ECC (European Communities. Commission), 2006a, Proposal for a Regulation of the European Parliament and of the Council on Classification and Labelling of Substances and Mixtures Based on the Globally Harmonised System, August 2006.

ECC (European Communities. Commission), 2006b, Analysis of the Potential Effects of the Proposed GHS Regulation on its EU Downstream Legislation, August 2006.

ECC (European Communities. Commission), 2007, Proposal for a Regulation of the European Parliament and of the Council on Classification and Labelling of Substances and Mixtures Based on the Globally Harmonised System, June 2007.

European Commission. EIDG & EDG (Enterprise and Industry Directorate General and Environment Directorate General), Feb 2007, REACH in Brief, http://europa.eu/enterprise/reach/docs/reach_intro.htm.

Hansen B.G., van Haelst A.G., van Leeuwen K. and van der Zandt P., 1999, Priority Setting for Existing Chemicals: European Union Risk Ranking Method, *Environ. Tox. and Chem.* **18** (4): 772–779.

Ruden C. and Hansson S.O, 2003, How Accurate are the European Union's Classifications of Chemical Substances, *Toxicology Letters* **144:** 159–172.

UN, 2005, Globally Harmonized System of Classification and Labelling of Chemicals (GHS) First Revised Edition, ST/SG/AC.10/30/Rev. 1, UN.

Wood M., Pichard A., Gundert-Remy U., de Rooij C. and Tissot S., 2006, The AETL Methodology as a Potential Solution to Current Challenges Associated with the Development and Use of Acute Exposure Levels in Seveso II Applications, *J. Haz. Mat.* **A133**: 8–15.

GLOSSARY

Aerosol (mist): liquid droplets of a substance or mixture suspended in a gas (usually air) [UN, 2005].
Particulate (dust): solid particles of a substance or mixture suspended in a gas (usually air) [UN, 2005].
ppmV: parts per million by volume (cm^3/m^3).
Vapour: the gaseous form of a substance or mixture released from its liquid or solid state [UN, 2005].

APPENDIX: ASSIGNMENT OF INFORMAL GHS CLASSIFICATIONS

For the purposes of this study, informal GHS classifications were assigned without carrying out the detailed checks and data gathering that would, for instance, form part of the work of EU harmonised classification assignment. For example: we did not check that data came from valid well-performed tests; where information was incomplete we did not request further details from the source; and we only considered data on the notional 'preferred test species' rather than carrying out a full evaluation of all available experimental animal data. For those substances that have an EU harmonised classification, we generally could not use the corresponding LC_{50} and LD_{50} as this information was not retained in the earlier years of the programme. (The need to publish the scientific motivations of classifications was proposed in [Ruden, 2003] which discusses the accuracy of harmonised classifications.) For inhalation exposures, assigning the physical state of the test atmosphere appropriately is non-trivial for substances that are liquids at ambient conditions: it may not be specified in the available account of the test, and testing may have been conducted using a mixture of physical states. Therefore, we considered all possible physical states. For instance, based on the manufacturer's toxicity data in IUCLID, tridemorph (CAS 24602-86-6), has overall GHS acute toxicity Cat 4, Cat 3, or Cat 2 according to whether the aerosol, vapour or gas state is assumed.

For the UK T, T^+ substances, the data sources used were toxicological reviews including: CICADS[11], ATSDR[12], OECD SIDS[13], EU Risk Assessment Reports on HPVCs under regulation 793/93/EEC [ECC, 1993], WHO Environmental Health Criteria reports[14], draft Technical Support Documents prepared by HSE for the EU ACUTEX project [Wood, 2006]; and UK Pesticide Safety Directorate substance evaluations carried

[11]CICADS: `Concise International Chemical Assessment Documents' International Programme on Chemical Safety IPCS Co-operative Programme of WHO/ILO/UNEP. See http://www.inchem.org/pages/cicads.html.
[12]ATSDR: US `Agency for Toxic Substances and Disease Registry', US Department of Health and Human Services. See http://www.atsdr.cdc.gov/2p-tox-substances.html.
[13]SIDS: `Screening Information Dataset for High Production Volume Chemicals', Organisation for Economic Co-operation and Development (OECD). See: http://www.inchem.org/pages/sids.html.
[14]See: http://www.who.int/ipcs/publications/ehc/ehc-numerical/en/index.html.

out[15] under the EU 'Pesticides' Directive 91/41/EEC on Plant Protection Products [ECC, 1991]. Additionally, confidential HSE records were accessed, discussions were held with HSE toxicologists involved in the EU harmonised classification process to clarify the basis of decisions taken for some substances and, where no other information was available, Manufacturer's Material Safety Data Sheets were used. Only 30% of the UK T, T^+ substances were considered. For the remainder either: data were not available for all exposure routes or were not in a format that would allow a comparison with classification criteria; the overall EU harmonised classification is not supported by the data regardless of the physical state assumed for inhalation classifications; or the EU harmonised classification is corrosive but not T or T^+ although there are data to support the latter. (Inconsistencies with corrosive substances can arise because of priorities, unrelated to Seveso II, within the administrative system for agreeing the EU harmonised classifications.)

[15]Evaluations carried out by the UK Pesticide Safety Directorate Advisory Committee on Pesticides, see: http//www.pesticides.gov.uk/psd_evaluations_all.asp.

PRACTICAL EXPERIENCE IN RADIO FREQUENCY INDUCED IGNITION RISK ASSESSMENT FOR COMAH/DSEAR COMPLIANCE†

Ian R Bradby, Senior Safety Consultant
ABB Engineering Services, Pavilion 9, Belasis Hall Technology Park, Billingham, Cleveland, TS23 4YS, UK

Is your site within 30km of a radio, TV or radar transmitter? If so, radio frequency induced ignition could pose a hazard to the assets on your site.

In order to comply with the Dangerous Substance and Explosive Atmosphere Regulations (DSEAR), companies handling substances capable of creating explosive atmospheres are required to carry out a formal risk assessment. This must consider the extent of foreseeable explosive atmospheres within and external to the process, and ensure that suitable equipment is installed to control all potential ignition sources. One potential ignition source arises from radio-frequency radiation, often identified during the preparation/review of company COMAH safety reports.

The radio-frequency environment is becoming increasingly severe, with the proliferation of transmitting sources, increased transmitter powers and the exploitation of new techniques. Sources for radio-frequency transmissions include radio and television broadcasts, radio communications, mobile phone communications, radar and navigational equipment. These transmission sources can affect an area of up to 30km and have the potential to impinge on most operating sites.

Electromagnetic waves produced by radio-frequency transmitters will induce electric currents and voltages in any conducting structure on which they impinge. The magnitude of the induced current and voltage depends upon the combination of the shape and size of the structure, the wavelength and the strength of the transmitted signal. A spark may occur if the induced voltage and currents are sufficiently large.

The latest standard for the assessment of inadvertent ignition of flammable atmospheres by radio-frequency radiation, BS 6656:2002, provides detailed guidance, but currently many companies are unclear about the level of risk posed by radio-frequency induced ignition.

INTRODUCTION

This paper describes the methodology and practical experiences of applying BS6656:2002 to the issue of inadvertent ignition of flammable atmospheres by radio frequency radiation.

Radio frequency (RF) induced ignition is a credible, but not well recognised mechanism for creating a source of sparks on operating plant structures. Much time, effort and expenditure is spent by companies in controlling sources of ignition, so that in the event of a release of flammable material, the potential for fire or explosion is minimised.

BACKGROUND

It is known that radio transmission sources can induce currents in metal structures. TV, radio, Radar systems, communication system (e.g. mobile phones) all fall within the radio frequency range of concern, as shown in Figure 1.

The types of RF transmission sources are numerous. Typical systems include; long and medium wave radio, ship communication & radar systems, radio beacons, amateur radio, FM & VHF/UHF radio, TETRA communication systems, radio telephones, civilian & military radar, satellite communications, television broadcasts, mobile phone networks and local site radio communication systems. Today's environment is a 'soup' of electromagnetic radiation which is a result of our modern technological society.

It is known that RF transmissions may produce spark ignition at a distance of up to 30km. Given the generous distribution and geographical location of RF transmission sources, it is probable that most industrial sites will be in range of relevant transmitters (see Figure 2 for a typical example).

The basic principle is that electromagnetic waves produced by radio-frequency transmitters (e.g. radio, television and radar) will induce electric currents and voltages in any conducting structure on which they impinge. These structures can include vessels, pipework, vent stacks and other equipment such as loading cranes. The magnitude of the induced current and voltages depends upon the shape and size of the structure relative to the wavelength of the transmitted signal and on the strength of the electromagnetic field.

In addition, parts of the plant structure (which are normally in contact) are caused to break or separate momentarily, this could be as a result of maintenance or vibration; a spark may occur if the induced voltage and current is sufficiently large. If this happens in a location where a potentially flammable atmosphere may be present a hazardous situation can occur (see Figure 3). However, the possibility of ignition will depend on many factors including whether the spark can deliver sufficient energy to ignite a particular flammable atmosphere.

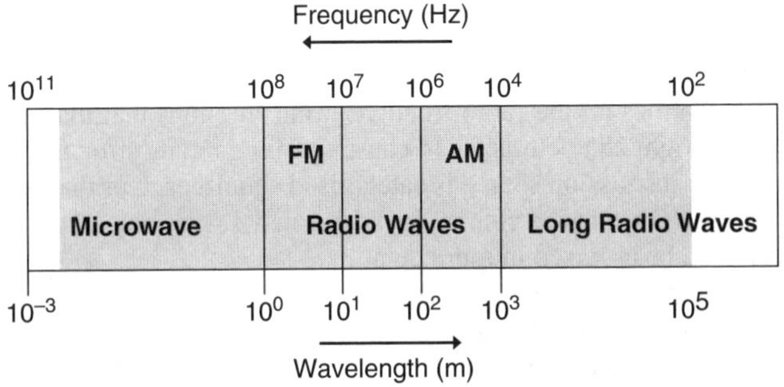

Figure 1. Electromagnetic spectrum, RF induced ignition risk

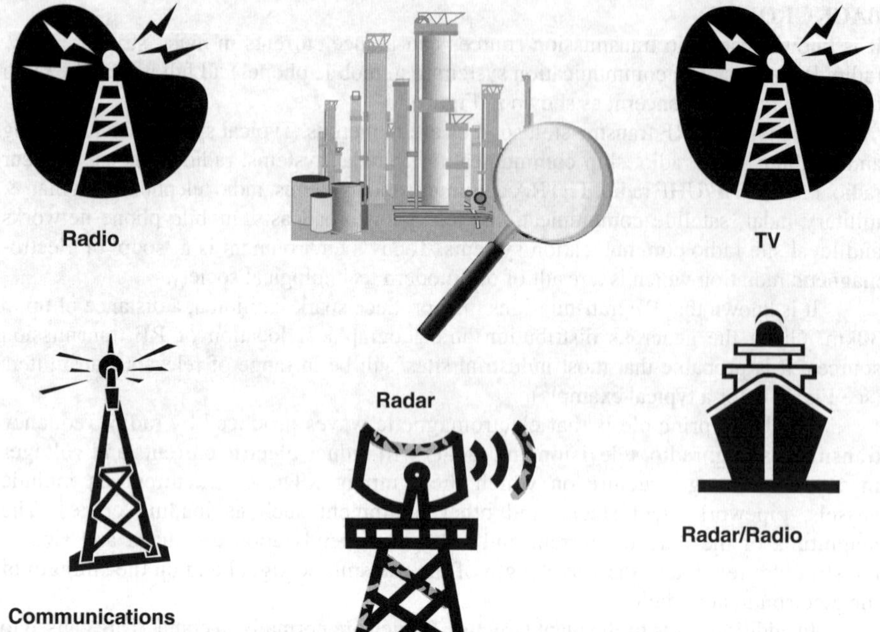

Figure 2. RF transmission sources that typically affect industrial sites

So how do we attempt to assess the risk of spark induced ignition from radio transmission sources? Fortunately there is a very pragmatic British Standard *BS 6656 (2002)*[1] that details how to assess the risks. The standard provides a systematic approach to the elimination of RF induced ignition hazards by assuming that realistic worst case conditions apply.

The standard outlines a 'screening study' or *initial assessment*. The *initial assessment* essentially identifies all the radio frequency transmissions that may be of sufficient power and have the right characteristics to cause sparking in any structures. The vulnerable zone from each transmission source is categorised against each of the gas groups I/IIA, IIB & IIC. Any radio transmission sources identified in the screening study that are a cause for concern, can then be assessed in more detail.

It also describes how to assess hazards in more detail. The 'detailed' study looks at the characteristics of the plant structures which are acting as an 'aerial', to determine whether sufficient energy can be extracted (by the structure) to exceed the threshold values to cause ignition of any flammable atmosphere that may be present. This can be achieved through either detailed analysis/calculations, or by carrying out practical on site tests to

Figure 3. Conditions required for RF induced hazardous events

determine signal strengths and the 'efficiency' of plant structures in acting as 'aerials'. Incendivity tests can also be used to establish the amount of energy required by a spark to ignite the flammable atmosphere under consideration.

The standard provides advice on mitigation measures where assessments indicate that a hazard exists. There are a number of possible solutions from; 'bonding' structures to prevent breaking any circulating currents that may be circulating as a result of RF transmission, to changing the configuration of the plant to reduce the 'efficiency of the aerial', to the use of insulation to prevent current circulating.

PRACTICAL EXPERIENCE

BS6656 (2002)[1] details the following methodical approach to assessing radio frequency induced ignition risk, as shown in Figure 3A. Steps 1, 2 & 3 form part of the initial assessment methodology as outlined in the standard.

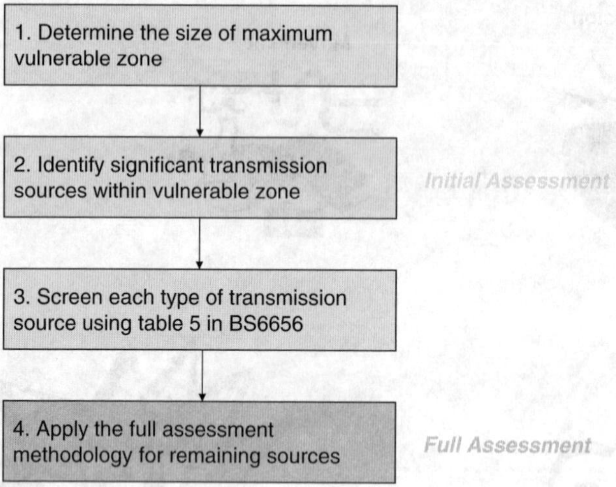
1. Determine the size of maximum vulnerable zone
2. Identify significant transmission sources within vulnerable zone
3. Screen each type of transmission source using table 5 in BS6656
4. Apply the full assessment methodology for remaining sources
Initial Assessment
Full Assessment

Figure 3A. BS6656 Assessment methodology

The following example is used to illustrate some of the issues involved in addressing radio frequency induced ignition risk from a practical point of view.

INITIAL ASSESSMENT

This simplified example describes a typical approach using the *Initial assessment* as outlined in section 10.2.2 of the standard. The *Initial assessment* is designed to eliminate from further consideration those locations from which it is highly unlikely that a hazard exists. It is based on realistic worst-case estimates of the radius of the zone around different classes of transmitter; therefore Table 5 in the standard has been used in preference to Table 6. (Tables 5 & 6 consider the effects of electromagnetic fields on differing plant geometries.) The initial screening is also based on gas group IIC (Hydrogen as the representative gas) as the flammable atmosphere, which again represents the worst-case scenario.

The first step (see item 1 in Figure 3A) is to determine the maximum vulnerable zone. This is taken from Table 7 in the standard, and in our example is 29.2 km which is based on gas group IIC. The extent of this zone is shown in Figure 4. Any transmission sources outside this zone do not require any further consideration and can be eliminated from the study.

The second step in our method (see Figure 3A) is to identify all the significant transmission sources within the vulnerable zone. The standard considers transmission sources operating in the 9 kHz to 60 GHz range as these present a potential ignition risk. In our example, these have been identified as shown in Figure 4.

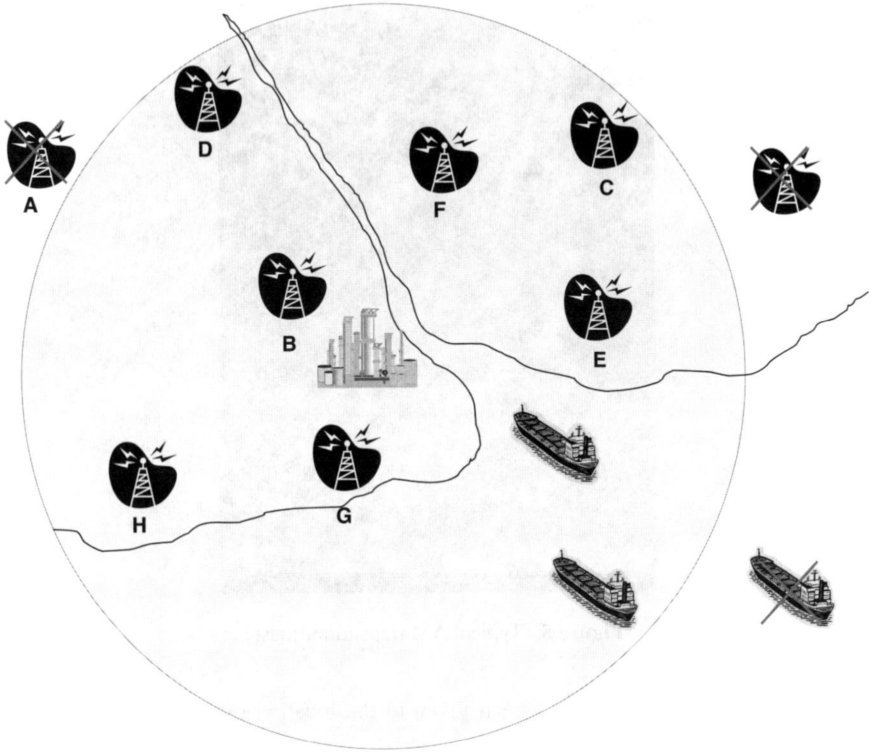

Figure 4. Example 29.2km vulnerable zone from radio frequency transmissions

Sites A, B & C are AM/FM radio transmission masts broadcasting on a number of frequencies. We have two transmitter masts at site D & E of unknown origin. There are mobile phone masts at sites F & G and a national TV mast at site H. We also have some shipping which is navigating the estuary near to our example site.

The third step in the method (see Figure 3A) is to take each of these identified sources in turn, and assess them against Table 5 in the standard, to determine whether any further action is required. Let us now consider each type of transmission source in our example.

Radio transmissions

From Table 5 in the standard, AM radio transmissions (typical example is shown in Figure 5) can have significant vulnerable zones measured in kilometers; whereas other transmission sources, at higher frequencies, tend to have smaller zones measured in

Figure 5. Typical AM transmitter mast

metres. Careful consideration has been given to the location of AM radio transmitter masts as listed in Table 1.

With the exception of the AM transmitter mast at site B (which is broadcasting three radio stations on frequencies 1063, 1179 & 1232 kHz), all the identified AM transmitters fall outside the 29.2km zone and can therefore be discounted from this study.

Using Table 5 in the standard, the transmitter mast at site B falls under item 17 and has an associated vulnerable zone of 4.3 km. The extent of this zone is shown in 06.

Table 1. Example AM radio transmitter mast locations

Frequency	Radio station	TX site	Tx power kW	Grid ref
613 kHz	BBC Radio	A	2	AB 274598
1073 kHz	Local Radio	A	1	AB 274598
1219 kHz	Local Radio	A	2.2	AB 274598
1451 kHz	BBC Regional	A	2	AB 274598
1063 kHz	*Local Radio*	*B*	*3*	*AB 420218*
1179 kHz	*Local Radio*	*B*	*3.3*	*AB 420218*
1232 kHz	*Local Radio*	*B*	*2*	*AB 420218*

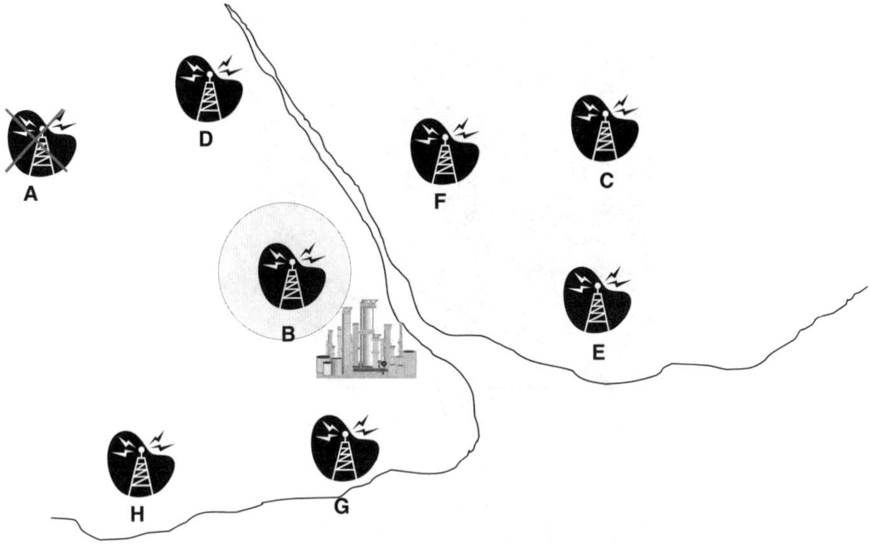

Figure 6. Size of vulnerable zone from AM radio transmitter mast at site B

This vulnerable zone does not directly impact the site under consideration and can therefore be discounted as a source of hazard in this study.

FM, DAB radio & TV transmissions

A similar approach is used to identify FM, DAB radio and TV transmissions. In this case two transmitter masts are within the 29.2 km zone at sites C & H.

From Table 5 in the standard, the FM transmitter mast at site C falls under item 50 and is considered to be non-hazardous, as the main radiation lobe of the transmitting antenna is so designed that it produces field strengths near the ground that are unlikely to exceed 1 V/m. Therefore this FM transmitter mast can be eliminated as a potential source of hazard. A similar argument can be used to assess the mast at site H.

Other transmission sources within vulnerable zone

Two further transmitter sites were located on the ordinance survey map within the 29.2km zone, as shown in Figure 4 at sites D & E.

The transmitter mast at site D (shown in Figure 7) and its ownership has been difficult to determine. This site is located more than 10km from the plant. From inspection, there appear to be 3 types of transmitting aerials on this mast. Two of the aerial types, which from an estimate of their dimensions would be suitable for operating around

Figure 7. Transmitter mast at site D

75 MHz & 130 MHz, would probably fall under item 54, Table 5 in the standard. This item indicates a maximum vulnerable zone of 6 metres from the transmitter location. The other aerial type appears from its dimensions to be suitable for operation at around 500MHz, and would probably fall under item 57 in Table 5 from the standard. This item indicates a maximum vulnerable zone of 1.3 metres. As these zones are very localised to the transmitter site, this transmitter mast can be eliminated as a potential source of hazard.

Identifying potential RF sources can be an interesting and entertaining activity. For example, the ordinance survey map for the area shows a transmission tower in close proximity to the site at location E (see Figure 8). After further investigation and a site visit to verify the status of the tower, the tower was used as part of the RAF's Second World War radar defence system to warn of attacking enemy aircraft. The tower has been decommissioned and now is more of a tourist attraction on a local industrial facility. This transmission source can therefore be discounted from the study.

Mobile phone masts

Mobile phone masts represent another potential hazardous source. The number of mobile phone masts has increased dramatically over the past few years, as this technology has

Figure 8. Transmitter mast at site E

been widely taken up by the general public (see Figure 9). There are many such sites situated within the 29.2 km zone. Typically these might be sites F and G in our example as shown in Figure 4.

Using a worst-case scenario from Table 5 in *BS 6656 (2002),* mobile phone transmitter masts fall under item 64. This gives a maximum vulnerable zone of 3 metres from the transmitter mast. As mobile phone transmitter masts are located outside the site boundary fences, and the operating plants under consideration are situated more than 3 metres from the site boundary, mobile phone transmitter masts can therefore also be eliminated as a potential source of hazard.

Shipping – radar & radio transmissions

Another local potential source of hazard can be found on shipping operating along the River estuary. Ship's radar and radio communication systems can use relatively high power transmitters. Using Table 5 in the standard the worst-case scenario for marine radio would give a vulnerable zone of 560 m under item 11. The worst-case scenario for marine radar falls under item 95, and gives a vulnerable zone of 420 m.

The conclusion from our example *initial assessment* is that Ship's radar and radio cannot be screened out and require further investigation. Many industrial sites are located

Figure 9. Typical mobile phone mast site

along major water ways where passing vessels have systems of sufficient RF power to create the potential for spark ignition. Radar systems need to be powerful enough to 'penetrate' poor weather conditions to allow the vessel to move in relative safety. Deep sea radio may need to transmit signals thousands of miles to communicate with the shore.

In our example we now apply step 4 of our method to the remaining transmission sources, as outlined in Figure 3A, and use the full assessment methodology as detailed in the standard.

FULL ASSESSMENT

In this example, the initial screening demonstrated that there may be a significant hazard from vessels navigating along the river estuary. Further investigation is required to examine the typical types of systems used onboard vessels navigating past the site.

Typical systems onboard vessels

The following systems are commonly used onboard ships, that propagate electromagnetic energy in the 9 kHz to 60kHz range; VHF radio telephone, MF/HF radio telephone, Satellite communications, Navigation radar and EPIRB emergency search and rescue systems.

VHF radio telephone
VHF radio is the primary means of voice communication between the ship and shore, capable of operating at a range of up to 50kms. When a vessel is being piloted to its berth, VHF radio is the system that is used to communicate with the Harbour Office. VHF marine radio operates from 150 Mhz to 165 MHz, with typical transmitter powers of 25 watts. When in port, these systems are required to operate on reduced power, typically around 1 watt, which reduces further any potential radio-frequency induced spark ignition risk. VHF radio remains operational at all times. Vessels also have independent additional VHF radio system(s) available in the event of a main VHF radio system failure.

Satellite communications
Satellite communication systems are used to provide computer-to-computer links for Internet, e-mail and other business systems. The satellite phone system also provides voice communications when the vessel is out of VHF radio range. Typical systems operate at 1.6 GHz with power ratings of around 150 watts. These systems operate continuously.

MF/HF radio telephone
MF/HF marine radio is essentially an independent backup communication system used to communicate with the shore. It is capable of operating over thousands of kilometres in the 1.6 MHz to 30 MHz range, with typical transmitter powers of around 400 watts. Current practice is for this system to remain in 'standby mode' until the vessel has docked. On docking, the antennae is switched to an 'earthed' position, to prevent any inadvertent transmissions. This is carried out by operating a switching device on the radio set. There are no recognized circumstances, other than planned maintenance, when the MF/HF radio system would be required in port.

EPIRB
Electric Position Indicating Radio Beacons are small self-contained, battery operated, low power systems used to assist in the location of survivors in search and rescue operations. These operate in the MHz frequency range with transmitter powers measured in milliwatts.

Navigation radar
Typically there are two independent radar systems on vessels; S-band and X-band.

X-band operates at 10 GHz with typical transmitter powers up to 50 kW and gives higher resolution than S-band. This is the system that is used by the Harbour Master's office to track ship movements. S-band operates at 3 GHz with typical transmitter powers up to 60 kW and can provide better definition in poor weather conditions e.g. ability to 'see through rain'. Both these systems are required to bring a vessel safely into port, and are actively used in poor visibility conditions. The radar systems are switched off once the vessel has safely docked.

ANALYSIS OF SYSTEMS ONBOARD SHIPPING

The geographical layout of the antennae relative to the shore, when the vessel is docked, is considered here as part of this example analysis.

Typical ship's beam again can vary considerably, however the smallest vessels have a beam of approximately 12 m. Radar and radio antennae are located on the highest points of vessels to eliminate potential 'blind spots' in communication. The lowest estimated antennae position would be at least 10 m above the deck.

Jetty loading arms are considered to be the most vulnerable pieces of land-based equipment to hazard sources from the ship, during unloading operations. They have the correct geometry to form an antenna capable of converting any received EM radiation into potential spark energy. However, due to concerns over static discharges when a ship is unloading at the jetty, loading arms incorporate an isolator and so cannot form a receiving loop with the ship. Jetty loading arms can therefore be discounted from the analysis.

VHF RADIO TELEPHONE

VHF radio is the primary means of voice communication between the ship and shore. VHF Radio Telephone falls under item 54 in Table 5 in the standard. In this example, if we take gas group IIB as the most demanding case for all the jetties, then this gives a maximum hazard range of 4.5 m. Taking the typical dimensions of the smallest 65 m vessel that docks by the site, then this hazard range is within the confines of the ship, and would not impact on shore operations (see Figure 10). Therefore VHF radio telephone can be discounted from this analysis. When in port, these VHF systems are also switched to operate on reduced power, typically around 1 watt, which would further reduce any spark ignition risk.

Satellite communications

Satellite communication systems are used to provide computer-to-computer links and operate continuously. Taking a conservative approach, Satellite Communications would fall under item 108 in Table 5, with an associated hazard zone of 4.5 m for gas group IIB. From Figure 10, this hazard range is within the confines of the ship and would not impact on shore operations. Therefore Satellite Communications can be discounted from this analysis as a source of hazard.

MF/HF radio telephone

MF/HF marine radio provides essentially an independent backup communication system that would not be used in port. Taking a conservative approach, MF/HF radio telephone falls under item 23 in table 5 in the standard. This would give a maximum hazardous range of up to 450 m for any shipping traffic navigating along the estuary either past the site or to the jetties. This could have a significant impact on site operations.

MF/HF marine radio is essentially a backup communications system for deep-sea operation, which is not required in port. The MF/HF marine radio is independent, and separate from, the VHF radio system used to communicate with the Harbour Master's

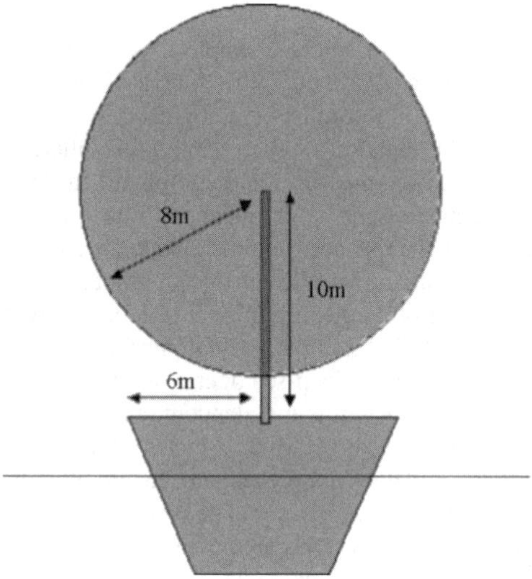

Figure 10. Position of hazard zone relative to the ship

Office in port. The MF/HF radio is in a 'standby mode' and does present any hazard other than accidental operation. Communication with the Harbour Authorities is via VHF channels, and so accidental operation is considered unlikely. When in port, these MF/HF systems are switched to an 'earthed' position, which eliminates any spark ignition risk. Additional procedural measures could be considered to further reduce the possibility of accidental operation while vessels are navigating along the estuary.

EPIRB

Electric Position Indicating Radio Beacons are small self-contained, battery operated, low power systems used to assist in the location of survivors in search and rescue operations. Therefore EPIRB can be discounted as a hazards source from this analysis.

Navigation radar

Typically there are two independent radar systems on vessels; S-band and X-band. Both these systems are required to bring a vessel safely into port, and are actively used in poor visibility conditions. The radar systems are switched off once the vessel has safely docked. S-Band radar represents the greater hazard and from item 93 in Table 5 has a range of up to 8 m for gas group IIB. From Figure 10, this hazard range is within the confines of the ship and would not impact on shore operations. Therefore Navigation Radar Systems can be discounted from this analysis as a source of hazard.

CONCLUSIONS

The conclusions from the full assessment example may be that RF transmissions from shipping could impinge on the site. Although this source of radio frequency energy represents a potential source of ignition, it must be recognised that safeguards may already in place on site that prevent the creation of flammable atmospheres and sparks, i.e. plant integrity, ATEX area classification measures and equipment earthing. Hence, the risk of a flammable atmosphere being present, co-incident with an ignition source due to radio frequency effects is small. As the risk of an explosion due to radio frequency induced ignition is small, then cost benefit analysis indicates that significant expenditure on further risk reduction measures could be grossly disproportionate. Hence, only low cost risk reduction measures should be considered. For example, MF/HF marine radio is essentially a 'backup' communications system intended for deep-sea operation, which is not required, or used in port. The MF/HF marine radio system is independent, and separate from, the VHF radio system used to communicate with the Harbour Master's Office. In port, MF/HF marine radio is in a 'standby mode', and does not present any hazard other than through accidental operation. Additional procedural measures could be considered to further reduce the possibility of accidental operation, while vessels are navigating along the estuary.

This paper has described in detail examples of initial and full assessments, in accordance with the guidance in *BS 6656 (2002)*. The assessment of the risk from RF induced ignition will be dependent on the location of RF sources in relation to the site under consideration. Each assessment will need to be tailored to the needs of the site. Long wave radio, radar and shipping represent the highest level of risk of RF induced ignition.

Many industrial sites are located along major water ways where passing vessels have systems of sufficient RF power to create the potential for spark ignition. Radar systems need to be powerful enough to 'penetrate' poor weather conditions to allow the vessel to move in relative safety. Deep sea radio may need to transmit signals thousands of miles to communicate with the shore.

REFERENCE

"BS 6656:2002 Assessment of inadvertent ignition of flammable atmospheres by radio-frequency radiation – Guide" – 30th October 2002 - ISBN 0 580 40595 8

LESSONS LEARNT FROM DECOMMISSIONING A TOP TIER COMAH SITE

Kevin Dixon-Jackson
Ciba Expert Services, Charter Way, Macclesfield, Cheshire, SK10 2NX, UK

Ciba completed the decommissioning of the Clayton site in 2007. The site has been in use for over one hundred and thirty years producing dyestuffs and intermediates mainly for the textile and allied industries and was a 'Top Tier' COMAH site.

Site staff were committed to maintaining high EHS standards until the decommissioning was completed and the site handed over to its new owners. A project of this size and scale had never been completed within Ciba before. Careful planning was required to manage the work. New risk analysis systems, safe systems of work and working procedures had to be developed. New relationships had to be developed with decommissioning and demolition contractors and the number of Ciba staff was reduced progressively during the project.

The paper shares Ciba's learning experiences, highlighting key project issues, the systems which were developed for managing each issue and some of the unexpected things that occurred during the project. Lessons learnt from the project are also highlighted.

KEYWORDS: COMAH, Decommissioning.

INTRODUCTION

THE CLAYTON ANILINE COMPANY

The Clayton Aniline Company (CAC) was set up over 130 years ago in 1876 as one of the first manufacturing sites for aniline textile dyes. Benzol was sourced from the coal gas industry and used as a raw material for producing nitrobenzol and then aniline. The alkalis and inorganic acids required for textile dye manufacture were sourced from the local chemical industry. At the height of its success, the company operated a site covering over 57 acres, one of the largest single factory sites in the Manchester area. 2,000 people worked at the site. Figure 1 shows an aerial view of the site from 1974.

Companies which would later become Ciba (part of the The Basel Community of Interests) acquired a financial interest in CAC in 1918. A major rebuilding program was completed in the 1960's. Most of these buildings and plants then served the site until its final closure in 2007. Ciba obtained a majority shareholding in CAC in 1971 (Abrahart, 1976).

THE CIBA CLAYTON SITE

The site lies in an industrial area to the east of Manchester, abutting the Manchester, Stockport and Ashton canal (see Figure 2). About 200 staff worked at the site in the years

Figure 1. Aerial view of Ciba Clayton site (1974)

leading to its closure. The site included process plant, tank farms, warehousing, a power station and effluent treatment units. There were two main types of production line:

1. Older plant fed from roof tanks using flexible hoses. This plant included about 130 vessels. The main process safety concern related to potential unknown build up of diazo compounds, which decompose violently when dried (Dixon-Jackson et al, 2002).
2. More modern plant, fed from outdoor tank farms using fixed hard piped transfer lines. This plant included 12 reactors and 20 receivers. The main process safety concerns were the potential to cause toluene or oleum leaks. Toluene is a flammable solvent and oleum is a corrosive liquid which liberates a toxic gas (SO_3) on contact with air.

The site handled a range of bulk hazardous chemicals and came within the scope of the COMAH Regulations (COMAH, 1999).

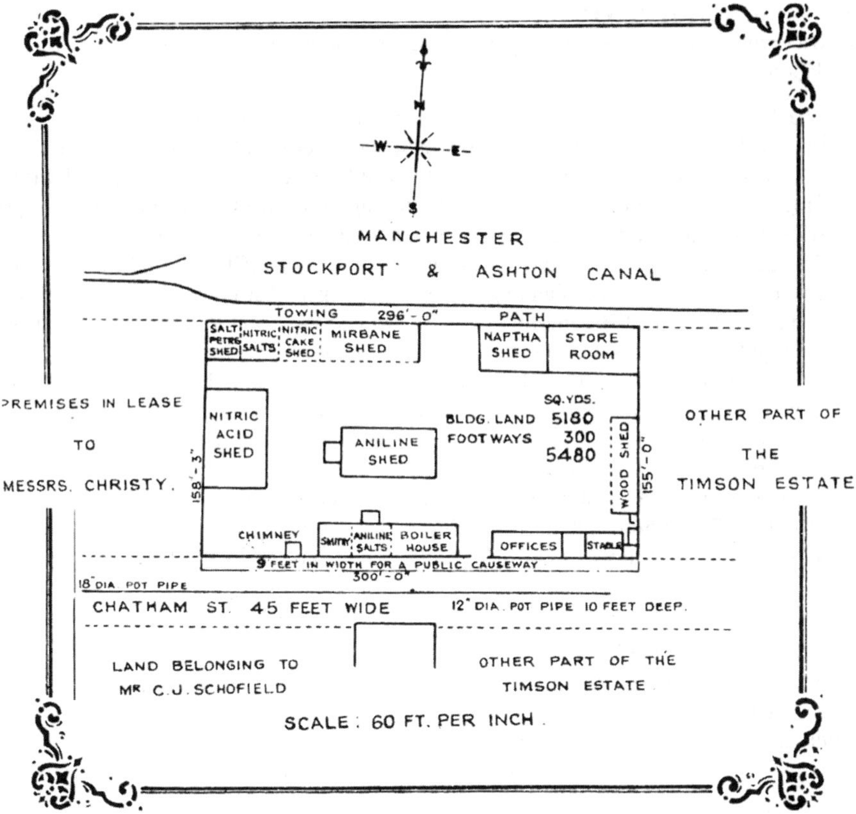

Figure 2. Site layout (extract from 1877 lease)

SITE CLOSURE

During the 1990's and the early part of the twenty first century, an acceleration occurred in the movement of the textile manufacturing industry from Europe to low cost countries, such as China, in Asia. The industry was subject to savage cost pressure as textile sales prices reduced. This price pressure was passed up the supply chain until it was uneconomic to produce textile dyes and intermediates in the UK. Over its 130 years, the Clayton site had specialised in products for the textile industry. There had been little diversification into other industry segments. The plant was old and the average age of the workforce was over 50. In 2003, it was therefore announced that the site would close. Detailed project planning then started, leading to final site closure in 2007. It is to the workforce's credit that the site was decommissioned with no serious safety incidents.

LEGAL REQUIREMENTS

The decommissioning project took place within the framework of four key pieces of EHS legislation:

1. The Health and Safety at Work Act (HASAWA, 1974), in particular covering the responsibilities of individuals for their own and other people's safety. Critical issues for the success of the project included the need for people to follow agreed written procedures, to challenge decisions and working procedures if people were unhappy with them and to comply with risk assessments and Permits-To-Work.
2. The Construction, Design and Management Regulations (CDM, 2007), in particular covering the role of the planning supervisor (a specialist contractor), the principal contractor (a specialist demolition contractor) and the client (Ciba) and the management of the interfaces between the three organisations.
3. The COMAH Regulations (COMAH, 1999), as the site was a 'Top Tier' COMAH site. In general, hazardous chemicals were removed from the site before decommissioning started, mainly by processing them into products, but also by removing them as waste. The project team was, however, aware that any unidentified residual quantities of hazardous chemicals could have led to chemical release, fire or explosion.
4. Pollution Prevention and Control (PPC, 1999), relating to releases of prescribed substances to air and water courses and the management of waste from the site.

PROJECT OBJECTIVES

Decommissioning the Clayton site represented the largest ever whole site demolition project in Ciba's history. As such, the project team had to adapt existing corporate standards and procedures and develop new procedures where none were present. Before developing detailed plans and procedures, the team set the following project objectives:

1. ***No accidents or incidents*** during the decommissioning project.
2. ***Equipment to be clean*** as far as practicable before decommissioning work was started.
3. ***Residual contamination*** to be identified and quantified prior to decommissioning.
4. Work to be completed using ***demonstrably safe systems***.
5. Plant to be handed over to demolition contractors safely, clearly ***identifying physical disconnections*** within a building and between buildings and low points where trapped chemicals could have accumulated.
6. ***Records to be provided***, proving how the decommissioning project was managed.
7. Project completion ***on time***.

DECOMMISSIONING STRATEGY

Considerable uncertainties exist when carrying out a decommissioning project on a site which was built in the nineteenth and twentieth centuries. Work which was completed many years ago may not have been documented or the documents could have been lost. Drawings may be inaccurate or may not exist. People will not remember all of the activities which have occurred on the site. This is especially true as the project progresses and site manning levels drop almost on a daily basis. Many of the people working at the site will be contractors and will be unaware of site operations and systems.

It was rapidly identified that robust systems would have to be used to manage the decommissioning project. Due to the large variety of plant and infrastructure which had to be decommissioned, it was decided that safe systems would have to be built around:

- Risk assessments.
- Permit-To-Work systems.
- Engineering method statements and risk assessments.
- Detailed step-by-step operating instructions.

This provided a framework to:

- Identify hazards, assess risks and identify appropriate risk controls.
- Control work in compliance with the requirements of each risk assessment.
- Ensure that everybody understood the detailed work requirements, working safely in line with agreed standard operating procedures.

This was considered to be the best way to minimise risk, accepting that some residual risks would always be present in a project of this complexity and novelty.

Many existing corporate and site standards and procedures fitted in well with the requirements of the decommissioning project. These included risk analysis methodologies, permit-to-work systems, confined space working, plant isolation and plant maintenance procedures. Indeed, safety management systems have to be suitable for decommissioning activities, as small scale decommissioning takes place regularly on most sites.

Three major gaps were, however, identified. Firstly, the corporate Ciba risk analysis methodology was too focused on chemical processes and did not have the required detailed guidewords which are suitable for decommissioning work. A special decommissioning risk analysis therefore had to be developed. Secondly, it was felt that workers and contractors required additional training about the decommissioning procedures for the site, supported by additional safety checks, which became known as 'transfer safety stops'. Thirdly, it was recognized that the demolition contractor needed to have assurance that plant was safe to decommission on handover. This was achieved by using a system of certificates to formally hand areas of plant over to the contractor.

PLANNING

Three main types of plant had to be decommissioned:

- Textile effects manufacturing plant, with feed tanks on the roof, linked to process vessels by flexible hoses. Battery limits for each part of the decommissioning work were easy to specify as the vessel.

Table 1. Typical activities covered by a work instruction

Decontamination	Decommissioning
• Receive washes.	• Drain oil.
• Transfer out washes.	• Open valves. Drain all pipes.
• Cleaning of nozzles.	• Electrical disconnection.
• Cleaning of routes not normally cleaned.	• Air isolation.
• Cleaning of reflux routes.	• Control isolation.
• Cleaning of receiver inlets and outlets.	• Pumps.
• Prevention of recontamination.	

- Printing chemicals (carbonless copying paper dyes) manufacturing plant, linked via pipes to tank farms. Battery limits for each part of the decommissioning work had to be defined on the P&ID documents.
- Site infrastructure.

The site was therefore split into sections and battery limits were defined for each section, supported by drawings. Drawings were not necessarily accurate due to the age of the plant, uncertainty about whether all modifications had been recorded correctly, difficulties in accessing all relevant drawings and because some plant and infrastructure (such as control systems, utilities and power supplies) may already have been removed or disconnected. It was therefore essential to walk the plant and adjust drawings to reflect actual plant conditions.

No activities were allowed to be completed without a risk assessment, which could be a permit-to-work or a specialised risk assessment. This team based activity identified the required controls involving the operators. The risk assessment was then incorporated into a detailed work instruction, known as a 'decontamination and decommissioning instruction'. Safe practices and required Personnel Protective Equipment (PPE) were identified in the detailed work instruction. Any permits were cross referenced to the relevant work instruction. Operators were briefed prior to starting each piece of work. The briefing included a walk of the job to confirm that the work was properly understood. The work instructions were signed off step by step to confirm that they had been completed correctly. Typical issues covered by a work instruction are illustrated in Table 1.

It was found that isometric drawings were particularly helpful for communicating the requirements within work instructions. A typical isometric is shown in Figure 3.

RISK ASSESSMENT

Ciba's normal risk analysis methodology is based around a checklist, using guidewords to identify possible process deviations. Each deviation is then assessed to estimate the frequency and severity of occurrence and to check that the required risk reduction measures are in place. The guidewords in the checklist are very effective for identifying process deviations

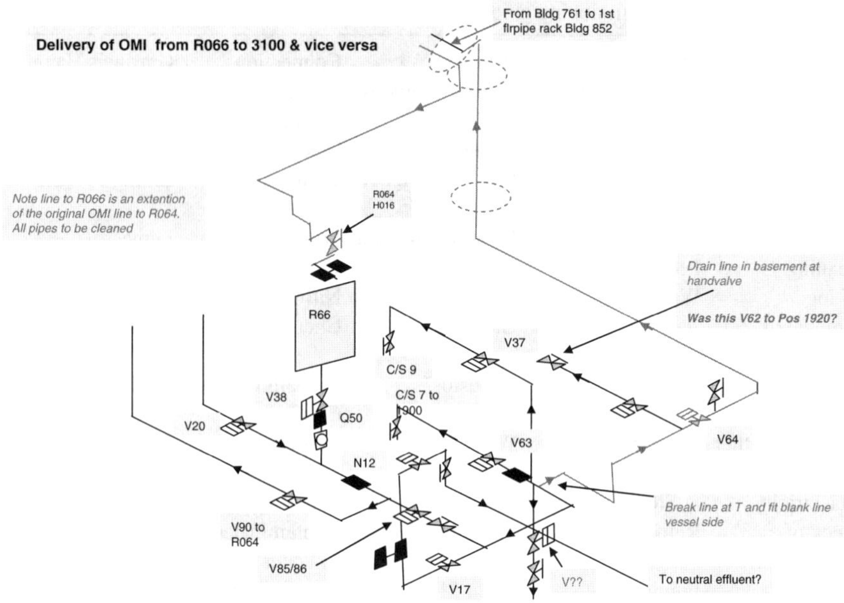

Figure 3. Typical isometric for supporting work instructions

from events such as power loss, overcharging, high temperature, loss of cooling etc but there are very few guidewords which specifically address decommissioning risks. The project team therefore produced a new set of guidewords, which were more relevant to decommissioning activities and incorporated them within the general framework of the existing Ciba risk analysis methodology. Table 2 summarises the general guidewords which were used for the decommissioning risk analysis. These guidewords were grouped by activity types for use in the risk analysis, based on the nature of each task being analysed.

PROJECT CONTROL

Risk assessment therefore lay at the heart of the decommissioning project as shown in Figure 4.

Risk assessment in itself will not guarantee that work is completed safely. It will provide a framework for safe operations. Safety at the plant level requires people to understand the risk assessments, communicate with colleagues and comply with the requirements of the risk assessments and the detailed work instructions which flow from them. For this reason, Ciba uses a system of 'safety stops' to confirm compliance before critical activities are started. Safety stops are really designed for controlling engineering building

Table 2. General decommissioning guidewords

Personnel Movements	Plant vehicle Movements	Paints lead/luminous	Connection lines/valves	Graviners/Hammers
Plant Activities	Burning	Draining	Steam	Cleaning agents
Confined spaces	Cutting	Access	Electricity	Blockages
Access equipment	Isolations	Work at Height	Computer	Lining of item
Manual Handling	Surround's	Recontamination	Waste gas	Fumes
Cleaning	Blowing	Lone working	Compressed air	Fume extraction
Lifting	Washing	Dust	Nitrogen	Biological
Removal of fittings	Jetting	Gases	Chemicals	Explosive
Welding	Breaking Flanges	Asbestos	Contamination	Radiation
Sharp Edges	Insulation/Lagging	Drains	Condensate	Mercury (thermos)
Adverse weather	Inspections	Water	Lubricating Oils	PCB
Supply of electricity	Heat/cool Media Stored Energy	Speed of agitation	Man-made Fibres	Mix-up chemicals
Check of equipment	Refrigerants	Sampling	Contaminated Water	Temperature
	Fumigation Agents	Open Manway	Hot Surfaces	Pressure pH Value
Charging/Dosing	Pathogens Treatments	Disposal		

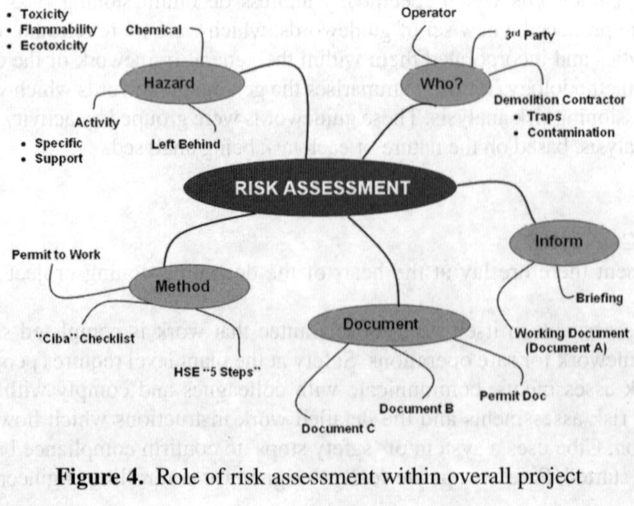

Figure 4. Role of risk assessment within overall project

projects and have limited use for controlling decommissioning work. It was therefore decided to evolve the existing safety stop system to cover decommissioning activities.

This led the project team to develop the 'transfer safety stop'. It is designed to ensure that risks are properly assessed before any material transfers are carried out. The transfer safety stop considers issues such as:

- Are the conditions (inerting, earthing etc) correct for the transfer?
- What is the consequence of any mixing which might take place?
- Is the resulting mixture safe (thermochemistry, flammability, combustability etc) and acceptable for disposal?
- Can effluent streams be safely sent to drain and will they be compliant with site discharge consents?
- What other transfers may be in progress at the same time?

This poses two practical problems:

1. Mixtures and residues inside vessels may have unknown safety properties. It is often necessary to consult experienced chemists and it will sometimes be necessary to conduct additional laboratory safety tests before an operation can proceed.
2. Sites often have extremely tight discharge consents for named chemicals. Normal operations are then managed using captive drainage systems. When the time comes to decommission plant, residual contamination cannot be released to the effluent system as this would cause a breach of the discharge consent. Careful thought and planning is therefore required and some wash streams will have to be sent for offsite disposal by road tanker. If this is the case, it is essential to miminise cross-contamination and cleaning liquid volumes, as this will massively increase waste disposal costs.

It was recognised at an early stage in the project that the project objective of no accidents or incidents could only be achieved if there was a strong link between Ciba and the demolition contractor. The selection of the demolition contractor was carefully considered and it was decided to use a contractor who had previous experience in the decommissioning of chemical sites. They had already worked at the Ciba Clayton site on other projects and their standards and performance had been audited.

Ciba staff then needed to have a system for formally handing over individual site areas to the contractor. This was achieved using a formal system of handover documents, which covered:

- Basic data about chemicals and equipment, supported by drawings and plans where they were available.
- A handover report for the building covering the chemical and engineering activities and hazards, bulk contaminations and photos, highlighting any disconnections. This report was typically about ten pages long.
- A walk round of the site area.
- Certificates for service electrical and control isolation, decontamination and decommissioning.

LEARNING EXPERIENCES

Ciba have gained a lot of valuable experience about decommissioning a large chemical manufacturing site. Six particularly important lessons have been learnt:

1. **The need for detailed risk assessment and systems to support the project.** The project was complex and involved people from different organisations. Detailed systems, procedures and paperwork were required to manage the project.
2. **Maintaining staff motivation as the site closes down.** The project can only succeed if the site staff are fully committed to the project. Effective consultation is required with all staff and key staff need to be retained until the project is finished. This can be difficult to manage as the project enters its final stages, with staff leaving the company on an almost daily basis. Detailed plant knowledge is essential for completing this type of project successfully.
3. **Ensuring site security prior to closure.** At the final stages of the project, the site will have a large perimeter and there will be few people present on the site. A lot of valuable scrap metal is often left at the site and is often cut and pre-loaded, ready for offsite disposal. The site was very vulnerable to theft and security breaches at this time.
4. **Additional safety testing is required before material can be removed for disposal.** Mixtures and unknown residues were found and collected during decommissioning. The hazardous properties (fire, explosion, thermal stability etc) of these materials were unknown. Material samples were taken and analysed in the Ciba Safety Testing Laboratory in Macclesfield before starting work.
5. **Chemicals must be used up before the production plants are shut down.** Commercial logic would suggest that valuable raw materials are converted into saleable finished products before the site is finally closed down. Inventory management plans must be in place well before closure operations start. Unforeseen problems can also occur with less valuable raw materials. In one case, an outline agreement was made with the supplier to buy back an inventory of chemical. Ciba planned on the basis of this agreement but last minute problems occurred when the liquid had to be removed. The supplier was not prepared to buy the material and as Ciba were not the producer of the raw material, it became classified as waste. This caused time delays whilst alternative disposal routes were found. Eventually, the liquid had to be sent offsite in road tankers for waste disposal, converting a small planned revenue into a large unplanned cost. Further practical problems occurred. Firstly, it was realised that the storage tank was designed for road tanker offloading into the tank and not for loading from the tank into a road tanker. Additional engineering work was required. Then, when this work had been completed and the transfer was behind schedule, it was discovered that the control system, instruments, power supplies, services and effluent connections had all been decommissioned. A job which appeared to be simple had become extremely complicated.
6. **Some operations will not go to plan for totally unforeseen reasons.** A storage tank of concentrated sulphuric acid had to be emptied and washed out. The tank was

constructed of mild steel. It was envisaged that the washing operation would be completed quickly. When the work was started, it was found that there was only one way to fill the tank with water because the control system was no longer operational. Unfortunately, this involved using a 2″ diameter water connection. It took a long time to fill the tank for washing and generated an unforeseen corrosion/reaction hazard as dilute sulphuric acid attacks mild steel, generating flammable hydrogen gas. The original procedure was safe but the practicalities of the operation changed the intent of the original procedure, generating a potential hazard.

CONCLUSIONS

The major decommissioning project was completed with no reportable accidents or incidents. This was achieved with a combination of the efforts and commitment of site and project team staff, the use of risk assessment and project control systems and careful briefing of plant staff before starting each item of work. Figure 5 summarises the links between the key planning and control elements which allowed the project to be completed successfully.

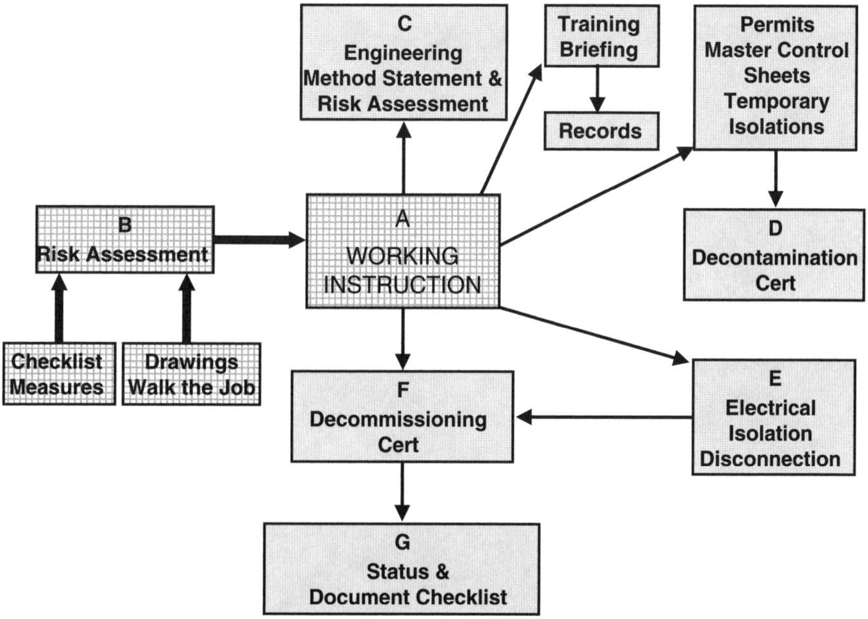

Figure 5. Overview of critical project planning and control elements

ACKNOWLEDGEMENT

The author gratefully acknowledges the underlying work and the help in producing this paper provided by Dave Sanderson, who has now retired from Ciba Clayton after many years service.

REFERENCES

(Abrahart, 1976) 'The Clayton Aniline Company Limited 1876–1976', E.N. Abrahart, published by the Clayton Aniline Company Limited in 1976.

(CDM, 2007) The Construction, Design and Management Regulations, 2007.

(COMAH, 1999) The Control of Major Accident Hazard Regulations, 1999.

(Dixon-Jackson et al, 2002) 'The manufacture of diazonium compounds…', K. Dixon-Jackson, F. Altorfer & R. Roper, Loss Prevention Bulletin Issue 164, April 2002.

(HASAWA, 1974) The Health and Safety At Work Etc. Act, 1974.

(PPC, 1999) Pollution Prevention and Control Regulations SI1973, 1999.

A SAFETY CULTURE TOOLKIT – AND KEY LESSONS LEARNED

Peter Ackroyd
Technical Head, Greenstreet Berman Ltd

INTRODUCTION

The recent Baker report (Ref 1) has again highlighted the importance of 'safety culture' on the actual effectiveness of safety management arrangements, and how degradation can readily occur even in 'mature' organisations. It also raises the challenge of safety culture assessment; what should be considered and how? Are there any pitfalls in undertaking safety culture assessments? The purpose of the paper is to prompt organisations to consider carefully how to make best use of safety culture assessments as part of an overall approach to safety management, without being seduced into false perceptions of their own strengths and limitations.

This paper has been prompted in part by recent work we have been undertaking. We have been developing a comprehensive 'web-enabled safety culture toolkit for an industry sector with our partners Enable Infomatrix. This toolkit is a significant development of safety culture assessment and guidance tools. This demonstrates that advanced tools can be developed that permit routine safety culture assessments to be easily undertaken and interpreted. It shows that simple to use, on-line tools can greatly aid process industry companies to assess key aspects of their safety culture and help to identify improvements. It raises the possibility for developing similar process industry toolkits that could be used to address 'process safety culture' and offer great opportunities to share good practices within an organisation or across the process industry.

The trials of the toolkit along with our collective experience of safety culture assessment has raised the whole issues of the objectives and appropriate use of safety culture assessment – i.e. for continual improvement not merely 'acceptance'. This paper considers the limitations and cautions that need to be placed on use of assessments to ensure that organisations do not mislead themselves and believe they are better than they are. It considers the key aspects of 'safety culture' that process industries should address to avoid the types of pitfall demonstrated so clearly by the Texas City incident. It draws on experience of other safety culture assessment approaches to highlight the potential for 'internal anchoring'. The paper concludes with thoughts for the process industry on effective use of safety culture assessment as a means to avoiding complacency. It relates back to the challenge of 'Organisational Drift' and the Baker panel recommendations following Texas City.

PART 1 THE SAFETY CULTURE TOOLKIT

Over the last two years Greenstreet Berman has been developing an advanced web-based safety culture toolkit. This permits companies readily to undertake safety culture assessments and conduct immediate analyses of their results. Importantly, the toolkit provides guidance on improvement strategies that are targeted in response to the assessed culture; it also provides a general source of useful information on safety culture and its improvement. Trials of the toolkit with companies have been very successful. However, the results from the two companies have pointed to potential limitations of any survey based approach to assessing safety culture, particularly via 'internal anchoring' i.e. that relative insularity means people can only judge on their limited experience and do not respond in an adequately absolute manner.

AN OVERVIEW OF THE TOOLKIT

The toolkit is web-based and has three main elements:

- A 'Useful Information' area – providing a source of reference on safety culture and generic advice on safety culture improvement
- The main safety culture assessment area – based around a questionnaire that incorporates an extensive automatic analysis and interpretation capability, including benchmarking against all or selected registered users
- A 'Good Practice' area – that currently contains around 70 examples and is formatted to allow exchange, inclusion of additional examples

The toolkit permits companies, once registered, to tailor their own confidential area within the overall toolkit by adding news items. Company confidentiality is ensured as no registered users are able to access the sites and results of other companies (other than via the benchmarking report). Similarly individual confidentiality is ensured via a variety of means.

The safety culture assessment survey is the core assessment tool and it comprises a questionnaire that has been developed using lessons gained from research into safety culture assessment approaches. It follows similar safety culture attitude surveys. The key advances are in its analysis and accompanying interpretation. The toolkit permits a very comprehensive set of analyses to be undertaken and provides both text and graphical outputs.

The toolkit comprises both generic and specific guidance; this has been developed from review of the latest safety culture developments. The guidance is based on literature and models that have gained considerable use and credibility; and that allow the users to gain maximum insights into understanding their safety culture and how to improve it.

PILOT TRIALS & LESSONS LEARNED

Two companies have undertaken full company surveys using the draft toolkit. These were very successful – both companies being impressed with the general ease of use of the toolkit and particularly how easy it was to undertake analyses. As the toolkit is web-based it permits rapid access to the results by all (with appropriate company authorisation) to run and view the results. This permits the toolkit to be used at a local level by managers as well as considering the overall company.

The results of both companies showed that the general safety culture in both organisations were good and enabled a variety of key issues to be identified, for example:

- Differences between departments
- A notable difference between one location and all others
- A noted difference in the general level of satisfaction in safety standards between front line staff and their immediate supervisors/managers

In detailed discussions, all of the results obtained reflected well on the companies safety professionals view of issues within the company.

The results generated by the two companies did highlight some key issues that are important when undertaking any attitudinal safety culture assessment. The main issues being:

- What should safety culture assessments really be used for, and what should they measure?
- What are the hidden issues to be aware of?
- How important is an external perspective on any safety culture assessment?

These are raised and discussed in the second part of this paper.

THE IMPLICATIONS

The development of this type of safety culture toolkit shows that some very effective safety culture assessments tools can be developed for industry sectors, or organisations, to use without requiring extensive external support. Further more such web-based tools provide many additional benefits to the user organisations. In particular, analyses and use of the assessments can be given to a much wider range of line managers and employees to help local improvements. Additional benefits include:

- The capability of sharing 'good practices' quickly and effectively
- Benchmarking across an organisation or industry sector
- Tailoring the questionnaire to meet specific industry sector/organisation needs

The challenge for safety culture assessment, particularly if such 'self-help' comprehensive tools become available is to ensure that they are used appropriate to aid improvement and do not unwittingly lead organisations astray.

PART 2 SAFETY CULTURE ASSESSMENT – HOW TO USE EFFECTIVELY

Our experience over many years of differing approaches to measure 'safety culture' in some guise or other is that they can be very useful and provide very valuable insights IF used appropriately. They can be positively mis-leading or even dangerous if used inappropriately. Key issues from our experience, and re-highlighted from the safety culture toolkit project include the following.

WHAT'S THE PURPOSE?

Safety culture assessments (based on attitudinal questionnaires and workshops) can provide some very useful insights into the relative strengths and weaknesses within an organisation. Hence this can provide a very useful platform for continuous improvement. However, attitudinal approaches particularly questionnaires, should not be used on simple pass/fail or acceptable/unacceptable basis. Responses to questionnaires can be open to many influences including other issues affecting staff within the organisation (e.g. pay & conditions; immediate manager; 'internal anchoring') and this can significantly limit their ability to be used as an absolute measure of performance. The judgements based against fixed or absolute values are likely to be mis-leading, and it is far more useful to use the results to identify comparative differences in responses to help identify strengths and weaknesses. This places limitations on the usefulness of benchmarking between surveys.

WHAT TO MEASURE?

The safety culture toolkit questionnaire is similar to others including the HSE CST questionnaire in that, arguably, it tends to address 'general safety' i.e. primarily relating to the H&S of employees. For many organisations this is likely to be the most appropriate topic to consider. However, the BP experiences at Grangemouth (2000) and Texas City (2006) and findings of the Baker report show that management can become excessively focussed on, and mis-led by simple employee LTA type indicators. So any assessment of safety culture needs to start from considering what aspects of safety this organisation/sector should be really concerned with.

The answer may well be that there is more than one aspect of H&S that needs to be considered. Consequently for some organisations/sectors having 'safety culture' questionnaires to tackle each key area may be required. The Baker report includes a 'process safety' safety culture questionnaire used to provide a greater focus on key issues for this area. The author's experience in the nuclear industry is that several non-nuclear safety measurement approaches were adopted, which caused unease amongst many technical specialists as they did not appear to provide sufficient focus on nuclear safety issues. Interestingly, companies won several safety awards at the same time as having a series of incidents on nuclear related safety issues.

Attitudinal surveys by themselves are not sufficient to identify all key issues; it is imperative to use them to complement the insights gained from other safety performance measures and audits. They can reveal 'hidden' issues that may otherwise be missed.

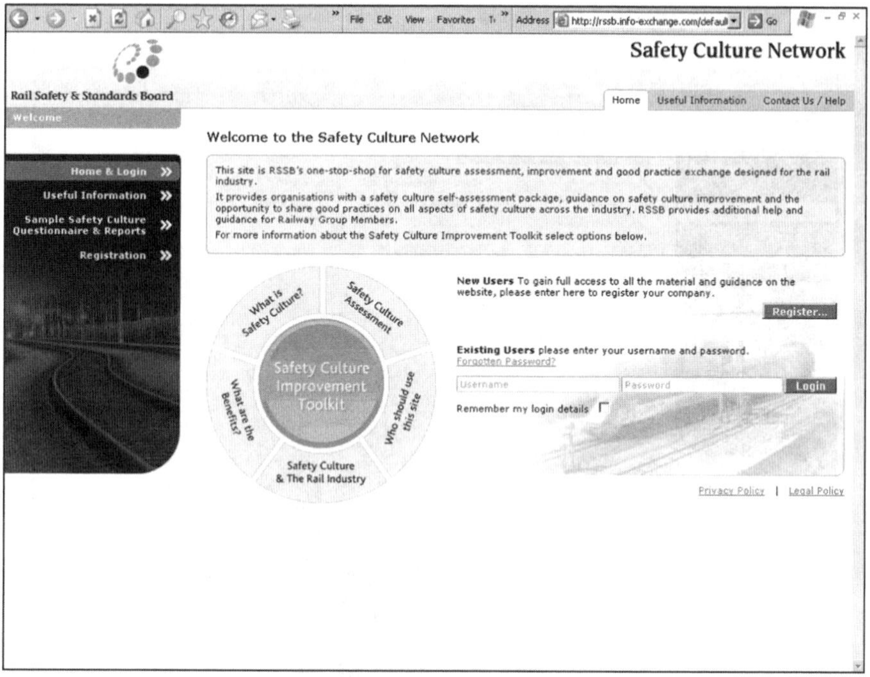

ISSUES TO CONSIDER

Internal anchoring – the recent toolkit survey results were very good, indeed better than envisaged based on other knowledge on the safety performance and culture. Detailed discussions revealed that in comparison with other types of rail companies, these were likely to be amongst the best performing, and that staff were generally very content with the companies. Also most employees have little experience outside the rail industry. A likely explanation of the better than judged responses is 'internal anchoring' i.e. the respondents are making the judgments against their own experiences – but if these have (collectively) been very limited they do not represent judgement against an objective global or 'absolute' scale. Very similar experiences occur with other organisational assessment approaches. Peer evaluation processes in the nuclear industry often show that staff judge things to be acceptable, but an international peer assessment team has very different judgements on the standards. Research into the development of a new safety culture assessment approach (SCART[1]) in the nuclear sector (using behavioural descriptions on key topics)

[1]SCART = Safety Culture Assessment & Rating Tool – developed by British Nuclear Group (Reactor Sites)

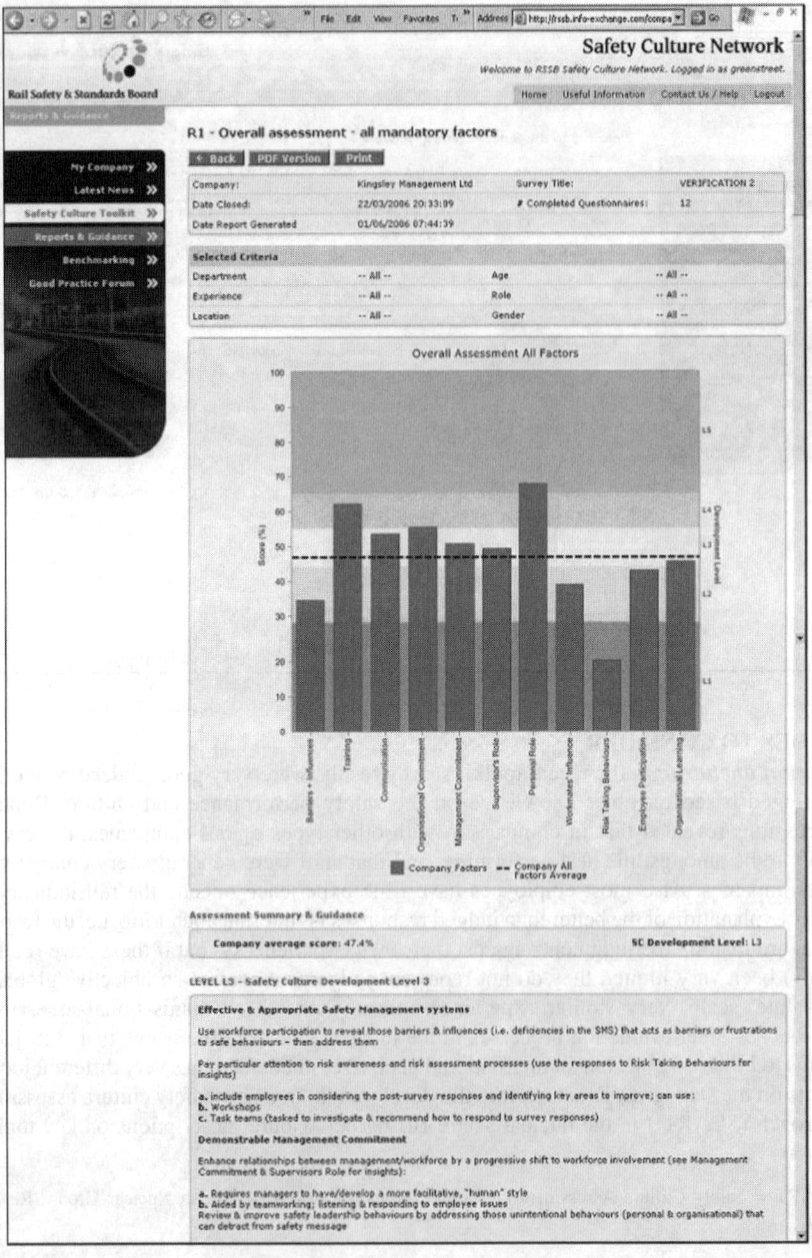
Safety Culture Network
Welcome to RSSB Safety Culture Network. Logged in as greenstreet.
Rail Safety & Standards Board
Home Useful Information Contact Us / Help Logout
Reports & Guidance
My Company
Latest News
Safety Culture Toolkit
Reports & Guidance
Benchmarking
Good Practice Forum
R1 - Overall assessment - all mandatory factors
Back PDF Version Print
Company: Kingsley Management Ltd
Survey Title: VERIFICATION 2
Date Closed: 22/03/2006 20:33:09
Completed Questionnaires: 12
Date Report Generated 01/06/2006 07:44:39
Selected Criteria
Department -- All --
Age -- All --
Experience -- All --
Role -- All --
Location -- All --
Gender -- All --
Overall Assessment All Factors
Score (%)
Development Level
L1
L2
L3
L4
L5
Barriers + Influences
Training
Communication
Organisational Commitment
Management Commitment
Supervisor's Role
Personal Role
Workmates' Influence
Risk Taking Behaviours
Employee Participation
Organisational Learning
Company Factors
Company All Factors Average
Assessment Summary & Guidance
Company average score: 47.4%
SC Development Level: L3
LEVEL L3 - Safety Culture Development Level 3
Effective & Appropriate Safety Management systems
Use workforce participation to reveal those barriers & influences (i.e. deficiencies in the SMS) that acts as barriers or frustrations to safe behaviours - then address them
Pay particular attention to risk awareness and risk assessment processes (use the responses to Risk Taking Behaviours for insights)
a. include employees in considering the post-survey responses and identifying key areas to improve; can use:
b. Workshops
c. Task teams (tasked to investigate & recommend how to respond to survey responses)
Demonstrable Management Commitment
Enhance relationships between management/workforce by a progressive shift to workforce involvement (see Management Commitment & Supervisors Role for insights):
a. Requires managers to have/develop a more facilitative, "human" style
b. Aided by teamworking; listening & responding to employee issues
Review & improve safety leadership behaviours by addressing those unintentional behaviours (personal & organisational) that can detract from safety message

showed very different assessments between station staff self-evaluation; corporate H&S staff evaluation, and that undertaken by an international peer review team. The corporate staff and international peers perceiving the 'safety culture' to be significantly lower than the internal self-evaluation.

This potential 'internal anchoring' is particularly relevant in industries that are relatively 'insular' and suggests that obtaining an external perspective on the results is likely to be very useful. It also reinforces the message that improvements should be driven from the comparative differences revealed by a survey rather than just judging on the absolute values.

Building on the existing culture – survey results allow greater insights to be gained into the existing culture, and help to identify relative strengths and weaknesses. Any improvement strategy should be based on building onto the existing culture, particularly its strengths. The relative success of many safety improvement initiatives (e.g. behavioural observation programmes) are dependent on whether the culture (or key aspects of it) is right.

Being prepared to respond – safety culture surveys tend to have a high profile within an organisation when implemented and create workforce expectations on the response. Doing a survey then not being seen to respond adequately is nearly always worse than not doing a survey at all. Consequently any organisation considering undertaking a safety culture assessment should already have planned and prepared as to how it intends to respond – and that it matches the expectations of the workforce.

Not just a measurement tool – any high profile safety culture assessment should not just be viewed as a means of measurement. This would be 'missing a trick' in the overall safety improvement strategy. Due to the high profile and expectations it creates, a safety culture assessment can help energise and create focus and interest in safety improvement. This is as much a benefit from the exercise and the insights it provides.

SUMMARY

The web-enabled safety culture toolkit represents a significant step forward in safety culture assessment and improvement. It permits organisations to 'self-help' with minimal external input and provides the organisation, or industry sector to allow much easier access and use of attitudinal based surveys. However its ease of use and capabilities also make it even easier for organisations to mis-use or be led astray by the results of safety culture surveys. If used appropriately such assessments are a significant additional 'weapon' to use to improve safety culture and prevent significant accidents occurring. The onus is on organisations to critically consider the key aspects of 'safety' that they need to address and tailor their safety culture assessments accordingly. Similarly effective use of the results and insights gained from such assessments needs organisations to be willing to use the tools in appropriate ways. As with any tool, it is only a tool and its users to use it as intended.

ACKNOWLEDGEMENTS

Enable Infomatrix have been a key partner in developing the toolkit; without their IT expertise it would not have been possible to create the toolkit. Thanks to all who have been involved in the development, particularly to Colin James all the hours of detailed discussions and development to help turn the ideas into a practical product.

REFERENCES

1. J.A. Baker et al January 2007 'The Report of the BP U.S. Refineries Independent Safety Review Panel'
2. HSE Heath & Safety Climate Survey tool, 1997

SAFETY MODEL WHICH INTEGRATES HUMAN FACTORS, SAFETY MANAGEMENT SYSTEMS AND ORGANISATIONAL ISSUES APPLIED TO CHEMICAL MAJOR ACCIDENTS

Linda J Bellamy[1], Tim A W Geyer[2], Joy I.H. Oh[3] and John Wilkinson[4]
[1]Managing Director, White Queen Safety Strategies BV, PO Box 712, 2130 AS Hoofddorp, The Netherlands, linda.bellamy@whitequeen.nl
[2]Partner, Environmental Resources Management Limited, 8 Cavendish Square, London W1G 0ER, UK, tim.geyer@erm.com
[3]Deputy Unit Head, Occupational Safety and Major Hazards Policy, Ministry of Social Affairs and Employment, Postbus 9080, 2509 LV Den Haag, The Netherlands, joh@minszw.nl
[4]HM Principal Specialist Inspector (Human Factors), Hazardous Installations Directorate, Health & Safety Executive, Redgrave Court, Merton Road, Bootle, Merseyside, L20 7HS, UK, John.Wilkinson@hse.gsi.gov.uk

This paper introduces a model called PyraMAP which is focused on human performance in hazardous systems. It was developed for the UK Health and Safety Executive. The model integrates the role of Human Factors in safety within a wider context of safety management and organisation to enable more cohesive and better structured approaches to analyzing the performance of Major Accident Prevention (MAP). The paper also looks at the model in the context of the Texas City refinery accident of 2005.

1. INTRODUCTION

Investigation and analysis of accidents and their underlying causes has a dependency on knowledge and models of cause and effect. In recent years analysis has targeted underlying causes of accidents, influenced by ideas such as the Swiss cheese model of Reason (1990, 1997) and the concept of the organisational accident. Failures of front line operators in hazardous systems are no longer to be regarded as sufficient single cause for an accident.

This paper describes a structured approach to integrating the different levels of the human contribution to accidents and their prevention, in particular concentrating on major hazard chemical and petrochemical accidents. The approach is based on extensive previous work (Bellamy et al, 1989, 1999, 2006a, 2006b, 2007a, 2007b; Bellamy & Geyer, 1992; Bellamy & Brouwer, 1999; Baksteen et al 2007; White Queen, 2003) and is the next step in structuring and defining the role of Human Factors in Major Accident analysis and prevention.

2. TEXAS CITY ACCIDENT

For illustrative purposes the paper looks at a recent accident, the BP Texas City refinery accident (U.S. Chemical Safety and Hazard Investigation Board, 2007) where 15 people were killed and 180 injured. According to the investigation report the accident in BP's refinery was caused by overfilling of a raffinate splitter tower resulting in the release of a flammable liquid from a blowdown stack that was not equipped with a flare. During start-up of the tower operators were unaware that it was overfilled because the level transmitter was inaccurate and the redundant high level alarm failed to activate. In addition the tower level sight glass was dirty and unreadable. The control board display did not provide adequate information on the imbalance of flows in and out of the tower to alert the operators to the dangerously high level. "The Board Operator truly had no functional and accurate measure of tower level on March 23, 2005" says the report.

The accident triggered an enormous amount of interest in the fundamentals of safety organisation and management, leading to a review in the US of BP's refineries by an independent safety review panel documented in the so-called Baker panel report (Baker 2007). In this paper we will analyse the conclusions of this report along the lines of the PyraMAP model.

3. PYRAMAP MODEL – PYRAMID OF MAJOR ACCIDENT PREVENTION (MAP)

The PyraMAP model was developed for the Health and Safety Executive in the UK with the purpose of having a framework for a more integrated approach to safety (Bellamy & Geyer 2007). In addition the Deputy Unit Head for major hazards policy at the Ministry of Social Affairs and Employment in The Netherlands provided independent review and advice of the work. In particular he was critical of current applications of human factors science in assisting with major accident prevention.

The purpose was to have a generic safety model that integrated human factors within its wider context of organisation and safety management. The model should be applicable to different hazards and industries but the initial focus was on major hazards. PyraMAP stands for *Pyramid of Major Accident Prevention*.

The generic pyramid is shown in Figure 1.

The socio-technical issues comprise 3 taxonomies – organisation, safety management and human factors – and any aspect of risk control, such as a measure or a barrier or a procedure. A theme is a set of items from the 3 taxonomies built round the hub of risk control. The theme defines a the selection of the elements from the taxonomies which are connected together in some way.

The idea of having taxonomies is to make it clear what is being addressed in each area. In particular there was a need to define human factors because non specialists had difficulties understanding what it meant. Good human factors in practice is about optimising the relationships between demands and capacities in considering human and system performance. Whether there is good fit or there is mismatch will be reflected in behavioural outcomes. The human factors taxonomy therefore focuses on these demands, capacities and outcomes. The organisational taxonomy came from literature identifying organisational aspects of accidents (Bellamy, Leathley & Gibson, 1995). However, what makes the model major hazard specific is the safety management taxonomy, the technical aspects of

Figure 1. Generic socio-technical pyramid (Bellamy & Geyer 2007) with its three taxonomies – Organisation, SMS and human factors

the system and the hazards. In the major hazards application the safety management taxonomy used the UK COMAH regulation Safety Report Assessment Manual criteria series 4 as a basis (Health and Safety Executive, 2003) The detailed taxonomies can be found in Bellamy & Geyer 2007). An overview is shown in Figure 2.

4. MAJOR HAZARD THEMES DERIVED FROM ACCIDENTS

The taxonomies of the generic PyraMAP are used to make specific warning triangles which function to highlight socio-technical aspects that come together to create strengths or weaknesses in the system. This coinciding of factors is called a theme. The major hazards pyramid has 4 themes. Each theme describes the recurrence in major accidents of a specific group of socio-technical issues identified from the taxonomies. These 4 themes are coming from the analysis of 8 major accidents with detailed accident reports. They are:

1. Failure by people with major hazard responsibilities to understand the risks and the risk controls in MAP, particularly involving the information derived from risk assessment and the allocation of roles and responsibilities where understanding of the risks is key.
2. Failure to competently perform tasks related to the integrity of MAP risk control measures because of failures to deliver appropriate competences to persons in the organisation carrying out MAP tasks

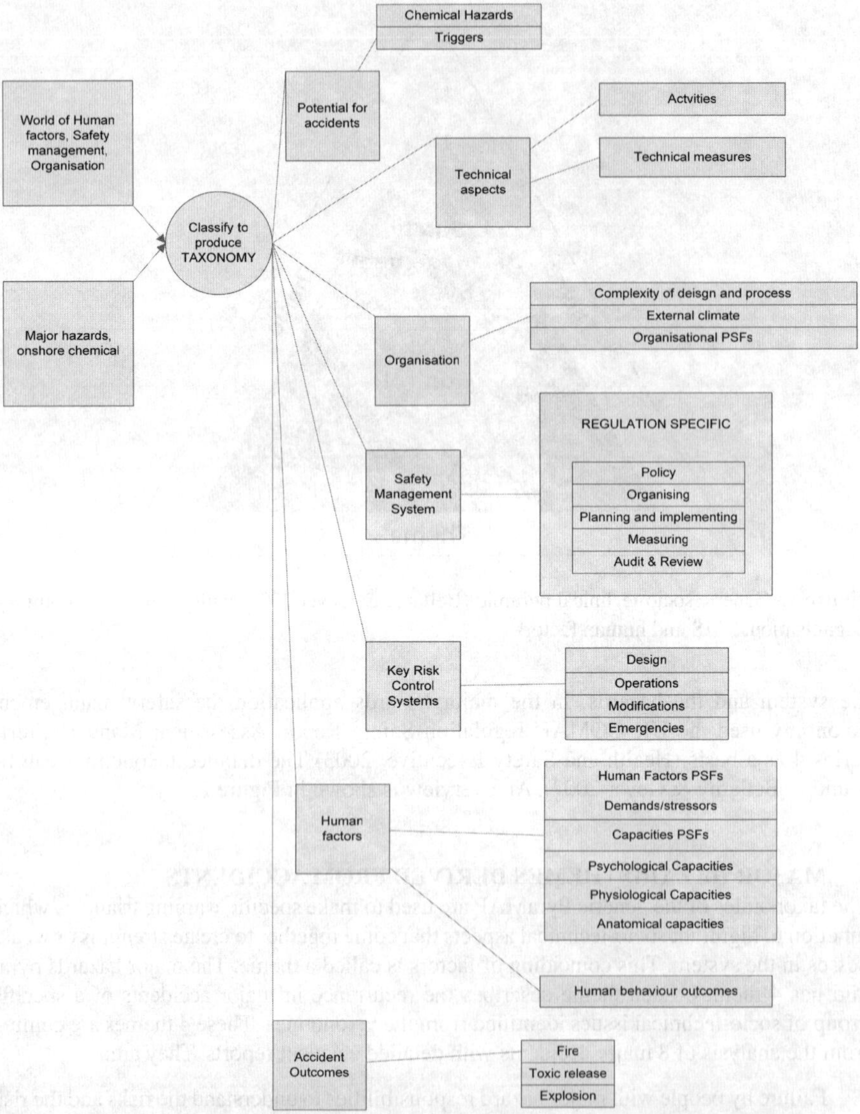

Figure 2. Basic taxonomy structure of the major hazards PyraMAP emphasising human factors, safety management, organisation and regulation

3. Failure to prioritise and give due attention to resolving demands on human performance capacities which conflict with MAP particularly through communications and workforce involvement
4. Failure to give assurance that there is a knowledgeable, learning organisation where behavior in relation to the MAP goals and procedures is being measured and improved.

The 4 themes of the PyraMAP, the 4 "warning triangles" are shown in Figure 3. Accident contributors identified in the reports were identified in the taxonomies such that each accident then had a list of taxonomy elements whose failures were related to causing the accident. The accidents analyzed were:

– Flixborough (UK, 1974): Explosion due to release from a temporary bypass assembly of inadequate design operated by insufficiently competent people

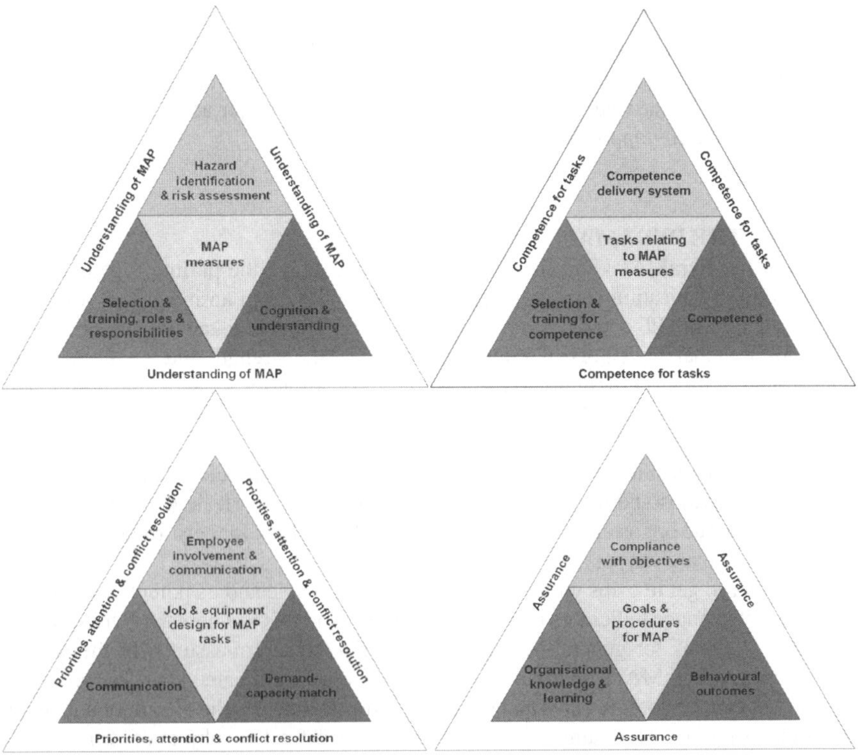

Figure 3. The 4 themes of the PyraMAP for major accidents (Bellamy & Geyer 2007) as derived from the analysis of 8 major hazard accidents

- Grangemouth (UK, 13 March 1987): Fire due to passing valve (poor design) and inadequate isolation procedures
- Allied Colloids (UK, 1992): Fire following misclassification of chemicals and failure to segregate incompatible substances in storage
- Hickson and Welch (UK, 1992): Jet fire following runaway reaction during non routine vessel cleaning due to lack of awareness of risks and inadequate precautions (Health and Safety Executive, 1994).
- Cindu (The Netherlands, 1992): Explosion due to runaway reaction in a batch processing plant. Trainee using wrong recipe in an old poorly designed plant
- Associated Octel (UK, 1994): Fire due to poor awareness of risks in complex poorly maintained plant
- Texaco (UK, 1994): Explosion and fires due to incorrect control instruments, poor MMI and alarm system and a lack of management overview
- Longford (Australia, 1998): Failure to identify hazards and properly train operators. Insufficient understanding led to a critical incorrect valve operation

Analysis of major accidents using showed the four dominant socio-technical themes mentioned earlier contributing to ***crucial mistakes*** that triggered those accidents. Failures in these four are considered to be archetypical of chemical major accidents and possibly can be more generally applicable.

5. USING THE PYRAMAPS

The use of the PyraMAPs and their themes is to encourage the pulling together of key aspects of an organisation surrounding a risk control system in an integrated way. From Figure 3 the PyraMAPS could be combined to make a 3-D pyramid, an organising structure for pulling together indicators of safety performance, the strength and weaknesses. The point of having the 3-D pyramid concept is to reinforce the idea that all the components combine to create a new whole. The purpose is to encourage holistic thinking in preventing major hazard accidents.

When laid out side by side the hallmarks of the dominant socio-technical archetypes become very obvious and predictable in major accident reports. If there are patterns then that might offer the opportunity to use the predictability to look for performance indicators which will fit the archetypes.

In order to get to grips with the organisation as a whole and its potential for a major accident, performance indicators need to be generated and this is where the PyraMAPs can be used as domains for generating indicators for risk control measures. In the working method for the PyraMAPs the analyst starts with a selected safety barrier, procedure, job design factor, or goals & rules and generates indicators across the 3 components within the specified theme. An example is shown in Figure 4 for the subject of *Ability of the Organisation to Learn*. This theme was considered to be a good test of the model because it is a broad issue. It was possible to map onto the structure the key attributes and key issues at a high level using the domain expertise of HSE inspectors.

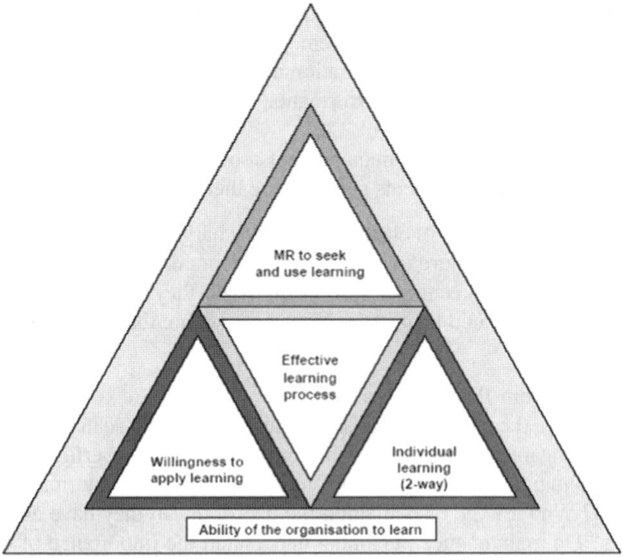

Figure 4. PyraMAP of the subject "ability of the organisation to learn" (from Bellamy & Geyer 2007)

Selected Risk Control Tasks: – Recognition of risk/danger from failure to learn – Identification of learning sources – Identification of learning tasks	**Safety Management System:** – Arrangements to identify and access sources of learning (internal/external & international and other sectors) – Assignment of responsibility and accountability for learning – Arrangements to assess and implement improvements from learning – Competence
Organisation: – Actively seeking learning opportunities (including external) – Willingness to apply learning – Commitment at senior level	**Human Factors:** – Use learning to train individuals (direct and awareness) – Individuals provide ideas for learning

The steps to go through in order to generate PyraMAP warning triangles are:

1. Identify measures specific to the risks. Analysis of near misses, incidents and accidents in context of the barriers model, can be used to iterate the model.

2. Develop PyraMAPs for (selected) measures. Sufficient measures have to be selected to reflect the whole sociotechnical system.
3. Specify relevant indicators. A combination of domain experts is needed.
4. Gather evidence of the level of performance for these indicators and make a set of warning triangles.
5. Use as an inspection or safety management tool to identify safety barriers and the important sociotechnical elements surrounding them.

For example the risk control chosen is flow discharge from a particular containment and associated indicators of failure/overfilling for that containment. Where are the strengths and weaknesses in the sociotechnical system of which they are a part? When these are identified can they be validated using other measures i.e. are they systematic. Taking each theme in turn:

1. Understanding: Were there criteria for inclusion of failure of this system in a risk assessment? Were the hazards and risks of failure in flow discharge included in training? Who understands its importance in terms of preventing overfilling e.g. in terms of recognition of maintenance and monitoring requirements, or in terms of what the indicators mean? Who are the ones making the decisions Do they have an understanding of the risks? Do maintenance personnel understand the importance of the measures? Do decision makers allocating personnel resources know what the knowledge requirements are for those job positions which have a role in MAP?
2. Competence: What are the associated tasks for provision, use, maintenance and monitoring of the overfilling prevention measures e.g. Were designers of the indicators for overfilling competent? Were competence requirements identified for identifying and responding to deviations? Do users get training in following procedures and in recognising and responding to the indicators for that particular containment? e.g. is it safe to start up if flow discharge is blocked? Were they trained in an appropriate way? Do they get refresher training? Are there training and performance criteria?
3. Priorities, attention and conflict resolution: Are the overfilling identification and response tasks within capacity, or are there competing demands, overload, distractions, insufficient manning, communication failures, etc that could conflict? Could capacities be reduced through fatigue or attention to other tasks? How could a person report problems? Do they? What kind of response would they get? Is safety being shown to be a priority? In effect is there workforce involvement in safety or is there emphasis on production?
4. Assurance: What are the goals & standards and rules & procedures of the organisation that apply to the overfilling prevention of the containment in question? What are they based on? Is it a sound basis? Have there been any symptoms of mismatch in the performance of people interacting with this system? Violations? Omissions? Fatigue? Have there been failures in flow discharge before? Is the organisation learning to do things better with respect to overfilling scenarios – better knowledge, training, interface?

The answers to these questions for a small part of the system can provide the start of a creative pattern identification process. This can be taken further by actively tracking the main line strengths and weaknesses in other systems.

6. APPLICATION OF THE PYRAMAP TO THE TEXAS CITY ACCIDENT

After elaborating on each PyraMAP theme, some examples of relevant CSB investigation and Baker panel findings are given below (US CSB 2007, Baker 2007).

1. UNDERSTANDING OF (MAJOR) ACCIDENT PREVENTION

The most important aspect of the technical system is the control measures themselves, the equipment and process controls which are the necessary measures of major accident prevention and the safe boundary of operation. This is where hazard identification and risk assessment comes in. The safety management processes make use of organisational resources and assessment criteria to undertake these risk assessment activities. An output of these processes is information on hazards and risks as criteria and inputs to other processes such as training. This provides an understanding of what the measures and safe boundaries are and why they are there. Processes such as selection and training and job allocation provide as outputs managers and supervisors in jobs of authority who understand the risks and the risk controls. These processes include providing criteria for manning specific activities and replacing absentees. The ultimate goal is to have the understanding of the risks and risk controls present whenever MAP measures could be affected by human intervention in any of the life cycle phases.

When contributors to this sociotechnical system fail people who do not understand the risk control measures could end up in a situation which demands a judgement or recognition which they do not have in order to keep the MAP measures in place.

The Texas City investigation report (US CSB, 2007) stated that:

"A lack of supervisory oversight and technically trained personnel during the startup, an especially hazardous period, was an omission contrary to BP safety guidelines."

"Occupied trailers were sited too close to a process unit handling highly hazardous materials. All fatalities occurred in or around the trailers."

In particular:

"BP had used a rigorous pre-startup procedure prior to the incident that required all startups after turnarounds to go through a PSSR26. While the PSSR had been applied to unit startups after turnarounds for two years prior to this incident, the process safety coordinator responsible for an area of the refinery that includes the ISOM was unfamiliar with its applicability, and therefore, no PSSR procedure was conducted. …. The PSSR required sign-off that all non-essential personnel had been removed from the unit and neighboring units and that the operations crew had reviewed the startup procedure."

The Baker panel (Baker 2007) believed that BP

– has active programs to analyze process hazards but " the system as a whole does not ensure adequate identification and rigorous analysis of those hazards. The Panel's examination also indicates that the extent and recurring nature of this deficiency is not isolated, but systemic." (PyraMAP 1)
– "have delegated substantial discretion to U.S. refinery plant managers without clearly defining process safety expectations, responsibilities, or accountabilities" (PyraMAP 1)

2. COMPETENCE

People undertake tasks which should keep the measures in place by making them available, by using them correctly, maintaining them and monitoring them so that the technical system remains within the safe envelope. It is important that people are competent to do these tasks.

People require both theoretical and practical training. Competence requirements are criteria for selection and training. Safety management makes use of organisational resources and criteria like selection and training systems, job and task analysis and job descriptions in processes which deliver competences to tasks which support the MAP measures. The workforce (which includes managers) needs the knowledge, procedures and skills to do their tasks competently.

The Texas City investigation report (US CSB, 2007) stated that:

"The operator training program was inadequate. The central training department staff had been reduced from 28 to eight, and simulators were unavailable for operators to practice handling abnormal situations, including infrequent and high hazard operations such as startups and unit upsets."

The Baker panel believed that BP:

– "has not effectively defined the level of process safety knowledge or competency required of executive management, line management above the refinery level, and refinery managers" (PyraMAP 2)
– "has not adequately ensured that its U.S. refinery personnel and contractors have sufficient process safety knowledge and competence." (PyraMAP 2)
– "over-reliance on BP's computer based training contributes to inadequate process safety training of refinery employees" (PyraMAP 2)

The competence PyraMAP applied to the Texas City accident is shown in Figure 5. Here the findings of the Baker panel are shown according to the triangle components. The point is that these weaknesses combine to weaken the barrier integrity, not just in competences to use barriers effectively, but underlying competences in the organisation from the leadership downwards through the line management indicating an organisational incompetence to manage process safety.

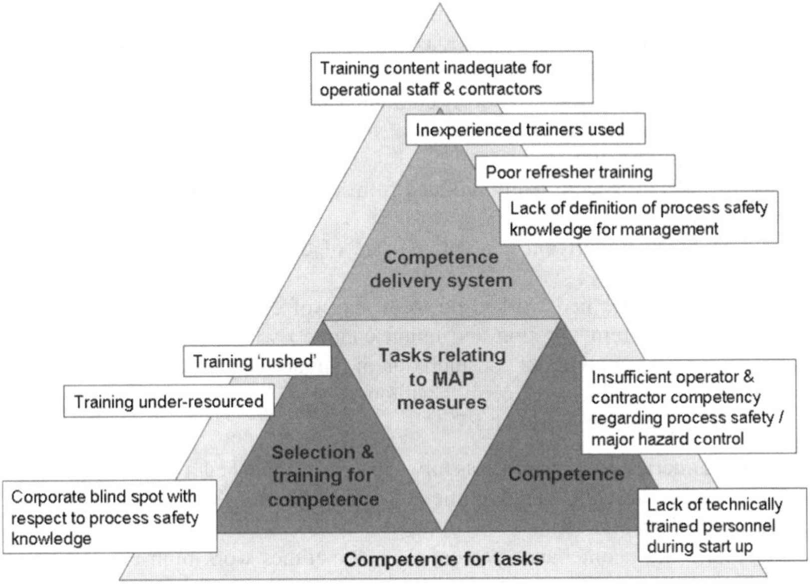

Figure 5. Baker panel and investigation report elements in the competence PyraMAP

3. PRIORITIES, ATTENTION & CONFLICT RESOLUTION

Performance on risk control related tasks should be supported by job and equipment design to prevent excessive demands which could lead to a demand-capacity mismatch. Mismatch means that a person is unable to perform psychologically, physically or physiologically in order to meet the task requirement like not being able to reach something because it is too high, being unable to analyse something because insufficient information is supplied or being unable to attend to something because it is lost in noise. These tasks should also be supported by information and communications that emphasize the criteria for what tasks should be given priority and attention. These communication systems should allow feedback and involvement of operators to indicate demand-capacity problems and help identify possible solutions as input to adjustment processes. Sometimes communications emphasize the wrong things because production pressures compete for time and attention or because the communication is badly designed and is giving the wrong message. Workload on operators, poor interface design, the stress of handling process deviations, insufficient procedural support can all cause attention and prioritizing problems in the use of resources.

The investigation report indicated that:

"An extra board operator was not assigned to assist, despite a staffing assessment that recommended an additional board operator for all ISOM startups."

" Supervisors and operators poorly communicated critical information regarding the startup during the shift turnover; BP did not have a shift turnover communication requirement for its operations staff."

"ISOM operators were likely fatigued from working 12-hour shifts for 29 or more consecutive days."

The Mogford Report cites fatigue as one of the root causes of the Texas City accident:

"Some employees had worked up to 30 days of consecutive 12-hour shifts. The reward system (staff remuneration and union contract) within the site encouraged this extended working period without consideration of fatigue. There were no clear limitations on the maximum allowable work periods without time off."

The Baker panel believed that BP

– in some refineries, including Texas city, "has not established a positive, trusting, and open environment with effective lines of communication between management and workforce" (PyraMAP 3)
– "operations and maintenance personnel ... sometimes work high rates of overtime, and this could impact their ability to perform their jobs safely and increases process safety risk" (PyraMAP 3)

4. ASSURANCE

How do the behavioural outcomes relate to the goals, objectives and rules (procedures) of the organisation? It is often said that what gets measured gets better or gets done. What is measured should reflect the objectives of the organisation. Are the objectives, the goals, the procedures and the standards of risk control being met and are they good enough? Are there deviations, use of wrong objectives? Are there symptoms of mismatch? It is important to have appropriate MAP objectives for risk control and a system that ensures these are being achieved including learning systems for improvement. Organisational change can influence the ability to meet objectives. Loss of memory or knowledge separating it from the risk control system can be disastrous. For this reason monitoring, learning and adjustment is required in all areas affecting the processes whose outputs impact on MAP.

The Texas City investigation report stated that:

"Outdated and ineffective procedures did not address recurring operational problems during startup, leading operators to believe that procedures could be altered or did not have to be followed during the startup process."

"The process unit was started despite previously reported malfunctions of the tower level indicator, level sight glass, and a pressure control valve."

"The BP Board of Directors did not provide effective oversight of BP's safety culture and major accident prevention programs. The Board did not have a member responsible for assessing and verifying the performance of BP's major accident hazard prevention programs."

The Baker panel believed that BP:

- "has not provided effective process safety leadership and has not adequately established process safety as a core value" (PyraMAP 4)
- " did not always ensure that adequate resources were effectively allocated to support or sustain a high level of process safety performance" (PyraMAP 4)
- "does not effectively translate corporate expectations into measurable criteria for management of process risk or define the appropriate role of qualitative and quantitative risk management criteria." (PyraMAP 4)
- "does not effectively measure and monitor process safety performance" (PyraMAP 4)
- "does not effectively use the results of its operating experiences, process hazard analyses, audits, near misses, or accident investigations to improve process operations and process safety management systems" (PyraMAP 4)
- exhibits "instances of a lack of operating discipline, toleration of serious deviations from safe operating practices, and apparent complacency toward serious process safety risks at each refinery" (PyraMAP 4)
- "corporate safety management system does not ensure timely compliance with internal process safety standards and programs" or "timely implementation of external good engineering practices that support and could improve process safety performance" (PyraMAP 4)

7. CONCLUSIONS

The PyraMAP model closely matched the conclusions drawn by the Baker Panel in the analysis of BP's refineries Therefore, PyraMAP could be a useful tool in the analysis of major accidents with respect to the contribution of human factors. Human factors are now better related to major hazards accidents. In general the relation between human factors and accidents has not been described in a systematic way. The PyraMAP model is a step towards bringing structure into this relation. Accident analysis at a detailed level, across a significant number of accidents, can increase understanding of the sociotechnical patterns which make up major accident prevention or causation. PyraMAPS provide a framework, with the 4 dominant themes of accidents described in this paper, for creative thinking "outside the box" in prevention of major accidents. They might provide a basis for generating indicators of safety on the wider organisational influences on human performance, whether at board, line management or operator level for a specific technical system. In the Human Factors context, based on this model, such indicators would be placed on the organisation and management aspects which influence the match between the capacities of and demands on front line operators with the end result of reducing the likelihood of the technical system failing.

REFERENCES

Baker J.A., 2007. The Report of the BP U.S. Refineries Independent Safety Review panel, January 2007

Baksteen, H., Mud, M., Bellamy, L.J., 2007. Accident Analysis using Storybuilder Illustrated with overfilling accidents including Buncefield, UK. Report prepared by RIVM, RPS Advies and White Queen for the Ministry of Social Affairs and Employment, The Netherlands, September 2007

Bellamy, L.J., Geyer, T.A.W., Astley, J.A., 1989. Evaluation of the human contribution to pipework and in-line equipment failure frequencies. Health and Safety Executive, Bootle: HSE, 1989. ISBN 0717603245. HSE Contract Research Report 15/1989.

Bellamy, L.J., Geyer, T.A.W., 1992. Organisational, management and human factors in quantified risk assessment – Report 1 – HSE Contract Research Report 33/92.

Bellamy, L.J., Leathley, B.A., Gibson, W.H., 1995. Organisational factors and safety in the process industry: inspection tool development. Establishing A Link Between Safety Performance and Organisation. Ministerie van Sociale Zaken en Werkgelegenheid, Den Haag, The Netherlands. ISBN 90-5250-976-X.

Bellamy, L.J., Brouwer, W.G.J., 1999. AVRIM2, a Dutch major hazard assessment and inspection tool. Journal of Hazardous Materials 65, 191–210.

Bellamy, L.J., Papazoglou, I.A., Hale, A.R., Aneziris, O.N., Ale, B.J.M., Morris, M.I., Oh, J.I.H. 1999. I-Risk: Development of an integrated technical and management risk control and monitoring methodology for managing and quantifying on-site and off-site risks. Contract ENVA-CT96-0243. Report to European Union. Ministry of Social Affairs and Employment. Den Haag.

Bellamy, L.J., Geyer, T.A.W., Wilkinson, J. 2006. Development of a functional model which integrates human factors, safety management systems and wider organisational issues. To be published in Safety Science, available on Science Direct:, doi:10.1016/j.ssci.2006.08.019

Bellamy, L.J., Ale B.J.M., Geyer T.A.W., Goossens L.H.J., Hale A.R., Oh J., Mud, M., Bloemhof A, Papazoglou I.A., Whiston J.Y., 2007a. Storybuilder—A tool for the analysis of accident reports, Reliability Engineering and System Safety 92 (2007) 735–744

Bellamy, L.J., Mud, M.L., Ale, B.J.M., Whiston, Y., Baksteen, H., Hale, A.R., Papazoglou, I.A., Anezeris, O., Bloemhoff, A., Oh, J.I.H., 2007b. The software tool Storybuilder and the analysis of the horrible stories of occupational accidents. To be published in Safety Science, available on Science Direct, doi:10.1016/j.ssci.2007.06.022

Bellamy, L.J., Geyer, T.A.W., 2007. Development of a working model of how human factors, safety management systems and wider organisational issues fit together. Contract research report RR543, Health and Safety Executive, HMSO 2007

Health and Safety Executive, 2003. COMAH safety report assessment manual. Issue date January 2003. http://www.hse.gov.uk/hid/land/comah2/

Reason, I.T., 1990. Human Error. Cambridge University Press, UK.

Reason, J., 1997. Managing the Risks of Organizational Accidents James Reason. Ashgate Publishing, ISBN 184014 105 0.

US Chemical Safety and Hazard Investigation Board, 2007. Investigation report Refinery Explosion and Fire BP Texas City March 23 2005. Report No. 2005-04-I-TX, March 2007

White Queen (2003) Seveso II Safety Report Information Method (SAVRIM2000) Handbook. White Queen BV, PO Box 712, 2130 AS, Hoofddorp, The Netherlands

CAN WE STILL USE LEARNINGS FROM PAST MAJOR INCIDENTS IN NON-PROCESS INDUSTRIES?

Frederic Gil[1] and John Atherton[2]
BP Process Safety & Fire Engineering Advisor, Refining Safety & Operations Excellence, Sunbury TW16 7LN UK
Process Safety Consultant, 8 Smolletts, East Grinstead, RH19 1TJ, UK

Major accidents from outside of the process industries, such as Titanic and Columbia, are reviewed against a 21st century process industry integrity management standard. Each such incident is then presented against accidents from the process industries to reinforce the learnings. For example, Titanic can be used to create learnings around the need for pre-start up safety reviews following an emergency shutdown; Columbia on institutional learning. The result is an in-house booklet that has been given wide circulation and subsequently offered to CCPS for publication. The paper will include extracts from non-process incidents and describe how they relate to the process industry.

INTRODUCTION

BP initiated a project in early 2004 to develop a process safety teaching aid in the form of a booklet containing analyses of non-process and process industries accidents against their Integrity Management programme. The first edition was published internally in early 2005, with a second edition containing lessons from BP Texas City March 2005 accident in 2006. BP has offered this booklet to CCPS for wider publication under the title "Incidents that define Process Safety – ISBN 978-0-470-12204-4".

In discussions with young graduates or even experienced technicians it has become clear that many had not heard of Flixborough, Bhopal or Piper Alpha, and if they had, few of them received enough information to be able to transfer lessons to their current activities. They had, however, heard of Titanic, Chernobyl, the NASA accidents (Challenger and Columbia), Concorde and other aviation accidents through the press and media. The objective of this booklet was to disseminate more widely the main lessons from major incidents from wherever they originated using a simple, user friendly, integrity management tool.

Jesse C. Ducommun was Vice-President, Manufacturing and a Director of American Oil Company in 1961 (joined in 1929) and Vice-President American Petroleum Institute in 1964. While working for Amoco, he conceived, inspired and co-authored the unique series of booklets on process safety in the 1960s that was then updated and complemented after 2002 by BP Refining Process Safety Community of Practices. In the preface, he stated:

> *"It should not be necessary for each generation to rediscover principles of process safety which the generation before discovered. We must learn from the*

experience of others rather than learn the hard way. We must pass on to the next generation a record of what we have learned."

So why focus on non-process accidents when there is plenty of fertile ground within our own oil and petrochemical industry? There are a number of reasons:

- Oil and petrochemical companies exist in a very similar environment to other commercial ventures, in that:
  - They operate in a highly competitive business environment.
  - They have hierarchal management structures.
  - They employ skilled people who have their own career progression aspirations, which can impact on an individual's behaviour.
  - They are strictly regulated.
  - They involve similar technologies or practices (for example board operators in a refinery control room and air traffic controllers in a control tower both have to work remotely and can have issues mentally visualising what is going on; replacing critical parts on an aircraft engine or a refinery compressor has to be done with similar rigorous procedures).
- It is very easy for individuals reading accident reports to focus on technical details, whereas the majority of root causes relate to softer, cultural and human behaviour issues. Lessons learned from an incident on a crude oil distillation unit (CDU), for example, may receive a technical review at another refinery which leads to a conclusion that a similar accident could not occur on their CDU as it has a different equipment arrangement or process conditions. In most cases, the root causes are likely to be around human factors and management protocols where meaningful parallels can be drawn. Lessons from non-process accidents do not have a technical equivalence and readers are forced to consider the full range of lessons learned.
- Some industrial sites, or locations, can build up a negative reputation for process safety if there is a history of accidents and incidents, and others may assume that they are better than the site suffering the incident, and that the same thing could not happen to them. Lessons from accidents such as Titanic and the aviation industry can excite the imagination, leading to the learning of valuable lessons that remain in the minds of those who had studied them.

Included in the booklet are a number of major process incidents that have been influential in shaping process safety, such as Flixborough, Bhopal, Phillips Pasadena, and Piper Alpha. These are used to illustrate synergies and because they are important in their own right.

BP'S INTEGRITY MANAGEMENT STANDARD

The model used to analyse the incidents in the booklet is BP's Integrity Management (IM) Standard, which sets priorities for process safety improvements in line with the best of industry practices. It sets out the requirements for IM to satisfy BP's Group Values, particularly

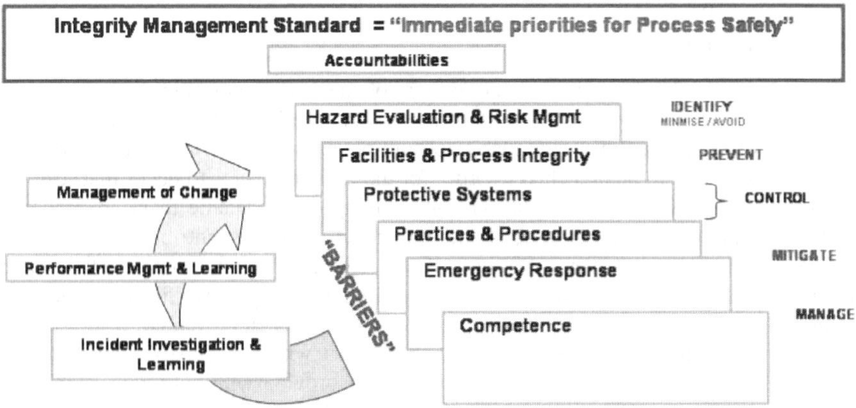

Figure 1. BP's integrity movement standard

those pertaining to Risk, Health and Safety, Environmentally sound operations and the achievement of internal targets as defined in BP's Management Framework. It requires the controlled application of BP's Major Accident Risk (MAR) assessment, process safety and engineering management, combined with internationally recognised industry standards and BP developed engineering and operating procedures. The objective is to sustain BP's license to operate by reducing the probability of uncontrolled release of hydrocarbons and chemicals, including catastrophic and chronic releases to the atmosphere water or ground, and to prevention of equipment failure and hence avoid serious harm to people, the environment and physical assets. Additional benefits are seen in improved operational integrity; enhanced HSE performance; increased life cycle value of assets and greater engineering standardisation and productivity. The IM Standard is shown schematically in Figure 1.

CATEGORISATION OF ACCIDENT SCENARIO'S STUDIED

The scenarios used in the booklet were chosen against a number of topic headings:

- Blind operations
- Design
- External causes
- Inspection & maintenance
- Knowledge and training
- Lack of HAZID (hazard identification)
- Management of Change
- Not learning from near misses
- Operating practices
- Permit to Work

- Emergency Response
- Human Factors

These headings reflect the view of the authors of the major causation of the respective incidents from which the most important lessons can be learned. However, as with all incidents, they are invariably attributable to multiple causes. These are often represented as barriers that have to be breached in order for the chain of events to occur that results in the eventual outcome. For example, with the sinking of the Titanic there are important design, operations and human factors, and emergency response contributors to the eventual tragedy that claimed over 1500 lives. If the original design to complete the vertical bulkheads up to the weather deck had been installed, the progressive filling of the ships' compartments as she went down by the bows would have been prevented; if the ship had been slowed down when ice was forecast, the impact with the iceberg may have been prevented or at least reduced; if the ship had been provide with sufficient lifeboats to enable all members of the crew and passengers to safely abandon it, the loss of life could have been far less.

THE INCIDENTS

The non-process incidents chosen represented a wide range, some of which are well known, e.g. Titanic and the NASA incidents; some less well known, such as the Royal Navy K Class submarines. Not all are recent incidents as, in addition to 1912 Titanic and 1916 K Class submarines, the 1937 Hindenburg, 1930 airship R101 and a 1906 mining incident are included. A significant number were drawn from the marine and aviation industry, such as Betelgeuse, Erika, HMS Glasgow, Concorde and Canada Air Flight 236. Non-oil and petrochemical industry incidents are included: Bhopal and Toulouse, together with nuclear power production incidents at Three Mile Island and Chernobyl. Also included are a number of important oil and petrochemical incidents, including: Feyzin, Phillips Pasadena, Texaco Milford Haven, Esso Longford, Total la Mede, Marathon HF release, Exxon Valdez, Seveso, and Tosco Hydrocracker. Other oil related incidents associated with exploration and production included the sinking of Platform P36 and Piper Alpha. Major BP incidents that are also included are Grangemouth Hydrocracker and Flare Line incidents of 1987, and BP Texas City incident of March 2005.

More details of incidents reviewed are included against each of the sections below.

THE LESSONS DERIVED FROM A REVIEW OF THESE INCIDENTS

BLIND OPERATIONS

In this section incidents are examined where those operating equipment were unaware of the actual situations they were in. Two of these are from the aircraft industry, and one from the nuclear industry:

- Incomplete information (Pan Am 1736/KLM 4805)

- Unclear information transfer (MD83/Shorts 330)
- Overwhelmed with too much information (Three Mile Island)

The lesson that runs through all of these incidents is that sometimes people create their own blindness or "mind set" either through lack of awareness of what is going on resulting from a lack of training or experience, or through external factors such as stress. It is good practice when faced with a situation that appears to be a bit out of the ordinary, to step back and take a thorough look at all the evidence, or ask a colleague to do that for you.

DESIGN

The five incidents described in this section demonstrate major areas of importance in design:

- Bhopal
- NASA Challenger Space Shuttle
- K Class Submarines (WW1 steam driven submarines built for the Royal Navy)
- TWA flight 800
- Hindenburg airship

These incidents demonstrate a range of important design issues: Bhopal and the steam driven K Class submarines – inherent safety, Challenger – not fully understanding the potential consequences of violating the true limits of a safe operating envelope (solid rocket booster seals), TWA 800 and Hindenburg – compromising a safe operating envelope as deterioration processes kicked in or where subtle changes were made to equipment.

Hazard identification and risk assessment are tools that are required to be used in all major jurisdictions. However, these can only be effective where those conducting these studies can do so knowing that their recommendations will be acted upon in a responsible manner. In more than one of the above incidents, political and commercial pressures either caused the risk assessment process to be shortened or ignored.

The importance of knowing the safe operating envelope and how this can be compromised by external factors or safety critical equipment malfunction is also demonstrated by some of these incidents.

EXTERNAL CAUSES

Two incidents are described:

- Mexico City – Pemex LPG terminal
- Tupras – Turkish Earthquake 1999

Even with the most sudden and extreme events, the use of traditional hazard identification, risk assessment and emergency planning techniques can still provide major benefits. As it is virtually impossible for every potential scenario to be assessed in depth, review team members need to be able to see the wider issues in order to make meaningful recommendations that will make a difference when the major event takes place.

INSPECTION & MAINTENANCE

Seven incidents are included:

- Explosion of the "Betelgeuse" at Bantry Bay
- Sinking of the "Erika" – fuel oil marine tanker in the Bay of Biscay
- Canadian Airways flight TS 236 – ran out of fuel over the Atlantic
- HMS Glasgow – fatalities during construction
- Marathon HF release at Texas City
- Flare line failure at Texaco Refinery, Milford Haven
- FCCU explosion at Total la Mede

The examples given here demonstrate a number of key points, for example the impact of the age of plant and equipment on integrity. Increasing age does not necessarily mean that something is unsafe, but more care needs to be taken to maintain its integrity as deterioration sets in and, for example, corrosion allowances are consumed as occurred in the cases of "Betelgeuse" and "Erika". Exceeding the corrosion allowance was also a major contributor of the Texaco Milford Haven flare line failure. The piping failure that led to Total la Mede also resulted from corrosion that occurred over a long period of time, >30 years. This corrosion mechanism was entirely predicable as it was due to the expected reaction of the process fluids with the materials of construction, but this section of piping had not been regularly inspected, most likely, because it was a bypass.

Failures to follow safe maintenance or construction procedures were a root cause of the remainder of these incidents.

Many major incidents have resulted from maintenance and inspection not being carried out correctly or not at all. Maintenance and inspection are, for the most part, costly and invasive and detract from production if not specified and planned correctly. The role of an Engineering Authority is vital here to ensure that the right risk based arguments on plant integrity are presented to the decision makers with ultimate accountability for safe operations, Directors and Senior Managers, to balance those being put forward by the commercial and production functions. The planning and execution of maintenance and inspection activities is, therefore, critical to ensuring all parts of the safe operation equation are properly balanced.

KNOWLEDGE AND TRAINING

In this section a number of incidents are reviewed where those most involved in the incidents were not aware of the hazards that they were exposed to:

- Feyzin LPG tank farm explosions 1966
- Ammonium nitrate explosions:
  - Oppau, Germany
  - SS Grandcamp, Texas City Port
  - Toulouse, France
- Dust explosion at Courrierres mine, France, 10 March 1906

Feyzin was one of the first post WW2 major process accidents in Europe. It occurred because the operator did not understand the importance of getting the valve sequence right when draining water from LPG storage vessels.

The lessons of the past are important to everybody, in every walk of life. Sadly, experience shows that the lessons learned many years ago become forgotten over time. The ammonium nitrate accidents are a case in point. The first occurred in 1921, the second in 1947 and the third in 2001. Nothing had changed in the fundamental properties of ammonium nitrate, and in some industries it was used as an explosive. However, those handling this material do not appear to have had an appreciation of its dangerous properties despite the major incidents that had occurred previously. Similar issues are demonstrated in the case of dust explosions, with 3 recent (1997 and two in 2003) incidents explained after the detailed description of the 1906 mining disaster.

LACK OF HAZARD IDENTIFICATION

The six incidents described in this section indicate the consequences of not having put in place some form of formal hazard identification, e.g. HAZID, and risk assessment process. The six cases demonstrate different scenarios:

- Titanic – compromising a safe design
- P-36 FPSO – invalidating a safe design
- Esso Longford not carrying out a formal HAZOP on the original plant design and any modifications
- Reactive chemicals:
  - Road tanker explosion, Teeside Rohm & Haas plant, UK, 3rd January 1976
  - Pesticide explosion during storage, Bartlo Packaging (BPS), Inc, 8 May 1997, where 3 died
  - Napp Technologies, Inc., Lodi, New Jersey 1995, where 5 died

In all cases, the consequences represented a worst-case situation, although the loss of life could certainly have been higher in all of them. The worst case concept has been repeatedly challenged over the years in favour of a more risk-based approach. There is no doubt that the concept of ALARP (as low as reasonably practicable) is valid, provided the risk assessments are based on well founded data. The very fact that a ship/platform floats means that it can sink. A process unit operating with materials that can flash to give very low temperatures means that it will be possible to achieve these temperatures under severe upset conditions. Prevention, control and mitigation measures need to be robust and not vulnerable to a common mode form of failure that renders them all ineffective simultaneously. Consideration of human factors where risk reduction is sought through procedural means must be carefully examined to ensure the right levels of checks and balances.

An unusual aspect of the Titanic incident can be related to Pre-Start Up Safety Reviews. In his book "Last Log of the Titanic" (1), David G. Brown describes how the Titanic was stopped after hitting the iceberg. An assessment of damage was carried out and a decision taken to restart the engines. Up to that point in time the water ingress into the ship was being matched by the pumps that had been rigged to pump out the damaged compartments. It is postulated that the ship could have remained afloat for a far longer time, in all probability enough time for the passengers to be rescued. This would have achieved one if its design intentions to be its own lifeboat. However, once restarted the damage to the hull plating was exacerbated by the forward motion of the ship to a point where the pumps could no longer cope. The parallel in our industry is that following an emergency shutdown a thorough pre-start up safety review needs to be carried out before recommissioning. Process plant will undergo extremes of pressure and temperature excursions in an emergency shutdown situation, demanding such a review. If this is not carried out, perhaps (as in the case of the Titanic) due to pressure from commercial interests to restart quickly, the consequences can be catastrophic and far outweigh any short term gain.

MANAGEMENT OF CHANGE

Two major accidents that have influenced the industry in major ways are described in this section:

- Flixborough explosion.
- Chernobyl radioactive cloud.

Flixborough describes the failure to create a safe design for a critical process modification, and Chernobyl where operations managers and technicians stray outside of a pre-defined safe operating envelope.

Change is inherent in the way in which we live and work. The most important aspect of change is being able to recognise when a change is being proposed or put into action. Even replacement of a machinery spare of apparently identical proportions with that manufactured by another supplier is a change that merits careful review of dimensions and materials of construction of all components to assure that there are no unintended harmful consequences. This applies equally to changes in staffing and organisation (no two human beings are identical), chemicals, feedstocks and all aspects of process plant operation.

Another critical lesson from these incidents is the ability for individuals to "know what they don't know". Competition for attractive jobs coupled with personal ambition has led to many cases of individuals being placed into roles with responsibilities they are not fully competent to manage. This in itself may not be problematical provided their limitations are recognised at the outset and those individuals are properly mentored and have no inhibitions in recognising when they need help and asking for it. However, many may see asking for help as running counter to their aspirations for advancement and coupled with the reduced numbers of experienced people available to support them, this may well conspire to major incidents being caused through inexperience and ignorance.

NOT LEARNING FROM NEAR MISSES

Three incidents are selected:

- NASA Columbia Space Shuttle: re-entry fatal incident

- Herald of Free Enterprise: capsized ferry
- Concorde – fatal crash at Paris

It is said that there is nothing new in safety – it has all happened before. These three incidents were all high profile with major loss of life. Had lessons from previous incidents been learned and applied, it is highly likely that they would not have happened.

In the case of Columbia, the time between lessons from Challenger incident being available and the second space shuttle incident was 17 years. This is around half of the career span of the average person, which implies that around 1/2 of the workforce were not around when the first accident happened. The only way they could have benefited from lessons from that far back is for these lessons to be assimilated into the culture of the company so that they remain "evergreen". Even then this is not a guarantee that they will remain in place as senior people evolve and move on, management theory and practices change, and commercial pressures take on different dimensions.

OPERATING PRACTICES

Despite the best design thinking with incorporation of lessons from the past, things can still go badly wrong if the people operating process plants and equipment do not follow proper operating procedures. In this section there are three incidents;

- R101 airship crash
- Tosco Hydrocracker fire
- BP Texas City ISOM explosion

Operating procedures are vital documents and need to be written by competent people. Training in these procedures must be accompanied by some form of testing regime to ensure that knowledge has been taken on board. Changes to procedures need to be reviewed against a robust management of change procedure with operators and others who need to know being properly informed and trained. The Texas City refinery Isomerisation incident on 23rd March 2005 highlights the importance of using operating procedures.

PERMIT TO WORK SYSTEMS

This section describes five major accidents where failure to comply with Permit to Work Procedures resulted in large-scale loss of life and/or property. These are all oil and petrochemical industry incidents, unlike the make up of the preceding sections:

- Motiva tank fire
- Phillips Pasadena process unit fire
- Piper Alpha platform fire
- Port Edouard Herriot tank farm fire
- BP Grangemouth flare line fatalities and fire

Work Permits or Permits to Work are an industry standard to ensuring that work of a non-routine manner can be safely controlled in hazardous areas, and in many jurisdictions are a legal requirement. However, experience has shown that no matter how simple or sophisticated the procedure and work permit forms are, it is by the strict application of the procedures and practices associated with the identification of hazards, assessment of risk and application of Permit to Work conditions that ensures that work can be safely carried out. A Permit to Work system can only be as good as the people implementing it are competent.

EMERGENCY RESPONSE

This section examines incidents where emergency preparedness was overlooked or undersized, communications were mishandled, inadequate tactical choices were made or those designing/operating/authorising installations or responding to the incident were unaware of the actual situations they were in.

Clarity of information is vital to understanding the situations in which we may find ourselves. When designing an installation, risks must be clearly assessed and communicated to all parties involved, so that everyone is clear on the potential consequences and adequate response strategy and resources needed the day things may take a bad turn.

Three types of incident have been selected to illustrate this section:

- The case of Seveso, the authorities did not know until late after the incident that Dioxin was involved.
- Warehouses incidents, such as the Sandoz Bale accident, show the importance of agreeing on tactical response long before an incident, with all parties that will respond (Operator, Local/National Authorities, etc.). The adequate strategy may not be the obvious one, nor the one that media pressure will put forward, and it's sometimes preferable to let a fire burn out than try to fight it to avoid putting firefighters at risk or contaminate huge quantities of water.
- The Tacoa boilover is another example of poor understanding of the risks involved and inadequate management of the incident.

On the face of it, it could be assumed that there was nothing anybody could have done about these major incidents. However, looking closely, there are three important areas that could have prevented or limited the consequences of the events:

- Learning lessons from outside the boundaries of ones own experience. There is a vast amount of information available to the manager, engineer and operator that can help them in times of crisis, but managing such a vast quantity of information is always a challenge.
- Designing installations that are inherently safer and that won't overcome the resources available to emergency responders. The Bhopal incident is a typical case where the hazardous material inventory could have been significantly reduced.
- Writing and exercising realistic major accident scenarios. In this way it is possible to agree and educate on the adequate tactics to respond to different events before they happen, and/or purchase additional equipment if necessary.

Many of the incidents in this book could also have been discussed in this section: from Piper Alpha to Bhopal and the Titanic. Emergency Response has often been perceived as an un-necessary cost centre that was considered disproportionately expensive when set against operational short term gains; why add life boats for all passengers when the ship is supposed to be unsinkable? It is probably part of human nature to think that scenarios like that "can't happen here" and that they are someone else business to deal with. It is the duty of the Managers and the Engineers to overcome these feelings and adopt a transparent and realistic approach to Emergency Preparedness.

HUMAN FACTORS

Two major accidents are described where failure to look at human factors in training and work organisation strongly contributed to large-scale loss of life and/or damage to the environment.

This Exxon Valdez incident report from the National Transportation Safety Board highlights the following human factors:

- drug & alcohol abuse;
- fatigue;
- excessive workload.

The Flash Airlines crash highlights the following human factors:

- training,
- organisational/hierarchical factors that impede communications.

Human factors are often overlooked, both at the design stage (including in the Management Of Change process: see specific section on MOC) and during investigation of incidents (the Egyptian report of the Flash Airlines crash is a typical example: as the NTSB comments document on this report diplomatically puts it: "the Egyptian Ministry of Civil Aviation's investigation of the operational and human factors related to the accident was minimal"). This may be because human factors seem difficult to grasp or deal with. However, incidents cannot be prevented or fully understood without them.

CONCLUSIONS

There is little doubt that the booklet has been well received. One senior BP manager wrote: "One resource that the Process Safety Specialists have made available for us all is a really great book entitled "Integrity Management - Learning from Past Major Industrial Incidents" (CCPS title "Incidents that define Process Safety"). This contains a large number of case studies both from within our industry and from other events. It is written in a very easy to read style and is a great resource for all of us to use in our facilities to enhance people's awareness of risk and to help us all reduce any sense we have of complacency. All the stories in the book involve people who did not set out to have the accident or incident - that is people just like us!"

Another shared the booklet with several of BP's manufacturing site safety representatives and the one comment that that was heard many times was: "You know, the circumstances that led to the incidents used for the examples could EASILY have existed at our facility".

Other examples of feedback received on the booklet from within BP included:

- The information was seen to be highly relevant to the readers' activities and was received with great interest. Many sites see this as a major additional "lessons learned" training resource that feeds into Knowledge Management.
- Many of those responding requested more additional copies. There were a number of cases where somebody saw a copy sitting on another person's desk and requested their own copy as they saw that the learnings would fit very well with their team's activities. One example was in connection with the supply of drilling and completions safety critical software.
- The professional way in which the case histories were presented was appreciated. The crisp language and detailed information in tabbed format makes it very user-friendly.
- One team intends to convert sections of the booklet into a CBT tool that can be easily shared via their intranet.
- Sections of the booklet have been translated into Dutch and German with PowerPoint presentations being created in other local languages.
- Some have translated the "sharing the experience" booklet and distributed it to operators during toolbox meetings. Others added it to basic operator training.
- One reader found the document excellent background reading for an Open University Post graduate project entitled 'Forensic Engineering'. He was sure his fellow students would benefit from reading the booklet.

REFERENCE

(1) The Last Log of the Titanic, David G. Brown, International Marine - McGraw-Hill, 2001, ISBN 0-07-136447-1

SOME OF THE SELECTED READING USED TO DRAFT THE VARIOUS SECTIONS OF THE BOOKLET

NUCLEAR INDUSTRY:

- Three Mile Island:- US Nuclear Regulatory Commission Fact sheet on the Accident at Three Mile Island, from: http://www.nrc.gov
- "United States President's Commission on the Accident at Three Mile Island." Vol. 9. U.S. Government Printing Office: Washington, D.C., 1979.
- Hodgson, Peter. "Chernobyl After Five Years". Contemporary Review. (July, 1992)
- Lewis, H.W. "The Accident at the Chernobyl Nuclear Power Plant and Its Consequences". Environment. (November, 1986).
- Space and Plane incidents:
- Challenger – NASA web-site: http://history.nasa.gov/sts51l.html: The Challenger Decision to Launch, Diane Vaughan, University of Chicago Press, ISBN: 0-226-85176-1

- MD83/Shorts 330 crash at Paris Charles de Gaulle airport: Bureau Enquêtes-Accidents Report "Accident on 25 May 2000 at Paris Charles de Gaulle (95) to aircraft F-GHED operated by Air Liberté and G-SSWN operated by Streamline Aviation.
- "US summary comments on draft final report of aircraft accident Flash Airlines flight 604, Boeing 737–300, SU-ZCF January 3, 2004, Red Sea near Sharm El-Sheikh, Egypt", National Transportation Safety Board.
- "Factual Report of Investigation of Accident, Flash Airlines flight 604, January 3, 2004, Boeing 737–300, SU-ZCF, Red Sea off Sharm El-Sheikh, Egypt", Egyptian Ministry of Civil Aviation.
- TWA 800 – National Transportation Safety Board Report "In-flight Break up Over The Atlantic Ocean Trans World Airlines Flight 800 Boeing 747–131, N93119 Near East Moriches, New York July 17, 1996" NTSB/AAR-00/03 Adopted August 23, 2000
- Hindenburg – Flight 100 years of Aviation, R.G. Grant, Smithsonian National Air and Space Museum/Duxford Imperial War Museum, ISBN 0-7513037323
- Columbia Accident Investigation Board – Volume 1, Report published by the CAIB on 26 August 2003 and subsequently printed and distributed by NASA and the Government Printing Office, Washington DC. Available on Board web-site: http://www.caib.us/
- Concorde – Bureau Enquêtes-Accidents Preliminary Report "Accident on 25 July 2000 at "La Patte d'Oie"in Gonesse (95), to the Concorde, registered F-BTSC, operated by Air France – English Version"
- R101: The Airship Disaster 1930, 1931, The Stationary Office, London, ISBN: 0 11 702407 4

SHIPPING INCIDENTS

- Disasters at Sea, Capt. Richard A. Cahil, Century, 1990, ISBN: 0-7126-3814
- Bantry Bay - "Disaster at Whiddy Island, Bantry, Co. Cork", Report of the Tribunal of Inquiry, published by the Stationary Office, Dublin
- "Report of the enquiry into the sinking of the "Erika" off the coast of Brittany on 12 December 1999" - Permanent Commission of Enquiry into Accidents at sea (Commission Permanente d'Enquetes sur les Evenements de Mer – CPEM)
- "Grounding of the US tankship Exxon Valdez on Bligh Reef, Prince William Sound near Valdez, Alaska. March 24, 1989" Marine Accident Report PB90-916405 NTSB/MAR-90/04, National Transportation Safety Board.
- "Fire on HMS Glasgow 23 September 1976", Health & Safety Executive/HM Factory Inspectorate, HMSO, 1978, ISBN: 0 11 883075 9
- K-Boats, Steam Powered Submarines in World War 1, Dan Everitt, Airlife Publishing Ltd, ISBN 1-84037-057-2
- MV Herald of Free Enterprise. Report of Court No. 8074 - Formal Investigation (Hon. Mr Justice Sheen - Wreck Commissioner), July 29th 1987. London, Her Majesty's Stationery Office, 1987, ISBN 0 11 550828 7

PETROCHEMICAL INDUSTRY:

- Tosco Hydrocracker explosion & fire - US Environmental Protection Agency Chemical Accident Investigation Report: EPA 550-R-98-009, November 1998
- Mexico City – Lees, F.P., 'Loss Prevention in the Process Industries – Hazard Identification, Assessment and Control', Volume 3, Appendix 4, Butterworth Heinemann, ISBN 0 7506 1547 8, 1996.
- Marsh and McLennan, 'Large Property Damage Losses in the Hydrocarbon-Chemical Industries a thirty-year Review', 16th Edition, Marsh and McLennan Protection Consultants, 1995.
- "Analysis of the LPG Disaster in Mexico City", C.M. Pietersen, TNO, Apeldoorn, Netherlands.
- Tuprus – US Geological Survey Circular 1193, Implications for Earthquake Risk Reduction in the United States from the Kocaeli, Turkey, Earthquake of August 17, 1999
- Bhopal:- "TED Case Studies: Bhopal Disaster," Trade and the Environment Database, 1996. Ashford, N.A.; Gobbell, J.G.; Lachman, J.; Matiesen, M.; Minzner, A.; Stone, R.
- Kalelkar, A.S.; Little, Arthur D., "Investigation of Large-Magnitude Incidents: Bhopal as a case study," Chemical Engineers Conference on Preventing Major Chemical Accidents, London, England, May 1988;
- Marathon HF release – UK Health & Safety Executive Hazardous Installations Directorate "Release of Hydrofluoric Acid from Marathon Petroleum Refinery, Texas, USA. 30th October 1987; website links directly to H&SE's Lifting Procedures guidance. http://www.hse.gov.uk/comah/sragtech/casemarathon87.htm"
- Health and Safety Executive, 'The explosion and fires at the Texaco Refinery', Milford Haven, 24 July 1994: A report of the investigation by the Health and Safety Executive into the explosion and fires on the Pembroke Cracking Company Plant at the Texaco Refinery, Milford Haven on 24 July 1994, ISBN 0 7176 1413 1, 1997.
- Total la Mede – "Etude de l'Accident de la Raffinerie "TOTAL" a la Mede", Dr Ir. P. Michaelis, Risk Manager Total Raffinage Distribution, presented at a meeting of GESIP at Tours on 12 June 1996 (in French).
- Conoco Phillips - Health and Safety Executive Public report of the fire and explosion at the Conoco Philips Humber refinery on 16 April 2001"
- Feyzin – (French) Ministère chargé de l'environnement – DPPR/SEI/BARPI – Inspection des installations classées : Septembre 2005
- P 36 Accident analysis, ANC/DPC Inquiry Commission Report, Agência Nacional do Petróleo/Diretoria de Portos e Costas, July 2001. (Brazilian National Petroleum Agency and Directorate of Port and Coasts joint report).
- Esso Longford – Report of the Royal Commission into the accident at Esso Longford.
- Lessons from Longford, Andrew Hopkins, CCH Australia Ltd., ISBN 1 86468 422 4
- The Fires and Explosion at BP Oil (Grangemouth) Refinery Ltd, a report of the investigations by the Health and Safety Executive into the fire and explosion at Grangemouth

and Dalmeny, Scotland, 13 March, 22 March and 11 June 1987, HSE Books 1989, ISBN: 0 1188 5493 3

- The Flixborough Disaster, Report of the Court of Inquiry, HMSO, 1975, ISBN 011 361075 0
- BP Texas City – web site http://www.bp.com/genericarticle.do?categoryId=9005029&contentId=7015905
- CSB Investigation Report No. 2005-04-I-TX, March 2007 "Refinery Explosion And Fire, Bp Texas City, Texas March 23, 2005" from http://www.csb.gov
- Motiva - US Chemical Safety and Hazard Investigation Report – Refinery Incident - Motiva Enterprises LLC, Delaware City Refinery, Delaware City, Delaware, July 17 2001; issue date October 2002.
- Phillips 66 Company Houston Chemical Complex Explosion and Fire, Implications for Safety and Health in the Petrochemical Industry, A Report to the President by US Department of Labor Occupational Safety and Health Administration, April 1990.
- Public Inquiry into the Piper Alpha Disaster, Volumes 1 & 2, November 1990, HMSO Publications Centre ISBN 0-10-113102-X
- Skikda: "The Incident at the Skikda Plant: Description and Preliminary Conclusions," LNG14, Session 1, March 21, 2004, DOHA-Qatar, Sonatrach.

AMMONIUM NITRATE

- Accident on the 21st of September 2001 at a factory belonging to the Grande Paroisse Company in Toulouse, Report Of The General Inspectorate For The Environment Ministry For Regional Development And The Environment (France), 24 October 2001
- Damages of the Toulouse Disaster 21st September 2001, Nicolas Dechy and Yvon Mouilleau, 11th International Symposium Loss Prevention 2004, Prague.

DUST EXPLOSIONS:

- "Courrières 10 mars 1906 : la terrible catastrophe" by Bruno Vouters (2006). Lille: Editions La Voix du Nord. ISBN 2-84393-100-2
- US Chemical Safety Board report 2003-07-I-NC Sept. 2004 (West plant)
- US Chemical Safety Board report 2003-09-I-KY February 2005 (CTA Acoustics plant)
- INERIS report "Explosion d'un silo de céréales, Blaye (33)", July 1998
- "Agricultural Dust Explosions in 1997 in the USA", February 27, 1998, University of Kansas State, Department of Grain.

REACTIVE CHEMICALS:

- "BPS Inc., West Helena Arkansas" EPA/OSHA chemical accident investigation report, EPA 550-R99-003, April 1999.

- "Napp Technologies Inc., Lodi New Jersey" EPA/OSHA chemical accident investigation report, EPA 550-R97-002, October 1997.
- "Morton International Inc.", Chemical Safety Board investigation report 1998-06-I-NJ.
- "The explosion at Concept Sciences: hazards of Hydroxilamine", Chemical Safety Board case study 1999-13-C-PA, March 2002.

WAREHOUSES:
- "Sherwin-Williams Paint Warehouse Fire, Dayton, Ohio (May 27, 1987) – With Supplement on Sandoz Chemical Plant Fire, Basel, Switzerland," United States Fire Administration, Federal Emergency Management Agency, Technical Report Series, 1987.
- "Flammable Liquid Warehouse Fire, Dayton, Ohio, May 27, 1987," NFPA Fire Investigations Report.

LIQUID MISTS AND SPRAYS FLAMMABLE BELOW THE FLASH POINT – THE PROBLEM OF PREVENTATIVE BASES OF SAFETY

Stephen Puttick
Process Hazards Section, Syngenta Huddersfield Manufacturing Centre, PO Box A38, Leeds Road, Huddersfield, HD2 1FF, UK
E-mail: Stephen.puttick@syngenta.com

Preventative bases of safety (BoS) (i.e. absence of flammable atmospheres or avoidance of ignition sources) are the most economic to establish, so there are clear drivers for using them where possible. However, where they are used they must be robust and maintainable. For example solvents can be used below flash point (with a safety margin) to avoid flammable atmospheres or dusts can handled where possible electrostatic discharges are below the MIE. To establish such a BoS the material must be well characterised relative to the possible ignition sources present.

Sprays and mists can be created deliberately in processing (e.g. cleaning vessels; and spraying materials onto substrates) as well as from leaks. A number of workers (e.g. Burgogne & Richardson, 1949) have identified that mists can be flammable below the flash point. A figure often quoted is that mists can be flammable as much as 60K below flash point. However, results from Syngenta's legacy organisation shows sprays to be flammable as much as 125K (Maddison, 1983) below the flash point; with no upper limit having been determined. There may be an upper limit but it will not help us for many materials. To use the BoS absence of viable ignition sources the sensitivity of the mist to ignition must be established in an analogy to dust MIE. Little work has been done in this field particularly for hazard assessment.

A rig has been built to spray materials at ambient temperature and perform ignition tests with pyrotechnic and electrostatic ignition sources. Electrostatic ignition presents some challenges in making reliable measurements: electrodes are wetted by the spray; the presence of droplets between the electrodes can lead to early breakdown and discharge below the desired voltage. Full scale nozzles use large quantities of fluid and ignitions can contaminate this. Characterisation of the spray will be necessary so that measurements are made at a relevant droplet size possibly with a much smaller nozzle. Several fluids have been tested.

FORMATION IN CHEMICAL MANUFACTURE

Mists can be formed accidentally, incidentally or deliberately during manufacturing processes. Mechanical formation can be from release through orifices, impingement, bubble collapse or disengagement from liquid surfaces. Condensation can results from change in temperature or pressure. Operations and activities leading to mist formation can include leaks and loss of containment; material transfer (tail end of blow transfers, splash filling, addition by tail pipes); cleaning (spraying) operations through jets, nozzles or spray-balls; and deliberate processes such as formulation where material may be sprayed on to substrates. This last case is of primary interest in this paper.

HAZARDS OF MISTS

The twofold hazards of mists are documented and fairly well known, and have been associated with a number of incidents (Bright *et al.* 1975; Kletz, 1988 & 1995; Kohlbrand, 1991;Owens & Hazeldean, 1995), some fatal. However, to re-iterate: mists of combustible liquids can be flammable even below flash-point; and the formation of mists is associated with charge separation processes that can lead to an electrostatically charged mist, and ultimately to incendive discharges. This paper will only look at the flammability hazards.

BASIS OF SAFE OPERATION

Chemical plants are (or at least should be operated) with a clearly stated basis for safe operation. By having a principle for safe operation it is possible to define the limits to that principle and consequently necessary precautions to maintain safe operation.

Bases of safe operation come under two categories

- **Preventative** where the basis of safety is designed to prevent the hazardous event
- **Protective** where the hazardous event may initiate, but the basis of safety is designed to protect people and plant from the consequences

When considering fire and explosion hazards, *Preventative Bases of Safety* are implemented by eliminating one element of the fire triangle, either by controlling flammable atmospheres or ignition sources. Controlling flammable atmospheres can be either by the fuel or oxidant concentrations, with appropriate safety factors. Obviously this requires detailed knowledge of the flammable boundaries of the system concerned. To control ignition sources it is vital to understand the sensitivity of the atmosphere to ignition.

As an example for liquids, control of fuel concentration is normally achieved by operating lower than the flash point less a safety margin (usually 5–10 K below the flash-point).

For *Protective Bases of Safety* it is necessary to understand the behaviour of the atmosphere once it has already ignited, particularly concerning flame speed and over-pressure.

MIST FLAMMABILITY

HISTORY OF FIELD

Sprays of liquid fuels have been in use for over 100 years (Williams, 1973) and as such there is a body of knowledge within the combustion literature. The formation of aerosols is an acknowledged way of making fuels of limited volatility easier to ignite for burners and the principle is also used in the auto-motive industry, particularly in diesel engines. Some work is associated with jet fuels and re-ignition (e.g. Ballal & Lefebvre, 1978).

Researchers as early as the 1920s are credited with measurement of mist flammability limits (Burgoyne, 1963), but the bulk of work started to be published after the second World War where some of these papers were directly concerned with safety (Burgoyne & Richardson, 1949; Sullivan *et al.*, 1947).

A certain amount of work has been associated with flame speeds and the possibility that for certain droplet size ranges flame speed may even be enhanced over vapour flame speeds (Polymeropoulos & Das, 1975). Although this is intrinsically interesting, and important for setting Protective Bases of safety, as well as other applications, it is of little relevance to setting Preventative Bases of Safety.

RELATIONSHIP TO MATERIAL FLASH POINT

As a measure of mist flammability the often quoted statistic is that mists can be flammable as much as 60 K below the flash point (Bowen & Cameron, 1999; Kletz, 1995) based on a 1982 report by the HSE. However, work conducted within this group (Maddison, 1983) has shown sprays to be flammable at ambient temperatures as much as 125 K below the material flash-point for solvent systems (see Table 1). Note that this work does not show limits of how far below the flash point a material may be flammable; rather the lower temperatures have been imposed by ambient temperature. Earlier work (Sullivan, 1947), which defined the flammability limits of sprays based on the limiting oxygen requirement, had an example where a fluid was shown to be flammable at ambient temperatures and oxygen concentration more than 200 K below the flash-point.

The interesting question raised here is "*is there a temperature (below the material flash point) at which the mist is no longer flammable?*" This question is difficult to definitively answer since most studies have been conducted at ambient conditions, and often only positive results have been reported. Another difficulty is the literature concerned with spray testing of hydraulic fluids, since this area uses quite high strength ignition sources compared with those that could be present on a chemical plant (e.g. BSI, 1979; Yule and Moodie, 1992). Yule and Moodie discuss how the strength of the propane burner ignition source needs to be high in order to create a stable flame. Ultimately many materials can be made to burn if provided with enough energy. Although for certain types of hazard assessment such ignition sources are important, they are largely irrelevant for much of the chemical industry, and would only lead to excessively conservative evaluations.

Data from Beattie (1988), shown in Table 2, on aqueous solvent mixtures shows that some are flammable a few degrees below the flash point, but at high water fractions they

Table 1. Ignition data from Maddison (1983)

Liquid	Droplet diameter (μm)	Flash point (°C)	Measured LEL (mg/l)	Temperature of mist at ignition (°C)
kerosene	42	47	30	20
tetralin	65	77	45	23
diphenyl ether	38	116	30	23
dimethyl phthalate	70	146	40	20
benzyl bezoate	86	148	60	23

Table 2. Ignition data for aqueous solvent mixes from Beattie (1988)

Solvent	Solvent concentration (% v/v)	Droplet diameter (μm)	Flash point (°C)	Ignition energy (mJ)	Combustion tube entry temperature (°C)
methanol	55	131	22.8	500	20
	50	119	28.3	5000	26
	50	119	28.3	no ignition	20
	25		41	no ignition	34–37
ethanol	50	134	25.6	110	20
	45	127	26.7	500	20
	40	137	27.8	no ignition	20
	25		35	5000	27–28
isopropanol	50		20	5000	17
	30	115	24.4	no ignition	20
	25		27.5	5000	26–27
	10		41	no ignition	33–36
THF	10	111	6	no ignition	20
acetone	7.5	108	22	no ignition	20

cease to be ignitable under the experimental conditions used. It is difficult to tell whether the water is having an inerting effect entirely due to its heat capacity, or whether the droplet sizes are also contributing to the limit, as there may be some mass transfer limitations within the droplets limiting the fuel concentration in the air surrounding the droplets.

The work from Sullivan and co-workers (1947) is one of the most comprehensive in terms of numbers of fluids studied. It was conducted at different oxygen concentrations some of them much above ambient showing that some fluids could only be made flammable at ambient temperature in oxygen enriched atmospheres, despite having measurable flash points.

The amount a mist may be flammable below its flash point will depend on physical properties as well as combustion parameters. Relevant parameters include: heat of combustion; combustion stoichiometry; vapour pressure; latent heat; heat capacity; and droplet size.

FLAMMABLE LIMITS

Since mists can settle out there is (in most cases) no useful upper flammability limit, since at some point it will cross back through the flammable range if left to itself. Although it is worth pointing out that droplets sized below about 20–30 μm will not settle out at any appreciable rate (Bowen & Shirvill, 1994).

For droplets less than approximately 10 μm in size then the mist will burn homogeneously much as vapour with the same flammability limits. Below this size range droplets will evaporate completely in advance of the flame. However, for droplet sizes above about 20 μm then LFL starts dropping and the burning is by individual droplets without complete evaporation. The flames will propagate from droplet to droplet (Burgoyne & Cohen, 1954). Local vapour concentrations may be high surrounding the droplets. Burgoyne (1963) suggested that heat transfer may be dominated by radiation rather than convection (as would be the case in homogeneous vapour combustion). Thermal radiation is absorbed better by the droplets than intervening air. This may explain some aspects of the apparent lower flammable limit, since there is no requirement for all the air, or even all the fuel to reach the same temperature. The other aspect is that the fuel concentration only needs to be high around the droplet. Burgoyne (1963) defined a dynamic concentration based on flame speed. LFL still drops with increasing droplet size, but much less dramatically.

IGNITION SENSITIVITY

The general trend of MIE is towards increasing MIE with increasing droplet size. (E.g. Chan, 1982; Law & Chung, 1980.) Ignition frequency is seen as the most appropriate approach since flammable region is not as clear cut as for vapour systems. However it has been developed for systems where ignition is wanted (such as jet re-ignition). Typically MIE is taken as 50% ignition (Danis *et al.*, 1980). Singh (1986) mapped out for 50% and 20% ignition frequencies for tetralin (unfortunately do not have pure vapour data for comparison).

In hazard assessment for dusts the MIE is mapped out across a range of parameters to be a minimum. However, the location of the minimum is rarely recorded; it is just important that there is one, and it has been determined. Ignition frequency will also be at a very low level, unlike automotive and jet systems which have often been studied. Many of these studies also make little reference to the ignition circuit which may be tuned to give minimum energy sparks, whereas we tend to use pure capacitative sparks as the model for discharges on plant.

TENDENCY TOWARDS AEROSOL FORMATION

Bowen and co-workers (Bowen & Shirvill, 1994; Bowen & Cameron, 1999; Maragkos & Bowen, 2002), and Krishna and others (Krishna *et al.*, 2003; 2003a; 2004) have both looked at the mechanical formation of aerosols through releases. Bowen & Shirvill (1994) reviewed the existing literature and looked at possible break-up mechanisms concluding that aerosol formation and consequent hazards were a real possibility. Krishna (2003) looked at 6 heat transfer fluids from an intrinsic safety point of view where fluid choice might be made on the basis of the fluid's likelihood to atomise. This was related as a dimensionless correlation for each fluid, and can be used to give a relative idea of better or worse. Unfortunately there are no ignition data to give an absolute measure of flammability.

LEAKS AND LOSS OF CONTAINMENT

The primary preventative basis of safety for mists external to process equipment is always containment, supported by normal plant controls over ignition sources (zoning etc.) to reduce the residual risk to acceptable levels. In some cases, such as for possible leaks from heat transfer fluid system flanges, it is possible to mitigate using gauze coalescers, which can trap spray and allow it to drip away in comparative safety (Bowen & Shirvill, 1994).

HOW TO TREAT COMBUSTIBLE MISTS FROM A HAZARD ASSESSMENT AND BASIS OF SAFETY POINT OF VIEW

Mists are most logically treated in a similar manner to combustible dusts i.e. there should be a primary assessment for combustibility or ambient temperature flammability with a relatively large ignition source, rather akin to the group A/B classifications for dusts. Then a secondary assessment for sensitivity to ignition, which would allow appropriate precautions to be defined. A potentially complicating aspect is that a mist is most analogous to a hybrid atmosphere: the droplets can be assumed to be analogous to dust particles; but there will also be some flammable vapour present too. The amount of flammable vapour present will depend on temperature, and the MIE of a mist should be expected to be much more sensitive to changes in temperature than a dust would. The work of Puttick & Gibbon (2004) on solvents in powders developed a criterion where 40 K or more below the flash point and the vapour contribution, hence hybrid behaviour, could be ignored.

Another issue will be droplet size, and droplet size distribution. In a polydisperse mist how will the MIE be affected by the fraction of 'small' droplets? Even so in a confined and un-drafted vessel it is possible that larger droplets will rain out anyway, leaving us with a possible worst case in the 20–30 μm and less range.

It would be interesting to develop a test methodology for mists with similar concepts to those used in the MIKE3 apparatus for dusts. This would require being able to vary spark energies and other discharge characteristics, and concentration ranges. Gibson and Harper (1988) developed an approach to MIE using ignition frequencies for dusts systems, where the flammable boundaries are not hard edged. The frequency approach limits the requirement for large numbers of repeat measurements; it does assume a form to the frequency distribution to allow extrapolation. In dust testing it is usual to sieve samples to less than 63 μm to obtain a reasonable worst case measurement, for a mist it may be necessary to determine droplet size distribution from a nozzle and perform experiments on a relatively mono-disperse mist at the lower end of the distribution.

EXPERIMENTAL WORK

A rig has been built to investigate the issues involved in testing sprays of solvents for MIE. The rig was used with nozzles similar to those used in formulation processes within the company (supplied by Spraying Systems Co). One was a single fluid nozzle (Figure 1), and the other a two fluid system (Figure 2). Both were tested at a range of solvent (and for the

Figure 1. Single fluid nozzle

two fluid nozzle air) pressures. The rig sprayed fluids at ambient temperature, but this could be easily adapted to spray heated fluids into an ambient atmosphere. The nozzles were also tested on a laser particle size analyser using water as the test fluid. The test bed used is normally used for testing agricultural spray nozzles and is not set up to deal with potential flammable hazards. Typical droplet size distributions are shown in Figures 3 and 4.

Two different solvents were used (kerosene and tetralin) with others planned for the future. Ignition tests were with 2 sizes of pyrotechnic igniters (84 and 250 J) and an electrostatic spark circuit.

ISSUES

Tests with the real nozzles use large quantities of solvent which is difficult to re-use since it becomes contaminated with burnt material. This is another driver towards doing tests on a relatively mono-disperse mist possibly generated by a much smaller ultrasonic or piezo-electric nozzle, to give a worst case.

Figure 2. Twin fluid nozzle

The hemi-spherical electrodes in the first spark system tended to bridge with solvent, and so stopped sparking. Another problem was that the mist seemed to encourage breakdown between the electrodes at much lower voltages than in air. This meant that the capacitor discharged at a lower voltage (that could not be accurately determined) than the intended 10,000 V. Hence some discharges (including some incendive discharges) were at lower energy than the nominal circuit energy. These problems meant that a moving electrode system was developed, using spare components from a MIKE 3 dust tester.

RESULTS

Single Fluid Nozzle

Testing the single fluid nozzle on kerosene ignitions were not obtained for 1 or 2 bar solvent pressure with the 84 J igniter, but there were positive results for the higher pressures. 1 bar corresponds to a Sauter Mean diameter (SMD) on water of 338 μm, 2 bar an SMD of 179 μm and 3 bar to 129 μm (or fractions below 100 μm of 3%, 10% and 25% respectively). Using a 250 J igniter and the 1 and 2 bar generated sprays both ignited,

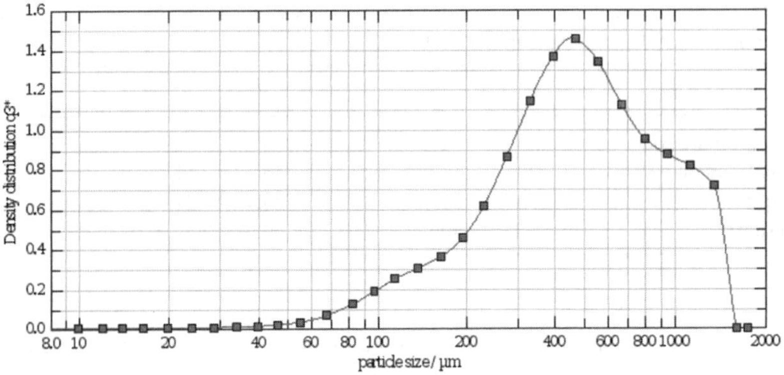

Figure 3. Droplet size distribution for single fluid nozzle at 1 bar using water as a model fluid

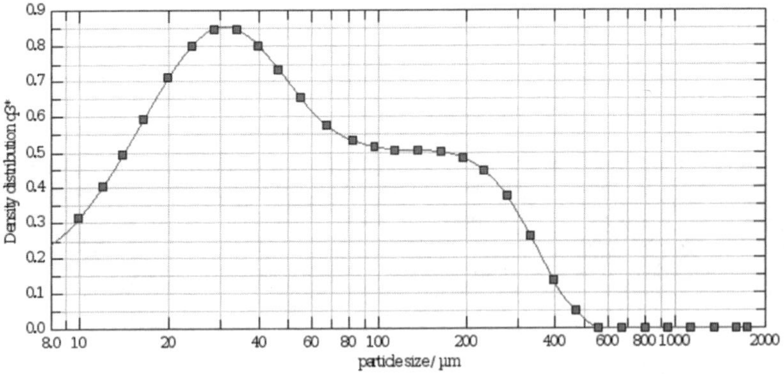

Figure 4. Droplet size distribution for twin fluid nozzle at 3 bar air and 4 bar water pressure using water as a model fluid

although the 1 bar spray was only at 50% frequency. The single fluid nozzle was also tested with the 84 J igniter in a variety of positions across the spray and towards the nozzle. There was some variation in ignition frequencies.

This bears out the observations that ignition energy trends down with droplet size, but that high strength sources can force ignition. The variation in ignition frequency across the spray indicates that there may be some variation in droplet size in different locations in the spray, and that concentration, which can affect ignition energy, probably varies across the spray.

Table 3. Ignition frequencies for kerosene using various settings for twin fluid nozzle with size distribution (water) for comparison (572 mJ spark)

Gas pressure (bar)	3		4			5	
Solvent pressure (bar)	3	4	3	4	5	4	5
Ignition frequency (%)	19	10	20	16	11	7	6
SMD (μm)	27.6	22.5	31.5	25.1	21.9	28.0	24.0
VMD (μm)	53.1	40.5	72.1	49.4	44.2	62.3	53.9

Table 4. Ignition frequencies for tetralin using various settings for twin fluid nozzle with size distribution (water) for comparison (281 mJ spark)

Gas pressure (bar)	3		4	5
Solvent pressure (bar)	4	5	4	3
Ignition frequency (%)	5	8	8	4
SMD(μm)	22.5	19.6	25.1	33.9
VMD(μm)	40.5	36.1	49.4	90.3

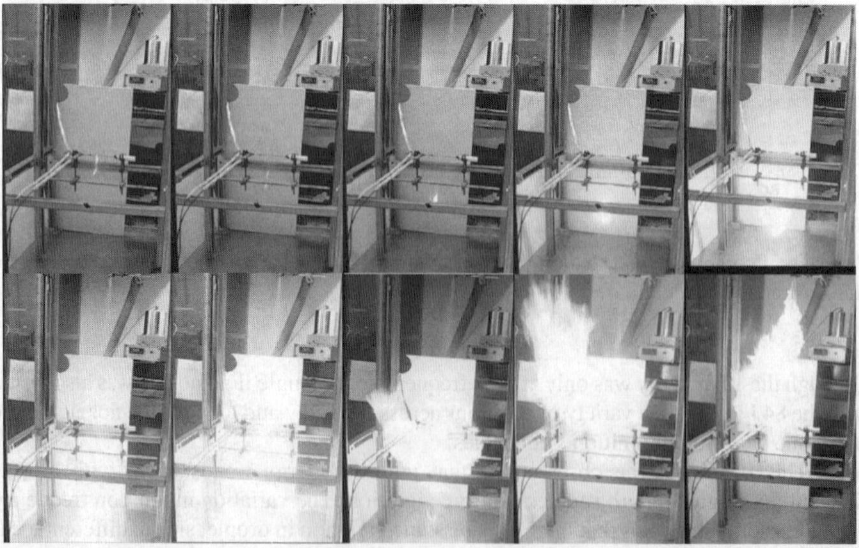

Figure 5. Kerosene ignition sequence by spark using twin fluid nozzle

Twin Fluid Nozzle

The twin fluid nozzle was tested with both kerosene and tetralin and a variety of conditions varying solvent and air pressure both between 1 and 5 bar. Some results are presented in Tables 3 and 4. There is no convincing trend for increasing ignition frequency with decreasing drop size. It is possible that there are simply too few tests to derive a trend, or more likely that issues of mist concentration and mist velocity are affecting the results. Hence a wider number of parameters might need to be measured.

An ignition can be seen in a series of video stills in Figure 5. It shows a small ignition near the spark, which drops and enlarges near the floor. This then spreads back up against the flow of the spray. Some ignitions near the spark did not enlarge and propagate; others enlarged a little, but did not propagate back against the flow.

CONCLUSIONS

There has been some development of a possible test method for the ignitability of sprays of high flash point solvents. Long-term the methodology requires smaller spray volumes to be practical. Some analysis of spray size distribution is also required to decide whether flammability can be dominated by a certain fraction of small droplets.

ACKNOWLEDGMENTS

The newer ignition results were generated by Ander Morelló Mentxaka of Universitat Ramon Llull, Barcelona who undertook a work placement at Syngenta through the ENGINE (Enterprises and New Graduates: International Network) part of the European Leonardo da Vinci scheme. Droplet sizing was carried out by Alan Cochran of Syngenta Jealotts Hill site. Thanks to Mike Bailey and Bob Mullins who gave support to construct the laboratory rig.

REFERENCES

Ballal, D. R. & Lefebvre, A. H. (1978) Ignition of liquid fuel sprays at sub-atmospheric pressures. *Combustion and Flame*, 31:115–126

Beattie, S. R. (1988) Electrostatic hazard during Batch transfers involving solvent. Technical Memo D96046A, ICI, Organics Division

Bowen, P. J. & Cameron, L. R. J. (1999) Hydrocarbon aerosol explosion hazards: A review. *Process Safety and Environmental Protection (Transactions of the Institution of Chemical Engineers, Part B)*, 77(1):22–30

Bowen, P. J. & Shirvill, L. C. (1994) Combustion Hazards Posed by the Pressurised Atomization of High-Flashpoint Liquids. *Journal of Loss Prevention in the Process Industries*, 7(3):233–241

Bright, A. W.; Hughes, J. F. & Makin, B. (1975) Research on electrostatic hazards associated with tank washing in very large crude carriers (Supertankers): I. Introduction and experiment modelling. *Journal of Electrostatics*, 1(1):37–45

Burgoyne, J. H. (1963) The Inflammability of Mists and Sprays. In *Second Symposium of Chemical Process Hazards with Special Reference to Plant Design.* IChemE, 2–4 April

Burgoyne, J. H. & Cohen, L. (1954) The Effect of Drop Size on Flame Propagation in Liquid Aerosols *Proceedings of the Royal Society of London. Series A, Mathematical and Physical Sciences*, 225, 375–39

Burgoyne, J. H. & Richardson, J. F. (1949) The inflammability of oil mists. *Fuel*, 28, 2–6

Cameron, L. R. J. & Bowen, P. J. (2001) Novel cloud chamber design for 'transition range' aerosol combustion studies. *Process Safety and Environmental Protection (Transactions of the Institution of Chemical Engineers, Part B)*, 79:197–205

Chan, K. K. (1982) *Experimental investigation of minimum ignition energy of monodisperse fuel sprays*. PhD thesis, Rutgers University

Danis, A. M.; Namer, I. & Cernansky, N. P. (1988) Droplet Size and Equivalence Ratio Effects on Spark Ignition of Monodisperse N-Heptane and Methanol Spray. *Combustion and Flame*, 74(3):285–294

Gibson, N. & Harper, D. J. (1988) Parameters for assessing electrostatic risk from non-conductors – a discussion. *Journal of Electrostatics*, 21(1):27–36

BSI (1979) Flammability Spray Test for Hydraulic Fluids: DD61. British Standard

Kletz, T. A. (1988) *What Went Wrong? Case Histories of Process Plant Disasters*, page 173. Gulf Publishing Co, 1988

Kletz, T. A. (1995) Some Loss Prevention Case Histories. *Process Safety Progress*, 14(4):271–275

Henry T. Kohlbrand. Case history of a deflagration involving an organic solvent/oxygen system below its flash point. *Plant/Operations Progress*, 10(1):52–54, 1991.

Krishna, K.; Kim, T. K.; Kihm, K. D.; Rogers, W. J. & Mannan, M. S. (2003) Predictive correlations for leaking heat transfer fluid aerosols in air. *Journal of Loss Prevention in the Process Industries*, 16, 1–8

Krishna, K.; Rogers, W. & Mannan, M. (2003) The use of aerosol formation, flammability, and explosion information for heat-transfer fluid selection. *Journal of Hazardous Materials*, 104(1):215–226

Krishna, K.; Rogers, W. & Mannan, M. (2004) Prediction of Aerosol Formation for Safe Utilization of Industrial Fluids. *Chemical Engineering Progress*, 100(7):25–28

Law, C. & Chung, S. (1980) An Ignition Criterion for Droplets in Sprays. *Combustion Science and Technology*, 22:17–26

Madison, N. (1983) Development of method to determine the Flammability Characteristics of Droplet Mist sprays. In ICI report D91044B.

Maragkos, A. & Bowen, P. (2002) Combustion hazards due to impingement of pressurised releases of high flashpoint liquid fuels. *Proceedings of the Combustion Institute*, 29:305–311

Owens, K. A. & Hazeldean, J. A. (1995) Fires, explosions and related incidents at work in 1992–1993. *Journal of Loss Prevention in the Process Industries*, 8(5):291–297

Polymeropoulos, C. E. & Das, S. (1975) The effect of droplet size on the burning velocity of kerosene-air sprays. *Combustion and Flame*, 25, 247–257

Puttick, S. & Gibbon, H. (2004) Solvents in Powder *Hazards XVIII: Process Safety – Sharing Best Practice*, 507–516

Singh, A. K. (1986) *Spark ignition of monodisperse aerosols*. PhD thesis, Rutgers University

Sullivan, M. V.; Wolfe, J. K. & Zisman, W. A. (1947) Flammability of the Higher Boiling Liquids and Their Mists. *Industrial and Engineering Chemistry*, 39, 1607–1614

Williams, A. (1973) Combustion of droplets of liquid fuels: A Review *Combustion and Flame*, 21, 1–31

Yule, A. J. & Moodie, K. (1992) A method for testing the flammability of sprays of hydraulic fluid. *Fire Safety Journal*, 18(3):273–302

MODELLING OF VENTED DUST EXPLOSIONS – EMPIRICAL FOUNDATION AND PROSPECTS FOR FUTURE VALIDATION OF CFD CODES

Trygve Skjold[1], Kees van Wingerden[1], Olav R. Hansen[1], and Rolf K. Eckhoff[2]
[1]GexCon AS; Fantoftvegen 38; P.O. Box 6015 Bergen Bedriftssenter; NO-5892 Bergen
[2]University of Bergen; Department of Physics and Technology; Allegaten 55; NO-5007 Bergen

Explosion venting is the most frequently used method for mitigating the effects from accidental dust explosions in the process industry. Optimal design of vent systems and credible execution of risk assessments in powder handling plants require practical and reliable ways of predicting the course and consequences of vented dust explosions. The main parameters of interest include flame propagation and pressure build-up inside the vented enclosure, the volume engulfed by the flame, and the magnitude of blast waves outside the enclosure. Extensive experimental work forms the empirical foundation for current standards on vent sizing, such as EN 14491 and NFPA 68, and various types of software for vent area calculations simply apply correlations from these standards. Other models aim at a more realistic description of the geometrical boundary conditions, as well as phenomena such as turbulent compressible particle-laden flow and heterogeneous combustion. The latter group include phenomenological tools such as EFFEX, and the CFD code DESC (Dust Explosion Simulation Code). This paper briefly reviews the empirical foundation behind modern guidelines for dust explosion venting, and explores current capabilities and limitations of the CFD code DESC with respect to reproducing results from one experimental study on vented dust explosions. The analysis emphasizes the influence of geometrical features of the enclosures, discrepancies between laboratory test conditions and actual process conditions, and inherent limitations in current modelling capabilities.

KEYWORDS: Dust explosions, explosion venting, deflagration venting, DESC

INTRODUCTION

Dust explosions pose a hazard whenever a sufficient amount of combustible material is present as fine powder, there is a possibility of dispersing the material forming an explosive dust cloud within a relatively confined volume, and there is an ignition source present. The materials involved in dust explosion accidents have evolved with the development of industry. The dust explosion hazard was first recognized in the handling of grain, feed, and flour, as well as in coal mining operations, but accidents can occur with all types of finely divided combustible solids: agricultural products, foodstuffs, pharmaceuticals, chemicals, plastics, rubber, wood, metals, etc. Typical process units involved in industrial powder handling operations include mills, dryers, hoppers, cyclones, filters, chain conveyors, bucket elevators, systems for pneumatic conveying, and storage silos. For practical and occupational health reasons, combustible suspensions of particulate matter are usually contained within closed systems during normal operation, and explosion venting is the conventional method of mitigating the damaging overpressures that could otherwise result from dust explosions in such systems. Blast waves, secondary dust explosions, collapse of buildings, flame burns, projectiles, and subsequent fires are some of the major hazards to personnel and equipment outside a vented enclosure.

Venting guidelines specify the required vent area A_v for an enclosure, for a given set of parameters describing the enclosure, the venting device, possible vent ducts, and properties of the potentially explosive atmosphere inside the enclosure (Table 1). Current standards that contain venting guidelines include VDI 3673 (2002), EN14491 (2006), and

Table 1. Summary of some frequently used parameters for vent area calculations

Symbol	Description
A_f	Effective vent area
A_S	Internal surface area of vessel/enclosure
A_v	Geometric vent area
A_v/V_v	Classical vent ratio (units m^{-1})
$A_v/V_v^{2/3}$	Non-dimensional vent ratio (e.g. Tamanini, 1990)
D	Diameter or equivalent diameter of vessel/enclosure
$(dp/dt)_{ex}$	Maximum constant volume rate of pressure rise at given arbitrary concentration
$(dp/dt)_{max}$	Maximum constant volume rate of pressure rise at optimum concentration
$(dp/dt)_{red}$	Maximum reduced rate of pressure rise in vented enclosure
E_f	Venting efficiency: $E_f = A_f/A_v$
K_{St}	Size corrected maximum rate of pressure rise: $K_{St} = (dp/dt)_{max} V_v^{1/3}$
l_{vd}	Length of vent duct
L	Longest dimension of vessel/enclosure
L/D	Length to diameter ratio for vessel/enclosure
L_F	Flame length
P_{bw}	External overpressure outside enclosure (blast wave)
P_{ds}	Design overpressure, i.e. design strength of enclosure
P_{ex}	Maximum constant volume explosion overpressure at given arbitrary concentration
P_{max}	Maximum constant volume explosion overpressure at optimum concentration
P_{red}	Maximum reduced explosion overpressure in vented enclosure
P_{stat}	Static activation overpressure, i.e. the overpressure required to activate the venting device
t_v	Ignition delay time
V_v	Volume of vessel/enclosure

NFPA 68 (2007). Other standards specify experimental procedures for determining dust specific properties such as P_{max} and K_{St}: EN 14034 (2004; 2006) and ASTM E 1226 (2000). Together, these standards form the methodology for explosion protection by pressure relief venting adopted in Europe and North America. The correlations found in current venting guidelines originate from extensive experimental work, and design according to this methodology provides acceptable levels of safety in most situations. However, their empirical origin limits the extent to which vent area correlations apply to the great variety of process conditions encountered in industrial practice.

Recent developments of more advanced methods for predicting the consequences of industrial dust explosions include both phenomenological tools (Proust, 2005) and methods based on computational fluid dynamics (CFD). With proper modelling of the relevant physical and chemical phenomena involved in dust explosions, the predictive capabilities of such methodologies should extend significantly beyond the limited range of scenarios covered by past and possible future experimental work. However, it is nevertheless necessary to adopt some simplifying assumptions, since detailed modelling of all aspects of dust explosions is currently not within reach for industrial applications. Hence, extensive experimental verification is essential for building confidence in the new methodologies. With the currently limited prospects for funding of further large-scale experimental work on dust explosions, the available validation option is to utilize experimental results obtained in earlier campaigns. Unfortunately, however, it is usually not straightforward to simulate the original experimental conditions. Reliable data for the dusts used are often missing, including chemical composition, particle size distribution, specific heats, heat of combustion, and experimental pressure-time characteristics such as P_{ex} and $(dp/dt)_{ex}$ for the combustible concentration range. Furthermore, generation of explosive dust clouds often involve transient particle-laden flows that are inherently difficult both to measure and to model, and the initial dust distributions and levels of turbulence are therefore significant sources of uncertainty in most cases. Finally, many written sources only report a limited number of explosion characteristics, typically P_{red} and $(dp/dt)_{red}$, omitting important details of the actual pressure development and flame propagation.

The CFD code DESC (Dust Explosion Simulation Code) was developed to simulate industrial dust explosions in complex geometries, but there are still unresolved issues concerning the modelling approach (Skjold, 2007). Some of the main limitations in the current version (DESC 1.0) include inherent shortcomings in available turbulence models suitable for engineering applications, a simplified modelling approach to particle-laden flows (equilibrium mixture assumption), uncertainties concerning the validity of the correlations used to describe turbulent burning velocity, and lack of reliable models for flame quenching phenomena. The validation work has nevertheless produced promising results for certain vented dust explosion scenarios (e.g. Skjold *et al.*, 2005; 2006). The following sections review the empirical foundation behind currently used guidelines for dust explosion venting, and illustrate current capabilities and limitations of the CFD code DESC when it comes to reproducing experimental results obtained for vented dust explosions in a 64 m^3 vented enclosure at various levels of initial turbulence.

VENTING GUIDELINES – THE EMPIRICAL FOUNDATION

Pressure relief by release of combustion products and still unburned dust cloud through vent openings is presumably the oldest method of explosion protection for enclosures such as buildings and process units. Nevertheless, a majority of the early publications in this field focused primarily on preventive rather than mitigation measures, and most guidelines for vent sizing were primarily of qualitative nature. Following a series of disastrous explosions on the British Isles in 1911, one of the recommendations given by Her Majesties Inspector of Factories was simply: *"The roof should be such as to offer little resistance in the event of an explosion"* (Price & Brown, 1922). Since then, venting guidelines have become increasingly quantitative in nature. The following paragraphs review some highlights from the extensive experimental work that forms the empirical foundation for modern venting guidelines, with a view to the applicability of the results for future validation of CFD codes. Some brief comments on the evolution of standards on venting and experimental characterization of dusts are also included.

Some of the first systematic large-scale investigations on the effect of vent size and ignition position on vented dust explosions include the contributions by Greenwald & Wheeler (1925) and Brown & Hanson (1933). Both investigations demonstrated clearly that vent openings positioned close to the point of ignition provide the most effective pressure relief. Wheeler (1935) reported on dust explosion experiments with rice meal in a vertical 37.5 m^3 silo, $L/D = 4$, illustrating the pronounced influence of the vent area on P_{red}: fully open cylinder produced barely measurable overpressures, 2/3 open 0.03-0.04 bar, 1/3 open 0.3-0.4 bar, and 1/9 open in excess of 1 bar.

Hartmann (1954) presented results from a 1 ft^3 test gallery, demonstrating how the effect of vent ratio on P_{red} differ for various types of dust. Hartmann & Nagy (1957) emphasized that *"results from relatively small explosion chambers can be useful for protecting equipment and also for larger commercial structures"*; this statement was supported by experiments in cubical galleries having volumes 1, 64, and 216 ft^3, with and without vent ducts. The first standardized test vessels for measuring the explosion-pressure characteristics of dusts, i.e. P_{max} and $(dp/dt)_{max}$, were closed cylindrical vessels of relatively small volume: a 1.2 litre bomb introduced by Hartmann (Dorsett *et al.*, 1960), and a similar 1.0 litre bomb developed in England (Raftery, 1968). These tests have later been replaced with standardized tests in 20-litre explosion vessels.

Early quantitative guidelines for the calculation of vent areas relied on the classical vent ratio A_v/V_v. A typical example is the preliminary guidelines provided by NFPA (1946): *"For mild explosion hazards – 1 ft² for each 100 ft³; for moderate explosion hazards – 1 ft² for each 50 ft³; for severe explosion hazards – 1 ft² for each 15 ft³; for extreme explosion hazards – maximum venting area obtainable"*. Palmer (1971) pointed out that the vent ratios method was limited to compact enclosures, i.e. enclosures having all three dimensions of the same order, and structures capable of withstanding pressures up to about 0.14 bar. Palmer mentioned the following guidelines for vent ratios, based on maximum rates of pressure rise determined in a 1.0 litre bomb: 1/6 m^{-1} (1/20 ft^{-1}) for $(dp/dt)_{max}$ less than 350 bar s^{-1}; 1/5 m^{-1} (1/15 ft^{-1}) for $(dp/dt)_{max}$ in the range 350 to 700 bar s^{-1}, and 1/3 m^{-1} (1/10 ft^{-1}) for $(dp/dt)_{max}$ exceeding 700 bar s^{-1}. Since A_v/V_v has dimensions $(length)^{-1}$, estimates for very large enclosures yield unnecessarily large vent openings (Eckhoff, 2003). The American *'Standard on explosion protection by deflagration venting'* was first released as a temporary standard in 1945, and then replaced with a *'Guide for Explosion Venting'* in 1954, before major revisions followed in 1974, 1978, 1988, 1994, 1998, 2002, and 2007 (NFPA, 2007). The 1954 edition of NFPA 68 provided vent area recommendations based on the size and bursting strength of the enclosure.

The currently used venting guidelines in Europe originate from the extensive amount of experimental work reported by Donat (1971) and Bartknecht (1971; 1974ab), as well as the theoretical analysis by Heinrich & Kowall (1971). These contributions established the concepts of the K_{St} value and the cube-root-law: under given assumptions, the K_{St} value is a material specific constant for a given particle size distribution and a certain level of turbulence in the dust clouds at the time of ignition (Bartknecht, 1986). Unfortunately, it is practically impossible to achieve experimental conditions fulfilling the underlying assumptions behind the cube-root-law, but the overall concept is nevertheless valuable for practical applications. This work culminated in the first venting guidelines based on nomographs from Verein Deutscher Ingenieure (VDI) in 1979. Vent area correlations were introduced later (Siwek, 1994). The current European standard *Dust explosion venting protective systems* (EN 14491, 2006) is based on the VDI guidelines (Moore & Siwek, 2002).

The monumental experimental contribution from Bartknecht in the field of dust explosion safety comprises a vast number of experiments performed in vessels covering a wide range of scales and shapes (e.g. Bartknecht, 1971; 1986; 1993). In 1966 he introduced the standard 1-m^3 ISO vessel for determining reference values of P_{max} and $(dp/dt)_{max}$ (Bartknecht, 1971). A pneumatic dispersion system produced the dust cloud, and ignition after a specified time delay secured reproducible levels of turbulence. In the 1-m^3 vessel, dust is injected from a 5 litre container pressurized to 20 bar*g*, and the resulting dust cloud is ignited by two 5 kJ chemical igniters after 0.6 s. The larger enclosures were fitted with similar dispersion systems, but the number and volume of pressurized dust containers were chosen to achieve satisfactory distribution of the dust and sufficiently high levels of turbulent in the flow prior to ignition. Siwek (1977; 1988) demonstrated good agreement between results obtained in the 1-m^3 ISO vessel and a 20-litre spherical vessel. Several researchers have studied the dispersion induced flow and transient combustion phenomena in this 20-litre vessel, including Pu *et al.* (1990; 2007) and Dahoe *et al.* (1996; 2001abc), and the Siwek sphere is currently used for most of the experimental characterization of industrial dust samples. However, results presented by Proust *et al.* (2007) show that there can be significant differences between results obtained in the Siwek sphere and the standard ISO vessel. Regarding the validation of CFD codes, there are several challenges associated with the use of data from this type of experiments: the dust dispersion process involves transient turbulent particle-laden flow, and the turbulent combustion process takes place in a flow field characterized by rapidly decaying turbulence. The next section illustrates some of the challenges associated with such validation work.

Palmer (1975/76) summarised the status on dust explosion venting, emphasizing the need to strengthen the theoretical foundation. Other relevant contributions from this period include the ones by Rust (1979), Field (1984), Lunn (1989), and Siwek (1994). Several researchers presented experimental work where the dust was introduced into the enclosure by constant rate pneumatic conveying, e.g. Siwek (1989), Eckhoff *et al.* (1987), and Hauert *et al.* (1996). Such scenarios are more straightforward to model with modern CFD codes, compared to experiments with transient dust injection from a pressurized container (Skjold *et al.*, 2005; 2006). Eckhoff (1986, 1990) emphasized the need for a differentiated approach to vent sizing based on a risk assessments. Tamanini & Chaffee (1989) and Tamanini (1990) investigated the effect of turbulence on explosion severity in a vented 64-m^3 enclosure, and the next section illustrates the process of simulating these experiments with the CFD code DESC.

DECS SIMULATIONS – DUST EXPLOSIONS IN VENTED ENCLOSURE

Experiments reported by Tamanini (1990) and Tamanini & Chaffee (1989) demonstrate that the initial turbulent flow conditions influence the reduced overpressure from dust explosions in vented enclosures. The dusts used in the experiments were either maize starch or a blend of bituminous coal and carbon. Results from these tests inspired the vent design requirements for turbulent operating conditions included in the 2007 edition of the NFPA 68 guidelines (Zalosh, 2007). This section describes CFD modelling with DESC of the tests with maize starch, nominal dust concentration 250 g m^{-3}, vent ratio $A_v/V_v^{2/3} = 0.35$, ignition by a 5 kJ chemical igniter in the centre of the enclosure, and ignition delays in the range 0.5–1.1 s; these conditions cover seven of the totality of 21 original experimental tests. Results from experiments and simulations are compared, and the discussion focuses on the key assumptions, and hence the inevitable uncertainties, inherent in this type of validation work.

The experiments were conducted in a 64 m^3 vented enclosure with dimensions 4.6 × 4.4 × 3.0 m, design pressure 0.7 bar*g*, and an open 2.4 × 2.4 m vent door in one wall. Figure 1 illustrates the implemented geometry and computational grid used in the simulations. The grid inside the enclosure consisted of 0.1 m cubical grid cells for the dispersion simulation (cell size dictated by the maximum pseudo diameter of the transient release), and 0.1 or 0.2 m cubical grid cells for the explosion simulations (to illustrate the effect of grid resolution on the simulation results).

The explosion overpressures and K_{St} values for the maize starch used in the experiments were 6.1 bar and 144 bar m s^{-1} for a nominal dust concentration of 250 g m^{-3}, and 7.4 bar and 178 bar m s^{-1} for the optimum concentration of 800 g m^{-3}; the corresponding values for the sample used to generate the empirical combustion model for maize starch in DESC were 6.3 bar and 75 bar m s^{-1} at 250 g m^{-3}, and 8.6 bar and 150 bar m s^{-1} at 800 g m^{-3}. To compensate for the lower reactivity of the model sample, especially at 250 g m^{-3}, the estimated laminar burning velocity for the model dust was multiplied by a factor 1.75. As for most studies of this type, the reactivity of the model dust is a major source of uncertainty.

Tamanini and co-workers used a pneumatic injection system for dispersing the dust inside the enclosure: operation of fast acting valves discharged four 0.33 m^3 air tanks, initially charged to 8.3 bar overpressure, and the resulting airflow entrained and dispersed

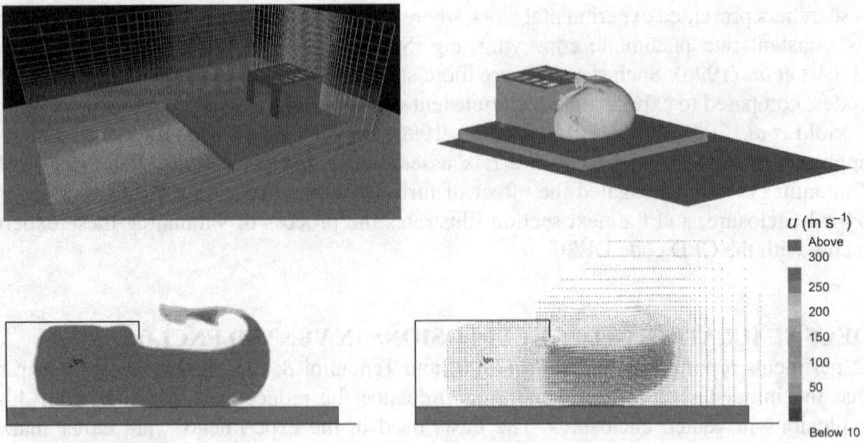

Figure 1. Geometry in DESC (above): grid (left) and flame ball exiting from the vent opening (right); the ten smaller vent openings on the top of the enclosure were blocked in both experiments and simulations. Cross sections of a part of the calculation domain (below), illustrating flame propagation visualised as mass fraction of combustion products (left) and velocity vectors (right)

dust from separate dust canisters through four perforated nozzles (i.e. 16 nozzles in total). The discharge stopped when the overpressure in the air tanks reached 1.4 bars. Several bi-directional velocity probes measured instantaneous velocities at various positions inside the enclosure during the injection process. Average and root-mean-square (RMS) turbulent velocities derived from these measurements indicated a high degree of non-uniformity of the flow field. In view of the importance of large-scale flow structures in the flow, *'the RMS of the instantaneous velocity was judged to be a more appropriate quantity to characterize the intensity of the turbulence inside the chamber'* (Tamanini, 1990). The RMS of the turbulent velocity fluctuations was roughly a constant fraction (60%) of the RMS of the instantaneous velocity fluctuations. The reported expected accuracy of the velocity measurements was 10-20% for velocity fluctuations with frequencies up to 400–500 Hz.

The DESC simulations imitate the actual dispersion process by introducing 16 transient leaks, releasing the same total amount of dust and air as in the experiments, from the same positions as the 16 original nozzles. It was not possible to resolve geometrical details of the perforated nozzles on the computational grid used here, but porous panels placed a few grid cells downstream of the leaks produced some spread in the flow from the nozzles. Figure 2 shows a comparison between experimental (with and without dust) and simulated turbulence intensities near the centre of the enclosure. A 0.1 s time shift in the simulated data accounts for the delay in opening the fast acting valves and charging the line between

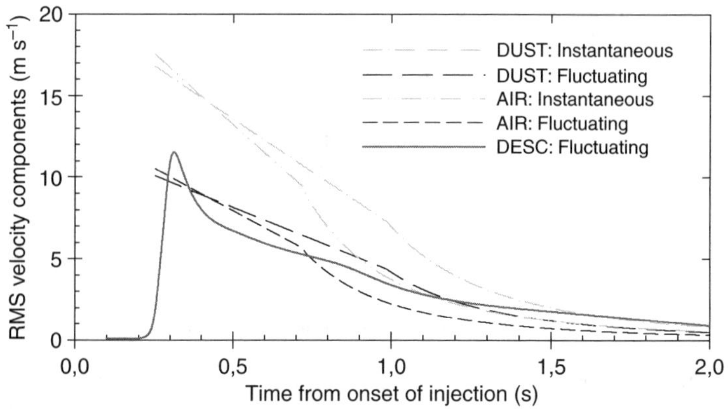

Figure 2. Measured instantaneous and estimated fluctuating velocity components, with and without dust, during the injection process in the 64 m^3 vented enclosure, and simulated fluctuating component (DESC)

the air tanks and the dispersion nozzles. Note that only the RMS of the fluctuating velocity components is relevant for comparison with the simulated values. Although the build-up time of the turbulent flow field is too long in the simulations, the simulated results are in reasonable agreement with measured values from about 0.3 s and onwards. The uncertainties in both measured and simulated values are nevertheless considerable. Turbulence measurements obtained with intrusive methods, such as bi-directional velocity probes, are not optimal, and it is generally very difficult to calculate the production of turbulence during the transient outflow of air through that takes place when a valve from a high-pressure tank opens quickly. Turbulence production during the initial phase of dispersion is probably largely due to the baroclinic term in the vorticity equation (Dahoe, 2001c), and current turbulence models for engineering applications, including the k-ε model used in DESC, are not able to accurately reproduce this phenomenon.

Figure 3 shows the measured increase in P_{red} for higher values of the average RMS of the instantaneous velocity. These results influenced resent modifications regarding turbulent flow conditions in the 2007 edition of NFPA 68 (Zalosh, 2007). The same Figure also shows results for two tests with other ignition sources (tests 21 and 22), one test with a smaller vent opening (test 10), and one test with a higher nominal dust concentration (test 6); stars indicate the experimental pressure traces that contained a distinct double peak (tests 3, 7 and 21).

Figure 4 illustrates the effect of ignition delay time and estimated RMS of the fluctuating velocity component (at the time pressures reach 0.8 P_{red}) on P_{red} and the average rate of pressure rise $(\Delta p/\Delta t)_{av}$ taken from 0.2 P_{red} to 0.8 P_{red}. Ongoing injection of dust at the

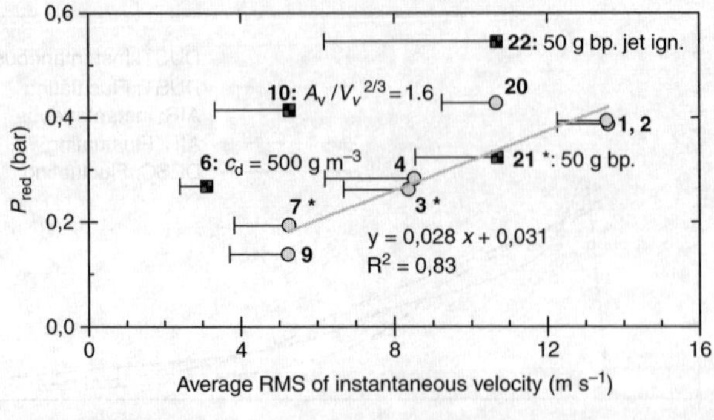

Figure 3. Effect of turbulence intensity at nominal time of ignition on P_{red} in 64 m^3 vented enclosure; original test numbers indicated for each data point, stars indicate double pressure peaks; error bars indicate RMS velocity at the time pressures reach 0.8 P_{red}; results from four tests with deviating experimental conditions included (bp. = black powder)

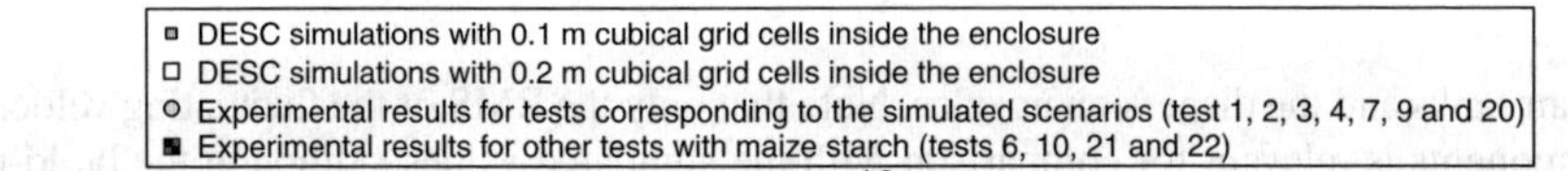

Figure 4. Experimental and simulated P_{red} and $(\Delta p/\Delta t)_{av}$ in the 64 m^3 vented enclosure for various ignition delay times and estimated RMS of the fluctuating velocity component at the time the pressure reaches 0.8 P_{red}

time of ignition results in the slight decrease in both P_{red} and $(\Delta p/\Delta t)_{av}$ for the shortest ignition delays and the highest turbulence intensities. The simulated values of both P_{red} and $(\Delta p/\Delta t)_{av}$ are too low at the shortest ignition delays, compared to experimental values, most likely due to the inability of the turbulence model in reproducing the high initial rates of turbulence production. For longer ignition delays, the experimental and simulated results are in better agreement. However, Figure 4 also reveals a significant effect of the computational grid on the simulation results. The flame model in the current version of DESC yields a flame that is three grid cells thick, and one reason for the higher rates of energy release inside the enclosure for the smallest grid cells is the fact that a finer grid resolution results in a thinner flame and a larger flame area. In future versions of DESC, local properties of the flow field, such as the RMS of the turbulent velocity fluctuations and the turbulent integral length scale, as well as dust specific parameters such as laminar burning velocity and laminar flame thickness, should determine both the turbulent burning velocity and the turbulent flame thickness. Turbulent combustion in dust clouds exhibits a high degree of volumetric combustion, and a fuel and flow dependent flame thickness that accounts for this phenomenon should produce less grid dependent results. The difference in rates of combustion between a coarse and a fine grid is enhanced for scenarios where ignition takes place in a flow field characterized by rapidly decaying turbulence. This effect is due to a too low rate of combustion during the initial phase of combustion, where a subgrid model governs the rate of growth of the flame ball up to a flame radius of about three grid cells. On a finer grid, it takes shorter time for the initial flame ball to reach a size where the subgrid model no longer governs the further rate of flame growth. Hence, due to the rapid decay of turbulence, the flame on the finer grid propagates through a flow field characterized by somewhat higher turbulence intensity, as compared to the flame on the coarser grid.

Enclosures with moderate L/D ratios yield the most pronounced effect of dispersion-generated turbulence on P_{red}. For enclosures with larger L/D, the influence of ignition position and explosion-generated turbulence dominates (e.g. Eckhoff, 1992).

CONCLUSIONS

The experimental work invested in developing, validating, and improving guidelines for explosion protection by venting represents a vast amount of information on flame propagation in dust clouds. Hence, in principle there is no doubt that such data represents a unique possibility for validating modern CFD codes. However, modelling of typical large-scale dust explosion experiments is a challenging task, not only because of the inherent complexity of particle-laden flows and turbulent combustion, but also due to the transient nature of the experimental procedures often adopted in this type of dust explosion research. Furthermore, the level of details included in descriptions of original experimental equipment, procedures, and results varies significantly, and some sources report only selected variables such as the maximum reduced explosion pressure and the maximum reduced rate of pressure rise. Future validation work for CFD codes is nevertheless likely to benefit significantly from previous experimental work, and it seems inevitable that our understanding of dust explosions will increase with improved modelling capabilities in the future.

REFERENCES

ASTM E 1226, 2000, *Standard test method for pressure and rate of pressure rise for combustible dusts*, ASTM International, PA, March 2000.

Bartknecht, W., 1971, *Brenngas- und Staubexplosionen*, Forschungsbericht F45, Bundesinstitut für Arbeitsschutz, Koblenz.

Bartknecht, W., 1974a, Bericht über Untersuchungen zur Frage der Explosionsdruckenlastung brennbarer Stäube in Behältern: Teil I, *Staub Reinhaltung der Luft*, 34: 381–391.

Bartknecht, W., 1974b, Bericht über Untersuchungen zur Frage der Explosionsdruckenlastung brennbarer Stäube in Behältern: Teil II, *Staub Reinhaltung der Luft*, 34: 456–459.

Bartknecht, W., 1986, Pressure venting of dust explosions in large vessels, *Plant/Operation Progress*, 5: 196–204.

Bartknecht, W., 1993, *Explosionsschutz – Grundlagen und Anwendun*, Springer Verlag, Berlin.

Brown, H.R. & Hanson, R.L., 1933, Venting Dust Explosions, *NFPA Quarterly*, 26: 328–341.

Dahoe, A.E., Cant, R.S. & Scarlett, B., 2001a, On the decay of turbulence in the 20–litre explosion sphere, *Flow, Turbulence and Combustion*, 67: 159–184.

Dahoe, A.E., Cant, R.S., Pegg, R.S. & Scarlett, B., 2001b, On the transient flow in the 20–litre explosion sphere, *Journal of Loss Prevention in the Process Industries*, 14: 475–487.

Dahoe, A.E., van der Nat, K., Braithwaite, M. & Scarlett, B., 2001c, On the sensitivity of the maximum explosion pressure of a dust deflagration to turbulence, *KONA*, 19: 178–195.

Dahoe, A.E., Zevenbergen, J.F., Lemkowitz, S.M. & Scarlett, B., 1996, Dust explosions in spherical vessels: the role of flame thickness in the validity of the 'cube-root law', *Journal of Loss Prevention in the Process Industries*, 9: 33–44.

Donat, C., 1971, Auswahl und Bemessung von Druckentlastungseinrichtungen für Staubexplosionen, *Staub Reinhaltung der Luft*, 31: 154–160.

Dorsett, H.G., Jacobson, M., Nagy, J. & Williams, R.P., 1960, *Laboratory equipment and test procedures for evaluating explosibility of dusts,* Report of investigation 5624, US Bureau of Mines.

Eckhoff, R.K., 1986, Sizing dust explosion vents – the need for a new approach based on risk assessment, *Bulk Solids Handling*, 6.

Eckhoff, R.K., 1990, Sizing of dust explosion vents in the process industries – advances made during the 1980s, *Journal of Loss Prevention in the Process Industries,* 3: 268–279.

Eckhoff, R.K., 1992, Influence of initial and explosion-induced turbulence on dust explosions in closed and vented vessels, *Powder Technology*, 71: 181–187.

Eckhoff, R.K., 2003, *Dust explosions in the process industries*, 3rd ed., Gulf Professional Publishing, Amsterdam.

Eckhoff, R.K., Fuhre, K. & Pedersen, G.H., 1987, Dust explosion experiments in a vented 236 m^3 silo cell, *Journal of Occupational Accidents*, 9: 161–175.

EN 14034-1, 2004, *Determination of explosion characteristics of dust clouds – Part 1: Determination of the maximum explosion pressure* p_{max} *of dust clouds*, CEN, Brussels, September 2004.

EN 14034-2, 2006, *Determination of explosion characteristics of dust clouds – Part 2: Determination of the maximum rate of explosion pressure rise* $(dp/dt)_{max}$ *of dust clouds*, CEN, Brussels, May 2006.

EN 14491, 2006, *Dust explosion venting protective systems*, CEN, Brussels, March 2006.

Field, P., 1984, Dust explosion protection: a comparative study of selected methods for sizing explosion relief vents, *Journal of Hazardous Materials*, 8: 223–238.

Greenwald, H.P. & Wheeler, R.V., 1925, *Coal dust explosions: the effect of release of pressure on their development*, Safety in Mines Research Board Paper No. 14: 3–12.

Hartmann, I., & Nagy, J., 1957, Venting dust explosions, *Industrial and engineering chemistry*, 49: 1734–1740.

Hartmann, I., 1954, Dust explosions in coal mines and industry, *The scientific monthly*, 79: 97–108.

Hauert, F., Vogl, A. & Radant, S., 1996, Dust cloud characterization and the influence on the pressure-time histories in silos, *Process Safety Progress*, 15: 178–184.

Heinrich, H.-J. & Kowall, R., 1971, Ergebnisse neuerer Untersuchunhungen zur Druckenlastung bei Staubexplosionen, *Staub Reinhaltung der Luft*, 31: 149–153.

Lunn, G.A., 1989, Methods for sizing explosion vent areas: a comparison when reduced explosion pressures are low, *Journal of Loss Prevention in the Process Industries*, 2: 200–208.

Moore, P.E. & Siwek, R., 2002, An update on the European explosion suppression and explosion venting standards, *Process Safety Progress*, 21: 74–84.

NFPA 68, 2007, *Standard on explosion protection by deflagration venting – 2007 edition*, NFPA, Quincy MA.

NFPA, 1946, *National Fire Codes, Vol. II, The prevention of dust explosions*, NFPA, Boston.

Palmer, K.N., 1971, The relief venting of dust explosions in process plant, *IChemE Symposium Series*, 34: 142–147.

Palmer, K.N., 1975/76, Recent advances in the relief venting of dust explosions, *Journal of Hazardous Materials*, 1: 97–111.

Price, D.J. & Brown, H.H., 1922, *Dust explosions – theory, and nature of, phenomena, causes and methods of prevention*, NFPA, Boston.

Proust, Ch., 2005, The usefulness of phenomenological tools to simulate the consequences of dust explosions – the experience of EFFEX, In *International European Safety Management Group (ESMG) Symposium*, October 11–13 2005, Nürnberg, Germany, 17 pp.

Proust, Ch., Accorsi, A. & Dupont, L., 2007, Measuring the violence of dust explosions with the "20 l sphere" and with the standard "ISO 1 m^3 vessel": Systematic comparison and analysis of the discrepancies, *Journal of Loss Prevention in the Process Industries*, 20: 599–606.

Pu, Y.K., Jarosinski, J., Johnson, V.G. & Kauffman, C.W., 1990, Turbulence effects on dust explosions in the 20–liter spherical vessel, *Twenty-third Symposium (Int.) on Combustion*: 843–849.

Pu, Y.K., Jia, F., Wang, S.F. & Skjold, T., 2007, Determination of the maximum effective burning velocity of dust-air mixtures in constant volume combustion, *Journal of Loss Prevention in the Process Industries*, 20: 462–469.

Raftery, M.M., 1968, *Explosibility tests for industrial dusts*, Fire research technical paper, no. 21.

Rust, E.A., 1979, Explosion venting for low-pressure equipment, *Chemical Engineering*, November 5: 102–110.

Siwek, R., 1977, *20-L Laborapparatur für die Bestimmung der Explosionskenngrößen brennbarer Stäube*, Thesis, HTL Winterthur, Switzerland.

Siwek, R., 1988, Reliable determination of safety characteristics in the 20–litre apparatus, In *Conference on Flammable Dust Explosions*, November 2–4, St. Louis.

Siwek, R., 1989, Dust explosion venting for dusts pneumatically conveyed into vessels, *Plant/Operation Progress*, 8: 126–140.

Siwek, R., 1994, New revised VDI guideline 3673 "Pressure release of dust explosions", *Process Safety Progress*, 13: 190–201.

Skjold, T, Arntzen, B.J., Hansen, O.R., Taraldset, O.J., Storvik, I.E. & Eckhoff, R.K., 2005, Simulating dust explosions with the first version of DESC, *Process Safety and Environmental Protection*, 83: 151–160.

Skjold, T., 2007, Review of the DESC project, *Journal of Loss Prevention in the Process Industries*, 20: 291–302.

Skjold, T., Arntzen, B.J., Hansen, O.J., Storvik, I.E. & Eckhoff, R.K., 2006, Simulation of dust explosions in complex geometries with experimental input from standardized tests, *Journal of Loss Prevention in the Process Industries*, 19: 210–217.

Tamanini, F. & Chaffee, J.L., 1989, *Dust explosion research program report no. 3: Large-scale vented dust explosions – effect of turbulence on explosion severity*, Factory Mutual Research Corporation, Technical Report FMRC J.I. 0Q2E2.RK, April 1989.

Tamanini, F., 1990, Turbulence effects on dust explosion venting, *Plant/Operations Progress*, 9: 52–60.

VDI 3673, 2002, *Guideline VDI 3673 Part 1: Pressure venting of dust explosions*, VDI, November 2002.

Wheeler, R.V., 1935, *Report on experiments into the means of preventing the spread of explosions of carbonaceous dust*, His Majesty's Stationery Office, London.

Zalosh, R., 2007, New dust explosion venting requirements for turbulent operating conditions, *Journal of Loss Prevention in the Process Industries*, 20: 530–535.

BRINGING RISK ASSESSMENTS TO LIFE BY INTEGRATING WITH PROCESS MAPS

Gordon Sellers[1], Alan Webb[2] and Chris Thornton[3]
[1]Safety Management Consultant, 34 Westbury Road, Northwood HA6 3BX, UK.
E-mail: gordon.sellers@dsl.pipex.com
[2]Poplars Landfill General Manager, Biffa Waste Services Ltd
[3]Milton Keynes Depot Manager, Biffa Waste Services Ltd

As one of the leading integrated waste management businesses in the UK and operating from over 160 locations nationwide, Biffa has many thousands of task-specific risk assessments. These are used by managers as input to local operating procedures and meet statutory requirements.

Following a tragic accident in 2006, Biffa decided to develop process maps for all its locations, focussing on vehicle-pedestrian interactions. The methodology was embryonic, however most managers found that their process maps gave a much better overview of operations than their existing risk assessments, which are not very 'user-friendly' and are slow to search.

Therefore Biffa initiated a project to integrate the best features of risk assessments and process maps. The resulting 'Process & Risk Assessments' (PRAs) used Excel spreadsheets, which are widely used throughout the company, and included thumbnail photographs to make them more readable. Following development of the methodology at one location by a safety professional, six location managers 'volunteered' to develop pilot PRAs for their own locations. The pilot site results were reviewed and found to be a significant improvement on the existing separate risk assessments and process maps.

Phase 2 of the Process & Risk Assessment project is underway at the time of writing this paper, and is developing standard best practice modules which each location will then use to indicate where physical constraints make it impossible to implement the full best practice – and therefore what additional control measures have been put in place. By the time of Hazards XX, we expect to be able to report the results of phase 2.

EXISTING RISK ASSESSMENTS

Biffa Waste Services employs more than 5,000 employees, operates over 1,500 vehicles and 160 operating locations including collection depots, recycling facilities and landfill sites. In the early 1990s, three-page risk assessments were introduced to assist managers to control the risks in their operations and to meet the company's statutory requirements under the Management of Health and Safety Regulations, subsequently an online one-page version was introduced to simplify the process. A typical location has 80 to 180 task-specific risk assessments depending on the complexity of its operations and, in total, there are around 30,000 risk assessments on the company intranet. Each responsible manager is required to review and update his or her risk assessments at regular intervals, normally annually.

PROCESS MAPPING – FIRST STEPS

Following a tragic fatal incident in 2006, Biffa's Chief Executive decided that the company should develop process maps for activities at its locations. These Excel-based process maps included risk assessments and, where appropriate, proposals for actions to reduce risk. The initial focus was on vehicle-pedestrian interactions but the scope was later widened to all activities at each location. The methodology was embryonic and was used in some haste, however the resulting process maps were generally felt to give a more complete risk picture than existing task-based risk assessments and to have led to worthwhile improvements.

DEVELOPING INTEGRATED PROCESS MAPS AND RISK ASSESSMENTS

Biffa now had two parallel systems of risk assessment:

- The 30,000+ existing 'text & tables' risk assessments on the company intranet.
- Around 150 Excel files each containing 5-10 process maps, duplicating some of the information from the existing risk assessments, uploaded onto the company intranet.

This was obviously unsustainable in the medium term so the Board decided to investigate merging the two systems into one. A project team was set up comprising a health & safety project manager, along with representatives from the operating divisions. We first reviewed the existing risk assessment process:

✓ Linked into the company's system for tracking actions and review dates (known as the 'Compliance Database').
✓ Well over 500 generic risk assessments have been developed for common activities across the company and each location is required to adapt the relevant ones to their own situation, but …
× … there is no simple way to ascertain if a specific location has implemented 'best practice'.
× Slow to download from the company intranet so it is a time-consuming task for managers to review and update them.
× No links between the task-specific risk assessments so it is difficult to check that no significant risks have been overlooked 'in the gaps', especially interactions between separate activities in adjacent areas.
× Output is not 'user-friendly' and is therefore unsuitable for use as briefing documents or toolbox talks for the majority of the workforce.

Similarly we reviewed the process maps:

✓ Visually show links between different activities so easier to review for any gaps and to discuss with operatives whether or not they accurately represent the activities carried out in practice.
× Not standardised across the company so no facility to promote best practice.
× Not linked into the Compliance Database so no automatic tracking of actions and review dates.

Therefore we set out to develop a system of Process & Risk Assessments to combine the best features of risk assessments and process maps. If successful, this would then replace both systems.

We developed the first draft based on a location which has three business units–an industrial/commercial collection depot (effectively a heavy vehicle park), a heavy vehicle maintenance workshop, and a transfer station (here Biffa vehicles and other customers from the local area tip their loads, from which recyclables are reclaimed and the residual waste is loaded into 44 tonne articulated trucks for despatch to landfill). During this period, Biffa had organised a series of IOSH 'Managing Safely' courses so that was an excellent opportunity to present the early drafts and get very useful comments from the managers who were attending, also the drafts were reviewed with Biffa's health and safety professionals. As a result of these consultations, we made significant improvements to the drafts, probably the most significant being to include thumbnail photographs.

From this first draft, we developed a template process & risk assessment which was issued for pilot implementations to be carried out by 'volunteer' managers at six representative locations – two industrial/commercial collection depots with vehicle workshops, one of which included a secure waste recycling facility; a municipal collection depot and workshop; a major landfill with composting and a municipal depot; a special waste treatment plant and transfer station; and an integrated waste management facility operated by Biffa at a customer site.

At the end of the pilot implementations, we held a review meeting at which the six location managers presented their findings to the responsible operating directors. They reported both the positives and negatives:

- ✓ Easy to follow and understand
- ✓ Logical process that makes you think of everything
- ✓ Pictures aid discussion with team – and involve them
- ✓ Comprehensive overview of activities on site, great for training
- ✓ Customer positive
- ✓ Improved understanding of ancillary plant operations
- ✓ HSE was positive about the approach
- ✓ Only ~ ½ day required to tailor the template at one similar location
- ✓ Key Safe Behaviours emphasise need for reinforcement
- ✓ Minimal IT training needs as all managers familiar with Excel
- × Excel file too big and cumbersome to edit and eMail (largely because the photograph files had not been reduced in size)
- × Significant time needed to develop initially
- × Not structured to print out relevant sub-sections e.g. for project pack or for contractor working in small area of site
- × Without an index, not easy to find way around
- × Repetitive e.g. PPE, slips & trips come up in many tasks
- × Concern that the format might not be acceptable to lawyers handling an injury claim
- × Not as good for a special waste treatment plant as the 'complex risk assessments' which had been developed locally

Following the presentations and discussions, the operating directors recommended two important improvements:

- To avoid different practices at different locations with similar operations (partly due to historical reasons following business acquisitions), the process & risk assessments should positively promote a standard best practice approach at locations.
- As far as possible, the process & risk assessments should use non-technical language so that they can be used directly for initial induction training and refresher training without needing further materials to be developed.

PHASE 2 PROCESS & RISK ASSESSMENTS

This work is currently in progress and 'best practice' process & risk assessment modules are being developed for the activities and tasks covered in the pilot implementations. These will be reviewed by the divisional best practice working parties and rolled out for a phase 2 series of pilot implementations at a different set of representative locations than used for the phase 1 pilot implementations, before being rolled out across the company.

Important differences from the phase 1 template are:

- Formal language has been replaced by more commonly used language (e.g. 'Hazards' has become 'Look out for', 'Make eye contact with mobile plant operative' has become 'Eyeball the driver'). This may sound trivial but is expected to have a major impact for use in training sessions. See Figure 1.
- For the same reason as above, the technical parts of each process & risk assessment are now on a manager's page – risk ranking; lists of key safe behaviours; references to legislation, company standards and training materials. See Figure 2.
- There is a formal statement of company best practices on the manager's page, with a requirement to state any issues where the local situation prevents use of the best practice and what other measures are in place to control risks.

By the time of HAZARDS XX we expect to report on experience of rolling out the process & risk assessments across all of Biffa's locations.

To roll out the process & risk assessments company wide, we anticipate that each business unit manager will probably spend one or two days adapting the relevant modules to his or her location (including pasting in local photographs and discussing with supervisors and workforce representatives). These will replace the existing risk assessments and process maps, and will also be useable for induction of new employees and toolbox talks for experienced employees. Thereafter routine reviews (typically annual) will take less time than the typical ½-day for reviewing the existing risk assessments and process maps – and will be more effective.

001 Baler - demo.xls | Large Baling Machine | Best Practice Module 001 Issue d1

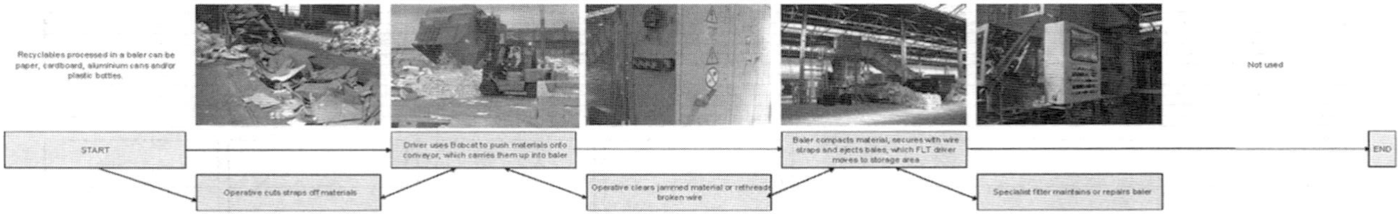

Who	Operative helping Bobcat driver	Bobcat driver	Baler operative	FLT driver	Specialist fitter	Not used
Risk level	Low Risk	Medium Risk	High Risk	Medium Risk	Medium Risk	Not assessed
Look out for	Falling onto conveyor, being carried up into hopper and getting crushed. Being hit by Bobcat or FLT.	Hitting colleague who is working with you. Hitting someone else who wanders into the area. Toppling Bobcat onto conveyor and getting hurt.	Falling into feed hopper when clearing blockage and getting crushed. Falling off conveyor when climbing up to clear blockage in feed hopper. Climbing into needle pit or reaching inside baler when it is running or is started up, and getting crushed.	Hitting someone. Bale stack collapsing.	Fitter being trapped or crushed inside baler.	Not used
Do's ✓	Check your Safetech belt before starting work and always wear it. Eyeball Bobcat driver when starting work with him. Stand in safety zone whenever Bobcat or FLT is moving.	When operating Bobcat, make sure that any operatives are in safety zone. If any one else wanders into the area, stop then ask them to leave. Wear your seatbelt.	Get PTW for agreed low frequency high risk activities. Use your padlock to lock off power at main isolator then lock off control panel and remove Castell key. If the problem is on the conveyor, use a pole to pull as much material off conveyor as possible. Visually inspect your safety harness before putting it on, climb up conveyor, clip onto the anchor point at the top. Use your pole to clear the jam. If you have to go inside the baler, fix open the access door.	When operating FLT, make sure that any operatives are in safety zone. Make sure that bales are secure in the clamps before moving. Keep clamps low or drive in reverse so you have a good view, Check bale stack is stable before adding more bales.	Obtain PTW. As far as possible, do same procedure as operative clearing a jam. If you have to keep the baler/conveyor powered up for a specific task, follow written SSoW, which should include safety/control measures for work at height, work in restricted spaces, work on machinery, electrical work, hot work etc.	Not used
Don'ts ✗	Don't stand on conveyor unless it has been locked off and controller key removed. Never go outside safety zone while Bobcat or FLT is working close to baler.	Don't carry on working if you can't see your colleague. Don't drive too close to feed conveyor pit.	Don't try to clear jam without locking off baler and if necessary PTW. Don't climb down into hopper. Don't climb onto the conveyor without it being locked off.	Don't move more than 2 bales at a time. Don't stack bales more than 4 high.	Don't try to repair baler without getting PTW, locking off baler and fixing open the access door.	Not used
What Biffa will do for normal & emergency situations	Test emergency grab cords every week. Test Safetech security belt system every week by sending one belt up conveyor to check operation. If Safetech is tripped by a retail security tag, make sure PTW & procedure is used to bypass Safetech.	Fit roll bar or cage to all mobile plant. Regularly check that green seat belt beacon is flashing when bobcat is being used.	Annual check by specialist engineer that power switch and Castell key isolate all power & hydraulics. Six monthly checks of safety harnesses and anchorages.	Fit roll bar or cage to all mobile plant. Regularly check that green seat belt beacon is flashing when FLT is being used.	Check fitter's competence (qualifications and experience). Review safe systems of work applicable to the task. Conduct & record comprehensive site induction for 3rd Party Employees.	Not used
	Provide efficient and effective Permit-to-Work (PTW) system, Maintain COSHH files / Regularly review & revise Traffic Management Plans and Pedestrian Access Routes as a result of consultation and operational changes Conduct and record regular inspections of mobile plant, static plant, safety harnesses, E-stop systems and Safetech device / Ensure effective defect reporting system / Conduct and record training/assessment of operatives. Regularly review / revise this risk assessment following consultation and/or changes in working practices.					

Hazards can change at any time - keep your eyes open and think "What if?" not "If only ..."

Figure 1. Process & risk assessment top page

001 Baler - demo.xls | Large Baling Machine | Best Practice Module 001 Issue d1

For manager's use only

Task-specific equipment	Safety knife to cut straps off materials	Bobcat to push recyclables onto conveyor	Personal padlocks	FLT to move bales to storage area	Personal padlocks	
Residual Risk						
Outcome	3 Lost Time Accident	5 Fatality	5 Fatality	5 Fatality	5 Fatality	0 SELECT CREDIBLE OUTCOME
Likelihood	1 Less than once in 40 years in Biffa	2 Once in 1 to 40 years in Biffa	3 Once in 10 to 40 years at this location	2 Once in 1 to 40 years in Biffa	2 Once in 1 to 40 years in Biffa	0 SELECT LIKELIHOOD OF SPECIFIED OUTCOME
Risk	Low Risk	Medium Risk	High Risk	Medium Risk	Medium Risk	Not assessed

Brief summary of any actions to reduce risk (details to be entered into Compliance Database)		Business Unit Responsible	Due Date
Action 1	When CDM2007 competence checks are issued company-wide, check competence of specialist fitters maintaining baler	Melchester	31-Jan-08
Action 2			
Action 3			

Key Safe Behaviours	PPE requirements	HSE Guidance	Company Standards & Guidance Notes	Company Best Practice	If best practice is not possible, what control measures are in place
Never operate plant without training & authorisation	Hi vis vest / jacket	INDG 392 - Recover Paper Safely	CS042 - Control of noise at work	Each conveyor belt to be fitted with a trip wire running the length of the conveyor with pull cords suspended at regular intervals that can be reached easily	Best practice in place
Be aware of mobile plant	Hard Hat	Waste09 - Safe transport in waste management and recycling facilities	CS057 - General/Hot Permit to Work	Each operative working in proximity of a conveyor to wear a belt which trips the conveyor if he is carried up it	Best practice in place
Challenge unauthorised persons in working area	Boots (toe and midsole protection)	Waste18 - Hand sorting of recyclables ('totting') with vehicle assistance	GN009 - Pedestrian Vehicle Segregation	All access points through which an operative could reach upper body into a hazardous machine are protected by a key-exchange or electrical interlock system	Best practice in place
Lock off & isolate before working on/inside conveyor / baler	Gloves	Minimum mechanical safety requirements for the design, construction and use of compacting and baling machines used in the waste-, recycling-and retailing- industries - dra Apr 07	GN024 - Guidance on seatbelts	All opening of access points into a hazardous machine are under permit to work which requires power to be locked off as well as the key system	Best practice in place
Safe stacking of bales	Safety glasses/goggles if dusty		GN044 - Working at Height Using Harnesses	Mobile plant to be fitted with audible reversing alarm and, if rear vision is restricted, with CCTV	Best practice in place
Cut away from yourself	No trainers, shorts, loose clothing		GNR04 - Safe System of Work: Safetech Personal Protection System	Maintenance to hazardous machines to be carried out only by specialist fitters whose competence has been checked	The maintenance company is experienced specialist, Biffa now developing formal CDM 2007 competence checks
	Hearing protection and dust masks as required		GNR08 - Safe System of Work: Baling Presses	Mobile plant operatives to wear seat belt and this is indicated by flashing green light	Best practice in place for all Biffa plant, exception is some plant on short term rental where facility is not available
			GNR09 - Safe System of Work: Conveyor Belt Operations	Mobile plant to be fitted with audible reversing alarm and, if rear vision is restricted, with CCTV	Best practice in place
			GNR10 - Safe System of Work: Emergency Stop Systems For Inclined Conveyors	Emergency procedure 'Rescue from inside baler feed chute' in place (written procedure, operatives trained, drill conducted every six months)	Best practice in place
			GNR14 - Stored Energy in Large Horizontal Baling Machines		

Date	Issued by	Reason for issue	Personnel involved in risk assessment
10-Dec-07	G.Sellers	Draft	

Figure 2. Process & risk assessment manager's page

CONCLUSIONS

For Biffa, which has large numbers of drivers and operatives operating from many locations and undertaking many similar activities, changing from the existing risk assessment system to an integration of process maps with risk assessments offers the prospect of a system in which hazards are less likely to be overlooked, which reinforces standard best practices, which is immediately usable for training purposes – and which is less cumbersome to review and update.

SIMPLIFIED FLAMMABLE GAS VOLUME METHODS FOR GAS EXPLOSION MODELLING FROM PRESSURIZED GAS RELEASES: A COMPARISON WITH LARGE SCALE EXPERIMENTAL DATA

V.H.Y. Tam[1], M. Wang[1], C.N. Savvides[1], E. Tunc[2], S. Ferraris[2], and J.X. Wen[2]
[1]EPTG, BP Exploration, Chertsey Road, Sunbury-on-Thames, TW16 7LN, UK
[2]Faculty of Engineering, Kingston University, Friars Avenue, London, SW15 3DW, UK

Gas explosion modelling is used widely to assess explosion loading due to overpressure or drag forces. These loadings depend upon the conditions when ignition is assumed to occur. We consider one of these conditions, namely the volume of a flammable gas cloud, specifically from a release of pressurized gas; and the application to the most commonly used commercial explosion code, FLACS.

There is a number of ways this volume is defined. The three main methods are (i) volume enclosed by the LFL contour surface, (ii) volume bounded by LFL and UFL contour surfaces and (iii) burning velocity weighted volume (Q9). These methods give vastly different flammable volumes. This can result in inconsistent overpressure prediction. As all these methods are currently being used, this situation is not satisfactory.

In this paper, we compared the prediction using different methods against data from the large scale explosion experiments (Phase 3b). Our results showed that the burning velocity weighted volume Q9 significantly underpredicted loads for explosion load higher than 0.1 bar compared with the other two measures which gave neutral to slightly over-conservative prediction of explosion loads.

1 BACKGROUND

Gas explosion is recognized to be one of the key major hazard accident risks in chemical, oil and gas processing facilities. Explosion modelling is used widely to assess these risks. Modelling techniques can vary from simple methods (e.g. multi-energy methods proposed by TNO) to more complex CFD codes (e.g. FLACS). The former is used widely to assess offsite risks, while the latter for on site or assessing potential explosion consequences close to or within the body of the congested facilities.

The consequence of a potential gas explosion varies depending on many factors. One of these is the volume of flammable gas cloud, usually within the volume of the production facilities where density of pipework and equipment form significant congestion.

In explosion consequence assessment, the flammable gas cloud volume definition varies between (a) the volume occupied by congestion, i.e. the body of the processing plant, and (b) the volume to represent the varying concentration field produced by a pressurized gas release within the body of the processing plant. The latter is referred to as "realistic release scenario".

There are a number of "realistic release scenarios" methods being used and these methods could give vastly different flammable volumes. This results in inconsistent explosion load estimations. As gas explosion overpressure loading is one of the key design input to the design of structures and buildings on onshore and offshore production facilities, the use of inappropriate methods could lead to over-design, or worse, a structure not fit for purpose. Similarly, risk would either be over or underestimated.

The subject of this paper is the various methods used in calculating flammable gas cloud volumes in the "realistic release scenario" approach. Specifically, we focus on methods used in the most widely used explosion code in the process industry: FLACS which was developed by Christian Michelsen Research (and latterly by GexCon) with over two decades of research funded by the major oil and gas companies including bp. The comparison with other explosion code will be addressed in future papers.

In this paper, we present our finding on the evaluation of these methods against data from the large scale experiments on explosion under realistic release scenarios, Phase 3b which will be described later.

2 FLAMMABLE GAS CLOUD VOLUME

There are a number of commonly used methods to derive the volume of a flammable gas cloud in a "realistic release scenario". One approach is to use the concentration and turbulence fields calculated for the dispersion phase of the gas release prior to ignition and applied them directly as input to gas explosion calculation. Leaving aside the issue of collecting data to verify this approach, the drawback of this is that it requires a large number of simulations that cannot be realistic completed within reasonable timescale (e.g. that of a typical design project).

A common approach that has been evolved and practiced today is to reduce the flammable gas cloud to a volume which is uniform and at stoichiometric concentration, in order to reduce the number of simulations to a manageable level. This still leave the issue of location and size of the volume. We examine three methods of representing this volume.

2.1 "> LFL": VOLUME BOUNDED BY LFL

This is the net volume of flammable gas above the lower flammability limits (LFL). This method has been used to estimate flammability distance for many years on onshore facilities and at bp for offshore facilities till the completion of the joint industry project on "jet dispersion" (Cleaver 1999).

As this measure includes volume above upper flammability limit (UFL) which is too rich to burn, the view of many is that this measure is over-conservative. This view is supported by previous work (Savvides 2001) indicated that this measure tend to give larger flammable volume than measured.

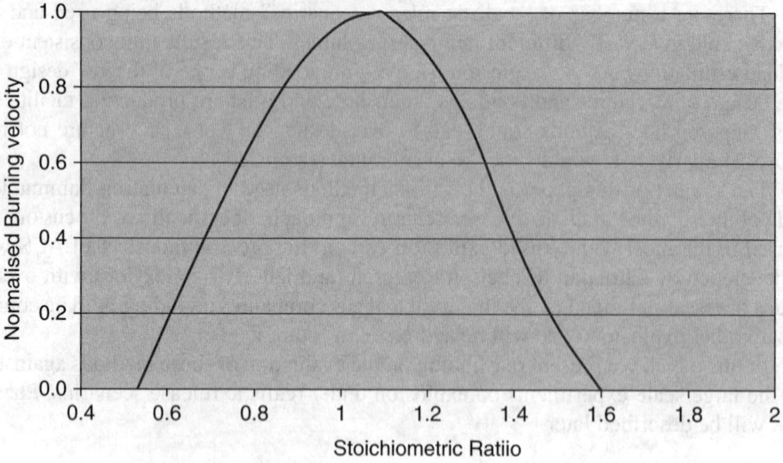

Figure 1. Variation of burning velocity with concentration of methane in stoichiometric ratio. This is taken from the correlation used in FLACS

2.2 "ΔFL": VOLUME BOUNDED BY UFL AND LFL

This is a logical development of the ">LFL" measure above, removing the volume with concentration >UFL, leading to a measure which is not seen to be "overly" conservative.

2.3 "Q9": WEIGHTED EQUIVALENT STOICHIOMETRIC VOLUME

In a pressurized gas release, the resultant flammable gas cloud would have a variation of concentration within it. Q9 is a volume measure which accounts for the effects of gas concentration by weighting the volume with the effect of burning velocity and expansion ratio. Prior to Gexcon introducing Q9 recently, a slightly different quantity, Q5, was used. To all intents and purposes, Q5 and Q9 give identical results. Here, we will only use Q9.

Experiments showed that burning velocity varies with concentration of flammable gas in air. For hydrocarbon, burning velocity is maximum at or near stoichiometric concentration of 1 and dropping off rapidly as gas concentration is rich (stoichiometric ratio > 1) or lean (stoichiometric ratio < 1), reaching zero at UFL and LFL (see Figure 1). Further, the expansion ratio follow similar pattern (Law 2006).

Q9 reduce the contribution of a volume of a parcel of gas to the total flammable volume by the product of these two effects, summarized in equation (1).

$$Q9 = \sum_{i=1}^{i=n} \frac{Vol_i(SE)}{S_{max} E_{max}} \qquad \text{Equation (1)}$$

Where

Vol_i = volume of a volume element i

S = burning velocity at the concentration of the Volume element i

E = Expansion ratio at the concentration of the volume element i

S_{max} = Maximum burning velocity of the flammable gas.

E_{max} = Maximum expansion ratio of the flammable gas.

Q9, effectively, puts a heavy weighting for flammable volume close to stoichiometric ratio of 1.

2.4 GENERAL BEHAVIOUR OF THESE MEASURES

These measures give different flammable gas cloud volumes. It can be seen that ">LFL" gives the largest volume, followed by ΔFL, then Q9. Magnitude of explosion loading would follow the same pattern.

We found that Q9 measures are being used increasingly by consultants. We are concerned that there has not been work to verify that this approach is indeed correct. Our observation is that there appears to be little fundamental understanding of the Q9 measures by consultants we encountered. Its application is based on a belief that since there is a varying gas concentration in a gas cloud formed from a pressurized gas release, assuming a uniform gas cloud concentration is thus 'over-conservative', and using Q9 would remove this perceived 'over-conservatism'. As we shall see later, this is not necessarily so.

3 METHODOLOGY

3.1 PHASE 3B – THE FIRE AND BLAST FOR TOP SIDE STRUCTURE PROJECT

We used data produced by the Phase 3b of the Fire and Blast for Topside Structures Project for this exercise.

The objective of the Phase 3b project was to study explosions resulted from the release of high pressure natural gas into a large scale model of an offshore production module. This work was funded by a consortium of international oil companies, including bp, and the UK regulator.

A brief history of the Fire and Blast for Topside Structure Project is as follows. Phase 1 of this project started in 1990 in response to the Piper Alpha accident in the UK sector of the North Sea. Phase 1 (SCI 1992) provided interim guidance to the industry and a review of knowledge in the fire and explosion area. This was followed by Phase 2, which consisted of a series of experiments to obtain data in full scale geometries representative of the offshore environment (SCI 1998). The results of Phase 2 indicated that high explosion overpressure could be generated. As a consequence, Phase 3a was commissioned by the UK Health and Safety Executive to study methods of reducing the severity of gas explosions (Al Hassan 1998).

The Phase 3b (Johnson 2002) tests consisted of laboratory, medium and large scale. In this paper, we employed the large scale data only. Figure 2 shows the experimental test rig which measured about 28 m long, 12 m wide and 8 m high. Natural gas was released within the module and was held constant for each test with release rates varying between 2.1 kg s^{-1} to 11.7 kg s^{-1}, and in direction of one of the three coordinate axes of the test rig. In total, twenty tests were carried out. Gas concentrations were measured prior to ignition and overpressure measured at locations distributed inside the module.

3.2 PREVIOUS WORK ON FLAMMABLE VOLUME

The issue of flammable volume was of concern and this was addressed by a joint industry project, called "Dispersion JIP" which studied the dispersion of releases of pressurized natural gas in a large scale module. This JIP (Cleaver et.al. 1999) took place between the completion of the Phase 3a programme and the start of the Phase 3b programme. The ">LFL" and ΔFL volume measures were evaluated and compared with predictions from FLUENT and FLACS (Savvides et.al 2001). These papers showed that FLACS was able to estimate ">LFL" and ΔFL with little bias. Q9 was not part of the evaluation. The remaining of this paper focuses on the investigation of the estimation of explosion overpressures by using the three volume measures described above.

3.3 MV DIAGRAM

An MV diagram shows the geometric bias and geometric variance of model predictions on a single diagram. It shows systematic overprediction or underprediction (bias), and the degree of scatter (variance).

Figure 2A. A picture of the Phase 3b test rig. It measures about 28 m long, 12 m wide and 8 m high and was a large scale model of a process module on an offshore platform

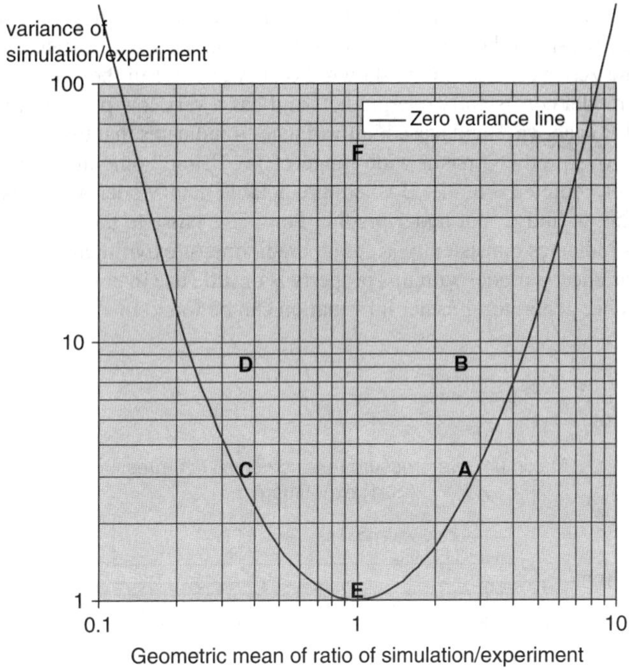

Figure 2B. An MV diagram which shows the behaviour of model predictions when it locates at various locations on the diagram (this supports the main text in the paper)

Hanna (Hanna et.al. 1991) developed the MV diagram for showing dispersion model performance when compared with a large range of data. This was later adopted by the scientific working group in the Blast and Fire for Topside Structure Phase 2 (SCI 1998) and in MEGGE protocol (SCI 1995) as a standard way to show model performance.

Bias indicates on average how much calculation over or under predicts experimental data. If we assume bias is zero, variance gives a measure of the scatter of prediction about the data. A high variance indicates poor consistency; one cannot be sure whether prediction grossly over or under predict. The bias and variance scale on the diagram are defined as:

$$\text{Mean Bias} = \exp\left(\langle \ln(P/O) \rangle\right)$$

$$\text{Variance} = \exp\left(\langle (\ln(P/O))^2 \rangle\right)$$

Where:

P = predicted overpressure

O = observed or measured overpressure

<X> denotes expectation value of X

The MV diagram consists of a number of parallel parabolas which gives line of constant variance. The lowest parabola is the zero variance line. This is illustrated in Figure 3 in which only the zero variance line is shown. Point A is close to the zero variance line: It indicates the model consistently overpredicts and has a very low probability of underprediction. Point B is above A and has a high variance; It indicates that the model, though has a tendency to overpredicts, has a wide range of prediction some are overpredicted and many are underpredicted – model B is less predictable than Model A. Points C and D are similar to Points A and B, but under-predict. Point E is close to the bottom of the lowest parabola; this indicates consistently accurate prediction of experimental data. Point F has a very high variance; a model with this property is of little use in practice as it behaves like a random number generator. Further information can be found in Tam (1998).

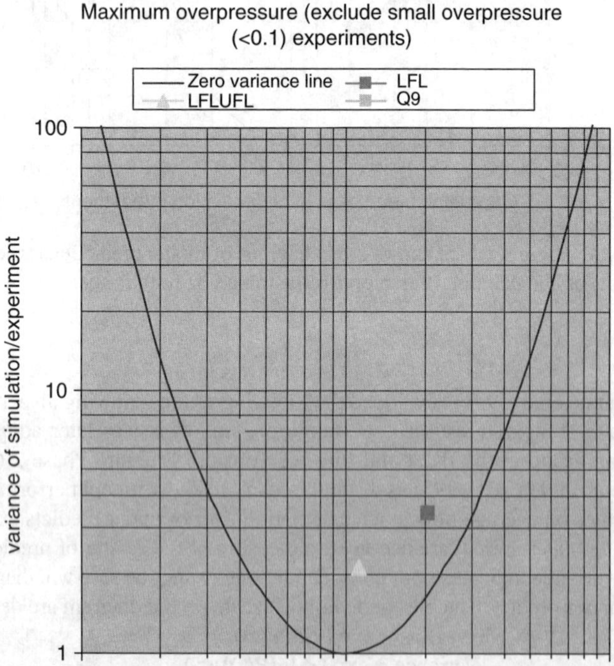

Figure 3. Bias and variance of the three volume measures. It shows that the Q9 measure underpredicts, ">LFL" is conservative and ΔFL is roughly neutral

4 RESULTS

Figure 4 gives a summary of predictions using the three volume measures for maximum overpressure anywhere within the module. The measured pressure data were processed by the experimental team: a rolling 1.5 ms time average was applied to data to remove fast transient effects, e.g. instrument noise. The rolling time-averaged data was used in this exercise.

For this exercise, we have excluded tests which had measured maximum overpressure of less than 0.1 bar. This is because there was large scatter of data at this low pressure. This aspect is discussed later. The trend is as follows: the prediction bias for ">LFL" is higher than ΔFL which is higher than Q9. Specifically, Q9 has a bias towards significant underprediction whereas ">LFL" and ΔFL tend to have a small overprediction or neutral respectively.

When we included the data with overpressure less than 0.1 bar, the relative position of the three measures remained the same. The bias of the Q9 measure is nearly zero and the other two biased towards overprediction. However, Q9 has a very high variance indicating a very wide scatter between predicted and observed values. This result is

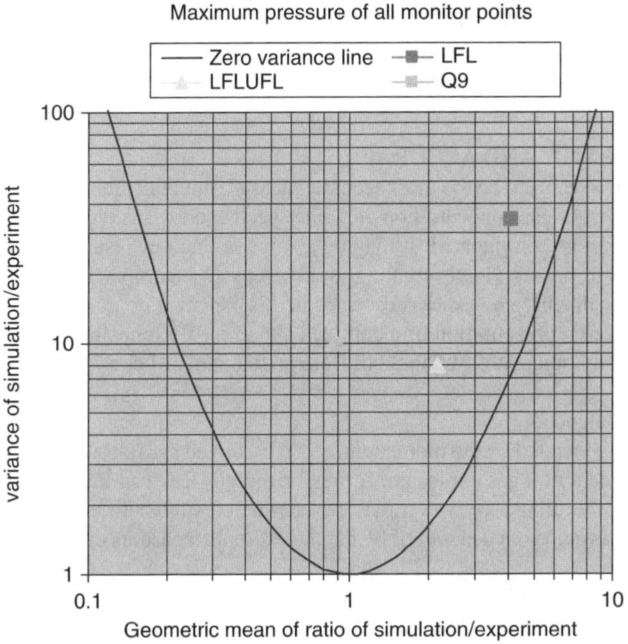

Figure 4. Bias and variance of the three volume measures when all data including those cases with a maximum of <0.1 bar recorded. It shows that the Q9 measure is unbiased but with a larger variance, both ">LFL" and ΔFL are conservative

summarized in Figure 4. The other two measures also have higher variance and showed significant bias towards overprediction.

Taking Figures 2a and 2b together, our results show that all three measures have significant variance for weak explosions (< 0.1 bar). Q9 showed significant bias for underprediction for explosions more than 0.1 bar, and the other two measures exhibit more consistent prediction behaviour (roughly neutral in bias and least variance). The reliability of overpressure predictions is important for structural design and becomes more important as the magnitude of the overpressure increases.

We also considered other ways to compare overpressures, e.g. average overpressure within the gas cloud, within the module. They all demonstrated similar trends. The major difference is that they all show Q9 has a negative bias (i.e. consistent underprediction) even for cases where maximum measured overpressure was less than 0.1 bar.

5 DISCUSSIONS

5.1 GENERAL COMMENTS OF THESE MEASURES

Use of Q9 in preference to the other two measures is widespread. We found that evidence supporting the use of Q9 is based on theoretical argument, sensitivity calculations and 'experience'. However, we found no supporting evidence against data presented even though data had been openly available for a number of years.

5.2 COMPLEXITY AND ACCURACY

Superficially, Q9 seems to be the most accurate measure out of the three as it accounts for the well known effect of gas concentration on flame speed and expansion ratio. The Q9 measure is certainly the most complicated. It may be a surprise that our results showed that the Q9 measure performs poorly. However, one should not confuse complexity and accuracy.

In reality, the three measures described in this paper are no more than a much simplified and idealized representation of a complex situation. By focusing on a couple of obvious factors, many others are overlooked. Here are a couple of examples:

A Initial turbulence

The turbulence generated by the momentum of a pressurized gas release is an important factor in the development of a gas explosion. A pressurized gas leak can impart a large amount of and high intensity turbulence which may be considered similar to turbulence produced by obstacles. This is not taken account of by the three simple measures discussed here.

B Size of the gas cloud

Limiting the flammable gas cloud to a smaller effective volume reduces the effect of flame acceleration over a larger distance and over longer period of time than that produced by larger cloud volumes and could lead to lower and the wrong distribution of overpressure.

If we take a hypothetical case of a gas cloud which is made up of two equal halves: one half at close to LFL and one half close to UFL. It can be seen from Figure 1 that Q9 would give a near zero flammable volume, whereas ΔFL would produce the whole volume.

Another reason for possible underestimation of flammable volume is that volumes with rich gas mixtures can be diluted with air or with lean gas mixtures during the course of a gas explosion, rendering the rich mixture closer to the stoichiometric ratio of 1.

Applying the Q9 method blindly, it is possible to reach a conclusion that a very large leak of flammable gas would not pose an explosion hazard.

5.3 SENSITIVITY TO POSITION

Gas cloud location can affect calculated overpressure predictions. For a gas cloud volume which is small compared with the process area, the choice of the gas cloud location within the process area becomes important. As Q9 produces the smallest effective gas cloud size for a given scenario, the results will be sensitive to this effect.

We compared the differences of overpressure prediction between the cases where a gas cloud is placed at the centre of the test module and where it is located at the edge for all the cases in the Phase 3b test programme. The mean ratio of overpressure at the centre to that at the edge are 1.3, 1.8 and 3.0 for ">LFL", ΔFL and Q9 respectively. It shows that the Q9 measure is the most sensitive followed by ΔFL, then by ">LFL".

5.4 VERIFICATION WITH DATA – A REQUIREMENT

Following the completion of the Phase 3b exercise and publishing of the results, there was no published validation of the methodology for calculating effective cloud volume measures for use with CFD codes.

Any methods used should be verified against experimental data as far as possible. It should be the duty of the model developer or user of the model to verify any new methods against available data. This requirement is stated in the MEGGE protocol (SCI 1995).

6 CONCLUSION

Based on the results of this work, we recommend that the non-conservative flammable cloud volume measure ΔFL be adopted as the basis for FLACS at mid to late stage of an engineering project definition. ">LFL" is a conservative measure which may be appropriate during the early stage of design of a process facility where uncertainties in the design is high. This work does not support the use of Q9.

REFERENCES

Al-Hassan, T. and Johnson, M., Gas Explosions in Large Scale Offshore Module Geometries: Overpressures, Mitigation and Repeatability, OMAE 98, Lisbon, May 1998.

Cleaver, R.P., Burgess, S., Buss, G.Y., Savvides, V., Tam, V., Connolly, S. and Britter, R.E., Analysis of Gas Build-up from High Pressure Natural Gas Releases in Naturally

Ventilated Offshore Modules, 8th Annual Conference on Offshore Installations: Fire and Explosion Engineering, Lord's Conference and Banqueting Centre, London, 30 November 1999.

Hanna, S.R., Strimaitis, D.G. and Chang, J.C., Evaluation of Commonly-used Hazardous Gas Dispersion Models, Washington 1991.

Johnson, D.M., Cleaver, R.P., Puttock, P.S. and Van Wingerden, C.J.M., Investigation of Gas Dispersion and Explosions in Offshore Modules, Offshore Technology Conference, Paper number: 14134, Houston, Texas U.S.A., 6–9 May 2002.

Law, C.K., Combustion Physics, Cambridge University Press (2006), ISBN-10: 0521870526.

Savvides, C.S., Tam, V., and Kinnear, D., Dispersion of Fuel in Offshore Modules: Comparison of Predictions Using FLUENT and Full Scale Experiments, Major Hazards Offshore Conference Proceedings, 27–28 Nov 2001, London, ERA Report 2001-0575

Savvides, C.S., Tam, V., Jon Erik, O., Hansen, O.R. and van-Wingerden, K., Dispersion of Fuel in Offshore Modules: Comparison of Predictions Using FLACS and Full Scale Experiments, Major Hazards Offshore Conference Proceedings, 27-28 Nov 2001, London, ERA Report 2001-0575.

Tam, V., Explosion Model Evaluation, Fire and Blast Information Group, Article R320, Newsletter Issue No 22, May 1998.

The Steel Construction Institute, "Interim Guidance Notes for the Design and protection of Topside Structures Against Explosion and Fire", Report No: SCI-P112, 1992.

The Steel Construction Institute, "Blast and Fire Engineering for Topside Structures, Phase 2, Final Summary Report, SCI Publication Number 253, 1998.

The Steel Construction Institute, Gas Explosion Model Evaluation protocol, Model Evaluation Group for Gas Explosions, European Communities Report, September 1995.

PRESSURISED CO_2 PIPELINE RUPTURE

Haroun Mahgerefteh[1], Garfield Denton[1], and Yuri Rykov[2]
[1]Department of Chemical Engineering, University College London WC1E 7JE
[2]Keldysh Institute of Applied Mathematics, 125047, Moscow, Russia

*Corresponding author (h.mahgerefteh@ucl.ac.uk)

Outflow data using a validated CFD model for the hypothetical full bore rupture of a pressurised pipeline transporting CO_2 are presented. For the sake of an example, the selected pipeline operating pressure of 117bara, 54 km long and 0.42 m dia. are the same as those for the main gas riser connecting the Piper Alpha to the MCP which ruptured during the Piper Alpha tragedy. Comparison of the CO_2 discharge data with those for the actual Piper Alpha natural gas composition indicate significantly greater amount of CO_2 released. Although both pipelines exhibit very similar depressurisation rates, almost 250,000 kg of CO_2 corresponding to only 3.7% of the total inventory is released in the first 300s following rupture. This compares with 125,000 kg of natural gas (9.7% of the total inventory) released for the same time duration. The temperature profile data indicate a significant drop in the temperature of CO_2 at the rupture plane corresponding to solid discharge at – 62°C and 4.1bara some 900s following pipeline failure. The combination of the massive amount of CO_2 released in a relatively short period of time, the resulting dense cloud followed by solid discharge and its slow sublimation will pose a major challenge to safety practitioners when dealing with the hazards associated with the failure of pressurised CO_2 pipelines.

INTRODUCTION

It is now well established that increasing amounts of CO_2 in the earth's atmosphere is leading to changes in the climate. Global use of fossil fuel which is the most significant source of CO_2 currently results in an annual emission of 32Gt of CO_2 to the atmosphere. The concentration now stands at about 375ppm by volume compared with a stable, pre-industrial level of around 280ppm, maintained for at least the last 6,000 years (UK Department of Trade and Industry report, 2002). UK is responsible for 2.3% of CO_2 emissions, despite the fact that it accounts for only 0.8% of the world population. It is the 6th largest producer of CO_2 per capita amongst the world (World Population Prospects, 2002).

In order to stabilise CO_2 concentrations or reduce them, global emissions of CO_2 would need to decrease dramatically.

Given this a portfolio of approaches is needed to drive CO_2 emissions down without impeding economic growth. For fossil fuels, this will mean ultimately the capture, transportation and long terms sequestration (CCS) of CO_2.

Bulk gaseous transport of CO_2 may be undertaken by tanker or pipeline. In view of the large volumes involved, pressurised pipelines are considered to be the most practical, and possibly the only option for many fossil fired generation plant. This has significant implications for the UK since more than 70% of its electricity is fossil fuel power generated (Energy Review, 2002). Additionally, given that most electricity generation plants are built close to energy consumers, the number of people potentially exposed to risks from CO_2 transportation facilities will be greater than the corresponding number exposed to potential risks from CO_2 capture and storage facilities.

Ironically (in line with its abbreviation), CCS and related legislation generally focus on the Capture and Sequestration of CO_2 and not on its Transportation. This is despite Intergovernmental Panel on Climate Change (IPCC, 2004) concluding 'public concerns about CO_2 transportation may form a significant barrier to large-scale use of CCS'. An especially commissioned study by the US congress in April 2007 states (Order Code RL33971, 2007) 'there are important unanswered questions about CO_2 pipeline safety'. It goes on to say that 'policy decisions affecting CO_2 pipelines take on an urgency that is, perhaps, unrecognized by many'.

It is noteworthy that CO_2 pipelines have been in operation in the US for over 30 year for enhanced oil recovery (Order Code RL33971, 2007). However, these are either confined to low populated areas, and/or operate below the proposed supercritical conditions (73.3 bar and 31.18 °C) that make CO_2 pipeline transportation economically viable thus representing significantly less safety issues. Additionally, due to their small number, it is not possible to draw a meaningful statistical representation of the risk. The US report predicts 'statistically, the number of incidents involving CO_2 should be similar to those for natural gas transmission'. It is noteworthy that the rupture of a natural gas pipeline during the Piper Alpha tragedy (Cullen, 1990) ultimately lead to the collapse of the platform onto the sea bed, the loss of 167 lives and a cost of £2 billion.

Despite all this, UK has no standards specific to CO_2 pipelines. Furthermore, CO_2 is not recognised as a dangerous fluid (Encyclopaedia of Occupational Health and Safety, 1989).

THE CHALLENGE

'A transportation infrastructure that carries carbon dioxide in large enough quantities to make a significant contribution to climate change mitigation will require a large network of pipelines spanning over hundreds of kilometres (IPCC, 2004)'. Putting this in perspective, a typical 100km, 0.8m dia. pipeline transporting CO_2 at room temperature and 170bara would contain approximately 9m tons of gas.

The near adiabatic expansion process following pipeline rupture could lead to a massive and rapid release. Depending on its discharge temperature, the escaping fluid could either form a very cold jet denser than the surrounding air covering distances of several kilometres or a solid discharge with its own characteristics hazards such as delayed sublimation and impact erosion of surrounding equipment.

In both circumstances, the resulting plume is the most dangerous with regard to toxic gases due to its poor mixing with the surrounding air. Connolly and Cusco (2007) provide an excellent review of the hazards associated with the accidental release of pressurised CO_2. At a concentration of 10%, an exposed individual would lapse into unconsciousness in 1minute (Lees, 1996). Furthermore, if the concentration is 20% or more, the gas is instantaneously fatal (Pohanish et al., 1996). The ability of CO_2 to collect in depressions in the land, in basements and in other low-lying areas such as valleys near the pipeline route, presents a significant hazard if leaks continue undetected. Hydrocarbons will eventually ignite or explode in such areas if, and when, conditions are "right", but CO_2 can remain undetected for a very long time.

Unlike other toxic gases that operate as chemical asphyxiants, CO_2 has no choking or distinctive odour and this attribute adds to its potency as a toxic gas. In 1986 in Cameroon a cloud of naturally-occurring CO_2 spontaneously released from Lake Nyos killed 1,800 people in nearby villages (Krajick, 2003).

It is clear that the hazards associated with CO_2 pipelines are quite different compared to those posed by hydrocarbon pipelines, presenting a new set of challenges. As such any confidence that existing experience with operating hydrocarbon pipelines can be wholly extended to CO_2 pipelines is dangerously misplaced.

Two key areas that will need to be demonstrated to gain public acceptance CO_2 pipelines are that such mode of transport is safe, and its environmental impact is limited. Pivotal to this is the estimation of the flow rate and its variation with time following pipeline rupture.

In this paper we employ our previously validated CFD model, PipeTech to report and compare outflow data for the rupture of hypothetical but nevertheless realistic of two identical pressurised pipelines each containing CO_2 and natural gas. Given the critical importance of the correct prediction of fluid density on the accurate prediction of outflow data, the efficacy of PipeTech in predicting CO_2 densities over an extensive range of temperatures and pressures is examined first.

BACKGROUND THEORY

PipeTech's background theory is extensively presented in previous publications (see for example Mahgerefteh et al, 2000, Mahgerefteh et al., 2006a,b, Mahgerefteh and Abbasi, 2007). Its formulation is rigorous with its predictions having been extensively validated against available field data (see for example Mahgerefteh et al, 2006a).

Briefly, the modelling involves the numerical solution of the mass, energy and momentum conservation equations assuming 1D flow using a suitable technique such as the Method of Characteristics (MOC).

PipeTech accounts for real fluid behaviour as well as flow and phase dependent heat transfer and frictional effects. It is applicable to both isolated and un-isolated flows where pumping at the high-pressure end continues despite pipeline failure. Liquid and vapour phases are assumed to be at thermodynamic and phase equilibrium. This assumption is found to be generally valid in the case of rupture of long pipelines (Chen et al., 1995).

Peng-Robinson equation of state (Peng and Robison, 1976) coupled with appropriate mixing rules is used for obtaining the relevant thermodynamic and phase equilibrium data. The speed of sound for real multi-component single-phase fluids is obtained using standard expressions (Picard and Bishno, 1987). In the absence of an analytical solution, the speed of sound for two-phase mixtures is calculated numerically.

RESULTS AND DISCUSSION

APPLICABILITY OF PR EOS IN PREDICTING CO_2 DATA

Although the PR EoS has been found to be particularly applicable to high-pressure hydrocarbon mixtures, its suitability in predicting CO_2 properties, particularly density covering an extensive range of pressures and temperatures has not been fully investigated. This is important since the accurate prediction of the discharge rate following pipeline rupture is critically affected by the efficacy of the EoS in predicting density data.

Tables 1–3 show the results of such analysis in the pressure and temperature range of 1–500 bar and 250–1100 K respectively. The corresponding fluid state is given in each table. The experimental data are those reported by Span and Wagner (1996). The tables also shows the predictions using the Bender EoS (Bender, 1975), specifically developed for CO_2.

Based on the comparison with the experimental data in the gaseous region (tables 1 and 2), it is clear that both EoS produce remarkably good agreement with the experimental data. The maximum discrepancy produced by PR EoS is 1.9%. The corresponding value using the Bender EoS is 1.2%.

Reasonably good density predictions are also obtained in the supercritical region (>31.9°C and >71.9 bar; table 3) with the Bender EoS (1.7% discrepancy) performing better than the PR EoS (4.25% discrepancy).

CO_2 PIPELINE RUPTURE OUTFLOW DATA

Figures 1–3 show the simulated discharge data following the full bore rupture of a hypothetical 54km long and 0.419m i.d pipeline transporting pressurised CO_2 at 117 bara and 283 K. For the sake of an example, these pipeline dimensions and the prevailing conditions are the same as those for the sub-sea natural gas line from Piper-Alpha to MCP-01 platform which ruptured during the Piper Alpha tragedy (Cullen, 1990). In the absence of reported values for the heat transfer coefficient, pipe wall thickness and pipe wall roughness corresponding values for a partially insulated mild steel pipeline are assumed. The corresponding simulated data for the actual natural gas inventory transported in the gas riser prior to its rupture are superimposed on the same graphs for comparison. For credibility, we chose the Piper Alpha conditions since PipeTech's output has been previously successfully validated by comparison against the actual pipeline intact end pressure data recorded during the night of the tragedy (Mahgerefteh et al, 1997).

Returning to figure 1, the data show the variation of discharge pressure with time for the first 300s following full bore pipeline rupture. Curve A shows the Piper Alpha data (natural gas). The CO_2 data are presented by Curve B. As it may be observed, pipeline failure is signified by a rapid instantaneous drop from the line pressure of 117bara to 10bara in approximately 25s followed by a gradual reduction. This type of hyperbolic behaviour is synonymous with full bore rupture (Mahgerefteh et al., 2006a,b).

Table 1. Comparison of the performance of various equations of state in predicting CO_2 densities in the gaseous state

		Density (kg/m^3)			% Difference	
Pressure (Bar)	Temperature (K)	PR EOS	Span & Wagner (1996)	Bender EOS	PR EOS	Bender EOS
			Gas			
1.01325	250	2.165	2.165	2.164	0.02	−0.03
	300	1.798	1.797	1.796	0.06	−0.03
	350	1.538	1.537	1.537	0.04	−0.03
	400	1.344	1.343	1.343	0.03	−0.02
	450	1.194	1.193	1.193	0.03	−0.02
	500	1.074	1.074	1.073	0.02	−0.01
	600	0.894	0.894	0.894	0.02	−0.02
	700	0.766	0.766	0.766	0.02	−0.02
	800	0.670	0.670	0.670	0.01	−0.02
	900	0.596	0.596	0.596	0.01	0.11
	1000	0.536	0.536	0.536	0.01	−0.02
	1100	0.488	0.487	0.487	0.01	−0.02
			Triple point			
	216	13.201	13.282	13.251	−0.61	−0.23
			Gas			
5	250	11.109	11.097	11.093	0.11	−0.04
	300	9.068	9.046	9.044	0.25	−0.02
	350	7.690	7.674	7.671	0.21	−0.03
	400	6.688	6.677	6.675	0.16	−0.03
	450	5.923	5.915	5.913	0.13	−0.03
	500	5.318	5.313	5.311	0.11	−0.03
	600	4.421	4.417	4.416	0.09	−0.03
	700	3.784	3.781	3.781	0.07	−0.02
	800	3.309	3.307	3.306	0.07	−0.02
	900	2.940	2.938	2.938	0.06	−0.02
	1000	2.646	2.644	2.644	0.06	−0.02
	1100	2.405	2.403	2.403	0.06	−0.01

It is interesting to note that both the natural gas and the CO_2 pipelines exhibit very similar depressurisation behaviour with the former demonstrating a marginally more rapid drop during the first 40s following rupture.

Figure 2 shows the corresponding discharge rate data for both pipelines. As it may be observed, the initial discharge rate upon rupture for the CO_2 pipeline is approximately

Table 2. Comparison of the performance of various equations of state in predicting CO_2 densities in the gaseous state

		Density (kg/m³)			% Difference	
Pressure (Bar)	Temperature (K)	PR EOS	Span & Wagner (1996)	Bender EOS	PR EOS	Bender EOS
			Gas			
10	250	23.464	23.435	23.409	0.12	−0.11
	300	18.672	18.579	18.341	0.50	−1.28
	350	15.645	15.581	15.575	0.41	−0.04
	400	13.521	13.477	13.470	0.32	−0.05
	450	11.930	11.899	11.894	0.26	−0.04
	500	10.687	10.664	10.659	0.22	−0.05
	600	8.860	8.845	8.842	0.17	−0.03
	700	7.575	7.564	7.562	0.15	−0.02
	900	5.879	5.872	5.871	0.12	−0.01
	1000	5.289	5.283	5.282	0.12	−0.01
	1100	4.807	4.801	4.801	0.11	−0.01
50	350	91.326	89.619	89.383	1.90	−0.26
	400	73.836	72.804	72.609	1.42	−0.27
	450	63.001	62.295	62.154	1.13	−0.23
	500	55.352	54.826	54.728	0.96	−0.18
	600	44.967	44.621	44.577	0.78	−0.10
	700	38.082	37.823	37.805	0.68	−0.05
	800	33.112	32.904	32.901	0.63	−0.01
	900	29.331	29.156	29.158	0.60	0.01
	1000	26.345	26.196	26.200	0.57	0.02
	1100	23.923	23.793	23.798	0.547	0.020

4500 kg/s as compared to 4150 kg/s for the natural gas pipeline. Thereafter the CO_2 pipeline maintains a noticeably higher discharge rate for the remainder of the discharge process under consideration.

The variation of the cumulative mass discharged with time results for the two pipelines is shown in figure 3. The data show that at any given time following rupture, a significantly larger amount of CO_2 is released as compared to natural gas. Almost 260000kg of CO_2 accounting for only 4% of the inventory (figure 4, curve B) escapes from the pipeline in the first 300s following rupture. Although significantly less than the amount release during the Lake Nyos irruption, nevertheless such huge amount of CO_2 released in such a short period of time would lead to catastrophic consequences where it to occur in a populated area.

Table 3. Comparison of the performance of various equations of state in predicting CO_2 densities in the supercritical state

		Density (kg/m³)			% Difference	
Pressure (Bar)	Temperature (K)	PR EOS	Span & Wagner (1996)	Bender EOS	PR EOS	Bender EOS
			Super Critical			
200	400	378.302	380.500	379.813	−0.58	−0.18
	450	288.499	285.140	280.201	1.18	−1.73
	500	239.343	235.240	231.913	1.74	−1.41
	600	184.222	180.500	179.114	2.06	−0.77
	700	152.430	149.270	148.681	2.12	−0.39
	800	131.033	128.340	128.096	2.10	−0.19
	900	115.378	113.040	112.969	2.07	−0.06
	1000	103.308	101.270	101.269	2.01	0.00
	1100	93.660	91.857	91.895	1.96	0.04
500	500	548.974	534.420	539.975	2.72	1.04
	600	430.109	414.840	411.227	3.68	−0.87
	700	357.326	343.270	340.460	4.09	−0.82
	800	307.839	295.340	293.585	4.23	−0.59
	900	271.622	260.550	259.497	4.25	−0.40
	1000	243.734	233.890	233.263	4.21	−0.27
	1100	221.462	212.660	212.288	4.14	−0.18

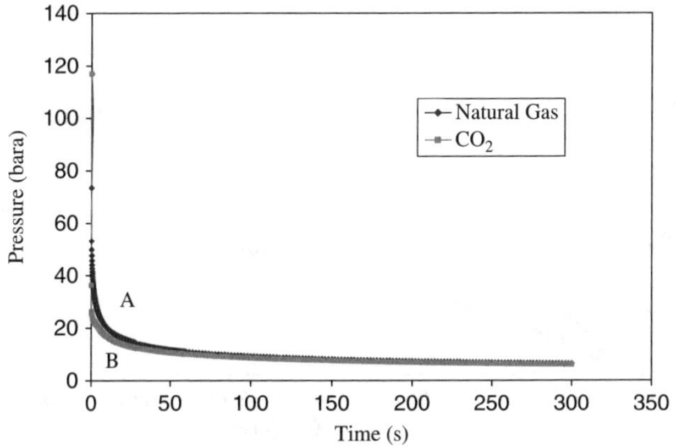

Figure 1. The variation of discharge pressure with time following full bore pipeline rupture. Curve A: Natural Gas (Piper Alpha). Curve B: CO_2

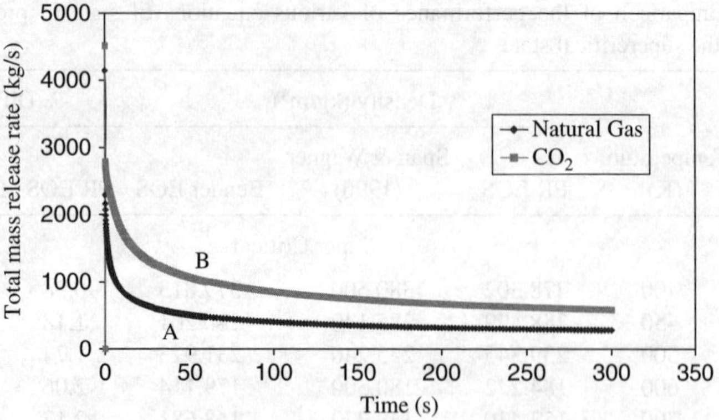

Figure 2. The variation of mass release rate with time following full bore pipeline rupture. Curve A: Natural Gas (Piper Alpha). Curve B: CO_2

The corresponding mass loss for the natural gas pipeline is approximately half of this value (125000 kg) representing a much higher percentage (10 %; figure 4, curve A) of the inventory lost.

Figure 5 shows the variation of the discharge temperature with time for the CO_2 pipeline. As it is clear, the initial gaseous inventory undergoes a significant drop in temperature

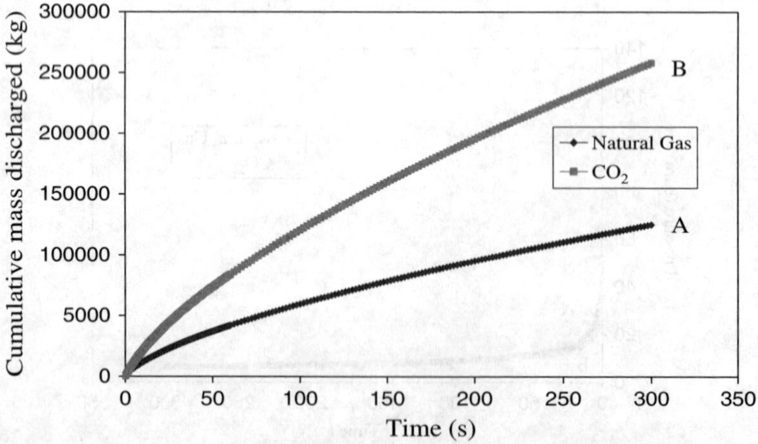

Figure 3. The variation of cumulative mass discharged with time following full bore pipeline rupture. Curve A: Natural Gas (Piper Alpha). Curve B: CO_2

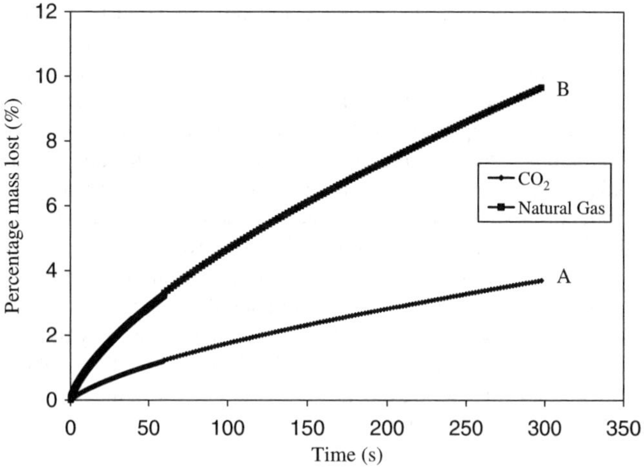

Figure 4. The variation of % mass lost with time following full bore pipeline rupture. Curve A: Natural Gas (Piper Alpha). Curve B: CO_2

reaching –212K (–62°C) at 4.1bara some 900s following failure corresponding to solid discharge. CO_2 triple point is –56.5 °C and 5.1bara.

CONCLUSION

In this paper we present transient outflow predictions following the full bore rupture of a pressurised CO_2 pipeline. This data is central to assessing all the hazards associated with such type of failure.

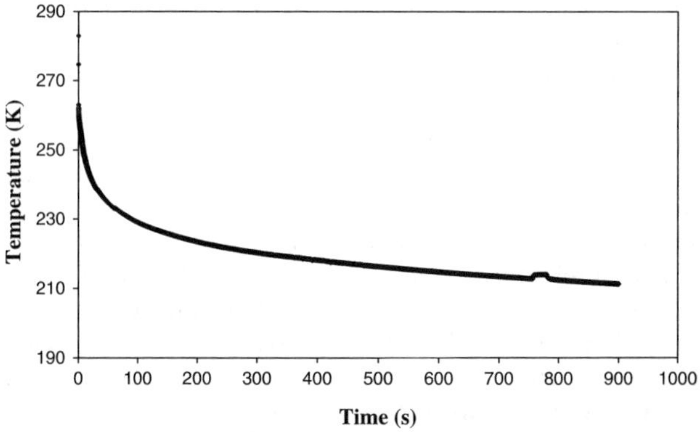

Figure 5. The discharge CO_2 temperature with time following full bore pipeline rupture

The simulated predictions, generated using our validated CFD model, PipeTech demonstrate a hyperbolic variation in the discharge rate with time characterised by a massive amount of inventory released in a relatively short period of time following pipeline failure. This type of release behaviour is the most catastrophic, significantly limiting the emergency response time available. Comparison of the outflow data with those for the rupture of the same pipeline containing natural gas indicates a significantly greater amount of CO_2 released representing only a fraction of the initial inventory. The tracking of the temperature/ pressure data of the discharged CO_2 at the rupture plane indicates cold dense vapour cloud discharge for the first 900s following rupture. This is followed by solid release at $-62^{o}C$ and 4.1bara. The released CO_2 would cover large distances remaining at lethal concentrations for a protracted period of time prior to sublimation and dilution to safe levels.

In conclusion, the hyperbolic release behaviour characterised by the massive burst of inventory coupled with its significant cooling clearly highlight the challenges faced by safety practitioners when considering the hazards associated with the rupture of pressurised CO_2 pipelines. The type of data presented in this paper is pivotal to the quantification of such hazards.

REFERENCES

Chen, J. R., S. M. Richardson, and G. Saville, 1995a, Modelling of two-phase blowdown from pipelines – I. A hyperbolic model based on variational principles, *Chem Eng Sci*, **50:** 695.

Connolly, S and Cusco, L, Hazards from high pressure carbon dioxide releases during carbon dioxide sequestration processes, Loss Prevention 2007, 12th International Symposium on Loss Prevention and Safety Promotion in the Process Industries.

Cullen, W. D., 1990, The public inquiry into the Piper Alpha disaster. *Dept of Energy*, HMSO.

Doctor, R. and Palmer, A., 2004, Transporting CO_2, Chapter 4.

Encyclopaedia of Occupational Health and Safety/Technical. 3 ed., 3 impr. Geneva: International Labour Office, 1989.

Intergovernmental Panel on Climate Change, IPCC, Carbon Capture & Storage ISBN 92-9169-119.

Krajick, K., 2003, Defusing Africa's Killer Lakes. *Smithsonian*, **34(6)**: 46 – 55

Lees, F. P., 1996, Safety and Loss Prevention in the Process Industries, Hazard Identification, Assessment and Control. *Butterworth-Heinemann,* Vol. 1(15): 75, 102-104.

Lees, F. P., 1996, Safety and Loss Prevention in the Process Industries, Hazard Identification, Assessment and Control. *Butterworth-Heinemann*. Vol. 2, pp 16/87-88.

Mahgerefteh, H., Saha, P. and Economou, I., 2000, Modelling fluid phase transition effects on the dynamic behaviour of ESDV. *AIChE Journal*, **46(5):** 997 – 1006.

Mahgerefteh, H., Oke, A. and Rykov. Y., 2006a, Efficient numerical simulation for highly transient flows. *Chem Eng Sci,* **61(15):** 5049-5056.

Mahgerefteh, H., Oke, A. and Atti, O., 2006b, Modelling outflow following rupture in pipeline networks. *Chem Eng Sci*, **61(6):** 1811-1818.

Mahgerefteh, H. and Abbasi, U., 2007, Modeling blowdown of pipelines under fire attack. *AIChE Journal*, **53(9):** 2443-2450.

Mahgerefteh, H., Saha, P. and Economou, I., 1997, A study of the dynamic response of emergency shut-down valves following full bore rupture of long pipelines. *Trans I ChemE: Process Safety and Environmental Protection*, **75(B4):** 201-209

Parfomak, P. and Folger, P., 2007, Carbon Dioxide (CO_2) Pipelines for Carbon Sequestration: Emerging Policy. CRS Report for Congress.

Peng, D. Y., and Robinson, D. B., 1976, A new two-constant equation of state. *Ind Eng Chem Fund*, **15:** 59-65.

Picard, D. J., and Bishnoi, P. R., 1987, Calculation of the thermodynamic sound velocity in two phase multi-component fluids. *Int J Multiphase Flow*, **13(3):** 295-308.

Span, R. and Wagner, W., **1996**, *J Phys Chem Ref Data*, **25:** 1509-1596.

Pohanish, P. R., and Greene, S.A., 1996, Hazardous Materials Handbook, Carbon Dioxide. *Van Nostrand Reinhold,* 330-331.

UK Department of Trade and Industry, 2002, Report on Carbon Dioxide Capture and Storage, *Pub URN 00/108.*

World Population Prospects: The 2002 Revision, 2003, New York: United Nations.

Cabinet Office PIU, 2002, Energy Review.

VERIFICATION AND VALIDATION OF CONSEQUENCE MODELS FOR ACCIDENTAL RELEASES OF HAZARDOUS CHEMICALS TO THE ATMOSPHERE

Henk W.M. Witlox and Adeyemi Oke
DNV Software, London, UK

This paper considers the "verification" and "validation" of consequence and risk models for accidental releases of toxic or flammable chemicals to the atmosphere. These models typically include a "physical description" of the phenomenon (e.g. cloud moving with the wind with air entraining into the cloud because of turbulence, etc.), a formulation of the corresponding "mathematical model", an "algorithm" for solution of the above mathematical model, and "implementation" of this algorithm into code.

Testing of the resulting software program should ideally include "verification" that the code correctly solves the mathematical model (i.e. that the calculated variables are a correct solution of the equations), "validation" against experimental data to show how closely the mathematical model agrees with the experimental results, and a "sensitivity analysis" including a large number of input parameter variations to ensure overall robustness of the code, and to understand the effect of parameter variations on the model predictions.

The current paper includes an overview on how the above verification and validation could be carried out for discharge, atmospheric dispersion (including pools) and flammable effects (e.g. pool fires, jet fires, explosions). A wide range of release scenarios is considered including sub-cooled liquid releases, superheated liquid releases, vapour releases, un-pressurised and pressurised releases, and a wide range of hazardous chemicals is considered (e.g. water, LNG, propane, ammonia, HF etc). Reference is made to the literature for the availability of experimental data and the verification and validation is illustrated by means of application to the consequence models in the hazard assessment package Phast and the risk analysis package Phast Risk (formerly known as SAFETI).

1. INTRODUCTION

Typical release scenarios involve liquid, two-phase or gas releases from vessel or pipe work attached to vessels. Consequence modelling first involves discharge modelling. Secondly a cloud forms which moves in the downwind direction, and atmospheric dispersion calculations are carried out to calculate the cloud concentrations. In case of two-phase releases rainout may occur, and pool formation/spreading and re-evaporation needs to be modelled. For flammable materials modelling is required of jet fires or fireballs in case of immediate ignition, pool fires in case of ignition of a pool formed following rainout, and explosions or vapour cloud fires (flash fires) in case of delayed ignition; Figure 1 illustrates the example case of a continuous release with rainout.

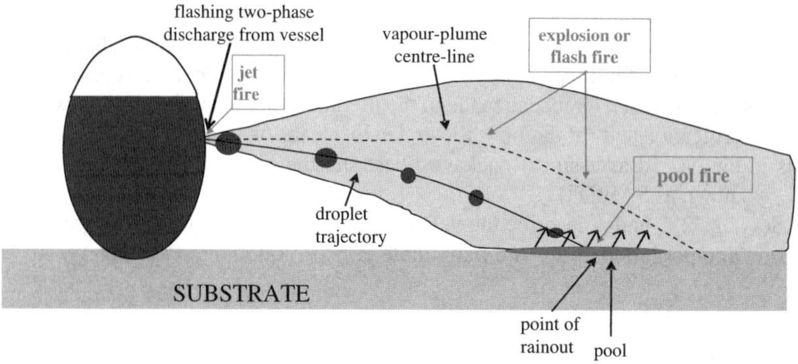

Figure 1. Continuous two-phase release of flammable material with rainout

To ensure the quality of consequence-modelling software thorough testing is paramount. This is ideally carried out by means of the following subsequent phases:

1. Verification that the code correctly solves the mathematical model, i.e. that the calculated variables are a correct solution of the equations. In case of a 'simple' mathematical model (e.g. not using differential equations but non-linear equations for unknown variables only), it can often be directly verified by insertion of the solved variables (calculated from the code) in the original equations, and checking that the equations are indeed satisfied. This is usually most expediently done by writing a 'verification' Excel spreadsheet in parallel with the code. In case of a more complex model expressed by a number of differential equations, the model can sometimes be solved analytically for some specific cases. Verification then consists of checking that the analytical solution is identical to the numerical solution. For a more general case, the more complex model can no longer be solved analytically. The only way of verifying the model is by comparing it with another model that solves the same (type of) equations.
2. Validation against experimental data. After, as shown above, the code has been verified to correctly solve the mathematical model, validation against experimental data will show how closely the mathematical model agrees with the experimental results. This provides a justification for the simplified assumptions made to derive the mathematical model.
3. Sensitivity analysis. This involves carrying out a large number of input parameter variations (e.g. hole diameter, ambient temperature, etc.) for a number of base cases (e.g. continuous vertical methane jet release, instantaneous ground-level propane un-pressurised release, etc.). Its purpose is to ensure overall robustness of the code, and to understand the effect of parameter variations on the model predictions.

This paper includes a brief overview of the "verification" and "validation" of consequence and risk models for accidental releases of toxic or flammable chemicals to the atmosphere. A limited number of key scenarios are considered, while reference is made to key papers for details. Reference is made to the literature for the availability of experimental data. The verification and validation is illustrated by means of application to the consequence models in the hazard assessment package Phast and the risk analysis package Phast Risk (formerly known as SAFETI).

Sections 2, 3 and 4 describe the verification and validation for discharge modelling, dispersion and pool modelling, and flammable effects modelling, respectively.

2. DISCHARGE

For releases of hazardous materials a wide range of scenarios can occur including instantaneous releases (catastrophic vessel rupture), and continuous and time-varying releases (leak from vessel, short pipe or long pipe). The stored material could be a sub-cooled liquid, a (flashing) superheated liquid, or a gas. As shown in Figure 2, the discharge model calculates both the expansion from the initial storage conditions to the orifice conditions, as well as the subsequent expansion from orifice conditions to atmospheric conditions. For superheated liquid releases, liquid break-up into droplets occurs along the expansion zone. It is typically assumed that the length of the expansion zone is very small with negligible air entrainment.

Key output data of the discharge model are flow rate, orifice data [velocity, liquid fraction] and post-expansion data [velocity, liquid fraction, initial droplet size (distribution)]. The post-expansion data are the starting point ("source term") of the subsequent dispersion calculations.

In the literature numerous discharge models can be found. Key literature including description of discharge models and experimental data include Perry's handbook (Perry et al.,

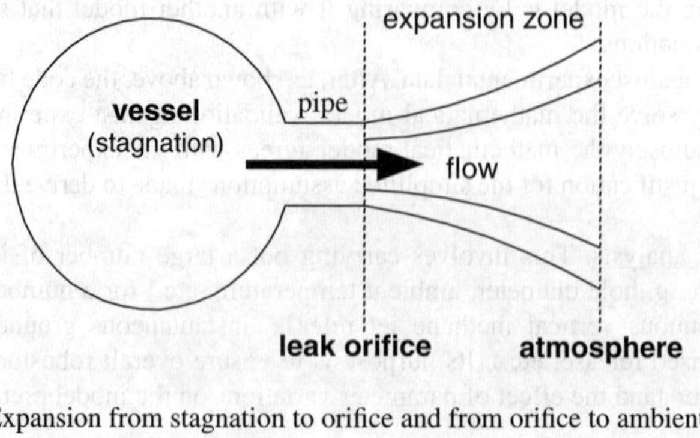

Figure 2. Expansion from stagnation to orifice and from orifice to ambient conditions

1999), the DIERS project manual (Fisher et al., 1992), CCPS QRA guidelines (CCPS, 2000), Sections 15.1–15.9 in Lees (Lees, 1996), and Chapter 2 in the TNO Yellow Book (TNO, 1997). The author did not find an up-to-date published overview of key experiments (benchmark tests for discharge models; input data and experimental results), in conjunction with a systematic evaluation of discharge models.

Key verification tests include comparison of the model against well-established analytical flow-rate equations for incompressible liquid (Bernoulli equation) and ideal gases. In addition verification could be considered between different discharge models and verification against results from process simulators (e.g. HYSIS or PROII).

Key validation tests include sub-cooled and saturated pipe and orifice releases of water (Sozzi and Sutherland, 1975; Uchida and Narai, 1966), and also data for hydrocarbon releases.

A detailed verification and validation has recently been carried out for the Phast discharge model for releases from vessels and/or short pipes including amongst others the above cases. Figure 3 illustrates the comparison for the Phast 6.53 model against sub-cooled water jets. The Phast long pipeline model has been validated for propane two-phase releases [Isle of Grain experiments (Cowley and Tam, 1988; Webber et al., 1999)].

Detailed validation of droplet modelling for two-phase releases was carried out by Witlox et al. (2007) using a range of droplet-size correlations accounting for both mechanical and flashing break-up of the droplets. This includes validation of initial droplet size against recently published experiments [STEP experiments (flashing propane jets), experiments by the Von Karman Institute (flashing R134-A jets), and water and butane experiments carried out by Ecole des Mines and INERIS]. It also includes validation of the rainout against the CCPS experiments (flashing jets of water, CFC-11, chlorine, cyclohexane, monomethylamine).

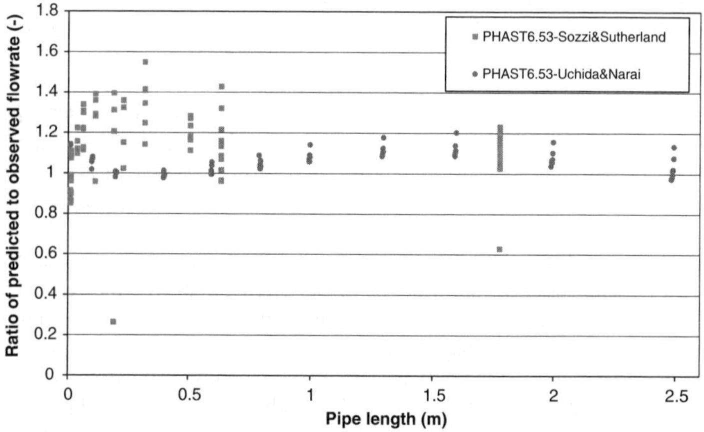

Figure 3. Phast 6.53 validation of flow rate for sub-cooled water release

3. DISPERSION AND POOL SPREADING/EVAPORATION

For dispersion modelling a very wide range of scenarios can be considered. Distinction can be made between momentum (un-pressurised or pressurised releases), time-dependency (steady-state, finite-duration, instantaneous or time-varying dispersion), buoyancy (buoyant rising cloud, passive dispersion or heavy-gas-dispersion), thermodynamic behaviour (isothermal or cold or hot plume, vapour or liquid or solid or multiple-phase, reactions or no reactions), ground effects (soil or water, flat terrain with uniform surface roughness, variable surface roughness, non-flat terrain, obstacles), and ambient conditions (e.g. stable, neutral or unstable conditions).

In the literature numerous text books and articles on dispersion can be found. Key literature including description of models and experimental data include Chapter 4 in the TNO yellow book (TNO, 1997), Sections 15.11–15.54 in Lees (Lees, 1996), and the CCPS dispersion guidelines (CCPS, 1996). Key experiments (benchmark tests for dispersion; input data and experimental results) have been stored in the MDA database by Hanna et al. (1993) in conjunction with comparison and validation of a wide range of models. Likewise data are stored in the REDIPHEM database partly as part of the EU project SMEDIS (Daish et al., 1999). The SMEDIS project has also produced a protocol for evaluating heavy gas dispersion models, which has also recently been proposed for application to LNG (Ivings et al., 2007).

Model verification and validation for dispersion models is illustrated below for the Phast dispersion model UDM (Witlox and Holt, 1999, 2007). This is an integral model, which can account for all the above type of releases except for effects of obstacles and non-flat terrain. The verification and validation for the UDM can be summarised as follows [see Witlox and Holt (2007) for full details and a detailed list of references]:

1. Jet and near-field passive dispersion. For an elevated horizontal continuous jet (of air), the UDM numerical results are shown to be identical to the results obtained by an analytical solution. For vertical jets very good agreement has been obtained against both the "Pratte and Baines" and "Briggs" plume rise correlations.
2. Heavy-gas dispersion. The UDM numerical results are shown to be in identical agreement against an analytical solution for a 2-D isothermal ground-level plume. The UDM has been validated against the set of three 2-D wind-tunnel experiments of McQuaid (1976). The new formulation has also been validated against the HTAG wind tunnel experiments (Petersen and Ratcliff, 1988). Furthermore the UDM model was verified against the HGSYSTEM model HEGADAS.
3. Far-field passive dispersion. For purely (far-field) passive continuous dispersion, the UDM numerical results are shown to be in close agreement with the vertical and cross-wind dispersion coefficients and concentrations obtained from the commonly adopted analytical Gaussian passive dispersion formula. The same agreement has been obtained for the case of purely (far-field) passive instantaneous dispersion, while assuming along-wind spreading equal to cross-wind spreading in the analytical profile.
4. Finite-duration releases. The UDM "Finite-duration-correction" module has been verified against the HGSYSTEM/SLAB steady-state results, and shown to lead to

finite-duration corrections virtually identical to the latter programs. Furthermore excellent agreement was obtained using this module for validation against the Kit Fox experiments (20-second releases of CO_2 during both neutral and stable conditions; see Figure 4).

5. Thermodynamics. The UDM dispersion model invokes the thermodynamics module while solving the dispersion equations in the downwind direction. This module describes the mixing of the released component with moist air, and may take into account water-vapour and heat transfer from the substrate to the cloud. The module calculates the phase distribution [component (vapour, liquid), water (vapour, liquid, ice)], vapour and liquid cloud temperature, and cloud density. Thus separate water (liquid or ice) and component (liquid) aerosols may form. The liquid component in the aerosol is considered to consist of spherical droplets and additional droplet equations may be solved to determine the droplet trajectories, droplet mass and droplet temperature. Rainout of the liquid component occurs if the droplet size is sufficiently large. The thermodynamics module also allows for more rigorous multi-component modelling (Witlox et al., 2006). The UDM homogeneous equilibrium model has been verified for both single-component and multi-component materials against the HEGADAS model. The UDM HF thermodynamics model (including effects of aqueous fog formation and polymerisation) was validated against the experiments by Schotte (1987).
6. Pool spreading/evaporation. If the droplet reaches the ground, rainout occurs, i.e. removal of the liquid component from the cloud. This produces a liquid pool which

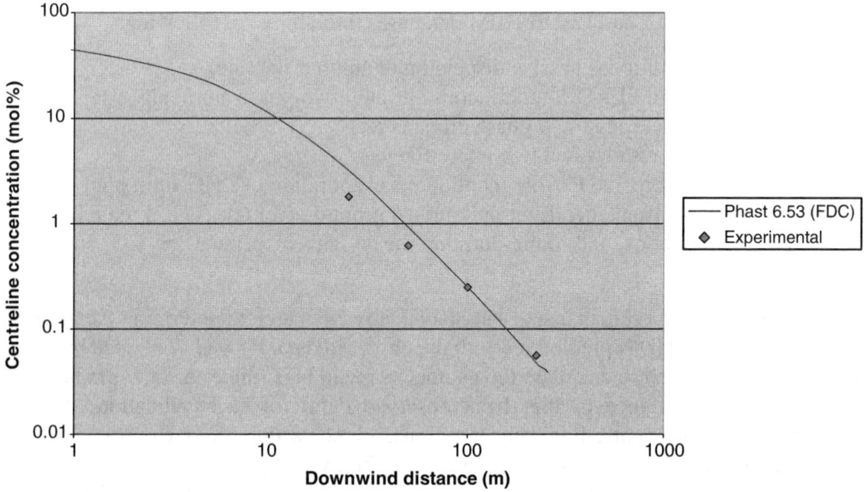

Figure 4. UDM dispersion results for Kit Fox experiment KF0706 (20 second release)

spreads and vaporises (see Figure 1). Vapour is added back into the cloud and allowance is made for this additional vapour flow to vary with time. The UDM source term model PVAP calculates the spreading and vapour flow rate from the pool. Different models are adopted depending whether the spill is on land or water, and whether it is an instantaneous or a continuous release. The pool spreads until it reaches a bund or a minimum pool thickness. The pool may either boil or evaporate while simultaneously spreading. For spills on land, the model takes into account heat conduction from the ground, ambient convection form the air, radiation and vapour diffusion. These are usually the main mechanisms for boiling and evaporation. Solution and possible reaction of the liquid in water are also included for spills on water, these being important for some chemicals. These effects are modelled numerically, maintaining mass and heat balances for both boiling and evaporating pools. This allows the pool temperature to vary as heat is either absorbed by the liquid or lost during evaporation.

PVAP was verified by David Webber against the SRD/HSE model GASP for a range of scenarios with the aim of testing the various sub-modules, and overall good agreement was obtained. The PVAP spreading logic was first validated against experimental data for spreading of non-volatile materials. Subsequently the PVAP evaporation logic was validated against experimental data in confined areas where spreading does not take place. Finally comparisons were made for simultaneously spreading and vaporising pools. The above validation was carried out for both spills on water and land, and a wide range of materials was included [LNG, propane, butane, pentane, hexane, cyclo-hexane, toluene, ammonia, nitrogen, water, Freon-11)].

The above covers the verification and the validation for the individual UDM modules. The validation of the overall model was carried out against large-scale field experiments selected from the MDA and REDIPHEM databases, including the following:

- Prairie Grass (continuous passive dispersion of sulphur dioxide).
- Desert Tortoise and FLADIS (continuous elevated two-phase ammonia jet)
- EEC (continuous elevated two-phase propane jet)
- Goldfish (continuous elevated two-phase HF jet)
- Maplin Sands, Burro and Coyote (continuous evaporation of LNG from pool)
- Thorney Island (instantaneous un-pressurised ground-level release of Freon-12)
- Kit Fox (continuous and finite-duration heavy-gas dispersion of CO_2 from area source)

Each of the above experimental sets was statistically evaluated to determine the accuracy and precision of the UDM predictions with the observed data. Formulas adopted by Hanna et al. (1993) were used to calculate the geometric mean bias (under or over-prediction of mean) and mean variance (scatter from observed data) for each validation run. This was carried out for centre-line concentrations, cloud widths, and (for the SMEDIS experiments) also off centre-line concentrations. The overall performance of the UDM in predicting both peak centreline concentration and cloud widths was found to be good for the above experiments.

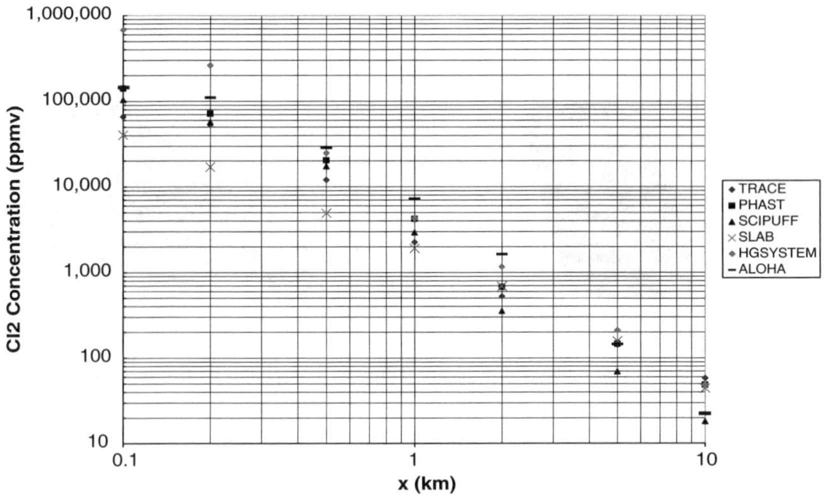

Figure 5. UDM (PHAST) verification against other models for Graniteville Chlorine accident

The overall UDM model was also recently verified by means of comparison against other models for three US chlorine accidents involving elevated two-phase chlorine jet releases. This is illustrated by Figure 5 for the case of the Graniteville accident; see Hanna et al. (2007) for full details.

4. FLAMMABLE EFFECTS

This section deals with the verification and validation of flammable effect models (fireballs, pool fires, jet fires and explosions, vapour cloud fires). Furthermore the most-established empirical models are considered only. Key literature including description of these models and experimental data include Chapters 5–6 of the TNO yellow book (TNO, 1997), Sections 16–17 in Lees (1996) and the CCPS guidelines (CCPS, 1994).

FIREBALLS, JET FIRES AND POOL FIRES

Empirical models for these fires include empirical correlations describing the fire geometry (most commonly a sphere for a fireball, a tilted cylinder for pool fire, and a cone for the jet fire) and the surface emissive power (radiation per unit of area emitted from the fire surface area); see Figure 6.

The radiation intensity (W/m^2) for a observer with given position and orientation is set as the product of the surface emissive power and the view factor. The view factor including the effects of atmospheric absorption is derived by means of integration over the

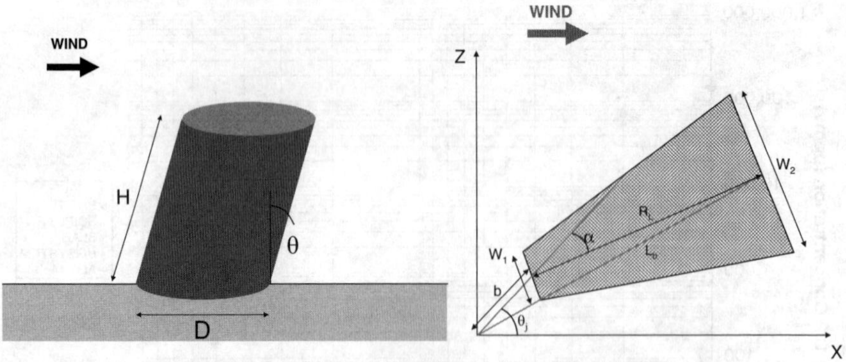

Figure 6. Geometry for pool fire (tilted cylinder) and jet fire (cone)

flame surface. In Phast this integration is carried out numerically, while other models adopt analytical expressions for specific fire geometries.

The fireball model from Martinsen and Marx (1999) is based on extensive literature, detailed tests and also allows for lift-off. More simplistic models are included in the above general references. The latter models can easily be verified by simple hand calculations.

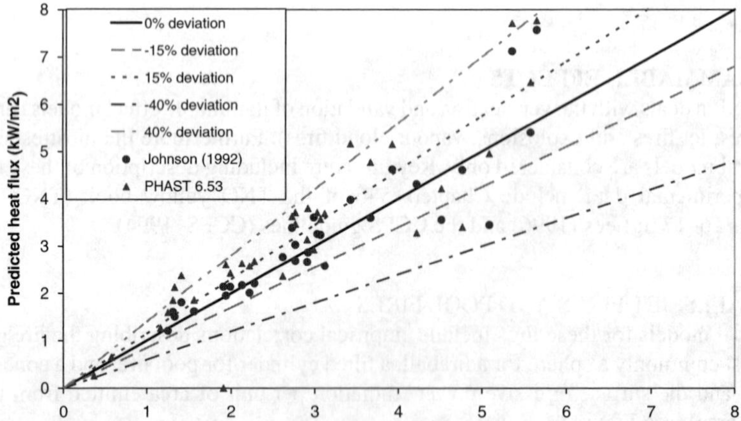

Figure 7. Predicted against measured incident radiation at different observer positions and orientations using the Phast 6.53 and Johnson pool fire models

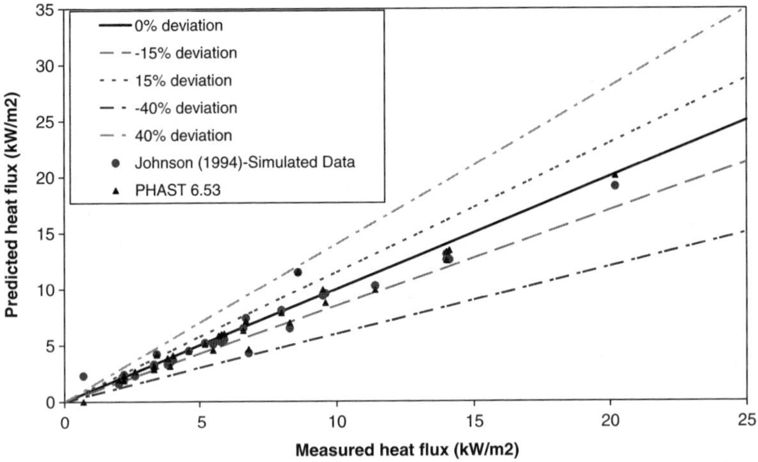

Figure 8. Predicted against measured incident radiation at different observer positions and orientations using the Phast 6.53 and Johnson jet fire models

The Phast pool fire model has been validated against data for LNG pool fires (Johnson, 1992); see Figure 7 which also includes verification against model predictions by Johnson (1992). Furthermore it has been validated against the Montoir LNG tests (Nedelka et al., 1990) and hexane tests (Lois and Swithenbank, 1979).

The Phast jet fire model has been validated against vertical natural-gas releases (Chamberlain, 1987), horizontal natural-gas and two-phase LPG releases (Bennett et al., 1991), and horizontal liquid-phase crude oil releases (Selby and Burgan, 1998). It has also been verified against model predictions by Johnson (Johnson et al., 1994) in the case of the horizontal natural-gas releases; see Figure 8.

EXPLOSION

Fitzgerald (2001) includes a detailed comparison of the TNO multi-energy (1988), Baker-Strehlow (1999) and CAM models (1999). This includes information of the latest versions of these models and comparison against experimental data (EMERGE experiments by TNO (EMERGE, 1998) and BFETS experiments by SCI (Selby and Burgan, 1998)). Clear conclusions are provided indicating under which conditions which model is best on over-pressure prediction. The latest available versions of the multi-energy (MULT) and Baker-Strehlow (BSEX) models have been implemented into Phast. They have been validated against the above EMERGE and BFETS experiments; see Figure 9 for the predictions of

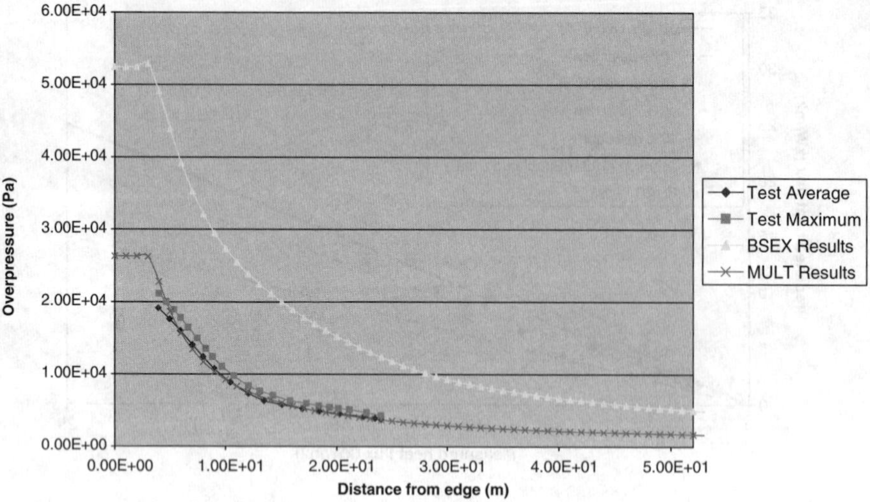

Figure 9. Validation of Phast models MULT and BSEX against EMERGE 6

overpressure (as function of distance from the edge of the congestion zone) for the case of the EMERGE 6 propane experiment (medium-scale 3D medium-congestion).

REFERENCES

Bennett, J. F., Cowley, L. T., Davenport, J. N., and Rowson, J. J., 1991, "Large scale natural gas and LPG jet fires - final report to the CEC", TNER 91.022

Center for Chemical Process Safety of the American Institute of Chemical Engineers (CCPS), 1994, "Guidelines for Evaluating the Characteristics of Vapor Cloud Explosions, Flash Fires and Bleves", American Institute of Chemical Engineers, New York

CCPS, 1996, "Guidelines for use of vapor cloud dispersion models", Second Edition, CCPS, New York

CCPS, 2000, "Guidelines for chemical process quantitative risk analysis", Second Edition, CCPS, New York, Section 2.1.1 – discharge rate models

Chamberlain, G.A., 1987, "Developments in design methods for predicting thermal radiation from flares", *Chem. Eng. Res. Des.*, 65: 299–309

Cowley, L.T. and Tam, V.H.Y., 1988, "Consequences of pressurised LPG releases: the Isle of Grain full scale experiments", *GASTECH 88, 13th International LNG/LPG Conference*, Kuala Lumpur

Daish, N.C, Britter, R.E., Linden, P.F., Jagger, S.F. and Carissimo, B., 1999, "SMEDIS: Scientific Model Evaluation Techniques Applied to Dense Gas Dispersion models in complex situations"., *Int. Conf. and workshop on modelling the consequences of accidental releases of hazardous materials*, San Francisco, California, CCPS, New York, 345–372

EMERGE, 1998, "Extended Modelling and Experimental Research into Gas Explosions", Final Summary Report for the project EMERGE, CEC Contract EV5V-CT93-0274

Fisher, H.G., Forrest, H.S., Grossel, S.S., Huff, J.E., Muller, A.R., Noronha, J.A., Shaw, D.A., and Tilley, B.J., 1992, "Emergency Relief System Design using DIERS technology", DIERS project manual, ISBN No. 0-8169-0568-1, Pub. No. X-123, AICHE, New York

Fitzgerald, G., 2001, 'A comparison of Simple Vapor Cloud Explosion Prediction Methodologies", *Second Annual Symposium, Mary Kay O'Connor Process Safety Center, "Beyond Regulatory Compliance: Making Safety Second Nature"*, Reed Arena, Texas A&M University, College Station, Texas

Hanna, S.R., Chang, J.C. and Strimaitis, D.G., 1993, "Hazardous gas model evaluation with field observations", *Atm. Env.*, 27a: 2265–2285

Hanna, S., Dharmavaram, S., Zhang, J., Sykes, I., Witlox, H. W. M., Khajehnajafi, S. and Koslan, K., 2007, "Comparison of six widely-used dense gas dispersion models for three actual chlorine railcar accidents", *Proceedings of 29th NATO/SPS International Technical Meeting on Air Pollution Modelling and its Application*, 24–28 September 2007, Aveiro, Portugal

Ivings, M.J., Jagger, S.F., Lea, C.J. and Webber, D.M., 2007, "Evaluating vapor dispersion models for safety analysis of LNG facilities", Contract by HSL for Fire Protection Research Foundation, Quincy, Massachusetts

Johnson, A.D., 1992, "A model for predicting thermal radiation hazards from large-scale LNG pool fires", *IChemE Symp. Series*, 130: 507–524

Johnson, A.D., Brightwell, H.M., and Carsley, A.J., 1994, "A model for predicting the thermal radiation hazard from large scale horizontally released natural gas jet fires", *Trans. IChemE.*, 72B:157–166

Lees, F.P., 1996, "Loss Prevention in the process industries: hazard identification, assessment and control", Second Edition, Butterworth-Heinemann, Oxford

Lois, E., and Swithenbank, J., 1979, "Fire hazards in oil tank arrays in a wind", *17th Symposium (Int.) on Combustion*, Leeds, Combustion Institute, Pittsburgh, PA, 1087–1098

Martinsen, W.E. and Marx, J.D., 1999, "An improved model for the prediction of radiant heat from fireballs", *International Conference and Workshop on Modelling the Consequences of Accidental Releases of Hazardous Materials*, CCPS, San Francisco, California, September 28–October 1, 605–621

McQuaid, J., 1976, "Some experiments on the structure of stably stratified shear flows", Technical Paper P21, Safety in Mines Research Establishment, Sheffield, UK

Nedelka, D., Moorhouse, J., and Tucker, R. F., 1990, "The Montoir 35m diameter LNG pool fire experiments", *Proc. 9th Intl. Cong and Exposition on LNG*, LNG9, Nice, 17–20 October 1989, Published by Institute of Gas technology, Chicago, 2-III-3: 1–23

Perry, R.H, Green, D.W. and Maloney, J.D., (eds.), 1999, "Perry Chemicals Engineering Handbook", 7th Edition, McGrawhill, Section 26 "Process safety"

Petersen, R.L. and Ratcliff, M.A., 1988, "Effect of homogeneous and heterogeneous surface roughness on HTAG dispersion", CPP Incorporated, Colorado. Contract for API, Draft Report CPP-87-0417

Schotte, W., 1987, "Fog formation of hydrogen fluoride in air", *Ind. Eng. Chem. Res.*, 26: 300–306; see also Schotte, W., "Thermodynamic model for HF formation", 31 August 1988, Letter from Schotte to Soczek, E.I. Du Pont de Nemours & Company, Du Pont Experimental Station, Engineering Department, Wilmington, Delawere 19898

Selby, C.A., and Burgan, B.A., 1998, "Blast and fire engineering for topside structures - phase 2: final summary report", SCI Publication No. 253, Steel Construction Institute, UK

Sozzi, G. L. and Sutherland, W. A., 1975, "Critical flow of saturated and sub-cooled water at high pressure", General Electric Co. Report No. NEDO-13418

TNO, 1997, "Methods for the calculation of physical effects" (TNO Yellow Book), CPR14E, SDU, The Hague

Uchida, H. and Narai, H., 1966, "Discharge of saturated water through pipes and orifices", *Proceedings 3day International Heat Transfer Conference, ASME*, Chicago, 5: 1–12

Webber, D.M., Fanneløp, T.K. and Witlox, H.W.M., 1999, Source terms for two-phase flow in long pipelines following an accidental breach, *International Conference and Workshop on Modelling the Consequences of Accidental Releases of Hazardous Materials*, CCPS, San Francisco, California, September 28–October 1, 145–168

Witlox, H.W.M. and Holt, A., 1999, "A unified model for jet, heavy and passive dispersion including droplet rainout and re-evaporation", *International Conference and Workshop on Modelling the Consequences of Accidental Releases of Hazardous Materials*, CCPS, San Francisco, California, September 28–October 1, 315–344

Witlox, H.W.M., Harper, M., Topalis, P. and Wilkinson, S., 2006, "Modelling the consequence of hazardous multi-component two-phase releases to the atmosphere", *Hazards XIX Conference*, Manchester, 250–265

Witlox, H.W.M., Harper, M., Bowen, P.J. and Cleary, V.M., 2007, "Flashing liquid jets and two-phase dispersion – II. Comparison and validation of droplet size and rainout formulations", *Journal of Hazardous Materials,* 142: 797–809

Witlox, H.W.M. and Holt, A., 2007, "Unified Dispersion Model – Technical Reference Manual", UDM Version 6.53 (distributed on reference CD as part of Phast 6.53 software), Det Norske Veritas, London

AVOIDANCE OF IGNITION SOURCES AS A BASIS OF SAFETY – LIMITATIONS AND CHALLENGES

Stephen Puttick
Process Hazards Section, Syngenta Huddersfield Manufacturing Centre, PO Box A38, Leeds Road, Huddersfield, HD2 1FF, UK
E-mail: stephen.puttick@syngenta.com

When operating plants and processes it is important to establish a basis or principle of safe operation. With a clearly defined basis of safety then appropriate precautions can be implemented to maintain that basis. The link between what is done (precautions and procedures) and actually being safe is then explicit.

For fire and explosion hazard assessment flammable and potentially flammable atmospheres must be identified and compared with the potential ignition sources present. With knowledge of the possible flammable atmospheres, their sensitivity to ignition and the possible ignition sources present and the incendivity of these sources a robust basis of safety may be selected. Preventative bases of safety (absence of flammable atmosphere and avoidance of ignition sources) are the most economic and so there will always be a driver to chose them over protective bases of safety (venting, suppression and containment). It is not always possible to use absence of flammable atmosphere due to insufficient fuel. Inerting brings its own set of problems, as well as expense, and possible difficulty of implementation. Avoidance of ignition sources can then appear to be an attractive option, but it has limitations. There are also challenges for all involved in research: some 'rules' that exist are based on very limited data and as such may be conservative, but without further data to show where the ultimate limits are we cannot justify breaking these rules.

This paper will discuss the limitations to the applicability of '*avoidance of ignition sources*', and the challenges to extending the validity of existing safety rules.

WHAT IS BASIS OF SAFETY/BASIS OF SAFETY CONCEPT

The terminology of basis of safety is well used, but it is worth a digression to clarify and elucidate on its meaning before looking in detail at avoidance of ignition sources.

Each element of the plant or processing step should have a unique *Basis of Safety* (BoS), otherwise known as *Basis for Safe Operation*, to counteract each specific type of hazard (e.g. fire and explosion, chemical reaction or toxicity). This is the principle which protects people from harm and injury.

TYPES OF BASIS OF SAFETY

There are two general types of BoS: *Preventative*; and *Protective*. Preventative Bases of Safety work on avoiding incidents or events. Protective Bases of Safety work on limiting the magnitude of an event or controlling the consequences to prevent harm. It should be emphasised that this is an admission that an event cannot ultimately be ruled out, and therefore must be managed.

For fire and explosion hazards the preventative bases of safety deal with the elements of the fire triangle. They fall into three categories: absence of flammable atmospheres achieved by limited fuel quantities; Control/Avoidance of Ignition sources; and absence of flammable atmospheres by limiting the quantities of oxidant present. Safety measures are chosen to eliminate an element of the fire triangle whilst causing minimal interference to plant operation (Gibson & Lloyd, 1963).

Protective Bases of Safety (for fire and explosion hazards) yet again fall into three categories: venting the explosion (to a safe area) to prevent pressure in the equipment exceeding a given level; containing the explosion (and limiting its ability to propagate to other equipment); and explosion suppression – detecting and quenching the explosion before it exceeds a given pressure. Contrary to the opinion of some, it is not unreasonable to expect a protective basis of safety to be activated under process conditions at some stage, even if this is an infrequent event.

RELATIONSHIP BETWEEN BASIS OF SAFETY AND CONTROL MEASURES

A basis of safety is the principle or philosophy of operation that maintains safety rather than the specific measures required to implement it. The basis of safety must be clearly documented as such, this includes the limits to the scope and delineation of where it changes in the plant or process, and how the two elements are separated. The specific precautions to establish and maintain the basis of safety must also be clearly recorded. If the precautions are not clearly distinguished from the principle it is possible to lose sight of the intent.

It is this author's experience that often people are confused about what is keeping them safe, and can fixate on one control measure. For example basis of safety is often described as nitrogen; this is insufficient. Nitrogen is often supplied to process vessels for quality reasons (excluding atmospheric water vapour or eliminating minor oxidation), it is not necessarily supplied with sufficient reliability or rigour to constitute a protective measure. Inerting with nitrogen may not protect against some decompositions. And finally stating it as nitrogen does not sufficiently highlight the need to achieve a given oxygen level, and then maintain that. Whereas this should be avoidance of flammable atmospheres by restricting oxygen, then it will have an associated set of precautions aimed at establishing an inert atmosphere, and then maintaining it. This might sound pedantic, but clarifying this in documentation can facilitate understanding, and with that can come thought about changes to process and operations.

By keeping the basis of safety as a principle, the measures or precautions taken can be audited and examined to verify that they will actually achieve the desired end result, or even that they can be sustained with sufficient robustness. This is an important aspect of the iterative process of choosing an appropriate basis of safety.

Another reason for separating the basis of safety from the precautions is that on any chemical plant many precautions (in particular those associated with avoiding ignition sources) are taken as a matter of good practice, but do not necessarily implement the basis of safety. They may be there to reduce the frequency of events where a protective basis of safety is employed. Or may be there to support another preventative basis of safety which may not be infallible, and will have a given failure rate, e.g. absence of flammable atmospheres through insufficient oxygen achieved by inerting with nitrogen; any nitrogen supply system will have some frequency of non-availability or pressure failure during downtime. Another reason precautions may be implemented is to prevent *practice creep* or *creeping change*. Precautions such as excluding the use of non- conducting plastics are implemented everywhere on a plant to make sure plastics do not find their way from an area where it does not matter, into an area where they can pose an ignition hazard. This can be important where many units essentially look identical, even though they may be processing different materials with different hazards. Practice creep can be associated with several incidents where change control has failed for example Ackroyd & Newton (2002) where a plastic IBC had been used for aqueous waste, and was then used for a non-conducting and flammable waste, which led to an ignition.

LIMITATIONS OF BASIS OF SAFETY

The Basis of Safety will not necessarily protect the plant, or the materials being processed. It will also not necessarily address the economics of continued operation nor the risks of interruption to the business.

A basis of safety which protects against one type of hazard can lead to another hazard and violate a different type of basis of safety. For example protecting against toxicity an appropriate basis of safety might be containment, but for fire and explosion it could be explosion venting which would potentially violate the first basis of safety. The basis of safety against one hazard cannot be chosen independently and without reference to other hazards.

USING AVOIDANCE OF IGNITION SOURCES AS A BASIS OF SAFETY

TERMINOLOGY

Various workers use the terms avoidance, elimination, control or absence of ignition sources, and at least amongst the author's co-workers there is some debate as to the correct terminology. However, whatever word is preferred perhaps the full phrase ought to be avoidance of *viable or effective* ignition sources.

SETTING A BASIS OF SAFETY TO COUNTERACT FIRES AND EXPLOSIONS

Setting a basis of safety is always based on knowledge of the flammable or potentially flammable atmosphere characteristics. Different bases of safety require different data.

In preference order the preventative bases of safety are:

1. Avoidance of flammable atmospheres by limiting fuel concentrations
2. Avoidance of viable ignition sources
3. Avoidance of flammable atmospheres by limiting oxidant concentrations

If the fuel concentrations can be reliably kept below flammable ranges then avoidance of flammable atmospheres achieved by limited fuel concentrations is a viable basis of safety.

Failing this potential ignition sources can be identified and compared to the sensitivity of the atmosphere. If measures can be rigorously enforced to avoid viable ignition sources, without unduly affecting plant operation, then this could be a viable basis of safety.

Otherwise the minimum oxygen concentrations for combustion (MOC) must be determined. If levels of oxygen can reliably and economically be kept below this level then it could be a feasible basis of safety. However, inerting has its own drawbacks, and there are a large number of fatal incidents associated with nitrogen asphyxiation. Another problem can be emissions of volatiles in waste nitrogen. Inerting should not be regarded as an easy option compared with engineering appropriate control measures against ignition sources, particularly electrostatic sources which have many well defined control measures. Maintaining inert atmospheres can have its own engineering challenges.

Finally if these are not feasible or insufficiently reliable, then protective bases of safety must be considered, all of which require explosion violence characteristics.

Several publications deal with ignition sources, control measures and quantifying material properties and have a level of detail that will not be covered here (e.g. Barton, 2002; Dickens, 1996; Gibson *et al.*, 1985; Gibson & Rogers, 1980 and Walmsley, 1992). What should be particularly emphasised is Avoidance of Ignition sources requires intimate knowledge of the actual operations being carried out.

ILLUSTRATIONS OF THE LIMITATIONS OF AVOIDANCE OF IGNITION SOURCES

LEAKS

One particular example in Kletz (2001) illustrates several points about the avoidance of ignition as a basis of safety. In the particular plant concerned leaks of ethylene were tolerated because it was considered that all sources of ignition had been eliminated. There are two reasons why this is not a valid basis of safety.

a) Avoidance of ignition sources is only suitable for use within defined areas of process plant where ignition sources can be rigorously controlled; and to a more limited extent where materials are charged into and discharged from plant, yet again within well defined areas.
b) Ethylene has an MIE of 0.07 mJ (NFPA, 2000). This is more sensitive than normal solvent vapour and flammable gas atmospheres which tend to be 0.2 mJ and above. It is not usual to apply this basis of safety to atmospheres with a sensitivity below 0.1 mJ. Below this level electrostatic sources of ignition cannot be controlled with any degree of confidence.

The correct basis of safety in this case should be absence of flammable atmospheres by containment of fuel within pipework. This could then be supported by avoidance of ignition sources to reduce ignition frequency for the inevitable times when there is a loss of containment, but the fact that humans can be present in the area, and that they are potentially a very effective ignition source means that containment should be emphasised.

In the example the area was zoned. However, the existence of zoned areas is an acknowledgement that losses of containment with resultant flammable atmospheres occur; it is not a measure of the acceptability of losses of containment.

This also applies to powder layers which are dormant flammable atmospheres and should not be allowed to accumulate. Powder layers which have been disturbed have led to major secondary dust explosions.

CHARGE CHUTES

Charging solids down a charge chute into a flammable vapour atmosphere one can acceptably avoid ignition sources provided that the charge chute is less than 3 m long, and the vapour atmosphere has a sensitivity of 0.2 mJ or above. It is likely that the acceptable charge chute length may be longer but this has not been determined. However, if charging into more sensitive atmospheres that may contain hydrogen for example then an alternative basis of safety must be employed. This would apply to a solid such as Sodium borohydride which could emit hydrogen on addition to a solvent.

DRYERS

In spray dryers it is normal to control ignition sources as part of the operating regime such as by earthing and bonding, and keeping inlet temperatures below the Minimum Ignition Temperature (MIT) of the dust cloud, but this is not the basis of safety. There is sufficient uncertainty about the thermal stability of accumulated powder layers that thermal decomposition cannot be reliably excluded as a potential ignition source (Gibson & Schofield, 1977). Spray dryers should be operated with an alternative basis of safety – usually protective such as venting, or suppression. Absence of flammable atmospheres through inert gas blanketing can also be feasible, although this usually relies on recirculation of a portion of the spent gas (Gibson *et al.*, 1985).

VENT HEADERS

Vent headers are another area where there is a huge temptation to try and employ avoidance of ignition sources as the basis of safety especially when one considers the potential cost of inerting or the necessary air flowrates and consequent fan sizes for dilution. However, vent headers are often shared with multiple vessels and there is potential for interactions between the vented materials. For example in Anonymous (1995) there are records of fires stemming from reactions of amines and NOx gases. Solids can and often do accumulate, leading to potential thermal stability issues, and vessels sometimes foam over into the vents. In fact this author has witnessed common vents which were effectively dug out, with many years of accumulated material. Although avoiding ignition sources where possible in these cases is a good idea, it is not suitable as a basis of safety (Iqbal Essa & Ennis, 2001).

SOLIDS ACCUMULATION
Solids near their melting point can stick and accumulate, even when being transported. In one case material formed a non-conductive layer on the inside of a metal pipe (Perbal, 2005). This layer became electrostatically charged, leading to incendive propagating brush discharges and hence explosion and fire.

In another case a powder which was known to be thermally unstable at relatively low temperatures was charged to a vessel using avoidance of ignition sources as the basis of safety. This assumed that the powder would not accumulate in layers greater than about 5 mm thick. The 5 mm layer ignition temperature was comfortably above operating temperatures. In practice the powder deliquesced with atmospheric water and formed layers up to 10 cm in thickness, these thicker layers would have an onset close to ambient temperature. Although the build up of material was discovered by operational staff who cleaned it out, the significance for safe operation was not appreciated. Eventually, after a failure in the normal processing sequence there was an overpressure event which blew a bursting disk and showed evidence of burning.

SOLVENTS IN POWDER
Some old guidance for avoidance of ignition sources discounted the incendivity of brush discharges with powders and simultaneously treated powders with up to 0.5% solvent present as only being as sensitive to ignition as the powder. An incident occurred when powder with less than 0.5% solvent was being milled. The milling released solvent from the powder, and a vapour atmosphere built up, which was in turn ignited by a brush discharge from the powder (SUVA, 2005). See also Puttick and Gibbon (2004) for more on this topic.

FUTURE CHALLENGES

Avoidance of ignition sources relies on well defined materials and ignition sources so that the sensitivity of the flammable atmosphere can be matched against the incendivity of the ignition source. There are still some large gaps in our knowledge, and filling these could allow us to apply avoidance of ignition sources to a greater range of situations.

DUSTS AND BRUSH DISCHARGES
Modern processes are creating finer dusts which are in turn increasingly sensitive to electrostatic spark ignition. Current standard dust testing equipment can only create sparks down to 1 mJ. There are a number of powders with MIEs below 1 mJ, but how far below? It is generally accepted now that brush discharges despite containing up to 3 mJ of discharge energy cannot ignite powders with MIEs above 1 mJ, unless there is solvent present. But it has also been seen that brush discharges can be made to ignite sensitive powders under extreme conditions (such as enhanced oxygen). However, the ability to measure MIEs below 1 mJ is a recent development, and there is not yet much data available. The missing

part of this jigsaw is to then determine how sensitive a flammable dust would need to be to be ignited by a brush discharge.

FLAMMABLE MISTS

It has been known for a long time that mists of fuel can be flammable below the flash point (e.g. Burgogne & Richardson, 1949). Industrially mists can be created when formulating products; solvent borne materials are sprayed onto solid substrates, often in rotary mixers where electrostatic charges can be generated.

Some work has been undertaken on ignition energies, e.g. Singh (1986), but much of this work has been associated with automotive ignition and high altitude jet re-ignition, rather than hazards of handling within an industrial situation. The sensitivity to ignition will depend on the fuel and droplet size. In a spraying operation there will be a range of droplet sizes, and it is likely that the incendivity will be influenced strongly by the fraction at the lower end of the size range. Large scale spraying tests use very large volumes of fuel, which can be difficult to justify and expensive. However, useful worst case data might be obtained by characterising nozzles with water, then to use a small scale nozzle with a narrow distribution, and small drop size for ignition testing. This would be analogous with sieving dusts to less than 63 μm for MIE testing.

If a reliable MIE can be determined for a flammable mist, then it is possible to consider whether precautions against incendive electrostatic discharges can be implemented.

ELECTROSTATIC MISTS

There is a body of work associated with washing large tanks for crude oil and chemical transport (e.g. Hughes, 1972; van de Weerd, 1975; Jones & Bond, 1984 and Walmsley, 1987), and some work on much smaller vessels including spraying of solvents and two phase mixtures (Post et al., 1989). Much batch processing and formulation requires cleaning of vessels between products. The standards required can be extremely challenging to avoid cross contamination and product quality issues, and there is always a demand to clean with solvents.

Looking at vessel sizes and/or spray set-ups it is clear that much of this falls outside the existing guidance and experimental work from which the guidance was derived. The charge that will be generated by a nozzle and the consequent mist electric field are not clearly calculable *a priori*. Many of the proposed operations are probably safe, but we cannot prove this; inerting vessels ends up being the only justifiable safety measure, but this is unsatisfactory when many of these vessels are not routinely supplied with nitrogen systems.

MECHANICAL SPARKS AND FRICTION

What is required from an assessment perspective is to be able to characterise a flammable atmosphere with respect to its ignition sensitivity, and to identify potential ignition sources so the two can be compared. Although there are gaps in our knowledge of electrostatics discussed above, the gaps are more glaring in the case of mechanical ignition sources, in particular the issue of quantification.

MECHEX (Proust et al., 2007) has set about correcting this state of affairs, and has some tools now in place. Disconcertingly one the the *rules* previously used (friction between surfaces is acceptable at less than 1 m/s) has been shown to not be so absolute, and the situation is much more complex. If this work is continued hopefully it will yield quantifications for atmospheres and ignition sources.

CONCLUSIONS

Avoidance of Ignition sources can be a useful and reliable basis of safety in certain circumstances provided that it is restricted to the inside of chemical plants, and certain well defined charging and discharging areas. Its reliability depends on having relatively insensitive atmospheres and the main applicability will be counteracting electrostatic and some mechanical ignition sources. It is vital that potential ignition sources are identified, and there is feedback from operational experience back into the hazard assessment process to identify changes and deviations from original expectations. If avoidance of ignition sources is to be safely applied it is vital to be fully conversant with the details of plant and operations. It can also be important to be aware of material handling properties which are outside the scope of normal hazardous properties, but can affect what occurs on plant.

Future work should be around better defining of potential ignition sources, and better characterisation of the sensitivity of atmospheres.

REFERENCES

Ackroyd, G. & Newton, S. (2002), 'Flash Fire during Filling', *Loss Prevention Bulletin* 165, 13–14.

Anonymous (1995), 'Nitrous fumes/amine interaction', *Loss Prevention Bulletin* 124, 15.

Barton, J. (2002), Dust Explosion: Prevention and Protection – a Practical Guide, IChemE.

Burgoyne, J. H. & Richardson, J. F. (1949) The inflammability of oil mists *Fuel*, 28(1) pp 2–6.

Dickens, A. (1996) Dust explosibility testing: to what extent is it necessary in prescribing a basis of safety? *Loss Prevention Bulletin*, 129, 18–22.

Iqbal Essa, M. and Ennis, A. (2001) Thermal Oxidiser Fire and Explosion Hazards, in *Hazards XVI: Analysing the past, planning the future*, IChemE Symposium Series 148 pp 209-217.

Gibson, N.; Harper, D. J. & Rogers, R. L. (1985), 'Evaluation of the Fire and Explosion Risk in Drying Powders', *Plant/Operation Progress* 4 (3), 181–189.

Gibson, N. & Lloyd, F. C. (1963) Static Electricity in Chemical Plants and Processes, ICI Technical Memo, D62100A.

Gibson N. & Rogers R. L. (1980) Ignition and Combustion of Dust Clouds in Hot Environments – An Exploratory Study, In *Chemical Process Hazards VII* IChemE Symposium Series 58 pp 209–217.

Gibson, N. & Schofield, F. (1977) Fire and Explosion Hazards in Spray Dryers, In *Chemical Process Hazards VI* IChemE Symposium Series 49 pp 53–62.

Hughes, J. F. (1972) Electrostatic hazards in supertanker cleaning operations *Nature*, 235, 381–383.

Jones, M. R. O. & Bond, J. (1984) Electrostatic hazards associated with marine chemical tanker operations: Criteria of incendivity in tank cleaning operations. *Chemical Engineering Research and Design* (Transactions of the Institution of Chemical Engineers, Part A), 62, 327–333.

Kletz, T. (2001), *Learning from Accidents*, Gulf Professional Publishing, Chapter 4: A Gas Leak and Explosion–the Hazards of Insularity, pp. 40–51.

NFPA 77, Recommended Practice on Static Electricity, 2000 Edition.

Perbal, R. (2005), 'ABS dust explosion in a silo–unexpected electrostatic ignition hazards', *Loss Prevention Bulletin* 181, 3–9.

Post, L.; Glor, M.; Lütgens, G. & Maurer, B. (1989) The Avoidance of Ignition Hazards due to Electrostatic Charges occurring during Spraying of Liquids under High Pressure, *Journal of Electrostatics*, 23, 99–109.

Proust, C.; Hawksworth, S.; Rogers, R.; Beyer, M.; Lakic, D.; Raveau, D.; Herve, P.; Pina, V.; Petitfrere, C. & Lefebvre, X. (2007), 'Development of a method for predicting the ignition of explosive atmospheres by mechanical friction and impacts (MECHEX)', *Journal of Loss Prevention in the Process Industries* 20(4–6), 349–365.

Puttick, S. J. & Gibbon, H. J., (2004) Solvents in Powder, in *Hazards XVIII: Process Safety–Sharing Best Practice*, IChemE Symposium Series 150, 507-516.

Singh, A. K. (1986) Spark ignition of monodisperse aerosols, PhD Thesis Rutgers University.

SUVA (2005) Dust Explosion Incidents: Their Causes Effects and Prevention, ISSA Prevention Series No 2051.

van de Weerd, J. M. (1975) Electrostatic charge generation during tank washing. Spark mechanisms in tanks filled with charged mist, *Journal of Electrostatics*, 1, 295–309.

Walmsley, H. L. (1989) Electrostatic hazards from water slugs formed during the washing of ships tanks: spark energy calculations, *Journal of Physics D. Applied Physics* 20, 329–339.

Walmsley, H. L. (1992) The Avoidance of Electrostatic Hazards in the Petroleum Industry, *Journal of Electrostatics*, 27 (1&2).

ASSESSMENT OF FLAMMABLE GAS INGESTION AND MIXING IN OFFSHORE HVAC DUCTS: IMPLICATIONS FOR GAS DETECTION STRATEGIES

C J Lea[1], M Deevy[2], and K O'Donnell[3]
[1]Lea CFD Associates Ltd, 12 Sheraton Way, Buxton, Derbyshire, SK17 6FA
[2]Health & Safety Laboratory, Harpur Hill, Buxton, Derbyshire, SK17 9JN
[3]Offshore Safety Division, Health & Safety Executive, Redgrave Court, Merton Road, Bootle, Merseyside, L20 7HS.

An assessment of flammable gas ingestion and mixing in offshore HVAC ducts is presented as a basis for a set of initial recommendations on gas detection strategies. These recommendations are based on the findings of a literature review and Computational Fluid Dynamics modelling of gas releases, supported by a scoping study which examined technologies for the detection of hydrocarbons on offshore platforms.

The circumstances by which a non-uniform distribution of gas could be present immediately inside or outside of an HVAC inlet are of particular interest: if an HVAC inlet ingests a non-uniform distribution of gas then there is the possibility that this could be 'missed' by the detection system.

The overall aim is to provide a basis for advice to HSE inspectors and industry on the effectiveness of flammable gas detection strategies for offshore HVAC ducts. This paper concludes with a number of initial recommendations on such strategies.

1. INTRODUCTION

The accidental release of flammable gas on offshore installations can potentially lead to the build-up of an explosive mixture. Natural or forced ventilation can help to mitigate such incidents and gas detection systems play a key role in reducing the risks from releases by enabling early detection and subsequent interventions. The provision and siting of gas detectors for open areas and gas turbine enclosures has been studied over a number of years and is comparatively well documented. However, there is much less information available on the provision and siting of gas detection systems for HVAC (Heating, Ventilation and Air Conditioning) ducts supplying air to accommodation modules, temporary refuges or process areas on an installation.

This paper presents research funded by the UK Health and Safety Executive (HSE) to examine the ingestion of flammable gas releases into offshore HVAC inlets and the subsequent mixing of gas inside HVAC ducts. Full details can be found in Lea & Deevy

(2007). The circumstances by which a non-uniform distribution of gas could be present immediately inside or outside of an HVAC inlet are of particular interest: if an HVAC inlet ingests a non-uniform distribution of gas then there is the possibility that this could be 'missed' by the detection system. The overall aim of the research was to provide a basis for advice to HSE inspectors and industry on the effectiveness of flammable gas detection strategies for offshore HVAC ducts. A summary of gas detection technologies used on offshore platforms is provided in Section 2.

The research is based on a review of the literature and Computational Fluid Dynamics (CFD) modelling. It has, in part, been prompted by an incident on the Brae Alpha platform in 2004 when there was a delay in confirmed detection and shutdown of the HVAC system despite gas being ingested into the HVAC inlets. There is no suggestion that the detectors were not operating correctly at the time of this incident.

The literature review has been very wide-ranging. It draws heavily on relevant research from the nuclear industry on the sampling of gas distributions in exhaust stacks. The key findings are summarised in Section 3. CFD simulations of a high and low pressure gas release have been undertaken for idealised representations of an offshore platform, as well as a high pressure release for a more realistic geometry based loosely on the Brae Alpha incident. These CFD results are post-processed to gain insights into the likely effectiveness of a range of detector systems for HVAC ducts. A representative sample of the CFD simulations which have been undertaken in this research are described in Section 4. A discussion of the main findings and a set of initial recommendations on flammable gas detection strategies for offshore HVAC ducts are given in Section 5. Although the study is focused on offshore HVAC ducts, the findings are also likely to be relevant to onshore installations in which gas detection is required for HVAC ducts.

2. GAS DETECTION TECHNOLOGIES FOR OFFSHORE PLATFORMS

Walsh *et al.* (2005) divide technologies for detecting hydrocarbons on offshore platforms into three main categories (excluding acoustic systems which do not measure gas concentration):

- Catalytic (also known as pellistor) point detectors.
- Infrared point detectors, which are based on the absorption of infrared light at different wavelengths by flammable (and other) gases.
- Infrared open path (beam) detectors, which use the same measurement principle as the infrared point detectors but the beam traverses a long open path and absorption in the beam is detected as a gas concentration in the same way as for the infrared point detector.

A recent development in HVAC detection technology is the extended closed path point infrared detector, which measures an average concentration over a path length of typically around 1 m. Such systems have quoted minimum alarm levels of 5% of the Lower Explosive Limit (LEL), which is significantly lower than typical alarm levels of 20% LEL.

Detectors for offshore HVAC ducts are usually located immediately outside, or just inside, the HVAC inlet. For monitoring inside ducts, point detectors can be employed in two ways: in the duct itself or on the end of a sampling system which extracts gas from the duct (known as aspirated systems).

3. FLOW AND DISPERSION OF GAS IN A DUCT

The flow in offshore HVAC ducts will generally have a Reynolds number in the range 10^5 to 10^6, based on a typical duct velocity of 5 m/s (BS EN ISO 15138) and a range of duct hydraulic diameters from 0.5 to 5 m. Whilst this is high enough to ensure fully turbulent flow some distance downstream from the entrance to a duct, such conditions may not exist immediately inside the entrance. Usually there is a development region over which turbulent boundary layers on the walls of a duct grow and eventually merge, ultimately leading to a local equilibrium in which the flow no longer changes. It is then said to be fully-developed. As a rule of thumb, Hinze (1975) recommended that fully developed turbulent flow can be assumed to occur in straight pipes with a rounded inlet after a minimum development length of 40 pipe diameters. The flow in a straight square or rectangular duct behaves in a broadly similar manner, in that the distance to fully-developed conditions is not short. Melling & Whitelaw (1976) present data which show that fully-developed turbulent flow in a square duct is reached at about 25 duct widths from the inlet.

In the offshore environment the entrance to an HVAC duct will typically be sharp-edged. In addition, there are usually obstructions present just inside and at the entrance to ducts comprising louvres, grilles and fire dampers to provide isolation from fire and gas in the event of an incident. All of these features will generate turbulence in the entrance region of a duct. Furthermore, the flow conditions immediately outside of an HVAC duct may also be turbulent due to wind flow over obstructions on an offshore platform. However, whilst the flow may well be turbulent at the inlet to an HVAC duct, and certainly will be turbulent across its full cross-section some distance downstream from the inlet, it would be wrong to simply assume that mixing will, as a consequence, be so rapid that any non-uniformity in the distribution of gas at an HVAC inlet will very quickly be dispersed to give well-mixed uniform conditions. This has a significant impact on the siting of gas detectors in HVAC ducts.

There is a significant body of literature on the mixing of a tracer gas in circular, square and rectangular ducts. In the presence or absence of bends and mixing elements it shows that the distance before well-mixed conditions are obtained can be very long and comparable to the length of the development region for fully-developed turbulent flow. This literature stems from research on the sampling of exhaust duct stacks in the nuclear industry undertaken to support the improvement and updating of American standards on gaseous radionuclide emissions (Hampl *et al.* 1986, McFarland *et al.* 1999, Anand *et al.* 2003 and Seo *et al.* 2006). In most of this research a passive tracer was released from a single location on the axis of a duct. Note that at low gas concentrations in the flammable range and for the velocities typically encountered in offshore HVAC ducts, a natural gas mixture can be regarded as a passive contaminant since the Richardson number (Simpson, 1997) is likely to be at least an order of magnitude too low for any turbulence-modifying effects of a slightly buoyant gas to dominate over shear-induced turbulence. In these tracer releases, multiple point concentration measurements were made across the entire cross-section of the duct at a number of axial locations downstream from the release. To characterise the degree of mixing, a parameter known as the Coefficient of Variation (COV) was introduced. This is defined as:

$$COV = \frac{1}{C_{mean}} \sqrt{\frac{1}{N-1} \sum_{i=1}^{N} \left(C_i - C_{mean}\right)^2} \tag{1}$$

where N is the number of samples at a particular downstream location, C_i is the concentration of the ith sample and C_{mean} is the mean concentration over all samples at that location, defined as:

$$C_{mean} = \frac{1}{N} \sum_{i=1}^{N} C_i \tag{2}$$

The COV is simply the sample standard deviation divided by the sample mean. As an example, if the concentration distribution is such that across one half of a duct the concentration is a uniform 30% LEL whilst across the other half the concentration is a uniform 10% LEL, then the COV obtained from a large number of samples would be 0.5 (the sample mean is 20% LEL and the sample standard deviation is 10% LEL). The updated American standards (ANSI/HPS N13.1-1999) on the sampling of radioactive materials in stacks allow for single point sampling of gaseous contaminants in a duct if the COV for both velocity and concentration of a tracer gas are less than 0.2 over the central two-thirds of a duct. We do not suggest or comment on the practicality or appropriateness of these criteria for offshore HVAC ducts. However, the notion of a COV is helpful in quantifying the uniformity of mixing in a duct and it is readily computed from CFD results.

This body of research on stack sampling provides much useful data on how the COV is affected by a range of configurations. Anand *et al.* (2003) show that the distance from a point release of a tracer in a straight pipe to the position at which the COV is less than 0.2 depends on the upstream turbulence intensity. For a low turbulence intensity of 1.5%, the COV falls very slowly with distance and is still greater than unity at 30 duct diameters downstream from the point of release. Even with a high turbulence intensity of 10%, generated by passing the flow through an array of thick rods, a COV of 0.2 was still not reached after 25 duct diameters downstream.

McFarland *et al.* (1999) investigated the effect of bends and static mixing elements on the COV. They show that a single smooth 90° bend in a circular duct still requires a distance of nine diameters downstream from the bend before the 0.2 COV criterion is met. The performance of the static mixing elements was very variable; they all resulted in the 0.2 COV criterion being met within nine diameters downstream, but the most effective mixers were able to meet the criterion within three duct diameters of the mixing element. The most simple and effective mixing elements consisted of two large flow deflectors attached to opposite walls of a duct, giving a slot-like opening in the centre of the duct. Two or more of these mixing elements were used in series. The common characteristic of the most effective mixing elements appears to be the generation of large turbulent eddies which promote mixing across a duct (Seo *et al.*, 2006). Mixing elements which only introduced flow swirl were less effective. One disadvantage of these simple deflector mixing elements is a relatively large non-dimensional pressure coefficient (5.0, for two elements in series).

Seo *et al.* (2006) examined the behaviour of the COV in square and rectangular ducts (aspect ratio of 3:1) with and without bends. They report that the COV is similar for circular and square-section ducts at large distances downstream from the tracer release point both with and without bends. This implies that the main findings of the above work on circular-section ducts are largely likely to carry over to square-section ducts. However, a significant difference was found when the COV for the square and rectangular-section ducts were compared; typically the COV was much higher for the rectangular duct at any given distance downstream from the point of release, by about a factor of four. The COV for rectangular ducts with bends are also consistently higher than the same flow configuration in a square duct. Seo *et al.* (2006) speculate that this is because turbulent eddies in a wide duct have less opportunity to effectively transfer mass and momentum from one side of a duct to another.

Much of the above research on stack sampling is based on a flow which is well-controlled at the inlet to a duct, for example by use of a rounded entrance or other flow control devices. As already discussed, this will not be the case for offshore HVAC installations. Turbulence can be generated by large-scale flow separation at the sharp-edged entrance to a duct or may already be present in the ambient flow outside of the duct. McFarland *et al.* (1999) note that the work of Hampl *et al.* (1986) was based on a sharp-edged inlet and in comparison to their later research using a well-controlled approach flow the COV are found to be reduced by between a factor of two to three: flow separation at the inlet enhances mixing inside a duct. Nevertheless, Hampl *et al.* (1986) still suggest that up to 50 duct diameters may be needed for near-uniform mixing of a passive tracer in a straight pipe, even with a sharp-edged inlet.

The effect of grilles on the turbulence in a duct is relatively well understood. Laws & Livesey (1978) explain that the effect is either to suppress or enhance turbulence dependent on the geometry of the grille. Thus a very fine grille, or mesh, will tend to suppress turbulence and any turbulence which is introduced by the mesh decays quickly due to its small scale. A grid of relatively large diameter rods will enhance turbulence, although Laws & Livesey (1978) state that it is difficult to achieve a turbulence intensity of much higher than 10%. It is not clear whether grilles typically used to cover HVAC inlets will suppress or enhance turbulence. However, even if turbulence is significantly enhanced, Anand *et al.* (2003) show that the COV will remain high for long distances downstream. For an inlet turbulence intensity of 10% created by an array of thick rods, they found that the COV was still over 0.5 at 15 duct diameters downstream from the rod array. The reason for the relative ineffectiveness of such devices on mixing is that they introduce turbulence on too small a length-scale.

It is also significant that both Anand *et al.* (2003) and Seo *et al.* (2006) note that the COV is little-affected by the Reynolds number. Seo *et al.* (2006) state that for a square duct the Reynolds number over the range 25,000 to 150,000 has only a small effect on COV, whereas for a rectangular duct with a 3:1 aspect ratio the COV shows a significant dependence on Reynolds number below 50,000 but relatively little dependence at higher Reynolds number. Anand *et al.* (2003) and Seo *et al.* (2006) conclude that for fully turbulent flow, mixing is primarily dependent on geometry.

4. CFD MODELLING OF GAS INGESTION AND DISTRIBUTION INSIDE HVAC DUCTS

The distribution of gas concentration over the cross-section of a free jet or plume varies continuously with radius (Rodi, 1982). For a very high pressure release the pseudo-source approach of Ewan and Moodie (1986) can be used in conjunction with empirical data from Rodi (1982) to give a good indication of this concentration distribution. Thus for a release of pure methane at a stagnation pressure of 100 bar from a hole of 12 mm diameter, the concentration at approximately 10 m downstream from the release would be 100% LEL on the jet centreline but just 10% LEL at a radius of 1.9 m. This distance, over which concentration varies by a factor of ten, is broadly comparable to the dimensions of typical offshore HVAC inlets. If such a release were ingested into an HVAC inlet then significant non-uniformity in gas concentration could be expected outside and inside the HVAC duct.

Two idealised scenarios have been modelled: firstly a 2.5 kg/s high pressure release of pure methane across the underside of a platform for conditions similar to those described above and secondly a 0.55 kg/s low pressure release of pure methane in the wake of a platform. In each case the platform was modelled as a rough-walled cube of side 30 m located 25 m above sea level. Wind speeds of 1.5 to 2 m/s were simulated by imposing a neutral atmospheric boundary layer profile upstream of the platform. A large region of the atmosphere around the platform 120 m wide by 115 m high and 240 m long was modelled. Figure 1 shows the idealised platform. Only the low pressure release is presented in this paper. The initial trajectory of the release is indicated, being at the rear of a partially-obstructed module. Also shown is a high level horizontal HVAC duct of internal dimensions 23 m × 2.8 m × 1.8 m. Figure 2 shows the modelled geometry and mesh at the inlet to the duct comprising a set of 24 louvres and rectangular obstructions having a blockage equivalent to that of a set of open fire dampers with their supporting structure. A mass flow rate is imposed at the interior end of the duct equivalent to a uniform velocity of 6 m/s. The flow around the platform and inside the duct was simulated using the k-ε turbulence model. A total of 564,000 nodes (control volumes) are used.

A more realistic scenario has also been modelled which draws upon some elements of an incident on Brae Alpha in 2004 where a high pressure gas riser failed resulting in gas

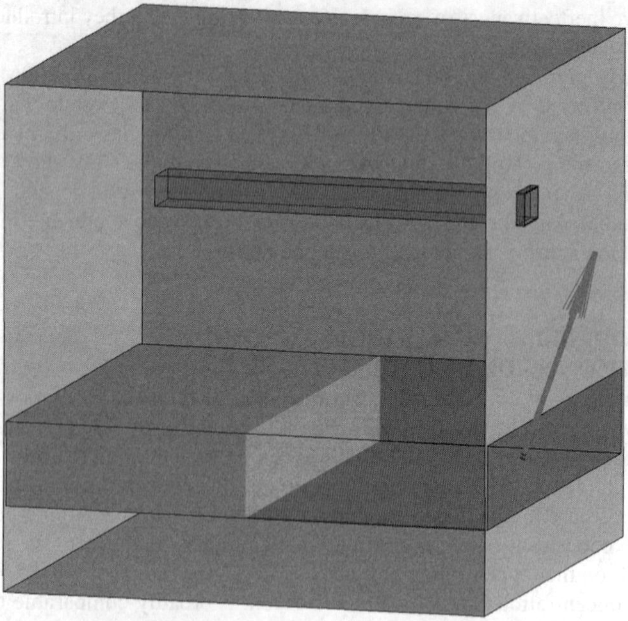

Figure 1. Idealised modelled geometry for an offshore platform, showing the location and orientation of a gas release from a partially-obstructed module. A HVAC duct is also shown

ingestion into Hazardous Modules via the Hazardous HVAC inlet duct. The intention has not been to replicate this incident but instead to devise a more realistic scenario than the idealised configurations outlined above so enabling more general conclusions to be drawn on the interaction between gas and HVAC inlets and ducts. A very simplified representation of the Brae Alpha platform has been modelled, see Figure 3. A large region of the atmosphere surrounding the platform has again been modelled. A wind speed of 12.3 m/s was imposed, with a wind direction chosen so as to direct the release towards the HVAC inlets located on the downwind side of the platform. The wind speed and direction is broadly consistent with that on the day of the incident. A high pressure gas release of pure methane at approximately 2 kg/s was modelled using the Ewan and Moodie (1986) approach. The release is initially directed vertically, but impinges on a horizontal pipe before being deflected by the wind. Credible flow rates are imposed through the three HVAC inlets shown in Figure 3. The geometry was meshed using a total of 665,000 nodes (control volumes).

All simulations have been undertaken using ANSYS CFX 10 software, in time-dependent mode.

The computed gas distribution for the low pressure idealised release scenario is shown in Figure 4. The location of the gas plume outside the HVAC duct is seen as a region

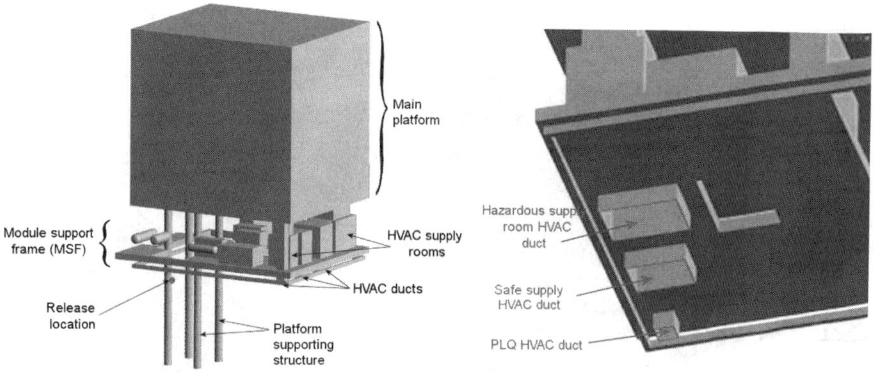

Figure 3. Geometry and location of HVAC inlets for the more realistic scenario

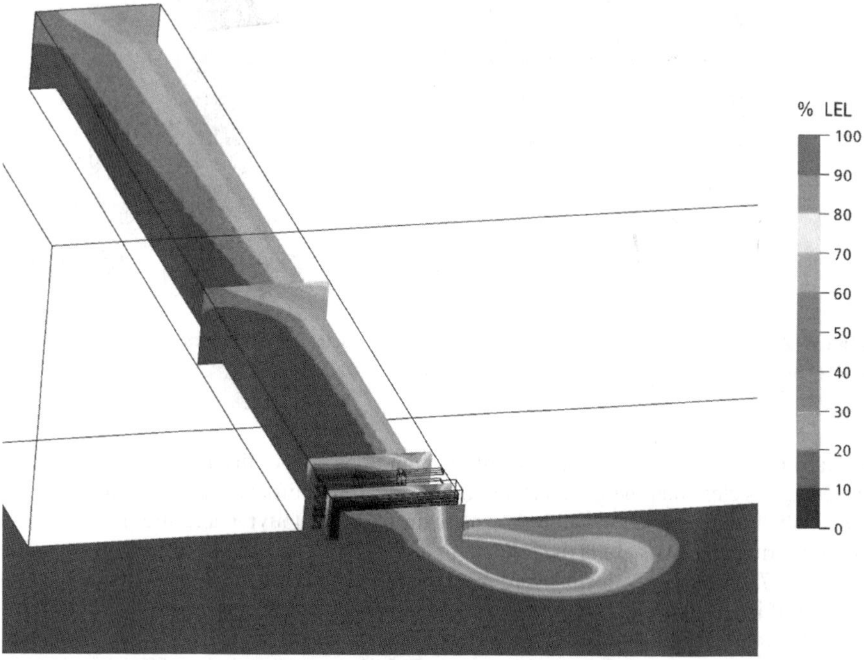

Figure 4. Gas concentration distribution inside and outside of an HVAC duct for a low pressure release, shown on a horizontal mid-plane of the duct

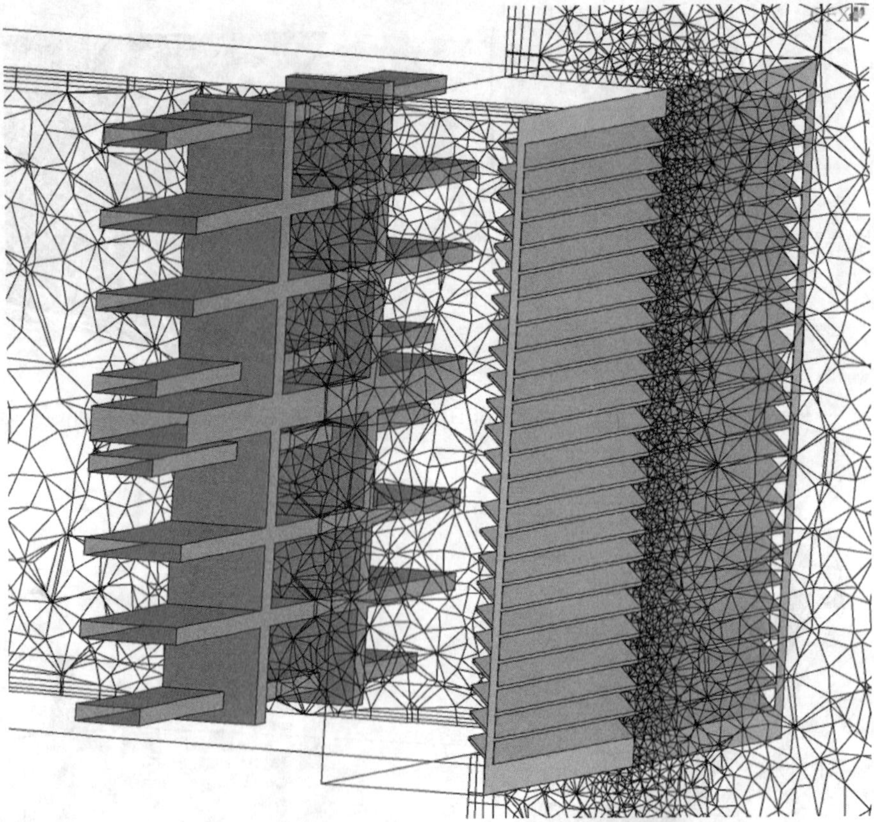

Figure 2. Geometry and meshing of louvres and open fire dampers at the inlet to an HVAC duct.

of high concentration close to one side of the inlet. A portion of this release is ingested into the duct. It is clear that the gas concentration is far from uniform at the inlet to the duct. It remains non-uniform immediately downstream from the louvres and fire dampers. This non-uniformity persists along the full length of the duct. Thus, the COV just outside of the duct is 1.27, immediately downstream from the fire dampers it is 0.97, whilst at the end of the duct it is still 0.67. Figure 5 shows more detail of the concentration distribution just inside the inlet. Although the average concentration in the duct is 21% LEL, in many locations the gas concentration is well below 10% LEL and often below 5% LEL. If point detectors were located on only one side of the duct then this release could potentially be missed by the detection system. Post-processing of these results shows that, in contrast, a

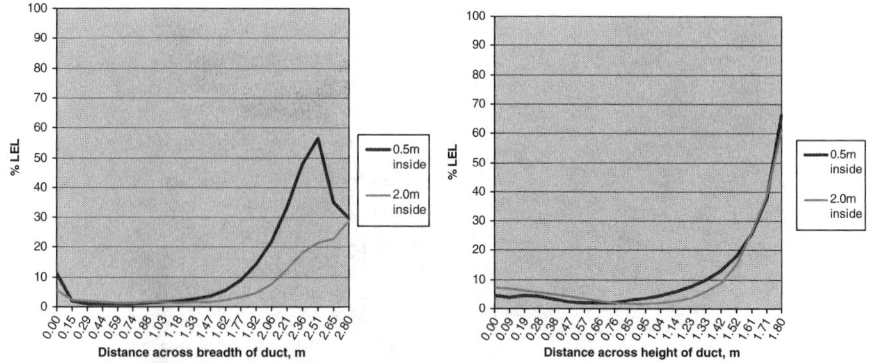

a) b)

Figure 5. Computed gas distribution inside an HVAC duct. The location '0.5 m inside' is immediately downstream from the modelled louvres, the location '2.0 m inside' is immediately downstream from the modelled fire dampers. Also see Figure 2. a) Across the breadth of the duct b) Across the height of the duct

beam detector located just upstream or downstream of the fire damper and oriented across the 2.8 m width of the duct would indicate a gas concentration of approximately 10% LEL per m.

Figure 6 shows an iso-surface of gas concentration at 10% LEL for the more realistic scenario. The gas can be seen to spread through and underneath the lowest parts of the platform and being ingested into the smaller of the three HVAC inlets. Inside the very large Hazardous HVAC inlet (~6 m × 4 m cross-section), the gas distribution is far from uniform, as shown in Figure 7. However, it is not just large ducts which can exhibit such non-uniformity in gas concentration. Figure 8 shows the gas distribution inside the smallest of the three ducts, with a 1.5 m square cross-section. The gas concentration is again very non-uniform; the COV is 0.43 at 2 m inside the duct. Although the average gas concentration in this smaller duct is approximately 16% LEL, it falls below 10% over a significant part of the cross-section. Post-processing of these results shows that beam detectors located 2 m inside the duct and oriented across the width of the duct would indicate a gas concentration of approximately 16% LEL per m which is comparable to the average concentration.

5. DISCUSSION AND INITIAL RECOMMENDATIONS ON FLAMMABLE GAS DETECTION STRATEGIES FOR OFFSHORE HVAC DUCTS

The most significant finding is that in all of the CFD simulations the distribution of gas at HVAC inlets is non-uniform: large variations in gas concentration are present over the cross-section of the modelled HVAC inlets. This implies a potential for gas releases to be

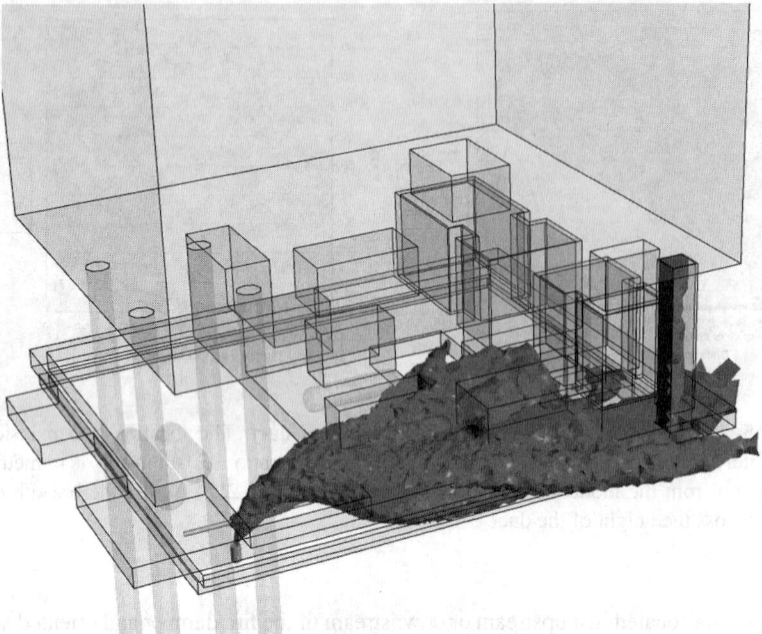

Figure 6. Iso-surface of gas at 10% LEL for the more realistic scenario

'missed' by detection systems unless this non-uniformity in gas concentration is anticipated in the selection and siting of gas detectors at HVAC inlets. The CFD results also show that a variation in gas concentration over a duct cross-section only reduces slowly with distance along a straight duct. These findings are consistent with theoretical considerations of the distribution of gas in a high pressure jet or low pressure buoyant plume, and the literature stemming from the sampling of gas distributions in the exhaust ducts of nuclear stacks.

The literature highlights that purpose-designed mixing elements and bends in a duct can be effective in creating well-mixed conditions but at the cost of increased pressure drop. It also suggests that relatively small-scale obstructions, such as louvres and fire dampers, are unlikely to significantly enhance mixing. This is borne out by CFD modelling of such obstructions in this study. The implications of the modelling work, substantiated by the literature, are that in the absence of purpose-designed mixing elements or a series of bends upstream from gas detectors, no significant benefit would be gained from siting detectors a significant distance downstream from an HVAC inlet. Also, no significant benefit can be expected to be gained from siting detectors inside an HVAC duct compared to locating them immediately outside the HVAC inlet.

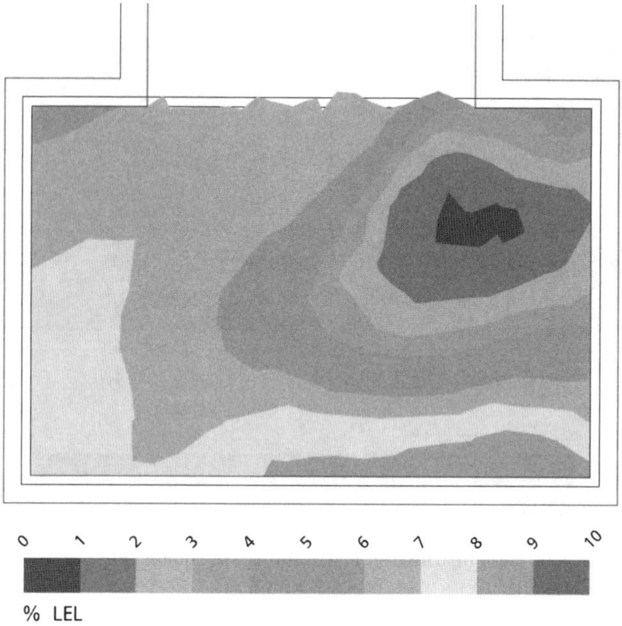

Figure 7. Gas concentration distribution 3.5 m inside the Hazardous HVAC inlet duct

Our initial recommendations on flammable gas detection strategies for offshore HVAC ducts are based on the outcomes of the research summarised in this paper and are listed below:

(a) Detector alarm levels should be set as low as reasonably practical: 10% LEL or less.

Justification: The possibility of significant non-uniformity in the distribution of gas which is ingested into an HVAC duct has been demonstrated by CFD modelling and is also indicated by theoretical considerations. The literature review has highlighted that, in the absence of purpose-designed mixing elements, an initial non-uniform distribution of gas in a duct requires a very long downstream distance before uniformity is approached. HVAC detectors are now available with a concentration range of 0 to 20% LEL and quoted minimum alarm levels of 5% LEL (Walsh *et al.*, 2005). Hence, to reduce the likelihood that detectors will 'miss' a non-uniform distribution of gas ingested into an HVAC duct, it is recommended that alarm levels be set no greater than 10% LEL. HSE have already provided information which states that although it is common practice for gas detector alarm levels to be set at 20% LEL, duty holders should explore the feasibility of reducing this alarm level to ~10% LEL (HSE, 2006). The low alarm levels have to be balanced with the minimisation of false alarms which arise from detector drift and transient operational activities.

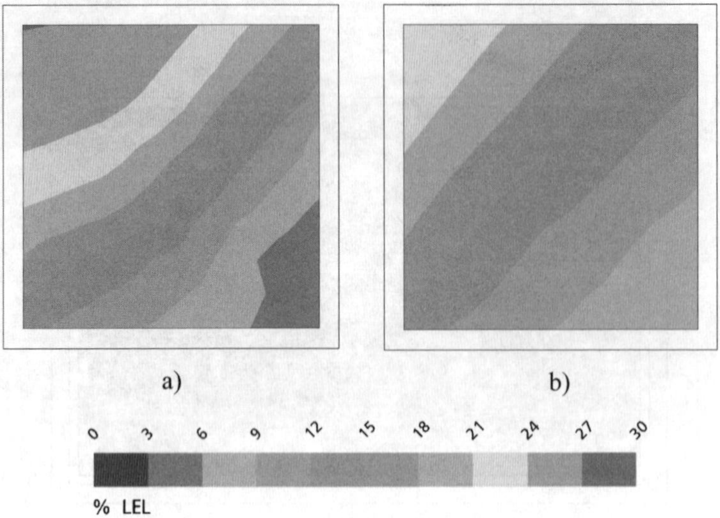

Figure 8. Gas concentration distribution inside the 1.5 m square 'PLQ' duct. a) 2 m inside the duct b) 6 m inside the duct

(b) Point catalytic, point infra-red, extended path point infra-red, cross-duct beam infra-red and aspirated point detector systems all have the potential to be effective in detecting non-uniform distributions of flammable gas in and around HVAC ducts provided that their sensitivity is sufficiently high (low detection limit) and that due regard is given to the possibility that gas will be distributed non-uniformly.
Justification: A range of detector types is available with high sensitivity, although there is some question as to whether all of the point and cross-duct beam infra-red systems have sufficiently-high sensitivity. Each detector type has its benefits and limitations, demanding differing siting requirements to ensure that non-uniform gas distributions are not 'missed'.
(c) Extended path point infra-red detector systems currently appear to offer the greatest sensitivity, but multiple detectors should be used and sited so as to anticipate non-uniform mixing.
Justification: Extended path point infra-red detector systems are available with a concentration range of 0 to 20% LEL and quoted minimum alarm levels of 5% LEL. In addition, specially designed point catalytic detectors (e.g. for gas turbine enclosures) are available with a similar sensitivity although typical catalytic detectors are usually not as sensitive or as reliable as infra-red types. However, whether either type operates reliably in the field with minimum false alarms is currently uncertain. The CFD modelling demonstrates that there is a possibility of significant non-uniformity in the distribution of gas inside and

around an HVAC inlet. The literature review indicates that this non-uniformity will reduce slowly with distance downstream in a duct. It is difficult to provide firm guidance on how many point or extended path detectors should be used since this depends on the size and shape of a duct. However, there should be good coverage of the cross-section of the duct. For large ducts this may mean that four detectors would be needed for systems which alarm upon two positive detections.

(d) Cross-duct beam infra-red, extended path or aspirated point detector systems should be based on two approximately orthogonal beams or lines of aspirated point probes.

Justification: As stated above, the CFD modelling demonstrates that there is a possibility of significant non-uniformity in the distribution of gas inside and around an HVAC inlet whilst the literature review indicates that this non-uniformity will reduce slowly with distance downstream in a duct. For these reasons there should be good coverage of the cross-section of a duct. This can be achieved by two infra-red beams arranged approximately orthogonally, either as open-path cross duct or extended path point infra-red, or lines of aspirated point probes.

(e) No significant benefit can be expected to be gained from siting detectors inside an HVAC duct compared to locating them immediately outside the HVAC inlet.

Justification: The literature review indicates that effective mixing in a duct is only achieved if large-scale turbulent eddies are introduced via purpose-designed mixing elements or bends. The CFD modelling indicates that louvres at the inlet to a duct or fire/gas dampers inside a duct will not, in themselves, be sufficient to rapidly ensure that well-mixed conditions exist in a duct. The literature review also indicates that grilles at the entrance to HVAC ducts are unlikely to significantly enhance mixing.

(f) In the absence of purpose-designed mixing elements or a series of bends upstream from gas detectors no significant benefit is to be gained from siting detectors a significant distance downstream from an HVAC inlet.

Justification: The literature review and the CFD modelling strongly indicate that, for a straight duct, well-mixed conditions are only achieved a very long way downstream from an HVAC inlet.

(g) Mixing elements have the potential to reduce any non-uniformity in the distribution of gas in a duct but their effectiveness should be proven by physical tests.

Justification: This is supported by the literature review. It should also be noted that mixing elements will result in an additional resistance to flow in a duct and that the resulting pressure drop may be significant.

The above recommendations are based on evidence from CFD modelling and the published literature. However, CFD modelling has inherent uncertainties and it is not certain that the findings from the literature are always directly relevant. These initial recommendations could be further substantiated by physical trials using real detectors.

6. ACKNOWLEDGEMENTS

The authors would like to thank Marathon Oil UK Ltd, Flamgard Engineering Ltd, Integrated Engineering Services Ltd, Groveley Detection Ltd and Honeywell Analytics for providing information and assistance during the course of this research. Dr Peter Walsh and Dr Mat Ivings, Health and Safety Laboratory, have also provided very helpful guidance.

REFERENCES

Anand N K, McFarland A R, Rajagopal K R, 2003, Gas mixing for achieving suitable conditions for single point aerosol sampling in a straight tube: experimental and numerical results, *Health Physics*, Vol. 84, No. 1, pp 82–91.

ANSI/HPS N13.1-1999, Guide to sampling airborne radioactive materials in stacks and ducts, American National Standards Institute/Health Physics Society.

BS EN 15138:2001, Offshore production installations – HVAC, British Standards Institute, London.

Ewan B C R, Moodie K, 1986, Structure and velocity measurements in under-expanded jets, *Combustion Science & Technology*, Vol. 45, pp 275–288.

Hampl V, Niemela R, Shulman S, Bartley D L, 1986, Use of tracer gas technique for industrial exhaust hood efficiency evaluation – where to sample? *American Industrial Hygiene Assoc J.*, Vol. 45, Part 5, pp 281–287.

Hinze J O, 1975, Turbulence, McGraw-Hill Book Company.

HSE, 2006, Testing regime for offshore TR-HVAC fire dampers & TR pressurisation requirements, Offshore Information Sheet No. 1/2006.

Laws E M, Livesey J L, 1978, Flow through screens, *Ann. Review of Fluid Mechanics*, Vol. 10, pp 247–266.

Lea C J, Deevy M, 2007, Assessment of gas detection strategies for offshore HVAC ducts based on CFD modelling, Health and Safety Executive Research Report RR602, http://www.hse.gov.uk/research/rrhtm/rr602.htm.

McFarland A R, Gupta R, Anand N K, 1999, Suitability of air sampling locations downstream of bends and static mixing elements, *Health Physics*, Vol. 77, No. 6, pp 703–712.

Melling A, Whitelaw J H, 1976, Turbulent flow in a rectangular duct, *J. Fluid Mechanics*, Vol. 78, Part 2, pp 289–315.

Rodi W, 1982, Turbulent buoyant jets and plumes, Pergamon Press.

Seo Y, McFarland A R, Ortiz C A, O'Neal D L, 2006, Mixing in a square and rectangular duct regarding selection of locations for extractive sampling of gaseous contaminants, *Health Physics*, Vol. 91, No. 1, pp 47–57.

Simpson J E, 1997, Gravity currents in the environment and the laboratory, 2nd edition, Cambridge University Press.

Walsh P, Johnson A, Ivings M, 2005, Effectiveness of gas detection in HVAC ducts: Scoping study, Health and Safety Laboratory Report FM/04/11.

GUIDANCE ON THE USE OF NON-CERTIFIED ELECTRICAL EQUIPMENT IN LABORATORY FUME CUPBOARDS

G.R. Astbury
Formerly Senior Engineer, Health & Safety Laboratory, Buxton, Derbyshire, SK17 9JN
E-mail: graham.astbury@gmail.com

This paper discusses a practical approach to realistic zoning of fume cupboards and the location of equipment with regard to the zoning within the fume cupboard is considered. A methodology is provided for the examination of laboratory equipment to determine its suitability for use in a potentially explosive atmosphere. Examples are given of the examination of several pieces of common laboratory electrical equipment, illustrating the potential areas of concern, and possible solutions to reduce the risk of fire or explosion from such equipment. A method is given to enable suitably competent persons to identify whether ignition hazards exist, whether modification to the equipment is practical or whether the equipment is unsuitable for use in the proposed location.

INTRODUCTION

In the Fine Chemicals industry, there is a tendency to manufacture short runs of differing products, generally requiring some prior research using laboratory facilities. As many products use flammable solvents for their manufacture, it is inevitable that flammable liquids are handled at Laboratory scale. Whilst there is good guidance available for the classification of hazardous areas in plant-scale environments, that for laboratory-scale operations is poor or non-existent. It is often assumed that fume cupboards, being well ventilated with a large throughput of air, are classed as non-hazardous but this can be erroneous as under some conditions, particularly loss of containment, explosive atmospheres can exist in fume cupboards for protracted periods.

In this paper, the practicalities of hazardous area classification are examined, and guidance is given on considering the location of equipment within a fume cupboard to minimise the risk of ignition of any explosive atmosphere that may exist. As little laboratory equipment is available which is certified as suitable for use in hazardous areas, a methodology for the examination of laboratory equipment is given, to determine whether the equipment can be classed as "unsuitable" for use in a hazardous area, or whether it can be classed as "not unsuitable". This is not an attempt at self-certifying equipment, and a clear distinction has to be understood in the definition of the term "unsuitable" and "not unsuitable". The former means what it says on the tin – it is definitely unsuitable for use in a potentially explosive atmosphere as there would be obvious parts of the equipment which would constitute a potential or permanent source of ignition. The term "not unsuitable" means that whilst it is not certified for use in a explosive atmosphere, it has been examined and there are no obvious indications that it would be a source of ignition if it were to be exposed to a explosive atmosphere. Three examples of the methodology are given to illustrate its application.

EXISTING STANDARDS

There are two well established standards for the classification of hazardous areas - the European Standard BS EN 60079-10:2003 (BSI, 2003) and the Institute of Energy Code of Practice Part 15 (Institute of Energy, 2005). Both of these standards are intended for industrial situations, rather than laboratory situations. However, each does include a small section giving advice for laboratory situations.

The European Standard BS EN 60079-10:2003 is aimed at large-scale industrial areas, and in Section 4.2, it suggests that if the inventory of flammable material is "small", then the methodology laid down in the standard may be inappropriate. Unfortunately, it does not define the quantity that would be considered as "small". Therefore it could be inferred from this statement that for a laboratory handling a few litres of flammable materials, no area classification would be required. The Institute of Energy guidance suggests that for laboratories and small-scale users where the inventory is less than 25 litres, the area can be classified as non-hazardous if it is well ventilated. Since there is no definition of "well ventilated" in the standard, it would be unwise to indiscriminately adopt a non-hazardous classification where the inventory is less than 25 litres.

From the first of these two standards, it would seem that a laboratory fume cupboard would not require an area classification exercise to be carried out, and that such a fume cupboard would be non-hazardous. From the second, it could be argued that since the inventory is less than 25 litres, and a fume cupboard is well ventilated, again the area could be classified as non-hazardous. A typical laboratory fume cupboard is illustrated in Figure 1.

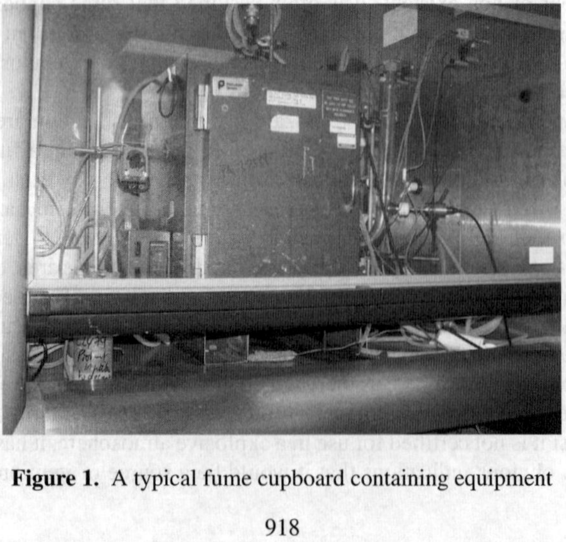

Figure 1. A typical fume cupboard containing equipment

Whilst it is possible for the methodology of the European Standard BS EN 60079-10:2003 to be applied to a fume cupboard, it would be necessary to be able to estimate release rates of any flammable vapours or gases to be able to determine the extent of any hazardous areas. Although typical releases such as vapour emanating from a boiling flask without a condenser can be determined from the heat input, it would be difficult to calculate a release rate for the foreseeable event of total loss of containment – the gross spillage of the contents of a beaker or flask. Consequently, the application of the method of European Standard BS EN 60079-10:2003 to a small laboratory fume cupboard is fraught with difficulty.

However, if the fume cupboard is classified as non-hazardous, the corollary is that sources of ignition would be acceptable. Clearly it would seem unacceptable to have a lit Bunsen burner in a fume cupboard containing a flask full of boiling solvent, so therefore good sense suggests that a fume cupboard containing solvents should at least have some sort of area classification exercise undertaken to establish the hazards present and the risk of an ignition occurring.

ZONING OF FUME CUPBOARDS

Whilst the above would suggest that fume cupboards are non-hazardous, this is far from realistic. Whilst a small enclosed process is being operated, this may be the case, but when the process is started or completed, the reality is one where spillages can occur quite easily. In the case of a spillage of flammable solvent, there is a considerable time whilst the solvent evaporates, and hence an explosive atmosphere will be present. In practice, the following need to be considered:

LARGE SCALE

Where the inventory is of a scale larger than about 5 litres, a walk-in type of fume cupboard in a pilot-scale laboratory would be more appropriate than a standard fume cupboard, and typically for such large-scale equipment, suitably certified electrical equipment is available and therefore should be used. Clearly the scale is such that it would be appropriate to use the methodology of BS EN 60079: Part 10, and further discussion is not necessary in this paper.

BENCH SCALE

This is more typical of a standard laboratory fume cupboard as illustrated in Figure 1. This is rather too small to be considered for treatment using the methodology of BS EN 60079: Part 10, and an alternative approach is required. Whilst it is easy to dismiss the risk of explosive atmospheres occurring as being insignificant or of being a trivial size, in practice, care is required to determine the likelihood of flammable material existing in the fume cupboard whilst not contained within the equipment. Operations such as pouring liquids into flasks, filtering, emptying equipment and cleaning out afterwards can all expect to form explosive atmospheres under normal and foreseeable circumstances. Clearly for very small inventories such as less than 1 gram, the extent and duration of any explosive atmosphere in a well-ventilated fume cupboard will be negligible.

VENTILATION QUALITY AND RELIABILITY

In most fume cupboards, the extraction rate is determined more by the need to maintain a minimum face velocity to prevent the escape of material from the cupboard than the dilution of emitted gases or vapours. However, for a typical face velocity of 1 m s^{-1}, and an access opening of 300 mm over a 1 m wide fume cupboard, the air throughput amounts to 0.3 m^3s^{-1}, which would safety dilute to less than 0.25% an evolved vapour rate of 0.075 m^3s^{-1}, which is far in excess of that likely to be released by an experiment which was considered safe to be undertaken within a standard laboratory fume cupboard. Where larger-than-bench scale experiments are to be undertaken, an alternative containment method may be required, and this is discussed briefly later.

The reliability of the ventilation also needs to be considered. Where an extraction system has a standby fan with automatic change-over on failure, the extraction can be deemed reliable, but where there is no standby, or the site is on an electricity supply with a known high rate of loss of power, then the reliability should be considered as low. In this case, consideration should be given to the use of standby supplies, automatic shut-offs for the experimental equipment and standby emergency cooling.

INVENTORIES

Gases present a particular hazard, as many flammable gases are heavy and spread along level surfaces, so explosive atmospheres at the base of the fume cupboard can easily form. Unless the gases are released during the experiment, they are often piped into the fume cupboard, and therefore the inventory is only limited by the supply, be it a cylinder external to the fume cupboard or natural gas piped in from the gas mains. Hydrogen and any other light gases present a hazard towards the top of the cupboard as they can accumulate in the top of the cupboard. Therefore the lighting needs to be either separate from the cupboard (such as shining through sealed glass partitions) or be suitably certified for the gases present.

Liquids present the greatest risk if spillage or other loss of containment occurs. If the liquid is above its flash point, then any spillage will accumulate on the base of the cupboard, and form a large pool which readily evaporates. Liquids approaching the boiling point will evaporate quickly initially until cooled by the base of the fume cupboard. Clearly immediately above the liquid surface, the atmosphere will be substantially pure vapour, but as the air flow through the cupboard mixes with the vapour, the concentration of the vapour will pass from over-rich, through stoichiometric, and the lower explosive limit, before being diluted to be non-explosive. This occurs over a relatively narrow height, so the base of the fume cupboard should therefore be classified probably as a Zone 1 area. Some estimate of the depth of the explosive zone can be gleaned from Hughes (1970) quoted by Lees (1996), who suggests that for spillage of 1 litre of petrol onto the ground resulted in explosive atmospheres being formed at least 6 m away, but the height of the explosive atmosphere was only a few centimetres.

Vapours present the same potential problems as gases, but can condense and form liquid pools, so both scenarios would have to be taken into account. Evaporation rates from small pools of liquid was investigated by Clancy (1974), and this may be used for small-scale spills such as occur in fume cupboards. There are alternative commercial software packages for modelling pool spread from spillages, but most are concerned with industrial scale quantities of a tonne upwards.

RELEASES

Any flammable releases of a continuous nature will be diluted constantly with the air flowing through the fume cupboard. Clearly if the vapour is pure, the concentration immediately around the point of release will be non-explosive as it is above the upper limit. Since there is an imposed flow through the fume cupboard, the vapour disperse and mixes with the air passing through the fume cupboard, and so the theoretical explosive zone will be a right circular cone extending from the point of release towards the rear extraction vent of the fume cupboard. The cone will be truncated at the base when it reaches the lower explosive limit. The volume of this zone (and hence its length) will depend on the dilution by the airflow. Since the airflow is usually much larger than the release rate, the resultant zone would be of negligible extent. However, sudden releases such as loss of containment or failure of a condenser will result in much larger zones. In practice, the zone is unlikely to be a cone - in fact due to the turbulence, the conical zone will vary in size, shape and location, and so an assumption will have to made on the likely size and approximate locations where it could be.

EQUIPMENT INSIDE FUME CUPBOARDS AND AIRFLOW MODIFICATION

The quality of the ventilation is difficult to determine in practice. Whilst it is simple to determine the total air flow through any one fume cupboard, the distribution of air within that fume cupboard is not straightforward. An empty fume cupboard behaves aerodynamically differently from one that is filled with equipment. Looking at the fume cupboard in Figure 1, it can be seen that the air flow will not be as designed with the cupboard empty. With the sash at the bottom in the normal position, the air enters through a horizontal slot, and passes through the cupboard to the rear, where it is extracted at both the top and bottom. Hence the air flow along the base of the cupboard is particularly prone to disturbance. The more equipment that is installed, the greater the air velocity through the remaining volume of the cupboard. Hence it is necessary to ensure that equipment is installed in a position where the airflow is disturbed as little as possible. This means locating it as centrally as possible, so it is not close to the front air inlet, the sides, or too close to the extraction points at the rear.

In Figure 1, it can be seen that the oven has been positioned on square hollow sections passing from front to rear, to allow the airflow to be disturbed as little as possible.

These are sometimes referred to as "technical bricks" as they do not impede the air flow to the same extent as a normal brick would do. Consequently, it is best to ensure that any equipment containing a potential ignition source is installed within fume cupboards in a position such that it is neither on the base nor at the top, and ideally it should be mid-way between the front and rear of the fume cupboard, so that any leakage of flammable vapours or gases will be carried towards the rear extraction port of the fume cupboard.

LABORATORY ELECTRICAL EQUIPMENT AND NON-CERTIFIED EQUIPMENT

Where electrical equipment is available in a certified form that is suitable for use in potentially explosive atmospheres, then this must be used. This applies even if the equipment is only available on a long delivery or is very expensive – cost or time cannot be used as mitigating circumstances for not using such equipment. If such equipment is not available, then the only choice is to asses the risk of the equipment that is available to be a potential source of ignition.

This is not an attempt to self-certify equipment as meeting the requirements of any standards for electrical equipment for use in explosive atmospheres, but is merely a methodology to examine the equipment to reject that equipment which has obvious potential ignition sources. This examination should be undertaken by a competent person, and not delegated to the Portable Apparatus Inspector, who is very unlikely to understand the hazards and risks involved. If necessary, both an Electrical Engineer and a person skilled in the art of Area Classification should examine the equipment together and come to an agreement on its suitability.

ITEMS PRESENTING RISK OF IGNITION

ELECTRICAL POWER SWITCHES AND INPUT CONNECTIONS

Any type of switch or connection that is likely to carry a heavy current or a high voltage will present a risk of arcing and the potential to ignite an explosive atmosphere. Where there is the potential for switches to be operated, the switch may be fitted with a small cover to prevent its inadvertent operation. Similarly, where mains connections are made using IEC-320 C14 appliance inlets, a simple clip can be fitted to prevent the withdrawal of the mains lead, as shown in Figure 2. The use of multi-way mains power distribution boards should be avoided within fume cupboards, as the switches can be operated inadvertently. Some types even have overload cut-outs which also could produce sparking, so boards of this type should not be used within fume cupboards.

MOTORS

These are often used in fume cupboards to drive stirrers, small pumps etc. As there are many types of motors used in laboratories, it is difficult to be specific, but there are several

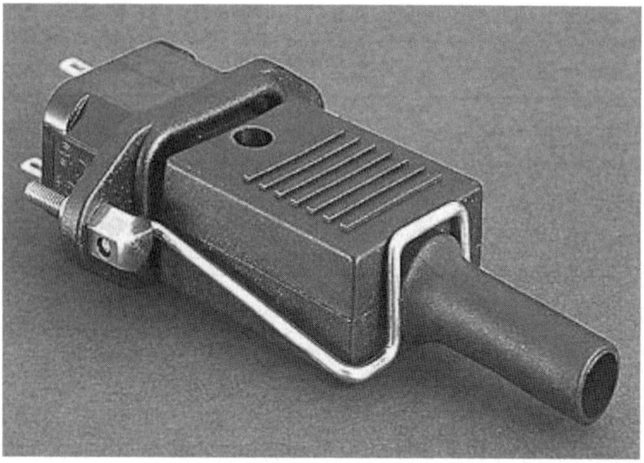

Figure 2. Clip to prevent withdrawal of mains lead (Courtesy of Bulgin components plc)

generic types most commonly seen in laboratory equipment. Small motors can be classified into three categories depending on their likelihood of sparking. These are:

- Types which do not spark under normal use, which include three-phase motors, permanent capacitor motors, stepper motors and shaded pole motors.
- Types which spark when starting, which include single-phase split-phase induction motors and capacitor start motors.
- Types prone to continuous sparking, which include motors with brushes and commutators or slip-rings.

Clearly, motors in the latter two categories would be unsuitable for use in a fume cupboard where an explosive atmosphere could exist. Where is it difficult to identify the type of motor, two series of articles describing different types of small motors were written many years ago by Philpott (1944) and Watts (1951), and include photographs and descriptions of the salient points to allow ready identification of small motor types. The later types of stepper motors are not described by these authors, but are recognisable by the use of a 4, 6 or 8 lead flat ribbon cable and a tendency to "cogging" when rotated by hand as the rotor aligns itself with the magnets within the motor.

As all motors tend to run hot, the maximum temperature that could be expected for the motor windings can be inferred from the motor temperature classification, as described in an IEC Standard (IEC, 1984). Typical temperature classes are given in Table 1 below. Note that sometimes a motor known to be used for short periods only may exceed the maximum service temperature determined by the insulation class. This is because the insulation class is selected to ensure a satisfactory service life assuming continuous isothermal operation. Some motors are also fitted with a re-settable thermal overload trip which would spark when opening under overload, and this can be identified by the reset button on the motor frame.

Table 1. Thermal classes according to IEC 85 (1984)

Thermal class	Temperature, °C
A	105
E	120
B	130
F	155
H	180

HEATERS AND OVENS

Where heating elements are used, there are two types that need to be considered. Firstly, there are those with essentially exposed elements which may well be coiled and supported in a ceramic insulation frame (similar to an electric toaster). Heaters of this type are the classic "Isomantle®" heater for round-bottomed flasks. In general, with electric heaters, the power input is constant, and the surface temperature will rise until it can dissipate all the input power. For a classic "Isomantle®" type of heater used to boil a small flask, the liquid level in the flask will slowly fall as the contents evaporate, and so the heat absorption capacity of the flask will diminish. As the power input is constant, the temperature of the element not in thermal contact with the flask's contents will rise, and may become red hot, thus acting as a potential ignition source.

Secondly, there are sheathed elements such as the mineral insulated heating elements as used on electric cooker hobs. The sheathed elements can also run red hot, but some are in such close contact with a large heat sink that their surface temperature may well be restricted simply by the extensive heat sink attached to them. A contact thermocouple or thermal imaging camera can be used to determine the maximum surface temperature. Providing that the maximum surface temperature does not exceed the autoignition temperature of the gases or vapours that are present, there is little risk of ignition. Where hot surfaces are present and are not enclosed, a higher surface temperature may be permissible as autoignition temperature determinations are carried out using enclosed volumes. Further information on the effect of volume on autoignition temperature is given by Coffee (1979).

Ovens present a different problem. Where an oven has no circulating fan and relies on convective heat transfer, the temperature of the heating element may reach red heat. This type of oven is likely to ignite any solvents which may be evaporated from the item being heated, and therefore should not be used. Where the oven is used for isothermal experiments, there is usually a circulating fan which circulates the air through the oven, and over the heating elements. This prevents the element surface temperature from reaching a high value under normal operation. For low power density elements, it may acceptable to use such an oven, providing that the oven is fitted with ventilation to allow solvent vapours to disperse. However, unless the maximum surface temperature is kept low, the same problem of potential ignition of vapours will exist.

In all types of electric heating, the control system requires some examination. Where the control is on and off using a classic thermostat, then the contacts are likely to spark, and such a controller should be located outside the fume cupboard. Where the control is electronic, then the controller may be satisfactory, providing the examination is acceptable as described above. There is also a need to check whether there is an over-temperature trip fitted within the heating system. This may be inaccessible and of the non-resettable thermal fuse type, in which case the heater manufacturer should be asked whether such a device is fitted. Otherwise, standard thermostats may be used as "policemen" in heaters, and these are likely to spark under over-temperature conditions.

RISK REDUCTION BY LOCATION

Location of equipment can play a significant part in risk reduction. Sometimes equipment does not actually need to be located within a fume cupboard - it is often just common practice that it is. An example of this is where a vacuum pump is located within the fume cupboard because of the potential of discharge of noxious materials from the outlet, but the simple modification of adding a flexible hose to the outlet and placing the open end within the fume cupboard with the pump outside is easy. Similarly, heating and cooling systems which circulate a heat transfer liquid are usually located within the fume cupboard as this reduces the length of tubing required to connect to the experimental equipment – but it could be relocated outside, standing on a suitable support so that the increase in tubing length is minimised.

OTHER POTENTIAL IGNITION SOURCES

Apart from electrostatic brush discharges from plastic equipment, there is the potential for frictional heating and impact sparks. Whilst the power, speed and load of much laboratory equipment is usually very small, and therefore unlikely to give rise to frictional heating (Hawksworth et al, 2004), consideration should be given to the potential for this to occur under abnormal conditions. For example, high speed laboratory grinding mills often run at speeds in excess of 8,000 rpm, and the stored energy of the rotor at such speeds can be sufficient to heat up small parts should the mill become choked. Other electrical equipment such as floor polishers and vacuum cleaners may be used in the vicinity of fume cupboards, and their use should be restricted to appropriate locations. If a torch is to be used in a fume cupboard, then an appropriately certified one should be used.

ITEMS NOT PRESENTING A RISK OF IGNITION

ELECTRONIC CIRCUITS AND COMPONENTS

In general, electronic circuits using solid state components exhibit a low risk of ignition. However, items such as resistors can run very hot, but as the surface area is relatively small, ignition is unlikely unless the resistor is enclosed in a small volume where there is little air flow. Whilst there is guidance on the risk of ignition from electronic circuits available in the European Standard EN 50020, (BSI, 1994), the main aspect of inspecting equipment is to identify any obvious high temperature large resistors or high current printed circuit boards.

Keyboards and keypads are often present on electronic laboratory equipment. The common types are membrane switches where there is a smooth plastic film over the keypad area and the keys are sealed within this to prevent water ingress; and computer type keyboards with individual keys. As both these types usually only handle low voltages and currents, typically less than 5 volts and less than 10 mA, they pose little risk of an incendive spark occurring. However, the membrane type may have an insulating plastic membrane which could become electrostatically charged, so the safety of this should be confirmed with published criteria (CENELEC, 2003).

Active electronic components such as transistors, thyristors, integrated circuits etc generally have a maximum operating temperature of about 150°C to avoid destruction of the silicon die within the encapsulation package, so the external temperature will always be less than 150°C. Passive components such as capacitors and inductors generally store energy, and therefore any mechanical switches associated with such components may well generate incendive sparks even though the circuit voltages and currents are low. Inductances can be identified by the core material, commonly steel laminations and ferrite pot cores. Capacitors can be identified by their physical size.

OTHER EQUIPMENT

This includes equipment not normally considered laboratory equipment. Typical examples include pocket calculators, personal data assistants (PDAs), MP3 players and other entertainment devices. Although such personal entertainment devices should not be present, many do find their way into laboratories, but the risk of ignition is low from solid state devices. However, motor driven CD players and cassette tape players may present a risk of ignition and appropriate controls should be put in place to prevent their use in and around fume cupboards.

EXAMPLE ASSESSMENTS OF EQUIPMENT

LABORATORY STIRRER MOTOR

A stirrer motor is shown in Figure 3, and this has been opened to reveal the type of motor. The speed variation is mechanical, and the motor has six leads, two for the main winding, two for the auxiliary winding, and the remaining two are for a thermistor sensor embedded in the windings for overload protection. Inspection of the printed circuit shows a large capacitor and a triac, so the motor is a permanent capacitor type and therefore non-sparking. The thermistor is wired to the triac in series with the supply, so that if the thermistor becomes hot, it inhibits firing of the triac and so cuts off the supply. Again, this is non-sparking so can be deemed acceptable for use in a fume cupboard.

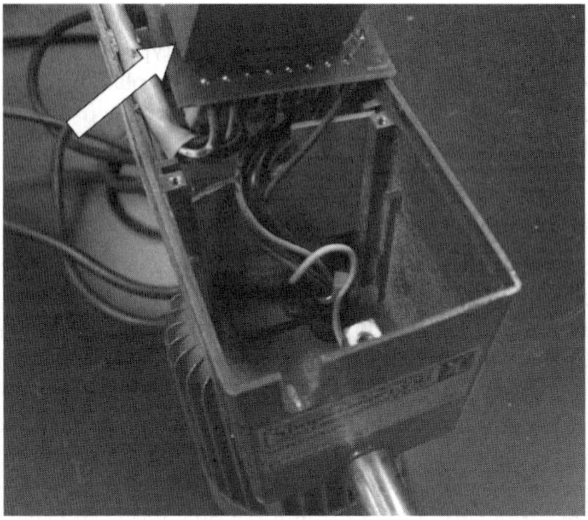

Figure 3. Internal view of stirrer motor showing six motor leads and capacitor (arrowed)

LABORATORY HOTPLATE

This is shown in Figure 4. Examination of the heating element shows that it is embedded in the metal plate, and therefore in good thermal contact, so there is little risk of hotspots occurring. The main temperature control is by thermocouple used to fire a triac to control the temperature. This control system does not have any sparking contacts, so it acceptable.

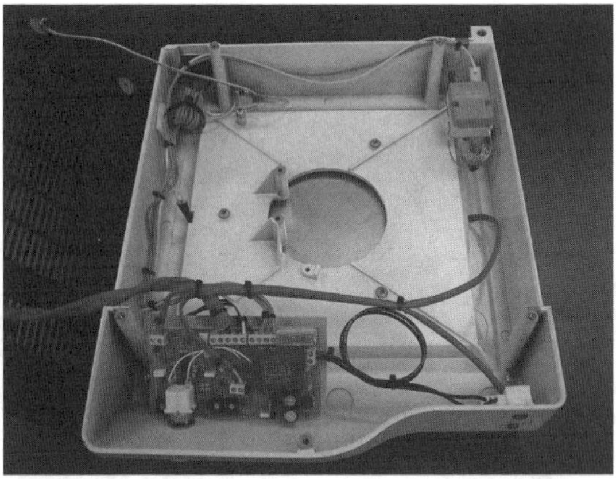

Figure 4. Hotplate showing thermal cut-out (arrowed)

However, the hotplate is fitted with a thermostatic "policeman" (arrowed) to cut off the supply if the thermocouple or triac controller fails. This type is an open snap type thermostat and will spark if it operates. Therefore this type of hotplate is probably unsuitable for use in the fume cupboard, particularly if located at the base of the cupboard where an explosive atmosphere could be anticipated.

ROTARY EVAPORATOR

In this piece of equipment, there is not only the heated water bath to consider, but also the variable speed motor. The hot water bath is similar to the hotplate above, but since there is no open-frame thermostat, this is acceptable. Removal of the cover as in Figure 5 shows that this is an induction motor, with a computer-style cooling fan. The cooling fan is powered by a solid-state switched three-phase motor with all the electronics encapsulated, so this will not give rise to sparks. Similarly, the main drive induction motor has no sparking contacts. Since there are only three wires to the main motor windings, it is a permanent capacitor two-phase motor, and is driven from a variable frequency supply with feedback from a speed sensor (arrowed). As the cooling fan operates at full speed all the time, the main motor has adequate cooling at low speeds.

AUTOMATIC CUT-OFF CONSIDERATIONS

Where loss of ventilation could result in the accumulation of flammable vapours, consideration should be given to the use of automatic cut-outs, mounted externally to the fume cupboard, to shut down the experiment safely. Typical shut-offs would include solenoid

Figure 5. Rotary evaporator motor and cooling fan, showing speed sensor (arrowed)

valves to shut off gas feeds, stopping feed pumps adding reagents and shutting off heater supplies. However, care should be taken to ensure that the cut-off does not revert to a hazardous condition. For example, shutting off a chiller unit used to cool an exothermic reaction would result in loss of cooling and a potential reaction runaway.

CONCLUSIONS

- The Area Classification Standards are not directly applicable to laboratory fume cupboards.
- Laboratory fume cupboards are should not be classified as non-hazardous.
- The most appropriate zoning for heavier-than-air vapours and gases would be that the bottom 100 mm of a fume cupboard should be classed as Zone 1 or Zone 2, with the remainder either Zone 2 or non-hazardous, depending on the inventory and equipment in use.
- The most appropriate zoning for lighter-than-air vapours and gases would be that the top 100 mm of a fume cupboard should be classed as Zone 1 or Zone 2, with the remainder either Zone 2 or non-hazardous, depending on the inventory and equipment in use.
- Where suitably certified electrical equipment is available, it should be used.
- Where no certified equipment is available, any electrical equipment to be used in fume cupboards should be examined to determine whether there is a risk of the equipment becoming a source of ignition.
- Minor modifications to equipment can be made to reduce the risk of ignition occurring.
- Equipment should not be located in positions where the occurrence of explosive atmospheres is foreseeable.

REFERENCES

BSI, 1994, European Standard BS EN 50020:1994 *Electrical apparatus for potentially explosive atmospheres - Intrinsic safety.*

BSI, 2003, BS EN 60079-10:2003 *Electrical Apparatus for explosive gas atmospheres - Part 10: Classification of hazardous areas.*

CENELEC, 2003, Technical Report CLC/TR 50404:2003, *Electrostatics - Code of practice for the avoidance of hazards due to static electricity.*

Clancy, V.J., 1974, *The Evaporation and Dispersion of Flammable Liquid Spillages*, Chemical Process Hazards, **5**, 80-89.

Coffee, R.D., (1979), *Cool Flames and Autoignitions: Two Oxidation Processes*, Chem. Eng. Prog. Loss Prev., 13, 74–82.

Hawksworth, S., Rogers, R.L., Proust, C., Beyer, M., Schenk, S., Gummer, J. & Raveau, D., 2004, *Mechanical Ignition Hazards in Potentially Explosive Atmospheres - EC Project MECHEX*, EMSG International Symposium on Process Safety and Industrial Explosion Protection, Nuremberg Trade Centre, 16-18 March.

Hughes, J.R., 1970, *Storage and Handling of Petroleum Liquids, Practice and Law, 2nd ed.* Griffin & Co., London, quoted by Lees (1996).

IEC,1984, IEC Standard 85, *Method for Determining the thermal classification of electrical insulation* (Note: Identical to British Standard BS 2757:1986).

Institute of Energy, 2005, *Area Classification Code For Installations Handling Flammable Fluids* Part 15 of the IP Model Code of Safe Practice in the Petroleum Industry, 3rd Edition, ISBN 0 85293 481 1.

Lees, Frank P., 1996, *Loss Prevention in the Process Industries*, Vol.3, A38, Butterworth Heinemann, 2nd Edition, ISBN 0 7506 1547 8.

Lettenmaier, T.A., Novotny D.W. & Lipo. T.A., (1991), *Single-Phase Induction Motor with an Electronically Controlled Capacitor*, IEEE Trans. Ind. Applic., **27**, No.1, 38-43.

Philpott, S.F., (1944), *Fractional Horse-Power Motors*, BEAMA Journal, September pp. 295-303 and October pp. 330-335.

Watts, J.A., (1951), *Small Motors*, Electrical Times, pp 325-330, 559-562, 699-703 and 1075-1080.

MAXIMISE THE USE OF YOUR EXISTING FLARE STRUCTURES

Neil Prophet[1], Georges A. Melhem[2], PhD and R. Peter Stickles[2], P.E.
[1]ioMosaic Corporation, 2650 Fountain View Drive, Houston, TX 77057
[2]ioMosaic Corporation, 93 Stiles Road, Salem, NH 03079

INTRODUCTION

Due to the design vintage of many petroleum refineries and petrochemical plants, existing pressure relief and flare systems may be overloaded because of:

- Prior unit expansions/upgrades which have increased the load on the flare for combined flaring scenarios beyond the original design intentions
- The desire to connect atmospheric relief valves to the flare for environmental and safety consideration and to eliminate blow down drums
- The addition of new process units that need access to flaring capacity

As a result, many petrochemical and chemical companies are engaged in comprehensive flare systems evaluation and upgrading projects to ensure continuing safe operations, to maximise the use of their existing flare systems, and to minimise the need for modifying existing flare structures or building new ones.

Achieving these goals presents several engineering challenges:

1. Which existing atmospheric relief devices present vapor cloud explosion and thermal radiation hazards and need to go to the flare?
2. What is the impact of the additional flaring loads on the existing flare header system and individual relief devices during combined flaring events (such as loss of power or cooling)?
3. Where and how many High Integrity Protection Systems (HIPS) should be employed to reduce the worst-case flaring load?
4. How should the HIPS components be configured to achieve the required safety integrity level (SIL)?

In order to properly and cost-effectively address these design questions, specialized expertise and tools for pressure relief systems design, risk analysis, and instrumentation are required:

- Dynamic simulation of relieving vessels and flare piping networks to identify capacity constraints
- Risk tolerability criteria related to vessel overpressure hazards
- Risk assessment and reliability analysis to properly select and configure the HIPS

This paper provides a general framework for evaluating and maximizing available flare systems capacity, and investigates criteria and approaches for determining a tolerable risk event for flare systems.

HIPS, SIS AND SILS: WHAT ARE THEY?

The ISA/ANSI Standard S84.01 96 defines a Safety Instrumented System (SIS) as a system composed of sensors, logic solvers, and final control elements for the purpose of taking the process to a safe state when predetermined conditions are violated. SISs act independent of the basic process control system (BPCS).

The term high integrity protective system is described in Annex E of API STD 521 Guide to Pressure-Relieving and Depressuring Systems, as an alternative in some scenarios for preventing overpressure conditions. A HIPS is a SIS that is designed to provide overpressure and over-temperature protection that is at least equivalent in reliability to a mechanical relief device.

HIPS have traditionally been used for rapid depressurization of Hydrocrackers and Acetylene Hydrogenators in runaway conditions, to simultaneously reduce pressure and remove heat, where a safety valve is ineffective. More recently, HIPS have been employed to remove the heating supply to fractionation columns to avoid activation of the pressure relief device and causing a release to atmosphere or a flare system. In this use it is a secondary overpressure protective system for the purpose of optimizing the design of the flare header system and connected pressure devices.

The Safety Integrity Level (SIL) is the discrete integrity level (SIL 1, SIL 2, SIL 3) of the SIS defined in terms of Probability of Failure on Demand (PFD) as presented in Table 1.

FLARE SYSTEM ANALYSIS

ESTABLISH GLOBAL OVERPRESSURE SCENARIOS

The first step is to establish worst-case global overpressure scenarios. Typically these are caused by failure of a utility system such as electric power (partial or total) or cooling water. Other typical potential causes are instrument air failure or fire. The global fire flaring load is often determined by applying a $232\,m^2$ ($2500\,ft^2$) fire circle based on API STD 521 (7.1.2), but does not usually define the worst case flaring load event.

Table 1. Safety integrity level

Safety integrity level	Probability of failure on demand average range (PFDavg)
1	10^{-1} to 10^{-2}
2	10^{-2} to 10^{-3}
3	10^{-3} to 10^{-4}

When developing global scenarios, consideration of basic process control systems (BPCS) and safeguards is also necessary to establish a credible event. For example credit can be given for some failure positions of control valves per API STD 521 (7.1.4.3). Credits or debits for other properly designed safeguarding systems may also be appropriate.

This review should conclude with an inventory of all the individual flare loads pertaining to each global scenario including relief devices, control valves, depressuring valves, etc. This will allow the establishment of a design flare load base case.

VERIFY RELIEF DEVICE CAPACITY

To complete the global scenario assessment, flow capacity information for different relief device contingencies is required. Depending on plant age and quality of relief systems documentation, this information may be incomplete or lacking for existing facilities. In most cases, it becomes necessary to verify the relief loads based on material and energy balance information and valve mechanical data. Other aspects that need to be considered when verifying the flows include:

- Multi-component representation of stream compositions
- Device inlet and outlet piping configuration
- Relief device flow and opening characteristics for accurate representation of peak flow
- The presence of multiphase, supercritical, high-viscosity, and/or reacting flows

CONSTRUCT FLARE NETWORKS MODEL

To cost-effectively analyze the flare system hydraulics requires constructing a network model of the flare collection system. This involves characterizing the geometric layout of the flare main header and sub-headers, including appropriate dimensional aspects. The individual design case flare loads are tied into the headers at their respective locations.

ANALYZE FLARE SYSTEMS HYDRAULICS

The flare network model is exercised to obtain a base-case flare system profile which establishes:

- Backpressure, flow reduction, pressure accumulation (%MAWP), and temperature accumulation (%MAWT) for protected equipment
- Sub-header, main header, and flare tip flow restrictions
- Exclusion zones for thermal radiation and noise restrictions

This base-case profile is used to identify sub-headers and individual relief devices that are deficient.

Many of these deficiencies are often associated with relief device instability caused by excessive inlet pressure loss or backpressure. Shelly (1999), confirms our experience that 30 to 40% of pressure relief valves in existence violate recommended guidelines for inlet pressure loss and backpressure. Excessive pressure loss can lead to valve instability and possibly valve failure. As a result, many operating companies are faced with significant upgrade or mitigation costs.

Typical flare system design and operating constraints are shown in Tables 2A and 2B. These design and operating constraints can differ depending on where the facility is located and who the operator/owner is.

At this point, an evaluation of options to correct the deficiencies is undertaken, with the purpose of maximizing the use of the existing flare collection system. Options that are usually considered include:

- Automate shutdowns and/or isolation systems currently requiring operator intervention
- Maximum use of bellows/pilot relief valves
- Account for actual timing of loads (e.g., automated de-pressuring systems)
- Make reasonable header and relief piping size adjustments to correct deficiencies, if possible
- Model vessel dynamics and establish actual pressure and temperature accumulation based on flare pressure profiles when using (a) reduced set points less than MAWP, and where (b) the required flow rate is less than the actual relief device rated capacity.

These aspects need to be thoroughly investigated and evaluated before consideration of HIPS as an alternative option. Flare systems mitigation can be costly. Careful analysis and

Table 2A. Typical flare system hydraulics design and operating constraints

Design criteria	Value	Description
Maximum Flow Velocity	Mach ≤ 0.6	Maximum value for header and sub-headers design
Flow rate	Rated Capacity	Value for sub-headers and relief discharge piping design
	Required Capacity	Value for main header design
Backpressure	≤0.1 P_{set}	Conventional relief valves
	≤0.3 P_{set}	Balanced relief valves. Balanced relief valves may be accepted for backpressures up to 0.5 P_{set} with prior consultation with manufacturer and ioMosaic Corporation
	≤0.5 P_{set}	Pilot operated valves. Pilot relief valves will be accepted for backpressures up to 0.7 P_{set} with prior consultation with manufacturer and ioMosaic Corporation

Table 2B. Typical flare system thermal radiation and noise design and operating constraints

Design criteria	Value		Description
Radiation Intensity	1.57 kW/m^2	500 BTU/h ft^2	Value at any location where personnel with appropriate clothing may be continuously exposed
Solar radiation component should be added and can be as high as 1 kW/m^2 in some locations	1.98 kW/m^2	630 BTU/h ft^2	Maximum value for pressured storage equipment
	3.15 kW/m^2	1000 BTU/h ft^2	Maximum value for atmospheric storage equipment
	4.72 kW/m^2	1500 BTU/h ft^2	Heat intensity in areas where emergency actions lasting several minutes may be required by personnel without shielding but with appropriate clothing. Maximum value for Process equipment.
	6.30 kW/m^2	2000 BTU/h ft^2	Heat intensity in areas where emergency actions up to 1 minute may be required by personnel without shielding but with appropriate clothing. Maximum value for Knock Out Drum.
	9.45 kW/m^2	3000 BTU/h ft^2	Heat intensity at any location to which people have access; exposure should be limited to a few seconds, sufficient for escape only.
Emergency Flaring Noise (working areas)	85 dBA		At maximum flaring load
Emergency Flaring Noise (residential areas)	80 dBA		At maximum flaring load
Normal operation Flaring Noise (residential areas)	68 dBA		At maximum flaring load

use of accurate and detailed simulation tools will ensure continued safety and a cost effective mitigation implementation where required. SuperChems™ Expert, or other flare network modeling software, can be used to produce more accurate answers for flow dynamics and flare sub-header optimization. This is crucial for effective selection of mitigation options where necessary.

HIPS EVALUATION

Typically HIPS are considered for de-bottlenecking existing flare collection systems in order to address one or more of the following conditions, without having to significantly modify the existing flare structures or building new ones:

- Header and/or sub-header connection Mach Number > 0.6
- Excessive relief device backpressure
- Excessive vessel accumulation/overpressure
- High flare thermal radiation levels on/off site
- High flare noise levels on/off site
- Adding atmospheric relief devices to the existing flare collection system

SELECT HIPS CANDIDATES

HIPS are generally applied to vessels that require external heat input, such as a distillation column. HIPS can also be applied to reactor vessels where crash cooling or isolation of feed may be required to prevent a runaway reaction. Quickly isolating the source of heat eliminates emergency venting for certain global scenarios. For petroleum refineries, HIPS are used on columns to eliminate power or cooling failure flare loads. The potential candidates are actually a result of the base design case global scenarios determination. Some potential candidates may be eliminated on the basis of a relatively small load that doesn't justify the cost of installing a HIPS system.

DEFINE HIPS CONFIGURATIONS

This activity focuses first on addressing the sub-header deficiencies. Using the base-case load information, a preliminary selection of HIPS equipment and identification of safety integrity levels (SILs) is established. This involves a risk-based analysis to determine the number of HIPS and the SILs required, and requires the establishment of a tolerable overpressure event risk criteria, which will be discussed later. These criteria are used to fix a tolerable event frequency target which is then utilized to evaluate different HIPS failure sequences to arrive at a possible design case.

CONFIRM HIPS DESIGN FLARE LOADS

A HIPS failure sequence and resulting flare loads that meet the target event frequency is run through the network simulation model to obtain new values for backpressure, accumulation, flow rates, Mach number and radiation/noise profiles from the flare. Depending on the results, HIPS configuration will be refined by adjusting the number of HIPS and SILs, and the simulations repeated. Several iterations may be performed to arrive at a cost-effective and tolerable risk solution.

VERIFY REQUIRED SIL

Once the HIPS design configuration is finalized, the next task is to analyze the proposed HIPS design to verify that the specified components and arrangement will meet the safety integrity level (SIL) requirement, which will be discussed later in this paper.

RISK CONCEPTS APPLIED TO FLARE SYSTEM DESIGN

DEFINE TOLERABILITY CRITERIA

A flare system which exerts excessive backpressure on relief devices poses a hazard to pressure vessels depending on the degree of overpressure. The risk tolerability of an overpressure condition in a vessel should be assigned based on:

- The consequences (effect) of the overpressure in terms of vessel integrity
- The frequency at which the severity of the overpressure can be tolerated

Effects of pressure accumulation on steel vessels designed to ASME VIII pressure vessel code are well documented and presented in Table 3. A set of risk criteria can be established using these overpressure effect characteristics.

In devising the criteria, one begins by deciding what level of overpressure is not acceptable and assigning a very low event frequency such as 1 in a million years (10^{-6}/yr). The probability of vessel failure becomes significant for any overpressure event that subjects a vessel to a pressure of 300% of the MAWP. No one should knowingly design for such an event. Hence, accumulations greater than this value are not considered. When setting the frequency for the 165–300% accumulation event, a value of 10^{-5}/yr is selected, which is an order of magnitude less than the unacceptable value for pressure accumulations

Table 3. Effect of pressure accumulation in carbon steel vessels

Accumulation (%)	Effects	Remarks
<135	None expected	None
135–165	Potential for slight permanent deformation	This range of pressure corresponds to the tensile limit of the vessel, and is both material- and code-dependent. The lower and upper limits correspond to ASME VII, Div. 2, and ASME VIII, Div. 1 (1998 edition and earlier) vessels, respectively. ASME VIII, Div. 1 (1998 edition with 1999 addenda) vessels fall in between these values. Therefore a representative value for this range is 150%.
165–300	Permanent deformation, possible small leak	Valid for remote contingencies, as more frequent overpressuring could weaken the vessel by fatigue
300–400	Same as above, but with a higher likelihood of a large leak or burst	Dangerous overpressuring
400–500	Burst	Typical for healthy ASME VIII code vessels

Table 4. Pressure accumulation frequency

Accumulation (%)	Frequency
<135	1 in 100 years (10^{-2}/yr)
135–165	1 in 1000 years (10^{-3}/yr)
165–200	1 in 10,000 years (10^{-4}/yr)
200–300	1 in 100,000 years (10^{-5}/yr)
>300	Not allowed

of greater than 300%. However, this pressure range spans a level that is barely above hydro-test at one extreme to a level above the yield point at the other. While a frequency of 10^{-5}/yr seems right for the upper end of the range, it is quite conservative at the lower end.

A better risk-consequence characterization is obtained by further dividing the 165 to 300 range into two ranges: 165–200 and 200–300; with frequencies of 10^{-4}/yr and 10^{-5}/yr respectively.

SELECT TARGET EVENT FREQUENCY

The target frequency for an overpressure event is determined from the matrix shown in Table 4 using the calculated vessel accumulations from the base-case network simulation. The process begins with analysis of each sub-header and associated loads. The HIPS candidate with the worst accumulation is used to establish the target frequency. Reducing flare loads in the sub-headers is often sufficient for achieving a satisfactory overall flare system design.

Combined scenarios involving HIPS failures on any device connected to the flare may need to be examined to complete the design. For example, failures occurring within the total HIPS population are considered when evaluating the radiation or noise effects from a global scenario. Also, the tolerable frequency target may be more relaxed for the radiation event than overpressure.

DETERMINE SAFETY INTEGRITY LEVEL

For each recommended HIPS, a design specification needs to be developed that details the actual configuration for the vessel being protected. The specified components and redundancy must be able to achieve the SIL requirement determined from the risk-based HIPS selection process. The application of fault tree analysis is an accepted method for determining the expected availability of a SIS or HIPS.

REFINE AND IMPROVE EQUIVALENT SIL BASED ON FUNCTIONAL TEST INTERVAL

The application of fault tree analysis has been shown effective in establishing the relative frequency of potential incidents associated with base-case and alternative HIPS design configurations. The technique has the versatility to handle equipment and control failures along with human errors. Examples of the application of fault tree and reliability analysis for evaluation of safety interlock systems have been reported elsewhere.

Since ISA is a performance based standard, it sets reliability performance requirements, rather than different integrity levels for an interlock based on configuration such as:

Type 3: Fully redundant components
Type 2: Partially redundant components
Type 1: No component redundancy

However, it may be possible to achieve a required SIL with lower reliability hardware through reduction of the test interval (i.e., more frequent testing).

Using appropriate component failure rates, the fractional dead times presented in Table 5 were calculated with incorporation of common cause failure. As Table 5 illustrates, this provides the decision-maker with a good picture of the reliability trade-offs for a given mission (testing interval) duration.

This information can also be utilized for determining reliability (availability) for different SIS configurations (e.g., Type 1 – fully redundant). For example, these data were used to determine the interlock reliability (1 – fractional dead time) for the three types of level interlock configurations as a function of functional testing interval (Table 6).

The reliability values account for common cause failures. Without considering common cause failures, the Type 3 system would meet SIL 3 criteria with monthly and quarterly testing. Analyzing the sources of common cause unreliability and if possible reducing its impact is also worth investigation before making a final select of SIS configuration.

As seen in Table 6, there is a trade-off between testing frequency, and the advantage gained by selecting the next higher SIL configuration. Combining these results with the

Table 5. Unreliability of level interlock systems with consideration of common cause failures

Test interval	Test interval (hours)	Unavailability type 1 design	Unavailability type 2 design	Unavailability type 3 design
1 shift	8	0.010%	0.007%	0.005%
1 day	24	0.029%	0.020%	0.016%
1 week	168	0.200%	0.140%	0.110%
1 month	720	0.870%	0.610%	0.490%
1 quarter	2,160	2.610%	1.840%	1.490%
6 months	4,320	5.220%	3.690%	3.030%
1 year	8,760	10.580%	7.540%	6.390%
18 months	12,960	15.660%	11.220%	9.780%
2 years	17,520	21.160%	15.270%	13.720%

Table 6. Reliability of different level interlock configurations

Configuration class	Redundancy	Test interval	Reliability, %	SIL
Type 3	Fully	Monthly	99.5	2
		Quarterly	98.5	1
		Annually	93.6	1
Type 2	Final Element	Monthly	99.4	2
		Quarterly	98.2	1
		Annually	92.5	1
Type 1	None	Monthly	99.1	2
		Quarterly	97.4	1
		Annually	89.4	0

ISA 84.01 SIL reliability requirements shown in Table 7 enables the designer to take into account cost-benefit considerations between initial capital cost and ongoing maintenance cost.

For example, a SIL 1 might be achieved using a Type 1 configuration with monthly function testing or a Type 2 configuration with annual testing.

A RECENT CASE STUDY

The methodology outlined in this paper was recently used to optimize a flare system in an operating large refinery. The refinery needed to add more than twenty large relief loads from atmospheric vents on several existing columns to the flare system. Additional flare loads from a new planned unit expansion needed to be connected to the existing flare system as well. The design plans called for relocating the flare stack and for expanding the additional new main header piping to 122 cm (48 inch) diameter. The refinery did not want to modify the existing main flare header or any of the existing seven sub-headers. A total of 340 relief devices were connected to the main flare system.

After careful optimization of two of the seven sub-headers connected to the main flare header, the main flare header calculated actual flow capacity was 890,000 kg/hr vs. a requirement of 1,340,000 kg/hr. At a flow capacity of 890,000 kg/hr several large vessels would exhibit pressures up to 1.7 times the maximum allowable working pressure.

Table 7. Combining results with the ISA 84.01 SIL

Safety integrity level	Availability range, %
1	90–99
2	99–99.9
3	99.9–99.99

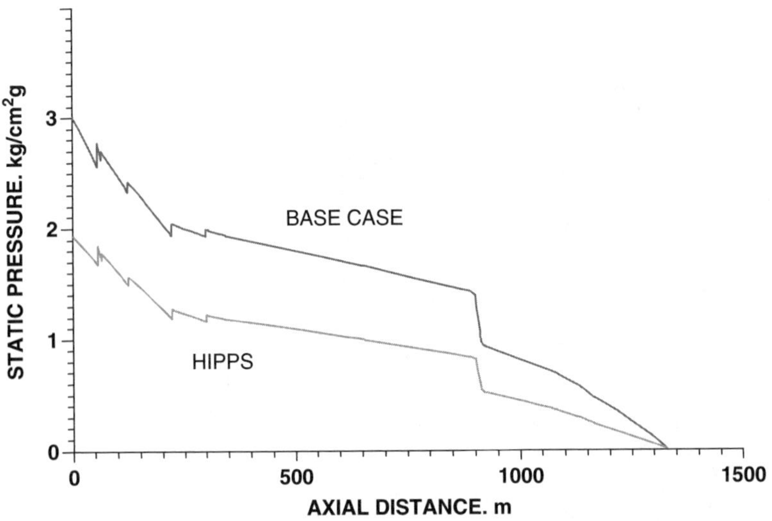

Figure 1. Main header pressure profile

Twelve HIPS systems with SIL levels of 1, 2, 2 + 2, and 3 were selected and optimized such that (a) all connected equipment comply with code requirements for pressure and temperature accumulation when ALL the HIPS function on demand, (b) it is not possible for any simultaneous failure of one or more HIPS to cause code violations at a frequency that exceeds the established target tolerability frequency, and (c) thermal radiation and noise criteria are met under both conditions a and b.

Profiles of pressure in the main header as well as the thermal radiation contours are shown in Figures 1 and 2 for the optimized flare system. Note the length of the main flare header. The HIPS solution enabled the refinery to maximise use of the existing flare structure and ensured continued safe operations with significant additional loads on the flare system. With HIPS, a cost optimal risk reduction was achieved easily and quickly.

CONCLUSIONS

The use of advanced pressure relief dynamics tools such as SuperChems™ Expert can provide accurate estimates of flaring loads and flare systems performance. When coupled with proper risk analysis techniques, accurate flow dynamics provide an optimal cost-risk reduction benefit of where and how to use safety instrumented systems (HIPS). This will yield a safe and cost effective design that meets code requirements for the best-case scenario (all systems working as designed) and that meets social and corporate risk tolerability criteria for worst-case scenarios (when one or more systems fail on demand).

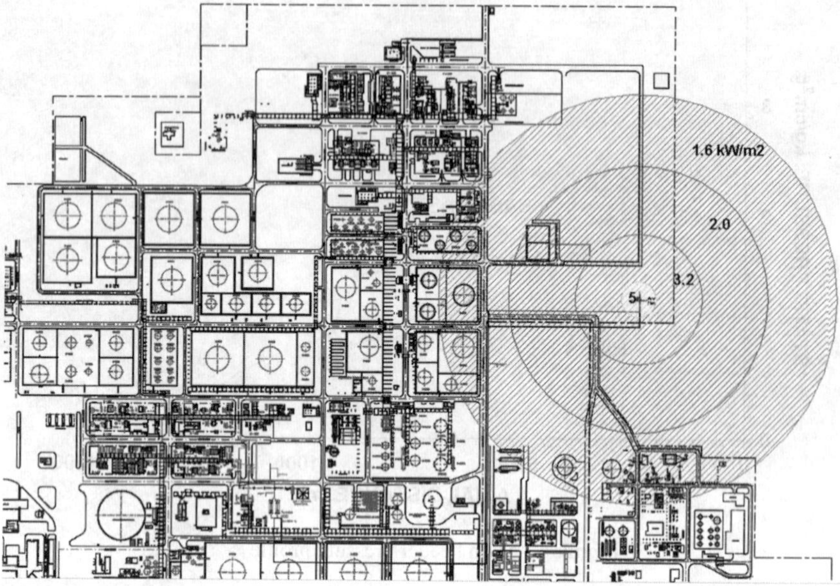

Figure 2. Flare system thermal radiation hazard zones at ground level

Risk tolerability criteria needs to account for the hazardous effects of accumulation on pressure vessels. Designs that result in a vessel accumulation >300% should not be allowed or considered. Note that SIL levels can be enhanced using shorter testing intervals.

The use of many existing flare structures can be maximized using the risk based approach outlined in this paper.

REFERENCES

Shelly, S., 1999, "*Beware: Your pressure protection may be inadequate*", Chemical Engineering, 106 (4), p. 58.

ISA, 1998, *Technical Report ISA-TR84.0.02*.

Freeman, R. A., 2004. "*Reliability of Interlocking Systems*", Process Safety Progress, Vol. 13, No.3 p.146–152

ANSI/API Standard 521 (ISO 23241), Fifth Edition, 2007, "*Pressure-relieving and Depressuring Systems*".

HEALTH AND SAFETY IN BIODIESEL MANUFACTURE

S W Harper, J C Etchells, A J Summerfield, and A Cockton
Health and Safety Executive

A simple process to convert vegetable and animal oils into an alternative and "green" fuel for diesel engines has resulted in a rapid increase in the number of producers of biodiesel in the last 3–4 years, particularly amongst small companies. However, the manufacture of biodiesel can be hazardous if suitable precautions are not taken, as it involves the storage, handling and use of several hazardous substances. Unfortunately, a significant number of new biodiesel producers have little or no experience of chemical processing. In addition, the benefits have prompted some equipment manufacturers to produce kits, for converting waste oils to biodiesel, that have become hazardous during use. In some cases the instructions have been found to be inadequate, so that the hazards have not been fully understood. A number of serious accidents and injuries have already occurred and there is concern that, as the number of producers continues to grow, this trend will increase.

In order to avoid this happening, HSE is producing free basic guidance on the measures necessary to achieve safe production of biodiesel through their website, and detailed advice for enforcement staff to assess such processes. This paper reviews the basic process, identifying the main physical and chemical hazards that need to be assessed to produce a safe process operation. Some of the incidents are also discussed. In addition, the paper examines some issues that have been addressed by inspectors.

INTRODUCTION

A simple process to convert vegetable and animal oils into an alternative and "green" fuel for diesel engines has led to a rapid increase in the number of producers of biodiesel in the last 3–4 years, particularly amongst small companies. However, the manufacture of biodiesel can be hazardous if suitable precautions are not taken, as it involves the storage, handling and use of hazardous substances. Unfortunately, a significant number of new biodiesel producers have little or no experience of chemical processing. In addition, the benefits have prompted some equipment manufacturers to produce kits, for converting waste oils to biodiesel, that have become hazardous during use. A number of serious accidents and injuries have already occurred and there is concern that, as the number of producers continues to grow, this trend may increase.

This paper reviews the basic process, identifying the main physical and chemical hazards that need to be assessed to ensure safe process operation. In addition, some typical issues that have been addressed by inspectors during the course of their inspections are outlined. Some of the incidents that have occurred are discussed, along with planned guidance in order to assist companies to manufacture the materials safely.

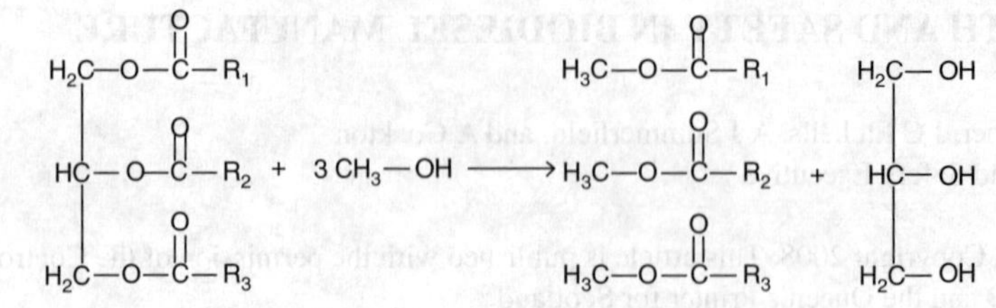

Figure 1. Transesterification reaction for the manufacture of biodiesel

OUTLINE OF MAIN PROCESSES INVOLVED

Feedstock oils are relatively unrefined, and although they can be used directly as fuel in diesel engines, they may eventually choke the engine. However, if the oil is broken down into smaller molecular units it can be used as a direct replacement for diesel.

The reaction commonly used to do this is given in Figure 1. An alkali metal hydroxide or methoxide catalyst is mixed with an excess of dry methanol, and added to the oil with agitation. Reaction times vary with temperature and efficiency of agitation, but are typically up to 8 hours at room temperature dropping to 1–2 hours at 55°C.

The methanol reacts with the oil to form a fatty acid methyl ester (FAME) and Glycerol, which is formed as a by-product. On standing, the two layers separate although, if the alkali ratio is incorrect, soaps may form which can lead to foaming and subsequent separation problems.

The densities of glycerol and biodiesel are approximately 1.3 and 0.9 respectively. The two layers are immiscible so that separation is fairly quick and clean. In larger continuous or batch systems, centrifuges are used to separate the two layers. Some producers choose to wash and dry their biodiesel product to bring it to the correct specification[1], whereas others use particulate filters and molecular sieves or similar. Material that does not meet this specification may damage engine components.

The waste products from the process include glycerol, water washings and excess catalyst, all contaminated with methanol. Specialist companies often remove the glycerol on contract for free as it can be used for a variety of purposes including cosmetics and food additives when cleaned. Direct disposal to the drain can pollute the public sewer.

INCIDENTS INVOLVING BIODIESEL

Several incidents are known to have occurred at biodiesel plants. Most were fires involving methanol or other combustible materials associated with biodiesel manufacture. A significant number were due to the corrosive properties of some of the materials involved. These include:

- A biodiesel plant was destroyed by fire when methanol was spilt during a transfer operation outside the plant building and ignited. Most of the inventory of raw material and product was saved, but the plant was entirely destroyed. The ignition source is unknown.
- An operator was burnt when caustic was being charged into the reaction vessel. The materials boiled over as the result of an uncontrolled exothermic reaction.
- Methanol/catalyst solution leaked onto unprotected electrical equipment when it was being charged into a reactor and ignited.
- Whilst oil/methanol/catalyst mix was being heated to the reaction temperature, vapour escaped from a poorly sealed lid on the reactor and was ignited by unprotected electrical equipment.
- Reaction mixture leaked from a large vertical reaction vessel (approximately 20 cubic metres) and ignited. The subsequent fire destroyed the plant and equipment. It is believed that material leaked either from a manhole cover near the base of the vessel or from a flange in the feed-pipe. There were a number of potential ignition sources in the vicinity, including a boiler and a portable inspection lamp.
- A homeowner, who was involved in domestic production of biodiesel, forgot to turn off the heating element when he left for the weekend. The element overheated which led to a huge fire and the shed and equipment being destroyed. The shed had also contained various materials used in the manufacture of biodiesel, along with some of the finished product, and these seeped into the ground during the fire.
- A company was pumping product biodiesel from a road tanker into IBCs which ignited. The subsequent fire killed 2 people. The ignition source is unknown.
- Approximately 7000 litres of biodiesel escaped into a bund from the lid on a filter pot on a transfer line at a biodiesel blending plant. This was found to have been caused by a vibrating pump.
- Electrical wiring faults have been blamed for other process fires.
- One UK company[2] was fined £24,000 and ordered to pay £5,282 in costs under environmental legislation. Used cooking oil escaped from the site and entered a local watercourse. Local residents reported a strong chip-shop like smell. It was found that they had failed to bund storage vessels and were fined for the release of the cooking oil.
- In the US[3], one person was killed whilst welding on top of a storage tank at a biodiesel plant. The storage tank contained biodiesel, and an explosion occurred followed by a fire. This fire spread to other storage tanks containing glycerine and other flammable materials, which exploded. A large part of the plant was destroyed in the fire, and a second person was injured whilst trying to rescue the man who died.
- Maintenance work involving a grinding operation on a vessel containing methanol is believed to have caused a separate explosion.
- An operator was injured by contact with solid potassium hydroxide on a biodiesel production facility whilst trying to clear a blockage in a reactor feed chute with a screwdriver.

This took about 2 hours. The blockage occurred due to the potassium hydroxide absorbing water from the atmosphere, and possibly methanol vapour being drawn up through the chute. Blockages had occurred in the past and a variety of methods were used to clear them. When the operator tried to remove his chemical suit and PVC gloves with a knitted cuff, skin from his wrist came away with the glove. A skin graft had to be performed on the affected area.

The accidents continue to occur.

HAZARDS

Regardless of the scale of operation the hazards are the same: a combination of flammable, toxic and corrosion hazards depending on the stage of the process. In particular:

Methanol

This is a highly flammable and toxic liquid. It will freely burn in the open air or explode if confined in a vessel or room and ignited. Whilst all precautions should be taken to avoid leaks of flammable vapours into the workroom, leaks and spillages may still occur and it is necessary to take further precautions to reduce the likelihood of their ignition. In areas where such materials are handled, companies are required to identify the areas where flammable atmospheres may exist, for example due to a leak, and determine their likely extent. Such areas are classed as hazardous and should be classified into zones, depending upon the likelihood of their occurrence. In such areas ignition sources such as naked flames should be excluded and only suitably protected electrical equipment should be used. Further information is given in reference 4. It should be noted that the presence of methanol can also render the product and any waste materials flammable, depending upon the way the biodiesel is manufactured.

The catalyst

This is normally potassium hydroxide, sodium hydroxide or sodium methoxide, sometimes in methanol solution, but often as a dry flake or prill. All are corrosive and sodium methoxide is violently water reactive and toxic. Powdered methoxides are a dust explosion hazard, and highly corrosive.

Feedstock oil

If clean and pure it should not be a health problem. However, if the source is unknown or of doubtful quality, then it should be treated as contaminated. Oils are a serious slipping hazard if spillage or contamination outside sealed vessels occurs. Oils can seep into lagging and many can self-combust following a period of chemical degradation. All oils are combustible and will add fuel to any developing fire.

Glycerol

This material is combustible (with a flash-point of 160°C). However, it may be contaminated with methanol and caustic, with their associated hazards, including a potential reduction in flash-point. Unless the initial by-product quality is reliably monitored, then prudence dictates that it should be regarded as contaminated until it has been suitably purified.

Biodiesel

If certified to BSEN 14241:2003[3] it may be regarded as combustible (its flash-point is approximately 150°C). It has rather unusual solvent properties, and will attack some common engineering polymers, including polyvinyls, natural rubber, some gasket and hose materials and metals, including copper, tin and zinc[5]. The effect can increase with heating and ageing of the biodiesel. It is also hygroscopic, and can absorb up to 1500 ppm water from the air. If even slightly contaminated with acid or alkali, biodiesel may be hydrolysed to fatty acids and methanol. This reaction also occurs more slowly in the absence of water, so material stored for more than a couple of weeks may show evidence of a different flash-point than anticipated (which may be significantly lower). Unless this can be reliably prevented, then the product should be used as quickly as possible after production, or reclassified and stored and handled accordingly. Unfortunately this is not always appreciated.

Wash water

This may be contaminated with acid, alkali and methanol. It should be treated as corrosive, toxic and flammable unless tests determine otherwise.

Reaction hazards

The main reaction hazards identified are in the preparation of the base catalyst, which can be by one or more of the following methods:

- The direct addition of sodium or potassium to methanol: this reaction is very exothermic and should be the subject of a rigorous risk assessment, particularly as molten sodium is spontaneously combustible in air and the reaction produces hydrogen as a by-product. Fortunately this route is not generally available to smaller/domestic producers.
- The addition of dried hydroxide or methoxide to methanol which is also very exothermic (heat of dilution)

A much gentler heat of reaction is produced if the catalyst is supplied in methanol solution (although this has to be manufactured safely elsewhere!), and further diluted to the required strength on plant. This is normally the preferred option for smaller companies.

A further reaction hazard occurs when concentrated mineral acid is mixed with water. Addition of water to acids often results in violent boiling and ejection of the acid from vessels. Acids should be added slowly to water with cooling and agitation.

General hazards

These include:

- Corrosion of processing equipment, building fabric, and supporting structures through exposure to caustic and acids leading to premature weakening and catastrophic failure. A by-product of the corrosion process is hydrogen;

- Biodiesel can soften and dissolve a variety of polymers commonly used in safety equipment, such as plastic aprons and rubber boots, causing premature failure.

EXPERIENCE AND TRAINING OF OPERATORS

One of the issues identified as a result of inspections is that practical experience and knowledge of handling hazardous chemicals amongst operators is extremely variable, and tends to drop with reducing scales of production. Larger producers tend to use experienced staff that have previously worked on chemical plant. Their products are invariably sold on to petrochemical companies for blending with various proportions of diesel derived from fossil sources and quality control is important. However some smaller producers may be relatively inexperienced. They include farmers, taxi drivers, and operators of small transport companies who may have little or no training in the hazards. Many of these small producers produce the fuel for their own use. HSE is not responsible for regulating health and safety issues at many of these premises, instead it will normally fall to the Local Authority.

PRODUCTION SCALES AND EQUIPMENT RANGES

Biodiesel output rates from individual producers ranges from DIY operators producing a few tonnes per year, small commercial producers manufacturing 1000 to 10,000 tonnes per year often in skid mounted batch units, and continuous production plants manufacturing hundreds of thousands of tonnes per year. Equipment is variable in cost, quality and construction, and ranges from plastic kitchen buckets and sieves costing a few pounds through small scale commercial mixers and separators made of HDPE costing less than a thousand pounds, then up the scale to the skid-mounted batch units costing many tens of thousands of pounds to the multi-million pound continuous production units. In addition there is a growing trend to use redundant batch chemical production plant at minimal cost and outlay. The output from these units is comparable with the skid mounted batch units.

ISSUES AND HOW THEY HAVE BEEN ADDRESSED SO FAR

As a relatively simple chemical process, biodiesel manufacture has given few problems to the established chemical industry. They have approached it using the same risk assessment procedures applied to other chemical processes, with due account of legislative requirements such as DSEAR, COSHH etc. Problems such as unwanted soap or emulsion formation are systematically dealt with through standard technology. However, small-scale operations sometimes operate less efficiently. They may follow basic instructions that they obtain from various sources, such as the internet, and may fail to fully appreciate the hazards. Consequently they may be unable to develop or provide a safe process. Some smaller scale operations declare their activity to their property insurers, who may give guidance to protect property, however this may not extend to personal safety which is outside their remit.

As a result of our inspections, HSE inspectors have identified a number of issues that have required enforcement action to be taken, ranging from advice to issuing improvement or prohibition notices. These mainly relate to small and medium sized enterprises. Issues that have been addressed at various installations include:

- No risk assessments were carried out on any part of the process;
- No written instructions for the operators, other than the brochure provided with the kit which was not sufficiently detailed;
- Untrained and unqualified operators;
- The operator (self-employed) had no chemical knowledge of the process and minimal understanding of the hazards associated with the chemicals he was using, consequently he had received serious burns from potassium hydroxide;
- The use of unprotected or incorrectly rated electrical equipment, for example the standard of electrical equipment on the reactor and its immediate vicinity was unsuitable for use in a potentially flammable atmosphere;
- Poorly ventilated production areas fitted with domestic electrics (lights, power connections, water heaters etc.);
- The manufacturing operation was run in a room, approximately 5 × 6 m, at the side of a woodworking factory with inadequate fire separation. The room was cluttered with equipment, oils and chemicals. On one side was a methanol mixing vessel, a 1000 litre reactor, a pump and a separation vessel. On the other side was a (domestic) fridge and a gas-fired boiler.
- Process equipment obstructed the means of access and egress.
- Materials of construction were not fire resistant, and no physical fire barrier was provided between the operation and non-industrial activities in the building;
- A number of sites where methanol was supplied in plastic drums, and stored indoors in non fire resisting and/or unventilated areas for security reasons.
- A laboratory (used for wet titrations) was in the same compartment as production, and heavily contaminated with reactants and products;
- Taking into account the raw waste oil, methanol, biodiesel product and by-products there was a considerable amount of combustible or flammable material in the building, resulting in a high fire loading;
- The method of addition of caustic soda to the reaction vessel was not effective in preventing spillage and there was a clear risk of corrosion to equipment on the top of the reactor, including a relief valve, isolation valve and a quick release coupling;
- Product and feedstock have leaked making the workplace slippery;
- Unwashed and untested biodiesel product was left for weeks in thin carbon steel unbunded tanks, which then corroded and leaked.
- A reaction vessel provided with a loosely fitting lid so that flammable vapours were able to escape into the workroom;
- No fire fighting equipment was provided;
- The local fire brigade was unaware of the production activity;
- Inappropriate personal protective equipment (PPE) was provided

- Contaminated PPE dissolving or melting due to biodiesel contamination;
- Wash water and glycerol waste were treated as non-hazardous, even though they were contaminated with methanol and had a low flash point;

Advice has already been issued to inspectors on the main hazards, their means of prevention control and the action that they should take in inspecting such premises which can be accessed on the internet. In addition, a number of equipment suppliers have been approached due to the supply of equipment for biodiesel manufacture, which was not suitable for use in a potentially flammable atmosphere.

NEED FOR GUIDANCE

There is much health and safety guidance readily available that addresses the hazards associated with biodiesel manufacture, for example references 4 and 6 to 10. However, small producers may not be aware of such guidance. In order to address this, HSE is producing an information leaflet that will be available on the Internet. The purpose of the leaflet will be to give information on the hazards of the main processes involved and the measures necessary for their safe design & operation. It will concentrate on the conventional manufacturing process and associated operations, rather than more novel production processes (such as hydrocracking) and will be particularly aimed at small and medium-sized biodiesel manufacturing operations, including chemical companies, farms and transport companies. Where more detailed guidance is available, such as standards for suitable electrical equipment, then these are referred to in the guidance.

By consolidating the relevant guidance into an industry specific document and making it freely available for download, we hope to influence biodiesel producers at all scales to work safely.

CONCLUSIONS

1. There has been a rapid increase in the number of biodiesel producers in the last few years, in particular the number of smaller companies who are manufacturing it.
2. The manufacture of bio-diesel can be hazardous if suitable precautions are not taken, as it involves the storage, handling and use of hazardous substances, in particular toxic, flammable and corrosive substances.
3. Whilst many of the larger producers tend to use experienced staff that have previously worked on chemical plant, unfortunately a significant number of new smaller companies producing biodiesel have little or no experience of chemical processing, and may be unaware of all the hazards that the process introduces. They include farmers, taxi drivers, and operators of small transport companies. Many of these small producers produce the fuel for their own use and HSE is not responsible for regulating health and safety issues at many of these premises.

4. Consequently, HSE has already identified a number of issues at such facilities, which have resulted in enforcement action. Advice to inspectors on the action they need to take has also been issued.
5. In addition, several incidents are known to have occurred at biodiesel plants both in Britain and overseas. Most were fires involving methanol or other combustible materials associated with biodiesel manufacture. A significant number were due to the corrosive properties of some of the materials involved, or from poorly controlled mixing of reactive chemicals.
6. In order to address these issues, HSE is producing guidance that will address the hazards of the various substances and process steps, and advise companies on how they should be controlled. The guidance will be aimed particularly at small and medium sized companies and will be freely available on the internet.

REFERENCES

1. EN14214:2003, Automotive fuels. Fatty acid methylesters (FAME) for diesel engines. Requirements and test methods
2. Incident description taken from The Environment Agency website
3. Incident description taken from website of Biodiesel Magazine, August 2006
4. BS EN 60079-10:2003, Electrical Apparatus for Explosive Gas Atmospheres – Part 10: Classification of Hazardous Areas
5. Private communication from Jeanne van Buren, Rotterdam Fire Brigade
6. HSE Publication HS(G) 51 The storage of flammable liquids in containers. ISBN 0717604810
7. HSE Publication HS(G) 140 Safe use and handling of flammable liquids. ISBN 0717609677
8. HSE Publication HS(G) 143 Designing and operating safe chemical reaction processes. ISBN 0717610519
9. HSE Publication L138 Dangerous Substances and Explosive Atmosphere Regulations 2002 Approved Code of Practice. ISBN 0717622037
10. HSE Publication L5 Control of Substances Hazardous to Health Regulations 2002 - Approved code of practice and guidance. ISBN 0717629813

A METHODOLOGY TO GUIDE INDUSTRIAL EXPLOSION SAFETY SYSTEM DESIGN

R J Lade[1] and P E Moore[2]
[1]Kidde Research, Mathisen Way, Poyle Road, Colnbrook, Berkshire, SL3 0HB, UK, Phone: +44 1753 689848. Fax: +44 1753 683810.
[2]UTC Fire & Security, Mathisen Way, Poyle Road, Colnbrook, Berkshire, SL3 0HB, UK, Phone: +44 1753 689848. Fax: +44 1753 683810.

Process plant owners/operators have an obligation to assess and ascribe the residual risk of an unmitigated explosion occurrence under the ATEX Directives, and in doing so are making a key decision on the acceptability of that residual risk. All safety systems have a residual risk that they fail to achieve their mission, whether it be via hardware failures, personnel errors, errors in the theoretical and design assumptions, or inadequacies in the quantification of the prevailing hazard.

It is essential that suppliers and users of explosion protection products and systems fully understand the efficacy and reliability upon demand of such products and systems. A systematic methodology for quantifying residual risk in the context of installed explosion mitigation has been described by the authors. This methodology explicitly accounts for the two principal mechanisms of failure:

a) failure of the hardware;
b) ineffective explosion protection (e.g, the reduced explosion pressure of a suppressed or vented explosion occurrence is greater than the pressure shock resistance of the vessel).

This paper considers the challenges faced in determining a meaningful residual risk datum for a processing plant. In particular it sets out the importance of the implicit assumptions and shows, by reference to process industry examples, the benefits of electing a systematic means of ascribing explosion protection security.

In order to quantify the residual risk of safety system failure in the practice, an overarching understanding of the efficacy of explosion mitigation means, system design, safety factors (both implicit and explicit) and the consequence of flame propagation between connected vessels, is of paramount importance. Existing explosion protection design guidance is invariably constructed around test data that have taken the premise that central ignition of a homogeneous and turbulent optimum fuel concentration in a closed vessel represents the worst case scenario. However this is not necessarily the most appropriate baseline for ascribing the risk of an unmitigated explosion occurrence.

This work demonstrates that explosion protection "trade off" decisions, design safety factors, and the design premise itself all contribute to the "relied upon" safety integrity of an industrial process. With this understanding and the adoption of a systematic methodology to determine a residual risk datum, practitioners can make more informed and cost effective design decisions, leading to enhanced overall process safety.

DESCRIPTION OF THE CALCULATION METHODOLOGY

A method for calculating residual risk of safety system failure has been set out previously by the authors [Ganguly, 2007], and the pertinent mathematical derivations are fully explained elsewhere [Date, 2008]. In this paper we present a brief description of the model and its implicit assumptions, using the same nomenclature as previously, for ease of reference.

Our intention is to demonstrate the value of such an approach to improve overall process safety. The process plant and its protection system are represented by a connected, bi-directional graph (West, 2001). In this architecture each plant item in the process is represented as a vertex, whereby edges between vertices represent possible flame paths (i.e. the connecting duct-work).

ASSUMPTIONS

- We use the probability of an unmitigated explosion in a given unit of time as a proxy for residual risk.
- All ignition locations within each plant item are equally probable.
- An unmitigated explosion (failure) is defined as any occurrence where the reduced explosion pressure of a suppressed or vented explosion is greater than the pressure shock resistance of a plant component.
- Given an ignition event, an unmitigated explosion is assumed to occur when any one component of the protection system fails, be it a vent panel, detector, suppressor or control panel. Consideration of component redundancy and the impact on residual risk is fully tractable, but is not addressed further here.
- We consider the consequence of all failures equally. In reality, not all failures will lead to a catastrophe, however by comparing all failures equally we are still able to compare different safety system designs.
- We only consider the probability of failure of the plant item that has the ignition event and those directly connected to it. The model is not bound by this assumption – extension to second order connectivity is tractable, but of negligible significance.

DEFINITION OF MODEL PARAMETERS

Each vessel or plant item *i* (vertex *i*) within the process plant, together with its associated explosion protection system is characterised by a set of parameters which are described in this section.

- $Q_E(i)$ is the ignition probability in vessel *i*. For a given process plant and over a given unit of time we assume that $\sum_i Q_E(i) = 1$, i.e. that there will be one ignition occurrence somewhere in the process plant.
- $P_{red}(i, j)$ is the reduced explosion pressure in vertex *i* following an ignition in vertex *j*. $P_s(i)$ is the pressure shock resistance of the vertex *i*. The values quoted for $P_{red}(i, j)$ and

$P_s(i)$ are intentionally very conservative to represent the worst case and to err on the side of safety. However, excessive safety factors will result in unrealistic values for the computed residual risk. We have elected to use a standard deviation of 10% of the mean value for both $P_{red}(i, j)$ and $P_s(i)$, and that the values specified are the two standard deviation limit values.

- $Q_{vessel}(i, j)$ represents the probability that the explosion protection hardware does not fail, but the reduced explosion pressure is still higher than the pressure shock resistance of the vessel:

$$Q_{vessel}(i, j) = P[P_{red}(i, j) - P_s(i) > 0] \qquad (1)$$

This allows us to represent the proximity of $P_{red}(i, j)$ to $P_s(i)$ in the system design and account for any intentional design safety factors in our computation of residual risk.

- In a similar manner, we can define a set of parameters which relate to the efficacy of explosion isolation barriers. We define $t_b(i, j)$ as the time taken from ignition for the isolation barrier to be established between vessels i and j. Implicit in $t_b(i, j)$ is the time taken to detect the explosion (whether via optical or pressure detection) and the actuation time of the isolation hardware such that flame cannot pass. $t_f(i, j)$ is the time taken for the flame front to arrive at the barrier location, and will be the summation of the time taken for the flame to enter the duct from the ignition location, and the time for the flame to transit the duct to the barrier position. Thus for efficacious explosion isolation $t_f(i, j) > t_b(i, j)$. Once again the specified values for these parameters are very conservative and we apply the same assumptions as with the pressure parameters, taking a standard deviation of 10% of the mean and that the specified values are the two standard deviation limit values.
- $Q_{barrier}(i, j)$ represents the probability that the isolation hardware is actuated and the barrier established, but the barrier is deployed too late and flame passes into the adjoining vessel:

$$Q_{barrier}(i, j) = p[t_b(i, j) - t_f(i) > 0] \qquad (2)$$

- $Q_f^s(i, j)$ is the probability of flame propagation between connected vessels i and j which then leads to an enhanced explosion in j. This of course will be sensitively dependent on the geometric configuration (relative vessel sizes, duct length and diameter, process flow direction and velocity) together with the fuel properties and the explosion mitigation means employed on both the source and connected vessels.
- The total flame propagation probability from vessel i to j, $Q^s(i, j)$, can be computed by summing the probability due to hardware failure ($Q_h(i, j)$) and the probability due to late activation of the barrier $1 - (Q_h(i, j)) \times Q_{barrier}(i, j)$) to give:

$$Q^s(i, j) = Q_f^s(i, j) \times [Q_h(i, j) + (1 - Q_h(i, j)) \times Q_{barrier}(i, j)] \qquad (3)$$

- $Q_h(i, j)$ can be calculated with knowledge of the mean-time-between-failure of the hardware components combined in an appropriate manner to represent the configuration of the protection system [Date, 2008, Ganguly, 2007].

When all of the above parameters have been specified for each vessel and connection, we have all the necessary information to compute the residual risk of safety system failure.

COMPUTATION OF RESIDUAL RISK

The risk of failure of any vessel i due to ignition in vessel j, is denoted $R_{i,j}$ and can be computed as the sum of the risk of hardware failure, $Q_h(i)$, and the risk of inadequate protection, $(1 - Q_h(i)) \times Q_{vessel}(i, j)$:

$$R_{i,j} = Q_h(i) + (1 - Q_h(i)) \times Q_{vessel}(i, j) \tag{4}$$

Once again, $Q_h(i)$ can be calculated with knowledge of the mean-time-between-failure of the pertinent hardware. We can now calculate the risk of failure in any vessel i due to an ignition in the same vessel or any vessel directly connected, ζ_i, as:

$$\zeta_i = Q_E(i) \times R_{i,i} + \sum_{i \in \Phi j} Q_E(j) \times (1 - R_{j,j}) \times Q^s(j,i) \times R_{i,j} \tag{5}$$

where Φ_i, denotes the set of vertices adjacent to vertex j.

EXAMPLE COMPUTATION OF RESIDUAL RISK

To illustrate this calculation methodology and demonstrate its use in guiding explosion safety system design, we consider the example of a simple milling and collection process (see Figure 1), where explosible dust represents the principal hazard [Eckhoff, 2003]. In this process a granulated chemical product is fed into a Grinder, and the product fines are pneumatically transported to a Storage Hopper. Residual dust from the Cyclone is extracted by a Bag-Filter before the process air is returned to the atmosphere. The Bag Filter and the Storage Hopper are protected by explosion suppression systems, whilst the Grinder and Cyclone are protected by appropriately sized explosion vent panels. In this example, a fast-acting explosion isolation valve has been installed to minimise the risk of flame propagation from the Grinder to the Cyclone.

First we must ascribe ignition probabilities ($Q_E(i)$) for the four vessels in our example process. This will of course be dependent on the material being processed (e.g. explosibility, concentration, minimum ignition energy etc.) and the nature of the process. In order to attain representative values we have taken literature data [Jeske, 1997] and organised it so as to be able to quote typical ignition probabilities for generic plant processes. Part of the organisation of this data involved excluding ignition sources that were external to the process, such as fire, and then grouping ignition sources that were pertinent to generic plant processes and then normalising these probabilities. Although this methodology is a simplification of the practice, it is based on real data and serves the purpose of providing representative ignition probabilities. The $Q_E(i)$ values determined for each vessel in our example are shown in Table 1. Also shown in Table 1 are the vessel strengths, $P_s(i)$, and the reduced

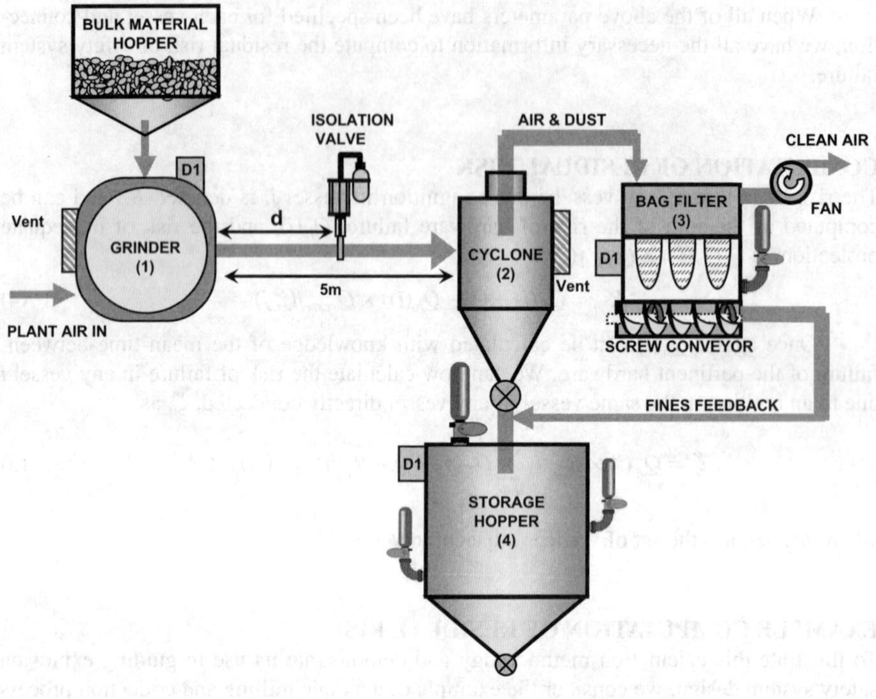

Figure 1. Schematic representation of an example milling and collection process. The grey arrows represent material flow through the plant. d represents the installed distance of the isolation barrier from the Grinder

Table 1. Ignition probabilities, $Q_E(i)$, pressure shock resistance, $P_s(i)$, reduced explosion pressure, $P_{red}(i, i)$, and the probability that the explosion protection hardware does not fail, but the reduced explosion pressure is still higher that pressure shock resistance of the vessel, $Q_{vessel}(i, i)$, (calculated using Equation 2) for each vertex in the example milling and collection process

Plant item	Vertex	$Q_E(i)$	$P_s(i)$/bar(g)	$P_{red}(i,i)$/bar(g)	$Q_{vessel}(i, i)$
Grinder	1	67%	0.55	0.50	2.34×10^{-4}
Cyclone	2	11%	0.45	0.42	1.29×10^{-4}
Bag filter	3	17%	0.40	0.40	6.52×10^{-3}
Storage hopper	4	5%	0.30	0.28	4.60×10^{-4}

explosion pressures from an ignition in i, $P_{red}(i, i)$, the latter being determined by using either proprietary software [Siwek, 2001] or in-house software packages [Moore, 2001]. Other means of calculating these pressures are equally valid.

We also need to determine $P_{red}(i, j)$ when $i \neq j$. This is the reduced explosion pressure following flame transfer from a connected vessel resulting in a flame jet ignition event. The resulting explosion incident is often more severe than the point ignition assumption that was used in designing the explosion protection on the connected plant item. The extent of the explosion enhancement due to flame jet ignition for our example has been estimated by referring to the literature data regarding this phenomenon, [Lunn,1996; Holbrow, 1996]. From these data the explosion enhancement was interpolated using the dust variant of an industry standard computational fluid dynamic (CFD) explosion modelling tool (FLACS,2005). Table 2 lists the $P_{red}(i, j)$ values for each connected vessel.

Next we need to represent the fast-acting explosion isolation valve fitted between the Grinder and the Cyclone at a distance, d = 3 m from the Grinder, see Figure 1. In this example the isolation valve relies upon a pressure detector fitted to the Grinder. The closing time of such a valve is typically 40 ms and we calculate $t_b(1,2) = 79$ ms, and $t_f(1,2) = 49$ ms using our in-house software package [Moore, 2005] with representative input parameters such as duct diameter, air flow, material explosibility etc. Other means of calculating these times are equally valid. This is **not** an explosion isolation solution since $t_b(1,2) > t_f(1,2)$ as a consequence of the valve being located too close to the Grinder, and therefore not prEN150089:2006 compliant.

Table 2 also lists $Q^s_f\,(i, j)$ for each flame path, together with the resulting $Q^s(i, j)$. $Q^s_f\,(i, j)$ has been determined from the large corpus of experimental data generated by Holbrow et al. [Holbrow, 1996] together with our own test data. A large proportion of these data sets are for explosions in connected vented vessels, therefore $Q^s_f\,(i, j)$ needs to be adjusted to represent configurations where either the source, connected or both vessels

Table 2. Reduced explosion pressure in vertex i following an ignition in vertex J, $P_{red}(i, j)$, the probability of flame propagation leading to an enhanced explosion in j, $Q^s_f(i, j)$ and total flame propagation probability, $Q^s(i, j)$, for each connection in the example milling and collection process

(i, j)	$P_{red}(i, j)$/bar(g)	Q^s_f, (i, j)	$Q^s(i, j)$
(1,2)	1.00	0.320	0.218
(2,1)	0.70	0.261	0.261
(2,3)	0.72	0.080	0.080
(3,2)	0.86	0.013	0.013
(3,4)	0.83	0	0
(4,3)	0.50	0	0
(4,2)	0.49	0.047	0.001
(2,4)	0.66	0.009	0.001

have explosion suppression systems fitted. The extent and form of this adjustment is work in-progress and so we have elected the following considered assumptions. Here V_1 refers to the source vessel where the ignition occurs and V_2 is the connected vessel.

- **V_1 Suppressed : V_2 Vented:** With the source vessel suppressed, only ignition locations close to the duct mouth will allow flame to enter the duct before the vessel is engulfed with suppressant. These ignition locations represent only a small fraction (~5%) of the vessel volume† and we have adjusted $Q_f^s(i, j)$ according to this criteria.
- **V_1 Vented : V_2 Suppressed:** If the source vessel is vented then flame transfer to V_2 is as probable as in the vented:vented case. However, in most configurations the pressure in the connected vessel will have risen sufficiently such that the suppression system will have actuated before the flame arrives at the vessel. The experimental data, supported by our CFD investigations show that on average only 25% of occurrences result in flame entry in V_2 before the suppressant has essentially engulfed the vessel volume. Once again we have elected this criteria to adjust $Q_f^s(i, j)$.
- **V_1 Suppressed : V_2 Suppressed:** With both vessels suppressed, it is difficult to envision a situation whereby an enhanced explosion in the connected vessel can occur, and we have therefore elected to set $Q_f^s(i, j)$ at zero for this scenario.

In our example, $Q^s(1,2)$ will of course include terms for the isolation hardware, $Q_h(1,2)$, and the probability due to late activation, $Q_{barrier}(1,2)$ according to Equation 3.

Finally we can now calculate the risk of failure in any vessel due to an ignition in the same vessel or any vessel directly connected, see Table 3.

IMPACT OF CHANGES IN EXPLOSION PROTECTION SYSTEM DESIGN

As we can see from Table 3, the Bag Filter is at greatest risk. This can be attributed to the proximity of $P_{red}(3,3)$ to $P_s(3)$ and the connection with the Cyclone and Storage Hopper

Table 3. Residual risk of safety system failure for each vertex in the grinding and milling example process

Plant item	Vertex	Risk of failure due to an ignition in the same vessel or any vessel directly connected
Grinder	1	$\zeta_1 = 2.79 \times 10^{-3}$
Cyclone	2	$\zeta_2 = 4.95 \times 10^{-3}$
Bag filter	3	$\zeta_3 = 7.10 \times 10^{-3}$
Storage hopper	4	$\zeta_4 = 5.43 \times 10^{-5}$

†This argument is similar to that presented later, and in more detail, regarding ignition location and isolation barrier placement when using pressure detection, see Figure 3(A)

which are both large vessels‡. Unless we have evidence to suggest that $P_{red}(3,2)$ or $P_{red}(4,2)$ has been overstated, and thus can be reduced, the best option is to change the suppression system design such that $P_{red}(3,3)$ is reduced somewhat. In this case, a simple reduction in the suppression actuation pressure from 100 mbar to 50 mbar is sufficient to reduce $P_{red}(3,3)$ to 0.33 bar and thus reduce ζ_1 to 1.16×10^{-3}. Of course, we must be thoughtful of the impact of any unnecessary or false actuations which may be greater with a reduced actuation pressure on the suppression system fitted to the Bag Filter.

With the risk in the Bag Filter reduced, the Cyclone now becomes greatest at risk with the largest contribution coming from $Q^s(1,2)$. This stems from the fact that the barrier is positioned too close to the Grinder and therefore does not allow enough time to establish the barrier before flame passes the barrier location. For efficacious explosion isolation using this hardware, the barrier must be placed at 8.0 m from the Grinder such that $t_b(1,2) = t_f(1,2)$. However, as shown in Figure 1, the duct-work between the Grinder and the Cyclone is only 5 m long. This represents a very real issue for explosion isolation systems in the practice.

One way to better understand this problem is to calculate the minimum barrier distance, d_{min}, such that $t_b(1,2) = t_f(1,2)$ as a function of ignition location in the source vessel. Figure 2 shows this for pressure, optical and dual (pressure AND optical) detection means to actuate the isolation barrier. With pressure detection, the worst case ignition location is close to the duct mouth, since the flame will have started propagating along the duct before the pressure in the source vessel has increased sufficiently to secure detection. Consequently the largest d_{min} is for ignition close to the duct mouth. For ignition locations far from the duct mouth, the pressure detector will have actuated long before the flame reaches the duct, and as Figure 2 shows, at ignition locations greater than 0.55 m from the duct mouth, the barrier can be place adjacent to the vessel ($d_{min} = 0$).

In our Grinder example with pressure detection, we can use Figure 2 to understand the consequences of barrier placement at 3 m. We see that with ignition locations further than 0.42 m from the duct mouth the calculated barrier distance, d_{min}, is less than our installed location of 3 m and we therefore predict efficacious isolation. However, for ignition locations less than 0.42 m from the duct mouth, d_{min} is greater than 3 m and we would expect flame passage. This ignition distance, $L_{press} = 0.42$ m, allows us to draw a locus and thus define a volume element in which if ignition were to occur the isolation barrier is likely to fail its mission, see Figure 3 (A). The volume of this hemispherical region close to the duct is $0.144\,m^3$ and so constitutes 7.22% of the total vessel volume. Since we assume that all ignition locations are equally probable, this volume element represents the percentage of ignition locations that will allow flame passage and is represented in our residual risk calculation by $Q_{barrier}(1,2)$.

Let us investigate changing the detection means for the isolation barrier to an optical detector located within a few duct diameters of the duct mouth. In this case ignition far from the duct is the most challenging case. Here the flame will be travelling very fast as it

‡The explosion enhancement from flame jet ignition is proportional to the connected vessel volume ratios (V_1/V_2) [Lunn,1996; Holbrow, 1996].

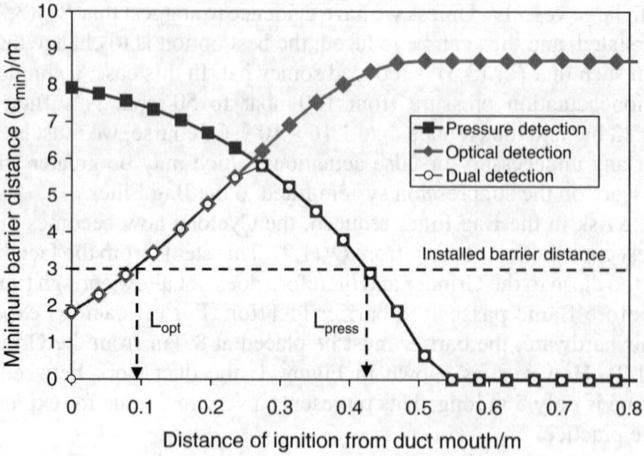

Figure 2. Minimum barrier distance (i.e. $t_b(1,2) = t_f(1,2)$) as a function of distance of ignition location from the duct mouth for pressure, optical and dual (pressure AND optical) detection means to actuate the fast-acting isolation valve. The Grinder has a volume of 2 m^3, with a air velocity through the DN300 duct of 10 m/s. The material has a fuel explosibility rate constant of 150 bar.m/s and the isolation system uses a 50 mbar detection pressure. L_{opt} and L_{press} are indicated for the installed barrier location of 3 m from the Grinder

enters the duct (when it is detected), and therefore requires a large barrier distance in order to establish the barrier. For optical detection, ignition close to the duct is the trivial case since the flame is moving very slowly as it propagates from the ignition kernel and will be detected immediately.

We see from Figure 2 that $L_{opt} = 0.09$ m and this represents a small volume element (1.52×10^{-3} m^3) whereby an ignition occurrence would result in efficacious explosion isolation. As shown schematically in Figure 3 (B), optical detection (in this example) is actually much worse than pressure detection, with 99.92% of ignition locations resulting in the isolation barrier failing its mission.

Employing dual detection takes the strengths of pressure and optical detectors, but avoids their respective weaknesses. This is shown in Figure 2 where at distances close to the duct, optical detection will actuate first, while at distances far from the duct pressure detection will actuate first. Using dual detection and locating the barrier at 5.8 m from the Grinder we would cover all possible ignition locations. However, in our example, we are still outside this design guidance and the resulting volume element in which an ignition would result in flame passage is shown schematically in Figure 3 (C) and is simply the difference in volume elements defined by L_{press} and L_{opt}. In this example, we only reduce the volume element to 7.14%, which is only marginally better than pressure detection alone (7.22%) because of the insignificant volume protected by the optical detector.

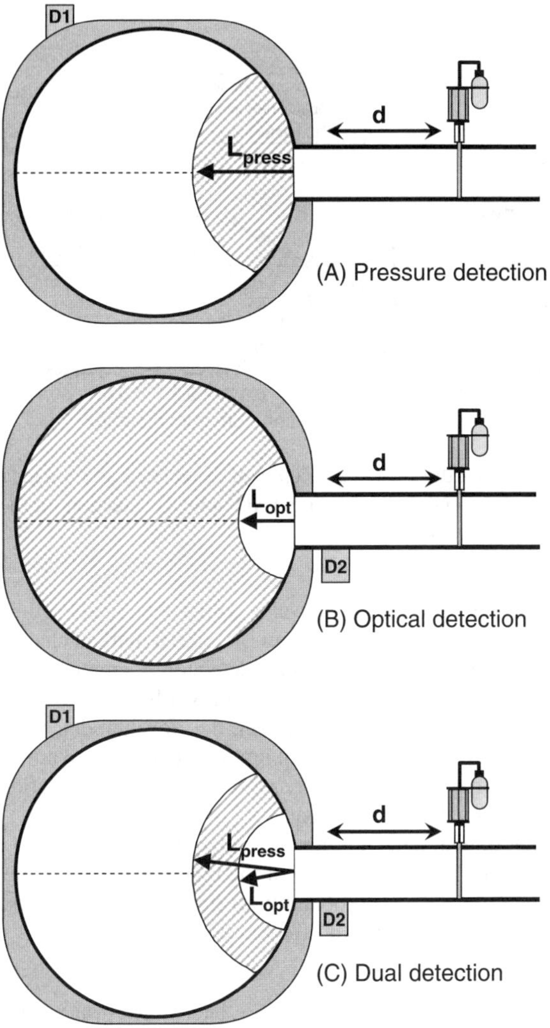

Figure 3. Schematic representation of the volume element (shaded region) in which if ignition occurred the isolation barrier would not prevent flame passage for (A) pressure, (B) optical and (C) dual detection means. L_{press} and L_{opt} represent the radius of the locus at which $t_f(i, j) = t_b(i, j)$ when using pressure or optical detection respectively

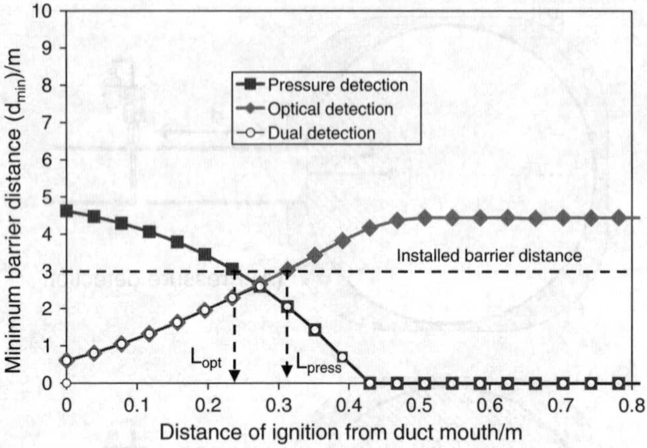

Figure 4. Minimum barrier distance as a function of distance of ignition location from the duct mouth for pressure, optical and dual (pressure AND optical) detection for the Grinder example using a chemically acting isolation barrier

If we were to replace the fast-acting valve with a chemically acting isolation barrier, the latter having a much faster deployment time, we see a marked change in the minimum barrier distances, see Figure 4. When pressure detection is employed, only 1.8% of ignition occurrences will allow flame to pass the barrier, whilst once again optical detection is a poor choice with 96.9% unprotected. However, d_{min} for dual detection is now less than our installed barrier distance meaning that all ignition locations are now protected. Table 4 shows the residual risk for the Cyclone when using a chemical isolation barrier actuated using either pressure or dual detection means.

Supposing that optical detection for the isolation barrier was not acceptable to the plant operator, maybe due to the frequency of the maintenance schedule of such a device in a dirty environment. Instead the probability of flame propagation and its consequence could be addressed to reduce the risk of safety systems failure.

Table 4. Residual risk of safety system failure in the Cyclone using either pressure or dual detection to actuate the chemical isolation barrier between the Grinder and the Cyclone. $t_f(1,2)$ varies for different detection means since the worst case ignition location is used [Moore, 2005]

Detection means	$t_b(1,2)$	$t_f(1,2)$	Risk of failure due to an ignition in the cyclone or any vessel directly connected
Pressure	61	49	$\zeta_2 = 8.25 \times 10^{-4}$
Dual	61	65	$\zeta_2 = 1.14 \times 10^{-4}$

Table 5. Residual risk of safety system failure for each vertex in the grinding and milling example process with explosion suppression fitted to the Grinder and the Cyclone. The tabulated risks are calculated using the changes in safety system design previously discussed, such as the reduced P_{red} (3,3) and a chemical isolation barrier fitted between the Grinder and the Cyclone actuated via pressure detection

Plant item	Vertex	Risk of failure due to an ignition in the same vessel or any vessel directly connected
Grinder	1	$\zeta_1 = 1.95 \times 10^{-4}$
Cyclone	2	$\zeta_2 = 3.93 \times 10^{-5}$
Bag Filter	3	$\zeta_3 = 5.19 \times 10^{-5}$
Storage Hopper	4	$\zeta_4 = 3.91 \times 10^{-5}$

Connected vented vessels (without efficacious explosion isolation) are much more likely to result in a flame transfer which can lead to an enhanced explosion occurrence in V_2. This stems from the fact that explosion venting simply mitigates against the rapid pressure rise and does not tackle the presence of flame which can lead to further ignition events. This is not the case for explosion suppression whereby the flame front itself is extinguished by the rapidly deployed suppressant agent.

In our example, fitting explosion suppression to the Grinder and the Cyclone, in place of explosion venting would significantly reduce the probability of flame transfer between the plant components. Table 5 lists the residual risk for each plant item with both the Grinder and the Cyclone fitted with explosion suppression as described above. In this case the residual risks have now been reduced by an order of magnitude for the Grinder and the Bag Filter, and by two orders of magnitude for the Cyclone.

CONCLUSIONS

- The benefit of using a systematic calculation tool to ascribe the residual risk of explosion safety system failure has been shown by reference to a practical process-industry example.
- By electing appropriate input assumptions and representative data to set out ignition probabilities and consequence of flame transfer, a meaningful measure of residual risk that an installed explosion protection measure will fail to mitigate an explosion incident can be determined.
- A software design support tool is clearly possible from the described calculation means.
- Residual risk determination is critically dependent on the detail of the elected explosion protection system. In this paper we have shown by reference to a practical example that the installation of an explosion isolation means, in this case a fast-acting valve

located at 3 m from the duct mouth rather than the prescribed 5.5 m, has important implications on overall process safety.

- Options to improve the design of explosion isolation include the incorporation of optical detection and electing a faster isolation means, and are shown to represent safety system enhancements for this example. However, the most profound change in the residual risk of failure was demonstrated by electing explosion suppression rather than explosion venting on connected vessels, and thereby significantly reducing the probability of flame transfer.
- We have shown that by determining the residual risk of failure, it is possible to select and quantify the safety integrity of explosion protection options. Through this process design, engineers and operators can make better and informed decisions, leading to enhanced safety integrity and cost effectiveness in delivering overall process safety.

ACKNOWLEDGEMENTS

The authors would like to thank Professor Gautam Mitra and Dr Paresh Date from the *Center for the Analysis of Risk and Optimisation Modelling Applications*, Brunel University for their work in devising the mathematical methodology employed for the computation of residual risk. We also would like to thank Robert Pallant from Kidde Research for his help in data analysis and its interpretation, together with his interest in this work.

NOMENCLATURE

$Q_E(i)$	Ignition probability in vessel *i*.
$P_{red}(i, j)$	Reduced explosion pressure in vertex *i* following an ignition in vertex *j*.
$P_s(i)$	Pressure shock resistance of vertex *i*.
$Q_{vessel}(i, j)$	Probability that the explosion protection hardware does not fail, but the reduced explosion pressure is still higher than the pressure shock resistance of the vessel
$t_b(i, j)$	Time from ignition for the isolation barrier to be established between vessels *i* and *j*
$t_f(i, j)$	Time taken for the flame front to arrive at the barrier location between vessels *i* and *j*
$Q_{barrier}(i, j)$	Probability that the isolation hardware is actuated and the barrier established, but the barrier is deployed too late and flame passes from vessel *i* to *j*
$Q_f^s(i, j)$	Probability of flame propagation between connected vessels *i* and *j* which then leads to an enhanced explosion in *j*.
$Q^s(i, j)$	Total flame propagation probability from vessel *i* to *j* which then leads to an enhanced explosion in *j*, taking into account any explosion isolation provision

$Q_h(i,j)$	Probability of explosion isolation hardware failure between vessels i and j
$Q_h(i)$	Probability of explosion protection hardware failure on vessel i
ζ_1	Residual risk of failure of vessel i due to an ignition in the same vessel or any vessel directly connected
Φj	The set of vertices adjacent to vertex i.
V_1	Source vessel where ignition occurs
V_2	Vessel connected to V_1
d	Distance of explosion isolation barrier from the source vessel
d_{min}	Minimum barrier distance from the source vessel such that $t_b(i,j) = t_f(i,j)$
L_{press}	radius of the ignition locus at which $t_f(i,j) = t_b(i,j)$ when using pressure detection
L_{opt}	radius of the ignition locus at which $t_f(i,j) = t_b(i,j)$ when using optical detection

REFERENCES

Date, P., Lade, R. J., Moore, P. E., Mitra, G., 2008, Modelling the Residual Risk of Safety System Failure, submitted to Operations Research, INFORMS.

Moore, P. E., Dunster, R. G., 2001, Improved Effectiveness in Explosion Suppression, VDI Berichte

Eckhoff, R., 2003, Dust Explosions in the Process Industries, Elsevier.

Ganguly, T., Date, P., Mitra, G., Lade, R. J. & Moore, P. E., 2007, A method for computing the residual risk of safety system failure, in `Proceedings of 12th International Symposium on Loss Prevention and Safety Promotion in the Process Industries'

Gexcon, 2005, FLACS: (FLame ACceleration Simulator), Gexcon AS, Bergen, Norway.

Holbrow, P., Andrews, S., Lunn, G. A., 1996, Dust Explosions in Interconnected Vented Vessels, J Loss Prevention Process Industries, 9(1): 91-103.

Jeske, A., Beck, H., 1997, Dokumentation Staubexplosionen, Analyse und Einzelfalldarstellung, BIA –Report NR. 4/82

Lunn, G. A., Holbrow, P., Andrews, S., Gummer, J., 1996, Dust Explosions in Totally Enclosed Interconnected Vessel Systems, J Loss Prevention Process Industries, 9(1): 45-58.

Moore, P., E., Spring, D., J., 2005, Design of Explosion Isolation Barriers, Trans IChemE, Part B, Process Safety and Environmental Protection, 83(B2): 161-170

Siwek, R., Cesana, C., 2001. Software for Explosion Protection, WinVent and ExTools, Safe Handling of Combustible Dusts, VDI Bewrichte 1601, Nuremberg

West, D., 2001, Introduction to Graph Theory, Prentice Hall.

APELL, SAFER PRODUCTION AND CORPORATE SOCIAL RESPONSIBILITY – LINKING THREE INITIATIVES TO IMPROVE CHEMICAL SAFETY IN THE THAI CHEMICAL INDUSTRY

Mark Hailwood
Belfortstrasse 8, 76133 Karlsruhe, Germany

INTRODUCTION

The APELL (Awareness and Preparedness of Emergencies at Local Level) Process is an initative of the United Nations Environment Programme (UNEP) which was initiated in the late 1980s in response to a number of major chemical accidents in industrialised and industrialising countries. The focus of the process is to enable the local community to identify the risks to which it is exposed and provide a mechanism for preparedness in dealing with such risks, coordinated between the relevant industries, local authorities, and the community. Safer Production is a term which is being applied to the tools, guidelines, and management principles implemented at site and local level to ensure the health and safety of the workers at the companies handling hazardous substances as well as preventing releases which may cause harm to the environment or endanger the health of the local community. Corporate Social Responsibility (CSR) is a more recent development. This initiative is a development of the environmental management issues which gained widespread acceptance in the 1990s to take in the interaction with other stakeholders, particularly local communities and other trading partners.

One of the aims of the project described in this paper is to use the resources and structures developed within these existing initiatives and apply them to the chemical industry in Thailand with the aim of improving chemical safety throughout the supply chain. This 2-year project started in October 2006 and is being run from within the UNEP Division of Technology, Industry and Economics (DTIE), based in Paris. Project activities in Thailand are being coordinated in conjunction with the local UNEP Regional Office for Asia and the Pacific, in Bangkok. At the national and local levels, the project is benefiting from cooperation and support by the Department of Industrial Works of the Thai Ministry of Industry, Mahidol University, the Thai Environment Institute, and by key stakeholders in the Thai chemical industry, such as the Responsible Care Management Committee of Thailand, and the Chemical Industry Club of the Federation of Thai Industries.

THE CHEMICAL INDUSTRY IN THAILAND

The chemical industry in Thailand is characterised by import, export and formulation activities. Most of the imports are made by traders, which play a key role in the chemical industry value-chain in the country. There is, according to the needs assessment report written in preparation for this project, little heavy chemical industry producing primary chemicals. In addition to reformulation and repackaging there is a large section of chemicals handling activities being carried out by the electroplating and electronics industry. The formulated chemicals cover the whole range of industrial chemicals, household chemicals, toiletries, cleaning agents and additives. According to the National Statistical Office there are 2241 registered chemical industries in Thailand. These industries have an official permit to operate issued by the Department of Industrial Works of the Thai Ministry of Industry, allowing the use and formulation of chemical products.

A number of multi-national (European and North American) chemical companies have subsidiaries in Thailand and the Thai government has developed a number of large industrial estates with central infrastructure.

This does however leave a large number of mainly indigenous Thai companies, mainly of the Small and Medium Enterprise (SME) scale, who handle a wide range of chemicals with a wide range of risks and hazards. In addition, if considering all the industries that use chemicals extensively in their manufacturing processes, the number of companies will be in the range of 10,000–12,000. These chemical industries are located all over Thailand.

Figure 1. The integration of industry, local authorities and community leaders in the APELL Process (Source: UNEP – DTIE)

THE APELL PROCESS

As the name "Awareness and Preparedness of Emergencies at Local Level" suggests the implementation of this process achieves a better level of preparedness by industry and local emergency services, and an understanding by local people of how to react to an emergency. This is attained by developing a meaningful dialogue between industry, local authorities and the local community leaders. The process is managed by a local co-ordinating group which includes representatives from all important stakeholders and is responsible for reviewing the hazard situation and then supervising some of the measures needed to address these hazards, towards the development of an integrated community emergency response plan .

The establishment of such a group can be initiated by anyone, however it has been found useful to request either industry or the local authorities to do this. Community groups can also convene a co-ordinating group if they are concerned that there are significant risks which have not been adequately addressed by official activities.

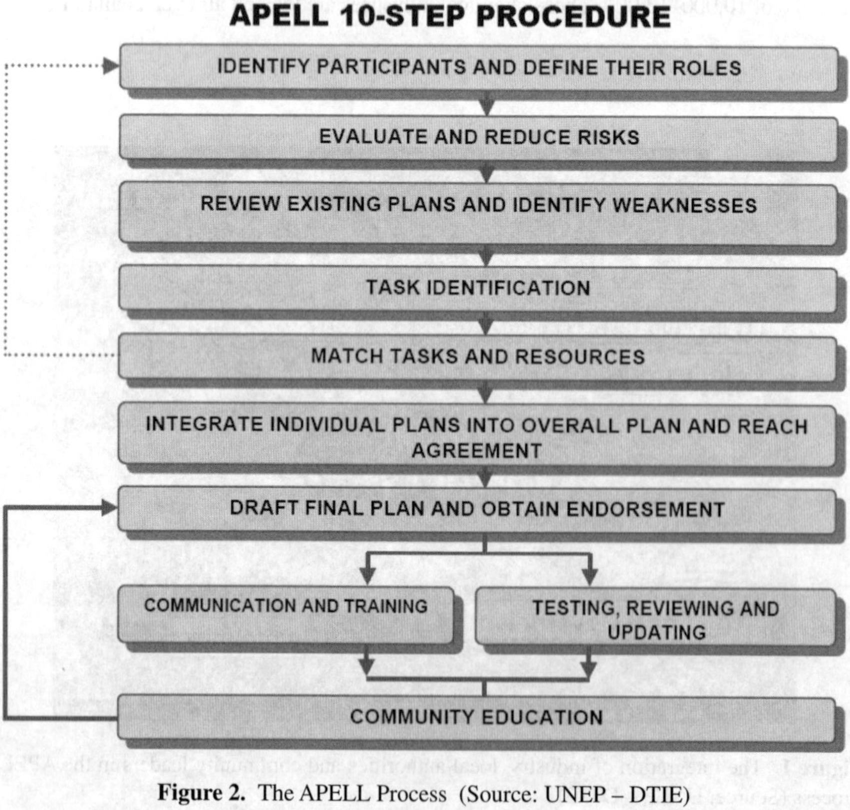
APELL 10-STEP PROCEDURE
IDENTIFY PARTICIPANTS AND DEFINE THEIR ROLES
EVALUATE AND REDUCE RISKS
REVIEW EXISTING PLANS AND IDENTIFY WEAKNESSES
TASK IDENTIFICATION
MATCH TASKS AND RESOURCES
INTEGRATE INDIVIDUAL PLANS INTO OVERALL PLAN AND REACH AGREEMENT
DRAFT FINAL PLAN AND OBTAIN ENDORSEMENT
COMMUNICATION AND TRAINING
TESTING, REVIEWING AND UPDATING
COMMUNITY EDUCATION

Figure 2. The APELL Process (Source: UNEP – DTIE)

The APELL Process is a ten step programme to identify hazards and assess their risks, whilst involving all necessary participants throughout [Balkau 2002].

SAFER PRODUCTION

Safer Production is a concept which comprises the tools guidelines and management principles which are implemented by a facility handling hazardous chemicals to ensure the health and safety of its workers and to minimise the risks of a release of hazardous chemicals which may harm the environment or endanger the local community. Within Europe it would be recognised that these aims fall within the scope of the Seveso II Directive (COMAH Regulations in the UK) for larger scale activities and other regulations on hazardous chemicals and machinery for those sites with smaller chemical inventories. Safer Production is based strongly on the approach of:

- identifying hazards
- assessing risk
- defining risk reduction measures
- setting out the safe modes of operation and the necessary competency
- defining inspection and control measures
- defining management structures and responsibilities

In addition to there being legal frameworks in which these aspects are regulated there are industry initiatives such as Responsible Care, which share many of the same elements. With the launch of the Responsible Care framework in Canada in 1985, the chemicals industry publicly declared a voluntary commitment to go beyond legal requirements. Under this framework, companies work together through their national associations in a continuous search to improve their health, safety and environmental performance, promoting communication with their stakeholders about products and processes.

In May 2003, the International Council of Chemical Associations (ICCA) announced it was undertaking a global review of the Responsible Care framework, which resulted in the new Responsible Care Global Charter, launched to the public at UNEP's International Conference on Chemicals Management in Dubai in February 2006, alongside with the industry's Global Product Strategy initiative (GPS). One of the important aspects linked to Responsible Care and the new GPS is Product Stewardship which moves the issue of product safety along the value chain from only involving the manufacturing unit to include customers and other users.

CORPORATE SOCIAL RESPONSIBILITY

Corporate Social Responsibility has been variously defined as:

> *"Corporate Social Responsibility (CSR) is the continuing commitment by business to behave ethically and contribute to economic development while improving the quality of life of the workforce and their families as well as of the local community and society at large."*

The World Business Council for Sustainable Development

"... a concept whereby companies integrate social and environmental concerns in their business operations and in their interaction with their stake holders on a voluntary basis."

The European Commission

"A values-based way of conducting business in a manner that advances sustainable development, seeking positive impact between business operations and society, aware of the close interrelation between business and society as well as of companies, like citizens, having basic rights and duties wherever they operate. Some, for example the negotiating texts of the World Summit on Sustainable Development, speak of corporate Environmental and Social Responsibility (CESR)."

UNEP Global Compact Environment Principles Training Package

Those used to working with or inside the chemical industry will recognise the high degree of congruence between the Responsible Care initiative and CSR. Responsible Care is one of a wide range of CSR frameworks and some companies and corporations assess their achievement in this area using more than one tool.

There are a number of reasons for implementing CSR measures within a company. However CSR has been shown to include the following benefits:

- it addresses economic, social and environmental considerations in an integrated manner
- it addresses ethical considerations
- it facilitates innovation and learning
- it provides employee motivation
- it improves risk management or risk reduction
- it can increase access to capital or increase shareholder value
- it can enhance reputation or brand
- it can enhance market position (market share) improvement
- it can strengthen supplier relationships
- it can result in cost savings and improved efficiency
- it can enable improved relationships with government authorities

It is thus apparent, that CSR provides the potential to bring about change in the way a corporation is perceived or even in its economic position; however it does not provide a guarantee. The main driving forces are the ethical issues and the role of the chemical company as a "citizen" and "good neighbour".

THE PROJECT

Project activities in Thailand have the central aim of building the capacity of SMEs, local and regional authorities towards improved chemicals safety, emergency preparedness and information on chemical risks along the value-chain. Regarding emergency preparedness, the "APELL-SP-CSR" project aims to build on the APELL approach which is very much orientated towards local communities and stake holder engagement. By integrating CSR components, in particular stakeholder engagement and site level reporting there is the opportunity to forge links between what happens on site and the local community, and between companies and their business partner partners and customers. By bringing the concepts described under the heading of Safer Production into the structure the project also gains a clear alignment towards the reduction of chemical risks, both at site level and off-site.

Originally the project was to be based around a cluster of SMEs located in Chonburi Province; however, following a request by the Department of Industrial Works of the Thai Ministry of Industry, the project is now centred on the Bangpoo Industrial Estate in Samutprakarn province. This will allow the DIW to ensure the contiguity of the capacity building activities activitites being promoted in Bangpoo since 1998 by the gtz (Gesellschaft für technische Zusammenarbeit – German society for technical cooperation) under the "Implementation of a System for the Safe Transport and Handling of Dangerous Goods" and the "Risk Management for the Handling of Hazardous Materials by SMEs in Bangpoo Area" projects.

Under these two capacity-building projects, gtz promoted the development of technical guidance and training sessions aimed at companies and governmental officials, and a risk assessment and profile of 88 factories dealing with hazardous materials in the Bangpoo area. Main conclusions of this risk assessment point to high exposure (through lack of Personal Protective Equipment – PPE and of understanding of the properties of the chemicals and hazards involved) and fire and explosion risks as the main problems to address for improving chemical safety management in Bangpoo.

As for the UNEP DTIE APELL-SP-CSR project, planned project activities include the development of practical tools for improving best practices and communicate risk in an effective way, and to assure that business partners (suppliers, transporters, distributors, customers, etc) and community stakeholders are clearly identified and engaged for raising their awareness to chemical hazards and risks. Within the scope of the project, companies' safety officers and government safety inspectors will be trained on chemical hazard identification, risk assessment and general best practice, for improving their understanding of the issues underlying improved chemical sasfety management. They will also be subject to awareness raising sessions on the economic benefits of improved chemical safety at site and in the companies they are doing business with.

For leading the training activitites under the APELL-SP-CSR Project, UNEP is involving a local technical institution – the Thai Environment Institute – and will seek to engage "Responsible Care Ambassadors" and "Code Champions" within the Thai chemical industry for additional support. At present, there are more than 70 companies in Thailand which have adopted Responsible Care. All of these companies are large scale companies or multinational companies. These companies have formed a group named as "Responsible care management Committee of Thailand (RCMCT)" under the Chemical Industry Club, Federation of Thai Industries.

"Responsible Care Ambassadors" and "Code Champions" are used by these industries to promote the Responsible Care programme by supporting smaller or less able companies through capacity building. In a similar way the Thai Chemical Manufacturers Association, which although it is not a member of ICCA is committed to the RC framework, could not only promote RC amongst its members but also champion the principles towards other trade associations such as transportation or associations which utilize the end products (e.g. electronics, surface treatment or household products).

One concept to developing the project further is not just considering chemical safety along the supply chain, but also across the whole chemical handling industries. This needs to involve transportation, storage and supply operations. There has to be a concerted effort not only to target individual enterprises but also the trade and industry associations. This is of particular importance, taking into account the role that traders play in the overall value chain. In this way, consistent paths of communication may be developed.

In considering the chemical supply-chain, chemicals may be sourced from companies which implement the ICCA "Responsible Care" programme together with its "Global Product Strategy" which aims to manage chemicals over their lifetime, or they may be sourced from companies which are outside of this scheme. Sourcing chemicals from outside of the RC Scheme introduces particular aspects for safety along the supply chain. Chemicals may be sourced from non-RC Thai companies or from outside of Thailand e.g. India or China. It is necessary to raise awareness for the minimum acceptable requirements for the packaging, labelling, documentation and transportation of the chemicals being supplied. The adoption of the Globaly Harmonised System (GHS) over the next few years (Thailand has an ambitious goal of implementing GHS by 2008/9) may go some way towards acheiving these aims; however it is to be expected that the time to reaching a high level of compliance will be long unless the principles found in RC and GPS are not understood and communicated up and down the supply chain and across the whole chemicals handling industries. Regardless of where from the chemicals are sourced the company or person handling or receiving them must be made aware of the risks involved at all stages of the supply-chain.

"Responsible Care" companies would be expected to supply the necessary information on the chemical hazards, for example as a Material Safety Data Sheet (MSDS) in the relevant languages for the supply chain, as a matter of course. Also contact details for dealing with spillages and emergencies relating to the chemicals supplied should be available. The chemicals should be packaged and labelled in a clear and suitable form for the conditions of transport and storage which can reasonably be expected over the entire transport route from the supplier to end user. Compliance with national and international legal requirements is a minimum of what is expected.

The "Responsible Care" companies should encourage the receiving partner to learn to recognise the value of achieving the standards of chemical management and safety of the "Responsible Care" programme and to adopt these values as far as possible themselves. In doing so the receiving partner would then place similar expectations upon those chemical suppliers which were outside of the "Responsible Care" Programme and which may not necessary hold these values automatically. This would lead automatically to an improvement of chemical safety across the whole chemical supply spectrum.

The chemical supply chain will necessarily involve transport systems by rail, road, ship (marine and inland waterways) or air. The parties involved in transportation must not only understand the special requirements relating to the transportation of hazardous materials (packaging, compatibility within loads, limits to load sizes, temperature, moisture, etc.); the requirements relating to general safe transport (speed restrictions, weight restrictions of loads, technical standard and integrity of the vehicle, etc.) must also be considered. The fact that the expectations and experience of companies used to operating in industrialised countries may be of a far higher standard than in many other countries needs to be considered when supplying chemicals to companies operating in these less industrialised countries. Appropriate robustness of packaging, information on handling or even restrictions on particular transport modes may therefore need to be considered.

The transportation route may not be direct from the supplier/manufacturer to the end user/final customer. The manufacturer may supply to distributors or other third parties. These may in turn combine deliveries from various suppliers to make up individual loads or even repackage the chemicals concerned (20 tonne lorry load into 25 kg sacks). Distributors and haulage firms need to understand the hazards and the risks involved in (intermediate) storage, packaging, commissioning, consignment management, loading and transportation of a wide range of chemicals. It is important that the information provided (MSDS) is comprehensible and is acted upon. Within the transportation link of the supply chain it is important that loads including hazardous chemicals are accompanied with appropriate documentation in the relevant languages to correctly identify the chemicals and also to be able to provide the correct emergency response in the event of a transportation accident or spillage. Contact details for advice and assistance are also of great value as in some cases special equipment or specialist knowledge may be required.

At the end of the supply chain the final customer/end user must be supplied with all of the necessary information to be able to carry out their activities safely. The end user is different to transportation in that they generally wish to utilize the chemical characteristics of those substances supplied (solvents, chemical reactions, pesticides, pigments, etc.) Therefore it is necessary for the end user to understand the chemical hazards in particular detail. Some chemicals require particular competence to be able to handle them safely, (e.g. explosives, pesticides). The supplier/manufacturer should consider the necessary requirements when making the chemicals available.

If the manufacturer/supplier is also supplying technology to handle the chemicals involved, then the supplier must consider the cultural and social differences as well as technological differences which may exist between the originating and destination countries. This may need to make particular allowances for irregular power supply, lack of maintenance and availability of spare parts, differing levels of manning, etc. The systematic identification of hazards and assessment of risks must be carried out for the location in which the chemicals or technology are going to be used. A direct transfer of results from previous assessments without considering the local situation will not reflect the particular risks to which the local employees and communities are exposed. This will lead to inadequate or incompatible emergency planning or land use planning and harm not only the local population but also the reputation of the supplying company. The loss of reputation may even be a bigger financial risk than the loss of plant or equipment.

Guidance on specific aspects of chemical accident prevention preparedness and response beyond those regulated directly in the international agreements or national legislation may be found in the OECD Guiding Principles on Chemical Accident prevention preparedness and response. Guidance for dealing with emergencies as a result of chemical accidents may be found in the documents of the UNEP APELL Program.

THE ISSUES TO BE PURSUED

Whilst the application of APELL, CSR and Safer Production to locations with multinational corporations is feasible and is already practice in differing degrees at various locations around the world, the situation changes enormously when moving to small and medium enterprises in developing countries. A number of questions arise which have as yet to be fully addressed. These include:

- The identification and assessment of hazards and risks. In societies where the day to day living is associated with a relatively high level of risk due to the environment and economic constraints (e.g. ranging from poisonous snakes to extreme poverty) the appreciation of the risks inherent with handling hazardous chemicals is different to that of western European or North American society. This poses obstacles to relatively simple issues such as the use of PPE, the design and operation of processes so as to minimise the need for PPE and also general good housekeeping to avoid spills, releases and emissions. It is not unusual to observe workers in chemical installations wearing flip-flops as shoe-work and makeshift, rag face masks as protection from vapours and dusts. However this must be seen in the context, that this is normal foot wear in this part of the world and that the air quality in Bangkok is very poor compared to European standards.
- A further issue is the communication of information on hazardous materials and processes. The information needs to be provided in a format that can be understood. It is probably not sufficient to provide MSDS in the Thai language. More efforts need to be made which take account of the general understanding of chemicals risks and also of the level of literacy in the workforce. For finding the best approach, it is of the utmost importance to not only involve local technical institutions, but also to engage successful companies that are experienced in risk communication and training.
- Thailand has agreed to adopt the GHS system by 2008, however there is likely to be a period in which there is dual use of both the GHS and the current Thai chemicals legislation.
- With the limits of risk understanding, literacy and the culture which abounds in Thailand there are constraints to the success of a paper based methodology. Whilst APELL, CSR and SP are all processes from the industrialized nations and thus are dependent on paper documentation, there needs to be a review of how much documentation is necessary.

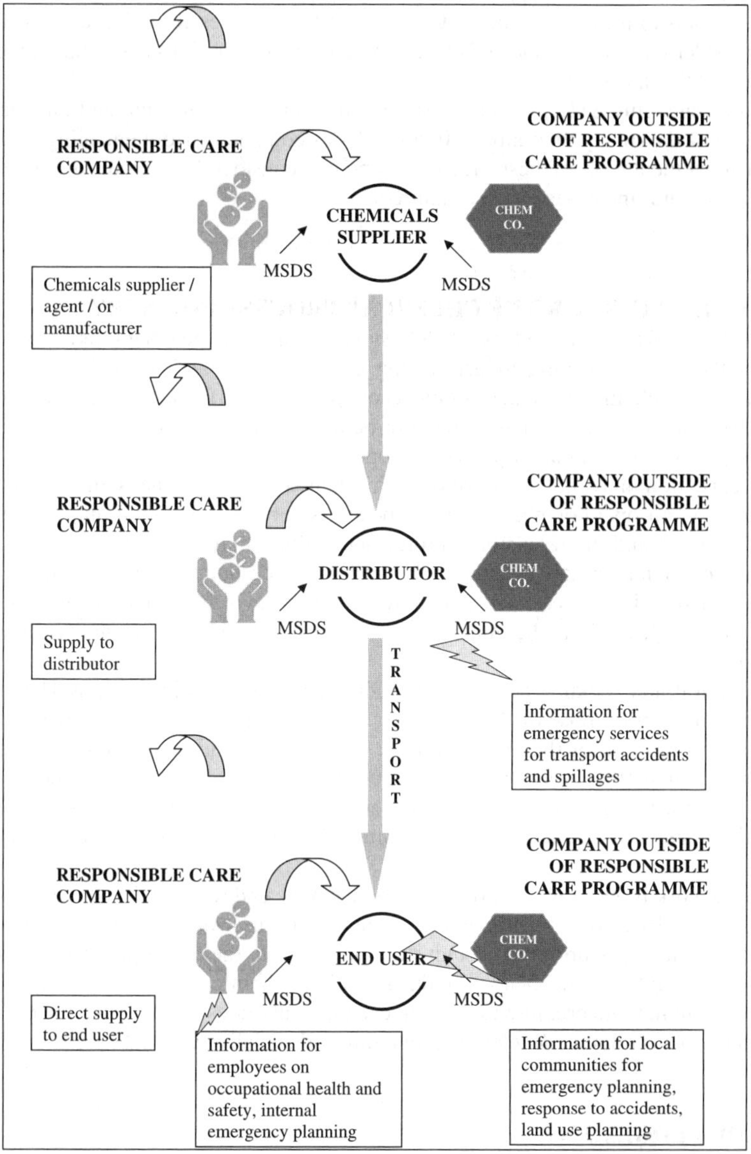

Figure 3. A model for the transfer of chemical safety competency along the supply chain and across the business sector

Of primary importance is an increase in the understanding of the risks, the hazards which leads to these risks and how to manage them successfully. It may be necessary to consider the use of visual aids including posters, illustrations and training videos to transport this message.

- Finally, engaging SMEs requires a strong effort in communication and outreach. The burden of day-today operations often precludes companies from engaging in training and implementation of best practices, where a considerable amount of time – and formal commitment – is usually required.

WIDER MEANING FOR THE CHEMICAL PROCESS INDUSTRY

In presenting this paper at the Hazards XX Symposium it is hoped that interest and awareness for the issues affecting industrializing countries such as Thailand can be raised. Companies which invest in developing economies, in particular those who maintain subsidiaries and export technology need to be aware that there is a major difference to operating in the UK or in other parts of Europe.

Legislation may not be developed in such a way that the same requirements are placed on an operator with regard to health, safety (occupational and process) and the environment and the regulatory enforcement culture may be weaker than that which they are used to in their native country. This means that corporate social responsibility and safer production initiatives have a real role to play in setting and maintaining high standards regardless of whether much lower standards would be sufficient to achieve legal compliance.

Cultural and social expectations in industrializing countries such as Thailand are also very different to those of Western Europe. In countries where the standard of living for a large proportion of the population is extremely low and day to day living poses a sizeable risk then other risks, such as those posed by chemical accidents or occupational exposure to hazardous chemicals will be seen by the local population as being tolerable at a higher level of risk if they perceive an immediate benefit than in more industrialized countries.

This work is very much work in progress and the end results will not be measurable for some time. However the situation is not unique in the world, and with an increasing globalization and pressure to move industrial processes to the developing world, there are increased demands on the chemical industry and the chemical safety community worldwide to ensure that the pressure to develop economically does not overtake the possibility to maintain this economic development in a safe and socially sustainable fashion.

ACKNOWLEDGEMENTS

This paper has been prepared with the kind assistance and co-operation of Tomas Marquez, Ruth Coutto and Sandra Averous of the UN Environment Programme (UNEP), Division of Technology, Industry and Economics (DTIE), to whom I am most grateful.

REFERENCES

Balkau, F. (2002) Industrial Disaster Preparedness – the APELL process for community information and awareness, APELL Seminar/CCA8, Odense, Denmark, 16 October 2002

Commission Green Paper 2001 "Promoting a European Framework for Corporate Social Responsibility", COM(2001)366 Final, cited at: http://ec.europa.eu/enterprise/csr/index_en.htm

UNEP (2005) United Nations Global Compact Environment Principles Training Package, http://www.unep.fr/outreach/compact/docs/Training-Package/pdf-trainer/UNGC_TRAIN_ALL.pdf

WBCSD (2000) Corporate Social Responsibility: Making good business sense. World Business Council for Sustainable Development. ISBN 2-94-024007-8

DEMONSTRATING CONTINUOUS RISK REDUCTION

Alastair Bird[1], Angus Lyon[1], and Volton Edwards[2]
[1]DNV
[2]BP Trinidad and Tobago

Measurement of risk at large industrial sites has generally been achieved through analysis of hazard scenarios which allows facilities to calculate theoretical individual risk and societal risk levels for comparison with specified criteria. This has provided plants with a measure of their 'inherent risk levels'.

Operations management are increasingly demanding that plants measure and report their risk levels on a regular basis with the rightful expectation that these facilities demonstrate a downward trend in risk levels as a function of time. 'Inherent risk' levels alone do not commonly allow facilities to make such a demonstration on a frequent basis, as they will only change significantly as a result of major engineering or personnel changes over the course of multiple years. Risk management personnel are therefore being required to employ alternative risk indicators, in addition to inherent risk, in order to construct multi-component risk measures that are better able to reflect short-term variations.

This paper describes the approach BP Trinidad and Tobago is taking to address this need based on the construction of a risk measure which combines indicators based on inherent risk (long term risk level), plant conditions (integrity status of plant and safety systems) and leading indicators (status of risk control measures in place to protect against identified hazards).

This paper first summarises the risk measures that are calculated to represent inherent risk. The shortcomings of these measures, and QRA in general, for helping management make decisions regarding risk reduction activity on a day to day basis is then discussed. The final section of the paper presents the authors thoughts on one approach that can be adopted to track risk reduction on a regular basis and thus better inform management regarding the effectiveness of risk reduction activity.

QUANTIFIED RISK ASSESSMENT MEASURES

Engineers and scientists employed as risk management professionals commonly express numerical measures of risk in one of two ways; either as Individual Risk or as Societal or Group Risk. The former is the risk experienced by an individual person, the latter is the risk experienced by the whole group of people exposed to the hazard. Both may be produced as an output of an analytical, quantitative risk assessment (QRA) for an industrial site or facility. Both may be regarded as a measure of the 'inherent risk' associated with operating that site or facility.

Individual risk has been formally defined by Jones (1992) as the frequency at which an individual may be expected to sustain a given level of harm from the realisation of specified hazards. It is usually taken to be the risk of death, and usually expressed on an annual basis. It is most commonly expressed in terms of the risk experienced by an individual worker (or member of the public), taking into account the amount of time they are likely to spend in the area impacted by the particular hazards and the approximate break-down of their movements within these 'hazard zones'. An alternative, less widely adopted means of representing individual risk is on the basis of location; i.e. the risk experienced by a theoretical person who remains in the same position for 24 hours, 365 days a year. This measure is usually employed for constructing isopleths for points of constant risk exposure (risk contours).

Societal risk is defined by Spouge (1999) as the risk experienced in a given time period by the whole group of personnel exposed. It reflects the severity of the hazard and the number of people in proximity to it. It is usually taken to refer to the risk of death, and usually expressed on an annual basis. Societal risk may be expressed in one of two ways. The simplest measure is an 'annual fatality rate' attributable to the site or facility. A more informative measure is a tabular or graphical representation of the relationship between the frequency and the number of people suffering a given level of harm from the realisation of specified hazards. The most common form of this second measure is the FN curve. FN curves are frequency-fatality plots, showing the cumulative frequencies (F) of events involving *N or more* fatalities.

Used collectively these QRA outputs are, potentially, a valuable tool for managing risk. QRAs require an analyst to build a numerical model of a site or facility. Failure rate data from historic archives is used to estimate how often the plant will experience a failure, this may be a leak of flammable or toxic substance, or other hazardous occurrence. Consequence modelling then assesses the results of a release, based on expected plant operating parameters and flammable, radioactive or toxic properties. The overall risk picture is completed by taking into account the geographic and chronologic distribution of vulnerable, populations and, in the case of flammable compounds, potential ignition sources.

This structured means of estimating exposure to particular risks is the greatest strength of a QRA. As a tool for making rational, informed judgements on potential risk-reduction measures it is unique. It identifies tangible linkages between plant operations and the risks experienced by vulnerable populations. However it does have its shortcomings.

SHORTCOMINGS WITH QRA AS A MONITOR OF CONTINUOUS RISK REDUCTION

Individual Risk and Societal Risk should be regarded as complementary. Used in isolation, they both have shortcomings. Individual risk does not give any indication of the scale of a particular incident that the person is exposed to. It is, by definition, a measure of the annual risk exposure of a single person. It makes no distinction between them being exposed to a relatively small incident that affects them, and them alone, or a relatively large incident that affects multiple other persons. Societal risk reflects the level of risk exposure for a population. It is not an effective tool as far as managing the risk exposure of particular worker groups is concerned, however. If, at a particular site, one worker group spends all their time in a blast-proof, fire-rated control room and another worker group spends all their time out amongst process vessels and pipe-work, societal risk would be representative of the exposure levels of the overall population and give no indication of the risk levels experienced by the most exposed individual.

A comparable view was expressed by the UK Health & Safety Executive (1989);

> *QRA is an element that cannot be ignored in decision making about risk since it is the only discipline capable, however imperfectly, of enabling a number to be applied and comparisons of a sort to be made, other than of a purely qualitative kind. That said, the numerical element must be viewed with great caution and treated as only one parameter in an essentially judgemental exercise.*

This endorsement is clearly not without reservations and the limitations of QRA, and the 'inherent risk' measures that result from it, have been widely discussed.

One major criticism is that it presents a theoretical 'snapshot' of risk at a 'point in time', and some practitioners will assert that this risk picture is valid for the entire lifetime of the plant or facility. The validity of the picture will vary dependent upon the quality of the inputs and the frequency with which the model is updated. As an example, in the UK, offshore operators must undertake a through review of the Operational Safety Case for their installations which details the findings of the QRA every 5 years or whenever they consider themselves to be completing major structural changes to their installations. This reflects the fact that the QRAs are not expected to be highly dynamic analysis tools.

Elsewhere in the world there is often no regulatory requirement for reviewing QRAs and the reports will often lie dormant for the duration of a facility's life. There may not even be a study completed for that phase of the platform life when risk is at its highest; during drilling and well intervention activity. Even if the model is frequently updated, the constraints placed upon the model by the necessarily generic input data must be taken into account. The statistical validity of failure rate frequencies employed in QRAs is dependent upon them being drawn from large sample populations. British offshore QRA practitioners will generally refer to analyses of the HSE's OIR12 database for their failure rate data, for instance. This records all leaks from platforms on the UK Continental Shelf since 1992. Whilst the extensive reporting of leaks in this database ensures that such analyses provide a good average leak rate for British facilities, across the entirety of their operating lifetimes they fail to successfully account for fluctuations in leak rate during different phases in a facility's operating life.

Reliability analysis have shown that plant failures are unevenly distributed throughout the lifetime of a facility, with a heightened probability of failure during commissioning, start-up and the initial months of a facility's operating life and, once again, during the final years of a facilities life. Similarly, QRAs are not suited to dynamically reflecting failures to carry out adequate fabric maintenance at a facility (data from a British source will inevitably reflect an average level of fabric maintenance associated with a British facility and levels of fabric maintenance will vary significantly across the UK sector and elsewhere in the world). Neither can they reflect failures to follow operating procedures that may have significant negative implications for Integrity Management (IM) at a particular facility. Even if QRAs were capable of reflecting such things, the time taken to go from acknowledging such a failure and reporting it and then working out how to demonstrate this failure as a statistical measure and incorporate it into the QRA would be very extensive, if it were even possible. To try and use QRA in this way would be to misinterpret it's purpose which, as already identified, is to provide a statistical measure of the 'inherent risk' associated with operating a particular facility.

THE NEED FOR A TECHNIQUE TO SUPPLEMENT QRA

Risk management professionals are increasingly being asked to identify tools which can be used to monitor and report 'major accident risk' levels (i.e. risks associated with incidents which may result in multiple fatalities) frequently throughout the lifetime of a facility, in order to demonstrate 'Continuous Risk Reduction'. Senior managers are familiar with such reporting for HSE risks, through the comprehensive recording and analysis of occupational risks, and may misguidedly believe a useful, dynamic record of inherent risk can be reported with comparable frequency. Such reporting of 'slips, trips and falls' is an effective measure of the success of HSE management programmes but this reactive reporting only highlights what is happening at one end of the 'risk spectrum'. It reports on the high-frequency, low-consequence events. Whilst no manager wants to report reactively on HSE risks they want, much less, to report on an event that might happen once every thousand years in the theoretical operating life of a facility, but that results in 10 or 20 fatalities. For an offshore facility, however, occupational risks will commonly only represent around a quarter of the platform risk. The remainder of the risk is associated with the major accident risks; the low-frequency, high consequence events.

As already indicated, 'Inherent Risk' measures such as Individual Risk and Societal Risk are one proactive means of measuring how effectively 'major accident risks' are being managed through good design, and QRA is, potentially, a very effective tool for allowing informed decisions to be made on the relatively significance of various different risks drivers. If a QRA is only to be valuably updated every five years or so, however, there is a demand for a more wide-ranging selection of dynamic, proactive indicators of major accident risk potential that managers can use to monitor 'Continuous Risk Reduction'. bpTT has identified a range of these measures that broadly fall into one or other of two categories. The first are measures of 'plant condition' (the integrity status of plant and safety systems) and the second are 'leading indicators' (measures of the status of risk control measures in place to protect against identified hazards).

'Continuous Risk Reduction' is a phrase increasingly being employed within risk management literature as a supplement to demonstrating ALARP. In the UK this demonstration of ALARP has generally been by means of cost benefit analysis. The cost of implementing potential safeguards will be balanced against their probabilistic potential to avert certain incidents with associated levels of plant damage. Alternatively, the comparison may be made against the expected number of fatalities associated with these incidents, using a formula which includes a suggested figure for the value of a human life. This method (if employed as intended) imposes a moral responsibility upon an operator to spend the calculated sum each year on mitigation measures to prevent the identified hazards being realised but certain safeguards will be shown to be disproportionately expensive. Whilst such analyses have been widely undertaken by risk management professionals in the UK they are increasingly passing out of favour as it has been used on occasions to justify not implementing safety measures which have previously been considered industry good practice.

As the ALARP demonstration falls out of favour operators are increasingly being asked to demonstrate 'Continuous Risk Reduction' instead. Relative to the established ALARP demonstration this still remains a rather nebulous concept. Nevertheless, risk management professionals are being asked to identify tools that can prove that such a process is in place. For the reasons identified above, QRA may not necessarily be the most appropriate tool for making this case, if applied in exclusion.

AN APPROACH TO DEMONSTRATING CONTINUOUS RISK REDUCTION ON A DAY TO DAY BASIS

In the long term Continuous Risk Reduction has to be demonstrated by an ongoing reduction in inherent risk measures such as Individual Risk and societal risk. However, these measures, in isolation, are insufficient as a management reporting tool to ensure appropriate focus on CRR. The reason is that the processes for generating these risk measures in most organisations only require updates infrequently, e.g. 3–5 years, or when a major design modification is implemented.

Additional measures are required which can be monitored on a more regular basis.

What should these measures be? Consider the following.

The management of hazards is based on a series of safeguards or barriers which either prevent a hazardous situation arising or mitigate against the consequences of the Hazardous situation should it arise, see Figure 1.

If barriers are in place to protect against all causes of hazardous events and their consequences, there should be zero residual inherent risk. In reality, there is a level of residual risk which is due to a degree of unreliability in each barrier. Conversely, improvements to the reliability of barriers will reduce the level of residual risk. Measuring the status of barriers is therefore a possible contributor to a CRR indicator. "Leading" indicators such as barrier performance status have the advantages that they can be measured and reported on a regular basis, say every month.

There is a large gap between the status of barriers and inherent risk levels. The harm, which the inherent risk values represent, is frequently preceded by a degradation of conditions, e.g. an increase in the amount of corrosion in hydrocarbon containing equipment. Monitoring a series of conditions or "lagging" indicators can also usefully feature in monitoring CRR.

Immediate improvements in the reliability of leading Indicators should translate in time to an improvement in lagging indicators which in time should translate to a reduction in inherent risk values, see Figure 2.

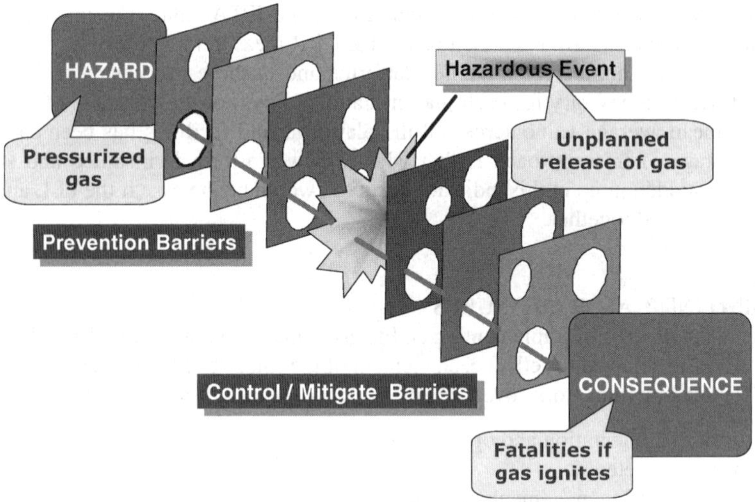

Figure 1. Hazard management barriers

A combination of inherent risk measures, leading and lagging indicators is considered as one possible approach to developing a CRR indicator capable of being updated on a regular basis. This approach has been adopted by bp Trinidad and Tobago.

BPTT CRR MEASUREMENT APPROACH

INHERENT RISK MEASURES

Two point measures of inherent risk are calculated within bpTT, The Potential Loss of Life (PLL) and the Individual Risk Per Annum (IRPA). Selecting either measure in isolation as a metric for demonstrating CRR could be misleading as discussed previously.

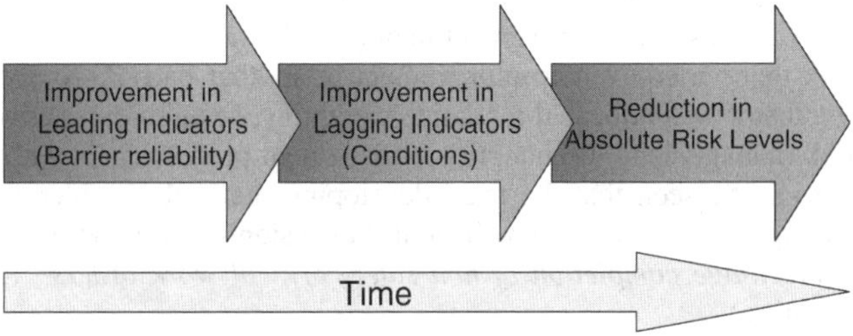

Figure 2. CRR factors timeline

CRR must demonstrate a reduction in both a PLL and IRPA value. For this reason both PLL and IRPA values have been included in the CRR measurement process.

bpTT has a number of offshore platforms and onshore processing facilities. The PLL and IRPA values vary for each platform and facility. For the purposes of measuring CRR the mean average value across all the platforms and facilities has been calculated. The advantage of this approach is that when a new platform is installed and starts to produce no sudden jump in this indicator occurs as would be the case if the PLL and IRPA values were added together.

LEADING INDICATOR MEASURES

bpTT is in the process of implementing an Integrity management Standard and the leading indicators were selected to reflect each element of the Integrity Management Standard.

The leading indicators cover the key aspects of a hazard management system:

- Status of documentation and operating procedures
- Status of competency
- Status of hazard management awareness
- Status of inspection and maintenance
- Status of management of change

The adopted list of leading indicators is listed in Table 1.

To demonstrate a trend it is important that the leading indicators are couched in terms that are comparable over time and measured on a consistent basis. For this reason where possible the leading indicators were based on % completion of an item rather than "number of tasks outstanding" etc. e.g. *"percentage of IM engineers and practitioners assessed as competent"*. In all cases a high percentage completion represents a larger contribution to risk reduction than a low percentage complete.

Where it was not possible to phrase a leading indicator in this form a scaling approach was adopted to translate, say, a number of outstanding items to an equivalent % complete. For example, the measure *"Number of outstanding critical work orders"* uses the scale listed in Table 2. If in a given reporting period the number of outstanding work orders is, say, 78 the equivalent percentage complete is recorded as 50% based on the scaling in Table 2.

Not all the barriers that the leading indicators measure necessarily have the same impact on risk levels. A small improvement in one barrier may have a much larger impact on reducing risk than an equivalent improvement in another barrier. To account for this each leading indicator was allocated a relative weighting, see Table 1, derived by a team approach of risk management and integrity management professionals within bpTT, e.g. from Table 1 it can be seen that the team developing the tool considered *"Number of outstanding critical work orders"*, if small, contributed significantly more to risk reduction than a high *"percentage completion of non safety critical work orders"* (20 weighting against a 5 weighting).

Translating the percentage completions and weightings to a risk reduction measure is described later in the paper.

Table 1. Leading indicators

Leading indicators	Weighting
Percent Design/Construction/Operations documents with appropriate level of EA/TA review	5
Percent Operations/Maintenance Technicians assessed as competent	20
Percent IM engineers and Practitioners assessed as competent	10
Percentage Contractors with approved competency management systems	20
Percent of assets/sites with current hazard registers	5
Percentage of (on time) closure of actions from risk assessments and hazard evaluations	20
Percentage of current hazard evaluations/risk assessment (completed according to schedule)	10
Percentage PM Plan Attainment (non SCE)	5
Number of outstanding PR1 WOs (non SCE) (20/100/100+)	20
Percentage Completion of inspection and tests (All Systems – non SCE)	5
Percentage completion SCE PMs	10
No of shortfalls against functional specification for safety critical protective systems during testing and actual demands on system (20/100/100+)	20
Percentage Compliance with high risk to STPs	15
Percent SOPs certified (by OPS Leadership) as up to date, accurate, accessible and being followed	15
Percent MOCs compliance	15
Percentage CM&ER plans current and in place	10
Percentage ER drills conducted according to schedule and lesson learned entered in Tr@ction	10

Table 2. Number of outstanding critical work orders scale

No of outstanding critical work orders	Equivalent % complete
<20	90%
21–99	50%
>100	20%

LAGGING INDICATOR MEASURES

Three lagging indicators representing conditions which have the greatest impact on risk levels were selected, again, based on the views of a team of specialists, see Table 3.

Hydrocarbon leaks pose one of the biggest threats to personnel on bpTT platforms and facilities. The status of the leading indicators as described previously has a large influence on this measure, hence its inclusion as a lagging indicator.

Excursions outside process design limits and Integrity management related incidents e.g. identification of pipes with excessive corrosion do not impact risk levels directly. However, they are "near misses" and as such are meaningful indicators of conditions, hence their inclusion as lagging indicators.

When a new platform comes on stream the number of unplanned releases may go up slightly because of the additional process facilities now operational, yet the "risk" posed by process releases on the existing facilities has not changed. To avoid these lagging indicators painting an unrealistic picture of conditions, the indicators were worded as average values per facility per time period, e.g. the total number of unplanned hydrocarbon releases divided by the number of facilities per reporting period.

As can be seen the lagging indicators are not percentage completed items as are the majority of the leading indicators. Scaling values had therefore to be developed for the lagging indicators, see Table 4. The scaling is used as follows. If, say, a particular asset has 6 platforms and there are 3 leaks during the reporting period the average number of leaks per platform is 0.5. This translates to a leaks measure of 5 using the scale in Table 4. Similarly, if there are 12 Integrity management (IM) related incidents in the same period (average 2 per facility) the IM measure, the IM measure is 4 using the scale in Table 4.

Table 3. Lagging indicators

Lagging indicators
Average number of unplanned hydrocarbon releases per facility per reporting period
Average number of excursions outside process design limits per facility per reporting period
Average number of Integrity management related incidents per facility per reporting period

Table 4. Lagging indicator scales

	Leaks		IM incidents		Process deviations	
	Scale	No.	Scale	No.	Scale	No.
Upper	1	10	1	10	1	10
Lower	0	0	0	0	0	0

COMBINING OF MEASURES TO PRODUCE THE CRR MEASURE

The inherent risk measure, leading and lagging indicators are then combined to produce a single CRR indicator which can be trended with time.

Each of the three indicators has been scaled to provide a 1–10 score. The scaling factors have been generated by using knowledge of good and bad indicators from within the business sector that bpTT operates, e.g. an IRPA value of 1×10^{-3}/year is considered the upper limit of acceptability with the UK offshore oil industry and an IRPA Value of 1×10^{-5}/year is considered extremely low within the same industry. These two values are used as two points on the 1–10 scale. The scaling factors have then been developed from them. A similar approach has been taken to the PLL values.

Scaling of the leading indicators was simpler to achieve as they were all couched in terms of % complete (or scaled to an equivalent). The scaling was then a case of setting 100% complete as a low value on the scale and 0% complete as the 10 value on the scale.

For the lagging indictors a similar approach to scaling as for the inherent risk measures was adopted by comparison with known good and bad performance within the industry for the lagging indicators, see Table 4.

Indicator	Target	Jun-07	Jul-07	Aug-07	Sep-07	Oct-07
Percent Design / Construction / Operations documents with appropriate level of EA/TA review	100%	0%	0%	0%	0%	0%
Percent Operations / Maintenance Technicians assessed as competent	100%	36%	37%	45%	53%	57%
Percent IM engineers and Practitioners assessed as competent	100%	0%	0%	0%	0%	0%
Percentage Contractors with approved competency management systems	100%	0%	0%	0%	0%	0%
Percentage of (on time) closure of actions from risk assessments and hazard evaluations	100%	0%	0%	0%	77%	77%
Percentage of current hazard evaluations / risk assessment (completred according to schedule)	100%	0%	0%	0%	82%	82%
Percentage PM Plan Attainment (non SCE)	100%	69%	85%	84%	88%	76%
Number of outstanding PR1 WOs (non SCE) (20 / 100 / 100+)	0	726	786	840	836	685
Percentage Completion of inspection and tests (All Systems- non SCE)	100%	41%	46%	22%	61%	30%
Percentage completion SCE PMs	100%	93%	92%	91%	95%	100%
Number of outstanding corrective Wos (SCE) (20 / 100 / 100+)	0	13	18	23	27	19
Percentage Compliance with high risk STPs	100%	0%	0%	0%	0%	0%
Percent SOPs certified (by OPS Leadership) as up to date, accurate, accessible and being followed.	100%	0%	0%	0%	0%	0%
Percent MOCs compliance	100%	65%	72%	68%	74%	78%
Percentage CM&ER plans current and in place	100%	0%	0%	0%	0%	0%
Percentage ER drills conducted according to schedule and lesson learned entered in Tr@ction	100%	0%	0%	0%	0%	0%

Indicator	Weighting	latest Value	Weighted Value
Percent Design / Construction / Operations documents with appropriate level of EA/TA review	5	0%	5.00
Percent Operations / Maintenance Technicians assessed as competent	20	57%	8.60
Percent IM engineers and Practitioners assessed as competent	10	0%	10.00
Percentage Contractors with approved competency management systems	20	0%	20.00
Percentage of (on time) closure of actions from risk assessments and hazard evaluations	20	77%	4.60
Percentage of current hazard evaluations / risk assessment (completred according to schedule)	10	82%	1.80
Percentage PM Plan Attainment (non SCE)	5	76%	1.20
Number of outstanding PR1 WOs (non SCE) (20 / 100 / 100+)	20	685	16.00
Percentage Completion of inspection and tests (All Systems- non SCE)	5	30%	3.50
Percentage completion SCE PMs	10	100%	0.00
Number of outstanding corrective Wos (SCE) (20 / 100 / 100+)	20	19	2.00

Date:	10/1/2007
KPI:	6.00

Return to Main

Figure 3. CRR Tool typical data entry sheet

Weighting									
	PLL Date:	Jul-07	Jul-07	Jul-07	Jul-07	Jul-07			
	IRPA Date:	Jul-07	Jul-07	Jul-07	Jul-07	Jul-07			
	Lagging Ind. Date:	Jun-07	Jul-07	Aug-07	Sep-07	Oct-07			
Weighting	Leading Ind. Date:	Jun-07	Jul-07	Aug-07	Sep-07	Oct-07			
0.5	PLL KPI:	4.113	4.113	4.113	4.113	4.113			
0.5	IRPA KPI:	7.584	7.584	7.584	7.584	7.584			
1	Lagging Ind. KPI:	4.818	4.626	4.485	4.237	4.020			
1	Leading Ind. KPI:	7.440	7.336	7.733	6.369	6.000			
3	Risk Reduction KPI:	6.036	5.937	6.022	5.485	5.290	0.000	0.000	0.
	Period	1	2	3	4	5	6	7	8

Return to Main

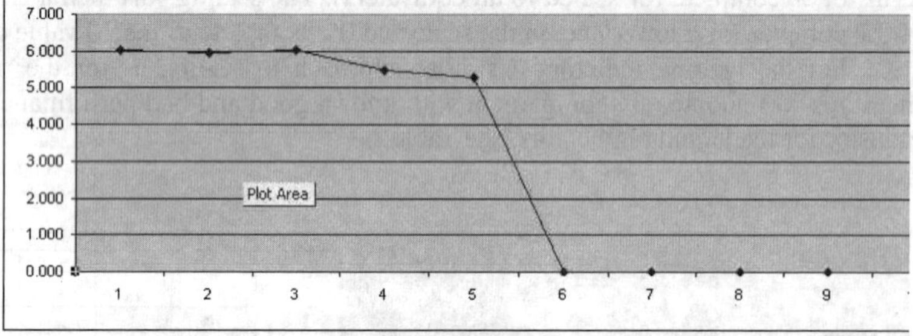

Figure 4. CRR trend line sheet

CRR TOOL

To facilitate the generation of the CRR indicator a spreadsheet template has been developed which generates a new point on the CRR trend graph each time a set of data is entered. The spreadsheet template consists of 4 sheets, one for each of the three sets of indicators, see Figure 3 as an example, and the fourth summarising the data and generating the CRR trend graph, see Figure 4.

CONCLUSION

The paper has discussed the need for a method of measuring continuous risk reduction on a regular basis. The shortcomings of the use QRA results in isolation have been discussed. A simple method of producing a continuous risk reduction indicator based on a combination of inherent risk values, leading and lagging indicators has been described when implemented gives management an indicator of status of their risk reduction journey. It is important to recognise that the indicator described is only an indicator; it is not an actual measure of risk.

The only actual measures are the PLL and IRPA values.

REFERENCES

Jones, D., 1992, Nomenclature for Hazard and Risk Assessment in the Process Industries, Institution of Chemical Engineers.

Spouge, J., 1999, A Guide to Quantitative Risk Assessment for Offshore Installations, CMPT.

HSE, 1989, Quantitative Risk Assessment: Its Input to Decision Making, Health & Safety Executive, HMSO.

HSE, 1992, The Tolerability of Risks from Nuclear Power Stations, HMSO.

BUSINESS CONTINUITY AND THE LINK TO INSURANCE: A PRAGMATIC APPROACH TO MITIGATE PRINCIPAL RISKS AND UNCERTAINTIES

Nicholas J.L. Gardener
Elementis plc

Accidents happen. Business profits and company reputation are at risk. Shareholders, and those who depend on the company for their livelihood, require reassurance that the business will survive and prosper if a major incident or crisis occurs.

Furthermore, most UK listed companies must now make forward looking statements in their Annual Report & Accounts, including, among other things, a description of the principal risks and uncertainties facing the company. Such statements should include what has been done to mitigate the potential impact of these risks.

This paper argues that good design and management of facilities must be backed by a business continuity plan (BCP). Insurance should be seen as a supplementary mitigation of risk not the first line of defence. Combining an effective risk management strategy with a suitable BCP linked to the insurance programme, has several benefits: it is an effective mitigation against known risks and uncertainties, and it creates the potential to reduce the total cost of risk borne by a company. The key is to convert the concept of a BCP into reality.

The paper describes how an international chemical company has worked on business continuity and insurance holistically. Coordinated collaboration within the company, partnership with insurers and brokers, and selective use of consultants provided the necessary expertise to distil what is required into a pragmatic BCP. As a result there is now greater confidence that profits and reputation would be protected against major hazards and other corporate risks.

KEYWORDS: Business Continuity, BCP, Crisis, Insurance, Reputation, Risks and Uncertainties, Total Cost of Risk

INTRODUCTION

A company is in business to trade. Having robust measures in place to prevent harm and damage is necessary but not sufficient. What happens when, despite best efforts, things go wrong? A well structured insurance programme will help mitigate some of the loss but cannot be expected to cover all eventualities, particularly loss of reputation (including brand). The business should be prepared to initiate effective business recovery rapidly to reassure shareholders, customers and other stakeholders that the company remains viable, and to protect the company's ability to recover insured losses. Reduction in risk accompanied by a business continuity plan provides a degree of confidence in the company's ability to handle a crisis[1] and minimise downtime.

[1]In this paper, the words crisis, major incident and catastrophe are used interchangeably.

This paper describes how Elementis plc[2] developed a cost effective business continuity plan (BCP) in conjunction with property and liability insurers, and with advice from insurance brokers, risk management consultants and a consultant with business continuity expertise.

Implementing a BCP does not, however, render insurance unnecessary. The paper also describes how business continuity planning integrates with insurance as part of a successful approach to optimising the company's total cost of risk. Success has been achieved through focussed risk reduction measures, attention to limits of cover, claims management, programme structure and insurance premiums.

RISK MANAGEMENT

DUTIES AND RESPONSIBILITIES

All companies have a moral and, in general, a statutory duty to protect people and the environment from the effects of their activities and external hazards. Chemical companies rightly spend much time identifying hazards and taking steps to reduce risks to an acceptable level[3]. Protection must be provided for routine and non-routine operations, and plant emergencies (from internal and external events). Nevertheless, accidents and natural disasters still occur, sometimes with devastating results.

At a financial level managers have a duty to protect the business and its profits. Companies must understand and deal effectively with operational and other risks they face, if they are to survive and prosper. Responsible companies also recognise their obligations to society, whether it is to the local economy or to global sustainable development. Success here comes from ethical prosperity.

CORPORATE GOVERNANCE

Sometimes we are lucky, but we should do all we can to reduce our dependence on chance. Good engineers and prudent commercial managers have always understood this, but thinking is often intuitive rather than structured. While luck will always play some role in business, because businesses operate in conditions of uncertainty, formal risk management is essential for effective corporate governance.

Knowing and reporting risks must not be an end in itself. Hazard identification and risk assessment must lead to an ability to continue in business – no matter what. Businesses should adopt the philosophy that: "the ruling interest in knowledge is the practical interest of action" (Lewis 1929). Appropriate mitigation is required.

[2]A UK specialty chemical company with manufacturing sites in the UK, US, The Netherlands and China.

[3]Hazard is used here to refer to the potential to cause harm; risk is a function of the likelihood and consequence of exposure to the hazard. Elsewhere the words: "risk" and "hazard" are sometimes used synonymously where such distinction is not necessary.

PRINCIPAL RISKS & UNCERTAINTIES

The Directors of most UK listed companies have a duty to promote the success of the company for the benefit of its members as a whole. As part of this duty they must make forward looking statements within a Business Review in their Annual Report & Accounts. This review must contain a description of the principal risks and uncertainties facing the company (The Companies Act 2006). The Company's auditors will, or should, be looking to verify that a suitably robust risk management process is in place to identify and manage these risks.

To provide balance and allow a fair assessment to be made, statements of risk should include what has been done to mitigate the potential impact of the event. For example, some mitigation efforts may include providing better detection and control, or seeking alternatives and redundancy in critical business processes and systems. Some risks may just have to be accepted. Evidence of an effective Business Continuity Plan (BCP) adds a strong measure of reassurance that, whatever the risks and uncertainties, the company is well prepared to respond positively.

BEYOND MANUFACTURING

It's not just an issue for manufacturing recovery: the knock-on effects must also be considered. A major process incident, fire or flood could damage a Company's IT systems or office functions. Loss of data, or destruction of hard copy records need to be considered. Without viable back-up systems it may be hard or even impossible to recover the business in a timely way. Even if you are not worried about a few lost records (and you should be) your auditors need to see full audit trails on financial data. Accidental loss of financial records could lead directly to difficulties in providing unqualified audited company accounts[4]; not having planned adequately in advance could compound the negative impact on reputation. Increasingly therefore, auditors are looking for evidence of business continuity planning, in addition to IT security.

BUSINESS CONTINUITY PLAN

DO NOT RELY SOLELY ON EMERGENCY SERVICES AND INSURANCE

Fire fighting and insurance are important of course, but the arguments for a BCP are clear – it could make all the difference to the speed and effectiveness of recovery following a crisis. Research by Templeton College, Oxford (Knight & Pretty 1996) went further. The researchers considered how companies are valued by the stock market following a catastrophe. Their research suggested that direct financial loss was a small part of the effect on shareholder value. The skill with which a company's management responded has a much

[4] In the US the consequences of losing financial data could be even more serious under the Sarbanes-Oxley Act 2002

greater effect. The implication is that an effective BCP is far more important as a first line of defence than insurance cover.

SO WHAT IS THE PROBLEM?

Why do many companies not have a BCP, or more to the point, an effective one?

It may be a cultural problem. Efforts are, understandably, put into making things safe, demonstrating (statistically) that risks are acceptable and so on. Having a BCP could be construed as planning for failure – not acceptable. Years of safe operation without a crisis can reinforce complacency. Practical people may need persuasion to accept that low likelihood events may still occur. Preventive action may not always be appropriate, but you should still know what you would do if the worst happens.

Another reason could be that the need is accepted but companies get bogged down in a complexity of worthy standards and software solutions. While these may have theoretical rigour, actual implementation can be lengthy and perhaps even impossible. There is also a risk that the BCP will become so comprehensive as to be unwieldy and unusable in an emergency – gathering dust on a shelf in a forgotten office. It may also be out of date or unavailable because it is "Confidential".

A PRAGMATIC SOLUTION

Elementis employs competent people who understand the business. They have empirical knowledge of potential crises from risk assessments. The BCP takes account of these factors and contains only those things required to prepare for and guide a comprehensive and timely business recovery in a crisis. Standards provide a framework for the BCP. The pragmatic concept is summarised in Figure 1.

The resulting utilitarian, tailored BCP reflects the fact that many of the required actions will be known, or be intuitive, to professional managers. In our experience, engineers as managers are good at handling internal recovery at a plant level. It is after all

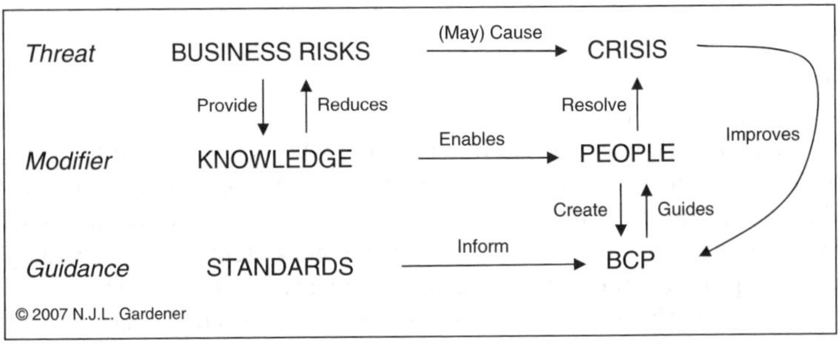

Figure 1. Pragmatic concept for business continuity planning

a project to be planned and managed - albeit an unwelcome one. It seems reasonable to assume that what these engineers and managers would need under the abnormal stress of a crisis is a clear definition of their roles & responsibilities in a crisis, prompts on actions to take, and key reference data (contact details etc). The impact of stress should not be underestimated. It is not just that some key individuals may not be available to help - consider how performance may be affected if there has been a major accident where staff or colleagues have been seriously injured or killed.

As in military operations (see for example US General Accounting Office 1997), the first 48 hours is a critical, and realistic, period to plan. After that, details on the ground will be clearer: the situation will have evolved; and there will have been time for detailed planning on implementing the next actions. Not taking decisive action in the first 48 hours risks harm to the company reputation.

It should be noted that the Elementis BCP is a product of the particular business, structure and intentions of the company. It is not an absolute solution. Alternatives would be possible and might suit other companies better.

AN EVOLUTIONARY APPROACH

Consider first the type of content. A BCP can serve many purposes, and progress through many iterations. For example: in the early days there may be sceptics who need to be convinced of the need for a BCP. If so, it will be important to provide a section showing justification for having a BCP. Others may agree with the concept of a BCP but will be unclear what a BCP contains and how it works. These people will need an explanation.

In the early stages, it is suggested, the BCP should contain justification and explanation as much as procedure and data. In time the balance can shift towards what is actually needed for a crisis and administrative details on how to be prepared.

GENERIC VS. SPECIFIC

Each crisis is different: different causes, different effects. This can be another cause of inaction on the BCP. Why create a BCP when you do not know, in advance, what will be required? The answer is to do all you can to predict, and then keep the BCP flexible. Plan for specific events and generalise for uncertainty. Fortunately, some tasks are generic, required immediately regardless of what has happened.

As an example, communication is a universal requirement. Good contact lists and communication channels are required showing who is responsible for what, how to contact them – and who to contact if they are not available, or cannot be contacted. You also need a well planned media strategy. Consistent, accurate and balanced statements are required from trained and authorised company representatives. Senior managers must be involved at appropriate times: what may appear to be a local crisis to those tackling the particular incident may have much wider ramifications. For example, the crisis may affect the trading outlook and hence share price, or customer confidence in the Company's products.

Crises can develop quickly. There may be legal implications and the Company's reputation needs to be protected from the outset. Questions may be asked by journalists, major investors, analysts, banks and other City institutions. The Board must know details urgently if they are to prepare an accurate and appropriate response to meet stock exchange regulations, and prevent unsubstantiated rumours. Regulators, local authorities and neighbours may also need to be informed promptly under local emergency plans. BCP communication protocols must accommodate these diverse requirements, and ensure a consistent message as the crisis evolves.

HOW MUCH CONTENT?

With time the BCP may grow in thickness as more and more useful content is added. For example content may include: hazards that might create a crisis, details of how to prepare for a crisis, critical equipment, roles and responsibilities, contact lists, maps, how to keep things up-to-date, and so on. Note however that the needs of users who would have to manage a crisis are different from those who have to prepare for one. Further, users' needs vary depending on whether they are considering strategic outward facing corporate issues or recovery operations at a tactical level.

SEPARATION OF CONTENT LEADS TO AN EFFECTIVE, INTEGRATED RESPONSE

The core of the BCP must provide the appropriate level of detail for the anticipated users, or their deputies. It must be in a user friendly and familiar format. It must be adequate to help them maintain the business and company reputation. One practical way to cater for the various needs is to divide the BCP into several parts for ease of use at different times. This is shown diagrammatically in Figure 2.

The lower part of Figure 2 shows diagrammatically the layered structure of the BCP concept for incidents at a manufacturing site. A local incident team takes actions to contain the incident using a local Emergency Plan (not in the BCP). The incident team is separate from, but interfaces with, a local business continuity team planning clean up and recovery actions. Simultaneously, if the crisis meets certain criteria, a corporate team coordinates external communications, provides leadership for business recovery and works with loss adjusters on insurance claims. In this way the company has a coordinated and optimised response to a crisis; local actions mesh neatly with corporate involvement.

The layered structure dictates what must be available in a crisis. Each site needs a local BCP linked to the corporate BCP. Each BCP collates the details and checklists that its users will need in a crisis. It is a document that can be carried around easily. It can be used quickly and unambiguously in a crisis. Appendices contain directories, maps, equipment lists, key material safety data sheets, and so on as appropriate. Functional groups such as IT and accounting have their own Disaster Recovery plans that interface with the corporate BCP.

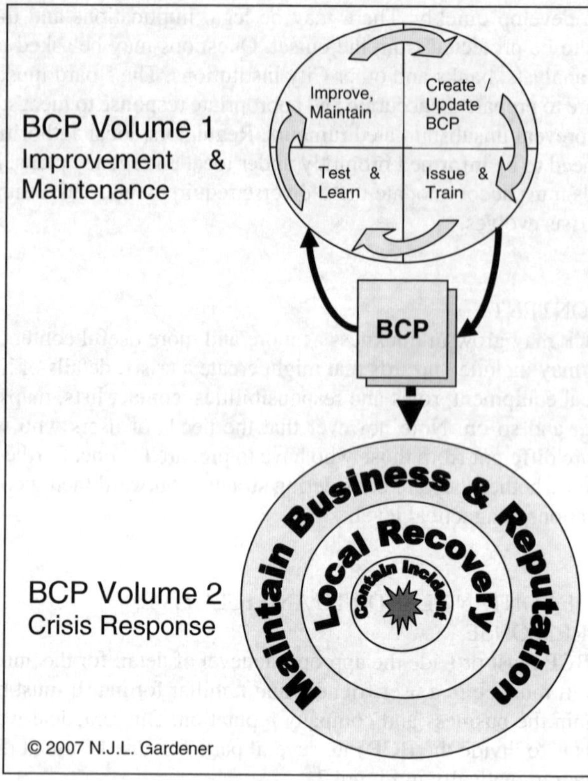

Figure 2. BCP improvement cycle maintaining effective layers of crisis response

A supporting volume of the BCP contains the policy on BCP, background information on perceived major risks, preparations for managing a crisis (training etc) and maintenance of the plan (updating, distribution etc) as a controlled document for improvement. The improvement cycle, shown at the top of Figure 2, is based on Deming's PDSA cycle[5] (Deming 1986). This administrative part of the BCP is to help prepare for a crisis. It can be used to provide explanation and guidance for education and training. It can also help demonstrate to interested parties[6] that the BCP is effective. However, the content is not required during a crisis so it can be held in a separate volume.

[5]Plan, Do, Study, Act – a flow diagram for learning, and for improvement of a product or of a process; credited by Dr Deming as the Shewhart cycle (Deming 1986)
[6]Such as auditors, insurers, customers who depend on your supply and investors

INTERNAL DRIVER HARNESSING EXTERNAL KNOWLEDGE

Elementis has achieved what it believes to be a workable BCP in a pragmatic way. Creation followed an evolutionary approach, with a blend of internal ideas and commitment guided by external advice. The property and liability insurer's head of risk management, with experience of property risks and implementing BCP's, provided considerable practical insight into what is required to minimise business interruption. A consultant with first-hand operational experience of business continuity management helped structure the BCP around the British Standard for Business Continuity Management[7] (BS25999-1:2006) in a way that recognised the maturity of the company. The BCP content is outlined in Appendix 1.

A small steering group facilitated the BCP project. The team comprised a corporate Financial Director[8], the VP Manufacturing and the HSE & Risk Manager. The HSE & Risk Manager, with wide knowledge of the company and its operations, led the development process in conjunction with insurers and the specialist BCP consultant. The corporate director acted as sponsor and provided a link to the Board of Directors. The senior manufacturing manager facilitated action at an operational level. As a result a BCP was produced and tested successfully in six months. The BCP was then refined over the following year. Some details of the phased introduction are shown in Appendix 2. Costs were limited to employees' time and travel, plus a few days of consulting. These costs were effectively financed by the reduction in insurance premiums obtained as a result of implementing a BCP.

NEVER ENDING PROCESS AND BENEFITS

The BCP needs to be maintained and kept up-to-date. The effectiveness of the BCP must be verified periodically by exercises at corporate and local level. Learning from these tests should also allow continual improvement to provide even greater reassurance that the company would maintain continuity despite a crisis. Company reputation will be enhanced both by providing evidence that a viable BCP exists and by the business continuity results it brings in the unfortunate event of a crisis.

Adding further benefits, the BCP has been developed so that it integrates with the company's insurance programme (see Figure 3).

Logically, risk management with a tested BCP reduces the likelihood, and mitigates the effects, of a major incident or crisis. It should also ensure that there will be efficient and effective claims management with insurers and loss adjusters in the event of a major incident. The resulting reduced exposure to risk and hence claims should then be reflected in insurance premiums.

But having an effective BCP is only one part of the equation. The total cost of preparing for, and recovering from, a major incident or crisis must also be borne in mind. The next section describes the role of insurance and the relationship with a viable BCP.

[7]This standard is broadly comparable to the US standard NFPA 1600 Standard on Disaster/Emergency Management and Business Continuity Programs:2004 Edition

[8]Now employed as a consultant from Capital Finance Advisory Services, 15 Moyleen Rise, Marlow Bucks SL7 2DP

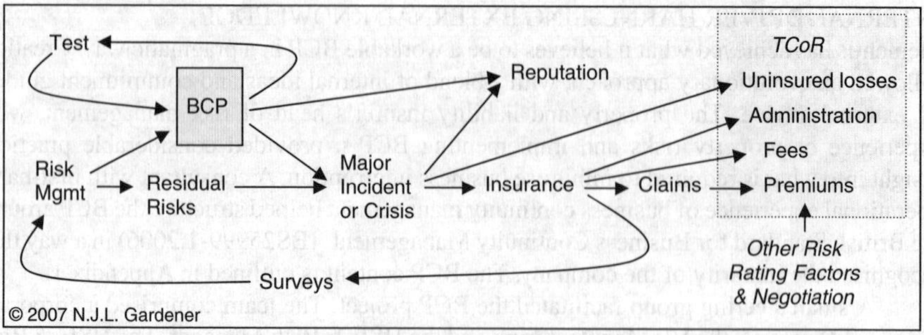

Figure 3. Linking business continuity planning to insurance and total cost of risk

INSURANCE & TOTAL COST OF RISK

INSURANCE MYTHS & LIMITS

There is sometimes a naïve belief that we will do our best and insurance is there to protect us when we fail. Others, more cynically, will say from bitter experience that insurance never pays! The truth often lies somewhere between the two.

Firstly, there may be an assumption that, whatever happens, you have insurance cover. Such complacency must be challenged. Most companies will have Property Damage insurance with additional cover for Business Interruption to recover loss of profit attributable to the property damage. The same may not hold for insurance for environmental incidents. Most policies are clear that cover is limited to "sudden and accidental" loss. Insurers, backed by the courts, are also defining "damage" more narrowly than had previously been assumed. Public Liability insurance may not provide cover for clean-up costs. You could be exposed to huge claims following an environmental incident and find yourself uninsured both in terms of your own and third party losses (unless you have taken out environmental liability insurance).

Second, even if you have insurance, there will be an upper limit (or cap) on the insurers' liability (the maximum claim that will be considered). The question is how to establish the limit and gain agreement with your insurers. Higher limits cost more in premium. Lower limits risk a shortfall of cover. For property, is the total value of your assets (buildings, plant and equipment, stock) an appropriate limit, is it the maximum foreseeable loss, or some other number plucked out of the air? For public and product liability the amount to be covered may be even more speculative. Whatever the method, limits should be reviewed with operational/business management and approved by the Board (or a body with delegated authority to make that decision). Limits should be reviewed annually to reflect changes in the company. Any uninsured or self-insured levels of retained risk should be adjusted in accordance with the Board's appetite for risk.

SUB-LIMITS, DEDUCTIBLES & PREMIUMS

Even when an upper limit is agreed, insurers may set sub-limits that are not so widely communicated within an organisation. For example there may be significant local restrictions, or even exclusions, to property insurance in areas vulnerable to natural perils such as earthquake, wind storm or flood.

Limits of cover are a major factor in setting a premium. But having agreed a limit, companies still have an opportunity to reduce their premiums in the way deductibles (excesses) are accepted. Insurance should be regarded as something to cover true catastrophic loss. A serious effort should be made to assess the probable level of future claims. This is what the insurer will be trying to do. If future claims are reasonably predictable, insurers will be charging you a premium to cover those, plus a share of the few catastrophic losses they suffer each year. They will also charge a share of their own administrative, marketing, and loss adjustment costs, together with an element for profit. If experience tells you, statistically, to expect a certain level of damage each year: budget for it, and negotiate a deductible based on accepting these losses. Do not insure for it – you will pay much more going the insurance route.

But that's not the end of the story. Budgeting for future losses should ring alarm bells. What are you doing to reduce the likelihood of them occurring? Insurers prefer companies to have a high deductible since it should encourage a company to act as if uninsured; protecting everyone's interests. Just as with the Heinrich triangle for the frequency of severe injuries as a proportion of all incidents[9] (Heinrich 1931), property damage can be assumed to follow some form of predictable distribution based on common causality. Statistically, for every major loss there will have been many minor losses from the same cause (for example fire). Attention to reducing the causes leading to the more frequent (and hence knowable) minor damage should reduce the frequency of major damage. Set your deductible with your insurer to optimise the cost of premium against potential claims.

Setting a limit (cap) and a level of deductible is ultimately subject to acceptance by the insurer. A further consideration is the non-linear nature of premiums (generally). Insurers may have a premium they wish to achieve. They can sometimes provide extra cover at competitive rates but will not reduce premium much, if at all, for a reduction in cover. These represent some of the "other factors" in Figure 3.

INSURED VALUES

While the required limits are, to a certain extent, at the discretion of the company, asset values, product characteristics, turnover, and so on are not. The duty of "utmost good faith" (*uberrimae fide*) applies to insurance contracts, requiring each party to be completely honest. So, for example, property values must be accurately declared. Declare too high and

[9]Heinrich proposed that major accidents have the same cause pathways as minor accidents. Analysis showed statistically that for every major injury occurring there were 29 minor injuries and 300 no-injury accidents. The implication is that working on the causes of lesser incidents reduces the likelihood of a major injury (or damage).

you will pay too much premium, too low and insurers may pro-rata a claim even if it is well below the limit – the insurance principle of "average" (or "co-insurance" in the US). If you are responsible for providing values to insurers – beware. Accountants may give you the written down book values. These are likely to be vastly different from replacement values, which is what you will want.

TOTAL COST OF RISK

A useful concept when optimising the structure of an insurance programme is to consider the Total Cost of Risk (TCoR) – defined here as the sum of: premiums, retained losses, fees and administrative costs (see Figure 3). TCoR is expressed per £1000 ($1000) of turnover to allow year on year comparison and a comparison against industry peers or other industries where data are available.

This is the approach taken by Risk International[10] – a professional risk management company who assisted Elementis to reduce substantially its TCoR in a sustainable way.

In particular, Elementis had a variety of separately placed policies to cover hazards such as construction, boiler & machinery and US flood[11]. Risk International was appointed by Elementis in 2003 to work with brokers to develop a more effective insurance programme. As a result exposures have been consolidated into various global policies with significant premium savings. In some cases, through effective risk assessment, it was determined that some cover was not required. This work not only eliminated the separate premium charge, but also reduced the administrative burden of maintaining separate programmes.

BROKERS

As the company began to focus on reducing TCoR it became clear that having a pro-active insurance broker was imperative, particularly in a hard market[12]. Unlike "placing brokers" – where your risk will be only one of a number of risks they are placing – a pro-active broker will not wait until renewal time to find the best price available in the market. A broker review catalysed a chain of events that led to a real engagement with the insurance market, and more competitive broking fees.

Close and regular collaboration with the brokers now ensures that the broker's Account Executive understands the company business, and finds insurers keen to work directly with us during the year before renewal – to give us an even better price than otherwise available. By developing this partnership type relationship with our brokers (Priest & Co[13]) and principal insurers (Allianz[14]) the mutual benefit of better understanding and, in

[10]Risk International Services Inc, 4199 Kinross Lakes Parkway, Suite 220, Richfield, OH 44286, USA

[11]Other examples included aviation products liability; US railroad liability, US-only marine cargo.

[12]In a hard market insurers coverage becomes more scarce (some insurers may even decline to quote), exclusions expand, and premiums grow

[13]Priest & Co, 8-9 Lovat Lane, London, EC3R 8DW, UK

[14]Allianz Global Corporate & Specialty AG, 27 Leadenhall Street, London EC3A1AA, UK

some areas, reducing the property and liability risk has led to greater insurer confidence in the company. This has been reflected in our premiums.

INSURANCE SURVEYS

A positive result of the TCoR approach has been greater engagement and dialogue on insurance surveys. Once, these might have been considered a nuisance: something done *to* the plant by insurers for underwriters' benefit when assessing a premium. All you got was a list of things to do – whether they were recommendations or, worse, requirements.

However, these surveys, undertaken by experienced professionals, are a rich source of information. Tapping into that resource can provide valuable advice on major loss scenarios based on the surveyor's access to industry-wide data on losses. They have the expertise to make judgements on different scenarios and hence the experience to project the estimated maximum loss (EML)[15] and maximum foreseeable loss (MFL)[16] at a location. Recommendations now feed back into the company's risk management process (see Figure 3).

Since property damage insurance often has associated cover for business interruption, the insurer, once a deductible limit is breached, can lose a great deal financially from plant downtime as well as damage. Having a viable BCP is therefore clearly in the interests of the insurer as well as the insured. Our insurers worked collaboratively with us helping to develop a pragmatic BCP based on experience of what really matters.

CONCLUSIONS

With the uncertainty and variability inherent in manufacturing - and business -companies needs to protect their ability to continue to trade whatever happens. Apart from that, others will need reassurance of this protection.

The first step is to know what risks the company faces. A formal process of identifying, analysing and managing major risks is essential. This process should allow some risks to be reduced or even eliminated.

If, despite best efforts, a crisis occurs, a well structured insurance programme will help, but cannot be expected to cover all eventualities, particularly loss of reputation and shareholder confidence. A company may have more to lose from the way it handles a crisis, through loss of reputation, than from pure financial loss.

[15]Estimated Maximum Loss is defined as the largest loss likely to occur as a result of a single incident under normal conditions. All existing physical protection measures and/or human factors are taken into account. The extraordinary circumstances likely to modify the nature of the risk are left out.

[16]Maximum Foreseeable Loss is the largest loss that may be expected from a single fire or explosion to any given property, when the most unfavorable circumstances are more or less exceptionally combined and when as a consequence, the fire is unsatisfactorily fought against and therefore is only stopped by impassable objects or lack of sustenance.

Insurance should not be relied on as a first line of defence. It is merely one method by which a company can transfer risk out of the organisation. An effective business continuity plan (BCP) should be the principal mitigation against the effects of the crisis.

Having an effective risk management process with a BCP not only gives the business the best chance of avoiding the adverse impacts of uncertainty: auditing the process and plans can provide confidence to others that, in the event of a crisis, measures are in place to maintain continuity from a range of events, with potentially positive impacts on reputation.

A company should also be working to reduce its Total Cost of Risk over the long term. An effective BCP is an integral part of this objective. The insurance programme should be designed with this in mind. Over a period of time Elementis has significantly reduced its Total Cost of Risk by attention to detail in the insurance programme and implementing a BCP.

However, implementing a BCP can be a major undertaking. The risk is that it is not available when needed. Alternative approaches exist but it is suggested here that an effective approach, certainly initially when starting to develop a BCP, is to take a pragmatic view of what is required. Actions taken in the first 48 hours are crucial, particularly where reputation is concerned. Develop and implement a plan that prompts people to do what they already know, but may miss under pressure, and provide the specific data that they will need to refer to. This approach should ensure that at least some guidance exists right from the start of the crisis.

Every BCP must recognise the specific risks faced by the company, and use the internal resources available. However, given the infrequent experience of such incidents, guidance from external experts in developing the BCP speeds up the implementation process – and provides a level of reassurance that the plan is sound. We may not be able to predict all shocks but we can provide resilience.

BCP is not difficult, but it takes wisdom and commitment to put a viable BCP in place. There should be no doubting the justification though. The end product has potentially huge intangible value – it can make the difference between a prosperous company and one that is facing financial ruin.

In the end there will always be some residual risk but, as The Combined Code on Corporate Governance states: "profits are, in part, the reward for successful risk-taking in business" (Financial Reporting Council 2003).

APPENDIX 1 BUSINESS CONTINUITY PLAN CONTENT

The corporate Business Continuity Plan (BCP) focuses on the essential details required to establish the organisation and actions necessary to help the company recover from a major incident or crisis in an efficient and structured way. The aim is to bring the business back to previous levels with minimum disruption. It includes roles and responsibilities, with clear definition of crisis control leadership team structure and communication protocols. Tables show the required notifications and consequent responses, according to type and severity of the event or incident. The BCP includes company contact details (with alternatives) and key external contacts (such as insurers and City contacts).

For ease of use in a crisis, only those parts needed in the crisis are included in the main BCP; the administrative sections for preparation and control are split into a separate volume. Specific persons are assigned to maintain the plan and to review it annually. Learning from regular exercises allows the BCP to be improved continually.

Using the corporate BCP as a template, a local BCP is prepared for each manufacturing location, off-site laboratory, major warehouse and corporate office. Each local BCP contains the specific details and checklists required for local recovery actions to interface seamlessly with the corporate BCP.

Local BCP's include checklists of required actions such as:

- local contacts, and alternatives, with call out details
- how to contact contractors and vendors equipped to repair or replace buildings and critical equipment, and the means to recover or replace essential utilities
- outline procedures for salvage of buildings, equipment and stock
- details of products, critical business records, engineering drawings, with copies to be stored at a safe off-site location
- how to contact alternate suppliers for critical raw materials and supplies and options to outsource operations to make up lost production and inventory

APPENDIX 2 BCP IMPLEMENTATION

Based on our experience, a phased approach, with clear milestones and reviews, is a successful model to follow.

PHASE 1: DEVELOP A CORPORATE BCP

The first step was to obtain senior management understanding and support by drafting a Business Continuity Policy for approval, supported by a draft BCP.

This phase drew widely on experience and expertise, wherever it existed. The Elementis BCP leader ensured that the BCP integrated with company culture and requirements - otherwise the BCP, however good, might have been implemented but not truly accepted. Our insurers with their range of experience and the trusted BCP consultant advised on best practice and shortened the development process.

The BCP was rolled out to key locations for desktop exercises to prove the BCP. During this phase the BCP grew in size and complexity but, since it was very much under the control of the project team, this was considered acceptable.

There were two key lessons learnt in this phase. Firstly, there was a tendency at a local level to mix up emergency response actions with business continuity actions. Fighting fires is important but it is not the same as planning for continuity of business. Second, external communications: at what point do you communicate, to whom, what message? How do you keep all the various interested parties adequately updated with a consistent, factually correct message as the crisis evolves? To expand on this: in a crisis the site team needs to communicate with corporate management. Corporate management then, amongst other things, may need to issue legally approved statements to the Stock Exchange on the trading impact of the incident. Meanwhile local press and television will be seeking interviews at the site, Regulators, and in some countries local politicians, will require detailed factual information – immediately. Major customers must be advised about the impact on supply – before they see the news and rush for alternative suppliers.

PHASE 2: REFINE AND EXTEND THE BCP

The purpose of this phase was to incorporate learning from Phase 1: dealing with the ullage, filling in the gaps, reorganising the content. The result transformed the BCP into a user friendly document that can, with training, be used without support. Phase 2 involved greater interaction with users at manufacturing sites, other locations[17] and functions such as IT and finance. This direct, and continuing, contact ensures that hazards are correctly identified and addressed in the BCP.

ACKNOWLEDGEMENTS

The work described represents collaboration over several years. I acknowledge with thanks the following for their contributions, critical reviews and comments:

Randall Davis and his colleagues at Risk International made several important contributions to content, clarity and accuracy on insurance and Total Cost of Risk;

Rob Hearn and his colleagues at Priest & Co for a comprehensive review of the content of the paper, which led to several improvements in the content and structure;

Clifford Perkins of Capital Risk Advisory Services Limited (formerly with Elementis plc) provided support and knowledge that led to real success with reducing TCoR and acceptance of the BCP at a senior level in Elementis.

Dennis Murphy from Allianz Global Corporate & Specialty AG provided wise, detailed advice and encouragement on how to produce a successful BCP in a relatively short space of time based on his experience and expertise in Business Continuity Planning.

Jeff Brooker provided specialist expertise and support on implementing the BCP based his own practical experience and authority.

Marty Neil provided ideas and support, particularly on operational aspects.

The author wishes to express his thanks to his employer, Elementis plc, for permission to publish this paper.

If you have any comments or questions on this paper, or ideas for improving BCP, please contact the author: nick.gardener@elementis-eu.com.

REFERENCES

BS 25999-1:2006 *Business Continuity Management – Part 1: Code of Practice*; UK, British Standards Institution

Deming, W.E., 1986, *Out of the Crisis* Cambridge MA: MIT

[17]Such as offices at non-manufacturing locations, major off-site warehouses and laboratories

Deming, W.E., 1993, *The New Economics* Cambridge MA: MIT
Financial Reporting Council, 2003, *The Combined Code on Corporate Governance*; London, FRC
US General Accounting Office, 1997, *Operation Desert Storm: Evaluation of the Air Campaign*, Letter Report, GAO/NSIAD-97-134, Appendix V
Heinrich H.W., 1931, *Industrial Accident Prevention* New York: McGraw-Hill
Knight R.F. and Pretty D.J., 1996, *The Impact of Catastrophes on Shareholder Value* Oxford Executive Research Briefings, Oxford: Templeton College
Lewis, C.I., [1929] 1956, *Mind and The World Order: Outline of a Theory of Knowledge* New York: Dover Publications, Inc.
The Companies Act 2006, 2006 Chapter 46, *Contents of directors' report: business review.* Part 15, Chapter 5, 417; UK, Office of Public Sector Information

A WORLD CLASS APPROACH TO PROCESS SAFETY MANAGEMENT (PSM) AFTER THE TEXAS CITY DISASTER

Dr E. Pape
DNV Energy, UK

There are many major facilities, such as Texas City which are aging with many legacy issues (e.g. atmospheric venting blowdown drums). These were built to conform to the industry practices of their day. Such facilities can be run safely provided more attention is focused on process safety than would be necessary in more modern facilities. Such sites need some guidance as to what operational safety programs they need to implement and monitor through life for their specific designs and safety challenges. Operational excellence benchmarking can be dangerous if this leads to under management of safety issues because peers operate safely with less.

The Baker and the Chemical Safety Board Reports following the Texas City Disaster have already had ramifications across the world. The major oil companies have each taken the report seriously and in their several ways are taking appropriate action. Many operators are creating a large program around Control of Operations and why many are so keen on the barrier approach.

DNV have compared the different industry approaches to Process Safety Management and have drawn some key lessons on the vital components to an effective system. These form a balance between the management of risk and the maintenance of effective barriers with the optimisation of performance. All robust systems have a component of measured improvement and DNV is providing just such a service to main players across the world with its isrs7 – PSM assessments.

This paper will conclude with some trends that are starting to emerge in the area of PSM which will dictate the Operators' activities over the next decade.

CONTENT

- The three areas of focus, investment, process optimisation and process safety management
  - Refining Investment pattern-growth
  - Optimising Performance
- The dilemma between declining frequency of occupational health events and regular losses through process events
- The Baker and CSB reports
- Trends from the main operators – comparing the approach between refining and offshore approaches
- What can we do to create a robust Process Safety Management approach
- Measured Improvement – one approach by DNV

1. INTRODUCTION

The refinery industry is making good margins today and we are seeing performance enhancement projects, expansion of existing refineries and new builds across the world. Loss statistics show an expensive dilemma over the last decade for the industry. While the frequency of occupational injuries and ill health have declined from around 2.5 to 1 per 200000 hours per year the trend of incident costs have risen from $200 to $300 million over the same period. Such events often mean a financial disaster as well as a physical event, that significantly affect the Profit & Loss Account and in extreme cases they can herald ruin.

The industry is feeling significant influences, the high oil prices, perceived shortage of capacity, increased demand for low sulphur fuels and the events at Texas City all of which have coincided to produce a decade of change. The continuing high oil price, currently $76 per barrel for Light Sweet Crude, combined with an increasing demand for environmentally friendly fuels has created some new refining trends. Three areas emerge

1. Investments, new builds and mergers
2. Production Optimisation – getting more from what you have
3. Minimising losses – by preventing major accidents, currently drawing on the learning's from Texas City and others

2. CURRENT INVESTMENT DEVELOPMENTS

For the last ten years we have seen mergers of the majors with the assets of Aramco, Elf, Fina, Mobil and others all being integrated into larger owners. This trend has continued and three typical developments illustrate the investment levels that are in play in the last few years:

- OMV's purchase of the Romania's oil industry Petrom (some 40,000 employees) to create the largest central European refining group
- Kuwait National Petroleum Company's planned construction of a 615k bpd refinery at Al Zour and consequent realignment of their existing 3 refineries
- Saudi Aramco's plans for several world scale refining and petrochemicals facilities

DNV believes that similar patterns of investment will continue for the next decade. This will accommodate the changing demand pattern for fossil fuels to optimise our carbon footprint.

However this is the subject for a different paper.

3. OPTIMISING PERFORMANCE – SOME DIFFICULT QUESTIONS

In parallel with investment pattern, asset owners want more margin from their current facilities. In times of high refining margins, maximising throughput and minimising downtime is critical. They are doing this by attempting to optimising performance. The operators are however faced with some difficult questions:

- Will we achieve target performance of 96% on-stream factor/availability?
- What is the optimum intermediate storage to protect against feed unit unavailability? Do we have too much?

– Do I have enough sulphur recovery capacity and redundancy?
– Do I have a reliable hydrogen supply?
– What is the impact of 1% increase in hydrotreater unit availability?
– What is the impact from crude import logistics issues (berthing, demurrage, storage)?; will there be feed shortfalls?

The challenge is how to reduce the production losses:

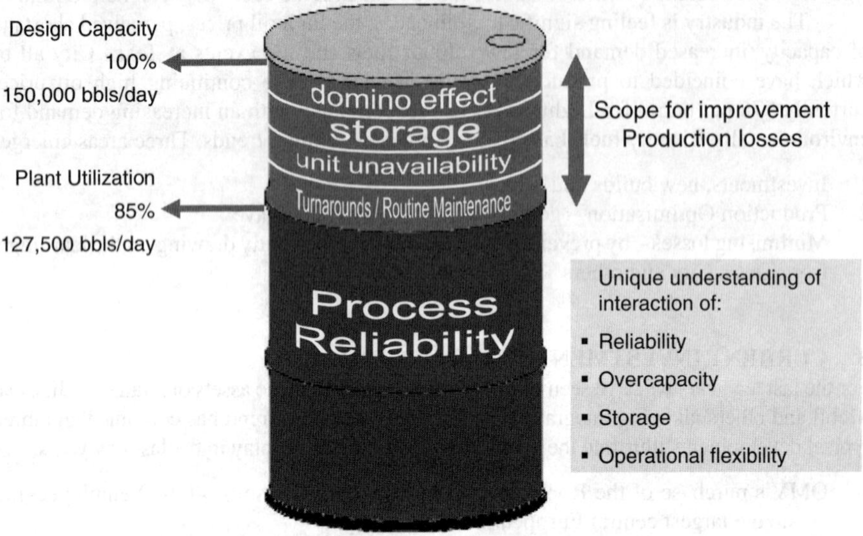

We at DNV are helping a range of Refineries address these issues through the active use of our TARO Plant Simulation models. TARO models such areas as process flows, current design envelopes, the equipment, units and their inter dependability, the storage volumes (for Feed, intermediate and product storage), the operations, maintenance and storage history, the refinery slate and material balance, the market forecast and demand profile.

TARO runs a series of simulations to establish the sensitivity of plant utilisation to a range of operational changes or investment strategies. The TARO model then provides

– Production Efficiency for the planned product streams
– Unit Criticality analysis and Unit Utilisation
– Storage Utilisation
– Production Improvement Opportunities
– Detailed forecasts for project and plant economics

In summary where to focus to achieve the clients' objectives

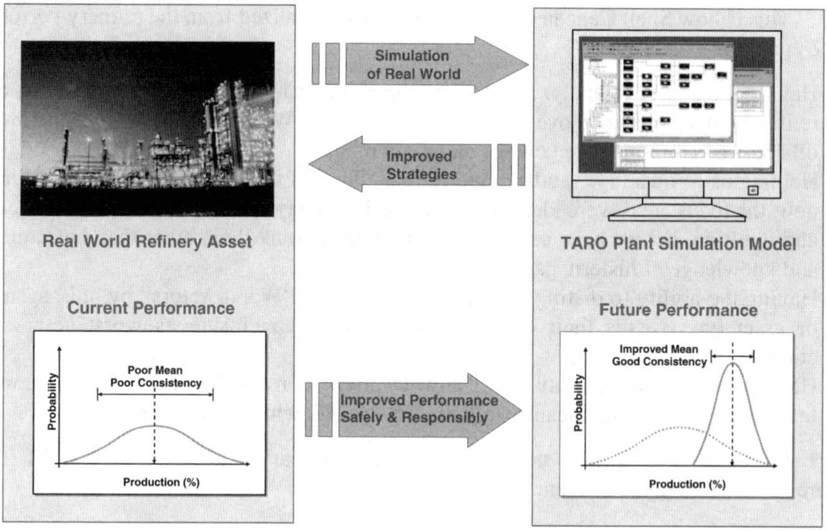

Refineries across the world are addressing these issues and benefiting accordingly. Here is a typical result:

Scenario	Proposed Design	Optimised Design
Gasoline Production Efficiency	88.30%	89.20%

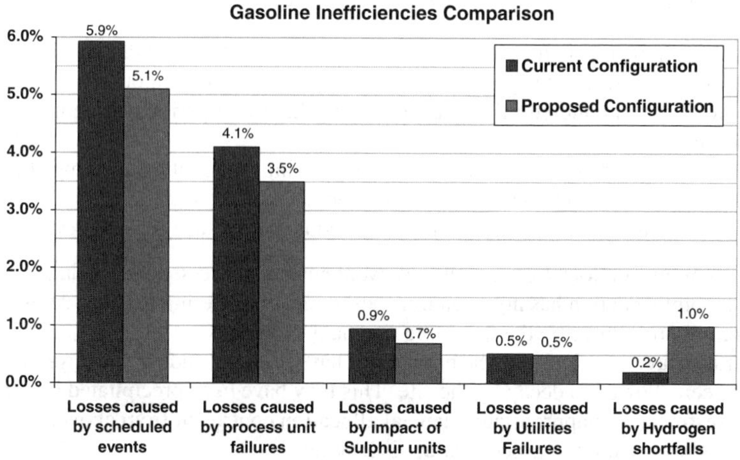

This is how Shell Canada described the benefits realized from the refinery performance analysis:

- Having the ability to analyze and study the affect that plant capacity, reliability, and refinery tankage have on overall refinery production. We are able to analyze the trade-offs between intermediate tankage and plant reliability for example.
- Helping us to quantify, validate, and rank and/or add to the initial focus areas. Up to now, the focus areas were identified and prioritized based on our collective knowledge and intuition, but we have no ability to analytically rank them other than by intuition and knowledge of historic gaps.
- Having the ability to distinguish critical equipment "Worst Actors" by unit, system, or asset type versus their contribution towards gap closure as worst actors are corrected.
- Having insight into the relationship between production and Solomon Utilization, what features contribute significantly to both and through what mechanism."

This subject deserves a more thorough coverage than can be achieved here and I refer interested readers to the referenced papers.

4. IMPROVING PROCESS SAFETY PERFORMANCE

The third area that refiners are focussing on is to minimise the unplanned events. These range from production upsets, through minor losses of containment to the significant event that can destroy the annual finances and damage your corporate reputation.

Plants built in the 1970's or earlier will be faced with aging aspects and legacy hardware no longer used in the industry – especially items which are prone to loss of containment – such as any system that vents to the atmosphere in a process area. Aging assets require more Process Safety care than current designs to provide a comparable safety performance.

Although the media focuses on the major events, the financial damage of a series of minor events should not be ignored these have been known to mount up to economic significance, several hundred million Euros over a three year period was compiled by one major refiner.

Headline grabbing events occur with depressing regularity indicating that a least some of the lessons from previous events have been ignored.

There are four reasons why the industry should suffer such repeated events:

- The recommendations from a previous event has not addressed the fundamental issue
- The recommendation has not been transposed into engineering barriers, rules or sound operating procedures by the operating company concerned
- The rules have been ignored, the procedures have been set aside or the physical barriers have been allowed to decay by the site. This may have been precipitated by restriction on investment or inappropriate financial allocation controlled from above.
- Leadership attention to process safety is lacking

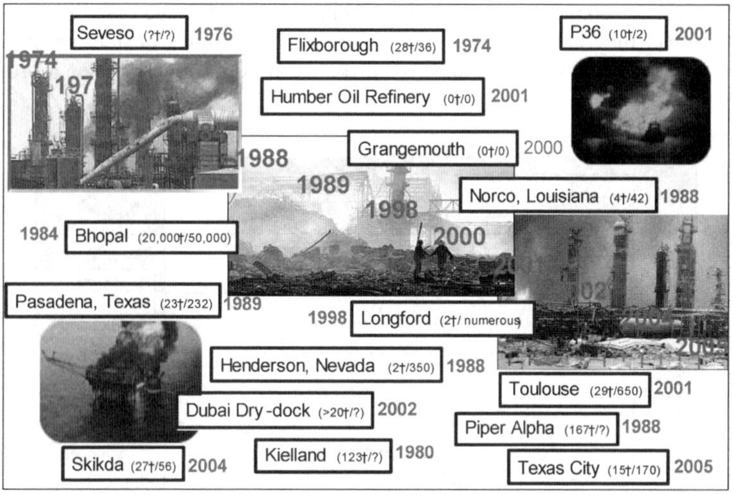

This brings us to the current dilemma in the process industry. Since the late 1980's the industry has focussed on HSE Management Systems and major advances have been made. Many companies have achieved a three times improvement, over 10 years even starting from a good initial performance, e.g. Conoco:

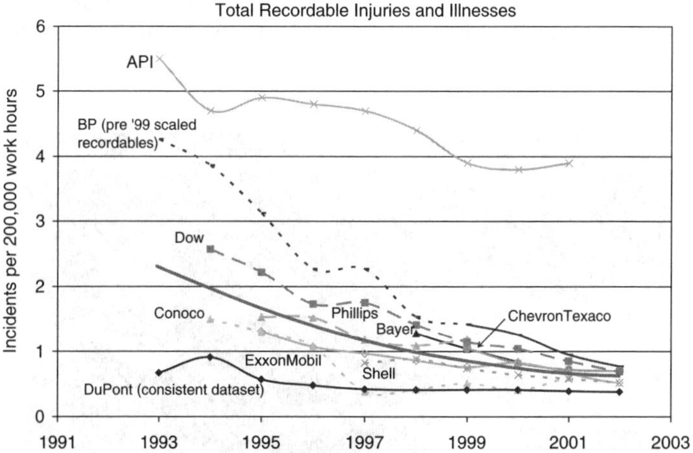

Most majors have deployed three approaches to achieve this result:

- Improving formal safety management systems
- Deploying risk assessment programs
- Most recently deploying safe behaviour programs

However, by contrast the trends in refinery material damage costs have not shown a decline curve, they have actually grown during this period. Considering the incident costs – \$ per 1000 bbls refinery capacity corrected to 2000 prices we see a progressive increase:

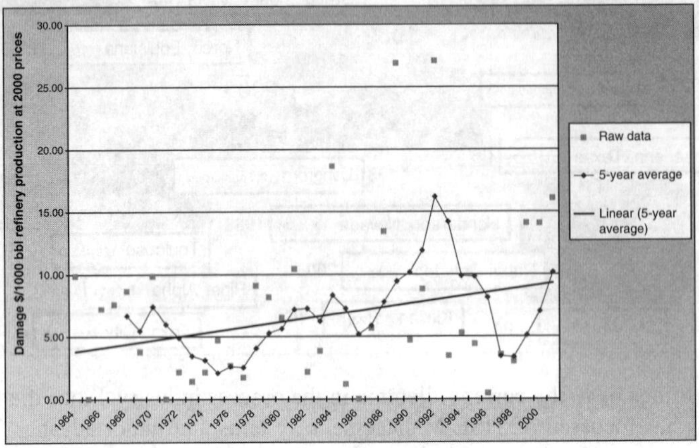

Most companies recognise the Bird triangle of losses and have been working on the HSE areas to reduce losses. However there is a parallel triangle related to outstanding work orders reflecting the PSM issues and always in the background stands the spectre of a major event.

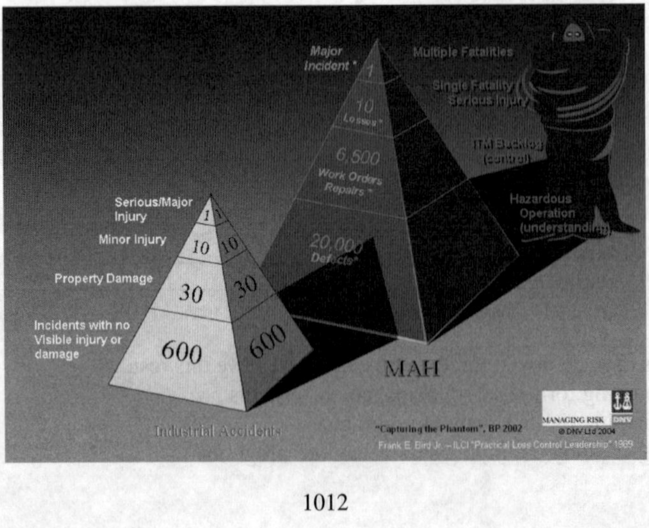

5. TEXAS CITY 2005

This was the largest accident to have occurred in the US in 20 years. It represented a wake-up call to the refining industry, suggesting that major assets are not completely protected by current safety methods and investment strategies.

The Baker Panel has done an excellent job assessing safety management system and culture issues in BP's US refining operations. The CSB investigation has also now been released. So what can the industry learn?

Michael Broadribb, BP's Lead Investigator, pointed out that at Texas City there were twelve layers of protection that could have prevented the event, each one failed.

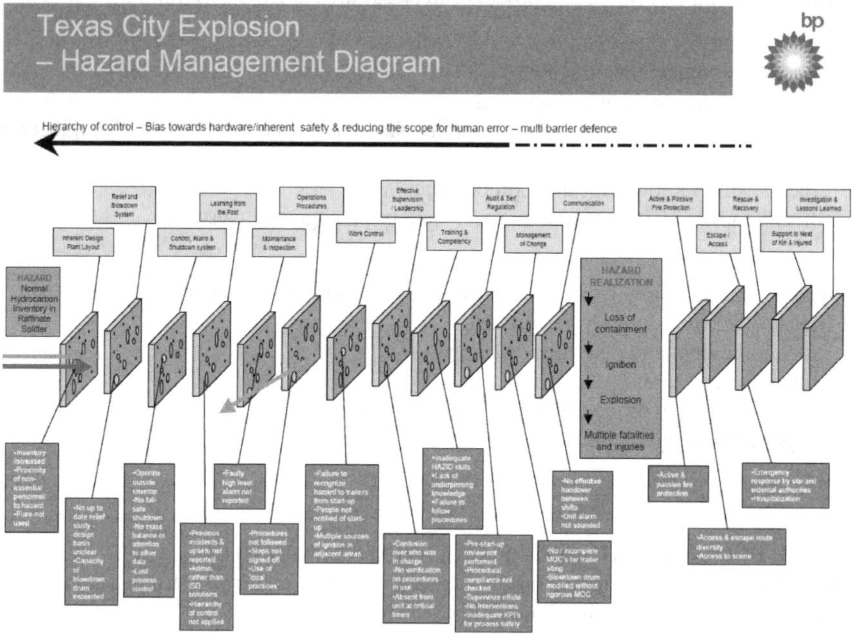

The pre-event failed controls included:

Management Control Related	**Asset Integrity**
Effective Supervision and Leadership	Inherent layout/design of the flare system
Work Control	Relief and Blow down System
Learning from Previous Events	Process Control, alarms and Shutdown System
Communications	Maintenance and Inspection
Training and Competence	Management of Engineering Change
Adherence to approved operational procedures	
Audits and Self Regulation	

However there were post event failures too. These included the Fire Protection Systems (Passive and Active), Escape and Access Routes and the Rescue and Recovery Arrangements.

These facts were also reflected by Baker who can be summarised as having concerns about:

Corporate Safety Culture
- Process safety leadership and management decisions for process safety
- Resources and positioning of process safety capabilities
- Employee empowerment and process safety cultures at BP's U.S. refineries

Process Safety Management Systems
- Process risk assessment and analysis
- Compliance with internal process safety standards and external good engineering practices
- Process safety knowledge and competence
- Effectiveness of BP's corporate process safety management system

Performance Evaluation, Corrective Action, and Corporate Oversight
- Measuring process safety performance
- Process incident and near miss investigation
- Process safety audits
- Timely correction of identified process safety deficiencies
- Corporate oversight

6. THE PROCESS INDUSTRY FOLLOWING TEXAS CITY

Many companies are reflecting on the Texas City event and considering whether these failures could equally describe some of their own operations:

- DNV are seeing the Majors reinforcing the deployment of sizeable asset review teams to each of their downstream operations in turn. In a number of cases, we are supporting these review activities. These teams test the robustness of their PSM philosophy, provide a structured verification of barriers and actions and develop suitable improvement programmes. These teams are spending between 12 and 45 man weeks on each site and so represent a significant investment in situation analysis. Most are using a Maturity Model Approach. Shell, BG, BP, C-P and parts of Total all undertake formal bi annual Asset Integrity reviews, reporting corporately.

This focus on analysis is likely to uncover significant requirements for additional investment.

- Several Majors are investing in enhancing PSM knowledge. They are running a series of Process Safety Management Workshops to ensure that the fundamentals of PSM are understood by sufficient of the operational staff to effectively lead their peers.
- Some companies are now forming/reinforcing PSM knowledge circles/experience exchange clinics to ensure best practise is shared across their group.
- DNV knows of two majors who are re-examining their performance metrics to create a better reflection of inherent risk, overall plant condition (through a collection of lagging PSM indicators) and the status of critical controls (through a selective collection of leading PSM indicators).

7. WHAT HAS WORKED FOR MAJOR ACCIDENTS

DNV believe that major hazards are as amenable to reduction as occupational health – provided companies do the necessary work. As described above occupational health has had a major improvement in the past 20 years (see Dow Chemical statistics showing a factor of 10 improvement). We know in the North Sea that HSE leak statistics show a 4 fold reduction since 1995 – and major accidents almost always start with a loss of containment.

Perhaps this is why many major upstream operators have exported the North Sea solutions across the world with good reason. The North Sea approach has been dramatically effective with the last headline grabbing event in the North Sea being Piper Alpha in 1988.

In UK – the safety case approach establishes performance standards for safety critical elements which are supported by an independent verification regime. In Norway there is a risk based process to demonstrate integrity. For Upstream assets there is more emphasis on Major Accident Control. They deploy:

- full safety management and mechanical integrity system driven by best practice risk assessment
- detailed holistic risk assessment, with a consolidated Hazard and Risk Register
- identification of the most important safety barriers/controls – hardware and systems (safety critical elements) and establishing performance standards for these controls
- valuing of every control, and independently verifying sufficient controls are in place for the level of risk and performance standards are achieved
- development of tracking metrics using recent data and real-time tools
- linking incident investigations back to the Hazard and Risk Register and the identified controls that failed

We are seeing that Numerical Techniques (QRA) being only selectively deployed with simpler Bowtie approaches growing in use and the demand for effective HAZOP/HAZID growing.

8. WHAT CAN BE DONE FOR REFINING PSM?

DNV is in the centre of much of this activity; we have drawn some fundamentals from our work and produced tools to assist companies evolve and demonstrate robust PSM.

We believe that an integrated approach to risk management is essential to ensure a robust system. The Refinery should:

- Design a sound layout and establish a robust operating envelope
- Understand the hazards at enterprise, site and unit level
- Work within the operating envelope and maintain its integrity
- Understand their critical controls
- State what they expect them to do
- Make sure they do it through effective inspection and maintenance
- Provide effective leadership in PSM, similar to the successful focus already applied to occupational safety issues
- Adequately resource the key functions
- Measure performance year on year ask and how they can make the situation better

	Plant	Processes	People
1. Leadership		✓	✓
2. Planning	✓	✓	
3. Risk Evaluation	✓	✓	
4. Human Resources		✓	✓
5. Compliance assurance	✓	✓	
6. Project Management	✓	✓	
7. Training and Competence		✓	✓
8. Communication and Promotion		✓	✓
9. Risk Control	✓	✓	
10. Asset Management	✓	✓	
11. Contractor management and purchasing	✓	✓	✓
12. Emergency Preparedness	✓	✓	
13. Learning from events	✓	✓	✓
14. Risk Monitoring	✓	✓	✓
15. Results and Review		✓	

We believe that a structured approach to managing the various layers of protection is essential:

DNV have incorporated world best PSM practise into a reference, measurement and development tool called isrs[7] PSM. This has emphasis on:

- Design Integrity
- Process Safety Information
- Management of Change
- Asset Management
- Process Hazard Analysis
- Operating Procedures
- Pre-start-up Safety Reviews
- Managing Proprietary Knowledge
- Technology Development

This PSM tool has incorporated OSHA 1910, MES, SEVESO, Baker Report and CSB Report from Texas City into the existing **isrs**[7] Omega tool. It uses the same quality control criteria, reviews the systems in place and the results they produce. It is supported by assessment guidance to ensure consistency in application. It can be used in a flexible way by supporting an integrated business management approach, by developing the PSM System or to develop the HS, E, or Q systems as required. It can focus on one specific area of PSM, Asset Integrity for instance or can delve into particular issues and develop the approach.

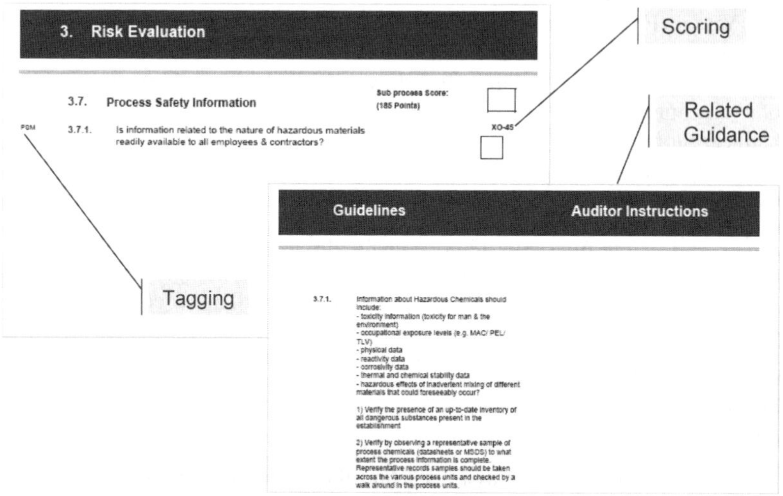

World majors are already using **isrs**[7] in a range of ways. Three are using it as a reference tool to develop their corporate PSM system. Two are using it as a Selective Improvement Tool. Three are using it as an Assurance Protocol mapped against the company's MS and two are using it as a full System Improvement Tool and Assurance Protocol.

However the PSM performance cannot be assured or evolved through good PSM systems alone. This must be balanced by focussed assessment of the Physical Conditions or barriers. DNV has developed a complimentary Physical Conditions Assessment (plant walkdown) that ensures that PSM systems are implemented through actual good management of physical assets. The PCA looks at typical areas of weakness known to DNV through its work with the industry. Many majors are using the barrier approach to PSM and isrs7 will link to this – through the field inspections. Isrs7's PCA will be looking at the status of critical safety barriers, safety defeat logs, etc and is integrated into scoring of overall assessment specifically focusing on:

- Bypasses and isolations
- Control system overrides
- Locked Close/Locked Open/Normally Close/Normally Open Valves
- Pipe, Flanges and Blinds
- MOV/ROV/Excess Flow (Depressurizing) Valve
- Small Bore Fittings (<2″) & Conduit
- Control Valves & Check Valves
- Relief Valves and Rupture Discs
- Culverts & Drainage
- Field Instrumentation
- Tankage and Bunds
- Fired Vessel (Furnace, Incinerator, Main Combustion Chamber, Package Boiler, Gas Turbines, etc.)
- Unfired Pressure Vessel (Column, Vessel, Reactor, Heat Exchangers, etc)
- Rotary Equipment
- Chemical Storage & Usage
- Emergency Equipment
- Electrical controls

DNV believes in Measured Improvement, and this structured tool will allow clear comparison of the health of the PSM system through benchmarking. Here is an example of a site being compared with a group of peers. The system will allow comparison with:

- Past performance
- Sister companies
- Market competitors
- World class performers

Companies who use the classic isrs find that it provides a basis for setting short and medium term objectives, calibrates internal measurement activities and sets a clear standard set so performance levels can be easily recognised and next steps understood.

It can be seen from the previous illustration that the approach makes the problem identification activity much clearer. Across a group the variations in performance indicate robustness of PSM implementation which will allow deployment of resources and avoid PSM events.

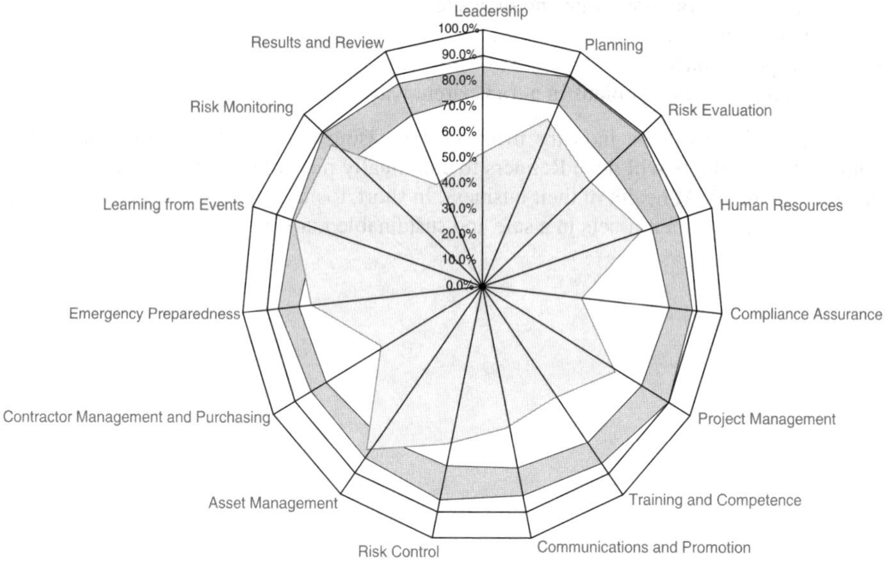

Nuclear Power stations are considered by many to have a high societal and installation risk profile, the consequence of a loss of containment are higher than a refinery. Many UK installations have used ISRS for the last decade to identify and manage their risk profile. This is what the Operations Director responsible for 14 UK stations said about the value of this approach:

> *"ISRS has given us the challenge and the recognition to continually and consistently improve in all aspects of efficient, safe and environmentally responsible operation of our sites.*
>
> *It has achieved strong staff participation and enthusiasm by measurable goals to drive overall site improvement plans."*

Magnox Electric – Chris Marchese Chief Operations Officer 2006

9. CONCLUSION

The industry is going through a period of significant change with tremendous opportunities for performance improvement. High oil prices and a shift in demand pattern are all driving investment while the spectre of a "Texas City in my company" is ensuring that companies take their PSM responsibilities seriously. DNV is working with many major players to assist in the optimisation of performance in the refinery through its TARO tools. However the work DNV is doing with isrs7 PSM is assisting companies to:

- Set standards, both corporate and on a site
- Implement change
- Assess performance
- Develop improvement plans in a transparent way.

This will assist the industry meet the increasing expectations of multiple stakeholders. Stakeholders will need Refiners to continually measure, improve and most of all demonstrate the PSM health of their business. In short, to ensure that companies maximise the performance of their assets in a safe and sustainable manner.

LEADING INDICATORS FOR THE MANAGEMENT OF MAINTENANCE PROGRAMMES; A JOINT INDUSTRY PROGRAMME

Keith Hart[1], Professor John Sharp[2], John Wintle[3], David Galbraith[4], and Ed Terry[5]
[1]Energy Institute
[2]Cranfield University
[3]TWI Ltd.
[4]Poseidon International Limited
[5]Sauf Consulting Limited

The Energy Institute has initiated and managed a Joint Industry Project (JIP) to deliver a Capability Maturity Model procedure to assist Duty Holders and contractors with the management of maintenance. The model is a generic tool which enables the assessment of the maturity of an organisation's capability to maintain facilities at offshore installations for improved safety and availability. It uses Capability Maturity Modelling as a means of measuring organisational behaviour and is aimed at the assessment of maintenance departments.

The first applications were projects for the Health & Safety Executive in the UK and the Petroleum Safety Agency in Norway and were specifically targeted at safety with elements of regulatory compliance. These have been the subject of papers in previous Hazards.

The JIP has now developed new applications with a more operational and commercial bias that have highlighted different emphases in the approaches used in the modelling technique; the modelling from the JIP describes the processes involved in the management of maintenance programmes and includes the application of the model to the different contracting strategies used in the UK oil and gas sector.

The paper will describe these most recent applications, the way in which these applications have modified the model development and how using workshops and case studies have added to our confidence in the effectiveness and applicability of the model.

BACKGROUND

The authors have been developing tools that can be used to identify and assess leading indicators; some of the applications have been presented in earlier Hazards Conferences. Through the Energy Institute, feedback from the oil and gas industry and the regulator (the UK Health & Safety Executive) has indicated that industry wanted a better mechanism to monitor and measure maintenance programmes

The industry is well aware of the impact of inefficient or ineffective maintenance on the reliability of systems essential for ensuring safety and production, and the costs and hazards of outage. This project has developed a model from which companies can assess the maturity of their capability to manage maintenance, and identified some of the steps needed for the companies to move to a different level. The project was set up as a JIP and attracted members from the oil & gas operators, the regulator and one of the new breed of contractors who operates assets on behalf of other owners.

As described in the earlier papers, the Capability Maturity Model (CMM) describes five levels of company culture and approach; in this project, it was the culture and approach adopted for the different processes of maintenance management that was subject to capability maturity assessment. This paper describes how the maintenance model was developed and summarises some of the outcomes.

MAINTENANCE MANAGEMENT PROCESSES

Maintenance management can be broken down into a number of discrete identifiable processes. Some processes are essential for the activity and apply to every installation undergoing maintenance, and these are called the Core Processes.

There are other processes which although common are not found in every application, and these have been called Complementary Processes.

In addition, there are management processes that affect the ability to carry out the Core and Complementary Processes in the longer term, these are called Supporting Processes.

The following sections describe the anticipated Core, Complementary and Supporting Processes which the authors consider form Maintenance Management Processes.

CORE PROCESSES

For almost every situation, the activity of maintenance management can be broken down into six Core Processes and the accuracy of the definition of, the effectiveness of their integration and communication between the processes is often as important as the conduct of the processes themselves. The six Core Processes are:

1. Setting policy, targets and controls over resources
2. Strategy and planning
3. Procurement of resources
4. Effecting maintenance on site
5. Data management and record keeping
6. Audit and assurance

These six Core Processes are continuous with inputs and outputs between processes at different points. This is shown in Figure 1. In organisations with multiple assets all the processes are in progress simultaneously.

Sometimes the processes will be hard to recognise, particularly if their maturity is not high, and sometimes responsibility for the processes will be divided between several different organisations. This is discussed later in the paper.

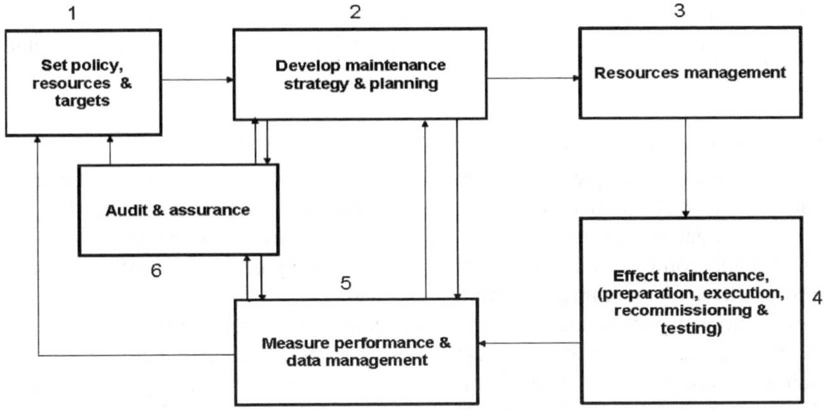

Figure 1. Core Processes for maintenance management

SETTING POLICY, TARGETS AND CONTROLS OVER RESOURCES

This process is normally undertaken at a senior level and may cover a range of assets. The policy for maintenance management would normally be expressed in general terms consistent with company objectives and values. These could cover aspects relating to maintaining the safety and reliability of the installation, the health and safety of undertaking the maintenance itself, change management, the requirement for data management and record keeping (both relating to the maintenance itself and the equipment performance), and the need for regular audit and assurance.

Targets may relate to production, availability of equipment, number of unplanned shutdowns for unscheduled maintenance, the time, duration or frequency of outages. The targets should take account of the life of the installation, the field and its characteristics, the market and the age and condition of the equipment.

A major part of this process is to define a consistent set of responsibilities for maintenance and allocate sufficient authority to purchase resources and equipment.

As part of this process, clarifying the divisions of responsibility between on-shore and off-shore staff, and the interactions between operations, maintenance and inspection, and between in-house and outsourced maintenance contractors is a key contributor to creating a positive culture.

DEVELOP MAINTENANCE STRATEGY, PLANNING AND SCHEDULES

Responsibility for this process is normally asset or installation specific and in the hands of a dedicated maintenance team or supervisor, the process should develop a general strategy or approach to be applied in planning maintenance for each asset consistent with meeting the policy and targets. The approach may depend on the type or function of the equipment or structure being considered, and the context in which it is operated. Possible maintenance strategies could range from routine, predictive, condition based, risk based, preventative, to breakdown only.

RESOURCE MANAGEMENT

This process gets the people, equipment, tools and parts to site ready for maintenance to be put into effect. It is concerned with resource management, procurement and logistics. Inputs include detailed planning, local constraints and context, and any business arrangements that the company may have made with its supply chain. The efficiency of this process can have a large impact on productivity and cost.

For more mature organisations resource management may involve risk considerations and a spares management strategy (considered as a Complementary Process). The availability, storage and location of equipment, tools and parts are key aspects, and for offshore installations there is the issue of the resources to be stored off-shore. Arrangements for rapid transport and delivery to site are key aspects of the process.

EFFECTING MAINTENANCE AT SITE

Management of this process is largely concerned with interfacing the maintenance task with the wider operations and safety context of the installation. It involves the preparation of the equipment, local isolation and access, Permit-to-Work, inspection and assurance, and finally re-commissioning and testing. While the normal work-breakdown for the task may have been specified from the computerised maintenance management system, the maintenance team needs to be alert to unexpected conditions of the equipment being maintained, when other actions may be needed.

More mature organisations will have a seamless interface between their own staff and subcontractors on site. Creating a positive work environment and culture is key to successful maintenance. There have been several recent industry guides on how a positive maintenance culture can be achieved. The level of supervision is important, particularly where the competencies required for a task are split.

Effective management of the schedule of maintenance tasks during an outage requires a level of maturity which is at least based on company standards (that is, level 3).

MEASURING PERFORMANCE AND DATA MANAGEMENT

To become a more mature organisation requires feedback from measuring performance and learning from experience. Here the performance measured refers to both the maintenance task itself and the subsequent operation of the equipment. For the maintenance task, typical areas to record are;

- How long did it take?
- What went well and what were the problems?
- Where did delays occur?
- Were the resources adequate?

More mature organisations will have a procedure for recording these data and updating the maintenance management system.

Information about failures and unscheduled shutdowns, abnormal operations, or from condition monitoring can be analysed and trended to provide insights to guide future maintenance strategy. Evidence of the beneficial or detrimental effects from maintenance approaches and practices can be gathered in this way.

ASSURANCE AND AUDIT

Within this model, assurance and audit demonstrate that maintenance and maintenance management have been done satisfactorily, and that appropriate reports are given to senior management. It is a management process in its own right and involves reviewing, reporting and gaining assurance on the performance of the maintenance and the condition and performance of the equipment. Within the UK offshore regulatory regime there are specific requirements for a safety management system and for verification by an independent competent person.

Assurance extends beyond quality control and compliance with maintenance procedures, and requires evidence that the functionality, availability, reliability and survivability of the equipment will meet the demands for production and safety with the maintenance that is being undertaken.

COMPLEMENTARY PROCESSES

A number of Complementary Processes have been identified and these are shown, together with Supporting Processes (discussed later) in Figure 2.

Complementary Processes are maintenance management processes that are common but, unlike the Core Processes, do not necessarily apply in every application.

RECOGNITION AND HANDLING OF NON-STANDARD EQUIPMENT AND COMPLEX TASKS

At most installations there will be items of equipment whose maintenance will (by the skills and knowledge required) need the involvement of specialists or the original manufacturer.

A mature company will have systems and a culture in place for recognising and managing such equipment and tasks efficiently. A less mature company will either not recognise that special measures are required or will have the work done ineffectively. The model highlights the attributes of different maturity levels and the interactions with specialist suppliers.

PLANNING FOR MAINTENANCE IN DESIGN AND BROWN FIELD DEVELOPMENTS

Considerable economies can be obtained if the requirements for in-service maintenance are considered when designing equipment for new installations or brown field developments.

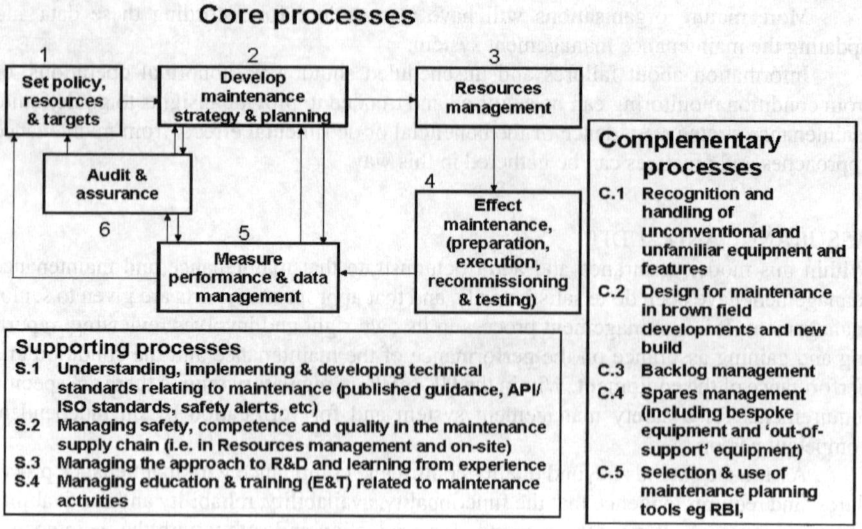
Core processes
1
Set policy, resources & targets
2
Develop maintenance strategy & planning
3
Resources management
Audit & assurance
6
4
Effect maintenance, (preparation, execution, recommissioning & testing)
5
Measure performance & data management
Complementary processes
C.1 Recognition and handling of unconventional and unfamiliar equipment and features
C.2 Design for maintenance in brown field developments and new build
C.3 Backlog management
C.4 Spares management (including bespoke equipment and 'out-of-support' equipment)
C.5 Selection & use of maintenance planning tools eg RBI,
Supporting processes
S.1 Understanding, implementing & developing technical standards relating to maintenance (published guidance, API/ ISO standards, safety alerts, etc)
S.2 Managing safety, competence and quality in the maintenance supply chain (i.e. in Resources management and on-site)
S.3 Managing the approach to R&D and learning from experience
S.4 Managing education & training (E&T) related to maintenance activities

Figure 2. Core Processes together with Complementary and Supporting Processes

The process involves determining the requirements and frequency of maintenance through life, and incorporating aspects of cost, risk and availability. Planning for maintenance also involves practical issues like access and isolation.

BACKLOG MANAGEMENT

Maintenance at most installations is based on information contained and co-ordinated within a computerised maintenance management system. The system will indicate when a maintenance task is due; when a task is not completed by the due date it is registered as a backlog item.

The significance and management of backlog items is a major preoccupation of operators and regulators. The measures that a company takes to reduce and eliminate backlogs require a specialist management process which is concerned with investigating whether the requirement for maintenance identified by the system is in fact appropriate, assessing the implications of not undertaking the maintenance, and managing the scheduling and resources required to tackle the backlog.

SPARES MANAGEMENT

The issues surrounding the holding of spare equipment, parts and consumables are particularly important for installations offshore and in remote places where transport and logistics are difficult.

Relationships with the supply chain are key to successful "just-in-time" ordering. Standardisation of equipment and parts and the sharing of depots between installations and assets can be beneficial, providing it does not reduce operational flexibility.

SELECTION AND USE OF MAINTENANCE PLANNING METHODS

Deciding on the optimum maintenance strategy for an item of equipment requires knowledge and application of the various maintenance planning methods that have been proposed (e.g. condition based maintenance, risk based maintenance and reliability centred maintenance).

For example, safety and production critical equipment like gas turbines may need a different approach to say a complex pipe work system or the painting of the structure.

SUPPORTING PROCESSES

Supporting Processes underwrite the Core Processes and affect the ability in which they can be carried out. The Supporting Processes take place over a longer period of time, and while failure to undertake them may not have an immediate impact, it will have a detrimental effect on the Core Processes in the longer term.

INTERACTION WITH TECHNICAL STANDARDS

The awareness and use of technical standards is a characteristic of organisations operating at maturity level three and above.

More mature organisations will understand the basis and limitations of applicability of the general industry standards, and will be developing their own internal standards based on experience of their particular equipment. They will also be active in supporting industry standards committees and gaining insight from a wider global experience base.

MANAGING SAFETY, COMPETENCE AND QUALITY IN THE SUPPLY CHAIN

Nowadays, it is common for operators and Duty Holders to sub-contract aspects of maintenance management and maintenance tasks to other organisations in their supply chain. In theory, the responsibility for undertaking these activities safely, competently and assuring quality should lie with the sub-contracting organisation, but in practice the Duty Holder cannot avoid sharing this responsibility for both legal and commercial reasons. The Duty Holder therefore has a duty to ensure the management of safety, competence and quality in the supply chain is fit for purpose.

The approach that Duty Holders take depends on the level of maturity. Lower maturity organisations will not consider these aspects at all or consider them through only general words in a commercial contract. Higher maturity organisations will insist on being involved with the supply chain. Depending on circumstances, this involvement may range from discussing the requirements for each job where the sub-contractor is weak to undertaking occasional audits of the sub-contractor where there is less cause for concern.

MANAGING THE APPROACH TO LEARNING FROM EXPERIENCE AND R&D

Learning from experience and wider longer term knowledge generation is at the heart of the maturity model. Within the field of maintenance management, more mature organisations will examine their processes and tasks and undertake work to make them safer, more reliable, and efficient. They will know that learning and process development do not always happen naturally, but need management drive and structure.

MANAGING EDUCATION AND TRAINING

Increasing the level of education and training of the maintenance workforce can have both short and long term benefits in terms of personnel having a better appreciation of the important aspects of and reasons for their work and increasing flexibility through multi-skilling.

Companies at maturity level three will ensure that their personnel have the education and training to meet the immediate needs of the job through a combination of on-the-job and formal training. Less mature companies will rely on close supervision on-the-job, where inexperienced personnel may be undertaking a task for the first time. More mature companies will anticipate the education and training needs to meet work requirements and would be interested in accreditation of their training and certification of their personnel to undertake specific maintenance tasks. They would seek to develop industry-wide standards for training and co-operate with their supply chain and other Duty Holders to provide the facilities required.

DIVISION OF RESPONSIBILITY

Traditionally, responsibility for all the processes of maintenance management lay largely within the organisation of the Duty Holder and operator. Nowadays, the oil and gas industry is more vertically segregated and responsibility for different Core Processes may be divided between different companies. Several scenarios can be identified as shown in Figures 3 to 6.

Scenario 1 (Figure 3) shows the traditional model where the Duty Holder was responsible for all the Core Processes and only called in specialist contractors for effecting certain maintenance tasks on-site.

In Scenario 2 (Figure 4), the Duty Holder retains responsibility for policy, planning/scheduling, but appoints a maintenance contractor to manage the resources required and implement the maintenance. The Duty Holder measures the performance of the maintenance and the (maintained) equipment and undertakes the audit function, but the maintenance contractor has no incentive other than to carry out the maintenance to the schedule and at minimum cost.

Scenario 3 (Figure 5) has the Duty Holder appointing a contractor to manage the asset against a policy and targets, but does not interfere other than to carry out audits against policy metrics. The asset management contactor takes the main responsibility for

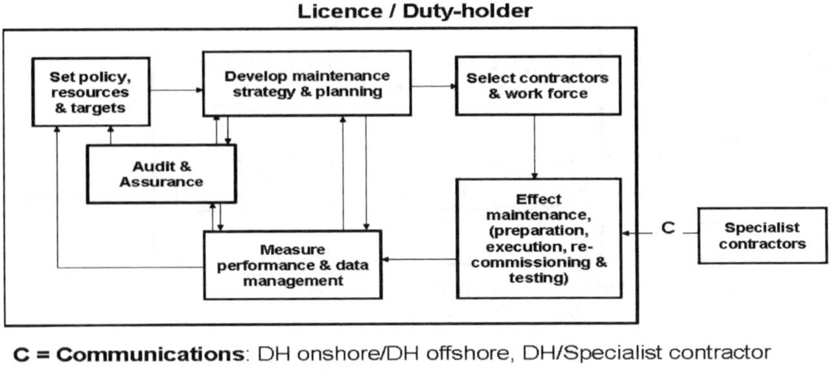

Figure 3. Scenario 1: Processes undertaken by licence/Duty Holder with specialist assistance

maintenance and can be motivated by incentives relating to the performance, availability, reliability and safety record of the equipment and the maintenance.

In Scenario 4 (Figure 6) the holder of the license (which may not be an oil company at all) has appointed an asset management contractor to undertake the role of Duty Holder and to operate and maintain the asset on its behalf. Policy and targets will be set in conjunction with the asset management contractor. The license holder's responsibility is one of due diligence to ensure that the organisation appointed has the necessary management, competence and track record to manage the assets effectively and meeting relevant

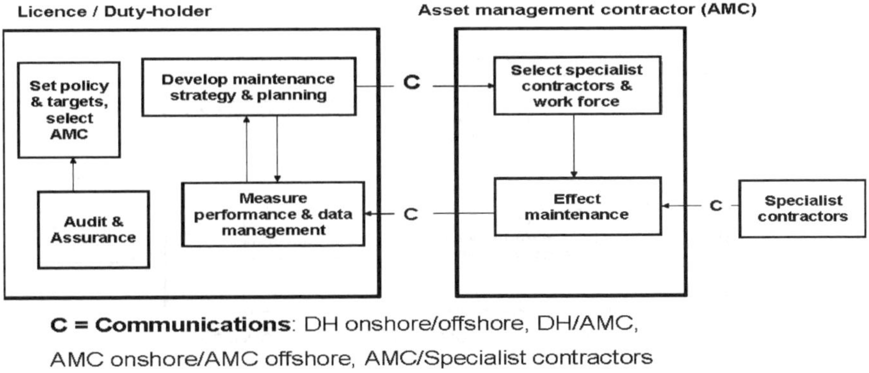

Figure 4. Scenario 2: Processes divided between licence/Duty Holder and asset management contractor

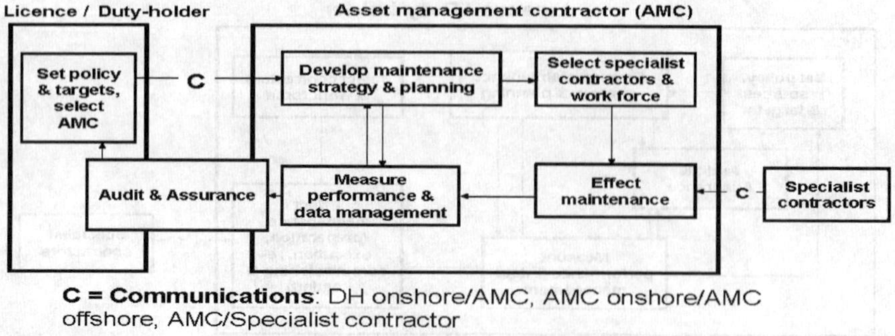

Figure 5. Scenario 3: Processes undertaken by asset management contractor except for policy and audit which are still undertaken by licence/Duty Holder

legislation. Responsibility for maintenance management is then the responsibility of the asset management contractor and the arrangement becomes similar to Scenario 1; dependent upon circumstances a separate maintenance contractor may be appointed as in Scenario 2.

It is important for personnel within companies to understand where responsibility lies. Whilst these divisions are contractually defined, training in the meaning of these agreements in practical terms is very beneficial. Long term partnering arrangements can generate understanding as experience of different organisations working together is gained.

Problems can sometimes arise when organisations within the supply chain are at different levels of maturity. Different expectations and management practices can cause

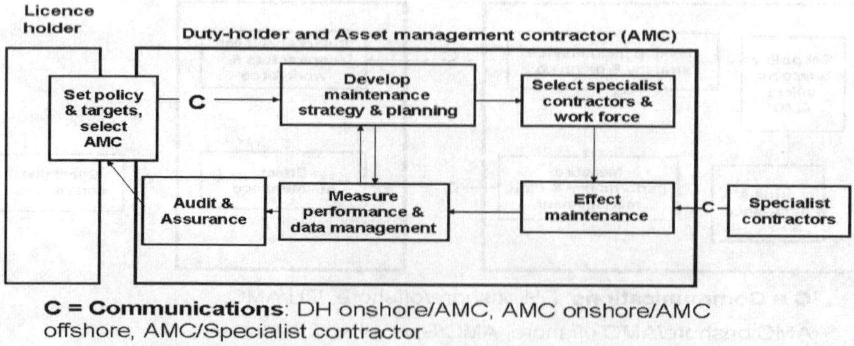

Figure 6. Scenario 4: Processes undertaken by asset management contractor except for policy which is shared with licence/Duty Holder

confusion and tension. This can occur both when the appointing company is at a higher and lower level of maturity than the supplier. In the former, the appointing company may feel a need to try to improve the maturity of the supplier through providing training as part of the supply agreement. It is more difficult for a more mature supplier to work with an uninformed Duty Holder, when the supplier must safeguard the integrity of its approach.

CAPABILITY MATURITY MODEL FOR MAINTENANCE MANAGEMENT

MODEL DEVELOPMENT AND TRIALS

The maintenance management maturity model was initially developed, based on descriptions for the five maturity levels for each of the six Core Processes. Following use of this model in a simple trial it was recognised that to enable more detail to be sought when a low maturity level was found sub-processes were needed and these were developed as indicated. Benefit was also taken from developing both Complementary Processes for particular situations where for example an installation had special new or unusual features, or where spares management had been a problem. It is not expected that all of the Complementary Processes would be included in every audit.

A more formal audit was undertaken with one of the sponsors, which involved interviews with four people with different responsibilities, ranging from the maintenance manager to the person responsible for maintenance data analysis. In general it was found that the six Core Processes worked well and were recognised by all those interviewed as representing the relevant management maintenance issues. It was also seen that inspection and integrity management are clearly relevant to several core maintenance processes and should be referred to within the model, although there is the difficulty of using the model when these two functions are separate, as they were in this case.

The trial also checked on some of the more complex Supporting Processes, but to a limited degree. Some limitations in the questions associated with maturity levels were recognised and there was a need to develop further sets of example questions. There was also a need to recognise the use and reliance placed on computer databases and management systems and to ensure that the model is consistent with their use. Overall these trials played an important part in developing and improving the model.

SPONSORS' WORKSHOP

A workshop for representatives of the sponsor's organisations was held in Aberdeen on April 20th 2007. The purpose of the exercise was to give the participants an insight into the concept and use of Capability Maturity Modelling for maintenance management by assessing the core maintenance processes. The exercise was based on an imaginary organisation which has recently taken over as operator of an early North Sea facility.

Teams of 4 or 5, who were part of an imaginary audit team, were each asked to interview a company representative (provided by members of the project team) and from the answers received identify the appropriate level of maturity for that process. After the exercise, feedback to the overall group was provided.

The exercise was well received and proved worthwhile in demonstrating the model. As a result several of those taking part felt that the model was worth trying within their own organisations.

GUIDANCE ON APPLICATION

The model is based on six Core Processes containing 17 sub-processes, together with 5 Complementary Processes and 4 Supporting Processes. These are illustrated in Figure 7. In addition there are maturity level descriptions which provide a single evaluation of maintenance management but are rather limited in the depth of analysis of the whole activity. Complexity increases with using more processes in the application of the model but this increases the value of the model in understanding the organisation's capability to perform.

The model can be used both for self assessment and for audit. In using the model it is necessary to identify the asset and identify the main organizations involved in the maintenance activities which could be the Duty Holder, asset maintenance contractor etc. From these organisations interviewees are selected, preferably at different levels in the organization. The interviews are carried out using the descriptions for each maturity level in the model.

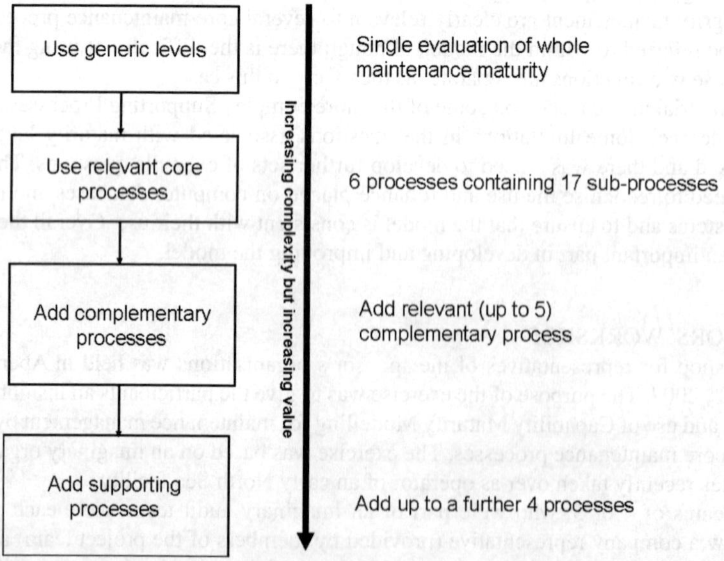

Figure 7. Outline of maturity model for asset maintenance management

The interviews would normally start with an assessment of the Core Processes. Sub-processes can be used when there is some level of concern or lack of understanding of the maturity level of a Core Process. Relevant Complementary Processes can also be included, depending on their relevance. There are also four Supporting Processes which in general are associated with longer term impact on the activity.

Information is collated for the different processes, and fed back to those interviewed, to establish the different maturity levels achieved. Depending on these results it may be necessary to recommend improvement steps to improve the maturity for some of the processes.

The project team have developed a number of audit check sheets to assist operators review their own approaches and procedures. The JIP report has a complete set of these audit sheets available to the JIP sponsors but a typical sheet is attached to this paper, see Table 1.

Table 1. Example audit sheet for Core Process 1

This table illustrates a check sheet which can be used for a maintenance management audit, this sheet is applicable to Cor e Process P1, other sheets have been developed for the Complementary Processes and Supporting Processes indicating maturity levels together with space for the auditor to record his observations.

Core Process P1	Set maintenance policy, resources & targets Provide the direction and wherewithal [Setting philosophy, policy & targets to be appropriate for condition of plant and field requirements and organization's needs, set resources {money, bed space, shut downs} to meet requirements, all based on capacity of, existing condition of and future requirements for existing plant]	
Maturity levels & description		Applicability and supporting evidence
1	Management does not have any maintenance policy or targets. It leaves decisions to local staff. It may impose unconsidered resource constraints or instructions. (E.g. "Spend as little as possible." "We have an outage from production – let's do some maintenance.")	
2	Management may have a maintenance policy and targets but these may be unwritten and do not necessarily address the current requirements of the plant. Resources for maintenance are determined following previous practice	
3	Policy and targets are documented and have been based on reservoir and facility condition but are not updated to reflect changing circumstances. Metrics exist for measuring performance against targets but could be out-of-date or inappropriate.	

(*Continued*)

Table 1. (Continued)

Maturity levels & description	Applicability and supporting evidence
Management is aware of the system design and the critical factors and equipment for production, safety etc Resourcing and organisation based on policy and targets but not necessarily updated to suit current requirements.	
4 Policy and targets are updated regularly to reflect the changing production and field characteristics including condition of plant and any life extension requirements. Resources allocated appropriately. Management is aware of system design and critical factors for production, safety etc, and develops its policy taking these into account. Training is introduced to support the development of policy and targets. Performance against metrics monitored and steps implemented to remedy any inadequacies. Maintenance management developed to meet integrity assurance and system reliability requirements.	
5 The approaches to determining maintenance policy, targets and resources are optimised making use of world-wide corporate and industry knowledge. Management aware of system design and critical factors for production, safety etc, and is taking steps to introduce redundancy and diversity to eliminate or mitigate the risks. Determine optimal maintenance requirements and resolve conflicts (e.g. breakdown vs. planned preventative maintenance vs. condition based). Optimise methods of delivering maintenance (in-house and/or outsourcing) through engaging efficient and incentivized management of the supply chain Business plan and forecasting embedded into best practice policy and targets including organisational changes/adaptations as required. Performance against metrics optimised by exchange of experience and utilisation of best available global practices. Develop and improve training for key staff and the supply chain to support policy and targets.	

CONCLUSIONS

GENERAL

The team have reported on applications of the maturity modelling approach before but this application has a particular emphasis that caused extra complexity. The previous applications have had more explicit elements of regulatory requirement embedded in them and this has helped to focus on a defined level of acceptability, previously this has tended to appear as a benchmark in the Core Processes at maturity level 3. There has been no obvious need for a clear benchmark to be defined in this project; this has allowed additional flexibility in the definition of the maturity levels.

The other key area of difference with this project has been the issue of the split of responsibilities. The maintenance sector in the oil and gas industry has evolved a large range of business arrangements, any of which work very well in the right context. However, when the team have been trying to define actions and responsibilities for each maturity level, the range of contract forms has influenced our thoughts considerably. It was therefore appropriate that the range of contract forms be discussed, and these form the basis of the 4 scenarios discussed earlier.

The other development used in this project was the workshop. The team have carried out trials in all of their maturity modelling applications but this dedicated workshop using role-playing is the largest we have yet applied to maturity modelling development. The workshop was deemed very successful and it seems likely that future applications will employ a similar technique.

RECOMMENDATIONS FOR SHARING EXPERIENCE

There are potential benefits for the continued capture of experience from companies' use of the maintenance model. An opportunity to collate and share feedback would help to refine the model, particularly in identifying further improvement steps, which would assist the transfer of best practice in maintenance management across the industry. There is currently limited experience of the practical application of the model and the question set, and there would be benefit in comparing experience from different types of application, (e.g. group versus individuals interviews), and setting down some guidelines for application.

DETAILS FOR ONSHORE APPLICATION

The development of maintenance model is generic and applicable to onshore and offshore process plant. The processes that have been defined are universally applicable, as are their attributes and the associated improvement steps. However, it should be noted that for an application to a specific onshore plant there are different emphases that should be understood and accounted for; in addition, due to historical developments in the downstream sector, there will often be different tools that may be adopted to implement the processes.

The elements requiring different emphases and which may affect some processes are;

- specific onshore legislative requirements
- onshore logistical arrangements, including;
  – the available transport options for both personnel and equipment
  – the accommodation arrangements for personnel
  – the increased opportunity for rapid response teams (Just in Time arrangements)
  – increased dependency on suppliers' warranty support
- options for increased flexibility of operations in larger onshore plants, with;
  – opportunities to maintain one train whilst continuing to operate others
  – increased redundancy within some processing trains, allowing running to continue with some plant down or bypassed with an acceptable short term impact on product quality

EXTENSION OF THE MODEL

The model had to be limited necessarily to accommodate the budget allocated in the JIP but the team foresee a range of other areas of maintenance management that should be considered at some future point. These extended areas should be able to follow the basics of the model developed but will necessarily need to account for additional drivers and yet another range of contract forms and agreements. Areas the team identified for further application are;

- Sub-sea/pipeline maintenance
- Maintenance of floaters/jack-ups
- Decommissioning, failure investigation
- Shut-down/production planning, management of upgrades

ACKNOWLEDGEMENTS

The authors of this paper would like to acknowledge the contributions to the work made by all the sponsoring organisations. The team would particularly like to thank the representatives of the sponsoring companies, that is Peter Elliot and Martin Hinchliffe from BP, Graham Walker from Talisman Energy, Quentin Davidson and Carl Everett from Shell, Neil Pickering, Andrew McGeachy and Bill Mclaren from Petrofac, and Rob Miles from the Health and Safety Executive, for the productive discussions at project and individual meetings and written material provided, and also to those at Talisman who took part in the trial of the model. Finally, the authors would like to thank The Energy Institute for co-ordinating the project and for permission to publish this summary paper of the project.

A PROGRESSIVE RISK ASSESSMENT PROCESS FOR A TYPICAL CHEMICAL COMPANY: HOW TO AVOID THE RUSH TO QRA

R. T. Gowland
European Process Safety Centre, Rugby

Safety Management Systems are a requirement for compliance with the Seveso 2 Directive. A very important part of any Safety Management System for a chemical company is the Process Safety Management System. Most companies have constructed their own approach to the task and in some cases these pre-dated any regulatory requirements. A fundamental part of the process Safety Management System is the risk management process and this in turn requires a risk assessment to determine the scale and likelihood of the operational risks of an establishment. Once this has been accomplished, decisions need to be made about acceptance, reduction or cessation on the basis of some risk tolerance criteria set by the operator or by the regulator. Figure 1 illustrates the risk management 'process'.

The assessment part of the process can range from the simple to the complex. As time goes by and technology advances, we seem to be moving in the direction of the complex. For small and medium enterprises and even for larger companies, this increasing complexity may not always bring a benefit in managing risk. Drawbacks include:

- A lack of full understanding of methods
- Reliance on experts who may be remote from the operation
- Undue reliance on third parties to evaluate risk
- Key operations staff being vaguely aware of the true risks of the production process and potential deviations
- A lack of appreciation of the impact of changes to the manufacturing process on the level of risk
- Illusory 'precision' in methods which use elegant models and mathematical expressions.

There is no doubt that the complex models are needed and have brought real benefits, but their application needs to be selective and consistent. In the past three years, several risk management systems have been observed where:

- Risk Assessments (including QRAs) have been done and no one was quite sure why they were done, what the results really meant and what was done with them. In extreme cases it appeared that little had been done to address risks which could have been reduced easily and economically. Furthermore, obvious risks had been overlooked.
- Several studies at European and National levels have shown that the consistency issue remains unresolved.

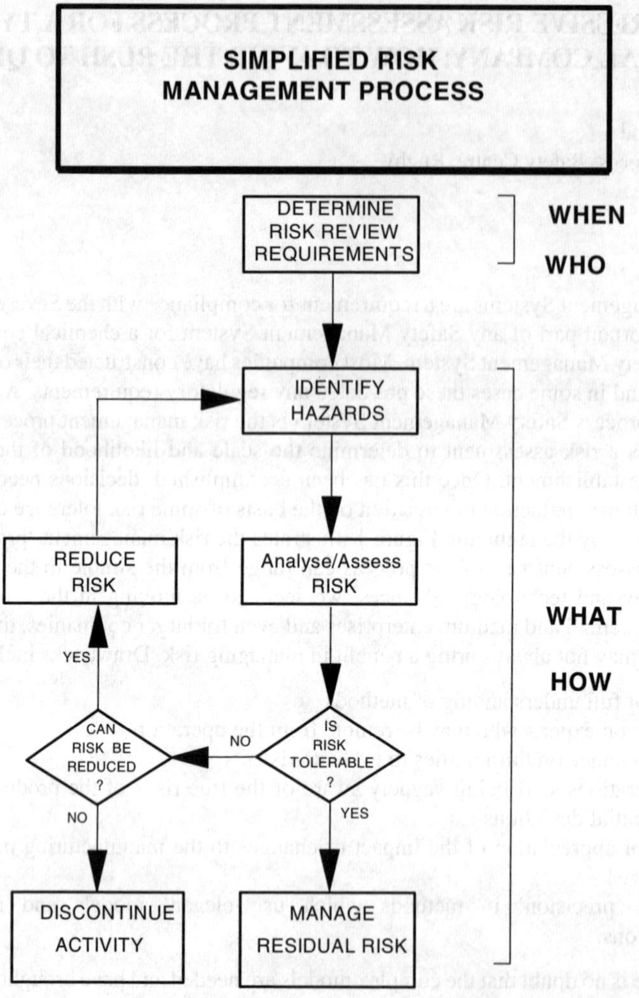

Figure 1.

We do not need to be reminded that it would be bad practice reject a method simply because we did not understand it. It would also be irresponsible to seek an assessment system which gives us the least provocative answer, or to manipulate the inputs or assumptions used in a given method to achieve the same thing. Whilst examples of this are rare, they do exist and are not always inspired by the establishment operator.

The following demonstrates a progressive approach which allows the use of simple tools to approximate the risk so that comparison with 'tolerance criteria' or 'risk review elevation criteria' can be made. Those cases which exceed these criteria are then taken to a deeper, more complex level of study. The example approach gives 4 levels of study, starting simple and progressing to complex. Each of these levels requires the use of tools, many of which are well known industry generic best practice and therefore allow some comparison establishment to establishment. Furthermore, some specific tools have appeared as examples in regulatory guidance from some member states in the European Union and other countries such as Thailand and Australia.

How it can be done:

Starting from the design phase of a plant it is normal to be able to identify the process materials and their physical and hazard properties

- process steps
- process conditions (temperature, phase, pressure etc.)
- basic reactive chemical issues (Thermal runaway potential etc.)

In the case of existing facilities, all these are well known and documented.

With such basic information it should be possible to make an estimate, at least in relative terms and in some cases in quantitative terms the consequences of major process deviations. In the case of Fire and Toxic releases, there are well established methods such as the Dow Fire and Explosion Index and the Dow Chemical Exposure Index.

For the case of vapour cloud explosion, there are simplified methods which can apply the TNO Multi Energy approach in a conservative way. These can all be used to quickly estimate the hazard consequences of deviations. A further resource is provided for worst case toxic release scenarios by the Environmental Protection Agency in the United States which supports the RMP plan.

It may help to have a standardised group of scenarios for the estimates being made. The ARAMIS project of the European Commission devotes some time to this topic. The referenced Dow methods include suggested scenarios and the explosion case is well covered in papers from Martin Goose of the United Kingdom Health and Safety Executive.

The whole risk management 'process' can be described graphically in the following diagram.

At each stage the work is done and the outcome compared with the elevation criteria to decide if the next level risk review needs to be done. The effect is that only the risks which have a high severity or about which there is most doubt reach the stage of Quantitative Risk Assessment. Furthermore, the scope of work for QRA will probably be narrowed because of all the knowledge gained from the lower level reviews.

The tools used range from the simple to the complex:

Stage 1 (simple)

- Formal Review to demonstrate operator knowledge of the production process
- Questionnaires designed to extract all process deviations and hazardous scenarios

LEVEL 1: PROCESS HAZARDS ANALYSIS
- Triggers : All plants, significant projects and changes
- Fire & Explosion Index (FEI)
- Chemical Exposure Index (CEI)
- Credible case scenarios and lines of defence (with frequency or LOPA target factors).
- Worst case scenarios and relationship to Emergency Plan (EPA RMPtool)
- Explosion Impact (Building Overpressure) evaluation*
- PHA Questionnaire

LEVEL 2: RISK REVIEW
- Triggers: F&EI >=110 or CEI = ERPG2 at fence line , LOPA Target Factor to be defined (check output from Level 1) e.g. fatality at freq > KNR governance criteria
- Cause-Consequence pair Identification* e.g. 'bow tie'
- HAZOP
- LOPA and Triggers: LOPA Target >= 5 or LOPA inappropriate.
  - Structured Hazard Analysis (Fault Tree analysis*, FMEA, Checklist, etc.)

LEVEL 3: ENHANCED RISK REVIEW
- Triggers: LOPA Protection Gap > 0 i.e. we are not meeting governance criteria
- More accurate Dose considerations e.g. AEGLs or AETLs
- Screen for QRA*

LEVEL 4: QUANTITATIVE RISK ASSESSMENT
- Triggers: Individual Risk contours in off-site population exceeds Business Governance Elevation Criteria
- Combination of Consequence Analysis, Frequency of Impact
- Focuses on highest risk activities

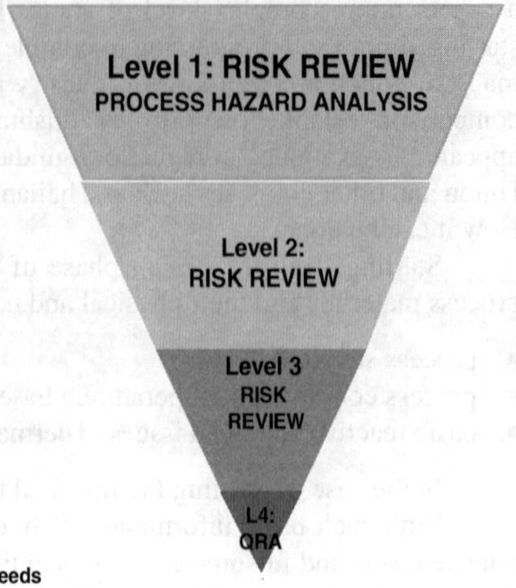

Figure 2. Summary diagram of a progressive system

- Semi Quantified Screening tools for Consequences of Fire, Explosion and Toxic substance releases. Examples are shown in Appendix 1, 2 and 3.
- Comparison of results with screening criteria set by the company

All these activities are carried out by **plant based production technical personnel** who have been trained by Process Safety expertise. The results are communicated to Process Safety Expertise for validation.

Stage 2 (intermediate)

- Detailed Hazard Identification (HAZOP)
- Layer of Protection Analysis
- Comparison of results with screening criteria set by the company

These activities are carried out by plant based production personnel under the facilitation of Process Safety expertise for the technology being used.

Stage 3 (Technology based Process Safety expertise)

Confirmation of severity and frequency of potential consequence by use of more complex modelling including

- Dispersion models e.g. DNV PHAST
- Potential exposure e.g. AEGLs, AETLs, for each predicted event
- Screening and scoping for QRA

These activities are carried out by Process Safety expertise for the technology being used with assistance of Technology Leaders and plant based production personnel.

Stage 4 (Corporate Process Safety Expertise/3rd party provider)

- Quantitative Risk Assessment

THE SIMPLE TOOLS EXPLAINED:

1) Toxic release from standardised credible case scenarios

First taking the case for toxic release, the Chemical Exposure Index can quickly provide the distance which is travelled by a specified toxic release. Criteria are based on Emergency Response Planning Guidelines. The estimate of 'hazard distances' from each case are made available. It seems logical that if the toxic release travels a distance which is less than the distance to the company's fence line it should be possible to protect site based persons from adverse effects. Various methods play a role from detection and response to Personal Protective Equipment and 'shelter in place' etc. These measures are not practical for application outside the site. In the scheme presented, it is suggested that any scenario which develops a toxic concentration of ERPG2 or greater outside the fence line should be subjected to deeper study.

The same approach can be made using the EPA dispersion model for catastrophic worst case scenarios. If these demonstrate a potential for ERPG3 concentrations at the fence line, it could be assumed that extra studies would be needed. These would be used to refine the on and off site emergency plans and any possible scale reduction or mitigation measures. The principle of Inherently Safer Design might be intensified or in extreme cases, the process changed. An example is shown in APPENDIX 1. The software is demonstrated in the presentation of the paper.

2) The worst case toxic release scenarios are studied via the EPA RMP approach

This assumes the catastrophic failure scenario with all inventory discharged in 10 minutes. This is available from the EPA website http://yosemite.epa.gov

3) For Fire Cases, the Dow Fire and Explosion Index can be used to assess fire effects

In this case the criteria for deeper study would be based on the criteria used by some companies and some legal authorities. For example – Fire and Explosion Index exceeds 110. An example is shown in APPENDIX 2: where the results are reproduced from a simple Excel spreadsheet with simple user inputs.

4) Occupied Buildings close to plants where flammable materials present an explosion risk

This is studied with the use of TNO Multi Energy Methodology. An example is illustrated in APPENDIX 3 where the results are reproduced from a simple Excel spreadsheet with simple user inputs.

APPLICATION

The Dow Chemical Company has based its approach on using these kind of criteria, its trigger criteria are developed, but the system may be used with user defined numbers.

CONCLUSION

The work of process risk assessment is aided by many tools. It seems reasonable that the greatest hazards should receive the deepest studies, possibly involving greater complexity and expense. The combining of simple publicly available indexing tools with screening or trigger criteria offers a way of adding efficiency to the activity. Most importantly, it ensures that the people closest to the risk and whose responsibility in managing it daily, understand their process and its hazards and the role they must carry out in preventing or responding to unsafe deviations.

APPENDIX 1 EXAMPLES:

Chemical Exposure Index Calculation	Make Entries in yellow cells only
Plant:	Chlorination
Unit of Process Plant:	store V101
Calculation by:	Gowland
Material in case studied	Chlorine in pressurised store
Date: January 4 2006	

Piping release or vessel nozzle release

Level 2 Risk Analysis is triggered	*(Liquid Release)*
Level 2 Risk Analysis is triggered	*(Gas Release)*

Distance travel by ERPG2 concentration Metres (Liquid Release)	16932
Distance travel by ERPG2 concentration Metres (Gas Release)	3571

The software also allows the cases of hoses, overflows and relief systems to be studied.

APPENDIX 2: FIRE AND EXPLOSION INDEX

	Solids pesticides warehouse	
	Fire and Explosion Index	41.78
	Material Factor (see Material Data)	14.00
	NFPA Health rating (Nh)	1
	NFPA Flammability rating (Nf)	2
	NFPA Instability rating (Ni)	1
	General Process Hazards	
	Base	1
1A	Exothermic Reaction (range of input 0.3–1.25)	0.00
1B	Endothermic Reaction (input range 0.2–0.4)	0.00
1C	Material Handling and Transfer (input range 0.25–0.8)	0.00
1D	Enclosed or Indoor Process or storage Units handling Flammable materials	0.00
1E	Ease of Access for Emergency Responders	0.35
1F	Drainage and Spill Control	0.50
	General Process Hazards Factor	1.85
	Base	1
2A	Toxicity of the material handled.	0.20
2B	Process or Storage operates at vacuum (<500 mmHg) – penalty 0.5	0.00
2C	Operation in or near the flammable range (input range 0.0–0.8)	0.00
2D	Dust Explosion (input range 0.0–2.0)	0.00
2E	Pressure Penalty	0.00
2F	Low Temperature Operation	0.00
2G1	Combustible and Flammable materials in Process	0.00
2G2	Liquids or gases in Storage	0.00
2G3	Solids in Storage or Process	0.41
2H	Corrosion and Erosion (input range 0.0–0.75)	0.00
2I	Leakage, Joints, packing, flexible joints	0.00
2J	Use of Fired Equipment (Fig 6)	0.00
2K	Hot Oil Heat Exchange Equipment (Table 5)	0.00
2L	Rotating Equipment	0.00
	Special Process Hazards Factor	1.61
	Fire and Explosion Index	41.78
	Level 2 Risk Analysis is not triggered	

Software calculation demonstration is included in the oral presentation.

APPENDIX 3: OCCUPIED BUILDING VULNERABILITY

Vapour Cloud Explosion Screening for: *Distillation Unit*

Date: 2/20/07

By Angel

Dimensions of the confined or semi-confined Zone which can be filled by a leak of flammable vapour above its flash point.

	Width (M)	Depth (M)	Height (M)	Volume	Standoff distance between source and potentially affected building (M)	Calculated Fuel mass (Kg)		
INPUTS in this row	15	15	5	1125	20	91.575		
TNO M.E.	Curve 1	Curve 2	Curve 3	Curve 4	Curve 5 Y	Curve 6 Y	Curve 7	Curve 8
Approximate Predicted Side on Overpressure at building (mBar)	10.00	20.00	52.10	109.25	201.70	612.93	1000.00	1641.29
Standard Brick or block construction without independent roof support	O.K.	O.K.	O.K.	Not Suitable	Not Suitable	Not Suitable	Not Suitable	Not Suitable
Conventional Windows allowed?	YES	NO	NO	NO	NO	NO	NO	NO

In this case, the screening criteria suggested is a peak side on overpressure of 70 mBar. If this pressure is exceeded, the building is vulnerable to major damage and consequent serious injury to occupants. When these criteria are met, the procedure outlined by HSE (Martin Goose) is very helpful. It proposes a system of estimating cumulative frequency of these events and assessing the result against ALARP (As Low As Reasonably Practicable) criteria frequency (e.g. frequency of serious injury meets tolerability criteria)

Software calculation demonstration is included in the oral presentation.

REFERENCES

The Dow Chemical Company Fire and Explosion Index Hazard Classification Guide Ed 7 1994

The Dow Chemical Company Chemical Exposure Index Guide Ed 2 1993

Location and Design of Occupied Buildings at Chemical Plants – Assessment Step by Step. Martin Goose – U.K. Health and Safety Executive

MANAGING BUSINESS RISKS FROM MAJOR CHEMICAL PROCESS ACCIDENTS

Mariana Bardy[1], Dr Luiz Fernando Oliveira[2], and Dr Nic Cavanagh[3]
[1]Head of Section, Risk Management Solutions Salvador, DNV Energy Solutions South America
[2]Head of Department, DNV Energy Solutions South America
[3]Head of Safeti Product Management, DNV Software, London, UK

Traditional risk management has focused on effects on people, particularly when using Quantitative Risk Assessment (QRA). Nowadays a much broader approach is being adopted as companies begin to consider potential accidental losses in their decision making process. This paper describes a study where business risks from major accident hazards were assessed for a large chemical process company.

Results are presented in the form of F-Cost curves, similar to traditional F-N curves, representing cumulative frequencies of events versus their cost. These are evaluated for various types of loss including property damage, business interruption, inventory loss, environmental loss, clean-up costs, legal costs, fatality and injury costs, amongst others. Adjusting insurance values according to risk is one of the benefits of such an analysis. Evaluating these risk criteria helps organisations to negotiate with their insurers to agree maximum insured loss values that will fit their strategy and satisfy their risk acceptance criteria. Also, values for civil liability insurance can be reviewed, based on the hazards that can result in fatalities and injuries, as well as property losses, for third parties.

This paper presents the results of this study which involved the evaluation of the company's maximum insured losses. The key driver was a board level initiative to implement a more objective approach to quantifying this metric which serves as the main basis for setting the level of insurance premiums. The results of the analysis are presented as F-Cost curves and we describe the risk-based decision making process and criteria adopted by the company to set the maximum loss value to be insured for their plant.

In conclusion, this paper presents results from a real life example of the application of the quantitative evaluation of business risks, illustrating that a financial or business risk approach can help management in day-to-day decisions when considering possible effects of major chemical process accidents.

INTRODUCTION AND BACKGROUND

Over the last 30 years, the management of risks associated with the operation of major accident hazard facilities has been high on the corporate agenda. This has been driven by a number of major accidents including Flixborough (1974), Bhopal (1984), Piper-Alpha (1988) and, more recently, Enschede (2000), Toulouse (2001), Fluxys (2004) and Texas City (2005). All these, and many more, have resulted in significant fatalities and injuries. The most recent three examples have resulted in a total of more than 60 fatalities and 3000 injuries. This has further driven legislation such as the Seveso directives in Europe and the

EPA Risk Management Plan regulations in the US (Fewtrell and Hirst, 1998). Legislation has generally been focused on reducing the risk of fatalities and injuries, and rightly so.

However, in this same period there have been many high profile accidents which have resulted in few, or even zero, fatalities and injuries, but enormous cost to business, both of the operator and more broadly in the vicinity of the plant concerned. Companies have suffered significant financial losses and entire countries have seen major disruption from single incidents involving relatively small direct asset loss and sometimes no fatalities.

The release of dioxin at Seveso, Italy, on 9th July 1976 resulted in no direct fatalities. However, this incident required the evacuation and decontamination of a wide area north of Milan. Although no immediate fatalities were reported, kilogram quantities of the substance, which can be lethal even in microgram doses, were widely dispersed. This resulted in an immediate contamination of some ten square miles of land and vegetation. More than 600 people had to be evacuated from their homes and as many as 2000 were treated for dioxin poisoning. This was a key driver in changing the regulation of major hazard facilities across Europe through the so-called Seveso directives, subsequently brought to the statute books in all member states of the European Community. From 3rd February 1999, the obligations under the Seveso II Directive have been mandatory for industry as well as the public authorities of the Member States responsible for the implementation and enforcement of the Directive. So, although causing no immediate fatalities, the Seveso incident had an enormous cost both to the operator and the environment as a whole.

In 1998 at Esso's Longford liquified petroleum gas processing plant in Australia there was a massive explosion, killing two workers and injuring eight. Although in comparison with some of the events described above, the number of fatalities and injuries were relatively small, gas supplies to the state of Victoria were severely affected for several months after the incident. Most of the state's gas supply was cut for almost two weeks with severe disruption for a further 2 months and a total estimated cost to industry of $1.3 billion.

The Exxon Valdez oil spill on 24th March 1989 resulted in no fatalities but in addition to the direct costs to Exxon, fines of around $150 million dollars were imposed along with a $900 million civil settlement. The lingering oil spill also had affects on the environment which are difficult to put a value on even now.

In recent years focus has moved from improved safety and compliance with legislation to a regime where companies need to look at improvements which can be shown to deliver benefits directly to their bottom line. In general, good safety means good business, as exemplified above, and the business risk concept is a way of demonstrating this to senior management. The techniques for QRA and Consequence analysis can be extended to assess the financial consequences of accidents and associated financial risk exposure. In today's competitive business environment key drivers are improved financial performance, maximised up-time, reduced insurance costs or reduced risk of interruption to business resulting from an accident.

The typical questions to be answered are:

- If I have an incident, what will it cost?
- What is the maximum loss I can incur as the result of an accident?

- How can I minimise the likelihood of an incident resulting in loss of production?
- What risks am I exposed to from a financial standpoint?
- How can I perform Cost Benefit Analysis on my operational business risks?

With greater global competition and much more challenging margins, there is now less cash available for activities that do not contribute directly to the bottom line, or are perceived as such. The "Q" of QRA need not only be fatalities, but effects on the environment, downtime or dollars. By quantifying impacts on people, operations or assets, analysts are better able to estimate the likely costs of an incident or incidents in terms of down-time, asset damage, personal injury and loss of life, brand damage, environmental clean-up, litigation and compensation, and so on.

Building on the methodologies and models used over many years for QRA and Hazard Analysis and built into many standard software tools (Worthington and Witlox, 2002 and Cavanagh, 2001), we have extended the classical QRA methodology to calculate financial risks and consequences in addition to the more traditional fatality risks and effects on life. This paper goes on to describe the methodology we have used in extending the classical QRA methodology to take account of other risk measures such as business interruption, environmental impact, loss of production and so on.

THE FINANCIAL RISK CONCEPT

QRA techniques have been used over many years to assist in the management of the safe operation of process plants. The focus has usually been on compliance with safety legislation and the approach taken dependant on whether a quantitative risk based approach is legislated (such as the Purple Book in the Netherlands) or a consequence based approach is used (such as RMP in the USA). Either way, both can be extended to assess the financial risks associated with a plant or the cost of a single event occurring (financial consequence).

The classical QRA methodology provides a technique for quantifying the risks associated with the activities involved in the production and processing of chemicals and petrochemicals. In order to quantify risks it is necessary to first identify all possible risk situations, quantify them in terms of event consequence and likelihood and compare them with acceptance criteria. The main questions to be answered by a QRA are what can go wrong, what are the potential effects if it does go wrong, how often is it likely to go wrong and is it important if it does.

Typical outputs of a QRA study are location specific risk contours, estimates of individual risk and the F-N curve for representation of societal risk. Individual risk can be defined as "the frequency at which an individual may be expected to sustain a level of harm from the realisation of specified hazards" and is usually taken to be the risk of death expressed as a risk per year. Societal Risk is defined as "the relationship between the frequency and the number of people suffering a given level of harm from the realisation of specified hazards". It is normally taken to refer to the risk of death expressed as a risk per year and displayed as F-N curves. These describe the cumulative frequency (F) of all event outcomes leading to N or more fatalities and this representation of societal risk highlights the potential for accidents involving a large number of fatalities.

Risk to life is only one of the risks inherent in the operation of a process plant which may be realised by the occurrence of an accident. Others include risk to the environment, risk to assets and equipment and risk to financial performance. Furthermore, all these "risks" can have a cost associated with them which can be calculated and integrated in the same way as fatality risk provided appropriate cost parameters are available for each cost category. The workflow and data required to extend the concepts of QRA to Financial Risk Analysis (FRA) are summarised in Figure 1.

Typical contributors to overall financial losses resulting from an accident may include:

- Impact on people in terms of fatalities and injuries
- Property damage including capital costs to repair or replace damaged equipment and damage to other assets
- Business interruption including lost production from original failures
- Cost of lost inventory, again from sources and other damaged equipment
- Environmental damage, including clean-up costs, fines, impact on animal and plant life
- Plus many other outcomes with financial impact including legal costs, fines, loss of reputation, brand damage, compensation, reduction of share prices and so on

Typical output from a financial risk analysis, looking purely at consequences of a single accident may be total cost of a single failure case, total cost per outcome or per cost category, or both, and cost ranking per scenario. This is of use in assessing areas of a plant where a single accident may result in unacceptable high risk of loss. Extending this to risk, measures such as Estimated Annual Average Loss and Estimated Maximum Loss may be calculated. Also, F-Cost curves (analogous with the F-N curves for societal risk from a traditional QRA) can be generated, along with other metrics (Cavanagh and Linn, 2006).

The Estimated Annual Average Loss (EAAL) helps to identify the factors contributing to the highest risks in financial terms. The rate of financial loss for a given event outcome is the product of the financial loss for that outcome and its frequency, and the total rate of financial loss is the sum over all outcomes. Each event in an FRA model has a likelihood (typically occurences per year) and consequence (typically in USD, GBP, Euro or other currency). The product of these is the EAAL for the event in USD/year, for example. By aggregating the EAAL for each event, the EAAL for the entire facility can be quantified. So the EAAL for a plant represents the average annual loss rate expected across the complete lifetime of the facility.

With respect to Estimated Maximum Loss (EML), this provides an estimate of the maximum loss that could be sustained at a facility due to a major accident. By assessing the financial consequences of all possible outcomes for all possible release scenarios and integrating these over all possible release directions, an accurate and defendable estimate of the maximum loss which could be sustained at the facility can be derived. This is also the position where the F-Cost curve crosses the cost axis on the F-Cost graph so from Figure 3, for example, we can see that the EML is close to $1billion.

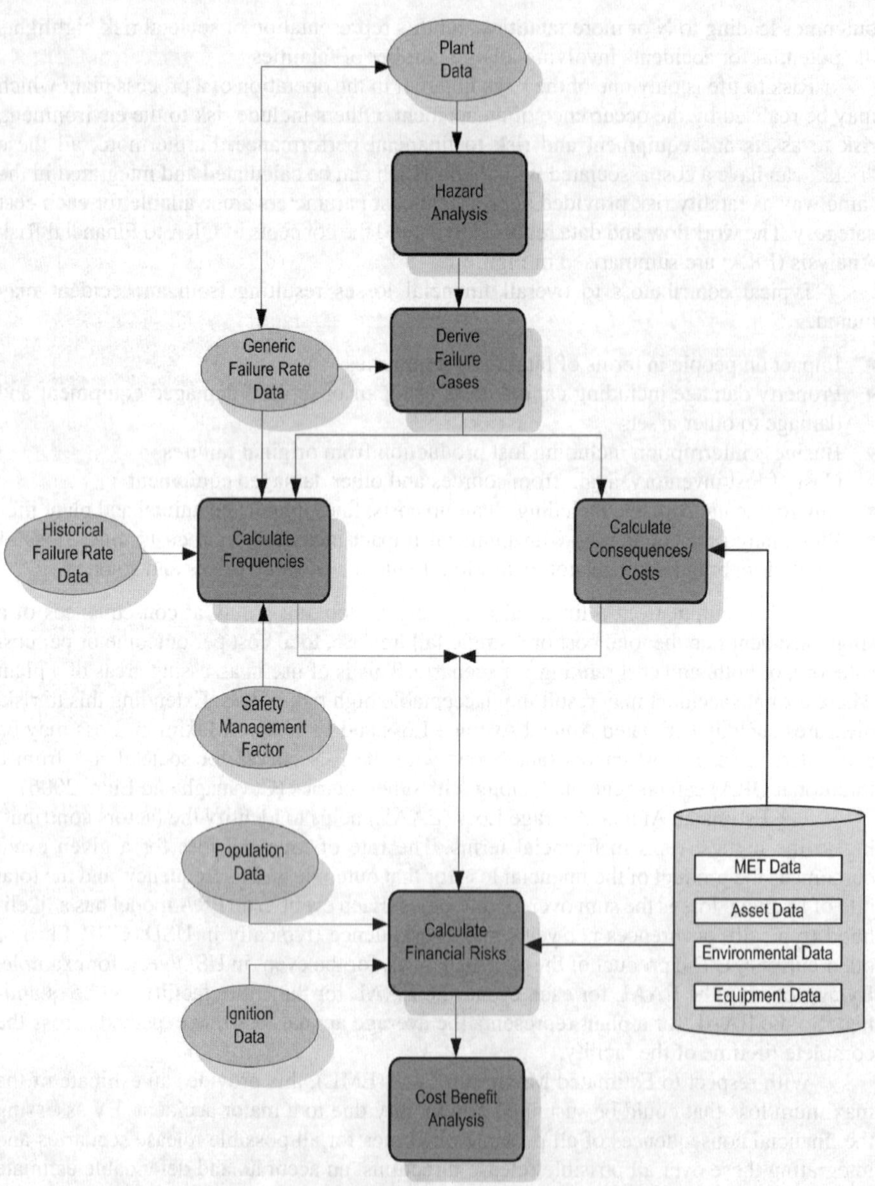

Figure 1. Classical risk analysis methodology extended to include financial risk calculations

Other approaches have defined EML as a "worst-case" or financial consequence measure rather than an absolute maximum possible loss with a given frequency. Marsh's SLAM model, for example, defines EML as "The largest loss that could result from a single incident in the plant" and uses the CAM explosion model for assessment of overpressure whilst SwissRe's ExTool focusses on damage to property only using the TNT model, determining-financial loss due to an individual explosion and extrapolating this to estimate Maximum Possible Loss (MPL), again without consideration of its likelihood of occurrence.

Chippindall and Butts (2004) adopted a similar approach to the current authors, using Phast as a consequence engine and performing the cost calculations and risk summations through a number of spreadsheet models. The advantage of the Safeti Financial model used in this study is that asset, equipment, source, population and ignition information is entered onto a map directly through the existing Safeti GIS and grouping and combination functionality enabling multiple combinations to be analysed easily.

For the model under consideration here, the key elements are:-

- A geographical model of the facility and surroundings including population, ignition and asset and equipment data sets
- A complete set of major accident hazard scenario failure cases for the facility
- A set of representative weather conditions and their directional probabilities (wind rose)
- Estimation of the likelihood of each failure case
- Modelling of the range of potential consequences for each failure case
- Assessment of the impact of each failure case on the plant, surrounding assets and population
- Calculation and assessment of the financial risks associated with these impacts reported in terms of F-Cost curves, total loss rates, Estimated Annual Average Losses (EAAL) and Estimated Maximum Losses (EML). (See for example Evans and Thakorlal, 2004)

Typical uses of this kind of financial risk analysis include

- Aiding the decision making process with risk reduction recommendations supported by cost benefit analysis techniques
- Reducing exposure to financial risk by assessing the relative benefits of different risk mitigation strategies
- Comparison of financial risk exposure for a range of process conditions
- Financial risk trends with time
- Direct assessment of financial risks from major process plant hazards
- Demonstrating a strong culture of corporate social responsibility (CSR) and adherence to the principles of triple bottom line (TBL) reporting which explicitly considers an organisation's economic, environmental and social performance (Elkington, 1994)
- Better understanding of appropriate levels of insurance in terms of both maximum insured losses and deductible levels

The next section goes on to describe in more detail the methodology we have adopted and its implementation in the Safeti Financial model.

METHODOLOGY AND MODEL

FINANCIAL RISK CALCULATIONS

The total financial risk is the summation of the total risk due to impacts on people (fatalities and injuries), impacts on equipment and other assets in terms of damage and replacement cost, cost of business interruption, cost of environmental impact and the cost of other outcomes such as legal costs, fines, brand damage, etc.

The methodology used in this model considers financial risk in terms of the following asset types:

- Population;
- Original Source Equipment;
- Other Specific Equipment;
- Other Assets (buildings, non specific plant, infrastructure, etc.);
- User Defined costs.

Cost contributors within each of these asset groups may include equipment damage, lost inventory, business interruption and environmental cleanup.

For population zone analysis, the cost of each release is calculated summing up the cost of a fatality and the cost due to injuries.

If the analysis involves specific equipment that constitutes a source of a release of hazardous material or that can be affected by the accident, the following different costs could be considered:

- Equipment repair: the cost of repairing the damage or replacing the equipment;
- Equipment business interruption cost: the financial loss incurred when production is halted because of the damage;
- Equipment lost inventory cost: the value of any inventory that had to be scrapped as a result of the damage.
- Equipment environmental: the cost of cleaning up any environmental effects associated with the damage;
- Or any other costs associated with the damage.

The same cost categories indicated for equipment can be applied to any asset zone (such as an industrial area, process area, adminstrative building, residencial area) that can be affected by a hazardous event.

The total cost related to any accidental release is calculated considering all the above individual costs and any aditional cost that can be defined by the analyst, as environmental fees and legal costs.

For cases with relatively low costs, total cost, $\mathbf{COST_T}$, is assumed to be given by a simple summation. For cases with larger costs, the value will be given by the power relation:

$$\mathbf{COST_{Total}} = \max\left(\mathbf{COST_T}, 0.5\ \mathbf{COST_T}^{1.05}\right)$$

This is already commonly used in Risk Based Inspection (RBI) calculations (Topalis and Cavanagh, 2001, Risk Based Inspection, 2000). The power relation accounts in a general way for the additional costs such as litigation, fines, etc., which increase with the size of the incident. This is commonly referred to as escalation.

DAMAGE LEVEL AND VULNERABILITY FACTOR CONCEPT

The simplest method for assessing the impact of each consequence on a receptor is to define a threshold value for each, above which the receptor suffers 100% damage and below which it suffers no damage. This concept will be familiar to Safeti users, where typically 2 threshold values are used for each outcome type. For explosions, jet-fires and pool fires an upper and lower damage level are defined, with a vulnerability factor between zero and 1 associated with each level. For flash fires a fraction of LFL is defined below which no damage occurs and above which maximum damage occurs, as defined by the appropriate vulnerability factor.

For toxic releases, a concentration level is provided below which no damage occurs and above which maximum damage occurs, again based on the relevant vulnerability factor. Toxic damage is intended primarily for use with assets where damage will result in some kind of pollution which has an environmental clean up cost associated with it.

A similar approach is taken for the effects on population and this is already well documented within the Safeti model (Worthington and Witlox, 2002). From the effect zone size and location determined using the consequence models available in Phast, the damage levels described above, the associated vulnerability factor and the overlap between assets and effect zones, the level of damage can be calculated, and converted to a total cost for each cost category.

In Figure 2, the bold inner and outer footprints represent typical upper and lower threshold damage levels respectively for a single release scenario and weather state. Also shown are typical population and asset zones and a single equipment item. Any receptor

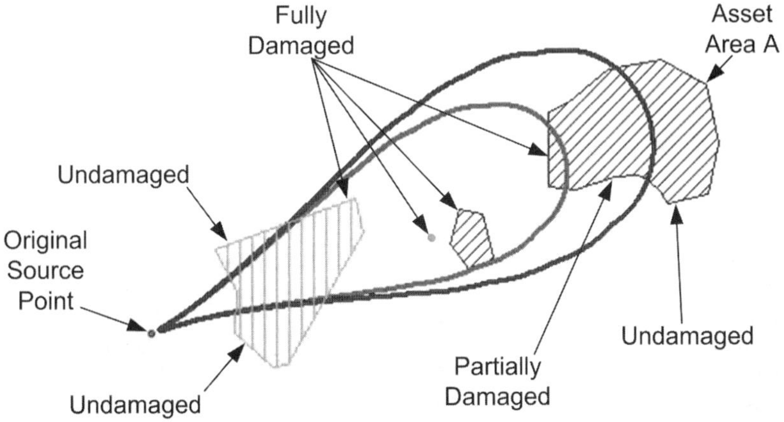

Figure 2. Definition sketch for asset damage due to upper and lower damage levels

outside the lower threshold value indicated by the bold outer line will be undamaged. Any receptor between the inner and outer threshold boundaries will be damaged to the degree indicated by the appropriate vulnerability factor for the lower threshold. Any receptor within the inner threshold boundary will be damaged to the degree indicated by the appropriate vulnerability factor shown for the upper threshold.

So, for example, in Figure 2, if upper and lower vulnerability factors are set to 1.0 and 0.5 respectively, then the section of asset area A outside the outer footprint will be unaffected, 50% of asset area A between the inner and outer footprints will be affected and 100% of asset area A within the inner footprint will be affected. Similarly, for the single equipment item illustrated, this will be 100% affected, since it is in the inner zone, as will the population which is also within the inner zone.

SOURCE COSTS

Sources or failure cases result in hazardous releases which in turn result in damage to other assets. However, each release will have a cost associated with it whether it impacts other assets or not. In order to calculate the source costs, the following information is required in addition to the release source term information. Each source is defined as a particular equipment type, which will have its own unique outage time and repair cost, based on the information in Table 1. The model also allows user defined equipment types to be added with their own outage time and repair costs.

Table 1. Outage time and repair cost look-up table for range of equipment types

Equipment type	Outage time (day)	Repair/replace cost ($)
Large Pipes	7	50,000
Medium Pipes	4	20,000
Small Pipes	2	5,000
Compressors	14	250,000
Exchangers	5	50,000
Vessels	7	40,000
Filters	1	10,000
Reactors	14	80,000
Tanks	7	80,000
Pumps	0	5,000
Heater	5	60,000
Column	21	10,000
Other/General	7	20,000
Mobile Buildings	5	25,000
Brick Buildings	15	100,000
Asset Zones	5	1/m²
None	0	0

In addition the following information is required to calculate costs in each of the cost categories described earlier:

- Cost of lost Production per day
- Cost of Environmental clean-up per kg
- Other Environmental costs for other effects per kg
- Value of lost inventory

The Inventory mass and spilled mass, which may not be the same, are also required, although the spilled mass can be calculated as part of the discharge calculation performed in the source term modelling if desired.

POPULATION ZONE COSTS

Population information can be entered through the GIS in the same way as in the Safeti model, as can ignition source information. However, for financial risk calculations based on population the following additional parameters are required:

- Cost of a Solitary Fatality
- Number of Injuries Per Fatality
- Cost of a Single Fatality Among Many Fatalities
- Cost of One Injury

EQUIPMENT COSTS

An equipment item in terms of financial risk is essentially the same as a source, merely acting as a point receptor to a hazard zone generated by a source. From that point of view an original source and an item of equipment require the same financial data. Typical examples of specific items of equipment which you may want to enter into the model as "equipment" rather than "sources" or "assets" are individual large capital value items of plant which do not represent a hazard themselves, or storage facilities for non hazardous material which have a high capital value.

ASSET SET COSTS

An asset is a generic receptor type, which will result in a cost being incurred if it is impacted by a hazard zone with a large enough damage level. An asset zone is treated in the same way as an equipment item in terms of financial risk, but with no contribution from environmental costs. Assets can be defined at a point, or over an area. Typical examples of assets may be on-site buildings like control rooms and offices, off site buildings and infrastructure including, for example, houses, commercial buildings, factories, warehouses, etc. Areas of a plant which do not contribute directly to risk as possible sources of hazardous releases may be defined by a series of assets, to account for background plant value.

FINANCIAL RISK PARAMETERS

Much of the data described above is available as a set of default parameters within the model which may be modified on a case by case basis for each receptor and cost category type. Data available as defaults includes:

- Damage levels and vulnerability factors for each outcome type, cost category and receptor type
- Cost of lost production per day, environmental clean-up per kg, environmental costs for other effects per kg and value of lost inventory
- Outage time and repair cost for each equipment type with ability to add new types
- Financial information relating to fatalities and injuries including cost of a solitary fatality, cost of an injury and number of injuries per fatality and cost of single fatality amongst many in order to deal with the concept of "risk aversion" (see for example Wiggins 1984) when dealing with multiple fatality events.

In addition, calculations and results may only be required for a sub-set of the complete set of cost categories available. For example, in certain situations it is undesirable to place a value on a life and therefore you may wish to exclude fatality and injury costs from the calculations. The complete set of cost categories is original source, fatality, injury, repair and replacement, lost inventory, business interruption and user defined. Any of these can be excluded at any level, either from defaults, so that they are not included in calculations and never appear in the results, at the study level so that they are not calculated for a particular study and don't appear in the results for that study, or at the results level so that they are calculated but not displayed individually in the results. The study described below was used to assess the financial risks associated with the business interruption cost category only.

PRACTICAL EXAMPLE OF MANAGING BUSINESS RISKS USING FRA

This real life study was performed by DNV Energy in order to support its client in assessing the adequacy of the maximum insurance values for their plant. The client wanted to review the maximum value for their insurance based on a quantitative risk assessment of accidents that could happen on their site. A typical QRA study was developed based on the classical approach supported by Safeti and the workflow illustrated in Figure 1. Subsequently equipment data was added to the model using the GIS capabilities of Safeti Financial and appropriate repair costs were defined for each asset zone within the plant.

Accident scenarios from eleven different systems on two of the client's Olefin plants were analysed. Frequencies were estimated for a range of typical failure case scenarios categorised within each of the systems. These were combined with the financial loss consequences calculated using Safeti Financial. The combined result from the frequencies per year and the financial consequences in US dollars based only on equipment damage costs were presented as F-Cost curves, thus representing cumulative frequency with which losses exceeding a particular dollar value would be expected to occur.

The major phases of the analysis were:

Phase 1: Identification of release events, based on material, release conditions, inventory, control systems, location and confinement, that could represent significant losses to the site.
Phase 2: Likelihood and consequence calculation for each selected event.
Phase 3: Financial loss calculation for each scenario and results assessment.

The processes on each plant which were included in the analysis were:

Plant Olefins A: Propylene refrigeration, Propylene separation, Ethylene refrigeration, Ethylene separation, Demethanizer and Deetanizer.
Plant Olefins B: Propylene refrigeration, Propylene separation, Ethylene separation, Deetanizer and Demetanizer,

All the calculations were performed using the Safeti Financial software, developed by DNV Software. Safeti Financial calculates the frequency of occurrence of each representative release scenario. This is based on individual equipment release frequencies factored by the number of pieces of equipment within the segment or area on the plant for which that scenario type could occur. This is achieved by "counting" equipment items for which similar releases may occur and factoring the individual release frequencies based on the equipment count.

These frequencies are then factored by probabilities associated with each possible weather condition (in terms of wind direction and velocity – the wind rose), ignition source probabilities for immediate and delayed ignition, as well as event trees providing the probabilities with which each possible outcome type is expected to occur (e.g. probability of ignition, flash fire, jet fire, VCE, etc.). This methodology is described in detail in the Safeti user manuals (Worthington and Witlox, 2002).

The financial consequences are estimated using the Phast modelling integrated in Safeti Financial. This takes account of the release conditions such as material, temperature, pressure, inventory and release hole size. These "source-terms" are then used to calculate the associated hazard zones from which the cost of repair of the equipment on the installation damaged by the release is calculated.

The cost of repair within each area, based on the damage for each release scenario, are then calculated considering the hazard zone, as follows:

$$F_{Repair,h,o} = v_{Repair,h}\, C_{Repair}\, a_{h,0}$$

where $v_{Repair,h}$ is the **repair vulnerability** for hazard zone h that is defined for each outcome type, as shown in Table 2 below, C_{Repair} is the cost of repairing or replacing the entire area and $a_{h,o}$ is the fraction of the area that is covered by hazard zone h for outcome o.

The result from the analysis are used to create the F-Cost curves shown in Figure 3, from which we can see the estimated repair costs we can expect to incur with frequencies between 0.2 and 10^{-6} (or 1 in a million) occurrences per year. The blue curve shows expected losses considering all possible consequences from the eleven events analysed, considering day weather conditions and the pink curve is related to night weather conditions.

Table 2. Repair vulnerability per outcome (based on API RP 581 (2000))

Outcome	Threshold		Maximum	
Pool Fire	12,5 kW/m^2	50% damage	37,5 kW/m^2	100% damage
Jet Fire				
Fireball				
Flash Fire	–	–	100% LFL	100% damage
Explosion	0,3 bar	50% damage	0,5 bar	100% damage

Based on the above, the client decided to use an accumulated frequency of 10^{-5}/year as acceptable for their maximum loss, which implies a total insured loss of USD 1 billion. The main contributors from each system to the maximum expected losses are shown in Table 3 below. From this we can see that the 3 largest contributors to expected maximum loss are the Propylene Refrigeration in Plant A, Propylene Separation in Plant A and the Propylene Separation unit within Plant B. It should also be noted that all scenarios with a frequency of occurrence less than 10^{-9}/year were neglected and that all

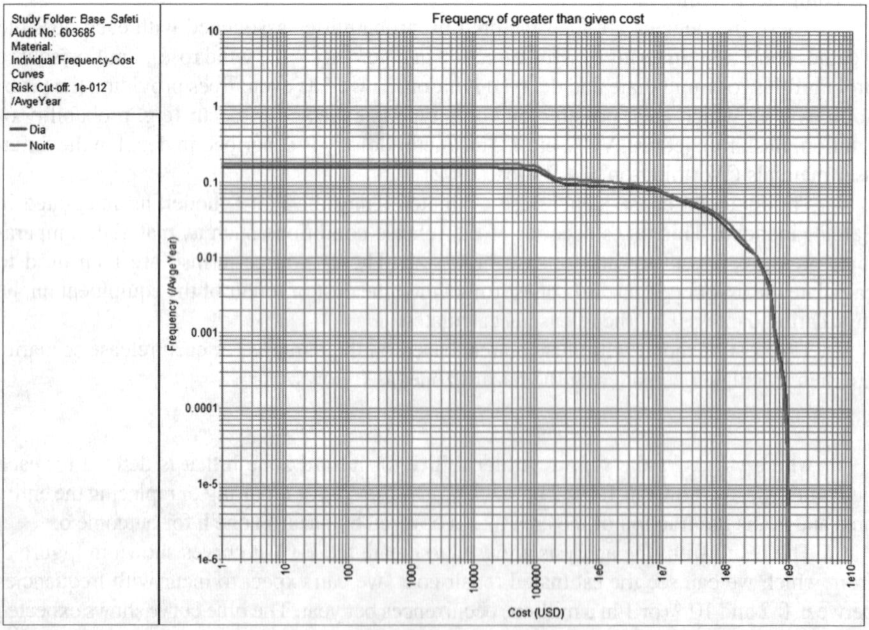

Figure 3 F-Cost curve

Table 3. Maximum loss by system

System	Event number	Maximum cost (USD)
Plant Olefins A		
Propylene Separation	6	$\mathbf{1.32 \times 10^9}$
Propylene Refrigeration	2	$\mathbf{1.20 \times 10^9}$
Ethylene Separation	1	9.30×10^8
Demethaniser	5	8.30×10^8
Ethylene Refrigeration	3	8.05×10^8
Deethaniser	4	8.00×10^8
Plant Olefins B		
Propylene Separation	10	$\mathbf{1.15 \times 10^9}$
Deethaniser	11	6.50×10^8
Propylene Refrigeration	7	6.30×10^8
Ethylene Separation	9	5.00×10^8
Demethaniser	8	4.00×10^8

scenarios with losses above USD 1 billion had frequencies of occurrence of less than 10^{-8}/year.

Drilling down further we can use the facilities of Safeti Financial to identify the scenarios contributing most to the overall financial risk in terms of Expected Maximum Loss (EML). This enables us to isolate the key failure cases we should focus on should we wish to recommend mitigating measures for reducing overall financial risk exposure. In this study the following were found to be the largest contributors.

EML above USD 1 billion

- Propylene Refrigeration – Plant Olefins A
- Propylene Separation – Plant Olefins A
- Propylene Separation – Plant Olefins B

EML above USD 900 millions

- Ethylene Separation – Plant Olefins A

We have also looked at individual F-Cost curves for each system in order to assess more accurately which system are contributing most to the financial risk exposure at different frequencies of occurrence and these are illustrated in Figure 4.

From here we can see that the main contributors at frequencies of 10^{-6}, 10^{-5} and 10^{-4} per year, in order of highest contributor are:

10^{-6}/year: Systems 6, 10, 2, 1 and 5
10^{-5}/year: Systems 10, 2, 6, 1 and 9
10^{-4}/year: Systems 10, 2, 9, 3 and 11

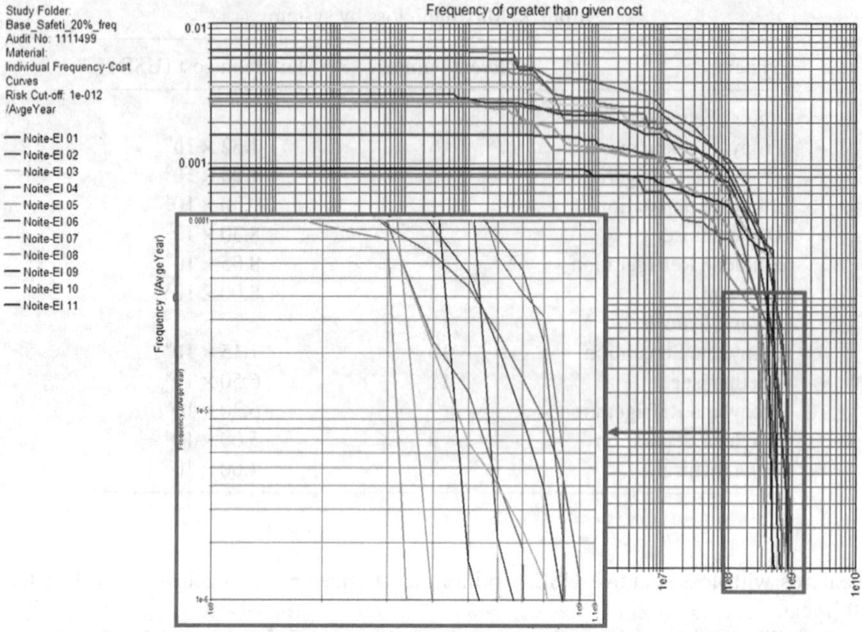

Figure 4. F-Cost curve per system

FUTURE POTENTIAL

The results presented above represent the actual conditions of the analysed systems. In order to reduce the estimated losses as well as the expected losses for each frequency, mitigation measures can be evaluated and the results reviewed. This mitigation should focus on inventory isolation to minimize consequences and reduction of frequency of release.

A cost benefit analysis can be performed, comparing the reduction on the insurance premium per year with the cost for the implementation of the mitigation proposed. This way a risk management approach based on the continuous reduction of the estimated losses can be used.

Also, as an extension to the study presented in this paper, other specific insurance products such as third party insurance can be analysed using the same approach taking account of property damage as well as business interruption and fatality/injury costs.

CONCLUSIONS

We have extended the classical approach to QRA to enable the calculation of broader financial or business risks and built this extended methodology into the Safeti QRA model.

We have then used this model to assess the financial risk exposure of an olefin production plant focusing on financial risk from accident scenarios resulting in damage or destruction of equipment and other assets. The Safeti Financial FRA model has been used to make recommendations on appropriate levels of insurance. We have also used the model to identify the main contributors to these broader business risks in order to propose measures to reduce our client's exposure to possible financial losses.

REFERENCES

Cavanagh, N.J., Calculating Risks, Hydrocarbon Engineering, Volume 6, Number 6, Palladian Publications, London, June 2001.

Cavanagh, N.J. and Linn, J., April 2006, Process Business Risk – A methodology for assessing and mitigating the financial impact of process plant accidents, AIChE Global Safety Symposium, Annual Conference of Centre for Chemical Process Safety, Orlando, Florida, pp 237–256.

Chippindal, L. and Butts, D., June 2004, Managing the Financial Risks of Major Accidents, Annual Conference of Centre for Chemical Process Safety, Emergency Planning: Preparedness, Prevention and Response, Orlando, Florida, pp 321–326.

Elkington, J., Towards the sustainable corporation: Win-win-win business strategies for sustainable development., California Management Review 36, no. 2: 90–100, 1984.

Evans, J. and Thakorlal, G., Total Loss Prevention – Developing Identification and Assessment Methods for Business Risks, 11th International Loss Prevention Symposium, Prague, 2004.

Fewtrell, P. and Hirst, I.L., A review of high cost chemical/petrochemical accidents since Flixborough 1974, IChemE Loss Prevention Bulletin, No. 140, April 1998.

Risk Based Inspection, Base Resource Document, API Publication 581, Second Edition, October 2000.

Topalis, P. and Cavanagh, N.J., 2001. "Optimising Inspection Regimes." Hydrocarbon Engineering, Volume 6, No.10. October 2001.

Wiggins, J. H., Risk Analysis in Public Policy, Proceedings of Victoria Division, Engineers Australia, Risk Engineering Symposium 1984: Engineering to avoid Business Interruption, 1984.

Worthington, D.R.E. and Witlox, H., 2002, SAFETI Risk Modelling Documentation – Impact of Toxic and Flammable Effects, DNV Software Risk Management Solutions, October 2002.

OVERVIEW OF HEALTH AND SAFETY IN CHINA

Hongyuan Wei[1], Leping Dang[1], and Mark Hoyle[2]
[1]School of Chemical Engineering, Tianjin University, Tianjin 300072, P R China,
E-mail: david.wei@tju.edu.cn
[2]AstraZeneca plc, Macclesfield, Cheshire, SK10 2NA, England

Current status and future development regarding health and safety production and legislation in China has been reviewed and analyzed in this paper. Since economic reform and opening, the entire society has noticed the importance of health and safety, as this could be a big barrier for the further development. Significant progress has been made over last 25 years with respect to health & safety research and practices. Meanwhile, health & safety legislation and environmental protection law has been established and operated in a much more stringent manner; however, safety is still a major issue which damages China's reputation and has significant societal as well as economic impact.

The Chinese government is paying increased attention to the areas of health and safety, such as establishing new legislations and applying new techniques to actively tackle key problem areas. The health and safety issue in China is not only confined by the legislation and regulation but also greatly influenced by its social and culture background. Therefore, it is very much an inter-disciplinary science and should have integration with local society. In order to maintain sustainable social and economic development the government has defined health and safety as a key area with interim and long-term plans for national development.

1. INTRODUCTION

With the major economic development and social reform in China over the last 25 years there has been an associated increase in living standards. The people and their government have paid more and more attention to health and safety issues. Safety management has become a central focus for scientific research and regulatory attention. The entire society has come to recognize that health and safety problems are critical to the welfare of each individual as well as to the sustainable social and the economic development of the country. Meanwhile, new technologies and innovations have been applied to health and safety issues along with the systematic establishment of legislation and regulation by the government.[1] It is a fundamental premise of conventional welfare economics for the public policy makers that public decisions should, as far as possible, reflect the preferences of those who will be affected by them. However, both technical and human errors are still seen, especially, in the field of safety management and accident prediction.[2] Safety and health is still a major problem, which will continue to damage society and the economy.

This paper will give an overall review and analysis of Health and Safety related legislations in China, under the five following subjects:

- Current status
- Background and origination of safety accidents

- Health & safety management system
- Establishment and application of legislations and regulations
- Development of health and safety in the near future

2. CURRENT STATUS

At present, the concept of safety is being underpinned and significant transformation is occurring across the entire society. Major progresses are being made in terms of safety theory, safety technique, and constructive safety culture. Nevertheless, many fatal accidents are still taking place, and with them come the associated damage to society with respect to people's lives and health and the underlying damage to the countries reputation and economy.

Figure 1 shows the distribution of accidents in different industries from November to December, 2006. Accidents during transportation were by far the major contributor, but as a subset explosions and accidents during mining activities made up another major proportion thereafter.

Based on the statistical data from the State Administration Work Safety of China, there were 627,158 accidents during production activities in 2006. Human casualty in these accidents totaled 112,822. 7 severe accidents (>30 fatalities each) resulted in 263 people losing their lives. The number of major accidents (>10 fatalities each) was 95, giving a total loss of life of 1570. 2357 accidents resulted in between 3 and 9 deaths (9065 fatalities in total).[3]

The analysis of major accidents in the last decades also calls for more systematic safety management systems (SMS) and regulatory frameworks governing systems applied in manufacturing facilities.

3. BACKGROUND AND CAUSATION OF SAFETY RELATED ACCIDENTS

Based on the survey of State Occupational Safety and Health Association (China), there are five main aspects with respect to the present background and causation of safety issues in China, illustrated as follows:[1, 4, 5]

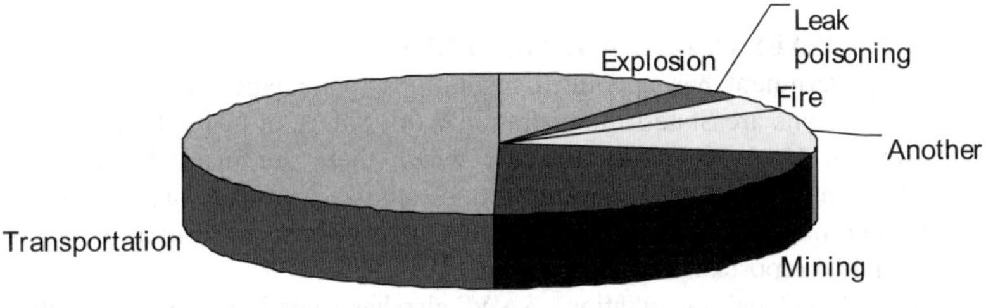

Figure 1. Accident Type Distribution from November to December, 2006

1. Limited resource is focused on safety during production. Inadequate supervision and management of safety is often tolerated in many production sites and limited finance/ funding is applied directly to safety. These are obviously major causal factors. It is estimated that only a third of the total cost of implementing safety procedures is implemented in China against that of other industrialised countries.
2. The cultural background in China is not one given over to safety consciousness. Although Chinese civilization is over 5,000 years old, China has mainly been an agricultural country. Industrialization in real terms only started 30 years ago when China began its economic reform.
3. Compared with the rate of economic development, health protection and accident prevention systems have developed at a much slower pace, especially, in mining and the chemical industries.
4. Means and monitoring of implementing the Safety in Production Act are very weak.
5. There is reluctance to use new techniques or equipment that will provide or improve safety during prod uction.

In fact, nearly all the accidents that occurred in the process industry were caused by improper process or equipment design or lack of the preventative control system. Nowadays safety management has become a fashionable term and the focus of study, but it could be dangerous to think it will solve all the safety issues.

Good engineering and behavioral safety (human factor analysis) are both equally important. One cannot simply substitute one for another in a developing economy. No matter how good safety management is, it cannot make up for poor design and lack of operating control. Safety and health effects are well studied and can be assessed according to science-based and commonly accepted methods of risk analysis.[6] In risk analysis terms we can formulate the task of a safety management system as a "common mode" influence that is designed to keep all failure probabilities in the fault tree at the lower ends of their intrinsic bounds. It also anticipates all significant risk scenarios and design measures to eliminate them, or at least to reduce and provide robust control of them. The question of interest is whether we know enough about how safety management operates in order to design, assess, and improve the systems applied in practice.

4. HEALTH & SAFETY MANAGEMENT SYSTEM

The Chinese government has appointed institutions to manage Health & Safety. The uppermost institutions are State Administration Work Safety of China (SAWS) and State Administration of Coal Mine Safety under which there are many local subordinate organizations. The function of local organizations are to ensure the enterprise's approach to safety during production and employee's rights, supervise the performance of legislation and regulation, and report accidents etc.

Besides all the local organizations, SAWS also has some branches such as the China Academy of Safety Sciences and Technology (CASST) and the State Occupational Safety and Health Association. The major roles of those branch organizations are: technological support to SAWS, promoting the development of safety science & technology, Health and Safety education, and trade, and providing services to enterprises related to safety technology, and information management.

CASST, as an example, is a research institute covering the areas of major accident prevention, major hazard identification, assessment and monitoring, risk assessment and management, building major accident & emergency systems and emergency planning, developing safety & hygiene engineering technology, work safety inspection and management systems, and basic theory of work safety and work safety economics.

5. ESTABLISHMENT AND APPLICATION OF LEGISLATION AND REGULATION

The Chinese government is seeking to establish new safety legislations revolutionarily to reduce the number of major accidents. Prescriptive laws and regulation aimed specially at technical preventive measures are to be implemented. There is an inevitability that these will potentially trail behind new developments in technology. The responsibility for devising the means of risk control will be taken not by the companies and industries that generated the technology, but by the regulator. The framework legislation requires companies to develop their own safety management system to specify, implement, and evaluate the detailed preventive measures. Therefore, more systematic management systems for safety are required to be further developed, structured, assessed, and improved. The Safety in Production Act was promulgated in 2002 in China. Consequently, it became a central task for 'safety' people to work on how to systematically and scientifically standardize the certification and accreditation of Occupational Safety and Health Management System, and how to establish an efficient and standardized mechanism. Objectivity and publicity in implementing the Safety in Production Act have also been given close attention by safety workers. In addition, the mechanism of market permission and inspection has been put forward in order to implement the Act effectively.

The constitution of national and international standards for safety management is in progress, modelled on the ISO 9000 series for quality management and the ISO 14000 series for environmental management systems.[7] Meanwhile, health & safety legislation and environmental protection law have been established and operated in a much more vigorous manner. The effective management system is identified as Occupational Health and Safety (OHS) Management System, which is a method of going about your business whilst reducing risks to your staff. An effective OHS Management System can also help establish the framework of compliance with the two fundamental elements of OHS legislation, that is:

- Employers providing and maintaining a working environment that is safe.
- Employees taking reasonable care for the health and safety of themselves and others.

6. SAFETY PRACTICES IN THE CHEMICAL INDUSTRY

China has been the manufacture centre for the world. The Chemical industry has also blossomed over the last 10 years; therefore, health and safety issue have become the major challenge for this industry. Although the chemical industry in China has far better performance in terms of safety than other industries, such as mining and transportation, there are still many problems that need to be solved. The main causes for chemical accidents are:

- Speed of the development without serious consideration of safety;
- Budget saving and cost reducing;
- Improper design and engineering work;
- Lack of technologies to control hazards;
- Mismanagement;
- Awareness for safety due to the culture background;

The good news is that the best practice of safety management system for the oil, petrochemical and coal chemical industries have been developed by the State Administration of Work Safety, aiming at establishing control systems for different safety classes, such as explosion-isolation or safety spark type, which ensures the production safety for chemical processes.

Current trends show that more and more chemical enterprises have taken initiatives to conduct Safety and Environment impact studies for their intended investment in China. Ecological consideration is becoming vital to the industry's sustainable development. Besides implementing novel technologies in chemical plants, 'Safe & Green Concept' for designing and engineering are widely applied in many middle size and large chemical companies. 'Responsible' chemical enterprises, which offer the moral, legal and financial necessities to ensure safety, are welcome by employees and the whole society.

Great effort has been successfully put on introducing advanced and safer chemical technologies from oversea during the periods of the "Tenth Five-Year Plan" (2001–2005), the "Eleventh Five-Year Plan" (2006–2010). Importing of technologies with emphasis on safety and environment has continued. There is also more opportunity for international collaborations in SHE areas.

Alongside safer technology implementations, an advanced management system for the chemical industry has been introduced. It consists of nine aspects:

1) Safety policy: Written safety policies for each factory shall be established and maintained with emphasis on protecting the safety, health and the general well being of every personnel working in the factory. This spells out management's commitment towards safety and the principles that govern safety and health decisions in the factory. Senior management shall ensure that the safety policy is communicated to all employees and relevant contractors and, shall establish a culture that safety and health are valued as basic and fundamental prerequisites for the business of the factory.

2) Safe work practices: Safe work practices shall be carried out in the factory through procedural and administrative control of work, critical operating steps and parameters,

pre-start up safety reviews for new and modified plant equipment and facilities, and management of change of plant equipment and process.

3) Training: Employees and contractors at all levels shall be equipped with the knowledge, skills and attitudes relating to the operation or maintenance of facilities so as to work in a safe and reliable manner. An effective training program shall be established to train employees at all levels. A scheme shall be devised to identify the safety and health training needs for each level of employees and contractors.

4) In-house safety rules and regulations: Written safety rules and regulations for all personnel in the factory shall be established to instill a common understanding of their safety and health obligations and responsibilities. A set of basic safety rules and regulations shall be formulated in the factory to regulate safety and health behaviour at the workplace. The rules shall cover the main work operation or process in the factory.

5) Safety promotion: The management's commitment and leadership shall be clearly demonstrated in promoting good safety and health behaviors and practices in the factory. Promotional programs shall be developed and conducted. Safety and health awareness shall be raised and maintained among all employees and contractors.

6) Safety inspection: A system shall be established to verify compliance with the relevant regulatory requirements, in-house safety rules and regulations and safe work practices. The factory shall develop and implement a written program for formal and planned safety inspections to be carried out in the factory. The program shall include safety committee inspection, routine safety inspection, plant and equipment inspection and other special or surprise inspection.

7) Hazard analysis: Hazards shall be methodically identified, evaluated and controlled in processes in a factory. The process of Hazard Analysis shall be documented and shall include measures over serious scenarios that could occur in a factory. Necessary corrective or control measures shall be incorporated to mitigate the frequency or severity of the risks associated with the scenario event.

8) Occupational health programs: Occupational health hazards shall be identified, evaluated and controlled to protect all personnel from developing occupational diseases or illnesses arising from their exposure. The program is to specific occupational health hazards presented in the work environment and establish policies and plans to protect all workers from them.

9) Emergency preparedness: A program of drills and exercises shall be established to assess the preparedness of the factory for prompt and effective response to emergency situations. Factories shall establish effective first-aid programs to provide first-aid and emergency treatment to victims of an accident, which include provision of adequate first-aid facilities and trained first-aiders.

People in the chemical industry start to understand that SHE challenges cannot be overlooked as we strive for a better life and a favourable economic achievement for the country while keeping in line with the goal of the global sustainable development. The effort on SHE issues has enabled us to make this industry a safe place to work despite the misperception by many that associate 'chemicals' with 'fear'. The chemical industry in

China will continue to push forth its efforts on SHE to manage the life-cycle of chemicals through collaborating closely with the world.

7. DEVELOPMENT OF HEALTH AND SAFETY IN THE NEAR FUTURE

The Chinese government has recently made high-profile efforts to build a harmonious society on an all-round, well-off basis, a goal which has given full expression to the people-oriented development perspective. To regulate industrial practices, in a sense to truly protect workers' safety and health, is a logical step amid efforts to build a harmonious society. That is a society built on the foundations of democracy, the rule of law, justice, sincerity, amity, vitality, and stability, harmony between man and nature, and coordinated economic and social development. These continuous improvements can benefit society as a whole. By taking major measures such as institutional reform, embedding safety mechanisms/techniques into production, increasing funding on safety issues, etc., then continuous improvement of workplace safety and working conditions nationwide will follow. This will drive improvements in the current rising trend of total accidents, whilst maintaining the sustainable and rapid growth of the Chinese economy.

Whilst China, the world's largest developing country, is going through a major industrialisation process, the work safety infrastructure remains weak, indicated by the reoccurrence of major accidents in certain industries (and in certain areas of the country). The economy of the country is relatively unbalanced and of relatively low productivity when looked upon as a whole. China's work safety status is so far incompatible with what a 'well-off' harmonious society demands, and in this connection there is still a lot of work to be done to improve it.

The Government needs to continue to give top priority to safety issues, promoting a preventive safety culture within workplaces and overseeing that the primary responsibility of enterprises is given to workplace safety. Government regulators need to fulfil their role in ensuring this is rigorously enforced and audited effectively. Sustained efforts have to be made in promoting safety culture, building a sound legal framework governing work safety, implementing the safety responsibility system, advancing safety science and technology, and increasing the funding of safety measures.

It is the Chinese government's aim to enhance safety management at source, eliminate hazards that stand in the way of improved safety conditions, and ensure safety at an intrinsic level. To better fulfil this mission, the Chinese government ought to broaden their horizons and continue to further embrace the outside world in the field of health and safety. The Chinese government is eager to enhance cooperation with the relevant intergovernmental organizations, foreign governments, intermediaries, multinational companies and enterprises of different categories, to examine and learn safety philosophies, sciences, technologies and management approaches that originate in other countries and regions.[8] Hopefully, assisted with international wisdom, China will blaze a new trail to put work safety on a solid basis and achieve steady and rapid improvement of its work safety situation. As a responsible player in the international community, China can make due contributions in implementing the Global Strategy on Occupational Safety and Health and creating for mankind a safe, harmonious environment with less potential for major accidents.

REFERENCES

[1] Luo Yun, The status analysis of safety production in China, Development and observation in China. 2005, 5, 33–37.

[2] Graham Loomes, (How) Can we value health, safety and the environment, Journal of Economic Psychology. 2006, 27, 713–736.

[3] Wang Yajun, Huang Ping, Statistics of industrial accidents in China during the period from November to December in 2006, Journal of Safety and Environment. 2007, 7(1), 154–157.

[4] Xu Deshu, Chinese Safety Culture Construction-Research and Discovery, Sichuan Safe Science and Technology. 1998, 16 (3), 26–34.

[5] Xu Deshu, Times Background on the Safety Culture Construction in China, China Safety Science Journal. 1999, (4), 1–7.

[6] Guntram Koller, Ulrich Fischer, and Konrad Hungerbu1 hler, Assessing Safety, Health, and Environmental Impact Early during Process Development, Ind. Eng. Chem. Res. 2000, 39, 960–972.

[7] Hale A.R., Safety Management in Production, Human Factors and Ergonomics in Manufacturing. 2003, 13 (3), 185–201.

[8] Jin Lei, Xu Deshu, Luo Yun, Strategy of Safety and Reducing Disasters in China for the 21st Century, Kaifeng, Henan University Publishing House, Dec. 1998.

ANALYSIS OF PAST INCIDENTS IN THE PROCESS INDUSTRIES

I. M. Duguid
Consultant

This article is the culmination of 15 years work on the quantitative analysis of major hazards incidents (Ref. 1 Duguid August 1998, Ref. 2 Duguid July 2001, Ref. 3 Duguid December 2005). It will only deal with items which are new. However to make use of this in a safety system to minimize future incidents it is necessary to include previous studies as well. The source article doing this is available from the author.

Table 1 is the first sheet of the 1000 item database. The format is unchanged from the earlier versions. However it is now an EXCEL database and so much easier to analyse. One point not made fully clear in the previous articles is that "Causes" are physical items such as corrosion and "Responsibilities" cover the requirement for provision of adequate safety equipment and procedures by management and for the workers to make proper use of these. There the lack of safety alarms and shutdowns which are in use elsewhere is an example. Also the total reliance on such alarms and shutdowns by operators when they are only meant as a second line of defense. Some extra codes, such as PTWF – Permit to Work Failure have been added where the new data made this appropriate.

One of the major new points in this analysis is the quantitative assessment of the Pareto effect (Table 2). For responsibilities and causes it is confirmed that the 20% of most frequent items are involved in close to 80% of the total incidents. This makes possible a very cost effective concentration of the safety effort on these items. Purists might say that all problems should be tackled equally. However until the top 20% have been dealt with this would slow progress.

Table 3 and Figure 1 including new data covering the past 10 years demonstrates the lack of progress in addressing the top 20% of problems. On a statistical basis the frequency with which these are involved in incidents has dropped little in the last 50 years. Note that the total incidents in the periods before 1970 and after 1999 contain less than half the numbers of incidents of those in each of the intervening decades. They are thus less significant statistically. There are two exceptions to this lack of progress. Firstly the major drop in frequency in HAZOP. Incidents involving a missing or inadequate HAZOP have dropped continuously from around 60 to 30% since the introduction of HAZOP's. Secondly the reduction in incidents involving poor design for safety. In view of recent interest in "Management of Change" it is disappointing that there is no indication yet of a fall in code MODI – Incidents involving modifications to plant or operations.

Safety audits for the most frequent problems have been proposed in earlier articles (Ref. 4 Duguid April 2004, Ref. 3 Duguid December 2005). However these can now be substantially reduced by limiting them to the Pareto top 20% items for causes and responsibilities and a few other important items (Table 4). The answers required are simply yes/no and in a majority of cases someone in the organization should be able to answer each

Table 1. Database of major hazards incidents

Item	Decade	ACT/POT	Industry	Process	Substance	Ignition	Equipment	Responsibilities			Causes			Hazop	Mode
12005	7	A	O	ALK	LPG	NON	VALV	OPER	PROC	PTWF	OPEN	PLUG		S	SU
12005	7	A	O	ALK	HYF	EXP	VESL	INST			INCO	OFIL	SVAP	U	NR
12006	7	A	O	ALK	LPG	NON	HEEX	PROC	SAFE		PRES	SHUT		U	MN
12006	7	A	O	ALK	LPG	FLA	FURN	PROS			CORE			S	AB
12008	7	A	O	ALK	LPG	NON	TANK	OPER	PROC		RUNA			U	AB
12009	7	A	O	ALK	LPG	FLA	TANK	OPER	PROC		DRVT			U	SD
12016	7	A	O	CRD	WAT	NON	VESL	OPER	PROC		SVAP			S	SU
12017	7	P	O	CRD	CRD	NON	HEEX	MECH			PLUG			U	NR
12017	7	A	O	CRD	CRD	HOT	PVRV	INST	SAFE		DRVT	MODI		U	AB
12018	7	A	O	CRD	CRD	HOT	FLAN	MECH			MATR			U	SU
12019	7	A	O	CRD	DIE	NON	VESL	OPER			SVAP	VLPS		S	SU
12019	7	A	O	CRD	WAT	NON	VESL	OPER	PROC		SVAP	VLPS		S	SU
12022	7	P	O	CRD	CRD	FLA	FURN	INSC			EROS	MATR		S	SU
12023	7	A	O	CRD	DIE	AUT	PUMP	INSM	MAIN		SEAL	VIBR		S	MN
12023	7	A	O	CRD	CRD	SPA	PUMP	MAIN	SAFE		SEAL	VIBR		S	MN
12027	7	P	O	FCC	GAS	NON	REAC	MECH			FATI	TSHK		S	SU
12029	7	A	O	FCC	GAS	AUT	REAC	OPER	PROC		VFLR			S	SD
12030	7	A	O	FCC	GAS	AUT	PIPE	OPER	PROC		SUPL			S	AB
12031	7	P	O	FCC	GAS	AUT	VALV	INSM	MAIN		EROS			S	NR
12033	7	A	O	FCC	DIE	ELE	VESL	PROC			VLPS			S	AB
12035	7	A	O	FCC	HVY	AUT	PIPE	OPER			CORI			S	NR
12036	7	P	O	FCC	GAS	AUT	VESL	MECH			FATI	TSHK		S	NR
12037	7	P	O	FCC	OTH	NON	PIPE	OPER	PROC		SUPL	VLPS		S	SU
12040	7	A	O	FCC	GAS	HOT	FURN	SAFE			FLOU			U	SD
12041	7	A	O	FCC	GAS	AUT	VALV	INSM	MAIN		FAIL			U	AB

Table 2. Pareto effect analysis

A) Responsibilities
1. The total number of categories is 14
2. Of these three: OPER – Operator error, PROC – Inadequate written procedures, and SAFE – Inadequate design for safety were involved in 823 or 82% of the incidents
3. Three out of fourteen is 25% of the categories

B) Causes
1. The total number of categories is 45
2. Of these nine: RUNA – Runaway reaction, TEMP – Above design temperature MODI – Plant or operation modified, VFLR – Flammable vapours in enclosed space, DRVT – Uncontrolled flow through drain or vent, MATR – Material of construction unsuitable, SAIN – Safety instrument fails, CORI – Corrosion internal, and OPEN – Equipment under pressure opened up, were involved in 833 or 83% of the incidents
3. Nine out of 45 is 20% of the categories

C) Other

Other items which were involved in a high percentage of incidents include:
1. HAZOP U – Unsatisfactory or absent hazard and operability study – 38%
2. STB – Storage and Blending – 24%
3. MODE – Plant not in normal operation – 45%

D) Conclusion
1. The figures confirm the Pareto effect which says that the 20% of problems which occur most often are responsible for 80% of the incidents

Table 3. Variation of frequency of problems over time

	Percentage of incidents with this problem				
Problem	1940–1969	1970–1979	1980–1989	1990–1999	2000–2005
Responsibilities					
Operator error	31	23	23	27	27
Written procedures inadequate	31	34	38	34	44
Design for safety inadequate	49	30	37	30	22
Causes					
Runaway reaction	25	8	18	23	25
Above design temperatures	10	13	9	12	5
Modification to plant or operations	4	13	8	9	11
Flammable vapours in enclosed space	5	7	9	10	17

(*Continued*)

Table 3. Continued

Problem	Percentage of incidents with this problem				
	1940–1969	1970–1979	1980–1989	1990–1999	2000–2005
Uncontrolled flow through drain or event	11	9	6	6	7
Material of construction unsuitable	6	7	6	9	8
Safety instrument failure	10	8	8	7	5
Corrosion (internal)	3	8	9	7	9
Equipment under pressure opened up	1	2	9	9	13
Other					
Storage and blending	26	28	20	23	24
HAZOP unsatisfactory or absent	60	40	37	33	27
Incidents not during normal operation	37	50	47	42	47

Note: The number of incidents in the first and last columns are less than half those for the intervening decades. In the case of 1940–1969 this was due to the paucity of incidents for which adequate information was available to permit inclusion. After 1999 only five years data was available at the time the database was finalised. This makes the figures in these columns less significant statistically than the rest.

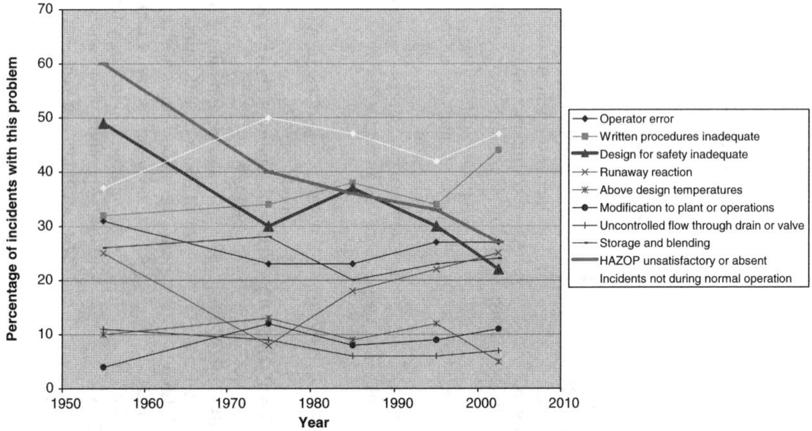

Figure 1.

Table 4. Proposed annual audit covering pareto top 20% problems

No	Code	Questions	Reply yes/no
		Responsibilities	
1	OPER	Are Operators trained on top 20% items which involve them (Table 6)	
2	OPER	Does Operator training include start up, shutdown and likely abnormal operations	
3	OPER	Do Operators attend a presentation such as "Remember Charlie" stressing their vulnerability in an incident	
4	PROC	Are operating procedure manuals standardized or audited to ensure consistent quality	
5	PROC (MODE)	Do these manuals cover shutdown and start up and likely abnormal operations fully. One off cases require temporary instructions	
6	PROC (MODE)	Do these manuals warn Operators not to go outside written operating procedures without advising supervision	
7	PROC	Are these procedures designed to eliminate split responsibility	
8	SAFE (HAZOP)	Are design safety features fully up to best industry standards (Table 6)	
		Causes	
1	RUNA (HAZOP)	Have all possibilities for runaway reactions to occur been checked and action taken where a problem was found (Table 6)	
2	TEMP (HAZOP)	Have all causes of above design temperatures been checked and action taken where problems were found (Table 6)	
3	MODI (HAZOP)	Are all substantial modifications and/or changes to operations covered by a HAZOP	
4	MODI	Are all minor modifications to plant and/or changes to operations covered by a safety check	
5	VFLR	Are the vapour spaces of all cone roof and internal floating roof tanks checked for flammable vapours regularly	
6	VFLR	Are the vapour spaces of tanks and vessels being cleared of sludge and/or scale during maintenance monitored continuously for flammable vapours	
7	DRVT	Do all operating drains and vents discharge to safe locations. (Including relief valve discharges)	
8	DRVT	Are all drains and vents not required for normal operation plugged off	
9	DRVT	Do operating drains with an open discharge which could release hydrocarbons/chemicals have two valves one of which is spring closed.	
		Where under automatic interface level control do they have an independent low interface level shutdown	
10	MATR	Is there a system for spot checking that the right materials are installed, particularly during maintenance	

11	SAIN	Is all safety instrumentation, including relief valves, checked for correct operation at least on each planned shutdown
12	SAIN (MODI)	If any safety instrumentation has been taken out of service has it been reinstated or replaced by a more reliable system performing the same function
13	SAIN	When any safety equipment, including relief valves, fails is every effort made to determine the cause of the failure and eliminate it.
14	CORI	Is corrosion monitoring concentrated on the most likely places for to it to occur (Table 6)
15	CORI	Where high corrosion rates are found, is the use of more corrosion resistant material evaluated
16	OPEN	Is there a system in place where an operator must demonstrate physically to the maintenance crew that the equipment is fully depressured and drained directly before they work on it
17	OPEN	At the same time does he make it clear exactly what items they can work on
		Other
1	STB	Have likely problems in tank farms been fully investigated and action taken where required. (This includes carrying out a HAZOP.) (Table 6)
2	HAZOP	Have HAZOPs been carried out on all parts of the site including utilities and other ancillaries
3	HAZOP	Have HAZOPs been updated when any substantial changes have been made to plant or operations
4	–	Has the five yearly audit been carried out as scheduled. (Ref. 4)
5	–	Have the actions recommended in previous safety audits and HAZOPs been completed or where not have reasons for not doing so been documented

Total Yes Answers	a
Total Questions	30
Safety audit score %	$\frac{a \times 100}{30}$

Notes

1. Where questions call for a check on all possible problems, details of the most frequent ones are given in Table 6. Each such question concerned is tagged Table 6 here.
2. See Table 2 for the meaning of the codes given here
3. As my old boss used to say "To make as if" is worse than useless. The safety audits and the HAZOPs must be carried out thoroughly by experienced personnel. At least one in each team should have studied the source article so that they are aware of the more frequent problems and actions required.
4. The reply to each question should be initialed by the auditor who made it. He is responsible if it proves to be wrong.

one without further investigation. Thus the percent of "yes" replies is a simple metric for progress in major hazards safety. Where there are multiple frequent problems related to a particular question, these are listed in Table 4. This will help ensure that the answers cover the main problems. This information could be of use in deciding on Key Performance Indicators for major hazards.

New information has been used to revise the cost/benefit analysis (Table 5) previously given in Ref. 5 Duguid March 2006. This analysis covers the nine petroleum refineries in the UK over a period of 18 years. That should be long enough to provide a significant average figure. The benefits now include a figure for the fifth major incident in the period. It had been suggested that for some companies who are less far down the road of implementing safety measures than others, the previous figure for the cost of correcting problems found by the first audit was too low. This has therefore been increased. Overall the cost/benefit ratio is now 1:2.2. Thus in the long term there is a good payout on the cost of the new safety measures. In addition this figure should be quite conservative because:

1) Data on loss of profits due to long plant shutdowns for repair were not always available.
2) It was assumed that only one of the five incidents could have been prevented by the proposed safety measures. Surely we can do better than that.

The substantial increase in the number of total and particularly chemical industry incidents in the database has made a quantitative analysis of the differences between the chemical and petroleum industries practical. Most of the common problems are the same for the two industries. Significant differences are as follows:

1) On HAZOP's the involvement of unsatisfactory or absent ones in major incidents dropped from 40% to 15% in the petroleum industry since the 1970's. In the chemical industry the corresponding figures are 45% to 43%. This indicates how effective HAZOP can be and how much catching up the chemical industry has to do.
2) In the petroleum industry a full third of the major incidents occurred in storage, that is tank farms. This indicates too much concentration of safety effort on process units compared with ancillaries, which is a mistake. Table 6 details the more frequent problems. It should be noted that all incidents occurring at petroleum distribution terminals are included in these figures. It was not always possible to determine whether or not they were located at refineries. They do form a significant proportion of the total. The much smaller use of storage tanks in the chemical industry means that this is less of a problem there. Only a sixth of the incidents there are involved. However it is replaced by incidents in warehouse storage.
3) In the chemical industry major incidents involving runaway reactions actually rose from 32% in the 1940's to 1980's to 47% after 1990. Clearly this industry needs to concentrate a substantial part of their safety efforts on tackling that problem. This is underlined by the fact that of 90 incidents involving unstable chemicals 73 resulted from runaway reactions occurring. For obvious reasons this is not a sserious problem in petroleum refining. Details of the frequent problems involved are given in Table 6.

Table 5. Cost/benefit analysis on proposed safety measures

A) Basis

1. This analysis is based on the record of the nine petroleum refineries in the UK over a period of 18 years
2. It covers the five very major incidents in this period
3. The costs for these incidents come mainly from Ref. 7. The benefits are the reduction in these costs by preventing only one of the five incidents by use of the proposed safety measures.
4. It is known that not all the figures available for these incidents included loss of profits due to the long plant shutdowns for repairs
5. The costs for the proposed safety measures are estimates based on experience. They allow for the fact that a substantial portion of the safety measures required, such as HAZOPs will already be in place.

B) Benefits

		$US
1.	Costs for all five incidents	600,000,000
2.	Average cost per incident	120,000,000
	Benefit	120,000,000

C) Costs

1.	These include the cost of the audit itself and the cost of correcting any deficiencies found	
2.	Initial five year safety audit per refinery (See Ref. 4)	2,000,000
3.	Subsequent two five year safety audits per refinery	1,000,000
4.	Annual safety audit per refinery	200,000
5.	Total for 15 annual audits per refinery	3,000,000
6.	Total cost of audits per refinery	6,000,000
7.	Total cost of audits for nine refineries	54,000,000

D) Cost Benefit

1. From the above figures the benefits are more than twice the costs in the long term
2. These figures are felt to be very conservative. It would be a sad comment on the thoroughness of the audits and consequential improvements of safety measures if at least two of the incidents could not have been prevented. This would raise the benefits to costs ratio from 2.2:1 to 4.4:1. This still does not take into account the missing loss of profits due to long plant shutdowns.
3. For the Chemical Industry the figures could be poorer due to the larger numbers of smaller plants. However, it is difficult to believe that they would fail to break even.

Table 6. Most frequent problems relating to Table 6 questions

Question – Operator Training

1) Are they made aware that over half all incidents occur during shutdown, startup, maintenance, and abnormal operations
2) Are they made aware of the risks of runaway reaction due to changes in operation such as charging an incorrect reactant, changing the order or rate of charging, temporary shutdowns, or loss of cooling
3) Are they made aware that they should not depart from written operating procedures without advising supervision. In emergency situations this may have to be after the event.
4) Are they aware of the need to only use alarms/shutdowns as a backup, rather than relying on them completely
5) Are they made aware of the need to physically check that equipment is depressured and drained in the presence of the maintenance crew directly before work on it starts. Also that the crew know exactly what they may work on
6) Are they made aware of the dangers that their operations can cause in tank farms. See Code STB (Table 6) & Storage tankage below.

Question – Design for safety inadequate

70 % of the design errors here would have been picked up by an adequate HAZOP. The most frequent specific problems are:

1) Alarms to detect high temperatures at critical locations or loss of cooling, particularly where runaway reactions are possible. Also lack of skin thermocouples to detect high temperatures in fired heater tubes.
2) High level Alarms/Shutdowns totally independent of the normal level control system to minimize the risk of overflow from tanks or vessels.
3) Spring closed valves to remind operators to remain in attendance while draining water bottoms from equipment containing flammables or toxics to an open drain. Also installing a totally independent low interface level shutdown system where such drainage is under automatic control.
4) Over reliance on check valves to prevent reverse flow.
5) Inadequate attention to design for safety when implementing modifications to plant or operation.

Question – Runaway reaction

1) Use of an incorrect reactant.
2) Sending a chemical to the wrong storage tank.
3) Temperatures not measured where maximums are likely.
4) Loss of cooling or mixing.
5) Feeding reactants at the wrong rate, ratio, or order.
6) Inadequate checking that safety instrumentation is reliable.
7) Inadequate precautions to prevent carryover or reverse flow mixing incompatible chemicals (e.g. H_2SO_4 and NaOH).
8) Inadequate testing to confirm whether runaway reaction is possible.

(*Continued*)

Table 6. Continued

Question – Above design temperature
1) Inadequate design for safety, including lack of effective alarms/trips and temperatures not measured where maximums are likely.
2) Inadequate procedures for handling likely abnormal situations.
3) See Runaway Reaction above

Question – Corrosion Internal
1) Monitor downstream of chemical and water injection points.
2) Monitor where initial condensation of water occurs from mixed process streams
3) Monitor low points, dead ends, and drop legs in piping.
4) Monitor bends and other maximum velocity points, particularly where solids are present.

Question – Storage tankage
1) See Code VFLR (Table 4)
2) Have the risks of sending high temperature streams to tankage been addressed, particularly where water bottoms are present.
3) See also Codes MODI, OFIL, RUNA, and DRVT (Table 4).

It must be emphasized that this presentation only deals with major hazards incidents. Lost time accidents (LTA) require a different approach. Recent reports on the Longford gas recovery plant, Texas City refinery, and Buncefield storage terminal incidents have highlighted the fact that concentration on achieving a good LTA performance is not a good measure of protection against major hazards incidents. It should be added that the causes involved in these three incidents were all highlighted in the earlier articles on analysis of past incidents. This further reinforces the case for carrying out the safety audits proposed here. Finally Oscar Wilde has a pithy saying for most occasions. The following one encapsulates what this article is about. "To lose one parent, Mr Worthing, could be regarded as a misfortune, to lose two seems like carelessness".

REFERENCES

1) Duguid I.M., 1998, Analysis of past incidents in the oil, petrochemical, and chemical industries, *Loss Prevention Bulletin*, Issues 142 pages 3–6, 143 pages 3–7, and 144 pages 26–30.
2) Duguid I.M., 2001, Preventing plant mishaps, *Chemical Engineering,* July 2001 issue, pages 80–84.
3) Duguid I.M., 2005, A framework for the use of key performance indicators of major hazards in petroleum refining, *IP Research report,* December 2005.
4) Duguid I.M., 2004, Proposal for a process plant safety audit, *Loss Prevention Bulletin,* Issue 176 pages 13–16.
5) Duguid I.M., 2006, Lessons from the past, *Petroleum Review,* March 2006 Issue pages 36–38 & 42.

6) European Commission Major Accidents Bureau, 2006, MARS Database of major hazards incidents, *http://mahbserv.jrc.it/mars/default.html*, Version dated January 2006
7) Marsh's Risk Consulting Practice, 2003, The 100 largest losses 1972 to 2001. *MAHB 20th edition. Also 11th and 16th editions.*
8) American Petroleum Institute, 1979, Safety digest of lessons learned, *API Publication 758.* Sections 2 to 9.
9) Institute of Chemical Engineers, *Loss Prevention Bulletin, Issues* 80 to 176.
10) Lees F., 1983, Loss prevention in the process industries, *Butterworth & Co Ltd,* Appendix 3.
11) Kletz T., 1999, What went wrong, *Gulf Professional Publishing,* 4th Edition.

HAZARDS OF AN EXPERT WITNESS – AN AUSTRALIAN EXPERIENCE

R. Ward
Visiting Fellow, Faculty of Engineering, University of New South Wales,
Sydney, N.S.W., Australia

INTRODUCTION

This paper is not, strictly, about process safety *per se*, it's about what can happen after a failure of process safety, that is, an inquiry into what has happened. Then it concerns the hazards faced by someone who becomes involved in the investigation, namely, the expert witness, in which role the author of this paper has acted for several years, investigating over a hundred incidents. The paper based on this abstract will outline some example cases of interest which illustrate the investigation process, how an answer was found for each, what can be learned from them, and how they demonstrate hazards of performing the expert witness function.

WHAT IS AN EXPERT WITNESS?

The author has found the expert witness is defined in both an English reference (well, one from part of the UK) and one from the USA. The first is brief and is quoted here in full (Stewart, 2001):

> *Expert witness in the law of evidence, a witness who is allowed to give opinion evidence as opposed to his evidence of his perception. This is the case only if the witness is indeed skilled in some appropriate discipline. An exception to the usual rule of practice whereby witnesses are heard one after another and do not hear the evidence of the preceding witness is made in relation to competing experts. The term* skilled *witness is preferred in Scotland.*

The American definition covers much the same ground and with more detail, hence is worth reviewing also in full (Gifis, 1996):

> *Expert witness a witness having "special knowledge of the subject about which he is to testify; that knowledge must be such as is not normally possessed by the average person. The expert witness is thus able to afford the tribunal having the matter under consideration a special assistance". This experience may derive from either study and education, or from experience and observation. An expert witness must be qualified by the court to testify as such. To qualify, he or she need not have formal training but the court must be satisfied that the testimony presented is of a kind which in fact requires special knowledge, skill or experience. Such testimony, given by an expert witness,*

> *constitutes EXPERT EVIDENCE or EXPERT TESTIMONY. Hypothetical questions [asking the witness to assume certain stated facts] may be asked of an expert witness as a way of educating the* ***trier of fact*** *in the area of the expert's knowledge or experience.*

(The above reference also has, within its text, references to USA statutes, which have not been included in this extract.)

Summing up: an expert witness is a person with special knowledge related to the matter under consideration, must be qualified as having that special knowledge, the court must be satisfied the person has the qualifications and special knowledge, and based on that the person can give opinion testimony rather than observed or factual evidence.

The essential feature of the above definitions is that although the expert witness may be hired by one of the parties he is not truly responsible to them, his loyalty is to the Court. His function is that of an *amicus curiae*, a "friend of the Court" (also defined by both Stewart and Gifis), and as such, if he discovers something contrary to either party he should, indeed, must disclose that.

The acceptance of opinion evidence is an interesting feature of expert witness testimony as it's almost contrary to the general rules of evidence in British and similar legal systems, which "normally" accept only factual, observed, recorded, physical evidence.

AND WHAT IS AN "EXPERT"?

It would have been reasonable, before proceeding to defining "expert witness", to have defined "expert" more generally, and we are fortunate in having available a reference which discusses the expert and its position in society very thoroughly, though with some satire and tongue-in-cheek cynicism, with a definition (Ford, 1982):

> *Experts are unassailable and superior individuals who use a language of their own to cloak their inner whims in a spurious aura of authority.*

That does not *quite* mesh with our understanding of the above definitions of an expert witness, only demonstrating how we can find humour in serious matters. However, let's proceed …

THE NEW SOUTH WALES CODE OF CONDUCT

The Australian judicial system is as complex as the English, with Local Courts presided over by a magistrate (named, years ago, as "Petty Sessions") in city suburbs and country towns, then there are the District Courts in major cities, the Supreme Court which is at the State level and hence in the state capitals, and the High Court of Australia which is nominally in Canberra but has branches elsewhere such as in Sydney. The District, Supreme, and High Courts are all presided over by judges. It's still possible to go one step higher to the Privy Council, but these days that happens very rarely.

One result of these divisions is different details in different states, and in New South Wales (where this author has worked) an expert witness must abide by a "Code of Conduct". Until recently this was District Court Procedures Part 28, rule 9c and Part 28A, rule 2, then in 2005 the Civil Procedure Act was proclaimed which led to the Uniform Civil Procedure Rules (Amendment No 12) in which there's Schedule 7 Expert Witness Code of Conduct (Rule 31.23) (cf SCR Schedule K). The full statement of this Code is attached here as Appendix A.

The essential features of the Code are: an expert witness is responsible to the Court, not to a party; must work or confer with other experts if so directed; and must include in the report his qualifications, plus the facts and assumptions on which his opinions are based, and reasons for those opinions.

THE DIFFICULTY OF AN OPINION BASED ON JUDGEMENT

The general rule followed in legal argument is that witnesses are only allowed to give facts as evidence, and these must be facts as they know them, not hearsay, what's been told to them by others, which is excluded as evidence. In similar manner a witness cannot in general give a conclusion, an inference, or an opinion in evidence, the exception being an expert witness who is permitted to give accepted evidence based on his opinion. This is a convention which has developed in English law through several centuries, progressively accepted by the courts (Jones, 1994).

The use of the word "opinion" immediately introduces one of the hazards, simply because an opinion is based on human judgement. In general, we humans try to make decisions objectively, but some decisions cannot be made entirely by use of objective techniques, and require use of the mysterious element we term "judgement", which can be faulty. The type of decision-making in which an individual's judgement is required has been given the title "judgement call", originally, it seems, by the USA military, but the term has by now been taken up by business. Expressing the need for such a type of decision is often indicated by another person saying to the decision-maker: "It's your call."

The characteristics or parameters of judgement calls are such that some risk is usually involved because there is a choice, in a high stakes environment, between two or more poorly-identified options, and the choice must be based on lack of accurate information, perhaps ambiguous information, while facing conflicting goals, often with time constraints and a close time-horizon, finally with a background awareness that getting it wrong may have serious consequences. All *that* is why a judgement-call-decision needs to be a correctly-made decision. There is, almost always, no second chance, to allow going over and correcting what was done. (For a good coverage of this see Mowen, 1993.)

An expert witness's opinions are certainly based on knowledge of the particular matter under review, but in practice one must qualify that by terming it *some* knowledge, for it's very rare the expert witness knows, intimately and in complete detail, all the facts and background applicable to the case. So the mind makes an agile chamois-like "leap of faith" from what's *known* to what can be *speculated* from that knowledge, and that's expressed as an opinion.

WHAT DOES AN EXPERT WITNESS DO? ON WHAT DOES HE DEPEND?

For continuity of work the expert witness depends, of course, on getting cases come to him. Most of the cases on which this author has worked have come from agents, the majority having come from the commercial departments of two Sydney universities, a few from independent agents, and a small number direct from law firms. The agents act as intermediaries between law firms and the experts, and the law firms are of course acting for an injured party or a defendant. So we have a problem: for whom is the expert working? Not for the agent, who usually sets operating procedures, "rules of engagement". Strictly, business-wise, for the law firm. But, morally and ethically, for the party who has hired the lawyers. Resolving this distinction is sometimes tricky.

First impressions suggest that what an expert witness does is *to find answers*. That's undoubtedly true, but recognizing that leads back to asking: how does one find those answers? The obvious answer is: look at the facts. But facts are, sometimes, indeed very often, hard to find, they may be hidden by a lack of expert knowledge, they may be buried under a mountain of contradictory statements, they may be obscured by irrelevant information, or they may simply be confused by statements made by persons directly involved in the matter under investigation. Or the persons providing the facts may be avoiding the truth, or distorting the truth, or providing disinformation, or, putting it simply, telling lies.

Sometimes the expert witness has specialised knowledge, which is the trick Sherlock Holmes used so often he has been accused of cheating when solving some of his cases. In real life such knowledge is often picked up through previous work situations, then remembered. The author's experience contains two examples of that, both under the title: The Cases of Spontaneous Combustion.

But without such specialised knowledge there is a simple procedure: *ask questions.* Having said that there is the further problem of asking the right questions, to excavate deep into contradictions, to throw away the irrelevancies, and to come as close as possible to what really happened. The examples illustrating use of those questions are The Case of the Chinese Restaurant, The Case of the 45 kg LPG Cylinder and The Case of the Pumpkin Soup.[1]

These cases, all now well in the past, illustrate what an expert witness has to do to serve the client's interests. They also illustrate how frustrating the task can be, when faced with contradictions, irrelevancies, and statements provided by others. But, most of all, they reinforce the need to ask questions. We don't know the answers, so we may not know what questions to ask when we begin, but there's no alternative to hunting through question after question, perhaps groping blindly to begin, in seeking "the most probable truth".

THE HAZARDS OF THE EXPERT WITNESS

The paragraphs above have outlined in narrative the hazards of the expert witness; here, now, is a summary.

[1]Although none of these cases are still active we believe the full reference details of these cases should not be given, that is, names of those involved in them, dates, etc. We can only assure the reader they are real cases, not at all fictional.

The need to give an opinion based on professional judgement
Lack of "the best evidence", which is seeing what's happened.
Contradictory statements.
Irrelevant information.
Avoidance or distortion of the truth, or complete fabrication of non-truth.

But experience has shown there's been an inherent hazard in all the investigations this author has performed: the length of time between the event and being requested to look into the matter. Two years is uncommonly short, four or five years is frequent, more has happened. By which time footprints in the garden outside the French windows have been raked over, the parlour maid has wiped fingerprints from the door handles, the kitchen has washed all the wine glasses, and the butler has moved all the chairs away from the card table. In addition memories have become frayed into disjointed fragments. The expert has to deal with what little remains, enough to puzzle Hercule Poirot into a frenzy.

THE CASES OF SPONTANEOUS COMBUSTION

The first of these was a small factory located in part of an old building in Sydney, and while operating a fire started and burned down the whole building. The small firm's product was plastic shavings, the material used to surround and protect packaged items, and the fire apparently started in the production system which used air flow to transport the plastic through ducts. Unfortunately for the owner of the business another part of the building was being used as a store by a major airline, which lost several million dollars of stock and of course sued the business in which the fire occurred. The owner tried to cover losses via insurance, and the insurance company's lawyers hired the author as expert witness.

The case came to this author months after the fire, indeed several months after the building had been demolished, a common and frustrating occurrence, meaning no direct evidence was available. Not even photographs. Only statements, of severe indignation by the airline and "it was purely an accident" from the business in which the fire started. We asked questions about the design and construction of the air flow conveyor system and it appeared to have been put together by relative amateurs, leading to asking: "Was the system earthed?" The only available answer was: "Apparently not."

That question came from experience in synthetic resin plants, in particular design of resin solution tanks, into which solvent is delivered to thin the resin. One of the design rules is the hydrocarbon solvent should not be allowed to fall through the air space in the tank, which would generate static electricity, but must be directed onto the side so the stream is earthed, promptly. This memory led to *the opinion* that blowing the pieces of plastic around the ducts generated static electricity, eventually building up a strong enough charge to ignite small particles, which led to a major fire. The result of the case was not disclosed (they very rarely are) but it seems likely the insurance company would have refused to cover losses.

The second case occurred in a warehouse, located in a relatively isolated industrial area well away from the city. Some vegetable oil was spilled, and an absorbent material was spread to soak it up, then left over a long weekend. The fire started in the spill, spread to discarded packaging, then to other stored materials, and the isolated location meant there was some delay before a fire crew arrived. Damage was extensive, to both the building and stock. Once again the case came to this author from the insurance company.

Several years earlier a fire had occurred in a store building where this author worked at that time. The source of the fire was traced to a wastepaper bin into which rags had been thrown after having been used to wipe up a vegetable oil spill, and company chemists agreed that the combination of *vegetable oil* plus *air* plus *dispersion through material such as cloth* plus *time can possibly* lead to ignition. (Quaint reflection: the person who threw the rags in the bin was never identified.)

In the case of the warehouse fire the absorbent material served the same purpose as the rags, and the memory of the earlier fire led to *the opinion* that leaving the oil-soaked absorbent material on the floor, that is, not collecting it immediately for disposal, led to the fire.

Both of those opinions were reached by the "Sherlock Holmes" method, of using specialised knowledge. Were they correct? No-one knows for sure. Probably, yes. Anyway, they were accepted as evidence. But the inherent hazard was they could have been wrong.

CASES WHEN THE RIGHT QUESTIONS WERE ASKED

In the Case of the Chinese Restaurant a worker sued the owner because he alleged he had been injured by hot cooking oil from a deep fryer being spilled over him when he was sweeping the floor. He claimed the broom had knocked off one of the fryer's front legs so that a front corner of the fryer had fallen over onto the floor and the oil had overflowed, spilling onto him. He sued his employer, who put the matter in the hands of a city law firm who hired this author to investigate the defence.

This was an annoying case for several reasons, the first of which was that the law firm hiring the author would not pay for a visit to the country town where the incident occurred, over three hundred kilometres from Sydney. Next, some details in reports from the relevant statutory authority and a doctor did not agree with photographs and a video taken in the restaurant. Third, correspondence with the manufacturer (in a neighboring state) and inspection of an identical deep fryer at a Sydney firm which sold them showed the photographs and video were correct, hence we concluded the reports were not correct. Furthermore, the design of the leg fitting was such that knocking it with a broom could not possibly disturb it. Finally, an experiment with the identical deep fryer showed that if a front leg was removed (by undoing bolts with a spanner) the fryer was still stable, and even if the fryer were pushed down so the frame touched the floor the oil would not overflow.

In the report this author pointed out all these discrepancies and discoveries as tactfully as possible, hoping the lawyers would agree to a more thorough investigation. But, no.

Several months later there was opportunity to pass through the country town when returning from a visit to relations further north, so we visited the restaurant which we found had been vacated, was being stripped, and was for sale. The agent handling the sale was almost directly opposite and was prepared to chat (in country town manner) about what he had heard (by country town gossip) of what had happened, and essentially there seemed to be an understanding the whole affair was some sort of scam involving the owner and the worker. The worker was still living in the country town but in the meantime the original owner of the restaurant (a Chinese woman) had disappeared into Hong Kong.

So, although what was reported to the lawyers was incomplete and rather inconclusive there was considerable personal satisfaction in what was discovered by going to the site and asking the right questions … answers given to the lawyer concerned, purely for personal satisfaction, but with no payment for them.

The author has been involved in a few LPG cases, and the Case of the 45 kg LPG Cylinder is selected for comment here because one might say it's "process engineering related". Briefly, a family had an LPG-fuelled stove on the back porch of a house in a Sydney suburb, and several were injured in what was poorly described as a fire but appears to have been a vapour cloud explosion, said (by both the Police and the Fire Brigade) to have been caused by gas from a 45 kg cylinder delivered just after midday. The incident occurred in mid-afternoon and the inspecting authorities reported the cylinder must have became hot by being left in the sun for about three hours, the relief valve lifted, and the gas ignited. The family sued the gas supplier, a small local hardware and general materials firm, for damages caused by the gas.

The case came to the author months later from the lawyers acting for the supplier. The first question was: well, where is the cylinder? Ah, it had been picked up late the same afternoon by the supplier and returned to the store. Which one is it? There was shrugging of shoulders, one 45 kg cylinder is the same as the next, by the time it went back into the stock we can't tell one from another, by now it could be anywhere. Why was it returned to stock? Well, it seemed to be full, so it could still be sold. The actual truck driver was questioned: when you picked it up, did it seem to be full? Answer, yes, it seemed to be full weight. Was the paint showing signs of being in a fire? Burned? Discoloured? No, it was clean. Sign of a leak? No. All that seemed to indicate no gas had left the cylinder.

The next set questions related to seeing the location of the incident. No, you can't go there, the people were renting and they have moved, anyway there's nothing to see, and you can't talk to them. Well, what about the LPG stove which was said to have been in use and provided the source of ignition? No-one knows where it is, it seems to have been thrown away. Well, where was it getting its gas? The records showed the 45 kg cylinder wasn't in use that day, only delivered for future use. Oh, yes, the stove had a 9 kg bottle attached to it. Someone had said the family refilled that 9 kg bottle from the 45 kg cylinders they had delivered from time to time. The only conclusion which seemed to fit the facts was that the explosion had been caused by leakage from the 9 kg bottle, maybe from the hose or a connection fitting. Incidentally, even if the whole 9 kg had been released it seemed to be insufficient for a VCE to occur, hence that was argued as being unlikely, even though it appeared, from neighbour's reported remarks and a broken window, that *something* had happened.

After delivering the report the case became even more interesting. At that time the number of injury cases had led to many being settled by arbitration performed by appointed lawyers, not in front of a judge, and this case went that way. The second interesting feature was the injured family also hired an expert, whose opinion was that the cylinder had been overfilled and expansion of the liquid, caused by the heat of the afternoon sun, had lifted the relief valve. The arguments against that came: first, from the depot where the cylinders were filled; that organisation had procedures and a quality control system which made overfilling virtually impossible, second, if that had been so there had been sun-exposure time for such a leak to have occurred days earlier, and third, by checking the location of the house in a street directory which showed the rear of the house would have been in shade after midday.

Altogether, the claim against the supplier of the gas seemed to collapse, but the arbitration result was a surprise. After the final meeting with the arbitrator the lawyer with whom this author had worked came out and said the family had been awarded some damages; when asked: "How much?" she gave a curious smile and replied: "Not much" without quantifying that. One could only assume the payment was to make the matter "go away".

The Case of the Pumpkin Soup is an example of asking the right questions by performing experiments, and has something of a process engineering flavour. The background was a young woman who had a blender, given to her by her sister as a wedding present, and when using it for the first time to prepare pumpkin soup the lid blew off and splashed her with hot liquid. The provided photographs showed serious scald injuries along her arm and body front. She sued the manufacturer.

This was one occasion when the vital piece of evidence was available: the lawyer had the blender and handed it over for inspection. We (this author and wife) made pumpkin soup, following the same recipe and procedure, poured it into the blender, and the switch was flicked very briefly. The lid was held down firmly, but in the one or two seconds of running the hot liquid spurted out from under the lid. Repeated tests under different conditions, different levels and temperatures, confirmed this was indeed a physically hazardous process!

The manufacturer's instructions were reviewed; they did not warn this could happen. Fourteen other brands of blender were inspected and reviewed, most only fairly tight labyrinth seals, some had loose lids, and some had lock-down lids which might leak but wouldn't spray as the one tested did.

The conclusion suggested that the firm was negligent in not advising users to hold the lid down, not to blend hot liquids, and not to fill above a certain safe level. Sadly, we were not informed what compensation was made by the blender manufacturer.

ANOTHER JUDGEMENT FACTOR – EMOTIONAL IMPACT?

Many injury cases have come to this author, but no fatalities, which might have, by their emotional impact, influenced expert opinion. But there was one injury case which did have an emotional impact, due not only to the nature of the injury but also because of the class of person injured.

A woman, an Asian migrant, employed as a worker in a factory in a Sydney suburb, was allowed by the management to bring her ten-year-old daughter to work during the school holidays, presumably because there was no-one at home to look after her and to save child-mining costs. The woman was operating a strange machine which was, in effect, a horizontal press, built to insert cloth materials into containers for advertising purposes. Somehow, the child put her hand between the ram and the container and her hand was crushed beyond repair.

After being prosecuted by the government safety authority the employer sued the manufacturer of the press, claiming they were responsible for the injury, and this author was hired to investigate and prepare a defence statement. Briefly, what came out was that the press was designed and built to the *owner's* specification, first a pilot version, then a final one, which actually had a three-position rotating die-holder so with two people working (one on each side of the machine) one could load, rotate the loaded die to in line with the ram and actuate the drive, then rotate the die-holder to the third position for the second person to unload the pressed item. There was a change of ownership, and the rotating die-holder was removed and replaced with a single die, which removed the original safety feature: the rotating system meant hands were never in line with the ram. Now, with the single die, a hand had to pass in line with the ram to load the die.

Even though the damaged hand was not seen, either in the person or in photographs, the description of the injury made objective reasoning rather difficult, but the history of the machine led to suggesting the manufacturer had little responsibility for what happened, the suggested proportion was five percent. This author is left with a niggling feeling that the factory's appearance may have had some effect on him; it was crowded, poorly set out, poorly lit and ventilated (strong chemical odour), with stacks of cardboard packaging (fire hazard), but no fire extinguishers, somewhat disgusting amenities, altogether a very poor quality workplace, what one names as a sweat-shop.

REVIEWING THE HAZARDS OF THE EXPERT WITNESS

The hazards of acting as an expert witness have been stated above, and the above cases illustrate those very well. However, we can now not only review but also add to the earlier statements.

The Cases of Spontaneous Combustion show how knowledge previously obtained can be used to solve the problem of what caused something, in these cases, fires. But there's an inherent hazard simply phrased by these questions. Is what is remembered from the past experiences relevant, truly applicable. to this present situation? Are the circumstances sufficiently alike to make that information applicable? Is an excessive level of speculation needed to apply the information? These questions are only resolved by using judgement, and one person's full-daylight judgement may be another person's blind leap in the dark. Both are hazardous for one's reputation, and one must decide which one is facing when applying judgement.

The Case of the Chinese Restaurant and The Case of the 45 kg LPG Cylinder involved lack of the best evidence, contradictory and irrelevant information, and we believe some disinformation. Not being able to see the actual location was a serious handicap. However, understanding the first was improved by seeing an identical deep fryer and testing it, and giving a satisfying report on the second depended on having been involved in the LPG industry. At the end both reports only gave opinions, believed to be correct, but of course one doesn't know whether those opinions were in fact correct.

The Case of the 45 kg LPG Cylinder was the only one in which this author was opposed by another expert. The arbitrator did not require the two to confer and agree, indeed, it's doubtful we would have or could have.

The Case of the Pumpkin Soup was a rare example of feeling, when the report was written, that we had the best evidence, there was no contradictions, irrelevancies, or lack of true information. We were able to perform experiments which confirmed the plaintiff's claims and, we hoped, led to compensation payment. The hazard, of course, was the possibility of being injured as the young woman was.

CONCLUSION

Being an expert witness is hard work, though not consistent week by week, sometimes there's been three or four cases in the one month, then a break of several months, and when a case turns up a report is usually needed in a week or two. Of the hundred–plus cases investigated and reported through some twenty years almost all have been frustrating, one way or another, some have been difficult, some have been related to tragedy, some have queried whether human nature is as good as it's supposed to be. But altogether, the experience has been interesting and in fact enjoyable.

Some lessons have been learned. Keep asking questions. Don't accept the first answers received. Don't put in writing to the lawyers anything found out which argues against the lawyer's client, written material can be "discovered" by the other side, talk it over first.

The ultimate hazard is "getting it wrong".

The present state of play has been reduced by our state government's introducing legislation about seven years ago which severely limited injury claims, which, inevitably, led to an equally severe drop in cases for investigation. However, there seems to have been, in the last year or so, a return to near what existed those years ago.

REFERENCES

Ford, B. J. The Cult of the Expert. Corgi Books, Transworld Publishers Ltd, London. 1982.

Gifis, S. H. Law Dictionary. Barron's Educational Series, Inc. Hauppauge, New York. 1996.

Jones, C. A. G. Expert Witnesses. Clarendon Press, Oxford University Press, Oxford. 1994.

Mowen, J. C. Judgement Calls. Making Good Decisions in Difficult Circumstances. Simon and Schuster, New York. 1993.

Stewart, W. J. Law (Collins Dictionary). Second edition. HarperCollins Publishers, Glasgow. 2001.

APPENDIX A
Expert Witness Code of Conduct (New South Wales)

Uniform Civil Procedure Rules (Amendment No 12) 2006
under the
Civil Procedure Act 2005
The Uniform Rules Committee made the following rules of court under the *Civil Procedure Act* 2005 on 4 December 2006.
Jennifer Atkinson
Secretary of the Uniform Rules Committee

Schedule 7 Expert witness code of conduct
(Rule 31.23)
(cf SCR Schedule K)

1 Application of Code
This code of conduct applies to any expert witness engaged or appointed:

(a) to provide an expert's report for use as evidence in proceedings or proposed proceedings, or
(b) to give opinion evidence in proceedings or proposed proceedings.

2 General duty to the court
(1) An expert witness has an overriding duty to assist the court impartially on matters relevant to the expert witness's area of expertise.
(2) An expert witness's paramount duty is to the court and not to any party to the proceedings (including the person retaining the expert witness).
(3) An expert witness is not an advocate for a party.

3 Duty to comply with court's directions
An expert witness must abide by any direction of the court.

4 Duty to work co-operatively with other expert witnesses
An expert witness, when complying with any direction of the court to confer with another expert witness or to prepare a parties' expert's report with another expert witness in relation to any issue:

(a) must exercise his or her independent, professional judgment in relation to that issue, and
(b) must endeavour to reach agreement with the other expert witness on that issue, and
(c) must not act on any instruction or request to withhold or avoid agreement with the other expert witness.

5 Experts' reports

(1) An expert's report must (in the body of the report or in an annexure to it) include the following:
   (a) the expert's qualifications as an expert on the issue the subject of the report,
   (b) the facts, and assumptions of fact, on which the opinions in the report are based (a letter of instructions may be annexed),
   (c) the expert's reasons for each opinion expressed,
   (d) if applicable, that a particular issue falls outside the expert's field of expertise,
   (e) any literature or other materials utilised in support of the opinions,
   (f) any examinations, tests or other investigations on which the expert has relied, including details of the qualifications of the person who carried them out,
   (g) in the case of a report that is lengthy or complex, a brief summary of the report (to be located at the beginning of the report).

(2) If an expert witness who prepares an expert's report believes that it may be incomplete or inaccurate without some qualification, the qualification must be stated in the report.

(3) If an expert witness considers that his or her opinion is not a concluded opinion because of insufficient research or insufficient data or for any other reason, this must be stated when the opinion is expressed.

(4) If an expert witness changes his or her opinion on a material matter after providing an expert's report to the party engaging him or her (or that party's legal representative), the expert witness must forthwith provide the engaging party (or that party's legal representative) with a supplementary report to that effect containing such of the information referred to in subclause (1) as is appropriate.

6 Experts' conference

(1) Without limiting clause 3, an expert witness must abide by any direction of the court:
   (a) to confer with any other expert witness, or
   (b) to endeavour to reach agreement on any matters in issue, or
   (c) to prepare a joint report, specifying matters agreed and matters not agreed and reasons for any disagreement, or
   (d) to base any joint report on specified facts or assumptions of fact.

(2) An expert witness must exercise his or her independent, professional judgment in relation to such a conference and joint report, and must not act on any instruction or request to withhold or avoid agreement.

The full text of the **Uniform Civil Procedure Rules (Amendment No 12) 2006** can be found at:- http://www.advertising.nswp.commerce.nsw.gov.au/NR/rdonlyres/er4z6bl6syffgusdj73aowiq2k3ajpqdrlc3bhh4z3ow5ff5a5yth4egnnxcojjbysr762ol6epgeh/Government+Gazette+8+December+2006+-+Part+A.pdf

'INCREDIBLE'

Glenn Sibbick
Operations Director, Centrica Storage

On Thursday 16th February 2006, the Rough 47/3 Bravo offshore platform was evacuated having suffered a catastrophic vessel failure and a resultant very large hydrocarbon fire.

'Incredible' is a video providing a clear account told by those directly involved, of the incident, the investigation and the platform's swift reinstatement.

The Objectives of the session will be to:

(i) Share the broader learning's from the incident
- Highlight the root cause (heat exchanger design).
- Safety critical system performance.
- Effective restoration and response.
- The human impact.
- Plant inherent risk.

(ii) Describe what really happened:
- Present a complete picture, pulling together all aspects of the event and follow-up activities.
- Demonstrate the professional approach of Centrica Storage Limited and all the individuals involved.

The session will:

- Demonstrate Centrica Storage's commitment to share lessons and its desire to achieve high levels of safety performance.
- Challenge the historical approach to communicating and sharing lessons from major accidents in Hazardous Industry.

In summary the DVD will provide broader lessons for the oil and gas, and supporting parts of the industry. This emphasis however must be placed on the audience deciding what this means to their organisations.

The following safety alert was distributed to operators to draw their attention to the root cause of the incident which was identified during forensic investigation.

HSE

Health and Safety Executive

CATASTROPHIC FAILURE OF SHELL AND TUBE PRODUCTION COOLER

Safety alert: 01/2006
Issue date: May 2006

INTRODUCTION

1. A recent serious incident occurred that involved the catastrophic failure of a shell and tube heat exchanger, and there is a potential risk of failure to heat exchangers of the same, or similar, design. This notice describes the incident and outlines the action that should be taken by duty holders.

BACKGROUND

2. The incident on an offshore gas production platform occurred when a shell and tube production cooler suffered a catastrophic failure. Seawater was being used to cool High Pressure (HP) hydrocarbon gas. The shell, tubes and titanium cladding sheet were torn from the steel tube sheet and propelled across the deck with sufficient force to rupture an adjacent exchanger. The cooling water pipe work and vent pipe work were torn off the shell and the tube sheet and channel end were ripped off the supports. There was a significant and immediate gas release followed by ignition and an explosion. Fortuitously there were only two relatively minor injuries, but under slightly different circumstances there could have been significantly more serious casualties.

HEAT EXCHANGER INFORMATION

3. The heat exchanger operates with sea water on the shell side and HP hydrocarbon gas plus condensates and water on the tube side. The tube sheet had been manufactured from carbon steel with an explosively bonded, 13 mm thick titanium cladding plate on the sea water side to provide corrosion protection. A key feature of this design is that it allows the HP hydrocarbon gas, condensates and water to come into contact with the steel tube sheet/titanium interface, thus creating the potential for in-service degradation mechanisms as detailed in paragraph 4 below.

Tube to Tube sheet and Titanium cladding to Steel tube sheet interface detail.

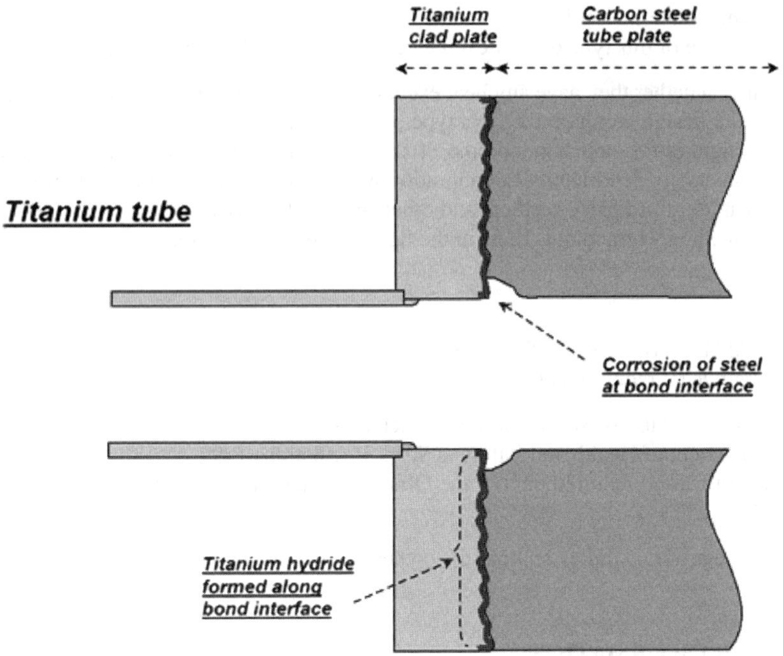

REASONS FOR FAILURE

4. Based on the results of an ongoing forensic investigation, the current evidence suggests the following sequence of events:

- That the explosively formed bond between the titanium cladding and the tube sheet was fit for purpose at the time of manufacture.
- It is probable that liquid in the process stream led to galvanic corrosion of the steel tube sheet at the intersection of the tube holes and the steel/titanium interface (bond interface).
- Hydrogen released by the corrosion process had reacted with the titanium and resulted in the progressive formation of brittle titanium hydrides at the bond interface.
- The growth of titanium hydrides weakened the bond interface sufficiently to cause a sudden, dynamic failure of the bond.
- Failure of the bond resulted in rapid pressurisation of the bond interface, complete separation of the titanium cladding plate from the steel tube sheet, followed by over pressurisation of the shell and catastrophic failure of the cooler.

ACTION REQUIRED

5. HSE considers that the emerging findings of this investigation are significant enough to require immediate action by duty holders in the light of the sudden and possibly catastrophic nature of this type of heat exchanger failure. Therefore, duty holders should:

- Identify whether they have any heat exchangers manufactured to the same design or to a similar design incorporating this type of steel-titanium interface.
- In the light of this notice, undertake a risk assessment of the continuing use of any heat exchangers so identified, in collaboration with the plant manufacturer (if still in business) and/or third party verifiers and other specialist advisers.
- Implement any remedial actions in the light of the risk assessment.

RELEVANT LEGAL REQUIREMENTS

6. The main legal requirements include:

- Provision and use of Work Equipment Regulations 1998.
- Management of Health and Safety at Work Regulations 1999.
- The verification requirements of the Offshore Installations (Safety Case) Regulations 2005.

There are legal duties on both the owner and the operator of the plant.

FURTHER INFORMATION

Any queries relating to this notice should be addressed to:

Team Leader – OSD3.4 Mechanical Systems
Health and Safety Executive
Hazardous Installations Directorate
Offshore Division
5N.2 Redgrave Court
Merton Road
Bootle\Merseyside L20 7HS
Tel: 0151 951 4036

This guidance is issued by the Health and Safety Executive. Following the guidance is not compulsory and you are free to take other action. But if you do follow the guidance you will normally be doing enough to comply with the law. Health and safety inspectors seek to secure compliance with the law and may refer to this guidance as illustrating good practice.

DUST EXPLOSION IN SUGAR SILO TOWER: INVESTIGATION AND LESSONS LEARNT

M Westran[1], F Sykes[2], S Hawksworth[3] and G Eaton[3]
[1]British Sugar, Sugar Way, Peterborough
[2]Health and Safety Executive, Old Chapel Way, Norwich
[3]Health and Safety Laboratory, Harpur Hill, Buxton

On 21 July 2003 a dust explosion occurred in the Sugar Silo Facility at British Sugar Refinery in Cantley, Norfolk. The plant was not operational at the time of the explosion and was actually undergoing maintenance. This paper is produced by HSE/HSL and British Sugar and describes:

i. The circumstances of the explosion in terms of the formation of the explosive atmosphere, the source of ignition and development of the explosion.
ii. The physical consequences of the explosion in terms of damage to the plant, the travel of the explosion through the plant, and the subsequent fire.
iii. The consequences for British Sugar during the days following the explosion and their close cooperation with HSE during the subsequent investigation.

The paper will also discuss lessons learnt from this incident, in terms of practical considerations of explosion prevention, the practicalities of complying with DSEAR, and the protection of industrial systems such as bucket elevators from the effects of such explosions.

1. INTRODUCTION

On 21 July 2003, a dust explosion occurred in the Sugar Silo facility at the British Sugar Refinery in Cantley, Norfolk. The explosion propagated through large parts of the facility, having its greatest impact in the bucket elevators and dust extraction system. At the time of the explosion, the plant was not operational but was undergoing maintenance and modifications which involved welding on one of the bucket elevators known as the Silo Feed bucket elevator, to attach support frames onto the elevator casing for out of alignment detectors. Incidents such as this clearly cause extensive damage, injury and even death and when they occur and it is essential to learn as much as possible from them. It is for this reason that HSE/HSL and British Sugar decided to produce this paper, which describes key features of the explosion and its impact on the site.

The Sugar Silo facility at Cantley is an extensive facility consisting of six silos, a bucket elevator tower and conveyers top and bottom for transferring sugar into and out of the silos. The initial explosion appears to have occurred in the bucket elevator tower at the east end of the Sugar Silos, which is the area where the welding work was taking place at the time. The photograph in Figure 1 shows the bucket elevator tower and adjacent silos.

Figure 1. Photograph showing bucket elevator and adjacent silos

The elevator tower is approximately 52 m high and the transfer of sugar from the base of the tower to the top, uses two linked elevators known as the Production and Silo Feed elevators.

Within the tower, the lower Production/Bulk Export elevator was a single casing that housed the two separate bucket elevators to carry crystallised sugar from ground level to a height of approximately 20 m. At this point, sugar from the Production Elevator transferred to a second elevator known as the Silo Feed elevator via a chute and sugar from the Bulk Export Elevator transferred to the bulk export area of the plant in an adjacent building. Although the two elevators shared the same case, they operated in independent channels separated by a central wall. The combined elevator was fitted with two explosion relief vents to protect it in the event of an explosion. The lower explosion vent was in a vertical orientation, fitted approximately 5 m above ground level on the up leg of the elevator. It was designed to operate at 0.1 bar with a tolerance of 25% and was actually made up of 4 individual vent panels. The top explosion vent was horizontal and was directed out through the tower wall via a right-angled duct. Again the vent was made up of four individual panels of the same size as the lower vent.

The Silo Feed elevator took crystallised sugar from the top of the Lower Production Elevator to the top of the storage silos discharging into the silo feed scroll and then onto a series of belt conveyors feeding the storage silos. This elevator was 38.4 m in height. This elevator was again fitted with explosion vents, the lower of which was a vertically orientated approximately 5.7 m from the base of the up leg of the elevator. This vent was made up of four individual panels, designed to operate at 0.1 barg with a tolerance of 25%. The second vent was horizontal at the top of elevator and feed out through the tower wall via a right-angled duct and was made up of two individual vents.

In addition to propagating through the two bucket elevators, the explosion also propagated through other parts of the plant, including the dust extraction system ducting which terminated at an Airmaster bag filter/dust collector system. The Airmaster filter body was fitted with two explosion relief vents, which used a membrane of 'Flexotalic Glingerite' with an estimated static bursting pressure of 0.05 barg. The dust collector was housed in a separate building. The ducting system was complex, connecting almost all parts of the plant. It was badly damaged by the explosion which propagated through the ductwork.

2. THE CAUSE OF THE INCIDENT

As already described, the work underway involved the welding of metal frames to the outside of the casing of the Silo Feed Elevator. Figure 2 shows a close up of a typical weld. This work was carried out under a permit and a risk assessment had been carried out. However, this risk assessment assumed that there was no potential for an explosive atmosphere because the plant was not operational, and hadn't been for some time, and the work was been carried out on vertical sections of the elevator well away from areas where any material may have settled.

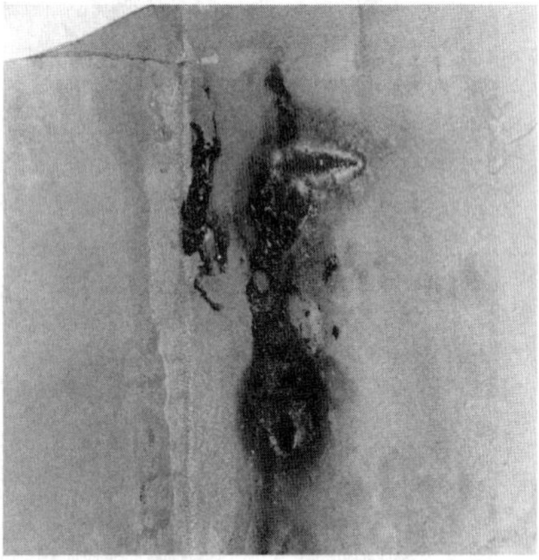

Figures 2 & 3. Close up of weld on outside and inside of case showing burnt sugar residue around weld and second area away from weld

Figure 3 shows a photograph of a typical weld viewed from reverse side, inside the elevator, which clearly shows that the heat of the welding process had transmitted through to the inside of the casing. In addition, the photograph shows black residues of charred sugar. Almost every weld examined showed similar charred residues.

A metallurgical examination was performed on the section of the elevator being worked on at the time which showed that during the welding, molten metal had penetrated through onto the inside of the casing producing temperatures approaching 1500°C. This penetration (stainless steel through mild steel) is clear from the section through one of the welds shown in Figure 4.

For thoroughness, given the serious nature of the explosion, it was important to establish that the explosive atmosphere was in fact sugar dust. This involved firstly establishing that the sugar that was likely to be in the system could have formed an explosive

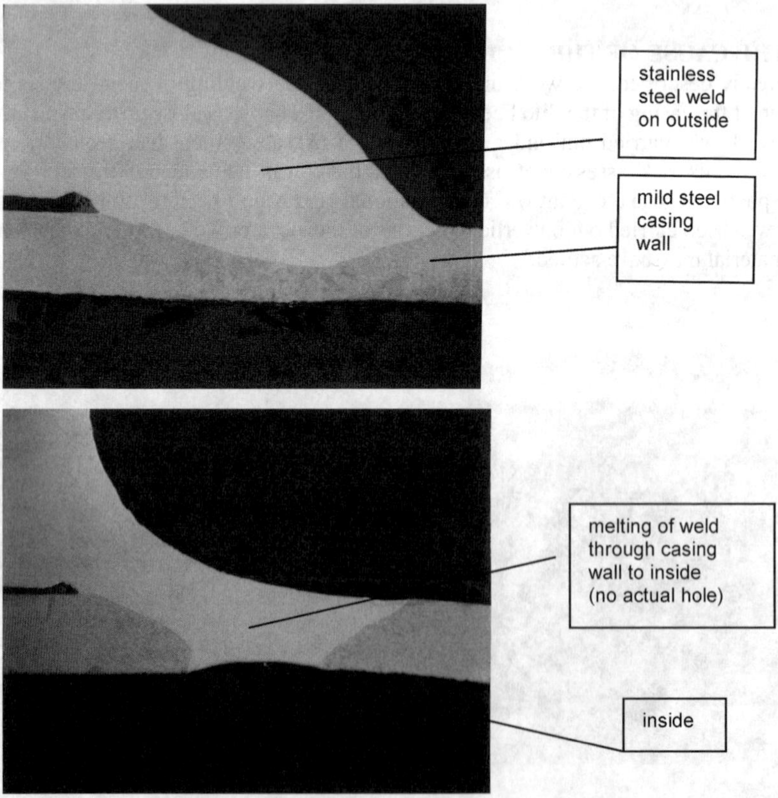

Figure 4. Sections through welds on sample

Table 1. Results of 20 litre sphere & BAM oven test

Maximum explosion pressure	8.6 bar
Maximum rate of pressure rise	650 bar/s
Specific material constant (K_{st})	176 m.bar/s
Dust classification	ST1
Lowest oven temperature for ignition	372°C
Lowest surface temperature for ignition	335°C

atmosphere and had typical characteristics of such an explosive atmosphere. Representative samples were taken from the material filtered out by the dust extraction system as these were unaffected by the explosion and were representative of the fine material that could have become airborne during transfer through the bucket elevator systems. The results of the 20L sphere and BAM oven test are given in Table 1.

At the time of the explosion the plant was not operational. As already mentioned, there was evidence of sugar on the vertical surfaces inside the elevator system. The Lower Explosive Limit (LEL) for icing sugar/sugar fines is around 60 to 100 g/m^3 (according to the BIA Report)[1]. In this incident it was assumed that explosive atmosphere was provided from the sugar coating the interior of the walls of the elevator, ignoring any sugar dust on the elevator belt or in the buckets. Based on the dimensions for the elevator section the amount of sugar required was calculated to be approximately 12 g/m^2 of surface area. Using a volume packing fraction of 50%, this gives a layer thickness of approximately 0.5 mm to achieve the LEL. Figure 5 shows a photograph taken looking into the up leg of Import/Export elevator which was not involved in the explosion. Clearly, there is considerably more than 0.5 mm of sugar coating the inside of the elevator, and so if the Silo Feed Elevator were similar the LEL would easily have been achieved. Figure 6 shows evidence of similar deposits after the explosion in the Silo Feed elevator.

Figure 7 shows a series of frames from a video recording of a test heating a sugar sample on a 1.5 mm thick metal plate. The plate is heated below with an oxy-acetylene welding torch. The sequence of events shown is (1) the heating chars the sugar and produces flammable vapours, (2) When the metal beneath reaches red heat (as would have the case in the incident) the vapours ignite and cause sustained burning of the layer even when the welding torch beneath is removed. The third frame in the sequence (bottom left) shows the moment of ignition, a small flame kernel starting around the hot spot on the plate.

3. THE DAMAGE CAUSED

The damage to the plant was extensive and the route taken by the explosion complex. It seems, as already discussed that the explosion started in the Silo Feed Elevator propagating up and then across the tops of the silos through the Silo Feed Scroll and Silo Feed

Figures 5 & 6. Photographs taken looking into the hatch on the Import/Export Elevator showing heavy build up of sugar on walls and up-leg of Silo Feed Elevator through hatch at Level 2.5 showing residual sugar fines on casing wall and bucket and less evidence of caramelisation

Conveyers. It also propagated down into and through the Production and Bulk Export elevators and to the Import/Export area of the plant and the Under Silo Elevators. The sections below summarise the explosion damage in the plant following the most likely routes for the explosion to progress from the Silo Feed Elevator, concentrating on the equipment in the elevator tower as the main areas of interest.

3.1 SILO FEED ELEVATOR TO SILO FEED SCROLL AND SILO TOP FEED BELTS

Although the Silo Feed Elevator was the area where the explosion appears to have been initiated, it had suffered comparatively little damage. Caramelised sugar could be seen inside the elevator, appearing more severe on the down leg than the up leg.

The explosion propagated out of the top of the elevator through the Silo Feed Scroll and along the conveyers blowing off the side cases and venting into the space outside. Caramelised sugar could be seen lying on the bottom of the conveyors. Explosion damage, such as displaced room/compartment doors and wall cladding material was observed along the full length of the walkway spanning the top of the elevators. On the top level of the tower the heavy steel door to the outside (access to the top of the silos and an emergency exit) which had been secured by a flat mild steel bolt of approximate size 25 mm × 5 mm, had been blown open, bending the bolt and distorting the door.

Figure 7. Frame grabs from video of heating of sugar deposits leading to ignition

3.2 PRODUCTION ELEVATOR/BULK EXPORT ELEVATOR

These two elevators are contained in a single casing, but are separated by a central steel wall. The most direct route for the explosion to have propagated into these elevators is via the chute connecting the Production Elevator to the Silo Feed Elevator. The explosion may well have then developed in the production elevator before linking into the adjacent Bulk Export Elevator.

The explosion damage to the combined elevator becomes progressively more severe towards the base. At the top there was little, if any, damage apparent to the top (head) of the elevator. Moving down the elevator at level 2 where the damage is more apparent, both legs are clearly distorted and some panels have become partially detached on the up leg.

Figures 8 & 9. Damage to Production/Bulk Export Elevator on level 2 to Production/Bulk Export Elevator just above ground level

Figure 8 show the elevator one floor level above the base where the bottom panel shown has opened up on the up and down legs. Figure 9 shows the damage to this elevator just above ground level. The elevator case opened up here and vented the explosion into the area at the base of the tower destroying the wooden partitions, entering adjacent rooms, blowing the exterior door off in this area causing injury to a worker who was close by. It also pushed down the exterior wall, caused other structural damage indicating that the flat roof in this area lifted and then resettled.

3.3 UNDER-SILO AREA

The door into the silo discharge area had been blown off its hinges and was lying on the floor at a distance of approximately 10 metres. The side and top panels on the conveyor from the silos had been blown off. The top covers had also been blown off the conveyor leading to the bagging shed, hitting the roof in several places. The fusible link on the drop down fire door on the conveyor had operated allowing the door to close part way where it had jammed, presumably as a result of the explosion.

3.4 DUCT SYSTEM AND AIRMASTER DUST COLLECTOR

The duct system on the plant is extensive and complex linking to most of the items of equipment. As such it provided a very effective means for the explosion to propagate through the plant.

The key areas of the ductwork were in the Import/Export area where the damage was severe. In terms of actual explosion damage to other items in this area, it seemed fairly limited. However, the explosion did lead to a fire in this area.

Figure 10. Building housing Airmaster filter. Note damaged ductwork at 3-way junction

At the end of the duct system is the Airmaster dust collection system. Both explosion vents on this system had operated and the filters and filter housing were disrupted. The doors of the building housing the dust collector had been blown off their hinges (see Figure 10). The ducting in this area was severely damaged. At the rear of the building a hole had been blown in the cladding where a piece of the ducting from inside had been blown through the wall and projected for tens of metres across the area outside.

4. ASSESSMENT

4.1 SOURCE OF IGNITION AND EXPLOSIVE ATMOSPHERE FORMATION

The source of ignition in this incident was the welding that was taking place on the outside of the Silo Feed Elevator in the tower. Evidence indicates that molten metal penetrated through to the inside of the elevator casing. This in theory could have directly ignited any explosive sugar dust cloud present.

However, it is also possible that the explosion did not commence immediately. A viable alternative sequence of events is that the sugar in contact with the heated region charred and then began to flame as demonstrated in the tests shown in Figure 6. This flaming could have persisted for several seconds and then ignited a cloud produced after a delay. This would have provided time for the welding work to finish and further work to begin which then disturbed the sugar to produce the explosive cloud, e.g. an impact on the elevator casing. Based on work by Gummer and Lunn (2003)[3], for this second mechanism to be viable, the sugar would almost certainly need to have been flaming, as simply smouldering material does not appear to provide such an effective ignition source.

It is clear that this work should never have been attempted without first ensuring there was no potential for an explosive atmosphere inside the elevator. However, there would clearly be problems in achieving this with the present situation, as any activity involving dismantling the bucket elevator would disturb the sugar coating within the elevator, causing an explosive dust cloud inside it and possibly outside. Care would therefore be required to ensure that ignition sources were not present. The action was and still is to ensure a flammable atmosphere does not exist, this can be by washing or cleaning a section and installing fire blanket barriers etc.

A better approach for consideration, which would be much more in keeping with the ALARP principle and Section 6 of the DSEAR regulations (2002)[7], would be to try and eliminate the potential for an explosive atmosphere to occur in the first place. To achieve this would require a means to prevent or at least limit the build of sugar fines on the inside of the casing. It may be that this has already been considered, but dismissed due to the limited occurrence of an explosive atmosphere because activities that disturb the sugar fines are so infrequent. This is partially achieved through dust extraction, but the nature of transporting sugar does result in fines formation because crystal damage occurs.

4.2 EXPLOSION DEVELOPMENT AND DAMAGE

Once ignited, the explosion was a classical self-feeding dust explosion in the sense that the pressure wave/air flow and vibration produced by the explosion disturbed dust to propagate it through the plant (See Eckoff, 2003)[2]. The initial explosive cloud was probably produced by disturbance of the coating in the area where the modification work was being carried out.

Evidence suggests that from the very early stages the explosion progressed through the system in both directions. Because of the linking chute, the pressure in the Production/Bulk Export Elevators would also have increased during these early stages and airflow and vibration would begin to stir up sugar fines to form an explosive cloud.

Work done by Holbrow et al (2002)[4] gives guidance on the venting requirements for dust explosions in bucket elevators. As a basic requirement they recommend vents (100 mbar static opening pressure) within 6 m of the boot (base) and in the head (top) and of area equal to the cross-sectional area of the elevator. This would appear to be the case here.

However, other important findings reported by Holbrow were:

i. The K_{st} value is of key importance to determine if and how the dust explosion develops. K_{st} values greater than 150 bar m s^{-1} will propagate explosions, and vents additional to those in the head and boot are required on elevators (discussed further below) to limit the reduced explosion pressures* obtained. No extra vents were incorporated on either elevator.

ii. The operation of the buckets had no significant effect on the reduced explosion pressure.
iii. Ignition position has little effect on measured reduced explosion pressures.
iv. The bucket spacing has an important effect on the reduced explosion pressures obtained. A smaller spacing tends to result in lower reduced explosion pressures.

From the data presented by Holbrow et al (2002)[4] it is possible to estimate that a fully developed explosion in the Silo Feed elevator would have produced a reduced explosion pressure of approximately 1 bar. This is based on the K_{st} of 176 bar m s^{-1} reported in Section 2.2.3 and a vent spacing greater than 10 m. An explosion pressure of 1 bar might be expected to cause some deformation of a structure such as this elevator casing, but generally does not appear to have done so. This suggests that the explosion did not develop fully in all parts of the elevator as indicated by the reduced level of caramelised sugar/heating on the up leg. Reasons for this may be that the concentration of sugar fines was not ideal, the relatively close spacing of the buckets may have impeded early development and also that the explosion was able to vent into the Production/Bulk Feed Elevator below thus limiting the pressure.

Considering the Production/Bulk Feed Elevator, clearly much greater damage occurred suggesting a greater reduced explosion pressure was achieved than in the Silo Feed Elevator. From Holbrow et al (2002)[4], an explosion in this elevator in isolation could again be expected to produce a reduced explosion pressure of approximately 1 bar using the same approach as above.

However, the damage to this elevator was much more severe because:

i. The effects of the explosion in the Silo Elevator vented into and so pre-pressurised the interior of the Production/Bulk Export Elevator and produced a turbulent dust cloud. Although the authors know of no work on the effects of dust explosions in linked bucket elevators, the general principles of explosions in linked vessels are described by Holbrow et al (1999)[5] entitled "Dust explosion protection in linked vessels: guidance for containment and venting". The high turbulence and higher pre-explosion pressurisation in the second vessel (in this case the second bucket elevator) combine to give a much higher K_{st} value. This in turn impacts on the effectiveness of the vents, much larger areas being required. This effect is illustrated by reference back to Holbrow (2002)[4], which shows that for a Kst of 211 bar ms^{-1} (20% increase) the reduced explosion pressure for a 12 m vent spacing is 3 bar and increases for greater vent separations. This in itself explains the severity of the damage. Note that secondary

*Reduced explosion pressure is the pressure in the system even with the limiting effects of the vents. Without the explosion vents the pressure inside the elevator could reach the maximum explosion pressure reported section 3.2.3 assuming the elevator was strong enough.

explosions in totally enclosed linked vessels have been shown to easily lead to a doubling of the Kst value (Lunn et al, 1996).

ii. The bucket spacing on the Production Elevator is much greater than for the Silo Feed Elevator, which would lead to greater explosion pressures according to Holbrow (2002)[4].
iii. Because of the wider casing of this elevator and apparently similar bolt spacing to the Silo Feed elevator, the former is probably weaker and so more susceptible to damage.

Because the casing of the Production/Bulk Export Elevator failed, the explosion vented in to the space at the base of the tower causing damage and injury. In addition, the explosion entered the Import/Export area of the plant causing damage and starting a fire. The explosion then propagated through to the Airmaster filter system where again pressure-piling and enhanced turbulence would have led to higher Kst, so that explosion vents were inadequate and reduced explosion pressure was greater than allowed and damage resulted.

5. SUMMARY

While this was clearly a very unfortunate incident, it was a good reminder of the dangers posed by explosive dusts, and in particular the hazards they pose even in plant that is not operational. It demonstrates the need to have a thorough risk assessment based on a very good understanding of the plant in question, and the potential hazards. More specifically, this incident also demonstrates the importance of a means of preventing explosions from propagating between different pieces of equipment. Such equipment could include rotary valves and fast acting valves. Suppression systems are an option, but consideration may be needed to ensure that such systems do not project a dust cloud outside the equipment/elevator at junctions in the conveyers etc. If the plant had been fitted with some means of isolation, then this explosion would have been confined to the Silo Feed Elevator. The Silo Feed Elevator may have suffered more severe damage, but this could be allowed for by adding further explosion vents.

If isolation is not fitted, then vent sizes should be increased, to allow for the higher K_{st} that should be expected during secondary explosions. There may be practical limits to this approach as reference to Holbrow et al (2002)[4] shows that for a K_{st} of 211 bar m s^{-1} vents at 3 m spacing (meeting other criteria described in Section 4.2) would only limit the reduced explosion pressure to 0.6 bar.

Specific recommendations that came out of this incident, but which probably have wider applications, are:

1. Greater care is required when carrying out hot work to avoid a repeat of this incident. In particular, consideration should be given to the presence or creation of a flammable/explosive atmosphere inside equipment when performing work on the outside.
2. Based on the K_{st} value measured for the sugar dust and guidance in the literature (Holbrow 2002)[4], the level of explosion venting on the elevators needs to be increased.

3. In plant where the risk of explosions is recognised, extra measures must be taken to prevent or limit the effects of propagation between interconnected items of equipment. Preferably this should be achieved using explosion isolation. However, if explosion venting is to remain as the means of explosion prevention, vent sizes should be increased to allow for the enhanced rates of pressure that will occur in secondary explosions.
4. Consideration should be given to prevent or limit the build-up of sugar inside the elevators at source (general principle set out in section 6 of DSEAR regulations, 2002) rather than having to deal with the dust build-up prior to maintenance with the associated risk of creating an explosive atmosphere.

Following the incident and during various reviews with explosion protection suppliers, the authors would also recommend the following points:-

1. Ensure that evaluations are undertaken of the strength of equipment that is to be protected by explosion protection equipment. The guidance available in the public domain for applying protection systems assumes strength data, which therefore could result in the incorrect level of protection being applied. For the correct application, actual strength data is required.
2. The design of explosion vents are also critical, ducts with bends or long ducts can render the vent ineffective.
3. Use recognised experts to advise on practical methods of explosion propagation protection. British Sugar believed they were being supplied with suitable systems for conveying to and from bucket elevators. For new installations they now have recognised industry consultants to sign off their designs/installations.
4. A team approach to high hazard activities is recommended, it avoids assumptions and ensures the right people are involved. They now also ensure that Safety Personnel, Explosion Risk Trained Personnel, Plant Managers and the Supervisor involved with Hot Work activities are present during Risk Assessments, Method Statement Generation as opposed to just Permit Issue.

6 REFERENCES

1. Combustion and Explosion characteristics of dusts, BIA-Report 13/97, ISBN 3-88383 469-6
2. Eckoff R K "Dust Explosions in the Process Industries – Third Edition", ISBN 0-7506-7602-7, Elsevier 2003, Chapter 3.
3. Gummer J and Lunn GA (2003) "Ignitions of explosive dust clouds by smouldering and flaming agglomerate" Journal of Loss Prevention in the Process Industries 16 (2003) p 27–32.
4. Holbrow P, Lunn G A and Tyldesley A (2003) "Explosion Venting of Bucket Elevators", Journal of Loss Prevention in the Process Industries 15 (2002) p 373–383.

5. Holbrow P, Lunn G A and Tyldesley A (1999) "Dust explosion protection in linked vessels: guidance for containment and venting", Journal of Loss Prevention in the Process Industries 12 (1999) p 227–234.
6. Lunn G A, Holbrow P, Andrews S and Gummer J (1996) "Dust Explosion in totally enclosed interconnected vessel systems" Journal of Loss Prevention in the Process Industries 9, p 45–58.
7. DSEAR 2002 – Statutory Instrument 2002 No. 2776 "The Dangerous Substances and Explosive Atmospheres Regulations 2002" Section 6 Paragraphs 4 a to e.

THE USE OF PROCESS SAFETY PERFORMANCE INDICATORS TO ENSURE THE EFFECTIVE MANAGEMENT OF MAJOR ACCIDENT HAZARD RISKS – THE HEALTH AND SAFETY EXECUTIVE'S EXPERIENCE

Ian Travers[1], Viki Beckett[1] and Jonathan Carter[2]
[1]HSE HID CI 4
[2]HDE HID CI 3

INTRODUCTION

In 2006 HSE, in collaboration with the Chemical Industries Association published a guidance document on the development of process safety performance indicators, "Developing Process Safety Indicators: A step by step guide for chemical and major hazard industries", HSE publication HSG 254[1]. In this paper we will discuss the approaches that HSE has undertaken to continue to promote the application of metrics in the chemical and major hazards industries, and the developing experience in the practical application of metrics at major accident hazard establishments.

PROMOTING THE USE OF PROCESS SAFETY PERFORMANCE INDICATORS

There has been considerable support for the step-by-step approach to the development of process safety performance metrics outlined in HSG 254. Most notably, the report of the BP US Refineries independent review panel, established on the request of the US Chemical Safety and Hazard Investigation Board following the BP Texas City disaster[2] (and commonly known as the Baker panel), recommends that BP develop, implement, maintain and periodically modify an integrated set of leading and lagging performance indicators to measure how the BP US refineries are performing with regard to process safety. In this recommendation BP are advised to refer to the guidance developed by bodies such as the Health and Safety Executive, and consider the example of leading and lagging indicators provided in this guidance.

[1]"Developing process safety indicators: a step-by-step guide for chemical and major hazard industries" HSE Books 2006 HSG 254 ISBN 0717661806
[2]"The Report of the BP US Refineries Independent Safety Review Panel" James A Baker III et al. January 2007. http://www.chemsafety.gov/index.cfm?folder=completed_investigations&page=info&INV_ID=52

The recommendation from the Baker panel has been very influential in increasing the interest and understanding of the importance of process safety metrics. Following the publication of HSG 254, HSE had a plan of work to publicise the approach and encourage its adoption among operators of chemical and major accident hazard sites, with the expectation that all operators of top tier COMAH establishments should in time develop appropriate metrics to monitor the effectiveness of the process safety management system on their site. The approach has been a combination of direct contact with duty holders on their sites, and the engagement with intermediaries such as industry trade associations to encourage the development, where appropriate, of sector indicators, to enable duty holders to benchmark their process safety performance against their peers.

The Baker panel findings, following on from the finding of the investigation into the major incidents at BP Grangemouth in 2000[3] and subsequent findings by the Buncefield Major Incident Investigation Board (MIIB)[4] have confirmed that performance metrics on key risk control measures are a vital component of the safety management system for any operation managing major accident hazard risks. The conclusions of these studies into the occurrence of major accidents demonstrate that the probability of the major accidents occurring could have been significantly reduced had the appropriate management systems been in place to ensure that senior decision makers within the relevant organisations received regular targeted information on the health of key components to prevent and mitigate major accidents. For this reason HSE HID CI is expecting operators of Top Tier COMAH establishments to have systems in place to measure the performance of key process safety risk control systems, or to be carrying out a programme of work to develop such a set of metrics.

One very effective mechanism that we have adopted in the development and application of the guidance on process safety performance metrics has been carrying out site specific workshops with operators of major accident hazard plant, using part of their major accident hazard plant as an example to work through the process outlined in the guidance document, to build up a suite of leading and lagging indicators for the key aspects of the major accident hazard safety management system in place for that particular piece of process. This approach, developing site specific leading and lagging metrics in collaboration with major accident plant operators, to enable them to work with the developed metrics, and to judge their practicability and efficacy, has been most effective in promoting the application and training those responsible for their development in how to undertake the process.

[3]"Major Incident Investigation Report – BP Grangemouth Scotland: A Public Report Prepared by the HSE on behalf of the Competent Authority" Health and Safety Executive. August 2003.

[4]"Recommendations on the design and operation of fuel storage sites" Buncefield Major Incident Investigation Board. March 2007. http://www.buncefieldinvestigation.gov.uk/reports/index.htm

In addition, we have worked directly with trade bodies to encourage the development of suites of sector indicators. For example, we have worked closely with the Chemical Industries Association Chlorine Users Group in the development of an extensive suite of process safety performance indicators of particular relevance to operators of sites where the principal major accident hazard is associated with their use of large quantities of chlorine. Representatives from the sector worked with HSE, identified the principal sources of hazard on chlorine handling sites, the key risk control systems (RCSs) relied upon to prevent and mitigate major accident hazards, and some example metrics for which site operators could collect data to monitor the effectiveness of these key risk control systems. The result is an extensive list of potential metrics. From this list site operators can choose a smaller set of particular relevance to their operations and for which they can collect data to monitor their major accident hazard safety management system.

The most important step in the implementation of an effective process safety performance indicators programme at major hazard establishments is the senior management commitment to, and support of the process. Another major recommendation from the Baker panel report, and also a recommendation from the Buncefield MIIB report into the design and operation of fuel storage sites is that senior managers should demonstrate leadership in the area of process safety. The Buncefield MIIB report states "Implementing our recommendations will require the sector to show clear leadership in setting high standards of process safety and environmental protection and in pursuing excellence in operations." Senior management involvement in the establishment and implementation of a PSPI programme is a clear demonstration of their commitment to leadership in the area of process safety, and the continued use of performance indicators to inform decision making continues the demonstration of that commitment.

TYPES OF INDICATORS

In our work with duty holder in the development of site specific process safety performance indicators we have found a number of common themes. In general, the metrics developed can be identified into one of three categories: process specific **operational control process safety indicators**; site **generic risk control system process safety indicators**; and **programme work process safety indicators**. These cover in turn the specific measures, engineering and human that are relied upon at the Unit level to prevent and mitigate major accidents; site wide common safety management system processes in place to ensure the prevention of major accidents; and site-wide or corporate programmes of work undertaken to underpin the delivery of effective process safety management.

There are a number of key aspects that any major accident hazard safety management system would expect to include. An effective process safety performance indicators system would be expected to have in place some metrics to enable the effectiveness of the delivery of all these elements of the SMS to be monitored and their effectiveness judged. Depending on the nature of the operations undertaken, it should be possible on most establishments to collect data from indicators in the three identified categories (operational control indicators, generic risk control system indicators and programme indicators) to give information on the effectiveness of the delivery of all the key components of the safety management system.

OPERATIONAL CONTROL INDICATORS

Operational control process safety indicators are those collected at the Unit or process level that relate directly to the measures in place to prevent loss of containment. These may typically relate to key pieces of process plant being inspected and maintained to the appropriate schedule, process controls and operating envelopes being set correctly, and operators undertaking key steps in safety critical activities at the correct time and in the correct manner.

Lagging indicators for operational control indicators should be set to identify challenges to the integrity of the process, and if possible set at the earliest point of deviation.

Leading indicators may, for example, be determined from data collected on:

- Operator actions to set process controls;
- Operator actions to maintain routine operations;
- Operator effectiveness to take remedial action if the process deviates from the identified safe operating envelope; and/or
- Inspection and maintenance of process control instruments and alarms.

GENERIC RISK CONTROL SYSTEM INDICATORS

Generic risk control system process safety indicators are those relating to site wide risk control systems that are applied similarly across all units on an establishment. These may be, for example the permit-to-work system, management of change system or emergency response arrangements.

Lagging indicators relating to the generic risk control systems will be incidents that are identified on investigation to be as a result of failures of the generic RCS. These may be loss of containment incidents, process upsets or any particular demands on critical safety systems.

Leading indicators in this category will be identified by checking that the critical tasks required by the RSC are undertaken correctly, for example by audit of isolations under the Permit-to-Work system, or of risk assessments undertaken as part of the management of change system.

PROGRAMME WORK INDICATORS

Programme work process safety indicators are those relating to statutory or other programme demands that are undertaken across the establishments. These are generally scheduled programmes of work that are expected to be delivered to ensure that the relevant safety management process can be relied upon. This covers key aspects of the safety management system, such as statutory inspections, audits, training of staff in safety critical tasks and development and maintenance of procedures.

The indicators for programme work will predominantly be leading rather than lagging, as it is difficult to identify failure of programme delivery as a single cause of failure of a component of the management system. The leading indicators could be for example:

- Percentage of statutory inspections completed on time;
- Percentage of staff trained to a specific competence; or
- Percentage of the procedures that are up-to-date.

LEADING AND LAGGING PROCESS SAFETY PERFORMANCE INDICATORS

When discussing the issue of process safety performance indicators, they are regularly referred to as being either leading or lagging metrics, and in the guidance document HSG 254 this terminology is used extensively. In particular, the guidance document promotes the use of leading and lagging indicators in conjunction to give "dual assurance" for each risk control system under examination.

LAGGING INDICATORS

The lagging indicators are derived to show whether or not the desired outcome for the relevant RCS is being delivered. In many examples of the use of process safety performance indicators, lagging indicators are metrics of loss of containment, fires or other releases that could be precursors to major accidents. However, it is possible to define lagging indicators as being any early indication that the relevant risk control system is failing to deliver its desired outcome, and depending upon the data collection and potential metrics available around the RCS, lagging indicators can be set to give early warnings of the deterioration of the RCS before loss of containment occurs. Lagging indicators should help identify challenges to the integrity of the RCS, and wherever possible should be selected to identify the lowest reasonably detectable event that demonstrates a compromise of the identified safe operating envelope. Depending on the system under consideration, this may be identified by determining that there is overfilling of a vessel, overpressure in a reactor, over-temperature in a furnace, low-flow in a pipeline or excessive corrosion of the pressure containment envelope. The indication point for the relevant metric should be set at the earliest point of deviation, so that reasonable quantities of data can be collected on the number of occasions that the RCS has failed to deliver its desired outcome, resulting in the detectable event occurring. Lagging indicators give no intelligence on why the deviation has occurred; why the RCS has on this occasion failed to deliver its desired outcome. The continuous collection of data of this nature about the key risk control systems that are relied upon to deliver process safety enables data trending to be undertaken and presented, to give evidence of whether the RCS is continuing to deliver the level of protection that is expected of it, whether that level of protection is improving as modifications and enhancements are made to the RCS, or whether the effectiveness of the RCS is deteriorating, indicated by an increasing rate of the relevant lagging metric. The guidance document recommends the establishment of tolerances around the metrics; setting values for acceptable ranges of the metric with a threshold value such that if the metric reaches a certain level such that it has deviated beyond the acceptable level then there is upwards reporting of this information to enable decisions to be made at an appropriate level as to whether intervention is necessary to address the reduction in the effectiveness of the delivery of the relevant RCS.

LEADING INDICATORS

The leading indicators derived using the methodology described in HSG 254 relate to the most important parts of the relevant risk control system under consideration; the activities that must be undertaken correctly every time, the aspects of the system that are liable to deterioration or the key activities that are undertaken most frequently. This must obviously be related to some aspect of the RCS that can be measured in some way to enable the collection of appropriate metric data. The leading metrics should be selected to show that for each relevant key RCS the critical controls to reduce and prevent major accidents are working as intended. If selected appropriately, leading indicators should highlight whether the risk control systems in place to deliver process safety are operating as designed. Similarly to the lagging indicators, tolerances should be set such that when deviation of the delivery of the key activity is beyond some acceptable level, there is upwards reporting to enable appropriate remedial action to be initiated.

However, it is recommended that duty holders don't get too concerned about determining whether the indicators that they are developing are leading or lagging. In some cases it is not immediately obvious, and a lot of time can be spent determining into which category a particular indicators sits. If duty holders are able to collect useful, informative data about the performance of the key risk control systems that they are relying on, such that they will get early indication that the RCS is deteriorating, and some indication what needs to be done to improve its performance, then that is valuable information that should be collected, analysed, reported and used.

LINKING METRICS TO THE PROCESS SAFETY MANAGEMENT SYSTEM

A typical major accident hazard safety management system would be expected to contain the following components:

- Inspection/Maintenance
- Process instrumentation, controls and alarms
- Permit-to-Work system
- Management of Change process
- Safe Plant Design
- Risk Assessments
- Emergency Arrangements
- Competence Framework
- Control of Contractors
- Operational Procedures

- Effective Communications
- Audit and Review

Ideally, each of these components of the SMS should be subject to monitoring of their effectiveness by the use of performance metrics derived from one or more levels within the organisation. In this way, managers at the unit, plant or corporate level, can determine the level of reliance that can be placed on these components to deliver process safety, and can receive early indication of any developing deterioration of the effectiveness of these key components, in advance of any serious loss of containment events occurring.

The information supplied to site managers, and upwards to corporate managers should be a representative sample of the current status of all of the relevant components of the safety management system. These metrics may come from any of the three levels, depending upon the nature of the risk control system, and should include a balance of leading and lagging metrics. Operators should regularly review and revise the metrics chosen for upwards reporting to ensure that the snapshot that is being examined is representative of operations across the establishment, and to ensure that some areas are not being dealt with in a different way because of the collection of particular metrics data.

REVIEW AND REVISION OF METRICS

It is recommended in the guidance document HSG 254 that all performance indicators are subject to periodic review. A review of the safety management system should be undertaken, based on the metrics information collected to confirm that it is delivering the critical outcomes that it is expected to. The metrics should be updated in the event of any modifications to the plant or process, or to the associated safety management system and risk control systems. In addition, as metrics are developed and used, they can be tailored and improved. Initially metric data may be collected to confirm that a particular aspect of a risk control system is in place, but as experience of the collection and use of metrics develops, and managers require more specific confirmation, data can be collected to determine the degree that a risk control system meets its functional demand. For example, an initial metric to determine the level of staff competence to undertake a safety critical activity may be based on the percentage of staff that have undertaken training in the safety critical activity. This will give some information on the receipt of training, but it will not confirm understanding of the training and translation of the training into working in a safe way. Therefore, a second level, quality metric relating to staff competence may be based on intermittent testing or auditing of those undertaking the safety critical task, and be percentage of staff who, on questioning, understand how to carry out the task in a safe way, or percentage of staff who when observed undertake the task in a safe way.

EXPERIENCE

The informal feedback so far received from organisations that have developed and adopted process safety performance metrics is very positive. Potential weaknesses have been

Types of Process Safety Indicators

Operational Control Indicators			Generic RCS e.g. PTW, Plant change, Emergency Arrangements		Programme Indicators
Lagging Indicators	Leading Indicators		Lagging Indicators	Leading Indicators	Leading Indicators
Identify challenges to integrity Select the lowest detectable event (breach of process control envelope Set indicator at earliest point of deviation	Critical operator actions to: Set process controls / operating envelope correctly Take remedial action if process deviates from operating envelope Routine operation control – monitoring actions	Inspection and maintenance of process control instruments and alarms	Loss of containment incidents Process upsets Demands on safety systems	Critical tasks undertaken correctly e.g. PTW isolations, plant change, risk assessment	% statutory inspections completed to time % audits completed to time % audit actions closed out on time % staff trained to specified competence % procedures up to date

Figure 1.

identified and important issues prioritised. A major part of this benefit has been delivered during the process of developing the metrics, where potential weaknesses in the layers of protection being relied upon can be identified. HSE needs to continue to work with organisations that have developed and adopted metrics to determine the benefits that can be gained from going through this process, so that this can be judged against the level of work required to implement such a programme to confirm scale of the benefit and the practicability of applying this process.

CONCLUSIONS

HSG 254 has been very influential in the development and adoption of process safety performance indicators on major accident hazard sites. Even so, there is still the need for continued promotion of the adoption of metrics, as they are very effective in the identification of potential failure in major accident hazard safety management systems.

HSE HID CI is expecting operators of Top Tier COMAH establishments to have systems in place to measure the performance of key process safety risk control systems, or to be carrying out a programme of work to develop such a set of metrics.

HSE will undertake further work with organisations that have adopted metrics to confirm the scale of the benefits to be gained.

It important that senior managers are committed to the application and use of performance metrics to ensure that they continue to be collected, and that the information that they reveal is understood and acted upon.

Data should be collected at the unit and site wide levels, and metrics from the delivery of key safety programmes should also be collected. These metrics should be selected and analysed to allow senior managers to receive brief summary data on the effectiveness of the key aspects of the safety management system on a regular basis. In this way senior managers can demonstrate understanding of the measures being relied upon to prevent and mitigate serious accidents, and be able to take informed decisions on the need for improvements in the management system as a result of the metrics information.

Both leading and lagging indicators should be developed to test the effectiveness of the safety management system, but in some cases it will not be possible to develop both, or it will be difficult to determine if an indicator is leading and lagging. If the metric is one that can be collected and that will give valuable information on the performance of the safety management system, then it is not necessary to determine whether it is leading or lagging, just that the information that it supplies is used effectively.

Index

C

D

E

H

I

J

M

N

O

P

Q

R

S

Y

V

W